Chemistry
FACTS, PATTERNS, AND PRINCIPLES

Dr W. R. Kneen studied natural sciences at the University of Cambridge before teaching at Dulwich College. He then took a PhD and did research in inorganic chemistry at University College, London. Latterly he lectured at Kingston Polytechnic. At present he is in the education department of Bedfordshire County Council.

M. J. W. Rogers studied natural sciences at the University of Cambridge before teaching chemistry at Westminster School, where he became senior chemistry master and then master of the Queen's scholars. From 1962–4, he was on the headquarters team of the Nuffield O-Level Chemistry Project. Mr Rogers has written and edited several chemistry texts. At present he is headmaster of Malvern College.

Dr P. Simpson studied natural sciences at the University of Cambridge and also took a PhD in organic chemistry there. He then taught chemistry at King's School, Worcester, before moving to the School of Molecular Sciences, University of Sussex, as lecturer in chemistry. Dr Simpson has written a background reader for sixth forms on organometallic chemistry. He is an examiner in A-Level chemistry for one of the examination boards. His main research interest is in organophosphorus chemistry.

Chemistry
FACTS, PATTERNS, AND PRINCIPLES

W. R. KNEEN Bedfordshire County Council
M. J. W. ROGERS Malvern College
P. SIMPSON University of Sussex

Contributing editors: Professor D. J. MILLEN
Professor SIR RONALD NYHOLM
Department of Chemistry, University College, London

ADDISON-WESLEY PUBLISHERS LIMITED
London · Reading, Massachusetts
Menlo Park, California · Don Mills, Ontario

Copyright © 1972 by Addison-Wesley Publishers Limited.
Philippines copyright 1972 by Addison-Wesley Publishers Limited.

All rights reserved. No part of this publication may be reproduced, stored in a retrieval system, or transmitted, in any form or by any means, electronic, mechanical, photocopying, recording or otherwise, without prior written permission of the publisher. Printed in Malta.

Preface

It is becoming increasingly difficult for a student to obtain an overall grasp of chemistry because of the continual expansion of chemical knowledge. This book has been written in an attempt to meet this need. It aims to provide a map of important areas and to help the student in learning how to get about. While we have attempted to give an overall view for a wide range of students, we hope to interest some in making a more formal and extensive study of the subject.

In writing the book, we have had in mind a student starting on an A-level course but it should continue to serve him well beyond that. Being an outline of facts, patterns, and principles it could provide a basis for a wide variety of courses in universities, polytechnics, colleges of education, colleges of technology, and technical colleges. Emphasis has been put on understanding, but the understanding of patterns in chemistry requires a certain breadth of knowledge. A wide range of factual material has been given from which the student may select.

We have not attached importance to formal mathematical development of ideas but have tried to develop a feeling for the way in which variables under discussion are related. Our experience has been that for many students, too early a development of mathematical formulation obscures rather than clarifies the subject. In treating quantitative areas of chemistry, we have tried to develop a feeling for the kind of relationships that might be expected, or to show what is plausible. For this reason certain material has been put into appendices, to make it available without interrupting the main development, which can be made in a less formal way.

As an aid to forming patterns and condensing materials we have often included summaries, tables, and charts. Problems at the end of each chapter are intended to assist the student to reinforce his understanding. Frequent references are made to technological and other applications of chemistry.

The first chapter attempts to give an outline answer to the question: "What is chemistry all about?" The following chapters develop the major themes of structure, bonding, energetics, equilibria, and rates, which are widely applied throughout the book. The aim has been to develop the required principles simply and not to take up the student's time in mastering principles which after all will not be used in an integrated study of chemistry. Thus, ideas on thermodynamics are introduced in relation to the way in which they will be applied. The central aim here has been to relate energy changes to bonding and to equilibria. The thermodynamic ideas required in relation to bonding are in fact developed in the chapter on bonding. We have introduced entropy through a simple molecular picture, our experience being that it is easier for students to follow this approach than more abstract ones. The approach leads up to the Boltzmann distribution law. Then the essential principles underlying the interpretation of the direction of change are developed as a step towards understanding chemical equilibrium. This leads to a discussion of electromotive forces of cells and to oxidation–reduction potentials generally. A chapter on rates and mechanisms completes the development of principles.

The integrating themes which are developed in the early part of the book are applied as occasion demands. In inorganic chemistry structural and thermodynamic considerations are of central importance. In applying electronic theory much use is made of electron configurations, electron shielding, ionization energies, electron affinities, and electronegativity. Whereas in inorganic chemistry the product of a reaction is often, though not always, controlled by thermodynamic factors, in organic chemistry reactions are frequently kinetically controlled and thermodynamic considerations are less important. Structure and mechanism are important themes here and, by contrast with inorganic chemistry, mechanisms are sufficiently well understood for important patterns to emerge. In developing inorganic chemistry much use is made of the Periodic Table, and in organic chemistry a systematic introduction is made in terms of functional groups. In the earlier chapters on organic chemistry the emphasis is on patterns of behaviour which ultimately relate to principles. The idea of mechanism, in the organic context, is introduced after this and is used throughout the later chapters with increasing sophistication. In neither area have we attempted to make much use of quantum mechanical pictures of bonding such as orbital theories, since while it is often comparatively easy to get a picture of a poor quantum mechanical treatment it is often difficult to get even a poor picture of a good quantum mechanical treatment. The difficulty stems from the complexity of the quantum mechanical problem of treating molecules. In the hands of those who appreciate how the visual models have been derived from the mathematical formulation they provide important ways of thinking about bonding. But we have taken the view that it is not for beginners; without such an appreciation, the use of such models easily becomes little more than using a jargon. Nevertheless, recognizing that in the particular area of hybridization the terms used have become a part of the language of chemistry we have treated this in an appendix and have used the terminology in some places where it has very important advantages of convenience.

We do not suggest that the order in which topics are presented is necessarily an appropriate learning or teaching sequence, rather it is merely a convenient classification of material within a book. This has the advantage that individual teachers and students have some freedom in the order in which material is studied. However, in an integrated study of the subject the degree of freedom is less than that in a non-integrated study, since necessarily most of the conceptual framework must be studied (to some extent) first. Certainly the student would be well advised to read Chapters 1 to 5 before reading any other section of the book. To understand even these chapters, however, it will be found necessary to dip into the later chapters. For example, the concept of oxidation number is used in the early chapters, but it is conveniently defined and discussed in the chapter dealing with the chemistry of oxygen. Similarly, some knowledge of organic reactions will help to clarify the material of Chapter 8, where concepts of reaction rates and pathways are introduced. We have not thought it necessary to write a stoichiometric equation for each reaction which is mentioned in the book. However, it is clearly desirable that each student should be able to formulate such equations and the ability to write down equations for redox reactions is also assumed early in the book, though the explanation is conveniently presented in the section on oxidation number. Many cross references are given, but the student is urged to use the index when he comes across an unfamiliar concept; the most important reference is given in heavy type. Even within a single chapter it may be thought desirable to change the order of approach depending upon the background knowledge of the student. For example, oxidation–reduction diagrams have been widely used in the book as convenient summaries of the comparative chemistry of the elements in a particular group of the Periodic Table. Until students have familiarized themselves with this method of presentation by constructing their own diagrams from reduction potential data, it might be desirable to deal with the chemistry of the elements first, and have a comparative interpretation afterwards. It must not be supposed, however, that we have not given thought to the order of presentation. On the contrary, during the study of inorganic chemistry, Chapters 11, 12, and 13 and the early parts of Chapter 10, are best read early on, but in Chapters 15 to 20 the order is less important. Similarly, during the study of organic chemistry, Chapters 21 to 25 are probably best read before Chapter 26 on organic reaction mechanisms, and an understanding of the material of Chapter 26 will be almost essential to the student who hopes to get the most out of Chapters 27 and 28. Where ideas have been introduced and taken up at a later stage, some overlap has often been provided, for brief recall, before making further development or application of the idea. We have assumed that the reader is undertaking a laboratory course and have not devoted space to experimental details, although the basic principles underlying separation and purification are treated in the final chapter. The book is intended to be complementary to an experiment-based approach.

Nomenclature has posed problems. We have attempted to strike a balance between a move towards increasing systematization and current usage. A glossary of systematic and trivial names of organic compounds is included as an appendix together with a glossary of symbols. State symbols have normally been included in equations where they characterize the conditions, but there are occasions, particularly in organic chemistry, where the conditions of reaction (such as mixed solvents, or a two-phase liquid system) are not readily described in this way and state symbols have been omitted. Recognizing that many students may be using Nuffield Foundation materials, we have taken standard thermodynamic data from the Nuffield Advanced Science, *Book of Data*, for consistency. We are grateful to the Nuffield Foundation for allowing us to use the data given in Appendices 15 and 16.

We are grateful to Mr. G. R. Maitland for the spectra for Figs. 2.4 and 2.10, to Dr. B. Beagley for an electron diffraction photograph (Fig. 3.8), and to Dr. P. J. Wheatley for an X-ray diffraction photograph (Fig. 3.6). We are grateful also to Mrs. M. G. Millen for assistance in preparing the indexes.

It is a pleasure to express our appreciation of the helpful criticisms received from Professor E. H. Coulson, Mrs. C. Groves, Professor M. J. Frazer, Professor A. K. Holliday, Dr. M. A. Jenson, Dr. J. Mason, and Dr. J. J. Thompson, all of whom were good enough to comment on the manuscript.

December 1971

W. R. Kneen
M. J. W. Rogers
P. Simpson

D. J. Millen
R. S. Nyholm

Contents

What is Chemistry?

Isolation, preparation, and synthesis	1
Structure and properties	2
Chemical reactions: how far and how fast?	3
Energetics in chemistry	3
Molecular theory	3
Hypotheses, theories, models, and laws: explanation and understanding	4
The dual character of science: classification and the scientific method	4
Using concepts and models in chemistry	5
Investigation, learning, and discovery	7
Pure and applied chemistry	8
Conservation and pollution	9
Looking to the future	10

1 The Mole Concept

1.1	Atomic and molecular weights	11
1.2	Relative atomic and molecular masses	13
1.3	Ions, electrons, and equations	13
1.4	Molar quantities	14
1.5	Amount of substance	15
1.6	Molality and molarity	16

2 Atomic Structure

2.1	Matter: continuous or discontinuous?	18
2.2	Evidence for the structure of the atom	19
2.3	Electrons in atoms: spectra and quantum theory	25
2.4	Energy levels of atoms other than hydrogen	41
2.5	Electron configurations of multi-electron atoms	48
2.6	Summary	55

3 Structure of Solids, Liquids, and Gases

3.1	Structure: evidence from diffraction	59
3.2	Structure of solids	64
3.3	Packing of spheres and correlation of crystal structures	66
3.4	Structure of liquids	71
3.5	Gases	72

4 Structure, Bonding, and Energy

4.1	Structure and bonding: types of bonding	77
4.2	Electronegativity	80
4.3	Fajans's rules	81
4.4	Dipole moments	82
4.5	The octet rule	83
4.6	Dative or co-ordinate bonding	85
4.7	Metallic bonding	85
4.8	Hydrogen bonding	86
4.9	Intermolecular forces	87
4.10	The electron-pair model and the shapes of simple molecules	88
4.11	Energy and bonding: internal energy and enthalpy	90
4.12	Bond energies and bonding	95
4.13	Ions in crystals and solutions	103

5 Gases, Solids, Liquids, and Solutions in Equilibrium

5.1	Equilibrium	111
5.2	Single-component systems	112
5.3	Two-component systems	114
5.4	Lowering of vapour pressure and depression of freezing point	119
5.5	Lowering of vapour pressure and elevation of boiling point	119
5.6	Osmotic pressure	120

6 Energy and Chemical Change: Why Do Changes Take Place?

- 6.1 Changes of mixing 124
- 6.2 Changes of falling 125
- 6.3 Changes involving heat flow 125
- 6.4 Chemical changes 127
- 6.5 Exothermic and endothermic changes . 127
- 6.6 Probability and change 128
- 6.7 How atoms and molecules hold their energies 130
- 6.8 Energy distributions 132
- 6.9 Entropy 134
- 6.10 Entropy and change 135
- 6.11 Measurement and tabulation of entropies . 137
- 6.12 Entropy changes for chemical reactions . 140
- 6.13 Changes at constant temperature and pressure 141
- 6.14 Entropy: summary 143

7 Equilibrium

- 7.1 Equilibria: distribution between two liquids 146
- 7.2 Equilibria: reactions in gases and solids . 147
- 7.3 The effect of pressure on equilibria . . 150
- 7.4 The effect of temperature on equilibria . 151
- 7.5 Free energy and equilibria 152
- 7.6 Free energy, cells, and redox reactions . 156
- 7.7 E.m.f. and free energy change . . . 158
- 7.8 Half-cells and electrode potentials . . 159
- 7.9 The sign of the e.m.f. 160
- 7.10 Ion–ion redox reactions 161
- 7.11 Variation of cell e.m.f. with concentration and temperature 162
- 7.12 Using the Second Law 163
- 7.13 Chemical change: summary 172

8 Rates and Mechanisms of Reactions

- 8.1 Concentration and rate of reaction . . 177
- 8.2 The effect of temperature on rates of reactions; molecular collisions 179
- 8.3 Activation energy 180
- 8.4 The molecularity of a reaction . . . 182
- 8.5 Order and molecularity 183
- 8.6 Encounters and collisions 186
- 8.7 Transition state theory 187
- 8.8 Catalysis 189
- 8.9 Kinetic and thermodynamic control of reactions 192
- 8.10 Kinetics and mechanism 194

9 Solutions and Acid–Base Equilibria

- 9.1 Solutions 198
- 9.2 Conductance of solutions 201
- 9.3 Acid–base equilibria 203

10 Periodicity I: The Periodic Properties of the Elements

- 10.1 Periodic tables and the periodic law . . 214
- 10.2 The classification of elements as metals and non-metals 218
- 10.3 The physical properties of the elements . 221
- 10.4 The chemical properties of the elements . 256
- 10.5 Classification of elements: summary . . 261

11 Periodicity II: The Periodic Properties of Compounds

- 11.1 The properties of selected types of compounds 269
- 11.2 Thermal decomposition of salts . . . 280
- 11.3 Salts 286
- 11.4 Summary of the main relationships in the Periodic Table 292
- 11.5 Metallurgy: the science of metals . . 294

12 The Noble Gases: Group 0

- 12.1 Occurrence, physical properties, and uses . 308
- 12.2 Importance in the history of bonding theory 309
- 12.3 The rationalization of the inertness of the noble gases, and its failings 311
- 12.4 Preparation and properties of XeF_4, a typical compound 312
- 12.5 Clathrate "compounds" or enclosure "compounds" 313
- 12.6 The molecular shapes of noble gas compounds 313
- 12.7 The Gillespie–Nyholm theory 313

13 Hydrogen

- 13.1 Isotopes 319
- 13.2 Occurrence, preparation, properties, and uses. Atomic hydrogen 320
- 13.3 Types of compound formed 322
- 13.4 Types of acid in aqueous solution . . 323
- 13.5 Complex hydrides 328
- 13.6 Water 329

14 The s-Block Elements: Groups IA and IIA

- 14.1 General properties 336
- 14.2 Group IA: the alkali metals 339
- 14.3 Group IIA: the alkaline earth metals . . 348

15 The Elements of the p-Block: Group IIIB

- 15.1 General trends: summary 360
- 15.2 Group IIIB 363

16 The Carbon Group: Group IVB

16.1	Group trends and types of compound formed. The inert pair effect	374
16.2	Occurrence, allotropes, and extraction	378
16.3	Properties of the elements	380
16.4	Peculiarities of carbon	381
16.5	Binary carbon compounds	387
16.6	Oxides of group IVB	388
16.7	Hydrides of group IVB	396
16.8	Halides of group IVB	396
16.9	Sulphides of group IVB	398
16.10	Properties of aqueous lead(II) and tin(II) ions	399
16.11	Organometallic compounds	399
16.12	Oxalates	400

17 The Nitrogen Group: Group VB

17.1	General chemistry	403
17.2	Oxidation state diagrams: stability of oxidation states	405
17.3	Group trends	407
17.4	The anomalous nature of nitrogen	408
17.5	Occurrence, preparation, and uses	412
17.6	Allotropes	413
17.7	Reactions of the elements	414
17.8	Nitrides and phosphides	414
17.9	Hydrides	415
17.10	Halides	418
17.11	Oxides	420
17.12	Sulphides	424
17.13	The oxyacids of nitrogen	424
17.14	The oxyacids of phosphorus	426
17.15	Oxyacids, oxyanions, and solution chemistry of arsenic, etc.	429

18 The Oxygen Group: Group VIB

18.1	Features of the oxidation state diagram	431
18.2	General chemistry	433
18.3	Group trends	434
18.4	The anomalous nature of oxygen	434
18.5	Occurrence, preparation, and allotropes	436
18.6	The reactivity of the elements	440
18.7	Oxides	440
18.8	Hydrogen peroxide	441
18.9	Hydrides	442
18.10	Sulphides	443
18.11	Halogen compounds	445
18.12	Oxides	446
18.13	Oxyacids	447
18.14	Oxidation and reduction	453

19 The Halogens: Group VIIB

19.1	General chemistry and group trends	463
19.2	Occurrence and preparation of the elements	468
19.3	Properties of the elements	469
19.4	Chemical properties of the elements	470
19.5	Uses	471
19.6	The anomalous nature of fluorine	472
19.7	Preparation of hydrogen halides	473
19.8	Hydrogen fluoride	474
19.9	Properties of hydrogen chloride, bromide, and iodide	475
19.10	Halides	476
19.11	Interhalogen compounds and polyhalides	478
19.12	The oxides and oxyacids of the halogens	480

20 *d*-Block Elements

20.1	Definition of a transition metal ion	487
20.2	Periodic trends	487
20.3	General properties	489
20.4	Transition metal complexes	491
20.5	Stereochemistry	492
20.6	Isomerism	492
20.7	Rules for the nomenclature of complexes	493
20.8	The uses of complex compounds	494
20.9	Trends among transition elements	495
20.10	Differences between the first and the last two transition series	497
20.11	Summary of general properties of transition elements	499
20.12	The scandium group and the lanthanides	501
20.13	The titanium group, group IVA	502
20.14	The vanadium group, group VA	504
20.15	The chromium group, group VIA	506
20.16	The manganese group, group VIIA	509
20.17	The iron group, group VIII	511
20.18	The cobalt group, group VIII	516
20.19	The nickel group, group VIII	518
20.20	The copper group, group IB	520
20.21	Zinc, cadmium, and mercury, group IIB	525

21 Hydrocarbons: Structure and Geometry

21.1	Chain structures	536
21.2	Nomenclature	539
21.3	The geometry of four-co-ordinate carbon: optical isomerism	540
21.4	Unsaturation and geometrical isomerism	543
21.5	Ring structures	550
21.6	Summary	553

22 Properties of Hydrocarbons

22.1	Physical properties of hydrocarbons	558

| 22.2 | Chemical behaviour | 559 |
| 22.3 | Important methods of hydrocarbon synthesis | 572 |

23 Compounds with Functional Groups containing Oxygen

23.1	The hydroxyl group	577
23.2	Carbon–oxygen–carbon linkages	597
23.3	The carbonyl group	604

24 Compounds with Functional Groups containing Nitrogen

24.1	Primary amino groups	629
24.2	Secondary amino groups	638
24.3	Tertiary amino groups	640
24.4	Quaternary ammonium compounds	642
24.5	Imino groups	645
24.6	Nitriles and isonitriles	646
24.7	The nitro group	648
24.8	Diazonium group	650

25 Compounds with Halogens as Functional Groups

25.1	Physical properties	658
25.2	Chemical behaviour	659
25.3	Introduction of the groups	664

26 Organic Reactions as the Making and Breaking of Chemical Bonds

26.1	Introduction	669
26.2	Classification of reactions	671
26.3	Nucleophilic substitution at tetrahedral carbon	672
26.4	Reactions of nucleophiles with multiply bound carbon	677
26.5	Reactions of electrophiles with multiply bound carbon	679
26.6	Elimination reactions	682

27 Organometallic Compounds

27.1	Introduction	693
27.2	Methods of forming carbon–metal bonds	695
27.3	Chemical character and synthetic uses of organometallic compounds	698

28 Molecules Containing Two or More Functional Groups

28.1	α-Amino acids	705
28.2	Carbohydrates	708
28.3	Compounds containing two carbonyl groups attached to the same carbon atom	716
28.4	α,β-Unsaturated aldehydes and ketones	721
28.5	Interacting functional groups in aromatic molecules	723

29 Separation and Purification of Compounds and the Determination of Molecular Formulae

29.1	Recrystallization and related techniques	739
29.2	Distillation and related techniques	743
29.3	Chromatography and related techniques	748
29.4	Chemical methods of separation	758
29.5	Automation	759
29.6	Determination of molecular formulae	760

Postscript: Organic Synthesis 764

Appendices

A.1	The covalent bond	771
A.2	Some further aspects of bonding and structure	772
A.3	Energy distributions and entropy	777
A.4	Entropy: molecular interpretation and calorimetric measurement	779
A.5	Determination and evaluation of entropy changes	780
A.6	Determination of standard entropies: an example	782
A.7	Chemical equilibrium	782
A.8	Transition state theory	784
A.9	Crystal field theory	785
A.10	Ionization energies and electron affinities	792
A.11	Bond energies and bond distances	794
A.12	Enthalpies of hydration	795
A.13	Radii, electronegativities, enthalpies of atomization, and atomic volumes of the elements	796
A.14	Boiling points, melting points, densities, enthalpies of fusion and vaporization, reduction potentials, thermal conductivities, and atomic conductances of the elements	798
A.15	Thermodynamic properties of selected inorganic substances	800
A.16	Thermodynamic properties of selected organic substances	811
A.17	Values of physical constants and conversion factors	813
A.18	Glossary of symbols	814
A.19	Glossary of organic nomenclature	818

General Bibliography 820

Answers to Selected Problems 823

Index of Named Substances 843

General Index 850

What is Chemistry?

Chemistry is about "stuffs". It involves making different substances, finding out their properties, and how they react with one another; finding patterns in this knowledge and attempting to understand properties and phenomena in terms of atoms and molecules, and their electronic structures; and using this knowledge and understanding to make predictions.

Chemistry is now so diverse in its methods, and ranges over such wide fields, that it would be pointless to attempt to define its scope precisely. Either things would slip through the definition or it would become so general that it would have little meaning. In any case, chemistry is not staying still; it is changing and expanding. It overlaps with other fields, such as biology and the more recently developed field of materials science, without any sharp dividing line. Chemistry can be expected to bring to these fields its wide experience in the handling of molecular models, and so contribute to a better understanding in complex areas.

ISOLATION, PREPARATION, AND SYNTHESIS

Chemistry is not only a body of knowledge it is also an activity. Chemistry, it has been said, is what chemists do. Three closely related activities have long attracted their attention: isolation, preparation, and synthesis—these are the three ways in which chemists obtain their materials. They isolate materials from plant and animal matter, or directly from the earth, and according to the origin of a particular material, it was once assigned to organic or inorganic chemistry. Isolation, in either branch, involves various techniques of purification, as, for example, in obtaining a substance such as morphine. Secondly, the chemist may prepare substances using fairly well established chemical techniques. Sulphuric acid, for example, is obtained from a variety of sources such as natural sulphur, anhydrite, iron pyrites, and zinc sulphide. Thirdly, we have synthesis. Here the chemist devises a route to obtain a substance perhaps previously unknown, or known only as a natural product; or he may design a molecule and then set about making it. Synthesis has changed vastly since the appearance of the first synthetic materials. Two major factors have been responsible for the change, one theoretical and the other experimental: developments in molecular theory; and the discovery of physical techniques. Both have increased the scope and power of chemistry.

These two developments were in fact so important in the growth of chemistry that they led essentially to a new branch: physical chemistry. With this addition there came about the traditional classification of the subject as inorganic, organic, and physical chemistry. The classification reflects the way in which the subject grew; it denotes two areas of study and adds to it techniques and a particular discipline which came later. The classification emphasizes first the extent of carbon chemistry, and secondly that there are areas of the subject having rigour and precision. While it is still possible to point to work which falls fairly neatly into one or other of these categories, the old boundaries have broken down. Outstanding among the factors which have contributed to this have been: the increasing use of physical methods throughout the entire subject; the development of a vast range of organometallic compounds; the use of structural ideas to understand properties; and the use of mechanistic ideas to understand chemical reactions. Not surprisingly attempts are being made to restructure chemistry. The classification "synthesis, structure, and dynamics", for example, has the advantage of providing a general framework and focusing attention on current general interests. Classifications however are useful only in so far as they help our understanding; there is no one unique classification of the subject which is ultimately to be achieved.

Developments in chemistry have also had the result that the classification of the sources of the chemists' materials (isolation, preparation, and synthesis) is itself in need of qualification in one very important respect. Today it is no longer necessary, in order to characterize a new chemical species, to isolate and purify it first. It is possible by modern techniques to study it in its environment. There are now an enormous number of examples. Some chemical species cannot in fact be isolated pure; simple examples are nitrous and hypochlorous acids. Many other species are known in electrical discharges in gases, or under other low pressure experimental conditions. The study of such ions or molecules can make important contributions to understanding chemistry. A simple example, the ion H_2^+, is obviously of interest, as the simplest of all molecules. There are other examples in which chemical species have been identified and studied, even though their lifetimes may be only minute fractions of a second. Again, they can be important in understanding chemistry, for example as intermediates in chemical reactions. Modern techniques have made it possible to study many chemical species which cannot be isolated and put into bottles, and frequently in fact structures have been obtained for such chemical species as precisely as for "stable" compounds.

STRUCTURE AND PROPERTIES

Structure is of very wide interest in chemistry. A great deal of the interest stems from the increasing understanding of properties of substances in terms of their structures. The aim of a structural investigation is to find out how the atoms are arranged with respect to one another and how they are bonded together. Even when a structure has been determined it is often desirable to infer by analogy more about the bonding than experiment reveals. The justification for this is that structure is so important in understanding properties that one wants to be able to visualize the structure in order to think about the properties and relate them to general patterns. A great deal of interest in chemistry is centred around trying to get a better understanding of bonding in molecules and crystals and so in turn to understand better the properties of chemical substances.

The properties of interest to the chemist include mechanical, optical, electrical, and magnetic properties; but the chemist is particularly interested in the properties that a substance shows in reacting with other substances. The great range of substances provides chemistry with an immense number of chemical reactions. The chemist is helped in this situation by patterns of increasing generality which are being built up from studies of properties and reactions.

Patterns in chemical knowledge have not only great value in correlating products of different reactions but also in allowing attempts to predict the products that one expects will be formed. One important application is the use of knowledge in this form in devising synthetic routes to new compounds or providing better ways of making known compounds. A great deal of preparative and synthetic chemistry relies upon such applications.

CHEMICAL REACTIONS: HOW FAR AND HOW FAST?

Two further questions arise which seek to extend quantitatively these ideas about products of chemical reactions: how much product will be formed, and how fast will it be formed? For example, when an acid and an alcohol are allowed to react, how much ester will be formed before equilibrium is reached, and how fast will equilibrium be approached? Providing a quantitative understanding of much of chemistry rests on answering these questions, and opportunities for applying these ideas, so far as they are developed, are widespread throughout chemistry. A research worker attempting to prepare some required compound can avoid fruitless attempts if it can be shown that the yields from certain proposed preparations will be inadequate. In industry, an important objective is to produce desirable products in the most efficient way from readily available materials (oil, natural gas, minerals, water, air, and so on). Time, money, and effort can be saved in the search for routes to new products, or alternative processes for already marketed products, if yields of certain routes can be shown to be unacceptably low. If knowledge is available which can be used to decide if particular reactions are not feasible, pointless effort can be avoided. Chemical thermodynamics provides a basis for understanding what determines the equilibrium position for a chemical reaction. As far as rate of reaction is concerned, recent years have seen considerable growth in understanding of the ways in which bonds are broken and new bonds formed.

ENERGETICS IN CHEMISTRY

Chemical changes are almost always accompanied by energy changes. The study of these energy changes proves invaluable in understanding large areas of chemistry, for example, understanding chemical bonding, and understanding the factors controlling the direction and extent of chemical changes. A large section of the chemical industry uses energy to bring about chemical transformations leading to desirable products; another section brings about transformations to provide convenient sources of energy. Success in either area requires an understanding of how and under what circumstances energy can be transferred, as heat, or as mechanical or electrical work.

MOLECULAR THEORY

While much of chemistry is performed on bulk matter in beakers, flasks, or reactors in industry, the interpretation of what is going on is nearly always made in terms of atomic and molecular theory, perhaps the most powerful theory ever introduced into science. While the theory is widely applied in this way, its development owes much to the study of phenomena which are directly attributable to atoms or molecules, such as spectra and X-ray diffraction, to mention two of the most important examples. Such studies are leading increasingly to wider knowledge and better understanding of structures of molecules and crystals. Furthermore spectroscopy hand-in-hand with quantum mechanics leads to a knowledge and understanding of molecular energy levels. This provides a basis for one of the long-term aims of chemistry, namely that of obtaining not only a qualitative but also a quantitative understanding of the properties of chemical systems in terms of the properties of their constituent molecules. The basic procedure is to consider an assembly of atoms or molecules whose properties are known, and then by a statistical treatment to attempt to link up with the properties observed for bulk matter. In particular, there is the attempt to calculate entropies and free energies of systems, and so to understand the position of chemical equilibria in terms of the properties of atoms and molecules. Only relatively simple systems can be treated with precision in this way, but the ideas are important in determining semi-quantitative thinking about more complex systems.

HYPOTHESES, THEORIES, MODELS, AND LAWS: EXPLANATION AND UNDERSTANDING

The terms "laws", "theories", and "hypotheses" continually recur in any treatment of chemistry. There is no essential difference between hypotheses, theories, models, and laws; they all refer to models of natural behaviour. Frequently "hypothesis" is used for an early model of the particular phenomenon, which on "verification" (when found to be a useful model) may become a "theory", and later, on further verification, may achieve the status of "law". Thus the difference is one of degree, not kind. A "hypothesis" has the flavour of something to be put to the test, whereas a "model" is a working model which is kept in mind and applied to the next appropriate situation, not necessarily expecting to test it, but rather to gain some insight into the chemical phenomenon through applying it. In addition, "model" is frequently reserved for a pictorial theory, for example the very useful billiard ball model of gas molecules, and "law" for a (more abstract) mathematical theory, such as Newton's law of gravity. Often a law refers to an "ideal" from which deviations may be measured for real systems. When a complex phenomenon, such as the movement of the planets around the sun, can be rationalized in terms of a simple law or model, for example Newton's law of gravity, it is said that the phenomenon is "understood", or that it has been "explained". This does not mean that everything about the system is known. Since it is unlikely that "everything" could ever be known about any system, it would be foolish to restrict the words "understanding" and "explanation" in that way.

THE DUAL CHARACTER OF SCIENCE: CLASSIFICATION AND THE SCIENTIFIC METHOD

The hypotheses, theories, models, and laws of science fulfil two related functions: they systematize or classify a vast body of knowledge, and they lead to predictions, the attempted verification of which suggests new experiments. Thus, the amount of knowledge increases, necessitating further systematization in the form of new or improved models or theories, and so on. This is an unending process. It can be summarized diagrammatically as shown:

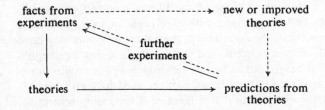

This model of the scientific method is useful, but like all models it is an imperfect over-simplification, and must be used with care. For example, it has sometimes been interpreted to mean that scientists indulge in a series of automatic methodical activities, with the implication that imagination plays little or no part in science. This is not so. It is not always obvious what experiments should be performed, especially in new fields of activity, nor do theories automatically spring to mind when some of the facts are known. The formulation of theories is a process usually requiring great imagination.

Science then is essentially dual in character: to the research scientist it is primarily a method of extending our knowledge; to the student and teacher it is, to a large extent, a number of useful classifications of the already known large body of facts and underlying theory, which gives the subject a structure, relieves the memory of much useless work, and offers intellectual rewards. All model building, classification, and theorizing are dangerous, however, since it is very easy to confuse the model with the real thing. Every student should pin a notice on his bedroom wall with a message such as: "Atoms are not very small billiard balls", or "No two elements are the same", or "Entropy is not randomness".

USING CONCEPTS AND MODELS IN CHEMISTRY

Chemistry uses a wide range of different kinds of concepts. There are areas of chemistry which employ precisely defined concepts as in physics, for example in thermodynamics and quantum mechanics. As examples of such concepts we have pressure, energy, electromotive force, and entropy. They are used in structuring knowledge of a wide range of phenomena, for example in relating equilibrium constants to calorimetric measurements. A large structure of this kind has been built up which gives precise relationships between precisely defined quantities. A simple example is the use of the Born–Haber cycle to evaluate the lattice energy of sodium chloride (Section 4.13.1). Precise concepts are also employed in another way. Chemists make much use of models both visually and mathematically: frequently they make "thought experiments" on these models, which suggest a mathematical treatment of the model or, alternatively may suggest directly a laboratory experiment as a test of the idea coming from the thought experiment. For example, in considering the bonding in sodium chloride, we may choose as a model an array of positively and negatively charged spheres and attempt to calculate the bonding energy in terms of interionic forces. The calculated value may then be compared with that derived from experimental measurements via the Born–Haber cycle. In such procedures it is recognized that the model is approximate, that there are assumptions involved, and that in treating the model mathematically further approximations are frequently necessary. But if the comparison of expectation from the model and experiment is good then this encourages the view that the model is a good one. The model may then be modified, or adapted, in an attempt to get even better agreement with experiment or to apply it to a wider area. On the other hand, if the agreement is poor then the model may have to be drastically changed, or even rejected, and the search for a new one begun. Almost always the model is recognized as being oversimplified because the knowledge of its constituents is inadequate, or because of the need to keep it readily visualizable, or to avoid the mathematical treatment of it becoming intractable. The aim is to refine the model by a trial-and-error process, being guided in this by comparison with experiment. Such approximate models can still use precisely defined concepts.

Besides precise concepts chemistry also makes much use of somewhat fuzzy concepts, such as electronegativity and solvation number, for example. Although numerical values may be quoted for these terms, it has to be realized that this does not mean such concepts are precisely defined. Electronegativity differs fundamentally from a concept such as electron affinity which is precisely defined and may be measured as precisely as experiment will allow. Electronegativity on the other hand, which is intended to be a measure of the electron-attracting power of an atom in a molecule, does not have this definite character; it will be influenced to some extent by the environment of the atom within the molecule. Such concepts even though they are only loosely defined are of great value in chemistry. They allow a rationalization of chemical phenomena in large areas of the subject which chemists have explored and which are too complex for a treatment in terms of precise concepts at the present time. Furthermore, such treatments frequently suggest new ideas and experiments, though one should perhaps be careful of using the term prediction. This does not imply that chemistry is a rather informal study, in a hurry, given to loose applications of ideas and fairly unrestrained speculation. Almost always the application of such ideas has to satisfy observations on a series of compounds or reactions, or even several series. Attempts to apply many of the ideas used in chemistry to one or two compounds would be quite fruitless. But when series of compounds are available, severe restraints are imposed and the procedure can be immensely powerful. It may be noted in passing that it is important to have a wide chemical knowledge to be able to use ideas in this way. By using such ideas chemical theory is forging ahead into areas where precise physical models are not yet available. Procedures of this kind run through much of chemistry, though a significant area of the subject now has a clear mathematical framework employing precise concepts. Chemists value both intensive and extensive approaches, which are complementary.

There are other areas of chemistry which are not formulated quantitatively in terms of either precise or fuzzy concepts, but which are qualitative and employ the verbal or the visual. In such areas patterns in knowledge play an important role. Thus we start from a single experimental observation that on mixing aqueous solutions of silver nitrate and sodium chloride a precipitate of silver chloride is formed. Then this is broadened to: alkali metal chloride solutions with silver nitrate give a precipitate. And then extended to: solutions of chlorides, bromides, and iodides with solutions of silver salts give precipitates. The interesting question arises as to whether the pattern can be extended to all halides including fluorides.

The recognition, extension, and interlinking of patterns are very important aspects of chemistry. Firstly, the building of such patterns relieves the burden on memory; all the facts no longer have to be remembered individually. Secondly, it paves the way for the development of theory, since a theory which covers a generalization includes all the specific cases. Thirdly, the pattern enables the knowledge to be grasped in such a way that it can be applied. For example, consider the synthesis of a compound B starting from a compound A. The pattern allows the chemist to propose, let us say, half-a-dozen routes which might lead from A to B, each perhaps involving several stages. He can go on to consider the snags in the various stages and to evaluate the different routes in order to decide which is to be preferred. Undeniably to succeed here the chemist needs a large body of knowledge. It is no good saying he can go and look up the facts. If the pattern of his knowledge is not large enough, the best routes may not even occur to him and the question of looking up some details before judging which is the preferred route does not arise.

Finally, there are throughout chemistry isolated facts and loose ends. The student should be aware of these if he is to have a realistic view of chemistry; it is not all in neat and tidy patterns.

INVESTIGATION, LEARNING, AND DISCOVERY

The processes of discovery and learning have in common the recognition and extension of patterns in knowledge, one in the general sense and the other for the individual. The usual account of scientific advance given in introductory books tells of scientists beginning by making experiments and collecting facts through observation. The rational correlation of facts leads eventually to hypotheses, theories, and finally scientific laws. The process, it is emphasized, is inductive rather than deductive. The account could be misleading; it neglects important features and it appears to make the process thoroughly systematized and almost mechanical. The fact is that scientists operate in a variety of ways, though doubtless there is much in common in their approaches. But it would be rare to find someone starting, as the account implies, simply with an experiment; just doing an experiment, quite in the dark, to see what happens. Everyone begins with a body of knowledge, having both patterns in it and loose ends. Much more frequently science proceeds through the attempt to extend patterns of knowledge, to tie loose ends together, or to test a hypothesis. The extension of factual knowledge is an important part of scientific activity but it leads of itself to no breakthroughs. It is the hypothesis which is the way to new ground. Putting forward a hypothesis is not something for which we can set out rules and systematize in a rational way: it calls for imagination. There is a vital creative element in science which accounts of scientific work have sometimes underrated.

A hypothesis is speculative of what may be. The next step is to test the hypothesis, a step calling for a quite different quality: critical power. Hypotheses take many forms; some, for example, may relate to patterns in factual knowledge, while others may refer to models. The hypothesis may emerge from the model, first in visual terms, and then have to be translated and developed in mathematical terms. Whatever the form of the hypothesis, there is the need to devise an experiment or experiments which will provide as severe a test of the hypothesis as possible. Then there follows the need to analyse the results of experiment, calling again for critical and analytical powers.

Now of course it may well happen, and not infrequently does happen, that after performing an experiment suggested by a hypothesis, the experimenter finds something quite unexpected, which may or may not be related to testing the original hypothesis. It may even be of greater interest than the original hypothesis which suggested the experiment. We cannot expect exploration to run systematically, what we can be sure of is that it will throw up new challenges to both imaginative and critical powers.

PURE AND APPLIED CHEMISTRY

It has been customary to talk of pure and applied chemistry as though there were some clear dividing line between them. Yet one merges into the other and there is much that is common to the approaches used in both areas. One seeks a better understanding of nature and the other a better control of nature. One is open-ended, the other has a defined objective, but there is a strong interaction between the two. No one today seeks to exploit crude oil without an understanding of organic chemistry. Research workers, even in the "purest" areas, depend increasingly on the availability of materials, instrumentation, and techniques that stem from applied research. Differences in approach spring from the different aims in the two areas. Applied research will generally be oriented towards objectives that are known to have, or are expected to have commercial importance. Some of the objectives may be long term, others may have to be achieved in a limited time and empirical solutions may have to be accepted rather than time being spent pressing for a deeper understanding. The challenge presented in such areas of research may be to find the best acceptable solution in the time available. Applied chemistry has to be seen in the wider context of meeting economic needs as well as in its relationship to gaining knowledge and understanding.

The increasing influence of chemistry on social changes is now seen in everyday life. Clothing, for example, is undergoing a revolution partly through the development of new materials such as rayon, nylon, and Terylene, and partly through the development of inexpensive dyestuffs and detergents. Plastics, particularly polythene and polyvinyl chloride, are now commonplace and steadily becoming cheaper. For example ethylene cost £200 a ton in the thirties when polyethylene was first prepared; it now costs £20 a ton. As the use of polyvinyl chloride steadily increases it becomes cheaper. It is of interest to note that the fluorine analogue, polyvinyl fluoride, is also known. It forms a remarkably stable film which resists sunlight-breakdown with ultra-violet radiation. It may well be an important material in the year 2000 providing coverings which are resistant to attack and breakdown in sunlight. New fuels appear for domestic use: coal gas, town gas, natural gas, and oil. The production of foodstuffs today owes much not only to the use of chemical fertilizers but also to the development of selective weedkillers and pesticides. In the field of health there are a whole range of pharmaceuticals in everyday use, analgesics, anaesthetics, antibiotics, and so on, not to mention the developments in understanding of hormones, vitamins, and enzymes. Underlying the development of new materials is an industry whose size is illustrated by sulphuric acid production which for Britain alone is measured in millions of tons. Over and above this is the production of materials such as steel, aluminium, and carbon fibres. So important is chemical industry that the production of sulphuric acid per capita is often used as a measure of the industrial development of a country.

All such materials are won from available raw materials (coal, natural gas, oil, ores, water, air, and so on) through the application of chemical understanding. At the present time petroleum is in fact the basis for the preparation of about 90 percent of all chemicals. Many questions besides purely chemical ones have to be answered by the industrial chemist before any particular process can be considered as a commercially viable proposition. For example, there is the availability and cost of raw materials, the supply of energy, the existence or development of suitable chemical plant, the scale of operation, and whether the technology is available to exploit the method. Advances in chemistry and technology, and the discovery of new materials, can all lead to marked changes in industrial processes. For example, before 1939 dehydration of ethanol (produced by fermentation) was the main source of ethylene. In the United States in 1971 ethylene was produced largely by cracking ethane and propane which are available there in natural gas. By contrast, North Sea gas (which is largely methane) is not a suitable raw material for ethylene production. In fact, ethylene in Europe is produced largely by cracking naphtha, a low-boiling fraction obtained from crude oil.

CONSERVATION AND POLLUTION

The increasing use of raw materials raises the question of exhausting world supplies. It may be, for example, that it will become desirable to conserve supplies of such materials as zinc, lead, and copper (which have hitherto been widely used) for the more essential uses and replace them for many purposes by aluminium or its alloys with magnesium, both of these elements being relatively abundant. Alternatively, plastics may increasingly replace metals in many applications. Already the total tonnage of plastics has surpassed that of all but a small number of metals. If the discovery of new oil and gas deposits does not keep pace with demands it may be that underground gasification of coal, or carbon dioxide from limestone, may become an important starting point in chemical industry. The key to such developments is likely to be the use of cheap energy. It would be rash to attempt to predict the future, but the changes in availability of raw materials will inevitably present new challenges to chemical industry in serving society.

But besides serving society's needs industry produces among its by-products some which are undesirable. Pollution is not a new problem, but one which takes different forms in different societies. We no longer have open drains running along our streets, but we do have to face problems not only of sewage but also of industrial waste. Industry makes heavy demands on water; it takes many gallons of water to make a pint of beer, and tens of thousands of gallons to make a ton of paper. The problem of river pollution is accentuated when several industries discharge effluent along the length of the river, even though each industry may be treating the effluent to some extent. Likewise there are pollutants of the atmosphere from several sources. Among these pollutants is sulphur dioxide; millions of tons of it are put into the atmosphere each year at the present time.

Generally techniques are available for reducing pollution by treating effluent, by extracting sulphur from flues, and so on. But the cost of applying pollution control raises production costs, sometimes sufficiently, it is argued, to make the product no longer competitive. Often a compromise can be reached between cost and an acceptable level of pollution control. The problem of pollution control is not basically a technological one; it is for the most part a question of affording the cost. Nevertheless in certain cases there is a technological challenge to provide alternative and cheaper techniques of pollution control, as was the case for example in the development of biologically degradable detergents.

Science and technology have sometimes been unfairly criticised for causing pollution. In fact, there are examples of enormous decreases in pollution in recent years. Thus in London, for example, the amount of smoke has fallen by over 70 percent in the last twenty years; the dirty fogs are things of the past. These changes owe much to the Clean Air Acts of 1956 and 1968, illustrating the point that the solution of pollution problems is frequently legal rather than chemical. Good laws need to be introduced and rigidly enforced; certainly this has been the key to the enormous clean-up of the atmosphere in the United Kingdom.

LOOKING TO THE FUTURE

In the scientific revolution of the past two hundred years chemistry has played an increasingly prominent part. It seems likely that in the future it will contribute to the development of new and improved materials for construction, medicine, clothes, and foodstuffs, to relate it to the basic needs of health, shelter, food, and clothing. But it is more likely that the social benefits will be greater than a simple extension of present applications. In the first place, looking back suggests that we can look forward to unexpected developments. Secondly, the interaction between the technologies of engineering, agriculture, and medicine, and those based on chemistry, physics, and the biological sciences, may lead to developments that might not be expected by extrapolation from any of these areas alone.

Let us not, however, in thinking of the material benefits and ills that science brings, neglect to remember its power to free us from ignorance.

BIBLIOGRAPHY

Caldin, E. F. *The Structure of Chemistry*, Sheed and Ward (1961).

>   An easily readable slim volume giving an introductory account of the nature of chemistry.

Medawar, P. B. *The Art of the Soluble*, Methuen (1967).

>   Contains two articles: "Two Conceptions of Science" and "Hypothesis and Imagination", which are suggested for further reading.
>
>   Introductory accounts of applied chemistry include the following:

Raitt, J. G. *Modern Chemistry: Applied and Social Aspects*, Arnold (1966).

Samuel, D. M. *Industrial Chemistry—Organic*, Monographs for Teachers No. 11, Royal Institute of Chemistry (1966).

Modern Chemistry in Industry. Proceedings of an International Union of Pure and Applied Chemistry Symposium held at Eastbourne. Society of Chemical Industry (1968).

CHAPTER 1

The Mole Concept

It is appropriate that the first concept to be discussed in this book should be one of the great unifying ideas in chemistry: the mole concept. The concept was born when Avogadro in 1811 first put forward the hypothesis that equal volumes of gases at the same temperature and pressure contain equal numbers of molecules. In doing so he described the molecule as an independent group of atoms. Dalton had suggested "compound atoms" between atoms of different elements but Avogadro saw the need for molecules in gaseous elements. Although Avogadro's hypothesis only works approximately in practice, the idea has proved to be extremely useful. It means that the space occupied by a molecule—which must not be confused with the volume of the molecule—is independent, or nearly so, of its size. The volume of a gas is a function of the number of molecules rather than of their size. This is less surprising when one realizes that at room temperature and pressure the distance between molecules is about ten times their size. The oxygen molecule is about 4 ångströms in diameter (1 ångström, symbol Å, is 10^{-10} m) and the average distance between centres of molecules at room temperature and pressure is about 30 Å. The diameters of gaseous molecules may differ (e.g. hydrogen 3 Å, and butane 8 Å) but the distance between their centres remains much the same.

1.1 Atomic and molecular weights
1.2 Relative atomic and molecular masses
1.3 Ions, electrons, and equations
1.4 Molar quantities
1.5 Amount of substance
1.6 Molality and molarity

1.1 ATOMIC AND MOLECULAR WEIGHTS

Another idea which is related to the mole concept is that of atomic weight. Dalton suggested that the relative weights of atoms could be determined if only the ratios of the numbers of atoms of each element in a compound were known. For his first scale of atomic weights, he guessed these ratios and, as can be seen in Table 1.1, Dalton's guesses were largely, though not entirely, wrong.

Dalton based his list of atomic weights on the weight of one atom of hydrogen. He defined atomic weight as the weight of one atom of a substance compared with the weight of one hydrogen atom. As relatively few elements combine with hydrogen, this definition was soon dropped as a basis for comparison in favour of oxygen which combines with almost all elements. The oxygen atom is nearly 16 times as heavy as the hydrogen atom. The oxygen standard for atomic weights gave oxygen the value of exactly 16. The atomic weight was then the weight of one atom of an element compared with one sixteenth of the weight of an atom of oxygen. Molecular weight was defined on the same scale. By the end of the nine-

Table 1.1

	Dalton's atomic weights	Present atomic weights for comparison
Hydrogen	1	1
Azote (nitrogen)	5	14
Carbon	5.4	12
Oxygen	7	16
Phosphorus	9	31
Sulphur	13	32.1
Iron	50	55.9
Zinc	56	65.4
Copper	56	63.5
Lead	90	207.2
Silver	190	107.9
Gold	190	197.0
Platina (platinum)	190	195.1
Mercury	167	200.6

teenth century, with the notable help of Avogadro and Cannizzaro, the atomic weights of all known elements had been determined. This is not the place for a full discussion of this remarkable achievement.

From the concepts of atomic and molecular weight were developed the ideas of the gram-atom—the atomic weight expressed in grams—and the gram-molecule—the molecular weight expressed in grams. The volume of a gram-molecule of a gas at standard temperature and pressure (s.t.p., 0 °C and 760 mmHg) is known as the gram-molecular volume. Actual values for the gram-molecular volume of some common gases are given below.

oxygen	22.40 litres
sulphur dioxide	21.90 litres
methane	22.40 litres
nitrogen	22.42 litres

These values are all very similar—a fact which agrees with Avogadro's hypothesis. The gram-molecular volume of an ideal gas is taken to be 22.4 litres at s.t.p.

As Avogadro stated that equal volumes of gases contain equal numbers of molecules, it follows that there must be the same number of molecules in 22.4 litres of any gas. If we take one molecule of hydrogen, relative weight 2, and compare it with a molecule of carbon dioxide, relative weight 44, the ratio of the two weights is 1 to 22. If we take 50 molecules of hydrogen and 50 molecules of carbon dioxide the ratio of the weights of this number of molecules is still 1 to 22. Indeed if we take the same number of molecules of each substance the relative weights will always be 1 to 22. Put the other way round, if we take 1 gram of hydrogen and 22 grams of carbon dioxide, we will have the same number of molecules of each gas. Similarly, if we take 2 grams of hydrogen (1 gram-molecule of hydrogen) and 44 grams of carbon dioxide (1 gram-molecule of carbon dioxide) we will have the same number of molecules of each substance. It can be seen from this that a gram-molecule of any substance (not only gases) will contain the same number of molecules. In the special case of a gas this number of molecules always occupies the same volume, but this is not so for liquids or solids. The actual number of molecules in a gram-molecule became known as the Avogadro number. This is no longer the definition which we use (see below) but it explains the origin of the term. As both atomic and molecular weights are based on the same scale, the number of atoms in a gram-atom of any element will be the same as the number of molecules in a gram-molecule. It will become clear that the Avogadro number is of fundamental importance. This is an appropriate place at which to look at it from a modern point of view.

1.2 RELATIVE ATOMIC AND MOLECULAR MASSES

Atomic weights are now determined by means of a mass spectrometer (Section 2.2). Indeed they are more properly called **relative atomic masses,** or for short, atomic masses. The mass spectrometer measures the relative masses of the atoms, and oxygen is no longer of any special value as a standard. For technical reasons, carbon has been chosen as the element most suitable as a standard. Carbon atoms occur in three different forms or isotopes (Section 2.2). As the mass spectrometer separates isotopes from each other, it is necessary to specify a particular isotope as the standard. Carbon has one stable and plentiful isotope, carbon twelve, written $^{12}_{6}C$. The nucleus of $^{12}_{6}C$ consists of six protons and six neutrons (Section 2.2). Natural carbon also contains a small percentage of $^{13}_{6}C$ (six protons, seven neutrons), and $^{14}_{6}C$ (six protons, eight neutrons) which is radioactive and is incidentally used in an important system of archaeological dating.

The modern system of atomic masses uses for its reference standard, the $^{12}_{6}C$ isotope which is given the value of 12. We can therefore define the Avogadro number as the number of atoms in 12 grams of $^{12}_{6}C$. The latest experimental value for the Avogadro number (1960) is $6.02252 (\pm 0.00028) \times 10^{23}$. The value used in most simple calculations is 6×10^{23}. The term used for an Avogadro number of any particular particle is the **mole**. It is a unit like a dozen, a gross, or a score. Starting from the above definition of the Avogadro number, we can define a **gram-atom** of an element as the mass, in grams, of the Avogadro number of the atoms concerned, or the mass in grams of a mole of the atoms. Similarly **gram-molecule** becomes the mass in grams of a mole of the molecules concerned. Strictly speaking, the term gram-molecule can only refer to a substance with a molecular structure. A compound like sodium chloride which has a giant structure of ions (Section 3.3) should not be referred to in this way, although this has been done. We need an expression to denote the same thing without making any statement about its structure. The phrase used is the **gram-formula**. When the expression is used the formula of the substance must be given. Thus we can talk about a gram-formula of NaCl or a gram-formula of $CuSO_4.5H_2O$. Only then do we know precisely what we are talking about. It will be seen that in a gram-formula of sodium chloride we have a mole of sodium ions and a mole of chloride ions. The expression **gram-ion** is also used, meaning the mass in grams of a mole of the ions concerned.

1.3 IONS, ELECTRONS, AND EQUATIONS

Ions and the gram-ion lead to the subject of electrolysis. The quantity of electricity required to deposit a gram-atom of silver from a solution of one of its salts is by definition a **Faraday** (96,487 coulombs). No element requires less than one Faraday of electricity to deposit one gram-atom of it. The silver ion has a positive charge of one (Ag^+) and it requires one electron which is negatively charged to make a silver ion into a neutral atom:

$$Ag^+(aq) + e^- \rightarrow Ag(s)$$

To deposit a gram-atom of silver, or a mole of silver atoms, a mole of electrons will be needed. A Faraday is the quantity of electricity equal to that carried by a mole of electrons.

In the equation just given a single ion of silver becomes an atom of silver. It could also be written:

$$Ag^+(aq) + 1 \text{ Faraday} \rightarrow Ag(s)$$

It would then mean that a mole of silver ions reacts with a mole of electrons to form a mole of silver atoms. The same symbols are used for two different statements. It is important to be clear about this. Zn can refer to one atom of zinc, or to one gram-atom of zinc. Usually when it refers to one gram-atom of zinc the state symbol is included; Zn(s) means one gram-atom of zinc in the solid phase.

Sometimes it is necessary to refer to the number of gram-formulae which react together in a chemical reaction. For instance, when discussing the heat evolved when ethanol is burnt (Section 4.11.2) we want to measure the heat given off when the number of gram-formulae in the equation,

$$C_2H_5OH(l) + 3O_2(g) \rightarrow 2CO_2(g) + 3H_2O(l),$$

react together. To avoid a cumbersome phrase the term **gram-equation** is used. Thus the heat given of in the above reaction is sometimes expressed in kilojoules per gram-equation.

1.4 MOLAR QUANTITIES

The common denominator which links all these terms together—gram-atom, gram-ion, gram-molecule, gram-formula, gram-equation, Faraday—is the mole. It is important because it helps us to see that when we are comparing quantities in these units we are comparing equal numbers of particles. There are the same number of atoms in a gram-atom as molecules in a gram-molecule, as electrons in a Faraday, and so on. And this is important because in many of the phenomena which chemists study it is the number of particles present that matters rather than their size. We have already come across one instance of this: the volume occupied by a gas. There are many others. Properties which depend on the number of particles present are called colligative properties.

Comparing the properties of similar numbers of particles sometimes gives an insight into the nature of phenomena which would not otherwise be apparent. Let us take one example. The specific heats of elements give no particular pattern when expressed in terms of joules per gram, but when expressed in kilojoules per gram-atom, or mole, a definite picture emerges (Table 1.2). The unit kilojoules per mole is abbreviated to $kJ \text{ mol}^{-1}$.

It is clear from Fig. 1.1 that many elements require approximately the same amount of heat per gram-atom to raise their temperature by a given extent. This is very significant and is known as Dulong and Petit's law—a law which was of great help in determining atomic masses in the nineteenth century.

Table 1.2. Specific and atomic heats of some solid elements.

Element	Specific heat (J g^{-1} K^{-1} at 25 °C)	Atomic heat (kJ mol^{-1} K^{-1} at 25 °C)
Li	3.39	0.0235
Be	1.85	0.0164
B	1.025	0.0111
C	0.711	0.0085
Na	1.235	0.0284
Mg	1.030	0.0250
Al	0.899	0.0242
Si	0.711	0.0200
P	0.670	0.0207
S	0.732	0.0235
K	0.754	0.0284
Ca	0.654	0.0262
Sc	0.566	0.0255
Ti	0.524	0.0251
V	0.481	0.0246
Cr	0.448	0.0233
Mn	0.477	0.0262
Fe	0.448	0.0250
Co	0.436	0.0257
Ni	0.439	0.0258
Zn	0.385	0.0251
Cu	0.435	0.0276
Ga	0.381	0.0266
Ge	0.322	0.0234
As	0.326	0.0244
Se	0.322	0.0254
Rb	0.360	0.0306
Sr	0.284	0.0249

1.5 AMOUNT OF SUBSTANCE

The ideas introduced so far have been in common use for some time. Nevertheless, in recent years another way of looking at things has been put forward. It makes use of the idea of **amount of substance**. This is regarded as a physical quantity like length, time, mass, and so on. Each physical quantity is measured in terms of certain units, for example, length in metres, mass in grams, and time in seconds. The choice of unit is a matter of convenience. The question arises then, of choice of a unit for amount of substance. In comparing amounts of substances the chemist wants to compare equal numbers of particles, and equal amounts of substance are chosen to contain the same number of particles. That is to say, the amount of substance is proportional to the number of specified particles. The unit of amount of substance is conveniently taken to be the mole, which is defined in terms of number of particles in the way already described. The constant of proportionality between amount of substance and number of particles (the same for all substances) is called the **Avogadro constant** (symbol N). In this scheme the Avogadro constant has units of mol^{-1}. Similarly in this scheme the **Faraday constant** (symbol F) is the constant of proportionality between electric charge and amount of singly charged ions. In SI units its value is 96,487 C mol^{-1}.

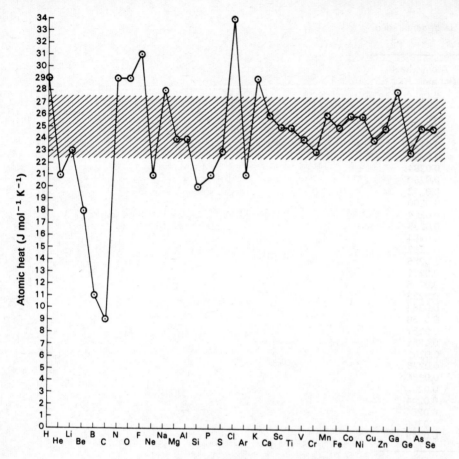

Fig. 1.1 Atomic heats

It is, of course, necessary to specify the particles being considered when one talks of a mole of some substance. This is readily seen if we consider, for example, one mole of hydrogen atoms and then one mole of hydrogen molecules. Some examples are given to show the use of the mole as a unit.

1 mole of hydrogen atoms (H) has a mass equal to 1.008 g.
1 mole of hydrogen molecules (H_2) has a mass equal to 2.016 g.
1 mole of methane (CH_4) has a mass equal to 16.04 g.
1 mole of chloride ions (Cl^-) and 1 mole of sodium ions (Na^+) together have a mass equal to 58.44 g.
1 mole of electrons has a charge equal to 96,487 C.

It is evident from the examples that there is no need to use the terms, gram-atom, gram-molecule, or gram-formula. These terms, although widely used, are no longer strictly necessary. Nevertheless, the student is likely to encounter them. Only time can tell how usage will change.

1.6 MOLALITY AND MOLARITY

It is worth noting also that the terms molar and molal are often used in referring to solutions. The unit of *molality* is mol kg^{-1}. A solution of molality equal to 0.1 mol kg^{-1} is often called an 0.1 molal solution. A solution with a molality of 1 mol kg^{-1} is prepared by dissolving one mole in one kilogram of solvent. A *molar* solution is prepared by dissolving one mole of substance in sufficient solvent to make one litre of solution. A 0.1 molal solution is often written 0.1m, and a 0.1 molar solution, 0.1M.

PROBLEMS

1. Write down in your own words what you mean by:
   a) relative atomic mass;
   b) relative molecular mass (molecular weight);
   c) gram-atom;
   d) gram-formula.

2. The following simple calculations are for you to check whether you understand and can use the mole concept.
   a) What is the mass of 10 moles of sodium atoms?
   b) What is the mass of 0.02 moles of hydrogen molecules?
   c) What is the mass of 2.5 moles of ethanol molecules?

3. a) How many Faradays of electricity are required to deposit 216 grams of silver from a solution of silver nitrate?
   b) How many electrons are required to deposit 6.35 grams of copper from a copper(II) sulphate solution?
   c) What weight of tin would be deposited from a solution of a tin(II) salt by 965 coulombs of electricity?

4. How many grams of sodium chloride are needed to make 500 cm^3 of molar solution?

5. How many grams of sodium chloride must be dissolved in 500 grams of water to prepare a molal solution?

6. Which of the following contains the largest number of atoms:
   a) 1 mole of hydrogen molecules;
   b) 1 mole of oxygen molecules;
   c) 1 mole of water molecules?

7. What volume of oxygen at s.t.p. is required for the complete combustion of one mole of each of the following molecules:
   a) H_2;
   b) CH_4;
   c) C_6H_6?

BIBLIOGRAPHY

McGlashan, M. L. *Physico-Chemical Quantities and Units. The Grammar and Spelling of Physical Chemistry*. Monographs for Teachers No. 15, Royal Institute of Chemistry (1971).

   This contains a general discussion of quantities, units, and notation. The ideas of amount of substance and the mole are introduced. It will also be found a useful reference book in making physico-chemical calculations.

Whiffen, D. H. *School Science Review*, **145**, 368 (1960).

   This contains a discussion of the ^{12}C standard for atomic masses.

CHAPTER 2

Atomic Structure

2.1 Matter: continuous or discontinuous?
2.2 Evidence for the structure of the atom
2.3 Electrons in atoms: spectra and quantum theory
2.4 Energy levels of atoms other than hydrogen
2.5 Electron configurations of multi-electron atoms
2.6 Summary

2.1 MATTER: CONTINUOUS OR DISCONTINUOUS?

From early times philosophers have debated the divisibility of matter. Could a piece of material be divided again and again into infinitely small particles—the theory Aristotle favoured—or was there a limit to this divisibility as Lucretius and, earlier, Democritus (born about 460 B.C.) had proposed? The latter school, the atomists, have generally been considered to be the "winners" of this debate. After all it was their word which Dalton used as he developed his atomic theory for chemical reactions. But were the Greek atomists right? It depends what is meant by indivisible. It is true that the particles which we call atoms are of fundamental importance. But it is certainly not true that atoms are indivisible as the atomists proposed and as Dalton himself at first suggested. It is interesting to note that Dalton changed his mind before the end of his life and in a remarkable prophesy suggested that a great deal of energy might be liberated if atoms could be split. We have no evidence then that matter is not infinitely divisible. It becomes increasingly difficult to divide particles the smaller they become. The limiting factor may be the energy required to divide the particle. It is impossible to "prove" the truth of any general statement since this would involve testing every possible case. But if it is generally true the statement is useful. It is however possible to disprove a general statement by a particular case.

2.1.1 The Size of Particles

We can take almost any solid substance and grind it into a fine powder. The energy required to divide up matter in this way depends on the toughness of the substance but in most cases is not great. The separation of solid particles from air (dust) or from water presents no great problems; the two methods used are filtration and centrifuging. Filter papers are made to filter out particles of various sizes, and the finest filter papers available will filter out particles of about 2×10^{-4} cm in diameter.

There are very small particles of solids suspended in liquids which can pass through the finest filter papers. These suspensions, called colloidal suspensions, have special properties. The size of colloidal particles varies between 10^{-7} and 10^{-4} cm approximately.

2.1.2 The Nature of the Particles

The separation of matter into smaller particles than those of powders or of colloidal suspensions can be performed in two simple ways: the substance may be vaporized or it may be dissolved. Since the process of dissolving may be a two-stage process, involving both separation and a reaction with the solvent (solvation), it is more profitable to look at vaporization. The process of vaporization is one in which a solid turns into a gas. The gas has a much larger volume than the solid. The process is clearly one in which the particles of the solid are pulled apart. How much energy is required to do this? An examination of molar heats of vaporization (that is the heat required to change a mole of a solid into vapour) shows that substances are roughly of two sorts: those with low molar heats of vaporization, and those with high molar heats of vaporization. It is simpler to consider elements only at this stage. It turns out that those with low molar heats of vaporization consist, in the vapour phase, of groups of atoms (molecules) and only occasionally, as in the noble gases, of single atoms. Those with high molar heats of vaporization almost always consist, in the vapour phase, of single atoms. Similarly, X-ray crystallography (Section 3.1.1) shows us that in the solid phase those substances with low molar heats of vaporization consist of groups of atoms with spaces between them (molecules again), and those with high molar heats of vaporization show a consistent regular pattern of atoms without such groupings (called giant structures). Solid structures are discussed in detail in Chapter 3.

The forces holding molecules together in solids (and in liquids) are called van der Waals forces and are relatively small (Section 4.9). The forces binding atoms together into giant structures are called chemical bonds and can be very large. In the case of a diamond for instance, the molar heat of vaporization is 718 kJ mol^{-1} (kilojoules per mole). The energy required to vaporize 12 g of diamond is sufficient to lift a kilogram over 70,000 metres! Compare this with the molar heat of vaporization of iodine which is 26.8 kJ mol^{-1}.

2.1.3 The Structure of Atoms

Looking further into the structure of atoms themselves, we find that the atoms consist of a nucleus surrounded by electrons. The nucleus contains two types of particles—protons and neutrons. The energy required to strip all the electrons from an atom is extremely high in the case of the larger atoms; oxygen atoms, for example, need about 2×10^5 kJ mol^{-1}. Of an even higher order is the energy required to split a stable nucleus. It would require 10^{10} kJ mol^{-1} to split the nitrogen nucleus into separate protons and neutrons. The proton has not yet been divided. We cannot say with any certainty that it will eventually be split but we can predict that if it is a very great deal of energy will be required. Protons, neutrons, and electrons are called **fundamental particles**. They are fundamental in the sense that they are found in all matter on earth. But if we assume that they are indivisible we may be making the same mistake that Dalton made in his first definition of the atom without the justification of making a major advance in theory.

2.2 EVIDENCE FOR THE STRUCTURE OF THE ATOM

The history of the discovery of the atom's structure is a fascinating but complicated subject. It is outside the scope of this book, and readers who are interested should consult Partington's *History of Chemistry* or some other specialist work. Potted histories are generally of little value. On the other hand, the evidence on which theories are based is of fundamental importance; accurately observed facts remain "true" for all time, although the theories that seek to explain these facts change. It is important to distinguish between the observation and the explanation.

2.2.1 Models

In many cases explanations take the form of models. A model can become more and more sophisticated but it can never be the real thing. Take as an example the model of a boat. A child playing in a stream is satisfied with a stick of wood as a model boat. Later a boy sails a boat with a real sail, keel, and rudder over a pond. Men, whose hobby it is, construct models of extreme complexity in which every part of a four-masted barque is exactly to scale. But even if every part of the model worked perfectly, it would still by its very nature be a model and not the real thing. The case of scientific models is very much the same, with the important distinction that the subject of the model is unseen and in one sense unknown. A model is used in this case to account for observations. For instance Dalton explained the constant composition of compounds by suggesting that matter was composed of atoms and that compounds consisted of combinations of atoms of elements in certain fixed ratios. He was proposing the atom as a model, and his model of the atom was extremely simple. Atoms, he suggested, were small, hard, discrete, indivisible particles. Atoms of the same element were all alike, atoms of different elements were different. Compounds were formed by the combination of atoms of different elements to form what Dalton called "compound atoms". Dalton also suggested that if the ratios of the numbers of atoms combining together could be determined it would be possible to work out the relative weights of the atoms of different elements. It was a remarkable feat of imagination to see that the relative weights of atoms might be found, and it led during the course of the nineteenth century to the fairly accurate determination of the atomic weights of all known elements. However, at this moment we are concerned less with that fact than with the "Daltonian atom" as a model. Eventually this model was found to be inadequate, as indeed, in time, all models become inadequate.

The experimental evidence which led to the replacement, in stages, of Dalton's simple model by more sophisticated (but still in some ways inadequate) models that are in general use today, will now be examined.

2.2.2 The Evidence of Radioactivity

Our knowledge of radioactivity dates from the discovery by a French scientist, Becquerel, in 1895. He kept photographic plates covered with black paper in a drawer which also contained some uranium salts and found that the plates became fogged. It seemed probable that some sort of radiation capable of activating photographic emulsion had come from the uranium salt and penetrated the paper covering the plates.

Subsequent investigations have shown that three types of radiation can be detected from radioactive substances. They are called α-, β-, and γ-rays. α-rays are found to consist of a stream of particles whose mass is equal to that of the helium nucleus. They travel for a distance of a few centimetres in air but have little penetrating power. The effect of a magnetic field shows that they are positively charged. Rutherford devised an experiment in which the radioactive gas, radon, was sealed in a tube with thin glass walls through which α-particles, but not radon, could pass. A tube with thicker walls was placed round the first tube. After a time this outer tube became filled with a gas which was identified by its spectrum as helium. This evidence finally demonstrated that α-particles were helium atoms with a double positive charge, or simply helium nuclei.

β-rays have a negative charge and can be identified by their charge to mass ratio as streams of electrons. β-rays have a greater penetrating power than α-rays, of the same energy, and can pass through thin sheets of metal. The denser the metal the greater its "stopping-power".

γ-rays have a very considerable penetrating power: they can pass through 15 cm of lead. They are like X-rays but with even shorter wavelength. γ-rays vary in wavelength, the longest being of the order of X-rays, the shortest being too short to measure by standard crystallographic means.

The evidence of radioactivity shows that the atom is far from being an indivisible particle. If it can emit electrons and helium nuclei, it must have a substructure of its own.

2.2.3 The Evidence of the Geiger and Marsden Experiment

Geiger and Marsden studied the effect of bombarding thin metal sheets with α-particles. They found in the case of gold foil for instance, that a small amount of scattering occurred. That is, most of the particles went straight through the gold leaf but a few were deflected. Most surprising of all at the time was the fact that a very few particles bounced right back towards the α-particle "gun". The thicker the gold foil, the more particles bounced back, so this was no surface effect. Rutherford suggested a new model of the atom to account for this phenomenon. He proposed that the atom consisted of a small, very dense nucleus surrounded by electrons. He suggested that the electrons rotated round the nucleus like planets round the sun. The nuclear theory has proved to be highly successful but the idea of planetary electrons, attractive as it is, has been as misleading as it has been helpful.

2.2.4 The Evidence of X-rays

X-rays have played a vital part in the determination of structure at sub-atomic and supra-atomic levels. At the sub-atomic level, evidence was gathered by H. G. J. Moseley in 1912. He bombarded elements with β-rays (electrons) in a discharge tube. It had been discovered that X-rays could be produced in this way. Moseley used a crystal as a diffraction grating to investigate the spectra of the X-rays given off by different elements. He discovered that the X-ray spectra were simple and that they varied in accordance with the position of the elements in the Periodic Table, showing for the first time that the number of an element in the Periodic Table has physical significance.

It can be seen from Fig. 2.1 that the frequency of the X-rays is related to the position of the element in the Periodic Table. There are three main sets of spectral lines in the X-rays given off by elements. The K-lines are given by all elements, the L-lines by elements after neon, and the M-lines only by the very heavy elements.

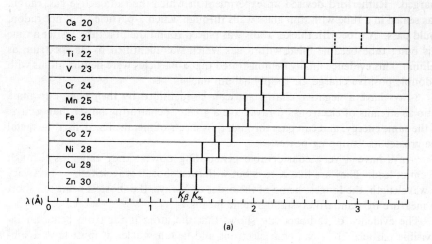

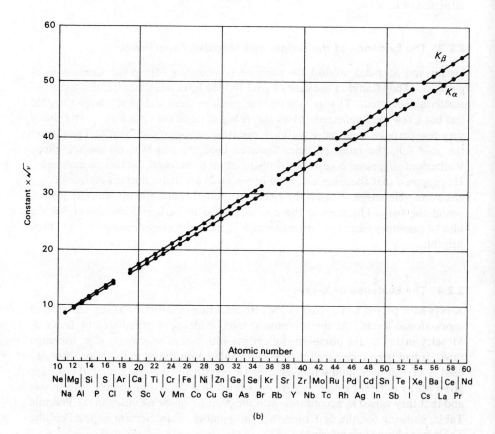

Fig. 2.1 (a) X-ray spectra. (b) Graphical illustration of Moseley's law. Adapted from E. A. Moelwyn-Hughes, *Physical Chemistry*, Cambridge University Press (1940). Reproduced by permission of the publisher.

The frequency, ν, of a line in the spectrum is given by

$$\nu = a(Z - b)^2,$$

where a and b are constants, and Z is the ordinal number of the element in the Periodic Table (hydrogen 1, helium 2, lithium 3, etc.). Subsequent work related this number to the charge on the nucleus. And so for the first time the Periodic Table was given a physical basis.

2.2.5 Evidence from the Mass Spectrometer

One of the most important instruments used in finding out about the atom has been the mass spectrometer. J. J. Thomson performed the first experiments on "positive rays" between 1910 and 1912, which led to the development of the first mass spectrograph by Aston in 1913. The modern mass spectrometer is a vastly more sophisticated instrument but the principle remains the same (Fig. 2.2).

The substance to be examined is first vaporized and then ionized. The positive ions thus formed are accelerated by an electric field and then deflected by a powerful magnet. The extent to which they are deflected will depend on their mass and their charge. The stream of ions also has to be focused at this stage. Finally the ions must be detected in some way. There are various methods of doing this: Aston used a photographic plate; modern instruments employ electronic techniques.

It was found in early experiments with neon that two lines were given on the photographic plate. One represented a particle of relative mass 20, the other of 22. The only satisfactory explanation of this is that there are two forms of neon of different atomic masses but otherwise with almost identical properties. Such forms are called **isotopes** and it is found that almost all elements exist in several isotopic forms. It was the existence of isotopes together with other evidence which led Rutherford to suggest that the nucleus consisted of two types of particles, protons which were positively charged, and neutrons which had no charge, but a mass equal to that of the proton.

The model of the atom, derived from the kind of evidence given in this section, can be described simply in the following terms. All atoms consist of a positively

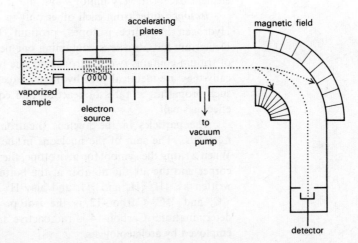

Fig. 2.2 Mass spectrometer.

charged nucleus surrounded by negatively charged electrons. The nucleus is composed of two types of particle—the **proton** and the **neutron**. The nucleus of an atom is very dense and contains almost all of its mass. The electrons around the atom, especially the less tightly bound outer electrons, are responsible for its chemical properties. The charge of the proton and electron are equal but opposite. Atoms contain equal numbers of protons and electrons and are themselves electronically neutral. The mass and charge properties of the electron, proton, and neutron can be summed up as in Table 2.1.

The simplest atom is that of hydrogen which consists of one proton with one electron. Each successive atom in the Periodic Table has one more proton in the nucleus. The number of protons in the nucleus equals the **atomic number**, Z, of the element and corresponds to its position in the Periodic Table. The number of electrons around the nucleus must be equal to the number of protons in the nucleus as atoms have no residual charge. The number of protons therefore controls the number of electrons. When atoms react to form compounds it is the outer electrons which form bonds and determine chemical properties. Thus the number of protons in the nucleus of an atom controls its chemical properties. An element is defined by the number of protons in its nucleus, that is, by its atomic number.

Table 2.1

	Relative charge	Relative mass	Absolute charge (C)	Absolute mass (g)
Proton	+1	1	1.602×10^{-19}	1.6725×10^{-24}
Neutron	0	1	0	1.6748×10^{-24}
Electron	−1	$\frac{1}{1840}$	1.602×10^{-19}	9.109×10^{-28}

The atomic masses of elements are usually about twice their atomic number. The difference is due to the presence of extra or fewer neutrons in the nucleus. The number of neutrons in the nucleus does not affect the chemical properties of the element except in very minor ways, but it does affect the stability of the nucleus.

Isotopes differ from each other only in the number of neutrons in the nucleus. Hydrogen has three isotopes, protium (H), deuterium (D), and tritium (T). Deuterium has a nucleus containing one neutron and one proton, and the nucleus of tritium contains two neutrons and one proton. Tritium is radioactive. Chemically they are identical with hydrogen, except when their physical differences affect such properties as rates of reaction and equilibrium constants, but generally the effect is small.

The particles in the nucleus (neutrons and protons) are sometimes called nucleons. The sum of the nucleons in the nucleus is called the **mass number**, A. When writing the symbol for an isotope, the mass number is put at the top left hand corner and the atomic number at the bottom left. The isotopes of hydrogen are written as, $^{1}_{1}H$, $^{2}_{1}H$, and $^{3}_{1}H$, and also H, D, and T. Those of carbon are, $^{12}_{6}C$, $^{13}_{6}C$, and $^{14}_{6}C$. Carbon-12 is the isotope used as a standard for atomic mass determination. Carbon-14 is radioactive and is used in the carbon dating system employed by archaeologists.

2.2.6 The Stability of Nuclei

Figure 2.3 shows the ratios of neutrons and protons for which a nucleus is stable. Isotopes whose neutron to proton ratio is outside the stable region are radioactive. They decay either by emitting an electron (β-decay) or by emitting an α-particle (α-decay). An electron can be emitted by neutron decay, that is to say a neutron turning into a proton which remains in the nucleus and emitting an electron. In this case, the atomic number goes up by one and the mass number remains the same. If an α-particle (helium nucleus $_2^4\text{He}^{2+}$) is emitted, the atomic number goes down by two and the mass number by four. Some isotopes inside the stable region are also radioactive.

2.3 ELECTRONS IN ATOMS: SPECTRA AND QUANTUM THEORY

Our knowledge of the way in which electrons are distributed in atoms comes largely from the evidence of spectra. Evidence from electron impact experiments has added to this knowledge but to a lesser extent. The latter is discussed in Section 2.4.1.

2.3.1 Spectra

Newton demonstrated that white light can be split up by means of a prism into its constituent colours. The spectrum produced by the splitting of white light is a continuous series of colours merging into each other starting with blue and ending

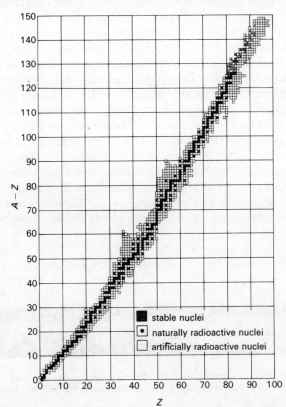

Fig. 2.3 Stability of nuclei. Number of neutrons $A-Z$, plotted against Z, the number of protons (A is the mass number). From I. Kaplan, *Nuclear Physics* (first edition), Addison-Wesley (1955).

with red. Such a spectrum is called a **continuous spectrum**. It is given not only by sunlight but also by the light from any hot solid or liquid body. The colour of light depends on its wavelength. Blue light has the shortest wavelength (about 4000 Å), and red light the longest (about 7000 Å). Light of a single wavelength is called monochromatic.

A different kind of spectrum may be obtained when light from a gas source is passed through a prism. In order to emit light, the gas must be "excited" in some way. A common way of doing this is by passing an electric current through the gas at low pressure. The neon lights used in advertisements make use of this method for producing light, and so do sodium vapour street lights. The latter produce a characteristic yellow light which is restful to the eyes but unattractive in other ways. Most coloured objects appear to lose their colours in this light. Analysis of this light through a prism or diffraction grating gives a spectrum which is not continuous but consists of a series of lines. The dominant lines in this case are the two yellow lines which give sodium vapour lamps their characteristic colour. The other lines are relatively faint, so this light is nearly monochromatic. Such a spectrum is called a **line spectrum**. It was Bunsen, the German chemist, who first noticed that each element has its own specific line spectrum.

Both continuous and line spectra are produced by the emission of light from atoms and are called **emission spectra**. A third type of atomic spectrum is formed when white light shines through a gas. Dark lines are found in the otherwise continuous spectrum at exactly the same places where some of the bright lines would be formed from the emission spectrum of the gas. These spectra are called **absorption spectra**. We will consider two examples. The first is the absorption spectrum of sodium (Fig. 2.4) which is closely related to the emission spectrum of sodium vapour. The second example is that seen when the spectrum of sunlight itself is observed carefully. Dark lines corresponding to the absorption spectra of hydrogen and of another element can be seen. These lines are called Fraunhofer lines after the man who first noticed them. The second series of lines did not correspond with the spectrum of any element then known on earth (1868). The element was therefore called helium after the Greek *helios*, the sun. Helium was, of course, later discovered on earth by Ramsay in 1895. It is a measure of the fundamental importance of spectra that their study enabled an element to be discovered in the sun before it was discovered on earth. But for chemists, the evidence from spectra has led to even more important advances in the understanding of electron distributions in atoms.

A study of part of the spectrum of hydrogen (Fig. 2.5) shows that there are three groups of lines only one of which is in the visible region. Long before the

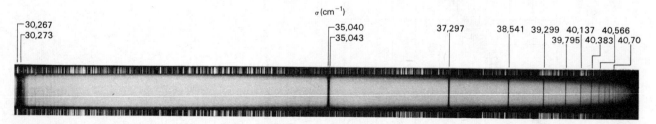

Fig. 2.4 Part of the absorption spectrum of sodium vapour. The figure is a print and absorption lines appear as dark lines. The wave number for each line is given in cm^{-1}. Above and below is seen the iron spectrum which is often used for calibration purposes in spectroscopy.

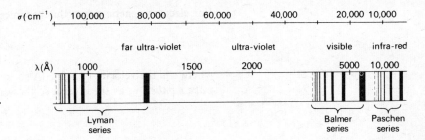

Fig. 2.5 Diagram of hydrogen atom spectrum.

significance of these lines for determining atomic structure was understood, a mathematical relationship between the wavelengths of the lines was discovered. The first series to be represented in this way was that in the visible region. It is called the **Balmer series** after Balmer (1885) who first fitted the lines to a mathematical formula. Mathematically, the series is very simple:

$$\frac{1}{\lambda} = R\left(\frac{1}{2^2} - \frac{1}{n^2}\right),$$

where λ is the wavelength, R is a constant known as the "Rydberg constant", and n is a whole number, in this case having values 3, 4, 5, etc.

The two other series are the **Lyman series** in the ultra-violet range for which

$$\frac{1}{\lambda} = R\left(\frac{1}{1^2} - \frac{1}{n^2}\right),$$

and the **Paschen series** in the infra-red range for which

$$\frac{1}{\lambda} = R\left(\frac{1}{3^2} - \frac{1}{n^2}\right).$$

For the Lyman series n has values 2, 3, 4, 5, etc. and for the Paschen series, 4, 5, 6, etc. What explanation is there for this remarkable regularity?

2.3.2 Particle and Wave Models

In order to understand the answer to this problem, we must look at two of the traditional models for understanding scientific phenomena. These are the particle and the wave models. The power of these two models is that they are expressed in terms which are familiar in our everyday lives, and which we can easily visualize. It is also incidentally their limitation. A golf ball, a stone, the moon, and the planets all behave in a similar way. They have mass, are subject to gravity, and their behaviour at low velocities can be explained or described by Newton's laws of motion. At velocities near the speed of light, different relationships apply, but the basic principles of the particle remain unaltered. The waves of the sea, the ripples on a pond, and sound waves also have a great deal in common. They can all be described in terms of wavelengths, amplitudes, and velocities (Fig. 2.6). For all waves we can relate the frequency, ν, to the wavelength, λ, and velocity, v:

$$\nu = \frac{v}{\lambda}.$$

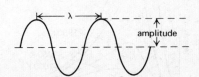

Fig. 2.6 Representation of a wave, giving definition of wavelength and amplitude.

For electromagnetic waves this becomes

$$\nu = \frac{c}{\lambda},$$

where c is the velocity of light. Also for electromagnetic waves we can define the *wave number*, σ, as the inverse of the wavelength:

$$\sigma = \frac{1}{\lambda}.$$

When we consider a phenomenon such as light, we have traditionally asked: is it a wave or a particle phenomenon? What we should ask is: which model is more appropriate or more useful? Light is light; it is in essence neither a particle nor a wave phenomenon. It can however usefully be described using both models. It has taken a long time to come to this conclusion and much effort was expended during the eighteenth and nineteenth centuries on arguing about which model represented a "true" description of light. One problem was that at first sight there seemed to be no relationship between wave and particle models. It was the bridging of the gap between the two models which led to the next great advance in understanding the atom.

2.3.3 The Quantum Theory

The first breakthrough was made by the German physicist, Max Planck, who proposed the quantum theory. Before discussing the theory one of the most important pieces of evidence supporting it must be described: it is the evidence of **black-body radiation**. Black-body radiation is the radiation emitted by a non-reflecting solid body. When a piece of iron, for instance, is heated it first radiates energy in the infra-red, non-visible, region. Anyone who has made the mistake of picking up a piece of metal which is hot, but not red hot, knows this to his cost. As the iron is heated it becomes progressively dull red, bright red, orange, yellow, and white hot. That is to say, the radiation emitted contains more radiation of shorter wavelengths. Figure 2.7 shows the relative intensities emitted at different temperatures. According to the nineteenth-century theory of Rayleigh and Jeans, there is no reason for the fall off in intensity towards shorter wavelengths. The black body ought to radiate over the whole range and as the body gets hotter the radiation emitted should get uniformly more intense. In other words the iron should radiate a little in the visible range even at room temperature. When the evidence of our eyes and fingers does not fit the predictions of the theory it is time for the theory to be modified. Planck's new theory was quite revolutionary. His suggestion led to the idea that energy could not be held by atoms, or other particles, in any arbitrary quantity, but only in specified amounts which he called **quanta**. According to this theory, a particle cannot change its energy continuously but only by a series of steps. This idea will later be shown to be consistent with the observed facts of black-body radiation. It is easier to understand why this is so if Einstein's extension to Planck's theory is also taken into account.

The phenomenon which Einstein was able to explain was the **photoelectric effect**. It had long been known that metals give off electrons when light shines on them. This can readily be observed when light shines on a clean metal surface in a

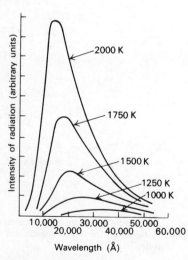

Fig. 2.7 Emission of radiation from a black body at different temperatures.

vacuum. There are interesting facts about this emission of electrons. The first is that no electrons are emitted unless the frequency of the radiation reaches a certain level, however intense the radiation is. This is like saying that something that happens depends not on how loud a sound is, but only on its pitch, which must be higher than a certain note. The second fact is that the kinetic energy of the electrons coming off the metal surface varies with the frequency of the radiation but not with its intensity. The higher the frequency the faster the electrons that come off are moving. Once again there was nothing in the nineteenth-century view of light as a wave phenomenon to explain all this. Einstein took Planck's quantum theory, which had been applied to the changes in energy of particles, and applied it to light and all other electromagnetic radiation. He proposed that energy is not only accepted and given out in quanta but also transmitted in quanta. The photoelectric effect is clear evidence that the energy of electromagnetic radiation is in some way dependent upon its frequency (or wavelength): the higher the frequency the greater the energy. Planck expressed the relationship between a quantum of energy, ϵ, and frequency, ν, as

$$\epsilon = h\nu,$$

where h is a constant known as **Planck's constant**. In the case of the electron emission of metals, the minimum frequency ν_0 required to produce quanta of energy sufficient to provide the work w to remove an electron from the metal surface is given in the expression

$$w = h\nu_0.$$

It follows from the quantum theory that electromagnetic radiation of a lower frequency than ν_0 will be unable to remove any electrons at all. Higher frequencies will simply result in the electrons having more than sufficient energy to leave the metal; they will move off faster. The phenomenon of black-body radiation can also be explained by quantum theory. According to this theory radiation is emitted by oscillators (atoms or groups of atoms) giving up energy quanta ϵ and emitting radiation of frequency ν, related to ϵ through Planck's equation

$$\epsilon = h\nu.$$

The energy of the black body is distributed among the oscillators according to Boltzmann's law (Section 6.8). The number of oscillators with energy ϵ is proportional to $e^{-\epsilon/kT}$. With Planck's assumption this means that the number is proportional to $e^{-h\nu/kT}$. This fraction falls off as ν increases and so the number of oscillators capable of emitting a frequency ν falls off rapidly with increasing ν. Consequently the intensity of black-body radiation falls off rapidly at high frequencies, that is towards shorter wavelengths. Also, the fraction will increase if the temperature of the black body is raised, that is more radiation of shorter wavelengths will be emitted under these conditions.

The significance of the expression $\epsilon = h\nu$ should be considered carefully. ϵ is largest when the frequency is highest. The spectrum of electromagnetic radiation and its associated energy is given in Fig. 2.8. On one end of the scale are the energy rich and highly dangerous γ-rays, while on the other are the comparatively harmless low energy long wavelength radio waves. It can be seen that an actual energy (expressed in kilojoules per mole or any other energy units) is associated with a particular wavelength of electromagnetic radiation. The total energy of

Spectrum		microwave			infra-red			visible	ultra-violet		X-rays	
Wavelength	(cm)	10	1	10^{-1}	10^{-2}	10^{-3}	10^{-4}	10^{-5}	10^{-6}	10^{-7}	10^{-8}
	(Å)	10^9	10^8	10^7	10^6	10^5	10^4	10^3	10^2	10	1
Frequency (Hz)		3×10^9	3×10^{10}	3×10^{11}	3×10^{12}	3×10^{13}	3×10^{14}	3×10^{15}	3×10^{16}	3×10^{17}	3×10^{18}
Wave number (cm^{-1})		10^{-1}	1	10	10^2	10^3	10^4	10^5	10^6	10^7	10^8
Energy (kJ mol^{-1})		1.2×10^{-3}	1.2×10^{-2}	1.2×10^{-1}	1.2	1.2×10	1.2×10^2	1.2×10^3	1.2×10^4	1.2×10^5	1.2×10^6

Fig. 2.8 Electromagnetic spectrum.

electromagnetic radiation is dependent both on frequency and the number of quanta. The size of the quanta are determined by the frequency. More intense radiation of the same frequency has more quanta of the same size. This must be borne in mind when relationships between energy and frequency of vibration are considered.

Perhaps the most remarkable fact about Planck's theory is that it provides the link between "particle properties" and "wave properties", or more accurately the particle model and the wave model. The relationship between energy, E, and mass was given by Einstein as

$$E = mc^2,$$

where m is mass and c is the velocity of light. Incorporating Planck's relationship we have, for a quantum of radiation of frequency, ν,

$$mc^2 = h\nu,$$

and hence

$$mc = \frac{h\nu}{c},$$

but

$$c = \nu\lambda$$

therefore

$$mc = \frac{h}{\lambda}.$$

If the idea can be extended to a particle with velocity v, as was suggested by de Broglie, then

$$mv = \frac{h}{\lambda},$$

which is an expression relating the momentum of a particle to its equivalent wavelength. The equation in fact proves to be of great importance. Rearranging this equation we have the wavelength

$$\text{"wave property"} \rightarrow \lambda = \frac{h}{mv} \leftarrow \text{"particle property"}$$

ATOMIC STRUCTURE

Put differently we can say that all momenta can be expressed either in terms of mass and velocity or in terms of wavelength (or frequency). In all energy phenomena either the wave or the particle model can be used. Which model we use depends only on which is most useful. As has been said, it was Einstein who first showed that the particle model could be usefully applied to light. He suggested that light could be described as a stream of particles, called "photons", whose momentum is

$$mv = \frac{h}{\lambda} = \frac{h\nu}{c}.$$

Later (in 1923) the same sort of reasoning was applied to a stream of electrons by de Broglie. Cathode rays had previously been described in terms of a stream of particulate electrons. De Broglie used the expression $\lambda = h/mv$ to find a wavelength for a stream of electrons. The wavelength depends on the velocity of the electrons (as does their momentum) and a typical electron beam, with an accelerating potential of about 200 volts, is found to have a wavelength of about 1 Å. A striking application of this is the electron microscope which has proved to be a most effective scientific instrument in recent years.

It is important to see that there are considerable differences between a stream of electrons and a light beam. All electromagnetic radiation travels in space with the same velocity; electron beams travel with variable velocity according to their energy. Nevertheless, in both cases, both the particulate and wave models may be used to describe their properties. Both models are useful but it would be a mistake to think of particles moving along in some kind of wave motion. We can still think in terms of a "particle" model. The "wave properties" imply that the probability of finding the particle at a given point can be determined by a mathematical relationship similar to that used for describing wave motion.

2.3.4 Quantum Theory Applied to the Hydrogen Atom

The quantum theory was first applied to the spectrum of the hydrogen atom by the Danish physicist, Niels Bohr, who published his results in 1913. His model of the hydrogen atom, called the "Bohr atom", has certain drawbacks but it does successfully explain the hydrogen atom spectrum. As a model it is not often used now but we will consider it carefully as it is helpful to the further understanding of atomic structure and because it is an excellent example of the usefulness and limitations of a model in science.

Bohr proposed that the electrons in the hydrogen atom could only exist in certain defined energy states. He suggested that the electron could move from a lower to a higher energy state and vice versa by absorbing or emitting electromagnetic radiation. It follows from Planck's hypothesis that if the energy levels for the electron in the hydrogen atom are defined then the frequencies of the radiation emitted or absorbed are also defined. That is to say hydrogen can only emit or absorb electromagnetic radiation of certain frequencies. This explains the remarkable line spectrum of hydrogen (and other elements). Furthermore, Bohr was able to show that each line in the spectrum was related to a specific change in energy of the electron. In the Bohr atom the electron is envisaged as rotating round the nucleus in a circular orbit. The potential energy of the electron increases as it is drawn further from the nucleus. Work has to be done against the electrostatic attraction between the positively charged proton and the negatively charged electron. But because of the quantized nature of energy, only a quantum of energy of a certain size can be absorbed by the electron. Hence there is a set series of radii at which the electron can orbit the nucleus. The orbits are given numbers, $n = 1, 2, 3, 4$, etc. The smallest radius represents the lowest energy the electron can have and is called the **ground state**. Now consider what happens when the atom is made to absorb energy. This may be done by heating to very high temperatures by electrical discharge or by other forms of bombardment. The electron moves from its ground state ($n = 1$) up to a higher energy level, an orbit of larger radius. It will then tend to fall back to a lower energy level, an orbit of smaller radius, giving off energy in the form of electromagnetic radiation of a particular wavelength, a quantum of energy. If the electron falls back to the ground state, we can see that the energy emitted, ΔE, could be ΔE_1, ΔE_2, ΔE_3, ΔE_4, etc., depending upon whether the electron falls from $n = 2$ to $n = 1$, $n = 3$ to $n = 1$, $n = 4$ to $n = 1$, etc. The size of the fall will define the frequency of the radiation emitted as ν_1, ν_2, ν_3, ν_4, etc. The relationship is shown diagrammatically in Fig. 2.9. With a large number of atoms, all of which may radiate, radiation will

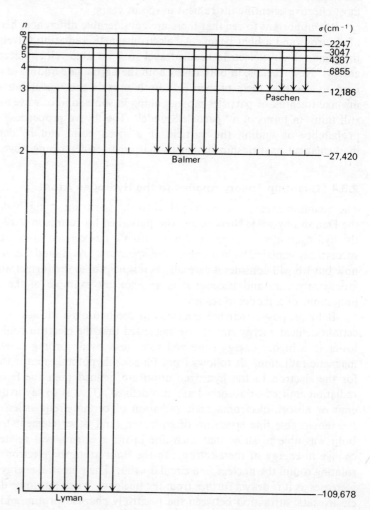

Fig. 2.9 Energy levels for the hydrogen atom and transitions giving rise to series. (The wave number for any line in the spectrum is given directly by subtraction.)

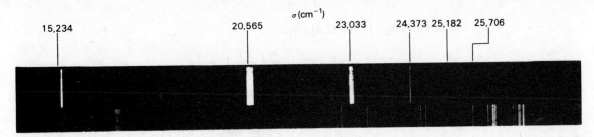

Fig. 2.10 Balmer series in the spectrum of the hydrogen atom. The figure is a print, and emission lines appear as light lines. The wave number for each line is given in cm^{-1}.

be emitted at all these frequencies and hence series of lines will be seen in the spectrum, each series resulting from a series of electron moves. The Lyman series is in the ultra-violet range and corresponds to the electron moving from higher levels down to $n = 1$. The Balmer series in the visible range represents the less energetic changes down to $n = 2$, and the Paschen series in the infra-red represents the changes down to $n = 3$. The Balmer series can be seen in Fig. 2.10, which shows part of the emission spectrum of the hydrogen atom.

2.3.5 The Mathematics of the Bohr Atom

Some readers may be interested to follow the relatively simple mathematics of the Bohr atom. For their sakes it is given below. Those who are less enthusiastic about mathematics may omit this section.

Two equations define the movement of the electron in Bohr's model of the atom. The first represents the application of Planck's hypothesis and has been discussed above. The energy change between two states is given by

$$\Delta E = h\nu, \tag{2.1}$$

where h is Planck's constant and ν is the frequency of the radiation absorbed or emitted.

The second equation defines the angular momentum of the electron:

$$mvr = n\frac{h}{2\pi}, \tag{2.2}$$

where m is the mass of the electron, v is its velocity, and r is the radius of its orbit, h is Planck's constant, and n a positive whole number (1, 2, 3, etc.). This was a postulate made by Bohr, namely that the angular momentum can have only certain definite values, given by Eq. (2.2).

Now if Newtonian mechanics are applied to the electron in orbit we can see that the centrifugal force must equal the electrostatic attraction between the electron and the proton:

$$\frac{mv^2}{r} = \frac{e^2}{r^2}, \tag{2.3}$$

where e is the charge on the electron (and the proton) in electrostatic units.

From Eq. (2.2):

$$v = \frac{nh}{2\pi mr}.$$

Substituting v in Eq. (2.3) gives:

$$\frac{m}{r}\left(\frac{nh}{2\pi mr}\right)^2 = \frac{e^2}{r^2}.$$

Simplifying this expression in terms of the radius gives:

$$r = \frac{n^2 h^2}{4\pi^2 m e^2} = n^2 a_0. \tag{2.4}$$

This expression defines the various possible radii of the electron orbits. Substituting the known values for h, π, m, and e, when $n = 1$ gives a value for r of 0.53×10^{-8} cm, about half an ångström. This distance is often called the **Bohr radius** and denoted by a_0.

To find the frequency of the radiation the total energy of the electron must be considered. This is the sum of its kinetic energy $\frac{1}{2}mv^2$ and its potential energy $-e^2/r$ (for hydrogen). The potential energy is negative if we take it to be zero when r is infinity, which is the usual convention. Hence the total energy,

$$E = \tfrac{1}{2}mv^2 - \frac{e^2}{r}.$$

Substituting for mv^2 from Eq. (2.3):

$$E = \frac{1}{2}\frac{e^2}{r} - \frac{e^2}{r}$$

$$E = -\frac{1}{2}\frac{e^2}{r}. \tag{2.5}$$

Substituting for r from Eq. (2.4):

$$E = -\tfrac{1}{2}e^2\left(\frac{4\pi^2 m e^2}{n^2 h^2}\right)$$

$$E = -\frac{2\pi^2 m e^4}{n^2 h^2}. \tag{2.6}$$

Now if we consider the energy change from one level E_1, corresponding to $n = n_1$, to another level E_2 corresponding to $n = n_2$, where n_2 is greater than n_1, the change of energy

$$\Delta E = E_2 - E_1$$
$$\Delta E = \frac{2\pi^2 m e^4}{h^2}\left(\frac{1}{n_1^2} - \frac{1}{n_2^2}\right), \tag{2.7}$$

but $\Delta E = h\nu$ from Eq. (2.1), so

$$\nu = \frac{2\pi^2 m e^4}{h^3}\left(\frac{1}{n_1^2} - \frac{1}{n_2^2}\right). \tag{2.8}$$

The most remarkable success of the Bohr atom is that substitution in this final equation gives values for the frequencies of the spectrum of hydrogen which correspond very accurately to those actually found. The Bohr model can also be used successfully to predict the spectra of other single-electron systems such as He^+ and Li^{2+}, both of course unstable ions in chemistry, but readily obtainable in electric discharges.

The limitations of the Bohr atom. The Bohr atom is outstandingly successful as a model for explaining the spectrum of hydrogen. It does not prove successful in explaining the spectra of other elements. It also has the disadvantage that it postulates an electron moving in a circular path, creating a changing electromagnetic field. According to classical theory, this should result in a loss of energy as radiation, causing the electron to slow down and be drawn in towards the nucleus. If this were to happen all matter would collapse. Fortunately it does not! In fact, further understanding of the nature of the atom depends on treating the electron not as a particle but as a wave. This idea is described in the next section.

2.3.6 Wave Description of the Electron

We have seen that a stream of electrons can be thought of in terms of a wave model. A wavelength can be attributed to it and we will see later that a stream of electrons can be used like a beam of X-rays to produce diffraction patterns (Section 3.1) and like a light beam as in the electron microscope. But what of the electrons in an atom? Where an electron is not completely free but restricted, an analogy can be made with a standing wave. A standing wave is one whose nodes do not move, as found for example in a vibrating violin string. In quantum theory, as we have seen, wavelength is very important for it determines the momentum of the system through the de Broglie relationship

$$mv = \frac{h}{\lambda},$$

and the kinetic energy follows from the equation

$$\tfrac{1}{2}mv^2 = \frac{h^2}{2m\lambda^2}.$$

A very important result emerges from these simple considerations. The shorter the wavelength associated with an electron the higher the kinetic energy. Now the smaller the space in which an electron is confined the shorter the associated wavelength becomes and the higher its energy.

In an atom there is also potential energy to consider. The attraction of the nucleus for the electron tends to draw the electron close to the nucleus leading to a low potential energy (that is, a larger negative value). On the other hand, the more closely the electron is confined the higher its kinetic energy. The lowest energy state of the hydrogen atom represents a compromise between these potential and kinetic energy considerations.

Mathematically the wave description of the electron is not simple and in any case it is a three-dimensional problem. The de Broglie relationship which applies to a freely moving electron was generalized by Schrödinger, whose equation can be applied to bound electrons as in atoms, where there are potential and kinetic energy terms. For the record, the Schrödinger equation which describes a particle in this way is

$$\frac{\partial^2 \psi}{\partial x^2} + \frac{\partial^2 \psi}{\partial y^2} + \frac{\partial^2 \psi}{\partial x^2} + \frac{8\pi^2 m}{h^2}(E - V)\psi = 0,$$

where ψ is the wave function, m is the mass of the particle, E is the total energy of the particle, V is the potential energy of the particle, and h is Planck's constant.

It is not easy to solve this equation but the significance of the solutions found will be easier to follow if the analogy of the vibrating string is used again. A wave equation can be written to represent the movement of a vibrating string. If such an equation is solved two results emerge: firstly there are solutions for only certain wavelengths; and secondly the solutions give values for the amplitude of the wave in particular places. Similarly, physically acceptable solutions to the Schrödinger equation exist only for certain energies, and the solutions also give amplitudes, providing a wave description for electronic systems. The wave description is consistent with the Uncertainty Principle (Section 2.3.8), according to which the position of an electron in a given energy state cannot be known exactly. The wave function, ψ, itself, has no physical meaning but the square of the wave function, ψ^2, which varies with the position in space, represents the probability of finding the electron in a particular place. Where ψ^2 is large, the probability of finding an electron is high. Ideally we need a three-dimensional graph. The problem of representing ψ^2 in this may be solved by breaking it down into two parts, one a radial part and the other an angular part. Figure 2.11 shows how ψ^2 varies with r, the distance from the nucleus of the hydrogen atom, when the electron is in its ground state. The probability can also be expressed in terms of **electron density**, that is the average number of electrons (a fraction) to be found per unit volume (usually a cubic ångström). Figure 2.11 shows that ψ^2 is finite from r equals zero to infinity, though in fact it falls off rapidly with increasing value of r. The description is quite different from that of the Bohr theory. A useful way of looking at the probability description is to indicate a volume, by a contour of constant ψ^2, within which there is, say, a ninety percent chance of finding the electron. For the hydrogen electron in its lowest energy state, the so-called ground state, this is done by drawing a sphere.

Wave functions describing single electrons are called **orbitals**. The name is intended to be similar to the term orbit used in Bohr theory but also to be different. We speak of an electron being in an orbital but this does not mean moving along a path. Instead we have a probability description of where the electron is likely to be found. While for the lowest energy level the orbital is spherical, this is not so of all orbitals. The electron for higher energy levels sometimes shows a directional

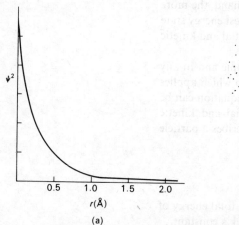

Fig. 2.11 The hydrogen atom 1s-orbital. (a) ψ^2 as a function of r. (b) Diagram in which density of dots shows how ψ^2 depends on r.

preference, the probability of finding the electron being higher along a particular direction. The energy levels will be discussed in greater detail in the next section.

The energy levels of the hydrogen atom. The evidence provided by the hydrogen spectrum leads to the idea that the hydrogen atom can exist in a number of different energy levels. In the Bohr model each energy level corresponds to a circular orbit at a set distance from the nucleus. The lowest energy level is that which is nearest to the nucleus. A change in energy of the electron may be brought about by the absorption or emission of energy according to whether the energy of the electron goes up or down. The frequency of the electromagnetic radiation absorbed or emitted depends, according to Eq. (2.1), on the size of the energy change. The lines in the Balmer series of the hydrogen spectrum can be linked to the energy levels of the hydrogen atom as shown in Fig. 2.12. The link between the energy of an electron and its distance from the nucleus is not so simple in the probability model of the atom as in the Bohr model (Eq. 2.5). Even so, it is true to say that the average distance of the electron from the nucleus increases as the energy levels become higher.

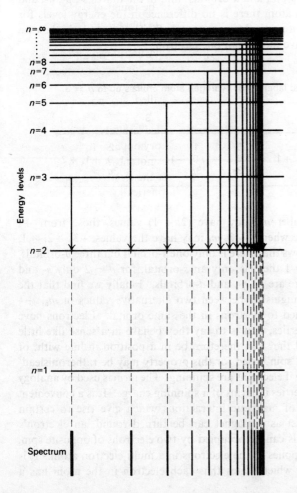

Fig. 2.12 Energy levels and transitions showing origin of the Balmer series.

2.3.7 Quantum Numbers

The simple picture of hydrogen atom levels given in Fig. 2.9 does not give the whole picture. The solution to the Schrödinger equation defines the orbital which an electron can occupy in terms of four **quantum numbers**. The first, called the *principal quantum number*, n, corresponds to the numbering of the energy levels. The second, called the *angular momentum quantum number*, l, represents a subdivision of the states for a given principal quantum number. The third and fourth need not concern us for the moment, but are known as the *magnetic quantum number*, m_l, and the *spin quantum number*, m_s. The principal quantum number, n, can have values of integers 1, 2, 3, 4, and so on; l can have values limited by the value of n of 0, 1, 2 ... up to $n - 1$. Thus if $n = 1$, $l = 0$ and if $n = 2$, $l = 0$ or 1. The subdivision is described by a terminology which stems historically from characteristics of certain spectral lines which were labelled "sharp", "principal", "diffuse", and "fundamental", and led to the notation s, p, d, and f. s-orbitals are those for which $l = 0$, p-orbitals those for which $l = 1$, d-orbitals those for which $l = 2$, and f-orbitals those for which $l = 3$. It can be seen from this, that in the principal quantum "shell" $n = 1$ there is only an s-orbital; in the second shell $n = 2$ there are s- and p-orbitals; in the third s-, p-, and d-orbitals; and in the fourth s-, p-, d-, and f-orbitals. For the hydrogen atom there is no difference in the energy levels for the s-, p-, d-, and f-orbitals in a particular quantum shell. This is not so in the case of orbitals in atoms which have more than one electron.

Table 2.2. Quantum numbers and labelling of hydrogen atom orbitals up to $n = 3$.

n	1	2		3		
l	0	0	1	0	1	2
m_l	0	0	$-1, 0, +1$	0	$-1, 0, +1$	$-2, -1, 0, +1, +2$
Label	$1s$	$2s$	$2p$	$3s$	$3p$	$3d$

The third quantum number m_l may have $(2l + 1)$ values, those from $+l$ through 0 to $-l$; for example when $l = 3$, m_l may have the values $+3, +2, +1, 0, -1, -2, -3$. Table 2.2 shows that there is only one s-orbital but three p-orbitals, and five d-orbitals. For $n = 1$ there is only an s-orbital, for $n = 2$ only s- and p-orbitals, and for $n = 3$ there are s-, p-, and d-orbitals. Finally we find that the fourth quantum number distinguishes between two alternative values of m_s, $+\frac{1}{2}$ and $-\frac{1}{2}$, which can be assigned to electrons in the same orbital. Electrons have electrical and magnetic properties, that is to say they behave in a sense like little magnets. In a magnetic field they can therefore be in a position in line with, or against, the field. The word "spin" used for this property may be rather misleading. We have no way of seeing if electrons are spinning. The term is used by analogy to account for magnetic properties in terms of a spinning charge. It is a convenient way of regarding the origin of magnetic interactions which give rise to certain fine structure in spectra. Just as a magnet can be turned round, an electron's spin can be reversed. Orbitals can be occupied by two electrons of opposite spin.

An important principle applies to the electrons in a multi-electron atom. It is the **Pauli exclusion principle**, which states that each electron in the atom has a unique set of quantum numbers. No two electrons in an atom can have all four quantum numbers identical. Whatever the state of the atom and whatever its combination with other atoms, this principle is always found to hold. The orbitals occupied by electrons in some simple atoms are summarized in Table 2.3; a comprehensive set is given in Table 2.6. The shorthand form, a "1s-electron" is often used, meaning an electron in a 1s-orbital. All electrons are of course indistinguishable. The arrows represent spin. The notation $2p_x$, $2p_y$, $2p_z$ is explained in Section 2.3.8.

Table 2.3 Electron configurations of elements 1 to 5. A pair of boxes is used here to represent an orbital.

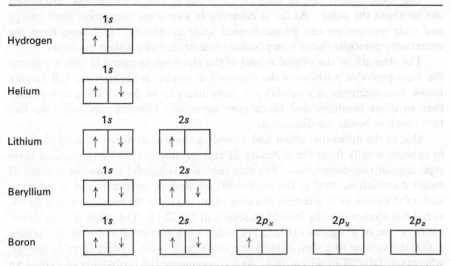

All this is based on the solution of the Schrödinger equation for the hydrogen atom. It can be extended usefully to atoms with more than one electron even though the actual orbitals for electrons in such atoms can never be directly worked out algebraically. We think of electrons in such atoms as occupying orbitals with quantum numbers as for the hydrogen atom even though repulsion between electrons means that the simplicity of the hydrogen atom is lost. One of the consequences of making this extension is that it provides a scheme which rationalizes a lot of knowledge related to chemical bonding. Multi-electron atoms are discussed in Section 2.5.

2.3.8 The Spatial Distribution of Electrons

In the previous section we stated that the square of the wave function (ψ^2) represents the probability of finding an electron of a given energy level at a particular point in space. The Bohr atom is a useful model for relating the energy changes in an atom to the wavelength of transmitted and absorbed quanta of energy. It tells us nothing of value about the distribution of the electron in space. As has already been mentioned, the picture of the electron rotating in an orbit round the nucleus is no more than a useful fiction. Bohr himself understood this and it has long been known that Newtonian mechanics, successfully applied to the movement of the planets, did not succeed in predicting the motion of electrons. Indeed it is not possible to predict *both* the position and the momentum of an electron (or any other particle for that matter) exactly. This hypothesis was first put into a precise form by Heisenberg in 1925 and is called the **Heisenberg uncertainty principle**. The principle states that the uncertainty of momentum (Δp) times the uncertainty of position (Δq) is equal to a constant. The constant is Planck's constant divided by 2π. Expressed as an equation this becomes

$$\Delta p \times \Delta q = \frac{h}{2\pi}. \qquad (2.9)$$

The more exactly one of the two quantities is measured the less certain one can be about the other. As far as electrons in atoms are concerned, their energy and their momentum can be determined quite accurately. It follows from the uncertainty principle that it is impossible to be precise about their positions.

The strength of the orbital model of the electrons in atoms is that it predicts the most probable position of the electrons in space. It suggests, as will be seen below, that electrons in *p*-orbitals are more likely to be found along certain axes than in other positions and hence goes some way towards explaining the fact that covalent bonds are directional.

One of the difficulties about understanding the spatial distribution of electrons in orbitals results from the difficulty of representing them in any physical form especially in two dimensions. For this reason it is helpful to use the concept of radial distribution, that is the probability of finding an electron in a spherical shell of thickness dr at a certain distance r (radius) from the nucleus. Consider the radial distribution for the 1*s*-orbital, shown in Fig. 2.13. The graph is quite different from that of ψ^2 against r (Fig. 2.11), which has a maximum at $r = 0$. The reason is that the volume of a shell of thickness dr increases with r, being given by $4\pi r^2 dr$, whereas ψ^2 falls off rapidly with r. As a consequence the probability of finding an electron at a distance r, irrespective of direction, has a maximum for some value of r. From the graph in Fig. 2.13, we can make a number of important points. The probability of finding the electron increases steeply with distance from the nucleus up to a maximum and then falls away gradually. The distance of maximum probability is equal to the Bohr radius, a_0. Strictly speaking there is some probability of finding the electron at great distances from the nucleus, but it is negligibly small.

In the case of the 1*s*-electron the distribution is symmetrical around the nucleus. The probability of finding the electron is the same for a given distance at all angles. How can we represent this in three dimensions? As the distribution is symmetrical, a sphere is the most suitable figure, but this sphere has no exact radius. The Bohr radius will not do as the electron is likely to be found outside the sphere of Bohr radius or inside it. In practice a radius is chosen so that the sphere contains the volume within which there is a ninety percent chance of finding the electron.

The case of the 2*s*-electron is similar, but also shows a different feature. From Fig. 2.13 it can be seen that there are *two* maxima and a minimum between them. This means that there are two radii of maximum probability and one radius at which there is no chance at all of finding the electron. A good way of representing the volume within which there is a ninety percent certainty of finding the electron is by means of a sphere with a thick outer "shell" around it. Between the sphere and

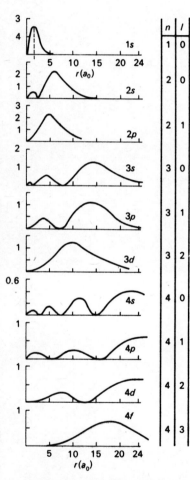

Fig. 2.13 Radial probability density for some orbitals of the hydrogen atom. All graphs have the same horizontal scale, r, in units of the Bohr radius, $a_0 = 0.529$ Å. The scale of the vertical axis has been varied to give plots of convenient size.

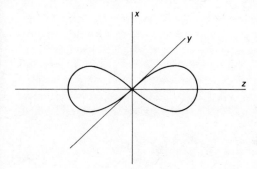

Fig. 2.14 Angular dependence of a 2p-orbital. The 2p_z-orbital is shown.

the shell is a space which represents the volume where there is little or no possibility of finding the electron.

Electrons in p-orbitals are a little more difficult to describe as the probability of finding them is not spherically symmetrical around the nucleus. From the radial distribution graph for the 2p-orbital in Fig. 2.13 it will be seen that it is rather more drawn out than the 1s-orbital and, unlike the 2s, has only one peak. Unlike an s-orbital, a p-orbital is not spherical; for a p-orbital the probability of finding an electron is concentrated along one direction. The angular dependence is shown in Fig. 2.14. For the p-orbital illustrated, there is a high probability of finding the electron along the x-direction, the probability falling off for points removed from this axis and becoming zero for the yz-plane. The figure is a three-dimensional one, and to visualize it one must imagine the diagram to be rotated about the x-axis by 360°. An s-orbital has spherical symmetry; a p-orbital has axial symmetry.

The three p-orbitals are at right angles to each other. The 2p_x-, 2p_y-, and 2p_z-orbitals, as they are called, can all be represented as in Fig. 2.15.

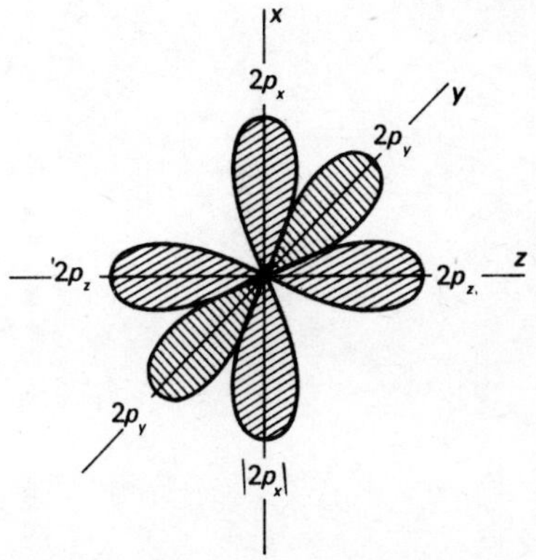

Fig. 2.15 The three 2p-orbitals are at right angles to one another.

2.4 ENERGY LEVELS OF ATOMS OTHER THAN HYDROGEN

2.4.1 The Evidence of Ionization Energies

The convergence of the lines in the hydrogen spectrum (Fig. 2.10) shows that there is a limit to the energy which can be put into the electron as part of a hydrogen atom. Eventually, when the energy absorbed by the electron reaches the limit shown by the converging lines of the spectrum, the electron leaves the atom altogether. The atom then becomes a positively charged ion. The energy required to remove an electron from an atom (to infinity) is called the **ionization energy**. From the hydrogen spectrum we find that the convergence limit of the series in the ultra-violet region of the spectrum (the Lyman series) is 109,678 cm^{-1}. This corresponds to an ionization energy of 1315 kJ mol^{-1}. Similarly the ionization energy of the sodium atom can be obtained from the convergence of the series shown in Fig. 2.4.

There are other methods of measuring ionization energies. A valve like that in Fig. 2.16 is filled with the vapour of an element at low pressure. Electrons are accelerated from the heated cathode towards the grid. The anode beyond the grid

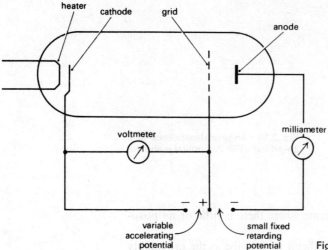

Fig. 2.16 Apparatus for electron impact experiments.

is at a small potential opposed to that of the grid. To put it in simple terms, the electrons accelerate towards the grid but slightly decelerate between the grid and the anode. The current between the cathode and anode is measured. The voltage between the cathode and the grid is raised and the current noted. At first the current rises, as would be expected, as the voltage rises. But then the rise in current ceases and at a certain voltage actually decreases. After this, the current rises again. The explanation for this phenomenon is that the electrons are at first deflected but not slowed down appreciably by hitting the atoms or molecules of the vapour in the valve. But as the energy of the electrons increases there comes a stage when they have sufficient energy to knock the electrons in the atoms up a few energy levels and finally to knock them off altogether so that ions are formed. This process absorbs sufficient energy from the electrons to prevent them from passing from the grid to the anode. In using this method care has to be taken to distinguish between collisions which lead to an atom being in an excited state (electrons in higher energy levels) and collisions which lead to ionization.

This method, called the **electron impact method**, has been further developed in recent years by the use of the mass spectrometer (Section 2.2). The voltage across the ionization chamber can be varied and the ions produced then analysed by the mass spectrometer. In this way ionization energies of different species including fairly unstable molecules have been found. This method can also be used to find the second, third, and further ionization energies of atoms.

Yet another method of determining ionization energies is to pass electromagnetic radiation, infra-red, visible, or ultra-violet as necessary, through the vapour of an element at low pressure. The wavelength is selected by the rotation of a prism or diffraction grating until the electromagnetic radiation reaches a wavelength of sufficient energy to ionize the vapour. Ions are detected in this case by measuring the current between two plates fixed in the vapour at right angles to the direction of the light ray. A high potential is placed across the plates and a current begins to flow as soon as ions are produced in the vapour. The ionization energy is obtained from the Planck relationship.

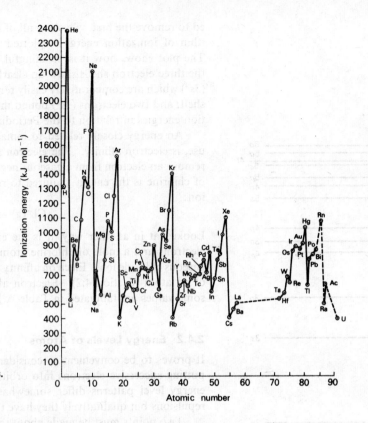

Fig. 2.17 First ionization energies of the elements.

The energy required to remove the first electron from an atom in the gaseous state is called the **first ionization energy**. The energies required to remove further electrons are called the second, third, etc. ionization energies as appropriate. Ionization energies inevitably become larger as each successive electron has a greater excess charge in the nucleus to hold it. It is shielded less from the nucleus by other electrons.

A graph of first ionization energies (Fig. 2.17) shows that they follow a periodic relationship. The inert gases have the highest first ionization energies, and the alkali metals the lowest. It would be a mistake to carry the analogy too far but it is interesting that the noble gases are the least reactive of the elements and the alkali metals amongst the most reactive. Ionization energies of the elements are given in Table A.10.1. Note that electrons with the smallest potential energy are those which are farthest from the nucleus and these are the ones which are most easily removed. They have the lowest ionization energy.

It is clear from the table that the differences between successive ionization energies in the same atom are not constant. Take magnesium for instance; there is a big jump between the second and third ionization energies. Similarly, with carbon there is a big difference between the fourth and fifth. With other atoms there are similar big jumps. It is not difficult to see that these jumps occur when electrons in an energy level of a lower principal quantum number are being removed. Successive ionization energies for magnesium are plotted in Fig. 2.18. The energy needed to remove the last electron is over two hundred times that need-

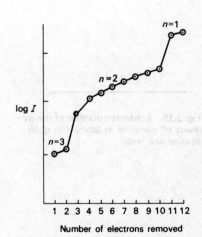

Fig. 2.18 Successive ionization energies for magnesium.

2.4 ENERGY LEVELS OF ATOMS OTHER THAN HYDROGEN

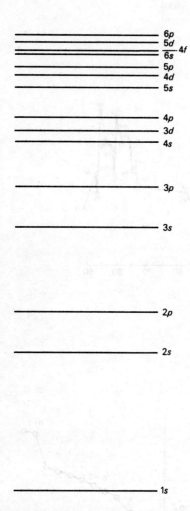

Fig. 2.19 Schematic diagram of energy levels of electrons in atoms (for qualification see text).

ed to remove the first. To show all of the successive ionization energies, the logarithm of ionization energy is plotted against the number of electrons removed. The plot shows how it is meaningful to talk of electron shells. For magnesium the three electron shells show up clearly: there are two electrons in the outer shell ($3s^2$) which are comparatively easily removed; eight electrons ($2s^2 2p^6$) for the $n = 2$ shell; and two electrons ($1s^2$) bound much more tightly in the $n = 1$ shell. Ionization energies in relation to the Periodic Table are discussed in Section 10.3.3.

An energy closely related to ionization energy, which we shall often want to use, is electron affinity. The **electron affinity** is the amount of energy required to remove an electron from a gaseous negative ion. For example, the electron affinity of chlorine is the energy required to remove an electron from a gaseous chloride ion:

$$Cl^-(g) \rightarrow Cl(g) + e^-.$$

Looked at in another way, it is the energy that would be liberated if a mole of electrons and a mole of chlorine atoms were brought together to form a mole of gaseous chloride ions. Electron affinity is a measure of the electron-attracting power of an atom (Section 4.2). Electron affinities are discussed in Section 10.3.4 and some values are tabulated in Table A.10.2.

2.4.2 Energy Levels of Atoms

It proves to be convenient to consider multi-electron atoms as being built up by putting successive electrons into orbitals like those for the hydrogen atom. The energy level patterns differ somewhat from atom to atom because of electron repulsions but qualitatively they have the form shown in Fig. 2.19.

Two points must be made about the energy level diagram. The first is that the size of the atom varies with the element. To quote an obvious example, the electron occupying the $1s$-orbital in, say, potassium, will be much closer to the nucleus (will have a lower potential energy, i.e. a larger negative value) than that of hydrogen. This is because the potassium nucleus contains 19 protons which attract the electron much more strongly than the one hydrogen proton does. The second point is that the actual order of energy levels can vary between elements. In the general diagram, Fig. 2.19, the $3d$-level is shown to be above the $4s$-level. This is the case for elements before scandium in the Periodic Table, but not for those after scandium. In one sense, neither of these points will make a great difference to the introductory understanding of this topic, but it would be most misleading to think of energy levels as in some way fixed like shelves on a wall. The ideas are taken up in more detail in the next section.

2.4.3 Penetration

1) Single-electron systems. Figure 2.13 shows the radial probability curves of the lowest energy orbitals of the hydrogen atom. It is obvious from these that *on average* an electron in a $3s$-orbital (called $3s$-electron for short) is farther away from the nucleus than a $2s$-electron. Therefore the $3s$-electron has a higher energy than a $2s$-electron, since more energy can be obtained by the "thought experiment" of allowing the $3s$-electron to fall into the nucleus than by allowing the $2s$-electron to fall into the nucleus. Similarly, energy must be provided to "excite" the $2s$-

electron into the 3s-orbital, and energy is emitted if the 3s-electron "drops" into the 2s-orbital. One feature of the mathematical treatment of these electron energies, that can be confusing at first sight, is that the maximum energy, where the electron is completely removed from the nucleus (at "infinity" as it is said), is given the value zero, and therefore all real energies have negative values. So when a system increases its energy by excitation of an electron into a higher orbital, say from the 2s- to the 3s-orbital, this is described by its energy becoming less negative rather than more positive as you might expect.

Now compare orbitals for the same value of n (say, 3) but different values of l. We see from Fig. 2.13, that there are different radial probability curves for 3s-, 3p-, and 3d-orbitals. This variation is very important for some points we are going to consider shortly, but does not, however, affect the energy for one-electron systems. This can be shown mathematically and can be looked at in the following way. In the Bohr theory the energy is simply obtained by halving the potential energy (Eq. 2.5):

$$E = -\frac{1}{2}\frac{e^2}{r}.$$

The reason for this, as the derivation of Eq. (2.5) shows, is that the kinetic energy is half the potential energy, but of opposite sign. This still holds in quantum mechanics but, by contrast with the Bohr theory, r no longer has a fixed value. In calculating the energy an average has to be taken over the probability distribution. For a one-electron, hydrogen-like system, consisting of a central charge Z, and one electron in an orbital with quantum numbers n and l, the average potential energy is given by

$$V_{av} = -Ze^2 \left(\frac{1}{r}\right)_{av}, \tag{2.10}$$

where $(1/r)_{av}$ is the average value of $1/r$. This can be obtained for each orbital using the mathematical expression for the radial distribution function. (This can be found in many standard works, but you could take it on trust at this stage!) The result is that

$$\left(\frac{1}{r}\right)_{av} = \frac{Z}{n^2 a_0} = \frac{Z}{n^2}\frac{4\pi^2 m e^2}{h^2}, \tag{2.11}$$

where a_0 is the Bohr radius (Eq. 2.4).

The important result is that the expression is dependent only on n; it is independent of l. From Eqs. (2.10) and (2.11) we have:

$$V_{av} = -Ze^2 \left(\frac{Z 4\pi^2 m e^2}{n^2 h^2}\right) = -\frac{4\pi^2 m Z^2 e^4}{n^2 h^2}. \tag{2.12}$$

Remembering that the total energy is one half of the average potential energy we have for the energy levels:

$$E_n = \tfrac{1}{2} V_{av} = -\frac{2\pi^2 m Z^2 e^4}{n^2 h^2}. \tag{2.13}$$

This is the same expression as Eq. (2.6), obtained by the Bohr theory, for the special case of $Z = 1$. The energy levels depend only on n and not on l, but we have one

2.4 ENERGY LEVELS OF ATOMS OTHER THAN HYDROGEN

piece of information, from Fig. 2.13, not obtainable from the simple Bohr model. This is that an *ns*-electron is more likely to be found very close to the nucleus than an *np*-electron (same value of *n*), which in turn is more likely to be found closer to the nucleus than an *nd*-electron. It is said that *s*-electrons *penetrate* closer to the nucleus than *p*-electrons (with same value of *n*), and so on. This is very important in connection with multi-electron systems (all atoms except hydrogen!).

2) Multi-electron systems. The methods of quantum mechanics have not yet successfully solved any multi-electron problem by exact algebraic means. This emphasizes the need for a useful model for these systems; such a model is developed below. It is assumed that the orbital description of electrons which is successful for hydrogen, can be qualitatively extrapolated for all other atoms, that is: electrons are associated with atomic orbitals; each orbital is described by a set of quantum numbers which are the same as those used in hydrogen; and the quantum numbers have similar "physical meanings" (*n* is important in determining the energy of the orbital, *l* correlates with shape, m_l with spatial direction, and m_s gives the spin). There are, however, important differences between single and multi-electron systems: whereas the angular distribution of electrons (the shape of the orbitals) in multi-electron systems is the same as in single-electron systems, the radial distribution (or how far the orbitals extend in space) differs (and cannot be calculated exactly), and therefore the energy of orbitals in multi-electron systems differs (and cannot be calculated exactly). But the most important difference is that the energy of an orbital in a multi-electron system now depends on the value of *l* as well as *n* (Fig. 2.19). Thus an electron in a 2*p*-orbital in a multi-electron system has a higher energy than a 2*s*-electron.

For a hydrogen-like system, consisting of a nucleus of charge *Z* with one orbital electron, the energy is described by Eq. (2.13); it depends upon the nuclear charge *Z* and $(1/r)_{av}$. Consider bringing up further electrons (assuming that *Z* is very large) into the available orbitals, so that each additional electron remains *completely* outside the previously added electrons, and the previously added electrons (say *N* in number) are a spherically symmetrical smear of negative charge surrounding the nucleus and shielding the outside (*N* + 1) electron from the charge on the nucleus. Then the energy of the (*N* + 1) electron (see Fig. 2.20) would be given by

$$E = -\frac{(Z-N)e^2}{2}\left(\frac{1}{r}\right)_{av}.$$

According to this picture the order of energy levels in a given atom would be the same as that in hydrogen, and the energy of each orbital would depend only on the quantum number *n* (or the associated average value $(1/r)_{av}$). This does not correspond even approximately to the real situation. The added outer electrons do not remain outside the cloud of previously added electrons, but penetrate this cloud to a greater or lesser extent depending on the two quantum numbers *n* and *l*. The more the "outer" electron penetrates the "inner" core of electrons, the greater the positive charge of the nucleus this electron "feels", and the lower its energy (Fig. 2.20). Thus whereas the energy of an electron is still largely determined by the value of *n* and the associated average value $(1/r)_{av}$, this is modified by the value of *l* which describes not only the shape of the orbital, but also the penetrating power of the electron in the orbital. For a given value of *n* the penetration decreases in the order

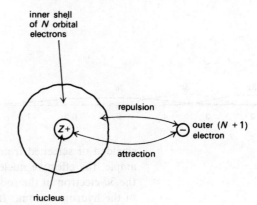

Fig. 2.20 A model of forces felt by an outer electron.

$s > p > d$, and so on, and since s-electrons have the greatest chance of being found near the nucleus, they have the lowest energies of electrons in the same shell. Thus in a given quantum shell (given value of n) the increasing order of orbital energies is ns, np, nd, nf, etc., and the energy of an electron in a multi-electron atom is determined, not only by n, but also by l (Fig. 2.19).

The mathematical model described is easily changed to accommodate this extension of the model. Instead of the term Z in the equations describing the one-electron system (or $Z - N$ in the multi-electron), a term Z_{eff} is included, the **effective nuclear charge**, so that the energy of an outer electron in a multi-electron system is given by

$$E = -\frac{Z_{eff} e^2}{2} \left(\frac{1}{r}\right)_{av}. \qquad (2.14)$$

Now the energy depends not only on $(1/r)_{av}$ (or n) but also on Z_{eff}. Z_{eff} is not the nuclear charge but should be regarded as the charge that a particular electron actually feels. This cannot be the same as the nuclear charge, because of the screening effect of the inner electrons. Nor can it be obtained simply by subtracting the number of inner electrons from the nuclear charge, since this assumes perfect shielding (see below) of the nucleus for the outer electrons by inner electrons. This is not the case, since different electrons penetrate the inner electron distribution to different extents, and different electrons shield the nucleus from outer electrons to differing extents. In fact the effective nuclear charge, Z_{eff}, depends upon: the nuclear charge Z; the net charge on the ion; the number and type of the inner screening electrons; and the type of outer electron being considered. This is discussed further in Sections 2.5 and 10.3.3.

The validity of this model is confirmed by the approximate calculation made of the radial distribution of electron density in the sodium atom (Fig. 2.21). This shows the distribution of electrons in the inner shells, "core electrons", and the $3s$-electron (ground state), and what the distribution would be if the $3s$-electron were excited to the $3p$- or $3d$-orbitals (excited state). The only difference between the three systems (described in the notation of Section 2.5 as $1s^2 2s^2 2p^6 3s^1$, $1s^2 2s^2 2p^6 3p^1$, and $1s^2 2s^2 2p^6 3d^1$) is the changing orbital of the outer electron. Clearly, in the excited state, $1s^2 2s^2 2p^6 3d^1$, the $3d$-electron spends virtually all its time far from the nucleus and well outside the regions of space where the central core of ten electrons spend most of their time. It is said that this electron is well

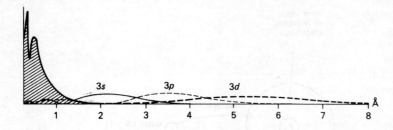

Fig. 2.21 The radial distribution of electron density in the sodium atom in its ground state and excited states. The shaded area represents the core electrons. The distribution of the valence electron when it is in the 3s-, 3p-, and 3d-orbitals is also shown. From B. H. Mahan, *University Chemistry* (second edition), Addison-Wesley (1969).

shielded or screened from the nucleus by the inner electrons. Indeed, in this example, the effective nuclear charge felt by this electron, $Z_{\text{eff}} \approx 11 - 10 = 1$, and the 3d-electron in the sodium atom has almost the same energy as the 3d-electron of the hydrogen atom. In the excited state $1s^2 2s^2 2p^6 3p^1$ however, the 3p-electron spends an appreciable amount of its time inside the inner core of electrons (it is said to "penetrate" the inner core) and therefore feels an effective nuclear charge, Z_{eff}, greater than one; thus the 3p-electron has a lower energy than the 3d-electron. The 3s-electron penetrates the inner core even more, feels an even greater effective nuclear charge, and has an even lower energy than the 3p-electron.

The shielding effect of inner electrons depends not only on their number and principal quantum number n, but also on the value of l. Thus the degree of penetration of an outer electron depends not only on the type of outer electron, but also on the type of inner electron. Compare the filled subshells ns^2 and np^6. The s-electrons spend more time close to the nucleus than do the p-electrons, and therefore it is more difficult for an outer electron to penetrate them than to penetrate the filled p^6-subshell; in other words the s-electrons shield the nucleus more successfully. Hence the shielding efficiency of inner electrons decreases in the order $s > p > d > f$, or conversely the tendency for inner electrons to be penetrated by outer electrons increases in the order $s < p < d < f$. This effect is enhanced by the fact that an s-subshell contains only two electrons, a p contains six, and a d contains ten. Thus the relatively easy penetration of a d-subshell by an outer electron reduces the effective shielding of ten electrons, whereas the relatively difficult penetration of an s-subshell reduces the effective shielding of only two electrons.

One very important result of this is that valence electrons which are immediately outside a full d-shell are much more strongly attracted to the nucleus (and therefore much less reactive) than valence electrons immediately underlain by a full p-subshell. Thus the group IA and IB metals appear to have similar electron configurations, but their properties are very different.

2.5 ELECTRON CONFIGURATIONS OF MULTI-ELECTRON ATOMS

First, it is convenient to summarize the nomenclature used in describing the energy levels (states) and the associated orbitals. An **electron shell** in a given atom refers to all of the energy states (and corresponding orbitals) which have the same value of the principal quantum number, n. The number of orbitals in a specified electron shell is given by n^2, and since two electrons can be accommodated in each orbital (see later) the maximum number of electrons which can be placed in a given shell is $2n^2$. An **electron subshell** of a certain atom refers to all the energy states (and corresponding orbitals) which have the same values of the two quantum numbers n and l. For example the 2p-subshell in which the values of the quantum numbers are $n = 2$ and $l = 1$. The number of orbitals in each subshell is $(2l + 1)$, and the maximum number of electrons which can be accommodated is $2(2l + 1)$. An **electron orbital** is defined by a given value for each of the three quantum numbers n, l, and m_l; for example, for a particular $2p$-orbital, $n = 2$, $l = 1$, $m_l = +1$, and it has a maximum capacity of two electrons. The **electron configuration** of an atom describes the way in which the orbital electrons are distributed among the available orbitals. Generally it is atoms in their lowest energy state (referred to as the *ground state* of the atom) that we shall be interested in. For example, the gaseous lithium atom in its ground state contains three electrons, two of which are in the lowest energy orbital (in this case the $1s$-orbital), and one of which is in the next lowest orbital (the $2s$-orbital), since the $1s$-orbital has a maximum capacity of two electrons. The shorthand notation for this electron configuration is $1s^2 2s^1$, where $1s$ and $2s$ describe the orbital, and the superscript gives the number of electrons in each orbital. If a sufficient amount of energy is given to the gaseous lithium atom, an electron can be **excited** from one of the occupied orbitals to an orbital with higher energy. For example, it is possible to obtain the **excited state configuration**, $1s^2 3s^1$. After a short time the electron in the $3s$-orbital will emit energy and "fall" into the $2s$-orbital, and the atom will again be in the ground state configuration. In all of the following discussion we are interested in ground state configurations only.

The importance of being able to write down the ground state configuration of atoms arises because the properties of the elements are related to the way in which the orbital electrons are distributed about the nucleus of the atom. The highest energy electrons (those in the outer shell) are particularly significant in this respect, and to a first degree of approximation the properties of any element are determined by the number and type of the highest energy electrons. Since the inner core of electrons is of only secondary importance, this is frequently ignored in describing the electron configuration, or is summarized in some way. For example, the ground state of potassium can be written as $[Ar]4s^1$ (Table 2.6), where $[Ar]$ is shorthand for the ground state of argon, $1s^2 2s^2 2p^6 3s^2 3p^6$. This shorthand description emphasizes the importance of the outer electrons in determining the properties of the elements. The Periodic Table is a model of this behaviour: the periodicity of the properties of the elements is interpreted in terms of recurring similar outer electron configurations (Section 10.3), and these are represented in vertical groups to emphasize their similarity. The elements in group IA, for example, all have similar properties because each element has the same outer electron configuration, ns^1.

2.5.1 The *Aufbau* Approach

As usual there are two approaches to the problem (in this case that of writing down the ground state electron configuration of each of the elements): calculation using the mathematical methods of wave mechanics; or by inventing a simple model. So far, it has not proved possible to solve this problem by calculation, but a very simple model exists which enables the student to write down the electron configuration of (almost) any atom for which the atomic number is specified. This model is known as the ***aufbau*** **approach** (*aufbau* means "building up").

The assumption is made that multi-electron atoms (atoms containing more than one extra-nuclear electron) contain electrons which are in orbitals similar to those of the hydrogen atom in that the orbitals are described by a set of quantum num-

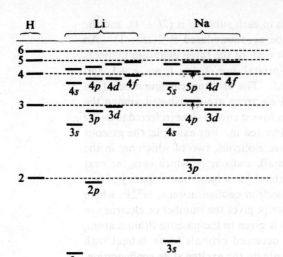

Fig. 2.22 Comparison of the energy levels of the hydrogen atom, lithium atom, and sodium atom. From B. H. Mahan, *University Chemistry* (second edition), Addison-Wesley (1969).

bers which are the same as those in the case of the hydrogen atom (and therefore the number of orbitals in each shell is the same), but which differ in that the energies of the orbitals are different from those in the hydrogen atom. Figure 2.22 compares the energy levels for the hydrogen and lithium atoms (why are the 1s-orbital energies missing from the diagram?). There are two major differences between the electronic energies of these two atoms, which are obvious from the diagram. First, the energy of a given orbital (for example the 2s-orbital) is lower in the case of the lithium atom than in that of the hydrogen atom. Secondly, whereas all orbitals in the same shell (with the same value of the principal quantum number n) have the same energy in the case of the hydrogen atom, this is no longer true for a multi-electron atom. Here the s-subshell is lower in energy than the associated p-subshell, which in turn is lower in energy than the associated d-subshell. Thus the quantum numbers n and l have slightly different meanings for multi-electron atoms compared with the hydrogen atom. Whereas in the case of the hydrogen atom the principal quantum number n alone determines the energy and the secondary quantum number l determines the shape of the orbital, in the case of a multi-electron atom the secondary quantum number l also determines the energy of the orbital to some extent. In both cases, however, the shapes of orbitals with the same value of l are the same but the orbitals are much less extensive in multi-electron atoms. In other words, in going from hydrogen to a multi-electron atom the angular distribution function remains the same, but the radial distribution function changes. The meanings of the magnetic and spin quantum numbers, m_l and m_s, are unchanged in going from the hydrogen atom to a multi-electron atom.

The only distinction it has been found necessary to make between any two electrons in a given atom is a difference in their quantum numbers. Clearly, therefore, no two electrons can have the same values for each of the four quantum numbers, otherwise they would be the same electron! This principle is known as the **Pauli exclusion principle** (Section 2.4). Another way of stating it is to say that no orbital can contain more than two electrons, and two electrons in the same orbital must have opposite spin (in other words have different values of the spin quantum number, $+\frac{1}{2}$ and $-\frac{1}{2}$). This follows from the first statement of the principle, since

| H | Ne → Ca | Sc |

4s 4p 4d 4f		4f
3s 3p 3d		4p 4d
2s 2p	3d	4s 3d
	4s	3s 3p
	3s 3p	
	2s 2p	2s 2p
1s		
	1s	
		1s

Fig. 2.23 Schematic representation of relative energies of orbitals.

if two electrons are in the same orbital, they must have the same values for the quantum numbers n, l, and m_l (by definition), and therefore the spin quantum number must be different (and there are only two possible values).

The *aufbau* **principle** states, that for any given atom the electron configuration with lowest energy (ground state) is found by assigning electrons to orbitals of lowest energy in sequence, two electrons to each orbital. This is another "thought experiment" in which, in order to obtain the electron configuration of an atom, it is imagined that the particular atomic nucleus is surrounded by empty orbitals into which the electrons are fed in order of increasing orbital energy, starting with the 1s-orbital. To do this it seems necessary, at first glance, to know the relative energies of the orbitals for each atom (or ion). Unfortunately, the relative energies of orbitals often change from one atom to another, and even from one atom to its ions. In other words, each atom or ion has a unique set of orbital energies determined by its nuclear charge and the number of orbitals occupied by electrons. Moreover, the problem is too complicated in general to be solved by the methods of wave mechanics. Fortunately, a simple model exists which, although it is oversimplified (as usual), predicts the correct ground state configuration in almost every case. The relative energies of the lowest energy orbitals of the lightest elements in the Periodic Table fall into three categories: (a) hydrogen; (b) the elements neon to calcium (inclusive); and (c) the elements from scandium onwards. These are represented in Fig. 2.23. The most important points to note are: (a) the energy of all orbitals decreases as the charge on the nucleus increases (but much more rapidly than the figure indicates); (b) all orbitals in the same shell have the same energy in the case of the hydrogen atom, but not in the case of a multi-electron atom; and (c) the main difference between the elements from neon to calcium, and those heavier than calcium, is that in the former case the 4s-orbital is lower in energy than the 3d-orbital, and in the latter case this order is reversed. This is made clear in Fig. 2.24. As the nuclear charge increases, the energy of all orbitals decreases sharply, but the 3d-orbitals are decreasing in energy more rapidly than the 4s-orbital, and the cross-over occurs between calcium and scandium. For elements from He up to Ne the energy levels are qualitatively like those for Li in Fig. 2.22.

The student should now be able to write down the ground state electron configuration of any atom. Consider nitrogen for example (atomic number 7), which has a nuclear charge of seven, and seven extra-nuclear electrons. The lowest

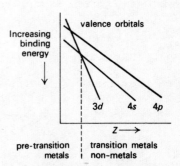

Fig. 2.24 Relative energies of orbitals shown diagrammatically.

energy orbitals, in order of increasing energy are: $1s$, $2s$, $2p$, and $3s$. The number of electrons that can be accommodated in any subshell is $2(l+1)$. Therefore two electrons can be placed in the $1s$-orbital, two in the $2s$-orbital, and three in the $2p$-orbitals (which could accommodate six altogether); and the electron configuration is $1s^2 2s^2 2p^3$. Note that $2p$ represents the $2p$-subshell (which consists of three $2p$-orbitals), not a particular $2p$-orbital. As another example, consider calcium; the order of subshells in increasing energy is $1s$, $2s$, $2p$, $3s$, $3p$, $4s$, $3d$ (Fig. 2.23). The twenty electrons therefore form the configuration $1s^2 2s^2 2p^6 3s^2 3p^6 4s^2$. The configuration of the other elements can be formulated in a similar manner.

A related process to that of writing down the electron configuration of any individual element, is the "thought experiment" of building up the elements, starting at hydrogen, by adding one proton to the nucleus (and perhaps one or more neutrons), and one electron to the atomic orbitals available. There are two useful mnemonics which help in remembering the order in which the orbitals are filled (note: this is not necessarily the order of energy levels in any one particular atom). The first of these is given in Fig. 2.25. From this figure the order of filling the orbitals during the thought experiment of building up the elements is given as $1s$, $2s$, $2p$, $3s$, $3p$, $4s$, $3d$, $4p$, $5s$, $4d$, $5p$, $6s$, $4f$, $5d$, $6p$, $7s$, etc. Another way of remembering this is the $(n+l)$ rule: in "building up the elements" the subshell with the lowest value of $(n+l)$ fills first; when two subshells have the same value of $(n+l)$, the orbital with the lower value of n fills first. This useful rule is a reminder that the energy of the subshells of multi-electron atoms depends upon the value of both the quantum numbers n and l, but mainly on the value of n. For example, which fills first, the $5s$- or the $4p$-subshell? For the $4p$-subshell ($n=4$, $l=1$) the value of $(n+l)=5$; for the $5s$-subshell (or orbital) the value of $(n+l)=(5+0)=5$ also; but the $4p$-subshell has the lower value of the principal quantum number n and therefore fills first.

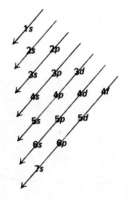

Fig. 2.25 Order of filling orbitals.

One further question which could be asked is: How are the electrons distributed among the orbitals in the subshells? For example, in oxygen (electron configuration $1s^2 2s^2 2p^4$) the four electrons could be arranged among the three $2p$-orbitals ($2p_x$, $2p_y$, and $2p_z$) in several ways. Generally a box is drawn to represent an orbital and the boxes are joined together to represent a subshell. Within each box an arrow is drawn to represent an electron, usually pointing upwards to represent one allowed value of the spin quantum number (say $+\frac{1}{2}$), and downwards to represent the other allowed value ($-\frac{1}{2}$) (Table 2.3). In order to represent the lowest energy state an additional rule is required; this is referred to as **Hund's rule of maximum multiplicity**, and states that, in a given atom in its lowest energy state, electrons in the same subshell will occupy different orbitals with the same spin, as far as possible. Nitrogen, for example, with the configuration $1s^2 2s^2 2p^3$ could have the electrons in the $2p$-subshell arranged in several ways, but the actual arrangement is:

Tables 2.4 and 2.5 give the more detailed electron configuration of some of the elements.

Table 2.4. Electron configurations of elements 1 to 10.

Atomic number	Element	1s	2s	$2p_x$	$2p_y$	$2p_z$	Electron configuration
1	H	↑					$1s^1$
2	He	↑↓					$1s^2$
3	Li	↑↓	↑				$1s^2 2s^1$
4	Be	↑↓	↑↓				$1s^2 2s^2$
5	B	↑↓	↑↓	↑			$1s^2 2s^2 2p^1$
6	C	↑↓	↑↓	↑	↑		$1s^2 2s^2 2p^2$
7	N	↑↓	↑↓	↑	↑	↑	$1s^2 2s^2 2p^3$
8	O	↑↓	↑↓	↑↓	↑	↑	$1s^2 2s^2 2p^4$
9	F	↑↓	↑↓	↑↓	↑↓	↑	$1s^2 2s^2 2p^5$
10	Ne	↑↓	↑↓	↑↓	↑↓	↑↓	$1s^2 2s^2 2p^6$

In Table 2.4, the electron configuration is shown in the *sp* notation. In Table 2.5 the number of electrons in each successive shell (*K, L, M, N*) is shown. A comprehensive set of electron configurations is given in Table 2.6.

Table 2.5. Electron configurations of elements 19 to 30. (× represents an electron pair ↑↓.)

Atomic number	Element	1s	2s	2p	3s	3p	3d					4s	$n=1$ K	2 L	3 M	4 N
19	K	2	2	6	2	6						↑	2	8	8	1
20	Ca	2	2	6	2	6						×	2	8	8	2
21	Sc	2	2	6	2	6	↑					×	2	8	9	2
22	Ti	2	2	6	2	6	↑	↑				×	2	8	10	2
23	V	2	2	6	2	6	↑	↑	↑			×	2	8	11	2
24	Cr	2	2	6	2	6	↑	↑	↑	↑	↑	↑	2	8	13	1
25	Mn	2	2	6	2	6	↑	↑	↑	↑	↑	×	2	8	13	2
26	Fe	2	2	6	2	6	×	↑	↑	↑	↑	×	2	8	14	2
27	Co	2	2	6	2	6	×	×	↑	↑	↑	×	2	8	15	2
28	Ni	2	2	6	2	6	×	×	×	↑	↑	×	2	8	16	2
29	Cu	2	2	6	2	6	×	×	×	×	×	↑	2	8	18	1
30	Zn	2	2	6	2	6	×	×	×	×	×	×	2	8	18	2

Whereas the Pauli principle is a basic principle and cannot be explained, the origin of the Hund rule can be traced farther back. Since electrons are negatively charged, they repel one another and, all things being equal, will keep as far apart as possible if they have the choice. Consequently if there are orbitals of similar energy available, electrons will spread themselves out in the different orbitals in order to be as far apart as possible. Similarly it can be calculated from wave mechanics that electrons with the same spin are slightly further apart on average, than electrons with opposite spin; therefore, if electrons have the choice, they keep their spins the same. Do not waste time trying to understand this last point, since it can only be understood in terms of detailed calculation.

Table 2.6. Electron configurations of gaseous atoms.

Atomic number	Element	Electron configuration	Atomic number	Element	Electron configuration
1	H	$1s$	53	I	$-4d^{10}5s^25p^5$
2	He	$1s^2$	54	Xe	$-4d^{10}5s^25p^6$
3	Li	[He] $2s$	55	Cs	[Xe] $6s$
4	Be	$-2s^2$	56	Ba	$-6s^2$
5	B	$-2s^22p$	57	La	$-5d6s^2$
6	C	$-2s^22p^2$	58	Ce	$-4f^26s^2$
7	N	$-2s^22p^3$	59	Pr	$-4f^36s^2$
8	O	$-2s^22p^4$	60	Nd	$-4f^46s^2$
9	F	$-2s^22p^5$	61	Pm	$-4f^56s^2$
10	Ne	$-2s^22p^6$	62	Sm	$-4f^66s^2$
11	Na	[Ne] $3s$	63	Eu	$-4f^76s^2$
12	Mg	$-3s^2$	64	Gd	$-4f^75d6s^2$
13	Al	$-3s^23p$	65	Tb	$-4f^96s^2$
14	Si	$-3s^23p^2$	66	Dy	$-4f^{10}6s^2$
15	P	$-3s^23p^3$	67	Ho	$-4f^{11}6s^2$
16	S	$-3s^23p^4$	68	Er	$-4f^{12}6s^2$
17	Cl	$-3s^23p^5$	69	Tm	$-4f^{13}6s^2$
18	Ar	$-3s^23p^6$	70	Yb	$-4f^{14}6s^2$
19	K	[Ar] $4s$	71	Lu	$-4f^{14}5d6s^2$
20	Ca	$-4s^2$	72	Hf	$-4f^{14}5d^26s^2$
21	Sc	$-3d4s^2$	73	Ta	$-4f^{14}5d^36s^2$
22	Ti	$-3d^24s^2$	74	W	$-4f^{14}5d^46s^2$
23	V	$-3d^34s^2$	75	Re	$-4f^{14}5d^56s^2$
24	Cr	$-3d^54s$	76	Os	$-4f^{14}5d^66s^2$
25	Mn	$-3d^54s^2$	77	Ir	$-4f^{14}5d^76s^2$
26	Fe	$-3d^64s^2$	78	Pt	$-4f^{14}5d^96s$
27	Co	$-3d^74s^2$	79	Au	$-4f^{14}5d^{10}6s$
28	Ni	$-3d^84s^2$	80	Hg	$-4f^{14}5d^{10}6s^2$
29	Cu	$-3d^{10}4s$	81	Tl	$-4f^{14}5d^{10}6s^26p$
30	Zn	$-3d^{10}4s^2$	82	Pb	$-4f^{14}5d^{10}6s^26p^2$
31	Ga	$-3d^{10}4s^24p$	83	Bi	$-4f^{14}5d^{10}6s^26p^3$
32	Ge	$-3d^{10}4s^24p^2$	84	Po	$-4f^{14}5d^{10}6s^26p^4$
33	As	$-3d^{10}4s^24p^3$	85	At	$-4f^{14}5d^{10}6s^26p^5$
34	Se	$-3d^{10}4s^24p^4$	86	Rn	$-4f^{14}5d^{10}6s^26p^6$
35	Br	$-3d^{10}4s^24p^5$	87	Fr	[Rn] $7s$
36	Kr	$-3d^{10}4s^24p^6$	88	Ra	$-7s^2$
37	Rb	[Kr] $5s$	89	Ac	$-6d7s^2$
38	Sr	$-5s^2$	90	Th	$-6d^27s^2$
39	Y	$-4d5s^2$	91	Pa	$-5f^26d7s^2$
40	Zr	$-4d^25s^2$	92	U	$-5f^36d7s^2$
41	Nb	$-4d^45s$	93	Np	$-5f^46d7s^2$
42	Mo	$-4d^55s$	94	Pu	$-5f^67s^2$
43	Tc	$-4d^55s^2$	95	Am	$-5f^77s^2$
44	Ru	$-4d^75s$	96	Cm	$-5f^76d7s^2$
45	Rh	$-4d^85s$	97	Bk	$-5f^97s^2$
46	Pd	$-4d^{10}$	98	Cf	$-5f^{10}7s^2$
47	Ag	$-4d^{10}5s$	99	Es	$-5f^{11}7s^2$
48	Cd	$-4d^{10}5s^2$	100	Fm	$-5f^{12}7s^2$
49	In	$-4d^{10}5s^25p$	101	Md	$-5f^{13}7s^2$
50	Sn	$-4d^{10}5s^25p^2$	102	No	$-5f^{14}7s^2$
51	Sb	$-4d^{10}5s^25p^3$	103	Lr	$-5f^{14}6d7s^2$
52	Te	$-4d^{10}5s^25p^4$			

The number of electrons in an orbital is given by a superscript. Where no superscript is given it is understood that there is one electron in this orbital.

2.5.2 The *Aufbau* Approach: Summary

The *aufbau* approach is a model for determining the ground state configuration of a multi-electron atom. In order to use the model, five pieces of information are necessary and these are summarized in Fig. 2.26. The orbital energy scheme is based on that of the hydrogen atom; for example, the nomenclature and the number of orbitals (n^2), and the maximum number of electrons ($2n^2$) that can be accommodated in each shell are the same, but in the case of a multi-electron atom an orbital energy depends not only upon the principal quantum number n, but also on the secondary quantum number l.

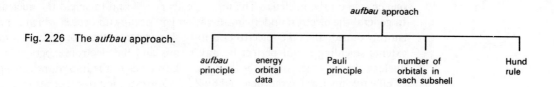

Fig. 2.26 The *aufbau* approach.

An exercise related to the determination of the ground state configuration of a particular atom is the "thought experiment" of building up the elements one by one by successively adding a proton to the nucleus (and perhaps one or more neutrons) and an electron to the next lightest element. Two mnemonics were given earlier for remembering the order in which the orbitals are filled. This order of filling for all the elements should not be confused with the order of the energy levels in any one element, though it may well be the same as that for any particular element.

Consider the following problems:

1. Why do all the levels decrease in energy as the nuclear charge increases?
2. Why do the *s*-levels decrease more than the *p*-, *d*-, and *f*-levels?

2.6 SUMMARY

It may be helpful to summarize briefly some of the more important ideas introduced in this chapter since they will often be used throughout the book.

1. The electron is an entity whose properties have to be found by examining its behaviour. It is found to have mass, charge, and magnetic moment, which is conveniently interpreted in terms of spin. The way in which it occupies space is accurately given by a wave function, ψ. In order to think about electrons in molecules we need to translate the mathematical statement into a visual picture. When we do this we find that the electron in a hydrogen atom cannot be thought of as moving along a defined path. Instead we have a probability picture, and the value of ψ^2 at a particular point tells us the probability of finding the electron at that point. The electron is said to occupy an orbital.
2. For the hydrogen atom the orbitals are characterized by four quantum numbers, n, l, m_l, and m_s. Only certain values are allowed for these quantum numbers as illustrated in Table 2.2. For the hydrogen atom the energy levels depend only on n.

3. The idea of orbitals, introduced for the hydrogen atom, can be taken over and used in describing multi-electron atoms. The new feature is that there is repulsion between the two or more electrons. An important consequence is that we have to consider orbitals for such atoms which have energies depending not only on n but also on l as illustrated in Fig. 2.23.
4. The repulsion between electrons which complicates the treatment of multi-electron atoms can be looked at in terms of shielding, which proves to be a useful concept for many purposes. An electron in such an atom experiences attraction by the nucleus of charge Z and repulsion from $(Z-1)$ electrons. If the electron considered were removed well from the atom it would experience a positive charge $Z - (Z-1)$, the $(Z-1)$ electrons cancelling all but one charge on the nucleus. The inner electrons are said to shield the nucleus. In general the electron under consideration will occupy an orbital which is not entirely outside the other electron orbitals. It penetrates these orbitals to some extent; shielding is not perfect by any means and the electron experiences a nuclear charge which is greater than one. This leads to the important concept of effective nuclear charge, denoted by Z_{eff}. Consideration of penetration leads to an understanding of relative energies of orbitals.
5. Electron configurations for atoms in their lowest energy states can be built up by putting successive electrons into orbitals according to the *aufbau* approach, which has already been summarized. Electron configurations of the elements in their ground states as gaseous atoms are given in Table 2.6.

PROBLEMS

1. After listening to a lecture by Rutherford a questioner said that he still did not believe there were such things as atoms. Exasperated, Rutherford replied "I can see the little b......s!" Is there any sense in which atoms can be seen? What do you consider is the most important evidence for the atomic theory today?
2. The structure of matter is described in terms of models. Choose two models of the atom and describe the way in which the models are useful. In each case point out the limitations of the model.
3. Describe three key experiments which contributed to the development of the modern theory of the atom. Point out the precise importance of each experiment.
4. This question illustrates some principles in the determination of ionization energy from spectra. We begin with the information given in Fig. 2.10 for the Balmer series. The first step is to find the convergence limit of the series of lines.
    a) Try plotting the wave number for each line against the number of the line and extrapolating to $n = \infty$. Does this prove to be a good method?
    b) Look at the expression for the Balmer series: $\nu = R(1/2^2 - 1/n^2)$. This suggests plotting ν against $1/n^2$ and extrapolating to $n = \infty$. Does this prove to be a good method? (Often in science it is important to arrange graphical work in the most convenient way. If a method which will give a straight line can be found it will usually be advantageous.) When you have found the convergence limit, what must be added to this to give the limit

for the Lyman series (see Fig. 2.9)? From the value you obtain in cm^{-1}, obtain the ionization energy in kJ mol^{-1} (see Section A.17).

5. This question extends the procedure developed in Problem 4 to determine the ionization energy of the sodium atom. Frequencies observed in the absorption spectrum are given in Fig. 2.4.

    Draw a schematic diagram labelling the levels with the principal quantum number, n. Remembering that in absorption the transitions all begin from the lowest level draw in the transitions. (Remember also for sodium that the electron which is being excited occupies the 3s-orbital and so the value of n for the ground state is 3.)

    The energy levels are not given, in this case, by a simple expression as for the Balmer series, because of the complication of electron repulsion. On the other hand, a method like 4(a) will not be very accurate. A better method is to plot ν against $1/n^2$ and extrapolate. Hence obtain the ionization energy. The plot will not be a straight line. Nevertheless, as the electron is excited to higher and higher orbitals, the effective nuclear charge becomes closer and closer to 1; that is for large distances the outer electrons shield the nucleus almost perfectly. Consequently, the plot approaches linearity in the limit as $n \to \infty$ and extrapolation is relatively straightforward. Wave numbers for some further lines besides those given in Fig. 2.4 are as follows: 40,383, 40,566, 40,706, 40,814, 40,901, 40,971, 41,029, 41,076, 41,116, 41,150 cm^{-1}.

6. What is the de Broglie relationship? Calculate the de Broglie wavelength for an electron accelerated through a potential of 1000 V. What are the wavelengths for (a) an α-particle emitted from radium (4.8 × 10^6 eV) and (b) a neutron, from an atomic pile, having energy 0.05 eV? Which of these are suitable for investigating the atomic structure of matter?

7. Which of the following are not permissible sets of quantum numbers for an electron in an atom?

	n	l	m_l	m_s
a)	3	2	−1	$\tfrac{1}{2}$
b)	2	3	−1	$\tfrac{1}{2}$
c)	3	2	−3	$\tfrac{1}{2}$
d)	4	1	1	$\tfrac{3}{2}$
e)	3	3	0	$\tfrac{1}{2}$

8. Make a table showing the various combinations of quantum numbers that are possible for 3d-orbitals. What is the maximum number of electrons that can be accommodated in the 3d-electron subshell?

9. Which of the following are not possible electron configurations for an atom:
    a) $1s^2 2s$;
    b) $1s$;
    c) $2s$;
    d) $1s^2 3s$;
    e) $1s^2 2s^2 2p^8 3s$;
    f) $1s^2 2s^2 2p^6 2d^2$?
    Which of the acceptable configurations represent (a) ground, and (b) excited states?

10. Arrange each of the following series of orbitals in order of increasing energy (a) for sodium and (b) for scandium:

   a) 1s, 2s, 4s, 3s, 5s; b) 2s, 3s, 2p, 3p, 3d; c) 3s, 3p, 3d, 4s, 4p.

11. Write down electron configurations for the ground states of each of the following:

   a) Na ($Z = 11$);
   b) Ca ($Z = 20$);
   c) Sc ($Z = 21$);
   d) Mg^{2+} ($Z = 12$);
   e) C^{1-} ($Z = 17$).

12. Make a plot of the sum of the first two ionization energies of the elements for elements 1 to 21. Compare this with the plot of first ionization energies (Fig. 2.17) and comment.

13. Elements X, Y, and Z have the following ionization energies (kJ mol^{-1}).

	1	2	3	4
X	738	1450	7730	10550
Y	495	4563	6912	9540
Z	800	2427	3658	25024

   From this information which of X, Y, Z do you select for the following:
   a) an element forming a covalent chloride;
   b) an element whose common oxidation state in chemistry is $+2$;
   c) an element forming an ionic univalent chloride?

BIBLIOGRAPHY

Partington, J. R. *A History of Chemistry*, Macmillan (1961–64).

    A four-volume work for reference, on history of chemistry generally, including development of ideas of atomic theory mentioned in this chapter.

    For a brief account see:

Mellor, D. P. "A Historical Approach to Atomic Molecular Theory," Topic I in *Approach to Chemistry*, edited by G. H. Aylward and I. K. Gregor, University of New South Wales, Sydney (1966).

Romer, A. *The Restless Atom*, Heinemann (1961).

    Easily readable account of study of radioactivity and structure of the atom.

Chadwick, Sir James (editor). *Collected Papers of Lord Rutherford of Nelson*, Vols 1,2,3. Allen and Unwin (1962–5).

    Original papers on scattering of α-particles, and other evidence from study of radioactivity.

Mahan, B. H. *University Chemistry* (second edition), Chapter 10, Addison-Wesley (1969).

    This gives an account of electronic structure of atoms at a somewhat more advanced level than in the present chapter.

Hochstrasser, R. M. *Behaviour of Electrons in Atoms*, Benjamin (1965).

    A book for further reference.

Herzberg, G. *Atomic Spectra and Atomic Structure*, Dover (1944).

    A standard work to dip into or for reference purposes.

CHAPTER 3

Structure of Solids, Liquids, and Gases

3.1 STRUCTURE: EVIDENCE FROM DIFFRACTION

3.1.1 X-ray Crystallography

One of the most useful techniques for scientific research developed during the twentieth century is X-ray crystallography. Through X-ray crystallography the first determination of the simple cubic structure of sodium chloride was made by Lawrence Bragg in 1913. Over fifty years later the structure of an extremely complex protein, lysozyme, was discovered by the same technique, much refined. It was fitting that this work was performed in the Royal Institution, London, under Sir Lawrence Bragg, its director at the time. It is obviously important to understand the principles of X-ray crystallography.

The principle on which X-ray crystallography is founded is that of diffraction, essentially a wave phenomenon. Looking through a handkerchief at a point source of light, one sees a pattern of dots. This pattern has been determined by the arrangement of the holes in the handkerchief, but is not directly related to the holes, as can be seen if the handkerchief is tilted towards and away from the light. The regular arrangement of holes in the handkerchief has caused the light from the point source to be diffracted. Similar diffraction patterns can be seen by looking at a light through a net curtain. Without going into the phenomenon of diffraction in detail, it is possible to get an idea of the way in which X-ray diffraction occurs in crystals. The first point is that X-rays have a very short wavelength (Fig. 2.8), of the order of 1 Å, and that this is of the same order as the distance between the nuclei of atoms in solids. Consider the case of a beam of X-rays reflecting off a row of atoms or ions in a crystal (Fig. 3.1). As with light the angle of incidence will equal the angle of reflection. Now consider the effect of a beam being reflected off the row of atoms below the surface at the same time. Taking θ as the angle of approach and d the distance between two layers of atoms, what is the extra distance that the ray reflected off the second layer has to travel? In Fig. 3.1 it is represented by $2x$. From this figure it can also be seen that

$$x = d \sin \theta;$$

therefore,

$$2x = 2d \sin \theta.$$

- 3.1 Structure: Evidence from diffraction
- 3.2 Structures of solids
- 3.3 Packing of spheres and correlation of crystal structures
- 3.4 Structure of liquids
- 3.5 Gases

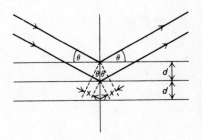

Fig. 3.1 Reflection of X-rays from layers of atoms in a crystal.

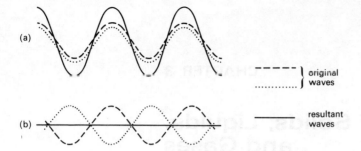

Fig. 3.2 Interference between waves. (a) Two waves in phase lead to reinforcement. (b) Two waves out of phase lead to cancellation.

The difference in path length determines whether the two waves cancel each other out or reinforce each other: two wave fronts which are in phase reinforce each other (Fig. 3.2a); two waves which are out of phase cancel each other out (Fig. 3.2b). The condition for this is that one wave should be displaced with respect to the other by $\lambda/2$. The condition for waves to cancel each other out is that the path difference should be $\lambda/2$ or any whole number of wavelengths plus $\lambda/2$. Similarly for waves to reinforce each other, there must be either no path difference or a path difference of a whole number of wavelengths, $n\lambda$. When light is undergoing diffraction, reinforcement causes a bright spot or line and cancellation causes a dark spot or line. In between these extremes, there will be intermediate brightness. The condition for reinforcement in the situation represented in Fig. 3.2 is that

$$n\lambda = 2d \sin \theta. \tag{3.1}$$

This simple relationship is known as **Bragg's law** after Sir Lawrence Bragg.

It can be seen that there will be a certain value of θ which will satisfy the equation. If the angle of incidence, $90 - \theta$, of the X-rays is changed slowly from 90° to 0° then an observer looking at the reflection (in the case of X-rays this would be a photographic plate) would "see" a series of dark and bright reflections according to whether reinforcement or cancellation was occurring at that particular angle. The first reinforcement is known as the first order diffraction. If the angle of incidence is measured for the bright spots and the wavelength of the X-rays is also known then

$$n\lambda = 2d \sin \theta$$

can be used to determine d. It is not difficult to see which is the first order diffraction, that for which $n = 1$, and so there is no need to do experiments to determine n.

Bragg's law enables us to determine the distance between rows of atoms or ions in a crystal. It cannot be used unless the compound is in a crystalline form. The example used to explain this principle is of course an oversimplification. Two further points must be made. The first is that X-rays are reflected not only from the first and second layers but also, with diminishing intensity, from all the parallel layers of atoms. The second is that as the angle θ changes there may be many parallel layers, even in a simple crystal like sodium chloride, which are capable of reflecting. Figure 3.3 shows this possibility.

The X-ray goniometer. The X-ray goniometer is the apparatus used to determine crystal structures in the way just described. It is illustrated in Fig. 3.4.

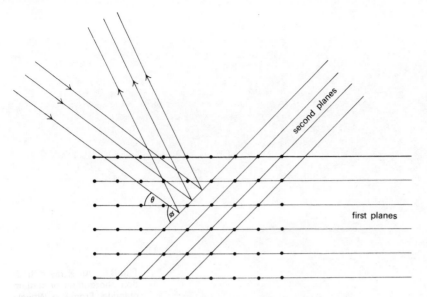

Fig. 3.3 Reflections are shown from a second set of parallel planes of atoms in a crystal.

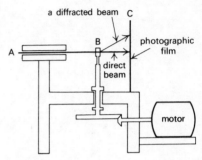

Fig. 3.4 Diagram of apparatus used in rotating crystal method.

X-rays are produced at A and diffracted by the crystal at B. The photographic film C is curved round cylindrically and the crystal (not the X-ray beam) is turned on its table. After a complete rotation, the pattern on the photographic film is developed and analysed. A typical photograph is shown in Fig. 3.6; this is the diffraction pattern of sodium chloride.

3.1.2 X-ray Diffraction by Powders

It is not always possible to produce a crystal good enough for the method of analysis described above. It is much easier to produce a powder which is in effect a large number of very small crystals in random positions. In a sample of powder there will be many crystals facing every direction.

The diffraction pattern produced is like the pattern which would be produced if a single crystal were rotated not in one plane only, as is the case in normal X-ray analysis, but at all possible angles in all three dimensions. If the normal X-ray diffraction pattern is rotated about a central point, the dots become circles. The characteristic of an X-ray powder photograph is that it consists of segments of circles. Figure 3.5 shows the scheme and the kind of pattern obtained. The diameters of the circles are used for the analysis of the dimensions of the unit cells of crystals.

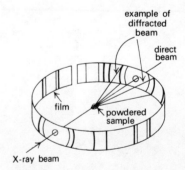

Fig. 3.5 X-rays are diffracted by a powdered sample and the diffraction pattern is photographed on a strip of film arranged cylindrically.

3.1.3 Electron Density Maps

A more refined technique has been developed which considers not only the position of the dots in the diffraction pattern but also their intensity. A series of diffraction patterns is analysed by a computer, and the results expressed in terms of electron densities. The electron density is an expression of the charge density at positions

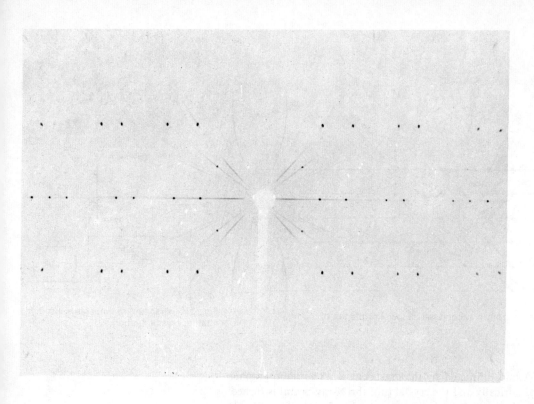

Fig. 3.6 An X-ray diffraction photograph of sodium chloride. From P. J. Wheatley, *Determination of Molecular Structure* (second edition), Oxford University Press (1959). Reproduced by permission of the publisher.

between nuclei; it is measured in electrons per cubic ångström. Points of equal electron density are joined by contours, to produce an electron density "map" of the type shown in Fig. 3.7. Such maps can be made in three dimensions by mounting them on layers of perspex. They have proved extremely useful in getting an overall picture of a molecule and seeing how it fits in the crystal. Notice that the nucleus itself is not shown but can be deduced from the position of the contours.

3.1.4 Electron Diffraction

X-rays are reflected by electrons rather than atomic nuclei. The effect on X-rays is most marked where electron density is greatest. The disadvantage of this is that the hydrogen atom scarcely shows up at all, as the electron density around the atom is too small to affect the X-rays. Compare for example, carbon and hydrogen in Fig. 3.7. It has already been shown that a beam of electrons has an associated wavelength (Section 2.3); it can be used to form diffraction patterns in the same way as X-rays. Its penetrating power is less than that of X-rays but it shows up the hydrogen atoms better as electrons are diffracted by protons as well as electrons. Perhaps the most useful contribution of electron beam diffraction has been with the structures of molecules in gases. A gas behaves rather like a powder in that the molecules (rather than the crystals) are in random orientations. There are always some molecules facing in every direction. The diffraction patterns formed by electron beams for gases are rings like those of powder X-ray diffraction. Figure 3.8 shows such a pattern for trifluorotriazine vapour. In this way, direct evidence for the diatomicity of the simple gases, such as hydrogen and nitrogen, has been

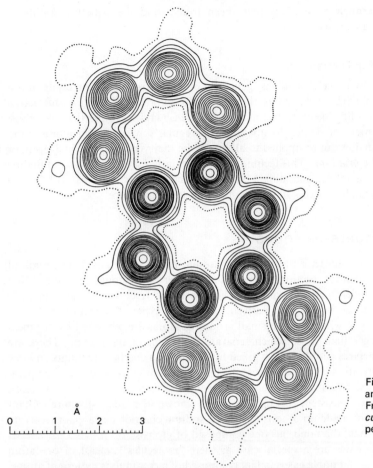

Fig. 3.7 Electron density contour diagram for the anthracene molecule.
From J. Monteath Robertson, *Organic Crystals and Molecules*, Cornell University Press (1953). Reproduced by permission of the publisher.

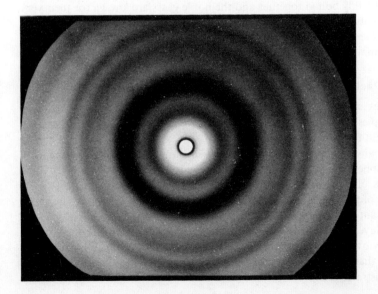

Fig. 3.8 Electron diffraction pattern for gaseous trifluorotriazine ($C_3N_3F_3$). From an original photograph kindly supplied by Dr B. Beagley, University of Manchester Institute of Science and Technology.

3.1 STRUCTURE: EVIDENCE FROM DIFFRACTION

obtained, internuclear distances have been found, and the structures of more complicated gases worked out.

3.1.5 Neutron Diffraction

A beam of slow moving neutrons, generally obtained from an atomic pile, has a wavelength of about 1 Å and is therefore also suitable for making diffraction patterns from solids. Neutron beams are diffracted by nuclei rather than electrons. Scattering powers of nuclei do not increase regularly with atomic number and neutron diffraction can be applied to all nuclei, be they very light like hydrogen, or very heavy like uranium. This technique is particularly useful, therefore, in finding the structures of solids containing a large number of hydrogen atoms. It has been used with particular success in determining the structure of ice.

3.2 STRUCTURES OF SOLIDS

The chemist sees matter as an array of atomic nuclei surrounded by clouds of electrons. In chemical reactions the nuclei are conserved, although rearranged, but the electrons undergo changes of energy and position. In crystalline structures the nuclei are arranged in a regular pattern. Two main types of structure are found. There are those in which distinct small groups of atoms are separated by distances which are larger than those which separate the atoms in the group. These are **molecular structures** and they are typified by low melting and boiling points and low heats of vaporization. All substances which are liquid or gaseous at room temperature are molecular or atomic. The second group shows no division into molecules. The structure is consistent throughout and is known as a **giant structure**. Giant structures can be made up of ions or atoms: sodium chloride has a giant structure composed of ions; diamond has one composed of atoms.

The noble gases are a special case as their "molecules" consist of one atom only. Hence a solid noble gas is in fact composed of a continuous pattern of atoms. They are very weakly bound together and the noble gases are known for their very low melting points (Section 12.1).

A relatively small number of basic structures are repeated in a large number of crystalline compounds. These structures will now be reviewed.

3.2.1 Close-Packed Structures

Giant structures may be conveniently divided into two types, close-packed and open. The simplest forms of close-packed structures are found in elements, arising from the fact that the atoms are all the same size. An experiment in packing apples, tennis balls, or any spherically shaped objects will show that in the closest type of packing, each sphere has six other spheres touching it in any one layer.

The centres of the six spheres if joined together make a hexagon (Fig. 3.9). When this type of packing is considered in three dimensions, it turns out that there are two important variations. In Fig. 3.10 a second layer of spheres is placed on top of the first layer. There are two ways in which a third layer can be added: it can either fit exactly over the first layer—called an *aba* structure; or it can be displaced by one hole so that it is not directly above the first layer—called an *abc* structure. Figure 3.10 illustrates this.

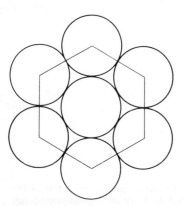

Fig. 3.9 Part of a layer of close-packed spheres.

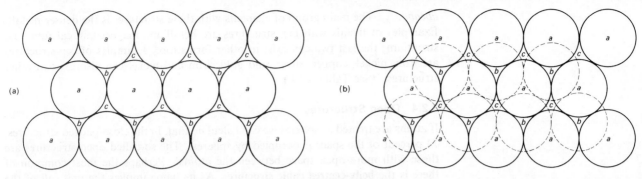

Fig. 3.10 (a) Layer of close-packed spheres. (b) A second layer can be placed to fill the spaces marked *b*, but not those marked *c* as well. A third layer can be placed either directly over *a* (hcp) or over *c* (ccp).

Viewed from a different angle, the *abc* structure presents the arrangement shown in Fig. 3.11. The square pattern in the diagram gives rise to the term **cubic close-packed** (ccp) to describe the *abc* structure. The *aba* structure is called **hexagonal close-packed** (hcp).

3.2.2 Co-ordination Number

It can be seen that in both these close-packed structures each atom is "touching" six atoms in its own layer, three atoms in the layer above, and three atoms in the layer below, making a total of 12. The number of atoms adjacent to a single atom is called the **co-ordination number**. The co-ordination number of both hexagonal close-packed and cubic close-packed structures is therefore 12.

3.2.3 The Unit Cell

The unit cell for simple crystals of this kind is the simplest arrangement of spheres which, when repeated, will reproduce the whole structure. The unit cell of the cubic close-packed structure is shown in Fig. 3.12(a); it is known as a **face-centred cube**.

It is interesting to note that although it would be possible to have an almost infinite number of variations of the close-packed structure, only the two described are commonly found naturally; that is, the common structures are *ababa*... and

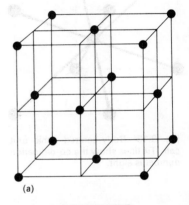

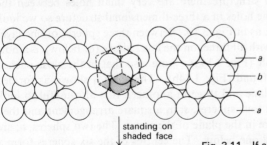

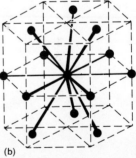

Fig. 3.12 (a) Face-centred cubic unit cell (*abc*). (b) Hexagonal close-packed structure (*aba*). The co-ordination number is twelve in both cases.

Fig. 3.11 If a part of the structure *abc* is rotated faces can be seen which show the cubic symmetry of the structure. The open diagram shows how this corresponds to the face-centred cubic structure. From Nuffield Advanced Science, *Chemistry*, Penguin (1970). Reproduced by permission of The Nuffield Foundation.

abcabc.... The main group of elements with these structures is the denser metals. Examples of metals with hcp structures are beryllium, magnesium, calcium, and strontium; the last two can exist in other forms too. Examples of ccp structures are iron, nickel, copper, silver, and gold; iron and nickel can also exist in other structures. (See Table A.14.)

3.2.4 Open Structures

If atoms are treated as spheres we can calculate that, in the close-packed structures, 74 percent of the space is occupied by spheres. The so-called open structures are those with more open space between the atoms. Perhaps the most common of these is the **body-centred cubic structure**. As its name implies the unit cell of the structure is a cube with one atom at its centre (Fig. 3.13). In the body-centred cubic structure 68 percent of the space is occupied by spheres. Examples of this structure are found in the lighter metals, in particular lithium, sodium, potassium, rubidium, and caesium. But the body-centred cubic structure is by no means restricted to light metals: it is also found in chromium, vanadium, and tungsten.

Another way of thinking of the body-centred cubic structure is to imagine a layer which is like that of the close-packed structure except that the spheres are further apart and therefore not touching. On top of this first layer it is possible to lay a second layer as before except that the sphere above touches four below. These spheres are in contact with the first layer spheres but are not in contact with each other. Finally imagine a third layer, like the first, on top of the second layer. The spheres touch the second layer but do not touch each other or the spheres of the first layer. Each sphere in the second layer touches four in the first layer and four in the third layer, giving a co-ordination number of eight. All the spheres in the structure are equivalent.

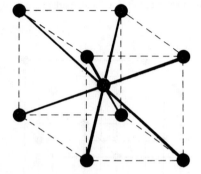

Fig. 3.13 Body-centred cubic unit cell. Tie lines show the co-ordination number is eight.

3.3 PACKING OF SPHERES AND CORRELATION OF CRYSTAL STRUCTURES

One of the simplest ways of describing the structures of a large number of solid compounds is to use a system based on the close-packed structures. In this system the "holes" in the close-packed structure are divided into two types. It is best to work this out with a pile of spheres of some sort in front of you. It can be seen that in any layer of a close-packed structure there are very small holes between the spheres. We are interested in the holes in a three-dimensional structure so we look at the holes between two layers, as in Fig. 3.10. Between three spheres of one layer and a fourth sphere in contact with all three, on the layer above (or below), is a hole which is called a **tetrahedral hole** (Fig. 3.14a). Note also that in the close-packed structure, a larger hole still is found (Fig. 3.14b). This hole is shown as formed between three spheres in one plane and three in another. It is also useful to look at it another way. Look at the four spheres, s, in a square arrangement and then imagine a tilt until these four are in the plane of the paper. The two spheres, o, are out of the plane, one above and one below. The centres of the six spheres form a regular octahedron and this hole is therefore called an **octahedral hole**. Many structures can be envisaged by assuming that these holes, or some of them, are filled with atoms or ions. It should be remembered that cations are often smaller, in some cases much smaller, than the anions they combine with (Table 3.1). In

Table 3.1. Ionic radii of some ions (Å).

Ion	Radius
Li^+	0.60
Na^+	0.95
K^+	1.33
Rb^+	1.48
Cs^+	1.69
F^-	1.36
Cl^-	1.81
Br^-	1.95
I^-	2.16

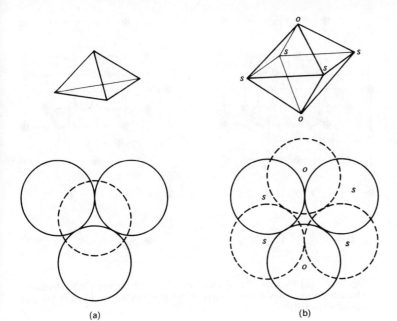

Fig. 3.14 (a) A tetrahedral hole formed between three spheres of one layer and one from the next layer. (b) An octahedral hole formed between two layers. It can also be viewed as four spheres, *s*, in a plane and two others, *o*, one above and one below, giving a regular octahedron.

certain cases it is possible for a small cation to enter the holes, or interstices as they are sometimes called, in a close-packed structure without distorting it at all.

It can be seen that in cubic close-packed structures each sphere is surrounded by eight tetrahedral holes and six octahedral holes. The total number of octahedral holes equals the number of spheres, but there are twice as many tetrahedral holes. Some structures of common binary compounds will now be described using the concept of tetrahedral and octahedral holes.

Sodium chloride, NaCl (rock-salt structure). In the sodium chloride structure the chloride ions can be thought of as having a cubic close-packed structure. The sodium ions occupy the six octahedral holes round each chloride ion. The sodium ion is bigger than the octahedral holes between the chloride ions and this means that the chloride ions must be pushed out so they do not actually touch one another. Each chloride ion "touches" six sodium ions and each sodium ion "touches" six chloride ions. This is called 6:6 co-ordination. The sodium chloride structure is a common one and is known as the "rock-salt structure" (Fig. 3.15). Lithium iodide is similar to sodium chloride but the lithium ions are so small (0.6 Å radius) that they can fit in the octahedral holes round the iodide ions without disturbing their cubic close-packed structure. The spacing between the ions in the crystal is in this case determined by the radius of the anions. This provides one way of working out a radius for an ion. Although ions do not have sharply defined outer surfaces the idea of ionic radius proves a good one. A set of **ionic radii** of alkali metals and halide ions is given in Table 3.1. The sums of the ionic radii give the distances that are found experimentally between the planes of alkali metal halides. The idea can be extended to many ions and crystals. It should be noted that generally the distances in ionic crystals are determined by the ionic radii of both cation and anion. In looking at crystal structures in the way just described we

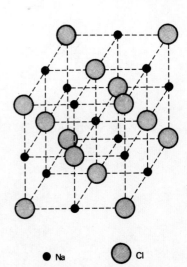

Fig. 3.15 Sodium chloride (rock-salt) structure.

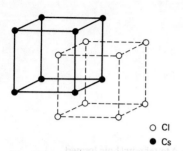

Fig. 3.16 Structure of caesium chloride.

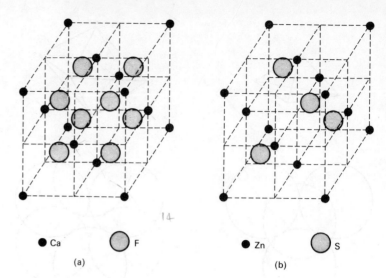

Fig. 3.17 (a) Fluorite (CaF_2) structure. (b) Zinc blende (ZnS) structure. The tetrahedral holes at the centres of the small cubes are filled in (a) and half filled in (b).

must remember in sodium chloride, for example, that the anions are not close-packed in the sense of touching spheres because of the need to fit in the cations. The anions form an expanded face-centred cubic structure.

Caesium chloride, CsCl. It can be seen from the discussion of the alkali metal halide (NaCl) structures that the relative sizes of the two ions forming the structure are important in that it requires more or less expansion of the close-packed sphere structure. Caesium chloride is interesting because the caesium and chloride ions are much the same size (caesium, 1.69 Å, and chloride, 1.81 Å). It is not possible for these to form a close-packed structure as such an arrangement cannot satisfy the need to have alternate caesium and chloride ions three dimensionally. The simplest way of achieving this is by having a simple cubic structure for each ion. The two cubic structures then "inter-penetrate" symmetrically, as shown in Fig. 3.16.

Each ion is in contact with eight of the other ions giving rise to 8:8 co-ordination. In general more anions can be fitted round a cation when it is large than when it is small. This leads to a larger number of close attractions between opposite charges and energetically to a more stable structure.

Calcium fluoride, CaF_2 (fluorite structure). A rather different situation arises when the ratio of ions is 1:2, as in calcium fluoride. Calcium ions can be thought of as lying on a face-centred cubic lattice. The fluoride ions occupy all eight *tetrahedral holes* round the calcium ions, as shown in Fig. 3.17(a). Effectively the layers of fluoride ions contain twice as many ions as the layers of calcium ions. Each calcium ion is in contact with eight fluoride ions and each fluoride ion is in contact with four calcium ions. The co-ordination is 8:4. This is known as the "fluorite structure".

68 STRUCTURE OF SOLIDS, LIQUIDS, AND GASES

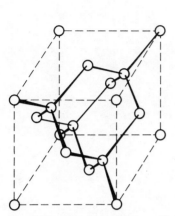

Fig. 3.18 Diamond structure. Unit cell with tie lines to show tetrahedral co-ordination.

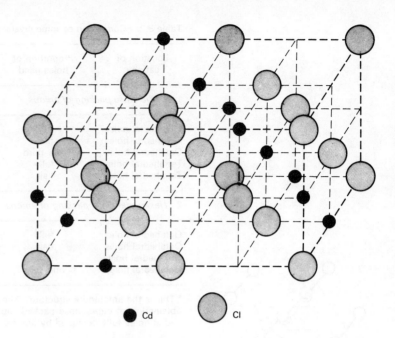

Fig. 3.19 Cadmium chloride structure showing relationship with sodium chloride structure (Fig. 3.15).

Zinc sulphide, ZnS (zinc blende structure). Another common structure is that for which zinc blende (zinc sulphide) is taken as typical. It can be thought of as similar to that of fluorite but with only half the tetrahedral holes round the zinc ions being filled, as shown in Fig. 3.17(b). This means that each zinc ion is in contact with only four sulphide ions. The co-ordination is therefore 4:4. Both the sulphide and the zinc ions are arranged tetrahedrally with regard to each other. The ions are in equivalent positions in this case and either can be considered to be filling holes in a structure formed by the other.

Diamond. If every atom in the zinc blende structure is replaced by carbon, the result is the diamond structure (Fig. 3.18). It is the same as that in Fig. 3.17(b) except that all atoms are equivalent. Tie-lines have been drawn in to show the tetrahedral co-ordination. The structure is contrasted with that of graphite in Section 16.2.

Cadmium chloride. By filling only one-half of the octahedral holes for a face-centred cubic structure, the structure shown in Fig. 3.19 may be obtained; this is the structure of cadmium chloride. It is often called a layer structure, and a layer can in fact be seen in the figure. The layer structure is often represented as in Fig. 3.20 which is a useful representation in that it emphasizes the layer character of the crystal. However, the approach in terms of close-packed spheres relates the structure to sodium chloride which would not be evident from Fig. 3.20.

Chromium(III) chloride. The structure of chromium(III) chloride can be obtained in a similar way. In this case, one-third of the octahedral holes are filled

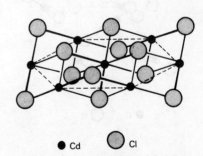

Fig. 3.20 Cadmium chloride structure. Alternative representation emphasizing layer structure.

3.3 PACKING OF SPHERES AND CORRELATION OF CRYSTAL STRUCTURES

Table 3.2. Correlation of some crystal structures.

Location of cations	Proportion of holes filled	Structure
1. *Cubic close packing of anions*		
Octahedral holes	fully	NaCl
Octahedral holes	half	$CdCl_2$
Octahedral holes	one-third	$CrCl_3$
Tetrahedral holes	fully	Na_2O*
Tetrahedral holes	half	ZnS (blende)
2. *Hexagonal close packing of anions*		
Octahedral holes	fully	FeS
Octahedral holes	half	CdI_2
Tetrahedral holes	half	ZnS (wurtzite)
Octahedral holes	one-third	$FeCl_3$

* This is the antifluorite structure. The fluorite structure is obtained from cubic close-packed cations with the tetrahedral holes fully occupied by anions.

by Cr(III) ions. Again another set of structures may be obtained by filling the tetrahedral holes.

By beginning with the hexagonal close packing the whole procedure of filling holes may be repeated to encompass even more structures. Table 3.2 gives examples of a number of common structures which fall within the scheme.

Molecular crystals. Where the crystal consists of molecules, X-ray diffraction studies lead to two types of information. The first is the way the atoms are arranged in the molecule and the distances between them. The second is the way the molecules are arranged in the crystal and the distances between the individual molecules. Thus for iodine two quite different distances come out of the analysis of the diffraction pattern. One is 1.66 Å, and the other is much larger, about 4.3 Å. The first corresponds to the distance between two iodine atoms tightly bound together in a molecule. The second is the nearest distance between the centres of two iodine atoms in adjacent molecules. These are held together by weak van der Waals forces (Section 4.9) and the distance apart, 4.3 Å, is much larger than that for two covalently bonded atoms in a molecule. The way in which the iodine molecules are arranged in the crystal is shown in Fig. 3.21(a). The structure is based on the familiar face-centred cubic structure, except that the positions are occupied not by single atoms but by diatomic iodine molecules.

It is of interest to note that such measurements as these lead to *covalent bond radii* and to *van der Waals radii*. To a good approximation the bond length in a molecule can be divided into two parts: the covalent radii for the two atoms involved. Bond lengths for many molecules can be obtained by adding together the two radii. Similarly, we may assign a van der Waals, or non-bonded, radius to iodine, of about 2.2 Å. Values for the halogens are given in Table 3.3. (See Section 10.3.2.)

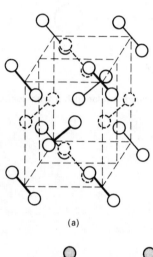

(a)

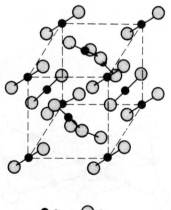

● C   ○ O

(b)

Fig. 3.21 Molecular crystals. (a) Structure of iodine. (b) Structure of carbon dioxide.

Table 3.3. Covalent and van der Waals radii.

	Covalent radius (Å)	Van der Waals radius (Å)
Fluorine	0.64	1.35
Chlorine	0.99	1.85
Bromine	1.14	1.95
Iodine	1.33	2.15

Another example of a simple molecular crystal is that of solid carbon dioxide. X-ray diffraction studies show that in this case also the structure is face-centred cubic but this time the sites are occupied by triatomic molecules as shown in Fig. 3.21(b).

3.4 STRUCTURE OF LIQUIDS

Liquids have some of the properties of solids and some of the properties of gases. Like gases, liquids flow rapidly and take up the shape of their containers, but they exhibit a surface. Also like gases, liquids show the properties of diffusion with the solids and liquids which are soluble in them.

As in solids, the atoms, ions, or molecules of a liquid are close together. In most cases they are less close together by about 10 percent than those of their corresponding solids. A well-known exception is water, which is *more* dense than ice; the structure of water is dealt with in Section 13.6. Like solids, liquids show a degree of structure: an X-ray diffraction picture of a liquid looks rather like that of a powdered solid (Fig. 3.6). In a powdered solid, tiny crystals are all in random positions, and the similarity of X-ray diffraction pictures indicates that the same is true of liquids. The X-ray diffraction photograph of a liquid has diffuse rather than sharp lines which suggests that the distance d in the Bragg equation (Eq. 3.1),

$$n\lambda = 2d \sin \theta,$$

is not constant. In other words the atoms are not held as rigidly in place as they are in solids. The picture of liquid structure which emerges from this and other evidence is that of local temporary order. Small localities of order exist but these localities are constantly changing. Figure 3.22(b) gives an idea of the structure of a liquid at a particular moment.

Not only are the atoms or molecules in liquids a little farther apart, there is also evidence that there are actually holes in the liquid structure. A few holes are found in solid structures but in liquid structures there are many more. The existence of these holes increases the general mobility of molecules and loosens the structure of the liquid. The number of holes in a liquid has been found to be dependent on the temperature of the liquid.

Liquids are held together primarily by van der Waals forces; in certain cases, such as that of water, hydrogen bonds play an important part. The forces between atoms and molecules are discussed in Chapter 4.

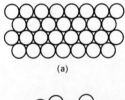

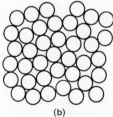

Fig. 3.22 Schematic view of structure in (a) a crystal and (b) a liquid. The latter changes with time. From G. W. Castellan, *Physical Chemistry* (first edition), Addison-Wesley (1964).

3.5 GASES

The structure of gases is in one sense extremely simple. The forces which hold liquids and solids together only have a minor influence in the case of gases. Most gases are molecular but ionic substances on heating produce what are often known as "ion pairs" in the vapour phase which behave as molecules. Na^+Cl^- is an example of an ion pair. We have already explained that the molecules of a gas are much farther apart than those of a solid or a liquid—the average distance between gas molecules at one atmosphere pressure is 20 to 30 Å.

The gas laws. This is an appropriate point at which to give a brief summary of the gas laws as a knowledge of these laws will be necessary later. **Boyle's law** states that at constant temperature the volume of a given mass of gas is inversely proportional to its pressure:

$$V_{T,m} \propto \frac{1}{P} \quad \text{or} \quad (PV)_{T,m} = \text{a constant}. \tag{3.2}$$

Charles's law states that the volume of a given mass of gas is proportional to its absolute temperature, if its pressure is constant:

$$V_{P,m} \propto T, \tag{3.3}$$

where V is the volume, P the pressure, T the temperature, and m the mass. The subscripts denote the variables which are being held constant. The two laws may be combined to give a relationship between temperature, pressure, and volume for a given mass of gas as

$$(PV)_m \propto T$$

or, for one mole,

$$PV = RT, \tag{3.4}$$

where R is a constant known as the gas constant. Its value is 8.3 J mol^{-1} K^{-1}. The relationship can be extended to cover all quantities of gases since for n moles we have:

$$PV = nRT. \tag{3.5}$$

This equation is known as the **ideal gas equation** and R as the **universal gas constant**. It is independent of the pressure, temperature, or the number of moles in the sample.

The gas law is extremely useful in this form but it must be remembered that it is only an approximation for real gases. Accuracy decreases as gases come near their boiling points for two reasons. The first is that the forces between molecules (van der Waals forces) become significant as the molecules become closer. The second is that the volumes of the molecules become important as they come closer together. Conversely the behaviour of a gas approximates to the "ideal" as its pressure tends to zero.

Kinetic theory of gases. Order in gases is the exception. Only in certain rather special cases is there structural grouping of molecules. In acetic acid vapour, for example, two molecules are hydrogen bonded to form a dimer which moves as a single entity subject to frequent dissociation and then association again with other monomers. But this is an example of exceptional behaviour. In general a gas is characterized by disorder, the molecules being distributed at random and in a continual state of random movement. Collisions transfer energy throughout the gas. Thus if the walls of a vessel containing a gas are heated collisions transfer energy throughout the gas until gas and walls have the same temperature. Collisions with the wall also provide us with a molecular picture of the origin of pressure exerted by a gas on the walls of the vessel. The impact of the molecules colliding with the walls leads to a force on the wall, and the pressure is simply the force on unit area. We can readily obtain a simple picture of the effect.

Consider a cube with sides of length l as shown in Fig. 3.23. Let us take a very simple model in which we suppose one-third of the molecules are travelling along the x-direction, one-third along the y-direction, and one-third along the z-direction. Let us suppose further that all the molecules are travelling with a velocity of v. Consider the impacts on face A. Each of the molecules moving along the x-direction will strike the wall A at intervals determined by the time to travel a distance $2l$. That is each molecule will collide with the wall $v/2l$ times in unit time. We have assumed collisions with the wall are perfectly elastic. Before a collision the velocity is v, and after it $-v$. Taking the mass of a molecule as m, then the momentum changes from mv to $-mv$, that is there is a change of $2mv$ for each collision at the wall. The change in momentum at the wall per second due to one molecule is

$$2mv \frac{v}{2l} = \frac{mv^2}{l}. \tag{3.6}$$

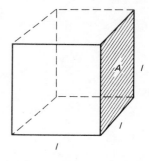

Fig. 3.23 Collisions on the face A of a cube, of sides, length l, are considered in calculating the pressure.

But there are $n/3$ molecules moving along the x-direction, where n is the total number of molecules in the vessel. The pressure, which is the total rate of change of momentum per second per unit area, is given by

$$P = \tfrac{1}{3}n \frac{mv^2}{l} \frac{1}{l^2} = \frac{1}{3} \frac{nmv^2}{V_b}, \tag{3.7}$$

where V_b is the volume of the box.

If we were to take a system containing one mole of the molecules considered, then we should have from Eq. (3.7):

$$PV = \tfrac{1}{3}Nmv^2 = \frac{2}{3}\left(\frac{Nmv^2}{2}\right), \tag{3.8}$$

where N is the Avogadro constant, V is the volume occupied by one mole of the gas, and $Nmv^2/2$ is the total kinetic energy of the molecules, which we shall call E. The equation reminds us of Boyle's law. Comparison of the model with experiment requires the total kinetic energy to be constant for a given temperature independent of P or V. We can take this further. For one mole of ideal gas we have

$$PV = RT.$$

The model leads to agreement with this equation if we put

$$E = \tfrac{3}{2} RT. \tag{3.9}$$

It leads to the remarkable result that temperature is a direct measure of the kinetic energy of molecules in a gas.

3.5 GASES

The model is an oversimplified one. In a more refined model, the molecules are not all regarded as moving with the same velocity. Even if we could start with all the molecules having the same velocity collisions would soon lead to some travelling with lower velocities and others with higher velocities. A model which allows for this and for random directions of motion leads to a similar result. It is in fact

$$PV = \frac{2}{3}\left(\frac{Nm\overline{v^2}}{2}\right), \quad (3.10)$$

where $\overline{v^2}$ is the mean of the squares of the velocities. We have from Eq. (3.10) and the ideal gas law (Eq. 3.4):

$$\frac{2}{3}\left(\frac{Nm\overline{v^2}}{2}\right) = RT. \quad (3.11)$$

From this we obtain the average kinetic energy of a molecule as

$$\frac{m\overline{v^2}}{2} = \frac{3}{2}\frac{R}{N}T = \tfrac{3}{2}kT, \quad (3.12)$$

where $k = R/N$ is known as the **Boltzmann constant**, the gas constant per molecule.

An important result can be seen, namely that the average kinetic energy of a molecule is $\tfrac{3}{2}kT$, independent of m, and so is the same for all gases. This means that if we compare two different gases at the same temperature heavy molecules move on average more slowly than light molecules. In fact if the molecules in one gas have mass m_1 and mean square velocity $\overline{v_1^2}$, and in the other m_2 and $\overline{v_2^2}$, then we have

$$m_1\overline{v_1^2} = \tfrac{3}{2}kT$$
$$m_2\overline{v_2^2} = \tfrac{3}{2}kT,$$

and therefore

$$\frac{\overline{v_1^2}}{\overline{v_2^2}} = \frac{m_2}{m_1} = \frac{M_2}{M_1}, \quad (3.13)$$

where M_1 and M_2 are the molecular masses of the two gases. The ratio of the root mean square velocities is given by

$$\frac{(\overline{v_1^2})^{1/2}}{(\overline{v_2^2})^{1/2}} = \left(\frac{M_2}{M_1}\right)^{1/2}. \quad (3.14)$$

Some values of root mean square velocities are given in Table 3.4. It is of interest to note the order of magnitude; for hydrogen the value is nearing 4000 miles per hour.

Table 3.4. Some values of root mean square velocities, $(\overline{v^2})^{1/2}$, for molecules in gases at 298 K.

	m sec^{-1}		m sec^{-1}
Hydrogen	1.93×10^3	Helium	1.37×10^3
Oxygen	4.84×10^2	Neon	6.07×10^2
Water	6.43×10^2	Argon	4.31×10^2
Carbon dioxide	2.38×10^2	Xenon	2.38×10^2

Finally we return to the point that in a gas there is a distribution of velocities, some molecules moving much faster than others. A graph may be plotted to show this distribution. An example is given in Fig. 6.4. We see that there is a whole range of velocities, the probability of finding a particular value being given by the height of the curve at that particular velocity. The effect of temperature on the distribution is of interest. We see that, as expected from Eq. (3.9), high velocities become more likely. The number with velocities in excess of a certain minimum is proportional to the areas of the shaded portions. Note how the proportion changes with temperature. The point is taken up in Section 8.2.

PROBLEMS

1. What information does X-ray diffraction give about the structure of matter?
2. Describe in your own words, with the aid of a diagram, the X-ray goniometer.
3. What methods would be suitable to find the structures for the following:
    a) crystalline sodium bromide;
    b) nitrogen gas;
    c) ice?
4. If the nearest distance between two atoms in Fig. 3.3 is 3 Å, what is the distance between the second series of planes shown in the diagram? If X-rays of wavelength $\lambda = 1$ Å are used, calculate the value of θ for first order diffraction from each of the two planes.
5. a) Calculate the radius of (i) the tetrahedral hole, and (ii) the octahedral hole, for close-packed spheres of radius 1 Å (see Fig. 3.14).
    b) Considering close-packed halide ions, in which of (i) or (ii) above can the lithium ion fit without producing distortion?
    c) Consider next the close-packed ions: (i) chloride ions, and (ii) sulphide ions. Compare fitting sodium ions in the octahedral and tetrahedral holes of (i), with fitting zinc ions in both kinds of holes in (ii).
6. Taking the density of sodium to be approximately 1 g cm^{-3}, devise a simple model and make a rough estimate of the distance between centres of sodium atoms in the metal. (Take any model you can think of that will allow you to make a simple calculation—the simpler the better.) How does the Avogadro constant enter your calculation? Do you think that with a better model, the calculation could be reversed to allow the Avogadro constant to be calculated?
7. Calculate the Avogadro constant from the information that the side of the cube for the diamond structure shown in Fig. 3.18, is 3.56696 Å. (In counting how many carbon atoms effectively belong to this cube, remember that those which are shared by one other cube count as halves, and those shared by seven others count as eighths.)
8. Twenty-two grams of a certain gas occupy 11.2 litres at s.t.p. Assume the gas behaves ideally.
    a) How many moles of gas are there in the system?
    b) What is the molecular weight of the gas?
    c) How many molecules are there in the system?
    d) How many molecules are there in 1 cm^3 of the gas?

e) What will be the volume occupied by the gas if it is heated to a temperature of 546 K and the pressure is simultaneously reduced to half of an atmosphere?

9. Make rough estimates (say $\pm 10\%$) of:
   a) the number of molecules in a radio valve with a volume of 25 cm^3, if the pressure of residual gas is 10^{-6} mmHg at room temperature;
   b) the weight of oxygen in a cylinder of 10 litres capacity, if the pressure at room temperature is 120 atmospheres;
   c) the mass of air in a room 6 × 3 × 3 m.

10. What volume of hydrogen at s.t.p. would be required to:
    a) reduce 1 kg of zinc oxide to zinc;
    b) form 1 kg of water on burning in oxygen;
    c) form 1 kg of lithium hydride (LiH)?

11. Trace out Fig. 6.4, and by counting squares, weighing cut-out areas, or any other suitable method, find for each temperature the proportion of molecules with speeds in excess of:
    a) 2.5×10^3 m sec^{-1};
    b) 3.0×10^3 m sec^{-1};
    c) 3.5×10^3 m sec^{-1};
    d) 4.0×10^3 m sec^{-1}.

    Make a plot and extrapolate to find the proportion, for each temperature, with speeds in excess of 10^4 m sec^{-1}. [*Hint*: consider a log plot.]

BIBLIOGRAPHY

Wheatley, P. J. *The Determination of Molecular Structure* (second edition), Oxford University Press (1959).

   This gives an introductory survey of the main physico-chemical methods that have been devised for the determination of molecular structure. The various diffraction and spectroscopic methods are discussed in turn and other physical methods are also examined.

Wells, A. F. *Structural Inorganic Chemistry* (second edition), Oxford University Press (1962).

   A valuable reference work for structures of inorganic compounds.

Chemical Society. *Interatomic Distances* (1958). *Supplement* (1965).

   A collection of structural information, interatomic distances, and bond angles.

Barclay, G. A. "Solid State Chemistry," in *Approach to Chemistry*, edited by G. H. Aylward and T. J. V. Findlay, University of New South Wales, Sydney (1965).

   An account of packing of spheres, and ionic and covalent radii, at about the same level as the present chapter.

CHAPTER 4

Structure, Bonding, and Energy

4.1 STRUCTURE AND BONDING: TYPES OF BONDING

In the last chapter the arrangement of atoms in substances was discussed but little mention was made of the forces which hold them together. What stops compounds from falling apart? The answer basically is very simple. Electrons are negatively charged and the nucleus of an atom, by virtue of the protons in it, is positively charged. When two atoms bond together to form a molecule, negatively charged electrons are sandwiched between two positively charged nuclei. The forces which keep them together are fundamentally electrostatic. But the situation is not quite as simple as that or all atoms would be capable of bonding with each other equally well. In fact atoms like those of the noble gases scarcely attract one another at all whereas hydrogen atoms have a very strong tendency indeed to join together in pairs. Two moles of hydrogen atoms will combine to form one mole of hydrogen with the evolution of 436 kJ:

$$2H(g) \rightarrow H_2(g); \quad \Delta H = -436 \text{ kJ}.$$

(ΔH is the enthalpy change of a process—see Section 4.11.1.) Everything depends on the number and distribution of electrons round the nucleus. The distribution of electrons round the nuclei of atoms is best seen by studying electron density maps (Section 3.1). Figures 4.1 and 4.2 show electron density maps of sodium chloride and calcium fluoride and figure 4.3 shows an electron density map of the hydrogen molecule. The lines on the maps join points of equal electron density just as the contours on a map join points of equal height above sea-level. Using the geographical analogy the sodium chloride and calcium fluoride maps consist of isolated, steep-sided hills. In terms of electrons, the electron density is high near the nucleus and falls off rapidly. In the hydrogen map the two peaks are linked with a shoulder of land. The sides of the hill slope gently away. In terms of electrons there is a relatively high concentration of electrons between the two nuclei. In this case the electrons are less confined to the individual nuclei. Clearly there are two rather different types of bonding here. Sodium chloride and calcium fluoride are well-known ionic compounds and hydrogen is molecular. The ionic compound caesium fluoride shows the ionic electron density pattern even more clearly than sodium chloride and calcium fluoride. Its bond type can be taken to be the extreme

4.1	Structure and bonding: types of bonding
4.2	Electronegativity
4.3	Fajans's rules
4.4	Dipole moments
4.5	The octet rule
4.6	Dative or co-ordinate bonding
4.7	Metallic bonding
4.8	Hydrogen bonding
4.9	Intermolecular forces
4.10	The electron-pair model and the shapes of simple molecules
4.11	Energy and bonding: internal energy and enthalpy
4.12	Bond energies and bonding
4.13	Ions in crystals and solutions

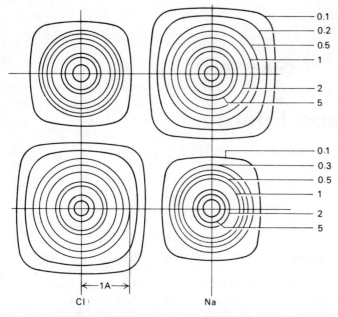

Fig. 4.1 Electron density map for sodium chloride. (Electron densities in electrons per cubic ångström.) After H. Witte and E. Wolfel, *Rev. Mod. Phys.*, **30**, 51–5 (1958). Reproduced by permission of the publisher.

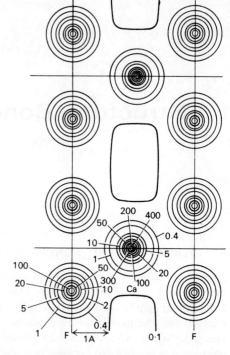

Fig. 4.2 Electron density map for calcium fluoride. (Electron densities in electrons per cubic ångström.) After H. Witte and E. Wolfel, *Rev. Mod. Phys.*, **30**, 51–5 (1958). Reproduced by permission of the publisher.

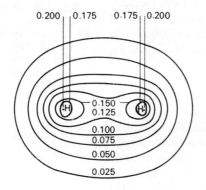

Fig. 4.3 Electron density map for hydrogen. (Electron densities in electrons per cubic atomic unit of length. The atomic unit of length is equal to the Bohr radius.) After C. A. Coulson, *Proc. Cam. Phil. Soc.*, **34**, 210 (1938). Reproduced by permission of the publisher.

example of ionic bonding. At the other end of the spectrum lies hydrogen. All chemical bonds lie between these two extremes.

In the case of ionic compounds the ions can be represented as separate species. The electron density maps show this separation well. The sodium atom loses an electron to form a sodium ion, Na^+, and the chlorine atom loses an electron to form the chloride ion, Cl^-. Treating ions as separate entities it is easy to see that ions of opposite charge will attract one another and that such attractions will hold giant structures of ions together. But nuclei all exert repulsive forces on each other since like charges repel and electrons also repel one another. However, these repulsive forces only become important when the ions approach each other closely.

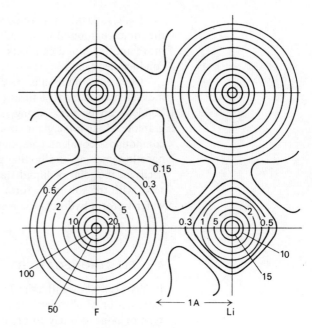

Fig. 4.4 Electron density map for lithium fluoride. (Electron densities in electrons per cubic ångström.) After H. Witte and E. Wolfel, *Rev. Mod. Phys.*, **30**, 51–5 (1958). Reproduced by permission of the publisher.

They lead to much smaller energies than those from the attractive forces due to charges on the ions and may be neglected in this simple model. Evidence for this lies in the large amount of energy needed to vaporize a mole of an ionic compound. The type of bonding which exists here is called **ionic bonding**. One additional point must be made here. The contours in the calcium fluoride map are perfectly circular, but those in the sodium chloride are squared off at the edges. This is a sign that the electrons do influence each other to some extent. A similar map of lithium fluoride (Fig. 4.4) shows this happening to a greater extent. It is important to realize that even when considering compounds commonly called ionic there is a tendency for them to show some of the characteristics of the hydrogen–hydrogen type of bond, though only very weakly.

The extreme represented by the hydrogen–hydrogen bond makes use of a different electron distribution. Here the two electrons are shared by the whole molecule. As the electrons spend much of their time between the two nuclei their negative charge acts as a bond between the two positively charged nuclei. This type of bonding is called **covalent bonding**. (See Section 12.2.)

Properties Associated with Bond Types

Compounds which are ionically bonded, that is, which consist of ions, are always found in the solid state at room temperature. The ions form a giant structure. A great deal of energy is required to break these bonds (Section 4.13.1) and consequently the melting and boiling points of ionic compounds are high. The giant structure is rigid and it can be seen from the diagrammatic representation in Fig. 4.5 that a shearing force on an ionic crystal will result in the ions being moved so that they are opposite like charges. The two planes of ions will then repel rather than attract one another and the crystal will break. This is a simplified picture but it accounts for the brittleness of ionic crystals.

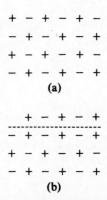

Fig. 4.5 Shearing an ionic crystal.

Ionic solids do not conduct electricity as the ions are not free to move. Once the ionic compound is melted (the term fused is often used) the ions are free to move and the liquid conducts electricity. Similarly, solutions of ionic compounds conduct electricity.

Covalent compounds vary in their properties according to whether they are molecular or giant structures (Section 3.2). Giant structures which are covalently bonded can be very strong, like diamond, or relatively weak and brittle, like graphite. Like ionic giant structures, they have high melting points. Unlike ionic compounds, covalent compounds are commonly molecular in structure. They have low melting and boiling points. Covalent compounds do not conduct electricity in the solid or liquid states, and conduct electricity in solution only if they react with the solvent to form ions. Hydrogen chloride, for instance, reacts with water to form hydrochloric acid which is an ionic solution.

4.2 ELECTRONEGATIVITY

It has been shown that most bonds are intermediate in character between the extremes of wholly electrovalent and wholly covalent. How can we predict which type of bond is likely to predominate in a compound between two elements? It would obviously be useful to have a quantity or property of each element to do this, but despite the fact that many attempts have been made, no simple and exactly measurable quantity has been found. Nevertheless the concept of electronegativity is very useful in a general way. Atoms which tend to attract electrons strongly are said to be strongly electronegative; fluorine is the most electronegative element. Elements which have a very weak tendency to attract electrons are weakly electronegative. Atoms which have a very weak electron affinity usually have relatively low first ionization energies; they lose one electron relatively easily. Caesium is the least electronegative element.

Attempts to quantify the concept of electronegativity and to give it a definite value for each element have been made independently by Mulliken and Pauling, and also by Allred and Rochow. Mulliken defined the electronegativity of an atom as the average of its ionization energy and electron affinity (Section 10.3.4). Ionization energies are quite easy to measure but electron affinities are more difficult and are not known for many elements. Pauling based his scale on bond energies (Section 4.12). A striking fact to emerge from the determination of bond energies is that for any pair of atoms A and B, the A—B bond energy is usually greater than the average of the A—A and B—B bond energies. A few examples are given in Table 4.1. Consistent with the generalization just made about relative bond strengths, reactions of the type

$$A\text{—}A + B\text{—}B \rightarrow 2A\text{—}B$$

are generally exothermic. The hydrogen–iodine reaction is an example of an exception. It can also be seen from Table 4.1 that the greater the difference in electronegativity the larger the heat of reaction. Pauling in fact used the ΔH values for such reactions to work out a quantitative scale of electronegativities, with the alkali metals, for example, having small values and the halogens large values.

Table 4.1. Enthalpy changes for some reactions:
A—A + B—B → 2A—B.

Reaction	ΔH (kJ)
H—H(g) + F—F(g) → 2H—F(g)	−542.3
H—H(g) + Cl—Cl(g) → 2H—Cl(g)	−184.6
H—H(g) + Br—Br(g) → 2H—Br(g)	−72.4
H—H(g) + I—I(g) → 2H—I(g)	+26.5
H—H(g) + H$_3$C—CH$_3$(g) → 2CH$_4$(g)	−65.0
Cl—Cl(g) + H$_3$C—CH$_3$(g) → 2CH$_3$Cl(g)	−79.4
Br—Br(g) + H$_3$C—CH$_3$(g) → 2CH$_3$Br(g)	+13.4

A more recent attempt to work out a scale of electronegativities which has been used a good deal is that due to Allred and Rochow. They look at **electronegativity** as "the force of attraction between an atom and an electron separated by a distance equal to the covalent radius of the atom". The force of attraction is expressed according to Coulomb's law as

$$F = \frac{Z_{\text{eff}} e^2}{r^2},$$

where Z_{eff} is the effective nuclear charge (Section 2.4), e the electronic charge, and r the covalent radius (Section 3.3). Z_{eff} can be calculated using certain rules about electron screening by electrons in different orbitals. Using this approach Allred and Rochow were able to obtain electronegativities in good agreement with Pauling's values. A major advantage of their method is that it may be applied to elements for which there is no information on ΔH values (required for the Pauling scale to be used) or on electron affinities (required for the Mulliken scale to be applied). Electronegativities on the Allred and Rochow scale are given in Table A.13. See Section 10.3.5.

4.3 FAJANS'S RULES

Fajans set out the series of "rules" which bear his name to guide the prediction of bond types between known atoms. According to these rules bonds will tend to be ionic if:

1. The stable ion of the atom has a small charge. For instance sodium (Na$^+$) is more likely to form ionic bonds than is aluminium (Al^{3+}), and chlorine (Cl$^-$) than is sulphur (S^{2-}). The explanation for this can be seen by studying ionization energies (Sections 2.4.1 and 10.3.3).
2. The atomic volume of the metal is large (like the alkali metals) and that of the non-metal is small.

The converse is true for covalent bonding.

A third rule may be added which is of particular value in discussing transition metal chemistry.

3. Where the ion has a d^{10} configuration, compounds formed from such atoms show more tendency to covalency. The explanation is that a d^{10} configuration is less shielding than a p^6 configuration (Section 2.4.3).

The second rule can be explained in terms of the radius (or volume) of the *cations* and *anions* that would be formed if an ionic bond were produced. The smallest cations are those with the largest charge: Al^{3+} with 0.5 Å radius; Mg^{2+}, 0.65 Å; Na^+, 0.95 Å. They also have the highest charge density both because of their higher total charge and their smaller volume. For this reason they are said to have a high polarizing power. Polarizing power means the ability to distort the electron distribution in a neighbouring atom, molecule, or ion. The smaller the ion the closer it can approach, and the greater the charge the greater the force it will exert electrically on a neighbouring electronic system. The electron distribution will be distorted in such a way that the part nearer the positive ion will become relatively negative (increased electron density) and the part farther away will be relatively positive. The atom, ion, or molecule under these conditions is said to be polarized. In the present context we consider cations attracting electrons and tending to distort the electron patterns of nearby atoms or ions in such a way that the bonds become more covalent. The exact opposite is true of anions. Those with a large charge are the largest in volume. Thus the radius of P^{3-} is 2.12 Å, of S^{2-} 1.84 Å, and of Cl^- 1.81 Å. Whereas cations are smaller than the atom from which they are formed, anions are larger. Anions are thus relatively polarizable and the larger the anion the more easily it is polarized. Looking at it from the point of view of polarizing power of cation and polarizability of anion, the bond between two atoms is more likely to be covalent if the possible cation is small and the possible anion large. (See Fig. 14.5.)

4.4 DIPOLE MOMENTS

Electronegativity and polarizing power have been presented as qualitative concepts. There is useful related concept which is quantifiable: the dipole moment. Consider a diatomic molecule consisting of two atoms of different electronegativities. The electrons forming the bond will be attracted more towards the nucleus of the atom with higher electronegativity. Such a molecule is said to be polar. The charge distribution in the molecule is such that one end is relatively negative and the other relatively positive. We can regard the molecule as having oppositely charged poles at its ends. Examples of polar diatomic molecules are HCl and HF, while H_2, Cl_2, and F_2 are non-polar molecules. Among polyatomic molecules, SO_2 and PCl_3 are polar molecules whereas CO_2 and CCl_4 are non-polar molecules. For these last two molecules the bonds are polar but the symmetry of the molecule is such that one end of the molecule cannot be relatively negative and the other positive, although this is the case for each bond considered separately. The symmetry of CCl_4 gives it a dipole moment of zero in spite of its having four polar bonds. The position is different for SO_2 which is a bent molecule and PCl_3 which is pyramidal. These molecules are themselves polar as well as having polar bonds. If the molecule is polar it behaves in the same kind of way as a small magnet: a field tends to align it. It has a small electric moment which is given by the charge at each pole times the distance between the charges, which for a diatomic molecule is approximately equal to the bond length. Bond lengths are of the order of 1 Å (10^{-8}cm) and the charges are of the order of 10^{-10} electrostatic units (e.s.u.). The units of dipole moment chosen by the Dutch physicist Debye, who introduced the theory, are 10^{-18} e.s.u. cm. The unit is known as the debye unit, D (the SI unit is coulomb meter; $D = 3.3356 \times 10^{-30}$ C m).

Dipole moments can be measured by the effect of a compound on the capacity of a condenser. The capacity of a condenser is increased if certain substances are placed between the plates. The ratio of the capacity with and without the substance between the plates is called the dielectric constant for the substance. If a compound with polar molecules is placed between the plates of a condenser and the plates are charged, the molecules will tend to line up in the electric field. They will do this against the tendency of the thermal motion of the molecules to knock them out of line. It is this that increases the capacity of the condenser. The greater the dipole moment the greater the dielectric constant. Values of dipole moments (μ) for some typical molecules are given in Table 4.2.

Table 4.2. Dipole moments of some typical molecules. (All values refer to molecules in the gas phase.)

Molecule	$\mu(D)$
CO	0.11
HCl	1.12
HBr	0.83
NaCl	8.5
CsCl	10.4
NH_3	1.47
PH_3	0.58
AsH_3	0.22
CH_3Cl	1.87
CH_3Br	1.80
CH_3CN	3.91
CH_3NO_2	3.46
SO_2	1.61
$(CH_3)_2CO$	2.93
$CH_3CH_2CH_3$	0.08

The alkali metal halides when vaporized give rise as might be expected to highly polar molecules. The cyanide group and the nitro-group are among the most polar groups in organic chemistry. Likewise carbon–halogen bonds are polar while carbon–hydrogen bonds are only weakly polar. Many molecules have zero dipole moment on account of symmetry: examples are CO_2, CH_4, BF_3, and PCl_5. (See Figs. 13.6, 16.5, and 19.6.)

4.5 THE OCTET RULE

A very simple rule allows us to predict the likely composition of ionic or covalent compounds. This is the so-called octet rule (Sections 12.2 and 4.10). The rule points to the stability of the noble gas atoms and to the fact that each of them (except helium) has eight electrons in the outer shell. No distinction is made between s- and p-orbitals in this case. The very frequent occurrence of the octet of electrons suggests that most atoms achieve a relatively stable state in compound formation by gaining or losing electrons so as to achieve this number of electrons in the outer shell. The rule works well for ionic compounds. The common cations for elements, other than the transition metals for which there are exceptions, are those in which all the electrons in the outer shell have been removed leaving an ion with a complete outer shell. This is the case for all the group I and group II cations and for many others. Similarly non-metals achieve an octet of electrons by accepting electrons to become anions. Groups VI and VII give particularly good examples.

Covalent compounds often achieve an octet of electrons but there are many exceptions. The way in which this is done is different from that of ionic compounds, as in covalent bonds electrons are shared between two nuclei. The case of chlorine provides a simple example. Each isolated neutral chlorine atom has seven electrons in the outer shell, the total electron configuration being $1s^2 2s^2 2p^6 3s^2 3p^5$. In ionic compounds it receives one electron which can be regarded as coming from the cation making Cl^-, $1s^2 2s^2 2p^6 3s^2 3p^6$. But in covalent compounds such as hydrogen chloride one of the outer electrons is shared with an outer electron of another atom, in this case hydrogen. The two bonding electrons are considered to belong to both nuclei and hence both achieve a noble gas structure. Note that in the case of hydrogen chloride the hydrogen atom achieves not an octet but the two-electron structure of helium. Hydrogen chloride has an appreciable dipole moment, 1.12 D, as might be expected for a compound formed between the highly electronegative chlorine atom and the much less electronegative hydrogen atom. Given the chance, as when the hydrogen chloride dissolves in water, the compound ionizes (Section 9.3.1).

Electron counting is a useful if occasionally misleading procedure. Diagrams called "dot and cross" diagrams are sometimes used to show the number of electrons involved in each bond. Usually the outer electrons only are considered. The electrons of one atom are given a dot as a symbol, those of the other a cross. Hence hydrogen H· combines with chlorine ×C̈l× to form hydrogen chloride:

$$H {\,\substack{\cdot \\ \times}\,} \ddot{Cl}{\,}^{\times\times}_{\times\times}$$

Similarly the structure of methane can be written as:

$$\begin{array}{c} H \\ {\scriptstyle \times | \cdot} \\ H {\,\substack{\cdot \\ \times}\,} C {\,\substack{\times \\ \cdot}\,} H \\ {\scriptstyle \cdot | \times} \\ H \end{array}$$

and carbon dioxide:

$$O {\,\substack{\times \\ \times}\,} C {\,\substack{\times \\ \times}\,} O$$

All models have their limitations and this one is certainly of very limited value. It is most important to understand that there are not two or more sorts of electrons in a molecule. There is no distinction at all between electrons which theoretically originate from different atoms. The electrons themselves are indistinguishable, though it may be convenient to denote them by dots and crosses for counting purposes, until one is familiar with the model.

4.6 DATIVE OR CO-ORDINATE BONDING

Some other types of chemical bonding must now be considered. It is well known that both water and ammonia can combine with copper(II) ions to form the blue $Cu(H_2O)_4^{2+}$ and the darker blue $Cu(NH_3)_4^{2+}$ ions. Neither the water nor the ammonia molecule has any unpaired electrons and so the bond cannot be formed according to the scheme described in the previous section. In fact it is the non-bonded electron pairs in water and ammonia molecules which are involved. A bond which is formed by a pair of electrons from one atom being shared between two nuclei is called a **dative** or **co-ordinate bond**. It is worth noting that the result is indistinguishable from any other two-electron bond; the difference lies in the way in which we view the formation of the bond. They are sometimes referred to as co-ordinate covalent bonds. A simple example is the formation of NH_3BF_3:

$$\begin{array}{ccccc} & H & & F & \\ & \cdot\cdot & & \cdot\cdot & \\ H:\!\!\!&N\!\!\!&: \;+\; &B\!\!\!&:F \\ & \cdot\cdot & & \cdot\cdot & \\ & H & & F & \end{array} \;\rightarrow\; \begin{array}{cc} H & F \\ \cdot\cdot & \cdot\cdot \\ H:N:B:F \\ \cdot\cdot & \cdot\cdot \\ H & F \end{array}$$

The octet rule is satisfied for both nitrogen and boron. If the electrons are assumed to be shared equally, then nitrogen has a half share of eight, that is effectively four electrons, instead of five as in an isolated neutral nitrogen atom. Boron on the other hand has effectively four instead of three. Therefore, the structure is sometimes written as:

$$\begin{array}{cc} H & F \\ | & | \\ H\!-\!\!N^+\!\!-\!B^-\!\!-\!F \\ | & | \\ H & F \end{array}$$

However, the electrons in the bond will not be shared equally and poles having full electronic charges will not be developed. Nevertheless such bonds are frequently strongly polar.

4.7 METALLIC BONDING

Metals have been recognized from the earliest times by their special physical properties. They are shiny, ductile, and malleable, and hence can be drawn and beaten into shape without breaking. They are above all good conductors of electricity. Chemically their properties depend on the ready formation of cations. But why do metals have their peculiar physical properties? Basically the structure shown by X-ray diffraction is a giant structure similar to non-metal covalent giant structures such as diamond and ionic giant structures such as sodium chloride. And yet only metals have the properties described above. As the structures are similar, differences in properties must be due to a different sort of bonding.

The outer electrons in metal atoms do not form pairs in the same way that covalently bonded non-metals do. They are not located between specific nuclei but are free to move around the whole structure. This is not a totally random movement as the free electrons all repel one another. These free electrons are said to be **delocalized**. A useful image of a metal structure is that of a sea of electrons around a giant structure of positively charged ions. This model accounts for the electrical conductivity and, since the bonds formed are non-directional, for the flexibility and consequently the malleability of metals (Section 10.2).

A comparison with electrons in atoms and molecules can be made. Electrons in atoms occupy orbitals about the nucleus. In a bond between two atoms electrons occupy orbitals encompassing both nuclei. In a molecule such as benzene, electron orbitals encompass six atoms (Section 4.12). In a metal the orbitals are giant orbitals, extending throughout the crystal structure of the metal, and there are a very large number of orbitals available to the electrons. The energy levels are so close together that they can be regarded as a band. As might be expected by comparison with the filling of atomic orbitals, electrons occupy the lowest levels first. If the whole band is fully occupied and the next band is much higher in energy the material is an insulator. If the band is incompletely filled or if it overlaps another higher band these additional orbitals are readily available to the electrons. Under these conditions the electrons are mobile and the material is a good conductor of electricity.

4.8 HYDROGEN BONDING

The anomalous properties of water can be understood in terms of another type of bonding which is sufficiently important to be considered on its own. The fact that water is an unusual liquid can be seen from the boiling point versus molecular weight graph of the hydrides of group VI in Fig. 4.6. Water behaves as if it had a much greater molecular weight than 18. In addition it has a high surface tension and heat of vaporization. Everything points to the fact that water molecules are bound together in some way. None of the bonding types mentioned above can account for this.

The electron pair which bonds the hydrogen and oxygen atoms together will be attracted more strongly by the more electronegative oxygen than by the less electronegative hydrogen. The electrons will spend more of their time near the oxygen nucleus than the hydrogen nucleus. This results in an unevenness of charge over the water molecule: the oxygen atom is more negatively charged than the two hydrogen atoms. As the molecule is V-shaped this results in the molecule having an electric dipole moment, the oxygen atom being the negative end and the hydrogen atoms the positive. This alone would cause water molecules to attract one another but there is more to it than this. A lone pair from the oxygen atom is able to form a weak bond with the slightly positively charged hydrogen atom of another water molecule as a result of the attractive forces between them. This bond is called a **hydrogen bond**. In water, since there are two hydrogen atoms each capable of forming a hydrogen bond with another water molecule, the structure built up is three dimensional. In ice the bonds hold the whole structure together. While —O—H---O hydrogen bonds are common, many other cases are known, for example:

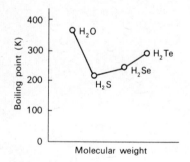

Fig. 4.6 Boiling points of hydrides of group VI elements.

The essential part of the hydrogen bond is the formation of the bond between a hydrogen atom attached to an electronegative atom and the lone pair of another electronegative atom. The more electronegative the atom is the stronger the hydrogen bond will be. It is interesting that only hydrogen forms this type of bond. When a hydrogen atom is bonded to a very electronegative atom, as in the HF molecule, electron withdrawal towards this atom exposes a neighbouring molecule to a less screened proton. The small size of the hydrogen atom allows close approach to the neighbouring molecule and there is consequently an interaction between the two molecules (Section 13.6.2).

4.9 INTERMOLECULAR FORCES

Dipole–Dipole Attractions

Hydrogen bonds are weak compared with covalent bonds, but there are still weaker intermolecular forces. These forces are important in determining fundamental physical properties such as boiling and melting points. One such force is the electrostatic attraction between molecules that are polar—called **dipole–dipole attraction**. Acetone, for example, a compound with a dipole moment of 2.9 D, has the boiling point 56.1 °C, compared with 0.8 °C for n-butane, a compound with a similar molecular weight but with a small or zero dipole moment.

We can readily envisage attraction between dipoles arising through electrostatic attraction between opposite charges. Polar molecules tend to orient others as shown diagrammatically in Fig. 4.7 and as a result there are attractive forces between the molecules.

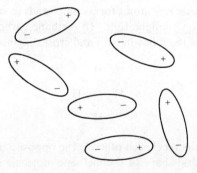

Fig. 4.7 Polar molecules tend to orient one another.

Van der Waals Forces

Weak attractive forces exist between all atoms and molecules. It is easy to see that in acetone vapour for example, the electric dipoles of two molecules will exert forces on each other. Similarly sulphur dioxide molecules interact through their electric dipoles. There are also weak forces between non-polar molecules such as methane or even between noble gas atoms. These forces are known as **van der Waals forces**. Their origin can be understood in the following way. At any particular instant we can imagine that the electron distribution in an atom may not be perfectly spherical, even though it is so on average. The instantaneous unsymmetrical distribution, with more electronic charge to one side than to the other, means the atom has at that instant an electric dipole moment. It will as a result induce a dipole moment in a neighbouring atom. Imagine that the positive end of the first dipole is pointing towards the second atom, then the induced dipole will have its negative end pointing towards the positive pole of the first. The result is a weak attraction between the atoms. Because the effect arises in this way between dipole and induced dipole it will always be attractive. Consequently, even though the atoms are on average spherical, the resultant forces do not average out to zero. Such forces fall off rapidly as the atoms or molecules become further apart. The greater the number of electrons in a molecule and the more weakly they are held by the nucleus, the greater the van der Waals forces between the molecules. It is these forces which determine the latent heats of sublimation of crystals of noble gases or substances composed of non-polar molecules. As the molecular size increases the number of electrons increases, and as a consequence heats of sublimation of substances composed of large molecules are large and vapour pressures are small.

4.10 THE ELECTRON-PAIR MODEL AND THE SHAPES OF SIMPLE MOLECULES

Although covalent bonds do not always consist of localized pairs of electrons (Section 4.12) the fact that many bonds are two-electron bonds leads to a useful model for correlating and predicting the shapes of simple molecules. In this model no distinction is made between s- and p-electron orbitals. It is assumed that all electrons are paired, whether they are involved in bonds or not, and that the pairs of electrons repel each other so that they take up a position of minimum potential energy.

According to the *octet rule* atoms form compounds in such a way as to achieve a noble gas like electron configuration. In methane carbon forms four covalent bonds with hydrogen as the following dot and cross diagram shows:

$$\begin{array}{c} H \\ {\scriptstyle \times |\, \cdot} \\ H \genfrac{}{}{0pt}{}{\cdot}{\times} C \genfrac{}{}{0pt}{}{\cdot}{\times} H \\ {\scriptstyle \cdot\, |\, \times} \\ H \end{array}$$

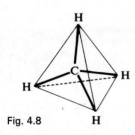

Fig. 4.8

These four electron pairs repel each other to the opposite corners of a tetrahedron and hence the tetrahedral shape of the methane molecule is accounted for in this model by bond-pair repulsions (Fig. 4.8).

Consider next the ammonia molecule. The octet of electrons around the nitrogen is derived from five for the nitrogen atom and one for each hydrogen atom. The electron dot diagram is therefore:

$$\begin{array}{c} {\scriptstyle ..} \\ H \genfrac{}{}{0pt}{}{\cdot}{\times} N \genfrac{}{}{0pt}{}{\times}{\cdot} H \\ {\scriptstyle \cdot\, |\, \times} \\ H \end{array}$$

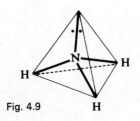

Fig. 4.9

If account is taken of the non-bonded electron pair the tetrahedral model can again be used. The shape predicted for the ammonia molecule will be a pyramid (Fig. 4.9).

In the case of water there are two non-bonded pairs of electrons:

$$:\!\overset{..}{\underset{H}{O}}\!\!\overset{.}{\underset{\times}{-}}\!H$$

and the predicted shape is shown in Fig. 4.10. Notice in this case that as the lone pairs can be involved in hydrogen bonding they determine the tetrahedral structure of ice and other structures in which hydrogen bonds are involved.

Finally it can be seen that the group VII hydrides are inevitably linear since there are only two atoms:

$$H\overset{.}{\underset{\times}{-}}\overset{..}{\underset{..}{F}}:$$

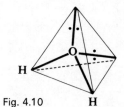

Fig. 4.10

The internal angle of a tetrahedron is 109.5°. How does the electron-pair repulsion model compare with the shapes of molecules found from X-ray diffraction and other experimental methods? Methane is found to have bond angles of 109.5° as predicted, but in ammonia the bond angle H—N—H is 107.3°, and in water the angle H—O—H is 104.5°. Why should the angles be smaller than those predicted? A simple explanation suggests that the non-bonding pairs of electrons are on average closer to the central nucleus than bonded pairs. This is reasonable since there is no positively charged nucleus to pull them out as there is for bonded pairs. As they are close to the nucleus they tend to repel the bonding electron pairs more effectively than these repel each other and so distort the tetrahedral shape. It follows that where there are non-bonded pairs of electrons, as in ammonia and water, bond angles will be smaller (Section 12.7).

The model can be extended to compounds which do not obey the octet rule. For instance boron trifluoride

$$\underset{\times\times}{\overset{\times\times}{\times F\times}}$$
$$\underset{\times\times}{\overset{\times\times}{\times F}}\!\overset{.}{\underset{\times}{-}}B\overset{.}{\underset{\times}{-}}\underset{\times\times}{\overset{\times\times}{F\times}}$$

is predicted to be a planar triangle. Mutual repulsion of bond pairs leads to these adopting positions as far apart as possible:

This is confirmed by experiment which establishes a regular planar structure with angles of 120°. Phosphorous pentafluoride, PF_5,

has the shape of a trigonal bipyramid (Fig. 4.11). Other examples are given under the chemistry of the elements. (See also Section 12.7.)

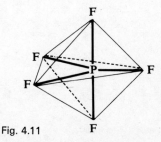

Fig. 4.11

4.10 ELECTRON-PAIR MODEL AND THE SHAPES OF SIMPLE MOLECULES

4.11 ENERGY AND BONDING: INTERNAL ENERGY AND ENTHALPY

The concept of energy is more difficult to define than to describe. "Energy is that which causes things to move or to heat up" is a reasonable generalization to start with. What is remarkable is that the concept of energy which started in men's minds from insignificant beginnings has turned out to be one of the most useful scientific ideas in existence.

Energy and its conservation. The first point to be made about energy concerns its conservation. We say that energy is conserved. This means that whatever the change or process we are considering the total amount of energy before and after the change is the same. The conservation of energy is the cornerstone of all energy theory. Heat and work are both ways of changing the energy of a system. No matter how a given change is brought about the algebraic sum of heat and work is always the same. This statement is made as a matter of experience. The physicist J. P. Joule demonstrated this and it is in recognition of this that the fundamental unit of energy is now the joule. The **joule** is defined as the energy transferred when a force of one newton acts through a distance of one metre. A newton is the force required to accelerate a mass of one kilogram by one metre per second per second. In electrical units one joule is the energy transferred when one coulomb of electricity is passed through a potential difference of one volt.

4.11.1 Internal Energy and Enthalpy

In this book we are mainly concerned with the application of energy concepts to chemical systems. By a system we mean the object of our discussion. We distinguish the system from its surroundings and focus attention on the system. The system, for example, might be contained in a flask, or it could be a single crystal or one mole of some substance, or a mixture of substances. When a chemical change takes place in a system, heat is usually either absorbed from the surroundings or given out to the surroundings, and sometimes a volume change takes place. Volume change represents an energy change as work must be done against pressure (usually atmospheric pressure) if the volume of the system is to expand. If the pressure P is constant, and the volume change is ΔV, then the work done is $P\Delta V$. That is to say

$$w = P\Delta V,$$

where w is the work done by the system.

We can associate with a system an **internal energy**, U. The internal energy of the system will change if the system absorbs heat or does work. The change in internal energy ΔU is given by

$$\Delta U = q - w, \qquad (4.1)$$

where q is the heat absorbed and w the work done by the system. The negative sign arises because by the convention used, the work done on the system is $-w$. The equation is a statement of the **First Law of Thermodynamics**. It also shows how ΔU may be obtained from experimentally measurable quantities. There is an important distinction to be made here. ΔU depends only on the final and initial states of the system; it does not matter how the change is made. On the other hand we cannot make such a statement about q or w; these will in general depend on how the change is made. The internal energy change of the system may arise through more or less absorption of heat provided the work done also changes in a compensatory way. U is said to be a state function; it has a value which depends only on the state of the system, independent of how that state was reached. For a change between two states, ΔU has a definite value independent of the way the change is brought about. One big advantage of using the concept of internal energy is seen straight away: we may tabulate values of ΔU for chemical changes whereas it would not be meaningful to attempt to do the same for q or w.

It is convenient to define another function of state which also proves to be of particular value for tabulation purposes in chemistry. The **enthalpy**, H, of a system is defined as

$$H = U + PV. \tag{4.2}$$

For any defined state of a system, P and V have definite values, and we have already seen U has a definite value, independent of how the state of the system was reached. It follows that H is a function of state. For a change between specified initial and final states of a system ΔH has a value independent of how the change is brought about. The question arises as to how to obtain values of ΔH from experimental measurements. Consider a change to be made at constant pressure, P; let the initial and final volumes be V_1 and V_2. Then from Eq. (4.2)

$$\begin{aligned}\Delta H &= \Delta U + P(V_2 - V_1) \\ &= \Delta U + P\Delta V \\ &= \Delta U + w.\end{aligned}$$

Hence from the First Law, using Eq. (4.1), we have

$$\Delta H = q_P, \tag{4.3}$$

where q_P is the heat absorbed by the system for the change at constant pressure. The enthalpy change is equal to the heat absorbed when the change is made at constant pressure.

This provides a basis for the experimental measurement of enthalpy changes. The determination of enthalpies, their tabulation, and use is considered in the following pages. We note here that the enthalpy change for a reaction depends to some extent on the temperature at which it occurs. Because of this the temperature at which ΔH has been measured is also given. Thus ΔH_{298} refers to a process occurring at 298 K.

We may also note that for a change at constant pressure we have

$$\Delta U = \Delta H - P\Delta V. \tag{4.4}$$

In practice for many purposes there is only a significant change in volume of a chemical system when a gas is evolved or absorbed, so that in many cases

$$P\Delta V = 0$$

and

$$\Delta U = \Delta H.$$

Even if a gas is evolved the energy used in "pushing back the atmosphere" is not very large. Consider the reaction of zinc in hydrochloric acid:

$$Zn(s) + 2HCl(aq) \rightarrow ZnCl_2(aq) + H_2(g)$$

i.e.

$$Zn(s) + 2H^+(aq) \rightarrow Zn^{2+}(aq) + H_2(g)$$

If a mole of hydrogen is produced at s.t.p. the work done against the atmosphere is 1×22.4 atmospheres litres, that is 2.5 kJ. The enthalpy change in this reaction is given by -152.3 kJ mol^{-1}. It can be seen that even in this case the contribution of $P\Delta V$ is not great. To simplify the application of energetics to chemistry the difference between ΔU and ΔH is often neglected. This will often be the case for applications of these ideas in this book. It is satisfactory where there is a large energy change resulting from chemical changes but would be quite wrong for a simple process such as expansion of a gas.

4.11.2 Measuring Changes in Enthalpy

Enthalpy changes are comparatively easy to measure. If the chemical change is allowed to proceed to completion in an insulated vessel the total heat change will take place entirely in the vessel itself. Provided the heat capacity of the system is known the enthalpy change can be worked out. The change measured in this way is obviously equal to the heat that would be transferred to the surroundings if the temperature of the system were allowed to fall to its initial value. Approximate enthalpy changes can be found by mixing reactants in a polystyrene cup and measuring the change in temperature.

Energy measurements are stated so that they refer to change in the system considered. An **exothermic reaction** is one in which heat is evolved. The system loses energy, and the surroundings (the environment) gain energy. The enthalpy change is therefore by definition negative; the ΔH value has a minus sign. For **endothermic changes**, in which heat is transferred from the environment to the system, ΔH has a positive value.

A number of terms are used in connection with enthalpy changes which must be understood. From Eq. (4.3), the **heat of reaction**, or more precisely the enthalpy change for a reaction, is the heat absorbed when the number of moles shown in the equation react completely at constant pressure, and the final temperature is the same as the initial value. The **heat of combustion** refers to the complete combustion of one mole of the substance in oxygen. The **heat of neutralization** of an acid, similarly, refers to complete neutralization of one mole of acid. Heats of reaction are frequently given as positive values when heat is evolved. Enthalpy changes for processes in which heat is evolved have, of course, negative values. The term **standard enthalpy of formation**, $\Delta H_f^\ominus$ is particularly important; it is the enthalpy change when one mole of the compound in its standard state is formed from its elements in their standard states. The **standard state** of a substance is the most stable form at one atmosphere pressure and at a temperature which is usually specified as 298 K. Standard enthalpies of formation are tabulated in Tables A.15 and A.16. Their use is described in Section 4.11.4.

Another quantity which we shall often want to use is the **enthalpy of atomization**, ΔH_{atom}. The enthalpy of atomization is the enthalpy change for the formation of one mole of gaseous atoms starting with the element in its standard state. For example the enthalpy of atomization of carbon refers to the change:

$$C(graphite) \rightarrow C(g); \quad \Delta H_{atom} = 715 \text{ kJ}.$$

But notice for a halogen it refers to:

$$\tfrac{1}{2}Cl_2(g) \rightarrow Cl(g); \quad \Delta H_{atom} = 122 \text{ kJ},$$

that is the formation of one mole of atoms from half a mole of molecules. Values are tabulated in Table A.13 and are discussed in Section 10.3.15 in relation to the Periodic Table.

Before considering enthalpy changes further, a consequence of the First Law called Hess's law of constant heat summation will be considered.

4.11.3 Hess's Law

There are many chemical reactions which cannot be studied directly by calorimetry. Hess's law enables information to be obtained for such cases. According to Hess's law heats of reaction are additive: a reaction which is performed in two or more separate stages gives the same final heat of reaction as that performed in one. More formally we may say the heat evolved or absorbed for a chemical change at constant pressure is independent of the steps taken in bringing about the change.

For example, the formation of sodium carbonate may be studied as a single-step process,

$$2NaOH(aq) + CO_2(g) \rightarrow Na_2CO_3(aq) + H_2O(l),$$

or in two stages via sodium hydrogen carbonate:

$$NaOH(aq) + CO_2(g) \rightarrow NaHCO_3(aq);$$

and

$$NaHCO_3(aq) + NaOH(aq) \rightarrow Na_2CO_3(aq) + H_2O(l).$$

For the first reaction $\Delta H_1 = -109$ kJ, and for the other two, $\Delta H_2 = +37$ kJ, and $\Delta H_3 = -146$ kJ. Hence $\Delta H_1 = \Delta H_2 + \Delta H_3$. This illustrates how heats of reactions may be added algebraically, which means that *heats of reaction which cannot be measured directly may be calculated indirectly.*

For instance the heat of formation of methane represented by the equation

$$C(graphite) + 2H_2(g) \rightarrow CH_4(g)$$

cannot be found directly as hydrogen and graphite do not readily react. The heat of formation, ΔH_f, is found by measuring the enthalpy changes of the following reactions:

$$CH_4(g) + 2O_2(g) \rightarrow CO_2(g) + 2H_2O(l); \quad \Delta H_1 = x$$
$$C(graphite) + O_2(g) \rightarrow CO_2(g); \quad \Delta H_2 = y$$
$$H_2(g) + \tfrac{1}{2}O_2(g) \rightarrow H_2O(l); \quad \Delta H_3 = z.$$

The enthalpy of formation of methane is $y + 2z - x$. The symbol for enthalpy of formation is ΔH_f. Under standard conditions and at a given temperature T the

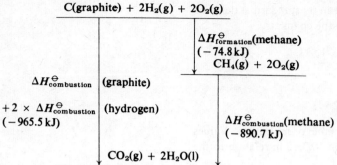

Fig. 4.12 Enthalpy diagram for formation of methane.

standard enthalpy of formation is given the symbol $\Delta H_{f,T}^{\ominus}$. The usual temperature is 25 °C (298 K). The actual values are:

enthalpy of combustion of graphite = -393.7 kJ mol^{-1}
enthalpy of combustion of methane = -890.7 kJ mol^{-1}
enthalpy of combustion of hydrogen = -285.9 kJ mol^{-1}.

Hence the enthalpy of formation of methane

$$\Delta H_{f,298}^{\ominus} = -74.8 \text{ kJ mol}^{-1}.$$

4.11.4 Enthalpy Changes and Enthalpy Diagrams

Information about energy changes can be usefully expressed in energy diagrams. In these diagrams the vertical axis represents enthalpy and the vertical distance between enthalpies for two systems represents the difference in enthalpy between them. The information used in finding the heat of formation of methane can be expressed as in Fig. 4.12. It is seen that the enthalpy of each system can be stated with respect to the enthalpy of a system consisting of the elements. This in fact provides a general basis for a systematic procedure for calculating enthalpy changes of chemical reactions.

The use of standard enthalpies of formation ($\Delta H_f^{\ominus}$) offers just such a systematic procedure. The standard state of a substance is chosen as the most stable form at one atmosphere pressure and a specified temperature. The elements in their standard states are chosen as reference points and by convention their standard enthalpies of formation are zero. The temperature is usually taken to be 298 K. For example we have:

$$\Delta H_f^{\ominus}(\text{graphite, 298 K, 1 atm}) = 0$$
$$\Delta H_f^{\ominus}(\text{hydrogen, 298 K, 1 atm}) = 0$$
$$\Delta H_f^{\ominus}(\text{methane, 298 K, 1 atm}) = -74.8 \text{ kJ}.$$

Some values of standard enthalpies of formation are given in Tables A.15 and A.16. From such tables heats of reactions may be readily calculated. For example, consider

$$C_2H_4(g) + H_2O(l) \rightarrow C_2H_5OH(l).$$

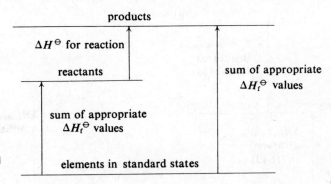

Fig. 4.13 Relation between ΔH for reaction and standard enthalpies of formation of products and reactants.

From the tables the enthalpy of formation of the reactants is -233.6 kJ and that of the product -277.0 kJ. Hence $\Delta H^\ominus$ for the reaction is given by the difference as -43.4 kJ. The procedure may be applied quite generally. It is summarized diagrammatically in Fig. 4.13. Written as an equation, $\Delta H^\ominus$ for a reaction is given by

$$\Delta H^\ominus = \sum \Delta H_f^\ominus(\text{products}) - \sum \Delta H_f^\ominus(\text{reactants}),$$

where the number of moles of products and reactants must be allowed for. The validity of such calculations rests directly on the First Law.

4.12 BOND ENERGIES AND BONDING

Examination of the enthalpies of combustion of the hydrocarbons given in Table 4.3 shows that there is a relationship between the heats of combustion and the number of carbon and hydrogen atoms present.

The graph in Fig. 4.14 shows that there is an approximately linear relationship between the heat of combustion and the number of carbon atoms. The conversion of methane by combustion to carbon dioxide and water may be regarded as the breaking of carbon–hydrogen and oxygen–oxygen bonds followed by the formation of carbon–oxygen and hydrogen–oxygen bonds. The increase in heats of combustion can be explained by the fact that the heat of combustion can be interpreted in terms of bond breaking and bond forming. Going along the series of

Table 4.3. Enthalpies of combustion.

Hydrocarbon	$-\Delta H$ (kJ mol^{-1})
Methane	890
Ethane	1560
Propane	2220
Butane	2880
Pentane	3510
Hexane	4195

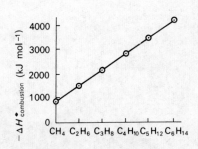

Fig. 4.14 Heats of combustion of hydrocarbons.

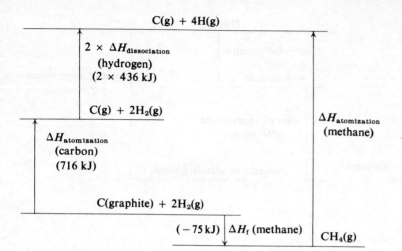

Fig. 4.15 Determination of C—H bond energy.

hydrocarbons each step involves the breaking of the bonds in one additional CH_2 group forming carbon dioxide and water. The approximately constant increase for each step suggests that the energy to break the carbon–hydrogen bond is approximately the same for each CH_2 group and for each carbon–carbon bond. This in fact proves to be a good approximation which can be quite widely used. Each bond is associated with a particular bond energy; this is the energy evolved when it is made, and absorbed when it is broken. Bond energies are not quite as easy to find as might perhaps be supposed however.

The enthalpy of formation does not lead directly to the bond energies. The definition of enthalpy of formation being "from the elements in their standard states" means that enthalpy of formation includes bond breaking of the elements in their standard states as well as the making of the bonds in the compound. Take the case of methane again. To make a mole of methane, a mole of graphite has to be "atomized" and two moles of hydrogen molecules divided into four moles of hydrogen atoms. In effect this means that the enthalpy of atomization of graphite must be found,

$$C(graphite) \rightarrow C(g),$$

and the bond energy of two moles of hydrogen molecules,

$$2H_2(g) \rightarrow 4H(g).$$

The bond energy of hydrogen can be found spectroscopically as the energy required to dissociate the molecule. The heat of atomization of graphite is difficult to determine and is in fact found through the temperature dependence of its vapour pressure (Eq. 5.5). The bond energy of carbon–hydrogen bonds in methane can be found by adding the heat of atomization of graphite and twice the bond energy of hydrogen to the heat of formation of methane. As can be seen from Fig. 4.15 this gives the heat of atomization of methane as $+1663$ kJ. The average bond energy for the carbon–hydrogen bond in methane is therefore 416 kJ mol^{-1}.

The bond energy for the carbon–carbon bond in ethane can now be calculated assuming that the carbon–hydrogen bonds have the same strength. Alternatively, by comparing heats of combustion of ethane and methane a value for the C—C and C—H bonds can again be found, there being two unknowns and two equations. It is found that this gives slightly different results. In fact the values for C—H bonds and all other bonds vary a little according to the molecule in which they are found. In order to obtain a value for a bond energy which can be used generally, an average value is taken for the bonds in a large number of compounds. This average value for the bond energy is called the **bond energy term**, but often for short simply bond energy.

It should be recognized that a bond energy term is not the same as a bond dissociation energy, except for a diatomic molecule where there is no distinction. For methane the first bond dissociation energy is the energy change for

$$CH_4 \rightarrow CH_3 + H,$$

the second dissociation energy relates to the process

$$CH_3 \rightarrow CH_2 + H,$$

and similar equations refer to third and fourth bond dissociation energies. The four have different values; they are in fact 423, 480, 425, and 335 kJ respectively. The average is of course equal to the bond energy term, since the sum of all four is

$$CH_4 \rightarrow C + 4H; \quad \Delta H_{298} = 1663 \text{ kJ},$$

and one quarter of ΔH for this process gives the required bond energy term.

Bond energy terms are average values and can be usefully applied to chemical changes in which products and reactants all have their usual valencies (Section 16.1), but not to processes such as those just considered. Under the former conditions the idea of bond energies being additive and transferable from molecule to molecule holds reasonably well. Nevertheless small deviations are common and must be expected. A number of bond energies are collected in Table 4.4. A more extensive set of values will be found in Table A.11.

Table 4.4. Some bond energy terms (kJ mol^{-1}). (Values for diatomic molecules are bond dissociation energies.)

H—H	436	N—H	389
C—C	347	O—H	464
C=C	611	F—H	565
C≡C	837	Si—H	318
C—H	414	P—H	322
C—Cl	326	S—H	368
C—Br	272	Cl—H	431
C—I	238	Br—H	364
F—F	158	I—H	299
Cl—Cl	244	As—H	247
Br—Br	192	C—O	360
I—I	151	C=O	736

An important application of bond energy terms is their use in the calculation of enthalpies of reactions. Unlike calculations from standard enthalpies of formation, such calculations are not precise because the idea of a bond energy term is not an exact one. Nevertheless, such calculations are often sufficiently accurate for many purposes. For example ΔH for the process

$$CH_4(g) + Cl_2(g) \rightarrow CH_3Cl(g) + HCl(g)$$

may be obtained from bond energies in Table 4.4 in the following way.

Bonds broken:		Bonds formed:	
Cl—Cl	244	H—Cl	431
C—H	414	C—Cl	326
	658 kJ		757 kJ

Estimated $\Delta H = 658 - 757 = -99$ kJ.

The value is in fact in this case very close to the experimental value of -99.5 kJ. To a good approximation bond energies are additive and to this extent they may be applied in the way illustrated above. Calculations of this kind are useful when insufficient calorimetric measurements have been made to allow the calculation from Hess's law. Another example would be the prediction of ΔH for a reaction involving a compound which has not yet been prepared or isolated in a pure condition. Bond energies are also useful in understanding bonding. Cases where the use of bond energies break down, for example in small rings or aromatic compounds, are of special interest for the further light they throw on bonding. For small rings the deviation has been attributed to strain, that is, to bond angles smaller than tetrahedral, and for aromatic compounds it provides evidence supporting the need for a modification of bonding theories for such molecules (Sections 4.12.1 and 21.5).

It is interesting to compare bond lengths and energies with the number of electrons involved in each bond. The carbon–carbon bonds in ethane, ethylene, and acetylene (Section 21.4) are compared in Table 4.5.

Table 4.5

	Bond length (Å)	Bond energy (kJ mol^{-1})	Number of bonding electrons
Ethane	1.54	346	2
Ethylene	1.34	611	4
Acetylene	1.20	837	6

There is a rough relationship between the number of bonding electrons and the length and strength of the bond. It is not surprising that the more electrons involved the stronger and shorter the bond. The electrostatic attraction is greater with more electrons and consequently the nuclei are drawn closer together. (While this is generally true, there are some exceptions. See for example, the oxygen molecule, Section 18.5.)

We shall see in the next section that although a large number of bonds have the expected lengths for single, double, or triple bonds, there are other cases where bonds have lengths which are intermediate between these values. Thus some bond lengths are found which are intermediate between the usual single and double bond values. This suggests the idea of fractional *bond order*, where the number of bonding electrons is effectively intermediate between two and four for example. We will see how this can arise in the next section. One way of looking at bond order is to draw a smooth curve through a plot of lengths of single, double, and triple bonds against bond orders 1, 2, and 3. From the curve a bond order can be assigned for bonds of intermediate length. For example, using such a curve, the centre bond in 1,3-butadiene, which has a bond length of 1.48 Å, is found to have a bond order of 1.2. The bonds in benzene have a bond order of 1.5.

4.12.1 Delocalization

For a very large number of chemical reactions the values of ΔH are in reasonably close agreement with the values calculated from bond energies. It is this that establishes the value of the idea of bond energies. Once an idea is well established it is often the exceptions, rather than further examples that follow the rule, which are of greater scientific interest. The exceptions call for a re-examination of ideas and may lead to a new development in our understanding.

An outstanding exception to the general success of using bond energies in calculating ΔH values for chemical reactions is provided by benzene and its compounds. The heat of hydrogenation of cyclohexene is -120 kJ mol^{-1}:

$$\text{cyclohexene} + H_2 \rightarrow \text{cyclohexane} \qquad \Delta H_{298} = -120 \text{ kJ}$$

The value is close to that typically found for double bonds. However, for benzene the value found is not $3 \times -120 = -360$ kJ but instead is approximately -205 kJ:

$$\text{benzene} + 3H_2 \rightarrow \text{cyclohexane} \qquad \Delta H_{298} = -205 \text{ kJ}$$

We see from the experimental value that the benzene molecule is energetically more stable by 155 kJ than we should expect from a structure with alternate single and double bonds.

The benzene molecule shows another deviation from expectations based on the structure with alternate single and double bonds. X-ray and electron diffraction studies lead to the conclusion that all six bonds are equal in length with a value of 1.40 Å whereas typical single bonds have a length of 1.54 Å and double bonds 1.34 Å. There are numerous examples of this outside the chemistry of aromatic rings. Simple examples are the nitrate and carbonate ions. A conventional structure for the nitrate ion would be:

The lines indicate electron-pair bonds in the usual way and the charges are those which result from assuming that bond pairs are shared equally between the bonded atoms (Section 4.6). Again, experiment shows not two long bonds and one short, but three bonds of equal length, at 120° in a planar structure.

Another example is the carbonate ion for which a conventional structure is:

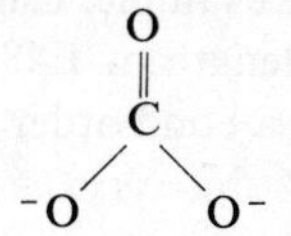

$$\begin{array}{c} O \\ \parallel \\ C \\ \diagup \quad \diagdown \\ {}^-O \quad\quad O^- \end{array}$$

It has a regular planar shape with all three bonds of equal length. Other examples are provided by the formate ion where both C—O bonds have the same length, and nitromethane where both N—O bonds are equivalent:

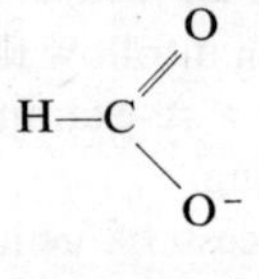

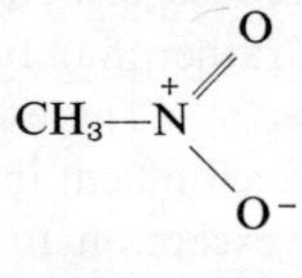

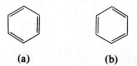

The idea of an electron-pair bond is inadequate to account for bonding in such molecules. Two important and widely used approaches to bonding have been extended to account for these effects. In one the idea of an orbital is extended so that it no longer refers simply to two atoms as in the usual electron-pair bond. Instead, orbitals are envisaged which extend over several atoms, six in the case of benzene. Such orbitals are called delocalized orbitals to distinguish them from orbitals localized on two atoms in the familiar electron-pair bond between two atoms. This description is discussed further in Section A.2. The other approach keeps more closely to the valence-bond structural formulae which chemists conventionally use. It describes the bonding in terms of conventional structures but with an additional feature. The Kekulé structure for benzene can be written in two ways:

(a)   (b)

Neither structure accounts for the stability or for the bond lengths of benzene. In the valence-bond account, the electronic structure of a molecule is regarded as intermediate between the structures represented by conventional valence-bond structures. The formate ion for example is regarded as intermediate between (a) and (b):

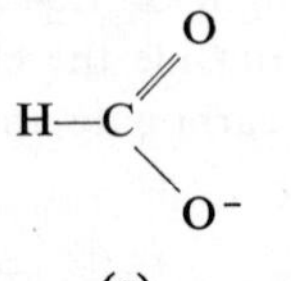

(a) 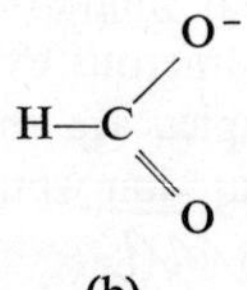   (b)

The carbon–oxygen bonds are equivalent, having a bond length intermediate between that for single and double bonds. Similarly in benzene all the bonds are equivalent being intermediate between single and double bonds, both in strength and in length.

The molecular-orbital and valence-bond approaches should not be regarded as rival theories awaiting a decision as to which is right. They begin from different starting points in an attempt to give an account of bonding and so make different approaches towards the same goal of describing the real state of affairs in molecules. The ideas they introduce are particularly important when the delocalization energy of electrons is large, as in benzene where localized electron pair bonding between adjacent atoms fails to provide a satisfactory description even of the regular hexagonal shape. The valence-bond approximation of describing the electronic structure as intermediate between valence-bond structures has the advantage of retaining structures with which the chemist is familiar, but it must not be thought that these individual structures have physical significance. They represent together a delocalization of electrons in the molecule. Thus the carbonate ion can be represented as intermediate in structure between (a), (b), and (c):

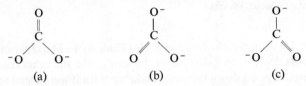

(a), (b), and (c) do not have separate physical significance; they are mental constructs. They are all called canonical forms. The actual structure of the molecule is said to be a resonance hybrid of the canonical forms. The term resonance is often used to refer to this type of description of bonding. The name can be misleading. Nothing moves, the molecule concerned does not fluctuate between the different canonical forms. Resonance refers to a situation where we cannot represent the electronic structure of a molecule by a single conventional structural formula, and where we attempt to overcome this by writing two or more such formulae which between them provide a better representation of the actual electronic structure of the molecule.

4.12.2 Bond Energies, Ionic Character, and Electronegativity

In Section 4.2 we saw that A—B bonds are usually stronger than the average of the strengths of A—A and B—B bonds. That is to say, the A—B bond energy is greater than the average of the A—A and B—B bond energies. It was also seen that the greater the difference between the electronegativities of A and B the greater the effect. In this section we examine the interpretation of this effect in terms of bonding.

At one extreme we take as an example the chlorine molecule, Cl_2, where the electronegativities are of course equal, and at the other extreme sodium chloride, NaCl, where there is a large difference in the electronegativities. The first of these is clearly an example of covalent bonding and the second an example of ionic bonding. As expected the first has zero dipole moment and the second a very large dipole moment, 8.5 D (Section 4.4). There are many molecules with dipole moments between these two extremes as Table 4.2 shows. The question arises as to how the bonding changes as we pass along a series of molecules between these two extremes. The answer is that there is no abrupt change but a gradual change from covalent to ionic bonding. These are two extreme forms of bonding but for most molecules the bonding is intermediate between these two extremes. These molecules have polar bonds, and bond energies which are greater than would be expected from the average of the two corresponding covalent bonds (Section 4.2).

Table 4.6. Bond energies and polarities.

	Dipole moment (D)	Bond length (Å)	Bond energy observed (kJ)	Bond energy calculated* (kJ)	Difference (kJ)
HF	1.90	0.92	565	297	268
HCl	1.03	1.27	431	340	91
HBr	0.78	1.41	364	319	45
HI	0.38	1.61	299	294	5

* For method of calculation, see text.

Consider, for example, the halogen acids (Table 4.6). The table shows that the dipole moments decrease in going down the series. On the other hand, the bond length increases in going down the series and this of itself would lead to an increase in dipole moment for the same partial charge developed on the hydrogen and halogen atoms in the molecules. It follows that the partial electric charges developed must be largest for hydrogen fluoride and fall off down the series. Hydrogen fluoride has the greatest tendency towards ionic bonding H^+F^-. We can write two canonical forms for the molecule, H—F and H^+F^- (Section 4.12.1). We may describe the molecule as a resonance hybrid of these two forms, the former being the more important. Because the latter has some importance we say that the bond has some ionic character. The greater the electronegativity difference between the atoms bonded together then generally the greater the ionic character.

The trend in bond energies is also of interest. The calculated values are the average of hydrogen and halogen molecule bond energies. For hydrogen iodide the calculated and observed are not very different, but as the ionic character of the bond increases so the difference between observed and calculated values increases. As the ionic character increases so the bond becomes increasingly stronger than would be expected on the basis of the averages.

The increase in partial ionic character means that the ionic structure H^+X^- is more important in describing the bond for HF than for HI. The reason for this is that electron affinity of iodine being small, H^+I^- has a relatively high energy, whereas H^+F^- has a relatively low energy. As a consequence both bonds have some ionic character but that for hydrogen fluoride is considerably more. The situation has analogies with the description of bonding in benzene (Section 4.12.1). In that case the bonding is described in terms of equal contributions from two equivalent structures leading to a structure which is energetically more stable than either. In the present case the two structures are not equivalent but as they become energetically closer so the ionic character increases and the more energetically stable the molecule becomes.

In the general case of A—A, B—B, and A—B, stability will be conferred if one of ionic structure A^+B^- or A^-B^+ has comparatively low energy. The requirement is that one atom should have a low ionization potential and the other a large electron affinity. The greater the electronegativity difference the more important will be the ionic structure in describing the real structure of the molecule and the greater its stability by comparison with A—A and B—B molecules. In fact as discussed in Section 4.2 the Pauling scale of electronegativity is based on a comparison of the bond strength of A—B and those of A—A and B—B.

Finally we note that there are molecules such as CCl_4 which have polar bonds but do not have dipole moments because of symmetry. Such bonds are also described as having partial ionic character, C^+Cl^-, and again the ionic character depends on electronegativity difference.

4.13 IONS IN CRYSTALS AND SOLUTIONS

4.13.1 Lattice Energies: The Born–Haber Cycle

Taking sodium chloride as an example, the energy change for the process

$$NaCl(s) \rightarrow Na^+(g) + Cl^-(g),$$

which is known as the **lattice energy**, ΔH_{latt}, is of fundamental importance in understanding the energetics of bonding. It cannot be determined calorimetrically or spectroscopically by direct experiment but can be obtained indirectly.

It is well known that strongly electronegative non-metals such as chlorine react with electropositive metals such as sodium with the evolution of a great deal of heat. If the process is analysed the following energy steps can be isolated. It must be emphasized that the reaction does not actually take place in this way. Hess's law enables us to break down the process into steps which can be conveniently studied experimentally. We begin with one mole of sodium and half a mole of chlorine.

1. The sodium is atomized and ionized; heat of atomization and ionization energy are required to bring these two steps about.
2. The chlorine is atomized; the heat of atomization of chlorine is required. Addition of an electron liberates the energy equal to the electron affinity of chlorine, forming the chloride ion.
3. The gaseous sodium and chloride are then combined. This last energy change, which is large, is the lattice energy. Looked at the other way round it is the energy required to convert a mole of an ionic compound into separate ions.

This scheme, which is called a Born–Haber cycle, can be conveniently illustrated in an energy diagram, such as Fig. 4.16. An alternative diagrammatic representation is given in Section 14.2.4. It is evident from Fig. 4.16 that the lattice energy for MX is given by

$$\Delta H_{latt} = \Delta H_{atom}(M) + \Delta H_{atom}(X_2) + I(M) - E(X) - \Delta H_f(MX), \quad (4.5)$$

where I and E refer to ionization energy and electron affinity. The expression allows the calculation of lattice energies which are a direct measure of the stability of ionic crystals. We consider next a simple model for ionic bonding and calculate the lattice energy in terms of the model. The extent of agreement with experiment is an indication of whether the model is a good one.

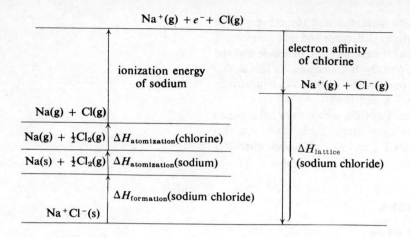

Fig. 4.16 Energy diagram for formation of sodium chloride. Born–Haber cycle.

The basis of such a calculation in terms of a model is outlined below and then comparison is made with experiment. If we represent two ions by point charges distance r apart,

|←r→|
+   −

we have according to Coulomb's law a force f between them given by

$$f = \frac{e^2}{r^2}, \qquad (4.6)$$

where e represents the electronic charge. The potential energy, V, that is the energy required to separate them to infinite distance, is given by the equation

$$V = \frac{-e^2}{r}. \qquad (4.7)$$

The negative sign arises because it is convenient to take the potential energy to be zero when the ions are separated to infinity. If we consider a linear array with a regular separation r:

|←r→|←r→|←r→|←r→|←r→|←r→|
−   +   −   +   −   +   −
            I

then the energy of a given ion I, due to the presence of the others which exert coulombic forces on it, is

$$V_I = -e^2\left(\frac{2}{r} - \frac{2}{2r} + \frac{2}{3r} - \frac{2}{4r}\cdots\right).$$

More realistically as shown in the three-dimensional model of face-centred cubic sodium chloride (Fig. 4.17), a given ion A has six nearest neighbours, twelve next nearest and so on. Putting in the numbers at the various distances we have, for the potential energy of ion A due to coulombic forces, the expression

$$V_A = \frac{-e^2}{r}\left(\frac{6}{1} - \frac{12}{2} + \frac{8}{3} - \frac{6}{4} + \cdots\right) = \frac{-Me^2}{r}, \qquad (4.8)$$

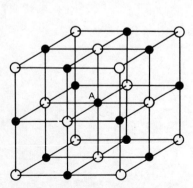

Fig. 4.17 Environment for ion A in a sodium chloride structure.

where r is the distance between centres of nearest neighbours. The quantity in brackets needs to be summed to infinity. It depends only on the geometry and not on the chemistry; it will have the same value for all face-centred cubic crystals. There will be such a constant for each crystal structure, with a value for all crystals having the fluorite structure and another for all crystals with the zinc blende structure. Such constants are generally called Madelung constants and denoted by the symbol M. Some typical values are listed in Table 4.7.

Table 4.7. Madelung constants.

Structure	Madelung constant
NaCl	1.748
CsCl	1.763
CaF$_2$	2.519
Zinc blende	1.638
Wurtzite	1.641

The generalized energy expression is written

$$\frac{-Me^2 z_+ z_-}{r},$$

where z_+ and z_- are the number of charges on the positive and negative ions, and $r = r_+ + r_-$.

So far we have considered only coulombic forces; these alone would lead to collapse of the model to a point. We need to include interionic repulsive forces which become very important when two ions approach closely, even though they may have opposite charges. As the electron shells of the two ions begin to overlap the forces rise rapidly, as the ions are brought closer together, and so does the interaction energy due to this repulsion. Figure 4.18 indicates the effect qualitatively. The energy term is of opposite sign to the coulombic attractive term and rises very rapidly at short distances of approach while it is negligible for larger distances. The curve for the coulombic term, according to the expression above, varies as $1/r$. The curve for the repulsion terms can be described by a much higher power function of the form k/r^n where n is large. A value of $n = 12$ is commonly used though present knowledge of the form of the repulsive energy term is in need of further development. For the present calculation it proves to be the case that the lattice energy is fortunately not particularly sensitive to n. When the repulsive term is allowed for, the general expression for the lattice energy per mole is found to be as follows:

$$\text{lattice energy} = \frac{MNz^2 e^2}{r}\left(1 - \frac{1}{n}\right), \qquad (4.9)$$

where $z^2 = z_+ z_-$, and e is the electronic charge in electrostatic units.

This expression is commonly used but other similar expressions will also be found depending on the functional form that is adopted for the repulsive energy term. Inspection of the above expression shows that it is not sensitive to the value

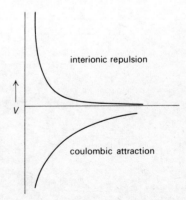

Fig. 4.18 Comparison of coulombic attraction and interionic repulsion potentials.

of n. For example, using 11 or 13 does not make a marked change in the calculated lattice energy.

It is of interest to compare theoretical values obtained by using the model (Eq. 4.9) with values obtained from experimental measurements using the Born–Haber cycle (Eq. 4.5). A comparison is made for some typical compounds in Table 4.8.

Table 4.8. Comparison of theoretical and experimental values (kJ mol^{-1}) for lattice energies

Compound	Theoretical value (from model)	Experimental value (from Born–Haber cycle)
NaCl	766	788
NaBr	731	719
NaI	686	670
KCl	692	718
KBr	667	656
KI	631	615
AgCl	769	921
AgBr	759	876
AgI	736	862
ZnS	3430	3739

Comparison of the theoretical values calculated from a simple model with experiment shows good agreement with experiment for the alkali halides, considering the simplicity of the model. Attempts have been made to improve the model to obtain better agreement, but for many purposes the important point is that the good agreement justifies the use of the simple ionic model for alkali metal halides.

Exceptions from the agreement are of interest in that they show the need to modify the model at least for some compounds. Table 4.8 shows that for silver halides and zinc sulphide, for example, the agreement is not nearly as good as for the alkali metal halides. The simple ionic model is not as satisfactory for these compounds; in these cases it needs modification. Where the electronegativity differences are large the ionic model gives good agreement with experiment. On the other hand where the difference in electronegativity becomes smaller, as for example between zinc and sulphur, there is a marked deviation from the theoretical value. The bonding is found to be stronger than the ionic model predicts.

The interpretation is that the bonding in such cases is not purely ionic but is partially covalent in character. The situation is reminiscent of that encountered in Section 4.12.2, where it was seen that covalent bonds A—B were stronger than might be expected and the additional strength was seen to parallel the electronegativity difference between A and B. The interpretation was seen in that case to be that the bonds are partially ionic in character. We see again that ionic and covalent bonds are extreme types and that bonds of intermediate character are found spanning the whole range (Section 10.3.7).

4.13.2 Solvation Energies: Heats of Solution

The heat of solution is the enthalpy change when one mole of solute is dissolved in a solvent. Generally it refers to the formation of an infinitely dilute solution, unless otherwise stated. In some cases heat is evolved and in others heat is absorbed in the process of dissolution. This is readily understood in qualitative terms. The process of separating the molecules of a solute always requires energy, and of itself leads to a positive enthalpy change, or the absorption of heat. In addition to this process there is also solvation to consider, that is the interaction of solute molecules with solvent molecules. This always leads to the liberation of heat. Whether dissolution is an endothermic or an exothermic process depends on which of the two terms is the larger. If the solvation energy is too small then the substance will not dissolve in the particular solute. Thus water solvates ions well and is a good solvent for ionic substances. Carbon tetrachloride molecules are only weakly solvating for ions, and ionic substances are generally insoluble in carbon tetrachloride. Solubilities of ionic substances are discussed in Section 10.3.17.

For an ionic substance such as sodium chloride dissolving in water we can readily relate these energy changes. We see from Table 4.8 that the lattice energy is 788 kJ and so we may write:

$$NaCl(s) \rightarrow Na^+(g) + Cl^-(g); \quad \Delta H_{latt} = 788 \text{ kJ}.$$

The heat of solution of sodium chloride is readily measured to give the **enthalpy of solution**, ΔH_{soln} and we have:

$$NaCl(s) + aq \rightarrow Na^+(aq) + Cl^-(aq); \quad \Delta H_{soln} = 4.5 \text{ kJ}.$$

By combining the two equations we find:

$$Na^+(aq) + Cl^-(aq) \rightarrow Na^+(g) + Cl^-(g) + aq;$$
$$\Delta H = 784 \text{ kJ} = \Delta H_{hyd}(Na^+) + \Delta H_{hyd}(Cl^-).$$

The last equation gives the sum of solvation energies, in this case **hydration energies** ΔH_{hyd}, for sodium and chloride ions. It is a quantity exactly analogous to lattice energy. It represents the energy required to break all the bonds due to solvation and separate the ions to infinity. In this example the bonding in the crystal and in solution are nearly equivalent energetically and the heat of solution is consequently small. It is generally the case for ionic substances that the heat of solution is obtained as the difference of two large quantities and in some cases is positive and in other cases negative. The relationship between the three quantities is illustrated in Fig. 4.19.

It should be noted that the result of an experiment leads to the sum of two ionic solvation energies and that the separate solvation energies, of Na^+ and Cl^- for example, cannot be obtained from experiment. However it is often convenient to attempt to obtain single ion solvation energies (Section 10.3.16). As might be

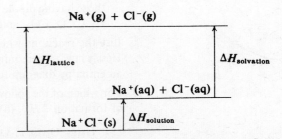

Fig. 4.19 Relationship between lattice energy, solvation energy, and heat of solution.

expected solvation energies of small positive ions are generally large. We can envisage strong electrostatic interaction between a cation and the lone pairs of water molecules.

PROBLEMS

1. Describe the two main types of chemical bond. What chemical and physical properties are associated with each bond type?

2. What is meant by electronegativity? What predictions can be made about a binary compound if the electronegativities of the two constituent elements are known? Look up and list the electronegativities of H, Li, B, C, N, O, and F.
   a) How does electronegativity change in (i) passing along a row, and (ii) passing down a group of the Periodic Table?
   b) Which element in the Periodic Table has the lowest electronegativity?
   c) Arrange the following into two groups—essentially ionic and essentially covalent: LiH, CH_4, O_2, N_2, Li_2O, and Na_2S.
   d) Arrange the following bonds in order of expected increasing ionic character: N—H, F—H, B—H, C—H, O—H, and S—H.

3. Describe the way in which atoms (or ions) are bound together in:
   a) calcium;
   b) ethane;
   c) potassium chloride.
   How are the properties of these substances dependent upon the bonds formed between their constituent atoms?

4. What are the most important intermolecular forces existing between molecules in each of the following compounds:
   a) propane;
   b) water;
   c) acetone?

5. How can the theory of "electron-pair repulsion" be used to rationalize the shapes of simple molecules?
   a) Which of the following molecules do you expect on the basis of the theory to be non-linear: H_2S, HOCl, $BeCl_2$, OCl_2?
   b) Which of the following are expected to be pyramidal: PH_3, NF_3, BF_3, PF_3?
   c) Pick out one pyramidal, one tetrahedral, and one trigonal bipyramidal molecule from the following gaseous molecules: BCl_3, PCl_3, PCl_5, $SiCl_4$, and OCl_2.
   d) Divide the following molecules into two classes, polar and non-polar (i.e. those having dipole moments and those without): NF_3, BF_3, CF_4, H_2, Cl_2, HCl, CO, CO_2, and H_2O.

6. List the reactions which could be studied experimentally and used, through Hess's law, to determine the standard enthalpy of formation of ethane. Draw an enthalpy diagram for the scheme.

7. For which of the following, at 298 K, are the values for the standard enthalpy of formation $\Delta H_f^\ominus$, (a) zero, (b) positive, (c) negative:

   $H_2(g)$, $O_2(g)$, $H_2O(l)$, $Cl_2(g)$, $Cl(g)$, $Br_2(g)$, C(diamond), $CO_2(g)$?

8. The following are graded examples of the use of tables of thermodynamic data. Find the standard enthalpy change, $\Delta H^\ominus$, for each of the following processes:
   a) $H_2(g) + \tfrac{1}{2}O_2(g) \rightarrow H_2O(l)$;
   b) $H_2(g) + Cl_2(g) \rightarrow 2HCl(g)$;
   c) $C_3H_8(g) + 5O_2(g) \rightarrow 3CO_2(g) + 4H_2O(l)$;
   d) $MgF_2(s) \rightarrow Mg^{2+}(g) + 2F^-(g)$;
   e) $Mg^{2+}(aq) + 2F^-(aq) \rightarrow Mg^{2+}(g) + 2F^-(g) + aq$.

9. a) Estimate the enthalpy of combustion of heptane (C_7H_{16}) from: (i) Fig. 4.10 and Table 4.3, and (ii) bond energy terms.
   b) Evaluate the enthalpy of combustion of heptane from the tables of standard enthalpies of formation.

10. Calculate $\Delta H^\ominus$ for the following processes:
    a) $3Mg(s) + Fe_2O_3(s) \rightarrow 3MgO(s) + 2Fe(s)$;
    b) $2Al(s) + 3ZnO(s) \rightarrow Al_2O_3(s) + 3Zn(s)$;
    c) $C_2H_4(g) + HCl(g) \rightarrow C_2H_5Cl(g)$;
    d) $Ag^+(aq) + Cl^-(aq) \rightarrow AgCl(s) + aq$;
    e) $Zn(s) + 2H^+(aq) \rightarrow Zn^{2+}(aq) + H_2(g)$;
    f) $NaCl(s) + aq \rightarrow Na^+(aq) + Cl^-(aq)$;
    g) $Fe(s) + 2Ag^+(aq) \rightarrow Fe^{2+}(aq) + 2Ag(s)$;
    h) $2Fe^{3+}(aq) + Zn(s) \rightarrow 2Fe^{2+}(aq) + Zn^{2+}(aq)$.

11. Using the bond energy terms in Table 4.4 find the expected standard enthalpies of formation of:

    a) cyclopropane ;
    b) naphthalene.

    In making the calculation assume that the bonds are as represented by the usual structural formulae. Compare with the tabulated values derived from experiment and comment.

12. From the tables of thermodynamic data find values for the following:
    a) the heat of combustion of methanol;
    b) the bond dissociation energy of chlorine;
    c) the heat of atomization of chlorine;
    d) the heat of neutralization of hydrochloric acid by aqueous sodium hydroxide;
    e) the heat of solution of potassium nitrate;
    f) the heat evolved in the precipitation of barium sulphate.

BIBLIOGRAPHY

Spice, J. E. *Chemical Binding and Structure*, Pergamon (1966).

    This gives a general introductory account.

Strong, L. E., and W. J. Stratton. *Chemical Energy*, Chapman and Hall (1966).

    This includes an introduction to the ideas of enthalpy and internal energy.

Ross, R. A. "Bond Energies," Chapter 2 in *Chemical Energetics and the Curriculum*, edited by D. J. Millen, Collins (1969).

This gives an account of bond energies at an introductory level. The same book also contains an introductory account of lattice energy and solvation energy:

Tobe, M. L. "Energetics of Ionic Crystals and Solutions," Chapter 5 in *Chemical Energetics and the Curriculum*, edited by D. J. Millen, Collins (1969).

Barclay, G. A. "Bond Energies and Bond Lengths," in *Approach to Chemistry*, edited by G. H. Aylward and I. K. Gregor, University of New South Wales, Sydney (1966).

Another introductory account which covers bond energies, bond lengths, and bond orders.

Coulson, C. A. *Valence* (second edition), Oxford University Press (1961).

A general account of quantum mechanical treatment of chemical bonding for dipping into, for further reading, and reference purposes.

Thompson, J. J. *An Introduction to Chemical Energetics*, Longman (1969).

An introduction written for sixth-formers.

Addison, W. E. *Structural Principles in Inorganic Chemistry*, Constable (1961).

An introductory account of bonding and structure, written for undergraduates but much of which could be read and understood by sixth-form students.

Dasent, W. E. *Inorganic Energetics*, Penguin (1970).

An introduction to the application of the ideas of energetics in chemistry. It makes much use of the Born–Haber cycle to elucidate various factors which contribute to energy changes for a wide variety of reactions in inorganic chemistry.

Harvey, K. B., and G. B. Porter. *Introduction to Physical Inorganic Chemistry*, Addison-Wesley (1963).

A book written for undergraduates, for further reading on bonding, structure, energetics, and equilibria in inorganic chemistry.

CHAPTER 5

Gases, Solids, Liquids, and Solutions in Equilibrium

Traditionally, changes have been divided into two classes, *chemical changes* and *physical changes*. In fact, such divisions are of limited value and in the following chapters all forms of structural change will be considered. The simplest of such changes involve the change of state of an element or compound between the solid, liquid, and gaseous phases.

The nature of the change itself, in the case of an element or compound changing phase, is relatively simple. Of more concern at this point is the balance between phases which we call *phase equilibria*. Before this is discussed in detail we must consider what is meant by equilibrium.

5.1 Equilibrium
5.2 Single-component systems
5.3 Two-component systems
5.4 Lowering of vapour pressure and depression of freezing point
5.5 Lowering of vapour pressure and elevation of boiling point
5.6 Osmotic pressure

5.1 EQUILIBRIUM

One might say that any object or system which is not moving is in a state of equilibrium. A lamp resting on a table is in a state of equilibrium, the force of gravity being opposed by the reaction from the table. If the lamp is pushed gently it will tip (unless it slides) but when the pressure is removed the lamp will return to its original position. This is an example of **stable equilibrium**. If the lamp is balancing precariously on the edge of its base, the slightest push will send it toppling over. In this case the lamp is in a state of **unstable equilibrium**. Both types of equilibrium are essentially static. In the equilibrium position, nothing is in motion.

Dynamic equilibrium. Chemical equilibria are of a different kind, a kind in which the molecules are in a perpetual state of movement and change. This type of equilibrium, dynamic equilibrium, is best explained in terms of a simple analogy.

London and Birmingham are connected by the M1 motorway. Suppose that traffic is limited to travelling between London and Birmingham on the motorway. If there are, say, 1,000,000 cars in London and 500,000 cars in Birmingham and *if an equal number of cars travel at equal speeds in each direction* along the motorway the total number of cars in London and Birmingham will always be the same. (We ignore the number actually on the motorway.) The point is that the number of cars leaving and entering London per hour is the same. The total number of cars in London therefore remains the same although the actual cars are different. The situation alters radically if we imagine that the London to Birmingham lane of the

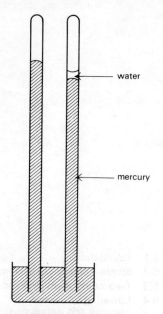

Fig. 5.1 Simple demonstration of vapour pressure above a liquid.

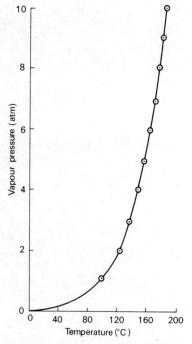

Fig. 5.2 Vapour pressure of water from 40 to 200 °C.

motorway is blocked at the border of London. Cars can now enter but not leave London. The number of cars in London goes up, the number in Birmingham goes down. Equilibrium is upset. In this simple model all the cars will eventually end up in London. Almost exactly the same principles hold for dynamic equilibria in chemical systems. A simple example is also found in phase equilibria.

5.2 SINGLE-COMPONENT SYSTEMS

5.2.1 Liquid–Vapour Equilibria

Liquid–vapour equilibria can be studied simply with a barometer, that is a tube full of mercury inverted over a bowl of mercury (Fig. 5.1). If a little water is injected into the bottom of the tube with a dropping pipette, it will rise to the top and evaporate. As a result the level of the mercury will go down. The vapour exerts a pressure which can be measured by the extent to which the mercury level is depressed. If more water is added the liquid will eventually cease to evaporate and both liquid and vapour will be in the tube together. If more liquid is added the level of the mercury will not be further depressed (except by a very small amount owing to the weight of the liquid). An equilibrium has now been established between the water and its vapour. Molecules of water are constantly leaving the water and entering the vapour phase. At the same time and at the same rate molecules of water in the vapour phase are entering the liquid phase. The result is a dynamic equilibrium.

The equilibrium can be upset by changing the temperature. Raising the temperature of the system means increasing the kinetic energy of the molecules in both phases. More molecules in the liquid phase have sufficient energy to leave the liquid and enter the vapour. According to Boltzmann's law (Section 6.8), the number of molecules leaving the liquid increases rapidly with temperature, and so the number in the vapour phase increases. This leads to an increase in the number of molecules entering the liquid from the vapour. A new equilibrium position is eventually reached in which a greater number of molecules is in the vapour phase. The pressure exerted by these molecules is greater than at the lower temperature. Thus the vapour pressure increases with a rise in temperature. A graph of vapour pressure against temperature is given in Fig. 5.2.

Note that the line in the diagram represents the temperatures and pressures at which water and water vapour are in equilibrium. At temperatures and pressures above the line, water can exist only on its own. At temperatures and pressures below the line only water vapour can exist.

5.2.2 Solid–Liquid Equilibria

It is a well-known fact that ice and water are in equilibrium at 0 °C and 760 mm of mercury (mmHg). As there is only a small change in volume when water freezes, the effect of pressure is relatively small. Water is unusual in that its solid phase (ice) is less dense than its liquid phase. The effect of pressure is therefore also unusual. The higher the pressure the lower the temperature at which ice and water are in equilibrium (see Le Chatelier's principle, Section 7.3), a fact which is useful for ice skaters.

5.2.3 Solid–Vapour Equilibria

It is not always realized that an equilibrium also exists between a solid and vapour. The vapour pressure in this case is so small that it is quite difficult to measure, but is nevertheless very significant. The vapour pressure of ice, for instance is 4 mmHg at 0 °C falling to 1 mmHg at −20 °C. Some solids have high vapour pressures, carbon dioxide and iodine being two well-known examples. If the vapour pressure of the solid reaches atmospheric before its melting point is reached, the solid will turn into the vapour phase without passing through the liquid phase. Both solid carbon dioxide (dry ice) and iodine will do this. The solid is said to *sublime*. By increasing the pressure the sublimation temperature is increased. If the pressure is increased sufficiently the sublimation temperature will rise above the melting point and the liquid can be obtained.

5.2.4 Phase Diagrams

Information about the equilibria between the three phases can be combined in the form of a phase diagram. Figure 5.3 gives the phase diagram for water. The diagram summarizes information about the stability of the three phases. At temperatures and pressures represented in the area A only water vapour is stable, in B water, and in C ice. The lines represent the temperatures and pressures at which two phases are in equilibrium. The "triple point" represents the only temperature and pressure values at which all three phases can exist together.

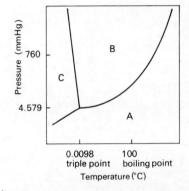

Fig. 5.3 Phase diagram for water.

5.2.5 Phase Changes and Energy Changes

During phase changes, changes in energy also take place. Solids and liquids are formed by atoms and molecules held together by forces of different kinds (Section 4.1). In the case of molecular substances, the forces (van der Waals forces, dipole–dipole attraction, and those of hydrogen bonds) are relatively weak. In the case of giant structures the ionic and covalent bonds are strong. Energy is required to break these bonds when solids are changed into liquids and gases. The heat absorbed when a mole of a solid element or compound is changed into a liquid is the **enthalpy of fusion**, ΔH_{fus}; that absorbed when a mole of liquid is changed into a gas is the **enthalpy of vaporization**, ΔH_{vap}. The **enthalpy of sublimation**, ΔH_{sub}, is the heat absorbed when a mole of a solid is changed into a vapour. Such quantities are often referred to as heats (e.g. heat of fusion) but strictly speaking they should be referred to as enthalpies.

Some examples of enthalpies of fusion and vaporization are given in Table 5.1.

5.2.6 Vapour Pressure and Temperature

Figure 5.2 shows that the vapour pressure curve becomes steeper and steeper as the temperature rises. To find the mathematical relationship between absolute temperature and vapour pressure, plot the logarithm of the vapour pressure against the inverse of the temperature. The result for water is shown in Fig. 5.4.

It is clear from this graph that a linear relationship exists between the logarithm of vapour pressure, P, and the inverse of the temperature, T:

$$\log P \propto \frac{1}{T},$$

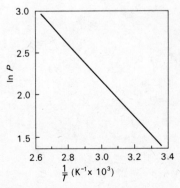

Fig. 5.4 Plot of ln P against $1/T$ for water.

Table 5.1. Some enthalpies of fusion and vaporization.

Substance (formula)	ΔH_{fus} (kJ mol^{-1})	ΔH_{vap} (kJ mol^{-1})
H$_2$	0.0586	0.452
He	0.0209	0.0836
Na	2.60	98
Ca	8.7	161
C	(sublimes)	718
N$_2$	0.360	2.79
O$_2$	0.222	3.39
Cl$_2$	3.22	10.2
NaCl	28.9	171
HCl	2.01	16.2
C$_2$H$_5$OH	4.60	42.5
NH$_3$	5.65	23.4
PCl$_3$	4.52	30.5
H$_2$S	2.47	18.7
SO$_2$	7.40	25.0
H$_2$O	6.03	41.5

or
$$\ln P = A \times \frac{1}{T} + B, \tag{5.1}$$

or
$$2.303 \log P = \frac{A}{T} + B, \tag{5.2}$$

where A and B are constants for a given substance. It is found that A is related to the heat of vaporization and to the gas constant R:

$$A = -\frac{\Delta H_{vap}}{R}. \tag{5.3}$$

Thus the equation may be written as

$$\ln P = -\frac{\Delta H_{vap}}{RT} + B. \tag{5.4}$$

A similar equation exists for sublimation of a solid:

$$\ln P = -\frac{\Delta H_{sub}}{RT} + B'. \tag{5.5}$$

5.3 TWO-COMPONENT SYSTEMS

Two-component systems in the liquid phase are of two types: those in which the liquids do not dissolve in each other, called immiscible; and those in which they do, called miscible. In fact there are many cases of partial miscibility and, as in the case of solids dissolving in liquids, no two liquids are really completely immiscible. Two immiscible liquids influence each other little and apart from their use in extraction by partition (Section 7.1 and Chapter 29), there is not much about them that concerns us here. But miscible liquids will be studied in some detail.

5.3.1 Raoult's Law

The most important pattern of behaviour which is found in miscible liquids is described by Raoult's law. This states that the partial vapour pressure of one liquid in a mixture of miscible liquids is directly proportional to its mole fraction. The mole fraction is the number of moles of a substance divided by the total number of moles in the mixture. If there are two miscible liquids A and B and a mixture contains a moles of A and b moles of B then the mole fraction of A, x_A, is given as

$$x_A = \frac{a}{a+b}. \quad (5.6)$$

Raoult's law gives the vapour pressure P_A due to A in this mixture as

$$P_A = P_A^\circ x_A, \quad (5.7)$$

where P_A° is the vapour pressure of pure A at that temperature. Similarly the vapour pressure P_B of B is given by

$$P_B = P_B^\circ x_B. \quad (5.8)$$

Figure 5.5 illustrates in graphical terms the vapour pressure of a mixture which obeys Raoult's law. The vapour pressure of the mixture is always the average according to the proportion of the number of moles of each component. That is, the total vapour pressure exerted by the mixture gives a straight line when plotted against mole fraction for the liquid. The partial vapour pressures of each of A and B are (according to Eqs. 5.7 and 5.8) directly proportional to mole fractions of A and B respectively. Plots of P_A and P_B against x_A and x_B give straight lines (Fig. 5.6). Raoult's law can be understood in terms of a simple picture of dynamic equilibrium (Section 5.1). The rate of escape of A molecules from a solution into the vapour is less than from the pure liquid. If we look at the surface, then if the molecules of the two compounds are the same size and arrange themselves randomly, the proportion of sites in the surface occupied by molecules of substance A will be given by the mole fraction of A. The number escaping is consequently reduced below that for the pure liquid. But of course the rate of return of molecules into the surface is not dependent on the molecular composition of the surface. The consequence is that the vapour pressure is reduced until a new dynamic equilibrium is set up with $P_A = P_A^\circ x_A$. (See Problem 5.)

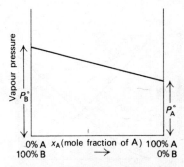

Fig. 5.5 Vapour pressure curve for a mixture obeying Raoult's law. Total pressure is plotted against mole fraction.

5.3.2 Exceptions to Raoult's Law. Why are Liquids Miscible?

There are many exceptions to Raoult's law and to understand why this is so we must consider briefly the factors which determine whether liquids are miscible or immiscible. In general terms liquids which are alike dissolve in one another. Alkanes for instance, are so alike that their molecules naturally intermingle and they are consequently miscible in all proportions. But alkanes do not dissolve in water. The explanation is that the water molecules are fairly firmly held together by hydrogen bonds (Section 13.6.2). Alkane molecules are incapable of breaking these bonds and hence cannot penetrate into the water. Similarly, the hydrogen bonds prevent the water molecules from entering the alkane. The result is that the liquids form two layers.

Ethanol has the property of dissolving both in organic solvents such as the

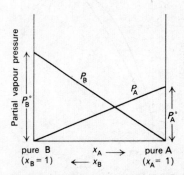

Fig. 5.6 Partial vapour pressures, P_A and P_B are directly proportional to x_A and x_B respectively.

5.3 TWO-COMPONENT SYSTEMS

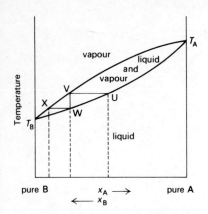

Fig. 5.7 Boiling point of mixtures of A and B as a function of composition.

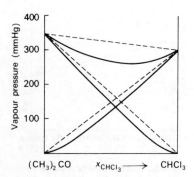

Fig. 5.8 Vapour pressure curves for trichloromethane-acetone system.

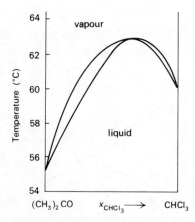

Fig. 5.9 Boiling point / composition curve for a mixture showing a maximum in the boiling point curve.

alkanes and in water. The ethyl group is responsible for dissolving in alkanes; it is an "alkyl" group and similar to an alkane molecule. The hydrogen bonding between the hydroxyl groups of ethanol molecules, although weaker than that in water, is strong enough to allow ethanol molecules to replace water molecules and thus dissolve in water with the formation of new hydrogen bonds between water and ethanol molecules.

Dipole–dipole attraction also plays a part in the dissolving of some liquids. Another factor is size: molecules of liquids which dissolve in each other are often of approximately the same size; substances with very large molecules are often insoluble.

5.3.3 Liquids which Obey Raoult's Law

Mixtures of liquids which obey Raoult's law are always of similar types of molecules. Examples are the alkanes, for instance hexane and heptane.

If a graph of *boiling point* against composition is drawn (Fig. 5.7) the result is a smooth curve but its slope is in the opposite direction to the plot of vapour pressure against composition. The liquid with the highest vapour pressure has the lowest boiling point and vice versa. Such mixtures can readily be separated by fractional distillation if the boiling points of the components are far enough apart. The high vapour pressure / low boiling point liquid comes off at the top of the column first and the low vapour pressure / high boiling point liquid will collect in the distillation flask and may be boiled off second (Section 29.2). The composition of the vapour in equilibrium with a liquid of composition U is represented by the point V. Drawing in a curve for points obtained in a similar manner to V completes the phase diagram in Fig. 5.7. If the vapour V is cooled, the resulting liquid will have composition W, and if reheated it will come into equilibrium with vapour of composition X. The process in a fractionating column can be regarded as equivalent to repeated condensation and re-evaporation of this kind. As the diagram shows the vapour becomes richer in the more volatile component.

5.3.4 Mixtures of Liquids with Negative Deviations from Raoult's Law

A number of mixtures show negative deviations from Raoult's law. An example is trichloromethane (chloroform) and acetone, and the vapour pressure against composition curve for this system is shown in Fig. 5.8. The partial vapour pressure of each component is less than expected from Raoult's law. If we consider a solution of trichloromethane in acetone (or vice versa), it is as if the apparent molecular weight of the solute has increased. This can be explained in terms of interaction or complex formation between the components. In this case the hydrogen atom in the trichloromethane makes a hydrogen bond with the oxygen atom in acetone. Evidence of bond making is provided by the fact that when the two compounds are mixed there is a temperature rise.

A boiling point / composition curve for a mixture of this kind (Fig. 5.9) shows that the mixture has a boiling point which falls above the straight line joining the boiling points of the pure components. In some cases the maximum boiling point is higher than the boiling point of either of the liquids. The mixture with the composition at this point is called the "maximum boiling point mixture" or "constant boiling mixture".

If a mixture of A and B with a composition different from the constant boiling mixture composition is boiled in an open beaker, the vapour boiled off will be richer than the liquid in either A or B until the liquid remaining has the composition of the constant boiling mixture. At this point, when the boiling point has risen to its maximum, the vapour and liquid have the same composition. The composition of the liquid mixture therefore remains constant as does its boiling point.

Such mixtures cannot be separated by fractional distillation. If any mixture of such a system is fractionated, either pure A or pure B will come out of the top of the column, depending on whether the mixture being separated is richer in A or B than the constant boiling mixture. The constant boiling mixture will remain in the flask until it boils over at a constant temperature as if it were a compound. Indeed, at one time constant boiling mixtures were thought to be compounds.

5.3.5 Positive Deviations from Raoult's Law

Some mixtures of liquids have vapour pressures which fall above the straight line joining the vapour pressures of the pure components. At first this seems surprising. How can the apparent molecular weight be smaller in the mixture than in either compound? If actual examples are taken the reason for this is readily understood. Ethanol and ethyl acetate is a good example. A vapour pressure / composition curve of this form is shown in Fig. 5.10.

Ethanol molecules are held together in the pure liquid by hydrogen bonds. Evidence of this is the fact that ethanol has a higher boiling point than ethyl acetate although its molecular weight (46) is lower than that of ethyl acetate (88). The addition of ethyl acetate breaks the hydrogen bonds between ethanol molecules and thus increases its apparent volatility. When ethanol and ethyl acetate are mixed, a drop in temperature of the mixture is recorded. Indeed a similar graph to the boiling point / composition curve can be drawn by plotting drop in temperature against composition (Fig. 5.11). The examples of negative and positive deviations we have considered both involve hydrogen bonding, and the deviations are large. There are many examples where smaller deviations occur and in which the intermolecular forces are weaker than those of hydrogen bonding.

As with the negative deviation a minimum in the curves in Fig. 5.11 means that the mixture cannot be separated by fractional distillation. If the mixture is fractionated the maximum vapour pressure / minimum boiling point mixture comes over first and pure A or B is left in the flask.

A well-known example of this is an ethanol–water mixture. The mixture has a maximum vapour pressure / minimum boiling point. When the mixture is fractionated, pure ethanol does not come over, but a 96 percent ethanol / 4 percent water mixture is obtained. In order to obtain pure ethanol another method must be used. A chemical solution is to add magnesium and iodine. The latter initiates a reaction to form magnesium ethoxide and hydrogen, and the water is ultimately converted to magnesium hydroxide.

5.3.6 Boiling Point and Vapour Pressure / Composition Curves

It is important to be quite clear about the distinction between these two types of graph. The boiling point curves show temperature against composition *at constant pressure*. The vapour pressure curves show pressure against composition *at*

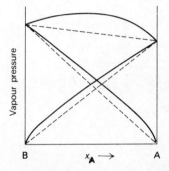

Fig. 5.10 Vapour pressures for system showing positive deviations. Ethanol–ethyl acetate is an example of this class of system.

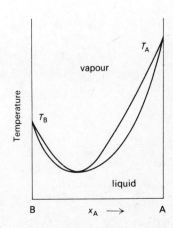

Fig. 5.11 Boiling point / composition curve for a mixture showing a minimum in the boiling point curve.

constant temperature. Although the one is often roughly the inverse of the other this is not necessarily so. At higher temperatures the vapour pressure / composition curves tend to flatten out. It is therefore possible that when a vapour pressure / composition curve at room temperature shows a maximum, the boiling point / composition curve does not.

5.3.7 Raoult's Law and the Lowering of Vapour Pressure

The vapour pressure of a solution of a non-volatile solute (e.g. a solid) in a solvent is less than that of the pure solvent. For dilute solutions it is found that the lowering of vapour pressure is directly proportional to the concentration of the solution. The lowering of vapour pressure of any solution is more generally expressed in terms of Raoult's law. It was shown earlier in this section that according to Raoult's law the vapour pressure of a component A in a mixture of A and B is given by Eq. (5.7) as

$$P_A = P_A^\circ x_A,$$

where P_A is the vapour pressure of pure A, and x_A is the mole fraction of A in the mixture.

The lowering of the vapour pressure of A is the vapour pressure of pure A minus the vapour pressure of A for a solution, that is

$$P_A^\circ - P_A.$$

The sum of the mole fractions of A and B is of course 1, from Eq. (5.6).

$$x_A + x_B = 1$$

or

$$x_A = 1 - x_B.$$

Substituting in the first equation:

$$P_A = P_A^\circ(1 - x_B)$$
$$P_A = P_A^\circ - P_A^\circ x_B$$
$$P_A^\circ - P_A = P_A^\circ x_B$$

or

$$\frac{P_A^\circ - P_A}{P_A^\circ} = x_B. \tag{5.9}$$

That is, the lowering of vapour pressure of A, divided by the vapour pressure of pure A is equal to the mole fraction of B, where B is the non-volatile solute.

5.3.8 Raoult's Law and the Determination of Molecular Weight

The lowering of vapour pressure can be used as a means of determining molecular weight. The mole fraction of B (Eq. 5.6) is given as $x_B = b/(a + b)$ where a and b are the number of moles of A and B respectively. Now the number of moles of a substance is given by its weight W divided by its molecular weight M. Thus:

$$b = \frac{W_B}{M_B}$$

$$a = \frac{W_A}{M_A}.$$

Raoult's law can then be rewritten from Eq. (5.9) as

$$\frac{P_A^\circ - P_A}{P_A^\circ} = x_B = \frac{b}{a+b},$$

or

$$\frac{P_A^\circ - P_A}{P_A^\circ} = \frac{W_B/M_B}{W_A/M_A + W_B/M_B}. \qquad (5.10)$$

If the lowering of vapour pressure is found for a solution of known concentration, and the molecular weight of A is known, then the molecular weight of B can be worked out from the expression above.

5.4 LOWERING OF VAPOUR PRESSURE AND DEPRESSION OF FREEZING POINT

The lowering of vapour pressure is closely related to the depression of freezing point of a solution. The vapour pressure / temperature diagram in Fig. 5.12 illustrates the point. For an aqueous solution, ice is in equilibrium with the solution for the temperature when both are in equilibrium with the same pressure of water vapour.

The depression of freezing point is found to be proportional to the concentration of the solution for a particular solution (Blagden's law), and in general proportional to the mole fraction of the dissolved solid. This holds for dilute solutions only.

The depression of freezing point can be used to determine molecular weight. A substance of known molecular weight is first used to find the *molal depression of freezing point*, that is, the depression of freezing point when a mole of solute is dissolved in a 1000 g of solvent. The molal depression of freezing point varies considerably from solvent to solvent: for water it is 1.85 °C per 1000 g solvent; for camphor it is 40 °C per 1000 g solvent. A solution of known concentration is then made up with the substance of unknown molecular weight. If M_x is the unknown molecular weight and a solution of W g per 1000 g solvent gives a depression of ΔT °C,

$$\frac{W}{M_x} = \frac{\Delta T}{K_f}, \qquad (5.11)$$

where K_f is the molal depression of the freezing point constant for the particular solvent.

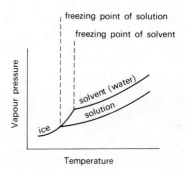

Fig. 5.12 Lowering of the freezing point of a solution.

5.5 LOWERING OF VAPOUR PRESSURE AND ELEVATION OF BOILING POINT

The elevation of the boiling point of a solution over that of the pure solvent is a direct consequence of the lowering of the vapour pressure of the solution, as can be seen from Fig. 5.13. The solution boils when the vapour pressure reaches one atmosphere.

Raoult's law predicts that the depression of vapour pressure is proportional to the molal concentration of the solute in a given solvent for dilute solutions. The molal elevation of boiling point may be found by determining the elevation of

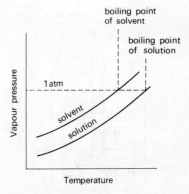

Fig. 5.13 Elevation of the boiling point of a solution.

boiling point of a substance of known molecular weight. For water the molal elevation of boiling point is 0.52 °C; for benzene it is 2.57 °C. Once the molal elevation is determined the molecular weight of a substance is found by determining the elevation of boiling point of a solution of known concentration (see depression of freezing point, Eq. 5.11).

Note that for both cases, the treatment holds only for dilute solutions, and so the accuracy of these methods of determining molecular weights depends on the accuracy of the means of measuring temperature. The limitation of a mercury glass thermometer is about 0.005 °C.

5.6 OSMOTIC PRESSURE

Properties of substances which depend on the *number* of particles present rather than the *weight* are called **colligative properties**. The lowering of vapour pressure, depression of freezing point, and raising of boiling point of solutions discussed in the last section are all examples of colligative properties. So is the phenomenon of osmotic pressure discussed in this section.

Osmosis and the Semi-Permeable Membrane

Certain membranes allow the passage of pure solvent molecules but do not allow solute molecules or ions to pass. These are called semi-permeable membranes. They are found frequently in living systems and may be made artificially. The exact mechanism of how the membranes operate is not understood. The membrane acts as a kind of filter, but cannot be described in these terms as the solute molecules which cannot pass through the membrane are sometimes smaller than the solvent molecules which do pass through.

When a semi-permeable membrane separates a solvent from a solution, solvent molecules tend to pass from the solvent into the solution. This process is called **osmosis**. In general, it is the tendency of solvent molecules to pass through a semi-permeable membrane from an area of lower concentration to an area of higher concentration.

It is possible to prevent the process of osmosis from occurring by applying a pressure to the solution of higher concentration. The pressure required to prevent osmosis from occurring is called the **osmotic pressure**. It is found to be proportional to the concentration of a solute in a particular solvent, or in general, proportional to the molal concentration of any solute in a given solvent. Equimolal concentrations of any solute in a given solvent have the same osmotic pressure. Thus the measurement of osmotic pressure can also be used for the determination of molecular weights. Osmotic pressures can be extremely high, of the order of 200 atmospheres, and this is a relatively sensitive way of measuring molecular weights. It is particularly useful in the determination of the molecular weights of substances with very large molecules.

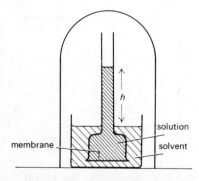

Fig. 5.14 An example of osmosis.

Lowering of Vapour Pressure and Osmotic Pressures

It is clear that there will be some relationship between the lowering of vapour pressure and osmotic pressure. Figure 5.14 illustrates a system in which osmosis has led to a column of solution of height h being in equilibrium with the solvent in the beaker. Let us call the solvent A and the solute B, with molecular weights M_A and M_B. Let the vapour pressure of the solution be P_A. This must be equal to the vapour pressure of the solvent *at that point*. Otherwise the system would be unstable. If P_A° is the vapour pressure of pure solvent, then the difference between P_A° and P_A must be equal to the pressure of a column of vapour of height h. Thus

$$P_A^\circ - P_A = hdg, \tag{5.12}$$

where d is the density of the vapour, and g the acceleration due to gravity. The gas law

$$PV = RT$$

leads to a re-expression of density. If the molecular weight of the solvent is M_A and we take one mole of solvent vapour, the volume V_V it occupies will be given by

$$V_V = \frac{RT}{P_A^\circ},$$

and the density by

$$d = \frac{M_A}{V_V} = \frac{M_A P_A^\circ}{RT}. \tag{5.13}$$

If it is assumed that the density of the solution is equal to the density of the solvent, D, then the osmotic pressure can be related to the height of the liquid column. We have the osmotic pressure Π which is given by

$$\Pi = hDg.$$

Since

$$hdg = P_A^\circ - P_A, \quad \text{from Eq. (5.12)},$$

and

$$d = \frac{M_A P_A^\circ}{RT}, \quad \text{from Eq. (5.13)},$$

we have

$$\frac{hg M_A P_A^\circ}{RT} = P_A^\circ - P_A,$$

which becomes

$$\frac{\Pi M_A}{DRT} = \frac{P_A^\circ - P_A}{P_A^\circ}. \tag{5.14}$$

Equation (5.14) shows that for a particular solvent for which M and D are constants, the osmotic pressure is directly proportional to the lowering of the vapour pressure. We have already seen that molecular weights can be calculated from the lowering of the vapour pressure, and so it follows that molecular weights can be calculated from a measured osmotic pressure. By combining Eqs. (5.14) and (5.9) we have

$$\frac{\Pi M_A}{DRT} = x_B, \tag{5.15}$$

and knowing x_B, the mole fraction of solute, we can calculate M_B from Eq. (5.10). Remembering that $M_A/D = V$, the volume of one mole of the solvent A, we have

$$\Pi V = x_B RT. \tag{5.16}$$

This is a convenient form for calculations. (See also Problem 7.)

Osmotic pressures can be large even for solutions of moderate concentration. A 0.1 molar aqueous solution has an osmotic pressure of about two and a half atmospheres. It is seen to be a powerful effect and is of great significance in biological systems. As a method of molecular weight determination osmosis has advantages for substances of high molecular weight. Such substances often form only dilute solutions on a molal scale, partly because they are often not very soluble and partly because of their high molecular weight. The consequence is that freezing point depressions and boiling point elevations of such solutions are small and it is difficult if not impossible to measure them reliably. On the other hand, such solutions frequently have readily measurable osmotic pressures. As a consequence osmotic pressure measurements are of importance for determining molecular weights of synthetic polymers and naturally occurring substances of high molecular weight.

PROBLEMS

1. Describe the phenomenon of dynamic equilibrium. How could you demonstrate that there is in fact a dynamic equilibrium between a liquid and its vapour?

2. What energy changes are associated with changes in phase? What molecular picture can be put forward to account for these energy changes?

3. Sketch a phase diagram for:
   a) ammonia;
   b) carbon dioxide.

4. What is the effect of temperature upon vapour pressure? Describe, in terms of the kinetic theory, what happens when a system of liquid and vapour in equilibrium is (a) heated, and (b) cooled down.

5. How can Raoult's law be accounted for in terms of a molecular picture? Consider a system of A and B with mole fractions x_A and x_B, and suppose each kind of molecule occupies the same area in the surface.
   a) What fraction of the surface is occupied by A molecules?
   b) Write an equation showing how the rate of escape of A molecules depends on the mole fraction x_A.
   c) Write an equation for the rate of return of A molecules into the surface.
   d) What is the condition for dynamic equilibrium?
   e) Write the corresponding equation for the pure liquid.
   f) What is the result of combining the equations from (d) and (e)?

6. Determine the molar enthalpy of vaporization of water, given the vapour pressure measurements as follows:

T(°C)	P(mmHg)	T(°C)	P(mmHg)
20	17.5	60	149
30	31.8	70	234
40	55.3	80	355
50	92.5	90	633

7. The equation relating osmotic pressure to concentration of solute is often written for dilute solutions as $\Pi V' = RT$, where V' is the volume which contains 1 mole of solute. Derive the above expression from Eq. (5.16) in Section 5.6. (*Hint*: In writing the expression for the mole fraction of a dilute solution, remember the number of moles of solvent is much larger than the number of moles of solute.)

8. The osmotic pressure of blood is 7 atm at 30 °C. It is desired to make a solution of sodium chloride of the same osmotic pressure. What should its concentration be? Make the calculation assuming the solution to be an ideal solution of sodium and chloride ions. (In fact because of non-ideality the factor to allow for ionization is 1.9 rather than 2.0.)

9. The triple point for water occurs at: temperature 0.0099 °C; and pressure 4.58 mmHg. The molar enthalpies of vaporization and sublimation are:

$$\Delta H_{vap} = 6.03 \text{ kJ mol}^{-1};$$
$$\Delta H_{sub} = 41.5 \text{ kJ mol}^{-1}.$$

Calculate the vapour pressure at -10 °C and $+10$ °C.

10. A solution of sucrose ($M = 342$) is prepared by dissolving 68.4 grams in 1000 grams of water. What is:
   a) the vapour pressure of the solution at 20 °C;
   b) the osmotic pressure at 20 °C;
   c) the boiling point of the solution;
   d) the freezing point of the solution?
   The vapour pressure of water at 20 °C is 17.5 mmHg. Take any of the information required from the text, and assume the solution to behave ideally.

BIBLIOGRAPHY

Mahan, B. H. *University Chemistry* (second edition), Chapter 4, Addison-Wesley (1969).

Barrow, G. M. *Physical Chemistry*, Chapter 18, McGraw-Hill (1966).

   Both of these provide general treatments which offer further reading following on from the present chapter.

Most of the books on physical chemistry in the general list contain material on phase equilibria.

CHAPTER 6

Energy and Chemical Change: Why Do Changes Take Place?

6.1 Changes of mixing
6.2 Changes of falling
6.3 Changes involving heat flow
6.4 Chemical changes
6.5 Exothermic and endothermic changes
6.6 Probability and change
6.7 How atoms and molecules hold their energies
6.8 Energy distributions
6.9 Entropy
6.10 Entropy and change
6.11 Measurement and tabulation of entropies
6.12 Entropy changes for chemical reactions
6.13 Changes at constant temperature and pressure
6.14 Entropy: summary

This chapter examines the question: why do changes take place? The patterns or "laws" which govern chemical changes are essentially the same as those that govern all changes. Indeed they are among the most fundamental laws of science, the laws of thermodynamics. The difference between chemical changes and other changes is that in chemical changes there is a redistribution of the structure of the substances on an atomic or molecular level. This brings in some new factors but does not affect the overall pattern. When a substance changes phase, from a liquid, say, to a gas, is this a chemical change? The answer depends only on your definition. Nature exists independently of our imposed definitions. We need not therefore concern ourselves with the relatively useless question of when a change is not a chemical change, but rather with the extremely interesting question of why changes take place.

Let us for a start look at a number of changes which we can see in everyday life.

6.1 CHANGES OF MIXING

If we fill a box carefully with a layer of white billiard balls covered by a layer of red billiard balls and shake it we will not be surprised to find that after shaking they are mixed up. That is to say that there are red and white billiard balls throughout instead of two layers. Similarly we should be surprised if we put our hand into a bag of mixed coins and pulled out all the pennies before all the 10 p pieces. It is easy to sort out coins or billiard balls, but if a bag of salt were mixed with a bag of sugar picking out the salt crystals from the sugar crystals would be no joke. The difference is not only in the size of the particles to be sorted out but in the number. There is a natural tendency for things to get mixed up. The more things there are the more mixed up they can become and the more surprised we would be if they sorted themselves out without our aid. If a good shake of the bowl produced separate layers of sugar and salt we should probably think that a miracle had taken place. To take an example from even smaller particles we can consider what happens in a laboratory or classroom when a gas-jar of hydrogen sulphide is opened on the demonstration bench. Gradually molecules of hydrogen sulphide mix with the molecules in the air, to everyone's discomfort. Would it be possible for all the hydrogen sulphide to get back into the jar? It has never happened in our experience and certainly seems most improbable. This general tendency for things to mix up can be expressed more precisely, but before doing so let us consider some other changes.

6.2 CHANGES OF FALLING

There seems to be a general tendency for things to fall down or run down. To take falling first, we notice that if there is nothing to stop it a ball tends to roll downhill. Books fall off shelves, water runs down to sea-level, mountains and hills are slowly eroded away, everything becomes smoother and flatter. Now this is not as simple as it seems at first. It is true that the ball runs downhill. Why does it not run on up the next hill? The fact is that if there were no "friction" it would do so. The question is not why does a ball roll downhill but why does it stop there? Why does a rubber ball stop bouncing up and down on a hard floor? And the answer is that the potential energy (energy of position) and the kinetic energy (energy of motion) are converted into thermal energy (heat). It is interesting that potential and kinetic energy can be converted in both directions without difficulty, as in the pendulum, but that kinetic and thermal energy cannot. The idea that when a gas is heated its molecules move faster is familiar. So is the idea that the molecules of a gas are moving about in a disorganized or random way. When a ball rolls down a hill only to stop at the bottom, the simple unidirectional motion of the one ball has been changed into the complex multidirectional movement of many, many molecules of the air (and perhaps the ground). Presumably it is possible that the molecules could gang up and all push the ball back to the top of the hill using the energy they acquired from the ball in the first place. But it seems very, very unlikely and we have never known it to happen. It looks as if the ball rolling downhill is not so very different from the change of mixing.

6.3 CHANGES INVOLVING HEAT FLOW

Changes which take place in thermal energy are on the face of it more obvious because they are such everyday occurrences: hot food out of the oven gets cold on the table; cold ice cream from the refrigerator warms up when it is removed. Hot things cool down to room temperature. Cold things warm up to room temperature. "Heat flows from the hotter body to the colder body" has been used as a succinct definition of the Second Law of Thermodynamics.

How do we recognize a hot body when we see one? We can touch it, or use a thermometer. If two substances are at the same temperature we know that when we put them in contact with each other there will be no net flow of heat from one to the other. They will in fact stay at the same temperature as each other. This could be (and is) used as the basis for a definition of temperature. The question is, why does heat pass from a hotter body to a colder body? We shall answer this question later on. Meanwhile it is worth looking a little more closely into what happens when we heat something up.

What happens when a given quantity of heat is absorbed by a body, say a block of metal? It will, of course, rise in temperature. How much it rises in temperature will depend on the mass of the block. The more metal, the more heat will be required to raise its temperature by a given amount. It depends also on the nature of the metal itself, that is on its heat capacity. Thus copper has a heat capacity at 0 °C of 0.436 joules per gram per degree (J g^{-1} deg^{-1}), and iron of 0.450 J g^{-1} deg^{-1}. If we compare the heat capacity of equal numbers of atoms of each metal instead of equal masses we come up against an interesting fact: they are almost the same. The heat capacity of copper is 27.7 J mol^{-1} deg^{-1}, whilst that of iron is 25.1 J mol^{-1} deg^{-1}. Indeed this is the case for most solids at room temperature. It is not the case for liquids or gases. Ice has a heat capacity of 1.75 J g^{-1} deg^{-1}, water of 4.134 J g^{-1} deg^{-1}, and steam of 1.51 J g^{-1} deg^{-1}. For chemical purposes we shall usually be interested in one mole rather than one gram, and we shall generally use the units J mol^{-1} K^{-1} or kJ mol^{-1} K^{-1}. The symbol K here stands for the kelvin, formerly called the degree Kelvin. It is in fact identical with the interval called the degree Celsius, °C. (The symbol K is also used to denote temperature, thus the freezing point of water is 273.15 K.)

Different substances have different heat capacities. The heat capacities of substances in the same phase do not vary very greatly at different temperatures, but

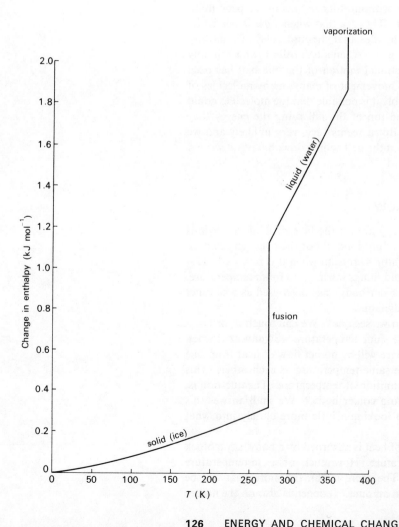

Fig. 6.1 Change in enthalpy of water with temperature.

they do vary greatly in different phases. The heat capacity tells us how much heat will have to flow in to produce a particular temperature rise. Its units are therefore of energy divided by temperature (e.g. J deg^{-1}). The temperature tells us which way heat will flow if a block of one substance is put in contact with a block of another substance at a different temperature.

It is instructive to look at a graph of the heat put into a mole of a substance against the temperature rise for a given pressure. It follows from Eq. (4.3) that such a plot is a plot of enthalpy against temperature. Figure 6.1 shows the plot for water. The two breaks correspond to phase changes, that from liquid to vapour being the larger. Similar graphs can be plotted for other substances.

6.4 CHEMICAL CHANGES

Suppose a mole of copper at 0 °C is put into a vacuum flask containing a litre of water at 20 °C. The temperature of the water drops, and that of the copper rises. Knowing the heat capacities of copper and water and assuming that they do not change appreciably over this temperature change, it is easy to calculate what the final temperature will be. If copper at 20 °C is put into the water at 20 °C there will be no change of temperature. However, if a mole of ammonium chloride at 20 °C is placed in a litre of water at 20 °C the temperature of the water will go down by 4 °C. What is the difference? The difference is that a change in the structure of the ammonium chloride and water has occurred. We may say if we wish, that a chemical change has taken place. Instead of neatly organized ammonium and chloride ions in solid ammonium chloride and the semi-structured water, we have a system which allows a more random movement of hydrated ammonium and chloride ions. The solution of ammonium chloride is quite different in structure from solid ammonium chloride and water, and has a different heat capacity.

A more obviously "chemical" reaction occurs when paper burns to form carbon dioxide and water. Heat is evolved, new substances are formed, and there is no doubt about this being a one-way change.

What are the underlying principles behind all these changes?

6.5 EXOTHERMIC AND ENDOTHERMIC CHANGES

Changes which evolve or absorb energy can both take place spontaneously. Most of the spontaneous physical changes we see, like the ball rolling downhill, evolve heat. One of the very few "physical" changes which takes place endothermically is the shooting of a stone with a catapult. The rubber of the catapult gets colder as it contracts. You can easily test this yourself with a thick rubber band. Stretch it and put it quickly to your lips; it feels warm. Let it cool while taut and then release it; put it to your lips again and it will feel cold. On the chemical level endothermic changes are more common. If ammonium carbonate dissolves in acetic acid a marked temperature drop occurs. There are many other examples, although exothermic changes are more familiar. Total energy is conserved, and therefore there is never any change in total energy when a transfer of heat takes place: what the system loses the environment gains; what the environment loses the system gains. Heat flow from a cold block to a hot block of metal is consistent with the Law of Conservation of Energy. So we should never expect the direction of heat

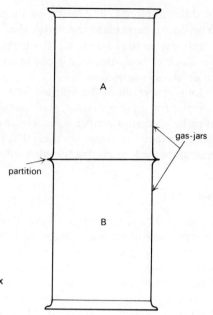

Fig. 6.2 Two gases A and B tend to mix spontaneously.

transfer in itself to be a guide as to whether a change can happen or not. In other words the endo- or exothermic nature of a change cannot be used as a criterion of whether or not the change will occur.

6.6 PROBABILITY AND CHANGE

One of the most successful models for explaining why things happen is the probability model. A certain event occurs because it is statistically most likely. We are well aware that the probability of our winning the football pools is small. But it is not impossibly small. It does happen. The probability of winning every week for three months is extremely small. It has never happened. The probability of a man winning the pools every week of his life is so improbable that we can virtually (but not accurately) say that it is impossible.

Now look at the case of two gases, A and B, in gas-jars arranged as in Fig. 6.2. We know from experience that when the partition is removed the two gases would tend to mix. What is the actual probability that they would do so? It is easier to take an even simpler case first, that of diffusion. The two flasks in Fig. 6.3 are separated by a tap. What is the tendency of a gas in one flask to spread into the empty flask when the tap is opened? To work this out consider the analogy of a number of counters being placed in two squares. If there is one counter there are two ways in which it can be placed in two squares—in one square or in the other:

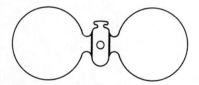

Fig. 6.3 Molecules of a gas initially in one flask distribute themselves equally between the two flasks when the tap is opened.

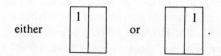

128   ENERGY AND CHEMICAL CHANGE: WHY DO CHANGES TAKE PLACE?

If there are two counters there are four different ways, and for three counters there are eight ways, and so on:

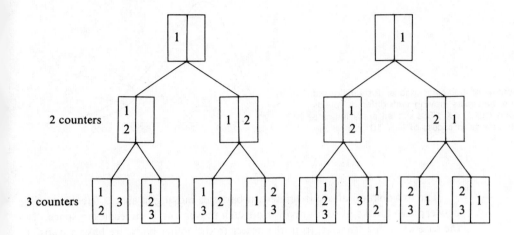

The total number of ways of dividing the counters between two squares will depend on the number of counters and will be 2^n if there are n counters.

With three counters there is one way, out of $2^3 = 8$, in which the counters can all be in one particular square. If there are six counters there is one way out of $2^6 = 64$ in which they can be all in one particular square. If the square into which a counter goes is purely a matter of chance, then the probability of it happening is dependent on the number of ways in which it can happen. The number of counters therefore makes a big difference. With three counters there is a probability of 1 in 8 that they will be in one particular square; with six counters it is 1 in 64. What of the problem of molecules in a flask? Consider the case of a litre of hydrogen diffusing into another litre flask. At 0 °C (273 K) the number of molecules in one litre of hydrogen

$$= \frac{6 \times 10^{23}}{22.4}$$
$$= 2.7 \times 10^{22}.$$

For simplicity call this 10^{22} molecules. Once the tap is turned and the hydrogen is allowed to spread into a second litre flask the chance of finding all the molecules back in the first flask is 1 in $2^{10^{22}}$. This is an extremely large number! It is equal to $2^{10000000000000000000000}$. Even the logarithm of it is difficult enough to comprehend:

$$\log 2^{10^{22}} = 10^{22} \log 2$$
$$= 10^{22} \times 0.3030$$
$$= 3 \times 10^{21}.$$

It becomes easy enough to see why we do not in practice find all the molecules in one flask. In the case of two gases diffusing into each other (assuming no interaction) the chances of the two separating out are even smaller. Probabilities are multiplicative.

6.6 PROBABILITY AND CHANGE

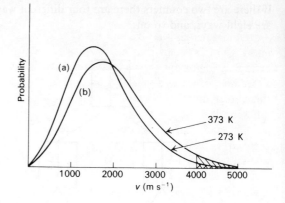

Fig. 6.4 Distribution of molecular speeds. Probabilities of finding hydrogen molecules moving with different speeds for (a) 273 K and (b) 373 K. The shaded areas indicate proportions with speeds in excess of 4×10^3 metres per second.

The explanation above is limited to particles moving in space. Heat transfer between two blocks of solid involves no such movement of molecules in space. In the case of such a transfer, from the hotter to the colder body, we have a transfer of heat. Before the transfer there is more energy in one block than in the other (in a particular sense); after the transfer the energy is spread evenly throughout the two blocks. As before, a more even distribution of energy is more likely. We must find out how atoms and molecules can hold their energy.

6.7 HOW ATOMS AND MOLECULES HOLD THEIR ENERGIES

Once again it is easier to consider this with reference to a gas. When the molecules of a gas in an enclosed vessel at a fixed temperature are banging around, bouncing backwards and forwards off each other and off the walls of the vessel, we would not expect them all to have the same kinetic energy (Section 3.5). A few will have only a very small velocity, that is a small kinetic energy, a few will have a large amount of energy, but many will be near the average. The distribution of molecular speeds was worked out by James Clerk Maxwell and has the form shown in Fig. 6.4. For higher temperatures this distribution spreads over a wider range.

So far we have talked about kinetic energy and have implied that molecules only have translational kinetic energy. That is to say energy of simple movement from one position to another. But there are other forms of molecular motion. All molecules can rotate and therefore have energy of rotation. Molecules can also vibrate. If we look into the atom itself we find that the electrons can move between different energy levels (Section 2.4.2). Looking at the three phases of matter we expect that the translation of molecules in a gas is freer than in a liquid, and that the translational energy for molecules in a solid is zero. Rotation is free in a gas, somewhat restricted in a liquid, and generally but not always absent in solids.

In the section on structure (Section 2.3.3) Planck's hypothesis that energy could only be increased or decreased discontinuously in quanta was discussed. This idea applies not only to electrons but also to molecules. Molecules in rotation and vibration are subjected to the same limitations as electrons in atoms. The molecule of hydrogen, for instance, can only have certain vibrational energies as shown in Fig. 6.5. The energy is quantized and the molecule cannot have inter-

Fig. 6.5 Vibrational energy levels for a diatomic molecule. To a good approximation the lower levels are equally spaced. Only the first seven levels are shown.

mediate vibrational energies. The same situation applies for more complicated molecules and crystals.

Unfortunately the size and number of these "energy steps" is not known for many of the systems that are of interest in chemistry. Nevertheless the idea is extremely important. It leads as illustrated below to the conclusion that there are a "number of ways" in which a number of molecules can hold their energies just as there are a "number of ways" in which a number of molecules can be placed in two different flasks.

The statistical implication of this can be illustrated by a simple model. Suppose there are five shelves on a wall, spaced at one metre intervals. You have four one-kilogram weights and have to give them a total of four kilogram metres of potential energy. There are various ways of doing this. For example, as shown on the left of Fig. 6.6(a), one weight could be on the top shelf and all the rest on the bottom shelf. Alternatively, all four weights could be on the shelf numbered one. There is only one way of doing the latter but four for the former. You will see that there are other ways, besides these, of distributing the total energy of four kilogram metres, as for example on the right of Fig. 6.6(a), and you may find it instructive to draw diagrams and find the total number of different ways. There are in fact 35 ways. Suppose now that the total energy to distribute for the same set up of four weights and five shelves is three kilogram metres. In how many ways can this be done? You may like to show that there are 17 ways.

It is of interest next to examine the effect of changing the spacing between the shelves. Suppose this time we have three shelves two metres apart and again four

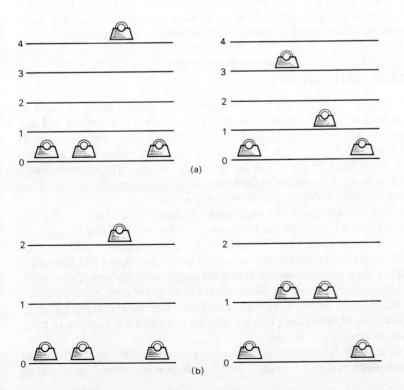

Fig. 6.6 Model showing examples of distribution of a total of four kilogram metres of potential energy. (a) Examples with shelves one metre apart. (b) Examples with shelves two metres apart.

6.7 HOW ATOMS AND MOLECULES HOLD THEIR ENERGIES

kilogram weights as in Fig. 6.6(b). Consider a total energy of four kilogram metres again. The number of ways of distributing this amount of energy is ten. You may like to extend the idea to consider shelves half a metre apart.

Two important ideas emerge from these considerations. Firstly, for a given spacing of the shelves, the more energy there is to distribute the more ways there are of distributing it. Secondly, for a given amount of energy the greater the number of shelves available, the greater the number of ways of distributing the energy.

A system of four atoms can be compared to the model; the shelves correspond to energy levels. The model needs to be refined since in the atomic or molecular system there are many energy levels. It turns out, however, that as might be expected, the more energy there is to distribute the more ways there are of distributing it, and for a given amount of energy the closer the spacing between the levels the greater the number of ways.

In molecular systems we find that for a gas the levels for translational energy are extremely close together, effectively a continuum of levels. Distribution of energy of translation makes a large contribution to the total number of ways. The average translational energy per molecule is $3RT/2N$ (Eq. 3.12). That is, the average energy per molecule is $3kT/2$ where k, Boltzmann's constant, is given by R/N. For high temperatures the average energy of molecular vibration is also closely related to kT. But for low temperatures the vibrational quantum is greater than kT, and for still lower temperatures the spacing between rotational energy levels is also greater than kT. Under these conditions the average energy of vibration and rotation may fall much below kT. There is less energy to distribute and a smaller number of ways of doing it. For many molecules in fact vibration makes only a small contribution to the number of ways at room temperature, and for lower temperatures the contribution of rotation falls off relative to translation.

6.8 ENERGY DISTRIBUTIONS

Suppose we consider a crystal made up of atoms. Each atom can vibrate about its equilibrium position in the crystal. The energy of vibration is quantized and to a good approximation the energy levels for each of the vibrating atoms are equally spaced as shown in Fig. 6.7. At a given temperature the crystal has a certain amount of energy held by the atoms as vibrational energy. The question under discussion is how many atoms are in the lowest state, how many with one quantum above this, how many with two quanta and so on.

Let us take a simple model to begin with. Consider 36 atoms arranged in a square as in Fig. 6.8. Suppose the total energy is 36ϵ, where ϵ is a quantum of vibrational energy. A game, described in Section A.3, illustrates the different ways in which the energy can be distributed among the atoms, and shows that some ways of distributing the energy among the atoms are much more probable than others. Although the numbers are small it illustrates the form of the most probable distribution of the energy. At the same time it reveals the extremely large number of ways that there are of distributing energy. Playing this game will help you to grasp the ideas introduced in this section.

Returning to Fig. 6.8, we can consider a few ways of distributing the energy, or **energy distributions** as they are called. One distribution is obtained by allocating

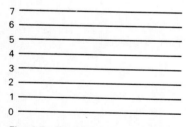

Fig. 6.7 Energy levels for vibration of atom about its equilibrium position in a crystal. A model with equally spaced levels is a good approximation. Only the first eight levels are shown.

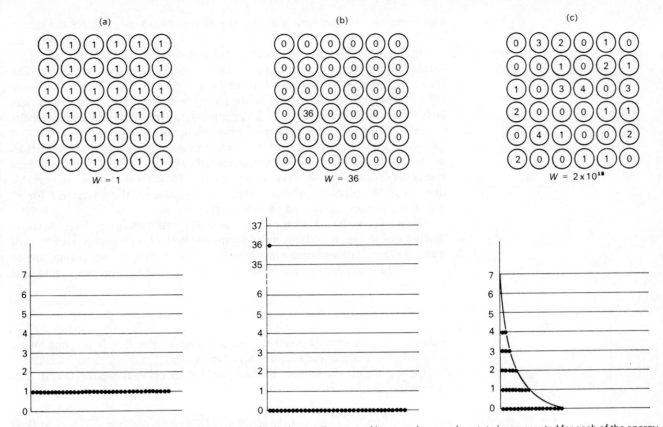

Fig. 6.8 Examples of energy distributions are shown in the lower diagrams. Above each, one microstate is represented for each of the energy distributions. W gives the number of microstates for each of the energy distributions.

one quantum to each molecule (Fig. 6.8a). Another way would be to give all the energy to one molecule, leaving the others with none (Fig. 6.8b). Figure 6.8(c) shows yet another distribution. Inspection of the diagram shows that there is only one way in which the first energy distribution can be obtained, while there are 36 ways of obtaining the second distribution. That is to say we can draw 36 different diagrams all having the energy distribution shown in the lower half of Fig. 6.8(b). For the third distribution there are in fact about 2×10^{18} different diagrams that can be drawn for the energy distribution shown in the lower half of Fig. 6.8(c). It is obviously convenient to have a term for the state of affairs described by a single diagram. It is called a **microstate**. That is there are 2×10^{18} microstates for the energy distribution $n_0 = 17, n_1 = 9, n_2 = 5, n_3 = 3, n_4 = 2$ where n_0 is the number of atoms with no quanta, n_1 the number with one quantum, and so on. We see that for some energy distributions there are many distinguishable microstates. There appears to be no *a priori* reason why any one of these distinguishable microstates should be preferred to the others. Statistical treatments of matter take as a basic assumption that all distinguishable ways in which a system can hold its energy are equally probable. That is, all microstates are equally probable. This is an assumption; it cannot be verified directly. Its justification is the success of treatments

6.8 ENERGY DISTRIBUTIONS

based upon it. In this particular example it would lead us to expect the energy distribution of Fig. 6.8(c) 2×10^{18} times more often than the energy distribution of Fig. 6.8(a). It is clearly highly improbable that under random distribution we should find the energy distribution in which all atoms have just one quantum. The form of the very probable energy distribution is of interest. The numbers occupying the various levels fall off rapidly as the energy increases. For systems with very large numbers of molecules, which correspond to typical atomic and molecular systems, it is found that for the most probable distribution of energy, the fall off follows an exponential function. For such large numbers of atoms and molecules as are usually encountered in normal systems the most probable distribution of the energy (or distributions which are undetectably different) is so much more probable than all of the other distributions that the properties of the system are for all practical purposes determined by this distribution.

We now consider a system of n atoms in a crystal with energy levels having a spacing ϵ as in Fig. 6.5. How many atoms are in the lowest energy state? How many are there with one quantum above this, how many with two quanta, and so on? Let these numbers be n_0, n_1, n_2, etc. The relationship between n_0 and n_1 is given by

$$\frac{n_1}{n_0} = e^{-\epsilon/kT}, \tag{6.1}$$

where ϵ is the spacing between the levels 0 and 1, $k = R/N$ is the Boltzmann constant, and T is the temperature (K) of the system.

Quite generally the number in the ith level at energy ϵ_i above the lowest state is given by

$$\frac{n_i}{n_0} = e^{-\epsilon_i/kT}. \tag{6.2}$$

This is a statement of the **Boltzmann distribution law**. It is quite general and applies to assemblies of identical units (atoms, molecules, ions) with any spacing between the energy levels. It is a fundamental law of great importance and turns up frequently in discussions of atomic and molecular phenomena. Not only is it important for the kind of treatment we are developing here, but also in many other areas such as interpreting rates of reaction or understanding intensities in spectra. (The equation needs some modification when two or more energy levels happen to coincide.)

6.9 ENTROPY

It is evident from the discussion in the previous section that the number of ways in which a system of atoms, ions, or molecules can hold its energy is an important quantity. The most probable distribution of energy in a system is the one with the greatest number of distinguishable ways, that is the one with the largest number of microstates. If a system is in a state with some less probable distribution of energy, change is possible in that system and that change can continue until the number of ways of holding the energy has a maximum value. A simple example, which we shall consider further in the next section, is a system of two blocks of metal at different temperatures which have been brought into contact. Change is expected to occur if it will lead to an increase in the number of microstates. This is a more probable state of the combined system, and heat flow is expected in this direction. We see here a basis of a criterion to determine in which direction change is possible.

Before taking these considerations further it is convenient to consider using the logarithm of W (number of microstates), instead of using the number of ways, W, itself. This proves to be advantageous since handling W is inconvenient. In the first place it is a very large number, and secondly, a much more serious inconvenience is that it is a multiplicative property. If the number of microstates for one mole is W, then that for two moles is W^2, since every microstate for one mole of copper can occur with every microstate for the second mole, if we are considering two copper blocks of one mole each. This is clearly inconvenient particularly as discussion of changes will frequently involve the properties U and H (internal energy and enthalpy of a system) which are extensive and thus additive.

The concept of entropy, given the symbol S, which is related logarithmically to W through the equation

$$S = k \ln W, \qquad (6.3)$$

has these advantages. The constant k could in principle be chosen arbitrarily. It is chosen, again for convenience, as the Boltzmann constant $k = R/N$. The importance of entropy stems from the criterion it provides of the direction of change. Quite simply, it can be stated that change continues in a system until the entropy has a maximum value. The condition refers to changes within an isolated system, which, as we have seen, continue until the number of microstates, W, has a maximum value.

The concept is a more powerful one than might appear from this preliminary account. Firstly, as we shall see later it provides a basis for discussing changes more generally than in isolated systems. Secondly, entropy changes can be measured experimentally and so the idea can be applied in systems which are too complicated to allow W to be calculated from a knowledge of atomic and molecular energy levels. In fact it is only for the simplest of systems that W can be calculated from a knowledge of atomic and molecular energy levels. An equation which allows entropy changes to be obtained from experimental measurements will be given in Section 6.11. This, together with the statement that *changes may continue in an isolated system until the entropy has a maximum value*, constitute a statement of the **Second Law of Thermodynamics**.

6.10 ENTROPY AND CHANGE

We saw in the last section that a change can occur if it leads to an increase in entropy. Now we are going to apply this to heat flow between two metal blocks which are identical except for a difference in temperature. Figure 6.9 illustrates the situation when the two blocks are brought into contact, but isolated from their surroundings so that no heat flows to or from the outside. One block initially has a temperature T_1 and the other T_2. The energy distribution in each is given by the Boltzmann law (Eq. 6.2). As illustrated in Figs. 6.9(a) and (b), the one with the higher temperature has the steeper distribution curve. Consider the two blocks together at the instant of contact (Fig. 6.9c). The distribution numbers for the combined system are obtained simply by adding the two numbers for each level as

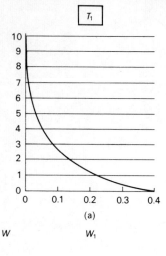
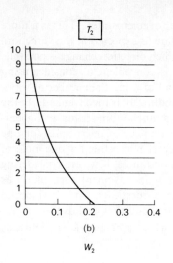
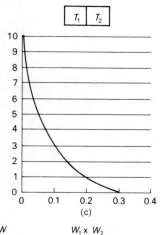
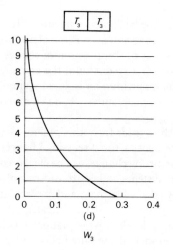

Fig. 6.9 Bringing two identical metal blocks at temperatures T_1 and T_2 into contact. (a) Energy distribution for temperature T_1. (b) Energy distribution for temperature T_2 where $T_2 > T_1$. (c) Energy distribution at moment of contact. (d) Energy distribution when equilibrium is achieved at temperature T_3.

plotted in (a) and (b). Each of the original distributions was a Boltzmann distribution, one at temperature T_1 and the other at temperature T_2. The new distribution shown in (c) is not a Boltzmann distribution for any temperature. There is a more probable distribution of the energy which will be a Boltzmann distribution at some new temperature T_3 (Fig. 6.9d).

The number of ways of distributing the energy at the instant of contact is given by $W = W_1 \times W_2$, since each microstate for block 1 can occur with each microstate for block 2. On contact we have

$$\ln W = \ln W_1 + \ln W_2, \tag{6.4}$$

or

$$S = S_1 + S_2, \tag{6.5}$$

the sum of the entropies of the two blocks. Heat transfer from block 2 leads to a decrease in the number of ways of distributing the energy in block 2: $\ln W_2$ and S_2

decrease. Heat transfer to block 1 operates in the opposite direction. There is more energy to distribute and there are more ways of doing it: $\ln W_1$ and S_1 increase. The process continues while the increase for block 1 exceeds the decrease for block 2. For as long as this is the case, W is increasing and the system of the combined blocks is moving to a probable state, that is one with a larger number of microstates and a larger entropy. The change continues until the entropy has a maximum value.

We now look briefly at a chemical process occurring in an isolated system. The system is isolated so that no energy transfer can occur between it and the surroundings (Fig. 6.10). Suppose for example the system, at a moderately high temperature, contains some gaseous 1-chlorobutane. We know that it will come into equilibrium with but-1-ene and hydrogen chloride:

$$C_4H_9Cl(g) \rightleftharpoons C_4H_8(g) + HCl(g).$$

How much but-1-ene and hydrogen chloride will be formed? What determines where equilibrium is reached? Change will continue until the entropy has a maximum value. We recognize certain problems in taking the answer further. Firstly, we need to know how to obtain entropies of the different substances. Secondly, the temperature may change as change occurs which means we would need to have entropies for different temperatures. This could be met but it is not very convenient. In any case chemists much more frequently work with systems at constant temperature rather than with isolated systems. It would clearly be helpful to remove the restriction on the treatment which prevents consideration of other than isolated systems. Both of these points will be examined in the Sections 6.11, 6.12, and 6.13.

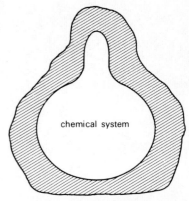

Fig. 6.10 An isolated chemical system.

6.11 MEASUREMENT AND TABULATION OF ENTROPIES

We have seen that the concept of entropy provides a basis for understanding the direction of change. For a number of comparatively simple systems entropies may be calculated from Eq. (6.3):

$$S = k \ln W.$$

Frequently in chemistry it is not possible to evaluate W for a system; the problem is far too complicated. An alternative and more generally applicable method for determining entropy changes is by making experimental measurements on the system.

Consider a simple system of some single substance. It is not difficult to see that if the system absorbs heat there will be an increase in W. There is more energy to distribute and there are more ways of doing it. The game described in Section A.3 can be used to see if, for the model, successive quantities of heat q always produce the same increase in $\ln W$, that is in the entropy. In fact one finds that each successive quantity q produces smaller and smaller changes in $\ln W$. It is as if at first disorder increases rapidly but, as the disorder increases, additions of successive quantities of heat do not produce as much additional disorder. This suggests that the change in $\ln W$ brought about by addition of a given quantity of heat falls off as the temperature of the system rises.

The relationship between entropy change and heat absorbed for an infinitesimal change is in fact

$$dS = \frac{dq}{T}. \tag{6.6}$$

This gives a way of expressing entropy changes which does not call for working out the number of microstates. It is in fact the way in which the idea of entropy was originally introduced, before a molecular understanding of the concept grew up. The form of the expression for dS, giving proportionality to dq and inverse proportionality to T (K), is obtained in Section A.4 for the system considered in the discussion of energy distributions.

To apply the equation to finite changes we need to note that the temperature will in general change as heat is added and so we have

$$\Delta S = \int_1^2 \frac{dq}{T}, \tag{6.7}$$

where 1 and 2 denote the initial and final states, and $\Delta S = S_2 - S_1$. There is one important condition to be satisfied in making the change. The heat must be transferred into the system sufficiently slowly that the system never departs significantly from equilibrium conditions. As heat flows in (or out) the system must be allowed to redistribute energy to give the distribution with the new maximum value of W, at the new equilibrium state of affairs. Such changes are called **reversible changes**.

A simple application would be the determination of the entropy change as the temperature of one mole of a substance is changed, at constant pressure, from T_1 to T_2. Under these conditions

$$dq = C_P \, dT,$$

where C_P is the molar heat capacity at constant pressure. Thus

$$\Delta S = S_2 - S_1 = \int_1^2 \frac{C_P \, dT}{T} = \int_1^2 C_P \, d \ln T. \tag{6.8}$$

The integration is readily made graphically by plotting C_P/T against T or alternatively C_P against $\ln T$.

Another example is the entropy change for a phase change. Suppose one mole of a substance is converted from liquid to vapour at its boiling point. In this case the temperature is constant throughout the process and

$$\Delta S = \frac{q}{T} = \frac{\Delta H_{\text{vap}}}{T}, \tag{6.9}$$

where ΔH_{vap} is molar heat of vaporization.

These equations provide a basis for measuring entropy changes. The zero to which entropy measurements may be referred remains to be considered. If we take a perfectly ordered crystalline solid at the absolute zero of temperature then there is no energy to distribute and there is only one spatial arrangement. Under these conditions there is only one microstate, $\ln W = 0$ and $S = 0$. For many substances, crystals are obtained sufficiently near to being perfectly ordered that it is entirely satisfactory to take the entropy to be zero for the crystalline solid at the absolute zero of temperature. There are some substances, however, for which there is disorder in the structure and there is small residual entropy at the absolute zero. The few exceptions are mostly substances with almost symmetrical molecules and the effect is well understood.

By using the principles outlined for measuring entropy changes, entropies of substances can be measured and tabulated. Experimentally, it calls for the measurement of heats of phase changes, and heat capacities down to low temperatures. The principles underlying the experimental measurements have been indicated. We shall be mainly interested in applying the ideas rather than in the measurements. For the interested reader Section A.5 contains further discussion of measuring entropy changes; and an example of the calculation of the entropy of a substance from experimental measurements is given in Section A.6. Entropies calculated in this way are often called absolute entropies. For tabulation purposes a standard state is chosen. Entropies are normally tabulated for 298 K and refer to a pressure of one atmosphere. A number of values of standard molar entropies, $S^\ominus$, are collected in Table 6.1. The values for gases are corrected for non-ideality to a kind of idealized state at one atmosphere pressure. The correction is generally small and we shall not discuss it further. For solutions a correction is applied to give the entropy for a kind of idealized concentration of unit molality. The corrections can be quite large but we shall not discuss them further. The values invite comparison. Those for iodine illustrate the general point that there is a large change in entropy in passing from the solid to the gas phase, as do values for carbon and sodium. Entropies for gases are generally large because of the contribution from translational freedom. The dependence of this contribution on mass is quite small as shown by the noble gases where a change in atomic mass from 4 to 222 results only in a 40 percent increase in the entropy, the entropies of all the noble gases falling in the range 150 ± 25 J mol^{-1} K^{-1}. The entropies of the halogens are

Table 6.1. Some standard molar entropies at 298 K (J mol^{-1} K^{-1}).

Li(s)	28.0	Zn(s)	41.6	F_2(g)	203
Na(s)	51.0	Cd(s)	51.8	Cl_2(g)	223
K(s)	64.2	Hg(l)	76.1	Br_2(g)	245
Rb(s)	76.2			I_2(g)	261
Cs(s)	82.8				
C(diamond)	2.4	Br_2(l)	152	H_2O(l)	70
Si(s)	19.0	Br_2(g)	245	H_2O(g)	189
Ge(s)	31.1	I_2(s)	117		
Sn(white)	51.4	I_2(g)	261	C(diamond)	2.4
Pb(s)	64.8			C(g)	158
				Na(s)	51.0
				Na(g)	153.6
CH_4(g)	186	He(g)	126	HF(g)	174
C_2H_6(g)	230	Ne(g)	146	HCl(g)	187
C_2H_4(g)	220	Ar(g)	155	HBr(g)	199
C_2H_2(g)	203	Kr(g)	164	HI(g)	207
		Rn(g)	176		
N_2(g)	131	O_2(g)	205	CH_3OH(l)	127
N(g)	115	N_2(g)	191	C_2H_5OH(l)	161
		CO_2(g)	214	CH_3OH(g)	238
				C_2H_5OH(g)	282

rather larger because of some contribution from rotation and a smaller one from vibration, but the translational contribution is still the one of major importance. Values of standard molar entropies are given in Tables A.15 and A.16.

6.12 ENTROPY CHANGES FOR CHEMICAL REACTIONS

Entropy changes for chemical reactions are of major interest in the application of the idea of entropy in chemistry. But, entropy changes for a chemical reaction cannot normally be measured directly by calorimetry. If we consider a system at some particular temperature in which a chemical reaction is taking place, we have by definition a system which is not in equilibrium and the course of events is not following a reversible path. The equations developed previously are not applicable.

For example, suppose we wish to know $\Delta S^\ominus$, the standard entropy changes, for the reactions

$$H_2(g) + Cl_2(g) \rightarrow 2HCl(g),$$
$$NH_3(g) + HCl(g) \rightarrow NH_4Cl(s),$$

or

$$C_2H_4(g) + H_2(g) \rightarrow C_2H_6(g).$$

That is, we want to know the entropy changes for the conversions at 25 °C, considering in each case one mole of each reactant at one atmosphere pressure. Suppose hydrogen and chlorine react together starting at 25 °C and we bring the resulting hydrogen chloride to 25 °C. The process has certainly not taken place reversibly in the thermodynamic sense of not having departed significantly from equilibrium conditions. Even for the ammonia and hydrogen chloride reaction, the mixture will start far from equilibrium conditions and again the process is not taking place reversibly, that is, it is not passing continuously through equilibrium conditions. The idea of direct measurement by calorimetry using Eq. (6.7) is not helpful.

Nevertheless the entropy change is readily obtained from tables of entropies. The standard entropies of hydrogen, chlorine, and hydrogen chloride are respectively 131, 223, and 187 J mol^{-1} K^{-1}. Hence $\Delta S^\ominus$ for the chemical change is 20 J K^{-1}. The entropy changes for other chemical reactions can be obtained equally readily from the equation

$$\Delta S^\ominus = \sum S^\ominus(\text{products}) - \sum S^\ominus(\text{reactants}), \qquad (6.10)$$

in which the number of moles of each product and reactant is to be allowed for.

Finally we examine typical entropy changes for chemical reactions and look for generalizations. Values of $\Delta S^\ominus$ for some typical reactions are collected in Table 6.2. Also included in the table are values of $\Delta H^\ominus$ and $T\Delta S^\ominus$ for reasons we shall encounter in the next section.

The examples of different types of reactions illustrate a general rule. There is an appreciable increase in the entropy of a system if the number of moles of gas increases during the reaction, and conversely there is a decrease in entropy when the number of moles of gas decreases. Similarly, it is found that dissociations in solution, as for example that of dinitrogen tetroxide, are also accompanied by an increase in the entropy of the system. The major contribution to the entropy change in these cases arises because of a change in the number of particles which can move independently: the greater the number of microstates, the larger the entropy.

Table 6.2. Entropy changes for various types of chemical reactions at 298 K. Comparison of values of $\Delta H^\ominus$ and $T\Delta S^\ominus$.

Type of reaction	Example	$\Delta S^\ominus$ (J K^{-1})	$T\Delta S^\ominus$ (kJ)	$\Delta H^\ominus$ (kJ)
solid + solid → solid	Fe(s) + S(s) → FeS(s)	8.3	2.5	−95.1
gas + gas → gas (without change in number of moles)	H$_2$(g) + Cl$_2$(g) → 2HCl(g)	19.8	5.9	−185
gas + gas → gas (with decrease in number of moles)	H$_2$(g) + C$_2$H$_4$(g) → C$_2$H$_6$(g)	−121	−36	−137
gas → gas + gas (with increase in number of moles)	N$_2$O$_4$(g) → 2NO$_2$(g)	176	52.5	57.1
solid + gas → solid	Ca(s) + Cl$_2$(g) → CaCl$_2$(s)	−151	−45.0	−795
salt + solvent → solution	KNO$_3$(s) + aq → K$^+$(aq) + NO$_3^-$(aq)	116	35	33
salt + solvent → solution	MgSO$_4$(s) + aq → Mg^{2+}(aq) + SO$_4^{2-}$(aq)	−189	−56.3	−93.0

The tabulated quantities refer to the amount of chemical reaction indicated by the equation.

The dissociation of a salt in water producing ions free to move independently throughout the solution, might, on the same lines, be expected to lead to an increase in entropy. An example where this is the case is provided by potassium nitrate, as reference to Table 6.2 will show. Many other examples could be quoted of salts consisting of relatively large univalent ions where there is an increase in entropy. But the last entry in the table, that for magnesium sulphate, shows the need for care in dealing with ionic solutions. Here there is a decrease in entropy. Other examples could be quoted to show that there are two factors which lead to this difference in behaviour. The effect arises for small ions and for ions carrying two or more charges. For such cases the entropy change may be negative. A simple qualitative interpretation can be given. We have to consider not only the effect on the entropy of the freedom of motion of the ions, but also the effect of forming a solution on the freedom of the water molecules (Section 9.1.1). So far as these are concerned solvation works in the opposite direction, and the smaller the ion and the larger its charge the more powerful the effect (Section 10.3.17).

Considerations of this kind allow patterns to be found in the stabilities, solubilities, and reactions of different compounds. The ideas will often be employed in subsequent chapters.

6.13 CHANGES AT CONSTANT TEMPERATURE AND PRESSURE

We have seen that for a change in an isolated system the entropy of the system increases. Change continues until the entropy has a maximum value. On the other hand, we have recognized that chemists often work not with isolated systems but with systems at constant temperature and pressure, when heat will generally flow into or out of the system as change occurs. This takes place in order to satisfy the condition of constant temperature, and is generally secured by a thermostat.

Let us now examine the system and its environment, represented in Fig. 6.11. We take the environment to be sufficiently large that any change in the system does not produce effects outside that environment. There are two contributions to the

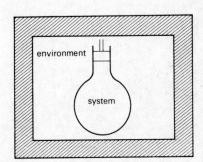

Fig. 6.11 A system and its environment.

total entropy change: the entropy change of the system, and the entropy change of the environment. We can write

$$\Delta S_{\text{total}} = \Delta S_{\text{system}} + \Delta S_{\text{environment}}. \qquad (6.11)$$

For any change the total entropy change must be positive; we can view the system and environment together as an isolated system. We have therefore

$$\Delta S_{\text{system}} + \Delta S_{\text{environment}} > 0. \qquad (6.12)$$

It is very inconvenient to have the second term in this form when we wish to fix attention on considerations of a particular chemical change in the system and not on the environment. A convenient substitution is readily made. If for the reaction the enthalpy change is written ΔH_{system}, then we have

$$\Delta S_{\text{environment}} = \frac{q}{T}$$
$$= \frac{-\Delta H_{\text{system}}}{T}, \qquad (6.13)$$

where q is the heat absorbed by the environment. (We have $-\Delta H_{\text{system}}$ because the heat absorbed by the environment is equal and opposite to that absorbed by the system.) Thus for a change to proceed at constant temperature and pressure the condition to be satisfied may be rewritten:

$$\Delta S_{\text{system}} - \frac{\Delta H_{\text{system}}}{T} > 0. \qquad (6.14)$$

The equation may also be put in the form

$$\Delta H - T\Delta S < 0, \qquad (6.15)$$

where the subscript "system" is understood, this being the normal usage of ΔH and ΔS.

There are two factors to be examined if we want to predict whether a chemical (or any other) change is feasible under given conditions of temperature and pressure. One is the enthalpy of the reaction, ΔH, which determines the entropy change of the environment. The other is the entropy change of the system itself. A vital consideration is also the temperature, because this affects the relative importance of the two factors. At high temperatures it is the change of entropy of the system which is most important; at low temperatures it is the entropy change of the environment that becomes important. It may be noted that for many chemical reactions ΔH and ΔS themselves do not change very much with temperature.

Put differently, ΔH is more influential at low temperatures and ΔS at high temperatures. It is clear that the balance between ΔH and $T\Delta S$ is of fundamental importance in determining whether a particular chemical reaction is feasible under certain conditions of temperature and pressure.

It is convenient to introduce a new term, the **Gibbs free energy** of the system, named after the American scientist J. Willard Gibbs. It is given the symbol G and defined by

$$G = H - TS. \qquad (6.16)$$

For a change at constant temperature we have

$$\Delta G = \Delta H - T\Delta S, \qquad (6.17)$$

and so the condition that a change in a system may occur may be written simply

$$\Delta G < 0. \qquad (6.18)$$

Change may continue until the Gibbs free energy has a minimum value. At this stage equilibrium will be reached and no further change will occur at that temperature and pressure. Applications of Gibbs free energy will be taken up in the next chapter which is devoted to equilibria. The ideas will be further developed in the next chapter and widely employed in subsequent chapters. Some of the more important ideas developed so far are briefly summarized in the next section.

6.14 ENTROPY: SUMMARY

At the beginning of the chapter we asked if it is possible to decide under what conditions change can occur. We found that the concept of entropy enables us to do so. Some important ideas which were developed are now briefly summarized.

1. Change is possible in an isolated system if it leads to an increase in entropy. Change can continue until the entropy has a maximum value.
2. A molecular interpretation of entropy leads to the relationship (Eq. 6.3):

$$S = k \ln W.$$

3. Entropy changes can be measured for certain changes experimentally by making use of the expression (Eq. 6.6):

$$dS = \frac{dq}{T}.$$

4. The entropy of a perfect crystalline substance is zero at absolute zero temperature. This, together with (3), provides a basis for the measurement and tabulation of entropies.
5. Tabulated values of entropies allow the calculation of entropy changes for chemical reactions (Eq. 6.10).
6. $\Delta S^{\ominus}$ has an appreciable positive value for:
   a) a chemical reaction in which the number of moles of gaseous products exceeds that of the reactants;
   b) a change of phase in which a liquid or solid is converted to a gas.

For chemical reactions in solution, entropy increases are expected where the number of solute molecules increases. Provided the interaction of the solute with the solvent is weak this is to be expected by analogy with gas phase reactions. For ionic solutions, interaction becomes strong. Nevertheless, it was seen that for the examples of dissolution of an ionic salt of relatively large univalent ions there is an entropy increase. For small ions or ions with two or more charges the interaction with the solvent water operates in the opposite direction and there may be an entropy decrease in such cases.

PROBLEMS

1. A flask contains an equal number of red and white balls. What is the probability if the balls are poured out one by one at random, that all the red will come first followed by all the white, if there are:

a) two balls of each kind;
b) three balls of each kind;
c) four balls of each kind;
d) n balls of each kind?

2. This question provides an opportunity to recall and test your understanding of concepts introduced in discussing energy distributions. Consider a simple model of three atoms arranged in a line at points X, Y, and Z.

Suppose each atom can vibrate about its mean position and that the energy levels for each atom are equally spaced, as in Fig. 6.7, and that the energy spacing between successive levels is ϵ. Suppose the system of three atoms has a total energy 3ϵ.
a) Draw a diagram for each microstate of the system. One example is shown below.

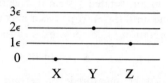

b) How many microstates are there altogether?
c) How many different energy distributions are there?
d) How many microstates are there for each energy distribution?
e) What is the most probable distribution of the energy?

3. Consider a system of a large number of particles each having energy levels as in Fig. 6.7, with a spacing of 4.14×10^{-21} J. Calculate the population of the first five levels relative to the lowest levels, for a temperature of 300 K.

4. What are the different forms of motion which contribute to the total energy of a diatomic molecule in the gas phase? For vibration, the energy levels of a diatomic molecule are equally spaced to a good approximation. From spectroscopy it is known that for the iodine molecule the transition between two adjacent levels has a wave number, $\sigma = 215$ cm^{-1} approximately. Find the ratio, with respect to the lowest state, of the number of molecules in each of the first three excited states at 298 K.

5. Calculate the entropy of fusion of ice at 0 °C and the entropy of vaporization of water at 100 °C and one atmosphere pressure (data from Table 5.1).

6. Arrange each of the following series in order of increasing molar entropy. Unless otherwise stated the conditions refer to standard states.
a) $H_2O(l)$, $H_2O(s)$, $H_2O(g)$ all at the triple point.
b) Ar(g), He(g), Ne(g), Xe(g).
c) $Br_2(g)$, $Cl_2(g)$, $F_2(g)$, $I_2(g)$.

7. The following graded examples are provided to offer experience in using tables of standard entropies to calculate standard entropy changes. Calculate standard entropy changes for the following processes:
a) $Br_2(s) \rightarrow Br_2(l)$;
b) $C_2H_4(g) + H_2(g) \rightarrow C_2H_6(g)$;

c) $H_2(g) + Cl_2(g) \rightarrow 2HCl(g)$;
d) $2N_2(g) + 3H_2(g) \rightarrow 2NH_3(g)$;
e) $H_2(g) + \frac{1}{2}O_2(g) \rightarrow H_2O(l)$.

8. From the changes noted below select three in each case, for which you would expect the standard entropy change (i) to have a positive value, (ii) to have a negative value, and (iii) to be small (positive or negative):
   a) $Cu(s) + S(s) \rightarrow CuS(s)$;
   b) $2Cu(s) + O_2(g) \rightarrow 2CuO(s)$;
   c) $CuO(s) + H_2(g) \rightarrow Cu(s) + H_2O(l)$;
   d) $2CO(g) + O_2(g) \rightarrow 2CO_2(g)$;
   e) $CH_4(g) + 2O_2(g) \rightarrow CO_2(g) + 2H_2O(g)$;
   f) $2NO_2(g) \rightarrow N_2O_4(g)$;
   g) $C_4H_9Cl(g) \rightarrow C_4H_8(g) + HCl(g)$;
   h) $2N_2O_5(g) \rightarrow 4NO_2(g) + O_2(g)$;
   i) $NH_4Cl(s) \rightarrow NH_3(g) + HCl(g)$;
   j) $2SO_2(g) + O_2(g) \rightarrow 2SO_3(g)$;
   k) $2Ag(s) + I_2(s) \rightarrow 2AgI(s)$.

BIBLIOGRAPHY

Nuffield Advanced Physics. Unit 9, *Change and Chance*, Penguin (1972).

> This provides a discussion of energy distributions with emphasis on underlying ideas. The treatment would fit in well with that adopted here. Illustrative film material also available: a 16 mm film, "Change and chance: a model of thermal equilibrium in a solid", 13 minutes, black and white, silent XX 1673, Penguin.

Series of films entitled "The Laws of Disorder" featuring Professor G. Porter. Imperial Chemical Industries.

> Three films which cover the ideas of energy distributions, the direction of change, and chemical equilibrium. 16 mm, colour, sound.

Gurney, R. W. *Introduction to Statistical Mechanics*, McGraw-Hill (1949).

> A standard work on statistical thermodynamics, which includes a more than usually detailed discussion of underlying ideas.

Bent, H. A. *The Second Law*, Oxford University Press (1965).

> A treatment of chemical thermodynamics, including some discussion of entropies in relation to structure, which was touched on in the present chapter. It provides insight in many other areas of thermodynamics.

CHAPTER 7

Equilibrium

7.1 Equilibria: distribution between two liquids
7.2 Equilibria: reactions in gases and solutions
7.3 The effect of pressure on equilibria
7.4 The effect of temperature on equilibria
7.5 Free energy and equilibria
7.6 Free energy, cells, and redox reactions
7.7 E.m.f. and free energy change
7.8 Half-cells and electrode potentials
7.9 The sign of the e.m.f.
7.10 Ion-ion redox reactions
7.11 Variation of cell e.m.f. with concentration and temperature
7.12 Using the Second Law
7.13 Chemical change: summary

In the last chapter (Section 6.13) we considered change occurring until there is a balance in two entropy changes. How far is this a special case? Most changes come to an end sometime; most chemical reactions slow down and eventually no further change is observed. How far a reaction goes is almost as important to the chemist as whether it goes at all. In this chapter we examine equilibrium and the factors which govern it.

7.1 EQUILIBRIA: DISTRIBUTION BETWEEN TWO LIQUIDS

Equilibria between phases were discussed in Chapter 5. As a first step towards understanding chemical equilibria we can examine the way in which a substance dissolves in two immiscible liquids (liquids that are insoluble in each other and therefore form two layers in a beaker). Two such liquids are trichloromethane (chloroform) and water. Ammonia dissolves in both trichloromethane and water. It reacts with the water to some extent forming ammonium ions and hydroxyl ions, but this does not affect the general principle illustrated. If trichloromethane and water are placed in a beaker and ammonia is dissolved in them, or alternatively trichloromethane is shaken with a solution of ammonia, the dynamic equilibrium

$$NH_3(aq) \rightleftharpoons NH_3(chloroform)$$

is set up between the ammonia dissolved in the water and that dissolved in the trichloromethane. If the amount of ammonia is varied and the concentrations in each solvent are measured the values given in Table 7.1 are obtained.

The equilibrium is established across the liquid interface and it is the concentrations which affect it, not the total amounts. Examination of the figures in the table shows that the ratios of the concentrations remain approximately constant. Note that the symbol [A] denotes concentration of A in moles per litre. For an equilibrium between A in one phase and B in another [B]/[A] is constant.

If the temperature is changed a new value for [B]/[A] will be found, but it will still be a constant for that temperature. This ratio for a particular temperature is called the **partition coefficient** or **distribution coefficient**.

Table 7.1. Distribution of ammonia between water and trichloromethane. From Nuffield Advanced Science, *Chemistry*, Penguin (1970). Reproduced by permission of The Nuffield Foundation.

Concentration of ammonia in trichloromethane, $[NH_3(CHCl_3)]$ (mol l^{-1})	Concentration of ammonia in water, $[NH_3(H_2O)]$ (mol l^{-1})	$\dfrac{[NH_3(CHCl_3)]}{[NH_3(H_2O)]}$
0.0163	0.405	24.8
0.0208	0.513	24.7
0.0261	0.648	24.8
0.0348	0.825	23.7
0.0429	1.05	24.5
0.0574	1.35	23.5

Square brackets [ ] denote concentrations of ammonia in each layer in moles per litre.

7.2 EQUILIBRIA: REACTIONS IN GASES AND SOLUTIONS

As might be expected, some of the simplest equilibrium reactions are found in gaseous systems. The reaction between hydrogen and iodine has been studied a great deal. The reaction is

$$H_2(g) + I_2(g) \rightarrow 2HI(g),$$

for which

$$\Delta H^\ominus_{298} = 51.9 \text{ kJ mol}^{-1},$$

and

$$\Delta S^\ominus_{298} = 21.8 \text{ J mol}^{-1} \text{ K}^{-1}.$$

If hydrogen and iodine are placed in a sealed vessel at a few hundred °C, after a time hydrogen iodide, hydrogen, and iodine are all found to be present. The concentration of hydrogen iodide increases with time up to a maximum (Fig. 7.1).

If a sample of hydrogen iodide is treated similarly the concentration of HI decreases, hydrogen and iodine being formed. The same equilibrium is reached

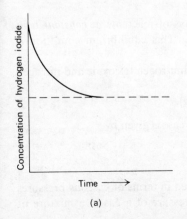
(a)

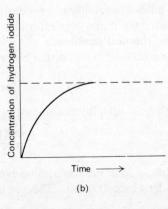
(b)

Fig. 7.1 The same equilibrium is reached whether the starting materials are (a) two moles of hydrogen iodide or (b) one mole of hydrogen and one mole of iodine. In (a) the concentration of hydrogen iodide decreases with time until equilibrium is attained; in (b) the concentration of hydrogen iodide increases until equilibrium is attained.

Table 7.2. Equilibrium between hydrogen, iodine, and hydrogen iodide at 698 K.

$[I_2]$ (mol l^{-1})	$[H_2]$ (mol l^{-1})	$[HI]$ (mol l^{-1})	$K_c = \dfrac{[HI]^2}{[H_2][I_2]}$
0.4789×10^{-3}	0.4789×10^{-3}	3.531×10^{-3}	54.35
1.1409×10^{-3}	1.1409×10^{-3}	8.410×10^{-3}	54.35
0.4953×10^{-3}	0.4953×10^{-3}	3.655×10^{-3}	54.58
1.7069×10^{-3}	2.9070×10^{-3}	16.482×10^{-3}	54.73
1.2500×10^{-3}	3.5600×10^{-3}	15.588×10^{-3}	54.61
0.7378×10^{-3}	4.5647×10^{-3}	13.544×10^{-3}	54.49
2.3360×10^{-3}	2.2523×10^{-3}	16.850×10^{-3}	53.96
3.1292×10^{-3}	1.8313×10^{-3}	17.671×10^{-3}	54.49

whether we start with one mole of hydrogen and one mole of iodine or with two moles of hydrogen iodide. Clearly an equilibrium is being reached, that is

$$H_2(g) + I_2(g) \rightleftharpoons 2HI(g).$$

The concentrations of the three species at equilibrium starting from different concentrations of hydrogen and iodine are given in Table 7.2. This time it can be seen that there is not such a simple relationship between the concentrations as with the distribution of ammonia between water and trichloromethane. Trying out various combinations we find that

$$\frac{[HI(g)]}{[H_2(g)][I_2(g)]}$$

is not constant; but

$$\frac{[HI(g)]^2}{[H_2(g)][I_2(g)]},$$

as Table 7.2 shows, is very nearly constant. This constant is called the **equilibrium constant** and is given the symbol K_c. It is found by experiment that for a reaction

$$vA + xB \rightleftharpoons yC + zD,$$

$$K_c = \frac{[C]^y[D]^z}{[A]^v[B]^x}. \tag{7.1}$$

This equation holds for equilibrium concentrations of reactants *at constant temperature*, whatever the initial concentrations are. This equilibrium condition is called the **law of chemical equilibrium**.

Another gaseous equilibrium is that between dinitrogen tetroxide and nitrogen dioxide:

$$N_2O_4(g) \rightleftharpoons 2NO_2(g).$$

It can be seen that the equilibrium constant in this case is given by

$$K_c = \frac{[NO_2]^2}{[N_2O_4]}.$$

The equilibrium constant may also be expressed in terms of gaseous pressures rather than molar concentrations. The partial pressure of a gas in a mixture of gases is defined as the pressure that the gas would exert if it occupied the space alone. **Dalton's law of partial pressures** states that in a mixture of gases the total pressure is the sum of the partial pressures. This means that in a mixture two gases do not interfere with each other; they each make their own independent contribution to the total pressure.

The symbol used for the partial pressure of a gas X is P_X. The equilibrium constant expressed in terms of partial pressures is denoted by K_P. Thus for the dinitrogen tetroxide equilibrium,

$$K_P = \frac{(P_{NO_2})^2}{P_{N_2O_4}},$$

K_P and K_c are closely related.

The ideal gas equation (Section 3.5) states that

$$PV = nRT.$$

The concentration of the gas A is given by:

$$[A] = \frac{n}{V}.$$

From the ideal gas equation

$$\frac{n}{V} = \frac{P}{RT};$$

hence

$$[A] = \frac{P}{RT}.$$

If we apply this to the equilibrium $N_2O_4 \rightleftharpoons 2NO_2$, we have

$$K_c = \frac{[NO_2]^2}{N_2O_4},$$

$$K_c = \frac{(P_{NO_2})^2/RT}{P_{N_2O_4}/RT}.$$

The pressure of each gas in the equilibrium mixture is of course its partial pressure, therefore:

$$K_c = \frac{(P_{NO_2})^2}{P_{N_2O_4}} \times \frac{1}{RT};$$

$$K_c = K_P \times \frac{1}{RT};$$

$$K_P = RTK_c.$$

It can be seen that the relationship depends on the relative numbers of moles on the left and right hand sides of the equations. If there is no change in the number of moles as in

$$H_2(g) + I_2(g) \rightleftharpoons 2HI(g),$$

then

$$K_P = K_c.$$

Table 7.3. Dissociation of dinitrogen tetroxide in trichloromethane at 281.3 K.

$[N_2O_4]$ (mol l^{-1})	$[NO_2]$ (mol l^{-1})	$K_c = \dfrac{[NO_2]^2}{[N_2O_4]}$ (mol l^{-1})
0.129	1.17×10^{-3}	1.07×10^{-5}
0.227	1.61×10^{-3}	1.14×10^{-5}
0.324	1.85×10^{-3}	1.05×10^{-5}
0.405	2.13×10^{-3}	1.13×10^{-5}
0.778	2.84×10^{-3}	1.04×10^{-5}

In general

$$K_P = (RT)^{\Delta\nu} K_c, \qquad (7.2)$$

where $\Delta\nu$ is the number of moles on the right hand side of the equation minus the number of moles on the left hand side.

We may also note that equilibria in solutions obey the equilibrium law in the same way. In this case we are, of course, concerned only with the equilibrium constant K_c. As an example some figures are quoted in Table 7.3 for the equilibrium between dinitrogen tetroxide and nitrogen dioxide in trichloromethane solution. The equilibrium constant for a reaction in solution does not in general have the same value as for the equilibrium in the gas phase.

It can be seen that the entries in the third column remain nearly constant over a six-fold change in concentration of the tetroxide.

7.3 THE EFFECT OF PRESSURE ON EQUILIBRIA

In a gaseous equilibrium the equilibrium position is found to be influenced by pressure if there is a change in volume (i.e. the number of moles of gaseous reactants) during the reaction. A rise in pressure tends to push the reaction in the direction of smaller volume. A lowering of pressure tends to push the reaction in the direction of increasing volume. Thus:

$$N_2O_4(g) \rightleftharpoons 2NO_2(g)$$
smaller volume $\qquad$ larger volume
$\leftarrow$ higher pressure
lower pressure $\rightarrow$

The new position of equilibrium is such that the equilibrium equation holds for the same equilibrium constant. This is in general accordance with the principle expounded by Henry Le Chatelier (1888). **Le Chatelier's principle** states that where a stress is applied to a system in equilibrium the equilibrium tends to react in such a way as to absorb the stress. The principle should be treated with some caution; it has its limitations, but it is a useful general guide to the behaviour of chemical (and of stable physical) equilibria.

7.4 THE EFFECT OF TEMPERATURE ON EQUILIBRIA

If the equilibrium mixture

$$N_2O_4(g) \rightleftharpoons 2NO_2(g)$$

is heated, the equilibrium position moves to the right—more $NO_2(g)$ is formed. The reaction is endothermic in this direction ($\Delta H^\ominus_{298} = 57.2$ kJ mol^{-1}). Similarly if the temperature is lowered the equilibrium position moves to the left—more $N_2O_4(g)$ is formed. The reaction $2NO_2(g) \rightarrow N_2O_4(g)$ is exothermic ($\Delta H = -57.2$ kJ mol^{-1}). This is in general agreement with Le Chatelier's principle. The reaction moves in such a way as to absorb heat when the temperature is raised and to evolve heat when the temperature is lowered.

In terms of entropy we can see that the entropy of two moles of NO_2 is greater than that of one mole of N_2O_4. The two moles have twice as many translational energy levels as the one mole of N_2O_4. The mole of N_2O_4 has more ways of vibrating but this does not compensate for the additional translational entropy of the NO_2. The extent to which the reaction will go will depend on $\Delta H^\ominus - T\Delta S^\ominus$ (Sections 6.12, 6.13, and 7.5). For positive values equilibrium will be over to the left and for negative values to the right. At high temperatures $T\Delta S$ will become the most important factor. Thus high temperatures favour the decomposition of N_2O_4. For low temperatures the term $T\Delta S^\ominus$ will become less important and the equilibrium moves in favour of formation of the tetroxide. In this discussion we have assumed that $\Delta S^\ominus$ and $\Delta H^\ominus$ are independent of temperature. Reference to Table 7.4 where data for this equilibrium are collected will show that this is a good approximation. *It is in fact the case for most chemical reactions that $\Delta H^\ominus$ and $\Delta S^\ominus$ change only slowly with temperature.*

For a more precise examination of the effect of temperature on the equilibrium constant we must look at values obtained experimentally. No obvious relationship can be seen by just looking at the figures in the table, but it is clear that K_P varies very rapidly with temperature. If we plot the logarithm of K_P against $1/T$ we get

Table 7.4. Thermodynamic data for the dinitrogen tetroxide and nitrogen dioxide equilibrium.

T (K)	$NO_2(g)$ $S^\ominus$ (J mol^{-1} K^{-1})	$N_2O_4(g)$ $S^\ominus$ (J mol^{-1} K^{-1})	$N_2O_4(g) \rightleftharpoons 2NO_2(g)$ $\Delta S^\ominus$ (J K^{-1})	$\Delta H^\ominus$ (kJ)	$\Delta G^\ominus$ (kJ)	K_P (atm)
100	202.3	239.1	165.5	55.73	39.54	3.61×10^{-21}
200	225.6	276.3	174.9	56.98	21.92	1.86×10^{-6}
250	233.4	291.2	175.6	57.15	13.10	1.78×10^{-3}
275	236.8	298.2	175.4	57.11	8.75	2.16×10^{-2}
298	239.9	304.2	175.6	57.07	4.69	1.51×10^{-1}
300	240.0	304.8	175.2	57.07	4.35	1.74×10^{-1}
325	243.0	311.1	174.9	56.98	-0.04	1.01
350	245.9	317.2	174.6	56.86	-4.39	4.52
400	251.2	328.7	173.7	56.61	-13.14	5.1×10
500	260.5	349.4	171.6	55.61	-30.42	1.51×10^3
600	268.7	367.7	169.7	54.52	-47.53	1.38×10^4

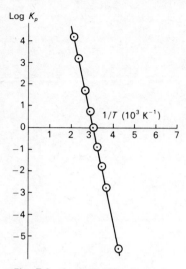

Fig. 7.2 A plot of log K_P against $1/T$ for the dinitrogen tetroxide and nitrogen dioxide equilibrium.

the graph in Fig. 7.2. The straight line shows that over the temperature range we may write

$$\ln K_P = C + \frac{D}{T}, \tag{7.3}$$

where C and D are constants.

We shall not develop here a full account of the way in which the relationship between K_P and temperature can be derived. Basically, the constant D is related to $\Delta H^\ominus$ for the reaction. Because $\Delta H^\ominus$ is not exactly constant (that is, independent of temperature) for precision the equation may be written in differential form. The equation in this form is

$$\frac{d \ln K_P}{dT} = -\frac{\Delta H^\ominus}{RT^2}. \tag{7.4}$$

Assuming that $\Delta H^\ominus$ is constant, by integrating we obtain the expression

$$\ln K_P = \text{constant} - \frac{\Delta H^\ominus}{RT}. \tag{7.5}$$

The graph in Fig. 7.2 illustrates this relationship between log K_P and $1/T$. The integrated form is not exact since $\Delta H^\ominus$ varies slightly with temperature but for many purposes it is entirely satisfactory. The slope is $-\Delta H^\ominus/2.303RT$.

This equation is useful as it allows us to calculate K_P at different temperatures. It can also be used as an indirect way of determining $\Delta H^\ominus$ for a reaction. This is often useful when calorimetry cannot be applied as for example when a reaction does not go to completion. It is of interest to compare the discussion of the temperature dependence of equilibrium constants with that for vapour pressure. Equation (7.5) can be regarded as a special case of Eq. (5.4), the vapour pressure P being regarded as equivalent to K_P for liquid–vapour equilibrium.

7.5 FREE ENERGY AND EQUILIBRIA

The importance of Gibbs free energy in the discussion of equilibria is apparent from its introduction in Section 6.13. Free energy was introduced through a discussion of the total change of entropy, considering a system and its environment.

In writing $\Delta G = \Delta H - T\Delta S$ for a change at constant temperature and pressure we are equating a number of energy terms: the free energy, the enthalpy, and the entropy change times T. The units of entropy may be joules per mole per kelvin, in which case the units of entropy times T are joules per mole, that is, the units of energy per mole.

If we divided the whole equation by T we should have an equation relating entropy terms:

$$\underset{\substack{\text{total change} \\ \text{in entropy}}}{\frac{\Delta G}{T}} = \underset{\substack{\text{decrease in} \\ \text{entropy of} \\ \text{environment}}}{\frac{\Delta H}{T}} \quad \underset{\substack{\text{decrease in} \\ \text{entropy of} \\ \text{system}}}{(-\Delta S)}.$$

Change may occur if it leads to an increase in total entropy, that is, if ΔG has a negative value. The extent to which a change takes place depends on the balance of entropies. Change can continue until, for any further change, the two terms just balance.

If
$$\frac{\Delta H}{T} = \Delta S$$
for a small displacement, or
$$\Delta H = T\Delta S,$$
then no change will take place in a system at constant temperature and pressure, and a state of equilibrium will exist. Under these conditions
$$\Delta G = 0$$
for a small displacement from equilibrium.

At this stage we must emphasize the difference between standard heats of reaction $\Delta H^\ominus$ (Section 4.11.4) standard entropy changes $\Delta S^\ominus$ and standard free energy changes $\Delta G^\ominus$, and the general changes ΔH, ΔS, and ΔG.

Consider a system in which change is occurring as the system moves towards equilibrium. Let us make the system so large that conversion of one mole of some reactant to products will not change the concentrations significantly. Then, for such a change, there will be a certain ΔG which will depend on the concentrations of reactants and products. As the change continues and the concentrations do change, ΔG changes. Eventually a stage is reached where for any further change $\Delta G = 0$, that is equilibrium has been reached.

On the other hand, in quantitative expressions we are often concerned with the standard free energy change. This is what the change would be if all the reactants were completely turned into products under standard conditions. It is this that is called the **standard free energy change** and is given the symbol $\Delta G^\ominus$. It refers to one g-equation's worth of change, starting with reactants in their standard states and ending with products in their standard states (Section 4.11.2). Similarly, $\Delta H^\ominus$ and $\Delta S^\ominus$ are changes in standard enthalpy and entropy.

A number of points in this discussion are conveniently illustrated by a diagram. Figure 7.3 shows the free energy relationships for the reaction
$$N_2O_4(g) \rightleftharpoons 2NO_2(g).$$

The graph shows the free energy of the system at different concentrations of the dioxide and tetroxide. It shows that the free energy of the system decreases whether we start with pure dioxide or tetroxide. Equilibrium is established when a minimum value is reached. For an infinitesimal change here in either direction $dG = 0$. For a finite change ΔG would clearly be positive, and so such a change will not occur. Notice that free energy change $\Delta G^\ominus$ refers to the overall change from reactant in standard state to product in standard state.

Now we might expect there to be some relationship between $\Delta G^\ominus$ and K, the equilibrium constant. Examination of actual values shows that, for many simple reactions, $\Delta G^\ominus$ is usually in about the range ± 500 kJ mol^{-1} and K can be as great as 10^{50}. It is not surprising therefore that the relationship is logarithmic. It can be derived, but we shall be more interested in applications than in the derivation. However, for the interested reader a treatment is given in Section A.7. The relationship is

$$\Delta G^\ominus = -RT \ln K, \qquad (7.6)$$

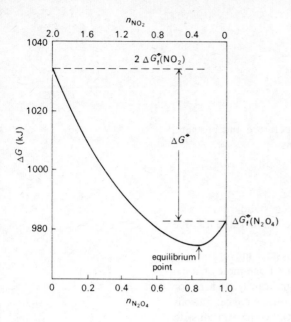

Fig. 7.3 Diagrammatic representation of free energy changes for the dinitrogen tetroxide and nitrogen dioxide equilibrium.

or in terms of logarithm to base 10,

$$\Delta G^\ominus = -2.303\, RT \log K. \tag{7.7}$$

The temperature is 298 K (25 °C) (unless otherwise specified), and R is about 8.3 J mol^{-1} K^{-1}, so the constant of proportionality is $2.303 \times 298 \times 8.3 \times 10^{-3} = 5.7$ kJ mol^{-1}. At this temperature $\Delta G^\ominus$ is roughly $6 \log K$. This is in fact useful in getting a rough idea of the relative values of K and $\Delta G^\ominus$.

If a reaction $A + B \rightleftharpoons C + D$ goes 90 percent to completion, an idea of the equilibrium constant can be found by assuming one mole of A is mixed with one mole of B in one litre of solution. At equilibrium there will be 0.9 moles of each of C and D and 0.1 moles of A and B. Therefore

$$K = \frac{0.9 \times 0.9}{0.1 \times 0.1}$$
$$= 81.$$

It is not unreasonable therefore to say that reactions for which K is not more than 100 or less than $\frac{1}{100}$ are in the range we usually call equilibrium reactions in the laboratory. Thus for $\Delta G^\ominus$ between $-6 \log 100 = -12$ kJ and $-6 \log 10^{-2} = +12$ kJ, reactions are in this equilibrium range. Many reactions appear to go to completion or not to go at all. For 1 percent conversion or 99 percent completion the limits are roughly 25 kJ or -25 kJ. If $\Delta G^\ominus$ is outside these limits the reaction is unlikely to be of interest for preparative work unless some way can be found of removing one of the products as it is formed. In fact all reactions can theoretically reach equilibrium. Those which appear to go to completion have very large equilibrium constants of the order of 10^{10} or more. Those that appear not to start have equilibrium constants of the order of 10^{-10} or less. Just where one draws the line depends on one's technique for measuring small concentrations.

The importance of Eq. (7.6) connecting equilibrium constant with standard free energy change can be recognized by considering how standard free energy values may be obtained. Equation (6.17) allows us to write for a given temperature:

$$\Delta G^\ominus = \Delta H^\ominus - T\Delta S^\ominus. \tag{7.8}$$

This allows values to be obtained from calorimetric measurements. For ease of calculation $\Delta H^\ominus$ may be obtained from tables of $\Delta H_f^\ominus$ as discussed in Section 4.11.4. Similarly, $\Delta S^\ominus$ may be obtained, as discussed in Section 6.12, from tables of standard entropies using Eq. (6.10). In this way it is possible using Eq. (7.8) to predict equilibrium constants from data obtained from calorimetric measurements.

The use of different tables for $\Delta H^\ominus$ and $\Delta S^\ominus$ to calculate $\Delta G^\ominus$ can be avoided by using tables of **standard free energies of formation**, $\Delta G_f^\ominus$, defined analogously to $\Delta H_f^\ominus$ (Section 4.11.2). Values for elements are taken by convention to be zero. Standard free energy values for compounds refer to the formation of compounds in their standard states from elements in their standard states. Values of $\Delta G_f^\ominus$ to form compounds from elements at one atmosphere pressure and 298 K are given in Table A.15. The standard free energy change for a chemical reaction is given by

$$\Delta G^\ominus = \sum \Delta G_f^\ominus(\text{products}) - \sum \Delta G_f^\ominus(\text{reactants}), \tag{7.9}$$

making allowance for the number of moles of each product and reactant.

An example will illustrate the meaning and use of standard free energies. The standard free energy of formation of nitrogen dioxide refers to the formation of nitrogen dioxide at one atmosphere pressure from nitrogen and oxygen at one atmosphere pressure:

$$\tfrac{1}{2}N_2(g) + O_2(g) \rightarrow NO_2(g); \quad \Delta G_f^\ominus(NO_2) = 51.3 \text{ kJ}.$$

For 298 K the value is obtained from Table A.15 as shown. Similarly, for dinitrogen tetroxide,

$$N_2(g) + 2O_2(g) \rightarrow N_2O_4(g); \quad \Delta G_f^\ominus(N_2O_4) = 97.8 \text{ kJ}.$$

The standard free energy change $\Delta G^\ominus$ for the process

$$N_2O_4(g) \rightarrow 2NO_2(g)$$

at 298 K can be obtained by applying Eq. (7.9):

$$\begin{aligned}\Delta G^\ominus &= 2\Delta G_f^\ominus(NO_2) - \Delta G_f^\ominus(N_2O_4) \\ &= 2 \times 51.3 - 97.8 \\ &= 4.8 \text{ kJ}.\end{aligned}$$

It should be added that use of Eq. (7.8) is not restricted to calorimetric values of entropies. Values calculated statistically as mentioned in Section 6.9 can also be used. For some purposes, such as high temperature reactions, statistical values are important since calorimetric values may not be available or may be inaccurate. Finally, the calculation may be inverted and measured equilibrium constants used to calculate $\Delta H^\ominus$ and/or $\Delta S^\ominus$. $\Delta H^\ominus$ may be obtained from the temperature dependence of an equilibrium constant through Eq. (7.5), and then $\Delta S^\ominus$ by using Eq. (7.8). Thus we have an extremely useful method of relating knowledge derived in quite different ways. In the next section we shall see how electrical measurements can also be tied up with thermodynamic quantities.

7.6 FREE ENERGY, CELLS, AND REDOX REACTIONS

Equilibrium constants and hence free energies can be measured, as we saw in the last section, by finding the concentration of reactants and products in the equilibrium mixture. We have also seen that the standard free energy change, $\Delta G^\ominus$, can be found if the enthalpy and entropy changes are known for a reaction. There is a third experimental way of finding $\Delta G^\ominus$. In some ways this is the most satisfactory way but it suffers from one drawback: it can only be applied to reactions in which there is electron transfer, that is to say redox reactions. Redox reactions have been used for a long time in batteries as sources of energy. More recently they have begun to be used in fuel cells. The energy produced is electrical energy and the output of energy can be measured very precisely. Electrical energy is measured in terms of the quantity of electricity (charge) which passes at a particular potential difference.

Any redox reaction can theoretically be made to generate an electric current. The familiar reaction between zinc and a copper sulphate solution is a redox reaction:

$$Zn(s) + Cu^{2+}(aq) \rightleftharpoons Zn^{2+}(aq) + Cu(s);$$
$$\Delta H^\ominus_{298} = -216.7 \text{ kJ mol}^{-1}$$
$$\Delta S^\ominus_{298} = -7.8 \text{ J mol}^{-1} \text{ K}^{-1}.$$

The zinc ion donates two electrons:

$$Zn(s) \rightarrow Zn^{2+}(aq) + 2e^-.$$

The copper ion accepts two electrons:

$$Cu^{2+}(aq) + 2e^- \rightarrow Cu(s).$$

In terms of oxidation states zinc goes from 0 to +2, copper from +2 to 0. The reaction can be brought about simply by mixing zinc dust with copper sulphate

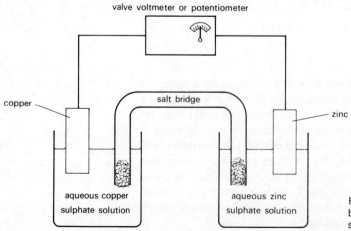

Fig. 7.4 A zinc-copper cell. The salt-bridge is closed with some material such as glass wool to prevent mixing.

solution. Alternatively, since electrons can be transferred through a metal conductor it is possible for the two halves of the above reaction to take place separately, but not quite separately for it is not possible to take electrons from a chemical system or add them to a system without a balancing of charges. In the case of cells the balancing charges are those of the ions which must be allowed to diffuse from one half-cell to the other. A cell with zinc and copper electrodes is shown in Fig. 7.4.

If the cell is short circuited the reaction will proceed until an equilibrium is reached, in this case well over to the right hand side of the equation. The solution will get hot producing the same amount of heat, equal to $-\Delta H$, as in the case of direct mixing. If the wire connecting the zinc and copper electrodes has an appreciable resistance some heat will be evolved in the wire instead of in the solution. This point will be taken up later. If the cell is left on "open circuit", a potential difference will exist between the two electrodes. This potential difference can be measured on a bridge circuit without any current being taken. Alternatively, the potential difference across the cell can be measured as a current is taken from the cell using the circuit of Fig. 7.5. If this is done the graph of current given by the cell against potential difference between the electrodes looks like the curve in Fig. 7.6. It can be seen that the maximum voltage is given with zero current. As soon as a small current is taken the voltage goes down. Notice also that as soon as the external voltage exceeds that of the cell, the current is reversed. The chemical reaction is then pushed backwards. This is the equivalent of the recharging of a battery.

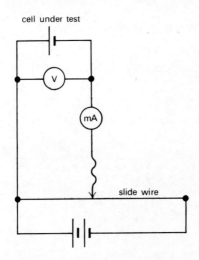

Fig. 7.5 Circuit diagram for measurement of potential difference across a cell as current through the cell is changed.

All the energy from the reaction can be released as thermal energy within the system. How much energy can be removed from the system as electrical energy? For the complete reaction to take place one mole of zinc must displace one mole of copper. Since the charge on each ion is $+2$, two moles of electrons must flow round the circuit. This quantity of charge is of course constant for the amount of chemical reaction. The amount of energy produced (potential × charge) will therefore depend only on the potential difference through which this charge moves, that is the voltage of the cell. To get the maximum amount of electrical energy from the cell it will be necessary to have the highest voltage, therefore the smallest current. That is to say the energy will have to be taken out very slowly indeed. If it is taken out infinitely slowly the voltage will be that on open circuit. The maximum electrical work, w_{max}, that can theoretically be obtained from the cell is thus given by the voltage of the cell on open circuit times the charge which flows, that is,

$$w_{max} = E \times z \times F, \qquad (7.10)$$

where E is the electromotive force (e.m.f.) of the cell, z is the number of moles of electrons, and F is the Faraday constant.

As will be seen in the next section the e.m.f. is determined by the free energy change of the process occurring in the cell through the equation

$$\Delta G = -zFE. \qquad (7.11)$$

If the reactants and products are in their standard states then the e.m.f. of the cell is called the standard e.m.f. for the process and is given by

$$\Delta G^\ominus = -zFE^\ominus. \qquad (7.12)$$

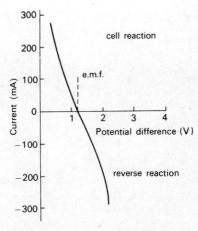

Fig. 7.6 Plot of current through cell against potential difference between electrodes.

7.6 FREE ENERGY, CELLS, AND REDOX REACTIONS

7.7 E.M.F. AND FREE ENERGY CHANGE

Consider the process,

$$H_2(g) + \tfrac{1}{2}O_2(g) \rightarrow H_2O(l);$$
$$\Delta H^\ominus = -285.9 \text{ kJ}$$
$$\Delta S^\ominus = -163.1 \text{ J K}^{-1}.$$

The reaction could be brought about by burning hydrogen, or alternatively it could occur in a fuel cell. We shall consider and compare the two processes. The combustion under standard conditions of one atmosphere pressure leads to the evolution of 286 kJ, giving us $\Delta H^\ominus$. The standard entropy change is readily obtained as discussed in Section 6.12, from Table A.15, using Eq. (6.10). We have

$$\Delta S^\ominus = S^\ominus(H_2O(l)) - S^\ominus(H_2(g)) - \tfrac{1}{2}S^\ominus(O_2(g))$$
$$= 70.0 - 130.6 - 102.5 = -163.1 \text{ J K}^{-1}.$$

Making the same kind of considerations as discussed in Section 6.13 we may calculate the total entropy change for the formation of water by combustion. We have, from Eq. (6.11),

$$\Delta S_{\text{total}} = \Delta S_{\text{system}} + \Delta S_{\text{surroundings}}$$
$$= \Delta S_{\text{system}} - \frac{\Delta H_{\text{system}}}{T}$$
$$= -163 + 956 = 793 \text{ J K}^{-1}.$$

In the combustion there is a total entropy increase of 793 J K^{-1}, and during the process there is a heat transfer to the surroundings equal to $-\Delta H$, in this case 286 kJ. But we see from the equation that it is not essential to transfer all of this quantity. Enough must be transferred to prevent the total entropy decreasing. To compensate for the change in entropy of the system by -163 J K^{-1} the entropy of the surroundings needs to be increased by $+163$ J K^{-1}. By using the equation $\Delta S = q/T$, we find that this could be achieved by transferring a quantity of heat $q = T\Delta S$. The result in this particular case is that the Second Law of Thermodynamics requires only a heat transfer of 48 kJ (298 × 163 J) to compensate for the entropy decrease of the system. The remainder of the energy transfer could be made available as work. In this case this amounts to 238 kJ; this is the maximum work that can be made available as a result of the reaction (286 − 48 = 238 kJ).

In the general case for a process at constant temperature and pressure the maximum work available is given by

$$-(\Delta H - T\Delta S),$$

that is

$$w_{\max} = -\Delta G.$$

It can in fact be shown that when a cell is operated under the conditions described in Section 7.6 the maximum work available from the chemical reaction is obtained electrically. That is,

$$w_{\max} = zFE = -\Delta G. \tag{7.13}$$

During the operation of the cell the maximum work obtainable is 238 kJ and at the same time 48 kJ is transferred as heat. In general, a quantity of heat $q = T\Delta S$ is transferred. It may be noted that if ΔS is negative then q will be negative and heat will be transferred to the cell during the course of the cell reaction, and in this case the maximum work done is greater than $-\Delta H$.

7.8 HALF-CELLS AND ELECTRODE POTENTIALS

We saw in Section 7.6 that the tendency for a reaction to go in a particular direction can be measured electrically. It proves convenient in fact to consider each part of the reaction, each half-cell, as having its own equilibrium situation, which can be treated independently. Consider again the copper rod surrounded by copper sulphate solution. There is a tendency for copper to form copper ions, giving off electrons, and there is the tendency of copper ions to accept electrons and become metallic copper:

$$Cu(s) \rightleftharpoons Cu^{2+}(aq) + 2e^-.$$

The equilibrium position will be dependent upon the concentration of copper ions, the temperature, and in the general case, on the tendencies for the ion to gain electrons and to hydrate. The potential difference between the copper and the surrounding solution will vary therefore with concentration and temperature. Standard conditions must be defined on which to base the **standard electrode potential** of the half-cell. The standard conditions chosen are a molar concentration of ions, unless otherwise stated, a temperature of 298 K, and atmospheric pressure (only important in the case of a gas). This standard electrode potential is known as the $E^\ominus$ value for the half-cell. Strictly rather than molar concentration we should specify unit activity, to allow for the non-ideality of the solution, but we shall not consider effects of non-ideality further.

Unfortunately a single electrode potential itself can never be measured directly since as soon as another metal conductor is put into the solution another half-cell is set up. A standard half-cell must therefore be produced against which to measure all other half-cells. The standard half-cell chosen is that of hydrogen. The obvious disadvantage that hydrogen is not a solid is overcome by using a black platinum electrode at which the hydrogen can undergo electron transfer, bringing hydrogen gas and hydrogen ions into equilibrium. The experimental arrangement is illustrated in Fig. 7.7. Frequently the junction of the two electrolyte solutions of the two half-cells is made via a salt bridge. Each half-cell is connected by a capillary or a sintered glass disc to a solution of potassium chloride which forms a bridge between the two electrolyte solutions. The $E^\ominus$ values given in tables are always given with reference to the hydrogen electrode. That is to say *the standard electrode potential for hydrogen is by convention taken to be zero.*

The standard potentials listed in Table 10.6 are often referred to as standard reduction potentials. Thus for copper,

$$Cu^{2+}(aq) + 2e^- \rightarrow Cu(s),$$

we have a standard reduction potential of $+0.34$ V. The term oxidation potential refers to the reverse process,

$$Cu(s) \rightarrow Cu^{2+}(aq) + 2e^-.$$

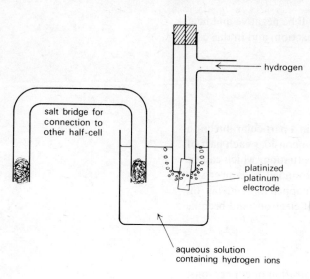

Fig. 7.7 Experimental arrangement for hydrogen electrode.

In this case the standard value is -0.34 V. Oxidation potentials are simply obtained from reduction potentials by reversing the sign.

7.9 THE SIGN OF THE E.M.F.

It is obviously possible to define the difference in potential between the metal and its surrounding solution in two ways: either from the solution looking at the electrode, or from the electrode looking at the solution. In fact the sign convention used in Europe, but not (yet) in the U.S.A., is that the potential of the metal is looked at from the solution. That means that the most reactive metals like sodium which are most likely to form ions and leave electrons behind on the metal have negative $E^{\ominus}$ values.

Special methods are conventionally used for representing cells and it is as well to be quite clear about their use. The electrode is referred to thus:

$$\text{Cu(s)} \mid \quad \text{or} \quad \text{Zn(s)} \mid$$

The junction is represented by a vertical broken line. Thus, the copper–zinc cell is written:

$$\text{Cu(s)} \mid \text{Cu}^{2+}\text{(aq)} \vdots \text{Zn}^{2+}\text{(aq)} \mid \text{Zn(s)}; \qquad E^{\ominus} = -1.1 \text{ V}.$$

Notice that the sign given to the $E^{\ominus}$ value is decided on the basis of the potential of the right hand electrode (zinc) with respect to the left hand (copper). The cell could be written the other way round:

$$\text{Zn(s)} \mid \text{Zn}^{2+}\text{(aq)} \vdots \text{Cu}^{2+}\text{(aq)} \mid \text{Cu(s)}; \qquad E^{\ominus} = 1.1 \text{ V}.$$

Factors governing standard electrode potentials are discussed in Section 10.3.18. A **standard electrode potential** is defined as the e.m.f. of a cell in which the standard electrode is on the right hand side, and in which the left hand electrode is the standard hydrogen electrode. Thus the standard electrode potential for copper refers to the e.m.f. of the cell:

$$\text{Pt } H_2(g) \mid H^+(aq) \vdots Cu^{2+}(aq) \mid Cu(s); \quad E^\ominus = +0.34 \text{ V}.$$

It can be seen that from Table 10.6 it is possible to predict what the e.m.f. of any particular cell will be. For instance, the e.m.f. of the copper–zinc cell under standard conditions can be obtained from the values:

$$\text{for zinc,} \quad E^\ominus = -0.76 \text{ volts};$$
$$\text{for copper,} \quad E^\ominus = 0.34 \text{ volts}.$$

The e.m.f. for the cell = 1.10 volts, since both $E^\ominus$ values are referred to the standard hydrogen electrode.

To obtain the sign of the e.m.f. the procedure is as follows. Write the cell diagram, reverse the sign of the $E^\ominus$ value for the left hand electrode (because it is then written as an oxidation), and add this to the value of the right hand electrode. This gives the polarity of the right hand electrode. If $E^\ominus$ is positive it means that the cell reaction proceeds spontaneously from left to right. For example for the cell:

$$\text{Zn(s)} \mid Zn^{2+}(aq) \vdots Cu^{2+}(aq) \mid Cu(s); \quad E^\ominus = +1.1 \text{ V}$$
$$\text{Zn(s)} \rightarrow Zn^{2+}(aq) \quad Cu^{2+}(aq) \rightarrow Cu(s)$$

the processes occurring are given under the cell diagram. The positive value of $E^\ominus$ means that if the cell were short-circuited, electrons would flow spontaneously from zinc to copper through the external circuit, that is, the polarity of copper is positive. A positive value of $E^\ominus$ leads to a negative value of $\Delta G^\ominus$, according to Eq. (7.12), consistent with a spontaneous process.

7.10 ION–ION REDOX REACTIONS (See Section 18.14.1.)

Many redox reactions are simply between ions in solutions. No metals are involved. To investigate these cases the electrodes must be of an inert metal. Platinum is the metal usually chosen. For instance, it can be seen from Table 10.6 that the redox potential of a half-cell of iron(II) and iron(III) ions is $E^\ominus = +0.77$ V. In this case the concentration of both ions is molar, ignoring effects of non-ideality. It can also be seen that for a half-cell of permanganate and manganese(II) ions with hydrogen ions present $E^\ominus = +1.52$ V. We can predict therefore that the cell formed between the two, written

$$\text{Pt} \mid Fe^{2+}(aq), Fe^{3+}(aq) \vdots MnO_4^-(aq), Mn^{2+}(aq), H^+(aq) \mid \text{Pt},$$

has $E^\ominus = +0.75$ V. This means that the $Fe^{2+}(aq), Fe^{3+}(aq)$ electrode is negative with respect to the $MnO_4^-(aq), Mn^{2+}(aq)$ electrode. The electrode which is negative is the electron-producing electrode. The fact that the iron ion electrode produces electrons shows that the direction of the half-reaction is:

$$Fe^{2+}(aq) \rightarrow Fe^{3+}(aq) + e^-.$$

Similarly the manganese ion half-reaction absorbs electrons:

$$MnO_4^-(aq) + 8H^+(aq) + 5e^- \rightarrow Mn^{2+}(aq) + 4H_2O(l).$$

The reaction which takes place when the ions are mixed *in the concentrations in which they are in the cell* is therefore:

$$5Fe^{2+}(aq) + MnO_4^-(aq) + 8H^+(aq) \rightarrow 5Fe^{3+}(aq) + Mn^{2+}(aq) + 4H_2O(l).$$

If the reaction went in the opposite direction, then $E^\ominus$ would be of opposite sign.

The $E^\ominus$ values give the e.m.f. of cells under standard conditions. How does the e.m.f. change with changes of concentration?

Before examining this we summarize the conventions about cells in the form of rules.

1. Solid lines represent interfaces between solutions and solid electrodes.
2. Vertical broken lines represent salt bridge, porous partition, or other junction between two half-cells.
3. The sign given to the e.m.f. indicates the polarity of the right hand electrode.
4. The least oxidized species in each half are all written next to the solid electrode. For example $Pt|Fe^{2+}, Fe^{3+}$ and $H^+, MnO_4^- Mn^{2+}|Pt$, etc.
5. The sign of a given standard electrode potential is that assigned to the cell formed by the standard electrode and the standard hydrogen electrode when the latter is written on the left hand side of the cell diagram.

7.11 VARIATION OF CELL E.M.F. WITH CONCENTRATION AND TEMPERATURE

It is important to make a clear distinction between standard values of E and ΔG (i.e. $E^\ominus$ and $\Delta G^\ominus$) and the general use of the terms under non-standard conditions. $E^\ominus$ is the e.m.f. of the cell under standard conditions of concentration, temperature, and pressure. E represents its e.m.f. quite generally under any conditions. E changes with concentration, temperature, and pressure (if a gas is involved). Similarly $\Delta G^\ominus$ refers to the total change in free energy when the reactants under standard conditions change completely to the products under standard conditions. But ΔG refers to the change in free energy at any concentration. Consider how these values change during the course of a reaction.

If a current is taken from a cell which initially contains the reactants at standard concentration the voltage is gradually found to decrease. The reason is that the two half-reactions are taking place and that the concentration of the reactants is decreasing. The reaction continues until E is zero. At this stage the concentrations of the reactants and products are those of the equilibrium mixture. If the contents of the two half-cells are mixed there will be no reaction. Since, from Eq. (7.11), $\Delta G = -zFE$, where $E = 0$, $\Delta G = 0$.

An experiment can be performed with a silver–copper cell:

$$Ag(s) \mid Ag^+(aq) \vdots Cu^{2+}(aq) \mid Cu(s).$$

To find exactly how the e.m.f. varies with concentration, the concentration of one ion, say the silver ion, is varied and the e.m.f. measured in each case. As the concentration of the silver ion is reduced so the voltage decreases until it reaches zero and is then actually reversed. It is clear that when the cell is set up under these conditions the silver ion concentration is so small that it is actually less than that in the equilibrium mixture. The reaction is now tending to go back towards the equilibrium position. If E is plotted against the logarithm of the silver ion concentration the result shown in Fig. 7.8 is obtained. It is clear that there is a linear

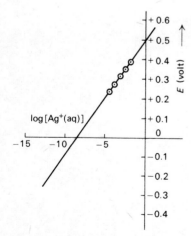

Fig. 7.8 Dependence of E on concentration.

relation between E and log [Ag$^+$]. In fact the complete relationship which was discovered by Nernst is that

$$E = E^\ominus + \frac{2.303RT}{zF} \log [\text{ion}], \qquad (7.14)$$

where R is the gas constant, T is the absolute temperature, z the charge on the ion, F the Faraday constant, and [ion] the concentration of the ion. Notice that although T is in this equation it holds good only at constant temperature.

For ion–ion reactions such as the iron(III)–iron(II) half-cell we have

$$E = E^\ominus + \frac{RT}{zF} \ln \frac{[\text{oxidized form}]}{[\text{reduced form}]}. \qquad (7.15)$$

For iron the ratio would be

$$\frac{[\text{Fe}^{3+}(\text{aq})]}{[\text{Fe}^{2+}(\text{aq})]}.$$

This is a way of expressing the equilibrium constant K, but upside down, hence

$$E = E^\ominus - \frac{RT}{zF} \ln K. \qquad (7.16)$$

Multiplying through by zF,

$$zFE = zFE^\ominus - RT \ln K.$$

But

$$\Delta G = -zFE$$

from Eq. (7.11), so

$$-\Delta G = -\Delta G^\ominus - RT \ln K.$$

If $\Delta G = 0$, i.e. equilibrium conditions obtain, then

$$\Delta G^\ominus = -RT \ln K,$$

an expression which was met previously in Eq. (7.6).

This is not a derivation of the relationship between $\Delta G^\ominus$ and K, but it does show that the Nernst equation is closely related to that expression. Strictly speaking the use of concentrations in Eq. (7.16) applies only to ideal solutions. For non-ideal solutions, the activity which allows for non-ideality should be used.

It can easily be shown that the e.m.f. of a cell is affected by temperature. This is to be expected as we have seen that E and ΔG are closely related and we also have from Eq. (6.17):

$$\Delta G = \Delta H - T\Delta S.$$

Although ΔH and ΔS are almost constant with temperature, $T\Delta S$ is certainly not. Hence ΔG varies greatly with temperature.

7.12 USING THE SECOND LAW

The expression

$$\Delta G = \Delta H - T\Delta S,$$

which holds at a particular temperature, stems from the Second Law of Thermodynamics. In the last chapter all three of the concepts which are linked by this equation were discussed in some detail. The best way to understand an idea is to use it. Throughout this book use will be made of the concepts of free energy, enthalpy, and entropy when these are helpful. While many of the reactions encountered in inorganic chemistry are thermodynamically controlled, many of the reactions met in organic chemistry are kinetically controlled. There are many exceptions, but nevertheless the major applications of thermodynamic considerations are in inorganic chemistry.

One way in which the use of the ideas can be appreciated is by looking at various reactions at various temperatures, looking at $\Delta G^\ominus$ values, and seeing how $\Delta G^\ominus$ varies with temperature. Why does calcium carbonate dissociate appreciably at 1000 °C but not at room temperature? Is hydrogen or carbon the better reducing agent for zinc oxide, and why? One way of examining such changes is to look at graphs of $\Delta G^\ominus$ against T. Such plots are frequently straight lines. This is readily understood from the relationship

$$\Delta G^\ominus = \Delta H^\ominus - T\Delta S^\ominus,$$

when it is recalled (Section 7.4) that for many chemical reactions $\Delta H^\ominus$ and $\Delta S^\ominus$ are only slightly dependent on temperature. Under these conditions the plot is linear (or nearly so) and has a slope $-\Delta S^\ominus$. An understanding of this relationship often proves valuable in the application of such diagrams (Sections 7.12.1, 11.2, and 11.5.1).

7.12.1 Ellingham Diagrams: Stabilities of Oxides and Salts

Look at the reaction

$$\text{CaCO}_3(s) \rightarrow \text{CaO}(s) + \text{CO}_2(g);$$
$$\Delta H^\ominus_{298} = 178 \text{ kJ mol}^{-1}$$
$$\Delta S^\ominus_{298} = 161 \text{ J mol}^{-1} \text{ K}^{-1}.$$

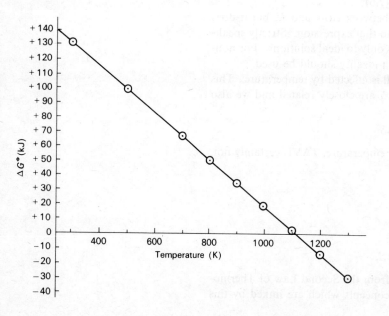

Fig. 7.9 Variation of $\triangle G^\ominus$ with temperature is often nearly linear. For positive values of $\triangle G^\ominus$, $K < 1$, and for negative values, $K > 1$. The values refer to the decomposition of calcium carbonate.

The reaction is endothermic as one might expect for a decomposition process in which bonds are being broken. But it is not for this reason that the reaction proceeds at high temperature only. There is an increase in entropy for the reaction. Again this seems likely as a gas and a solid are being produced from a solid (Section 6.12). The fact that the reaction is endothermic will work against its happening spontaneously (decrease in the entropy of the environment) but the increase in the entropy of the system will act in its favour (Section 6.13). At room temperature ΔH is dominant but at 1000 °C $T\Delta S$ is dominant.

From Fig. 7.9 it can be seen that at 1100 K $\Delta G^\ominus = 0$. Above this temperature $\Delta G^\ominus$ is negative and the equilibrium position is in favour of calcium oxide and carbon dioxide. At this temperature the equilibrium pressure is 1 atm. Below, it is less, and at higher temperatures greater, than 1 atm. For further discussion see Section 11.2.

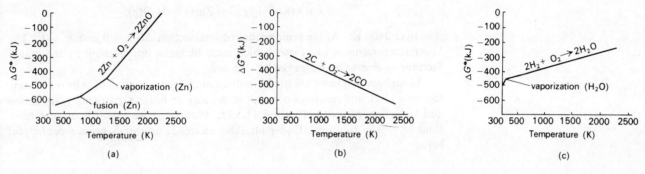

Fig. 7.10 Ellingham diagrams for (a) zinc oxide, (b) carbon monoxide, and (c) water vapour. In each case $\Delta G^\ominus$ has been plotted against T. The change to which the standard free energy change refers is also indicated.

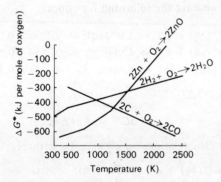

Fig. 7.11 Ellingham diagrams for reduction of zinc oxide by carbon and hydrogen.

Graphs of $\Delta G^\ominus$ against temperature are known after their originator as **Ellingham diagrams**. In an Ellingham diagram the variation with temperature of standard free energy change for the formation of a number of compounds may be compared to see which of two or more possible reactions is more feasible. Suppose for instance that we want to compare hydrogen and carbon as reducing agents for zinc oxide. Which is the better reaction in practice,

$$\text{ZnO(s)} + \text{C(s)} \rightarrow \text{Zn(s)} + \text{CO(g)}$$

or

$$\text{ZnO(s)} + \text{H}_2(\text{g}) \rightarrow \text{Zn(s)} + \text{H}_2\text{O(g)}?$$

Many different factors are involved in such a decision—the cost of carbon and hydrogen, the cost of the plant, and so on. One factor is the temperature at which the reaction will have to be carried out and this is where the Ellingham diagram helps. In Fig. 7.10 diagrams of standard free energy changes for formation of compounds against temperature are given separately for three reactions. Notice that the slope changes when there is a change in phase. This is because there is an entropy change at constant temperature where a change of phase takes place (Section 6.11 and Eq. A.4).

To find out at what temperature a reaction is feasible we superimpose these diagrams on each other as in Fig. 7.11. It can now be seen that for

$$ZnO(s) + C(s) \rightarrow Zn(s) + CO(g),$$

the temperature at which ΔG becomes negative is 1200 K whereas for

$$ZnO(s) + H_2(g) \rightarrow Zn(s) + H_2O(g),$$

it is over 1400 K. At the temperature of intersection $\Delta G^\ominus = 0$ and $K = 1$. The lower temperature is one factor which must be taken into account by the manufacturer of zinc when he makes his decision.

Ellingham diagrams are useful in discussing decomposition of salts on heating (Section 11.2), and reactions of solids with gases on heating, particularly oxidation and reduction of metals (Section 11.5.1). Frequently we wish to consider reactions of ions in solution. Understandably electrode potentials prove most helpful here.

7.12.2 Relative Stabilities of Different Oxidation States in Solution

From Table 10.6 we have the following for copper:

$$Cu^{2+}(aq) + e^- \rightarrow Cu^+(aq); \qquad E^\ominus = +0.16 \text{ V}$$
$$Cu^+(aq) + e^- \rightarrow Cu(s); \qquad E^\ominus = +0.52 \text{ V}$$
$$Cu^{2+}(aq) + 2e^- \rightarrow Cu(s); \qquad E^\ominus = +0.34 \text{ V}.$$

Making use of Eq. (7.12),

$$\Delta G^\ominus = -zFE^\ominus,$$

we can understand the relationship between these three. We have:

$$Cu^{2+}(aq) + e^- \rightarrow Cu^+(aq); \qquad \Delta G^\ominus = -0.16F$$
$$Cu^+(aq) + e^- \rightarrow Cu(s); \qquad \Delta G^\ominus = -0.52F$$
$$Cu^{2+}(aq) + 2e^- \rightarrow Cu(s); \qquad \Delta G^\ominus = -0.68F.$$

The third equation has been obtained from the first two, making use of the fact that $\Delta G^\ominus$ values are additive. It follows that for the third electrode reaction

$$E^\ominus = -\frac{\Delta G^\ominus}{2F} = +\frac{0.68F}{2F} = +0.34 \text{ V},$$

and so we may write

$$Cu^{2+}(aq) + 2e^- \rightarrow Cu(s); \qquad E^\ominus = +0.34 \text{ V}.$$

The example illustrates the way in which, for oxidation reduction systems, electrode potentials may be calculated from other electrode potentials. In such calculations it is important to remember free energies can always be added (or subtracted), but that this is not necessarily true of electrode potentials themselves, as the example shows. Another example for the MnO_4^-/MnO_2 reduction potential is given later in this section, and other examples are given in Problems 15.7, 19.18, and 20.6.

The meaning of the equations is evident when it is recalled that electrode potentials are referred to the standard hydrogen electrode which is by convention assigned a value of zero (Section 7.8). The first of the above equations thus gives $\Delta G^\ominus$ for the change

$$Cu^{2+}(aq) + \tfrac{1}{2}H_2(g) \rightarrow H^+(aq) + Cu^+(aq).$$

That is to say e^- in the above equations is shorthand for

$$\tfrac{1}{2}H_2 \rightarrow H^+(aq) + e^-.$$

A negative value for $\Delta G^\ominus$ implies that the equilibrium lies over to the right in such equations. This follows from Eq. (7.6),

$$\Delta G^\ominus = -RT \ln K,$$

K being equal to 1 for $\Delta G^\ominus = 0$. The fact that $\Delta G^\ominus$ and $E^\ominus$ have opposite signs means that for positive electrode potentials the equilibrium lies to the right, and for negative potentials to the left. This refers to the potential expressed as a reduction potential as in Table 10.6 and for the equation written as a reduction from left to right. In the present examples copper is stable in acid solution with respect to both copper(I) and copper(II) ions.

It is convenient to set out electrode potentials diagrammatically when a number of different oxidation states (Section 18.14.2) are being considered; Fig. 7.12 gives the diagram for manganese. Reduction potentials are shown for the relevant half-reactions in acid solution. For example

$$MnO_4^-(aq) + 8H^+(aq) + 5e^- \rightarrow Mn^{2+}(aq) + 4H_2O;$$
$$E^\ominus = +1.51 \text{ V}$$
$$\Delta G^\ominus = -5 \times 1.51F.$$

This is equivalent to

$$MnO_4^-(aq) + 8H^+(aq) + \tfrac{5}{2}H_2 \rightarrow Mn^{2+}(aq) + 4H_2O(l) + 5H^+(aq),$$

which when simplified is

$$MnO_4^-(aq) + 3H^+(aq) + \tfrac{5}{2}H_2 \rightarrow Mn^{2+}(aq) + 4H_2O(l).$$

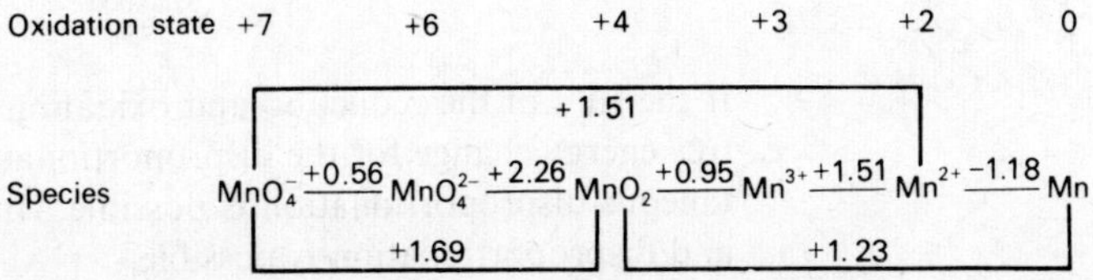

Fig. 7.12 Reduction potentials for manganese couples in acid solution.

Oxidation state: +7, +6, +4, +3, +2, 0

Species: $MnO_4^- \xrightarrow{+0.56} MnO_4^{2-} \xrightarrow{+2.26} MnO_2 \xrightarrow{+0.95} Mn^{3+} \xrightarrow{+1.51} Mn^{2+} \xrightarrow{-1.18} Mn$

$MnO_4^- \xrightarrow{+1.69} MnO_2$, overall $MnO_4^- \xrightarrow{+1.51} Mn^{2+}$, $MnO_2 \xrightarrow{+1.23} Mn^{2+}$

Only the middle line values are strictly necessary, the others can be calculated. For example, consider the calculation of the MnO_4^-/MnO_2 reduction potential.

We have:

1. $MnO_4^-(aq) + e^- \rightarrow MnO_4^{2-}(aq);\ E^\ominus = +0.56\ V,\ \Delta G^\ominus = -zFE^\ominus = -0.56F.$
2. $MnO_4^{2-}(aq) + 4H^+(aq) + 2e^- \rightarrow MnO_2(s) + 2H_2O(l);$
$$E^\ominus = +2.26\ V,\ \Delta G^\ominus = -zFE^\ominus = -4.52F.$$

Adding (1) and (2) we obtain

$$MnO_4^-(aq) + 4H^+(aq) + 3e^- \rightarrow MnO_2(s) + 2H_2O(l);\quad \Delta G^\ominus = -5.08F.$$

Hence $E^\ominus$ for MnO_4^-/MnO_2 is obtained as $\Delta G^\ominus/3F$. That is, the required $E^\ominus = +1.69\ V$.

From this diagram it is possible to tell which states are:

a) Powerfully oxidizing (relative to lower states): these have large positive $E^\ominus$ values, e.g. MnO_4^-.
b) Powerfully reducing: these have large negative $E^\ominus$ values, e.g. Mn^{2+}/Mn.
c) Thermodynamically unstable in regard to disproportionation (see the end of this section). Consider Mn(VI) as MnO_4^{2-}. We have

1. $MnO_4^{2-}(aq) + 4H^+(aq) + 2e^- \rightarrow MnO_2(s) + 2H_2O(l);$
$$E^\ominus = +2.26\ V,\ \Delta G^\ominus = -4.52F.$$
2. $MnO_4^-(aq) + e^- \rightarrow MnO_4^{2-}(aq);\qquad E^\ominus = +0.56\ V,\ \Delta G^\ominus = -0.56F.$

Taking (1) and subtracting (2) × 2 gives

$$3MnO_4^{2-}(aq) + 4H^+(aq) \rightarrow 2MnO_4^-(aq) + MnO_2(s) + 2H_2O(l);$$
$$\Delta G^\ominus = [-4.52 - (2 \times -0.56)]F$$
$$= -3.40F$$
$$E^\ominus = 1.70\ V.$$

$\Delta G^\ominus$ is found to be negative for the disproportionation reaction, so the reaction is expected to "go" in the direction written. This conclusion could equally well be reached by considering simply the electrode potentials. Inspection shows that what has been done above is to multiply each electrode potential by two, then add them algebraically, and divide by two. A little thought will show this is general for disproportionation reactions. The free energy change is negative if the "sum" of the potentials is positive. In short, if the algebraic sum of the electrode potentials is positive the state is unstable with respect to disproportionation. Alternatively, one can look at this in the following way. Drawing out the relevant piece of the diagram and remembering an oxidation potential is just the negative of the reduction potential, we can write:

$$MnO_4^- \xrightarrow{+0.56} MnO_4^{2-} \xrightarrow{+2.26} MnO_2$$
$$\underset{\substack{\text{oxidation}\\-0.56}}{\xleftarrow{\hspace{2cm}}} \quad \underset{\substack{\text{reduction}\\+2.26}}{\xrightarrow{\hspace{2cm}}}$$

If the sum of the reduction and oxidation potentials is positive, then the standard free energy change for the disproportionation will have a negative value and spontaneous disproportionation is possible. In this case we have $2.26 - 0.56 = +1.7$ and disproportionation is possible.

All this may be summarized as a simple rule. Disproportionation of an intermediate oxidation state is possible (thermodynamically) if the reduction potential of the couple linking the intermediate to the lower oxidation state is greater (more positive) than that linking the intermediate to the higher oxidation state (here $+2.26 > +0.56$). In general, oxidation states being considered need not be adjacent but they frequently are.

The term **disproportionation** refers to a special case of an oxidation–reduction process in which a single substance acts as both oxidant and reductant. Part of it is oxidized and part is reduced in the process. A familiar example is the decomposition of hydrogen peroxide which can be written as

$$H_2O_2(l) + H_2O_2(l) \rightarrow H_2O(l) + O_2(g).$$

Another example is provided by the Mn^{3+} ion. In aqueous solution the oxidation–reduction reaction

$$2Mn^{3+}(aq) + 2H_2O(l) \rightarrow MnO_2(s) + Mn^{2+}(aq) + 4H^+(aq)$$

occurs. Reactions in which a single substance undergoes change to produce products, one of which is in a higher oxidation state and the other in a lower oxidation state, are called disproportionation reactions. It can happen that a substance is unstable because it readily disproportionates. Thus it is shown in Section 14.3.3 that salts of Mg^+ would be expected to disproportionate according to:

$$2Mg^+ \rightarrow Mg + Mg^{2+}.$$

The equation represents clearly the meaning of the term disproportionation; there are two products, one in a higher oxidation state and the other in a lower oxidation state.

7.12.3 Free Energy Diagrams and Relative Stabilities of Oxidation States

Discussion of relative stabilities of oxidation states usually involves considerations of reduction potentials as in Section 7.12.2. However, many people find data easier to understand if presented in diagrammatic rather than tabular form, as for example in Ellingham diagrams. Diagrams can be constructed to display information about different oxidation states. The conventions used are listed below.

1. Free energy is plotted in units of electronvolts along the vertical axis. Electronvolts (eV) although not an SI unit are exceedingly convenient in this instance. The free energy differences between oxidation states are readily calculated from reduction potentials (Table 10.6); the procedure is illustrated in the example in Fig. 7.13. Consider

$$Mn^{2+} + 2e^- \rightarrow Mn^0; \quad E^\ominus = -1.18 \text{ V}; \quad \frac{\Delta G^\ominus}{F} = -zE^\ominus = 2.36.$$

The difference in free energy between the $+2$ and 0 states of manganese is given by $\Delta G^\ominus = -zFE^\ominus = -2 \times (-1.18) \times 96{,}484$ joules. It is inconvenient to keep multiplying by 96,484 to convert electronvolts ($-zE^\ominus$) into SI units (joules) so this is not done and instead of calculating $\Delta G^\ominus$ we calculate $\Delta G^\ominus/F = -zE^\ominus$ electronvolts. Our justification for this is that we are concerned only with the relative positions of oxidation states on the diagram, and

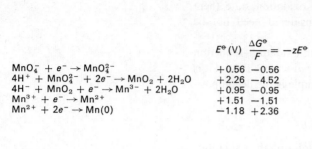
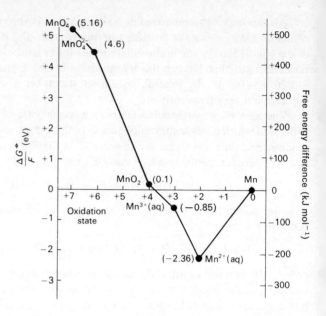

Fig. 7.13 Oxidation state diagram for manganese.

these are unaltered by multiplying all ordinate positions by the same (inconvenient) constant. To show that this is unimportant, on this diagram the ordinate is also given in kilojoules. In our example, therefore, the diagram must convey the information that the standard free energy difference between $Mn^{2+}(aq)$ and Mn (1 mole) is

$$\frac{\Delta G^\ominus}{F} = -zE^\ominus = -2 \times (-1.18) = +2.36 \text{ eV}.$$

2. The horizontal axis represents the oxidation state of the element, in this case Mn. The change from left to right is represented as a reduction to conform to previous conventions.
3. Arbitrarily (again to conform to usual conventions), the element in oxidation state zero is considered to be of zero free energy (Section 7.5).
4. Usually lines are drawn connecting adjacent oxidation states though these have no more (or less) theoretical significance than those lines connecting non-adjacent states.
5. The lower on the diagram an oxidation state is, the more stable it is; for example, Mn^{3+} is a stable oxidation state relative to MnO_4^- etc. but not to Mn^{2+}.

Construction of diagram. Start at Mn(0). Consider the reduction potential linking manganese(0) with the lowest oxidation state Mn(II).

$$Mn^{2+}(aq) \rightarrow Mn(0); \qquad \Delta G^\ominus/F = +2.36 \text{ eV}.$$

Clearly the difference between the two states is $+2.36$ eV on the diagram, and the change $Mn^{2+} \rightarrow Mn(0)$ is not spontaneous (ΔG is positive). But $Mn(0) \rightarrow Mn^{2+}$ is spontaneous, therefore Mn^{2+} must appear lower in the diagram than Mn(0) and thus it appears at -2.36 eV (in order that the change $Mn^{2+} \rightarrow Mn(0)$ gives $\Delta G^\ominus/F = +2.36$ eV). We now consider the couple Mn^{3+}/Mn^{2+}, $\Delta G^\ominus/F =$

$-zE^\ominus = -1 \times (+1.51) = -1.51$. The minus sign tells us that $Mn^{3+} \to Mn^{2+}$ is spontaneous. Therefore, Mn^{3+} appears higher in the diagram, at such a position that it is 1.51 eV above Mn^{2+}, that is at $-2.36 + (1.51) = -0.85$ eV. And so on. Obviously the diagram is internally consistent; for example, consider the couple MnO_4^-/Mn^{2+}, $\Delta G^\ominus/F = -zE^\ominus = -5 \times (+1.51) = -7.5$ eV. The negative sign tells us that the change, $MnO_4^- \to Mn^{2+}$, is spontaneous; therefore MnO_4^- appears at 7.5 eV, higher than Mn^{2+} in the diagram. Finally it is useful to consider the slopes of lines connecting oxidation states.

$$\text{Slope} = \frac{\Delta y}{\Delta x} = \frac{\Delta G^\ominus/F}{n} = \frac{\Delta G^\ominus}{nF},$$

where $n =$ change in oxidation number of the two states which equals z the number of electrons involved in the half-reaction relating them. But $\Delta G^\ominus = -zFE^\ominus$, therefore

$$\frac{\Delta G^\ominus}{nF} = -E^\ominus.$$

Thus the slopes of the lines are equal to minus one times the appropriate potential. This knowledge is often useful in checking the free energy diagram.

7.12.4 Interpretation of Oxidation State Diagrams: Summary

1. The lower in the diagram, the more stable the state is relative to those states above it. Thus changes "downhill", such as $Mn^{3+}(aq) \to Mn^{2+}(aq)$, are favourable thermodynamically (and vice versa). Although only adjacent states are linked by lines (whose slope equals $-E^\ominus$), this is significant only in that more lines would complicate the figure beyond comprehension.
2. A particular state may be represented by a point which is:
    a) A minimum, e.g. Mn^{2+}. This is stable relative to its two neighbouring states (in this case it is stable relative to all states).
    b) A maximum—unstable relative to its two neighbouring states and will disproportionate unless kinetically controlled (Section 8.9).
    c) On a convex curve, e.g. Mn^{3+}. Such states are relatively unstable to disproportionation and if equilibrium is achieved (i.e. not kinetically controlled) there will be less of this in the equilibrium mixture than of the products. It is easy to see why this is a disproportionating state. Consider the Mn^{3+} case in more detail by reference to Fig. 7.14. The disproportionation reaction can be considered as follows:

$$Mn^{3+}(aq) \to MnO_2(s); \quad \frac{\Delta G^\ominus}{F} = +0.95 \text{ eV}$$

$$Mn^{3+}(aq) \to Mn^{2+}(aq); \quad \frac{\Delta G^\ominus}{F} = -1.5 \text{ eV}.$$

Hence by addition

$$2Mn^{3+}(aq) \to MnO_2(s) + Mn^{2+}(aq); \quad \frac{\Delta G^\ominus}{F} = -0.55 \text{ eV}.$$

$\Delta G^\ominus$ is negative; the reaction "goes".

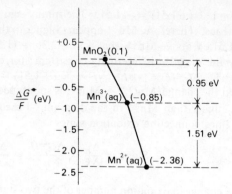

Fig. 7.14 Oxidation state diagram for discussion of disproportionation of Mn^{3+}(aq).

d) On a concave curve, e.g. MnO_2 relative to MnO_4^{2-} and Mn^{3+}. This disproportion does not tend to occur since $\Delta G^\ominus$ is positive.

e) On a straight line. When the line connecting three points is a straight line (or almost), as for example in the nitrogen or chlorine diagrams, Figs. 16.1 and 18.1, then assuming thermodynamic control an equilibrium mixture containing about equal amounts of product and reactant will be obtained.

It must be realized that disproportionation reactions occur without the need for other reagents; for example, Mn^{3+}(aq) at pH = 0 in the absence of other species decomposes to MnO_2(s) + Mn^{2+}(aq). However oxidizing agents (states high in diagram e.g. MnO_4^- on left hand side) and reducing agents (states high in diagram on right hand side, e.g. Ti^{2+}, Table 20.2) need something to oxidize and reduce (though water may provide this if the agent is sufficiently powerful). Thus $K^+MnO_4^-$ (solid or solution) is reasonably stable.

In interpreting these diagrams the following points should be borne in mind:

1. The diagram represents thermodynamic properties, and the rate of attaining the equilibrium position for a particular reaction may well be small or negligible, that is the reaction may be kinetically controlled; for example, MnO_4^- should oxidize water, but does not. MnO_2 is a lot more stable than would be expected from such a diagram because of low solubility.
2. Frequently the data depend on pH: MnO_4^- is much less oxidizing in alkaline solution (generally true for oxyanions, and therefore all are made more easily in alkali). Since many states precipitate in alkali as hydroxides, hydrated oxides, or oxides, the system is less useful.
3. Despite the fact that these diagrams refer to a particular set of solution conditions they are nevertheless generally useful under other conditions such as solid state. Thus Mn^{2+} is generally a stable state and forms compounds with all stable anions and ligands.

7.13 CHEMICAL CHANGE: SUMMARY

The value of the methods developed in this chapter stems from their ability to relate equilibrium, e.m.f., and calorimetric measurements. There is a network of ideas here and applications can be made in practice as occasion demands. For example equilibrium constants can be calculated from e.m.f. measurements or calorimetric measurements. Or equilibrium measurements can be used to add to the table of $\Delta G_f^\ominus$ values and so make them more widely useful. The ideas will be applied in later chapters as required for the particular discussion and it may be helpful to summarize here some of the main ideas.

1. Change can continue in a closed system at constant temperature and pressure until the Gibbs free energy has a minimum value (Sections 6.13 and 7.5).
2. The equilibrium constant, K, for a chemical reaction is related to $\Delta G^\ominus$, the free energy change for the complete conversion of reactants to products under standard conditions. The relationship is given by Eq. (7.6):

$$\Delta G^\ominus = -RT \ln K.$$

3. The standard free energy change can be calculated from Eq. (7.8),

$$\Delta G^\ominus = \Delta H^\ominus - T\Delta S^\ominus,$$

using Table A.15. It is generally more convenient to use Eq. (7.9):

$$\Delta G^\ominus = \sum \Delta G_f^\ominus(\text{products}) - \sum \Delta G_f^\ominus(\text{reactants}).$$

Values of $\Delta G_f^\ominus$ are tabulated in Table A.15.

4. In Eq. (7.8) given above, $\Delta H^\ominus$ and $\Delta S^\ominus$ are found for many chemical reactions to change only slightly with temperature. As a consequence plots of $\Delta G^\ominus$ against T are frequently linear (or nearly so). Such plots form the basis of Ellingham diagrams (Section 7.12.1) which prove useful in discussions when it is important to compare the effect of temperature on related chemical reactions.

5. Equation (7.12) which relates standard e.m.f. to $\Delta G^\ominus$ (or equilibrium constant),

$$\Delta G^\ominus = -zFE^\ominus,$$

proves particularly useful in the discussion of redox equilibria. A major application to relative stabilities of different oxidation states has already been summarized (Section 7.12.3).

PROBLEMS

1. Why do you think that equilibrium constants involve concentrations expressed in mol l^{-1} rather than in g l^{-1}?
2. What are the effects of changing in turn each of the three variables, (i) concentration of a reactant, (ii) temperature, and (iii) total pressure, on:
   a) the equilibrium position;
   b) the equilibrium constant for a gaseous equilibrium?
3. How can the enthalpy change for a chemical change be obtained from equilibrium studies? Using the information given in Table 7.4, plot $\log K_P$ against $1/T$. Use the graph to obtain $\Delta H^\ominus$ for the process:

$$N_2O_4(g) \rightarrow 2NO_2(g).$$

Calculate $\Delta S^\ominus$ at 225 K and 310 K.

4. Consider the following equilibria:
   i) $2NO_2(g) \rightleftharpoons N_2O_4(g)$;
   ii) $H_2(g) + I_2(g) \rightleftharpoons 2HI(g)$;
   iii) $2O_3(g) \rightleftharpoons 3O_2(g)$;
   iv) $2SO_2(g) + O_2(g) \rightleftharpoons 2SO_3(g)$.
   a) Evaluate $\Delta H^\ominus$ for each process from tabulated data.
   b) In which direction will each equilibrium change as the temperature is raised?
   c) Which of the equilibria will not be affected by a change in total pressure?

5. One mole of hydrogen iodide is introduced into a vessel held at constant temperature. When equilibrium is reached it is found that 0.1 moles of iodine have been formed. Calculate the equilibrium constant K_c for $2HI(g) = I_2(g) + H_2(g)$ at the temperature of the experiment.

6. Given K_P for $N_2O_4(g) \rightleftharpoons 2NO_2(g)$ at 300 K is 1.74×10^{-1} atm, calculate K_c.

7. The following are graded examples in the use of tables for the calculation of standard free energy changes. Evaluate from the tables the standard free energy change for each of the following processes:
   a) $Br_2(l) \rightarrow Br_2(g)$;
   b) $\frac{1}{2}H_2(g) + \frac{1}{2}Cl_2(g) \rightarrow HCl(g)$;
   c) $H_2(g) + \frac{1}{2}O_2(g) \rightarrow H_2O(l)$;
   d) $C_2H_4(g) + H_2(g) \rightarrow C_2H_6(g)$;
   e) $2SO_2(g) + O_2(g) \rightarrow 2SO_3(g)$;
   f) $2ZnS(s) + 3O_2(g) \rightarrow 2ZnO(s) + 2SO_2(g)$;
   g) $5CuO(s) + 2NH_3(g) \rightarrow 5Cu(s) + 2NO(g) + 3H_2O(l)$.

8. Write representations, for cells made from the following pairs of elements or compounds, using the conventional symbols:
   a) iron and zinc;
   b) hydrogen and copper;
   c) potassium permanganate and iron(II) chloride.

9. A cell is set up between copper and silver $Cu(s) | Cu^{2+}(aq) \vdots Ag^+(aq) | Ag(s)$. The reaction is: $Cu(s) + 2Ag^+(aq) \rightleftharpoons Cu^{2+}(aq) + 2Ag(s)$. A valve voltmeter (a very high resistance voltmeter) is placed across the cell. What do you think would happen if:
   a) the concentration of $Cu^{2+}(aq)$ was increased;
   b) the silver electrode was exchanged for one with twice the surface area;
   c) the temperature of the cell was raised?

10. What are the standard electrode potentials for cells in which the following reactions occur:
    a) $Cu^{2+}(aq) + H_2(g) \rightarrow 2H^+(aq) + Cu(s)$;
    b) $Cu^{2+}(aq) + Zn(s) \rightarrow Cu(s) + Zn^{2+}(aq)$;
    c) $2Cr(s) + 3Cu^{2+}(aq) \rightarrow 2Cr^{3+}(aq) + 3Cu(s)$?
    (Standard electrode potentials are given in Table 10.6.)

11. In an Ellingham diagram $\Delta G^\ominus$ is plotted against temperature. We have also: $\Delta G^\ominus = \Delta H^\ominus - T\Delta S^\ominus$. For many processes $\Delta H^\ominus$ and $\Delta S^\ominus$ are to a good approximation independent of temperature. Under these conditions the equation affords a simple method of obtaining the Ellingham diagram, since it shows the slope of the line is $-\Delta S^\ominus$. The diagram can thus be drawn from a

knowledge of $\Delta G^{\ominus}$ at one temperature and $\Delta S^{\ominus}$, both of which can be obtained from tabulated thermodynamic data. Plot Ellingham diagrams on this basis for:

$$2C(graphite) + O_2(g) \rightarrow 2CO(g); \text{ and}$$
$$2Fe(s) + O_2(g) \rightarrow 2FeO(s).$$

State in your own words the physical significance of the intersection of these two lines.

BIBLIOGRAPHY

Allen, J. A. *Energy Changes in Chemistry*, Blackie (1965).

    A general introductory account of energy changes in chemistry, introducing chemical equilibria and rates of reaction.

Ashmore, P. G. "Energetics and Equilibrium," Chapter 4 in *Chemical Energetics and the Curriculum*, edited by D. J. Millen, Collins (1969).

    This gives an introductory treatment of equilibria and free energy.

    A fuller account at an introductory level is also available by the same author:

Ashmore, P. G. *Principles of Chemical Equilibrium*, Monographs for Teachers No. 6, Royal Institute of Chemistry (1961).

Guggenheim, E. A. *Elements of Chemical Thermodynamics*, Monographs for Teachers No. 12, Royal Institute of Chemistry (1966).

    This gives an introduction to a more formal treatment of chemical thermodynamics.

Bent, H. A. *The Second Law*, Oxford University Press (1965).

    A stimulating discussion of the Second Law of Thermodynamics and applications.

    Two books which make use of the application of Ellingham diagrams are:

Ives, D. J. G. *Principles of Extraction of Metals*, Monographs for Teachers No. 3, Royal Institute of Chemistry (1966).

Warn, J. R. W. *Concise Chemical Thermodynamics*, Van Nostrand Reinhold (1969).

    Standard works for reference:

Kauzmann, W. *Thermodynamics and Statistics*, Benjamin (1967).

Klotz, I. *Chemical Thermodynamics*, Benjamin (1964).

Denbigh, K. *The Principles of Chemical Equilibrium*, Cambridge University Press (1966).

    For further reading especially on applications in inorganic chemistry:

Ives, D. J. G. *Chemical Thermodynamics with Special Application to Inorganic Chemistry*, Macdonald (1971).

CHAPTER 8

Rates and Mechanisms of Reactions

8.1 Concentration and rate of reaction
8.2 The effect of temperature on rates of reactions; molecular collisions
8.3 Activation energy
8.4 The molecularity of a reaction
8.5 Order and molecularity
8.6 Encounters and collisions
8.7 Transition state theory
8.8 Catalysis
8.9 Kinetic and thermodynamic control of reactions
8.10 Kinetics and mechanism

Important as the consideration of the thermodynamics of a chemical reaction is, the rate at which a reaction proceeds can be even more important. The first, perhaps surprising, thing is that the rate at which a reaction proceeds seems to have little to do with either the enthalpy or the entropy change for the reaction.

Take a reaction such as that between hydrogen and oxygen:

$$2H_2(g) + O_2(g) \rightarrow 2H_2O(l); \quad \begin{aligned} \Delta H^\ominus_{298} &= -572 \text{ kJ} \\ \Delta S^\ominus_{298} &= -326 \text{ J K}^{-1} \\ \Delta G^\ominus_{298} &= -475 \text{ kJ}. \end{aligned}$$

But at room temperature the reaction does not take place. Put a match to the mixture and there is an explosion. An energy barrier exists which must be surmounted if the reaction is to take place. This energy is called the activation energy. The size of the activation energy depends on the way in which the reaction takes place and the nature of the activated complex (Section 8.3) which is formed as the reactants change into the products. Reaction mechanisms, as they are called, are discussed in the next section.

Simple experiments with reactions such as the decomposition of calcium carbonate with hydrochloric acid,

$$CaCO_3(s) + 2HCl(aq) \rightarrow CaCl_2(aq) + CO_2(g) + H_2O(l); \\ \Delta H^\ominus_{298} = -15.7 \text{ kJ} \\ \Delta S^\ominus_{298} = 135 \text{ J K}^{-1},$$

or the decomposition of hydrogen peroxide,

$$H_2O_2(aq) \rightarrow H_2O(l) + \tfrac{1}{2}O_2(g); \quad \begin{aligned} \Delta H^\ominus_{298} &= -94.5 \text{ kJ} \\ \Delta S^\ominus_{298} &= 70 \text{ J K}^{-1}, \end{aligned}$$

show that the rate at which a reaction proceeds depends upon (a) the temperature (the hotter, the faster), (b) the concentration (the more concentrated, the faster), (c) the state of division (the smaller, the faster), and (d) the presence of a catalyst.

8.1 CONCENTRATION AND RATE OF REACTION

When quantitative measurements of rates of reaction are made it is found that most reactions slow down as the reactants are used up. A typical graph of quantity of product versus time looks like that in Fig. 8.1. The rate depends on some function of the concentration of the reactants, and sometimes also on the concentration of a substance, such as an acid or alkali, which is present in the reaction mixture but is not represented in the equation. It is quite impossible to deduce what these functions are from the stoichiometric equation.

The equation which relates the rate of a reaction, v, to the concentration of the reactants A and B, is called the **rate equation**. It can be expressed as

$$v = k[A]^x[B]^y. \tag{8.1}$$

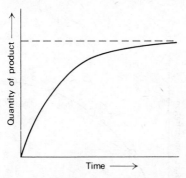

Fig. 8.1 Increase in concentration of product with time.

k is called the **rate constant**, or more accurately the **rate coefficient**. x and y can be any small whole numbers, and even fractional values are known. Some examples of rate equations are given in Table 8.1. The **order of a reaction** refers to the powers x and y in the rate equation. The order with respect to A is x and with respect to B is y. Sometimes the order of a reaction is expressed with reference to the whole rate equation. It then refers to the sum of x and y. Or to put it more precisely but rather heavily, the order of a reaction is the sum of the indices of the concentration terms in the rate equation. What does this really mean?

To take a simple example, it is sometimes found that the rate of a reaction does not vary with the concentration of a particular reactant. Suppose we call this reactant A and keep the other reactants in excess so that they are effectively at constant concentration throughout the experiment. The graph of product versus time is shown in Fig. 8.2. The graph is a straight line. Its slope represents the rate of the reaction (product/time), and as it is a straight line the rate is clearly not changing, despite the fact that the concentration of the reactant A is decreasing. For such a reaction the rate equation could be written:

$$v = k[A]^0 \tag{8.2}$$
$$v = k.$$

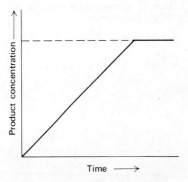

Fig. 8.2 Increase in concentration of product with time for a zero order reaction.

Table 8.1. Experimental rate laws.

Reaction	Rate law
$H_2(g) + I_2(g) \rightarrow 2HI(g)$	$-d[I_2]/dt = k[H_2][I_2]$
$H_2(g) + Br_2(g) \rightarrow 2HBr(g)$	$-d[Br_2]/dt = k[H_2][Br_2]^{3/2}/([Br_2] + k'[HBr])$
$2NO(g) + O_2(g) \rightarrow 2NO_2(g)$	$-d[O_2]/dt = k[NO]^2[O_2]$
$2N_2O_5 \rightarrow 4NO_2(g) + O_2(g)$	$-d[N_2O_5]/dt = k[N_2O_5]$
$H_2O_2(aq) + 2H^+(aq) + 2I^-(aq) \rightarrow 2H_2O(l) + I_2(aq)$	$-d[H_2O_2]/dt = k[H_2O_2][I^-](1 + k'[H^+])$
$S_2O_8^{2-}(aq) + 2I^-(aq) \rightarrow 2SO_4^{2-}(aq) + I_2(aq)$	$-d[S_2O_8^{2-}]/dt = k[S_2O_8^{2-}][I^-]$
$BrO_3^-(aq) + 5Br^-(aq) + 6H^+(aq) \rightarrow 3Br_2(aq) + 3H_2O(l)$	$-d[BrO_3^-]/dt = k[BrO_3^-][Br^-][H^+]$
$CH_3Br + OH^- \rightarrow CH_3OH + Br^-$	$-d[CH_3Br]/dt = k[CH_3Br][OH^-]$
$(CH_3)_3CCl + H_2O \rightarrow (CH_3)_3COH + Cl^-$	$-d[(CH_3)_3CCl]/dt = k[(CH_3)_3CCl]$
$CH_3COOH + C_2H_5OH \rightarrow CH_3CO_2C_2H_5 + H_2O$	$-d[CH_3COOH]/dt = k[CH_3COOH][C_2H_5OH]$
$CH_3COOC_2H_5 + OH^- \rightarrow CH_3COO^- + C_2H_5OH$	$-d[CH_3CO_2C_2H_5]/dt = k[CH_3CO_2C_2H_5][OH^-]$
$I_2 + CH_3COCH_3 \rightarrow CH_2ICOCH_3 + HI$	$-d[I_2]/dt = k[CH_3COCH_3][H^+]$

Such a reaction is said to be of **zero order** with respect to A. An example of zero order kinetics is found for the nitration of toluene under certain conditions (Section 8.5).

In the case of many decompositions of single substances, the product/time graph looks like that in Fig. 8.3. It is found then that, if the substance decomposing is A, the rate is simply proportional to the molar concentration of A:

$$v = k[A]. \tag{8.3}$$

Such a reaction is called a **first order** reaction. An example is the decomposition of dinitrogen pentoxide

$$N_2O_5 \rightarrow N_2O_4 + \tfrac{1}{2}O_2.$$

A reaction whose rate depends on the product of two concentrations, as does the reaction between ethanol and acetic acid,

$$C_2H_5OH + CH_3CO_2H \rightarrow CH_3CO_2C_2H_5 + H_2O,$$

has a rate equation:

$$v = k[C_2H_5OH][CH_3CO_2H].$$

This is called a **second order** reaction overall. It is first order with respect to alcohol and to acetic acid.

Clearly a **third order** reaction is one with a rate equation of the type

$$v = k[A]^3, \tag{1}$$

or

$$v = k[A]^2[B], \tag{2}$$

or

$$v = k[A][B][C]. \tag{3}$$

(8.4)

The other way of expressing the order of a reaction is to specify it with respect to a particular reactant. Referring to the above equations, we could say that (1) is third order *with respect to A*, and zero order for B, (2) is second order *with respect to A*, and (3) is first order *with respect to A*. In some ways this is a more practical approach. Each reaction remains of course a third order reaction overall.

Finding Orders of Reaction

Finding an order of reaction is basically a matter of determining the concentration of the reactants or products at set intervals of time. In solution this can be done by taking samples, stopping the reaction, and titrating, or by using some other method of analysis such as spectroscopy.

Complications can arise if the reaction is greater than first order. In these cases the art is to keep one reactant at constant concentration while allowing the other to vary. The order is then found with respect to the varying reactant. The conditions are then changed so that the concentration of the first reactant varies but the second remains constant. The two results put together give the rate equation. One way of keeping the concentration constant is to have a large excess of that reagent. Changes in its concentration are then relatively so small that they can be neglected. Such a method assumes that the reaction path does not change when there is a large excess of one reagent.

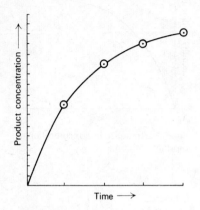

Fig. 8.3 Increase in concentration of product with time for a first order reaction.

8.2 THE EFFECT OF TEMPERATURE ON RATES OF REACTIONS; MOLECULAR COLLISIONS

It is clear that a reaction between two molecules or other species cannot take place unless the two actually meet or collide. Even reactions in which one molecule decomposes depend on collisions to gain enough vibrational energy to break apart the two fragments. Reactions are certainly dependent on collisions. What is interesting is whether the rate of reaction is directly dependent upon the number of collisions which take place or whether other factors are more important.

It is possible to work out the approximate number of collisions which take place in a given amount of a gas. For instance, if we consider the reaction

$$2HI(g) \rightleftharpoons H_2(g) + I_2(g)$$

at 500 °C at a concentration of 0.01 moles of hydrogen iodide per litre, we find that the rate of reaction is such that about 7.2×10^{17} effective collisions between hydrogen iodide molecules must be taking place per litre per second. But the total number of collisions which take place per litre per second is about 1.3×10^{32} under the same conditions. In other words only one collision in $(1.3 \times 10^{32})/(7.2 \times 10^{17}) = 2 \times 10^{14}$ is effective. The key to understanding this comes from studying the effect of temperature. Rates of reaction usually increase very rapidly with temperature as the rate constants in Table 8.2 for the hydrogen iodide decomposition show. What is the increase in the number of collisions? Between 550 K and 780 K the increase in the number of collisions is about 20 percent. The increase in the rate of reaction however is 3950/0.0352, that is, over 100,000 times. The increase in the average energy per molecule (Eq. 3.12) is only about 4 percent, so neither the number of collisions nor the *average* energy of the molecules can account for the increase of rate of reaction with temperature.

A closer look at the energy distribution, leads to the answer to the problem. Figure 6.4 shows how the speeds of the molecules in a gas are distributed (Section 6.7). The kinetic energy is closely related to molecular speed and has a similar distribution. In Fig. 6.4 the distributions have been plotted for hydrogen molecules in hydrogen gas at 298 K and 398 K. The shaded area in each case is proportional to the fraction of molecules moving with speeds in excess of 4×10^3 m s^{-1}. Notice how the proportion changes very markedly with temperature. A speed of 4×10^{-3} m s^{-1} corresponds in fact to a comparatively low energy (16 kJ mol^{-1}).

Table 8.2. Rate constants for decomposition of hydrogen iodide at different temperatures.

Temperature (K)	Rate constant (litre mol^{-1} s^{-1})
500	3.75×10^{-9}
600	6.65×10^{-6}
700	1.15×10^{-3}
800	7.75×10^{-2}

Had a higher value been chosen, the change with temperature would have been even greater, though less easy to show on the diagram (see Problem 11, Chapter 4).

The fraction of molecules having more than a certain amount of energy is given by

$$x = e^{-e_A/kT} = e^{-Ne_A/NkT} = e^{-E_A/RT},$$

where E_A is the energy per mole, and e_A is the energy per molecule. For the hydrogen iodide reaction E_A is about 209.5 kJ mol^{-1}. At 581 K

$$x_{581} = e^{-(209.5 \times 10^3)/(8.3 \times 581)}$$
$$= 2.11 \times 10^{-19}.$$

At 781 K

$$x_{781} = e^{-(209.5 \times 10^3)/(8.3 \times 781)}$$
$$= 1.26 \times 10^{-14}.$$

The ratio $x_{781}/x_{581} \approx 10^5$. The ratio is of the right order of magnitude to account for the change in rate. It appears that it is the number of molecules with sufficient energy that is the significant factor in determining rates of reaction. This energy is called the activation energy. The vast majority of collisions are ineffective in bringing about the reaction because they are insufficiently energetic.

8.3 ACTIVATION ENERGY

The subject of activation energy was introduced at the beginning of this chapter. It will now be described in terms of the mechanism of a reaction. The **mechanism of a reaction** is a term which covers the whole process by which a reaction occurs, including the number of steps and the way in which the various intermediate species react and dissociate. Reactions take place in a series of stages or steps, often called the **reaction pathway**. In each step a number of species react together. The term mechanism is used to cover not only a knowledge of all the individual steps but also includes a stereochemical picture of how the reaction occurs. The aim in studying mechanisms is to decide what species collide in each successive step, what the composition of the activated complex is for each step, and also what the geometry of each of the activated complexes is, if there is more than one (Section 26.1).

Take, for example, a simple case in which iodine and hydrogen react together. The two molecules meet and form a complex called an **activated complex** which has a very short life. Indeed it only exists during the time of the impact. The activated complex then breaks down, either back again to hydrogen and iodine or on to hydrogen iodide. A certain minimum energy of the molecules in collision is required to produce the activated complex. This is called the **activation energy** E_A. The idea of activation energy is best understood with reference to a diagram called a **reaction profile** (Fig. 8.4). In the figure, E_1 represents the activation energy of the forward reaction, E_{11} the activation energy of the reverse reaction. It can be seen that

$$\Delta H = E_1 - E_{11}. \tag{8.5}$$

The decomposition of hydrogen iodide as described is a one-stage reaction involving one activated complex. Two molecules are involved and so it is classed as

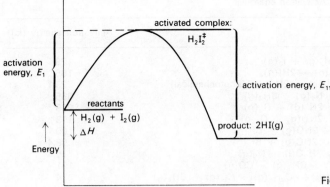

Fig. 8.4 Reaction profile.

bimolecular. The way in which mechanisms are related to kinetics is described in the next section. But the hypothetical nature of reaction mechanisms is well illustrated by the fact that for many years the mechanism described was thought to be the only one for this reaction. Recently it has been found that the reaction between hydrogen and iodine can also proceed by the following route:

$$I_2 \rightarrow I + I$$
$$I + H_2 \rightarrow HI + H$$
$$H + I_2 \rightarrow HI + I \text{ etc.}$$

Arrhenius studied the relationship between the rate constant of reaction and the temperature. He found that the relationship was exponential. It is usually expressed mathematically as

$$k = Ae^{-E_A/RT}, \tag{8.6}$$

where k is the rate constant, A is a constant, T is the absolute temperature, R is the gas constant, and E_A is the activation energy.

The expression may be unfamiliar to some students and it may be written in an alternative form as

$$\frac{d \ln k}{dT} = \frac{E_A}{RT^2}. \tag{8.7}$$

Note that this is similar to the expression for the temperature dependence of an equilibrium constant, Eq. (7.4). It means that the rate of change of the Naperian or natural logarithm of the reaction rate constant with absolute temperature is inversely proportional to the square of the temperature. The activation energy is nearly constant with temperature and R is as before the gas constant.

In its integrated form the above equation becomes

$$\ln k = \ln A - \frac{E_A}{RT}, \tag{8.8}$$

or in terms of log to base ten,

$$\log k = \log A - \frac{E_A}{2.303 RT}.$$

8.3 ACTIVATION ENERGY

Table 8.3. Activation energies.

	Activation energy (kJ)
$2HI(g) \rightarrow H_2(g) + I_2(g)$	179
$H_2(g) + I_2(g) \rightarrow 2HI(g)$	165
$H_2(g) + Cl_2(g) \rightarrow 2HCl(g)$ (photochemical)	25.1
$2NH_3(g) \rightarrow N_2(g) + 3H_2(g)$	335
$2N_2O(g) \rightarrow 2N_2(g) + O_2(g)$	245
$2NO_2(g) \rightarrow 2NO(g) + O_2(g)$	113
$2NOCl(g) \rightarrow 2NO(g) + Cl_2(g)$	100
$2NOBr(g) \rightarrow 2NO(g) + Br_2(g)$	58.4
$2H_2O_2(l) \rightarrow 2H_2O(l) + O_2(g)$	75.4
$CH_3CHO(g) \rightarrow CH_4(g) + CO(g)$	191
$C_2H_5OC_2H_5(g) \rightarrow C_2H_6(g) + CO(g) + CH_4(g)$	224

From this equation it can be seen that if log k is plotted against $1/T$, the result should be a straight line with gradient $-E_A/2.303RT$. This enables the activation energy of chemical reactions to be found. A list of some activation energies is given in Table 8.3.

8.4 THE MOLECULARITY OF A REACTION

We have seen that the rate of a reaction depends on the temperature because there is a minimum energy which the reacting molecules must have if they are to react. But what actually happens during the course of a reaction? Since it is impossible to see molecules or ions, such processes must be guessed at and then tested by experiment. When considering statements about mechanisms, therefore, it should be borne in mind that there is no deductive proof of a particular mechanism.

Most reactions are not simple one-step changes but they take place in a number of stages. In each stage a number of molecules or other particles take part. In some cases a step involves one molecule decomposing; in many cases two particles are involved; rarely three meet together. Reaction steps are categorized by the number of particles involved. A single-particle reaction step is called **unimolecular**, two-particle **bimolecular**, and three-particle **termolecular**. No reactions are known in which more than three particles meet together in one stage. It is obvious that the statistical chance of three particles with the right energy meeting together is much less than that of two. This accounts for the relative rarity of termolecular reactions. The chances of four particles meeting are extremely small and no such reaction has been found.

It must be emphasized that the order of a reaction can be found by direct measurement of the rate of the reaction but that its mechanism, including the molecularity of each step, is a matter of speculation. A reaction takes place in a number of steps. The rate of the whole reaction is often determined by the rate of the slowest step which acts as a "bottle neck" in the reaction. The slowest step is therefore called the **rate determining step**.

As far as each step is concerned the molecularity of the step determines the order. A bimolecular step will itself have second order kinetics. Unfortunately this does not enable the overall order to be predicted. Neither is it possible to assume that if the order is second order, say, the rate determining step will be bimolecular. Some examples will be given in the next section.

8.5 ORDER AND MOLECULARITY

8.5.1 Bimolecular and termolecular reactions

A bimolecular step in a reaction pathway is the most straightforward of steps to imagine. Two molecules collide and in the collision bonds break and new bonds form. We can readily envisage, for example, the reaction

$$H_2(g) + I_2(g) \rightarrow 2HI(g)$$

occurring through collision between hydrogen and iodine molecules. If the collision is favourable the two original bonds are broken and hydrogen–iodine bonds are formed. The same idea can readily be extended to suggest bimolecular mechanisms for many reactions, such as

$$CO_2(g) + NO(g) \rightarrow CO(g) + NO_2(g).$$

Whether these reactions actually proceed through bimolecular mechanisms is another matter; the idea must be tested by experiment. A simple bimolecular mechanism between reactants implies second order kinetics, the rate being proportional to the number of collisions which depends on the product of concentrations of the two reactants.

On the other hand, the reaction may involve several steps and still show second order kinetics. For example suppose the reaction

$$2A + B \rightarrow C + D$$

proceeds by the following steps:

$$A + B \rightarrow X \quad \text{slow}$$
$$X + A \rightarrow C + D \quad \text{fast}$$

Suppose that the first step is a slow bimolecular one, and the second a fast step, so that effectively all of X reacts with A as fast as it is formed. The rate of formation of C plus D under these conditions equals the rate of formation of X. This kind of situation may occur even when there are several steps, the rate of the reaction being determined by the slowest step. In the present case the rate determining step is bimolecular and the reaction will be second order, being first order with respect to both A and B.

A termolecular mechanism requires the collision of three molecules. Thus a termolecular mechanism for the reaction

$$2NO(g) + O_2(g) \rightarrow 2NO_2(g)$$

means reaction through collision of the three reactant molecules, indicated in the equation. The reaction will be a third order one. On the other hand, the observation of third order kinetics does not necessarily mean there is a termolecular step in the reaction pathway. This may be seen as follows. Suppose the reaction

$$2A + B \rightarrow C + D$$

proceeds by the following mechanism:

$$A + B \rightleftharpoons X \quad \text{fast}$$
$$X + A \rightarrow C + D \quad \text{slow}$$

Suppose that the first step is fast and achieves equilibrium, while the second is a slow bimolecular step.

The rate of the reaction
$$= k[X][A]$$
$$= kK[A][B][A],$$

where K is the equilibrium constant for the first step, that is

$$K = \frac{[X]}{[A][B]}.$$

Thus the rate of the reaction

$$= k'[A]^2[B],$$

where $k' = kK$. The kinetics are seen to be third order even though no termolecular step is involved. In the nitric oxide reaction, the following steps

$$2NO \rightleftharpoons N_2O_2$$
$$N_2O_2 + O_2 \rightarrow 2NO_2,$$

of which the first is postulated to be fast, and the second slow, would lead to third order kinetics. It is evident that it is not possible to decide between this mechanism and a termolecular one from the observed order alone.

8.5.2 Unimolecular reactions

The idea of a bimolecular step in a reaction, as we have seen, is readily grasped and can be extended without much difficulty to the idea of a termolecular mechanism, even though it may be rather improbable. In fact very few reactions have been shown to be termolecular. But the idea of a unimolecular mechanism, in which the rate is determined by single molecules decomposing or rearranging, is a little more difficult to grasp. It appears at first sight to disconnect the idea of reaction from that of collision. At the same time there is the awkward fact that the rate of reaction increases with temperature. Historically in fact, there was a good deal of uncertainty about how to account for first order reactions, which appeared to be unimolecular.

Amongst the examples we have many isomerization reactions (Section 21.4) such as

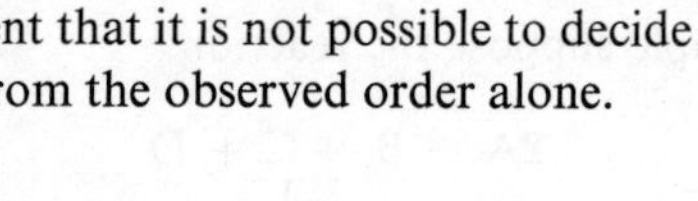

$$\begin{array}{c} H \\ \diagdown \\ Cl \end{array} C=C \begin{array}{c} Cl \\ \diagup \\ H \end{array} \rightarrow \begin{array}{c} H \\ \diagdown \\ Cl \end{array} C=C \begin{array}{c} H \\ \diagup \\ Cl \end{array}$$

as well as dissociation reactions, such as the following gas phase reactions,

$$C_2H_6 \rightarrow 2CH_3$$

and

$$CH_3-\underset{CH_3}{\overset{CH_3}{C}}-Cl \rightarrow CH_2=\underset{CH_3}{\overset{CH_3}{C}} + HCl$$

The origin of unimolecular reactions can be understood as follows. Take the reaction
$$A \rightarrow B.$$
Consider the mechanism
$$A + M \rightleftharpoons A^* + M \quad \text{fast}$$
$$A^* \rightarrow B \quad \text{slow}$$

where M represents a molecule that A may collide with; it could be another A molecule, or a B molecule, or a molecule of foreign gas. A^* represents an energized molecule, that is one which has achieved as a result of the collision the activation energy to undergo change into B. Suppose the first stage is fast and that an equilibrium is set up, while the second step is the rate determining step. We have then the rate
$$= k[A^*].$$
The equilibrium constant is given by
$$K = \frac{[A^*][M]}{[A][M]},$$
and so the rate
$$= kK[A]$$
$$= k'[A],$$
where $k' = kK$. So the mechanism leads to first order kinetics.

We can imagine the isomerization of dichloroethylene to proceed as follows. Suppose the activation energy for the isomerization is E_A, that is the energy barrier to be overcome for rotation about the double bond is E_A. Collisions will lead to a small fraction of molecules having this energy, which the molecule may hold as vibrational energy. Molecules may also loose this energy by collision and an equilibrium is set up. The energized molecules hold this energy as vibrational energy in many ways, for example by C—H and C—Cl bond stretching and bending vibrations. Only if the energy is transferred into the twisting vibration about the double bond can it be effective in leading to isomerization. Evidently this is a slow process. Only a fraction of molecules achieve this; the majority lose energy by collision before isomerizing. Collisions keep up a steady concentration of energized molecules, a small fraction going on to isomerize, but the majority are deactivated by collision before the energy is shuffled round the molecule to where it can lead to chemical change. The same kind of considerations apply to ethane where the energy may be distributed among the bonds as vibrational energy due to stretching, bending, and twisting. Only when sufficient energy happens to be transferred into the C—C bond will dissociation occur. Before this happens most molecules will be de-energized by a collision, only a small fraction dissociating before the energy is lost.

8.5.3 Zero Order Kinetics

That zero order kinetics should ever be observed seems odd at first sight, but is in fact readily understood. First of all it should be said that such a statement means that the reaction is zero order with respect to a particular reactant and may be first or second order with respect to other reactants. An example is provided by the iodination of acetone

$$I_2 + CH_3COCH_3 \rightarrow CH_3COCH_2I + HI$$

which is found to be first order with respect to acetone, and to hydrogen ion, but zero order with respect to iodine (Table 8.1). The observations are accounted for by the following mechanism.

$$CH_3COCH_3 + H^+ \rightarrow CH_3\underset{}{C}(OH)=CH_2 + H^+ \quad \text{slow}$$

$$CH_3\underset{}{C}(OH)=CH_2 + I_2 \rightarrow CH_3COCH_2I + HI \quad \text{fast}$$

The first step is slow and the second is fast, every enol molecule (Section 28.3) effectively reacting with iodine as fast as it is formed; thus the rate of formation of iodoacetone is the same as the rate of formation of enol, and independent of iodine concentration. The validity of the mechanism is supported by the observation that the rate of bromination is identical with that for iodination. Another example comes from the field of aromatic nitration where it is found under certain conditions that the rate of nitration (Section 26.5) is zero order with respect to the aromatic compound being nitrated. The following mechanism accounts for this behaviour.

$$2HNO_3 \rightarrow NO_2^+ + NO_3^- + H_2O \quad \text{slow}$$
$$NO_2^+ + ArH \rightarrow ArNO_2 + H^+ \quad \text{fast}$$

Again the first is the slow step and rate determining. The nitronium ion is sufficiently reactive that effectively it reacts as soon as it is formed and the rate of formation of the aromatic nitro compound is the same as the rate of formation of nitronium ion. This is the case for reactive aromatic compounds such as toluene; for less reactive aromatics such as benzene it is no longer the case that the nitronium ion reacts as soon as it is formed, and the rate depends on the concentration of such aromatic compounds.

8.6 ENCOUNTERS AND COLLISIONS

Most of the discussion of kinetics and mechanism has been given in terms of collision theory. This is a perfectly satisfactory picture for gas phase reactions. But many of the reactions of interest in chemistry occur in solution and the picture needs to be modified in these cases. It might be thought that the position would be much more complex. On the other hand, it turns out that for solution reactions rates are generally determined by a single step and the observed kinetics are comparatively simple, whereas for a gas phase reaction there are often several steps or even a chain reaction and correspondingly a more complicated kinetic order.

In solution we need to distinguish between encounters and collisions. The situations for gas and solution are compared diagrammatically in Fig. 8.5. We fix

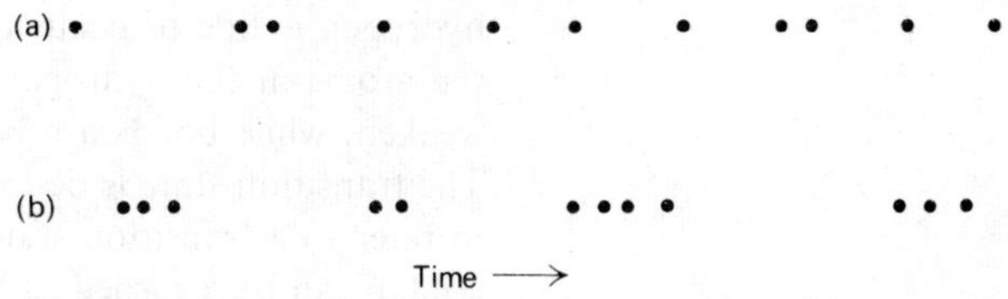

Fig. 8.5 Comparison of time elapse between collisions for (a) gas and (b) solution.

attention on a particular molecule and indicate in the diagram the time lapse between collisions. In the diagram for a solution (b) the collisions are bunched into groups. The total number of collisions may be the same for the two phases but their distribution is very different. In solution, our chosen molecule collides with a reactant molecule and cannot easily escape after a collision, being surrounded by solvent molecules. It continues to bump against the other molecule within a cage of solvent molecules for some time before it escapes. Each group of points represents an encounter and each individual point a collision. Clearly the rate of reaction cannot be greater than the encounter rate. Some of the fastest reactions in solution are accounted for by supposing that reaction occurs at each encounter. They are said to be diffusion controlled, their rates being determined by just the same factors as determine the diffusion of a solute in a solution. An example of such a reaction is the neutralization reaction in aqueous solution:

$$H_3O^+(aq) + OH^-(aq) \rightarrow 2H_2O(l).$$

Another distinction between gaseous and solution reactions is that many reactions in solution involve ions, or activated complexes which are ionic. Accordingly the rate may be affected by the dielectric constant of the solvent. In some reactions the solvent may act as a catalyst being consumed at one stage and regenerated at another stage of the reaction. In others, the solvent may actually enter into the reaction, being included in the stoichiometric equation.

8.7 TRANSITION STATE THEORY

So far we have considered what is known as collision theory of reactions. An alternative approach, known as transition state theory, has advantages for many purposes. It makes use of the concepts of equilibrium constant and free energy developed in earlier sections of the book. These concepts are applied to the activated complex (Section 8.3), or **transition state**, as it is often called. The activated complex is regarded for this purpose as being like a molecule even though its life time is very short, and even though it cannot be studied as a chemical species. In spite of this, the idea of thinking of the activated complex as being in equilibrium with the reactants proves useful. The idea is taken up in Section A.8, but we will make an introductory approach here. Consider as an example the reaction between hydrogen and iodine. We know that finally an equilibrium will be set up between hydroden, iodine, and hydrogen iodide:

$$H_2 + I_2 \rightleftharpoons 2HI \quad \text{standard free energy change, } \Delta G^\ominus.$$

The equilibrium position will be related to $\Delta G^\ominus$, the standard free energy difference between products and reactants (Section 7.5). We can imagine the formation of hydrogen iodide to occur through a transition state, in which the bonds between the atoms in the hydrogen molecule and in the iodine molecules have begun to weaken, while bonds are beginning to form between hydrogen and iodine atoms. The transition state is denoted by $(H_2I_2)^{\ddagger}$, the symbol $\ddagger$, by convention, being used to refer to a transition state. Making use of the idea that this activated complex $(H_2I_2)^{\ddagger}$ can be regarded as being in equilibrium with the reactants we can write

$$H_2 + I_2 \rightleftharpoons (H_2I_2)^{\ddagger} \quad \text{standard free energy change, } (\Delta G^{\ominus})^{\ddagger}.$$

From Section 7.5 we have $(\Delta G^{\ominus})^{\ddagger} = -RT \ln K^{\ddagger}$ where $K^{\ddagger}$ is the equilibrium constant for the above equilibrium. The activated complex may undergo unimolecular decomposition to produce the reaction products:

$$(H_2I_2)^{\ddagger} \rightarrow 2HI.$$

Alternatively, it may decompose to hydrogen and iodine which, of course, maintains the equilibrium with the reactants. Looked at from this point of view, it is seen that the rate of the reaction depends on the position of equilibrium for the formation of the activated complex, that is it depends on $(\Delta G^{\ominus})^{\ddagger}$, which is known as the **standard free energy of activation**. Other things being equal the larger the value of $K^{\ddagger}$ (that is the smaller $(\Delta G^{\ominus})^{\ddagger}$), the faster we shall expect the reaction to occur. The idea is discussed further in Section A.8, but the main point for the present discussion is that the rate of reaction can be related to $(\Delta G^{\ominus})^{\ddagger}$.

It is helpful to draw a diagram which shows the standard free energy of the reactants, the products, and the activated complex. In fact it proves useful to go even further and draw a complete curve showing how the free energy changes as the reaction proceeds from reactants, through the activated complex, to the products. Such a plot of free energy change is often called a **reaction free energy profile**. A profile for a simple bimolecular step such as we have been considering is shown in Fig. 8.6. It must be emphasized that this figure is not a graph, or plot, but a diagram, since although the free energies of reactants, products, and activated complex may be known from experiment, the general shape of the curve is normally unknown. The standard free energy change is being shown schematically as the reaction proceeds along a pathway from reactants, through activated complex, to products. In making such a plot the pathway is often called the **reaction co-ordinate**. Again the term is being used quite schematically to describe the change in molecular geometry (three-dimensional!) as the reaction proceeds. There are a number of features of interest in Fig. 8.6. In the reaction profile two important pieces of information are brought into one diagram. $(\Delta G^{\ominus})^{\ddagger}$ tells us how fast the forward reaction will go, and $\Delta G^{\ominus}$ tells us how far it will go. The diagram also shows a smooth change in the standard free energy as the reaction proceeds. It can be seen that the transition state corresponds to a point on this profile. It is special in that it corresponds to the maximum in that profile, but is like any other point on the profile in that it represents a state through which the reaction proceeds and is unlike the products or reactants which are stable entities.

As we have already seen many reactions take place in a number of steps (see also Section 26.1). The progress of a complete reaction can very conveniently be described by a reaction profile. An example will show how this can be done. We saw in Section 8.5.3, that the nitration of toluene under certain conditions takes

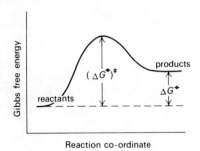

Fig. 8.6 Free energy profile for a reaction.

place in two steps. In the first step the nitronium ion is produced by dissociation of nitric acid:

$$2HNO_3 \rightarrow NO_2^+ + NO_3^- + H_2O.$$

This process has a large free energy of activation and its rate is slow. The second step

$$NO_2^+ + CH_3C_6H_5 \rightarrow CH_3C_6H_4NO_2 + H^+$$

has a much smaller free energy of activation, and proceeds rapidly. The situation is shown diagrammatically in Fig. 8.7.

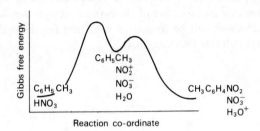

Fig. 8.7 Free energy reaction profile for nitration of toluene.

The second step is much faster than the first; as soon as NO_2^+ is produced it is converted into nitrotoluene. The first step is rate determining and the rate is consequently independent of the concentration of toluene. That is to say, the reaction is zero order with respect to toluene. Other reactions take place in more than two steps and the reaction profile has more than two maxima. Some examples will be considered in the next section.

8.8 CATALYSIS

The essential characteristic of a **catalyst** is that it increases the rate of a chemical reaction without itself being permanently changed by the reaction. Many reactions can be catalysed and the details of how the catalyst operates are often unknown, but essentially a catalyst provides an alternative reaction pathway, that is the mechanism is changed to one that proceeds more quickly. A new activated complex is formed as a result of the presence of the catalyst. This activated complex has a lower activation energy than the original one and hence allows the reaction to proceed more rapidly. As an example we may consider the hydrolysis of an ester, such as methyl acetate, in water (Section 23.3):

$$CH_3-\underset{\underset{O}{\parallel}}{C}-O-CH_3 + H_2O \rightarrow CH_3-\underset{\underset{O}{\parallel}}{C}-OH + CH_3OH.$$

The hydrolysis is slow but is catalysed by hydrogen ions. A lot of evidence has been accumulated which supports the following mechanism for the acid-catalysed reaction:

$$CH_3-\underset{\underset{O}{\parallel}}{C}-OCH_3 + H^+ \rightarrow CH_3-\underset{\underset{+}{\underset{O-CH_3}{|}}}{\overset{\overset{O\ H}{|}}{C}} \qquad \text{fast}$$

$$\text{CH}_3-\overset{\overset{\text{O}}{\|}}{\underset{+}{\text{C}}}-\overset{\overset{\text{H}}{|}}{\text{O}}-\text{CH}_3 + \text{H}_2\text{O} \rightarrow \text{CH}_3-\overset{\overset{-\text{O}}{|}}{\underset{\underset{+\text{OH}_2}{|}}{\text{C}}}-\overset{\overset{\text{H}}{|}}{\text{O}}-\text{R} \rightarrow \text{CH}_3-\overset{\overset{\text{O}}{\|}}{\text{C}}-\overset{+}{\text{O}}\text{H}_2 + \text{CH}_3\text{OH} \qquad \text{slow}$$

$$\text{CH}_3-\overset{\overset{\text{O}}{\|}}{\text{C}}-\overset{+}{\text{O}}\text{H}_2 \rightarrow \text{CH}_3-\overset{\overset{\text{O}}{\|}}{\text{C}}-\text{OH} + \text{H}^+ \qquad \text{fast}$$

A number of points can be noted. There are several steps in the reaction. The hydrogen ion takes an important part in the mechanism, but is reformed in the process so that it remains at the end of the reaction. The catalysis provides a route for hydrolysis that is not possible in its absence. Proton transfers to oxygen are generally fast; it is the breaking and forming of the carbon–oxygen bonds which is a slow process. The mechanism can be compared with that for the reaction between ammonia and acetyl chloride (Section 26.4).

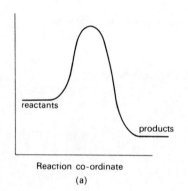

(a)

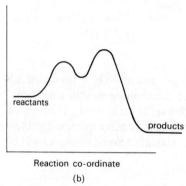

(b)

Fig. 8.8 Reaction profiles for (a) uncatalysed reaction, (b) catalysed reaction.

This kind of scheme for the operation of catalysts is a general one. It can be summarized diagrammatically by drawing qualitative reaction profiles as in Fig. 8.8. Here the uncatalysed reaction is represented as a single step with a high free energy of activation. The catalysed reaction is represented as a two-stage process with lower activation energy and is therefore faster. Of course it is often the case that the uncatalysed reaction consists of two or more steps and the catalysed reaction may consist of several steps. The important point is that for the latter there is no step with such a large free energy of activation as for the uncatalysed reaction. Even though the catalysed reaction frequently involves a longer route, in the sense of more steps, it does not have the steep hill of the uncatalysed reaction and so the reaction proceeds more readily.

As a second example of catalysis in solution we take an example from inorganic chemistry. The oxidation of thallium(I) by cerium(IV) is very slow although the oxidation potential is favourable:

$$2\text{Ce}^{4+}(\text{aq}) + \text{Tl}^+(\text{aq}) \rightarrow 2\text{Ce}^{3+}(\text{aq}) + \text{Tl}^{3+}(\text{aq}).$$

It has been suggested that the reason for the slow rate of reaction is that the mechanism is a termolecular one, there being no stable oxidation state (II) for thallium or cerium. The reaction is catalysed by Mn(II) ions. The pathway proposed is a three-stage process,

$$Ce^{4+}(aq) + Mn^{2+}(aq) \rightarrow Ce^{3+}(aq) + Mn^{3+}(aq)$$
$$Mn^{3+}(aq) + Ce^{4+}(aq) \rightarrow Ce^{3+}(aq) + Mn^{4+}(aq)$$
$$Mn^{4+}(aq) + Tl^{+}(aq) \rightarrow Mn^{2+}(aq) + Tl^{3+}(aq),$$

each step being bimolecular. The three-step process occurs more rapidly than the single-stage termolecular process. Again it will be noted that the catalyst, the Mn(II) ion, is regenerated during the reaction.

Besides homogeneous catalysis which we have considered so far, there are many examples of heterogeneous catalysis. Catalysis of gas reactions at solid surfaces is of immense importance in industrial processes. The manufacture of ammonia,

$$N_2(g) + 3H_2(g) \rightarrow 2NH_3(g),$$

depends on the use of a catalyst, usually iron, with nickel or vanadium. The cracking of hydrocarbons which is carried out on an enormous scale is a catalysed process; generally, silica or alumina is used. Metal oxide catalysts are widely used in dehydrogenation of saturated hydrocarbons to form unsaturated hydrocarbons, for example butadiene. The detailed mechanisms of heterogeneous catalysis are so complex that choice of catalysts for a particular process is largely a matter of trial and error. But some general ideas about factors controlling rates of such processes are fairly well accepted. The process can be subdivided into five different types of step: (a) diffusion of gas to catalyst; (b) absorption of reactants; (c) making and breaking of bonds; (d) desorption of products; and (e) diffusion of products away from catalyst. Any one of these may be rate determining though it is usually step (c).

As an example we may consider the catalysis of the hydrogenation of ethylene to ethane at a metal surface:

$$C_2H_4(g) + H_2(g) \rightarrow C_2H_6(g).$$

The reaction pathway proposed for the metal-catalysed reaction can be represented as a two-step process as follows:

$$2M + H_2 \rightarrow 2MH$$
$$2MH + C_2H_4 \rightarrow 2M + C_2H_6.$$

The equations refer to processes occurring at the surface, that is the hydrogen reacts with the metal M at its surface to form M—H bonds. The M—H bonds need to be of about the right strength, not too strong but not too weak. Two M—H bonds are formed in the first step, and provided the M—H bond energy is rather more than half the dissociation energy of the hydrogen molecule the process is a thermodynamically favourable one. Providing the M—H bond energy is not too large the second step will also be a feasible one. Besides these considerations, there may also be geometrical ones so that the hydrogen atoms are spaced correctly to react with the ethylene molecule. Even so the activation energies must also be favourable. It is not surprising that catalysts are very specific, and that it is still too difficult a matter to predict the choice of catalyst for a particular reaction from fundamental principles.

8.9 KINETIC AND THERMODYNAMIC CONTROL OF REACTIONS

In this section we look further at two important ideas about chemical reactions that have already been discussed. The first concerns thermodynamic criteria for a reaction to be feasible and the second the kinetic criteria which determine the rate of a chemical reaction. Suppose we consider an example. Methanol is used in the laboratory as a stable substance which may be poured from bottle to bottle through the air. But we know that thermodynamic considerations show a large free energy decrease accompanies the reaction:

$$2CH_3OH(l) + 3O_2(g) \rightarrow 2CO_2(g) + 4H_2O(l); \qquad \Delta G^\ominus = -1405 \text{ kJ}.$$

Oxidation by air is therefore feasible. Methanol is stable for kinetic reasons; there is a large activation energy and the thermodynamically stable products are not formed under normal atmospheric conditions, but change the conditions, and methanol can readily be burned to give the thermodynamically stable products.

Methanol can also undergo another reaction with oxygen to give formaldehyde as a product:

$$2CH_3OH(g) + O_2(g) \rightarrow 2CH_2O(g) + 2H_2O(l); \qquad \Delta G^\ominus = -368 \text{ kJ}.$$

Comparison of the free energy changes shows that the most thermodynamically stable product formed by oxidation of methanol is not formaldehyde but carbon dioxide. The way to obtain formaldehyde from the oxidation of methanol is to use conditions so that the second of these reactions proceeds rapidly, and the first, not at all or only negligibly. In this case the conditions can be obtained by using a suitable catalyst. Under these conditions the products are said to be kinetically controlled. On the other hand, when methanol is burned in oxygen, carbon dioxide and water are formed and the products are said to be thermodynamically controlled.

These ideas can be generalized and conveniently represented diagrammatically. Let A represent the reactants, and B and C the products of alternative reactions.

	Standard free energy change	Standard free energy of activation
B	$\Delta G_1^\ominus$	$(\Delta G_1^\ominus)^\ddagger$
A ↗ ↘		
C	$\Delta G_2^\ominus$	$(\Delta G_2^\ominus)^\ddagger$

The thermodynamically stable products are those for which there is the larger negative free energy change. Suppose that $\Delta G_1^\ominus$ has the larger negative value, then the products represented by B are the thermodynamically stable ones. On the other hand, the rate of formation of product is determined by the standard free energy of activation. If $(\Delta G_2^\ominus)^\ddagger$ is smaller than $(\Delta G_1^\ominus)^\ddagger$, the rate of formation of C will be greater, and if $(\Delta G_1^\ominus)^\ddagger$ is high enough, C may be the only product formed.

The position is summarized in Fig. 8.9. The lower free energy of B favours its formation thermodynamically, while the high free energy of activation for this profile means that the reaction may not proceed at a measurable rate. But C may be formed quite rapidly because of the low barrier. If the barrier for the formation of B from C is also high then the products will be C. The formation of C represents a kinetically determined product.

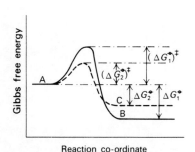

Fig. 8.9 Reaction profiles illustrating thermodynamic and kinetic control.

The sulphonation of naphthalene provides an interesting example of kinetic and thermodynamic control. Two products may be formed: the 1-sulphonic acid and the 2-sulphonic acid as shown:

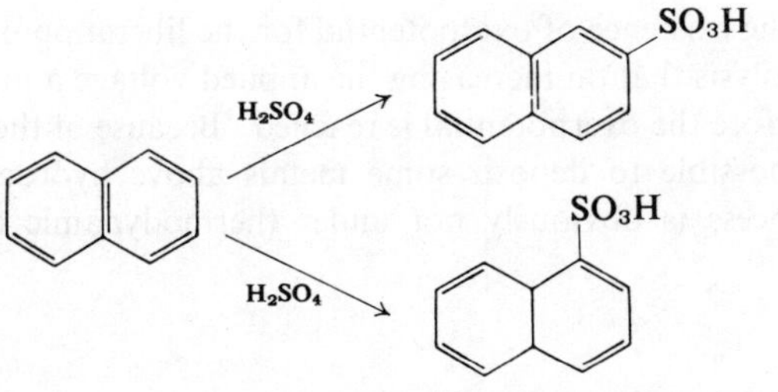

The product formed is found to depend on the conditions of the reaction. At a comparatively low temperature (80 °C) the product is almost completely the 1-sulphonic acid, whereas at a temperature of 160 °C the 2-sulphonic acid is the major, though not the only, product. The explanation is that at 80 °C the rate of formation of the 2-sulphonic acid is extremely slow and so the reaction produces almost entirely 1-sulphonic acid, that is the product is kinetically controlled. As the temperature is raised the activation energy for the formation of 2-sulphonic acid becomes less and less important and the thermodynamically stable 2-sulphonic acid is produced. At the higher temperature the reaction is thermodynamically controlled and this will generally be the case. The situation is summarized in Fig. 8.9; the 1-sulphonic acid corresponds to C having a low free energy of activation for its formation, and the 2-sulphonic acid corresponds to B being the thermodynamically stable product. In this example the barrier between B and C is small and if 1-sulphonic acid is heated with sulphuric acid to about 160 °C an equilibrium mixture, in which the 2-sulphonic acid predominates, is formed. Again thermodynamic control takes over at the higher temperatures.

Another example of kinetic and thermodynamic control can be taken from electrochemistry. It was seen in Section 7.6 that an electrode shows its maximum potential in the limit as the current drawn decreases to zero. Raising the potential above the maximum value leads to a reversal of the electrode reaction, metal being deposited instead of going into solution for example. The e.m.f. of the cell is given according to Eq. (7.11) by

$$E = \frac{-\Delta G}{zF}.$$

The reaction occurring at the electrode is reversible, and the process is thermodynamically controlled. A different situation is sometimes found; as the voltage applied to a cell is raised, the value reaches the reversible potential, rises above it, and the expected electrode reaction does not occur. Only when the applied voltage is significantly larger than the value expected on the basis of the reversible electrode potential does electrode reaction occur. The excess of applied voltage over the reversible potential is called the **overvoltage**. Under these conditions the process is no longer thermodynamically controlled. There is a slow step in the electrode process with an appreciable activation energy, that is to say kinetic factors are operating. As an example we may consider the evolution of hydrogen by electrolysis of dilute sulphuric acid at metal electrodes. For platinum black electrodes evolution occurs at the reversible electrode potential. But for copper an overvoltage of 0.23 volt is found and for a mercury electrode the overvoltage has the even larger value of 0.78 volt.

Because of the existence of overpotential for the liberation of hydrogen, it may happen in electrolysis that on increasing the applied voltage a metal is deposited at some voltage, before the overpotential is reached. Because of the existence of overpotential it is possible to deposit some metals above hydrogen in the activity series. The process is obviously not under thermodynamic control but under kinetic control.

8.10 KINETICS AND MECHANISM

It will be evident from the discussion of order and molecularity (Section 8.5) that the elucidation of reaction mechanisms is not something that can be systematized into a set of simple rules or into a formal scheme. Nevertheless certain generalizations are possible. The problem arises because it is not possible to deduce a mechanism simply from observed kinetics. We have to *imagine* a mechanism which fits the evidence, because only the reverse process of deducing the kinetics from the proposed mechanism is possible. But agreement with experiment does not prove the mechanism. It may be possible to think of more than one mechanism which fits the observed kinetics, and there is always the possibility of a mechanism which one has not thought of!

While the study of kinetics is central to the understanding of mechanisms, a variety of other evidence is also used in working out mechanisms for chemical reactions. This is particularly true for reactions in solution. Some of the more important ways of obtaining supplementary evidence for testing proposed mechanisms will be considered here briefly. These are outlined here to indicate that evidence of this kind is important, and to show the way in which it is used. We shall not in fact have occasion to use these methods at the level of treatment given in this book.

Firstly, there is the use of isotopic tracers, as a help in tracing reaction pathways. For example, by the use of ^{18}O in the hydrolysis of an ester it is possible, by examination of the products, to determine which bond to an oxygen atom is broken in the reaction.

$$R-C\overset{O}{\underset{^{18}O-R'}{\diagup}} + H_2O \longrightarrow \begin{cases} R-C\overset{O}{\underset{OH}{\diagup}} + R^{18}OH \\ R-C\overset{O}{\underset{^{18}O-H}{\diagup}} + R'OH \end{cases}$$

Secondly, stereochemistry of products can be of assistance. If, for example, the addition of bromine to a double bond leads to the *trans* product then a single-step mechanism is ruled out. In other cases the retention or loss of optical activity is a crucial mechanistic clue.

The effect of change of solvent can provide another clue. If, for example, the reaction involves development of a strongly polar transition state, with non-polar or only weakly polar reactants and products, then it may be expected that moving to a more polar solvent will stabilize the transition state relative to the ground state, that is lower the activation energy and so lead to a higher rate of reaction. If instead of ions the proposed reaction mechanism involves free radicals, then it may be tested by adding to the system substances such as nitric oxide or cyclohexene which are known to react rapidly with free radicals.

One of the most powerful guides in finding mechanisms is the use of series of compounds with varying reactivities. For example, there will be changes in rates for a given reaction, say hydrolysis, of a series of alkyl chlorides. Methyl chloride may undergo hydrolysis by a different mechanism from t-butyl chloride, but by studying the whole series of chlorides a test is provided by the need to obtain a consistent interpretation along the whole series. Then the ideas must be capable of extension to include observations made with a series of related substituents. That is, the suggested mechanism must be consistent with the variation in properties along a series of compounds.

In proposing a mechanism for a reaction a chemist will be guided by the mechanistic steps that are accepted for related reactions. Of course if it is not possible to fit the kinetic and other kinds of evidence in this way he is forced to introduce a different mechanism and this may be an important new step with wider implications. But for the most part it is the coherence of mechanisms with one another that gives strength to the subject. Certainly for understanding mechanisms of reactions in solution, a reaction cannot be considered in isolation. A feeling for the pattern that emerges is required.

PROBLEMS

1. Make a list of factors which affect the rates of chemical reactions. Describe the way in which each factor affects the rate of reaction. For each factor mention whether it is specific to a certain type of reaction, or whether it is general.

2. Dinitrogen pentoxide, N_2O_5, decomposes on standing to produce dinitrogen tetroxide, nitrogen dioxide, and oxygen. The decomposition is first order, that is:

$$\text{rate} = \frac{dc}{dt} = -kc,$$

where c is the concentration of dinitrogen pentoxide, and k the first order rate constant for the decomposition. Make a plot of c against time, t, for the following results:

t(sec)	0	184	319	526	867	1198	1877	2315	3144
c(mol l^{-1})	2.33	2.08	1.91	1.67	1.36	1.11	0.72	0.55	0.34

Take tangents to obtain dc/dt at $c = 2.0, 1.6, 1.4, 1.0, 0.6$, and 0.4 mol l^{-1}. For each value find k, the first order rate constant, from the above equation. How accurate do you think your results are? Is this a good method of finding a rate constant? As mentioned before it is often important to arrange the equation for calculation of results to the best advantage. If it can be arranged to give a straight line plot this will normally be desirable. In the present case this is readily done. We have:

$$\frac{1}{c}\frac{dc}{dt} = -k \quad \text{or} \quad \frac{d\ln c}{dt} = -k.$$

If a plot of $\log c$ against t is made, a straight line of slope $-k/2.303$ will be obtained. Use this method to obtain a better value for k. How good were your earlier results?

3. For the reaction $A + 2B \to C$ it was found that (i) when the initial concentration of A was doubled and B was held constant, the initial rate quadrupled; and (ii) when the initial concentration of B was doubled and A was held constant, the initial reaction rate was doubled. What is:
   a) the order of the reaction with respect to A;
   b) the order with respect to B;
   c) the overall order?

4. For the reaction $A + B \to C + D$ the initial rates for different initial concentrations were found as follows:

	Initial concentrations		Initial rate
	[A]	[B]	(mol l^{-1} sec^{-1})
i)	1.0	1.0	2.0×10^{-3}
ii)	2.0	1.0	4.0×10^{-3}
iii)	4.0	1.0	8.0×10^{-3}
iv)	1.0	2.0	2.0×10^{-3}
v)	1.0	4.0	2.0×10^{-3}

What is:
a) the order with respect to A;
b) the order with respect to B;
c) the form of the rate equation?
Suggest a possible mechanism for the reaction. Calculate the rate constant for the reaction. (Use result (i).)

5. For a reaction $A + 2B \to$ products, results are as follows:

	Initial concentrations		Initial rate
	[A]	[B]	(mol l^{-1} sec^{-1})
i)	4.0×10^{-2}	4.0×10^{-2}	6.5×10^{-5}
ii)	4.0×10^{-2}	8.0×10^{-2}	12.8×10^{-5}
iii)	8.0×10^{-2}	4.0×10^{-2}	25.6×10^{-5}
iv)	8.0×10^{-2}	8.0×10^{-2}	51.2×10^{-5}

a) What is the form of the rate equation?
b) What units could be used for the rate constant?
c) Calculate the rate constant (use result (i)).
d) Suggest a mechanism for the reaction.

6. First order rate constants for the decomposition of dinitrogen pentoxide at various temperatures are as follows:

T (°C)	0	25	35
k (sec^{-1})	7.9×10^{-7}	3.5×10^{-5}	1.35×10^{-4}

T (°C)	45	55	65
k (sec^{-1})	5.0×10^{-4}	1.5×10^{-3}	4.9×10^{-3}

Draw a graph and from it determine the activation energy for the reaction.

7. Draw a reaction profile for a reaction $A + B \rightarrow C + D$ for which the forward activation energy is 250 kJ and for which $\Delta H = -50$ kJ.
   a) What is the activation energy for the reverse reaction?
   b) What is the effect of raising the temperature from 300 K to 310 K on the rate of the forward reaction?
   c) Eventually an equilibrium is set up. By what factor does the equilibrium constant change when the temperature is changed from 300 K by increasing it to 310 K?

8. The mechanisms of five reactions are given below. In each case write down the form of the rate equation and state the order with respect to each reactant.
   a) $A + B \rightarrow C$
      $A + B \rightarrow C$ single bimolecular step
   b) $P + Q \rightarrow R$
      $P \rightleftharpoons S$ slow
      $S + Q \rightarrow R$ fast
   c) $L + 2M \rightarrow N$
      $2M \rightleftharpoons M_2$ fast
      $M_2 + L \rightarrow N$ slow
   d) $A + B \rightarrow C + D$
      $A \rightarrow M + D$ slow
      $M + B \rightarrow C$ fast
   e) $S + R \rightarrow P$
      $S + H^+ \rightleftharpoons SH^+$ fast
      $SH^+ + R \rightarrow P + H^+$ slow

Take the slow step in each case to be rate determining.

BIBLIOGRAPHY

Ashmore, P. G. *Principles of Reaction Kinetics*, Monographs for Teachers No. 9, Royal Institute of Chemistry (1967).

King, E. L. *How Chemical Reactions Occur*, Benjamin (1964).

    Both of these provide general introductory accounts.

Campbell, J. A. *Why Do Chemical Reactions Occur?* Prentice-Hall (1965).

    Gives an introductory account of collision theory, and breaking and making of chemical bonds. Rates and equilibrium are also discussed.

Daniels, F., and R. A. Alberty. *Physical Chemistry* (second edition), Chapter 12, Wiley (1962).

    For further reading, giving good coverage of subject within a single chapter.

Frost, A. A., and R. G. Pearson. *Kinetics and Mechanism*, Wiley (1961).

    A standard work for dipping into and for reference.

Most of the books on physical chemistry in the general list contain material on kinetics of reactions.

CHAPTER 9

Solutions and Acid–Base Equilibria

9.1 Solutions
9.2 Conductance of solutions
9.3 Acid–base equilibria

9.1 SOLUTIONS

Several types of solution may be prepared by using different combinations of gas, liquid, and solid for the solvent and solute. The more important types will be considered briefly. The most important class is that for which the solvent is a liquid; in this class we shall pay attention to ionic solutions, and consider acid–base equilibria in particular.

9.1.1 Types of Solutions

Gases

Gas–gas. All gases are miscible (soluble) in all proportions. The process is one of simple mixing in which the van der Waals forces between molecules are too small to prevent mixing taking place.

Gas–liquid. Many gases dissolve by a process which is essentially one of mixing. Thus oxygen and nitrogen are both soluble in water. But these solubilities are small; the intermolecular forces are weak. Other gases such as ammonia and carbon dioxide have very high solubilities. These gases react with water: carbon dioxide to form carbonic acid; hydrogen chloride to form hydrochloric acid.

Most gases dissolve exothermically and so, by the application of Le Chatelier's principle (Section 7.3), we would expect them to be less soluble at higher temperatures. This is generally the case, but there are exceptions.

The effect of pressure is similarly readily predictable. Solubility increases with increased pressure. For a gas which does not react with the solvent, the mass dissolved is found to be directly proportional to the pressure at constant temperature. This is called **Henry's law**. The *volume* which dissolves is independent of the pressure, since the volume is directly proportional to mass and inversely proportional to pressure at a given temperature.

Gas–solid. Gases also dissolve in solids, a fact which is of considerable concern to metallurgists. On the whole, the solubility of a gas in a solid is less than that in a liquid. When metals solidify the gases dissolved may come out of solution forming bubbles which weaken the metal. Metals which adsorb gases, which is not quite the same thing, may be used as catalysts in gaseous reactions. As an example of adsorption, palladium foil can adsorb almost 400 times its volume of hydrogen at room temperature. When gases "dissolve" in metals they are often found both to adsorb to the surface and to dissolve internally.

Liquids

Liquid–liquid. Liquids which are completely soluble in each other in all proportions are said to be completely miscible. Those which dissolve to some extent in one another are partially miscible and those which do not dissolve at all are immiscible.

Ethanol and water are miscible in all proportions. They are of course structurally similar, ethanol, C_2H_5OH, having an ethyl group where water has a hydrogen atom. Water molecules are quite firmly held to each other by hydrogen bonds. Another liquid can dissolve in it by interacting with the molecules or by forming similar hydrogen bonds. Ethanol itself is also hydrogen bonded in the liquid phase, but less strongly than water. When mixed, water molecules and ethanol molecules tend to bond with each other as well as with themselves. That a change in structure takes place can be seen by the change in volume and the change in temperature on mixing. The mixing is exothermic at room temperature (see also Section 5.3).

An example of partial miscibility is provided by phenol and water at temperatures below 67 °C. Above 67 °C phenol and water are completely miscible. Chloroform and water are examples of liquids which are generally regarded as immiscible. In fact very small quantities of water dissolve in chloroform and vice versa.

Liquid–solid. Of all inter-phase solutions it is those between solids and liquids which have received the most attention. Generally speaking, non-polar solids dissolve in non-polar solvents and polar solids in polar solvents. In the non-polar field, hydrocarbon waxes, such as greases, dissolve in hydrocarbon solvents such as the liquid alkanes. Organic solids generally dissolve in organic solvents. Such solvents as acetone and the chlorinated hydrocarbons such as chloroform (trichloromethane) and 1,1,1-trichloroethane are used industrially for this purpose. Polar solvents, of which water is the most common example, generally dissolve polar solids. Thus many salts dissolve in water but very few dissolve in the non-polar organic solvents. Acetone is an exception as, although organic, it is polar. Even so, salts which are soluble in water are generally very much less soluble in acetone.

A most interesting question is why some ionic solids dissolve in water and others do not. An important consideration is the entropy change for the process. It might be thought that the entropy of a solid in solution must be greater than that of separate solid and solvent, but this is not necessarily the case. When an ionic solid dissolves in water the ions become hydrated. That is to say, they become surrounded by water molecules. In the case of the cations it is the negatively charged oxygen in the molecule which is attracted to the ion. For anions the reverse is the case (Fig. 9.1).

The freedom of water molecules is reduced when they are bound in hydrated ions, and this effect of itself leads to a decrease in entropy. The greater the degree

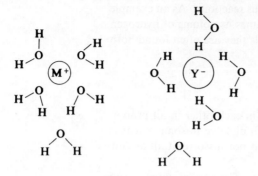

Fig. 9.1 Diagrammatic representation of hydration of ions.

of hydration the greater this effect becomes. Eventually it may overcome the gain in entropy resulting from the greater freedom of the dissolved ions (Section 6.12). Often, heats of solution are not very great and therefore do not have very much effect. For a general prediction of solubility, we must look for factors which govern the degree of hydration. The most obvious of these are the charge and size of the ion. The effect will be greater for the group II cations than those of group I. Similarly it will be greater for doubly charged anions such as the carbonate or sulphate ions than for singly charged ions such as the chlorides and bromides. The alkali metal salts (group I) are generally more soluble than those of the alkaline earth metals (group II), and chlorides and bromides are generally more soluble than sulphates and carbonates. The solubility of salts is discussed more fully in Section 10.3.17.

9.1.2 Solubility

Solubility curves. The variation of solubility with temperature has been widely studied. It has practical implications with respect to the separation of salts and compounds generally. Graphs drawn of solubility against temperature are called **solubility curves** (Fig. 9.2). The solubility is usually expressed as the number of grams of solid which dissolve in 100 g (or 1000 g) of solvent. The solubility of most solids increases with temperature but there are exceptions.

Solubility product. Consider the equilibrium of an ionic solid AB and its ions in solution:

$$A^+B^-(s) \rightleftharpoons A^+(aq) + B^-(aq).$$

The equilibrium constant is

$$K = \frac{[A^+(aq)][B^-(aq)]}{[A^+B^-(s)]}.$$

The concentration of solid in pure solid is constant and so $[A^+B^-(s)]$ can be taken as constant, and thus the ionic product $[A^+(aq)][B^-(aq)]$ is also constant. This product is known as the **solubility product**:

$$K_{sp} = [A^+(aq)][B^-(aq)].$$

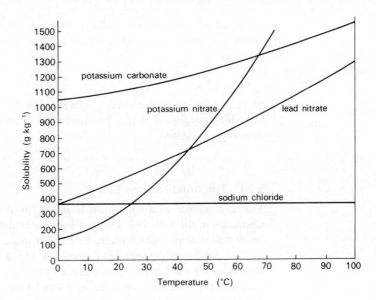

Fig. 9.2 Solubility curves.

A number of solubility products are given in Table 9.1.

Table 9.1. Some solubility products at 298 K.

	K_{sp} (mol² l⁻²)		K_{sp} (mol³ l⁻³)
AgCl	2.8×10^{-10}	CaF$_2$	1.7×10^{-10}
AgBr	5.2×10^{-13}	Mg(OH)$_2$	1.8×10^{-11}
FeS	4.5×10^{-19}	Ag$_2$CrO$_4$	1.9×10^{-12}
ZnS	4.5×10^{-24}	Zn(OH)$_2$	4.5×10^{-17}

9.2 CONDUCTANCE OF SOLUTIONS

The quantitative results of electrolysis, which occurs when an electric current is passed through a solution, are summarized by **Faraday's laws** which can be stated as follows:

1. When a solution in a cell undergoes electrolysis the mass of an element released is proportional to the quantity of electricity which has passed through the cell.
2. The quantities of electricity required to liberate one mole of different elements bear a simple relationship to one another.

The least quantity required is one Faraday. For example one mole of silver ions, Ag^+, is liberated by one Faraday. Two Faradays are required to liberate one mole of copper(II) ions, Cu^{2+}. In general z Faradays are required for one mole of X^{z+} or X^{z-} to be discharged (Section 1.3).

Ionic solutions like metallic conductors obey Ohm's law, that is to say the current, I, flowing through a solution of resistance, R, is related to the potential difference V by

$$V = IR.$$

More usually the **conductance**, $1/R$, is referred to for ionic solutions. Measurements of conductance have in fact provided important information about the nature of electrolyte solutions.

9.2.1 The Conductance Cell

The conductance of a solution is found by measuring the resistance between two electrodes in the solution, using a resistance bridge. Alternating current is used to avoid polarization, which increases the resistance.

A typical conductance cell is shown in Fig. 9.3. Two platinum (or platinized) electrodes are fixed firmly opposite each other as shown. They are insulated from each other and each is connected by a wire to the outside circuit, by which the conductance of the solution is to be measured.

Resistivity, ρ, (or specific resistance) is defined as the resistance across opposite faces of a centimetre cube of the material. That is, it is related to the resistance R, between two parallel faces of area A, distance l apart, by

$$\rho = \frac{RA}{l} \quad \text{or} \quad R = \frac{\rho l}{A}.$$

Conductivity, κ, (or specific conductance) is the reciprocal of resistivity:

$$\kappa = \frac{1}{\rho} = \frac{l}{RA}.$$

It is difficult to make a cell to exact dimensions so that conductivity can be measured directly. Usually a standard solution (generally 0.1 molar potassium chloride) of known conductivity is put into the cell and the conductance of the cell is measured. This gives the factor by which the conductance of the solution in the cell must be multiplied to give the conductivity, and is known as the cell constant.

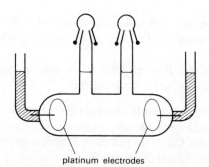

Fig. 9.3 Conductance cell.

platinum electrodes

9.2.2 Molar Conductance

The molar conductance of a solution is the conductivity (specific conductance) multiplied by the volume of the solution which contains a mole of the solute. It is interesting to compare the way in which molar conductance varies for strong (completely dissociated) and weak (partially dissociated) electrolytes with concentration.

It might be expected that molar conductance would not vary with dilution. Conductivity depends on the number of ions present, and so conductivity will decrease with dilution. The number of ions in a unit volume will decrease as the solution is diluted. But molar conductance is different. Multiplying κ by V, the volume to contain one mole, will compensate for this effect of dilution. So if there were no other changes the molar conductance, Λ, would be expected to remain constant. The **molar conductance**, Λ, is given by

$$\Lambda = \kappa V = \frac{\kappa}{c},$$

where c is the concentration of the solution.

Figure 9.4 compares the variation of molar conductance with concentration for a weak electrolyte (acetic acid) and a strong electrolyte (potassium chloride). For strong electrolytes the plot against $\sqrt{c}$ is approximately linear. It can be seen that the strong electrolyte does not actually reach a limiting value but clearly is likely to do so if the dilution is great enough. The weak electrolyte does not appear to be reaching any sort of limiting value.

These results can be explained as follows. The ions in a strong electrolyte are completely dissociated. Nevertheless, as the concentration increases they exert some influence on the movement of ions in an electric field. The result is some decrease in the molar conductance. In a weak electrolyte the molecules are only partly dissociated into ions. This dissociation increases as the substance is diluted, and so the molar conductance increases with dilution:

$$CH_3CO_2H(aq) \rightleftharpoons CH_3CO_2^-(aq) + H^+(aq).$$

The variation of molar conductance provides a method of determining the equilibrium constant for the dissociation.

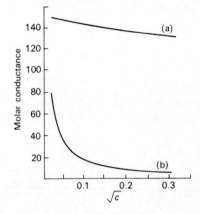

Fig. 9.4 Variation of molar conductances of potassium chloride (a) and acetic acid (b) with concentration.

9.3 ACID-BASE EQUILIBRIA

9.3.1 Acidity and Alkalinity

Acids and alkalis are characterized by certain chemical properties. A substance which is acidic has a sour taste, is generally corrosive, and changes the colour of indicators. It also reacts with carbonates to produce carbon dioxide and with more reactive metals to produce hydrogen.

Similarly substances described as alkaline are known for their soapy feel, which is usually attributed to their ability to hydrolyse the proteins in the skin, and for their ability to neutralize the properties of acidic substances. A number of theories on acids and bases have been developed and they will now be described.

Lavoisier and Davy. One of the first generalizations about acids (it is scarcely a theory) came from the great French scientist Antoine Lavoisier (1777). He proposed that all acids contained oxygen. Having demonstrated that oxygen was the principal agent of combustion he named the gas oxygen, meaning "acid-producer". This theory fell when it was found by Humphry Davy that the so-called "oxymuriatic acid" was in fact hydrochloric acid and contained no oxygen. Davy (1816) therefore proposed that all acids contain hydrogen, a better generalization.

Arrhenius. Neither Lavoisier nor Davy made any attempt to explain the nature of acidity on an atomic level. This was left to Arrhenius who, in 1887, proposed a theory of acidity which did so. He defined an acid as a compound which could produce hydrogen ions in water solution, and an alkali as a compound which could produce hydroxyl ions in water. A strong acid produced a lot of hydrogen ions and a weak acid, being only partially dissociated, produced only a few hydrogen ions. According to Arrhenius's theory, hydrochloric acid (a strong acid) is completely ionized,

$$HCl(aq) \rightarrow H^+(aq) + Cl^-(aq),$$

but acetic acid (a weak acid) is only partially ionized,

$$CH_3CO_2H(aq) \rightleftharpoons CH_3CO_2^-(aq) + H^+(aq).$$

Conductivity measurements confirm these results provided that the solutions are not too strong.

Strong bases such as sodium hydroxide are completely ionized in solution,

$$NaOH(s) + aq \rightarrow Na^+(aq) + OH^-(aq);$$

but weak bases such as aqueous ammonia are only partially ionized,

$$NH_3(g) + H_2O(l) \rightleftharpoons NH_3(aq) \rightleftharpoons NH_4^+(aq) + OH^-(aq).$$

The process of neutralization occurs by the reaction of hydrogen and hydroxyl ions:

$$H^+(aq) + OH^-(aq) = H_2O(l).$$

Arrhenius's theory marked a very great advance and is still useful in many situations. But it has weaknesses, one of which is that its application is limited to aqueous solutions.

Brönsted. In 1923 Brönsted published a new theory of acidity which has certain important advantages over the Arrhenius theory. Brönsted defined an acid as a proton donor and a base as a proton acceptor. This definition applies to all solvents, not only to aqueous solutions as does the Arrhenius definition. As far as acids are concerned, there are similarities between the Brönsted theory and the Arrhenius theory for aqueous solutions, though the Brönsted theory takes a broader view.

The Brönsted theory does not consider an acid or base in isolation. It readily interprets the different properties of pure acids and acids in solution. Pure dry liquid sulphuric acid or pure dry acetic acid does not change the colours of indicators nor does either react with carbonates or metals. As soon as water is added they do both these things.

In this scheme an acid is a species having a tendency to lose a proton and a base a species having a tendency to accept a proton. Every acid is related to a conjugate base and every base to a conjugate acid, in the Brönsted theory. For example when hydrogen chloride dissolves in water a reaction takes place and an equilibrium is established:

$$\text{HCl(aq)} + \text{H}_2\text{O(l)} \rightleftharpoons \text{H}_3\text{O}^+\text{(aq)} + \text{Cl}^-\text{(aq)}.$$
$$\text{acid 1} \quad \text{base 2} \quad \text{acid 2} \quad \text{base 1}$$

According to Brönsted's theory HCl is an acid for the forward reaction, but the hydroxonium ion, H_3O^+, is an acid in the reverse reaction. It is the conjugate acid (acid 2) of water (base 2). Similarly, the chloride ion (base 1) accepts protons in the reverse reaction to form its conjugate acid (acid 1). In this theory acids are not confined to neutral species or positive ions. The hydrogen sulphate ion can behave as an acid:

$$\text{HSO}_4^-\text{(aq)} + \text{H}_2\text{O(l)} \rightleftharpoons \text{H}_3\text{O}^+\text{(aq)} + \text{SO}_4^{2-}\text{(aq)}.$$

The hydroxonium ion. The hydrogen ion is exceedingly small consisting as it does of a proton only. Its diameter is about 10^{-13} cm as compared with other ions which have diameters of the order of 10^{-8} cm. The charge density of the proton is therefore very high indeed and it is almost inconceivable that it should exist on its own in water or any other solvent. There is evidence to suggest that the proton attaches itself to one water molecule via one of the lone pairs on the oxygen atom. The species is written as H_3O^+ and called the **hydroxonium ion**. But it seems probable that the hydroxonium ion itself is further hydrated. As with other ions, it is impossible at present to be precise about the number of water molecules which surround the ion. The hydroxonium ion is therefore written as $\text{H}_3\text{O}^+\text{(aq)}$. Sometimes it will be found written as $\text{H}^+\text{(aq)}$ to avoid complicating an equation. There is no other significance in this symbol. Note that the "lone pair" of electrons is characteristic of a base. This idea is taken up in relation to organic reaction mechanisms in Section 26.3.

Lewis. Another definition of acids and bases extends the theory yet further. This theory was proposed by G. N. Lewis in 1938. He defined bases as species which can donate an electron pair, and acids as species which can accept an electron pair. The hydrogen ion or proton is an acid under this definition and so are a great number of other species. For instance the reaction of ammonia with boron trifluoride is seen as an acid–base reaction:

$$\text{NH}_3 + \text{BF}_3 \rightleftharpoons \text{H}_3\overset{+}{\text{N}}{-}\overset{-}{\text{BF}}_3.$$

This theory greatly increases the number of reactions considered as acid–base reactions. It is less easily linked with the traditional qualities of acidity and will not be followed up here.

9.3.2 Strength of Acids

The formulation of acid–base theory in terms of an equilibrium suggests a way of measuring the strength of an acid.

Take an acid HA. In aqueous solution the equilibrium is:

$$\text{HA(aq)} + \text{H}_2\text{O(l)} \rightleftharpoons \text{H}_3\text{O}^+\text{(aq)} + \text{A}^-\text{(aq)}.$$

The equilibrium constant for this reaction is the dissociation constant for the acid, K_a, where

$$K_a = \frac{[H_3O^+(aq)][A^-(aq)]}{[HA(aq)]}.$$

The water is not included in the equilibrium expression as it is in great excess and its concentration does not noticeably alter with change in the other concentrations. Some typical dissociation constants for acids are given in Table 9.2.

Table 9.2. Some dissociation constants of acids at 298 K.

	K_a (mol l^{-1})
Hydrogen cyanide	4.8×10^{-10}
Formic acid	1.8×10^{-4}
Acetic acid	1.8×10^{-5}
Chloroacetic acid	1.4×10^{-3}
Hydrogen sulphate ion	1.2×10^{-2}

pH

A scale is needed for the day-to-day comparison of the acidity of solutions and for this purpose the pH scale is used. pH stands for the power (or puissance) of the hydrogen ion. It is defined as the negative logarithm to base 10 of the molar concentration of hydrogen ions:

$$pH = -\log[H_3O^+(aq)].$$

Clearly acidic solutions will have pH values which are low. A 0.01M solution of hydrochloric acid (fully dissociated) has a hydrogen ion concentration of 10^{-2} and hence a pH of 2.

The self-ionization of water can be written as an acid–base equilibrium:

$$H_2O(l) + H_2O(l) \rightleftharpoons H_3O^+(aq) + OH^-(aq).$$

For this equilibrium

$$K = \frac{[H_3O^+(aq)][OH^-(aq)]}{[H_2O(l)]^2}.$$

The equation still holds if an acid or base is added. The water concentration is so high compared with those of H_3O^+ and OH^- that it can be regarded as constant and the dissociation constant for water defined as

$$K_w = [H_3O^+(aq)][OH^-(aq)].$$

The value for this K_w at 25 °C is 10^{-14}:

$$K_w = [H_3O^+(aq)][OH^-(aq)] = 10^{-14};$$

hence

$$[H_3O^+(aq)] = 10^{-7},$$

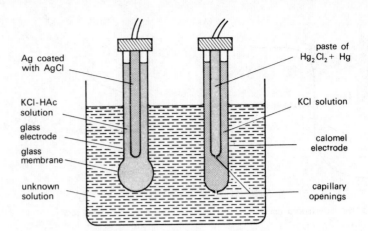

Fig. 9.5 Glass electrode as used in pH meter. A calomel electrode is being used as a reference electrode. From J. N. Butler, *Ionic Equilibrium*, Addison-Wesley (1964).

and the pH of water is 7. As water is so often used as a solvent a pH of 7 is defined as neutrality.

K_w enables the relationship between [H_3O^+(aq)] and [OH^-(aq)] to be worked out simply for any aqueous equilibrium. For instance a 0.1M solution of sodium hydroxide has a hydroxyl ion concentration of 10^{-1} moles per litre. Since

$$pH - \log [OH^-] = 14,$$
$$pH = 13 \text{ for this solution}.$$

Measuring pH. The most accurate way to measure pH is by using the hydrogen electrode described in Section 7.8. A more convenient method is provided by the pH meter. The essential part of this is a *glass electrode* (Fig. 9.5). A glass surface in contact with a solution gains a potential which is related to the hydrogen ion concentration in the solution. The glass electrode is a thin-walled glass bulb containing a standard acid solution of fixed pH. There is a reversible electrode such as an Ag|AgCl electrode in this solution. The bulb is inserted in the solution whose pH is to be found. This contains another reversible electrode and the difference in potential between the electrode in the unknown solution and that in the solution inside the bulb is measured. The pH meter has a scale of pH which gives a direct reading of the result. The other reversible electrode is frequently a calomel electrode ($Hg|Hg_2Cl_2$) as shown in Fig. 9.5.

9.3.3 Titration Curves

Acid–base titration is a fundamental method of analysis. The essence of a titration is the determination of the **equivalence point**, that is the point at which equivalent quantities of acid and base have reacted. In acid–base titrations this is done by means of an indicator, a pH meter, or by conductivity.

One of the remarkable things about an acid–base titration performed with an indicator is the fact that the colour of the indicator changes sharply when only one drop of acid or base is added at the equivalence point. This is particularly marked in the case of a strong acid and a strong base. To understand this phenomenon, and

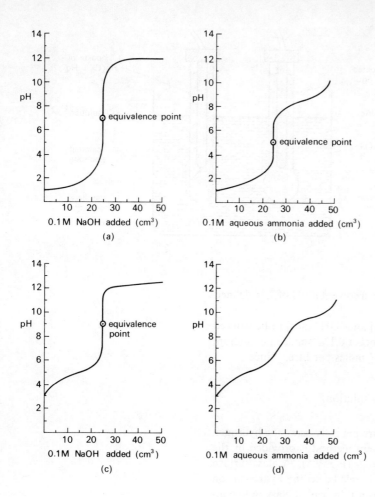

Fig. 9.6 Titration curves. (a) 0.1M HCl with 0.1M NaOH. (b) 0.1M HCl with 0.1M aqueous ammonia. (c) 0.1M CH_3CO_2H with 0.1M NaOH. (d) 0.1M CH_3CO_2H with 0.1M aqueous ammonia. 25 cm³ of acid has been used in each case.

indeed the whole of an acid–base titration, it is important to see how the pH of the reacting mixture varies during the course of a titration. For this experiment it is best to use a pH meter. The graphs produced are called **titration curves**. The series of titration curves in Fig. 9.6 show the change in pH when 25 cm³ of an acid are titrated with a base.

A typical result for a strong acid and a strong base is that given by the curve for 0.1M hydrochloric acid titrated with 0.1M sodium hydroxide (Fig. 9.6a). It can be seen that a very large change in pH takes place for a small addition of sodium hydroxide near the equivalence point. This accounts for the sharp change in colour of an indicator. Note also that in this case the equivalence point is at pH 7.

Figure 9.6(b) shows the result of a similar experiment performed with hydrochloric acid and aqueous ammonia. The result is typical for a strong acid and a weak base. In this case, the flat part of the curve is shorter and the equivalence point is at about pH 5.

The third curve (Fig. 9.6c) shows the change in pH when a strong base is titrated against a weak acid. The base is sodium hydroxide and the acid, acetic acid. As with the strong acid versus weak base, the flat point of the curve is shorter than that in Fig. 9.6(a). But in this case the equivalence point is on the alkaline side at about pH 9.

The final curve (Fig. 9.6d) shows the result of the titration of a weak acid against a weak base. The acid is acetic acid, the base aqueous ammonia. It can be seen in this case that there is no straight part to the curve, only a point of inflection. This means that there is no sudden change in pH when a small amount of base is added near the equivalence point. It is not surprising, therefore, that an indicator cannot be used to find the end point of such a titration.

9.3.4 Indicators

It is clear from the titration curves shown in the previous section that an effective indicator for a titration must change colour over the pH range covered by the flat part of the titration curve. That is the pH over which the reaction mixture changes rapidly at the end point of the titration. With a strong acid and a strong base (Fig. 9.6a) the pH range is about 3 to 11, for a strong acid and weak base 3 to 7, and for a weak acid and strong base about 7 to 11. Many indicators change colour over the wide range from 3 to 11 but fewer are found which correspond to 3 to 7 or 7 to 11. For a strong acid and weak base (pH 3 to 7) methyl orange, preferably "screened" with a blue dye to make the colour change easier to see, is often used. For a strong base and a weak acid (pH 7 to 11) phenolphthalein is commonly chosen; the change from colourless (acidic) to red (alkaline) is particularly clear making this a useful indicator even for the colour-blind! How can the behaviour of indicators be explained?

The behaviour of indicators can be explained by assuming that they are either weak acids or weak bases. This is not the whole story, but it provides a theory which, although simplified, is perfectly adequate for the present purpose. The important characteristic of an indicator is that the un-ionized molecule and the ions into which it dissociates must be of different colours. Alternatively, one may be coloured and the other colourless.

The theory is best described in terms of a particular indicator. Phenol red is a good example of an indicator which behaves as a weak acid. The full formula of phenol red and the products into which it partially dissociates in water are

[Structural formula of phenol red: left structure with OH group on phenol ring and SO_3^- group, $+ H_2O \rightleftharpoons$ right structure with O^- group and SO_3^- group, $+ H_3O^+$]

which, for short, can be written

$$HIn(aq) + H_2O(l) \rightleftharpoons H_3O^+(aq) + In^-(aq).$$

In this case the undissociated compound (HIn) is red and the anion (In^-) is yellow.

The dissociation represents a reaction in equilibrium and we can therefore say that the equilibrium constant for the reaction is given by:

$$K_a = \frac{[H^+(aq)][In^-(aq)]}{[HIn(aq)]}.$$

Taking the logarithm of both sides we have

$$\log K_a = \log [H^+(aq)] + \log [In^-(aq)] - \log [HIn(aq)].$$

Now $-\log [H^+(aq)]$ is, by definition, pH. So the equation can be written as

$$\log K_a = -pH + \log [In^-(aq)] - \log [HIn(aq)],$$

or as:

$$pH = -\log K_a + \log \frac{[In^-(aq)]}{[HIn(aq)]}.$$

It can be seen that the equilibrium position of the indicator will be affected by the hydrogen ion concentration. If the hydrogen ion concentration is high, the equilibrium position will move to the left and undissociated HIn will predominate; the solution will be red. If a base is added which reduces the hydrogen ion concentration the equilibrium position will move to the right and the solution will become yellow.

It is useful to know at what pH the indicator will be in its "neutral" position—the position in which there are equal concentrations of HIn and In^-. At this position the colour will be a mixture of red and yellow; it will appear to be orange.

If $[In^-(aq)] = [HIn(aq)]$ then

$$\frac{[In^-(aq)]}{[HIn(aq)]} = 1$$

and

$$\log \frac{[In^-(aq)]}{[HIn(aq)]} = 0.$$

Thus at the balance position

$$pH = -\log K_a.$$

$-\log K_a$ is sometimes called pK_a. The point is, that if the dissociation constant K_a for an indicator is known, it is very easy to work out the pH at which the indicator is in its balance position. Log K_a for phenol red is -7.6. The pH at which the indicator will show an orange colour is therefore 7.6.

9.3.5 Buffer Solutions

The control of acid–base equilibria is important both in nature and in chemical and biochemical research. In living systems the control of pH is essential for health. It is found, for instance, that the blood has a pH of about 7.4. The pH remains constant although the concentration of carbon dioxide, and hence carbonic acid, varies considerably. If acid is added to a sample of blood its pH is found to change very little unless an excess of acid is added. A solution which tends to resist changes in pH is called a buffer solution. To understand the ways in which the pH of a solution can be controlled, acid–base equilibria must first be examined.

For a strong acid such as hydrochloric acid the acid–base equilibrium

$$HCl(aq) + H_2O(l) \rightleftharpoons H_3O^+(aq) + Cl^-(aq)$$

lies far over to the right. We say that it is fully dissociated. A 0.01M solution therefore has a pH of $-\log 0.01 = 2$. If a drop of concentrated hydrochloric acid is added the change in acidity of the solution will be directly proportional to the amount of acid added. If the concentration of hydrochloric acid is doubled the concentration of hydrogen ions is doubled. The same is not true for weak acids.

The equilibrium for a weak acid such as acetic acid:

$$CH_3CO_2H(aq) + H_2O(l) \rightleftharpoons H_3O^+(aq) + CH_3CO_2^-(aq)$$

lies over to the left. The pH of 0.01M acetic acid is about 4 as it is only about 1 percent dissociated at this concentration. The effect of adding an equal amount of a strong acid is not in this case proportional to the amount added. When hydroxonium (hydrogen) ions are added the equilibrium tends to absorb them by moving to the left. Thus the rise in hydroxonium ion concentration is relatively less than it would be if the acetic acid was not present. The solution is tending to resist changes in pH and can be said to be acting as a weak buffer. Note that it will also tend to resist changes in an alkaline direction. If an alkali is added, hydroxyl ions will react with hydrogen ions to form water:

$$H_3O^+(aq) + OH^-(aq) \rightarrow H_2O(l).$$

This will make the equilibrium move to the right thus forming more hydrogen ions to replace those used up. The pH will change, but not as much as would be the case if no acetic acid were present.

To understand how more effective buffer solutions can be devised, consider the dissociation constant of the weak acid.

$$K_a = \frac{[H_3O^+(aq)][CH_3CO_2^-(aq)]}{[CH_3CO_2H(aq)]}$$

i.e.

$$[H_3O^+(aq)] = \frac{K_a\,[CH_3CO_2H(aq)]}{[CH_3CO_2^-(aq)]}.$$

It can be seen that if $[H_3O^+(aq)]$ is to remain constant, $[CH_3CO_2H(aq)]/[CH_3CO_2^-(aq)]$ must remain constant. In the example of acetic acid the concentration of undissociated acetic acid $[CH_3CO_2H(aq)]$ *does* remain approximately constant as the acid is a weak one. If the acid is only about 1 percent dissociated, large changes in hydrogen ion concentration will only make relatively small changes in concentration of the undissociated acetic acid. But the concentration of acetate ions $[CH_3CO_2^-(aq)]$ changes and hence the pH changes. How can the acetate ion concentration be kept constant?

The answer is to ensure that there is an excess of acetate ions by adding a salt such as sodium acetate. Sodium acetate is fully dissociated.

$$CH_3CO_2^-Na^+ + aq \rightarrow CH_3CO_2^-(aq) + Na^+(aq).$$

If it is added to acetic acid it will push the acetic acid equilibrium to the left thereby reducing the hydrogen ion concentration and increasing the pH. But this mixture will not change its pH appreciably when even quite large amounts of acid or alkali are added. In general a weak acid mixed with a solution of a salt of that weak acid with a strong base will form a good buffer solution with a pH less than 7. For an alkaline buffer solution a mixture of a weak base and a salt of that base with a strong acid may be used.

In practice more sophisticated buffer solutions have been developed for use, particularly in the field of biochemistry. The buffer in blood is the carbonic acid, itself a weak acid, in conjunction with protein molecules.

PROBLEMS

1. What is meant by solubility product?
   What are the units of K_{sp} for:
   a) AgCl;
   b) CaSO$_4$;
   c) Ag$_3$PO$_4$;
   when concentrations are expressed in mol l^{-1}?

2. Magnesium hydroxide, Mg(OH)$_2$, has a solubility of 1.6×10^{-4} mol l^{-1}.
   a) Calculate the solubility product.
   b) Calculate the solubility of magnesium hydroxide in 0.05 molar sodium hydroxide solution.

3. The solubility product of silver chloride is 2.8×10^{-10} mol^2 l^{-1}. What is the solubility of silver chloride in:
   a) pure water;
   b) 0.01 molar hydrochloric acid?

4. A sample of hard water has a calcium ion concentration of 10^{-3} mol l^{-1}. Calculate the maximum concentration of fluoride ion that can be obtained for this water. The solubility product of calcium fluoride, CaF$_2$, is 1.7×10^{-10} mol^3 l^{-3}.

5. Molar conductances of lithium chloride in water are as follows:

c(mol l^{-1})	0.0005	0.001	0.005	0.01	0.05
Λ(ohm^{-1} mol^{-1} cm^2)	113.2	112.4	109.4	107.3	100.1

   Find the limiting value of the molar conductance in the limit of infinitely dilute solution.

6. Which of the following cannot act as: (i) a Brönsted acid, or (ii) a Brönsted base:
   a) HSO$_4^-$;
   b) CO$_3^-$;
   c) NH$_3$;
   d) BF$_3$;
   e) HS$^-$;
   f) H$_3$O$^+$?

7. Which of the following are true statements? An acid and its conjugate base:
   a) neutralize one another;
   b) differ only by a proton;
   c) form a salt and water;
   d) carry opposite charges?

8. Complete the following equilibria between conjugate acids and bases, making the first named in each case a Brönsted acid.
   a) $HBr + H_2O \rightleftharpoons$
   b) $ H_2O \rightleftharpoons SO_4^{2-}$
   c) $NH_4^+ \rightleftharpoons H_2O$
   d) $HS^- \rightleftharpoons H_3O^+$
   e) $ \rightleftharpoons H_3O^+ + HS^-$
   f) $H_2O + CN^- \rightleftharpoons$

9. Write down the pH of the following:
   a) 0.01 molar hydrochloric acid;
   b) 0.002 molar hydrochloric acid;
   c) 0.001 molar sodium hydroxide solution;
   d) 0.001 molar barium hydroxide solution.

10. Calculate the pH of the following aqueous solutions (use the data in Table 9.2):
    a) 0.1 molar acetic acid;
    b) 0.1 molar chloroacetic acid;
    c) 0.01 molar acetic acid;
    d) 0.1 molar hydrogen cyanide.

11. What is the pH of each of the following solutions:
    a) 0.05 molar acetic acid;
    b) 0.2 molar sodium acetate;
    c) a solution of 0.05 molar acetic acid to which sodium acetate has been added to make a 0.2 molar solution of the salt? Sodium acetate is a strong electrolyte.

BIBLIOGRAPHY

Aylward, G. H. "Solution Equilibria" in *Approach to Chemistry*, edited by G. H. Aylward and T. J. V. Findlay, University of New South Wales, Sydney (1965).

   This gives an introductory treatment of equilibria in solution.

Butler, J. N. *Ionic Equilibrium*, Addison-Wesley (1964).

   A treatment of a wide range of equilibria in solution.

Bell, R. P. *Acids and Bases. Their Quantitative Behaviour* (second edition), Methuen (1969).

   A comprehensive discussion of the use of the concepts of acids and bases.

Stark, J. G. (Editor). *Modern Chemistry*, Penguin (1970).

   Contains introductory account of "Theories of Acids and Bases" by D. Nicholls and another on "Non-Aqueous Solvents" by A. K. Holliday.

CHAPTER 10

Periodicity I: The Periodic Properties of the Elements

10.1 Periodic tables and the periodic law
10.2 The classification of elements as metals and non-metals
10.3 The physical properties of the elements
10.4 The chemical properties of the elements
10.5 Classification of elements: summary

10.1 PERIODIC TABLES AND THE PERIODIC LAW

Periodic tables are arrangements of the elements in order of their atomic numbers in such a way as to demonstrate the periodic law, that is, to show clearly which elements have related properties. Earlier tables, such as that of Mendeleev (1869), were based on atomic weights measured as bulk properties from, for example, vapour densities, and on a combination of physical and chemical properties, especially valency relationships. Mendeleev (and others independently) found that when he wrote down the elements in order of increasing atomic weight (leaving gaps where necessary for elements then undiscovered) there was a periodic recurrence of elements with similar chemical and physical properties. On this basis he formulated his periodic law which stated that *the properties of the elements are a periodic function of their atomic weights*. This law fulfilled the dual function of theories described on page 4: firstly, it classified the elements into groups with similar properties, and secondly, it enabled predictions to be made, and directed relevant research. Originally the latter function was the more important: Mendeleev pointed out many wrongly determined atomic weights, and also predicted the existence of new elements and their properties, such as eka-silicon (germanium). But today the former function of the periodic model is more important. The Periodic Table is the usual method of classifying the elements in such a way that the most important periodic relationships between them are apparent. Elements are thus treated as members of families instead of as individuals. This gives the subject a structure and helps our understanding of chemistry. It must not be allowed to cloud the fact that elements are unique: *no one element is exactly like any other*.

Mendeleev's periodic law was formulated long before its fundamental basis was understood. He knew nothing of electronic structure. That his law amounted to almost the same thing as modern statements of the law (below) is due to the fact that atomic weight is almost a monotonic function of atomic number. However, owing to differences in the amounts of the heavier isotopes of cobalt and nickel present in the natural elements, the bulk atomic weights decrease from cobalt to nickel as their atomic numbers increase. Can you find any other pairs of similar elements for which this is also the case?

Modern chemists attempt to relate the bulk chemical and physical properties of elements (such as conductivity, melting point, and ease of decomposition of salts) to the atomic properties of elements (such as atomic radius, ionization energy, and so on). This is usually at least partly successful. Since it is found that the atomic properties of an element are related to its electron configuration, there is clearly a relationship between the bulk properties of elements and their electron configuration. This is why chemists pay so much attention to the electron configuration of the elements. Modern forms of the periodic law are: (a) *the properties of the elements are a periodic function of their atomic numbers;* and (b) *the properties of the elements depend upon their total electron configuration.* The equivalence of these statements is due to the fact that the repetitive filling of shells of electrons, one after the other, produces the same outer electron configuration time and time again (Section 2.5). To a first approximation, therefore, the bulk and atomic properties of an element depend upon the number of outer or valence electrons, that is, electrons in the outer or valence shell (Section 2.5). *Elements with the same outer electron configuration have similar properties.* In most forms of the Periodic Table this relationship is shown vertically by groups. But, the number of valence electrons is not the only major factor which determines properties: the relative energies of these electrons from one element to another are also important. Various useful models are employed to describe the force that a nucleus exerts on its outer electrons, for example, effective nuclear charge, shielding and penetration (Section 2.4.3), ionization energy, and electronegativity. The electronegativity of an atom is a convenient concept to use here (Section 4.2). The greater the electronegativity of an atom, the more the nucleus attracts the outer valence electrons, and therefore the lower their energy. The energy depends upon how far from the nucleus the outer electrons are (i.e. upon the principal and second quantum numbers primarily), and upon the nature of the other electrons screening the outer electrons from the nucleus (Section 2.5).

The overall result is that elements within a group or family show: (a) similar properties owing to similar outer electron configuration; and (b) a gradation of properties down the group owing to gradually changing electronegativities. In addition it will be found later that: (c) the first member of a typical group shows "anomalous" properties owing to abnormally high electronegativity, small size, and restriction to an octet of valence electrons (Section 14.2.7).

The periodic law is fundamental, but in the form given above it is of little use as an aid in rationalizing the multitude of chemical facts. Periodic Tables are models of the above law which help in this way. Many models or types of table are in existence, each of which shows in different ways some of the relationships existing between the elements. The minimum aims of any table are to show as *simply* as possible (therefore excluding three-dimensional tables): (a) similar elements grouped together as families; (b) the gradual changes in properties which occur as the electronegativity changes from element to element; and (c) dissimilar elements clearly distinguished. All two-dimensional forms have some disadvantages, and none can show completely satisfactorily all the relationships existing between the elements. Each classification or table (like all models) is useful for a particular purpose: there is no one form of the Periodic Table which is "right". The most satisfactory form for our purposes appears to be the separated "long form" shown in Fig. 10.1. This groups vertically elements consisting of atoms with similar outer electron shells, whose electronegativity changes relatively slowly in going from one

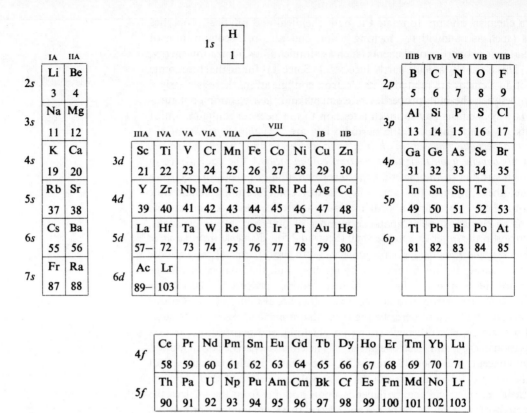

Fig. 10.1 The separated long form of the Periodic Table.

element to the next in atomic number order; the alkali metals lithium to caesium provide an example of such a group. Dissimilar elements with dissimilar outer electron shells in which the number of electrons increases monotonically, are shown as horizontal periods, as for example, lithium to fluorine. The table is broken either where no obvious relationship exists, or where a gradual transition is broken. For example, the series potassium, calcium, gallium, germanium, arsenic, selenium, bromine, shows the expected marked transition from the weakly electronegative potassium to the strongly electronegative bromine. However, the sequence is broken by the appearance of the transition series starting with scandium.

In most tables the elements fall into four obvious classes (Fig. 10.2) although subdivision into further classes is possible.

1. **Noble gases** (group 0). These are atoms in which the outer s- and p-subshells are filled. Their characteristic property is that they are chemically rather unreactive compared to all other elements, so much so that they were previously called the "inert" gases, when it was mistakenly believed they formed no chemical compounds.

2. **Major** or **representative elements**, or s- and p-block elements (groups IA, IIA, IIIB, IVB, VB, VIB, VIIB). In the atoms of these elements only the outer s-

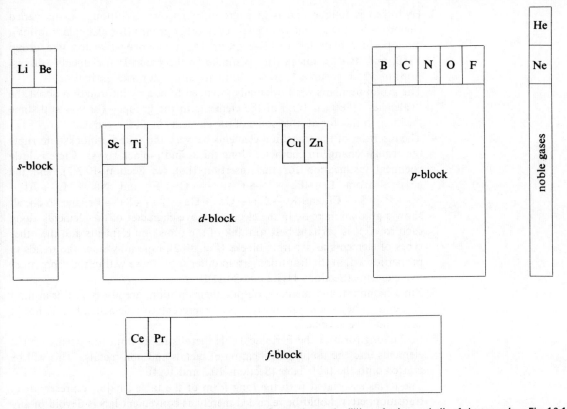

Fig. 10.2 Classification of the Periodic Table into blocks according to the filling of valence shells of electrons (*see* Fig. 10.1).

or *p*-electron subshell is incompletely filled. Their characteristic property is that they use only their outer shell electrons in bonding with other atoms.

3. **Transition metals,** or *d*-block elements (groups IIIA, IVA, VA, VIA, VIIA, VIII, IB). These are atoms in which there are two partially filled outer shells of electrons coexisting, ns and $(n-1)d$. Electrons from both shells may be used in bonding. The periodic relationships in these elements differ from those of the major elements since there is a marked horizontal similarity in properties as well as the usual vertical relationship.
4. **Lanthanides,** or rare earth elements, and **actinides,** or inner transition elements; or *f*-block elements. These atoms have three partially filled shells of electrons coexisting. The horizontal similarities existing in these groups are so great that difficulties were experienced in separating the lanthanides from one another.

The long form of the Periodic Table is an adequate representation of the periodic law as long as several misleading features are remembered. These are listed below:

1. The group nomenclature A and B is arbitrary and is often inverted for groups III to VII inclusive. To avoid confusion the first member of the group is often specified, for example group IVB is referred to as the "carbon group".

10.1 PERIODIC TABLES AND THE PERIODIC LAW   217

2. Hydrogen is unique and is in a group of its own. Although in its varied chemistry it shows some resemblances to other groups (for example it forms a monopositive hydrated ion like group IA, a mononegative ion like group VIIB, and its electronegativity is similar to the group IVB elements), discussions of which group it "should" be in are largely a waste of time.
3. The group numbers were originally intended to convey information about the "valencies" (Section 16.1) of the elements in the group. There is still some meaning in this for at least some of the elements in all the groups.
4. The insertion of the transition elements between the s- and p-blocks interrupts the regular change in character along the s- and p-block rows. There is little chemical justification for this insertion (but see Section 10.3.2), and the changes from $Li \rightarrow Be \rightarrow B \rightarrow C \rightarrow N \rightarrow O \rightarrow F$, and $Na \rightarrow Mg \rightarrow Al \rightarrow Si \rightarrow P \rightarrow S \rightarrow Cl$, and $K \rightarrow Ca \rightarrow Ga \rightarrow Ge \rightarrow As \rightarrow Se \rightarrow Br$, and so on, all show a gradual increase in the electronegative character of the elements along each row. It is perhaps best to think of the transition elements and the other types of elements as separate blocks (Fig. 10.2), especially since the trends in properties within the transition group differ from those within the representative elements (Sections 11.4.2 and 20.11).
5. Zinc, cadmium, and mercury, despite their position, are not typical transition elements. They are in fact fairly typical representative elements, but do not fit easily into the scheme.
6. In the long form of the Periodic Table there is no attempt to systematize the elements into the useful classification of metals and non-metals. This will be grafted onto the table later (Sections 10.2 and 10.4).
7. The terms associated with the long form of the table (major, representative, transition, etc.) should be regarded merely as convenient labels devoid of any associative meaning. The "transition" elements, for example, do not show "transitional" character more than any other series of elements.

10.2 THE CLASSIFICATION OF ELEMENTS AS METALS AND NON-METALS

The Periodic Table does not classify elements as metals and non-metals, but this can be done relatively easily. It is merely necessary to establish and apply a criterion of metallicity, that is, a definition of a metal. Any one of the periodic properties to be discussed later (Section 10.3), such as electronegativity or ionization energy, could be used as such a criterion. Thus many arbitrary classifications are possible, most of which, if chosen reasonably, would be similar but not necessarily identical. This is due to the fact that in passing from metals to non-metals (say along a row or period in the Periodic Table, for example from sodium to chlorine) there is a gradual change in the character of the elements, not a sudden one. It is not unlike attempting to classify all shades of grey as either black or white. There is a complete spectrum of shade in passing from black to white. If the shades are all to be classified either as black or white, then a definition or criterion is needed to assist in drawing the line between what is to be *called* black, and what is to be *called* white. Obviously many different reasonable definitions could be made. These would classify shades of mid-grey which are in the middle of the spectrum sometimes as one class, sometimes the other. But any reasonable definition would be in broad agreement with any other reasonable definition. Very dark grey would always be classified as black, light grey always as white. In classifying metals it is convenient to divide the spectrum from metals to non-metals into three classes: metals, semi-metals (or metalloids), and non-metals (cf. white, mid-grey, and black). Either a physical property (such as electrical conductivity, or structure of the elements), or a chemical property (such as the reaction of the elements with acids, or the chemical nature of the oxide) could be used. A convenient property to use as a criterion of metallicity is the atomic electrical conductivity, since this property can be expressed as an easily measurable accurate number. The **electrical conductivity** of a substance is usually defined as the reciprocal of the electrical resistance of a section of the substance one centimetre square in cross-section and one centimetre long, at a given temperature, usually 0–20 °C. The units are therefore $ohm^{-1}\,cm^{-1}$ at say, 20 °C. This compares equal volumes of substances. It is more appropriate to compare the conductivity of equal numbers of atoms (of the same cross-section of conductor), say one mole of atoms. This is called the **atomic electrical conductivity** (or atomic conductance) and is obtained in units of $ohm^{-1}\,cm^{-4}$ by dividing the above conductivity by the atomic volume. It is the conductivity of a block of the substance of cross-section one centimetre square, sufficiently long to include one mole of elemental atoms.

It is now possible to give a criterion which will allow all elements to be classified as either metals, non-metals, or semi-metals. (A criterion is merely a rule or definition which allows a certain pile of objects, which could be elements, cars, or people, etc., to be separated into sub-piles or classes. The relevance of the criterion can only be judged by the usefulness of the related classification. It will be found that the classification of elements in the way given below is very useful; therefore the criterion is relevant. A classification of elements according to the number of vowels in the name of the element would be found to be less useful; therefore the criterion is judged to be irrelevant).

1. **Metals** are good conductors of electricity with atomic electrical conductivities greater than $3 \times 10^{-4}\,ohm^{-1}\,cm^{-4}$, and their conductivity decreases slowly as their temperature is increased (Section 10.3.12).
2. **Non-metals** are insulators.
3. **Semi-metals**, or metalloids, have small but measurable conductivities, which increase as their temperature is increased and which are very sensitive to impurity.

Using these definitions, the elements are classified as in Fig. 10.3.

The main feature of this, and all other relevant classifications, is the separation of the metals from the non-metals by a diagonal band of semi-metals running from top left to bottom right of the *s*- and *p*-blocks in the table. The exact position of the boundary depends upon the particular criterion used. As another example, chemical this time, consider the acidic or basic nature (Section 11.1.1) of the oxides of the *s*- and *p*-block elements having their group oxidation numbers (Section 10.3.20); for example, Na_2O, MgO, Al_2O_3, SiO_2, P_2O_5, SO_3, Cl_2O_7 etc. A metal can be defined as a (representative) element of which the oxide having the group oxidation number is basic in character (Section 11.1.1). (N.B. The definition would have to be changed for the *d*- and *f*-block elements except for the zinc group.) Similarly non-metallic and semi-metallic oxides are acidic and amphoteric respec-

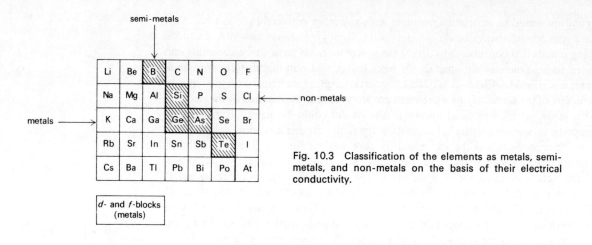

Fig. 10.3 Classification of the elements as metals, semi-metals, and non-metals on the basis of their electrical conductivity.

tively. Using these criteria the elements are classified as in Fig. 10.4. In this case the diagonal dividing line occurs in a slightly different place from that in Fig. 10.3.

The definition in terms of electrical conductivity is the usual one employed. Even here, however, where the measurement of the property and the application of the criterion are apparently simple, difficulties occur. Firstly, the conductivities of semi-metals are very sensitive to small amounts of impurities. Secondly, elements occur in different forms called allotropes (Section 10.3.1). This phenomenon is widespread throughout the whole Periodic Table, but in groups IVB, VB, and VIB the allotropes have different electrical properties. For example, carbon occurs in two forms: diamond, which is a non-metal and an insulator; and graphite which is a semi-metal if judged by its electrical conductivity. This is not a cause for alarm. Any reasonable but arbitrary classification of a continuous spectrum of properties may differ slightly from any other reasonable classification, or in other words, the dividing line may occur in a slightly different position. This idea is developed in Section 10.5. The important point here is that the application of the methods of science requires commonsense, not quick reflexes.

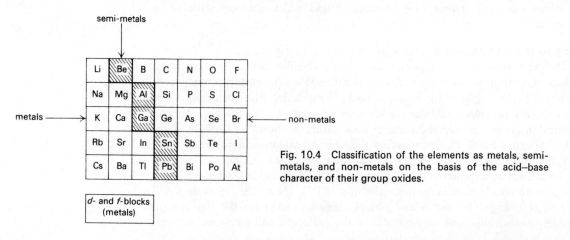

Fig. 10.4 Classification of the elements as metals, semi-metals, and non-metals on the basis of the acid–base character of their group oxides.

Summary. The classification into metals, semi-metals, and non-metals is for convenience, and it must not cloud the fact that each element differs from every other element (though as Orwell might have said, "some more than others"). Nevertheless, it would be unwise not to recognize the fact that the elements with broadly similar properties are placed in the same areas of the Periodic Table, leading to the classification above. The general behaviour is clear: metals are found in groups IA, IIA, and the bottom of groups IIIB, IVB, and VB (all the d- and f-block elements are metals also); the non-metals are the lighter elements of groups IVB, VB, VIB, and VIIB; the semi-metals occur as a diagonal block running from top left to bottom right of the s- and p-block elements, dividing the metals from the non-metals. The exact position of this dividing block depends upon the particular criterion used.

10.3 THE PHYSICAL PROPERTIES OF THE ELEMENTS

The bulk physical properties of the elements, such as melting point, depend upon their structure—metallic, molecular, etc.— and upon the type of bonding between the elemental atoms, which in turn depends upon the total electron configuration of the atom (Section 2.5). Problems often exist in attempting to relate the bulk properties to electron configuration, partly because of their indirect relationship; but the gross features of the problem are now understood. However, because of this indirect relationship between physical properties (and chemical of course) and electron configuration, it is convenient to discuss appropriate atomic properties, such as atomic radii, before discussing the physical properties. The latter can then be related to atomic properties when this relationship is clearer than that between physical properties and electron configuration, though of course, since atomic properties and electron configuration are directly related, in the end it amounts to the same thing.

10.3.1 Structure

Metals have metallic-bonded crystal structures (Section 4.7) with high co-ordination numbers (Section 3.2.2), usually either twelve (close-packed) or eight (body-centred cubic). The non-metals form covalently bonded molecules (O_2, N_2, etc.) with very small co-ordination numbers—one or two; they form volatile molecular solids in which the intermolecular bonding is due to weak van der Waals forces. The semi-metals form giant molecules (layer or long-chain lattices) in which the co-ordination number is small—less than or equal to four (Table 10.1).

Many chemical substances (elements or compounds) can exist in more than one form. The phenomenon is known as **polymorphism** in the case of compounds, or **allotropy** in the case of elements. Compounds exhibiting this property are said to be polymorphic, and elements allotropic. About one-half of the known elements are allotropic, and the number of allotropic forms of an element can be greater than two. Three types of allotropy are usually distinguished: monotropy, enantiotropy, and dynamic allotropy. Consider a solid element which can exist as two allotropes. At a particular temperature the difference in free energy between these two forms is given by $\Delta G = \Delta H - T\Delta S$ (Section 6.13). They can exist in equilibrium, for a given pressure, only at one particular temperature, T, when $\Delta G = 0$

Table 10.1. A summary of the differences in structure and bonding between metals, non-metals, and semi-metals.

	Non-metals	Semi-metals			Metals
Structure	molecular crystals	lattice structures			close-packed or body-centred cubic
		3-dimensional	2-dimensional	1-dimensional	
	discrete molecules	giant molecules e.g. diamond	infinite layers of molecules e.g. graphite	infinite chains of molecules e.g. selenium	
Bonding: a) intramolecular	covalent	covalent	covalent within layers	covalent within the chain	metallic
b) intermolecular	van der Waals	—	van der Waals between layers and chains		
Co-ordination number	very small, 1 or 2		small, 4 or less		large, 8 or 12

(condition for equilibrium, Section 7.5), that is when $\Delta H = T\Delta S$. Below this temperature one form only is thermodynamically stable, the low temperature form L (say); above this temperature the other form is stable, the high temperature form H (say). If the element is crystallized below the temperature T, the low temperature form L is obtained; above temperature T, H is obtained; at temperature T a mixture of the two forms H and L is obtained. The high temperature form H can be obtained from the low temperature form L by heating L above the temperature T; the process can be reversed by cooling below temperature T. This reversible relationship phenomenon is known as **enantiotropy**: each allotrope is stable over a limited temperature range only. The temperature T at which one allotrope reversibly changes into another is known as the transition temperature. Examples of enantiotropes are: rhombic and monoclinic sulphur ($T = 95.6\ °C$), and grey and white tin ($T = 13\ °C$).

Not all pairs of allotropes exhibit this type of relationship: red phosphorus cannot be changed to white phosphorus in the solid state; nor can graphite be changed to diamond in the solid state. In such cases one allotrope, for example white phosphorus, is thermodynamically unstable (the so-called metastable form) relative to the other form (the stable form) at all temperatures below the melting point; ΔG for the reaction, white phosphorus (s) $\rightarrow$ red phosphorus (s), is negative at all temperatures below the melting point. In other words, red phosphorus is the stable allotrope at all temperatures below the melting point. This phenomenon is called **monotropy**. The transition temperature for such a pair can be regarded as being higher than the melting point of the solids. The high-temperature or metastable form is usually obtained by rapid cooling from a high temperature. Although white phosphorus is thermodynamically unstable relative to red phosphorus, the change does not occur automatically. White phosphorus is kinetically stable (Section 8.9) or metastable. The reason is not difficult to see. In order to change form, quite strong phosphorus–phosphorus bonds must first be broken (Section 17.6) and the energy required to promote this change, the activation energy

(Section 8.3), is high and is not available at moderate temperatures. In **dynamic allotropy** the allotropes are capable of co-existence in dynamic equilibrium with one another over a wide range of physical conditions, the proportion of each in the mixture depending upon these conditions, especially the temperature. In this case the difference between allotropes is in the number of atoms in a molecule e.g. $I_2 \rightleftharpoons 2I$.

The structural change in passing along a period from metal to non-metal, for example from lithium to fluorine, is clearly periodic (Table A.14). Also there is some correlation between position in the Periodic Table and the structures adopted by the metals: metals in the same group tend to have the same structure. It is not possible at the present time to relate this closely to electron configuration. For example how can groups VA and IA be related? Nor is it known: (a) why metals usually crystallize in only two close-packed structures out of the very large number possible (Section 3.3); nor (b) why a given metal crystallizes with the cubic close-packed structure rather than the hexagonal close-packed structure, or vice versa. Clearly our close-packed, hard sphere model of metallic structure is insufficient at this level of sophistication. One structural change which does appear reasonable is the fairly common change from a close-packed to a body-centred cubic structure at high temperatures. It is generally true that, provided the nature of the bonding does not change, a structure of higher co-ordination number has a lower energy (is more stable) than a related form with lower co-ordination number, since there are fewer bonding interactions in the latter case. Consequently, in a given metal, the body-centred cubic structure (co-ordination number eight) is the high energy form and will be the stable form at high temperatures, whereas the close-packed forms (co-ordination number twelve) are the low energy forms and one or both of them will be the stable form(s) at lower temperatures. This expectation of enantiotropy is fulfilled in many cases, though iron is exceptional in showing the opposite behaviour. Presumably all metals which have body-centred cubic structures at room temperature can be converted to a close-packed structure at a lower temperature, like lithium and sodium. Since metallic transformations from one structure to another require only comparatively slight movement of layers of atoms relative to one another, they occur rapidly at the transition temperature (low activation energy). However, as discussed above, during the transformations of the allotropes of non-metals, where the change involves the rupture of strong covalent bonds, the changes are very slow except at high temperatures (high activation energy) and the thermodynamically unstable allotrope (e.g. diamond) is kinetically stable or metastable.

10.3.2 Atomic Radii (*see* Section 3.3)

These are very important since other properties can be related to them. The radius of an atom can be defined as the distance of closest approach to another atom in a given bonding situation. At this distance the internuclear repulsions and the interelectronic repulsions just balance the nuclear–electron attractions. Thus the sizes of atoms are not determined directly by the size of the nucleus, which is very small, but by the effective volume of the outer electrons. Although a few electron densities in compounds have been measured accurately (Sections 3.1 and 4.1) for the most part the only data available at this time are internuclear distances, frequently called "interatomic" distances. The radii of atoms have to be allocated from these

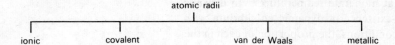

Fig. 10.5 A classification of atomic radii.

data. This is simple for covalent radii, but difficult for ionic radii, which have to be allocated somewhat arbitrarily (Section 3.3). The power of the concept, however, lies in the fact that *an individual bond distance is approximately* (*usually within 10 percent*) *the sum of the individual radii*, and this is as true for ionic as for covalent radii (though somewhat less true for van der Waals radii). In fact, if a particular bond distance deviates from the "expected" value (i.e. the sum of the atomic radii), then it is usually worth investigating, and the bonding is probably abnormal in some way; for example, if covalent, it may be a multiple bond. The additivity principle fails with compounds for which there are alternative methods of writing down the structure, such as, NO_3^-, CO_3^-, and C_6H_6 (Section 4.12.1).

The allocation of ionic radii is a problem and various empirical methods exist. If in the past it had been possible to measure electron densities as accurately as interatomic distances, then allocations based on electron densities would have been postulated. Such measurements are now possible but are unlikely to make a vast contribution to the concept of atomic size, because of the approximate nature of the model of ionic (and other atomic) sizes. The value of the concept lies in the almost additive nature of atomic sizes. But this is not sufficiently correct to make really accurate measurements of radii worth while, since the size of an atom (best regarded as the amount of space from which the atom excludes other atoms) depends upon many factors, all of which are themselves related to the total electron configuration. These include: the type of bonding; the multiplicity of the bond if covalent; the oxidation number of the atom; special interelectronic repulsions (e.g. fluorine); co-ordination number; and steric factors. Of these the most important factor determining the size of an atom with a given oxidation number is the type of bond. The other factors (apart from multiple bonding in the case of covalent bonds) are so relatively unimportant that the type of bond provides a classification of atomic size within which size is assumed to be constant. Thus the four main types of atomic radii (Section 3.3) coincide with the four major types of bonding: covalent, ionic, metallic, and van der Waals (Fig. 10.5). Briefly, what they tell us approximately is:

1. Ionic radius—the space occupied by an ion in any direction in the lattice of an ionic solid.
2. Covalent radius—the space occupied by an atom in a covalently bonded compound in the direction of the covalent bond.
3. Metallic radius—the space occupied in any direction by an atom of a metal in the metal lattice.
4. Van der Waals radius—the space occupied by an atom in a solid compound in any direction towards another atom with which it is not chemically bonded (that is, the force of attraction is due only to van der Waals forces). In this case the concept is not confined solely to the noble gases (the only elements forming solids with formally non-bonded atoms), but is extended to the non-

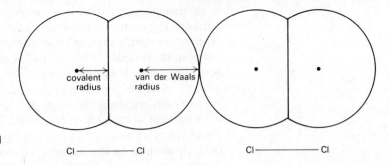

Fig. 10.6 Solid chlorine: covalent and van der Waals radii.

bonded interactions between molecules (Fig. 10.6). Its importance is that it governs steric effects in chemistry.

The approximate nature of the concept should be obvious from the classification in terms of bond type, which is itself an approximate classification. However, the approximately additive nature of the concept of atomic size, together with its direct relationship to electron configuration, makes it very useful in chemistry. Like all concepts it can be misused. For example, different types of radii are often included in the same table, inviting bogus comparisons between elements. In any odd-looking comparison it is first necessary to find out what radii are being compared. Often in tables of covalent radii the van der Waals radii of the noble gases are included, making it look as if there is an increase in radius from fluorine to neon!

Two general periodic trends are found for all four types of atomic radius: the size decreases along a period and increases down a group, in the long form of the Periodic Table (Fig. 10.1). Along any horizontal period of the table there is a decrease in atomic size as the outer electron shell is being filled. For example, in any period the covalent radius of the alkali metal is the largest and that of the halogen is the smallest (Table A.13). This general decrease in size is due to the effective nuclear charge (the charge actually felt by the outer electrons, Section 2.4) becoming greater along the period. From one element to another going along the period from left to right, the extra attraction of the nucleus for the outer electrons (owing to the increasing nuclear charge) is not neutralized exactly by the extra valence electron in the outer shell, since electrons in the same shell do not screen each other well from the nucleus (Section 2.4.3). Therefore all the outer electrons are pulled in towards the nucleus, and the radius decreases. The rate of decrease of size becomes smaller as the atoms become heavier: the change from indium to iodine is less than that from boron to fluorine.

The decrease along a period is particularly small when the difference in electron configuration from one element to the next is that of an additional inner electron. This added inner electron screens the size-determining outer electrons from the nucleus much better than an additional outer electron, and therefore the decrease in size is small from one element to another. Along the transition series for example (Table A.13), there is a relatively small decrease in atomic radius from one element to the next; but because there are ten such decreases, the decrease in atomic radius from calcium to gallium is much greater than that from magnesium to aluminium, where there is no intervening transition series. In other words the third row representative elements, gallium to krypton, are not as large as would be expected by simple extrapolation from the group elements of the first two rows, owing to the insertion of the transition elements—the so-called **transition contraction**. For example in descending the boron group from boron to aluminium to gallium, instead of the expected rise in covalent radius from aluminium to gallium, the two radii are about the same. It is important to emphasize that the screening of an outer $(n + 1)s$-electron by an nd- (or "inner") electron is more efficient than the screening of an $(n + 1)s$-electron by another $(n + 1)s$-electron (or that of an $(n + 1)p$-electron by another $(n + 1)p$-electron) in the same shell. This is why the decrease in atomic radius from sodium to chlorine is greater than that from scandium to copper. But this is not the same as saying that the so-called inner shell of electrons, nd^{10}, is a good electron shield for outer electrons in general. It is a good shield for $(n + 1)s$-electrons compared to the shielding effect of electrons for other electrons in the same shell, for example $(n + 1)s$-electrons for other $(n + 1)s$-electrons. But the nd^{10}-subshell is a poor shield for $(n + 1)$-electrons compared to the shielding effect of the ns^2- or np^6-subshells (Section 2.4.3). This is in fact the usual comparison made in chemistry (for example in the comparison of the chemistries of the group IA and IB metals) and therefore the d^{10}-subshell is usually called a poor electron shield. The question of whether or not the nd^{10}-subshell is a poor or good shield for outer $(n + 1)$-electrons is similar to the question "is ten a large or a small number?". Of course it depends on what it is being compared with. To the young child "ten" is a large number because he is used to dealing with very small numbers; to the statistician in his work it is a very small number. Thus in chemistry, where the usual comparison is between the shielding of the nd^{10} and the ns^2 (or np^6) subshells for the electrons in the $(n + 1)$-shell, since the ns^2 (or np^6) electrons form a much better shield, the nd^{10}-subshell is generally described as a poor electron shield.

Another way of looking at the transition contraction uses the model of effective nuclear charge (above). Owing to the insertion of the transition elements, the nuclear charge between the elements calcium and gallium has increased by ten units. Although there are ten additional orbital electrons, these do not screen the outer s- and p-electrons completely from the increased nuclear charge, and therefore the effective nuclear charge of the elements gallium to krypton is greater than if the transition elements were not present in this row. This greater than expected effective nuclear charge (greater than expected for gallium to krypton by simple extrapolation from the elements of the first two rows) leads to a group of related "anomalous" properties of the third row p-block elements, of which smaller than expected size is one example. Others are high electronegativities, ionization energies, reduction potentials, and so on. Collectively these anomalous properties have been labelled the **middle row anomaly**. The word anomalous in this and other contexts merely means irregular and does not imply unexpected or inexplicable. The anomalies are largely understood. Towards the right hand side of the middle row (bromine) the effect is less marked, or as it is said, "tends to die away".

The rate of decrease in size along the lanthanide series is even less than in the transition series, since electronic differences between the lanthanides are in the numbers of electrons in the antepenultimate (last but two) shell of electrons, and these shield the size-determining outer electrons from the nucleus even more effectively. However, there are 14 of these small decreases, and this affects markedly

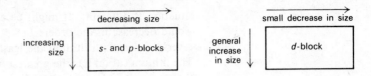

Fig. 10.7 Periodic changes in atomic radii.

the atomic sizes and therefore all the chemistry of the following elements—the third transition series—which are all much smaller than expected (Table A.13). Alternatively, as above, this can be regarded as due to the effect of the greater than expected effective nuclear charge of the third row transition series, hafnium to gold, owing to the insertion of the lanthanides. The small decrease of size along the lanthanides is referred to as the **lanthanide contraction**. The anomalous properties of the third row transition series (which are due to the presence of the lanthanides), for example smaller than expected size, can be referred to collectively as the **third row transition anomaly**. In this context the most striking result is that, unexpectedly, the atomic sizes and chemistry of the third row transition series closely resemble those of the second row series (Table A.13).

The second periodic regularity in size occurs within a given group, for example group IA. On descending any group of the Periodic Table, the number of outer electrons remains constant but the number of shells increases monotonically. Despite the increased nuclear charge this causes a large increase in size on descending the group, except in the case of the transition series. The rate of increase decreases as the atomic number increases. This is mainly due to the presence of the transition and lanthanide elements. For example, the change from boron to aluminium is much greater than that from indium to thallium. This means that the later elements in each group are more similar to one another than the earlier members are to each other. This is particularly true in groups IVA and VA: zirconium and hafnium, with similar electron configuration and size, and therefore similar chemical properties, were particularly difficult to separate. This similarity of the later elements in the groups lessens in traversing the Periodic Table from left to right, since owing to the absence of a lanthanide group in the fourth (rubidium) period, these elements decrease in size more rapidly than the fifth period elements.

The main periodic relationships discussed above are generally applicable to all atomic radii and are summarized in Fig. 10.7. One further general point is that the greater the oxidation number of the element, the smaller the atomic radius. Thus the ionic radius of iron(III) is smaller than that of iron(II), and the covalent radius of bromine in bromine trichloride is less than that in bromine monochloride.

1) Ionic radii. There are two types of ion: positively charged (cations) and negatively charged (anions). All common cations are smaller than all common anions (except that the rubidium and caesium cations are larger than the oxide and fluoride anions). This is not too surprising, since not only is there a loss of a partially filled outer shell on cation formation, but also there is an overall positive charge on the ion. Conversely, the addition of an electron to an atom increases its size (compare anion radii with corresponding covalent radii, Table A.13). In general there is a decrease in size going from anions, to covalent radii of corresponding atoms, to cations. It must be remembered that ionic radii are crystal radii. In solution the situation is different. It might be expected that the electrical mobilities in water would decrease in the order: $Li^+ > Na^+ > K^+ > Rb^+ > Cs^+$. In fact it is the reverse order, with lithium the least mobile owing to solvation effects (Fig. 9.1). The lithium cation has the greatest charge density (charge divided by atomic size), therefore attracts more polar solvent molecules around it than does the sodium ion, and is effectively larger and slower.

2) Covalent radii. In the case of covalent radii, values for the radii of hydrogen and fluorine are not taken as one-half the radius of the respective molecules; instead arbitrary values which give good agreement with other covalent molecules are accepted. This is at least in agreement with the unique bonding in the hydrogen molecule, and the unusual bonding in the fluorine molecule. Multiple covalent bonds are shorter than single bonds. In fact the bond order can often be deduced from the bond length.

3) Metallic radii. The metallic radii are larger than the corresponding covalent radii, although both involve a sharing of electrons. This is because the average bond order of an individual metal–metal bond is considerably less than one (Section 4.7), and therefore the individual bond is weaker and longer than a covalent bond. This does not mean that the overall bonding is weak, as there are a large number of these bonds, eight or twelve per metal atom. The metallic radii of the transition elements show a characteristic minimum in the iron group. In this type of bonding where electrons are shared between nuclei, the interatomic forces, and therefore the effective radii, depend not only on the effective nuclear charge, but also on the number of half-filled orbitals. This is a maximum at about group VIIA or VIII. Below this number of electrons there are too few electrons for optimum metal–metal bonding; above this, there are too many electrons in filled orbitals (cf. the noble gases).

10.3.3 Ionization Energy (see Section 2.4.1)

Ionization is the process of producing positive ions from neutral atoms or molecules by removing electrons. The ionization energy, I, of a gaseous atom (or ion) is defined as the molar internal energy change, ΔU, at 0 K which accompanies the particular electron-removing ionization process. The ionization energies of carbon are as follows:

		ΔU_0 (kJ mol^{-1})
first ionization energy:	$C(g) \rightarrow C^+(g) + e^-$	1086
second ionization energy:	$C^+(g) \rightarrow C^{2+}(g) + e^-$	2352
third ionization energy:	$C^{2+}(g) \rightarrow C^{3+}(g) + e^-$	4619, etc.

Clearly the second ionization energy of carbon is the same as the first ionization energy of the carbon cation $C^+(g)$. The ionization energy of the reaction $C(g) \rightarrow C^{2+}(g) + 2e^-$ is the sum of the first two ionization energies. Ionization energy is a measure of how firmly the nucleus of a gaseous atom or ion holds on to its outer electrons.

There is clearly a periodic trend in the first ionization energies of the elements (Fig. 2.17). To clarify this further the student should draw a long form of the Periodic Table containing the first ionization energies with data taken from Table

A.10. As usual in chemistry there are two possible approaches to the problem, which in this case is the rationalization of the first ionization energy data: calculation from theory (wave mechanics) which is impossible at the moment except for the simplest atoms; or qualitative discussion in terms of appropriate atomic parameters (concepts) such as effective nuclear charge, which are in turn related to electron configuration. The latter is a useful approach, and the trends in ionization energies can be rationalized in terms of the position of the elements in the Periodic Table and their electron configuration. According to a simple model (Eq. 2.13) the ionization energy of an outer electron is given by

$$I = \frac{Z_{\text{eff}} e^2}{2} \left(\frac{1}{r}\right)_{\text{av}},$$

where e is the charge on the electron, $(1/r)_{\text{av}}$ is the average value of $1/r$, r being the distance of the electron from the nucleus, and Z_{eff} is the effective nuclear charge (Section 2.4.3). The latter parameter is not the same as the nuclear charge, but is the charge that a particular outer electron actually feels (Section 10.3.2). This cannot be the same as the nuclear charge, owing to the screening effect of the inner electrons. Nor can it be obtained simply by subtracting the number of inner electrons from the nuclear charge, since this assumes perfect shielding of the nucleus by inner electrons. This is not the case, since outer electrons penetrate the inner core of electrons to a certain extent (Section 2.4.3).

The effective nuclear charge depends upon the nuclear charge, the net charge on the ion, the number and type of the inner screening electrons, and the type of outer electron being considered. Thus, in attempting to explain in detail how ionization energies change, there are a number of factors to consider. The distance of the outer electron from the nucleus is mainly determined by its principal quantum number n, which increases monotonically in descending any group in the Periodic Table. As n increases, the average distance r increases sharply and the ionization energy decreases. This is seen in the groups of the s- and p-block elements, where there is a general decrease in ionization energy in descending any group. This is not the case in the transition element groups, however, where irregular changes are found, but in which the general group trend is opposite to that in the representative elements. Clearly, more important factors are involved here. Certainly the very high ionization energies of the third row transition elements (third row anomaly) are consistent with the relatively small sizes of these atoms. This is due to the insertion of the lanthanides which causes the third row transition elements to have a greater than expected effective nuclear charge (lanthanide contraction). A similar effect is expected, and found, after the insertion of the transition elements in the third row of the table. Gallium in fact does not have a smaller first ionization energy than aluminium (middle row anomaly) owing to its greater than expected effective nuclear charge. Again there is the same correlation with atomic size (Section 10.3.2).

The concept of effective nuclear charge is vital to an understanding of ionization energies, and therefore to the chemistry of the elements in general. The factors influencing effective nuclear charge are:

1. The net charge on an atom or ion. For a given element the ionization energies increase in the order 1st < 2nd < 3rd < 4th, etc., since the pull of the nucleus for the outer electrons increases as the overall positive charge on the ion

increases, or as the number of (screening) electrons in the same valence shell decreases. The successive ionization energies for magnesium are given in Fig. 2.18, where they are used as evidence for shells of electrons. In addition to the general increase in successive ionization energies, there is an additional large increase whenever a new shell is broken into by removal of electrons. This is due to the poor shielding power that electrons in the same subshell have for one another (Section 2.4.3) resulting in strong attraction by the nucleus for each of them. This limits the possible oxidation states of elements in the same group in the same way, and results in *elements in the same group having similar stable oxidation states and similar chemistries.* For example, the prohibitively high third ionization energy of the beryllium group elements (Table A.10) limits the possible oxidation states of these elements to +1 or +2. It must be recognized that all gaseous atoms hold on to their valence electrons very firmly; even the most electropositive elements (those which form positive ions most readily), those in group IA, require as much energy as will break the strongest covalent bonds to cause them to ionize in the gas phase (about 400 kJ mol^{-1}) (Section 14.2.4).

2. The shielding effect of inner electrons (Section 2.4.3). All other things being equal, shells of smaller principal quantum number are more effective shields than shells of higher quantum number. Within a given shell the shielding efficiency of inner electrons decreases in the order $s > p > d > f$. The chemistries of the lithium and copper groups differ markedly from each other despite their similar outer electron configuration because the underlying nd^{10}-subshell of the copper group elements is a poor shield (compared to the underlying np^6-subshell of the lithium group) and is easily penetrated by the outer s-electron. This electron is therefore more firmly held by the nucleus than is that of the group IA elements which have the less easily penetrated underlying p^6-subshell. The result is that the group IA elements are much more electropositive (lose their outer electron more easily to form cations) than the group IB elements. The most obvious result is that the group IA elements have much lower first ionization energies than the group IB elements.

3. Penetrating power of the valence electron. In a given shell n, the penetrating power of the electrons decreases in the order $s > p > d > f$; therefore the ns-electrons are more firmly held by the nucleus than the np-electrons, and so on (Section 2.4.3). In other words, in a given element the ionization energies increase in the order $ns > np > nd$, etc. (Fig. 2.18).

Across any period, for example the lithium row (Fig. 10.8), there is a general increase from left to right in the first ionization energy, corresponding to the related decrease in atomic radius (Section 10.3.2). Both effects are due to the fact that in changing from element to element across the period electrons are being successively added to the same shell while the nuclear charge is increasing. Such electrons shield each other poorly from the increasing nuclear charge (Section 2.4) and the result is an increase in effective nuclear charge along the series. This causes a decrease in radius and an increase in ionization energy along the series. Boron and oxygen are anomalous, however, as can be seen in Fig. 10.8; clearly another factor is important in these cases. As usual chemists seek an explanation in terms of electron configuration. The electron configurations of the boron and oxygen atoms are $1s^2 2s^2 2p^1$ and $1s^2 2s^2 2p^4$ respectively. The outer electron is being taken

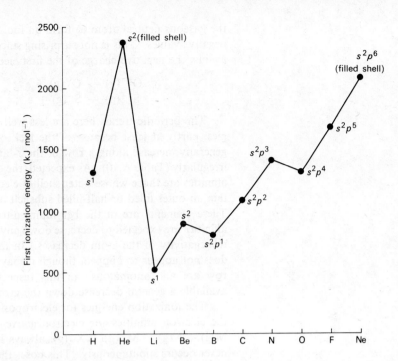

Fig. 10.8 Variation of the first ionization energy of the first row elements.

away (ionized) from an underlying filled ($2s^2$) or half-filled ($2p^3$) subshell. This is an aspect of the concept known as the *stability of the half-filled or filled shell* (Section 10.3.20). Both of these types of configuration are relatively good shields, and this causes the first ionization energies of boron and oxygen to be lower than expected. If this theory is correct (i.e. useful) then a drop in the first ionization energy is also to be expected between the group IIB and group IIIB elements. Is this found? The same general, but irregular, increase in ionization energy is found along the transition series.

10.3.4 Electron Affinity (see Table A.10)

The electron affinity, E, of a gaseous atom, A, is defined as the internal energy change per mole, at 0 K, which accompanies the process: $A^-(g) \rightarrow A(g) + e^-(g)$. For example, for

$$F^-(g) \rightarrow F(g) + e^-(g), \quad E = +333 \text{ kJ mol}^{-1}.$$

The electron affinity of fluorine is $+333$ kJ mol^{-1}. Clearly the electron affinity is the amount of energy required to remove an electron from a gaseous negative ion, in other words it is the ionization energy of the particular anion. It is called the electron affinity because it is a measure of the attraction or affinity of the atom for the extra electron. A positive electron affinity means energy is required to remove an electron from the gaseous anion. A negative electron affinity means that the gaseous anion spontaneously dissociates to the gaseous atom and the gaseous electron. Most elements appear to have positive electron affinities, but a few are known which have negative values; in these cases the electron must be forced onto the gaseous neutral atom to form an ion. All second electron affinities are large negative values. This is not surprising since the second electron must be forced on against the negative charge of the first electron. For example:

$$O^{2-}(g) \rightarrow O^-(g) + e^-; \quad E = -710 \text{ kJ mol}^{-1}$$
$$S^{2-}(g) \rightarrow S^-(g) + e^-; \quad E = -565 \text{ kJ mol}^{-1}.$$

The periodic trends here are less well defined than those for ionization energies, partly at least because of the lack of reliable data. The electron affinities generally increase along a row of the s- and p-block elements, though somewhat irregularly (Table A.10). As expected, the elements with small or negative electron affinities are those whose outer shell of electrons is a good shield, that is they contain an outer filled or half-filled subshell of electrons (s^2, p^3, or p^6 configuration). These elements are in the beryllium, nitrogen, and neon groups. The electron affinities are expected to decrease down any group of the Periodic Table as the electronegativity of the atom decreases. In fact, from the limited data available this does not appear to happen, though it may just be that the elements in the lithium row are all "anomalous" in this respect, and that when more data become available a general decrease down the groups will be seen as in group VIIB.

The ionization energies for electropositive atoms (M) are always greater than the electron affinities for electronegative elements (X). Therefore the process, $M(g) + X(g) \rightarrow M^+(g) + X^-(g)$, always has a positive free energy change, and never occurs spontaneously. This poses the question of why ionic compounds are stable (Section 14.2.4). It will be seen that this stability is largely due to the lattice energy. In the vapour phase the alkali halides dissociate at high temperatures not to the ions but to the atoms.

10.3.5 Electronegativity (see Section 4.2 and Table A.13)

Electronegativity, A/R, describes the tendency of an atom to become negatively charged in its covalent compounds. Like electron affinity and ionization energy, therefore, it describes the attraction of an element for electrons in a particular situation; but, unlike these, it is not a directly observable physical quantity. Several quantitative scales of electronegativity exist: Pauling's, based on bond energies; Mulliken's, based on ionization energies and electron affinities; and Allred and Rochow's, based on effective nuclear charges. All give similar values for the electronegativities of the elements when put on the same scale. This is massive verification of the general utility of the concept. It is very important since it describes with a single number the general chemical behaviour of an atom in a manner that (a) is self-consistent and (b) correlates well with chemists' general qualitative knowledge. It must be understood, however, that the values refer to an atom in an "average" environment; clearly the electron-attracting power of an element depends upon its oxidation state and its environment. Despite this, the concept is widely useful as a measure of electron-attracting power, and as a means of predicting bond type, dipole moments, and bond energies.

The electronegativities of elements increase sharply across a row of the s- and p-block elements (for example from lithium to fluorine), as expected, since the effective nuclear charge of these elements is increasing rapidly. This is because electrons are being added to the same electron shell, and are shielding each other badly from the extra nuclear charge. Along a transition row however, the increase in electronegativity is much slower because the additional electron is being added to an inner shell, which provides relatively good (but still not perfect!) shielding of the outer electron from the nucleus.

In the descent of a group of the representative elements (for example the lithium group), there is a general decrease in electronegativity which is relatively small except between the first two elements. The much greater electronegativity of the first row elements in any group correlates well with their anomalous behaviour (Section 14.2.7). As expected, the third row elements from gallium onwards have greater electronegativities than would be expected by extrapolation from the first two elements in the group (middle row anomaly). This is due to the insertion of the transition elements, which causes the effective nuclear charge of the third row elements from gallium onwards to be greater than if the transition elements were not there. Similarly the presence of the lanthanides is responsible for the electronegativities of the third row transition elements being greater than would be expected by extrapolation from the first two rows, in which the trend appears to be the same as in the s- and p-block elements.

10.3.6 Electropositivity

The quantitative measure of how well an atom attracts electrons in different situations is given jointly by electron affinity, ionization energy, and electronegativity. The word electropositivity is similarly used, but in a qualitative way, to describe how relatively easily atoms tend to give up electrons. Weakly electronegative elements such as the alkali metals are said to be strongly electropositive; more electronegative metals such as lead are said to be moderately electropositive. Even the most electropositive of metals, however, (in the gas phase) only give up their electrons to form gaseous ions reluctantly, as for example,

$$Cs(g) \rightarrow Cs^+(g) + e^-; \quad \Delta H = 420 \text{ kJ mol}^{-1},$$

and the term is used in a relative, not an absolute, sense. (See Section 12.2.)

10.3.7 Lattice (or Crystal) Energies (see Section 4.13.1)

The lattice energy, ΔH_{latt}, of a crystal, is strictly the change in internal energy when one mole of the solid is converted to one mole of the separated ions in the gas phase, at 0 K. This is not very different from the corresponding enthalpy change at 298 K and can be used in Born–Haber cycles (Section 4.13.1) without much error (within 5 kJ mol^{-1}). For example,

$$Na^+Cl^-(s) \rightarrow Na^+(g) + Cl^-(g); \quad \Delta H_{latt} = 787 \text{ kJ mol}^{-1}.$$

The lattice energy depends upon the charge, radius, and electron configuration of the constituent ions. These factors determine both the type of bonding (see Fajans's rules, Section 4.3), and the magnitude of the lattice energy if the bonding is mainly ionic. The lattice energy depends upon the reciprocal of the distance between the ions, $1/(r_+ + r_-)$. For any given positive ion, therefore, the lattice energy decreases as the size of the negative ion increases. The same is true for any given negative ion: the lattice energy decreases as the size of the positive ion increases (Table 10.2). Moreover, for large positive ions the size of the lattice energy is determined mainly by the size of this positive ion and is relatively in-

Table 10.2. Some lattice energies, ΔH_{latt} (kJ mol^{-1}).

(a) Univalent metal halides (MX).

	Li	Na	K	Rb	Cs	Cu	Ag	Au	Tl
F	1037^S	918^S	817^S	784^S	729^S	—	960^S	—	828
Cl	862^S	788^S	718^S	694^S	672^C	979^B	912^S	1042	736^C
Br	785^S	719^S	656^S	634^S	603^C	971^B	904^S	1042	724^C
I	729^S	670^S	615^S	596^S	568^C	958^B	891^B	1050	703

Structures: ^{C} caesium chloride; ^{S} sodium chloride; ^{B} zinc blende.

(b) Divalent metal halides (MX$_2$).

	Mg	Ca	Sr	Ba	Ti	V	Cr	Mn	Fe	Co	Ni	Cu	Zn	Cd	Hg	Pb
F	2908^R	2611^F	2460^F	2368^F	2749	2810	2879	2770^R	2912^R	2962^R	3046^R	3042^R	2971^R	2770^F	2740^F	2490^F
Cl	2528		2117^F													
Br	2406^A							2439^A	2540^A	2611^A						
I	2293^A	2058^A		2293^A				2356^A	2464^A	2531^A				2410^A		2134^A

Structures: ^{R} rutile; ^{F} fluorite; ^{A} cadmium iodide.

(c) Divalent metal chalconides (MX)

	Mg	Ca	Ti	V	Cr	Mn	Fe	Co	Ni	Cu	Zn	Cd	Hg	Pb
O	3580	3465^S	3882^S	3917^S		3813^S	3923^S	3992^S	4076^S		4035^W	3783^S		
S						3353^S					3617^Z 3604^W	3400^W	3571^Z	3098^S
Se						3303^S					3611^Z	3382^Z	3640^Z	

Structures: ^{S} sodium chloride; ^{W} wurtzite; ^{Z} zinc sulphide (sphalerite).

sensitive to changes in the anion. Compare, for example, the change in lattice energy from fluoride to iodide in the case of lithium, with the same change in the case of rubidium (Table 10.2). (In comparisons of this type it is always necessary to compare compounds of similar structures.) The change down the lithium series is much greater than that down the (larger) rubidium series.

Lattice energy is very dependent upon the charge on the ions; thus, magnesium oxide has a much larger lattice energy than lithium fluoride, despite the similar radii of the corresponding ions. The electron configuration is also an important factor. Ions with a d^{10} configuration, as in copper(I), zinc(II), silver(I), cadmium(II), gold(I), and mercury(II), have larger lattice energies than those predicted for them using an ionic model. This is due to their effective nuclear charge being greater than expected owing to the insertion of the transition elements giving them the poor shielding configuration, d^{10}. Except for zinc, and to a lesser extent cadmium, this causes their bonding to be largely covalent and increases the lattice energy, so that most of their compounds are sparingly soluble in water (Section 10.3.17).

Along the first row transition series for a typical salt, such as the difluoride MF$_2$, there is a general increase in lattice energy (Table 10.2), as would be expected from the decrease in ionic radius. However, the increase is not regular, and there is a small minimum, within the general increase, at manganese. This is associated with the relatively good shielding configuration, d^5 (half-filled shell), which reduces the effective nuclear charge and decreases the lattice energy.

10.3.8 Density (*see* Table A.14)

(Density is defined as mass per unit volume, for example g cm^{-3}.) For gases and liquids the density usually quoted is that of the liquid at its boiling point. For solids the density is a function of: the atomic weight, the volume of the separate atoms (which is not the same as the atomic volume, Section 10.3.9), and the structure of the solid which determines the closeness of the packing. Despite the fact that all three factors may vary from element to element, there are two obvious trends:

1. The density increases on descending any group, transition or representative. Since the elemental structures are often the same within any group, this means that the increase in atomic weight from element to element in passing down a group (which tends to increase the density) is more important in this respect than the increase in atomic volume (which tends to decrease the density), despite the fact that the atomic volume depends upon the third power of the radius. This underlines the surprisingly small increase in radius which occurs in descending a group. The increases in density within the d-block groups are greater than those within the s- and p-block groups. There are often surprising exceptions to this general trend among the representative elements, potassium for example. Can you find any others?

2. Along any period there is a general increase in density (while the elements are solids) related to the increasing atomic weight and decreasing radius. A maximum is reached somewhere in group VIII among the transition elements where the metallic radius is a minimum owing to the strength of bonding reaching a maximum value (Section 10.3.2). All the d-block elements have high densities owing to strong intermetallic bonding (many valence electrons), and the third row elements osmium, iridium, and platinum have the highest (twice as dense as lead), since they have high atomic weights, close-packed structures, and (less importantly) comparatively small metallic radii.

   The relatively high densities of the post-transition elements in group IIB and the other long period elements (gallium, etc.) are due to the insertion of the transition elements, causing a large increase in atomic weight and (less importantly) a small decrease in atomic radius.

10.3.9 Atomic Volume

This is the volume occupied by one mole of the element and is obtained by dividing the atomic weight by the density of the solid, or by the density of the liquid at the boiling point if the element is liquid or gaseous. Thus the atomic volumes of nitrogen and other gases appear large. It is a function of atomic radius and structure. Comparisons are difficult for various reasons:

1. The existence of allotropes of differing densities.
2. Densities in the solid state are not known for all elements, and liquid densities are often quoted, inviting misleading comparisons.
3. Not all densities are measured at the same temperature.
4. The atomic volume is very dependent on structure. For example, the covalent radius of silicon is greater than that of phosphorus, as expected, but the atomic volume is much less than that of phosphorus. This is a structural effect. Silicon is a tight covalently bonded giant molecule, whereas phosphorus contains P_4 molecules loosely bonded together. In other words, whereas it is the covalent radius of silicon that matters in determining its atomic volume, it is mainly the van der Waals radius that determines the atomic volume of phosphorus.

Despite these qualifications, periodic trends in atomic volumes are apparent if comparisons are confined to elements of similar structure. Discontinuities in these trends are almost invariably due to changes in structure from one element to the next. While the bonding remains metallic the trends are:

1. On descending any group within which the structures are generally the same, the atomic volume increases owing to the increase in atomic radius, except in the third row transition series.
2. Along any period there is a general decrease in size, superimposed upon which there is a minimum in group VIII, corresponding to the minimum metallic radius (Section 10.3.2).

10.3.10 Melting Point

Periodicity is evident here, but is a little less well defined than in some of the other properties. This is partly because of uncertain data, since there is difficulty both in obtaining very pure specimens and in measuring the melting point at high temperatures, and partly because, as usual, there is more than one factor involved in determining the melting point. The melting point is defined as the temperature at which pure solid is in equilibrium with pure liquid at atmospheric pressure. This temperature depends upon the size of the forces holding the particles together as a solid, and the extent to which these must be broken down to form a liquid. In a *metal*, for example, the bonding may be strong, but since similar forces are present in the liquid state the melting point may not be very high. In *giant molecules* such as silicon, however, although the covalent bonding may not be very strong, yet because many of the bonds have to be broken before melting can occur, a high temperature is required. In *molecular crystals*, where the intermolecular bonding is weak (van der Waals forces) as in iodine, relatively low melting points are observed. Melting points, therefore, depend both upon the type of structure and upon the strength of bonding. However, since both are periodic functions of electron configuration or atomic number, the melting points show quite well-marked periodicity. But, since structural changes are abrupt, not gradual, in going along a period, so are changes in melting point, as for example in going from the giant molecule carbon to the molecular crystal nitrogen.

In any given group where the structures of the elements are the same or similar, a gradual change will be observed which depends upon the relative strength of bonding. In the alkali metals, group IA, the atoms are held in close-packed structures by weak metallic bonding (since there is only one bonding electron per atom) which becomes weaker as the atomic size increases; therefore the melting point decreases down the group (note, other metallic groups show different behaviour). By comparison, the halogens are molecular crystals in the solid state, held together by weak van der Waals forces (Section 4.9.). These forces increase as size increases, and therefore the melting points increase down the group. Groups IIIB, IVB, VB, and VIB are more complicated because of the great structural changes which occur on descending the groups. But where similar structures are compared, the melting points depend upon the relative strength of bonding. This is also clearly seen in going along a period comparing elements with the same (metallic) structure. To a first approximation the strength of metallic bonding depends upon the number of valence electrons, and therefore increases from, for example, sodium to magnesium to aluminium. This is reflected in the consequent increase in the temperature of the melting points. The transition elements all have similar metallic structures and all have relatively high melting points due to very strong metallic bonding (all have more than two bonding electrons per atom, Section 20.1). Descending a transition group the strength of metallic bonding tends to increase, and therefore the melting points increase (cf. alkali metals). Along any row of the transition series a maximum is reached in the middle of the row, corresponding roughly to where the bonding is strongest. Manganese has an abnormally low melting point; this element often shows "anomalous" properties, associated with its half-filled d-electron shell.

The elements with the highest melting points are: carbon and silicon, which have giant covalent structures most of which must be broken down in order to melt the element; the transition elements, which have very strong metallic bonds; and beryllium and boron, which have giant structures with small atoms containing more than one valence electron (leading to strong bonding).

10.3.11 Boiling Point

The periodic trends are similar to those of melting point. When comparisons of elements with similar structures are possible, the relative boiling points depend upon the relative strength of bonding. In the case of metals the process of boiling requires almost complete breaking of the bonds, as there is very little association of atoms in the vapour phase. Since much metallic bonding exists in the liquid phase, and must be broken down for vaporization, high temperatures are necessary and the boiling points of metals are much higher than their melting points. The boiling of elements existing as giant molecules, carbon for example, also requires almost complete bond breaking. In this case, however, most of these are already broken at the melting point and so the boiling point, though high, is not much higher than the melting point. The boiling of molecular non-metals, which requires the complete breaking of very weak van der Waals forces, is easy and occurs at low temperatures. The boiling point is only slightly above the melting point. The elements of lowest boiling point and melting point are therefore molecular (or atomic) solids of low molecular (atomic) weight where van der Waals forces are minimized, such as hydrogen and helium. The element with the greatest liquid temperature range will probably be a metal with strong bonds, such as gallium.

10.3.12 Thermal and Electrical Conductivity (*see* Section 4.7)

It is known that heat is transferred more rapidly by "free" electrons than by atomic or molecular vibrations. The delocalized electrons present in metals conduct heat rapidly, and metals are therefore good thermal conductors (Table A.14). In contrast, non-metals have their high energy valence electrons localized in covalent bonds and are consequently poor thermal conductors (thermal insulators). Since "free" electrons are also responsible for conducting electricity, a good correlation between thermal and electrical conductivity is expected, and is found (Table A.14). In fact, a classification of metals and non-metals based on thermal conductivity is identical to that based on electrical conductivity (Section 10.2). The changes in conductivities as the atomic number increases are not entirely gradual owing to abrupt changes in bonding and structure which usually occur in or near the dividing band of semi-metals. Within the class of metals there is a wide variation in conductivity. The group IB metals have much larger conductivities than any other metals (Cu > Ag > Au), followed by aluminium, beryllium, and magnesium. In general, for *s*- and *p*-block elements *in which the bonding is metallic* the electrical and thermal conductivities (a) decrease as a given group is descended, and (b) increase along a given row or period, as the number of valence electrons increases. The same general trends are found in the transition series.

When the bonding changes from metallic to covalent there is a sharp drop in conductivity, as between aluminium and silicon, since there are no free electrons in covalent compounds. At high temperatures, however, the covalent bonds can be broken to provide "free" electrons for conduction, and therefore the conductivity of non-metals increases with temperature. An alternative method of breaking some bonds and causing conduction, is to subject the non-metal to γ-radiation. Valence electrons are liberated, and a pulse of current is produced if a potential difference exists between the ends of the "insulator" (photo-ionization). This is used to detect γ-rays in the diamond crystal counter. Another way to make an insulator conducting is to incorporate impurities into the lattice. For example, if a phosphorus atom is present in a silicon lattice, then it has a spare electron of high energy available for conduction. Elements which are semi-metals show similar behaviour. These are covalently bonded solids in which the bonding is weaker than in the case of non-metals, and can therefore be broken down more easily to provide "free" electrons, especially if impure. In contrast, impurities reduce the conductivity of metals, as does increasing the temperature. In fact, at sufficiently low temperatures where the atoms vibrate very little, metals have practically no resistance and are "superconducting".

10.3.13 Enthalpy of Fusion (Latent Heat of Fusion) (*see* Section 5.2 and Table A.14)

Enthalpy (or latent heat) of fusion, ΔH_{fus}, represents the difference in enthalpies between one mole of an element in the liquid phase and one mole in the solid phase, both phases being at the temperature and pressure at which the phase change takes place. In other words, it specifies the amount of bond energy that must be supplied to melt one mole of the solid at the melting point. Obviously it is expected that there will be some relationship between the enthalpy of fusion and the melting point, and in fact two roughly linear relationships are found, one for metals and one for non-metals (Fig. 10.9). For many non-metals the ratio given by dividing the enthalpy of fusion by the melting point is approximately 25 J mol^{-1} K^{-1}. The same ratio for most metals is 10 J mol^{-1} K^{-1}. Since $\Delta G = \Delta H - T\Delta S$ (Section 6.13), and at equilibrium $\Delta G = 0$ (Section 7.5), then at the melting point $\Delta H/T = \Delta S$. In other words ΔH_{fus}/m.p. measures the entropy of fusion. It is not surprising that this ratio is roughly similar for most metals, since melting involves a similar change in structure and a similar increase in the freedom of movement of the atoms in changing from the solid to the liquid state. That is, for most metals there is a similar change in entropy of fusion, ΔS_{fus}. We would not expect non-metals to show a similar pattern because of the wider variation in elemental structure, and in fact agreement with the rule, ΔH_{fus}/m.p. = constant, is much less for non-metals than for metals.

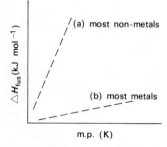

Fig. 10.9 The relationship between enthalpy of fusion and melting point.

Large changes in the enthalpy of fusion occur from one element to another when a change of elemental structure occurs, as in the changes from aluminium to silicon, silicon to phosphorus, gallium to germanium, and tin to lead. As usual, these dramatic changes occur on or near the diagonal dividing line between metals and non-metals. These large differences in enthalpies are not due to differences in total bond energies. They are due to the facts that: (a) in metals much metallic bonding remains in the liquid state and therefore the enthalpy of fusion is relatively low (relatively little bond breaking), and the entropy of fusion is relatively small (relatively little structural change); and (b) in the covalent giant structures of elements such as carbon, nearly all the bonds have to be broken down in order to melt the solid and therefore the enthalpy of fusion is relatively large, as is also the entropy of fusion. The enthalpies of fusion of covalent molecular crystals such as H_2, N_2, and O_2, are very small because fusion of such crystals does not involve the breaking of strong covalent bonds, but only the disruption of the much weaker van der Waals forces. Their entropies of fusion are moderately high (about 15 J mol^{-1} K^{-1}) since the structural change is greater than that for metals.

The periodic trends are similar to those observed for the melting points of the elements. In particular, although the transition elements are metals, and the bonding in the solid and liquid phases is similar, yet the enthalpies of fusion are large because of the very strong metallic bonding (many valence electrons). The value for carbon is unknown but is probably very high.

10.3.14 Enthalpy of Vaporization (Latent Heat of Vaporization) (see Section 5.2 and Table A.14)

Enthalpy (or latent heat) of vaporization, ΔH_{vap}, represents the difference in enthalpies between one mole of an element in the liquid phase and one mole in the gas phase, both being at the temperature and pressure at which the phase change takes place. In other words, it specifies the amount of bond energy that must be supplied to boil one mole of liquid at the boiling point. Usually the enthalpy of vaporization is greater than that of fusion since melting generally destroys only a small fraction of the bonds in the solid, whereas vaporization generally destroys (almost) all of these bonds. Just as the enthalpy of fusion is related to the melting point, so the enthalpy of vaporization is related to the boiling point. In fact there is again almost a linear relationship, this time for all elements (previously known as Trouton's rule):

$$\frac{\Delta H_{vap} \text{ (kJ mol}^{-1}\text{)}}{\text{b.p. (K)}} = 90 \text{ J K}^{-1} \text{ mol}^{-1} \quad \text{(entropy of vaporization)}.$$

10.3.15 Enthalpy of Atomization (see Section 4.11.2 and Table A.13)

The enthalpy of atomization, ΔH_{atom}, represents the difference in enthalpies between one mole (of atoms) of an element in the gaseous state, and one mole (of atoms) of the element in the standard state, both being at the temperature and pressure at which the change takes place. Where the standard state is a solid, and the gas obtained on vaporization is monatomic, this is the same as the **enthalpy of sublimation**, ΔH_{sub}. In the case of the halogens which consist of molecules, X_2, the enthalpy of atomization is one-half the **enthalpy of dissociation**, ΔH_{diss}, which is defined relative to one mole of molecules. The enthalpy of atomization specifies the total forces holding the element together in its standard state relative to dissociated elemental atoms. The periodicity is more obvious here than in the cases of enthalpies of fusion and vaporization, since the question of how many bonds must be broken to form a melt is not involved. Along a given row of the *s*- and *p*-block elements the enthalpy of atomization reaches a maximum in group IVB where the number of possible bonding electrons reaches a maximum (four). In the earlier groups the number of bonds per atom is restricted by the number of valence shell electrons, and in the groups beyond IVB by the number of unpaired electrons. The change in bonding from metallic to covalent along such a period is largely irrelevant here, since, although metallic bonding is delocalized, the total strength of the bonding is also limited by the number of valence electrons available, just as in the case of covalent compounds. Thus in changing from sodium to magnesium to aluminium the change in total bond energy is very like that in changing from one to two to three covalent bonds. A similar trend, for the same reason, is found along a row of the *d*-block elements, in which the enthalpy of atomization reaches a rough maximum in the middle (maximum number of unpaired electrons), except that manganese is exceptionally small (the usual exception). The group trend (that is, the change as a given group is descended) in the *s*- and *p*-block elements is towards weaker bonds and lower enthalpies of atomization as the atomic size increases (fluorine is exceptional, Section 19.6). The opposite trend is found in the transition metals, due in part to the lanthanide contraction.

10.3.16 Enthalpy of Hydration (*see* Sections 4.13.2 and 9.1.1, and Table A.12)

The electrolytic conductivity of electrolytes in aqueous solution is due, not to the transfer of electrons as in metal conduction (Section 10.3.12), but to the passage of mobile positive and negative hydrated ions. When, for example, solid sodium chloride is dissolved in water, the lattice of sodium and chloride ions is broken down, and hydrated ions are formed (Section 4.13.2). This process is called aquation, hydration, or (more generally) solvation:

$$Na^+Cl^-(s) \rightarrow Na^+(aq) + Cl^-(aq).$$

Clearly there must be strong forces of attraction between the ions and the solvent in order to break down the stable ionic lattice. In other words, hydration energies must be of the same order of magnitude as lattice energies.

The solution of sodium chloride is best considered by imagining that the process occurs in two steps and constructing a Born–Haber cycle (Section 4.13.1). This is not the mechanism of the reaction, but is energetically equivalent to it, and provides insight into the process. Assume that one mole of sodium chloride is sublimed to form *free gaseous ions*, and these are then plunged into water (Fig. 10.10a). The enthalpy change of the first step, may be taken as the lattice energy (ΔH_{latt}). Actually this is not quite true since the lattice energy is strictly defined as an internal energy change, not an enthalpy change, but the difference is too small to worry about in this calculation. It does raise the point, however, that in a Born–Haber cycle all the energy terms must be the same, for example all enthalpies or all free energies. The enthalpy change of the second step is "minus the enthalpy of hydration", $-\Delta H_{hyd}$. This represents the difference in enthalpy between one mole of the compound as separated ions in the gaseous state, and one mole of

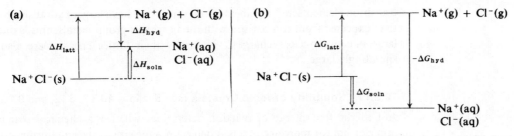

Fig. 10.10 Born–Haber cycles for the dissolution of sodium chloride: (a) enthalpy cycle, and (b) free energy cycle.

the compound as hydrated ions in aqueous solution. This cannot be measured, but is obtained from the equation (see Section 4.13.2):

$$\Delta H_{\text{soln}} = \Delta H_{\text{latt}} - \Delta H_{\text{hyd}}.$$

The heat of solution, ΔH_{soln}, is the enthalpy change when one mole of substance, for example sodium chloride, dissolves in water; it can be measured directly. The lattice energy can be found independently. Some values for these two quantities are given in Table 10.4. The most striking feature is the similarity of the values and the trends. This is due to the fact that both depend upon the same atomic properties: charge, radius, and the nature of the electron core of the ions (Section 10.3.2). In general the difference between them, the enthalpy of solution, is very small and this complicates solubility relations (Section 10.3.17).

The enthalpy of hydration for the whole "molecule" can be regarded as the sum of the constituent parts: the sum of the hydration enthalpies of one mole of sodium ions and one mole of chloride ions. These cannot be measured directly however, and we have the same problem as in the assignment of ionic radii, that of how to assign energies to individual ions. One such attempt takes the value of zero for the hydration enthalpy, ΔH_{hyd}, of the hydrogen ion; a different convention, the one used in this book, assigns the value of 1075 kJ mol^{-1} for the hydration enthalpy of the hydrogen ion, H^+, based on various indirect pieces of evidence. All the other enthalpies can then be calculated from the (total) hydration enthalpies of the compounds, and these are quoted in Table A.12. The hydration enthalpies of compounds are obtained by addition of the hydration enthalpies of the constituent ions. The most striking feature is that enthalpies of hydration are large because the ion–solvent bonds are strong and there are many of them.

There are several obvious trends. On a simple electrostatic view of the bonding (Section 4.1) the enthalpy of hydration of an ion is expected to depend on three factors, the charge on the ion, the radius of the ion (or, taken together, the charge density of the ion), and the nature of the electronic core; in other words, on the effective nuclear charge. This is largely substantiated by the data available. On descending any group, for example the lithium group, the hydration energy decreases as the size of the ion increases. Ions of the same charge have similar hydration enthalpies, which are much greater than those of lesser charge. In traversing the transition series, for ions of the same charge, there is a general slight increase in hydration enthalpy, correlating with the slight decrease in atomic radius. The hydrogen ion has a large hydration enthalpy due to its small size. The transition elements and members of the zinc group have larger hydration enthalpies than expected from comparison with the beryllium group metals; this is due to the large effective nuclear charges. Naturally their lattice energies are also correspondingly large.

10.3.17 Solubility of Ionic Crystals (*see* Sections 4.13.2, 6.12, and 9.1.1)

Just as the free energy of reaction determines whether a chemical reaction can occur or not (or more accurately it determines the size of the equilibrium constant) so the free energy of solution determines how soluble a compound is. It is related to the equilibrium constant in the usual way:

$$\Delta G^{\ominus}_{\text{soln}} = -RT \ln K_{\text{sp}}$$

or

$$\Delta G^{\ominus}_{\text{soln}} = -5.7 \log K_{\text{sp}} \text{ (kJ mol}^{-1}),$$

where $\Delta G^{\ominus}_{\text{soln}}$ is the standard free energy of solution; and K_{sp} is the solubility product, from which (owing to the non-ideal behaviour of solutions) only a rough value of the solubility of the salt can usually be obtained. Clearly a change of 5.7 kJ mol^{-1} in the free energy causes a change by the factor 10 in the solubility product. Thus, owing to the log relationship between free energy and solubility product, the relative solubility of two compounds is very sensitive to small differences in the free energy of solution between them. For example, the difference in the free energy of solution between potassium nitrate, conventionally regarded as a "soluble" salt, and potassium perchlorate, conventionally regarded as an "insoluble" salt, is only about 12.5 kJ mol^{-1}. This is not a hopeful start to a semi-quantitative view of solubility since even small errors in calculating $\Delta G^{\ominus}_{\text{soln}}$ values, owing to poor data or necessary approximations, may have large effects on calculated solubilities.

There is worse to come. In chemical reactions it is often possible to ignore changes in entropy and to obtain revealing information about the reaction in terms of bond strengths etc. by considering enthalpy changes only. But whereas the entropies of solid salts are small and vary little, those of hydrated ions are larger and vary greatly, and differences between them cannot be ignored. Solution is a complex process, and the calculation of entropy changes in terms of atomic and molecular motion is not possible at the present time. But two important contributions to it can be recognized: firstly, the ions in the crystal gain freedom of motion in going into solution; and, secondly, the water molecules lose freedom of motion, since they become associated with the ions. These two factors make opposing contributions to the entropy change. Entropies of solvation cannot be calculated, but they can be measured. As the charge density of an ion increases, say from sodium to aluminium (Fig. 10.11), the entropy of solution, ΔS_{soln}, (the difference in entropy between the ions in the lattice and in solution) decreases rapidly. This is because the greater the charge density of the ion, the more water molecules it attracts around it, and the less freedom of movement the water molecules have; and therefore the smaller the entropy change is in going from the solid to the liquid phase. Thus the entropy of solution associated with sodium ions is greater than that associated with aluminium ions, and in comparing the relative solubilities of sodium and aluminium salts, the differences in their entropies of solution,

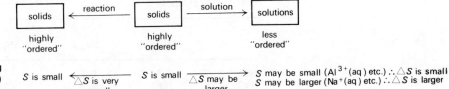

Fig. 10.11 Entropy changes during the processes of (a) reaction, and (b) solution (for qualification see text).

one large—the other smaller, cannot be ignored. *Free energy changes must generally be considered in considering solubility changes from compound to compound.*

There is yet another difficulty. Consider again the Born–Haber cycle for the dissolution of sodium chloride (Fig. 10.10); this time we are interested in the free energy cycle ($\Delta G^\ominus$) in Fig. 10.10(b) instead of the enthalpy cycle in Fig. 10.10(a). The standard free energy of solution, $\Delta G^\ominus_{soln}$, is determined by the free energy of sublimation, $\Delta G^\ominus_{sub}$, and the free energy of hydration, $\Delta G^\ominus_{hyd}$. These are both large numbers of about equal magnitude, the difference of which gives a small number, $\Delta G^\ominus_{soln}$:

$$\Delta G^\ominus_{soln} = \Delta G^\ominus_{sub} - \Delta G^\ominus_{hyd}.$$

This is an example of a typical dilemma of the chemist: the only way to obtain the relatively small numerical value of a certain parameter is frequently by subtraction of two large numbers, at least one of which is difficult to obtain accurately. This has been compared to obtaining the weight of the captain of the liner *Queen Elizabeth II* by weighing the liner first with him on board and then with him on shore, and subtracting the two weights. Even small percentage errors in the estimation of $\Delta G^\ominus_{sub}$ and $\Delta G^\ominus_{hyd}$ can lead to large percentage errors in $\Delta G^\ominus_{soln}$. $\Delta G^\ominus_{sub}$ here refers to the process of converting crystalline solid to *free gaseous ions*.

Entropies of solution can be decisive, therefore, in determining solubility. Most of the alkali metal halides, for example, have small positive enthalpies of solution, $\Delta H^\ominus_{soln}$, yet have negative free energies of solution and are freely soluble. This is due to the fact that their entropies of solution, $\Delta S^\ominus_{soln}$, are relatively large positive numbers on account of their small charge densities, small attraction for water, and relatively small restriction of the freedom of water molecules in solution (Fig. 10.11). Despite the fact that $\Delta H^\ominus_{soln}$ is positive, this makes the free energy of solution, $\Delta G^\ominus_{soln}$, negative at room temperature, and most alkali metal salts are soluble:

$$\Delta G^\ominus = \Delta H^\ominus - T\Delta S^\ominus.$$

Similarly, large singly charged anions tend to form soluble salts with simple cations, as in the cases of perchlorate and nitrate, whereas doubly and triply charged anions and cations tend to form less soluble salts, as for example phosphates, carbonates, sulphates, and the triply charged lanthanides.

Despite the qualifications discussed above, useful comparisons of solution enthalpy changes can be made and related to atomic parameters such as size, in carefully selected series of compounds. In the alkali metal halides, for example, the variations in entropy changes are small within the series, and can be ignored. Thus enthalpy changes can be discussed *as if* they alone determined solubility changes.

10.3 THE PHYSICAL PROPERTIES OF THE ELEMENTS

Table 10.3. Enthalpies of solution (ΔH_{soln}, kJ mol^{-1}) and solubilities of some chlorides.

	ΔH_{latt}	ΔH_{hyd}	ΔH_{soln}	Solubility (mol l^{-1})
LiCl	862	883	−21	19
NaCl	788	778	+10	6
KCl	718	695	+23	5
AgCl	912	845	+67	1.3×10^{-5}

In this way solubility can be considered, within a series of closely related compounds such as the alkali metal halides, to be the result of a competition between the enthalpy of converting the solid to free ions (the lattice energy) and the sum of the hydration energies of the gaseous ions. Notice that although both the lattice energy and the enthalpy of hydration may vary smoothly with increasing radius in a series of related compounds, the difference of these, the enthalpy of solution, need not do so. Table 10.3 gives figures for some chlorides.

The trend of decreasing solubility of the chlorides of lithium, sodium, and potassium, correlates with the increasingly positive enthalpies of solution, which are due to the enthalpy of hydration decreasing more rapidly than the enthalpy of sublimation as the size of the cation increases. Silver chloride does not fit into the series at all, and other factors are important, particularly the very large lattice energy which is due to a high proportion of covalent bonding, itself a result of the non-noble gas configuration (Section 10.3.7). Thus we can compare the enthalpies of the alkali halides, and say with much truth that lithium chloride is more soluble than sodium chloride because, although its lattice energy is higher than that of the sodium salt, tending to make it less soluble, its enthalpy of hydration is even larger, tending to make it more soluble. However, this cannot be done when comparing series of unrelated salts, particularly those containing different bond types or charges.

Since the enthalpy of solution (and therefore the solubility of related compounds) can be regarded in terms of a competition between the enthalpies of sublimation and hydration, we are interested in factors that affect ΔH_{latt} and ΔH_{hyd} unequally. The ionic radii of the ions is clearly important. Whereas in the case of lattice energy it is the distance apart of the ions $(r_+ + r_-)$ that is important (the sum of the ionic radii), in the case of the enthalpy of hydration the ions in solution contribute independently of the size of each other. Several deductions can be made on this basis, taking as examples the alkali metal halides. Consider the series of salts LiI, NaI, KI, RbI, and CsI, that is, changing the alkali metal from lithium to caesium in the case of a large anion such as iodide (Table 10.4). Descending the series, the size of the cation changes sharply and this causes a fairly sharp decrease in the enthalpy of hydration. The lattice energy, which depends on the sum of the ionic radii $(r_+ + r_-)$, does not change so rapidly since it is determined mainly by the size of the large anion. Therefore the trend in the enthalpies of solution, and therefore in the solubilities of a series of related metal salts with large anions, is determined by the trend in the enthalpy of hydration of the cations. This decreases as the radius increases: as group IA is descended, the solubilities of the

Table 10.4. Enthalpies of solution (ΔH_{soln}, kJ mol^{-1}) and solubilities of the alkali metal fluorides and iodides.

	LiF	NaF	KF	RbF	CsF	LiI	NaI	KI	RbI	CsI
ΔH_{latt}	1037	918	817	784	729	729	670	615	596	568
ΔH_{hyd}	958	853	770	744	711	803	698	615	589	556
ΔH_{soln}	+79	+65	+47	+40	+18	−74	−28	0	+7	+12
Solubility (g/100 g)	0.27[18]	4.22[18]	92[18]	130[18]	367[18]	165[20]	184[25]	127[0]	152[17]	44[0]

Superscript on solubility gives temperature (°C) at which solubility was measured.

iodides etc. decrease (Fig. 10.12). Consider the case of small anions, for example the series LiF, NaF, KF, RbF, CsF (Table 10.4). In this case the lattice energy is very sensitive to changes in the size of the cation and it is the lattice energy which changes most as the radius of the cation increases from compound to compound. Thus, as the size of the cation increases, the lattice energy decreases more quickly than the enthalpy of solution and the compounds become more soluble. Anions behaving like the iodide include perchlorates and bromides; those behaving like the fluorides include nitrites and hydroxides; chlorides and nitrates often show intermediate behaviour.

Similar considerations determine the trends in solubilities when the metal is kept constant and the anions are changed. In the series caesium fluoride to caesium iodide, the solubilities decrease since the most important factor is the change in enthalpies of hydration of the halide ions. In the series lithium fluoride to lithium iodide the solubilities increase since the lattice energies decrease dramatically down the series. Clearly intermediate situations arise in which the enthalpies of solution are small and vary irregularly, for example the sodium salts of the halides, and the chlorides of the alkali metals. The reasoning above implies that whenever there is a mis-match in size in ionic compounds, the salt is expected to be fairly soluble. This is because whenever there is a large ion and a small counter-ion the lattice energy will be relatively small since it depends on the sum of the ionic radii, and the hydration enthalpy will still be relatively large since the small counter-ion contributes independently to this. There are many examples: lithium and sodium perchlorate are very soluble in water, and are even soluble in solvents of low dielectric constant such as acetone; similarly lithium iodide is fairly soluble in ether. Obviously, this can also be associated with the covalent character of such compounds. If a given ion has to be dissolved in a given solvent, this can often be effected by choice of a suitable counter-ion. For example, very large organic groups tend to make ionic compounds more soluble in organic solvents since they both reduce the lattice energy and increase the solvation enthalpy.

Qualitatively much of the solubility data summarized in Table 10.5 can be understood in terms of the discussion above. The general solubility of the alkali metal ions and that of large singly charged anions such as nitrate and perchlorate has been explained on the basis of entropy changes. The smaller solution entropies of doubly charged ions, however, tend to make their ionic compounds insoluble (both lattice energies and hydration enthalpies increase by about the same amount

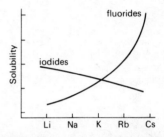

Fig. 10.12 The solubilities of the alkali metal fluorides and iodides.

Table 10.5. Solubility of common compounds in water.

Common anions	Common cations	Solubility of corresponding compounds
all	alkali metal ions: Li^+, Na^+, K^+, Rb^+, Cs^+	soluble
all	hydrogen ion: H^+	soluble
all	ammonium ion: NH_4^+	soluble
nitrate, NO_3^-	all	soluble
acetate, CH_3COO^-	all	soluble
chloride, Cl^- ⎫ bromide, Br^- ⎬ iodide, I^- ⎭	Ag^+, Pb^{2+}, Hg_2^{2+}, Cu^+ all others	low solubility soluble
sulphate, SO_4^{2-}	Ba^{2+}, Sr^{2+}, Pb^{2+} all others	low solubility soluble
sulphide, S^{2-}	alkali metal ions (Li^+, Na^+, K^+, Rb^+, Cs^+), H^+, NH_4^+, Be^{2+}, Mg^{2+}, Ca^{2+}, Sr^{2+}, Ba^{2+} all others	soluble low solubility
hydroxide, OH^-	alkali metal ions (Li^+, Na^+, K^+, Rb^+, Cs^+), H^+, NH_4^+, Sr^{2+}, Ba^{2+} all others	soluble low solubility
phosphate, PO_4^{3-} ⎫ carbonate, CO_3^{2-} ⎬ sulphite, SO_3^{2-} ⎭	alkali metal ions (Li^+, Na^+, K^+, Rb^+, Cs^+), H^+, NH_4^+ all others	soluble low solubility

A substance is said to be *soluble* if its saturated solution is greater than 0.1M at 20 °C.

for such compounds). Thus the phosphates and carbonates tend to be insoluble, except for those of the alkali metals. Sulphates are larger than carbonates and are therefore more soluble (smaller charge density). If both the anion and cation have high charge densities (a match in size) then the compounds tend to be insoluble; for example, phosphates of the lanthanides, and fluorides and hydroxides of lithium and triply charged cations.

10.3.18 Electrode Potential or Reduction Potential (*see* Section 7.8 and Table 10.6)

The standard electrode potential, $E^\ominus$, is a quantitative measure of the tendency of reactants in their standard states to form products in their standard states for a particular reaction. It is a convenient way of presenting equilibrium and free energy data, to which it is related by the equations, $\Delta G^\ominus = -zFE^\ominus = -RT \ln K$ (Section 7.5). Electrode potentials are presented in tables, not for complete reactions, but as half-reactions relative to the hydrogen electrode, in order to present the information as efficiently as possible. The half-cell reaction, which is always written as a reduction,

$$Ca^{2+}(aq) + 2e^- \rightleftharpoons Ca(s); \quad E^\ominus = -2.87 \text{ V},$$

is shorthand for the reaction,

$$Ca^{2+}(aq) + H_2(g) \rightleftharpoons Ca(s) + 2H^+(aq); \quad E^\ominus = -2.87 \text{ V},$$

or

$$\Delta G^\ominus = -2 \times (-2.87)F \text{ kJ},$$

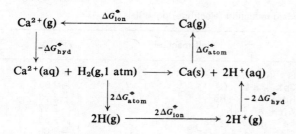

Fig. 10.13 A Born–Haber cycle representing the reduction potential of calcium.

all substances being in their standard states. The value of $E^\ominus$ gives the potential of the corresponding cell. The negative sign and the large numerical value (three is a very large number on this particular scale) indicate that there is a very strong tendency for the reaction to proceed from right to left in this case, or alternatively, that the equilibrium constant K is very small for the reaction as written. The free energy change for the reaction as written is given by

$$\Delta G^\ominus = -zFE^\ominus = 2 \times 96.5 \times (-2.87) = 554 \text{ kJ mol}^{-1};$$

and from $\Delta G^\ominus = -5.7 \log K \text{ kJ mol}^{-1}$, we obtain $K \approx 10^{-100}$, showing that the reduction of calcium ions by hydrogen under these conditions (standard states in aqueous solution) hardly proceeds at all.

The factors that influence the reduction potentials of metal ions can be investigated by considering the formation of the hydrated metal ion as the result of three processes in a Born–Haber cycle: atomization of the bulk metal; ionization of the gaseous metal atoms; and solvation of the gaseous metal ion. This of course is not the mechanism of the electrode process, but is its thermodynamic equivalent, since the initial and final states are the same in each case. The complete breakdown of the reaction is shown in Fig. 10.13.

The free energy change for the reaction, $\Delta G^\ominus$, is given by:

$$\Delta G^\ominus = 2[\Delta G^\ominus_{\text{atom}} + \Delta G^\ominus_{\text{ion}} - \Delta G^\ominus_{\text{hyd}}]_{\text{hydrogen}} - [\Delta G^\ominus_{\text{atom}} + \Delta G^\ominus_{\text{ion}} - \Delta G^\ominus_{\text{hyd}}]_{\text{calcium}};$$

and the standard electrode potential is given by:

$$E^\ominus = -\frac{\Delta G^\ominus}{zF} = \frac{1}{zF}[(\Delta G^\ominus_{\text{atom}} + \Delta G^\ominus_{\text{ion}} - \Delta G^\ominus_{\text{hyd}})]_{\text{calcium}} - \frac{1}{F}[(\Delta G^\ominus_{\text{atom}} + \Delta G^\ominus_{\text{ion}} - \Delta G^\ominus_{\text{hyd}})_{\text{hydrogen}}].$$

Since the contribution of the hydrogen is the same for all electrode potential systems, this can be ignored in qualitative discussion, though of course actually to calculate an electrode potential from free energy data it must be included. It works out that for hydrogen

$$\frac{1}{F}(\Delta G^\ominus_{\text{atom}} + \Delta G^\ominus_{\text{ion}} - \Delta G^\ominus_{\text{hyd}})$$

is equal to $+4.44$ volts, therefore

$$E^\ominus = -4.44 + \frac{1}{zF}(\Delta G^\ominus_{\text{atom}} + \Delta G^\ominus_{\text{ion}} - \Delta G^\ominus_{\text{hyd}})_{\text{calcium}}.$$

Table 10.6. Standard reduction (electrode) potentials.

	Half-reaction	$E^{\ominus}$(V)	
very weak oxidizing agents	$Li^+ + e^- \rightarrow Li$	−3.04	very strong reducing agents
	$Cs^+ + e^- \rightarrow Cs$	−2.95	
	$Rb^+ + e^- \rightarrow Rb$	−2.93	
	$K^+ + e^- \rightarrow K$	−2.92	
	$Ba^{2+} + 2e^- \rightarrow Ba$	−2.90	
	$Sr^{2+} + 2e^- \rightarrow Sr$	−2.89	
	$Ca^{2+} + 2e^- \rightarrow Ca$	−2.87	
	$Na^+ + e^- \rightarrow Na$	−2.71	
	$Mg^{2+} + 2e^- \rightarrow Mg$	−2.37	
	$Al^{3+} + 3e^- \rightarrow Al$	−1.66	
	$Mn^{2+} + 2e^- \rightarrow Mn$	−1.18	
	$2H_2O + 2e^- \rightarrow H_2(g) + 2OH^-$	−0.83	
	$Zn^{2+} + 2e^- \rightarrow Zn$	−0.76	
	$Cr^{3+} + 3e^- \rightarrow Cr$	−0.74	
	$Fe^{2+} + 2e^- \rightarrow Fe$	−0.44	
	$2H^+(10^{-7}M) + 2e^- \rightarrow H_2(g)$	−0.414*	
	$Cr^{3+} + e^- \rightarrow Cr^{2+}$	−0.41	
	$Co^{2+} + 2e^- \rightarrow Co$	−0.28	
	$Ni^{2+} + 2e^- \rightarrow Ni$	−0.25	
	$Sn^{2+} + 2e^- \rightarrow Sn$	−0.14	
	$Pb^{2+} + 2e^- \rightarrow Pb$	−0.13	
	$2H^+ + 2e^- \rightarrow H_2$	0.00	
	$Sn^{4+} + 2e^- \rightarrow Sn^{2+}$	+0.15	
	$Cu^{2+} + e^- \rightarrow Cu^+$	+0.15	
	$Cu^{2+} + 2e^- \rightarrow Cu$	+0.34	
	$Cu^+ + e^- \rightarrow Cu$	+0.52	
	$I_2 + 2e^- \rightarrow 2I^-$	+0.54	
	$O_2(g) + 2H^+ + 2e^- \rightarrow H_2O_2$	+0.68	
	$Fe^{3+} + e^- \rightarrow Fe^{2+}$	+0.77	
	$NO_3^- + 2H^+ + e^- \rightarrow NO_2(g) + H_2O$	+0.78	
	$Ag^+ + e^- \rightarrow Ag$	+0.80	
	$\frac{1}{2}O_2(g) + 2H^+(10^{-7}M) + 2e^- \rightarrow H_2O$	+0.815*	
	$NO_3^- + 4H^+ + 3e^- \rightarrow NO + H_2O$	+0.96	
	$Br_2 + 2e^- \rightarrow 2Br^-$	+1.07	
	$\frac{1}{2}O_2(g) + 2H^+ + 2e^- \rightarrow H_2O$	+1.23	
	$4H^+ + MnO_2 + 2e^- \rightarrow Mn^{2+} + 2H_2O$	+1.28	
	$Cr_2O_7^{2-} + 14H^+ + 6e^- \rightarrow 2Cr^{3+} + 7H_2O$	+1.33	
	$Cl_2 + 2e^- \rightarrow 2Cl^-$	+1.36	
	$Au^{3+} + 3e^- \rightarrow Au$	+1.50	
very strong oxidizing agents	$MnO_4^- + 8H^+ + 5e^- \rightarrow Mn^{2+} + 4H_2O$	+1.52	very weak reducing agents
	$H_2O_2 + 2H^+ + 2e^- \rightarrow 2H_2O$	+1.77	
	$F_2 + 2e^- \rightarrow 2F^-$	+2.87	

(left side: oxidizing strength decreasing ↓; right side: reducing strength decreasing ↓)

For simplicity, state symbols have been omitted from all these equations.
*These are not standard electrode potentials but are included for convenience.

We must emphasize that the value of +4.44 V is not a thermodynamic value. Single ion quantities cannot be obtained by thermodynamic arguments. It has been obtained on the basis of certain assumptions.

One further simplification which can be made in qualitative discussion is to ignore the differences between free energies and enthalpies. This is usually justifiable here because the entropy changes during these processes are relatively small compared to the enthalpy changes for these reactions. The final conclusion, therefore, is that electrode potentials depend mainly upon the relative sizes of the enthalpies of atomization, ionization, and hydration, from one metal to another. For large negative electrode potentials, the enthalpies of atomization and ionization (which are positive) must be small, and the enthalpy of hydration (which occurs with a negative sign) must be large. This model proves to be useful.

The standard electrode potentials are reproduced in Table 10.6. The metals with the highest negative electrode potentials are the alkali metals. This is due mainly to their relatively low enthalpies of atomization (Section 10.3.15) and low ionization energies, which cause the electrode potentials to be large and negative despite their low hydration energies (Section 10.3.16). However, within group IA the order is irregular, and unexpected from the viewpoint of their electronegativity. This is not totally surprising since, although electronegativity and electrode potential measure the broad affinity of an atom for electrons, they do so in two completely different ways, pertaining to two completely different situations. Broad agreement between the concepts would be expected, but not detailed agreement. It is an example of a typical difficulty in chemistry that the electrode potential depends on three factors whose sum need not vary smoothly from element to element, even if the factors themselves vary fairly smoothly, since their rate of change differs (Problem 25). In any comparison all three factors must be taken into account since no individual factor can explain all the facts. Lithium has the most negative reduction potential because its hydration energy is so much larger than that of any of the other alkali metals, owing to its very small size. The group IIA metals appear high in the series because, although their ionization energies are higher than those of the alkali metals, their enthalpies of atomization are fairly small and their hydration enthalpies (high charge density) are very large. In fact calcium, strontium, and barium have larger negative electrode potentials than sodium, which on nearly all other criteria would be defined as being more electropositive than these group IIA metals. The exceptionally high ionization energy of beryllium makes its electrode potential less negative than those of the other group IIA metals, despite its high hydration energy. Compared to the group IIA elements, the transition elements have fairly large ionization energies and very large atomization enthalpies which, despite their higher hydration energies, reduce their electrode potentials below those of the alkaline earth metals. Zinc, cadmium, and mercury have low enthalpies of atomization, and fairly large hydration energies, but the ionization energies are very large and this ensures fairly low electrode potentials.

In general the periodic trends in reduction potentials found for metals are: a sharp increase across the rows of the *s*- and *p*-blocks; and a smaller increase along the transition series. The behaviour down any group tends to be irregular.

A similar Born–Haber cycle can be written down for the anions, though it is generally less useful since there are fewer of them (Fig. 10.14). The hydrogen part

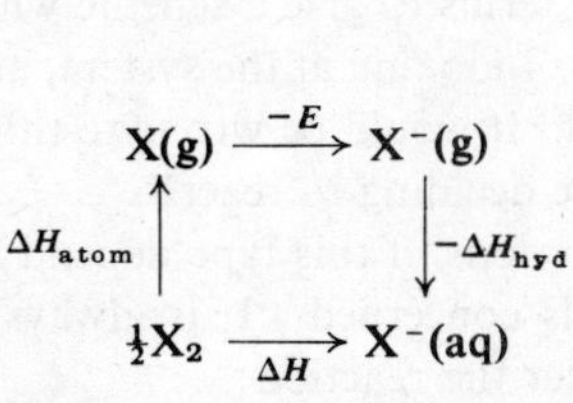

		F	Cl	Br	I
$\frac{1}{2}X_2 \to X(g)$	ΔH_{atom}	+79	+122	+111	+106
$e^-(g) + X(g) \to X^-(g)$	$-E$	−333	−348	−340	−297
$X^-(g) \to X^-(aq)$	$-\Delta H_{hyd}$	−460	−385	−351	−305
$e^-(g) + \frac{1}{2}X_2 \to X^-(aq)$	ΔH	−714	−611	−580	−496

Fig. 10.14 A Born–Haber cycle for the reduction potentials of the halogens.

Table 10.7

				Zn	Cu	Fe	(kJ mol^{-1})
(1)	$M^{2+}(aq) \to M^{2+}(g)$		ΔH_{hyd}	+2017	+2075	+1890	
(2)	$M^{2+}(g) + 2e^- \to M(g)$		$-(I_1 + I_2)$	−2638	−2703	−2323	
(3)	$M(g) \to M(s)$		$-\Delta H_{atom}$	−130	−341	−406	
(4)	$M^{2+}(aq) + 2e^- \to M(s)$		ΔH	−751	−969	−839	

I_1—first ionization energy
I_2—second ionization energy

of the cycle has been omitted (as for cations) and enthalpy data is quoted. The analysis in the figure reveals that the greater tendency of fluorine than chlorine gas to form aquated anions is primarily due to the greater hydration energy of the fluoride ion, and to a lesser extent the lower heat of atomization, which together outweigh the (surprising) greater tendency of gaseous chlorine atoms to form gaseous anions.

The data given in Table 10.7 provide satisfactory answers to the question: why does the reaction

$$Zn(s) + Cu^{2+}(aq) \to Zn^{2+}(aq) + Cu(s); \quad \Delta H = -218 \text{ kJ mol}^{-1}, \quad (5)$$

proceed from left to right with a large evolution of heat? Clearly this is associated with the greater positive electrode potential of the copper, for which element reaction (4), in Table 10.7, has a greater decrease in enthalpy. Reaction (5) can be obtained by subtracting the half-reaction (4) for zinc from that of copper:

(4) (a) $Cu^{2+}(aq) + 2e^- \to Cu(s);$ $\Delta H = -969$ kJ mol^{-1}
(4) (b) $Zn^{2+}(aq) + 2e^- \to Zn(s);$ $\Delta H = -751$ kJ mol^{-1}

(a) − (b) $Cu^{2+}(aq) + Zn(s) \to Cu(s) + Zn^{2+}(aq); \quad \Delta H = -218$ kJ mol^{-1}.

Reaction (4) has a greater negative enthalpy in the case of copper than in the case of zinc, mainly because of the greater enthalpy of atomization of copper than zinc. Therefore reaction (5) occurs with a decrease of enthalpy from left to right as written, mainly because the heat of atomization of copper is greater than that of zinc. This, surprisingly, is the most accurate statement that can be made. Since the ionization energies are very similar, changing electrons from one gaseous atom to the other gaseous positive ion changes the energy of the system by a relatively small amount, and we cannot say that gaseous copper ions attract electrons *much* more than gaseous iron ions. One further important general point: it is merely a matter of convenience to structure knowledge in this way. This is a particular breakdown of the energy terms to give a scheme which helps us to interrelate a lot of chemistry. It is a way of looking at the system, and it is not necessarily the only way. It is a useful model. It would be wrong to think that iron and copper do this type of calculation before deciding to react!

Of course not all reactions of this type depend upon the relative enthalpies of atomization of the metals concerned. It is always necessary to analyse the particular situation. Consider the reaction:

$$Fe(s) + Cu^{2+}(aq) \to Cu(s) + Fe^{2+}(aq); \quad \Delta H = -130 \text{ kJ mol}^{-1}.$$

Here the more negative enthalpy for the half-reaction (4) (Table 10.7), in the case of copper compared to iron, is due entirely to the greater ionization energy of copper, which swamps the opposing effects of the atomization enthalpy and the hydration enthalpy. In this case it is accurate to say that the reaction occurs mainly because the tendency for gaseous copper ions to acquire electrons is greater than that for gaseous iron(II) ions. Note that in both cases a statement such as: "aqueous copper ions acquire electrons from solid zinc more easily than aqueous zinc ions do from solid copper" merely restates the problem. *Chemistry is complicated* because there is more than one factor involved in even the simplest problems.

10.3.19 Activity Series

When the electrode potentials for ions are written in order of increasing (becoming more positive) electrode potentials (Table 10.6), we have the table known as the **electrochemical series**, which is a particular example of an **activity series** (Section 10.4) in which the criterion used to list the elements in a rank order is electrode potential. In such a table, *under the specified standard conditions* (Section 7.8), the reduced state of any couple, sodium in the couple Na^+/Na for example, is thermodynamically capable of reducing the oxidized state of any couple below it in the table. Zinc will reduce copper ions to copper, for example (Table 10.6):

$$\begin{array}{ll} & \text{oxidant or oxidized state} \qquad \text{reductant or reduced state} \\ (1) & Zn^{2+} + 2e^- \rightarrow Zn \\ (2) & Cu^{2+} + 2e^- \rightarrow Cu \\ \hline (2)-(1) & Zn(s) + Cu^{2+}(aq) \rightarrow Cu(s) + Zn^{2+}(aq). \end{array}$$

The values of $E^\ominus$, and the $\Delta G^\ominus$ values which can be derived from them, refer only to the specified standard conditions, though these can be extended by means of the Nernst equation (Section 7.11). Changing the conditions, however, may change the reaction which occurs. Chlorine is expected to oxidize manganese(II) ions to manganese dioxide in acid solution under standard conditions:

$$\begin{array}{ll} (1) & Cl_2(g) + 2e^- \rightarrow 2Cl^-(aq); \qquad E_1^\ominus = +1.36 \text{ V} \\ (2) & 4H^+(aq) + MnO_2(s) + 2e^- \rightarrow Mn^{2+}(aq) + 2H_2O(l); \\ & \qquad\qquad\qquad\qquad\qquad\qquad\qquad\qquad E_2^\ominus = +1.23 \text{ V} \\ \hline (1)-(2) & Cl_2(g) + Mn^{2+}(aq) + 2H_2O(l) \rightarrow 4H^+(aq) + 2Cl^-(aq) + MnO_2(s); \\ & \qquad\qquad\qquad\qquad\qquad E^\ominus = E_1^\ominus - E_2^\ominus = +0.13 \text{ V}. \end{array}$$

In fact, of course, chlorine is obtained by boiling manganese dioxide with concentrated hydrochloric acid (non-standard condition), and the volatile chlorine is removed from the system, causing the reaction to proceed further to the left. The second limitation on the use of electrode potentials in predicting chemical reactions is that the thermodynamic criterion for a reaction to occur ($E^\ominus$ is positive, or $\Delta G^\ominus$ is negative) is a necessary but not a sufficient criterion: some reactions may have high activation energies, and be kinetically controlled (Section 8.9). Electrode potential data predicts that magnesium can react with water spontaneously to form hydrogen and magnesium oxide. However, an impermeable oxide layer *effectively* raises the activation energy for the reaction and reduces the rate of reaction. If mercury is added, the coherence of the layer is destroyed by amalgamation of the magnesium and mercury, the activation energy is reduced, and the reaction becomes thermodynamically controlled and proceeds rapidly. Many other metals—aluminium and several transition elements—behave similarly. In general, reactions which, as written, appear to involve either only the transference of electrons (e.g. $Zn + Cu^{2+} \rightarrow$) or of charged species ($Ag^+ + Cl^- \rightarrow$), tend to be fast. Reactions which involve the breaking of covalent bonds may be fast or slow. Another snag is that although a reaction may be thermodynamically allowed and reasonably fast, another reaction may go even more quickly.

To summarize, explanations about general chemical reactivity based on the electrochemical series are only completely valid for thermodynamically controlled reactions under standard conditions in aqueous solutions. The series is very important for the understanding of ionic reactions in aqueous solution. But many important non-metals do not form ions in aqueous solutions, and in any case reactions can be performed in non-aqueous conditions—in the gas or solid phases, or in other solvents. For such reactions other activity series and other concepts are more useful in rationalizing the chemistry. Always, of course, the best guide for predicting the thermodynamic possibility of a reaction is the $\Delta G^\ominus$ value for the reaction, though often these data are not available.

An activity series, which we would expect to be more suitable for non-aqueous reactions generally, can be formulated using as a criterion the standard enthalpy of formation, $\Delta H_f^\ominus$, of the oxide of the element per mole of oxygen atoms (to make a fair comparison of elements with different oxidation numbers). Clearly there is a periodic trend (Fig. 10.15): the enthalpy of formation decreases along any period of the representative elements. On descending the groups there is no clear trend however, the enthalpy varying irregularly (cf. electrode potentials). The rank order of elements based on the above enthalpy change is found to be (in part):

Element	Ca	Be	Mg	Li	Sr	Al	Ba	B	Na	Mn	Cr	K	Ga	Zn
Rank order	1	2	3	4	5	6	7	8	9	10	11	12	13	14
$\Delta H_f^\ominus$ (kJ mol^{-1})	635	611	602	596	590	559	558	424	416	385	376	361	350	348

In general, oxygen forms the most stable compounds with elements of greatest electropositivity, though the correlation with electronegativity is by no means exact. Its compounds with the most electronegative elements, nitrogen, chlorine and fluorine, all have positive free energies of formation. This activity series, computed using free energy of formation of the oxide as the criterion of rank order, clearly differs from the activity series formulated from electrode potentials. This is not unexpected, since they refer to different situations. We would expect the former to be more useful in correlating solid state reactions, and the latter to be more appropriate for solution chemistry. When they are both written down fully, apart from the alkaline earth metals, a rough correspondence can be seen between them, and indeed between all activity series based on reasonable criteria (Section 10.2): electropositive metals are found near the top; weakly electropositive metals in the middle; and strongly electronegative non-metals at the bottom. For this reason a general activity series is formulated which is a rank order averaged over a number of such series based on reasonable criteria, often with the troublesome alkaline earth metals (particularly calcium) removed, and it is used to rationalize a wide range of data (Section 11.2).

The generic name "activity series" is an unfortunate one since "activity" is usually associated with rate of reaction which is dependent on activation energies,

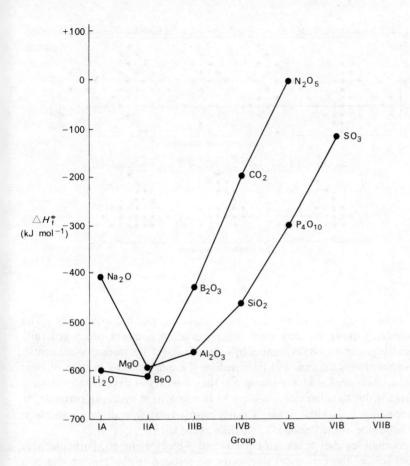

Fig. 10.15 The enthalpies of formation, per mole of oxygen atoms, of oxides of the elements for the first two rows of the Periodic Table.

whereas all activity series are in fact based on thermodynamic criteria. There are numerous examples in chemistry of historical nomenclature giving implications to concepts which are misleading, if not positively wrong. Rather than change much nomenclature, however, the chemist usually prefers to remember that such naming is merely a label. There are many examples of this in chemistry, particularly of concepts which were given the title of old theories, such as inert pair effect, first row anomaly, oxidation, reduction, and spin quantum number.

10.3.20 Oxidation States

Figure 10.16 shows that there is an obvious periodic relationship between the common oxidation states (Section 18.14) of the elements and their electron configuration (or atomic number). Consider the s- and p-block elements first. The maximum positive oxidation number shown by any representative element is equal to the total number of electrons, s and p, in the valence shell. This is called the group oxidation number (and is of course the group number with a positive sign), though it need not necessarily be the most common or the most stable oxidation state for a particular element; indeed, a particular element may have no compounds at all in this state, as in the case of fluorine. The s-block elements form

10.3 THE PHYSICAL PROPERTIES OF THE ELEMENTS

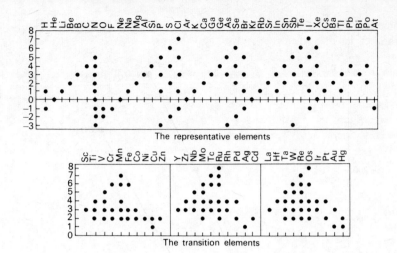

Fig. 10.16 The stable oxidation states of the elements.

compounds in which they exhibit the group oxidation number exclusively. The *p*-block elements, however, may show other oxidation numbers which generally differ from the group oxidation number by steps of two (cf. the transition elements). But this is not always the case and intermediate (i.e. not differing in steps of two) oxidation numbers are found, for example in the chemistry of oxygen and nitrogen, and are usually due to either element–element bonding, as in hydrogen peroxide, or multiple bonding, as in nitric oxide. Clearly more oxidation states are possible at the right hand side of the Periodic Table than at the left.

The common oxidation states of the *s*- and *p*-block elements, particularly in the lithium row of the Periodic Table, often correspond to the gain or loss of a sufficient number of electrons (either "real" in ionic compounds, or "formal" in covalent compounds) to achieve the noble gas configuration ns^2np^6. This tendency is particularly marked for elements with electron configurations near to those of the noble gases: groups IA, IIA, VIB, VIIB, boron (except for the hydrides), and aluminium. In fact groups IA and IIA only form compounds in which they obtain the noble gas configuration and display the group oxidation number. The heavier elements of groups IIIB and 0, together with groups IVB, VB, VIB, and VIIB, except for fluorine and to a lesser extent oxygen, form a variety of compounds in different oxidation states in many of which the law of the octet (Section 4.5) is not obeyed. Two common electron configurations are the *pseudo-noble gas configuration*, $ns^2np^6nd^{10}$, and the *inert pair configuration*, $ns^2np^6nd^{10}(n+1)s^2$. Fluorine, by definition (Section 18.14), shows only the oxidation state(−I), except in the fluorine molecule. The next most electronegative element, oxygen, shows only the oxidation state(−II) except in its fluorides and in compounds containing oxygen–oxygen bonds, as in the oxygen molecules (0), peroxides (−I), and superoxides (−½). The other electronegative elements, nitrogen, sulphur, and the halogens, while they do show stable negative oxidation states, as expected from their electronegativity, also display positive oxidation states in compounds in which they are combined with more electronegative elements than themselves. Again, as expected, in groups VB, VIB, and VIIB (to a lesser extent) the negative oxidation states become less important, and the positive oxidation states become more important for the heavier more electropositive members of the group.

In groups IIIB, IVB, and VB there is a characteristic change in the relative stabilities of compounds in which the group element shows the group oxidation number, and those in which it displays the oxidation number two less than the group oxidation number. Compounds in which the group element displays the group oxidation number are the more stable for the lighter elements of the group, but become increasingly less stable on descending the group. Thus in group IVB, lead(II) is the more stable state for lead, but silicon(IV) is the more stable state for silicon. This phenomenon is referred to as the *inert pair effect*, since in the lower oxidation state it was thought that the two ns^2-electrons had become "inert" (Section 16.1). In groups VIIB and VIB there is a wide variety of oxidation states, and the general trend for the lower positive oxidation states of the heavier elements to become relatively stable is less well defined.

Each transition element exists in several oxidation states, usually differing from one state to another by units of one (cf. the representative elements). For any first row transition element the maximum oxidation number possible can be obtained as follows. Write down the outer electron configuration as $d^m s^2$ (even though this is not the ground state in the cases of chromium and copper). The maximum oxidation number possible is equal to the sum of the number of *unpaired d*-electrons present (Section 20.9) and the two *s*-electrons. For example, in the case of nickel, $d^2 s^2$, this is two unpaired *d*-electrons plus two *s*-electrons equals four. This is the same as the group number up to and including group VIIA, the manganese group, but decreases in units of one thereafter. In fact for iron, cobalt, nickel, and copper, compounds containing these elements with oxidation numbers greater than three are uncommon. In general, the stability of compounds in which the group element displays its highest possible oxidation number decreases from left to right across the transition series, and the relative stability of the lower oxidation states increases correspondingly. Thus titanium(IV) is stable, whereas manganese(VII) is highly oxidizing, as are iron(VI), nickel(IV), and copper(III). If the relative stabilities of the (III) and (II) oxidation states from chromium to zinc are compared (Table 10.8) then the (II) oxidation state generally increases in stability (relative to the (III) oxidation state) along the series, except for manganese where the expected relative stabilities are reversed: manganese(II) is much more stable than manganese(III). This is attributed to the stability of the configuration d^5, and the concept is referred to as the *stability of the half-filled shell*. Many examples will be found of similar "anomalous" behaviour in the manganese group.

Table 10.8 The relative stabilities of the oxidation states (II) and (III) of the later first row transition series.

Cr	Mn	Fe	Co	Ni	Cu	Zn
3	(3)	3	(3)	(3)	(3)	
(2)	2	2	2	2	2	2

See Table 20.4, Section 20.9, for meanings of symbols.

One other general trend found among the transition elements is the following: on descending any transition metal group, the higher oxidation states increase in stability relative to the lower oxidation states, and the rule given above for the maximum oxidation numbers of the first row series does not always apply to the second and third rows. This trend is the opposite to that found in the representative elements, especially in the boron, carbon, and nitrogen groups. For example, whereas the chemistry of chromium is mainly that of the (II) and (III) oxidation states, and chromium(VI) is oxidizing, the chemistry of molybdenum and tungsten is mainly that of the non-oxidizing state (VI). More strikingly, whereas simple aquo-ions of the first row transition elements in oxidation states (II) and (III) exist, e.g. $Co(H_2O)_6^{2+}$, these are almost unknown for the second and third row elements.

For completeness, two other points about the oxidation states of transition elements can be mentioned here: negative oxidation numbers are found (Section 20.9); and oxidation states between the group oxidation number and oxidation numbers (II) and (III) tend to disproportionate (Section 20.9).

10.4 THE CHEMICAL PROPERTIES OF THE ELEMENTS

Metals tend to react as reducing agents (Section 18.14) and to form ionic compounds:

$$M(s) \rightarrow M^{n+}(s) + ne^-.$$

There are three common ways of expressing this relative tendency to lose electrons: electrode potential; ionization energy; and electropositivity (Section 10.3.6). The more electropositive a metal is (in other words the greater the tendency to lose electrons) then the more reactive it is (Table 11.9). Non-metals tend to react as oxidizing agents by gaining electrons in one of two ways, either by forming ionic compounds in which they form an anion, $X + ne^- \rightarrow X^{n-}$, or by sharing an electron to form a covalent bond, $X + Y \rightarrow X-Y$. There are also three ways of expressing the relative tendency of non-metals to gain electrons: electrode potential; electron affinity; and electronegativity. The more electronegative the non-metal, the more reactive it is. It is convenient to classify the chemical reactions of the elements by their reactions with: (a) air; (b) water, non-oxidizing acids, and other cations; (c) oxidizing acids; (d) alkalis; and (e) other elements. The tendency for the elements to undergo such reactions can be related to the appropriate "electron-gaining" concept such as electronegativity.

10.4.1 Action with Air

Nearly all elements burn in air or oxygen on heating, especially if they are finely divided. Air contains oxygen, nitrogen, water vapour, carbon dioxide, and traces of sulphur dioxide, hydrogen sulphide, and so on. On exposure to air at normal temperatures, the more electropositive elements form oxides (some elements form traces of nitrides—lithium, magnesium, calcium, strontium, and barium). In general, the more electropositive the metal, the more rapid the reaction. This statement is true, but is a dangerous correlation of thermodynamic (electropositivity) and kinetic (rate of reaction) properties, which are not necessarily related (Section 8.9). In this case the heat of reaction is related to the electropositivity of the metal (the more electropositive the metal, the greater the heat of reaction), and the greater heat of reaction may speed up further reaction. But this relation may be due to physical causes such as the relative impermeability of oxide layers. The alkali metals form oxide layers very quickly (lithium more slowly than the others), and they must be stored underneath non-aqueous aprotic solvents such as paraffins. The alkaline earth metals react more slowly, the rate increasing down the group: calcium, barium, and strontium are stored in paraffin; but magnesium and beryllium are not, since they form a layer of protective oxide.

The nature of the oxide layer which is initially formed decides whether further reaction can take place. Often this oxide layer is coherent and continuous and does not itself react with water, carbon dioxide, sulphur dioxide, etc. Such oxide layers protect the metal from further attack. This occurs for beryllium, magnesium, aluminium, zinc, and nearly all the transition elements except iron, which forms a permeable non-coherent oxide layer (rust). Often the oxide films are so thin they cannot be seen. Alternatively, the oxide layer may be permeable (e.g. iron), or may react with one of the constituents of air to form further products. The oxides of the alkali metals react with water to form hydroxides, which are deliquescent and give an aqueous solution of the alkali. In time this forms the carbonate by reacting with carbon dioxide. The oxides of the less electropositive alkaline earth metals react more slowly with atmospheric water, and the hydroxides formed are not very soluble and are not deliquescent, so eventually a crust of the basic carbonate (a mixture of hydroxide and carbonate) is formed. Copper roofs are green owing to the protective layer of basic copper salts formed by reaction with the air, for example the sulphate $3Cu(OH)_2 \cdot CuSO_4$. As the metals become less electropositive, they have to be heated to higher temperatures to form the oxides. Mercury, for example, forms the oxide at about 360 °C, but this dissociates just above this temperature. Neither platinum nor gold reacts with oxygen; presumably the temperature at which they would react is greater than the decomposition temperature of the oxide.

Non-metals, except for white phosphorus which smoulders in air to give a mixture of the trioxide and pentoxide, do not react with air at room temperature.

10.4.2 Displacement Reactions in Aqueous Solution

In aqueous solution the electrode potential is the most useful concept for classifying the reactions of the elements. For metals, the important principle is (Section 7.6)—a metal (or hydrogen) will reduce the ions of any metal (or hydrogen) lying below it in the electrochemical series:

$$Fe(s) + Cu^{2+}(aq) \rightarrow Fe^{2+}(aq) + Cu(s);$$

but not

$$Cu(s) + Zn^{2+}(aq) \not\rightarrow \text{(no reaction)}.$$

The analogous principle for non-metals is—a non-metal will oxidize the ions of any non-metallic element or radical lying above it in the electrochemical series:

$$F_2(g) + 2Cl^-(aq) \rightarrow 2F^-(aq) + Cl_2(g);$$
$$OH^-(aq) + Cl_2 \rightarrow Cl^-(aq) + HOCl(aq).$$

When a metal can exist in two oxidation states in aqueous solution, the situation is more complicated since alternative reactions can occur. Iron can form the cations $Fe^{2+}(aq)$ and $Fe^{3+}(aq)$. The position of iron in the electrochemical series is based on the couple Fe^{2+}/Fe. Thus the reaction $Fe^{2+}(aq) + Cu \not\rightarrow$ does not occur. However, the ion $Fe^{3+}(aq)$ does react with copper, but not to form iron metal:

$$Fe^{3+}(aq) + e^- \rightarrow Fe^{2+}(aq); \quad E^\ominus = +0.77 \text{ V}$$
$$Fe^{3+}(aq) + 3e^- \rightarrow Fe(s); \quad E^\ominus = -0.04 \text{ V}$$
$$Cu^{2+}(aq) + 2e^- \rightarrow Cu(s); \quad E^\ominus = +0.34 \text{ V}.$$

Therefore the reaction

$$2Fe^{3+}(aq) + Cu(s) \rightarrow 2Fe^{2+}(aq) + Cu^{2+}(aq); \quad E^\ominus = +0.43 \text{ V}$$

can occur; but the reaction

$$2Fe^{3+}(aq) + 3Cu(s) \rightarrow 2Fe(s) + 3Cu^{2+}(aq); \quad E^\ominus = -0.38 \text{ V},$$

cannot occur.

The reaction of metals with non-oxidizing acids, which contain the ion $H^+(aq)$, is a special case of displacement reactions—any metal lying above hydrogen in the electrochemical series will reduce hydrogen ions from dilute acid solutions and will liberate hydrogen:

$$Mg(s) + 2H^+(aq) \longrightarrow Mg^{2+}(aq) + H_2(g); \quad E^\ominus = +2.37 \text{ V}$$

but

$$Cu(s) + 2H^+(aq) \not\rightarrow \text{(no reaction)}; \quad E^\ominus = -0.34 \text{ V}.$$

The reaction with water is another special example of displacement reactions. Water contains a smaller concentration of hydrogen ions than do dilute acid solutions. This has two results: water reacts with metals less rapidly than do dilute acids (a kinetic result); and fewer metals react with water than with dilute acid (a thermodynamic result). The thermodynamic result is demonstrated by the value of the electrode potential of the hydrogen electrode in pure water, compared to that in acid solution:

$$\text{pH} = 7 \text{ (water)} \quad H^+(aq) + e^- \rightarrow \tfrac{1}{2}H_2(g); \quad E = -0.41 \text{ V}$$
$$\text{pH} = 0 \text{ (dilute acid)} \quad H^+(aq) + e^- \rightarrow \tfrac{1}{2}H_2(g); \quad E^\ominus = 0 \text{ V}.$$

The former reaction goes less readily, and only metals with electrode potentials less (more negative) than -0.41 V (iron and above) can reduce the hydrogen ions in water to hydrogen. Nickel, lead, and tin, can react with dilute acids but not with water. In general, the rate of reaction of metals with other metal ions or hydrogen ions (water and acids) is greater the further apart the reacting elements are in the series. This is the familiar correlation between kinetic and thermodynamic quantities, which is true in this context, but dangerous in general. Here the successful correlation may be due to the fact that both the reaction rate and the free energy change for the reaction are partly dependent upon the free energy of atomization of the metals, which is itself related to the position of the metal in the electrochemical series, being low for electropositive metals and becoming increasingly high down the series. Cold dilute hydrochloric acid, for example, reacts with sodium exceedingly quickly and explosively, very quickly but controllably with magnesium, and very slowly with tin. The slow reactions with metals which are low in the series can usually be made quicker by using more concentrated acid. Metals below hydrogen in the series never give off hydrogen with acids. The alkali metals (except lithium) react very vigorously with water to yield a solution of the hydroxide, and to liberate hydrogen. The rate of reaction and the enthalpy of reaction increase as the group is descended: lithium floats, gives off a steady stream of hydrogen but does not fuse; sodium fizzes about, melts, and sparks, but does not generally catch fire; the others react increasingly violently and the hydrogen evolved is ignited. The alkaline earth metals react more slowly with water than do the group IA metals. The initial reaction of magnesium is fairly rapid but the hydroxide is fairly insoluble and this coats the metal surface and prevents further reaction. Since the hydroxide is more soluble in boiling water, magnesium is slowly attacked by hot water. Since, as the metals become less electropositive, the hydroxides become less soluble, this factor is also important in slowing the rate of attack by water on the metals and is another reason why acids dissolve metals more rapidly than does water (hydroxides cannot form in strongly acidic solutions). Iron reacts reversibly with steam:

$$3Fe(s) + 4H_2O(g) \rightleftharpoons Fe_3O_4(s) + 4H_2(g).$$

Elements below iron in the electrochemical series do not react with water.

There are other factors influencing the rate of reaction. In general these are temperature, purity of metal, purity of acid, state of division of the metal, hydrogen overpotential of the metal, and solubility of the salt formed. Where the salt formed is insoluble, as in the reaction of calcium with sulphuric acid, the insoluble calcium sulphate coats the metal and slows down the reaction. Compare this with the reaction of copper on concentrated hydrochloric acid in the absence of air; the copper dissolves owing to the formation of the stable soluble complex ion, $CuCl_2^-$.

In their reactions with acids, metals are obtained in the lowest oxidation states which are stable in aqueous solutions:

$$Na^+(aq), Cr^{2+}(aq), Mn^{2+}(aq), Fe^{2+}(aq), Ti^{3+}(aq) \text{ etc.}$$

The only non-metals which react with water and dilute acids are the halogens (Section 19.4). In addition, displacement reactions can occur:

$$F_2(g) + 2Cl^-(aq) \rightarrow Cl_2(g) + 2F^-(aq).$$

Oxidizing acids. The noble metals, which do not react with dilute acids to displace hydrogen from solution, do react with oxidizing acids. They can be classified into two groups according to their electropositivity:

1. Metals with small positive electrode potentials, e.g. copper, silver, mercury. These react with warm concentrated sulphuric or nitric acid to give a mixture of sulphur and the oxides of sulphur, or ammonia and the oxides of nitrogen. No single equation can represent what happens and usually the major product is represented as if it were the only product; for example,

warm conc. HNO_3: $Cu(s) + 4HNO_3(aq) \rightarrow Cu(NO_3)_2(aq) + 2H_2O(l)$
$+ 2NO_2(g)$

warm 50% HNO_3: $3Cu(s) + 8HNO_3(aq) \rightarrow 3Cu(NO_3)_2(aq) + 4H_2O(l)$
$+ 2NO(g)$

warm conc. H_2SO_4: $Cu(s) + 2H_2SO_4(aq) \rightarrow CuSO_4(aq) + 2H_2O(l) + SO_2(g).$

Those metals just above hydrogen in the electrochemical series (lead and tin) react similarly under the same conditions. As the series is ascended, the proportion of hydrogen evolved and the violence of the reaction increase.

2. Metals with large positive electrode potentials, e.g. platinum, gold. These are not attacked by concentrated sulphuric or nitric acids but are dissolved by aqua regia. This is a mixture of concentrated nitric and hydrochloric acids in the ratio of one to three parts by volume. At first sight it appears that the reaction should not proceed:

(1) $\qquad Au^{3+}(aq) + 3e^- \rightarrow Au(s);\qquad E^\ominus = +1.5\ V$

(2) $\qquad 4H^+(aq) + 2NO_3^-(aq) + 2e^- \rightarrow N_2O_4(g) + 2H_2O(l);$
$$E^\ominus = +0.8\ V$$

(2) − (1) $2Au(s) + 12H^+(aq) + 6NO_3^-(aq) \rightarrow 2Au^{3+}(aq) + 3N_2O_4(g) + 6H_2O(l);$
$$E^\ominus = -0.7\ V.$$

The negative reduction potential shows that the reaction does not proceed. However, the presence of the chloride ions reduces the potential of reaction (1) below +0.8 V, by forming the stable complex $AuCl_4^-$, and the reaction proceeds:

$$AuCl_4^-(aq) + 3e^- \rightarrow Au(s) + 4Cl(aq);\qquad E^\ominus = 1.0\ V.$$

Certain metals are rendered passive by concentrated nitric acid (Be, Al, Cr, Ga, Fe, Ni, Co). The initial attack of the acid coats the metal with an impervious, coherent, unreactive layer of oxide, which prevents further attack.

Many non-metals react with oxidizing acids in the same way as metals do: in each case the element is oxidized and the anion of the acid is reduced:

$$C(s) + 2H_2SO_4(l) \rightarrow CO_2(g) + 2SO_2(g) + 2H_2O(l).$$

Aqueous alkalis. The electrode potential of the hydrogen electrode in 1M alkali is −0.83 V:

$$2H_2O(l) + 2e^- \rightarrow 2OH^-(aq) + H_2(g);\qquad E^\ominus = -0.83\ V.$$

The evolution of hydrogen by metals in alkaline solution is, therefore, more difficult than in water or in acid solution; only metals with electrode potentials lower than −0.83 V (manganese etc.) in the electrochemical series would be expected to release hydrogen from 1M alkaline solutions. It is somewhat surprising, therefore, to find that some metals below manganese react with alkalis to evolve hydrogen, for example zinc and tin (the amphoteric metals). This is due to the formation of stable complex ions between the zinc ions and hydroxyl ions to give zincates, stannites etc., moving the equilibrium to the right:

$$Zn(s) + 2OH^-(aq) + 2H_2O(l) \rightarrow Zn(OH)_4^{2-}(aq) + H_2(g).$$

Non-metals generally undergo disproportionation reactions with alkalis (simultaneous oxidation and reduction):

$$\text{White } P_4(s) + 3OH^-(aq) + 3H_2O(l) \rightarrow 3H_2PO_2^-(aq) + PH_3(g)$$

$$3S(s) + 6OH^-(aq) \rightarrow 3H_2O(l) + 2S^{2-}(aq) \xrightarrow{\text{excess S}} \text{polysulphides}$$
$$+\ 2SO_3^{2-} \xrightarrow{S} S_2O_4^{2-}$$

$$Cl_2(g) + 2OH^-(aq) \rightarrow Cl^-(aq) + OCl^-(aq) + H_2O(l).$$

The electrode potential for the O_2/OH^- couple is $+0.4$ V:

$$O_2(g) + 2H_2O(l) + 4e^- \rightarrow 4OH^-(aq); \quad E^\ominus = +0.4 \text{ V}.$$

Non-metals with electrode potentials greater than $+0.4$ V can displace oxygen from alkaline solutions. Fluorine is an example:

$$2F_2 + 4OH^- \rightarrow O_2 + 2H_2O + 4F^-$$

Metals cannot react in this way since in aqueous solution they form no stable compounds containing metals in negative oxidation states.

10.4.3 Reactions of Metals and Non-metals

Most metals combine with most non-metals under suitable conditions. In general, the greater the difference in electronegativity between the reacting elements, the more stable is the compound formed, and the easier the reaction is to perform. All metals combine with fluorine, normally to produce a compound in which the metal shows its highest oxidation state. All metals react with chlorine, most with oxygen, and most with the remaining non-metallic elements.

Just as hydrogen ions are reduced and displaced from water or aqueous solutions of alkalis or acids by some metals above hydrogen in the electrochemical series, so hydroxyl ions are displaced by the halogens which are below the hydroxyl ion in the series (since this reaction is an oxidation).

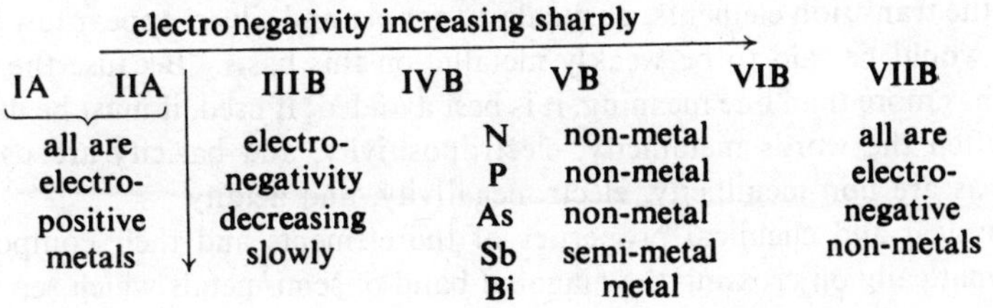

electronegativity increasing sharply →

IA IIA	IIIB	IVB	VB	VIB	VIIB
all are electro- positive metals	electro- negativity decreasing slowly ↓		N P As Sb Bi	non-metal non-metal non-metal semi-metal metal	all are electro- negative non-metals

Fig. 10.17 Trends of the character of the s- and p-block elements.

10.5 CLASSIFICATION OF ELEMENTS: SUMMARY

There are three (related) general methods of classifying the elements: (a) the Periodic Table (Section 10.1); (b) as metals, semi-metals (metalloids), and non-metals (Section 10.2); and (c) in an "activity" series such as the electrochemical series (Section 10.3.19). The Periodic Table presents elements in related groups or families, the modern basis (or criterion) of which is electron configuration. The trends in the character of the s- and p-block elements are given in Fig. 10.17. The electronegativity, non-metallicity, acidity of the oxide (Section 10.2), and oxidizing power of the element all increase in going across any period from left to right, or in ascending any s- and p-block group. Conversely, the electropositivity, metallicity (quality of being metallic), basicity of the oxide, and reducing power of the element all increase in going across any period from right to left, or on descending any group. Thus the Periodic Table, together with the concept of electronegativity or a related concept such as metallicity, emphasizes the trends in properties among related elements. In each group the elements show:

1. A general similarity of properties, especially valency relationships, which is greatest at the edges of the table (Fig. 10.17). This is related to the similar electron configuration of elements in the same group.
2. A gradation of chemical properties as the group is descended, which is attributable to decreasing electronegativity. The effect is most noticeable in the middle of the table among elements of moderate electronegativity. Among such elements on the border between metals and non-metals, small differences in electronegativity cause large differences in properties.
3. A gradation of physical properties (Section 10.3).
4. Anomalous behaviour of the first member of the group, owing to this member's being far smaller and more electronegative than succeeding members, and restricted to eight valence electrons in its outer shell (Section 14.2.7).

The classification of elements into metals and non-metals is discussed in Section 10.2, and activity series are discussed in Section 10.3.19. Any one of a number of criteria can be used to classify the elements into metals and non-metals, or to produce an activity series. For this reason the word "metallic" has more than one meaning, depending upon the context in which the word is used, or the property being used to determine it. If the criterion used to determine a rank order is, say, electrical conductivity, then the activity series obtained is very different from that obtained using the criterion of electropositivity (Section 10.3.6). Transition metals, particularly copper and silver, are near the top of an activity series based on electrical conductivity and would be said to be very metallic. But on the basis of electropositivity the group IA elements appear at the top of the associated activity series, and the transition elements, particularly copper and silver, appear low in the series, and would be said to be weakly metallic on this basis. Because the word "metallic" has more than one meaning, it is best avoided; if used, it must be defined exactly. Often the words metallicity, electropositivity, and basicity are used as synonyms; as are non-metallicity, electronegativity, and acidity.

The physical and chemical properties of the elements and their compounds change dramatically on crossing the diagonal band of semi-metals which separates the metals from the non-metals in the s- and p-block elements. A list of these periodic properties, any of which could be used as a criterion of metallicity, is given in Table 10.9. Semi-metals generally show intermediate behaviour and are not classified in the table. A completely satisfactory classification is not possible because of the fairly wide range of properties in each class, particularly in the case of non-metals; here the properties are very dependent on structure, which differs greatly from one non-metal to the next as the number of valence electrons changes. Carbon, for example, is a high melting point giant molecule; iodine forms a low melting point molecular crystal; and their properties differ greatly. Two factors are involved in determining the properties of non-metals: the type of structure and the strength of bonding. For this reason the range in properties of non-metals is wide and their general properties are qualified with words such as "often". Metals differ from non-metals in the relative ease with which the metals lose electrons, leading to low ionization energies and electronegativities, and high electropositivities. Moreover, since metals have similar structures to one another, their properties can be usefully correlated in terms of parameters such as strength of bonding (cohesive strength) and electropositivity. A useful comparison of the properties of weakly and strongly electropositive metals is given in Table 10.10.

Table 10.9. Comparison of the properties of metals and non-metals.

Metals	Non-metals
Structure:	
Close-packed structures	Molecular structures
High co-ordination number	Low co-ordination number
Metallic bonding	Intermolecular bonding: van der Waals
	Intramolecular bonding: covalent
	Many types of structure: giant and layer lattice, chains, molecular, atomic
"Free" electrons	No "free" electrons
Physical:	
Elastic, ductile, malleable	Brittle
Characteristic lustre	May be coloured
Good conductors	Poor
High density	Often low
Hard, high tensile strength	Often soft
High m.p. and b.p.	Often low
Ring when struck (sonority)	Do not
Chemical:	
Tend to form cations	Tend to form anions
Seldom form covalent bonds	Form many covalent bonds
Reducing agents	May be oxidizing agents
+Water: some give off hydrogen	Never give hydrogen. Only halogens react
+Dilute acids: some (more) give off hydrogen	Never give hydrogen
+Oxidizing acids: elements usually oxidized	Similar reaction
+Oxygen: ionic oxides formed	Covalent oxides formed
+Metals: give alloys	Ionic or interstitial compounds formed
+Non-metals: give ionic or interstitial compounds	Give covalent compounds
Compounds:	
Oxides: lower oxides are ionic and basic. Higher oxides are increasingly covalent and acidic	Covalent, acidic
Hydroxides: ditto	ditto
Chlorides: ionic, water soluble, not hydrolysed, involatile (increasing volatility F < Cl < Br < I)	Covalent, hydrolysed by water, soluble in organic solvents, (decreasing volatility F > Cl > Br > I)
Hydrides: basic, ionic, and stable, (or interstitial and often unstable), reducing agents	Seldom reducing agents, stable
Complexes: form complex cations	Form complex anions
Oxidation number: nearly always positive	Can be positive or negative
Ionization energy: relatively low	High
Electron affinity: very low	Higher
Covalent radii: large	Smaller
Enthalpies of fusion, atomization, vaporization } may be high	Often low
Electronegativity: low	High
Electropositivity: high	Low
Electrode potential: usually negative	Always positive
Bond energy of X_2 -homonuclear covalent molecule } low ∴ form largely monatomic vapours	Generally higher (except noble gases) ∴ usually molecular

Table 10.10. Comparison of the properties of strongly and weakly electropositive metals.

Strongly electropositive metals	Weakly electropositive metals
1. *Occurrence*	
s-block elements	p-, d-, f-block metals
2. *Cohesive properties*: properties of element associated with intermetallic bonding being relatively weak:	strong:
Low ΔH_{atom}, ΔH_{vap}, ΔH_{fus}	High
Low m.p. and b.p.	High
Low homonuclear bond energy	High
Low density	High
Soft	Hard
3. *Properties of metal associated with attraction for electrons being relatively low*:	*higher*:
Low electronegativity, reduction potential, and ionization energy	Higher
Large atomic radii, ∴ low density	Smaller / Higher
ΔH_{hyd} is low, but lattice energies generally even lower ∴ compounds are soluble	ΔH_{hyd} is higher, but lattice energies are generally even higher ∴ compounds are less soluble
Good reducing agents e.g. tarnish rapidly in air; react with water and acids rapidly	Poorer; tarnish less rapidly, often forming a protective layer, and react with water and acids slowly if at all
Dissolve in liquid ammonia to give blue reducing solutions	Do not
Ions have inert gas structure and only one oxidation state	Usually have non-inert gas structure and may have several oxidation states
Ions have small charge density and inert gas structure leading to:	Higher
a) low polarizing power of ion	Higher
b) high ionic character of compounds	Lower
Salts are highly ionic, and are stable with respect to dissociation into elements	Less ionic, less stable
Organometallic compounds are highly ionic and unstable e.g. $Na^+C_2H_5^-$	Covalent, more stable
Oxides and hydroxides are strong bases	Less strong, may be amphoteric
Form stable salts with large easily polarized anions (e.g. I_3^-, HCO_3^-, CO_3^{2-} etc.)	Less stable
Displace less electropositive cations from solution	
Oxides are difficult to reduce	Easier
Elements obtained by electrolysis	Reduction of oxide
Salts less hydrated	More
Salts less hydrolysed in aqueous solution, and anhydrous salts can be obtained by the evaporation of aqueous solutions	More, cannot—they form basic salts
Few complex compounds	More
4. *Property associated with the number of valence electrons being low* (1 or 2):	*high* (2 or more):
Low conductivities	High

These two classifications, metals and non-metals (Table 10.9), and weakly and strongly electropositive metals (Table 10.10), provide a framework for comparing elements or groups of elements with one another. For example, a comparison of the chemistry of potassium and copper is largely a question of the comparison

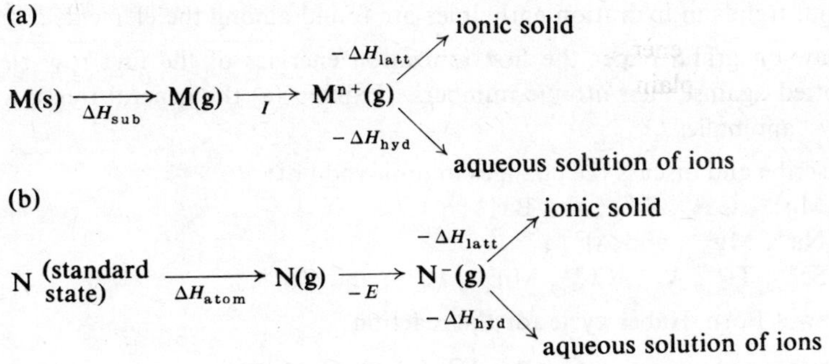

Fig. 10.18 Factors involved in (a) cation-forming ability of metals, and (b) anion-forming ability of non-metals.

of a weakly and a strongly electropositive metal, exemplified by the chemistry of these two elements. Whereas a comparison of the chemistry of sodium and chlorine is largely a question of comparing typical properties of metals and non-metals, exemplified by these two elements.

The more complex nature of non-metals compared to metals can be illustrated by comparing the properties of "cation-forming ability" (metals) and "anion-forming ability" (non-metals), using Born–Haber cycles to isolate the factors involved (Fig. 10.18). Each property depends upon a number of factors, as usual. For metals ΔH_{sub} depends upon: (a) the strength of elemental bonding; and (b) the number of element–element bonds, which is dependent upon the type of structure. Since the structure of all metals is somewhat similar, trends from one metal to another in cation-forming ability are relatively simple. For non-metals the equivalent property, ΔH_{atom}, depends upon the strength of bonding and the number of inter-elemental bonds. The latter depends upon the type of structure, and, since this changes dramatically from one non-metal to the next, this results in anion-forming ability being very dependent on structure.

In both tables an effort has been made to classify the properties. Electropositive metals have high negative reduction potentials because, compared to weakly electropositive metals, they have relatively low enthalpies of atomization (weak metallic bonding) and low ionization energies (low attraction for electrons) (Section 10.3.3). These are generally regarded as the fundamental properties to which other properties are related. It is convenient to list (Tables 10.9 and 10.10) the typical properties of metals and non-metals, and of weakly and strongly electropositive metals at this stage, even though some of these properties have not yet been explained.

PROBLEMS

1. Mulliken defines electronegativity as being proportional to the average of the ionization energy (which measures the attraction a gaseous atom has for its own electrons) and the electron affinity (which measures the attraction a gaseous atom has for other electrons). Assume that the electronegativity of iodine is 2.5 and calculate the electronegativities of the other halogens from the ionization energy and electron affinity data given in Table 19.1. Why is this simple scale not used more widely?

2. What trends in hydration enthalpies are found among the elements?
3. Draw on graph paper the first ionization energies of the first row elements plotted against their atomic numbers. Explain (a) the general trend, and (b) any "anomalies".
4. Describe and discuss the changes in ionic radii of:
   a) Mg^{2+}, Ca^{2+}, Sr^{2+}, and Ba^{2+};
   b) Na^+, Mg^{2+}, and Al^{3+};
   c) Sc^{3+}, Ti^{3+}, V^{3+}, Cr^{3+}, Mn^{3+}, Fe^{3+}, and Co^{3+}.
5. Draw a Born–Haber cycle for the reaction:
$$Cs(s) + \tfrac{1}{2}Cl_2(g) \rightarrow Cs^+Cl^-(s).$$
   Use this to calculate the lattice energy of caesium chloride (using data from Tables 14.1, 19.1, and A.13).
6. Discuss which of each of the following pairs of elements has the higher first ionization energy:
   a) Na and K;
   b) Na and Mg;
   c) Ne and Na;
   d) Cl and Br;
   e) O and F;
   f) F and Ne;
   g) Ca and Sc;
   h) Sc and Ti;
   i) Ca and Zn;
   j) Sn and Pb;
   k) C and N.
7. Draw a Born–Haber cycle for the reaction:
$$2K(s) + Cl_2(g) \rightarrow 2KCl(s).$$
   Given that the lattice energy is 718 kJ mol^{-1}, calculate the standard enthalpy of formation, $\Delta H_f^\ominus$ (using data from Tables 14.1 and 19.1).
8. The lattice energy of the unknown ionic compound $NaCl_2$, has been estimated (how?) as 2300 kJ mol^{-1}. Calculate the standard enthalpy of formation, $\Delta H_f^\ominus$, of this compound. Comment on the result.
9. The lattice energy of MgO has been calculated as 3850 kJ mol^{-1}. (How?) Calculate the electron affinity of the ion, O^{2-}. Comment on the result.
10. How are the properties of the following elements related to their structures: graphite, diamond, plastic sulphur, copper, and sodium?
11. Arrange the following groups of elements in order of their melting (and boiling) points, and explain the trends:
    a) Li, Na, and K;
    b) Be, Mg, and Ca;
    c) F, Cl, Br, and I;
    d) K, Ca, Sc, Ti, and V;
    e) Na, Mg, and Al;
    f) Br, Kr, Rb, and Sr;
    g) Si, P, S, Cl, and Ar.

12. Classify the elements according to the co-ordination numbers they show in their elemental forms.

13. In which group of the Periodic Table would you find the following elements (A, B, and C)?
    a) A forms a close-packed non-conducting solid.
    b) B shows a co-ordination of three in the solid state which consists of puckered sheets held together by metallic-type bonds.
    c) C forms a red solid consisting of puckered rings of atoms.

14. Discuss the general trend(s) that exist in the electron affinities of the elements. Why is the electron affinity of nitrogen much lower than that of the neighbouring elements oxygen and carbon?

15. Draw a Born–Haber cycle for the formation of sodium chloride. Calculate the electron affinity of chlorine using data given in this book.

$$2Na(s) + Cl_2(g) \rightarrow 2Na^+Cl^-(s)$$

(Lattice energy of Na^+Cl^- is 788 kJ mol^{-1}.)

16. The chlorides of lithium and potassium are more soluble in water than are the fluorides. Why? (All have the same structures.)

17. Calculate the enthalpy of formation of sodium iodide (lattice energy of Na^+I^- is 670 kJ mol^{-1}).

18. Place the fluorides of the group IA metals in order of decreasing lattice energy and discuss the result. Which of these compounds do you expect to be the least soluble in water?

19. Why are the changes in properties (melting point for example) so dramatic in going from carbon to nitrogen?

20. In 1829 Döbereiner, in one of the earliest ideas of periodicity, pointed out the existence of triads of elements such as Ca, Sr, and Ba in which a property of the middle element was (roughly) the average of the other two (melting point for example); do the other properties of these elements follow this pattern? Are there any other triads?

21. The only elements which do not have isotopes are: Be, F, Na, Al, P, Sc, Mn, Co, As, Y, Nb, Rh, I, Cs, Pr, Tb, Ho, Tm, Ta, Au, and Bi. What do they have in common? Which is the odd man out?

22. Comment on the fact that whereas the allotropes of phosphorus are metastable, those of metals change rapidly at the transition temperatures into the more stable allotrope.

23. Discuss the trends in the atomic sizes of the group IA metals.

24. Place the following isoelectronic species in order of decreasing atomic radii: Na^+, Mg^{2+}, Al^{3+}, Si^{4+}, N^{3-}, O^{2-}, F^-, Ne. Discuss.

25. On one sheet of graph paper plot against atomic number (a) the enthalpies of atomization (ΔH_{atom}), (b) the enthalpies of hydration of the ions M^+ (ΔH_{hyd}), and (c) the first ionization energies, of the group IA elements. Plot the sum of these three factors for each element, and comment on the result.

26. Discuss the anomalous density and atomic volume of phosphorus in the series aluminium to argon.

27. Discuss the trends in (a) ionic radii, and (b) ionization energies among the transition elements.
28. Which is the larger member of each of the following pairs of ions, and why:
    a) Tl^+ and Tl^{3+};
    b) O^{2-} and F^-;
    c) Ti^{3+} and Cr^{3+};
    d) S^{2-} and Se^{2-};
    e) Na^+ and K^+;
    f) Ca^{2+} and Zn^{2+};
    g) Zr^{4+} and Hf^{4+}?
29. Why is the ionic radius of zirconium(IV) about the same as that of hafnium(IV)?
30. Why do the elements of the transition series show more than one stable oxidation state; whereas those of the s-block elements do not? What trends in the relative stabilities of the oxidation states of the transition elements have been recognized?
31. Account for the trends in the enthalpy of vaporization, ΔH_{vap}, and the boiling point of the elements:
    a) Na, Mg, and Al;
    b) Si, P, S, Cl, and Ar.
32. Separately plot the following against atomic number:
    a) specific heat;
    b) melting point;
    c) boiling point;
    d) the difference between boiling point and melting point;
    e) enthalpy of fusion;
    f) enthalpy of vaporization.

    Which of these illustrate the concept of periodicity? Comment on any interesting features.

BIBLIOGRAPHY

Rich, R. L. *Periodic Correlations*, Benjamin (1965).
    Introductory account.

Sanderson, R. T. *Inorganic Chemistry*, Reinhold (1967).
    Slightly more advanced than the book by Rich with a vast amount of comparative data.

Sisler, H. H. *Electronic Structure, Properties, and the Periodic Law*, Reinhold (1963).

Wells, A. F. *Models in Structural Inorganic Chemistry*, Clarendon (1970).

Bassow, H. *The Construction and Use of Atomic and Molecular Models*, Pergamon (1968).

Addison, W. E. *Structural Principles in Inorganic Chemistry*, Constable (1961).
    Written for undergraduates, but very useful at sixth-form level.

Stark, J. G. (Editor). *Modern Chemistry*, Part 5: "Allotropy", article by W. E. Addison, Penguin (1970).

Addison, W. E. *The Allotropy of the Elements*, Oldbourne (1964).
    This is a university-level text, but the more descriptive sections can be understood at sixth-form level.

CHAPTER 11

Periodicity II: The Periodic Properties of Compounds

11.1 THE PROPERTIES OF SELECTED TYPES OF COMPOUNDS

The properties of various compounds of the elements can be used, as well as properties of the elements, to classify the elements into metals and non-metals.

11.1.1 Normal Oxides

These are oxides which contain oxygen in its normal oxidation state of -2. There are two related ways of classifying oxides: (1) acid–base character of the oxides; and (2) structure of the oxides.

1) Acid–base character of oxides. Acidic oxides are those oxides which either: (a) react with water to produce a solution containing a greater concentration of hydrogen ions than does pure water, e.g. $N_2O_5(s) + H_2O(l) \rightarrow 2H^+(aq) + 2NO_3^-(aq)$; or (b) if insoluble in water, react with a base to form a salt, e.g. $SiO_2(s) + Na_2O(s) \rightarrow Na_2SiO_3(s)$. Acidic oxides are often gases or liquids, and are highly covalent.

Basic oxides are those oxides which either: (a) react with water to produce a solution containing a greater concentration of hydroxyl ions than does pure water, e.g. $Na_2O(s) + H_2O(l) \rightarrow 2Na^+(aq) + 2OH^-(aq)$; or (b) if insoluble in water react with an acid to form a salt, e.g. $MnO(s) + 2HCl(aq) \rightarrow Mn^{2+}(aq) + 2Cl^-(aq) + H_2O$. The better the base (or acid) performs these functions, the stronger it is said to be. Although acidic and basic oxides often react directly, the product need not be a salt, but may be a mixed oxide. Amphoteric oxides are oxides with both acidic and basic properties, for example zinc oxide:

$$ZnO(s) + 2HCl(aq) \rightarrow Zn^{2+}(aq) + 2Cl^-(aq) + H_2O(l);$$
$$ZnO(s) + 2OH^-(aq) \rightarrow ZnO_2^-(aq) + H_2O(l).$$

When all the normal oxides of the s- and p-block elements in their group oxidation states are classified into acidic, basic, and amphoteric oxides, using the criteria above, the result given in Fig. 10.4 (Section 10.2) is obtained. The nature of the oxide is clearly related to the position in the Periodic Table of the element forming the oxide: metals form basic oxides; non-metals form acidic oxides; and semi-metals tend to form amphoteric oxides. In general: the more electropositive the metal, the more basic is its oxide; and the more electronegative the non-metal,

- 11.1 The properties of selected types of compounds
- 11.2 Thermal decomposition of salts
- 11.3 Salts
- 11.4 Summary of the main relationships in the Periodic Table
- 11.5 Metallurgy: the science of metals

Table 11.1. The changing acid–base character of normal oxides as the oxidation state of the element increases.

Compound	VO	V_2O_3	VO_2	"V_2O_5"	PbO	PbO_2
Structure type	NaCl	Al_2O_3	distorted rutile	P_4O_{10}	layer	rutile
Acid-base type	basic	basic	amphoteric	acidic	basic	amphoteric
Bond type	ionic	largely ionic	largely covalent	covalent	ionic	largely covalent

the more acidic is its oxide. Conversely, the concept can be used to classify elements as metals and non-metals; in this case the diagonal dividing line between them is not that of semi-conductors, as in the case of the electrical conductivity criterion (Section 10.2), but one of elements which form amphoteric oxides. In some cases high temperature ignition produces inert oxides which are strongly resistant to reaction, for example, aluminium oxide (α-form, Section 15.2.6).

Many elements form several normal oxides in which the oxidation state of the element differs from one compound to another. This is widespread among the transition elements, but also occurs among the p-block elements. In general when this occurs, the acidity of the normal oxides increases as the oxidation state of the element increases, and the compound becomes more covalent. Several examples are given in Table 11.1.

2) Bond type or structure. The oxides of vanadium (Table 11.1) change from an ionic structure (sodium chloride) to a molecular structure (phosphorus pentoxide) as the oxidation number of the vanadium, and the related acidity of the oxide, increase. This is generally true and provides an alternative classification of oxides: by bond type or structure. In the normal oxides of the s- and p-block elements in their group oxidation states, it is found that metals form ionic oxides (groups IA and IIA). Non-metals form oxides containing primarily covalent bonds. These may be: (a) small discrete molecules in the case of the more electronegative elements (carbon dioxide etc.); (b) partly polymerized molecules for less electronegative elements (phosphorus pentoxide etc.); or (c) completely polymerized molecules (silicon dioxide). In the case of the transition elements, the

Table 11.2. The relation between the acid–base character, bond type, and structure of the oxides of the elements.

electropositive elements	intermediate elements	electronegative elements
	tend to be	
basic	amphoteric	acidic
ionic	semi-ionic	covalent
	and to have	
ionic structures	infinite lattices	discrete molecules

lower oxides (normal oxides in which the transition metal has an oxidation number of +2 or +3) are ionic, and the higher oxides become increasingly covalent. The oxides of the heavier non-metals and those of the semi-metals tend to be solids of complex structure; for example, silicon dioxide, germanium dioxide, and tin dioxide contain three-dimensional lattices.

This classification of normal oxides in terms of structure and type of bonding is not identical to, but correlates fairly closely with, the classification in terms of acidity (Table 11.2). As the bond type of the oxide gradually changes from ionic to covalent in going from electropositive to electronegative elements, there is a corresponding change in the type of structure and the acid–base character of the oxide, as illustrated in Tables 11.2 and 11.3.

Table 11.3. The structure and acid–base character of oxides and sulphides.

Type of structure	Co-ordination number ratios	Name of structure	Oxides (examples)	Ionic and acid–base character	Sulphides (examples)
Infinite three-dimensional	4:8	anti-fluorite	Li, K, Rb, Na	basic	Li, K, Na, Rb
	6:6	sodium chloride	V, Ti, Mg, Ca, Sr, Ba, Cd, Fe		Mg, Ca, Sr, Ba, Mn, Pb
	6:3	cadmium iodide	Ti, Sn, Mn, Pb, V		
	4:4	zinc blende	Be ⎫		Cd, Hg, Zn, Be
	4:4	wurtzite	Zn ⎬ often amphoteric		Cd, Mn, Zn
	6:6	nickel arsenide	⎭		MS: Fe, Co, Ni, V, Ti
	6:6	pyrites			MS_2: Co, Ni, Mn, Os
Long chain			Sb_2O_3; SeO_2; CrO_3		Sb_2S_3, Bi_2S_3, SiS_2
Layer	6:3		As_2O_3; PbO; SnO		TiS_2, ZrS_2
Molecular			P_4O_6; As_4O_6; Sb_4O_6; P_4O_{10}; As_4O_{10}; Sb_4O_{10}; CO_2, etc. (all molecular oxides)	acidic	P_4S_b, P_4S_{10}, N_4S_4

As expected, the bond type and structure of the oxides determine their physical as well as their chemical properties. The ionic oxides have high melting and boiling points, conduct electricity in the molten state, are hard and brittle, and are more soluble in polar than non-polar solvents. Weak van der Waals forces are the only intermolecular forces in oxides consisting of discrete molecules (except in the case of hydrogen-bonded compounds), and therefore such oxides: have low melting and boiling points; when solid, have low tensile strength, low density, and are soft and brittle; and are more soluble in polar than non-polar solvents. These properties change as the degree of polymerization increases when the element becomes less electronegative.

A quantitative classification of acid–base character in the form of an activity series can be formulated by considering, as a criterion of acid or base strength, the change in free energy (or the related equilibrium constant, Section 7.5) for a series of related reactions. In order to compare base strengths, the reactions of basic oxides with a typical acid are contrasted; and in order to compare acid strengths, the reactions of acidic oxides with a typical base are contrasted. A few examples are:

$$MgO(s) + H_2O(l) \rightarrow Mg(OH)_2(s); \quad \Delta G^\ominus = -27.1 \text{ kJ}$$
$$CaO(s) + H_2O(l) \rightarrow Ca(OH)_2(s); \quad \Delta G^\ominus = -55.4 \text{ kJ}$$
$$BaO(s) + H_2O(l) \rightarrow Ba(OH)_2(s); \quad \Delta G^\ominus = -90.9 \text{ kJ}$$
$$CaO(s) + CO_2(g) \rightarrow CaCO_3(s); \quad \Delta G^\ominus = -130 \text{ kJ}$$
$$CaO(s) + N_2O_5(s) \rightarrow Ca(NO_3)_2(s); \quad \Delta G^\ominus = -251.4 \text{ kJ}$$

A small part of the activity series obtained is:

$$K_2O > BaO > CaO > MgO > CuO > H_2O > SiO_2 > CO_2 > N_2O_5 > SO_3.$$
basic　　　　　　　　　　　　　　　　　　　　　　　　　　　acidic

A summary of the trends in the acid–base behaviour and bond-type of the oxides of the representative elements is given in Fig. 11.1.

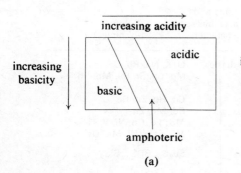
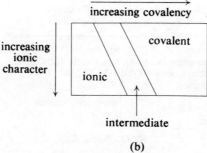

Fig. 11.1 (a) Acid–base properties of oxides. (b) Bond type of oxides.

11.1.2 Hydroxides

The properties of the hydroxides of the elements can also be classified either according to bonding and structure, or according to acid–base character. The results are similar to those obtained for oxides. Consider the hypothetical compound MOH. There are two extreme cases:

1. If M is a strongly electropositive element such as sodium, the compound is a solid with an ionic lattice, $M^+OH^-(s)$, and is a strong base (good source of hydroxyl ions).
2. If M is a strongly electronegative element such as chlorine, there is a strong covalent bond between the atom M and oxygen, for example in hypochlorous acid Cl—O—H; in aqueous solution the compound will dissociate to give protons and is an acid: $Cl—O—H + H_2O \rightarrow ClO^-(aq) + H_3O^+(aq)$.

Between these two extreme cases there is, as usual, a whole spectrum of compounds. If M is of intermediate electronegativity it often happens that the hydroxide can behave as either an acid or a base, and is amphoteric. Usually such compounds have polymeric structures. The hydroxides of transition elements in an oxidation state of three or greater, and the higher oxides of weakly electropositive p-block metals, usually exist as hydrous oxides: $M_2O_3 \cdot xH_2O$ etc.

11.1.3 Hydrides

The binary hydrides of the elements can be classified in the same manner as the oxides and hydroxides, according to (a) bond type and structure, and (b) acid–base character. The classification can also be related to the electronegativity of the element concerned and its position in the Periodic Table (Table A.13). The hydrides are conveniently divided into three main groups according to bond type (Fig. 11.2): (1) ionic hydrides; (2) covalent hydrides; and (3) metallic hydrides. As usual, borderline cases exist exactly where expected in the spectrum of bonding types: copper and zinc are metallic/covalent; beryllium, magnesium, and aluminium are ionic/covalent; and scandium, yttrium, and lanthanum are ionic/metallic.

1) Ionic hydrides. These are formed only by the group IA elements together with calcium, strontium, and barium, all of which have electronegativities of less than 1.1. A comparison of the enthalpies of formation of the halide and hydride ions reveals why there are fewer ionic hydrides than halides (Table 11.4). Whereas the enthalpies

Table 11.4. Enthalpies of formation of halide and hydride ions (kJ mol^{-1}).

		F	Cl	Br	I	H
ΔH_{atom}:	$\tfrac{1}{2}X_2 \rightarrow X(g)$	+79	+122	+111	+106	+218
$-E$:	$e^- + X(g) \rightarrow X^-(g)$	−333	−348	−340	−297	−73
	$\tfrac{1}{2}X_2 \rightarrow X^-(g)$	−254	−226	−229	−191	+145

of formation of the halide ions are very favourable, that of the hydride ion is unfavourable, and therefore the existence of ionic hydrides is much less widespread than that of the ionic halides: only the most electropositive metals form hydrides. These compounds are prepared by direct combination of the elements at high temperatures, and are white crystalline solids with high melting points. Except for lithium hydride, they decompose into their constituent elements when heated before reaching their melting points; the thermal stability of the hydrides decreases as the groups are descended. The compounds can be electrolysed by dissolution in the melt of a low-melting solid such as lithium chloride; hydrogen is evolved at the

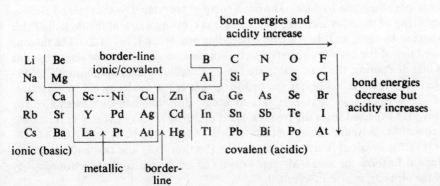

Fig. 11.2 Classification of hydrides.

anode, showing the presence of the hydride ion. The ionic hydrides react vigorously with proton donors such as water, acids, and alcohols, evolving hydrogen, and leaving an alkaline solution in the case of water:

$$Li^+H^-(s) + H_2O(l) \rightarrow Li^+(aq) + OH^-(aq) + H_2(g).$$

Since the compounds are a source of hydroxyl ions in water, they are basic. In general the reactivity of these hydrides increases down the groups. They are powerful reducing agents at high temperatures, their reducing power increasing down the groups. Lithium hydride can be used in metathetical preparations of other hydrides, such as silane, $4LiH + SiCl_4 \rightarrow SiH_4 + 4LiCl$ (the more soluble lithium aluminium hydride is more commonly used nowadays). Ketones are reduced to alcohols, sulphates to suphides, nitrates to ammonia, and so on. All the ionic hydrides react vigorously when heated with oxygen; caesium hydride in fact ignites spontaneously in air. The ionic hydrides all have negative enthalpies of formation, and are always stoichiometric. They are insoluble in all solvents with which they do not react. The size of the H^- anion is slightly larger than the F^- anion, and the lattice energies and, in the case of group IA, the structures of the hydrides are similar to those of the corresponding fluorides. The group IA hydrides all have the sodium chloride structure in which the hydride ions can be regarded as occupying the octahedral holes of a face-centred cubic lattice of sodium ions. In metallic hydrides the hydrogen is generally considered to be atomic hydrogen, which is considerably smaller than the hydride ion and occupies tetrahedral holes in the metallic lattice.

2) Covalent hydrides. These are formed by elements with electronegativities greater than 1.8. Since the electronegativity of hydrogen is 2.1, most of these compounds contain polar covalent bonds in which the hydrogen is somewhat positively charged. As the electronegativity of the other element increases, the covalent bonds become more polar until, with highly electronegative elements such as oxygen and fluorine, the bond is so highly polar that hydrogen bond formation is possible. Since the co-ordination number of hydrogen never exceeds two, and only rarely exceeds one, no three-dimensional lattices are possible and all the covalent hydrides consist of discrete molecules. The intermolecular forces can only be weak van der Waals forces, or in a few cases the stronger hydrogen bonds. Typically, therefore, they are gases or liquids of low boiling and melting points, are soft, are non-conductors, and are soluble in organic solvents. Some of the lighter elements form more than one hydride. The bond energies are related to the size and electronegativity of the other element, and increase in crossing a row of the Periodic Table from left to right, and decrease on descending any group (Fig. 11.2). The thermal stabilities of the hydrides correlate well with these trends: the most stable hydride is that of fluorine, the least stable are those of thallium, lead, and bismuth. However, although the acidity of the hydrides, that is the tendency for the reaction, $M—H + H_2O \rightarrow M^-(aq) + H_3O^+$, to occur, does increase on descending any group (as expected from the trend in bond energies) it also increases in traversing a row of the Periodic Table from left to right. Clearly the bond strength is not the only factor involved in determining acidity (Section 13.4) and the tendency of the element to form the anion $M^-(aq)$, which can be related to the electronegativity of the element, is also involved.

The boiling points and enthalpies of vaporization of the covalent hydrides show an interesting trend. The expected increases as the molecular weight increases, owing to increased van der Waals forces between the larger more polarizable molecules, is only found in the heavier hydrides. The first member of the group has an anomalously high boiling point and enthalpy of vaporization. This means there are greater than expected intermolecular forces for the first members of the groups. A "new" type of bonding was postulated to explain this: hydrogen bonding (Section 4.8).

The covalent hydrides are usually prepared in one of three ways: (a) direct combination by burning the element in hydrogen (groups VI and VII); (b) reduction of the corresponding halide using lithium aluminium hydride (groups III and IV); or (c) treatment of an "-ide" compound with water or acid (e.g. silicides, borides, carbides). Increasing acid strength is required for this method on going across a row from group IV to group VII; mixtures of compounds are often obtained.

3) Metallic hydrides. These are formed by many transition elements, lanthanides, and actinides, by heating the metal in hydrogen. Within this class there is a wide variation in the manner in which hydrogen is bonded to the metal, and a corresponding variation in properties. In general these hydrides are metallic in appearance, conduct electricity, and are usually non-stoichiometric; and the metal undergoes only a small volume change on absorbing hydrogen to form the hydride, in which the intermetallic distance is rarely different from that in the pure metal (but in which the structure of the metal may have changed). Previously these were regarded as "interstitial" compounds in which the hydrogen atoms were merely accommodated by the metallic lattice, and no form of bonding was postulated. This is obviously too simplified a picture. The hydrogen is clearly strongly bonded to the metal since: (a) energy to dissociate the hydrogen molecules must be provided; (b) the hydrides have high negative enthalpies of formation; and (c) the arrangement of the metal atoms in the hydride often differs from that in the pure metal lattice. A more satisfactory model is to consider these compounds as alloy phases, in which the hydrogen atoms are bonded in the holes in the lattice. This model can explain the bonding, the metallic, and the non-stoichiometric properties. Palladium is outstanding in its ability to absorb hydrogen; it takes up four hundred times its own volume of hydrogen at room temperature and one atmosphere pressure. At lower temperatures and higher pressures of hydrogen, up to one thousand times its own volume of hydrogen can be absorbed. The hydrogen can be expelled by heating in a vacuum. Transition metals, particularly palladium and platinum, are excellent catalysts for hydrogenation reactions and this is clearly related to their ability to absorb atomic hydrogen. (See Sections 8.8 and 22.2.4.)

11.1.4 Halides, Especially Chlorides

The properties of the chlorides of the elements also illustrate well the gradual change of character among related elements in the Periodic Table: (a) the sharp increase in electronegativity in traversing a row of the s- and p-block elements from left to right; (b) the small increase in electronegativity in crossing a row of the transition elements; (c) the decrease in electronegativity in descending a group of the s- and p-block elements; and (d) the small but irregular changes in electronegativity in descending a group of the transition metals. The gradual change in the bond type

Table 11.5. The chlorides of the second row elements.

Compound	NaCl	MgCl$_2$	AlCl$_3$	SiCl$_4$	PCl$_3$	SCl$_2$	Cl$_2$
Melting point (°C)	808	714	192(P)	−68	−92	−80	−101
Boiling point (°C)	1465	1418	180(s)	57	76	59	−34
ΔH_{atom} (kJ mol^{-1})	640	515	460	393	335	285	244
Structure	three-dimensional (NaCl)	layer (CdCl$_2$)	layer (CrCl$_3$)	←———————molecular———————→			
Bond type	ionic	semi-ionic	partly ionic	highly polar	←— decreasing polarity —→		non-polar

(s)—sublimes; (P)—under pressure.

and structure of the chlorides in crossing a row of the *s*- and *p*-block elements is illustrated in Table 11.5. The most electropositive elements form compounds in which the bonds are mainly ionic, and in which the three-dimensional lattices consist of essentially discrete ions, with no discrete molecules, as in the case of sodium chloride. Less electropositive elements such as magnesium form semi-ionic bonds and layer lattices, or chain lattices as in the case of palladium chloride (Figure 11.3) in which the halogen acts as a bridging group. Aluminium chloride also forms a layer lattice of intermediate type between ionic and covalent compounds. Silicon tetrachloride has a lattice which consists of discrete molecules, but in which the bonds are more polar than those of phosphorus trichloride and sulphur dichloride. In the chlorine molecule the bonds have no ionic character.

The melting points and boiling points are useful guides in such a classification since they both depend upon the strength of the intermolecular forces which are strong for the involatile ionic compounds, and which become increasingly weak as the bonding becomes more covalent, and the weak intermolecular van der Waals forces become increasingly important. The sudden change in melting point between aluminium chloride and silicon tetrachloride corresponds to a dramatic change in structure. In the volatile essentially covalent compounds the melting points generally decrease as the bonding becomes less polar. Among the essentially ionic compounds the fluorides, being the smallest of the halides, have the highest lattice energies and the highest melting points and boiling points, and these decrease in the order F$^-$ > Cl$^-$ > Br$^-$ > I$^-$. Among the essentially covalent compounds, however, this order is reversed, since the larger, more polarizable iodides have the strongest van der Waals intermolecular forces. The change down any *s*- or *p*-block group is less dramatic, but is present: beryllium chloride is mainly covalent and has a chain structure like palladium dichloride; barium chloride is mainly ionic. As usual, if a given element forms more than one chloride, then the chloride of the metal in its lower oxidation state tends to be ionic, that in the higher oxidation

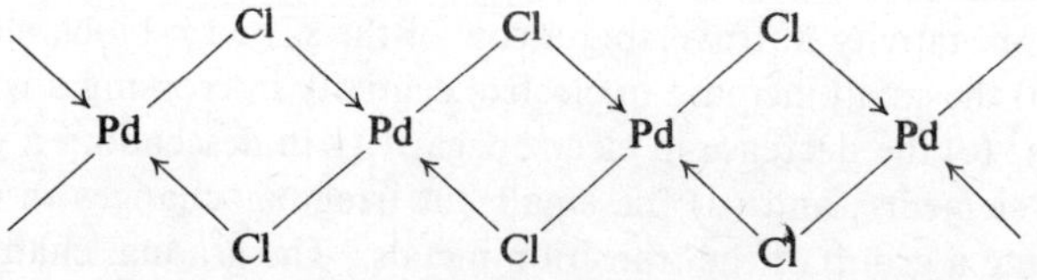

Fig. 11.3 The structure of palladium chloride (planar).

Table 11.6. The bond type of the aluminium halides.

Compound	AlF_3	$AlCl_3$	$AlBr_3$	AlI_3
Melting point (°C)	1200	192	97.5	191
ΔH_{atom} (kJ mol^{-1})	683	460	404	320
Conductivity of fused compound	conductor	conductor	non conductor	
Structure	semi-ionic	intermediate layer lattice	molecular lattice	

state tends to be covalent: lead dichloride is largely ionic, lead tetrachloride is largely covalent. This can be looked at from the point of view of polarizing power and Fajans's rules (Section 4.3). Clearly the polarizing power of the Pb^{4+} ion is greater than that of the Pb^{2+} ion, and its corresponding compounds are therefore more covalent. The same model predicts that the larger and more polarizable bromide ion will form compounds which are more covalent than the corresponding chlorides. It is found, in general, that for a given metal the order of decreasing ionic character is $F^- > Cl^- > Br^- > I^-$. Table 11.6 illustrates this for the compounds of aluminium.

All the transition elements are moderately electropositive; the structures of the dichlorides, MCl_2, are usually layer lattices, in which the metal is six-coordinated, except for a few examples of chain lattices, mainly for copper(II), platinum(II) and palladium(II) compounds. The bromides and iodides are similar, but the fluorides show more ionic three-dimensional structures: CuF_2, for example, has a distorted rutile structure. Oxides and fluorides of the same stoichiometry often have similar ionic structures, whereas the heavier halides often have more covalent structures similar to sulphides and selenides of the same stoichiometry. As the oxidation number of the metal increases, the halides become less ionic. The compound IrF_6, for example, is a molecular compound with octahedral molecules. A typical classification for the chlorides, which is actually based on electrical conductivity, is shown in Fig. 11.4.

An alternative classification can be based upon the reactions of the chlorides with water (cf. acid–base classification of oxides). The relevant criteria are: ionic chlorides dissolve in water (not organic solvents) to give hydrated metal and chloride

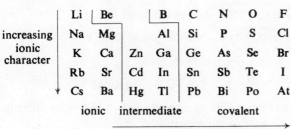

Fig. 11.4 Classification of chlorides based on electrical conductivity.

ions; covalent chlorides dissolve in organic solvents, but in water they hydrolyse (react with water) to give hydrochloric acid and the oxide, oxyacid, or oxychloride of the element involved.

$$Na^+Cl^-(s) + H_2O(l) \rightarrow Na^+(aq) + Cl^-(aq)$$
$$PbCl_4(l) + 2H_2O(l) \rightarrow PbO_2(s) + 4H^+(aq) + 4Cl^-(aq)$$
$$PCl_3(l) + 3H_2O(l) \rightarrow H_3PO_3(aq) + 3H^+(aq) + 3Cl^-(aq)$$
$$BiCl_3(l) + H_2O(l) \rightarrow BiOCl(s) + 2H^+(aq) + 2Cl^-(aq)$$

As usual this is not a sudden change from ionic to covalent compounds—intermediate cases occur. The partly covalent salts of the less electropositive metals undergo partial hydrolysis in solution, as in the case of zinc chlorides:

$$[Zn(H_2O)_4]^{2+} \rightleftharpoons [Zn(H_2O)_3OH]^+(aq) + H^+(aq),$$

and solutions of zinc salts in water are partly acidic, but less so than aqueous solutions of more covalent chlorides. This is a general feature of aqueous solutions of weakly electropositive metal ions. This hydrolysis occurs negligibly for the group IA and IIA metals. Even in the case of magnesium, however, when the hydrated chloride or an aqueous solution is heated, say in an attempt to form the anhydrous chloride, hydrogen chloride is lost, and the basic chloride is formed:

$$MgCl_2 \cdot 6H_2O(s) \xrightarrow{\Delta H} MgCl(OH)(s) + 5H_2O(s) + HCl(g).$$

Thus the anhydrous chlorides of weakly electropositive metals cannot be prepared by this method, and instead are prepared by the high temperature action of dry chlorine or hydrogen chloride on the metal. In the case of the hydrolysis of covalent chlorides the species present in solution are often complex, and not as simple as presented above. Some covalent halides, chloromethane for example, are kinetically fairly stable to hydrolysis by water, and alkali must be used to effect reaction. In general, for a given element there is a simple relationship between the strength of the covalent bond and the size and electronegativity of the halogen forming the bond: the strength of bonding decreases in the order, $F > Cl > Br > I$. Usually the weakly electropositive metals, which form intermediate types of layer and chain lattices, hydrolyse in water to yield aquo-ions, owing to the large hydration enthalpies involved:

$$AlCl_3(s) + H_2O(l) \rightarrow Al^{3+}(aq) + 3Cl^-(aq).$$

These aquo-ions further dissociate to yield an acidic solution (Section 13.4).

The properties of ionic chlorides are those of the chloride ion together with those of the metal ion. The most important properties of the chloride ion are the reactions with:

a) Concentrated sulphuric acid to yield hydrogen chloride:

$$NaCl(s) + H_2SO_4(l) \rightarrow NaHSO_4(s) + HCl(g).$$

At higher temperatures a further reaction takes place:

$$NaHSO_4(s) + NaCl(s) \rightarrow Na_2SO_4(s) + HCl(g).$$

b) Silver nitrate to yield a white precipitate of silver chloride, which is insoluble in dilute hydrochloric acid, but soluble in concentrated ammonia:

$$Cl^-(aq) + Ag^+(aq) \rightarrow AgCl(s) \xrightarrow{\text{conc. ammonia}} [Ag(NH_3)_2]^+(aq) + Cl^-(aq).$$

c) Lead acetate to yield a white precipitate of lead chloride which is fairly soluble in hot water:

$$2Cl^-(aq) + Pb^{2+}(aq) \rightarrow PbCl_2(s).$$

d) Concentrated sulphuric and manganese dioxide to yield chlorine:

$$2NaCl + MnO_2 + 3H_2SO_4 \rightarrow Cl_2 + MnSO_4 + 2NaHSO_4 + 2H_2O.$$

The occurrence of ionic halides is more widespread than that of hydrides (Section 11.1.3). There are no compounds corresponding to the alloy-like metallic hydrides. This is because the halides are too large to be accommodated in a metallic lattice.

11.1.5 Sulphides (Selenides and Tellurides)

Sulphur is less electronegative than oxygen, and this results in metal–sulphur bonds being more covalent than the corresponding metal–oxygen bonds. Thus the existence of ionic sulphides is less widespread than that of ionic oxides, and although many elements have oxides and sulphides of the same stoichiometry, they frequently have different structures (Table 11.3). Moreover, the less ionic sulphides have lower bond strengths than the corresponding oxides, and consequently have lower melting points, lower enthalpies of formation, and usually form the oxide on heating in air. The more covalent nature of the metal–sulphur bond is apparent from the structural chemistry of sulphides: whereas oxides (and the similar fluorides) rarely form layer or chain lattices, sulphides (and the similar heavier halides) do so commonly. Another result of the differing electronegativity of oxygen and sulphur is that oxygen produces higher oxidation states than sulphur in combination with other elements; compare, for example, the existence of the compounds AgO and RuO_4 with the non-existence of the compounds "AgS" and "RuS_4".

Like the oxides, the sulphides can be classified either according to bond type and structure, or by their acid–base behaviour. The classification in terms of acid–base properties is given in Fig. 11.5. A typical acid–base reaction is: K_2S(base) + CS_2(acid) $\rightarrow$ K_2CS_3(salt). Potassium thiocarbonate, K_2CS_3, is the salt of thiocarbonic acid. The lower sulphides of transition metals (those in the low oxidation state, $+2$) are basic, the higher sulphides are acidic. The presence of only three amphoteric sulphides in the classification may be due to the great insolubility of the metal sulphides, which prevents reaction with bases and results in their being classified as basic. The qualitative analysis scheme for identifying metal ions

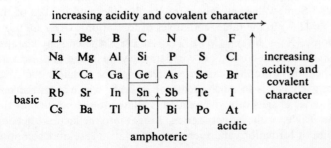

Fig. 11.5 Classification of sulphides in terms of acid–base properties.

depends upon the relative insolubilities of the metallic sulphides, and upon the solubility of the amphoteric sulphides in excess alkali or sulphide ions:

$$\underset{\text{acid}}{\text{SnS}_2(s)} + \underset{\text{base}}{\text{S}^{2-}(aq)} \rightarrow \underset{\text{salt}}{\text{SnS}_3^{2-}(aq)} \text{ (thiostannate).}$$

In general the basic sulphides react with acids to give a salt and hydrogen sulphide; the acidic sulphides react with bases; and the amphoteric sulphides react with both acids and bases:

$$K_2S(s) + 2HCl(aq) \rightarrow 2KCl(aq) + H_2S(g)$$
$$CS_2(l) + 6OH^-(aq) \rightarrow CO_3^{2-}(aq) + 3H_2O(l) + 2S^{2-}(aq).$$

An alternative classification is in terms of bond type and structure. The group IA and IIA metals form colourless, stoichiometric, non-conducting ionic sulphides with the antifluorite and sodium chloride structures respectively, formed by direct combination of the elements on heating, or by reduction of the sulphate by carbon. The more weakly electropositive transition elements form sulphides which may be non-stoichiometric, and which often show the nickel arsenide structure in which each metal atom is very close to two other metal atoms directly above and below it in the unit cell. This results in some metal–metal bonding being present, and these sulphides are all found as alloy phases: they have variable composition, are usually black with a metallic lustre, and they conduct electricity. Iron pyrites, FeS_2, is an example with a different structure, the pyrites structure, which is similar to the sodium chloride structure; iron atoms are in a face-centred cubic arrangement with S—S groups occupying the octahedral holes. There is a range of compounds and properties in the sulphides of the *p*-block elements, from the stoichiometric acidic molecular sulphides such as sulphur trioxide, to layer and chain lattices such as the compounds SnS_2 and SiS_2, and to alloy-type sulphides such as PbS. The general trends are: an increase in covalent character in traversing the Periodic Table from left to right, and a decrease in covalent character of the sulphide in descending a group of the *s*- and *p*-block elements. For example, carbon disulphide is a molecular compound, silicon disulphide shows a chain lattice, and tin disulphide has a layer lattice (cadmium iodide). Selenides and tellurides are often similar to sulphides, but are even more alloy-like.

11.2 THERMAL DECOMPOSITION OF SALTS

The ease of the thermal decomposition of binary salts of metals into their constituent elements is given by their free energies of formation, $\Delta G_f^\ominus$. The more negative the value of $\Delta G_f^\ominus$, the more stable is the salt relative to decomposition into its constituent elements. Some values are given for the fluorides and iodides of the alkali metals in Table 11.7. In the table, enthalpies of formation, $\Delta H_f^\ominus$, are tabulated for the reactions, $\frac{1}{2}X_2 + M(s) \rightarrow M^+F^-(s)$. At 298 K the entropy changes for these reactions in the present context are comparatively small and are also very similar to one another; therefore they can be ignored since the enthalpy contribution to the free energy change is so much greater than $T\Delta S$. The ease of thermal decomposition of a series of salts of a given anion is related to the position of the metal in the Periodic Table, or the activity series. Insight into the factors affecting the relative enthalpies of formation, and therefore the relative ease of decomposition into constituent elements, can be obtained by considering the Born–Haber cycle (Fig. 11.6), in this case for the alkali metal fluorides.

The enthalpy of formation, $\Delta H_f^\ominus$, of the salt MF is given by (Eq. 4.5):

$$\Delta H_f^\ominus = (\Delta H_{atom} + I)_{metal} + (\Delta H_{atom} - E)_{fluorine} - \Delta H_{latt},$$

where the terms stand for the enthalpies of atomization and ionization energy (metal), the enthalpies of atomization and electron affinity (fluorine), and the lattice energy respectively. Since the terms involving the anion alone are constant from metal fluoride to metal fluoride, the relative values of $\Delta H_f^\ominus$ for such compounds are dependent upon the sum of the enthalpy terms for the particular metal, $(\Delta H_{atom} + I)_{metal}$, and the lattice energy, ΔH_{latt}. The larger the lattice energy, the more negative the enthalpy of formation is and the more stable the compound; the larger the sum of the metal enthalpy terms, the smaller the enthalpy of formation and the less stable the compound. All of these terms become smaller on descending the series from lithium to caesium. The relative stability of the halides is decided by which decreases the more quickly: the lattice energy, or the sum of the metal enthalpies of atomization and ionization. For small anions of high charge density such as fluoride (also nitride, hydride, and oxide), the lattice energy (Section 10.3.7) is very dependent upon the size of cation and decreases rapidly on descending the group from lithium to caesium (Table 11.7).

Thus, as the cation size increases, the change in the lattice energy, which is greater than the corresponding change in the sum of the metal enthalpies, determines the change in the stabilities of the fluorides: the lattice energies decrease on descending the series, therefore the compounds become less stable. This is reflected in their enthalpies of formation, $\Delta H_f^\ominus$, which become less negative; lithium fluoride is the most stable alkali metal fluoride relative to dissociation into the constituent elements. This is also true for the hydrides, nitrides, and oxides of the group IA metals.

The opposite trend of stability is found for large anions of low charge density such as iodide and bromide. In such cases the lattice energy is relatively insensitive to changes in cation size, and the more rapid change in the decreasing ionization and atomization enthalpies of the metal on descending the group causes the stabilities of the iodides to increase from lithium to caesium, (Table 11.7).

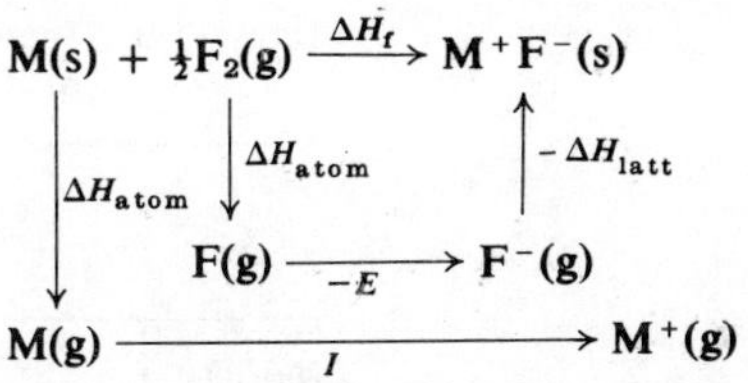

Fig. 11.6 Born–Haber cycle for the formation of alkali metal fluorides.

Table 11.7. Enthalpy changes for the alkali metal halides, $\Delta H_f^\ominus$ (kJ mol^{-1}).

	Li	Na	K	Rb	Cs	The change from Li → Cs
$M(s) \rightarrow M(g)$ $(\Delta H_{atom}^\ominus)$:	159	108	90	86	78	
$M(g) \rightarrow M^+(g)$ (I):	520	495	418	403	374	
$M(s) \rightarrow M^+(g)$ $(\Delta H_{atom} + I)$:	679	603	508	489	452	−227
fluoride: $M^+(g) + F^-(g) \rightarrow M^+F^-(s)$ $(-\Delta H_{latt})$:	−1039	−919	−817	−779	−730	−309
iodide: $M^+(g) + I^-(g) \rightarrow M^+I^-(s)$ $(-\Delta H_{latt})$:	−763	−703	−647	−624	−601	−162
	fluorides decrease in stability →					
fluoride: $M(s) + \frac{1}{2}F_2(g) \rightarrow M^+F^-(s)$ $(\Delta H_f^\ominus)$:	−612	−569	−563	−549	−531	−81
iodide: $M(s) + \frac{1}{2}I_2(s) \rightarrow M^+I^-(s)$ $(\Delta H_f^\ominus)$:	−271	−288	−328	−329	−337	+66
	iodides increase in stability →					

Table 11.8. Thermodynamic data for:
MCO$_3$(s) → MO(s) + CO$_2$(g).

	$\Delta H^\ominus_{298}$ (kJ mol^{-1})	$\Delta G^\ominus_{298}$ (kJ mol^{-1})	$T\Delta S^\ominus_{298}$ (kJ mol^{-1})	$\Delta S^\ominus_{298}$ (kJ mol^{-1} K^{-1})	T^* (°C)
MgCO$_3$	117	67	50	0.168	540
CaCO$_3$	176	130	44	0.148	900
SrCO$_3$	238	188	50	0.168	1280
BaCO$_3$	268	218	50	0.168	1360

T^* represents the temperature at which the pressure of carbon dioxide gas reaches one atmosphere.

For a given metal, the fluoride is always more stable than the iodide, owing to the greater lattice energy of compounds of the smaller fluoride ion.

Decomposition into the constituent elements is not the usual mode of decomposition of metal compounds (other than binary salts) at temperatures below about 1000 °C, however. Consider the decomposition of the alkaline earth carbonates, MCO$_3$: MCO$_3$(s) → MO(s) + CO$_2$(g) (Table 11.8). The stabilities of the compounds are given by the free energies of reaction at 25 °C, $\Delta G^\ominus_{298}$. The more positive the free energy of reaction, the more stable the compound; all the compounds are stable at room temperature (positive values of $\Delta G^\ominus$), and the relative stabilities at 25 °C decrease in the order Ba > Sr > Ca > Mg. Like all compounds they can be decomposed by heating to a sufficiently high temperature, and the question is: "Is the order of stability at higher temperatures the same as that at 25 °C?" The answer is "Yes." We have $\Delta G^\ominus = \Delta H^\ominus - T\Delta S^\ominus$, and $(\partial \Delta G^\ominus / \partial T)_P = -\Delta S^\ominus$. From the table it is observed that $\Delta S^\ominus$ is roughly constant for all the reactions; therefore $\Delta G^\ominus$ changes with temperature in about the same way for each reaction: $(\partial \Delta G / \partial T)_P = $ constant. As the temperature rises and the term $-T\Delta S$ becomes more negative, $\Delta G^\ominus$ becomes more negative at about the same rate for each reaction. In other words, although the absolute value of $\Delta G^\ominus$ for each reaction depends upon the temperature, the rank order of stability stays the same at all temperatures. This is confirmed by the higher temperature required for barium carbonate to obtain the same degree of dissociation (a decomposition pressure of one atmosphere of carbon dioxide above the metal carbonate) as the other carbonates. As usual we make the approximation that $\Delta S^\ominus$ does not vary significantly with temperature (Sections 7.12 and 11.5.1).

One other result of $\Delta S^\ominus$ for the decomposition reactions being fairly constant throughout the series is that the order of enthalpies of reaction, $\Delta H^\ominus$, is also a good guide to relative stabilities. This is often true when similar reactions are being compared, and is important because, not only is there more enthalpy data than free energy data available, but also the enthalpy data can more often be related to fundamental properties such as bond energies. However, the limitations and dangers of using enthalpy data as a rough criterion for the tendency for reactions to take place must always be remembered: it should really be free energy data that are used.

Insight into the factors involved in these changes can be obtained by considering the Born–Haber cycle for the process (Fig. 11.7). Clearly the enthalpy of reaction, $\Delta H^\ominus_{298}$, is equal to:

$$(\Delta H_{\text{latt}})_{\text{carbonate}} + \Delta H_{\text{diss}} - (\Delta H_{\text{latt}})_{\text{oxide}},$$

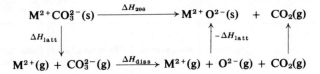

Fig. 11.7 Born–Haber cycle for the decomposition of a metal carbonate to the oxide.

where $(\Delta H_{latt})_{oxide}$ and $(\Delta H_{latt})_{carbonate}$ are the lattice energies of the oxide and carbonate respectively, and ΔH_{diss} is the dissociation enthalpy of the carbonate ion to form carbon dioxide and the oxide ion. Since ΔH_{diss} is involved in all cases, it can be ignored in comparing the decompositions with one another. As the size of the cation increases in descending the series, both lattice energies decrease in size. The trend of increasing $\Delta H^{\ominus}$ values (Table 11.8) and therefore increasing stabilities, is due to the lattice energies of MO and MCO₃ changing *unequally* as the cation size is increased (cf. Table 11.7). The oxide ion is much smaller than the carbonate ion. The lattice energy is inversely proportional to the sum of the radii of anion and cation $(r_- + r_+)$. For a large anion (carbonate) the lattice energy decreases only slowly as the cation size increases, since it is determined mainly by the large anion radius, r_-. For a small anion (oxide), however, the lattice energy decreases more quickly as the cation size increases. This results in the oxides of small cations being more stable relative to their carbonates than the oxides of larger cations are, and thermal decomposition is easier for the smaller cations.

By now it will be apparent that satisfying rationalizations of comparative chemistry are frequently difficult, and always time consuming. Fortunately, a simpler, but less accurate, model for the above phenomena exists which is frequently useful: the pictorial concept of **polarizing power** (Section 14.2.5). The simple idea is that as the charge density (or effective nuclear charge, or polarizing power) of the metal ion increases, the thermal stability of salts of large polarizable anions (such as carbonates) decreases, relative to some decomposition product. This is usually true, as seen above, because the decomposition product usually contains a smaller anion, as in the case of carbonate forming an oxide. It is at least a convenient label, and is used widely in chemistry (though its use implies the fuller discussion given above for the carbonates) because it is impossible to discuss each reaction correctly, since the time and space are rarely available. A few examples of the use of the concept are now briefly discussed. *In general, the least polarizing metal ions are those of the most electropositive metals, and these form the most stable salts of large anions*: the stability of the salts of any particular anion increases as the electropositivity of the metal ion increases, or as the polarizing power decreases. Large anions include: I_3^-, HCO_3^-, NO_3^-, NO_2^-, SO_4^{2-}, CO_3^{2-}, O_2^{2-}, O_2^-, HF_2^-, and even OH^- (relative to O^{2-}). For example, only the triiodides of large cations are stable, such as rubidium, caesium, and tetramethylammonium, and the stability increases down group IA of the Periodic Table. Examples of the decreasing stability of salts of large anions as the electropositivity of the metal decreases are given in Table 11.9. It is a typical property of weakly electropositive elements such as the transition metals that their highly polarizing cations form less stable oxyanions than those of the alkali metals (Table 11.9).

The so-called "anomalous" nature of lithium (Section 14.2.7) is due to its high polarizing power, since it is so much smaller than the other alkali metals.

Table 11.9. Properties of metals.

Ion	Electrode potential	Allred-Rochow electronegativity	Action of air (20 °C)	Combustion	Reaction of water	Reaction of dilute acids	Reaction of oxidizing acids	Reduction of oxides
K^+	−2.92	0.90	rapid reaction	burn on heating	react in cold very quickly	explosive	all react	↑ all can be reduced by carbon monoxide but increasing temperatures required as series is ascended
Na^+	−2.71	1.00						
Ca^{2+}	−2.87	1.05			slow in cold			
Li^+	−3.04	1.15						
Mg^{2+}	−2.37	1.25			very slow in cold	give off hydrogen		
Al^{3+}	−1.66	1.45						
Mn^{2+}	−1.18	1.60			react with steam at red heat			
Zn^{2+}	−0.76	1.65						
Cr^{3+}	−0.74	1.55	slow reaction rate decreasing down series ↓					
Cd^{2+}	−0.4	1.46						
Fe^{2+}	−0.44	1.65			reversible			
Ni^{2+}	−0.25	1.75						
Pb^{2+}	−0.13	1.55				very slow		can even be reduced by hydrogen
Sn^{2+}	−0.14	1.70						
$[H^+]$	[0]	[2.10]		slowly oxidized	no reaction			
Cu^{2+}	+0.34	1.75				no reaction		
Hg^{2+}	+0.85	1.45						
Ag^+	+0.80	1.40						decomposed to metal by heat alone
Pt^{2+}	+1.2	1.45	no reaction	no reaction			aqua regia only	
Au^+	+1.7	1.40						

This results in its compounds with small anions (hydride and nitride) being more stable owing to the high lattice energies, and its compounds with large anions being less stable "owing" to its high polarizing power. This fits into the trends observed for the alkali metals, but partly because of the larger percentage change between lithium and sodium than between any other two alkali metals, and partly because of chance ambient conditions of temperature and pressure on this planet, lithium appears very different from sodium. In fact its polarizing power resembles that of magnesium, and these two elements have many similar properties—the so-called "diagonal relationship" (Section 14.2.7). The alkaline earth metal ions are much smaller than those of the alkali metals and have a charge of +2. Consequently their charge density and polarizing power are greater, and their salts with large

Table 11.9. continued

Ion	Action of water on hydroxide	Action of heat on					Solubility of sulphides
		carbonates	nitrates	sulphates	hydroxides	bicarbonates	
K^+	soluble—strong bases	stable and water-soluble	form nitrites	stable	stable	form carbonates	soluble
Na^+							
Ca^{2+}				stable			
Li^+							
Mg^{2+}	sparingly soluble						
Al^{3+}							insoluble in water but soluble in dilute hydrochloric acid
Mn^{2+}							
Zn^{2+}		insoluble in water and decomposed by heat to the oxide	form oxides		form oxides		
Cr^{3+}							
Cd^{2+}				lose sulphur trioxide and form the oxide		do not form solid bicarbonates	
Fe^{2+}	insoluble—weak bases						
Ni^{2+}							
Pb^{2+}							
Sn^{2+}							
$[H^+]$							insoluble in water and dilute hydrochloric acid
Cu^{2+}							
Hg^{2+}							
Ag^+	hydroxides are unstable and are not formed	unstable	form metals	form metals	form metals		
Pt^{2+}							
Au^+							

anions are less stable to thermal decomposition than those of the alkali metals; the group IIA metal ions are said to be poorer "anion-stabilizers". The transition metal ions are even smaller than those of the corresponding group IIA metals and are even poorer anion-stabilizers: copper carbonate dissociates much more easily to the oxide than does calcium carbonate.

The relation between polarizing power and stability of salts is not too surprising, since the latter can be related to lattice energies (above), and both lattice energies and polarizing power depend upon the effective nuclear charge exerted by the nucleus of the cation upon electrons at bonding distances.

From the discussion above it is clear that a rough relation exists between the electropositivity of a metal, and the thermal stability of its salts. This is summarized in Table 11.9. together with certain other elemental properties. This is an activity series (Section 10.3.19) based upon the properties of the metals and their salts. In general the more electropositive the metal: (a) the more thermally stable are its salts with large anions; and (b) the more reactive the metal is to water, oxygen, acid, and so on. The dividing lines on the table must not be taken too seriously in most cases, since it is a spectrum of behaviour that is usually being discussed. For example, when it is said that the alkali metal sulphates are "heat-stable", we mean, of course, at temperatures normally obtainable in the laboratory, say up to 700 °C. Any compound can be decomposed at a sufficiently high temperature; in the case of sulphates the line is drawn above aluminium because aluminium sulphate begins to decompose at just below 700 °C. In other words, a continuous spectrum of properties, in this case the temperature required to decompose sulphates, has been taken, and divided using some arbitrary criterion usually related to conditions easily obtainable in the laboratory. The dividing lines are less important than the trends which the table illustrates: *as metals become more electropositive, their salts with large anions become more stable, and the metals become more reactive.* Many other properties could be classified in this way, for example the stabilities of other salts such as iodides. The table also shows that a very good correlation exists not only between electropositivity and chemical properties, but also between electrode potential and general chemical properties.

11.3 SALTS

11.3.1 Classification of Salts

It is convenient to summarize the properties of salts at this stage. A salt can be defined as a substance, other than water, which is formed as a result of the reaction between an acid and a base. Salts are often classified as: normal, acidic, basic, double, and complex. Polyprotic acids (those containing more than one replaceable hydrogen atom, Section 9.3.1) give rise to acid salts when they are incompletely neutralized. During this process a salt is formed which itself has a replaceable hydrogen atom or atoms (an acid salt); this hydrogen can be replaced by the addition of further base, for example:

$$\underset{\substack{\text{diprotic}\\\text{acid}}}{H_2SO_4} + \underset{\text{base}}{KOH} \longrightarrow H_2O + \underset{\substack{\text{acid}\\\text{salt}}}{K^+(HSO_4)^-} \xrightarrow{\text{excess KOH}} H_2O + \underset{\substack{\text{normal}\\\text{salt}}}{K_2SO_4}.$$

A normal salt is one in which all of the replaceable hydrogen has been replaced. Clearly monoprotic acids cannot form acid salts. Acid salts show the properties of both acids and salts, since in aqueous solution they form the constituent aquated ions:

$$Na^+HSO_4^-(s) + H_2O(l) \rightarrow Na^+(aq) + HSO_4^-(aq) \rightleftharpoons Na^+(aq) + H^+(aq) + SO_4^{2-}(aq).$$

Just as an acid salt is halfway between an acid and a salt, a basic salt can be regarded as being halfway between a base and a salt, for example basic magnesium chloride, Mg(OH)Cl. Thus there is a spectrum of acid–base character:

acid—acid salt—normal salt—basic salt—base.

The structure of acid salts is simple; potassium hydrogen sulphate, for example, consists of potassium ions and hydrogen sulphate (or bisulphate) ions in a lattice. The structure of basic salts, however, is often complex.

Generally, when a solution containing a mixture of ions is allowed to crystallize, the least soluble salt separates first. Sometimes, however, when a mixture of two simple salts is allowed to crystallize, they crystallize together as a **double salt**. The crystals of a double salt have a definite composition, and a definite crystal structure which is different from that of either of the original salts. For example, if equimolar portions of potassium sulphate and aluminium sulphate are dissolved in the minimum quantity of hot water, and then cooled, colourless crystals of potassium aluminium sulphate (common or potash alum) are formed: $K^+ \cdot Al^{3+}(SO_4)_2^{2-} \cdot 12H_2O$ or $K_2SO_4 \cdot Al_2(SO_4)_3 \cdot 24H_2O$. A series of such compounds is known, of general formula $M^+ \cdot M^{3+}(SO_4)_2^{2-} \cdot 12H_2O$, in which M^+ may be Na^+, K^+, or NH_4^+, and M^{3+} may be one of a large number of trivalent ions, mainly transition metal ions. The alums are isomorphous. In aqueous solution, double salts show the properties of the constituent aquated ions.

When two isomorphous compounds are mixed and crystallized, they frequently form crystals which have the same structure as the original salts, but which may contain a wide range of proportions of the original salts, depending on the proportion of the original salts added. These are called **mixed crystals**, and are best regarded as a solid solution of one substance in another. Some examples of isomorphous compounds which form mixed crystals are: potassium chlorate and potassium permanganate; sodium sulphate and sodium selenate; and the alums. Although isomorphous compounds usually form mixed crystals, it must be noted that some compounds which are not isomorphous form mixed crystals, and some compounds which are isomorphous do not form mixed crystals. Frequently, when a crystal of a compound is placed in a saturated solution of an isomorphous compound, the crystal will continue to grow: a colourless overgrowth of potash alum will grow on a violet chrome alum crystal, for example.

During the formation of double and mixed salts, no new ions are formed, and in aqueous solutions of such salts the constituent ions retain their identities. Another class of salts, **complex salts**, can also be formed by mixing together two simple salts. These, however, are characterized by the formation of a new ion, called a **complex ion** (Section 20.4), which displays characteristic properties of its own both in the solid state and in aqueous solution. The complex salt potassium hexacyanoferrate(II) (or potassium ferrocyanide), for example, can be prepared by mixing together an iron(II) salt and potassium cyanide:

$$4K^+(aq) + 6CN^-(aq) + Fe^{2+}(aq) \rightarrow K_4^+[Fe(CN)_6]^{4-} \cdot 3H_2O(s).$$

This compound shows none of the properties of ferrous ions or cyanide ions, but displays a new set of properties, which are due to the complex ion, $Fe(CN)_6^{4-}$. Not all complex ions are as stable as the ferrocyanide ion, however, and some dissociate wholly or partly in solution to give an equilibrium between the original ions and the complex ions, and consequently show the properties of both. Thus a whole range of stabilities of complex ions exists.

Two important types of complex salts are those formed by the hydroxyl ion and the water molecule. Weakly electropositive metals are frequently amphoteric

(Section 11.1.1): the metal, or the oxide or hydroxide of the metal, dissolves in alkali (as well as in acid) to give a salt. Zinc is such a metal:

$$Zn(s) + 2OH^-(aq) + 2H_2O(l) \rightarrow Zn(OH)_4^{2-}(aq) + H_2(g)$$
$$ZnO(s) + 2OH^-(aq) + H_2O(l) \rightarrow Zn(OH)_4^{2-}(aq).$$

The exact nature of the complex ion, the zincate ion, is not known and it is usually written as ZnO_2^{2-} or $Zn(OH)_4^{2-}$. Almost certainly a variety of aquated species exists in solution. Aluminium forms the aluminate ion, usually written $Al(OH)_4^-$ or AlO_2^-. In aqueous solution all metal ions are hydrated, though the exact nature of the species may not be known (Fig. 9.1). When such salts are crystallized from water, the solid crystal obtained may contain an exact amount of water, for example, $CuSO_4 \cdot 5H_2O$, which is an important part of the structure. These salts are said to be "hydrated", and the water is called "water of crystallization". It is usually possible to get rid of the water of crystallization by heating, but during this process the structure of the compound is altered (usually a powder is formed), and if the crystals are coloured there is a colour change, often to colourless; for example,

$$CuSO_4 \cdot 5H_2O \underset{H_2O}{\overset{heat}{\rightleftharpoons}} CuSO_4$$
$$\text{blue} \qquad\qquad \text{colourless}$$

Hydrated salts are usually complex salts and contain complex ions of the metal ion and water, similar to, but not necessarily the same as, the hydrated species in solution. Solid copper sulphate, for example, contains the complex ion $[Cu(H_2O)_4]^{2+}$, the sulphate ion, and a water molecule in the lattice, called "lattice water" (Fig. 11.8). The existence of different types of water in this structure is demonstrated by the action of heat on solid copper sulphate: four molecules of water of crystallization are easily removed on heating; but the fifth molecule requires temperatures above 200 °C to remove it. A useful model of the bonding in this type of complex ion is to consider it to be primarily ionic between the positive cation and the negative dipole of the oxygen atom of the water. By using this theory it can be predicted that the smallest cations with the greatest charges (i.e. the highest charge densities), for example the transition metal ions, will form the most stable complex ions with water. Conversely, the largest cations with the lowest charges (i.e. low charge density), for example the group IA cations, are expected to form the least stable complex ions with water. These predictions are amply justified: the transition

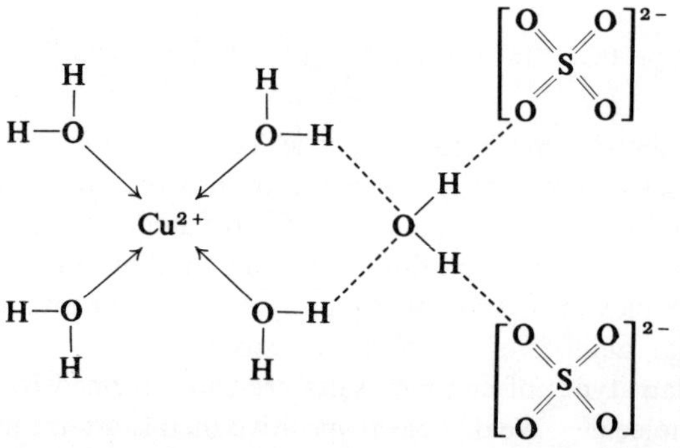

Fig. 11.8 A simplified representation of the copper(II) sulphate structure.

metals form stable hydrated metal salts; but large monovalent ions such as Rb^+, Cs^+, Au^+, Ag^+, and NH_4^+, tend to crystallize anhydrously. Moreover, in group IA, the small lithium ion forms many hydrated salts, sodium and potassium progressively fewer, and rubidium and caesium form hardly any at all. Cations with low charge densities are, of course, the strongly electropositive ions; those with high charge densities are weakly electropositive. In general, therefore, weakly electropositive metals form more stable complex ions than do strongly electropositive metals (Table 10.10).

11.3.2 Preparation of Salts

There are five general methods of preparing soluble salts: the reaction of the corresponding acid on the metal, metal oxide, metal hydroxide, or metal carbonate; and direct combination of the elements in the case of binary salts. Examples of these are:

$$Zn(s) + H_2SO_4(aq) \rightarrow ZnSO_4 \cdot 7H_2O(s) + H_2(g)$$
$$CuO(s) + H_2SO_4(aq) \rightarrow CuSO_4 \cdot 5H_2O(s) + H_2O(l)$$
$$2Fe(s) + 3Cl_2(g) \rightarrow 2FeCl_3(s).$$

Insoluble salts may be prepared by one of two methods: metathesis (double decomposition) by precipitation; or direct combination. Examples are:

$$AgNO_3(aq) + NaCl(aq) \rightarrow AgCl(s) + NaNO_3(aq),$$
or
$$Ag^+(aq) + Cl^-(aq) \rightarrow AgCl(s)$$
and
$$Fe(s) + S(s) \rightarrow FeS(s).$$

11.3.3 Properties of Salts

1) Action of water. Insoluble salts are those upon which water has no reaction. If the salt is soluble, the resulting solution can be either acidic, neutral, or basic. When the solution is acidic or basic, the salt is said to be hydrolysed (Section 13.6.3). Soluble salts can be classified as in Table 11.10. This is discussed in Section 9.3.3. A group of particularly interesting examples is that of salts of weak bases and strong acids, when the weak bases are those of weakly electropositive metals such as copper, aluminium, iron(III), and zinc. Salts such as iron(III) chloride dissolve in water to form acidic solutions, and eventually some iron(III) hydroxide is produced, unless the solution is acidified. Clearly the proton-donating capacity of a water molecule attached to such a cation is greater than that of an

Table 11.10. Hydrolysis of salts (see Section 13.4).

Type of acid and base from which the salt is derived	Example	Acid–base nature of aqueous solution	Ion hydrolysed by the water
strong base, weak acid	sodium acetate	alkaline	anion
weak base, strong acid	ammonium chloride	acidic	cation
strong base, strong acid	sodium chloride	neutral	neither
weak base, weak acid	ammonium acetate	approximately neutral	both

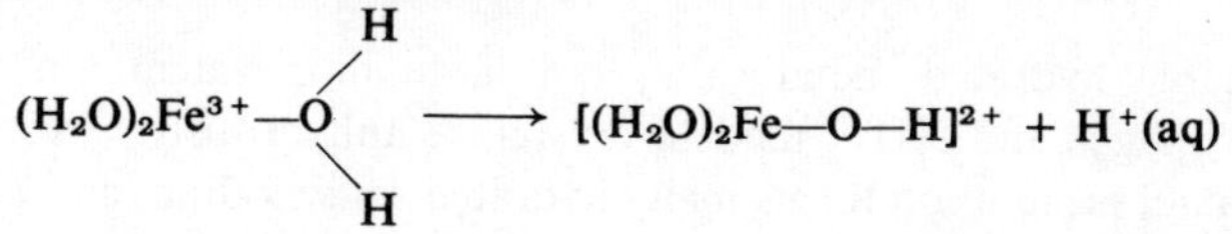

Fig. 11.9 The hydrolysis of the aquated iron(III) ion.

$$(H_2O)_2Fe^{3+}\!\!-\!\!O\!\!\begin{array}{c}H\\ \\ H\end{array} \longrightarrow [(H_2O)_2Fe\!-\!O\!-\!H]^{2+} + H^+(aq)$$

unattached water molecule. This is due to the positively charged cation attracting bonding electrons from the oxygen, which in turn attracts the electrons in the oxygen–hydrogen bond more powerfully, making it easier for the hydrogen to dissociate as a proton (Fig. 11.9). The result is that the aqueous solutions become acidic. This need not stop at this stage but can occur for another water molecule; eventually iron(III) hydroxide precipitates:

$$[Fe(H_2O)_6]^{3+} + H_2O \rightleftharpoons [Fe(H_2O)_5(OH)]^{2+} + H_3O^+$$
$$[Fe(H_2O)_5OH]^{2+} + H_2O \rightleftharpoons [Fe(H_2O)_4(OH)_2]^+ + H_3O^+$$
$$[Fe(H_2O)_4(OH)_2]^+ + H_2O \rightleftharpoons [Fe(H_2O)_3(OH)_3] + H_3O^+$$

or $Fe(OH)_3 \cdot 3H_2O(s)$.

Clearly each successive step becomes less likely since the charge on the cation is reduced and its attraction for electrons decreases. In fact, even in the first reaction, the equilibrium lies well to the left ($K = 6.3 \times 10^{-3}$). Nevertheless, some iron(III) hydroxide is precipitated eventually. The larger the charge density of the cation, the more readily the hydrolysis occurs. Alternatively, the more electropositive the metal, the less readily hydrolysis of its cation occurs. This was one of the differences between strongly and weakly electropositive metals quoted in Table 10.10. The alkali metals are hardly hydrolysed at all, and their salts with strong acids are neutral. Some equilibrium constants for the first hydrolysis step are: $Na(H_2O)_n^+$—very small; $Ca(H_2O)_n^{2+}$—10^{-13}; $Al(H_2O)_n^{3+}$—10^{-5}; and $Fe(H_2O)_6^{3+}$—6×10^{-3}. The addition of an acid moves the equilibria to the left, and precipitation of iron(III) hydroxide, etc., is prevented. (See Section 13.4.3.)

Some salts of very weakly electropositive elements, such as tin, bismuth, and antimony, precipitate the basic chloride on hydrolysis instead of the free base, for example,

$$SbCl_3 + H_2O \rightleftharpoons SbOCl(s) + 2HCl.$$

2) Action of acids. The addition of a strong acid to a salt will in general produce an equilibrium:

$$NaNO_3 + H_2SO_4 \rightleftharpoons NaHSO_4 + HNO_3.$$

In order for the reaction to form an appreciable amount of product, one of the products must be "removed" from the reaction (Le Chatelier, Section 7.3). This can be achieved: if one of the products is volatile; if one of the products is insoluble; or if the acid produced is weak, and largely undissociated (therefore $H^+(aq)$ is "removed" from the reaction). These are illustrated below:

$$NaCl + H_2SO_4 \rightarrow NaHSO_4 + HCl(g)$$
$$Na_2CO_3 + 2HNO_3 \rightarrow 2NaNO_3 + H_2CO_3 \rightarrow H_2O + CO_2(g)$$
$$AgNO_3 + HCl \rightarrow AgCl(s) + HNO_3$$
$$CH_3CO_2Na + HCl \rightarrow NaCl + CH_3CO_2H.$$

3) Action of alkalis. For a reaction to take place, one of the products must be either volatile, insoluble, or a stable complex ion:

$$NH_4^+Cl^-(s) + Na^+OH^-(s) \rightarrow Na^+Cl^-(s) + H_2O(l) + NH_3(g)$$
$$Pb^{2+}(aq) + 2OH^-(aq) \rightarrow Pb(OH)_2(s)$$
$$Zn^{2+}(aq) + 2OH^-(aq) \rightarrow Zn(OH)_2(s) \xrightarrow[OH^-]{excess} Zn(OH)_4^{2-}(aq)$$
$$Cu^{2+}(aq) + 2NH_3(aq) \rightarrow 2NH_4^+(aq) + Cu(OH)_2(s) \xrightarrow[NH_3]{excess}$$
$$[Cu(NH_3)_4]^{2+}(aq) + 2OH^-(aq).$$

4) Action of heat on salts. In general the more electropositive the metal, the more thermally stable are its salts with large anions (Section 11.2).

Nitrates all decompose on heating: those of the group IA metals except lithium form the nitrite; those of less electropositive metals (including lithium) form the oxide; those of very weakly electropositive metals, whose oxides are thermally unstable, form the metal; and ammonium nitrate forms dinitrogen oxide (nitrous oxide). These are illustrated below:

$$2NaNO_3(s) \longrightarrow 2NaNO_2(s) + O_2(g)$$
$$2Pb(NO_3)_2(s) \longrightarrow 2PbO(s) + 4NO_2(g) + O_2(g)$$
$$Hg(NO_3)_2(s) \longrightarrow Hg(l) + 2NO_2(g) + O_2(g)$$
$$NH_4NO_3(s) \longrightarrow 2H_2O(g) + N_2O(g).$$

Normal oxides are all stable except for those of very weakly electropositive metals, such as copper, which decompose on heating to the metal and oxygen.

Carbonates of highly electropositive metals (group IA, except lithium) are stable to heat. The other carbonates decompose. Those of moderately electropositive metals such as calcium give the oxide; those of weakly electropositive metals give the metal; and ammonium carbonate gives ammonia:

$$CaCO_3(s) \rightarrow CaO(s) + CO_2(g) \quad (800\ °C)$$
$$PbCO_3(s) \rightarrow PbO(s) + CO_2(g) \quad (300\ °C)$$
$$2Ag_2CO_3(s) \rightarrow 4Ag(s) + 2CO_2(g) + O_2(g)$$
$$(NH_4)_2CO_3(s) \rightarrow 2NH_3(g) + H_2O(g) + CO_2(g).$$

Sulphates are generally more thermally stable than carbonates and nitrates. Those of groups IA and IIA are stable except at very high temperatures; those of less electropositive metals lose sulphur trioxide and form the oxide; those of the very weakly electropositive metals form the metal; and ammonium sulphate forms the acid salt:

$$Fe_2(SO_4)_3(s) \rightarrow Fe_2O_3(s) + 3SO_3(g)$$
$$2HgSO_4(s) \rightarrow 2Hg(l) + O_2(g) + 2SO_3(g)$$
$$(NH_4)_2SO_4(s) \rightarrow NH_3(g) + NH_4(HSO_4)(s)$$

The **hydroxides** of group IA (except lithium) are thermally stable, the other hydroxides decompose. Those of the moderately electropositive metals decompose to the oxide and water, while those of the very weakly electropositive metals are so unstable they do not exist at normal temperatures:

$$Ca(OH)_2(s) \rightarrow CaO(s) + H_2O(g)$$
$$2Ag^+(aq) + 2OH^-(aq) \rightarrow Ag_2O(s) + H_2O(g).$$

Only the **bicarbonates** of the group IA metals (except lithium) are known in the solid state, and even these decompose on heating to the carbonate:

$$2NaHCO_3 \rightarrow Na_2CO_3 + H_2O + CO_2.$$

The bicarbonates of lithium and the group IIA metals exist in solution.

In general the **halides** are fairly stable to heat, in the absence of oxygen and water. Those of the weakly electropositive metals are the least stable:

$$AuCl_3(s) \longrightarrow AuCl(s) \longrightarrow Au(s) + \tfrac{1}{2}Cl_2(g).$$

In the presence of air the oxide may be formed and the hydrated chlorides of moderately electropositive metals tend to form the basic chlorides on heating:

$$4FeCl_3(s) + 3O_2(g) \rightarrow 2Fe_2O_3(s) + 6Cl_2(g)$$
$$MgCl_2 \cdot H_2O(s) \rightarrow Mg(OH)Cl(s) + HCl(g)$$

11.4 SUMMARY OF THE MAIN RELATIONSHIPS IN THE PERIODIC TABLE

11.4.1 *s*- and *p*-Block Elements

Across any row of the *s*- and *p*-block elements there is a general decrease in electropositivity (or metallic character) from element to element, or a general increase in electronegativity (or non-metallic character). This is illustrated in Table 11.11 with some typical properties of the row of elements from sodium to argon. These properties have already been discussed in the text, and the relationships between them will only be briefly developed here. There is a gradual change from metallic to non-metallic structures across the group. The two classes, metals and non-metals, are separated by a semi-metal, silicon, which has a three-dimensional covalent giant molecular structure (Section 16.2). The dramatic change of structure from aluminium through silicon to phosphorus is reflected in the properties of the elements. The expected increases in the melting and boiling points from sodium to silicon (Section 10.3.10) are followed by a sudden drop in these transition temperatures at phosphorus. This is due to the molecular structure of phosphorus in which the intermolecular forces holding the molecules together are weak van der Waals forces, instead of the strong metallic or covalent forces of the earlier elements in the period. The irregular changes in the transition temperatures of the last four elements in the period are due to the differing sizes of the molecular units. In general, the larger the molecule, the larger the van der Waals forces; and therefore the melting and boiling points decrease in the order $S_8 > P_4 > Cl_2 > Ar$. The atomic volumes of the metallic elements decrease from sodium to aluminium as expected (Section 10.3.9). There is a general increase in atomic volume for the subsequent elements except that sulphur is smaller than would be expected by interpolation from the other elements. This is due to the facts that: the van der Waals radius is much larger than the covalent radius for a given atom; and these atomic volumes depend upon both the covalent and van der Waals radii of the atoms, but to different extents from element to element. The general increase in atomic volume from phosphorus to argon, despite the fact that both types of radii decrease along the row, is due to the increasing importance of van der Waals radii, relative to covalent radii, in determining atomic volume as the molecular

Table 11.11. Comparison of the properties of the second row elements of the s- and p-blocks.

Property	Na	Mg	Al	Si	P	S	Cl	Ar
			decreasing metallicity or electropositivity →					
Atomic number	11	12	13	14	15	16	17	18
Atomic weight	23	24.3	27	28.1	31	32.1	35.5	39.9
Elemental structure	metal	metal	metal	giant molecule	P_4	S_8	Cl_2	Ar
Melting point (°C)	98	651	660	1410	44	119	−101	−189
Boiling point (°C)	892	1107	2467	2355	280	445	−34.5	−186
Electronegativity	1.0	1.25	1.45	1.74	2.05	2.45	2.85	
Ionization energy, 1st (kJ mol^{-1})	495	738	577	787	1060	1000	1255	1520
Electron affinity (kJ mol^{-1})	21	−67	26	135	60	196	348	
Atomic volume	23.7	14.0	10	12	17	15.4	22.7	28.5
Chloride: formula MCl_n	NaCl	$MgCl_2$	$AlCl_3$	$SiCl_4$	PCl_3 PCl_5	SCl_2 SCl_4	Cl_2	—
valency of M	1	2	3	4	3 or 5	2 or 4	1	0
				increasing covalent character →				
nature of bonding	ionic	ionic					covalent	
Oxide: formula MO_n	Na_2O	MgO	Al_2O_3	SiO_2	P_2O_5	SO_3	Cl_2O_7	
valency of M	1	2	3	4	5	6	7	
			increasing covalent character →					
nature of bonding	ionic	ionic				covalent	covalent	
acid–base character	base	base	amphoteric	acid	acid	acid	acid	
Hydride: formula MH_n	NaH	MgH_2	AlH_3	SiH_4	PH_3	SH_2	ClH	
valency	1	2	3	4	3	2	1	
			increasing covalent character →					
nature of bonding	ionic						covalent	
Classification of element		metals		semi-metal		non-metals		

size decreases. The irregular atomic volume of sulphur is due to its large molecule which increases the relative importance of covalent radii relative to van der Waals. The relative ease of formation of ions is illustrated by the values for the ionization energy, electron affinity, and electronegativity of the elements: the metals form cations more readily than do non-metals; the non-metals form anions more readily than do metals; and the semi-metals form neither anions nor cations. Three types of representative compounds are shown in the table: chlorides, oxides, and hydrides. There is a striking periodic trend in the valencies (Section 16.1) of the elements from row to row: in general, along any period of the s- and p-block elements, the valencies of the elements follow the pattern 1, 2, 3, 4, 5 or 3, 6 or 2, 7 or 1, with a few exceptions. In such compounds, as well as the gradual change in stoichiometry in traversing the period from left to right, there are also changes in the type of bonding, volatility, and acid–base character (Section 11.1.1) reflecting the changes in electropositivity across the row.

As well as the periodic relationship discussed above, there is also a group relationship among the s- and p-block elements: in passing down any main group there is a gradual increase in electropositivity or metallic character from element to element. This will become more obvious later in the book where the chemistry of the elements is discussed from the point of view of group relationships. The trend is most obvious in group IVB where the first element, carbon, is clearly a non-metal, and the last element, lead, is clearly a metal. Nevertheless, the trend is present even in group IA where all the elements are metals (for example ionization energies), and even in group VIIB where all the elements are non-metals. One of the most striking group relationships, however, is that of valency: for groups IA, IIA, IIIB, IVB, and 0, the most common valency state is the group valency (Section 10.3.20); for groups VB, VIB, and VIIB, the most common valencies are the group valency and (eight minus the group valency).

In addition to the group and period relationships in the s- and p-block elements, there is also a diagonal relationship between elements. The cause of this is the same as that which results in the line of semi-metals, which separates the metals from the non-metals, running diagonally from top left to bottom right across the Periodic Table: electronegativity increases from element to element in traversing the Periodic Table from left to right, and electropositivity increases in descending the Periodic Table from top to bottom; these effects tend to cancel each other in moving diagonally from top left to bottom right. Thus, the compounds of elements related diagonally in this way tend to have similar properties, despite their differing valencies. This diagonal relationship is particularly noticeable in the first two rows of the s- and p-block elements:

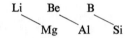

It will be discussed later (Section 14.2.7). This model of the diagonal relationship in terms of similar electropositivity or electronegativity is useful, but is hardly substantiated by the electronegativity data in Table A.13. A more satisfactory, but very similar, argument is in terms of the similar charge densities (charge to size ratios) of the diagonally related atoms (see Problem 2, Chapter 14).

11.4.2 Transition Elements

The relationships between elements in the transition series differ from those in the s- and p-block elements, and are discussed fully in Section 20.2. In brief, as well as a vertical relationship, there is a horizontal relationship between similar elements, and although there are vertical relationships, they differ from those in the s- and p-block elements.

11.5 METALLURGY: THE SCIENCE OF METALS

One important aspect of metallurgy is the extraction and refining of metals from their ores (naturally occurring compounds from which the metal can be extracted). All metals occur naturally as compounds, the most important ones being the oxides, sulphides, chlorides, and carbonates. In addition, the most inert metals are also found as the free elements: the noble metals copper, silver, and gold; and the platinum metals ruthenium, osmium, rhodium, iridium, palladium, and platinum.

11.5.1 Reduction of Oxide Ores by Carbon at High Temperatures

The reduction of oxides by carbon plays a central role in extractive metallurgy and will be discussed first, before giving an overall picture. The higher the temperature

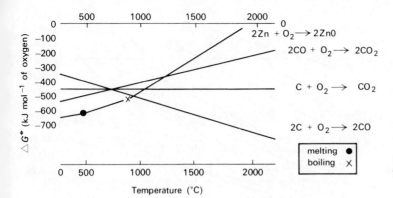

Fig. 11.10 Ellingham diagrams for oxides of carbon and zinc.

of an industrial process, the more costly is the process; therefore a knowledge of the effect of temperature on such reactions is important. At high temperatures the rates of reactions are high, and the reactions are under thermodynamic control (Section 8.9); therefore a knowledge of the change in free energy for the reaction, $\Delta G^\ominus$, will give a good indication of what actually happens. Consequently, the vital question is: how does $\Delta G^\ominus$ for a reaction change with temperature? Diagrams which show the variation of $\Delta G^\ominus$ with temperature are called *Ellingham diagrams* (Section 7.12). On these are plotted the $\Delta G^\ominus$ values at different temperatures for reactions between the elements and one mole of a common reagent such as oxygen, chlorine, or sulphur (Figs. 11.10, 11.11, 11.12, and 11.13).

Consider, for example, the $\Delta G^\ominus/T$ line for the reaction $2Zn(s) + O_2(g) \to 2ZnO(s)$ (Fig. 11.10). At 0 °C, the $\Delta G^\ominus$ value for this reaction is -640 kJ, and the reaction between zinc and oxygen at this temperature is highly exothermic, but slow. As the temperature rises, the $\Delta G^\ominus$ value becomes less negative; at 1000 °C, for example, it has increased to -400 kJ. By 1900 °C, the $\Delta G^\ominus$ value has become positive, and at this temperature zinc oxide spontaneously decomposes to its constituent elements. This behaviour is typical for all elements (except carbon): at a sufficiently high temperature the oxide becomes unstable relative to its constituent elements.

The slopes of these lines can be explained in terms of the changes in entropy during the reactions. Consider how the standard free energy change for a reaction varies with temperature (Section 6.13):

$$\Delta G^\ominus = \Delta H^\ominus - T\Delta S^\ominus \quad \text{and} \quad \left(\frac{\partial \Delta G^\ominus}{\partial T}\right)_P = -\Delta S^\ominus.$$

The rate of change of $\Delta G^\ominus$ with temperature (at constant pressure) is equal to the standard entropy change for the reaction with negative sign (called the temperature coefficient of the free energy change). In other words, since to a first approximation $\Delta H^\ominus$ and $\Delta S^\ominus$ do not change appreciably with temperature for the large majority of chemical processes, and for many purposes can be regarded as constant, $\Delta G^\ominus$ plotted against T gives a line of constant slope equal to $\Delta S^\ominus$ for the reaction, with negative sign. Thus, a knowledge of the standard entropy change, $\Delta S^\ominus$, for a reaction enables a prediction to be made about how the free energy change depends on temperature. Tabulations of standard entropies are clearly important in

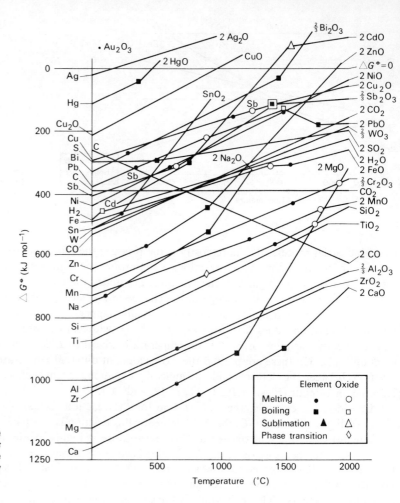

Fig. 11.11 $\Delta G^{\ominus}/T$ diagram for oxide formation. After D. J. G. Ives, *Principles of Extraction of Metals*, Royal Institute of Chemistry (1966). Reproduced by permission of the publisher.

this context. A few values are given in Table 11.12 (see Section 6.11 and Tables A.15 and A.16).

Table 11.12. Standard state entropies at 298 K, $S_{298}^{\ominus}$ (J mol^{-1} K^{-1}).

Solids	Zn, 41.6; ZnO, 43.9; Ca, 41.6; CaO, 39.7; C(gr) 5.7.
Liquids	Hg, 76.1
Gases	CO_2, 214; CO, 198; O_2, 205.

In general it is found that the standard entropies increase in the order solids < liquids < gases. Since for a reaction, $\Delta S^{\ominus} = \sum S^{\ominus}$ (products) $- \sum S^{\ominus}$ (reactants), or in other words, the standard entropy change for a reaction is equal to the sum of the standard entropies of the products less the sum of the standard entropies of the reactants, then it is obvious that the sign and approximate size of $\Delta S^{\ominus}$ can be

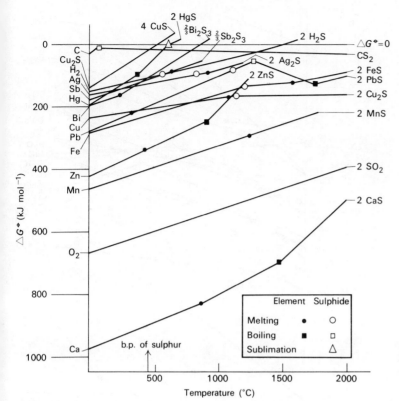

Fig. 11.12 $\Delta G^{\ominus}/T$ diagram for sulphide formation. After D. J. G. Ives, *Principles of Extraction of Metals*, Royal Institute of Chemistry (1966). Reproduced by permission of the publisher.

guessed by merely looking at the reaction. Consider the oxidation of zinc by molecular oxygen (above) (see Section 6.12):

$$2Zn(s) + O_2(g) \rightarrow 2ZnO(s)$$

$S^{\ominus}$ (J K^{-1}) 2×41.6 205 2×43.9
                small   large   small

The change in standard entropy [$\Delta S^{\ominus} = (2 \times 43.9) - 205 - (2 \times 41.6) = -200$ J K^{-1}] is a large negative quantity. This could be guessed: during the reaction, one mole of a disordered gas is converted into a more highly ordered solid; in other words, an increase in "order" has occurred, which is reflected in the large negative value of $\Delta S^{\ominus}$ for the reaction. Since $(\partial \Delta G^{\ominus}/\partial T)_P = -\Delta S^{\ominus} = -(-200) = +200$, the slope of the $\Delta G^{\ominus}/T$ line for this reaction is positive and slopes upward from left to right. At 0 °C the $\Delta G^{\ominus}$ value of the reaction is -640 kJ, but this becomes less negative as the temperature rises (positive slope), and eventually (at 1900 °C) the $\Delta G^{\ominus}$ value for this reaction becomes positive. At this temperature, where $\Delta G^{\ominus} = 0$, the equilibrium constant for the reaction equals one. This is typical of reactions which occur with a decrease in entropy, and this includes the reactions of all the metals with molecular oxygen (see Fig. 11.11). The general similarity of the entropy changes during these reactions leads to a generally similar slope of the $\Delta G^{\ominus}/T$ line. As the temperature is raised $\Delta G^{\ominus}$ becomes more positive

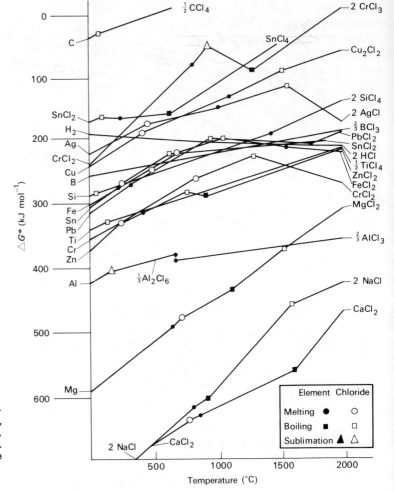

Fig. 11.13 $\Delta G^\ominus/T$ diagram for chloride formation. After D. J. G. Ives, *Principles of Extraction of Metals*, Royal Institute of Chemistry (1966). Reproduced by permission of the publisher.

or less negative, and the reaction has less thermodynamic tendency to occur than at a lower temperature (the equilibrium constant is smaller). (See Section 7.12.1.)

Another class of reactions is that in which the standard entropy change is very small, for example the oxidation of carbon to yield carbon dioxide:

$$\begin{array}{cccc} & C(s) & + O_2(g) & \rightarrow CO_2(g) \\ S^\ominus & 5.7 & 205 & 214 \\ & \text{small} & \text{large} & \text{large} \end{array} \bigg\} \begin{array}{l} \Delta S^\ominus = 214 - 205 - 5.7 \\ = +3 \text{ J K}^{-1} \end{array}$$

Since there is only one gaseous molecule on each side of the equation describing the reaction, it is expected that the change in entropy, $\Delta S^\ominus$, will be small, and that $\Delta G^\ominus$ will change little with temperature (Fig. 11.10).

A third class of reactions is that which takes place with a large increase in entropy:

$$\begin{array}{cccc} & 2C(s) & + O_2(g) & \rightarrow 2CO(g) \\ S^\ominus & 2 \times 5.7 & 214 & 2 \times 198 \\ & \text{small} & \text{large} & \text{large} \end{array} \bigg\} \begin{array}{l} \Delta S^\ominus = (2 \times 198) - 214 - (2 \times 5.7) \\ = +171.6 \text{ J K}^{-1} \end{array}$$

A large positive $\Delta S^\ominus$ is expected for this reaction since an extra gas molecule is produced during the reaction and the $\Delta G^\ominus/T$ line has a negative slope. The result is that, as the temperature rises, $\Delta G^\ominus$ becomes increasingly negative (Fig. 11.10), and the equilibrium constant for the reaction becomes larger. This is a general feature of reductions involving carbon. Consider the general reaction, where M is a metal:

$$C(s) + MO(s) \rightarrow M(s) + CO(g).$$

Clearly, since a gas is produced during the reaction, the change in entropy, $\Delta S^\ominus$, for the reaction is positive, therefore the $\Delta G^\ominus$ value for the reaction becomes increasingly negative as the temperature rises. In the case of zinc oxide (see below) the $\Delta G^\ominus$ value for the reduction is $+182.5$ kJ mol^{-1} at 25 °C, but has become negative by 1000 °C. In fact, *carbon can reduce the oxide of any other element at sufficiently high temperatures*, but various difficulties prevent the preparation of the more electropositive metals by this method: that of obtaining very high temperatures cheaply; that of preventing the back reaction occurring on cooling; and the formation of carbides.

Into which of the above three classes of reactions would you place the oxidation of carbon monoxide by oxygen, and what would you predict for the slope of the $\Delta G^\ominus/T$ line (Problem 15)?

These predictions are all based upon the hope that the temperature coefficient, $(\partial \Delta G^\ominus/\partial T)_P$, at 25 °C will be insensitive to changes in temperature, or in other words that plots of $\Delta G^\ominus$ against temperature for a given reaction give a linear slope. This hope is largely borne out in practice (Fig. 11.11)—the curves are roughly linear. The breaks in the lines are at temperatures at which substances involved in the reaction melt or boil. When a *reactant* (zinc, for example) boils or melts (907 °C and 419 °C), then its entropy increases and the change in entropy during the reaction, $\Delta S^\ominus = \sum S^\ominus$ (products) $- \sum S^\ominus$ (reactants), becomes more negative. This causes the temperature coefficient and the slope of the line to become more positive, and lessens the tendency for the reaction to go. The opposite effect occurs if one of the *products* melts or boils. Boiling has a larger effect than melting, since larger changes in entropy are involved.

The significance of Ellingham diagrams is not only that they illustrate and summarize much of the above discussion, but that, more importantly, free energy changes for a large number of reactions at different temperatures can be found from them. For example, we can read off from the diagram (albeit somewhat roughly because of the reduced scale) the standard free energy changes for the following reactions, (1) and (2), at (say) 25 °C (298 K):

(1) $2C(s) + O_2(g) \rightarrow 2CO(g);$ $\Delta G^\ominus_{298} = -275$ kJ
(2) $2Zn(s) + O_2(g) \rightarrow 2ZnO(s);$ $\Delta G^\ominus_{298} = -640$ kJ

(1) − (2) $2ZnO(s) + 2C(s) \rightarrow 2CO(g) + 2Zn(s);$ $\Delta G^\ominus_{298} = -275 - (-640)$ kJ
$= +365$ kJ

By subtracting the equations and their associated free energies, one from the other, a third reaction and its associated free energy change at that temperature is obtained, in this case the reduction of zinc oxide by carbon. The $\Delta G^\ominus$ value obtained for the third reaction in this case at 25 °C, is of course the difference between the ordinates of the two lines, measured at 25 °C. Clearly this reaction has little tendency to go at this temperature. However, since two moles of gaseous product are produced during the reaction, it is expected that $\Delta G^\ominus$ will become increasingly negative as the temperature increases, and that at a certain high temperature $\Delta G^\ominus$ will become zero (and just above this temperature $\Delta G^\ominus$ will become negative). At this temperature, about 900 °C in this case, the equilibrium constant equals one, and appreciable amounts of product are present at equilibrium. This point is where the two lines representing the relation between $\Delta G^\ominus$ and temperature for reactions (1) and (2) cut one another.

These diagrams are thus very useful for giving an indication of the temperature at which appreciable reaction occurs. The lower on Fig. 11.11 the $\Delta G^\ominus/T$ line of a given element is, the more stable its oxide is relative to dissociation into the element and oxygen; such elements reduce the oxides of other elements whose $\Delta G^\ominus/T$ line appears above them on the diagram at the given temperature. For example, below about 900 °C, zinc will reduce carbon monoxide to carbon, but above 900 °C carbon will reduce zinc oxide to zinc. At 900 °C where the two $\Delta G^\ominus/T$ lines cross, the equilibrium constant for the reaction is unity. Because the $\Delta G^\ominus/T$ line of carbon / carbon monoxide slopes downwards, it will eventually be below all other lines at sufficiently high temperatures. Theoretically, therefore, it is possible to reduce all oxides with carbon, but in practice these are limited as discussed above. The oxides of the most electropositive metals (those with the most stable oxides) are reduced either by using a more reactive element than carbon (aluminium for example), or by electrolytic reduction. Although this section is particularly concerned with reduction by carbon, the diagrams, of course, yield information about any metal–oxide system. For example, magnesium reduces silicon dioxide to silicon at temperatures below about 1700 °C, but above this temperature silicon reduces magnesium oxide to magnesium. This is used in the Pidgeon process:

$$2MgO(s) + Si(s) \rightarrow SiO_2(s) + 2Mg(s).$$

Actually, calcined dolomite, $CaO \cdot MgO$, is used as the raw material, and the calcium oxide reacts with the silica produced, forming calcium silicate. This has the effect of shifting the equilibrium to the right, and allowing the process to be conducted at a much lower temperature:

$$2MgO(s) + CaO(s) + Si(s) \rightarrow 2Mg(s) + CaSiO_3(s).$$

This illustrates a very important point: the diagrams refer to standard conditions, and these can often be changed to allow reactions to be conducted more easily. For example, gaseous products can be removed by sweeping them away, and liquid and solid products removed by combination. This is the process of slag formation: an oxide, calcium oxide in this case, is added to the reactants to react with an unwanted product, silica in this case, to remove it as an easily melted compound (the slag) which is immiscible with (therefore easily removed from) the molten metal produced. This causes the equilibrium to move to the right, and the process can be conducted at a lower temperature than would otherwise be the case. Since silica is an acidic oxide, a basic oxide is used. When basic oxides are the impurities, an acidic oxide, silica, is added. For example, in the extraction of copper from copper pyrites the oxides of iron present are removed as iron silicate slag. Another important point is that reactions are under thermodynamic control at high temperatures only. At 25 °C the reaction between zinc and oxygen is highly exothermic, and is accompanied by a large negative free energy change, but in fact does not occur under normal conditions since the activation energy is too high, and the reaction occurs very slowly.

Nothing has been said about the mechanism of the reduction of oxides by carbon. The oxidation of carbon produces an equilibrium mixture of carbon monoxide and carbon dioxide, with carbon monoxide predominating at higher temperatures. From Fig. 11.10 it is obvious that, from a thermodynamic point of view, carbon monoxide is the better reducing agent below 700 °C, whereas above 700 °C carbon is the better reductant (lower in the diagram). However, it is unlikely that in, for example, the reduction of zinc oxide by carbon, the equation

$$ZnO(s) + C(s) \rightarrow Zn(s) + CO(g)$$

represents the reaction mechanism, since this would be a reaction between two solids. Almost certainly the carbon monoxide does the actual reducing, even at temperatures above 700 °C, but the carbon dioxide produced is reduced by carbon to give more carbon monoxide, and so on:

$$
\begin{array}{ll}
(1) & CO(g) + ZnO(s) \rightarrow Zn(s) + CO_2(g) \\
(2) & CO_2(g) + C(s) \rightarrow 2CO(g) \\
\hline
(1) + (2) & ZnO(s) + C(s) \rightarrow Zn(s) + CO(g)
\end{array}
$$

By adding equations (1) and (2), it is seen that the thermodynamics of the reaction correspond to reduction by carbon.

The uses of gaseous hydrogen as a reductant are limited to the oxides of elements above it in the diagram. Unlike carbon, its versatility is not increased by raising the temperature. Moreover, its use is limited by the formation of stable hydrides by many metals.

The diagrams can also be used to define an activity series (Section 10.3.19) based upon the $\Delta G^\ominus$ values of the reactions of the elements with, for example, oxygen (or, conversely, upon the thermodynamic stability of the oxides with respect to dissociation into the constituent elements). If a temperature is selected, say 2000 °C, and a vertical line drawn on the diagram representing this temperature, then this line will cut each $\Delta G^\ominus/T$ line on the diagram. On ascending this line the metals become less reactive and the corresponding oxides become less stable. This defines an activity series under these conditions: the noble metals are at one end (the top in this case); the electropositive metals are at the other end. The limitations of such series are obvious here, where the order changes from one temperature to another. The oxides of all metals become unstable with respect to dissociation into the constituent elements at sufficiently high temperatures, but those of the noble metals become unstable at lower temperatures than those for the more electropositive elements. The noble metals are often found native.

The Ellingham diagram for sulphides is given in Fig. 11.12. In general, sulphide ores are roasted in air to form the oxides, and these are then reduced. That this is necessary is due to the ineffective nature of both hydrogen and carbon as reductants for sulphides (Problem 16).

The Ellingham diagram for chlorides is given in Fig. 11.13. Carbon is clearly useless as a reductant for chlorides (Problem 17); but hydrogen is better, and its versatility increases slightly as the temperature is raised (Problem 18). As usual the diagram refers to standard conditions, and the range of reductions can be widened by altering the conditions. The reduction of chromium(II) chloride by hydrogen looks an unlikely proposition (look at the $Cr/CrCl_2$ line, not the $CrCl_3$ line!), but in fact reduction can be effected at about 1000 °C by continuously sweeping fresh hydrogen over the salt, and removing the hydrogen chloride formed. The chlorides of the more electropositive metals are reduced electrolytically, or by replacement with a more reactive metal, for example, $2Mg(s) + TiCl_4(l) \rightarrow Ti(s) + 2MgCl_2(s)$. Both methods are clearly more expensive than carbon reduction.

11.5.2 Extractive Metallurgy

There are three main processes involved in obtaining pure metals from their ores: (1) concentration (or extraction) of the ore; (2) reduction of the ore to the impure metal; and (3) refining of the impure metal.

1) Concentration of the ore. Unwanted earth and siliceous material (gangue) can be removed by either physical or chemical processes (or both). "Floatation" or "differential wetting" is a common physical method, especially for sulphide ores. The ore is powdered, mixed with oil and water, and air is blown through. The oil "wets" the ore, and the water wets the gangue. The result is that the froth of oil contains mainly ore, and is skimmed off. Simpler methods are available for particular elements: the lighter gangue can be simply washed away from galena (lead sulphide), gold, and diamonds; magnetic separation may be possible (wolframite-tungsten). Chemical methods involve the leaching of ores with suitable aqueous solutions to extract the metal as soluble salts: sulphides of copper can be dissolved by dilute sulphuric acid in the presence of air to yield copper(II) sulphate; bauxite can be dissolved by sodium hydroxide to yield sodium aluminate; and silver and gold ores can be dissolved by aqueous cyanide solutions in the presence of air to yield soluble complex cyanides.

2) Reduction. The conversion of a compound of a metal to a metal is a reduction (Section 18.14). The method chosen will be the cheapest available. The ease of reduction of an ore depends upon the position of the relevant $\Delta G^\ominus/T$ line in the appropriate Ellingham diagram, or more generally upon the position of the metal in the activity series. The more electropositive the metal is, the more difficult it is to reduce one of its compounds to the metal. The reductive methods available can be classified into four groups, of generally increasing expense.

a) Roasting in air—for metals below hydrogen in the activity series (those with positive reduction potentials). When sulphide ores are roasted in air, the oxides which are formed in these cases are unstable relative to dissociation into the constituent elements at moderately high temperatures (Fig. 11.11). Take, for example, the red sulphide of mercury, cinnabar:

$$2HgS(s) + 3O_2(g) \longrightarrow 2SO_2(g) + 2HgO(s) \xrightarrow{500\,°C} 2Hg(l) + O_2(g).$$

In the case of copper, the stable copper(I) oxide (Fig. 11.11) formed is reduced by the excess sulphide present:

$$Cu_2S + 2Cu_2O \rightarrow 6Cu + SO_2.$$

b) High temperature reduction by carbon or carbon monoxide—for metals with low negative reduction potentials. The sulphide or carbonate ores are first converted to the oxide by heating, and the oxide is then reduced to the metal by heating with carbon or carbon monoxide (Section 11.5.1). Typical examples of this class are lead, tin, iron, and zinc:

galena $\quad 2PbS(s) + 3O_2(g) \longrightarrow 2SO_2(g) + 2PbO(s)$
$$C(s) + 2PbO(s) \longrightarrow 2Pb(s) + CO_2(g)$$
$$SnS_2(s) + 3O_2(g) \longrightarrow 2SO_2(g) + SnO_2(s)$$
$$C(s) + SnO_2(s) \longrightarrow Sn(s) + CO_2(g)$$
haematite $\quad 2Fe_2O_3(s) + 3C(s) \longrightarrow 4Fe(s) + 3CO_2(g)$
zinc blende $\quad 2ZnS(s) + 3O_2(g) \longrightarrow 2SO_2(g) + ZnO(s)$
$$C(s) + ZnO(s) \longrightarrow Zn(s) + CO(g).$$

c) Electrolysis of molten compounds of the metal—for metals with high negative reduction potentials for which carbon reduction is not suitable (the most electropositive metals). This process is called cathodic reduction since metal ions are forced to accept electrons at the cathode of a cell, and are reduced. The molten salt is itself the electrolyte in the cell, but often inert electrolytes are added to lower the melting point of the salt and increase the conductivity. Suitable inert electrodes are necessary, for example graphite, and often the container itself, or the container lining, is used as the cathode so that molten metal collects on the bottom of the cell from where it can be easily run off. Typical examples of this class are aluminium, magnesium, sodium, and zinc, for example:

$$Al^{3+} + 3e^- \longrightarrow Al(l).$$

Clearly aqueous solutions are useless for this purpose since, even if the metal were discharged at the cathode (actually hydrogen would be), it would react with water.

d) Reduction by a more reactive metal (one of more negative reduction potential). This is a general method since any metal cation can be reduced by a more reactive metal (one above it in the electrochemical series). However, since it involves the use of an already purified reactive metal, the method is clearly expensive and is used only when the above methods are unsuitable. One important example is the preparation of titanium from rutile (TiO_2)—Kroll process. Titanium forms a stable carbide with carbon, and also reacts with oxygen and hydrogen at high temperatures. It is therefore isolated by forming the tetrachloride, purifying this by fractional distillation, and reducing it with molten magnesium in an atmosphere of argon:

$$TiO_2(s) + C(s) + 2Cl_2(g) \xrightarrow{1000\,°C} TiCl_4(l) + CO_2(g)$$
$$TiCl_4(l) + 2Mg(s) \xrightarrow[800\,°C]{Ar} Ti(s) + 2MgCl_2(s).$$

The reduction of metal salts using aluminium is called the thermite or aluminothermic process, because the heat liberated is so great (owing to the great stability of aluminium oxide—Fig. 11.11). Chromium and manganese can both be produced by this method:

$$3MnO_2(s) + 4Al(s) \longrightarrow 2Al_2O_3(s) + 3Mn(s)$$
$$Cr_2O_3(s) + 2Al(s) \longrightarrow Al_2O_3(s) + 2Cr(s).$$

3) **Refining.** The purification of impure metal (Chapter 29).
Various methods are used, all of which have limited applicability.

a) *Electrolysis.* An electrolytic cell is used in which: the anode is made of the crude metal to be purified; the cathode is a strip of pure metal coated with a thin layer of graphite from which deposited pure metal can be removed; and the electrolyte is an aqueous solution of a salt of the metal. On electrolysis the impure metal goes into solution at the anode and pure metal ions are reduced and deposited at the cathode. Obviously, only weakly electropositive metals can be purified in this manner, for example, copper, tin, and lead, since they must not react with water, and must be easily oxidized (at anode), and reduced (at cathode) relative to hydrogen.

b) *Distillation.* Volatile metals can be purified by distillation, for example zinc, cadmium, and mercury. Zinc and cadmium in fact are separated from one another by this method.

c) *Zone-refining.* (See Section 29.1.) This method is used for elements required in a very high state of purity, for example the semi-conductors silicon, germanium, and gallium.

d) *Van Arkel process*—"*vapour phase refining*". In this process a volatile halide is formed by direct action of the metal and a halogen, the volatile halide is purified (for example by fractional distillation), and the halide is then decomposed to the metal and halogen at a higher temperature, for example:

$$Zr(s) + 2I_2(g) \longrightarrow ZrI_4(s) \xrightarrow[800\,°C]{filament} Zr(s) + 2I_2(g).$$

The elements Ti, Hf, Zr, V, W, Si, and Be have been purified in this way.

e) *Purification via the volatile carbonyl compounds* is similar to the van Arkel process, but has been used widely only for nickel:

$$Ni(s) + 4CO(g) \xrightarrow{50\,°C} Ni(CO)_4(g) \xrightarrow[\text{(b) heat to 180\,°C}]{\text{(a) distill}} Ni(s) + 4CO(g).$$

Summary

The more electropositive metals, groups IA and IIA, occur in nature combined with the more electronegative anions such as the singly charged chloride and nitrate in the case of the group IA metals, and the doubly charged sulphate and carbonate in the case of the group IIA metals. These metals are generally prepared by electrolysis of their molten salts. Various transition metals, which form stable carbides, are prepared by reduction with more reactive metals, for example Mn and Cr by the thermite process. Less electropositive metals commonly occur as silicates, sulphides, and oxides, and are manufactured by carbon reduction of their oxides, using slag formation to purify and move the position of equilibrium to the right.

PROBLEMS

1. How do the properties of the compounds $M(OH)_n$ (M is any element) depend on the position of M in the Periodic Table (e.g. NaOH, H_2SO_4, $Al(OH)_3$)?

2. Repeat Problem 1 for (a) chlorides, (b) oxides, and (c) sulphides.

3. Discuss the usefulness of the Periodic Table as a classification of the elements. What are its limitations?

4. Discuss the differences in properties between the following pairs of compounds:
    a) CCl_4 and $SiCl_4$;
    b) NH_3 and PH_3;
    c) HF and HCl;
    d) $B(OH)_3$ and $Al(OH)_3$.

5. Discuss the following misleading generalizations with reference to the substances diamond, methane, sodium chloride, and ammonium chloride: "A useful model of bonding and properties is: covalent compounds are low-melting, soft, and soluble in organic solvents; ionic compounds are high-melting, hard, and insoluble in organic solvents".

6. Place the following chlorides in order of increasing melting point: HCl, NaCl, Cl_2, and CCl_4. Discuss.

7. Discuss the statement: "The chemistry of the first member of any group is not typical of the group".

8. Outline the effect of heat on the salts of the oxyacids (sulphates, carbonates, and nitrates).

9. Comment on the possible existence of the following compounds: CCl_2, HeF_2, KCl_2, KCl_3, KO, and BaTe.

10. Choose from each of the pairs, (i) (ii) etc., that which forms:
    a) the stronger oxidant: (i) F_2 and Br_2; (ii) S and Cl_2.
    b) the stronger reductant: (i) Li and Be; (ii) Be and Mg; (iii) HF and HI.
    c) the stronger acid: (i) HF and HCl; (ii) HOCl and HOBr; (iii) HOCl and $HClO_2$; (iv) $Fe^{2+}(aq)$ and $Fe^{3+}(aq)$; (v) $Na^+(aq)$ and $Al^{3+}(aq)$.
    d) the stronger alkali: (i) LiOH and NaOH; (ii) NaOH and $Mg(OH)_2$; (iii) $Al(OH)_3$ and $Si(OH)_4$.
    e) the more metallic element: (i) I_2 and Br_2; (ii) Be and Ba; (iii) Be and B; (iv) Ca and Zn.

11. Describe the structures and nature of the bonding in the compounds:
    a) KO_2, TiO_2, CO_2, SiO_2, SO_2, ClO_2;
    b) BF_3, NF_3, SbF_3, AlF_3;
    c) MgF_2, CaF_2, OF_2, XeF_2;
    d) BeO, CaO, CO, NO, O_2.

12. Which of each of the following pairs of hydrides is the more stable with respect to dissociation:
    a) HCl and HI;
    b) PH_3 and SbH_3;
    c) NH_3 and H_2O?
    Discuss.

13. The enthalpy of formation of a typical metallic chloride, MCl_2, is usually negative:

$$M(s) + Cl_2(g) \rightarrow MCl_2(s); \quad \Delta H_f^\ominus < 0.$$

Write down a Born–Haber cycle for this reaction and use it to comment on how this enthalpy of formation is affected by:

a) the strength of the metallic bonding;
  b) the ionization energy of the metal;
  c) the size of the ion, M^{2+}.
14. Is there any relation between the ability of an element to form double-bonded compounds to oxygen and nitrogen, and its position in the Periodic Table?
15. Reactions can be classified into three groups (Section 11.5.1.) depending on the entropy change which occurs during the reaction. Into which group would you place the reaction:

$$CO(g) + \tfrac{1}{2}O_2(g) \rightarrow CO_2(g)?$$

16. Why are sulphide ores first roasted to oxides and then reduced to yield the metal, instead of the ore being reduced directly (with carbon for example)?
17. Carbon is a better reductant for oxides than is hydrogen, but the position is reversed for chlorides. Why?
18. Does hydrogen becomes a better reductant for chlorides at higher temperatures? Discuss.
19. What temperature is required to reduce the following oxides by (i) carbon and (ii) hydrogen:
    a) CaO;
    b) FeO;
    c) TiO_2?
20. Can hydrogen be used to reduce the oxides of:
    a) Ni;
    b) Na;
    c) Zn;
    d) Ag;
    e) Al?

21. Which of the following pairs of oxides has the highest melting point:
    a) MgO and Na_2O;
    b) Na_2O and Li_2O;
    c) Sb_2O_5 and N_2O_5;
    d) SeO_2 and SO_2?

BIBLIOGRAPHY

Rich, R. L. *Periodic Correlations*, Benjamin (1965).
> Introductory account.

Sanderson, R. T. *Inorganic Chemistry*, Reinhold (1967).
> Slightly more advanced than the book by Rich with a vast amount of comparative data.

Sisler, H. H. *Electronic Structure, Properties, and the Periodic Law*, Reinhold (1963).

Wells, A. F. *Models in Structural Inorganic Chemistry*, Clarendon (1970).

Bassow, H. *The Construction and Use of Atomic and Molecular Models*, Pergamon (1968).

Addison, W. E. *Structural Principles in Inorganic Chemistry*, Constable (1961).
> Written for undergraduates, but very useful at sixth-form level.

Clapp, L. B. *The Chemistry of the OH Group*, Prentice-Hall (1967).

    Most of this treatment of the chemistry of the OH group can be understood at sixth-form level.

Ives, D. J. G. *Principles of Extraction of Metals*, Monographs for Teachers No. 3, Royal Institute of Chemistry (1960).

MacKay, K. M. *Hydrogen Compounds of the Metallic Elements*, Spon (1966).

    A degree-level discussion, much of which can be understood at sixth-form level.

Shaw, B. L. *Inorganic Hydrides*, Pergamon (1967).

    A more qualitative treatment.

CHAPTER 12

The Noble Gases: Group 0

12.1 Occurrence, physical properties, and uses
12.2 Importance in the history of bonding theory
12.3 The rationalization of the inertness of the noble gases, and its failings
12.4 Preparation and properties of XeF$_4$, a typical compound
12.5 Clathrate "compounds" or enclosure "compounds"
12.6 The molecular shapes of noble gas compounds
12.7 The Gillespie–Nyholm theory

12.1 OCCURRENCE, PHYSICAL PROPERTIES, AND USES

All the noble gases, helium, argon, neon, krypton, xenon, and radon, had been identified by 1900 (by William Ramsay and others). They are all constituents of air except radon, which is radioactive and rare. All, except radon, are obtained commercially from the fractional distillation of liquid air. In addition, helium is found in high concentrations (up to 10 percent) in some natural gases, especially in the U.S.A. and Canada, and is separated by condensation of the other constituents. It has resulted from the decay of radioactive material. All are monatomic, colourless, odourless gases (at 25 °C), and are sparingly soluble in water owing to the formation of hydrates. Their properties are given in Table 12.1. There is a gradual increase in their very low boiling and melting points as the atomic number and size of the atoms increase. This behaviour is typical of atoms (or molecules) whose sole interatomic forces are the weak van der Waals forces which are proportional to the size of the atom (Section 4.9): the melting and boiling points are low (weak forces) and increase with atomic size (the forces are proportional to size). In fact, these transition temperatures (°C) are similar to those of small diatomic molecules such as hydrogen (-253, -259), nitrogen (-196, -210), fluorine (-188, -220), and oxygen (-183, -218). The intermolecular forces in these cases are also the weak van der Waals forces. Thus it is not their physical properties which distinguish these elements from other elements, but their lack of chemical reactivity (Section 4.5). The noble gases have a very narrow temperature range over which the liquid exists owing to the forces in the liquid being similar to those in the solid, another feature of van der Waals forces. The solids formed by the elements are regular close-packed arrangements (Section 3.3). The systematic variation of properties from one element to another, superimposed upon an essential similarity, which is expected for elements in the same group, is very obvious from Table 12.1. The first member of each group is usually "anomalous" (Section 14.2.7), and helium is outstanding among elements in this respect. The distinctive atomic volume and the impossibility of solidifying helium, except under pressure, are obvious anomalies. But helium is even stranger. The liquid isotope helium-3 behaves normally, but if helium-4 (which constitutes almost 100 percent of atmospheric helium) is cooled below 2.2 K at one atmosphere pressure, the normal liquid helium-4, called helium I, changes to an abnormal form called helium II.

Table 12.1. Properties of the noble gases, group 0.

Property	Helium He	Neon Ne	Argon Ar	Krypton Kr	Xenon Xe	Radon Rn
Atomic number	2	10	18	36	54	86
Electron configuration (outer)	$1s^2$	$2s^22p^6$	$3s^23p^6$	$3d^{10}4s^24p^6$	$4d^{10}5s^25p^6$	$5d^{10}6s^26p^6$
Isotopes (order of abundance) —naturally occurring	4, 3	20, 22, 21	40, 36, 38	84, 86, 82 83, 80, 78	132, 129, 131 134, 136, 130 128, 124, 126	
Atomic weight	4.0026	20.183	39.948	83.80	131.30	(222)
Van der Waals radius (Å)		1.31	1.74	1.89	2.1	2.15
Atomic volume (cm^3 mol^{-3})	31.7	16.8	28.5	32.3	42.9	50.5
Boiling point (°C)	−269	−246	−186	−152	−107	−62
Melting point (°C)	−272 (P) 26 atm	−249	−189	−157	−112	−71
Enthalpy of fusion (kJ mol^{-1})	0.02	0.33	1.18	1.64	2.3	2.9
Enthalpy of vaporization (kJ mol^{-1})	0.084	1.77	6.5	9	12.6	16.4
Density, liquid at b.p. (g cm^{-3})	0.126	1.20	1.40	2.6	3.06	4.4
Ionization energy (kJ mol^{-1})	2372	2081	1520	1350	1170	1040
Electron affinity (kJ mol^{-1})	−54	−99				
Percent by volume in dry air	5.2×10^{-4}	1.8×10^{-3}	0.93	1.1×10^{-4}	8.7×10^{-6}	0

P = under pressure; atm = atmospheres.

All its physical properties change, for example its thermal conductivity increases to 800 times that of copper (superconductivity) and its viscosity becomes very low (superfluidity).

The gases are widely used as inert atmospheres in welding and cutting (antioxidant), electric bulbs (prolongs filament life), and metallurgical processes (e.g. titanium). They are useful coolants for low temperature work. Helium or argon are the cheapest and most widely used. In addition each has particular uses: helium in deep sea diving (Problem 1) and in balloons; radon in cancer treatment; and neon, argon, krypton, and xenon, as fillings for coloured discharge tubes in the lighting industry.

12.2 IMPORTANCE IN THE HISTORY OF BONDING THEORY

The distinguishing characteristic of the noble gases is their comparative lack of chemical reactivity: for example, they are the only elements to exist as uncombined atoms. In fact, until 1962 they were thought to be totally unreactive, and were previously known as the "inert" gases. Even today, only the oxides and fluorides of the heavier elements xenon and krypton (to a lesser extent) are known (the compounds of radon are less studied because the element is rare and radioactive). However, it was their chemical stability which gave them a key place in the development of chemical bonding theory. Since chemists attempt to interpret reactivity in terms of electron configuration (Section 10.1), the stability of the noble gases was interpreted as a consequence of the number of electrons in the outer shell: the configuration with filled s- and p-subshells was obviously very stable. This reasoning led to an empirical rule known variously as the *noble gas rule*, the *law of the octet*, or the *rule of eight*. Early bonding theory was based on: (a) this rule; (b) two models of bonding which followed from it (the ionic and covalent models); and (c) a general understanding of "reactivity", all of which are briefly discussed below.

One form of the noble gas rule is: *every atom combines in such a way that its outer shell achieves the stable configuration of eight electrons, that is it tends to adopt the electron configuration of the nearest noble gas.* There are two general methods of achieving this desired configuration (Section 4.5):

1. Loss or gain of electrons to form ionic compounds; for example:

$$\overset{x}{Na} + \cdot \ddot{\underset{\cdot\cdot}{Cl}}: \rightarrow Na^+ \quad {}^x_{\cdot\cdot}\ddot{Cl}:^-$$

$$\underbrace{\underset{\text{electropositive element}}{} \quad \underset{\substack{\text{electro-}\\\text{negative}\\\text{element}}}{}}_{\substack{\text{tends to form a}\\\text{positive}\quad\text{negative}\\\text{ion}\qquad\text{ion}}} \quad \underset{\text{ionic compound}}{\text{cation}\quad\text{anion}}$$

2. Sharing of electrons (elements of similar electronegativities); for example:

$$H^x + \cdot\ddot{\underset{\cdot\cdot}{Cl}}: \rightarrow H{}^x_{\cdot}\ddot{\underset{\cdot\cdot}{Cl}}: \qquad \text{usually written H—Cl}$$

$$\underset{\substack{\text{covalent}\\\text{compound}}}{}$$

The most reactive elements are found to be those with electron configurations closest to those of the noble gases: (a) the alkali metals, which have one more electron than the nearest noble gas, and which react by losing this electron to form ionic compounds in which the alkali metal is present as a positive ion (above); and (b) the halogens, which have one electron less than the nearest noble gas, and which react either by gaining one electron to form ionic compounds in which the halogen is present as a negative ion, or by sharing electrons to form a covalent bond. The example of chlorine is given above.

This model of bonding is very useful for compounds of the elements in the lithium period, but often breaks down for heavier atoms of the subsequent periods; sulphur, for example, forms the stable compound sulphur hexafluoride, SF_6, in which the central sulphur atom has twelve electrons in its outer shell. Nevertheless, this is a useful model of bonding, but more should not be read into the rule and its consequences than that given above. It is not true to say, for example, that the "electropositive elements have a great tendency to lose electrons", since (on any reasonable interpretation of this statement) they have not. The first ionization energy of potassium, for example, is 418 kJ mol^{-1}. This means that to take a mole of electrons away from one mole of gaseous potassium ions requires about the same amount of energy as that required to break the bonds of one mole of hydrogen molecules (about the strongest covalent bond known):

$$K(g) \rightarrow K^+(g) + e^-(g); \quad I = 418 \text{ kJ mol}^{-1}$$
$$H\text{—}H(g) \rightarrow 2H(g); \quad \Delta H_{\text{diss}} = 435 \text{ kJ mol}^{-1}.$$

In fact, all neutral atoms resist losing an electron strongly, but the more electropositive elements resist less than the more electronegative elements (Section 10.3.6). The following statements are not equivalent:

1. Electropositive elements have a strong tendency to lose electrons (*false*);
2. Electropositive elements have a strong tendency to form ionic compounds in which they can be regarded as having lost an electron forming a cation (*true*).

These points are discussed further in Section 14.2.4. It is true, however, to say that the more electronegative atoms have a great tendency to gain an electron and form an anion. In fact, most atoms except the inert gases and other atoms with stable configurations (Section 10.3.4) have some tendency to gain an electron in the gaseous state. This is obvious from the electron affinities (Table A.10). The more electronegative elements, the halogens for example, have more tendency to gain an electron than the less electronegative elements, however, and the greater tendency of the halogen to form ionic compounds is associated with this.

12.3 THE RATIONALIZATION OF THE INERTNESS OF THE NOBLE GASES, AND ITS FAILINGS

The relatively inactive nature of the noble gases is easily rationalized. In each period of the Periodic Table they have the highest ionization energies and are very unlikely either to form cations or to form donor co-ordinate bonds, since they hold on to their outer electrons very firmly (Section 10.3.3). Their electron affinities are negative and they show the least attraction for electrons among members of their period in the Periodic Table; consequently, they would not be expected to form anions, nor to share electrons to form covalent bonds, nor to act as acceptors in co-ordinate bond formation. This comprehensive indictment of their bond-forming ability was generally accepted until 1962. Bartlett had prepared a red compound which he formulated as $O_2^+ \cdot PtF_6^-$, dioxygenyl hexafluoroplatinate(v). He realized that the first ionization energy of molecular oxygen (1180 kJ mol^{-1}) was almost identical with that of xenon (1170 kJ mol^{-1}), and he prepared another red compound by mixing PtF_6 and xenon, which he formulated as $Xe^+ \cdot PtF_6^-$. Soon afterwards, XeF_4 and other compounds were synthesized. The fluorides of krypton, xenon, and radon, are now known. The xenon compounds have been investigated more fully than the others because they are more thermally stable with respect to dissociation into the constituent elements than those of krypton, and because xenon is more available than radon. The main requirement for forming xenon fluorides appears to be the coexistence of xenon and fluorine atoms. In fact, the three fluorides are generally formed together, the proportions depending upon the exact conditions of temperature, pressure, reaction time, proportions of reactants, and so on. Purification is always necessary:

$$Xe + F_2 \xrightarrow{heat} \begin{array}{l} XeF_2 \xrightarrow{H_2O} Xe + \tfrac{1}{2}O_2 + 2HF \\ XeF_4 \longrightarrow XeO_3 + 2Xe + 1\tfrac{1}{2}O_2 + 12HF \\ XeF_6 \longrightarrow XeO_3 + 6HF \text{ (via } XeO_4F\text{)}. \end{array}$$

All three are white solids, and are stable with respect to dissociation into elements at ordinary temperatures (i.e. a negative free energy of formation at 25 °C).

The ordinary chemical nature of these compounds needs to be emphasized. No new theory has to be postulated in order to account for them. On the contrary, their existence was considered long before they were prepared, and they fit in with well-known periodic trends. If the thermodynamically more stable fluorides are

Table 12.2. The stable fluorides of groups IVB to 0.

CF_4	NF_3		OF_2		F_2			Ne
SiF_4	PF_3	PF_5	SF_4	SF_6	ClF	ClF_3		A
GeF_4	AsF_3	AsF_5	SeF_4	SeF_6	BrF_3	BrF_5	KrF_2	
SnF_4	SbF_3	SbF_5	TeF_4	TeF_6	IF_5	IF_7	XeF_2	XeF_4
PbF_4		BiF_5					RnF_6	

written down (Table 12.2) it can be seen that as we go towards the bottom right hand side of the Periodic Table, the more stable fluorides become the ones with an increasing number of fluorine atoms, that is, an increasing number of electrons in the outer shell of the central heavy atom (yet another diagonal relationship). If the ten, twelve, and fourteen, valence electron diagonal sequences: ClF_3, KrF_2; SF_4, BrF_3, XeF_2; SF_6, BrF_5, XeF_4; and IF_7, RnF_6 are considered, then not only does the existence of these compounds appear reasonable, but it is predictable that the more stable fluorides would be expected to be: KrF_2, XeF_4 and XeF_6, and RnF_6, at normal temperatures.

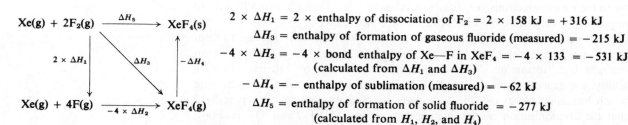

$2 \times \Delta H_1 = 2 \times$ enthalpy of dissociation of $F_2 = 2 \times 158$ kJ $= +316$ kJ

$\Delta H_3 =$ enthalpy of formation of gaseous fluoride (measured) $= -215$ kJ

$-4 \times \Delta H_2 = -4 \times$ bond enthalpy of Xe—F in $XeF_4 = -4 \times 133 = -531$ kJ (calculated from ΔH_1 and ΔH_3)

$-\Delta H_4 = -$ enthalpy of sublimation (measured) $= -62$ kJ

$\Delta H_5 =$ enthalpy of formation of solid fluoride $= -277$ kJ (calculated from H_1, H_2, and H_4)

Fig. 12.1 Stability of fluorides, e.g. XeF_4.

From the Born–Haber cycle (Fig. 12.1), the bond energy of the Xe—F bond is calculated as 133 kJ, and the enthalpy of formation of the solid fluoride as -277 kJ. The stability of this compound clearly depends upon the moderately high bond energy of Xe—F, and the low dissociation energy of the fluorine molecule. The corresponding chloride $XeCl_4$ has not been prepared, and the free energy of formation has been calculated to be positive, largely because of the larger dissociation energy of the chlorine and the smaller bond energies of chlorides, compared to fluorides (Table 19.2).

12.4 PREPARATION AND PROPERTIES OF XeF_4, A TYPICAL COMPOUND

This can be prepared by the action of an excess of fluorine (5:1) on xenon for one hour at 400 °C in a sealed nickel container at about six atmospheres. Some difluoride and hexafluoride are also formed (Section 12.3). The compound is a transparent crystalline solid, m.p. 114 °C, which is stable for indefinite periods when kept dry and pure, relative to dissociation into elements. It is a fairly strong fluorinating agent and oxidant, reacting with almost everything, and generally forming xenon, for example: $SF_4 \rightarrow SF_6$; $H_2 \rightarrow HF$; $I^- \rightarrow I_2$; $Xe \rightarrow XeF_2$;

$C_2H_4 \to C_2H_4F_2$; $Hg \to HgF_2$; $BCl_3 \to BF_3$. With water it forms xenon trioxide, which is stable in solution, but explosive in the solid state: $3XeF_4 + 6H_2O \to Xe + 2XeO_3 + 12HF$. The aqueous solution of xenon trioxide is acidic and is called *xenic acid*. It is a strong oxidant. In basic solution xenates are formed (solid salts are also known) but these disproportionate on standing to perxenates (salts of these are also known):

$$2OH^- + 2XeO_3 \longrightarrow 2HXeO_4^-$$
$$\text{xenate}$$

$$8OH^- + 4HXeO_4^- \longrightarrow Xe + 6H_2O + 3XeO_6^{4-}.$$
$$\text{perxenate}$$

Thus in acid or neutral solution xenates(VI) are stable; in alkali solution perxenates-(VIII) are stable. Perxenates are the strongest oxidants known and will oxidize manganese(II) to permanganate, etc. Some reduction potentials are:

$$XeO_6^{4-} \xrightarrow{0.9} HXeO_4^- \xrightarrow{0.96} Xe \quad \text{alkali}$$
$$H_4XeO_6 \xrightarrow{2.3} XeO_3 \xrightarrow{1.8} Xe \quad \text{acid}$$
$$1.6 \searrow \quad \nearrow 2.2$$
$$XeF_2$$

Xenon compounds are very useful oxidants and fluorinating agents since the reduction product is xenon, a gas, and they leave behind no contaminant.

12.5 CLATHRATE "COMPOUNDS" OR ENCLOSURE "COMPOUNDS"

Small molecules such as nitrogen, methanol, and argon can be trapped in the interstices of crystal lattices. For example, if hydroquinone (1,4-dihydroxybenzene) is crystallized from water or benzene, under argon at 20 atmospheres pressure, a clathrate "compound" $[C_6H_4(OH)_2]_3 \cdot 0.67Ar$ is obtained. This is not a true compound, but is a three-dimensional hydrogen-bonded structure of hydroquinone molecules, which contains cavities, two-thirds of which are occupied by trapped argon atoms. On heating or dissolution the trapped atoms escape. The cavities are never found completely filled. Helium and neon do not form clathrates, presumably because they are too small and can escape. This provides a method of separating helium and neon from the other noble gases. The crystallization of water containing dissolved argon, xenon, or krypton also results in the formation of clathrate compounds, called the noble gas hydrates. Again, helium and neon do not form such compounds. The limiting formula of such compounds is $5.75H_2O \cdot Xe$, but again the interstices are never completely filled.

12.6 THE MOLECULAR SHAPES OF NOBLE GAS COMPOUNDS

One of the more interesting aspects of the noble gas compounds is their shape. Figure 12.2 gives some examples. The shape of XeF_6 is unknown.

12.7 THE GILLESPIE–NYHOLM THEORY

The shapes of the noble gas compounds and most other simple covalent molecules, other than those of the transition elements, can be deduced by applying the rules

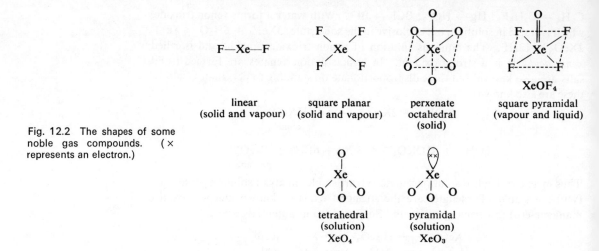

Fig. 12.2 The shapes of some noble gas compounds. (× represents an electron.)

below. These rules have been developed from the idea, put forward by Sidgwick and Powell, that electron-pair repulsions play an important role in determining molecular shape. (See Section 4.10.)

1. The shapes of covalent molecules are mainly determined by the number of electron pairs (bonding and non-bonding) in the outer shell of the central atom. First count this number. These pairs arrange themselves in space in such a way as to minimize electrostatic repulsion between the pairs (Table 12.3).
2. In the interactions between the pairs of electrons the repulsions decrease in the order: lone pair / lone pair > lone pair / bonding pair > bonding pair / bonding pair. This is due to the fact that non-bonded electron pairs tend to be closer to the central nucleus than bonded electron pairs which have been "pulled out" from the nucleus by the other bonding atom. Since it is repulsion in the congested space near this nucleus that is important, non-bonded electron lone pairs repel other electron pairs more than do bonding pairs.
3. It follows from this discussion that, in order to minimize electrostatic repulsion, whenever alternative structures are possible the one adopted is either (a) that in which the non-bonded pairs occupy the largest solid angles, or (b) if this is not possible (see later) the one in which the lone pairs are as far apart as possible.
4. Multiple bonds are considered as single bonds.
5. Single electrons repel other electrons less than pairs of electrons do. The repelling power of electrons for other electrons is therefore: lone pair > multiple bonds ⩾ bonding electron pair > single electron.

The application of these rules will be illustrated below. The simplest examples are those which contain single bonds only, that is, no lone pairs or multiple bonds. Some examples are given in Table 12.3. The theory predicts these shapes because in each case the particular shape gives the farthest distance between the bonds and the largest solid angles for each electron pair to occupy (Section 4.10).

Structures containing lone pairs are based upon the simple structures above, but the lone pairs occupy one of the positions. Tin(II) chloride, for example, contains three pairs of electrons around the central atom: two bonding pairs and

Table 12.3. Simple structures containing no lone pairs of electrons.

Bonds	Shape	Examples
2	linear	$HgCl_2$
3	triangular	BF_3
4	tetrahedral	CH_4, $SnCl_4$
5	trigonal bipyramidal	PF_5
6	octahedral	SF_6, $AlCl_6^{3-}$
7	pentagonal bipyramidal	IF_7

one non-bonding pair. The structure is therefore based on a triangular arrangement of the three pairs of electrons (rule 1), and this structure (Fig. 12.3) is known as *angular* (as opposed to linear). Since the lone-pair / bonding-pair repulsion is greater than the bonding-pair / bonding-pair repulsion (rule 2), the angle in tin(II) chloride is less than that "expected" for a triangular shape (120°). Similarly, in the case of water, based on a tetrahedral arrangement of the four pairs of electrons about the central atom, the bond angle is less than the tetrahedral angle (109° 28′). The bond angle in ammonia is also less than the tetrahedral angle, for the same reason. The shapes of the other molecules in Fig. 12.3., can be similarly rationalized (remember rule 4).

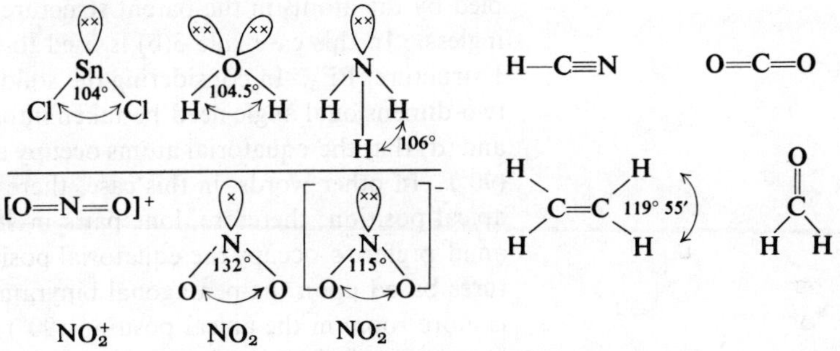

Fig. 12.3 The shapes of some molecules and ions.

The application of rule 3 requires a little skill and a further piece of knowledge. Consider the simple structures in Table 12.3. They can be classified into two groups: (a) those in which the non-central atoms are not all equivalent, e.g.

Table 12.4. Structures containing lone pairs of electrons.

Bonded pairs of electrons	Lone pairs of electrons	Total pairs	Structure is based on:	Examples
2	1	3	triangle	$SnCl_2$ (gaseous)
2	2	4	tetrahedron	H_2O
3	1	4	tetrahedron	NH_3
3	2	5	trigonal bipyramid	ClF_3
4	2	6	octahedron	XeF_4, ICl_4^-
5	1	6	octahedron	IF_5

12.7 THE GILLESPIE-NYHOLM THEORY

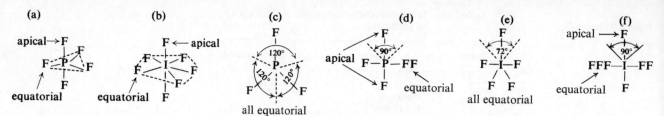

Fig. 12.4 Structures of PF_5 and IF_7: (a), (c), and (d) refer to PF_5; (b), (e), and (f) refer to IF_7; (c) shows only the equatorial groups in the PF_5 molecule. Angles between the dotted lines give the effective angle occupied by the enclosed atom.

PF_5, IF_7; and (b) those in which the non-central atoms are all equivalent, e.g. $HgCl_2$, BF_3, CH_4, SF_6. In order to check this classification imagine yourself seated on each of the non-central (peripheral) atoms in the molecule in turn, for example one of the fluorine atoms in PF_5. Look towards the central atom and remember the arrangement of atoms you see. Repeat this for each of the peripheral atoms. Either the same view is obtained in each case—class 2—or it is not—class 1. In the case of class 1 there are two types of atoms, which are in different chemical environments, referred to as the apical and equatorial atoms (Fig. 12.4). Rule 3(a) is only of value for compounds based upon class 1 structures. For compounds based upon class 2 structures (actually the difficulty only occurs for those based on the octahedron), since the positions are all identical, the solid bond angles occupied by the atoms in the parent structure are all the same, and rule 3(a) is meaningless. In this case rule 3(b) is used to predict the structure. Consider the class 1 structure, PF_5. In considering the solid angles occupied by each atom, only the two-dimensional angle need be taken into account. It is obvious from Fig. 12.4(c) and (d) that the equatorial atoms occupy a larger solid angle (120°) than the apical (90°). In other words, in this case, there is more room in the equatorial than the apical position; therefore, lone pairs in structures based upon the trigonal bipyramid prefer to occupy the equatorial position. This situation is reversed for structures based upon the pentagonal bipyramid. In this case (Fig. 12.4e and f), there is more room in the apical position (90°) than the equatorial (72°), and therefore lone pairs of electrons in structures based on the pentagonal bipyramid prefer to go into the apical position. Consider the structure of ClF_3. The central atom has five pairs of electrons around it and the structure is therefore based on the trigonal bipyramid (rule 1). There are three possibilities, Fig. 12.5: (a) the lone pairs can both be in apical positions; (b) one lone pair can be in an apical position, and the other in an equatorial position; or (c) both can be in equatorial positions. Since there is more room in the equatorial positions, the non-bonding electrons, which repel

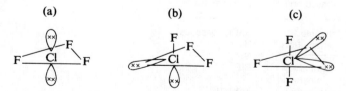

Fig. 12.5 The possible structures of chlorine trifluoride.

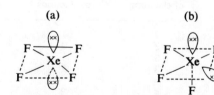

Fig. 12.6 The possible structures of xenon tetrafluoride.

other electrons the most, prefer to enter these in order to minimize electrostatic repulsions. For the same reason ICl_2^- and XeF_2 are linear molecules.

For structures based on the octahedron, class (b), the rule 3(a) is useless since all the positions are the same and all the solid bond angles are the same. However, application of rule 3(b) enables the structure to be predicted. Consider the structure of XeF_4. There are six pairs of electrons around the central atom, therefore the structure is based on an octahedron. There are two possibilities (Fig. 12.6). All "other" structures are identical with (can be superimposed upon) these. The structure adopted, (a), is that in which the lone pairs get as far apart as possible, rule 3(b). What is the structure of ICl_4^-?

There are, of course, cases in which the model predicts the "wrong" structure. In fact, there are surprisingly few of these for such a relatively simple model.

PROBLEMS

1. Deep sea divers may suffer from the "bends"; this is caused by the formation of gas bubbles in the veins, disrupting the flow of blood. This gas is dissolved in the blood at the high pressure necessary for deep sea diving and is released from the blood when the pressure is lowered. An "artificial air" composed of 80 percent helium and 20 percent oxygen is found to assist in preventing the bends. Why?

2. Why do compounds of xenon and fluorine exist, whereas compounds of (a) argon and fluorine, and (b) xenon and bromine, are unknown?

3. Why do the melting and boiling points of the noble gases increase on descending the group?

4. Describe the importance of the noble gases in the history of bonding theory.

5. Why is helium much less soluble in water than xenon?

6. What are the shapes of the following molecules and ions: XeF_2, XeF_4, XeF_6, XeO_6^{4-}, $XeOF_4$, XeO_4, XeO_3, ICl_4^-, ClF_3?

7. What are the shapes of the following molecules and ions: BF_3, BO_3^{3-}, CO_3^{2-}, NO_3^-, SO_3? A consideration of such series led to the formation of a rule known as the *isoelectronic rule*—what is this?

8. Write the Lewis Structures of the following ions: PO_4^{3-}, NO_2^-, NO_3^-, ClO_2^-, ClO_3^-, ClO_4^-, SO_3^{2-}, SO_4^{2-}, IO_6^{5-}. What are the shapes of these ions, and which have dipole moments?

9. What are the shapes of the following pairs of molecules, and which show dipole moments:
    a) CH_2Cl_2 and CCl_4;
    b) BF_3 and NF_3;
    c) IF_5 and PF_5;
    d) SO_2 and CO_2?

10. Predict the geometries of: XeF_2, PF_6^-, I_3^-, $TeCl_4$, BF_4^-, and $CuCl_2^-$.

11. Write "electron dot" structures and predict the shapes of the following molecules: BCl_3, CCl_4, NCl_3, PCl_5, SCl_4, SF_6, and IF_5.

12. Suggest three ions which are isoelectronic with SF_6. What is their shape?

BIBLIOGRAPHY

Holloway, J. H. *Noble Gas Chemistry*, Methuen (1968).
  A mainly qualitative review of the chemistry of the noble gases.
Stark, J. G. (Editor). *Modern Chemistry*, Part 2: "Stereochemistry," Penguin (1970).
  Articles by E. Sherwin and G. Baddeley at an introductory level.
Gillespie, R. J., and R. S. Nyholm. *Quarterly Reviews*, **11**, 339 (1957).
  A discussion of stereochemistry.

Most of the books on inorganic chemistry in the general list contain material on the chemistry of the noble gases.

CHAPTER 13

Hydrogen

13.1 ISOTOPES

Hydrogen is the first element in the Periodic Table and has the simplest atomic structure of all the elements: it consists of a nucleus with a charge of $+1$, and one electron. Three isotopes are known (Table 13.1): protium, ^1_1H; deuterium, ^2_1H or D; and tritium, ^3_1H or T. Tritium is radioactive and decays by emitting a particle (electron) from the nucleus: $^3_1H \rightarrow {}^3_2He + e^-$. The three isotopes of hydrogen contain 0, 1, and 2 neutrons respectively (Section 2.2). Since their electron configurations are identical, their chemical properties are largely the same; but owing to their very different masses their rates of reaction and equilibrium constants for corresponding reactions are slightly different. These differences allow them to be separated. The chemical equivalence of the isotopes of a given element (any element, not only hydrogen) led to their use as "labels" or "tracers". Isotopic substitution of an element by one of its isotopes can be detected through its difference in mass or, in some cases (tritium for example), by its radioactivity. The isotopes of an element are usually chemically indistinguishable from each other, and

- 13.1 Isotopes
- 13.2 Occurrence, preparation, properties, and uses. Atomic hydrogen
- 13.3 Types of compound formed
- 13.4 Types of acid in aqueous solution
- 13.5 Complex hydrides
- 13.6 Water

Table 13.1. Properties of atomic and molecular hydrogen.

	Hydrogen	Deuterium	Tritium
Boiling point (°C)	−252.6	−249.4	
Melting point (°C)	−259.2	−254.5	
Density of liquid at b.p., H_2 (g cm^{-3})	0.071		
Ionization energy, H (kJ mol^{-1})	1312		
Electron affinity, H (kJ mol^{-1})	73		
Electronegativity (A/R)	2.1		
Bond length, H_2(g) (Å)	0.749		
Bond energy (ΔH_{atom}), H_2 (kJ mol^{-1})	436		
Ionic radius, H^- (Å)	2.08		
Covalent radius (Å)	0.29		
Van der Waals radius (Å)	1.2		
ΔH_{fus} ⎫	117	219	
ΔH_{vap} ⎬ latent heats (J mol^{-1})	904	1226	
ΔH_{sub} ⎭	1030	1426	
Percent in normal hydrogen	99.985	0.015	10^{-15}
Atomic weight (C = 12)	1.008	2.014	3.016

Table 13.2. Properties of water and heavy water.

	H_2O	D_2O
Density at 20 °C (g cm^{-3})	0.9982	1.1059
Melting point (°C)	0	3.82
Boiling point (°C)	100	101.4
Temperature of maximum density (°C)	4	11.6
Dielectric constant at 25 °C	82	80.5
Solubility at 25 °C of NaCl per 100 g	3.6	3.0
Solubility at 25 °C of BaCl$_2$ per 100 g	36	29

undergo the same chemical reactions. Replacement of some of the naturally occurring atoms of an element by an isotope is therefore equivalent to labelling atoms, and it is possible to find out what happens to them in complicated processes. This is an excellent method for tracing reaction pathways. The isotopes of hydrogen are used in this manner as tracers. However, in this case the percentage weight difference between the isotopes is so great that it is possible that substitution of deuterium or (especially) tritium for hydrogen in a system could actually alter a reaction path: an alternative pathway may become more likely to occur owing to large changes in rates and relative rates of reactions. This is less likely to be true for isotopes of heavier atoms.

The large differences in some physical properties between hydrogen and deuterium, and between the compounds water and heavy water (deuterium oxide), are obvious from Table 13.2. In other elements corresponding differences are relatively trivial. The isolation of heavy water exemplifies these differences: it can be obtained from water, which contains 0.016 percent of deuterium oxide, either by fractional distillation or by electrolysis (in Norway, for example, where electricity is cheap). Hydrogen is liberated six times more quickly than deuterium at the cathode, and the residual liquid is therefore being continuously concentrated in heavy water during electrolysis.

Deuterium oxide is used as a moderator in nuclear reactors since it does not react with neutrons by absorbing them (and therefore reducing the number of neutrons) but it does slow them down and is itself heated. Heavy water undergoes all the reactions of water and is a useful source of deuterated compounds:

$$PCl_5(s) + D_2O(l) \rightarrow POCl_3(l) + 2DCl(g)$$
$$Mg_3N_2(s) + 3D_2O(l) \rightarrow 3MgO(s) + 2ND_3(g), \text{ etc.}$$

It also undergoes exchange reactions with hydrogen under suitable conditions:

$$NH_3(g) + D_2(g) \rightarrow NH_2D(g) + HD(g)$$
$$OH^-(aq) + D_2O(l) \rightleftharpoons OD^-(aq) + HOD(l), \text{ etc.}$$

See Problems 22 and 23 in Chapter 26.

13.2 OCCURRENCE, PREPARATION, PROPERTIES, AND USES. ATOMIC HYDROGEN

On earth, hydrogen occurs only in trace quantities as molecular hydrogen, but is abundant in its compounds. It forms more compounds than any other element, and forms compounds with all elements except the noble gases.

Hydrogen is the lightest gas known. It is colourless, odourless, and very insoluble in water. In the laboratory it is prepared by the action of dilute acids on metals such as zinc and iron. For example, granulated impure zinc reacts with aqueous acids:

$$Zn(s) + 2H^+(aq) \rightarrow Zn^{2+}(aq) + H_2(g).$$

Certain elements react with alkalis to yield hydrogen (e.g. aluminium), and hydrogen can also be obtained by the action of water on ionic hydrides (Section 11.1.3). On a large scale hydrogen is obtained by: the thermal cracking of hydrocarbons (Section 22.3); the reduction of water by carbon (Section 16.6); the electrolysis of water; or as a by-product during the electrolysis of brine (Section 19.2).

Hydrogen is thermally stable with respect to dissociation into atoms:

$$H_2(g) \rightarrow H(g) + H(g); \quad \Delta H = 436 \text{ kJ mol}^{-1}.$$

Atomic hydrogen can be produced from molecular hydrogen: in an electric arc; at high temperatures; or with ultra-violet radiation. The atomic hydrogen produced can recombine on a suitable surface (to conduct the heat of reaction away) giving off the bond energy. This has been used for welding metals, since the surrounding hydrogen gas prevents the hot metal from being oxidized. Atomic hydrogen is a powerful reductant (elements are reduced to hydrides, for example) and cannot exist except at high temperatures, for example in hot flames and the sun.

Despite the highly endothermic nature of (molecular) hydrogen gas, it is moderately reactive because hydrogen forms strong bonds with many other elements. The reactions are frequently slow however (kinetically controlled) because of the high activation energy for a simple bimolecular process. But once a chain reaction is initiated, perhaps by a spark, the reaction may proceed explosively. For example, hydrogen–oxygen mixtures are stable at room temperature in the dark, but if exposed to a flame such mixtures explode. Most non-metals react with hydrogen on heating to give hydrides; nitrogen forms ammonia, for example (Section 17.9). The hydrides of the elements are discussed in Section 11.1.3.

Hydrogen is a reducing agent:

1. The metal oxides of the less electropositive metals are reduced to the metals (see the Ellingham diagram, Section 11.5.1, for an explanation). This method is not used industrially (since carbon is more versatile and cheaper) unless either carbon cannot be used (the element may form carbides for example), or only small amounts of metal are required. Iron is the most electropositive metal that can be reduced in this manner:

$$Fe_3O_4(s) + 4H_2(g) \rightleftharpoons 3Fe(s) + 4H_2O(g); \quad \Delta G^\ominus = 0 \text{ at } 500\,°C.$$

The equilibrium can be moved to the right by continually providing fresh hydrogen and removing the steam by sweeping with hydrogen. Below iron in the activity series the $\Delta G^\ominus$ values for the oxide reductions by hydrogen become increasingly negative, and the reactions more favourable.

2. Unsaturated organic compounds are reduced to saturated compounds by hydrogen in the presence of catalysts such as finely divided platinum and nickel. An example is the hardening of fats and oils (Section 22.2.4):

$$\text{C=C (l)} + H_2(g) \xrightarrow{Pt} \text{H-C-C-H (s)}.$$

3. Carbon monoxide is reduced to methanol (Section 16.6.1).
4. Hydrogen often fails to cause reduction in aqueous solution when a reaction might be expected from the point of view of thermodynamics. For example, the half-cell potentials of the reductions of iron(III) ions and silver ions in aqueous solution are: Fe^{3+}/Fe^{2+}, $+0.77$ V; and Ag^{+}/Ag, $+0.8$ V. If hydrogen is passed through solutions, either of iron(III) ions or of silver ions, in the absence of catalyst, no reaction takes place. These reactions are kinetically controlled; the high activation energy is associated with the high bond energy of the hydrogen molecule (Section 8.9).

Homogeneous catalysts are now known for hydrogenation reactions. These are complex compounds of the transition metals, the ion $Co(CN)_5^{3-}$ for example.

Although a little hydrogen is used for special purposes such as filling balloons and oxy-hydrogen welding, most of the hydrogen produced is used in preparing other compounds: methanol; formaldehyde (plastics); and ammonia (fertilizer, nitric acid, and explosives) (Section 17.9).

13.3 TYPES OF COMPOUND FORMED

Hydrogen forms three general types of compound, in which it occurs as (1) the anion H^- (hydride ion), (2) the cation H^+ (proton), and (3) the covalently bonded atom as in H—X (X = halogen).

1. Ionic hydrides are expected to occur less widely than halides since, whereas the heat of formation of the gaseous halide ion for the least electronegative halogen, iodine, is -190 kJ mol^{-1}, that of hydrogen is $+145$ kJ mol^{-1}. In fact, hydrides occur only among the group IA and IIA elements (Section 11.1.3), the most electropositive elements.

$\frac{1}{2}X_2(g) \xrightarrow{\Delta H_{atom}} X(g) \xrightarrow{-E} X^-(g)$

	X = H:	X = I:
$\Delta H =$	$+145$ kJ mol^{-1}	-190 kJ mol^{-1}

The difference between hydrides and halides is clearly seen in the behaviour of the transition metals which form ionic halides but metallic hydrides (Section 11.1.3), and in the relative reactivity of typical ionic halides and hydrides.

2. Free protons exist only in discharge tubes, since owing to its uniquely small size the proton has a vast polarizing power (Section 11.2) and is therefore invariably found associated with other atoms. In water the proton is associated with water molecules, species such as H_3O^+ and $H_5O_2^+$, existing in solution. These are usually written, H^+(aq). The enthalpy of hydration, ΔH_{hyd}, for the reaction $H^+(g) + H_2O(l) \rightarrow H^+(aq)$ cannot be measured directly since only the enthalpies of hydration of *compounds* can be measured. Just as the electrode potential of the hydrogen electrode was arbitrarily taken to be zero owing to a similar problem of allocation of energies, the ΔH_{hyd} for the hydrogen ion is often taken to be zero (despite the fact it is obviously very large) in order to assign arbitrary, but relatively consistent, values for all other ions.

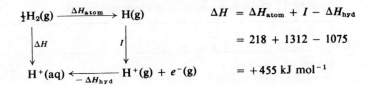

Fig. 13.1 The enthalpy of formation of the aquated proton.

Recently it has been estimated to be about -1075 kJ mol^{-1}. Using this value we can estimate the energy for the process: $\frac{1}{2}H_2(g) \rightarrow H^+(aq)$ (Fig. 13.1). The estimated value $+455$ kJ mol^{-1} makes the process seem unlikely, but it has not included the hydration of the electron, which converts ΔH for the process into a negative quantity. The large negative value for the enthalpy of hydration of the proton is the main reason why many covalent hydrides are acidic, that is exist in water as solutions containing H$^+$(aq), despite the fact that the H—X bond is often very strong.

3. Covalent bonds are formed between hydrogen and all the non-metals except the noble gases. Since hydrogen is of intermediate electronegativity, covalent bond formation is the normal mode of combination (Section 4.1). Usually, the bonds have a dipole moment (Section 4.4), and their strengths are large (Table A.11).

In addition to the above bond types, hydrogen forms metallic hydrides with the transition elements (Section 11.1.3), hydrogen bonds with highly electronegative elements (Section 4.8), and bridging hydrogen bonds in the hydrides of boron (Section 15.2.4).

13.4 TYPES OF ACID IN AQUEOUS SOLUTION

The ionization energy of hydrogen is greater than the first ionization energy of xenon, and this makes the existence of protonic hydrogen, H$^+$, unlikely. For this reason (see above) hydrogen forms mainly covalent compounds, and H$^+$ cannot exist except in special circumstances—only in solution when it is aquated. Compounds which dissolve in water to give a solution containing a greater concentration of hydrogen ions than does pure water are called acids. These can be classified into three groups: (1) hydro-acids, which are binary hydrides of the non-metals; (2) oxyacids, which are hydroxy compounds of non-metals; and (3) solvated cations.

13.4.1 Hydro-acids: Binary Hydrides

Consider the halogen acids for example. The problems are firstly, how to understand the relative acid strengths in terms of the free energy changes, and secondly, how to rationalize these in terms of a simple model. The acidity, pK_a, of a binary hydride is a measure of how far the reaction, $H—X(aq) \rightarrow H^+(aq) + X^-(aq)$, proceeds. The more negative the $\Delta G^\ominus$ value is for this reaction, the larger the equilibrium constant K_a, the more negative the pK_a value, and the more acidic the hydride. The Born–Haber cycle for this reaction, the free energy values $\Delta G^\ominus$ for each step, the $\Delta G^\ominus_{\text{diss}}$ value for the reaction, and the calculated pK_a values (which are close to those measured), are given in Table 13.3, and Fig. 13.2. Clearly, in water, the order of decreasing acid strength is HI > HBr > HCl ≫ HF; though

Table 13.3. The relative acidities of the hydrogen halides.

H		H—F	H—Cl	H—Br	H—I
H—X(aq) → H—X(g)	ΔG_1	23.8	−4.2	−4.2	−4.2
H—X(g) → H(g) + X(g)	ΔG_2	534.7	403.8	338.9	272
H(g) → H$^+$(g) + e$^-$	ΔG_3	1319	1319	1319	1319
e$^-$ + X(g) → X$^-$(g)	ΔG_4	−347.3	−366.5	−345.2	−315.1
H$^+$(g) + X$^-$(g) → H$^+$(aq) + X$^-$(aq)	$\Delta G_5 + \Delta G_6$	−1512.5	−1392.4	−1362.7	−1328.9
H—X(aq) → H$^+$(aq) + X$^-$(aq)	ΔG_{diss}	17.7	−40.8	−54.2	−57
$\dfrac{\Delta G}{5.69} = pK_a$		3.1	−7.2	−9.5	−10.2

in water all are classified as strong acids except hydrogen fluoride, which is weak. The principal factor in making hydrogen fluoride a much weaker acid than the others, is the large free energy of bond-breaking, ΔG_2. This, together with the oddly low electron affinity of fluorine, ΔG_4, more than compensates for the large hydration energy of the fluoride ion, which tends to make it a stronger acid than the others. Since the entropies of bond-breaking are similar for all the halides, the relative weakness of the acid H—F can be attributed to the strength of the H—F bond, an enthalpy term. This is the typical chemists' trick: in order to simplify discussion, and to avoid having to write out Born–Haber cycles the whole time, a trend in properties is attributed to the principal cause, in this case the bond enthalpies, and a simple model is formulated.

Consider the pK_a values for other hydrides in aqueous solution at 298 K (Fig. 13.3). The same trend of increasing acidity on descending any group is found, and this again correlates with decreasing bond enthalpies (or bond free energies). However, in going across a period the bond strength increases, but so also does the acid strength (Fig. 13.3). Clearly, whereas the changes in bond strength are the most important changes on descending a group from the point of view of acidity, in traversing a period some other factor is more important. Lack of data precludes quantitative discussion, but it can obviously be attributed to the massive increase in electronegativity which occurs along any period, compared to the relatively small increase in electronegativity down any group. The increasing electronegativity "causes" the free energy changes ΔG_5 and ΔG_6 (Fig. 13.2), to increase along the group and favour the dissociation of H—X(aq) to H$^+$(aq) and X$^-$(aq).

Our model is, therefore, that changes in acidity of the binary hydrides H—X from element to element are attributed to two main factors:

1. The electronegativity of X: the greater the electronegativity of X the greater is

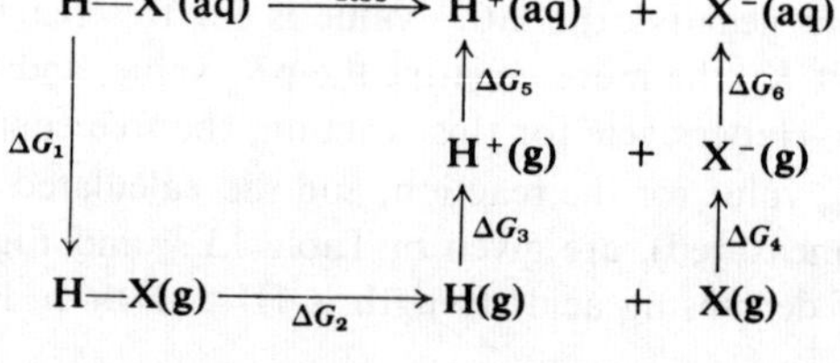

Fig. 13.2 Born–Haber cycle for the ionization and hydration of hydrogen halides.

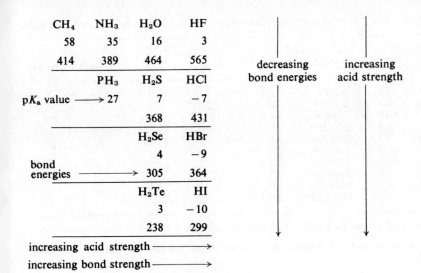

Fig. 13.3 The acidities of the binary hydrides of p-block elements.

the electron affinity of X(g) and the hydration energy of X⁻, therefore the more acidic is the hydride.
2. The bond strength of H—X: the greater this is, the less acidic is the hydride.

In descending any *p*-block group, the change in bond strength is greater than the change in the decreasing electronegativity; and the changes in bond strength determine the changes in acidity. In traversing any *p*-block group, the changes in electronegativity are greater than the changes in bond strength, and the trend in electronegativity determines the trend in acidity. In general, factors which tend to make acids strong are small (H—X) bond energies, large (X) electron affinities, and large (X⁻) hydration energies.

13.4.2 Oxyacids

Oxyacids of non-metals have the general formula $XO_p(OH)_q$. Sulphuric acid, for example (Section 18.13), has X = S, $p = 2$, and $q = 2$. If these acids are grouped into classes of equal values of p, it can be seen that such acids have similar pK_a values (Table 13.4). Thus the number of oxygen atoms attached to the non-metal, X, is the principal factor determining the acid strength. In fact, the structure of an unknown acid could be deduced from a knowledge of the molecular formula and the pK_a value. Consider phosphorous acid which is often misleadingly written $P(OH)_3$ or H_3PO_3. Its pK_a value is 1.8. Obviously, therefore, it must have one oxygen attached to the phosphorus (and to no other atom). The structure cannot be $P(OH)_3$ which would have a pK_a value of about 9, but is $OPH(OH)_2$ (Section 17.14). Another good example is the increasing acid strength of the oxyacids of chlorine: $Cl(OH) < OCl(OH) < O_2Cl(OH) < O_3Cl(OH)$. The effect can be attributed to the electronegativity or electron-withdrawing power of the oxygen atoms which leave the central atom positively charged. This in turn takes negative charge from the oxygen of the hydroxyl group, which in turn takes negative charge from the hydrogen atom, leaving it positively charged and making the oxygen–hydrogen bond more easily breakable, and therefore more acidic. The more

Table 13.4. The first pK_a values for some inorganic oxyacids in aqueous solution (298 K).

| $p = 0$ | | $p = 1$ | | $p = 2$ | | $p = 3$ | |
Acid	pK_a	Acid	pK_a	Acid	pK_a	Acid	pK_a
Cl(OH)	7.2	NO(OH)	3.3	NO_2(OH)	−1.4	ClO_3(OH)	(−10)
Br(OH)	8.7	ClO(OH)	2.0	ClO_2(OH)	−1		
I(OH)	10.0	CO(OH)$_2$	3.9*	IO_2(OH)	0.8		
B(OH)$_3$	9.2	SO(OH)$_2$	1.9	SO_2(OH)$_2$	(−3)		
As(OH)$_3$	9.2	SeO(OH)$_2$	2.6	SeO_2(OH)$_2$	(−3)		
Sb(OH)$_3$	11.0	TeO(OH)$_2$	2.7				
Si(OH)$_4$	10.0	PO(OH)$_3$	2.1				
Ge(OH)$_4$	8.6	AsO(OH)$_3$	2.3				
Te(OH)$_6$	8.8	IO(OH)$_5$	1.6				
		H_2PO(OH)	2.0				

* This is the true value for H_2CO_3, which takes into account the H_2CO_3–CO_2 equilibrium (Section 16.6).
(From K. B. Harvey and G. B. Porter *Introduction to Physical Inorganic Chemistry*, Addison-Wesley (1963).)

oxygen atoms there are attached to the central atom, the more effective is this withdrawal of charge and the more acidic is the compound. Within a given class, the differences in acidity depend on other, less important, factors. For example, the acidities of the halogen oxyacids, XOH, decrease in the order ClOH > BrOH > IOH (Problem 7).

The successive pK_a values of polyprotic acids are also interesting. Consider carbonic acid, H_2CO_3 (Fig. 13.4). The successive pK_a values (Table 13.5) give the ease of ionization of each proton. Clearly, after one has been removed, there will be a negative charge on the molecule and it will be much more difficult to remove the second proton. Thus there is a large difference between successive pK_a values; $\Delta pK_{12} = 10.32 - 3.88 = 6.4$. This is the case for all class I acids, those in which the hydroxyl groups that yield the protons are bonded to the same central atom. Consider the case of the class II acids (Table 13.5), that is those which have hydroxyl groups bonded to different central atoms, such as ethane-1,2-dicarboxylic acid (Fig. 13.4). For such acids there is a small difference between some successive pK_a values, and a large difference between others. Again this can be correlated with structure. In ethane-1,2-dicarboxylic acid the hydroxyl groups are so far apart that the ionization of one group hardly affects that of the other, and therefore the difference between the pK_a values, ΔpK_{12}, is relatively small. Moreover, the effect clearly depends on the distance apart of the hydroxyl groups (Table 13.5). Again, this phenomenon can be used to interpret structure. The acid $H_4P_2O_6$, hypophosphoric acid, has successive pK_a values of 2.2 and 2.81. The absolute values of

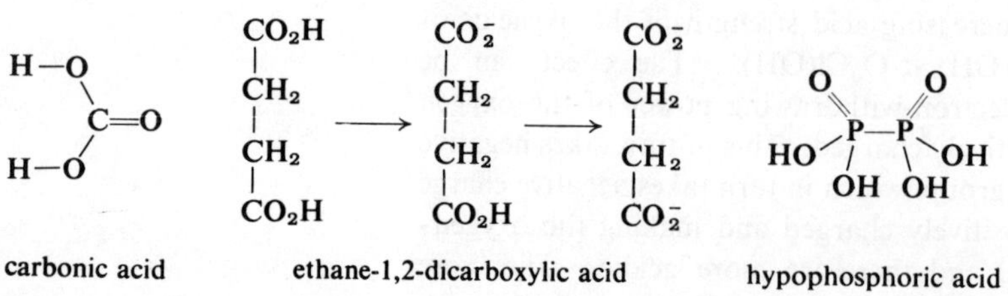

Fig. 13.4 The structures of some polyprotic acids.

carbonic acid

ethane-1,2-dicarboxylic acid

hypophosphoric acid

Table 13.5. Successive pK_a values of some polyprotic acids in aqueous solution (298 K).

Class I	pK_1	pK_2	pK_3	ΔpK_{12}	ΔpK_{23}
H_2CO_3	3.88	10.32	—	6.4	—
H_5IO_6	1.64	8.36	15.0	6.7	6.6
H_3PO_3	2.00	6.70	—	4.7	—
H_2S	7.00	12.92	—	5.9	—
H_3PO_4	2.12	7.21	12.0	5.1	4.8
H_2SO_3	1.76	7.21	—	5.5	—
H_2Se	3.80	11.0	—	7.2	—

Class II	pK_1	pK_2	pK_3	pK_4	ΔpK_{12}	ΔpK_{23}	ΔpK_{34}
$H_4P_2O_6$	2.2	2.81	7.27	10.03	0.6	4.5	2.8
$H_4P_2O_7$	0.85	1.96	6.54	8.46	1.1	4.6	1.9
Citric acid	3.13	4.75	6.39	—	1.6	1.6	—
$(COOH)_2$	1.23	4.19	—	—	3.0	—	—
$CH_2(COOH)_2$	2.83	5.69	—	—	2.9	—	—
$(CH_2)_2(COOH)_2$	4.19	5.48	—	—	1.3	—	—
$(CH_2)_3(COOH)_2$	4.34	5.27	—	—	0.9	—	—
$(CH_2)_7(COOH)_2$	4.55	5.41	—	—	0.9	—	—

(From K. B. Harvey and G. B. Porter *Introduction to Physical Inorganic Chemistry*, Addison-Wesley (1963).)

these tell us that there is one oxygen on each phosphorus (Table 13.5); and the small difference between them, ΔpK_{12}, tells us that there are hydroxyl groups on each phosphorus atom. The probable structure is therefore that in Fig. 13.4.

In transition metals, which exist in a variety of oxidation states, the lower hydroxides are basic, but hydroxides of higher oxidation states, where the metal has a high formal positive charge and is highly electronegative, are acidic; for example, $Mn(OH)_2$ is basic, $HOMnO_3$ is acidic.

13.4.3 Solvated Metal Ions

Many solvated metal ions are acidic, and can be regarded as oxyacids. Aqueous solutions of iron(III) salts, for example, contain the hydrated ion $[Fe(H_2O)_6]^{3+}$, and this dissociates in water (hydrolyses):

$$H_2O + [Fe(H_2O)_6]^{3+} \rightleftharpoons [Fe(H_2O)_5(OH)]^{2+} + H^+(aq); \quad pK_a = 4.$$

The rationalization of this effect is the same as in Section 13.4.2: the more electronegative or electron-withdrawing the central (metal) atom is, the more acidic is the solution. Thus solutions of the ions of weakly electropositive metals with high charges are more acidic than those of more strongly electropositive metals. Acidity decreases in the order:

$$Fe(H_2O)_6^{3+} > Al(H_2O)_6^{3+} > Ca(H_2O)_n^{2+} > Na(H_2O)_n^+$$

	acidic	less acidic	very weakly acidic	neutral
pK_a	4	5	13	very large

non-noble gas core ∴ polarizing   noble gas core ∴ less polarizing

13.4 TYPES OF ACID IN AQUEOUS SOLUTION

13.5 COMPLEX HYDRIDES

These can be regarded as compounds which contain hydride ions co-ordinated to metal ions. There are examples found in both transition and non-transition metal chemistry.

The most important examples in non-transition metal chemistry are formed by the group IIIB elements boron, aluminium, and to a lesser extent gallium. The most important examples are lithium aluminium hydride, and lithium and sodium borohydride. They are prepared by the reaction of ionic (saline) hydrides on appropriate compounds:

$$4Na^+H^- + B(OCH_3)_3 \xrightarrow{250°} Na^+BH_4^- + 3NaOCH_3;$$
$$(\text{excess}) \; 4Li^+H^- + AlCl_3 \xrightarrow[\text{ether}]{\text{dry}} LiAlH_4 + 3LiCl.$$

An excess of lithium hydride is used, otherwise aluminium hydride is formed. Other metal borohydrides are prepared from the sodium salt. The BH_4^- ion is tetrahedral, as expected (Section 12.7), and sodium borohydride is an ionic compound with a face-centred cubic lattice of borohydride ions and sodium ions. Lithium aluminium hydride is also a white crystalline solid and has a similar structure but is more covalent. The borohydrides are white ionic solids, and their reaction with water depends upon the cation present; the lithium salt reacts violently, the sodium salt only slightly, and the potassium salt not at all:

$$LiBH_4 + 2H_2O \rightarrow LiBO_2 + 4H_2.$$

This makes sodium borohydride useful for aqueous reactions. Lithium aluminium hydride is rapidly hydrolysed by water, and is used in dry ether solutions:

$$LiAlH_4 + 4H_2O \rightarrow Li^+OH^- + Al(OH)_3 + 4H_2.$$

These compounds are used as reducing agents, and sources of hydride ions (in general the AlH_4^- ion is more reactive than the BH_4^- ion).

1) **Reducing agents.** Sodium borohydride is used as a selective reducing agent since it will normally reduce aldehydes and ketones to alcohols, but does not react with acid and ester groups. Lithium aluminium hydride is more powerful and usually reduces, not only aldehydes and ketones, but also esters and acids to alcohols. In addition nitro compounds and cyanides are reduced to amines, but carbon–carbon double bonds are not normally reduced.

2) **Preparation of hydrides.** Both aluminium hydrides and borohydrides are used to prepare the hydrides of other elements. Some examples are:

$SiCl_4 + LiAlH_4 \xrightarrow{\text{ether}} SiH_4 + LiCl + AlCl_3$ (also $PCl_3 \rightarrow PH_3$; $BF_3 \rightarrow B_2H_6$);
$NaBH_4 + B_2H_6 \rightarrow$ mixture of hydroborate ions, $B_2H_7^-$, $B_3H_8^-$ etc.;
$2NaBH_4 + 2HPO_3$ (any acid) $\rightarrow B_2H_6 + 2H_2 + 2NaPO_3$;
$3BH_4^- + 4H_3AsO_3 + 3H^+ \rightarrow 3H_3BO_3 + 4AsH_3 + 3H_2O$ (also antimonite, stannite).

In addition, metal hydrides and transition metal hydrides can be prepared, for example: $LiAlH_4$: $MgEt_2 \rightarrow MgH_2$; $CuI \rightarrow (CuH)_4$; $NaBH_4$: $FeI_2(CO)_4 \rightarrow FeH_2(CO)_4$.

Table 13.6. Properties of some compounds related to water.

Compound	m.p. (°C)	ΔH_{fus} (kJ mol^{-1})	b.p. (°C)	ΔH_{vap} (kJ mol^{-1})	Dipole moment (D)	Dielectric constant	Surface tension (dyn cm^{-1} at b.p.)
a) Non-polar molecules							
H_2	−259	0.13	−253	0.9	0	1.2	
O_2	−219	0.44	−183	6.8	0	1.5	
Ne	−249	0.33	−246	1.77	0		
CH_4	−183	0.96	−161	8.2	0		
b) Polar molecules							
NH_3	−78	5.6	−33	23.4	1.45	22.4	
HF	−83	4.56	20	7.5	1.98		
H_2O	0	6.02	100	40.7	1.84	78.5	58.9
H_2S	−85	2.38	−60	18.6			28.7
H_2Se	−66	2.5	−41	19.3	0.90		28.9
H_2Te	−51	4.18	−2	23.2			30.0
CH_3OH	−98	2.18	64.7	35.2			
CH_3OCH_3	−138.5	4.94	23.7	21.5			

13.6 WATER

The human body contains about 65 percent by weight of water. Water is an essential constituent of all living tissue, and is present in the cells, blood, and all body fluids. It is both a reactant and the medium in which cell reactions take place. The average human total intake of water per day (in drink, food, and through oxidation processes) is about 2.5 litres.

13.6.1 Anomalous Properties of Water

The most interesting point about the chemistry of water is that nearly all its physical properties differ from those expected by a comparison with the properties of similar compounds. This is obvious from Table 13.6, particularly by comparison with the corresponding group VIB hydrides (H_2S etc.) and the substituted water molecules methanol (CH_3OH) and dimethyl ether (CH_3OCH_3). All of these anomalies are closely related to the structure of water.

13.6.2 The Structure of Water

The gas phase structures of the group VIB hydrides are commonly represented as in Fig. 13.5. The bond angle is less than the tetrahedral angle and decreases on descending the group from H_2O to H_2Te. Is this expected (Problem 4)? This model of the structure of water applied to the liquid and solid phases is somewhat misleading, however, since: (a) there is a permanent dipole in the molecule owing to the differing electronegativities of the oxygen and hydrogen atoms and the presence of lone pairs of electrons on the oxygen atom; and (b) the oxygen atom is much larger than the hydrogen atom. More satisfactory models of the molecule are shown in Fig. 13.6; Fig. 13.6(a) is the usual representation of the dipole nature of the molecule, but Fig. 13.6(b) is more useful for our purpose since it emphasizes

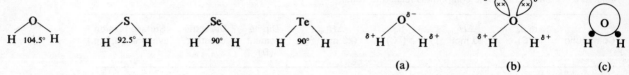

Fig. 13.5 The structures of the group VIB hydrides in the gas phase.

Fig. 13.6 Alternative representations of the water molecule.

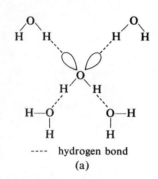

---- hydrogen bond
(a)

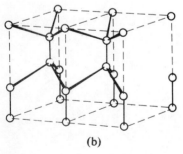

(b)

Fig. 13.7 Tetrahedral co-ordination in the oxygen skeleton of common ice. (a) The tetrahedral arrangement about each oxygen in ice, and (b) the open structure of ice. The hydrogen atoms have been omitted in (b), and are located on the lines joining the oxygen atoms.

the tetrahedral arrangement of the lone pairs of electrons and the hydrogen atoms about the central oxygen atom. Figure 13.6(c) shows the relative sizes of the atoms. It is known that molecular crystals formed by non-polar molecules such as methane (in which the intermolecular forces are the non-directional van der Waals forces only) form close-packed structures similar to those formed by metals, as far as the shape of the molecule will allow. Methane, for example, which is effectively spherical, forms a face-centred cubic close-packed structure in which each methane molecule is surrounded by twelve other methane molecules the same distance away. So do the inert gases. So also does solid hydrogen sulphide, in accordance with the essentially spherical nature of the molecule (Fig. 13.6c). If the density of ice is computed using a similar close-packed model, it is found that it "ought" to be about 1.9 g cm^{-3}, about twice the value found in practice. Clearly, therefore, ice is far from being a close-packed structure; it must have an open cage-like structure with very inefficient packing of the molecules. Moreover, forces stronger than van der Waals forces must be present in ice (but not in solid hydrogen sulphide) in order to hold this structure together (or rather apart!). The forces postulated to account for this phenomenon (an extension of the model of bonding) are the so-called hydrogen bonds (Section 4.8). The simplest model of hydrogen bonding is an electrostatic one, and this will be used here: in water the positively charged hydrogen from one water molecule is attracted to the area of high electron density (negative charge) of one of the lone pairs of electrons around the oxygen atom of another water molecule. This attraction constitutes a "hydrogen bond", since two oxygen atoms are joined together via a hydrogen atom bridge. It is important to realize that the bonding is directional, as expected from our model: in ice each oxygen atom is surrounded exactly tetrahedrally by four other water molecules (via hydrogen bonds), as in Fig. 13.7. Clearly the existence of hydrogen bonds affects the bond angles about each oxygen atom. A representation of the resultant open-cage structure of ice is given in Fig. 13.7(b). There is more than one way of packing tetrahedrally surrounded oxygen atoms together in a structure, and there is more than one form of ice. The common form of ice is related to the wurtzite structure of zinc sulphide, where each position is occupied by a water molecule.

Ammonia and hydrogen fluoride also have high dipole moments, form hydrogen bonds (Table 13.7), and show similar anomalous properties. However, neither can form a solid with a three-dimensional structure as in ice since ammonia has too many N—H bonds, and hydrogen fluoride too few H—F bonds. Despite the fact that hydrogen fluoride has a higher dipole moment than water, and forms a stronger hydrogen bond (Table 13.5), the boiling point (etc.) of hydrogen fluoride is lower than that of water because it cannot form such a large "molecule".

Just as the density of ice is "abnormally" low, so is the density of water, though not to such a great extent. Clearly, therefore, liquid water must retain some of the

Table 13.7. The energies of some hydrogen bonds (kJ mol^{-1}).

F—H---F	29
N—H---N	25
O—H---O	25
N—H---F	21
O—H---N	20
C—H---O	11
N—H---O	10

--- hydrogen bond

ice structure on melting. The usual behaviour of liquids on cooling is a contraction of volume and consequent increase in density (owing to the molecules vibrating less at the lower temperatures and occupying effectively smaller volumes) until the melting point is reached, when solidification occurs on further cooling with a sharp decrease in volume and increase in density. Water does not behave like this. When water is cooled from (say) 10 °C, the density increases until 4 °C when a maximum is reached; below this temperature the density decreases down to the melting point, and on solidification a further sharp decrease in density occurs. On the melting of ice the reverse changes occur.

These anomalous changes in water are the result of two conflicting tendencies: (a) the gradual breakdown of the hydrogen-bonded structure present in water as the temperature rises, tending to decrease the volume and increase the density; and (b) the usual increase in volume and decrease in density as the temperature of a liquid rises. Thus, when ice is melted to form a liquid at 0 °C, much of the hydrogen-bonded structure is broken down, the water molecules pack more closely together, and the density shows a sharp increase. As the temperature rises above 0 °C this process continues, and the density shows a gradual slow increase as the increasing kinetic energy of the molecules breaks down an increasing percentage of the hydrogen bonds. At first this effect is the more important, but above 4 °C the increasing vibration of the molecules (tending to increase the volume) becomes the major factor, and the density decreases. The maximum density of water has very important consequences for marine life and ice formation.

The unusually high enthalpy of fusion of ice is due to the fact that hydrogen bonds have to be broken to melt the solid. Similarly, the unusually high enthalpy of vaporization shows that much hydrogen bonding is present in water even at 100 °C. Thus there is abnormally high order in water even at 100 °C owing to the presence of hydrogen bonding, but this order is dynamic: the molecules which are hydrogen-bonded are continuously changing with those that are not, that is, a dynamic equilibrium is present. Clearly the high melting and boiling points are also related to the presence of hydrogen bonds, as is the high heat capacity, which makes water a useful coolant in the range 0–100 °C. For engines working beyond these temperatures, alcohols and glycols are useful.

The ions H^+ and OH^- have a much greater electrical mobility in water than any other cations or anions. This clearly cannot be explained by saying they are "smaller" than other ions and therefore move more quickly in an electric field. In fact, it is because the mechanism of their transference is quite unlike that of other ions, and more like that of electrons moving through a metal. They are "handed on" from one water molecule to the next (Fig. 13.8).

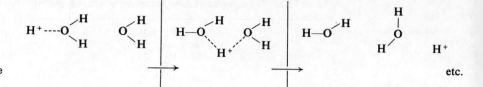

Fig. 13.8 The electrical mobility of the H^+ ion in water.

13.6.3 Classification of the Reactions of Water

1) Water as a solvent. (See Section 9.1.) Water is an excellent solvent for many ionic compounds. This is also associated with its polar nature and its consequent high dielectric constant. The enthalpies of hydration of ions, ΔH_{hyd}, are high and frequently the energy of hydration of the ions is greater than the high lattice energy of ionic compounds, and the compound dissolves in water. The high dielectric constant describes the way water weakens the electrostatic attraction between solute anions and cations in solution. Covalent compounds containing hydrogen bonds (alcohols, and carboxylic acids, for example) also dissolve in water since the water preferentially hydrogen-bonds to the covalent molecule. Hydration energies are so large that many covalent molecules react with water (hydrolyse) to form ions, for example, aluminium chloride, and hydrogen chloride.

Water is especially good as a medium for acid–base reactions, on account of its small but appreciable auto-ionization (self-ionization):

$$2H_2O(l) \rightleftharpoons H_3O^+(aq) + OH^-(aq).$$

Thus it can act as a source of either hydroxonium or hydroxyl ions, and is said to be amphoteric.

2) Water as oxidant and reductant. Water can act as either an oxidant or a reductant. In neutral solution (pH = 7, or $[H^+] = 10^{-7}$ M) the reducing action of water is summarized by the equation:

$$\tfrac{1}{2}O_2(g) + 2H^+(aq) + 2e^- \rightleftharpoons H_2O(l); \quad E = +0.815 \text{ V}.$$

Thus oxygen is liberated only by strong oxidizing agents. Actually the reduction potential of the oxidant must be greater than +0.815 V in practice, and the reaction is said to have an overvoltage (Section 8.9). The potential is dependent on pH and it is easier to release oxygen in alkali than in acid. The oxidizing action of water (again in neutral solution) is summarized by the equation:

$$2H_2O(l) + 2e^- \rightarrow H_2(g) + 2OH^-(aq); \quad E = -0.415 \text{ V}.$$

Thus water ought (thermodynamically) to yield hydrogen with reducing agents with reduction potentials in neutral solution more negative than this, for example Zn^{2+}/Zn, $E^\ominus = -0.76$. But the reaction is slow for most couples near to this value, and it is said that such reactions have high overpotentials. Again the reaction is dependent on pH (applying Le Chatelier's principle). A summary of the pH dependence of both reactions is given in Fig. 13.9.

3) Hydration: water of crystallization. In solution, both anions and cations are hydrated (see Table A.12) and the problem can be considered on the basis of an

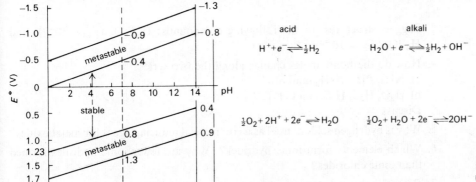

Fig. 13.9 The pH dependence of the reduction potentials of water.

ionic model of the bonding. In solid salts, however, the water of hydration (referred to as water of crystallization) is almost invariably bonded to the cation, as for example $[Fe(H_2O)_6]^{2+}SO_4^{2-}$, and it is difficult to reconcile this with a purely ionic view of the bonding. Moreover, if the bonding were purely ionic it would be expected that the co-ordination of water around the cations would be symmetrical. This is not always the case; for example, hydrated copper sulphate, $CuSO_4 \cdot 5H_2O$, contains a square planar arrangement of four of the water molecules around the copper cation, not tetrahedral (Fig. 11.8). Thus the bonding must be regarded as having at least some covalent character (Section A.9) involving donation of lone pairs of electrons into empty metal orbitals.

4) **Hydrolysis reactions.** These reactions can be classified into four groups; they include reactions of water with: (a) cations such as iron(III) (Section 13.4.3); (b) the anions of weak acids,

$$Ac^-(aq) + H_2O \rightleftharpoons OH^-(aq) + HAc(aq) \ (Ac = acetate),$$

yielding alkaline solutions (Section 11.3.3); (c) covalent molecular compounds such as PCl_3 (Section 11.1.4); and (d) salts such as halides, oxides, sulphides, nitrides, and hydrides (Section 11.1).

5) **Reactions with elements.** See Section 10.4.2.

6) **Catalytic reactions.** Many chemical and physical changes appear to require minute traces of water in order to proceed at reasonable rates, for example sodium does not appear to react with chlorine in the complete absence of water. The function of the water in such reactions is difficult to study.

PROBLEMS

1. What are the formulae and properties of the hydrides of: Na, Ge, S, Ne, Pd, and Cl?

2. How could the following compounds be prepared from heavy water (D_2O): NaD, DCl, ND_3, LiD, $LiAlD_4$, SiD_4, B_2D_6, D_2SO_4, C_2D_2, and $Ca(OD)_2$?

3. Suggest structures for the following compounds: H_6TeO_6 ($K_a = 10^{-9}$); and H_3PO_3 ($K_a = 10^{-2}$).

4. How do the bond angles change along the two series:
   a) NH_3, PH_3, AsH_3, and SbH_3;
   b) H_2O, H_2S, H_2Se, and H_2Te?
   Discuss.

5. Why is hydrogen seldom used as a commercial reducing agent for metal oxides?

6. Which elements form ionic hydrides? Why are fewer ionic hydrides formed than ionic chlorides?

7. Discuss how acid strength changes along the series:
   a) HF, HCl, HBr, and HI;
   b) H_4SiO_4, H_3PO_4, H_2SO_4, and $HClO_4$;
   c) ClOH, BrOH, and IOH;
   d) HClO, $HClO_2$, $HClO_3$, and $HClO_4$;
   e) $Fe(H_2O)_6^{3+}$, $Al(H_2O)_6^{3+}$, $Ca(H_2O)_n^{2+}$, and $Na(H_2O)_n^{+}$.

8. Draw a graph (qualitatively) of the change in density which occurs on cooling: (a) water; and (b) a normal liquid.

9. Describe at a molecular level what happens when water is cooled from (say) 10 °C.

10. Describe what happens to a pond when the air temperature falls well below 0 °C at night. What would happen if: (a) water did not show a maximum density; or (b) ice were more dense than water? Speculate on the effect on marine life.

11. Is the vapour pressure of water greater or less than expected from a comparison with that of the other group VIB hydrides?

12. What order would you expect for (a) the viscosity and (b) the Trouton constants of water, methanol, and dimethyl ether?

13. Explain why: (a) anions are not normally hydrated in solid salts; and (b) the transition metal ions form more stable hydrated salts than the s- and p-block metal ions.

14. What is the essential difference between "hydration" and "hydrolysis"?

BIBLIOGRAPHY

MacKay, K. M. *Hydrogen Compounds of the Metallic Elements*, Spon (1966).
   A degree-level discussion, but the qualitative material is understandable at the sixth-form level.

Shaw, B. L. *Inorganic Hydrides*, Pergamon (1967).
   A more qualitative treatment.

Lister, M. W. *Oxyacids*, Oldbourne (1965).
   Another undergraduate text, but one which the reader of this book can go on to.

Stark, J. G. (Editor). *Modern Chemistry*, Part 4: "Acids and Bases," article by D. Nicholls, Penguin (1970).

Clapp, L. B. *The Chemistry of the OH Group*, Prentice-Hall (1967).

> Most of this treatment of the OH group is qualitative and easily understood.

Bell, R. P. *Acids and Bases*, Methuen (1969).

> The "classic" of the modern treatment of acids and bases.

Drago, R. S., and N. A. Matwiyoff. *Acids and Bases*, Raytheon (1968).

> Introductory account.

Sanderson, R. T. *Inorganic Chemistry*, Reinhold (1967).

Most of the books on inorganic chemistry in the general list contain material on the chemistry of hydrogen.

CHAPTER 14

The s-Block Elements: Groups IA and IIA

14.1 General properties
14.2 Group IA: the alkali metals
14.3 Group IIA: the alkaline earth metals

14.1 GENERAL PROPERTIES

It is convenient to give a summary of the general properties and trends first, in order to gain perspective, and then to deal with each group separately. This method involves some repetition of material, and the difficulty of a completely satisfactory treatment is a reflection of the nature of the material: these elements of groups IA and IIA have broadly similar characteristics, but differ in detail.

The s-block elements have an outer electron configuration of $np^6(n + 1)s^1$ or $np^6(n + 1)s^2$, and are found in groups IA and IIA in the Periodic Table: the lithium and beryllium groups. In order to emphasize the gradual changes which take place in the properties of the elements, and to discourage the idea that the s- and p-block elements are *completely* different from each other, the properties of aluminium will be mentioned from time to time when it is appropriate, since this element shows many resemblances to the s-block elements, particularly the less electropositive members.

The elements (and their oxides) are too reactive to occur in the native state, and are found as cations combined with the most electronegative anions. The less electropositive elements in each group are found as silicates (Li, Be, Mg, and Al), the more electropositive elements in group IA and magnesium are found as chlorides, and the group IIA elements are found as the carbonates and sulphates (Ca, Sr, Ba, and Mg). All are prepared by the electrolysis of fused salts (Section 11.5.2).

These elements are the most electropositive elements in the Periodic Table; their properties are summarized in Tables 14.1 and 14.6, and many of them were discussed in Section 10.3. Some trends in these properties are illustrated in Figs. 14.1 and 14.2, and they are summarized in Table 14.2. The relatively large differences in these properties between the first and second members of the groups, and to a lesser extent between the second and third members, are reflected in their chemistry ("anomalous" behaviour of first member of the group, Section 14.2.7).

14.1.1 Physical Properties

The group IA elements have only one valence electron, and their intermetallic bonding is relatively weak (Section 10.3.2): they are soft metals of low density, with low melting and boiling points, and low enthalpies of fusion, vaporization, and

Table 14.1. Properties of the group IA metals.

Property	Lithium Li	Sodium Na	Potassium K	Rubidium Rb	Caesium Cs
Atomic number	3	11	19	37	55
Electron configuration (outer)	$1s^22s^1$	$2p^63s^1$	$3p^64s^1$	$4p^65s^1$	$5p^66s^1$
Isotopes (in order of abundance)	7, 6	23	39, 41, 40	85, 87	133
Atomic weight	6.939	22.898	39.102	85.47	132.905
Metallic radius (Å)	1.55	1.90	2.35	2.48	2.67
Ionic radius (Å)	0.60	0.95	1.33	1.48	1.69
Covalent radius (Å)	1.23	1.56	2.03	2.16	2.35
Atomic volume ($cm^3\ mol^{-1}$)	13.1	23.7	45.5	55.8	71
Boiling point (°C)	1317	892	774	688	690
Melting point (°C)	180	97.8	64	39	28.5
Enthalpy of fusion (kJ mol^{-1})	2.89	2.6	2.32	2.3	2.1
Enthalpy of vaporization (kJ mol^{-1})	135	98	79	76	68
Enthalpy of atomization (kJ mol^{-1})	159	108	90.0	85.8	78.2
Enthalpy of hydration (kJ mol^{-1})	−498	−393	−310	−284	−251
Density (g cm^{-3})	0.53	0.97	0.86	1.53	1.87
Electronegativity (A/R)	1.15	1.0	0.9	0.9	0.85
Ionization energy (kJ mol^{-1})	520	495	418	403	374
Electron affinity (kJ mol^{-1})	57	21			
Electrode potential (V)	−3.04	−2.71	−2.93	−2.99	−3.02
Ionic mobility, 20 °C	33.5	43.5	64.6	67.5	68
Common co-ordination numbers	4, 6	6	6	6	6, 8
Dissociation enthalpy of X_2 (kJ mol^{-1})	104.6	71.1	50.2	46.0	43.5
Atomic conductance × 10^3 ($ohm^{-1}\ cm^{-4}$)	8	10	3.5	1.4	0.7
Abundance, p.p.m.	65	28,300	25,900	310	7

atomization (Section 10.3). The group IIA metals, which have two valence electrons, have stronger intermetallic bonding and are harder, have greater densities, higher melting and boiling points, and higher enthalpies of fusion, and so on. Beryllium has the strongest intermetallic bonding among the *s*-block elements, and its properties are similar to those of aluminium (diagonal relationship). All except lithium, beryllium, and magnesium, impart characteristic colours to flames (Section 14.2.2).

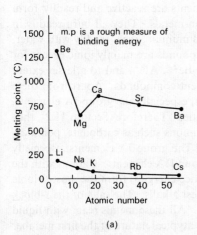
(a)

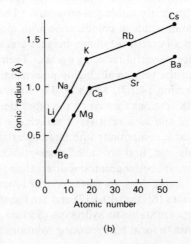
(b)

Fig. 14.1 Trends in properties of the *s*-block elements: (a) the melting point, and (b) the ionic radii.

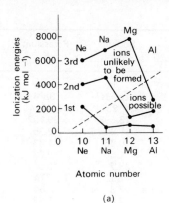
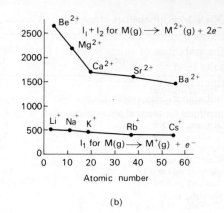

Fig. 14.2 Trends in ionization energies of the s-block elements.

(a) (b)

Table 14.2. The main trends in the properties of s-block elements.

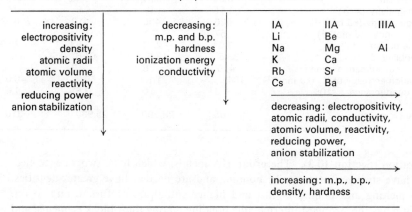

increasing:	decreasing:	IA	IIA	IIIA
electropositivity	m.p. and b.p.	Li	Be	
density	hardness	Na	Mg	Al
atomic radii	ionization energy	K	Ca	
atomic volume	conductivity	Rb	Sr	
reactivity		Cs	Ba	
reducing power				
anion stabilization				

decreasing: electropositivity, atomic radii, conductivity, atomic volume, reactivity, reducing power, anion stabilization →

increasing: m.p., b.p., density, hardness →

14.1.2 Chemical Properties

The most important property is that each of these elements (and aluminium) shows only one stable oxidation state (Section 10.3.20), that in which its cation has the appropriate noble gas structure. The elements are reactive and readily form ionic compounds in combination with most non-metals. They all burn readily in air, and all except beryllium, magnesium, and aluminium, react readily with water. They are powerful reducing agents. Their compounds are mainly ionic except for those of the very small, highly polarizing, ions Be^{2+}, Al^{3+}, and to a lesser extent Li^+ (Section 14.2.7), which tend to form covalent compounds. Apart from these three, the cations formed by the other elements, especially the group IA elements, are large and have relatively small charge densities. Therefore (Section 11.2), they form stable compounds with large polarizable anions such as carbonate, peroxide, triiodide, etc., and are called *anion-stabilizers*. The group IA elements, especially caesium, are better anion stabilizers than the group IIA elements. Cations with the lowest charge densities (therefore the lowest lattice energies) form the most soluble hydroxides (Section 10.3.17) and are the strongest alkalis. The salts of the s-block elements are stable to hydrolysis (Section 13.4). All these metals react with liquid ammonia to form blue reducing solutions. The atypical nature of the first member of each group will become more obvious later. Trends in properties, other than those in ionization energy etc. mentioned above, are dealt with later. These include the stabilities of salts (Section 14.2.4) and the solubilities of salts. The group IA salts tend to be soluble (low lattice energies) though solubility is very dependent upon the cation (Section 10.3.17). The group IIA salts have higher lattice energies, but also higher solvation energies. In general, the salts AB (calcium sulphate, etc.) are relatively insoluble (therefore occur naturally) and the salts AB_2 are relatively soluble.

14.2 GROUP IA: THE ALKALI METALS

In this group all the elements are electropositive metals (Section 11.4) and the resemblances between the elements, owing to their similar outer electron configuration, are more pronounced than the differences between them, owing to the electropositive character of the elements increasing as the lithium to caesium group is descended. As usual, the first member of the group, lithium, is much more electronegative than the other members, and the difference in character between the first and second members is much greater than that between any other two members of the group (the so-called "anomalous" nature of the first member, Section 14.2.7).

14.2.1 Occurrence and Extraction

Since these metals are reactive and highly reducing, they are never found native (in the elemental state), nor as their oxides, which are deliquescent, but they are found as other compounds. Sodium and potassium are relatively abundant (Table 14.1), the others relatively rare. Sodium is found as sodium chloride in sea-water, in solid deposits in the ground (dried-up seas?), and as sodium nitrate (Chile saltpetre) in the deserts of Chile (why there?). Potassium too, is found in sea-water, and also as Carnallite (Stassfurt), $KCl \cdot MgCl_2 \cdot 6H_2O$. Lithium, rubidium, and caesium, occur in a few rare aluminosilicates. All these elements are prepared by electrolysis of their fused salts (Section 11.5.2); sodium, for example, is prepared by the electrolysis of fused sodium chloride, the Downs process:

$$\text{(cathode)} \quad Na \xleftarrow{+e^-} Na^+Cl^- \xrightarrow{-e^-} \tfrac{1}{2}Cl_2 \quad \text{(anode)}.$$

On a smaller scale, the metals can be prepared by displacement (reduction) by the less volatile calcium, and the more volatile alkali metal distilled off:

$$Ca(s) + 2Cs^+Cl^-(s) \rightarrow CaCl_2(s) + 2Cs(g).$$

14.2.2 Physical Properties

All the metals have body-centred cubic structures at room temperature, which presumably change to close-packed structures at sufficiently low temperatures (Section 10.3.1). The melting and boiling points decrease as the atomic number increases (cf. the noble gases), a typical feature of metallic bonding as compared to van der Waals bonding (Section 10.3.10). Also, the liquids are stable over a much wider range of temperature. All have the typical properties of highly electropositive metals (lithium to a lesser extent), compared to non-metals (Table 10.9), and to less electropositive metals (Table 10.10). The trends in atomic radius, ionization energy, etc. (Tables 14.1 and 14.2), going down the Periodic Table, are just as expected (Section 10.3). Thus, compared with the less electropositive metals such as iron, they have low cohesive properties, and these become weaker from lithium to caesium. For example, only lithium is sufficiently hard to be difficult to cut with a knife, whereas rubidium and caesium are like putty. Owing to this softness, they are structurally useless. Their attraction for outer electrons is relatively low and decreases down the group; their ionization energies are relatively low and decrease down the group; their metallic bonding is weak and decreases down the group (Fig. 14.1a), and this is reflected in their small decreasing heats of atomization; and they show relatively little tendency to form covalent bonds. Just above the melting point the vapours contain only about 1 percent of covalent diatomic molecules, with the low bond energies decreasing down the group (Table 14.1). Their atomic volumes are relatively large (larger than the nearest noble gas, Section 10.3.9), but there is a large decrease in size on ionization to the cation $M^+(g)$, since a sole outer-shell electron is lost. As a result cations are almost always smaller than anions: only the largest cations, Rb^+ and Cs^+, are larger than the smallest anion, F^-. All their properties emphasize the relative ease with which an electron can be lost. For example, the heavier elements rubidium and caesium emit electrons when irradiated, and are used in photo-cells. If the outer electron is not completely removed (ionized), but merely excited to a higher energy level, then, when this electron falls down to the ground state, the corresponding energy is emitted; and for the alkali metals this energy is sufficiently low to appear in the visible part of the spectrum. Thus the alkali metals, and also calcium, strontium, and barium, have characteristic flame colours.

14.2.3 Uses

Formerly, the alkali metals were restricted to synthetic uses as alloys and reductants in the preparation of other chemicals, such as lead tetraethyl (Section 16.11) for example. Sodium and potassium have high specific heats, low melting points, and high thermal and electrical conductivities; they have recently found use as coolants and heat exchangers in nuclear reactors, and plastic-covered sodium wire is being used increasingly as an electrical conductor since it is quite cheap, very flexible, and surprisingly safe. Lithium is a useful organometallic reagent, and rubidium and caesium are used in photo-cells.

14.2.4 Chemical Properties

The most striking feature of the chemistry of the group IA metals is that, except when the anion is coloured, they form colourless ionic compounds such as sodium chloride, Na^+Cl^-, in which they show the $+1$ oxidation state exclusively. There are two related questions to be answered:

1. Why are these compounds stable with respect to dissociation into the constituent elements?
2. Why are compounds such as sodium dichloride, $Na^{2+}2Cl^-$ not formed?

Consider the thermodynamics of the reaction: $Na(s) + \frac{1}{2}Cl_2(g) \rightarrow Na^+Cl^-(s)$

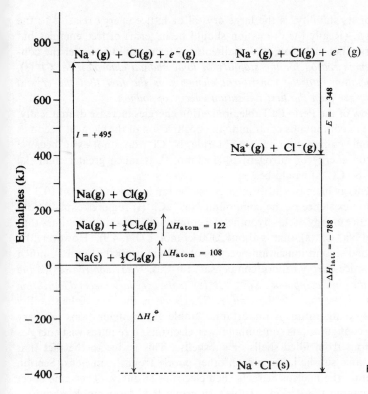
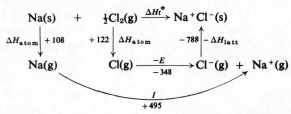

Fig. 14.3 A Born–Haber cycle for the enthalpy of formation of sodium chloride.

Fig. 14.4 An alternative presentation of Fig. 14.3.

(Figs. 14.3 and 14.4). The enthalpy of formation of sodium chloride, $\Delta H_f^\ominus$, is given, according to Eq. (4.5), by

$$\Delta H_f^\ominus = \Delta H_{atom}(Na) + \Delta H_{atom}(Cl_2) + I(Na) - E(Cl) - \Delta H_{latt},$$

where the symbols have their usual meanings (Section A.16). Therefore, and this should be obvious from the diagrams (which are merely two ways of presenting the same information):

$$\Delta H = 108 + 122 + 495 - 348 - 778 = -411 \text{ kJ mol}^{-1}.$$

The formation of sodium chloride from the constituent elements is highly exothermic, and therefore sodium chloride is stable with respect to dissociation into the elements in their standard states (what has been assumed in this argument?).

It is often stated (noble gas rule, Section 12.2) that atoms close to the noble gases in the Periodic Table "tend to" lose or gain electrons in order to form ions which are isoelectronic with the nearest noble gas (isoelectronic means having the same electron configuration). The rule is very useful, but it gives the wrong impression of the causes of the stability of these compounds. No atoms in isolation have any tendency to lose electrons. Even in the case of caesium, the most electropositive element known, it is necessary to supply 374 kJ mol^{-1} in order to remove the outer "loosely bound" electron. This is of the order of the strength of the strongest covalent bonds known. It is obvious from the diagram that the main reason for the large negative enthalpy of formation of sodium chloride, Na$^+$Cl$^-$(s), and therefore for its stability, is the large crystal or lattice energy relative to the ionization energy. (Really the discussion should be in terms of free energies but the ΔH_f is so large that inclusion of a small entropy term would not alter the conclusions.) *The main reason for the stability of ionic sodium chloride, $Na^+Cl^-(s)$, relative to dissociation into the constituent elements, is the fact that the crystal energy is much larger than the first ionization energy of sodium.*

Along any row of the Periodic Table, ionization energies increase dramatically, and the existence of compounds containing the positive ion of the element becomes less likely; the ionic compound nitrogen chloride, N^+Cl^-, does not exist, because the first ionization energy of nitrogen, 1403 kJ mol^{-1}, is much greater than the lattice energy of $N^+Cl^-(s)$ would be.

The crystal energy increases as the charge on the ion increases (Section 10.3.7). Therefore, the non-existence of the compound $Na^{2+}2Cl^-(s)$ requires explanation. The estimated lattice energy of this hypothetical compound is about three times that of the compound $Na^+Cl^-(s)$, that is about 2300 kJ mol^{-1} (Eq. 4.9). However, the sum of the first and second ionization energies of sodium is 5060 kJ mol^{-1}, much larger, and the lattice energy cannot compensate for this. *The main reason for the non-existence of the ionic compound, $Na^{2+}2Cl^-(s)$, is the very large second ionization energy of sodium relative to the lattice energy.* This is where the idea of the stability of the noble gas configuration is important. Noble gas configurations are very stable relative to configurations containing fewer electrons; it requires vast energies to remove electrons from filled shells or subshells. This is due to the fact that electrons in the same subshell screen each other poorly from the nucleus (Section 2.4.3), and therefore the nucleus attracts such electrons strongly. The same effect is observed throughout the Periodic Table. In group IIA, for example, the third ionization energy is very large, and this results in compounds of the type $Mg^{3+}\text{-}3Cl^-(s)$ being unstable with respect to dissociation into: (a) their elements; and (b) the more stable compounds, $MgCl_2$, etc. *High ionization energies always occur when noble gas configurations are broken into.*

In accordance with their highly electropositive character (Section 10.3.6), these metals (except lithium) are highly reactive, and are powerful reducing agents, reacting with water and most non-metals (Table 14.3). They form white crystalline ionic salts with high melting and boiling points. These salts are usually soluble in water (Section 10.3.17), giving conducting solutions, and are not hydrolysed. The

Table 14.3. The reactions of the group IA elements.

Reaction	Notes
$2Li + O_2(\text{excess}) \rightarrow Li_2O$	The higher metals form Na_2O_2, K_2O_2, KO_2, RbO_2, CsO_2.
$2M + S \rightarrow M_2S$	(Also Se, Te) Very vigorous reaction. Polysulphides also formed.
$M + H_2O \rightarrow MOH + \frac{1}{2}H_2$	Li fairly slow, K explodes. All are explosive with acids.
$M + ROH \rightarrow MOR + \frac{1}{2}H_2$	R = alkyl, aryl. Li fairly slow.
$M + \frac{1}{2}H_2 \rightarrow M^+H^-$	At high temperatures, ionic compounds are formed. LiH is the most stable.
$M + \frac{1}{2}X_2 \rightarrow M^+X^-$	X = halogen. The higher members can form polyhalides.
$3Li + \frac{1}{2}N_2 \rightarrow Li_3^+N^{3-}$	Room temperature.
$3M + Z \rightarrow M_3Z$	Z = P, As, Sb, Bi require heating.
$M + NH_3(l) \rightarrow [M(NH_3)_n]^+ + e^-(NH_3) \xrightarrow{\text{catalyst}} M^+NH_2^- + \frac{1}{2}H_2$	
$2M + C \text{ (or } C_2H_2) \rightarrow M_2C_2$ acetylides $\xrightarrow{H_2O} C_2H_2(g)$	
$M + Hg \rightarrow$ amalgams	

normal oxides and the hydroxides are basic, the chlorides are ionic and not hydrolysed, and the hydrides are ionic and basic. Only lithium and sodium form appreciable numbers of hydrated solids; potassium forms few, and rubidium and caesium none, except when the anion is hydrated. The table shows the high reactivity of these metals. In general, as expected, the order of reactivity of the metal, and the corresponding stability of the salt formed, is Cs > Rb > K > Na > Li. For example, caesium and rubidium catch fire in air and water, and also form the most stable salts with large anions. This order is reversed only for anions of very high charge density, H^-, N^{3-}, C_2^{2-} (Section 11.2), because of the exceptionally high lattice energy of these anions with cations of high charge density. The electrode potential of lithium is exceptionally large owing to the high enthalpy of hydration of the small lithium ion (Section 10.3.16). From this it would be expected that lithium would be the best reducing agent in aqueous solution. This is not true; for example, the reaction of lithium with water is fairly mild. This has been attributed to a high activation energy associated with the high enthalpy of atomization. As usual, whenever (correct) thermodynamic arguments predict the wrong conclusion, chemists take cover with kinetic arguments.

14.2.5 Compounds

1) Oxides. (It is convenient to discuss the group IIA oxides at the same time.) Three types of oxide are formed by the alkali metals (Table 14.4): normal oxides containing the ion O^{2-}, and peroxides containing the ion O_2^{2-}, both of which are diamagnetic (Section A.9) and colourless, and the paramagnetic (Section A.9) coloured superoxides, which contain the ion O_2^-. The large polarizable superoxides are formed directly by burning the metal in air, but only by those metals with the least polarizing cations: sodium (only a little, yellow), potassium (yellow), rubidium (orange), and caesium (red). These all have a distorted sodium chloride structure, and are powerful oxidizing agents. They react vigorously with water to yield hydrogen peroxide and oxygen:

$$2O_2^- + 2H_2O \rightarrow 2OH^- + H_2O_2 + O_2.$$

Peroxides are formed by all the group IA and group IIA metals except beryllium, though not in all cases by heating in air. These compounds react with water to give hydrogen peroxide,

$$O_2^{2-} + 2H_2O \rightarrow H_2O_2 + 2OH^-,$$

which was once manufactured this way from barium peroxide. The general tendency is for the most electropositive metals to form the compounds containing the

Table 14.4. Oxides formed by the group IA and IIA metals.

		Oxide formed in adequate supply of air	Oxides known (formed by other methods)
Normal oxide	O^{2-}	Li, Be, Mg, Ca, Sr, Ba	all group IA and IIA metals
Peroxide	O_2^{2-}	Na, Ba	all except Be
Superoxide	O_2^-	(Na), K, Rb, Cs	all except Be, Mg, Li

most oxygen. This is merely another way of stating the general rule that large polarizable anions form the most stable compounds with the most electropositive (least polarizing) cations (Section 11.2).

The normal oxides M_2O, except for lithium, are prepared by the reduction of nitrates or nitrites by the metal:

$$2KNO_3 + 10K \rightarrow N_2 + 6K_2O.$$

All have the antifluorite structure and react with water readily (in fact the IA oxides are deliquescent) to give alkalis:

$$O^{2-} + H_2O \rightarrow 2OH^-(aq).$$

2) Hydroxides. The group IA hydroxides are obtained by the electrolysis of aqueous solutions of the chlorides; hydrogen and chlorine gases are obtained at the cathode and anode. A technical problem is to prevent diffusion of the chlorine into the cell where it would react with the alkali formed. The solid hydroxides have the sodium chloride structure, and are all deliquescent (except LiOH). Solutions of these hydroxides pick up carbon dioxide from the air, and form carbonates.

The alkali metal hydroxides are soluble, the solubility decreasing in the order Cs > Rb > K > Na > Li, and are used as sources of hydroxyl ions in quantitative and qualitative analysis, and in synthesis.

A classification of the properties and uses of hydroxyl ions is given below:

a) Precipitation of insoluble hydroxides e.g. qualitative analysis of iron(III) ions:

$$Fe^{3+}(aq) + 3OH^-(aq) \rightarrow Fe(OH)_3(s).$$

b) Precipitation and redissolving of amphoteric hydroxides (Al, Sn, Pb, Cr, Be) e.g. extraction of aluminium, and analysis of aluminium:

$$Al^{3+}(aq) + 3OH^-(aq) \rightarrow Al(OH)_3(s) \xrightarrow[OH^-]{excess} Al(OH)_4^-(aq).$$

c) Neutralization of acids to form salts:

$$H^+(aq) + OH^-(aq) \rightarrow H_2O(l).$$

d) Hydrolysis of organic compounds such as esters:

$$CH_3COOC_2H_5(aq) + Na^+(aq) + OH^-(aq) \rightarrow$$
$$CH_3COO^-(aq) + Na^+(aq) + C_2H_5OH(aq).$$

e) Reaction with salts of weak bases to displace the base:

$$NH_4^+Cl^-(aq) + Na^+(aq) + OH^-(aq) \rightarrow$$
$$Na^+(aq) + Cl^-(aq) + NH_3(g) + H_2O(l).$$

f) Reaction with non-metals, chlorine for example, to cause a disproportionation reaction to occur (Section 7.12.2).

g) Reaction as catalysts (Section 8.8).

3) Salts. The interesting properties of the alkali metal salts are: (a) their thermal stabilities, discussed in Section 11.2; (b) their solubilities (Section 10.3.17); and (c) their ionic character. The solubilities show two general trends (though there are of course intermediate cases):

i) For salts of small anions of high charge density, such as fluoride and hydroxide (salts of weak acids), the (decreasing) change in lattice energy on descending the group from lithium to caesium is greater than the (decreasing) change in the heats of hydration, and therefore the salts become more soluble on descending the group.

ii) For salts of large anions of small charge density, such as sulphate, iodide, etc. (salts of strong acids), the reverse is the case.

Vaporization of alkali metal halides leads to a vapour containing strongly polar MX molecules. Their percentage ionic character can be calculated from their dipole moments and bond distances, and some figures are given in Table 14.5. The trends are exactly as expected. In all cases the percentage ionic character is less than 100. The more polarizing the cation, and the more polarizable the anion, the less ionic, and the more covalent character there is found for the bond. This polarization is commonly represented as in Fig. 14.5.

The salts of the alkali metals are the most ionic salts known. It is a useful model to consider the ionic salts as being composed of two independent ions, each exhibiting its own properties independently (principle of additivity). For example, to a first approximation, the distance between the sodium ions and chloride ions in the sodium chloride crystal can be considered to be a distance given by adding a contribution due to the sodium ion, to one due to the chloride ion. This approximation, or model, of additive properties is particularly useful for the general properties of aqueous solutions of ionic compounds. The properties of a solution of sodium chloride, for example, can be considered as the sum of the independent properties of the aquated sodium ion, and the aquated chloride ion (Section 10.3.16).

Table 14.5. Percentage ionic character of some alkali halides in the gas phase.

	F	Cl	Br	I
Li		73	60	54
Na		75		
K	83	78	75	
Cs			76	
Be	79	44	35	22
B	63	22		
C	44			

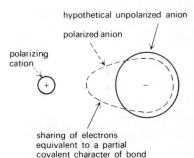

Fig. 14.5 A model of polarization.

Properties of salts containing sodium and potassium ions
a) All are colourless unless the anion is coloured, e.g. permanganate.
b) They have characteristic flame colours: sodium—intense yellow; potassium—lilac.
c) Most salts are soluble except: the zinc or magnesium uranyl acetate (yellow) of sodium; and the perchlorate (ClO_4^-), cobaltinitrite $[Co(NO_2)_6]^{3-}$ (yellow), and chloroplatinate ($PtCl_6^{2-}$) of potassium.

4) **Complex compounds.** (Section 20.4.) The group IA metal ions form fewer complex compounds than any other group of metal ions, and the complex-forming ability decreases down the group: Li > Na > K > Rb > Cs. These facts are consistent with the covalent-bond-forming capacity of these metal ions. Nevertheless, many stable complex compounds of the alkali metals are now known, mainly with chelating oxygen ligands (Section 20.6), such as salicylaldehyde, acetylacetone, and so on:

5) Organometallic compounds. Preparation:

$$2Li + C_2H_5Cl \xrightarrow{ether} C_2H_5Li + LiCl.$$

These are covalent compounds, and are low-melting solids or liquids, soluble in organic solvents. The lithium compounds are much more stable than those of the higher alkali metals, which are more ionic, as expected, and less stable.

14.2.6 Ammonia Solutions

The group IA metals (and calcium, barium, and strontium) dissolve in liquid ammonia and some amines to form blue solutions, when dilute. These solutions are thought to contain solvated electrons, since they are such good conductors of electricity:

$$Na(s) + NH_3(l) \rightarrow Na^+(NH_3) + e^-(NH_3).$$

At higher concentrations the solutions are copper-coloured, and are better regarded as liquid metal solutions. If no impurities are present, the alkali metal (except lithium) can be recovered unchanged by evaporation of the ammonia. However, in the presence of catalysts, such as iron, the amide is formed:

$$Na(s) + NH_3(l) \rightarrow Na^+NH_2^- + \tfrac{1}{2}H_2(g).$$

If the blue solutions contain free (solvated) electrons, they should behave as good reducing systems, and indeed they do, finding use in both organic and inorganic chemistry. With oxygen (correct amount) they form the oxides, peroxides, or superoxides (cf. Section 14.2.5); with sulphur, selenium, and tellurium, they form sulphides, polysulphides, etc.; metal halides are reduced to metals.

14.2.7 Anomalous Nature of Lithium (and other first row elements)

In descending any *s*- or *p*-block group of the Periodic Table there is a general decrease in electronegativity, or increase in electropositivity. The difference in electronegativity between the first and second elements of each group is much greater than that between any other two successive elements. This is reflected in the properties of the elements. Thus, not only is the first element more electronegative than the other elements in the group, but it is much more electronegative than expected by simple extrapolation from the heavier members of the group, and does not fit into the gradual changes in properties expected from such extrapolations. *Nevertheless, the difference between lithium and the other alkali metals is one of degree, not one of kind.* It is the "unexpectedly" large difference in electronegativity together with fortuitous ambient conditions in the laboratory which often make it appear different in kind. It is not. The differences merely reflect the trend towards decreasing electropositivity on ascending an *s*- or *p*-block group, albeit with a somewhat larger than expected decrease between the second and first elements. The differences, therefore, between lithium and the rest of the group IA metals (those given in Table 10.10) are those between a less electropositive metal and more electropositive metals. These are briefly summarized below. Lithium has relatively high cohesive properties associated with relatively strong intermetallic bonding. This is shown in its relatively high enthalpy of atomization etc., melting and boiling points, density, hardness, and homonuclear bond energy. The relatively high attraction of lithium for its outer electron results in relatively high electronegativity, ionization energy, hydration enthalpy, and electron affinity, and relatively low atomic radii. The high electrode potential which is due to its high hydration energy (Section 10.3.16) should result in its acting as an excellent reductant in aqueous solution, but this is not the case (Section 10.3.18). Its compounds have less ionic character than those of the other members of the group. It is clear that all these properties are part of regular trends. In fact, so are the anomalies in chemical properties, but the differences appear greater. This is generally due to fortuitous ambient laboratory conditions, and because small changes in free energies of reactions cause large differences in equilibrium constants (Section 7.5), not to sudden unexpected reversals of trends and stabilities. Some of the more important differences are listed below:

1. Lithium salts of large polarizable anions are less stable than those of other alkali metals, as expected (Section 11.2).

$$Li_2CO_3 \xrightarrow{700\,°C} Li_2O + CO_2 \qquad Na\ etc.,\ no\ reaction\ below\ 800\,°C$$

$$LiNO_3 \longrightarrow Li_2O + NO_2 \qquad NaNO_3 \longrightarrow NaNO_2 + \tfrac{1}{2}O_2$$

$$2LiOH \longrightarrow Li_2O + H_2O \qquad Na\ etc.,\ no\ reaction.$$

In each case, the more stable salts of the heavier alkali metals decompose at a temperature higher than that normally obtainable in the laboratory. Lithium forms no solid bicarbonate, triiodide, hydrosulphide, or superoxide. These are unstable at room temperature, whereas those of the other alkali metals require a higher temperature to effect their decomposition.

2. Solubility differences. The lithium salts of anions of high charge density are less soluble than those of the other alkali metals (Section 10.3.17), for example LiOH, LiF, Li_3PO_4, Li_2CO_3. The halides of lithium are more covalent than the other halides and are more soluble in organic solvents.

3. Complex formation. Lithium forms more stable covalent bonds than the other alkali metals and therefore forms more stable complex compounds (Section 20.4) not only with oxygen donor ligands, but also with nitrogen donors. For example, lithium cannot be recovered unchanged from its liquid ammonia solution, owing to the formation of $Li(NH_3)_4$.

4. Lithium reacts only very slowly with water (Section 10.4.2).

5. Lithium forms stable salts with anions of high charge density owing to their high lattice energy (Section 11.2). For example, in air lithium forms the normal oxide, whereas the others form higher oxides. Lithium reacts with nitrogen to form the nitride, Li_3N; the others do not react. Lithium hydride is more stable than the other hydrides, and lithium carbide forms more easily (with acetylene).

6. Lithium compounds are more covalent. Thus the halides are more soluble in organic solvents, and the alkyls and aryls are more stable than those of the other alkali metals.

The so-called "anomalous" properties of lithium occur because lithium is "unexpectedly" much less electropositive than sodium. Since the group IIA metals form another group of electropositive metals, each slightly less electropositive than its alkali metal neighbour, and also with the electropositivity increasing down the group, it would be surprising if there were not a member of this group whose properties resemble those of lithium. Magnesium is this element, and the properties above in which lithium differs from sodium, are also those in which magnesium resembles lithium—"the diagonal relationship". The two general trends in the s- and p-blocks of the Periodic Table, of increasing electropositivity down the group, and decreasing electropositivity along a period, clearly lead to diagonal relationships (Section 11.4.1).

This correlation of properties with the concepts of electronegativity and electropositivity is only one way of rationalizing the data. One other way of looking at the "anomalous" nature of lithium (and other first row elements) is by considering the variation of the reciprocal ionic radius, $1/r_+$, down the group IA metals. This determines the energy of interaction of the alkali ions with other (negative) ions and polar molecules, and determines such important factors as hydration enthalpies, lattice energies, and complexing power, and hence solubilities, stabilities, electrode potentials, etc. (Problem 18).

14.3 GROUP IIA: THE ALKALINE EARTH METALS

This is a group of fairly electropositive metallic elements (though less electropositive than the alkali metals) in which the electropositivity increases down the group. Their properties reflect these facts (Table 14.6 and 14.7). Like the group IA elements, they show a distinct group relationship in which the similarities between

Table 14.6. Properties of the group IIA metals.

Property	Beryllium Be	Magnesium Mg	Calcium Ca	Strontium Sr	Barium Ba
Atomic number	4	12	20	38	56
Electron configuration (outer)	$1s^2 2s^2$	$2p^6 3s^2$	$3p^6 4s^2$	$4p^6 5s^2$	$5p^6 6s^2$
Isotopes (in order of abundance)	9	24, 26, 25	40, 44, 42, 48, 43, 46	88, 86, 87, 84	138, 137, 136, 135, 134, 130, 132
Atomic weight	9.012	24.312	40.08	87.62	137.34
Metallic radius (Å)	1.12	1.60	1.97	2.15	2.22
Ionic radius (Å)	0.31	0.65	0.99	1.13	1.35
Covalent radius (Å)	0.89	1.36	1.74	1.91	1.98
Atomic volume (cm^3 mol^{-1})	4.9	14.0	26	33.7	39.3
Boiling point (°C)	2970	1107	1487	1334	1140
Melting point (°C)	1280	651	845	789	725
Enthalpy of fusion (kJ mol^{-1})	9.8	9.0	8.7	8.7	7.7
Enthalpy of vaporization (kJ mol^{-1})	310	132	161	141	149
Enthalpy of sublimation or atomization (kJ mol^{-1})	322	150	177	163	176
Enthalpy of hydration (kJ mol^{-1})	−2455	−1900	−1565	−1415	−1275
Density (g cm^{-3})	1.85	1.74	1.54	2.6	3.5
Electronegativity (A/R)	1.5	1.25	1.05	1.0	0.95
Ionization energy (kJ mol^{-1})	900	738	590	549	502
Electron affinity (kJ mol^{-1})	−66	−67			
Electrode potential (V)	−1.85	−2.37	−2.87	−2.89	−2.90
Common co-ordination numbers	2, 4	6	6	6	6
Atomic conductance (ohm^{-1} cm^{-4})	51	16	9.6	1.3	0.4
Abundance, p.p.m.	6	20,900	36,300	150	430

Table 14.7. Reactions of the group IIA metals.

$2M(s) + O_2(g) \rightarrow 2MO(s)$	All burn if heated. Some BaO_2 formed.
$M(s) + S(s) \rightarrow MS(s)$	The sulphides are insoluble, but hydrolyse if heated in water.
$M(s) + 2H_2O(l) \rightarrow M(OH)_2(s) + H_2(g)$	(Be and) Mg in steam only; others in water.
(acids) $M(s) + 2H^+(aq) \rightarrow M^{2+}(aq) + H_2(g)$	Be only slowly; others more quickly.
$M(s) + H_2(g) \rightarrow M^{2+}2H^-(s)$	Not Be. Others at high temperatures only (Mg under pressure).
$M(s) + X_2(g) \rightarrow MX_2(s)$	No polyhalides are formed.
$3M(s) + N_2(g) \rightarrow M_3N_2(s)$	At red heat. Stability: Be > Mg > Ca. Hydrolyse to NH_3.
$3M(s) + 2NH_3(g) \rightarrow M_3N_2(s) + 3H_2(g)$	In liquid ammonia, Ca, Sr, Ba give blue solutions.
$2OH^-(aq) + Be(s) \rightarrow BeO_2^{2-}(aq) + H_2(g)$	Not Mg etc.
$M(s) + 2C(s) \rightarrow MC_2(s)$	High temperatures. Ionic compounds; NaCl structure. Be_2C has antifluorite structure.

the elements are more pronounced than the differences between them, and in which the first member of the group, beryllium, is "anomalous". Their less electropositive nature shows clearly in a comparison of their properties with the alkali metals (Section 14.1); as a group they resemble the less electropositive elements of group IA, and the metal most resembling lithium turns out to be magnesium—the diagonal relationship (Section 11.4.1).

14.3.1 Occurrence, Extraction, and Uses

The elements are too reactive to occur native. Magnesium is the second most abundant metallic element in the sea, and also occurs as carnallite in the Stassfurt deposits, $KCl \cdot MgCl_2 \cdot 6H_2O$. Calcium occurs as various forms of calcium carbonate (marble, chalk, limestone, calcite, aragonite), and with magnesium as dolomite, $CaCO_3 \cdot MgCO_3$. It is also found as sulphates: anhydrite, $CaSO_4$; and gypsum, $CaSO_4 \cdot 2H_2O$. Strontium and barium are more rare, and are found as the carbonates and sulphates; beryllium is even rarer and is found as beryl, an aluminosilicate, $Be_3Al_2(SiO_3)_6$. All are prepared by the electrolysis of fused chlorides (Section 11.5.2), though magnesium has been manufactured by the carbon reduction of the oxide (Section 11.5.1) obtained from sea-water:

$$Mg^{2+}(aq) + Ca(OH)_2(s) \rightarrow Ca^{2+}(aq) + Mg(OH)_2(s) \longrightarrow MgO(s)$$
$$\xrightarrow{C} CO(g) + Mg(s).$$

At what temperature does this reduction occur? Small amounts of each can be obtained by aluminium reduction (Section 11.5.2) and the metals purified by vacuum distillation:

$$2Al(l) + 3BaO(s) \rightarrow Al_2O_3(s) + 3Ba(g).$$

Magnesium is used with aluminium as a light structural alloy, in aeroplanes, for example. Beryllium would be expected to be similarly useful, since it is light and less reactive than magnesium. However, it is costly, toxic, and extremely brittle unless very pure. All the elements find chemical uses: magnesium in Grignard reagents (Section 27.2.1); and calcium and magnesium as reductants in the preparation of pure metals (titanium, etc.) (Section 11.5.2). Their compounds find uses in cement and bleaching powder.

14.3.2 Physical Properties (see Section 10.3)

Except for barium the alkaline earths have close-packed metallic structures at room temperature. Since they each have two valence electrons, they have stronger metallic bonding and show higher cohesive properties than their group IA neighbours. Thus they are harder but become increasingly soft as the atomic number increases. Similar trends occur with the other cohesive properties (enthalpies of sublimation, boiling point, etc.), though magnesium is often anomalous. Similarly the properties associated with the relatively low attraction for outer electrons (though higher than that of the alkali metals) show the expected trends. For example, the electronegativity and ionization energies are relatively low (though higher than their group IA analogues), and within the group these decrease as the atomic weight increases. They are good reducing agents, but less reactive than the alkali metals; for example they also react with water, but more slowly. Similarly their ionic radii are smaller than those of their group IA neighbours; and since their ions are doubly charged, their polarizing power is much greater. This has several effects: first, their compounds are less ionic (Fig. 14.5); secondly, their compounds with large polarizable anions (triiodide, carbonate, bicarbonate, etc.) are less stable (Section 11.2); and thirdly, their enthalpies of hydration are much larger than those of the group IA ions. Superficially it would be expected that this would lead to their salts being generally more soluble than the corresponding group IA salts. But, of course, the lattice energies are also greater and the effect on the solubility is a delicate balance between these two (Section 10.3.17). Actually, the alkaline earth salts are usually less soluble than the corresponding alkali metal salts, since in general the lattice energy increases more than the enthalpy of hydration on changing the metal ion. A typical result is quoted below:

	Lattice energy (kJ mol^{-1})	ΔH_{hyd} (M^{n+}) (kJ mol^{-1})
MgF$_2$	2908	1900
NaF	918	393
difference	1990	1507

The group IIA salts AB, for example calcium carbonate, are less soluble than the salts AB$_2$, for example calcium chloride, since the lattice energy is much greater for the doubly charged anions, CO_3^{2-}, than for the singly charged anions, Cl$^-$, and the difference in the enthalpies of hydration is less marked. A final result of the high enthalpy of hydration is that the electrode potentials of the IIA metal ions are greater than expected. That of calcium is greater than that of sodium for example (Section 10.3.18).

14.3.3 Chemical Properties

The most important chemical fact about the group IIA elements is that they form colourless, largely ionic, compounds in the solid state in which they show the +2

Table 14.8. Enthalpy data for the formation of the compounds, $Mg^+Cl^-(s)$, $Mg^{2+}2Cl^-(s)$, and $Mg^{3+}3Cl^-(s)$ (all in kJ mol^{-1}).

Reaction step	Source of information				
$Mg(g) \rightarrow Mg^+(g) + e^-$ etc.	ionization energies	$I =$ 738$^{(1st)}$	1450$^{(2nd)}$	7730$^{(3rd)}$	2188$^{(1+2)}$
		9918$^{(1+2+3)}$			
$Mg(s) \rightarrow Mg(g)$	enthalpy of atomization	$\Delta H_{atom} = 150$			
$\tfrac{1}{2}Cl_2(g) \rightarrow Cl(g)$ etc.	enthalpy of atomization	$\Delta H_{atom} = 122$			
$Cl(g) + e^- \rightarrow Cl^-(g)$	electron affinity	$-E(Cl) = -348$			
$Mg^+(g) + Cl^-(g) \rightarrow Mg^+Cl^-(s)$	lattice energy (est.)	$-\Delta H_{latt} = -830$			
$Mg^{2+}(g) + 2Cl^-(g) \rightarrow Mg^{2+}Cl_2^-(s)$	lattice energy	$-\Delta H_{latt} = -2528$			
$Mg^{3+}(g) + 3Cl^-(g) \rightarrow Mg^{3+}Cl_3^-(s)$	lattice energy (est.)	$-\Delta H_{latt} = -5000$			
$Mg^+(g) + $ water $\rightarrow Mg^+(aq)$	enthalpy of hydration (est.)	$-\Delta H_{hyd} = -350$			
$Mg^{2+}(g) + $ water $\rightarrow Mg^{2+}(aq)$	enthalpy of hydration	$-\Delta H_{hyd} = -1900$			

oxidation state exclusively, for example compounds like $MgCl_2(s)$. Again (Section 14.2.4) there are two related questions to be answered: (1) why such compounds are formed at all; and (2) why the compounds $Mg^+Cl^-(s)$ and Mg^{3+}-$3Cl^-(s)$ are not formed. Table 14.8 presents the relevant enthalpy data, and Fig. 14.6 presents one of the typical Born–Haber cycles for the formation of these compounds, in this case for the compound $Mg^+Cl^-(s)$. The others are similar. In each case the enthalpy of formation of the solid is given by:

$$\Delta H_f^\ominus = \Delta H_{atom} + \Delta H_{atom} - E + I - \Delta H_{latt}$$

MgCl $\Delta H_f^\ominus = $ 150 $ + $ 122 $ - $ 348 $ + $ 738 $ - $ 830 $ = -168$ kJ mol^{-1}

MgCl$_2$ $\Delta H_f^\ominus = $ 150 $ + 2 \times 122 - 2 \times 348 + 2188 - 2528 = -642$ kJ mol^{-1}

MgCl$_3$ $\Delta H_f^\ominus = $ 150 $ + 3 \times 122 - 3 \times 348 + 9918 - 5000 = +4390$ kJ mol^{-1}

Although these are enthalpy data, and not free energy data, the numbers involved are large and the argument is unaltered (cf. Section 10.3.18). From these data it can be seen that both MgCl(s) and MgCl$_2$(s) are stable relative to dissociation into the constituent elements (negative enthalpies) but that MgCl$_3$ is not. The reason for the instability of MgCl$_3$ is, as expected, the very large third ionization energy, which is due to breaking into the noble gas electron structure (Section 10.3.3). Magnesium monochloride, $Mg^+Cl^-(s)$, is not obtained in the solid state, not because it is inherently unstable relative to dissociation into the constituent elements, but because it disproportionates to the more stable MgCl$_2$:

$$2Mg^+Cl^-(s) \rightarrow Mg^{2+}2Cl^-(s) + Mg(s); \quad \Delta H = -306 \text{ kJ}$$

$\Delta H_f^\ominus \quad -2 \times 168 \quad\quad -642 \quad\quad 0$

It could only exist if it were stabilized kinetically (in other words if the above reaction had a high activation energy) and this rarely occurs with ionic compounds,

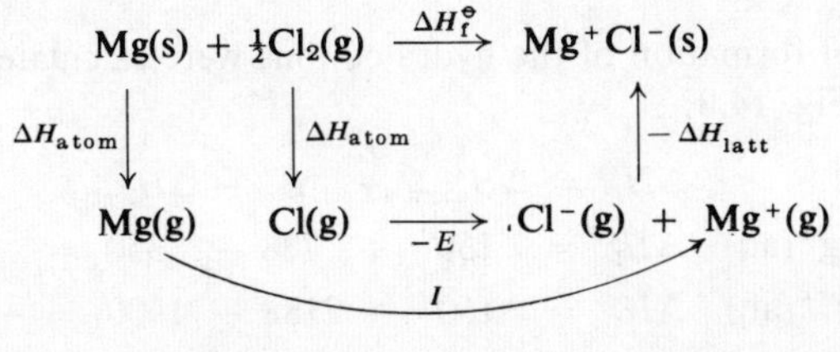

Fig. 14.6 Born–Haber cycle for enthalpy of formation of $Mg^+Cl^-(s)$.

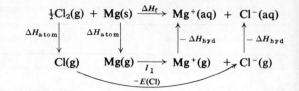

Fig. 14.7 The Born–Haber cycle for the enthalpy of formation of the aquated compound Mg^+Cl^-. The cycle for the compound $Mg^{2+}2Cl^-$ is similar.

though frequently with covalent. The thermodynamic stability of $MgCl_2$ relative to MgCl is principally due to the fact that the difference in lattice energy between them is much greater than the difference between the first and second ionization energies. The reverse is true when $MgCl_2$ is compared to $MgCl_3$. To summarize, *the compounds of the alkaline earth metals are ionic and show only one oxidation state (+2); consequently their chemistry is simply that of the constituent ions (principle of additivity).*

Similar considerations (but with hydration enthalpies replacing lattice energies in the argument) apply to the stability of the divalent ions in aqueous solution. But, consider the disproportionation of the ion $Mg^+(g)$ in the gas phase, where arguments concerning lattice energies, etc., are redundant.

(1) $Mg(g) \rightarrow Mg^+(g) + e^-$; $I_1 = +738$ kJ mol^{-1}
(2) $Mg^+(g) \rightarrow Mg^{2+}(g) + e^-$; $I_2 = +1450$ kJ mol^{-1}
(2) − (1) $2Mg^+(g) \rightarrow Mg^{2+}(g) + Mg(g)$; $\Delta H = +712$ kJ mol^{-1}

In the gas phase, therefore, the ion $Mg^+(g)$ is stable to disproportionation because the second ionization energy is greater than the first. However, in the solid state it disproportionates because of the greater lattice energy of the divalent ion (above). Similarly, in aqueous solution it also disproportionates owing to the greater enthalpy of hydration ΔH_{hyd} of the divalent ion. The Born–Haber cycle for the formation of the aqueous solution, $Mg^+(aq)Cl^-(aq)$, from the elements is shown in Fig. 14.7. That for the compound $MgCl_2$ is similar. Where necessary lattice energies can be estimated (see Eq. 4.5). In both cases the enthalpy of formation of the aqueous ions is given by $\Delta H_f^\ominus$, where:

$\Delta H_f^\ominus = \Delta H_{atom} + \Delta H_{atom} + I - E(Cl) - \Delta H_{hyd} - \Delta H_{hyd}$

MgCl $\Delta H_f^\ominus = 122 + 150 + 738 - 348 - 350 \text{(est.)} - 385 = -73$ kJ mol^{-1}

$MgCl_2$ $\Delta H_f^\ominus = 2 \times 122 + 150 + 2188 - 2 \times 348 - 1900 - 2 \times 385 = -784$ kJ mol^{-1}

The situation is the same as in the case of the solid compounds. MgCl in aqueous solution is unstable, not relative to decomposition into constituent elements, but relative to disproportionation (negative enthalpy):

$2Mg^+(aq) \rightarrow Mg(s) + Mg^{2+}(aq)$; $\Delta H = 438 - 2 \times 538$
ΔH_f 2×538 0   438 $= -638$ kJ

The enthalpies of formation of the hydrated ions were calculated from the Born–Haber cycle in Fig. 14.8:

$\Delta H_f^\ominus = \Delta H_{atom} + I - \Delta H_{hyd}$
$Mg^+(aq)$ $\Delta H_f^\ominus = 150 + 738 - 350 = +538$
$Mg^{2+}(aq)$ $\Delta H_f^\ominus = 150 + 2188 - 1900 = +438$.

$Mg(s) \xrightarrow{\Delta H_f} Mg^{2+}(aq)$
$\Delta H_{atom} \downarrow \quad \uparrow -\Delta H_{hyd}$
$Mg(g) \xrightarrow{I} Mg^{2+}(g)$

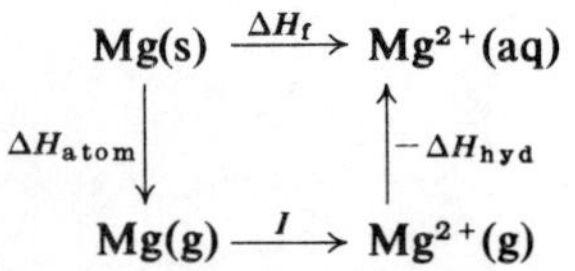

$\Delta H_f = \Delta H_{atom} + I + \Delta H_{hyd}$
(where $I = I_1 - I_2$)

Fig. 14.8 The Born–Haber cycle for the enthalpy of formation of the hydrated ion $Mg^{2+}(aq)$. The cycle for the ion $Mg^+(aq)$ is similar.

Table 14.9. Enthalpy data for the formation of the group IIA hydrated ions, M^{2+}(aq) (kJ mol^{-1}).

	Be	Mg	Ca	Sr	Ba
ΔH_{atom}	322	150	177	163	176
$I_1 + I_2$	2657	2188	1736	1613	1467
ΔH_{hyd}	2455	1900	1565	1415	1275
ΔH_f	+524	+438	+348	+361	+368

The major factor is clearly the high hydration enthalpy of the Mg^{2+} ion. Similar results are obtained with the other elements of the group.

Despite the higher heats of hydration of beryllium and magnesium they have lower electrode potentials than the other alkaline earth metals, and are poorer reducing agents in aqueous solution (and generally). This is obvious from the data below, based on the Born–Haber cycle in Fig. 14.8. The enthalpy of formation of the divalent ion is given by (Table 14.9):

$$\Delta H_f = \Delta H_{atom} + I - \Delta H_{hyd}.$$

The variations in electrode potential are obvious from the trends in these enthalpy data (remember, to actually calculate electrode potentials, free energy data must be used). The electrode potentials of the group IIA metals are generally smaller than those of the corresponding alkali metals because, although the hydration energies are high, the sum of the first two ionization enthalpies is even higher. The electrode potential of beryllium is low because of the very high ionization energies and the high enthalpy of atomization.

14.3.4 Compounds

1) Oxides and hydroxides. The group IIA metals burn in air to yield the normal oxides, except for the most electropositive element, barium, which also forms some barium peroxide (Section 14.2.5). The oxides react with water to form the hydroxides. The solubility of the hydroxides is low, and increases as the cation size increases (Section 10.3.17). As usual in the s- and p-block groups, the basic character of the oxide and hydroxide decreases as a group is ascended and in fact beryllium oxide is amphoteric:

$$H_2O(l) + BeO(s) + 2H_3O^+(aq) \text{ (very low pH)} \rightarrow Be(H_2O)_4^{2+} \quad \text{(slowly)}$$
$$H_2O(l) + BeO(s) + 2OH^-(aq) \rightarrow Be(OH)_4^{2-} \quad \text{beryllates}$$

The hydroxides are weaker bases than the alkali metal hydroxides because of their lower solubility; there are therefore fewer hydroxyl ions in solution. The oxides are more easily reduced than those of the alkali metals; in fact, magnesium has been manufactured this way (Section 11.5.1). As expected, the ease of reduction decreases as the atomic weight and electropositivity increase. The oxides all have the sodium chloride structure except for beryllium oxide, which has the wurtzite structure. The predicted (and found) co-ordination numbers, according to the radius ratio rules, are given in Table 14.10.

Table 14.10. Structures of the alkaline earth oxides.

Oxide	r_+/r_-	Co-ordination number predicted	Co-ordination number found	Structure
Beryllium	0.22	4	4	wurtzite
Magnesium	0.46	6	6	
Calcium	0.71	6	6	sodium chloride
Strontium	0.81	8	6	
Barium	0.96	8	6	

2) Chlorides, etc. The properties in Table 14.11 show clearly the increase in the ionic character of the bond as the electropositivity of the metal increases on descending the group. The trends in solubility are as expected: for very small anions of high charge density the changes in the lattice energy decide the trend in solubility, so that solubility increases as ionic size increases and lattice energy decreases (F^-, OH^-); for large anions, however, the changes in enthalpies of hydration of the cation decide the solubility trends, and solubility decreases as ionic size increases and hydration enthalpies decrease (Section 10.3.17). Their other properties are much as expected for a group of metals slightly less electropositive than the alkali metals. Their salts are ionic and fairly stable, but less so than the corresponding salts of the alkali metals. Similarly, their salts are more hydrated, e.g. $Ca(Mg)Cl_2 \cdot 6H_2O$, $BaCl_2 \cdot 2H_2O$, and more extensively hydrolysed (though still very little). Magnesium chloride hexahydrate, when heated, forms the basic chloride, $Mg(OH)_2 \cdot MgCl_2$, not the anhydrous chloride.

3) Thermal stability of salts. The thermal stabilities of the salts of the group IIA elements with large polarizable anions are less than those of the alkali metals, but increase as the electropositivity of the metal increases (Section 11.2): they are said to be less good "anion-stabilizers". Some examples are: ease of thermal decomposition of carbonates (Section 11.2); decomposition of nitrates to oxides (Section 11.3.3); easier decomposition of sulphates, nitrates, etc.; non-existence of solid bicarbonates and triiodides; and the less easy formation of peroxides and superoxides (Section 14.2.5). Their bicarbonates can exist only in solution (Section 16.6).

Table 14.11. Properties of the group IIA chlorides.

	Boiling point (°C)	Melting point (°C)	Equivalent conductivity at melting point	ΔH_f (kJ mol^{-1})	ΔH_{atom} (kJ mol^{-1})
$BeCl_2$	547	405	0.09	502	540
$MgCl_2$	1420	714	29	640	515
$CaCl_2$	1600	782	52	795	607
$SrCl_2$	1250	875		828	619
$BaCl_2$	1560	962		862	640

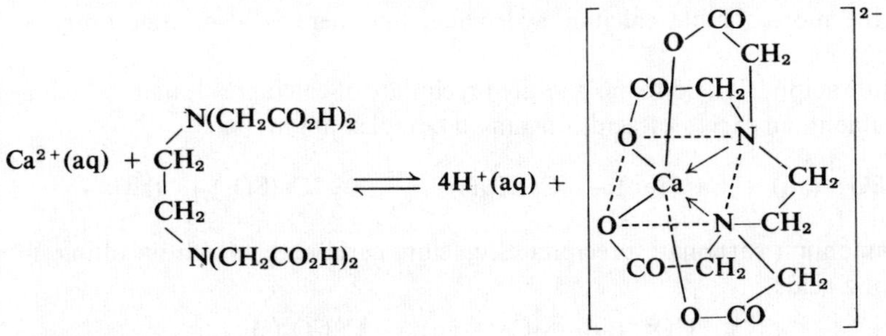

$$Ca^{2+}(aq) + \begin{matrix} & N(CH_2CO_2H)_2 \\ & CH_2 \\ & | \\ & CH_2 \\ & N(CH_2CO_2H)_2 \end{matrix} \rightleftharpoons 4H^+(aq) + [\text{EDTA-Ca complex}]^{2-}$$

Fig. 14.9 The EDTA complex of calcium.

4) Complexes. The alkali earth metal ions form more complexes (Section 20.4) than the alkali metal ions, and, like them, mainly with oxygen and nitrogen donors. As expected, in accordance with its greater tendency to form covalent bonds, beryllium forms the most stable complexes. Those of the larger cations are mainly confined to ring systems formed by chelating ligands—the "chelate effect". The ligand ethylenediaminetetra-acetic acid, abbreviated to EDTA, is particularly useful and is used in the volumetric determination of calcium and magnesium (Fig. 14.9). Actually the disodium salt of the acid is used in practice. Chlorophyll is a complicated complex compound of magnesium in which the ligand is a chelating ligand containing four nitrogen atoms arranged around magnesium in a square-planar arrangement (cf. haemoglobin, Section 20.17 and Fig. 21.10).

14.3.5 Reactions of Magnesium, Calcium, and Barium Ions

1. Reactions of *magnesium ions* (aquated), for example a solution of magnesium chloride in water:
   a) Aqueous sodium hydroxide and ammonium hydroxide (even in excess) precipitate magnesium hydroxide: $2OH^-(aq) + Mg^{2+}(aq) \rightarrow Mg(OH)_2(s)$. Magnesium ions in aqueous solution show little tendency to form ammines and therefore magnesium hydroxide is insoluble in an excess of ammonia. Ammonium hydroxide will not precipitate magnesium hydroxide in the presence of an excess of ammonium chloride (see below).
   b) Ammonium carbonate solution forms basic magnesium carbonate: $MgCO_3 \cdot Mg(OH)_2$. Again this reaction does not occur in the presence of ammonium chloride, since the concentration of carbonate ions is reduced below that necessary to exceed the solubility product of magnesium carbonate: $NH_4^+(aq) + CO_3^{2-}(aq) \rightarrow NH_3(aq) + HCO_3^-(aq)$.
   c) Ammonium hydroxide and ammonium chloride solutions, followed by the addition of sodium dihydrogen phosphate, give the insoluble salt $Mg(NH_4)PO_4$.

   Magnesium salts give no colour in flames.
2. Reactions of *calcium ions* (aquated), for example a solution of calcium chloride in water:
   a) Sodium hydroxide precipitates calcium hydroxide, which is insoluble in the presence of an excess of sodium hydroxide.
   b) Ammonium hydroxide does not precipitate calcium hydroxide. Not only is the concentration of hydroxyl ions too low to exceed the solubility product

of the more soluble calcium hydroxide, but there is also some ammine formation.

c) Dilute sulphuric acid forms a white precipitate of calcium sulphate, which is soluble in an excess of acid, forming a complex anion:

$$SO_4^{2-}(aq) + Ca^{2+}(aq) \rightarrow CaSO_4(s) \xrightarrow[SO_4^{2-}]{excess} [Ca(SO_4)_2]^{2-}(aq).$$

d) Ammonium carbonate precipitates calcium carbonate, which is soluble in strong acids:

$$CO_3^{2-}(aq) + Ca^{2+}(aq) \rightarrow CaCO_3(s).$$

e) Ammonium oxalate precipitates calcium oxalate, which is insoluble in acetic acid, but soluble in strong acids:

$$C_2O_4^{2-}(aq) + Ca^{2+}(aq) \rightarrow CaC_2O_4(s).$$

f) Potassium chromate in the presence of acetic acid, gives no precipitate.
Calcium salts colour flames brick red.

3. Reactions of *barium ions* (aquated), for example a solution of barium chloride in water:

a) Sodium hydroxide or ammonium hydroxide do not precipitate the fairly soluble barium hydroxide.

b) Dilute sulphuric acid precipitates the very insoluble barium sulphate.

c) Ammonium carbonate precipitates the insoluble barium carbonate, which is soluble in strong acids.

d) Ammonium oxalate precipitates the insoluble barium oxalate which is soluble in acetic acid, owing to the reduction in the concentration of the oxalate ions (oxalic acid is a weaker acid than acetic acid (Section 13.4)):

$$2H^+(aq) + C_2O_4^{2-}(aq) \rightarrow H_2C_2O_4(aq).$$

e) Potassium chromate gives the very insoluble barium chromate which is insoluble in acetic acid.

Barium salts colour flames green.

14.3.6 Anomalous Nature of Beryllium: Diagonal Relationship to Aluminium

Beryllium, the first member of the group, appears to be very different from the other members, in the same way as lithium differs from the other alkali metals, and for the same reasons (Section 14.2.7). In fact, the anomalous nature of the first member of the *s*- and *p*-block groups becomes more pronounced towards the middle of the table: beryllium differs more from magnesium than lithium does from sodium. Also, beryllium shows a diagonal resemblance to aluminium in the same way as lithium does to magnesium; and the properties in which beryllium differs from magnesium, it shares with aluminium (in general). The cohesive properties of beryllium are much greater than those of magnesium: beryllium has higher melting and boiling points, enthalpy of fusion, etc., and density, and it is much harder. Similarly its attraction for outer electrons is greater than that of magnesium, leading to much lower atomic radii, higher electron affinity and ionization energy, etc. (Section 10.3). Its higher polarizing power leads to all its compounds being largely covalent, with lower melting and boiling points, enthalpies of formation etc.,

(Table 14.6), and with greater solubility in organic solvents than the corresponding magnesium compounds. The hydration enthalpy of the small Be^{2+} ion is very high and its salts are among the most soluble known. Despite this, its electrode potential is not high, because of its very high second ionization energy (Table A.10). Nevertheless, it would be expected to react with water, and react vigorously with acids ($E^\ominus = -1.85$). In fact, it does not react with water, and is resistant to acid. This must be a kinetic effect: perhaps an oxide film protects the metal. Certainly this is one of the metals rendered passive by concentrated nitric acid (Section 10.4.2). The halides are hygroscopic and fume in air, and all soluble salts are largely hydrolysed and polymerized in water except in strong acid or strong alkali solutions (beryllium is amphoteric, unlike magnesium, etc.):

$$2[Be(H_2O)_4]^{2+} \underset{H_3O^+}{\overset{H_2O}{\rightleftharpoons}} 2H_3O^+ + [(H_2O)_3BeOBe(H_2O)_3]^{2+} \xrightarrow{OH^-}$$
$$Be(OH)_2(s) \xrightarrow{OH^-} [Be(OH)_4]^{2-}(aq).$$

Beryllium is a poor reducing agent, and does not dissolve in ammonia to give blue reducing solutions.

Beryllium forms very unstable salts with large anions, and those which are stable are hydrated, and hydrolyse in solution, for example $BeCO_3 \cdot 4H_2O$ and $BeSO_4 \cdot 4H_2O$; both decompose to the oxide on heating. Beryllium has a strong tendency to form complexes in which it almost invariably shows the tetrahedral co-ordination number of four (it cannot be greater, because of the small size of Be^{2+}). Examples are: $[Be(H_2O)_4]^{2+}$ in its salts (above); BeF_4^{2-}; the vapour $(BeCl_2)_2$ which contains bridging chlorines; and $Be(acac)_2$ where (acac) is acetylacetone (Fig. 14.10). In all of these there is a tetrahedral arrangement of ligand atoms around the beryllium. In the solid state the chloride is a chain polymer, and above 750 °C the dimeric vapour becomes a monomer. The hydride $(BeH_2)_x$ formed by the reaction of lithium aluminium hydride on beryllium dichloride is also a polymer. The highly poisonous nature of beryllium is obviously associated with its strong complexing power with oxygen and nitrogen ligands. The oxide of beryllium is high-melting and hard, and is amphoteric. It forms no peroxide or superoxide. Unlike magnesium etc., but like aluminium, it forms not the acetylide BeC_2, but the carbide Be_2C, when treated with acetylene.

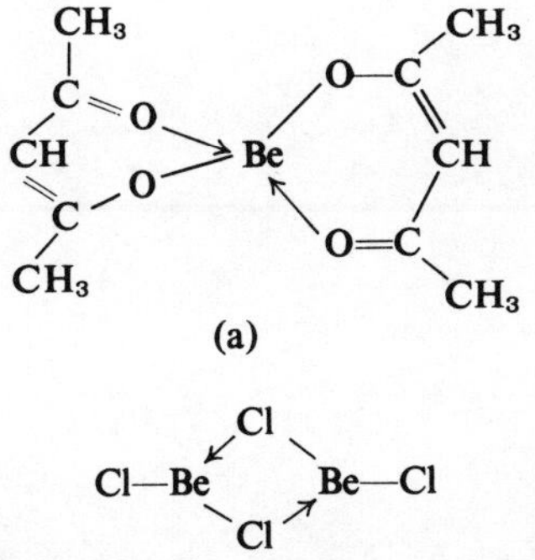

PROBLEMS

1. Account for the trends in the ionic radii of the group IA, IIA, and IIIA metals.

2. Charge density is a rough measure of polarizing power. Which ion in group IIA has a charge density most closely resembling that of (a) the lithium ion, and (b) the aluminium ion? Is the close relationship between charge densities paralleled by a similarly close relationship in chemical properties?

3. The lattice energy of unknown compounds can often be guessed by comparison with similar compounds. For example, the lattice energy of the unknown compound, calcium monofluoride, CaF, cannot be very different from that of potassium fluoride, KF. From data given in the book calculate the standard enthalpy of formation, $\Delta H_f^\ominus$, of the unknown compounds CaX (X = halogen). Comment on the results. Why has the compound CaF not been prepared?

Fig. 14.10 Compounds of beryllium: (a) $Be(acac)_2$ and (b) $[BeCl_2]_2$ vapour (compare Al_2Cl_6).

4. Comment on the following facts:
   a) Lithium iodide is quite soluble in non-polar solvents, the order of solubility in such solvents decreasing in the order:

   $$LiI > LiBr > LiCl > LiF.$$

   b) Lithium sulphate forms no alums.
   c) Lithium triiodide is much less stable (does not exist!) than caesium triiodide, CsI_3.
   d) The reduction potential of lithium is much the same as that of caesium, despite the fact that the ionization energy of lithium is much greater than that of caesium.

5. In the replacement of chlorine by fluorine using the reaction:

   $$R-Cl + M^+F^- \rightarrow R-F + M^+Cl^-,$$

   caesium fluoride is a more effective fluorinating agent that sodium fluoride. Why? [Hint: (a) What is the driving force for the reaction? (b) Why is CsF better?]

6. What properties do you expect for the metal francium of atomic number 87 (e.g. ionization energy, atomic radius, ionic radius, etc.). Has the lanthanide contraction any relevance?

7. Draw graphs of the melting (and boiling) points against atomic numbers of the group IA and group 0 elements. Comment on the differences and similarities.

8. Why are potassium and caesium, rather than lithium and sodium, used in photo-electric cells?

9. Explain why the group IA metals: are soft, electrically conducting, good reductants, and have low melting points; form colourless monovalent ions, with little tendency to form complex ions; form soluble salts, and form soluble hydroxides which become stronger alkalis on descending the group.

10. Explain why the group IIA metals are harder, have higher melting and boiling points, form hydroxides which are weaker alkalis, and form carbonates which are less stable, than those of the corresponding group IA metals.

11. Compare the melting points, densities, and enthalpies of vaporization of the group IIA metals. Comment on these.

12. Calculate the ratios of the second to the first ionization energies of the group IIA metals. Comment on these.

13. Compare the trends in the solubilities of the hydroxides and the melting points of the chlorides of the group IIA metals. Comment. What trend would you expect in the melting points of the halides of calcium?

14. Offer a qualitative explanation for the non-existence of calcium trifluoride, CaF_3.

15. The first ionization energy of beryllium is greater than that of lithium, but the position is reversed in the case of the second ionization energy. Why?

16. Why are the salts of the group IIA metals generally less soluble than the corresponding group IA salts? In general the salts of group IIA decrease in solubility as the group is descended. Why?

17. Which is the more stable to heat, beryllium carbonate or barium carbonate?
18. For the group IA metals plot against atomic number (on the x-axis): (a) the ionic radius, and (b) the reciprocal of the ionic radius, separately on the y-axis. How do these plots differ? What is the consequence?

BIBLIOGRAPHY

Most of the books on inorganic chemistry in the general list contain material on the chemistry of groups IA and IIA. A selection is listed at the end of Chapter 16.

CHAPTER 15

The Elements of the *p*-Block: Group IIIB

15.1 General trends: summary
15.2 Group IIIB

15.1 GENERAL TRENDS: SUMMARY

There are three related comparisons it is useful to consider together: the differences between *s*- and *p*-block elements; the differences between the first "anomalous" member of a group and the other members; and the trends in the chemistry of the *p*-block elements. These are summarized at the end of this section.

In groups IA and IIA (and also VIIB), the similar chemistry of the group (Section 10.5) is the most striking feature. The so-called "anomalous" behaviour of the first member of each of the *s*-block groups is due to the difference in size and electronegativity between the first and second members of the groups being much greater than that between any other successive group elements (Section 14.2.7). The same "anomalous" behaviour is found in the *p*-block elements and is attributed to the same cause. However, in groups IIIB, IVB, VB, and to a lesser extent VIB, the differences in the chemistry of all the group elements become more pronounced: on descending any of these groups the change in character from non-metallic to metallic is very obvious. Nevertheless, some points of similarity between the elements in a group, for example valency relations, are always present. One other general point can be made: in the *s*-block elements the change in character on descending a group is regular, with no sharp reversals in properties such as electronegativity; in the *p*-block elements, although there is a general increase in electropositivity on descending a group, the middle row element (Ga, Ge, As, Se, Br) is often anomalous—the so-called "middle row anomaly" (Section 10.3.2).

The *s*-block elements show only one stable oxidation state, that of the group number, whereas *p*-block elements usually show more than one stable oxidation number (Section 10.3.20). This complicates the chemistry of the *p*-block elements compared to that of the *s*-block elements. Oxidation states other than the group oxidation number generally differ from the group oxidation number in steps of two (Fig. 10.16) (compare the transition elements, Section 20.9). Oxidation states other than these are found, however, and are usually due to the presence of either multiple covalent bonding (nitric oxide, NO, for example, oxidation number of nitrogen = +2) or element–element bonding (peroxides, O_2^{2-}, for example, oxidation number of oxygen = −1). Only fluorine (by definition, since it is the most electronegative element) and the lower noble gases (since they form no known compounds) show only one oxidation state. In groups IIIB, IVB, and VB, the oxidation state two below the group oxidation state increases in stability relative to the group oxidation state, as the atomic number increases—the "inert-pair effect" (Section 16.1). (This trend is the opposite to that for the transition elements, Section 20.9.) Thus, in group IVB silicon forms no stable silicon(II) compounds, germanium(IV) compounds are more stable than germanium(II) compounds, the compounds of tin(II) and tin(IV) are of about equal stability, and those of lead(II) are more stable than those of lead(IV) which are oxidants. The situation is more complicated in the oxygen, fluorine, and noble gas groups, owing to the greater number of oxidation states shown by these elements. Usually, an element which is said to be in a stable oxidation state forms the sulphide, oxide, and all four halides in this oxidation state. If the higher oxidation state is unstable relative to a lower oxidation state (and is therefore oxidizing in nature, for example lead(IV)) then the polarizable (Section 14.2.5) sulphide and heavier halides of this highly polarizing cation (for example Pb^{4+}) will be unstable, but the less polarizable fluoride and oxide will be relatively stable. Conversely, when the lower oxidation state is relatively unstable (and is therefore reducing in nature, for example germanium(II)) then the sulphide and heavier halides will be stable, but the oxide and fluoride may not be.

The p-block elements are more electronegative than the s-block elements and form stronger covalent bonds. In general, as expected, on descending any group (as the electronegativity of the group elements decreases) the stability of a particular covalent bond decreases. For example, the larger more electropositive elements in group VB form the least stable covalent bonds to hydrogen in their group (Table A.11). The thermodynamic stability of the group VB hydrides decreases sharply on descending the group, mainly for this reason. The first row elements, nitrogen, oxygen, and fluorine, are anomalous, however, in the single covalent bonds they form with each other. The single bond strengths, such as N—N, are lower than expected by extrapolation from the heavier group members (Table A.11). This is attributed to high interelectronic repulsion of the non-bonding electrons, owing to the small bond lengths. Their double bond strengths in their compounds with each other (C=O, N=O, etc.), however, are very high. This is illustrated in Table 16.6 for the oxides of carbon and silicon; the first row elements carbon and oxygen form relatively weak single bonds with each other (C—O) and relatively strong double bonds (C=O). With the second row element, silicon, the reverse is the case for its oxides. The result is that carbon prefers to form one double bond to oxygen (C=O) rather than two single bonds, whereas silicon would prefer to form two single bonds (O—Si—O). This causes a great difference in the structure and properties of carbon dioxide and silicon dioxide (Section 16.4.3).

The first row element in each group cannot accommodate more than eight electrons in its outer shell, but this is not true for the subsequent elements in the group. This difference has a number of effects. Firstly, the co-ordination numbers of the first row elements cannot exceed four, whereas those of subsequent rows can. This causes dramatic differences between the chemistry of the first and other elements; for example, the pentafluorides of group VB all exist except for the nitrogen compound. Secondly, compounds of the second and subsequent rows are often more reactive than the corresponding compound of the first row element: carbon tetrachloride is less reactive than silicon tetrachloride. This is a kinetic effect, and is due to the fact that whereas the silicon compound can react by the central silicon atoms accepting a lone pair of electrons, the carbon compound cannot (Section

16.4.5). Similarly, whereas silicon tetrachloride can form acceptor complexes as for example with tertiary amines, carbon tetrachloride cannot.

The trends in properties on descending a p-block group are not smooth because of the insertion of the transition elements between the s- and p-block elements of the third and successive rows. This insertion results in the effective nuclear charge of the elements gallium, germanium, etc., in the third row being greater than that expected by simple extrapolation from the first and second row elements. In other words, the nuclei of these third row elements attract electrons more than expected, and this affects their properties. This has been called the *transition contraction* since one of the effects of the insertion of the transition elements is to make the third row p-block elements smaller than expected (Section 10.3.2). Thus the electronegativities of gallium, germanium, arsenic, and selenium are "unexpectedly" greater than those of the corresponding second row elements. However, the fourth and fifth rows of the p-block elements (the indium and thallium rows) also have transition series inserted before them, and therefore a gradual change in properties is found for the last three elements of each p-block. Actually this is somewhat modified by the presence of the *lanthanide contraction*, which has the same effect as the transition contraction, and results in the last two elements of the three (e.g. tin and lead) being more similar (in size for example) than expected. Two typical *middle row anomalies* are the (present) non-existence of the compound arsenic pentachloride, and the instability (and presumed high oxidizing power) of perbromic acid (previously thought not to exist). The typical p-block trend in properties, therefore, in descending a group is: an anomalous first row element; a second row element which is much less electronegative than the first row element; a middle row element which is more electronegative than expected (middle row anomaly); and a gradual increase in electropositivity for the last three elements, except that the last element is a little less electropositive than expected owing to the lanthanide contraction, and is more like the next-to-last element than expected.

Summary. The elements of an s-block group show: largely similar chemistries; regular changes in properties as a group is descended; and one stable oxidation state. The elements of a p-block group show: more dissimilar chemistries (halogens to a lesser extent); less regular changes in properties as a group is descended (see above); and more than one stable oxidation number. In both cases the first element of the group differs from the others in that it is much more electronegative; in the p-block group this results in a greater tendency for it to form anions and negative oxidation states than the succeeding members of the group. The first row elements nitrogen, oxygen, and fluorine form relatively weak single covalent bonds with one another, but stronger covalent bonds than the heavier members of the group with other elements. In addition (except fluorine) they form strong double covalent bonds with one another. This results in the oxides of carbon and nitrogen being volatile (Section 16.4.3). Finally, the first member of each group cannot expand its outer shell of electrons beyond eight, resulting in a maximum co-ordination number of four, and often in kinetic stabilization of the compounds (Section 16.4). In descending a p-block group there is a general increase in electropositivity, but the changes are not smooth as in the case of the s-block elements. In particular, the middle row element often shows anomalous properties, and the first element is much more electronegative than the others. The heavier elements show higher co-ordination numbers, and generally form weaker covalent bonds.

Table 15.1. Properties of the group IIIB elements.

Property	Boron B	Aluminium Al	Gallium Ga	Indium In	Thallium Tl
Atomic number	5	13	31	49	81
Electron configuration (outer)	$1s^22s^22p^1$	$3s^23p^1$	$4s^24p^1$	$5s^25p^1$	$6s^26p^1$
Isotopes (in order of abundance)	11, 10	27	69, 71	115, 113	205, 203
Atomic weight	10.81	26.981	69.72	114.82	204.37
Metallic radius (Å)	0.98	1.43	1.41	1.66	1.71
Ionic radius, M^{3+} (Å)	0.20	0.50	0.62	0.81	0.95
Covalent radius (Å)	0.82	1.25	1.25	1.44	1.55
Atomic volume ($cm^3\ mol^{-1}$)	4.6	10.0	11.8	15.7	17.3
Boiling point (°C)	2550	2467	2403	2000	1457
Melting point (°C)	2300	660	30	156	303
Enthalpy of fusion (kJ mol^{-1})	22.1	10.7	5.59	3.26	4.27
Enthalpy of vaporization (kJ mol^{-1})	536	284	256	225	162
Enthalpy of atomization (kJ mol^{-1})	562	326	277	243	182
Enthalpy of hydration (kJ mol^{-1})		4630	4645	4060	4140
Density (g cm^{-3})	2.34	2.7	5.91	7.3	11.8
Electronegativity (A/R)	2.0	1.45	1.8	1.5	1.45
Ionization energy (kJ mol^{-1}) 1st	800	577	579	558	589
2nd	2427	1816	1979	1820	1970
3rd	3658	2745	2962	2703	2879
4th	25024	11575	6192	5251	4895
Electron affinity (kJ mol^{-1})	15	26		−0.25(I)	−0.34(I)
Electrode potential (V)		−1.66(III)	−0.53(III)	−0.34(III)	+0.72(III)
Common co-ordination numbers	3, 4	3, 4, 6	3, 6	3, 6	3, 6
Common oxidation states	3	3	(1) 3	1, 3	1, (3)
Atomic conductance × 10^3 ($ohm^{-1}\ cm^{-4}$)		38	5	7.8	3.2
Abundance, p.p.m.	10	81300	15	1	0.3

15.2 GROUP IIIB

15.2.1 Group Trends

The trends in properties are much as expected (Section 11.4) for the change from the semi-metal boron (covalent polymer, very hard, black metallic appearance, semiconductor, high melting point, inert, etc.) to the metal thallium (soft, white, good conductor, low melting point, dissolves in dilute acids, oxidized by air, attacked by steam). The change from boron to aluminium is dramatic—aluminium shows mainly metallic properties. Thus, whereas the aquated ion B^{3+}(aq) does not exist, the less polarizing ion Al^{3+}(aq) is well known. In fact boron exists: as an electron acceptor in combination with metals as a wide variety of borides, which are generally hard and inert, such as AlB_2 and VB; and with non-metals as a variety of covalent compounds, such as BCl_3. This is clear non-metallic behaviour. Although aluminium is metallic, it is only moderately electropositive and shows the properties typical of such a metal (Table 15.1): its oxide is amphoteric, its chloride is covalent, and so on. In addition it is protected by a layer of aluminium oxide from, for example, further combination with air and attack by dilute acids. Gallium is very similar to aluminium, but despite its oxide layer it slowly dissolves in both alkali and acid to yield the ions $Ga(OH)_4^-$(aq) and $Ga(H_2O)_n^{3+}$ respectively. The latter ion, like that of $Al(aq)^{3+}$, is extensively hydrolysed in solution (Section 13.4.3).

Thallium differs in that it has no protective oxide layer, and reacts much more quickly with oxygen and dilute acids. It also differs in that, unlike the other elements, its most stable oxidation state is the $+1$ state, and it reacts to form compounds such as Tl_2O, $TlCl$, and $Tl(aq)^+$; the thallium(III) state is oxidizing. The increasing electropositive character down the group is well illustrated by the character of the oxides, which become more basic as the group is descended: boric oxide is amphoteric, mainly acidic; aluminium and gallium oxides are amphoteric; and those of indium and thallium are basic. Moreover, the bonding in the halides of these elements changes from largely covalent in the case of boron, to largely ionic in the case of thallium. The elements in this group form compounds in which there are fewer than eight electrons in the outer shell, called "electron-deficient" compounds.

15.2.2 Occurrence, Extraction, and Uses

Both aluminium and boron have a great affinity for oxygen, and neither element is found native. Boron occurs principally as boric acid, $B(OH)_3$, and as borates such as borax, $Na_2B_4O_7 \cdot 10H_2O$. One method of obtaining boron is first to treat borax with acid to yield boric acid, then heat this to obtain boric oxide, and finally reduce the oxide with magnesium (Section 11.5.2) to yield boron as an impure brown amorphous powder. Pure boron can be obtained by reducing boron tribromide with hydrogen, by passing the mixture over a tantalum filament at 1000 °C (Section 11.5.2).

Aluminium is the third most abundant element in the earth's crust and is the most abundant metal. It occurs widely in complex aluminosilicates such as those in clay, from which it can only be extracted at high cost. The most important ores are bauxite, $Al(OH)_3 \cdot xH_2O$, and cryolite, Na_3AlF_6. The metal cannot be obtained from the oxide, Al_2O_3, by carbon reduction (Section 11.5.1) and must be obtained electrolytically. However, electrolysis of the chloride cannot be used since it is a covalent solid, and sublimes at 180 °C. Nor is the oxide suitable for electrolysis since its melting point (2300 °C) is too high. Similar problems are encountered with all other salts. The method is to dissolve the purified oxide in fused cryolite at 800–1000 °C; electrolysis of the solution of aluminium and oxide ions produced, yields aluminium at the cathode and oxygen at the anode. The cell consists of an iron bath lined with graphite which acts as the cathode, and carbon rods suspended in the electrolyte acting as anodes. The aluminium discharged sinks to the bottom of the cell and is tapped off. Fresh alumina is added as required. The anode is slowly attacked by the oxygen to form carbon monoxide. The temperature of the cell is maintained by the current. A low voltage and a high current density are used, to avoid decomposition of the cryolite. The alumina (aluminium oxide) required by the process is obtained from bauxite. The ore contains iron(III) oxide and silica, among other impurities. These must be removed efficiently since the aluminium is difficult to purify once it has been prepared and must be produced pure. The ore is crushed and the amphoteric alumina is extracted by dissolution in hot sodium hydroxide under pressure:

$$Al_2O_3(s) + 2OH^-(aq) + 3H_2O(l) \rightarrow 2Al(OH)_4^-(aq);$$
$$SiO_2(s) + 4OH^-(aq) \rightarrow SiO_4^{4-}(aq) + 2H_2O(l).$$

The basic iron(III) hydroxide and other insoluble impurities are filtered off, the solution is cooled, and most of the aluminium hydroxide is precipitated either by passing in carbon dioxide, or by seeding with some freshly precipitated aluminium hydroxide:

$$2Al(OH)_4^-(aq) + CO_2(g) \rightarrow 2Al(OH)_3(s) + CO_3^{2-}(aq) + H_2O(l);$$
$$Al(OH)_4^-(aq) \rightarrow Al(OH)_3(s) + OH^-(aq).$$

The silicates remain in solution (silica is a more acidic oxide than alumina). The alumina is filtered, washed, heated to form the oxide, and reduced as above.

The other three elements are found only in trace quantities and are produced by electrolytic reduction in aqueous solution.

Crystalline boron is used in transistors. Boron is a good neutron absorber and is used as shields and control rods in nuclear reactors. Boranes are used as high energy fuels, for example in rockets. Aluminium is used extensively for structural purposes, either alone or alloyed, in aircraft, ships and buildings, as well as for cooking utensils, wrapping materials, and in electric cables. It is a good reducing agent and is used in the thermite process (Section 11.5.2). Aluminium forms many important alloys, which are generally harder and stronger than the parent metal, but which may be more or less resistant to corrosion, depending on the other metals. Duralumin, which contains 4 percent copper is more resistant to corrosion than aluminium. Other alloys containing more electropositive metals, such as magnesium and calcium, may be less resistant to corrosion, owing to the protecting oxide film being less firmly held. Such alloys can either be coated with aluminium, or have an oxide film coated electrolytically on them by using the metal as an anode during electrolysis—*anodization*. The thickness of the film on aluminium itself can also be increased by this method. This oxide film is very absorbent and is often coloured with a dye.

15.2.3 Properties

Boron is semi-metallic and its structure is that of a giant covalent polymer in which units of twelve boron atoms in an icosahedron are linked to each other by chains of boron atoms. The crystalline form is black, very hard, inert, semi-conducting, and high melting; it reacts with other elements only at high temperatures. The more common amorphous form is brown and more reactive than the black form. Aluminium is a light white metal which is very hard, but is malleable and ductile and can be shaped to any design and is therefore very useful for structural purposes. It is less reactive than expected from its electrode potential because of a layer of protective oxide (kinetic stabilization). Therefore, like boron, normally it reacts at high temperatures only. Mercury removes this oxide film, and rapid reaction occurs with either moist air to form aluminium hydroxide overgrowths, or with dilute acids to give hydrogen. Gallium, indium, and thallium are relatively soft metals which react with dilute acids to yield hydrogen, and are generally more reactive than boron and aluminium.

The elements all react when heated: with oxygen to form the oxide M_2O_3 (exception, Tl_2O); with halogens to form MX_3 (exception, TlX); with sulphur, selenium, and tellurium to form M_2S_3, etc. (exception, Tl_2S); and with dilute acids to yield hydrogen and M^{3+}(aq), except boron (not at all), aluminium (slowly unless amalgamated), and thallium (yields Tl^+aq). When heated in nitrogen, boron and aluminium form the nitrides BN and AlN. Boron nitride is isoelectronic with carbon, and occurs in two forms: one has a layer structure like graphite, and is soft and lubricating (but light in colour); the other, formed at high pressure, has the diamond-like tetrahedral structure and is very hard. Aluminium nitride has the diamond structure. The nitrides of gallium and indium can be formed by heating the elements with ammonia. Aluminium and boron react with carbon on heating to form the carbides B_4C and Al_4C_3 (Section 16.5). Boron reacts with many metals on heating to give a wide variety of binary compounds called borides, for example MgB_2, Fe_2B, and VB. These are usually hard inert compounds but magnesium boride is hydrolysed by dilute acids to yield boranes. Aluminium and the other group IIIB elements form alloys with other metals. Aluminium reduces the oxides of most other metals (Section 11.5.2). Even hot concentrated oxidizing acids attack boron only slowly to yield boric acid $B(OH)_3$ and the reduction products of the acids (sulphur dioxide from sulphuric, nitrogen dioxide from nitric). Oxidizing acids attack the metals of the group in the usual way (Section 10.4) to yield the aquated metal ions and the reduction products of the acids, except that concentrated nitric acid renders aluminium passive, probably owing to formation of a further oxide layer. Boron dissolves in fused alkali, to give the borate ion, BO_3^{3-}:

$$2B(s) + 6OH^-(aq) \rightarrow 2BO_3^{3-}(aq) + 3H_2(g).$$

Aluminium and gallium dissolve in aqueous alkali to yield the aluminate and gallate ions (both are amphoteric): $Al(OH)_4^-$ (aq) and $Ga(OH)_4^-$ (aq). A summary of these reactions is given in Table 15.2.

Table 15.2. A summary of the reactions of the group IIIB elements.

	Boron	Aluminium	Gallium	Indium	Thallium
Air (25 °C)	X	only if amalgamated	increasing reactivity →		
Air heating	———— all burn to a mixture of oxide and nitride ————				
N_2 (heat)	BN	AlN			
C (heat)	B_4C	Al_4C_3			
S, Se, Te (heat)	———— all form sulphides etc. ————				
X_2 (heat)	———— all form halides ————				
Metals (heat)	borides	alloys			
Steam	$H_3BO_3 + H_2$	only if amalgamated	increasing reactivity, $H_2(g)$ →		
Dilute acid	X	only if amalgamated			
Concentrated HCl	X	fast on heating (H_2)	fast, $H_2(g)$		
Concentrated H_2SO_4	$H_3BO_3 + SO_2$	Al^{3+}(aq) + SO_2	M^{3+}(aq) + SO_2		
Concentrated HNO_3	$H_3BO_3 + NO_2$ etc.	passive	M^{3+}(aq) + NO_2 etc.		
Hot concentrated alkali	BO_2^-(aq) + H_2	$Al(OH)_4^-$ (aq)	$Ga(OH)_4^-$ (aq)		

X = no reaction.

15.2.4 Boron Compounds

The most striking features of the chemistry of boron are: (a) the total absence of the cation B^{3+} (owing to the very high sum of the first three ionization energies of boron) either in the solid state or in aqueous solution; (b) the high bond strength of the B—O bond (523 kJ mol^{-1}); and (c) it is the only non-metal which contains fewer than four valency electrons, resulting in some interesting chemistry.

1) Boron trihalides. These are the most important (but not the only) halogen compounds (B_2Cl_4, etc., also exist). Some data for these are given in Table 15.3. Clearly these are covalent molecules.

Table 15.3. Some properties of the boron trihalides.

	Melting point (°C)	Boiling point (°C)	ΔH_f (kJ mol^{-1})
BF_3	−129	−100	−1145
BCl_3	−107	+12.5	−395
BBr_3	−46	90.5	−222
BI_3	49	210	−115

They are electron deficient since they contain only six electrons in the outer shell of the central atom, boron. Unlike aluminium, these molecules do not complete the octet by forming dimers containing halogen bridges. However, they are electron-acceptor molecules, behaving as powerful Lewis acids with molecules that can donate a lone pair of electrons (Lewis bases). Some examples are: (ethers) $R_2O \rightarrow BF_3$; (amines) $R_3N \rightarrow BF_3$; fluoride ions to form BF_4^-; and various chelating ligands, salicylaldehyde (Section 14.2.5) for example $[B(sal)_2]^+$. Write down the structures of these compounds. All the trihalides react with water to yield boric acid and the halogen acid:

$$BX_3 + 3H\text{---}OH \rightarrow B(OH)_3 + 3HX.$$

2) Boron–oxygen compounds: boric oxide, boric acid, borates. The bond energy of the B—O single bond is very high (523 kJ mol^{-1}) and therefore boron, unlike carbon and nitrogen, does not form stable B=O double bonds to oxygen. Thus, instead of forming small volatile covalent molecules and small anions, such as $B_2O_3(g)$ and BO_3^{3-}, it forms oxygen polymers with chains —B—O—B—O—B—O— etc. These can be based on either a tetrahedral or a triangular arrangement about the boron atom, in the form of rings, chains, or sheets. Some typical structures are shown in Fig. 15.3. One other result of the high B—O bond energy is that most boron compounds burn readily to boric oxide, and hydrolyse readily to boric acid or borates.

Boric oxide can be prepared as a glass-like solid by heating boric acid:

$$B(OH)_3 \xrightarrow{100°} HBO_2 \xrightarrow{>100°} B_2O_3 + H_2O$$
$$\text{orthoboric} \quad\quad \text{metaboric} \quad\quad \text{boric}$$
$$\text{(or boric) acid} \quad \text{acid} \quad\quad \text{oxide}$$

Boric oxide can exist in two forms, both of which are polymeric: as a glassy solid which has a random structure consisting of trigonal planar BO_3 units; and as a crystalline solid with a regular structure based upon tetrahedral BO_4 units. It is an acidic oxide; when added to water it slowly forms boric acid, and when heated with metal oxides it forms metaborates (borate glasses) which often have characteristic colours and which are used in qualitative analysis (borax bead test), for example:

$$CoO(s) + B_2O_3(s) \xrightarrow{heat} Co(BO_2)_2(s) \quad \text{deep blue}.$$

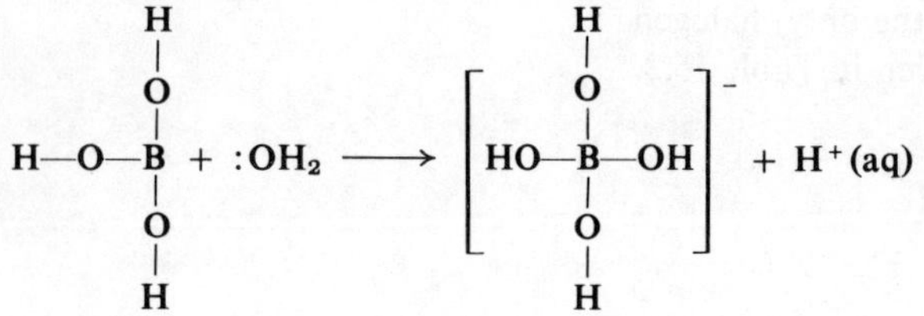

Fig. 15.1 The reaction of boric acid with water.

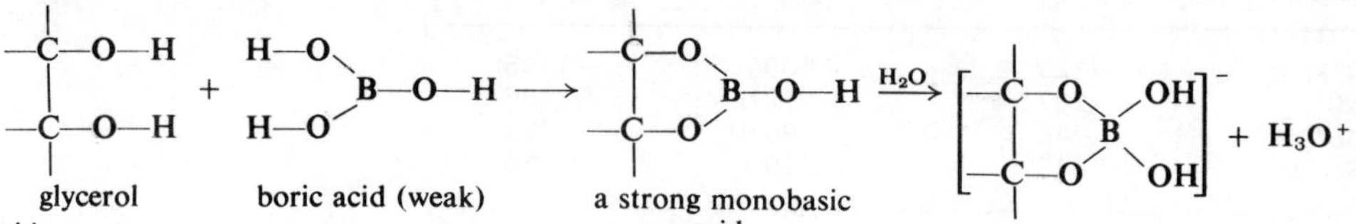

Fig. 15.2 The reaction of boric acid with glycerol.

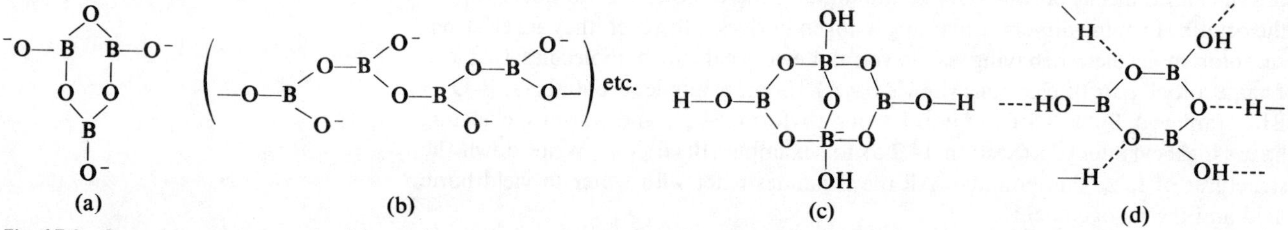

Fig. 15.3 Some boron compounds: (a) a cyclic metaborate, e.g. $[K(BO_2)]_3$; (b) a linear metaborate, e.g. $[Ca(BO_2)]_n$; (c) $[B_4O_5(OH)_4]^{2-}$ in borax $Na_2B_4O_7 \cdot 10H_2O$ or $Na_2[B_4O_5(OH)_4]8H_2O$; (d) the cyclic form of metaboric acid.

Orthoboric acid (called boric acid) is a white crystalline solid, moderately soluble in cold water, and is prepared by adding an excess of a strong acid to a hot saturated solution of borax; on cooling, boric acid crystallizes out. It consists of layers of planar triangular $B(OH)_3$ units, held together within the layers by hydrogen bonds. (What are the forces between the layers?—Section 4.9.) Boric acid dissolves in water to form the anion $B(OH)_4^-$ and a hydroxonium ion, H_3O^+ (Fig. 15.1). Thus its acidity is due to electron pair acceptance, not proton donation, and it is therefore a monoprotic acid, not triprotic as expected. It is a very weak acid, but its strength can be increased by the addition of polyhydroxy compounds which contain *cis* dihydroxy groups, such as glycerol and mannitol, and in their presence boric acid can be titrated with sodium hydroxide as if it were a strong monobasic acid (Fig. 15.2). The salts of the weak boric acids are extensively hydrolysed in solution. Sodium tetraborate, borax, $Na_2B_4O_7 \cdot 10H_2O$, is alkaline in aqueous solution and can be titrated using methyl red indicator; it can be used as a primary standard if kept in a well-stoppered bottle (it tends to effloresce to the pentahydrate):

$$5H_2O(l) + B_4O_7^{2-}(aq) + 2H^+(aq) \rightarrow 4H_3BO_3(aq).$$

The borates are rarely simple salts containing the BO_3^{3-} anion, but they contain polymeric units (Fig. 15.3). Nor is the BO_3^{3-}(aq) the only ion present in aqueous solution; like boric acid itself the hydrated borate ions are polymerized.

The borax molecule is interesting since it contains boron atoms with both tetrahedral and trigonal planar stereochemistry.

Perborates are formed by treating borates with hydrogen peroxide. These contain the anion $[B_2(OH)_4(O_2)_2]^{2-}$. In hot water perborates react to yield hydrogen peroxide, and they are used in washing powders as bleaches.

3) Hydrides: borohydrides, boranes, borazole. The borohydrides were discussed in Section 13.5. The action of dilute acids on magnesium boride yields a mixture of volatile binary compounds of boron and hydrogen, called the boranes: B_4H_{10}, B_5H_9, B_5H_{11}, B_6H_{10}, and $B_{10}H_{14}$. The lowest member of the series, diborane, B_2H_6, cannot be isolated by this method since it hydrolyses very quickly, but it can be prepared by the action of lithium hydride with boron trifluoride:

$$6LiH + 8BF_3 \rightarrow 6LiBF_4 + B_2H_6.$$

The other boranes can be prepared by heating diborane above 100 °C and separating the mixture of products. The formulae of the boranes are B_nH_{n+4} ($n = 2, 5, 6, 10$) and B_nH_{n+6} ($n = 4, 5, 9, 10$). B_2H_6 and B_4H_{10} are gases, the others are liquids or solids. They are decomposed by heat fairly readily, most are spontaneously inflammable in air, and they are readily hydrolysed to boric acid and hydrogen. The large bond energy of the B—O bond results in the oxidation of the boranes being highly exothermic, and they are used as rocket fuels:

$$C_2H_6 + 3\tfrac{1}{2}O_2 \rightarrow 2CO_2 + 3H_2O; \quad \Delta G = -1427 \text{ kJ}$$
$$B_2H_6 + 3O_2 \rightarrow B_2O_3 + 3H_2O; \quad \Delta G = -2020 \text{ kJ}.$$

All the boranes are electron-deficient molecules in the sense that they contain fewer electrons than is necessary to write the formula of each compound in such a way that there is a two-electron bond between each pair of bonded atoms (Fig. 15.4a, for diborane). Clearly, even if hydrogen could form two conventional covalent bonds (which it cannot) this formulation of the structure would require $8 \times 2 = 16$ valence electrons. However, there are only 12 electrons available (three from each boron, one from each hydrogen). A satisfactory model of the bonding is given in Fig. 15.4(b) and (c). Note: Figs. 15.4(b) and (c) represent the same description of the bonding in diborane and are merely two different ways of drawing it. The four terminal hydrogen atoms, which are in the same plane as the two boron atoms, are linked to the boron atoms by conventional two-electron covalent bonds. Each of the two central hydrogen atoms, which are above and below the plane, are part of a three-nucleus electron-pair bond, B—H—B. In this system, instead of a pair of electrons binding together two nuclei, they bind together three nuclei. The most probable average position of the electrons is shown in Fig. 15.4(b), one between each pair of nuclei. This is shown in a different way in Fig. 15.4(c), where the so-called banana bonds, above and below the plane,

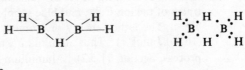

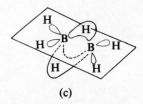

(a) (b) (c)

Fig. 15.4 The structure of diborane: there are insufficient electrons for the electronic formulation (a).

represent volumes within which there is a high probability of finding the bonding electrons. Such hydrogen atoms are referred to as "bridging" hydrogens. Interestingly, in the *immediate* vicinity of the boron atoms, the distribution of electron pairs is approximately tetrahedral as expected (Section 12.7).

The reaction of ammonia on diborane at 200 °C yields the compound borazole, $B_3N_3H_6$ (Fig. 15.5). This compound is isoelectronic with benzene and has many similar properties, but is generally more reactive.

In aluminium hydride the bonds are more ionic than in the boranes and the compound is an insoluble white polymer.

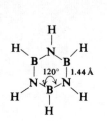

Fig. 15.5 Borazole.

15.2.5 Aluminium

The most striking features of the chemistry of aluminium are: (a) the absence of the ion Al^{3+} (owing to the high sum of the first three ionization energies) except when hydrated or as the very insoluble compounds of anions of very high charge density (therefore very high lattice energy), such as Al_2O_3, Al_4C_3, and AlF_3; (b) the lack of reactivity of the metal, which is due to the presence of a protective oxide layer; and (c) the high bond strength of the Al—O bond, and the consequent stability of Al_2O_3 and $Al(H_2O)_n^{3+}$.

The aluminium ion Al^{3+} is small and highly polarizing. This results in: most of its compounds being covalent; high lattice energies for ionic compounds such as the oxide and fluoride; and a high hydration energy, and therefore in a high electrode potential. It would seem that the hydration energy of B^{3+} would be even larger, and this would result in the existence of $B^{3+}(aq)$. In fact, the hydration energy of B^{3+} might not be greater since there is not much room around B^{3+} for water molecules; and even if the unknown hydration energy of the ion B^{3+} is larger than that of Al^{3+}, the non-existence of $B^{3+}(aq)$ can still be explained. Firstly, the sum of the first three ionization energies of boron is much larger than that of aluminium; and secondly, even if the reaction $BCl_3(s) + H_2O \rightarrow B^{3+}(aq) + 3Cl^-(aq)$ does have a negative ΔG value, the hydrolysis to boric acid has a free energy change which is even more negative.

15.2.6 Aluminium Compounds

1) Aluminium oxide. There are several different forms of aluminium oxide, all of which are white. γ-Al_2O_3 can be prepared by heating aluminium hydroxide in a vacuum at 500 °C. In this structure the aluminium ions are randomly distributed among the tetrahedral and octahedral sites formed by a face-centred cubic array of oxide ions. If γ-Al_2O_3 is heated at 1000 °C it forms α-Al_2O_3 (corundum), in which the aluminium ions are distributed symmetrically in two-thirds of the octahedral sites formed by a hexagonal close-packed array of oxide ions. This form is very hard and inert. It is insoluble in all solvents except fused alkali, whereas the more random γ-form is soluble in dilute acids. Rubies and sapphires are forms of alumina containing traces of transition metal ions which colour it: rubies contain traces of the ion Cr^{3+}, and blue sapphires contain traces of the ions Fe^{2+}, Fe^{3+}, or Ti^{4+}. The lattice energy of alumina is large, and the heat of formation is high (1676 kJ mol^{-1}). Thus aluminium will reduce the oxides of most metals (thermite process, Section 11.5.2). Aluminium hydroxide is amphoteric (Section 11.1.1):

$$Al(OH)_4^-(aq) \xleftarrow{OH^-} Al(OH)_3(s) \xrightarrow{H_3O^+} Al^{3+}(aq)$$

Table 15.4. Some properties of some group IIIB halides.

	Melting point (°C)	Boiling point (°C)	ΔH_{atom}	ΔH_f (kJ mol^{-1})	Character
AlF$_3$	very high	1270	2049	−1504	mainly ionic
AlCl$_3$	192	180(s)	1390	−704	covalent
AlBr$_3$	97.5	265	1212	−527	covalent
AlI$_3$	191	382	960	−315	covalent
BCl$_3$	−107	12.4	1346	−395	covalent

2) Aluminium halides. Some properties of these compounds are given in Table 15.4. The fluoride, AlF$_3$, exists in the solid state as a giant molecule in which the aluminium is six-co-ordinate. The bonding is more ionic than that in the other halides, and owing to its giant three-dimensional structure, its boiling point is higher than those of the other halides. Aluminium chloride has a layer structure and much lower melting and boiling points than the fluoride. The bromide and iodide have molecular structures in the solid state, discrete Al$_2$X$_6$ molecules being present in the lattice. The intermolecular forces in the latter two are due to van der Waals forces, which increase as the size of the molecule increases, and therefore the melting and boiling points increase from the bromide to the iodide. These four structures illustrate well the dependence of structure on bond type, and the dependence of other properties on structure. In the liquid and vapour phases the molecules Al$_2$X$_6$ (X = I, Br, Cl) consist of dimeric discrete molecules that have a halide bridge structure with a tetrahedral arrangement of atoms around each aluminium, as in diborane (Fig. 15.4). At high temperatures dissociation into planar trigonal monomers, AlX$_3$, occurs:

$$Al_2Cl_6 \rightleftharpoons 2AlCl_3$$

In water, aluminium chloride hydrolyses to yield the ion Al^{3+}(aq). This change, from a covalent compound to an aqueous solution of ions, is mainly due to the high hydration enthalpy of the aluminium ion. This is clear from the Born–Haber cycle in Fig. 15.6, where, despite the large energy involved in the ionization of the aluminium ion, the heat of formation of the aqueous solution of ions has a large negative enthalpy (−300 kJ), owing mainly to the high hydration energy of the aluminium. It might be thought that the same argument would apply to boron. In fact, the hydration energy necessary to make the corresponding ΔH_f negative is $\Delta H_{hyd} = -6009$ kJ. A hydration energy this large is unlikely, and in fact boron trichloride hydrolyses to yield, not B^{3+}(aq), but boric acid.

3) Salts of aluminium. Just as the aquated ion Al^{3+}(aq) exists in solution, so the ion Al(H$_2$O)$_6^{3+}$ can exist in the solid state. Aluminium salts form colourless hydrates of the type AlCl$_3 \cdot$6H$_2$O, Al(NO$_3$)$_3 \cdot$9H$_2$O, and the series of double salts, MAl(SO$_4$)$_2 \cdot$12H$_2$O, called alums, in which M can be any univalent cation. All these decompose to yield the oxide Al$_2$O$_3$ on heating, and are extensively hydrolysed in solution. Aluminium can be replaced in the alums by other trivalent cations—Fe^{3+}, Cr^{3+}, Co^{3+}, Ti^{3+}. Aqueous solutions of aluminium salts are acidic, owing to the high polarizing power of the aluminium ion (Section 14.2.5):

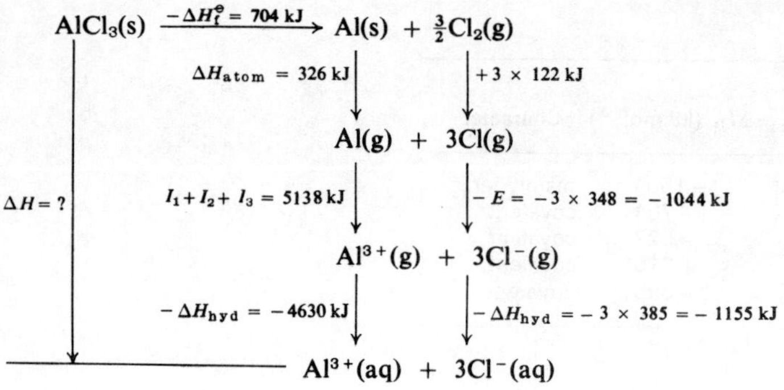

$$AlCl_3(s) \xrightarrow{-\Delta H_f^\ominus = 704 \text{ kJ}} Al(s) + \tfrac{3}{2}Cl_2(g)$$

$\Delta H_{atom} = 326 \text{ kJ} \downarrow \qquad \downarrow +3 \times 122 \text{ kJ}$

$$Al(g) + 3Cl(g)$$

$\Delta H = ? \qquad I_1+I_2+I_3 = 5138 \text{ kJ} \downarrow \qquad \downarrow -E = -3 \times 348 = -1044 \text{ kJ}$

$$Al^{3+}(g) + 3Cl^{-}(g)$$

$-\Delta H_{hyd} = -4630 \text{ kJ} \downarrow \qquad \downarrow -\Delta H_{hyd} = -3 \times 385 = -1155 \text{ kJ}$

$$Al^{3+}(aq) + 3Cl^{-}(aq)$$

$AlCl_3(s) \rightarrow Al^{3+}(aq) + 3Cl^{-}(aq);\ \Delta H = 704 + 326 + 366 + 5138 - 1044 - 4630 - 1155$
$\qquad\qquad\qquad\qquad\qquad\qquad\qquad\quad = -295 \text{ kJ}$

$BCl_3(g) \rightarrow B^{3+}(aq) + 3Cl^{-}(aq);\ \Delta H = +395 + 562 + 366 + 6885 - 1044 - \Delta H_{hyd}(B^{3+}) - 1155$
$\qquad\qquad\qquad\qquad\qquad\qquad\qquad\quad = (-6009 - \Delta H) \text{ kJ}$

Fig. 15.6 Born–Haber cycle for formation of hydrated Al^{3+} (aq) ion.

aluminium ions in water are almost as acidic as acetic acid. In order to prevent hydrolysis it is necessary to add mineral acids.

Reactions of Al^{3+}(aq), for example an aqueous solution of aluminium sulphate:

a) Sodium hydroxide precipitates a gelatinous white precipitate of aluminium hydroxide, which dissolves in an excess of hydroxide to form the aluminate ion, written $Al(OH)_4^-$(aq), which is probably a mixture of aquated anions such as $[Al(OH)_4(H_2O)_2]^{2-}$:

$$Al^{3+}(aq) + 3OH^-(aq) \rightarrow Al(OH)_3(s) \xrightarrow{OH^-} [Al(OH)_4]^-(aq).$$

b) Ammonium hydroxide precipitates aluminium hydroxide, which does not dissolve in an excess of ammonia, because the hydroxyl ion concentration is too low to form aluminates, and aluminium does not form ammines in aqueous solution.

c) Aluminium sulphide is not precipitated by hydrogen sulphide in aqueous solution.

4) Complexes. Both boron and aluminium show a greater tendency to complex formation than the s-group metals. Boron is invariably tetrahedrally co-ordinated in such compounds: borohydride, BH_4^-; tetrafluoroborate, BF_4^-; and tetraphenylborate, $B(C_6H_5)_4^-$, which forms insoluble compounds with potassium and the heavier alkali metals. Aluminium forms both tetrahedral ($AlCl_4^-$ and $AlBr_4^-$) and octahedral complex ions (AlF_6^{3-}). Both boron and aluminium form complexes with chelating oxygen and nitrogen ligands, for example the insoluble 8-hydroxyquinoline ("oxine") complex (Fig. 15.7), which is used to determine aluminium gravimetrically (aluminium "oxinate"). The formation of the ion $AlCl_4^-$ is important in the Friedel–Crafts reaction (Section 26.5).

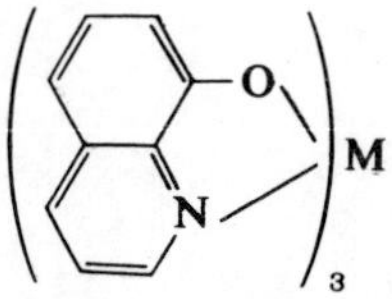

Fig. 15.7 The 8-hydroxyquinoline complex of aluminium.

5) Organometallic compounds (See Chapter 27.) All the members of the group form organic compounds of the types R_3M, R_2MX, and RMX_2, where R is an alkyl group and X an electronegative group such as a halogen. Whereas the boron compounds are monomers with a trigonal planar structure, BR_3, the other members of the group give polymers. An example is the compound $Al_2(CH_3)_6$, trimethyl aluminium dimer, which has an electron-deficient structure like that in diborane in which the carbons of the two central methyl groups act as bridging atoms.

PROBLEMS

1. The boiling points of the halides of aluminium are: "AlF_3", 1270 °C; "$AlCl_3$", 180 °C; "$AlBr_3$", 265 °C; "AlI_3", 382 °C. Account for these figures.

2. Write down the formulae for the oxides of thallium. Are either of these oxides coloured? Which oxide is more basic?

3. Explain why the melting point of boron is so high.

4. Why do the ions $Al(OH)_6^{3-}$ and AlF_6^{3-} exist, while $B(OH)_6^{3-}$ and BF_6^{3-} are unknown?

5. Account for the facts that aluminium bromide (molecular weight in the vapour phase, 530) is a poor conductor even when fused, but is a good conductor in aqueous solution, which is acidic.

6. Why is aluminium a better reductant than gallium?

7. What is the reduction potential for the couple Tl^{3+}/Tl^+, and the equilibrium constant for the disproportionation of the thallium(I) aqueous ion:

$$3Tl^+(aq) \rightarrow Tl^{3+}(aq) + 2Tl(s)$$

$$\begin{bmatrix} Tl^+(aq) + e^- \rightleftharpoons Tl(s); & E^\ominus = -0.34 \text{ V} \\ Tl^{3+}(aq) + 3e^- \rightleftharpoons Tl(s); & E^\ominus = +0.72 \text{ V} \end{bmatrix}$$

8. Account for the fact that thallium(III) trifluoride is known, but thallium(III) triiodide has not yet been prepared.

9. Thallium(I) and potassium salts are often isomorphous, but potassium salts are generally much more stable. Why?

10. Summarize the general trends which are found in the properties of the *p*-block elements.

11. The properties of the middle row element gallium are not always those expected by simple extrapolation of the properties of boron and aluminium (e.g. ionization energy, metallic radius). Discuss this anomaly.

12. Why are the oxygen compounds of boron so different from those of carbon and nitrogen?

13. Discuss the bonding in the diborane molecule.

BIBLIOGRAPHY

Most of the books on inorganic chemistry in the general list contain material on the chemistry of group IIIB. A selection is listed at the end of Chapter 16.

CHAPTER 16

The Carbon Group: Group IVB

16.1 Group trends and types of compound formed. The inert pair effect
16.2 Occurrence, allotropes, and extraction
16.3 Properties of the elements
16.4 Peculiarities of carbon
16.5 Binary carbon compounds
16.6 Oxides of group IVB
16.7 Hydrides of group IVB
16.8 Halides of group IVB
16.9 Sulphides of group IVB
16.10 Properties of lead(II) and tin(II) ions
16.11 Organometallic compounds
16.12 Oxalates

16.1 GROUP TRENDS AND TYPES OF COMPOUND FORMED. THE INERT PAIR EFFECT

In groups IA and IIA (and group VIIB), the most striking feature of the group chemistry (Table 16.1) is the similarity of the elements within the group—except for the first member—(Sections 10.1 and 14.2.7). By the time group IVB is reached, although similarities between the group members still exist (especially valency relationships), the most striking feature of the group chemistry is the dramatic increase in electropositivity in descending the group (Tables 16.2 and 16.3).

Looking at the chemistry in more detail (Table 16.1), we see that carbon and silicon differ from each other and from the other three elements; these last three form a series in which there is a gradual increase in electropositive character, except that lead is less electropositive than expected. Germanium is also more electronegative (and more like silicon) than expected. This trend is typical for a p-block group (Section 15.1) and is illustrated by properties such as the electronegativity and ionization energies of the elements, as well as being reflected in their general chemistry. There is also a strong diagonal relationship between silicon and boron (Section 16.3).

Before discussing the types of compound formed, the modern meaning of the word *valency* must be defined, and carefully distinguished from *oxidation number* (Section 18.14) and *co-ordination number* (Section 3.2.2). *Valency* refers to the number of valence electrons of an atom used in bonding. Consider the compound carbon tetrachloride, CCl_4 (Section 16.8); the carbon uses all four of its valence electrons in covalent bonding, and therefore its valency is four. In this example the co-ordination number of the carbon is also four, and its oxidation number is *plus* four. Thus, sometimes these numbers are the same, but not always. In ionic compounds the valency is the number of electrons gained or lost in forming the ion, for example Sn^{2+} has a valency of two, and C^{4-} has a valency of four. The word *valency* is particularly useful in carbon chemistry in describing, collectively, compounds such as those in the series: CCl_4, $CHCl_3$, CH_2Cl_2, CH_3Cl, and CH_4. These compounds are all very similar, but in them the carbon has the oxidation numbers $+4$, $+2$, 0, -2, and -4 respectively. It is convenient to describe them all as compounds in which carbon has a valency of four.

The main point of similarity between the elements in this group is the formation of compounds showing valencies of two and four, mainly those of oxidation

Table 16.1. Properties of the group IVB elements.

Property	Carbon C	Silicon Si	Germanium Ge	Tin Sn	Lead Pb
Atomic number	6	14	32	50	82
Electron configuration (outer)	$1s^2 2s^2 2p^2$	$3s^2 3p^2$	$3d^{10} 4s^2 4p^2$	$4d^{10} 5s^2 5p^2$	$4f^{14} 5d^{10} 6s^2 6p^2$
Isotopes (in order of abundance)	12, 13, 14	28, 29, 30	74, 72, 70, 76, 73	120, 118, 116, 119, 117, 124, 122, 112, 114, 115	208, 206, 207, 204
Atomic weight	12.01115	28.086	72.59	118.69	207.19
Metallic radius (Å)	0.91	1.32	1.37	1.62	1.75
Ionic radius, M^{2+} (Å)	—	—	0.93	1.12	1.20
Covalent radius (Å)	0.77	1.17	1.22	1.40	1.54
Atomic volume ($cm^3\ mol^{-1}$)	5.3	12.0	13.6	16.3	18.3
Boiling point (°C)	4827	2355	2830	2270	1744
Melting point (°C)	3550	1410	937	232	327
Enthalpy of fusion ($kJ\ mol^{-1}$)		46.5	31.8	7.2	4.78
Enthalpy of vaporization ($kJ\ mol^{-1}$)	718	297	285	291	178
Enthalpy of atomization ($kJ\ mol^{-1}$)	716	456	376	302	195
Enthalpy of hydration, M^{2+} ($kJ\ mol^{-1}$)				1520	1448
Density ($g\ cm^{-3}$)	2.22^{gr} 3.51^d	2.33	5.32	7.3	11.3
Electronegativity (A/R)	2.5	1.74	2.0	1.7	1.55
Ionization energy $1+2$ ($kJ\ mol^{-1}$)	3438	2364	2297	2119	2167
$1+2+3+4$	14277	9949	10008	8989	9330
Electron affinity ($kJ\ mol^{-1}$)	123	135			
Electrode potential, M^{2+}/M (V)				−0.136	−0.126
Common co-ordination numbers	2, 3, 4	4, 6	4, 6	4, 6	4, 6
Common oxidation numbers	IV	IV	(II), IV	II, IV	II, IV
Abundance, p.p.m.	320	277, 200	7	40	16
Atomic conductance ($ohm^{-1}\ cm^{-4}$)	0.14^{gr}			5.5	2.6
Conductivity	non-cond.^{d} fairly good^{gr}	semi-cond.	semi-cond.	semi-cond.^{g} good^w	good
Co-ordination numbers (solid)	$3^{gr}, 4^d$	4	4	$4^g, 6^w$	12
Ions	−4	−1		+2 (+4)	+2 (+4)

^{d} diamond; ^{gr} graphite; ^{g} grey; ^{w} white.

numbers +2 and +4, but not exclusively. The formation of compounds in which the elements show more than one stable oxidation number is characteristic of the *p*-block elements (Section 15.1). However, even here, a striking trend emerges: the stability of the +4 oxidation state relative to the +2 oxidation state decreases as the group is descended—the +4 oxidation state becomes more highly oxidizing, and the +2 oxidation state becomes more stable and less reducing. This is obvious from the oxidation state diagram, Fig. 16.1(a). It is the relative positions of the +2 and +4 oxidation states on the diagram for each element that is important in this context. Silicon(IV) dioxide is very stable (low on the diagram) and is not an oxidant; in fact no stable silicon(II) compound, SiO, is known under normal conditions. Germanium(IV) dioxide is also not an oxidant, since it is well below germanium(II) (a reducing state) on the diagram. Tin(IV) is a mild oxidant since it is just above tin(II) (a mild reductant); but lead(IV) is a strong oxidant, and lead(II) is a stable oxidation state. Carbon(II) and carbon(IV) have anomalous positions on the diagram, but nevertheless carbon dioxide is clearly stable and non-oxidizing, and carbon monoxide is a reductant. These facts are generally summarized in

Table 16.2. Electrode potentials of group IVB elements.

(a) Acid

O.N.	+4	+2	0	−4	+4	0	−4	+4	+2	0	−4	+4	+2	0	+4	+2	0
	CO_2	CO	C	CH_4	SiO_2	Si	SiH_4	GeO_2	Ge^{2+}	Ge	GeH_4	Sn^{4+}	Sn^{2+}	Sn	Pb^{4+}	Pb^{2+}	Pb
	−0.12	0.5	0.13		−0.86	0.1		(−0.3)	0	<−0.3		0.15	−0.14		1.46	−0.13	

(b) Alkali

O.N.	+4	+2	0	−4	+4	0	−4	+4	0	−4	+4	+2	0	+4	+2	0
	CO_3^{2-}	HCO_2^-	C	CH_4	SiO_3^{2-}	Si	SiH_4	$HGeO_3^-$	Ge	GeH_4	$Sn(OH)_6^{2-}$	$HSnO_2^-$	Sn	PbO_2	PbO	Pb
	−1.0	−0.52	−0.7		−1.73	−0.73		−1.0	<−1.1		−0.9	−0.9		+0.28	−0.54	

Table 16.3. The nature of the group IVB elements.

increasing electropositivity ↓	carbon	non-metal
	silicon	semi-metallic physical properties; non-metallic chemistry
	germanium	typical semi-metal
	tin ⎫	both have the physical and chemical properties
	lead ⎭	of a weakly electropositive metal

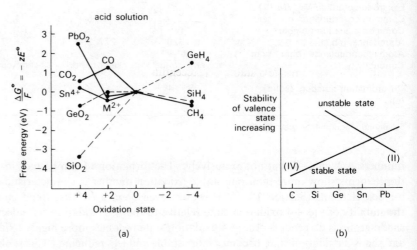

Fig. 16.1 Group IVB: (a) oxidation state diagram; and (b) simple representation of the relative stabilities of valence states.

diagrams such as Fig. 16.1(b). The decreasing stability of the valence state four on descending the group is true for the oxidation state −4, as well as the oxidation state +4. This is obvious from Fig. 16.1(a), where the hydrides appear higher on the diagram as the group is descended. Comment on the stability of germanium(II) compounds in aqueous solution at pH = 0 (25 °C) (Problem 16).

Since all the elements have four valence electrons, it is not surprising that the valency state four is common to the group. But no cations M^{4+} are found, owing to the high ionization energies involved (alternatively, view it in the sense of such ions being very highly polarizing, which would make the bonding covalent); thus, the bonding is always covalent (CCl_4, $SiCl_4$, $PbCl_4$, etc.). Partly ionic divalent compounds are formed by tin and lead, however. Only the most electronegative element in the group, carbon, forms anions: C^{4-} (methanides), and C_2^{2-} (acetylides). The low electron affinities of the other elements preclude the possibility of anion formation. In addition, carbon forms unstable ionic alkyls with electropositive elements, such as $Na^+CH_3^-$, and various other anions (carbanions), cations (carbonium ions), and radicals of moderate stability which involve double-bonded carbon. No stable divalent carbon compounds are known, but they are thought to occur as transient reaction intermediates in organic chemistry—carbenes—such as difluorocarbene, CF_2.

The increasing stability of the divalent state on descending the group was previously "explained" in terms of the "inert pair" of electrons, and called the *inert pair effect*. The idea was that as the atomic weight increased, the valence electrons, s^2, became "inert" and took less part in the chemistry of the elements. It would be expected therefore that a dramatic increase in the third and fourth ionization energies would be observed in descending the group. This is not found. Moreover, the "inert" pair of electrons is not inert stereochemically (Section 12.7). Nowadays, the term inert pair effect is used not as an explanation, but as a label for the phenomenon, and merely means that the lower valence state (two less than the group oxidation state) becomes more stable as the group is descended. This phenomenon, as by now you will have come to expect, is determined by a number of factors, including bond enthalpies for covalent compounds, and lattice energies for ionic compounds. The Born–Haber cycles required are too complex for this book, but the results can be quoted: methane is thermodynamically stable relative both to dissociation into its constituent elements, and to dissociation into carbene (CH_2) and hydrogen, mainly because of the formation of four strong C—H bonds (carbene only has two)—see Fig. 4.15. The problem of why the tetravalent covalent state becomes less stable on descending the group is more tricky, but one of the main causes is the fact that the strengths of the covalent bonds (M—H for example, Table 16.6) decrease down the group, and since the sum of the first two ionization energies also decreases (Table 16.1), it becomes increasingly thermodynamically favourable to form compounds of valency two, which are more ionic in character. To summarize, these elements form divalent and tetravalent compounds in which the bonding is mainly covalent, but which becomes increasingly ionic for the divalent compounds as the group is descended.

A wide variety of co-ordination numbers and shapes is shown by these compounds. Carbon compounds can be linear (CO_2), triangular (CO_3^{2-}), or tetrahedral (CH_4), but the co-ordination number never exceeds four, since the number of electrons in the valence shell cannot exceed eight. This restriction does not hold for the later elements in the group, and the co-ordination numbers of the divalent heavier group IV elements can be two to six inclusive, and those of the tetravalent heavier group IV elements can be four, five, or six. The shapes of those compounds, which are discrete covalent molecules, can generally be determined by the Gillespie–Nyholm rules (Section 12.7). The formation of complex compounds is a consequence of the ability of the heavier elements (silicon, etc.) "to expand the octet". Thus, whereas carbon forms no complex compounds, the other group IVB elements do: for example SiF_6^{2-}, $SnCl_6^{2-}$, $GeCl_6^{2-}$, $PbCl_4^{2-}$. Most of these can be regarded as adducts of halide ions, or nitrogen, oxygen, sulphur, or phosphorus, donor ligands to the tetrahalides which act as Lewis acids:

$$Cl^- + SnCl_4 \longrightarrow SnCl_5^- \xrightarrow{Cl^-} SnCl_6^{2-}.$$

The stability of these complexes generally increases as the central element becomes more electropositive: for silicon few complexes exist; for lead, many more. Most of the group IVB complexes are hydrolysed by water, however.

16.2 OCCURRENCE, ALLOTROPES, AND EXTRACTION

Carbon is the sixteenth most abundant element and occurs naturally in the form of diamonds (Brazil and South Africa) and graphite (Ceylon, Germany, U.S.A.), but much more abundantly in impure forms such as coal, and in the combined state in living matter, and as petrol, gaseous hydrocarbons, limestone, and so on. Both allotropic forms of carbon, graphite and diamond (Section 3.3), are giant molecules consisting of carbon atoms linked by a network of covalent bonds (Fig. 16.2). Their properties reflect this structure (Section 10.3.1): high melting and boiling points, high enthalpies of fusion, atomization, etc. In diamond each carbon atom is covalently bonded to four others located at the corners of a tetrahedron. The C—C bond enthalpy (355 kJ mol^{-1}) and bond distance (1.54 Å) are similar to those in other compounds containing carbon atoms linked by a single covalent bond, such as ethane. The unit cell can be described as fourteen atoms in the face-centred cubic arrangement, with one half of the tetrahedral holes filled symmetrically. Thus diamond has the same structure as zinc blende, with all the zinc and sulphur atoms replaced by carbon atoms. This interlocking arrangement of strong covalent bonds results in a very hard dense compound which is an insulator since there are no "free electrons" to conduct heat or electricity (Section 10.3.12); all the valence electrons are used in forming strong covalent bonds. Graphite has a very different structure. It is a two-dimensional sheet polymer of fused benzene-type rings which are regular hexagons. The bond length is 1.42 Å, suggesting some multiple bond character, since this is intermediate in length between a single and double carbon–carbon bond. It is not possible to draw a single valence-bond structure for a layer, which shows all the bonds equivalent. Three resonance hybrids are drawn in Fig. 16.3 to represent the bonding in a small part of a layer. Clearly, this model represents the bond order as being intermediate between a single and a double bond. The distance between the layers is 3.35 Å, much longer than a carbon–carbon single covalent bond. Thus the inter-layer bonding is of the weak van der Waals type, and the layers can slip over one another relatively easily. This partly accounts for the softness of graphite and its use as a lubricant; but also involved in this property is the ability of graphite to absorb small molecules such as oxygen. There are two forms of graphite, which differ in the order of the layers; neither has all the atoms in adjacent layers superposed. Just as in metals, these are the hexagonal *abab*... (the more stable), and the rhombic *abcabc*... modifications. Alternatively, the bonding in graphite can be viewed as three trigonal bonds formed by three of the four valence electrons of carbon, the fourth electron being delocalized over the whole planar structure. This delocalization of electrons, just as in the case of metals, results in graphite conducting electricity (but only in the plane of the layers, not at right angles to the layers), and in its characteristic metallic lustre. By comparison, diamond, which has no free electrons and no easy electronic transitions, is a transparent insulator.

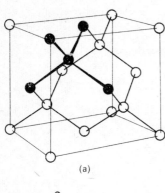
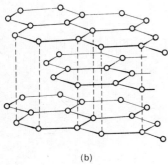

Fig. 16.2 The structures of (a) diamond, and (b) graphite.

Fig. 16.3 Three resonance structures for a graphite fragment (canonical forms).

Surprisingly, graphite is the (slightly) more stable allotrope at 25 °C and one atmosphere pressure:

$$C_{diamond} \rightarrow C_{graphite}; \quad \Delta H = -2.1 \text{ kJ mol}^{-1}.$$

However, since the change involves the initial breaking of a large number of strong covalent bonds, there is a large activation energy for this conversion to graphite (Section 10.3.1), and diamond is kinetically stable (called metastable) under normal conditions with respect to the change to graphite. Since diamond is more dense than graphite it follows that at higher pressures diamond will become increasingly stable relative to graphite (Le Chatelier), and in fact it has been estimated that the two allotropes are in equilibrium at 20 °C under a pressure of 15,000 atmospheres. Small diamonds, useful industrially, can therefore be obtained from graphite at high temperatures and pressures in the presence of transition metals (e.g. 2000 °C, 100,000 atmospheres, platinum). Graphite can be manufactured by the Acheson process in which sand and coke are heated electrically for one day:

$$SiO_2(s) + 3C(s) \rightarrow 2CO(g) + SiC(s) \rightarrow Si + C(graphite) + 2CO(g).$$

Despite their relative thermodynamic stabilities, graphite is more reactive than diamond. For example, diamond burns in air at 900 °C, graphite at 700 °C. This is therefore a kinetic effect: the "open" structure of graphite enables reagents to penetrate the layers and react; whereas the "closed" structure of diamond prohibits this. Therefore activation energies are lower, and corresponding reactions go more quickly for graphite, than for diamond, and we say graphite is more "reactive". Charcoal, soot, etc., are now thought to be, not different allotropes of carbon, but microcrystalline graphite. Their finely divided nature makes them more reactive than graphite.

Table 16.4. The enantiotropes of tin.

	(grey) α-tin	$\xrightleftharpoons{13.2\,°C}$	(white) β-tin	$\xrightleftharpoons{161\,°C}$	γ-tin	$\rightleftharpoons$	tin (liquid)
Structure	diamond		metallic		metallic		
Density (g cm^{-3})	5.75		7.31		both distorted close-packed		

Although silicon and germanium (no allotropes) and grey tin have the diamond structure, they are all less hard than diamond, the hardness decreasing down the group. This is due to the decreasing bond strength as the size of the atoms increases. Moreover, all except diamond are coloured (silicon is brown/black, the remainder are grey). The graphite structure is peculiar to carbon, and is associated with its unique ability, in this group, to form double bonds to itself (Section 16.4.2). As well as the diamond form, α-tin, tin forms complex metallic-type allotropes (Table 16.4): white, β-tin, which is stable above 13.2 °C, and which is a conductor; and another form, γ-tin. Lead exists in only one form, a distorted cubic close-packed metallic structure. The change in elemental character down the group is a good example of the increasing electropositive character of the elements (structure, melting point, electrical conductivity, etc.). During a hard winter β-tin exposed to the atmosphere changes to α-tin, increases in volume, and crumbles. Napoleon's retreat from Russia has been attributed, perhaps by a French historian, to this crumbling of tin equipment during the Russian winter.

Silicon is the second most abundant element, and occurs as silica (SiO_2) and many forms of silicates. By comparison, germanium, tin, and lead, are rare elements; but tin and lead are well known on account of their ease of extraction and their (related) technical importance as unreactive metals and alloys. Germanium is present in coal in very small amounts, and accumulates in flue dust as the oxide, GeO_2. This is treated with hydrochloric acid to yield the tetrachloride, which is volatile and easily purified by fractional distillation. It is then hydrolysed to the dioxide, now pure, reduced with hydrogen to the metal, and made ultra-pure by zone refining (Section 29.1). Pure silicon is obtained in almost the same way:

$$SiO_2 \xrightarrow{C} Si \xrightarrow{Cl_2} SiCl_4 \xrightarrow{distill} \text{pure } SiCl_4 \xrightarrow{H_2} Si \longrightarrow \text{zone refined.}$$

Lead is found as the ore galena (lead sulphide, PbS). The ore is concentrated from earth and zinc sulphide by froth flotation and is then roasted in air to yield the oxide. The oxide is then reduced to the metal with coke in a blast furnace (cf. iron). Some scrap iron is added to convert any unreacted sulphide to the metal:

$$2PbS + 3O_2 \rightarrow 2PbO + 2SO_2$$
$$PbO + C \rightarrow Pb + CO$$
$$PbS + Fe \rightarrow Pb + FeS.$$

The molten lead is tapped off from the bottom, and can be purified electrolytically (Section 11.5.2) using the complex salt, $PbSiF_6$, as electrolyte. Tin occurs as the dioxide, SnO_2, and is extracted and purified similarly.

Diamonds are used in jewellery on account of their high refractive index and dispersive power, and in industry as a hard material in drilling, cutting, and grinding instruments, and for bearings of precision instruments such as watches. Graphite is used in "lead" pencils, as a lubricant, as an inert conductor in cells (electrodes, lining), and for slowing down neutrons. Charcoal is used for absorbing gases and decolorizing liquids, and lamp black as a blackening agent in rubber, ink, boot polish, etc. Coal is used as a fuel; coke as a fuel or reductant. Silicon and germanium are used as semi-conductors in transistors by the electronics industry. Tin is used for tin-plating steel, and in various alloys: bronze (90 percent Cu); pewter (25 percent Pb); solder (50 percent Pb); and type metal (75 percent Pb, 15 percent Sb). Lead is used as an inert material to line less resistant materials, for example on roofs, in chemical vessels, gas and water pipes, and cable sheathing, as well as in batteries, in lead tetraethyl (anti-knock), in various alloys, and as a screen from radioactivity.

16.3 PROPERTIES OF THE ELEMENTS

There is an impressive group trend in nearly all the atomic and physical properties (if the graphite allotrope of carbon is ignored) corresponding to the general increase in metallicity down the group. The anomalies in some of these properties—electronegativity etc.—have been discussed previously (Section 10.3), and these are reflected in some of the chemical properties (Table 16.5). In general, the chemical reactivity increases down the group. Silicon is not very reactive, but is more so than carbon:

$$Si(s) + O_2(g) \rightarrow SiO_2(s); \quad \Delta G = -799 \text{ kJ mol}^{-1}; \text{ temp.} = 400 \text{ °C.}$$
$$C(gr) + O_2(g) \rightarrow CO_2(g); \quad \Delta G = -393 \text{ kJ mol}^{-1}; \text{ temp.} = 700 \text{ °C.}$$

Table 16.5. The chemical properties of the group IVB elements.

Reagent	Reaction	
Hot concentrated HCl	$M + 2H^+ \rightarrow M^{2+} + H_2$	Not C, Si, Ge. Pb is slow (insoluble $PbCl_2$).
Hot concentrated H_2SO_4	$C + 2H_2SO_4 \rightarrow CO_2 + 2SO_2 + 2H_2O$ $M + H_2SO_4 \rightarrow M^{n+} + SO_2$	Not Si. $M = Sn^{4+}, Pb^{2+}, Ge^{4+}$ (slow).
Concentrated HNO_3	$3M + 4H^+ + 4NO_3^- \rightarrow 3MO_2 + 4NO + 2H_2O$ $3Pb + 8H^+ + 2NO_3^- \rightarrow 3Pb^{2+} + 2NO + 4H_2O$ $Si + 6HF \rightarrow 2H_2 + H_2SiF_6$	Not Si, Pb. Ge and Sn give hydrated oxide. Fuming nitric renders lead passive. Only HF attacks Si—stability of SiF_6^{2-}.
Aqueous alkali	$Si + 2OH^- + H_2O \rightarrow SiO_3^{2-} + 2H_2$	Not C, Ge, Pb. SiO_3^{2-} fast. $Sn(OH)_6^{2-}$ very slow.
Molten alkali		Not C. $SiO_4^{4-}, GeO_4^{4-}, Sn(OH)_6^{2-}, Pb(OH)_4^{2-}$.
O_2 or air, heat	$M + O_2 \rightarrow MO_2$ molten $Pb + O_2 \rightarrow PbO \xrightarrow[O_2]{470°C} Pb_3O_4$	C, Si, Ge, Sn. Lead reacts slowly with cold air $\rightarrow PbO \xrightarrow{H_2O} Pb(OH)_2 \xrightarrow{CO_2}$ basic carbonate.
H_2O, 20 °C	lead + soft water $\xrightarrow{O_2} Pb(OH)_2$ lead + hard water $\xrightarrow{O_2}$ protective layer of insoluble $PbSO_4$ or $PbCO_3$	Poisonous.
Steam, heat (strongly)	$M + 2H_2O \rightarrow MO_2 + 2H_2$ $C + H_2O \rightarrow CO + H_2$	$M = Sn, Si$. Not Ge.
S, heat	$M + 2S \rightarrow MS_2$	All except lead, which gives PbS.
Cl_2, heat	$M + 2Cl_2 \rightarrow MCl_4$	All except lead which gives $PbCl_2$.
Metals, heat	carbides and silicides, lead and tin alloys.	

Silicon is attacked by alkalis at normal temperatures. Germanium is slightly more reactive than silicon, and tin and lead are even more reactive. Lead is less reactive than expected from its electrode potential. This is usually due either to its high overvoltage for hydrogen (high activation energy for reactions with dilute acids), or to the formation of insoluble lead compounds preventing further reaction.

As expected (Section 14.2.7), there is a diagonal relationship between boron and silicon. Examples of this are: the acidity of the oxides, which react with metal oxides to yield borates and silicates; formation of oxide glasses; hydrolysis of halides to boric and silicic acids; and the formation of volatile reactive hydrides (cf. aluminium).

16.4 PECULIARITIES OF CARBON

The differences between the chemistry of the first row elements and the succeeding members in the same s- or p-block group reach a maximum in group IVB. The main differences between carbon and the other group IVB members are discussed below.

16.4.1 Catenation

This is the property of direct bonding between atoms of the same element, and can be manifested in rings or linear chains, singly or multiply bonded, e.g. ethane, butadiene, benzene, and hexahydrobenzene. Carbon shows this property to a uniquely great extent among the elements of the Periodic Table: C ⋙ S ≫ Si. The tendency to catenation in group IVB decreases in the sequence C ⋙ Si > Ge ≈ Sn > Pb (little tendency). (See Chapter 21 *et seq.*)

Before discussing the reasons for the relative stability of the catenated group IVB compounds (or of any other compounds), two points require emphasis. First, to compare the corresponding reactions of two elements or their compounds rigorously, it is necessary to use the concept of the Born–Haber cycle in order to make it obvious which are the important factors concerned. Usually, space (and often lack of data) precludes this type of analysis, and only the results of such an analysis are quoted. When covalent compounds are being compared it frequently turns out that a difference in bond enthalpies is the critical factor. Secondly, the terms "stability", "lability", and "reactivity" are often used in chemistry without the qualifications which are, strictly speaking, necessary for clarification. Generally, the word stability refers to thermodynamic stability relative to a given change, which should be specified. Methane, for example, is thermodynamically stable relative to dissociation into the constituent elements, and this can be attributed mainly to the large bond enthalpy of the carbon–hydrogen bond (Section 4.12):

$$CH_4(g) \rightarrow C(s) + 2H_2(g); \quad \Delta G = +50.8 \text{ kJ mol}^{-1}.$$

However, methane is thermodynamically unstable relative to oxidation:

$$CH_4(g) + 2O_2(g) \rightarrow CO_2(g) + 2H_2O(l); \quad \Delta G^\ominus = -581 \text{ kJ mol}^{-1}.$$

This exposes the statement "methane is stable" as meaningless, unless the reaction to which it refers is specified. Thermodynamic instability relative to a certain change is a *necessary condition* for that change to occur ($\Delta G^\ominus$ to be negative), but it is not a *sufficient condition*. Methane does not react with oxygen (or air) at room temperature, despite the fact that $\Delta G^\ominus$ is negative for the reaction, but if a naked flame is applied to a suitable mixture a very rapid reaction takes place (an explosion). This is due to the fact that, for this reaction (and many others), a larger activation energy is required than is available to the molecules at normal (room) temperatures. It is said that the reaction is under kinetic control (Section 8.9). Molecules which react quickly in a given set of circumstances are said to be *reactive* or *labile* (low activation energies); those which react slowly are said to be *non-labile*, *unreactive*, or *inert*. These words refer to rates of reaction. Methane is inert with respect to oxidation at room temperature by molecular oxygen. To summarize: if a molecule is stable under a given set of conditions relative to a certain change, this may be due to either thermodynamic stability ("stability") or kinetic stability ("non-lability"). If the compound undergoes the given change quickly, it is said to be reactive.

In chemistry we are particularly interested in the three types of reaction most likely to occur on this planet: dissociation into the constituent elements; oxidation; and hydrolysis. Dissociation can always be achieved by heating the compound to a sufficiently high temperature, but of course we are mainly interested in what happens at room temperature and temperatures easily obtained in the laboratory and in industry, say up to 1000 °C. If the atmosphere contained appreciable amounts of hydrogen and the ambient temperature was 200 °C, then chemists would be very interested in hydrogenation reactions at this temperature. Since the

Table 16.6. Bond energies of the group IVB elements and sulphur (kJ).

Bond	C	Si	Ge	Sn	Pb	S
E—E	347	226	188	150		272
E=E	611	318 est.				431
E≡E	837					
E—H	414	318	285	251		368
E—O	360	464	360			335
E=O	736	640				523
E≡O	108	803				
E—S	272	293				
E—N	305	335	255			
E—F	489	598	473			
E—Cl	326	402	339	314		272
E—Br	272	331	280	268		213
E—I	238	234	213	197		
E—C	347	305	297	285		

E = element.

atmosphere contains oxygen and water vapour at 20 °C, the reactions of major interest are oxidations and hydrolyses.

The conditions necessary for catenation are: (a) the valency of the element must be greater than or equal to two; and (b) the element must form strong element–element bonds of strength equal to or greater than those to other elements, especially oxygen (to exclude the possibility of energetically favourable reactions, particularly hydrolysis and oxidation). From the bond energies (Table 16.6), it is clear that only carbon qualifies, and indeed the only catenated group IVB elements found in nature are those of carbon. It works out that all the catenated compounds of silicon, etc. are thermodynamically unstable to both hydrolysis and oxidation, as well as dissociation into the constituent elements. The catenated alkanes, C_nH_{2n+2} where $n \geqslant 2$, are stable to hydrolysis and dissociation but unstable to oxidation. Clearly (see above) a compound can be thermodynamically unstable, yet exist if it is kinetically inert. Most carbon compounds are in fact inert (see later), as indeed are many of the covalent inorganic compounds of the non-metals nitrogen, sulphur, etc. But many compounds of the metals are labile. Table 16.7 summarizes the catenated compounds formed by the group. The trend is obvious: the catenated compounds of silicon etc. become increasingly less stable and more reactive relative to all three changes as the number of element–element bonds increases. In Table 16.8 the properties of ethane and disilane are compared.

Table 16.7. Summary of group IVB catenated compounds.

Carbon	Many C_nH_{2n+2} (n without limit). Also halides.
Silicon	Si_nH_{2n+2} (n up to 6) Si_nCl_{2n+2} (n up to 10), Si_2F_6.
Germanium	Ge_nH_{2n+2} (n up to 9).
Tin	Sn_2H_6 and $(R_2Sn)_n$—cyclic and linear, R = alkyl.
Lead	$(R_2Pb)_n$

Table 16.8. Comparison of the properties of ethane and disilane.

Reaction: Compound	dissociation $\rightarrow$ 2M + 3H$_2$	oxidation $\rightarrow$ MO$_2$ + H$_2$O	hydrolysis $\rightarrow$ MO$_2$ + H$_2$
C$_2$H$_6$	$\Delta G^\ominus$ is positive ∴ stable	$\Delta G^\ominus$ is negative *but* kinetically inert	$\Delta G^\ominus$ is positive ∴ stable
Si$_2$H$_6$	$\Delta G^\ominus$ is negative	$\Delta G^\ominus$ is negative	$\Delta G^\ominus$ is negative: hydrolysed by alkali

To summarize: the thermodynamic stability of the alkanes, with respect to dissociation into the constituent elements and hydrolysis, is due partly to the high bond energy of the carbon–carbon bond. The inertness to oxidation at room temperature is due to a comparatively high activation energy for this reaction. The silanes, etc., are all thermodynamically unstable relative to dissociation, oxidation, and hydrolysis. This can be partly attributed to the low bond energies of the Si—Si, etc., bonds. Note that, although all bond energies decrease down the group, the element–element bond enthalpies decrease more rapidly than the others. Therefore compounds containing M—M bonds are destabilized relative to compounds containing M—O, etc., bonds as the group is descended. Thus no compounds containing Si—Si bonds, etc., are found in nature.

16.4.2 Multiple Bonding

Carbon is the only group IVB element which forms stable double (and triple) bonds to itself and to the other first row elements nitrogen and oxygen. As a result there are whole classes of carbon compounds—alkenes, ketones, nitriles, etc.—which have no analogues in the chemistry of silicon, germanium, tin, and lead. Stoichiometric analogues (the typical group valency relationship) often exist, but they are structurally different: for example, CO$_2$(g) and SiO$_2$(s); (CH$_3$)$_2$C=O and [(CH$_3$)$_2$SiO]$_n$. The heavier group IVB elements cannot form the graphite structure, nor any of the graphite compounds. Nor do the analogues of carbonium ions, carbanions, and methyl radicals exist. The apparently similar compounds (C$_6$H$_5$)$_3$M, where M = Si, Ge, Sn, Pb, are dimers containing element–element bonds, not radicals; and the compounds, (C$_6$H$_5$)$_3$MCl, are not ionic.

16.4.3 Formation of Gaseous Oxides by Carbon

Carbon forms gaseous oxides, whereas the other members of the group do not. This is a consequence of the multiple-bonding capacity of carbon (above). Consider the bond energies of the E—O, E=O, and E≡O bonds, formed by carbon and silicon (Table 16.6). *The single most important feature of the chemistry of silicon is the large bond strength of the* Si—O *bond*. Not only is it greater than that of the C—O bond, but it is two-thirds that of the S=O bond, and one-half that of the S≡O bond. In the case of carbon, the C—O bond strength is only about one-half that of the C=O bond, and one-third that of the C≡O bond strength. The result

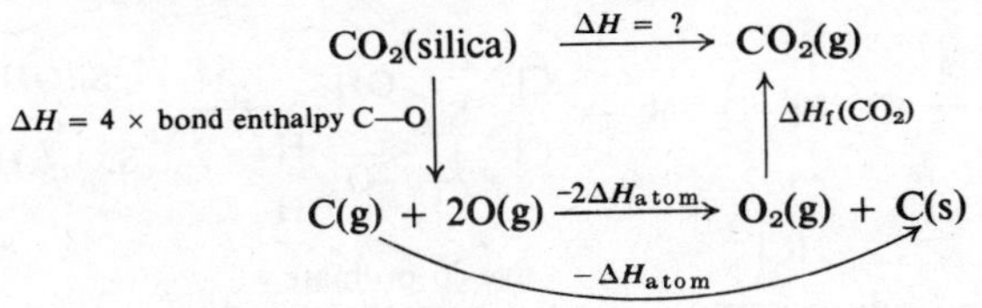

$$CO_2(\text{silica}) \xrightarrow{\Delta H = ?} CO_2(g)$$

$\Delta H = 4 \times$ bond enthalpy C—O ↓ ↑ $\Delta H_f(CO_2)$

$$C(g) + 2O(g) \xrightarrow{-2\Delta H_{atom}} O_2(g) + C(s)$$

$-\Delta H_{atom}$

$CO_2(\text{silica}) \longrightarrow C(g) + 2O(g);$	ΔH	$= +4 \times 360$	$= +1440$ kJ mol^{-1}
$C(g) \longrightarrow C(s);$	$-\Delta H_{atom}$		$= -716$ kJ mol^{-1}
$2O(g) \longrightarrow O_2(g);$	$-2\Delta H_{atom}$	$= -2 \times 250$	$= -500$ kJ mol^{-1}
$C(s) + O_2(g) \longrightarrow CO_2(g);$	ΔH_f		$= -394$ kJ mol^{-1}
(Adding) $CO_2(\text{silica}) \longrightarrow CO_2(g)$	ΔH		$= -170$ kJ mol^{-1}

Fig. 16.4 Born–Haber cycle for the stability of the "silica" form of carbon dioxide.

of this is that the presence of the single bond Si—O dominates the chemistry of silicon, whereas carbon forms many compounds containing C=O and C≡O bonds. To illustrate this, consider the reaction CO_2 (silica form, solid) → $CO_2(g)$. This hypothetical solid form of carbon dioxide is supposed here to have the same structure as silica (Section 16.6.5), with each silicon atom replaced by carbon. To estimate the enthalpy change for this reaction, the Born–Haber cycle in Fig. 16.4 is used. Clearly, although the data are enthalpy and not free energy data, the "silica" form of carbon dioxide is unstable relative to the gaseous form containing double bonds. (Would the free energy change for this hypothetical reaction be greater or less than -169 kJ mol^{-1}? Problem 13.) Therefore the "silica" form is not found. This is partly due to the fact that carbon prefers to form two double bonds to oxygen rather than four single bonds. Silicon, however, prefers to form four single bonds rather than two double bonds, owing to its high Si—O single bond strength. Therefore the solid form of SiO_2 is found, but not the gaseous form.

16.4.4 Carbon has a Maximum Covalency of Four

Carbon cannot have more than eight electrons in its outer shell (like the other first row elements). Since carbon already has four electrons in the outer shell, this means that the maximum covalency of carbon is four; and carbon forms compounds such as CCl_4, CH_4, etc. In such compounds carbon cannot act as a donor (no lone pairs), nor as an acceptor (cannot exceed eight outer electrons), and therefore forms no complex compounds. The heavier group IVB elements, however, can exceed eight outer electrons and can act as acceptors (Lewis acids), and they form complexes such as SiF_6^{2-}, $SnCl_6^{2-}$, etc.

16.4.5 Carbon Compounds are Relatively Inert

Another consequence of carbon obeying the "rule of eight" is that its compounds are relatively inert (kinetically stable), compared to the other group IVB elements. A good example is the hydrolysis of the chlorides:

$$CCl_4(l) + 2H_2O(l) \to CO_2(g) + 4HCl(aq); \quad \Delta G = -380 \text{ kJ mol}^{-1}$$
$$SiCl_4(l) + 2H_2O(l) \to SiO_2(g) + 4HCl(aq); \quad \Delta G = -289 \text{ kJ mol}^{-1}.$$

Clearly, carbon tetrachloride is less thermodynamically stable relative to hydrolysis than is silicon tetrachloride. The main reason for this is the high bond energy

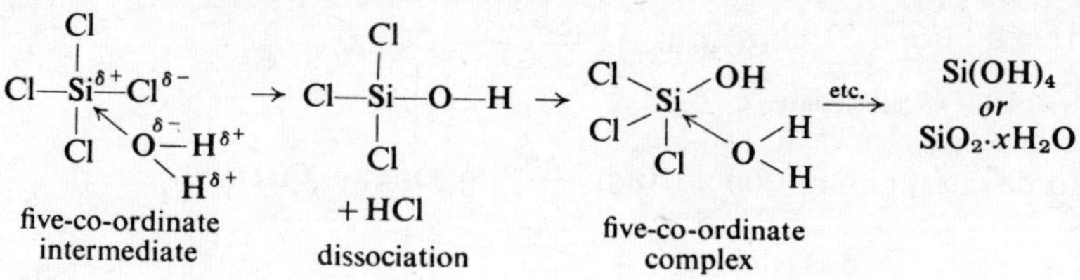

$$\underset{\substack{\text{five-co-ordinate}\\\text{intermediate}}}{\text{Cl}-\underset{\underset{\text{Cl}}{|}}{\overset{\overset{\text{Cl}}{|}}{\text{Si}^{\delta+}}}\text{Cl}^{\delta-} \underset{\overset{|}{\text{H}^{\delta+}}}{\overset{\delta-}{\text{O}-\text{H}^{\delta+}}}} \rightarrow \underset{\substack{\text{+ HCl}\\\text{dissociation}}}{\text{Cl}-\underset{\underset{\text{Cl}}{|}}{\overset{\overset{\text{Cl}}{|}}{\text{Si}}}-\text{O}-\text{H}} \rightarrow \underset{\substack{\text{five-co-ordinate}\\\text{complex}}}{\overset{\text{Cl}}{\underset{\text{Cl}}{\text{Si}}}\overset{\text{OH}}{\underset{\text{O}}{\diagdown}}\overset{}{\underset{\text{H}}{\diagup}}\text{H}} \xrightarrow{\text{etc.}} \begin{array}{c}\text{Si(OH)}_4\\\text{or}\\\text{SiO}_2\cdot x\text{H}_2\text{O}\end{array}$$

Fig. 16.5 Supposed mechanism of hydrolysis of chlorides.

of the Si—Cl bond compared to that of C—Cl (Table 16.6). However, whereas silicon tetrachloride hydrolyses readily, carbon tetrachloride is inert to hydrolysis. This is clearly a kinetic effect. A plausible mechanism of these hydrolytic reactions is that illustrated in Fig. 16.5. The positive polar central atom, silicon, is attacked by the negative polar oxygen atom of water forming a transient five-co-ordinate intermediate, $SiCl_4(\leftarrow OH_2)$, in which the water is co-ordinated to the silicon, and which then dissociates. This mechanism is repeated until eventually hydrated silica is formed. The five-co-ordinate intermediate contains a silicon atom with ten electrons in its outer shell. Assuming that this is a good model of the mechanism of the reaction, since carbon cannot expand its octet and form five-co-ordinate compounds, then the analogous reaction for carbon must have a high activation energy and proceed much more slowly. Alternatively, another (slower) mechanism of the reaction must occur. Because of the wide occurrence of such mechanisms (S_N—Section 26.3) most carbon compounds tend to be inert, whereas the analogous compounds of the heavier group IVB elements are less inert, and since they are often thermodynamically unstable, they are more reactive generally.

In general, organic (carbon chemistry) reactions tend to be under kinetic control (high activation energies), whereas in inorganic chemistry (the rest of chemistry) reactions are often under thermodynamic control (Section 8.9).

16.4.6 Carbon Shows No "Inert Pair Effect" (see Section 16.1)

There are no stable compounds in which carbon can be said to be divalent, as defined above. The complex reasons for this have been discussed briefly above for covalent compounds; for divalent ionic compounds to be stable, the lattice energy must be sufficient to make up for the sum of the first two ionization energies and the sublimation energy (Section 14.2.4), so that dissociation into the elements will not occur. Nor must disproportionation occur, e.g. $2 :CF_2 \rightarrow CF_4 + C$ (two very stable substances). But the cation C^{2+} is too small to form stable ionic compounds (since the anions, being much larger, would touch each other, lowering the lattice energy). Also, the sublimation enthalpy and ionization energies are large for carbon. In fact, only in the case of lead are stable ionic compounds formed in this group.

16.4.7 Carbon Has a Much Higher Electronegativity than Silicon, etc.

This is typical of the first member of the group, and has the usual associated trends in electron affinity, ionization energy, and so on. Thus, carbon is the only atom to form definite anions, C_2^{2-}, C^{4-}. Another example is the reactivity of the hydrides, where a useful complementary model to that above is to consider the polarity of the $Si^{\delta+}$—$H^{\delta-}$ and $C^{\delta-}$—$H^{\delta+}$ bonds. The silanes are more like the ionic hydrides: they are reactive, reducing, and attacked by nucleophilic reagents.

16.4.8 Bond Strength

Carbon, in general, forms stronger bonds to other elements than do the other group IVB elements. Therefore, its compounds tend to be more stable with respect to dissociation into elements, despite the stability of elemental carbon:

$$CH_4(g) \rightarrow C(s) + 2H_2(g); \quad \Delta H^\ominus = +74.8 \text{ kJ mol}^{-1}$$
$$SiH_4(g) \rightarrow Si(s) + 2H_2(g); \quad \Delta H^\ominus = -34.3 \text{ kJ mol}^{-1}.$$

(Do you think the free energy values are greater or less than the enthalpy values? cf. Problem 13.) In fact, all the free energy values for the dissociation of alkanes are positive, and those of the other group IVB hydrides are all negative. This does not mean, of course, that such compounds (silane, etc.) are impossible to make and keep. Their rates of decomposition depend upon activation energies and, though much smaller than those of the analogous carbon compounds, they are sufficiently large to allow many of these compounds to exist at room temperature. Typical decomposition temperatures, showing the typical trends, are: CH_4, 800°; SiH_4, 450°; GeH_4, 290°; SnH_4, 150°; PbH_4, 0°; and Si_3H_8, 20 °C. Even carbon compounds, however, decompose at sufficiently high temperatures.

The combination of the above effects, kinetic and thermodynamic, results in there being *more carbon compounds than compounds of any other element except hydrogen*.

16.5 BINARY CARBON COMPOUNDS

Carbon forms binary compounds with most elements and these can be classified as below.

1) Lamellar or graphitic compounds. The rather open structure of graphite allows many atoms and molecules to penetrate the lattice, forming two types of lamellar (layer) compounds. The first type are those in which the layers expand to accommodate the atoms or molecules, but in which the layers remain planar, and in which the essential graphite structure and properties are retained (e.g. conductivity). The compounds are non-stoichiometric and are formed by: the highly electropositive metals (K, Rb, and Cs); the highly electronegative non-metals (Cl_2 and Br_2); and by the halides, oxides, and sulphides of many metals (e.g., $FeCl_3$, FeS_2, and MoO_3). The second type, graphitic fluoride and graphitic oxide (formed by the two most electronegative elements), are again non-stoichiometric, but in them the carbon layers are no longer planar, but buckled. These compounds are non-conductors, and those "free" electrons present in graphite are considered to be used forming C—O and C—F bonds, reducing the conductivity and causing buckling of the planes.

2) Carbides. These are usually defined as binary compounds of carbon in which it is combined with an element (other than hydrogen) of about equal or lower electronegativity than itself. All carbides are high-melting solids, and they are classified according to the most satisfactory model of the bonding (below). All can be prepared by the action of carbon on the metal or metal oxide at a high temperature, except those of groups IB and IIB which are prepared by passing acetylene into a suitable solution of a metal derivative, e.g. aqueous ammoniacal solutions of silver nitrate or copper(I) chloride, or petrol solutions of zinc or cadmium dialkyls.

a) Ionic carbides are formed by the most electropositive elements: those in groups IA, IIA, IIIA, IB, IIB, and IIIB (except boron). They are colourless (except for copper(I) carbide, Cu_2C_2, and beryllium carbide, Be_2C, which are both dark red-brown), high-melting, ionic solids, and are decomposed by water and dilute acids into either acetylene (the acetylides) or methane (the methanides):

$$CaC_2(s) + 2H_2O(l) \rightarrow C_2H_2(g) + Ca(OH)_2(s)$$
$$Al_4C_3(s) + 12H_2O(l) \rightarrow 3CH_4(g) + 4Al(OH)_3(s).$$

The methanides (Be_2C and Al_4C_3) contain the anion C^{4-}; the acetylides (all ionic carbides except Be_2C and Al_4C_3) contain the anion C_2^{2-}. All are fairly stable to heat, except silver and copper(I) acetylides which are explosive when dry. The group IIA carbides have a distorted sodium chloride structure.

b) The transition metal carbides can themselves be classified into two types, depending upon the size of the metal atom. Interstitial carbides are formed by the larger transition metal atoms (Ti, Zr, Hf, V, Nb, Ta, Mo, W). In these the carbon atoms are "accommodated" in the octahedral sites (why not the tetrahedral?) without distorting the metal lattice, thus stabilizing the metallic bonding without altering it much. Therefore these compounds are alloys which are very hard and brittle, conducting, lustrous, have high melting and boiling points, and are chemically stable except to oxidizing agents. They are non-stoichiometric and approach the stoichiometry AB and the sodium chloride structure when nearly all the octahedral holes are filled; an example is tungsten carbide, WC, which is used as a cutting edge on tools. The octahedral sites of the smaller transition metals (Cr, Mn, Fe, Co, Ni) are too small to accommodate the carbon atom without distortion of the metal lattice. In the structures of these carbides the carbon atoms form long chains, and although they are still hard, high-melting, and conducting, they are softer, have lower melting and boiling points, and are more reactive than the interstitial type. With water and dilute acids, for example, they give a variety of hydrocarbons (the carbon atoms are already in chains in the structure). They are best viewed as intermediate in type between alloys and ionic carbides.

c) Covalent carbides are formed by two elements of almost equal electronegativity to carbon: boron and silicon. Both are prepared by the reduction of the oxides by carbon in an electric furnace. Both are covalent giant molecules, and are very hard (used as abrasives and for polishing and tool sharpening), high-melting, and chemically inert. The black shiny crystals of the semi-conducting boron carbide, B_4C, contain icosahedral B_{12} units linked by linear C_3 chains, the whole packed in a sodium chloride-like structure. The grey insulator silicon carbide, SiC, exists in three modifications, each with 4:4 co-ordination (tetrahedral arrangements about each atom). Two of these are related to the zinc sulphide and wurtzite structures in which the zinc and sulphur atoms are replaced by silicon and carbon atoms.

16.6 OXIDES OF GROUP IVB

The oxides of the group show a quite well-graded series of properties (Tables 16.9 and 16.10), except that the oxides of carbon are gaseous (Section 16.4). The basic character of the oxides increases: (a) on descending the group; and (b) on de-

Table 16.9. Properties of the group IVB dioxides.

Compound	$\Delta H_f^{\ominus}$ (kJ mol^{-1})	State at 25 °C	Melting point (°C)	Boiling point (°C)	Density (g cm^{-3})	Number of allotropes	Structure	Type of co-ordination	Bonding type
CO_2	−394	colourless gas	−56.5^a	−78.5^s	1.1			2:1	covalent
SiO_2	−910	white solid	1700	2590	2.6	3	quartz	4:2	covalent
							cristobalite	4:2	covalent
GeO_2	−551	white solid	1116	1200	4.7	2	quartz	4:2	covalent
							rutile	6:3	ionic
SnO_2	−581	white solid	1827	1900^s	6.7	3	rutile	6:3	ionic
PbO_2	−277	dark brown solid	752^d		9.0		rutile	6:3	ionic

^{s} sublimes; ^{a} under 5 atm pressure; ^{d} decomposes under 1 atm of oxygen.

Table 16.10. The acid–base character of the group IVB oxides.

	Oxidation state	Carbon	Silicon	Germanium	Tin	Lead
basic ↑ character increases	II	CO, weakly acidic	SiO, (unstable)	GeO, amphoteric	SnO, amphoteric	PbO, amphoteric mainly basic
	IV	CO_2, acidic	SiO_2, acidic	GeO_2, amphoteric mainly acidic	SnO_2, amphoteric	PbO_2, amphoteric

basic character increases →

creasing the oxidation number for a given element (Table 16.10). There are three well-characterized oxides of carbon: carbon monoxide, CO; carbon dioxide, CO_2; and carbon suboxide, C_3O_2 (O=C=C=C=O); as well as several less well characterized oxides.

16.6.1 Carbon Monoxide

This is prepared in the laboratory by the dehydration of formic or oxalic acids, or their sodium salts, by concentrated sulphuric acid. The gas is collected over water:

$$HCO_2H \xrightarrow{-H_2O} CO; \qquad (CO_2H)_2 \xrightarrow{-H_2O} CO + CO_2.$$

The pure gas is not manufactured on a large scale, but is an important constituent of water gas, producer gas, and coal gas. High temperature carbonization of coal produces, among other things, *coal gas*, which was used as a domestic and industrial fuel. The composition of the gas is approximately: 50 percent hydrogen; 35 percent methane; and 15 percent ethylene. Coke is also produced (about 90 percent carbon), and this is used to make producer gas. Air is passed over a bed of red hot coke at 1000 °C, when carbon monoxide is produced, via carbon dioxide. The product is called *producer gas*:

$$2C(\text{excess}) + O_2(\text{air}) \rightarrow 2CO; \qquad \Delta H = -221 \text{ kJ}.$$

It has a low calorific value since it contains only about one-third carbon monoxide and two-thirds nitrogen. Alternatively, if steam is passed over red hot coke a mixture called *water gas* is formed:

$$C(s) + H_2O(g) \rightleftharpoons CO(g) + H_2(g); \quad \Delta H = +121 \text{ kJ}.$$

This product is a much better fuel, but on its formation the coke cools and the reaction ceases. To avoid this, either air is blown through at intervals to raise the temperature, or a mixture of air and steam is blown through, giving a continuous supply of gas called *semi-water gas*.

	Composition (%)			
	H_2	CO	CH_4	CO_2, N_2 etc.
Water gas	50	40	1	9
Semi-water gas	17	25	8	50

These gases are used as domestic fuels (added to coal gas), and in the production of carbon dioxide, hydrogen, and carbon monoxide. Carbon monoxide is also produced during the incomplete combustion of carbon or carbon compounds, for example in the exhaust gases of petrol engines.

The electronic structure of carbon monoxide can be represented as:

$$:\!C\!:\!:\!O\!: \quad \text{usually written as} \quad :C\!\equiv\!\!\!\equiv\!O: \quad \text{or} \quad \overset{-}{C}\!\equiv\!\overset{+}{O}.$$

It might appear from this that the molecule should have a large dipole moment, but of course oxygen is more electronegative than carbon, and in fact the molecule has a small dipole moment (0.112 D). Carbon monoxide is one of the few molecules for which the *sign* of the dipole moment has been determined and found to be $C^{\delta -}O^{\delta +}$. Consistent with the triple bond is the fact that carbon monoxide is very stable relative to dissociation into its constituent elements, and has the highest bond energy of all diatomic molecules, 1071 kJ mol^{-1}. It is isoelectronic with nitrogen (941 kJ mol^{-1}). Carbon monoxide is a colourless, odourless gas, slightly less dense than air, only slightly soluble in water, is difficult to liquefy, and is highly poisonous (below). It is an important reducing agent (Section 18.14), reducing metallic oxides to the metals at high temperatures, forming carbon dioxide. In air it burns with a blue flame to yield carbon dioxide, $\Delta H = -284$ kJ mol^{-1}. It is weakly acidic, reacting with water under pressure to form formic acid, or with sodium hydroxide to form sodium formate. Thus it is an anhydride of formic acid. It acts as a Lewis base, performing as a ligand with certain transition metal atoms to form metal carbonyls and related compounds, e.g. Ni(s) + 4CO(g) → Ni(CO)$_4$(l) (Fig. 16.6). Other examples are iron pentacarbonyl, Fe(CO)$_5$, and chromium hexacarbonyl, Cr(CO)$_6$, both of which also obey the inert gas rule. These are typical covalent compounds: most carbonyls are fairly volatile and soluble in organic solvents. The reaction of carbon monoxide with the haemoglobin (Section 20.17) of the blood is similar. Oxygen is transported around the body as a weakly bonded complex with haemoglobin, called oxyhaemoglobin: $O_2 \rightarrow$ haemoglobin. Oxygen is a poor ligand, is only weakly bonded to the haemoglobin, and is therefore easily delivered to the required site (muscle tissue, etc.). Carbon monoxide is a much better ligand, and preferentially bonds to the haemoglobin forming the very stable pink complex carboxy-haemoglobin: CO → haemoglobin. This reaction is not reversible, and the body slowly destroys

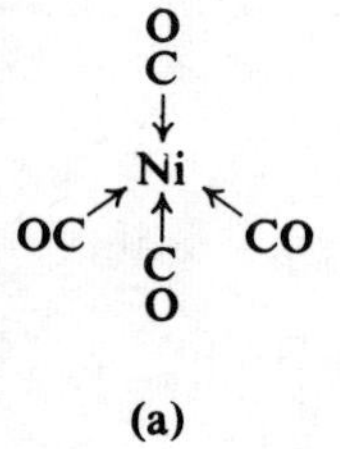

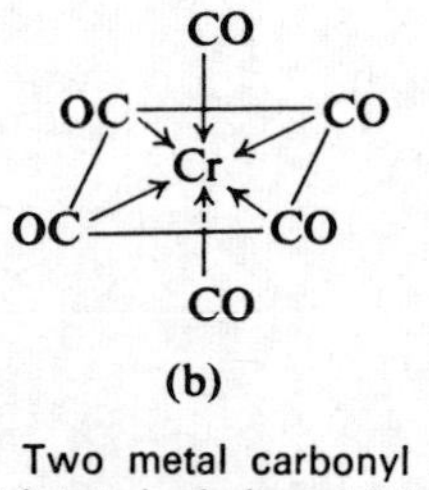

Fig. 16.6 Two metal carbonyl compounds: (a) tetrahedral, and (b) octahedral.

the complex and synthesizes fresh haemoglobin. This is what happens to the carbon monoxide we breathe in. However, above a certain small concentration (less than one part in 1000) the body cannot synthesize fresh haemoglobin sufficiently quickly. Another important complex is carbonyl copper(I) chloride (Fig. 20.16b) formed by passing the gas into a solution of copper(I) chloride in ammonia or hydrochloric acid. This reaction is used to absorb carbon monoxide, in gas analysis for example. Although carbon monoxide is very stable with respect to dissociation, it is very reactive and combines with many elements: with hydrogen at 200 atmospheres, 300 °C, and in the presence of catalysts such as alumina, it forms methanol; with chlorine in the presence of sunlight (or over a charcoal catalyst at 150 °C) it forms phosgene, $COCl_2$; with sulphur in a red-hot tube it forms carbonyl sulphide, COS; and with electropositive metals such as potassium it forms white salts, $K_2^+ \cdot C_2O_2^{2-}$. Carbon monoxide is used: for the preparation of phosgene, methanol, and sodium formate; in the reduction of ores; and in the preparation of pure nickel and iron, by decomposition of the carbonyls.

16.6.2 Carbon Dioxide

This is prepared in the laboratory by the action of dilute acids on carbonates or bicarbonates. A convenient method is to use marble chips in a Kipp's apparatus, the gas being purified from acid spray by passing through potassium bicarbonate, and from water by passing through concentrated sulphuric acid, and collected by downward delivery:

$$CaCO_3(s) + 2H^+(aq) \rightarrow Ca^{2+}(aq) + H_2O(l) + CO_2(g).$$

It is manufactured as a by-product in the production of: semi-water gas; quicklime; and during fermentation processes:

$$CaCO_3(s) \underset{}{\overset{800\,°C}{\rightleftharpoons}} CaO(s) + CO_2(g)$$

$$\text{glucose } C_6H_{12}O_6 \xrightarrow{\text{yeast, 30 °C}} 2C_2H_5OH + 2CO_2.$$

The structure of carbon dioxide can be represented by O=C=O. It is a colourless, odourless gas, denser than air, fairly soluble in water, easily liquefied at 60 atmospheres, and non-poisonous. Although it clearly does not support life, it stimulates respiration and, with oxygen, is used in reviving cases of drowning, gas poisoning, and electric shock (kiss of life). It is the anhydride of carbonic acid, H_2CO_3. This acid would be expected to be a fairly strong acid (Section 13.4.2). In fact the dissociation constants are low (Fig. 16.7), and it appears to be a weak acid. However, this assumes that all the carbon dioxide dissolved in solution is present as carbonic acid. This is not true. The total solubility of carbon dioxide in water is about 0.035M, but of this only 0.37 percent is present as carbonic acid; the rest is present in the form of hydrated carbon dioxide molecules, $CO_2(aq)$. Thus the dissociation constant is better written as $[H^+][HCO_3^-]/[CO_2(aq)] = 4.3 \times 10^{-7}$, and if the actual concentration of carbonic acid is taken into account, and the dissociation constant calculated, it works out to be $[H^+][HCO_3^-]/[H_2CO_3] = 2 \times 10^{-4}$ as

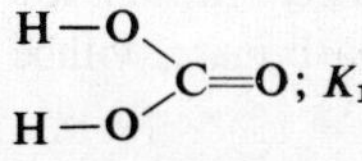

$$\begin{array}{c}H-O\\H-O\end{array}\!\!\!\!C=O;\ K_1 = \frac{[H^+][HCO_3^-]}{[H_2CO_3]} = 4.3 \times 10^{-7};\ K_2 = \frac{[H^+]^2[CO_3^{2-}]}{[H_2CO_3]} = 4.8 \times 10^{-11}$$

Fig. 16.7 The structure and apparent dissociation constants of carbonic acid.

expected (Section 13.4.2). The anhydrous acid is unstable but has been isolated as the etherate $H_2CO_3 \cdot O(C_2H_5)_2$, a white solid which decomposes above 5 °C.

If liquid carbon dioxide is allowed to expand rapidly, solid carbon dioxide is obtained. It is known as *dry ice*, since when it is used as a refrigerant it sublimes to carbon dioxide gas, leaving no trace (cf. ice). It is also used as a freezing mixture when mixed with different organic solvents to obtain different low temperatures. Its high enthalpy of vaporization (364 kJ mol^{-1}) makes it very suitable for these purposes.

Carbon dioxide is an acidic oxide. It reacts with bases to yield salts, and it dissolves in water to give carbonic acid which forms two series of salts, carbonates and bicarbonates:

$$NaOH(aq) + (excess) CO_2 \rightarrow NaHCO_3(aq) \text{ sodium bicarbonate}$$
$$(excess) 2NaOH(aq) + CO_2 \rightarrow Na_2CO_3(aq) \text{ sodium carbonate.}$$

One important example is the lime-water test for carbon dioxide. The gas reacts with a solution of barium hydroxide or calcium hydroxide (lime-water) to form a precipitate of (barium or) calcium carbonate. This precipitate dissolves if an excess of carbon dioxide is passed through, owing to the formation of the soluble bicarbonate:

$$Ca^{2+}(aq) + 2OH^-(aq) + CO_2(aq) \rightarrow CaCO_3(s) + H_2O(l)$$
$$CaCO_3(s) + CO_2(aq) + H_2O(l) \rightarrow Ca^{2+}(aq) + 2HCO_3^-(aq).$$

Carbon dioxide is much less reactive than carbon monoxide. The gas does not support combustion except that of substances burning at sufficiently high temperatures to decompose it, such as magnesium.

It reacts with ammonia to give a white solid, ammonium carbamate, from which urea can be prepared:

$$CO_2(g) + 2NH_3(g) \rightarrow NH_4O \cdot CO \cdot NH_2(s) \rightarrow H_2O + NH_2 \cdot CO \cdot NH_2(s) \text{ (urea).}$$

In the presence of sunlight (energy) and chlorophyll (catalyst), plants are able to build up carbohydrates from water and carbon dioxide:

$$6CO_2 + 6H_2O \xrightarrow{photosynthesis} 6O_2 + C_6H_{12}O_6 \begin{cases} \xrightarrow{respiration} 6CO_2 + 6H_2O + energy \\ \xrightarrow{fermentation} 2C_2H_5OH + 2CO_2 \end{cases}$$

This process is known as photosynthesis.

16.6.3 Properties of Carbonates (carbonate ion)

1. Ammonium carbonate and the carbonates of the group IA metals (except Li) are soluble in water; all the others are insoluble (Section 10.3.17).
2. All carbonates decompose on heating below 800°, except those of Na^+, K^+, Rb^+, and Cs^+. Calcium and barium carbonates require heating to about 1000 °C (Section 11.2).
3. All carbonates yield carbon dioxide on treatment with dilute acids, though the reaction of those acids forming insoluble salts, which coat the carbonate, will be slow.

4. Aqueous solutions of soluble carbonates treated with barium chloride yield a white precipitate of barium carbonate, which is soluble in dilute hydrochloric acid.
5. Aqueous solutions of soluble carbonates treated with silver nitrate yield a white precipitate of silver carbonate, which is soluble in both aqueous ammonia and nitric acid.
6. Solutions of soluble carbonates yield precipitates of basic carbonates (a mixture of hydroxide and carbonate) when treated with solutions containing cations of weakly electropositive metals (these form insoluble hydroxides and carbonates). This is due to the fact that aqueous solutions of carbonate ions contain hydroxyl ions owing to extensive hydrolysis:

$$CO_3^{2-}(aq) + H_2O(l) \rightleftharpoons HCO_3^-(aq) + OH^-(aq)$$
$$2Pb^{2+}(aq) + 3CO_3^{2-}(aq) + 2H_2O(l) \rightarrow PbCO_3 \cdot Pb(OH)_2(s) + 2HCO_3^-(aq).$$

Solutions of bicarbonates are much less extensively hydrolysed (to hydroxyl ion and carbonic acid), and these precipitate the carbonates of weakly electropositive metals: $Pb^{2+}(aq) + HCO_3^-(aq) \rightarrow PbCO_3(s) + H^+(aq)$.

7. The carbonate ion is planar trigonal. The calcite form of calcium carbonate, for example, has a deformed (owing to the size of CO_3^{2-}) sodium chloride structure, the calcium replacing sodium, and the carbonate anion replacing chloride.

16.6.4 Properties of Bicarbonates (Bicarbonate Ion)

Only those of the alkali metals (except Li^+) exist as solids; others exist only in solution. All decompose on heating to the carbonate:

$$2HCO_3^-(aq) \rightarrow CO_3^{2-}(aq) + H_2O(l) + CO_2(g).$$

All give carbon dioxide with dilute acids, and give no precipitate with barium chloride. The solid bicarbonates have structures containing infinite chains of hydrogen-bonded bicarbonate ions (cf. hydrogen fluoride, Section 19.8).

16.6.5 Silica, SiO_2

$$\beta\text{-quartz} \xrightleftharpoons{870} \beta\text{-tridymite} \xrightleftharpoons{1470} \beta\text{-cristobalite} \xrightarrow{1710} \text{liquid}$$
$$\updownarrow 573 \qquad \updownarrow 120\text{--}160 \qquad \updownarrow 200\text{--}275 \qquad \downarrow \text{cool}$$
$$\alpha\text{-quartz} \qquad \alpha\text{-tridymite} \qquad \alpha\text{-cristobalite} \qquad \text{glass}$$

Several forms of silica are known, both amorphous (e.g. flint) and crystalline (quartz, etc.). In all the naturally occurring crystalline forms of silica the silicon is tetrahedrally surrounded by four oxygen atoms, each bonded to the silicon by a single strong covalent Si—O bond, and each oxygen is covalently linked to two silicon atoms (i.e. 4:2 co-ordination). Another view of the structures is that each is built (differently) of SiO_4 tetrahedra, with each oxygen common to two tetrahedra, so that a giant molecule is formed in a symmetrical manner, with long-range order. Figure 16.8 gives a two-dimensional view of this. Quartz is the most stable crystalline form thermodynamically, but tridymite and cristobalite (metastable forms) are found in nature, because the inter-conversion involves the breaking of strong bonds, and therefore activation energies are high. On the other hand, the

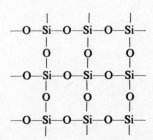

Fig. 16.8 A two-dimensional representation of a silica structure.

α → β conversions do not involve bond breaking, the differences being slight rotations of the tetrahedra relative to one another, and these occur at lower temperatures. In cristobalite, the silicon atoms occupy the same positions as the atoms in zinc sulphide or the carbon atoms in diamond, with an oxygen atom between each pair of (bonded) silicon atoms. Tridymite bears the same relation to the wurtzite structure. Quartz has the silicate tetrahedra arranged in a spiral around a vertical axis in the structure, and is therefore optically active (Section 21.3).

All varieties of silica begin to soften at about 1600 °C, and melt at 1710 °C to give a viscous liquid. On cooling, this liquid does not crystallize to form β-cristobalite, but forms a glass. This structure is again a giant molecule built of SiO_4 tetrahedra, but in this case the units are randomly arranged, with no long-range order as in a crystal. Silica glass has many desirable properties: it has a small coefficient of expansion, resists thermal shock, is transparent to infra-red and ultraviolet radiation, and is chemically inert. Very few oxides form glasses (e.g. B_2O_3, GeO_2, P_2O_5), since in order for a sufficiently random arrangement of the basic units to be possible: (a) the co-ordination number of the element to oxygen must be three or four (if two, the oxide can only form a chain, if greater than four it is too rigid); (b) only one oxygen atom must be shared, since if two oxygen atoms are shared by the same two units, too rigid a structure results; and (c) the element must be of intermediate electronegativity.

Silica is attacked only by: fluorine (when damp); hydrogen fluoride (on account of the great strength of the Si—F bond); and fused alkalis (it is an acidic oxide):

$$SiO_2 + 4HF \rightarrow 2H_2O + SiF_4$$
$$SiO_2 + 2NaOH \rightarrow Na_2SiO_3 + H_2O.$$

There is a mixture of silicates present in alkaline solutions, depending upon the amount of alkali present. If rich in alkali, the mixture is soluble and contains mainly simple species: SiO_3^{2-}(aq), SiO_4^{4-}(aq), $Si_2O_7^{6-}$(aq), and $Si_3O_9^{6-}$(aq) (cyclic). If low in alkali, relatively few Si—O bonds are broken and polymeric, less soluble, anions such as $(SiO_3)_n^{2n-}$, etc., are present. Silicate anions all have three structural points in common: (a) all structures are built from SiO_4 tetrahedra; (b) two tetrahedra never share an edge, only a corner; and (c) the charge on the anion depends upon the number of oxygens shared by each tetrahedron. For each shared corner the number of oxygens per silicon, compared with SiO_4^{4-}, is reduced by one-half, and the anionic charge per silicon is reduced by one. The silicates can be classified according to the number of shared corners per tetrahedron: (a) none shared—discrete anion SiO_4^{4-}; (b) one shared—$Si_2O_7^{6-}$; (c) two shared—chains or rings $[SiO_3]_n^{2n-}$ e.g. asbestos; (d) three shared—sheets or planes $[Si_2O_5]_n^{2n-}$ e.g. mica; (e) four shared—silica $[SiO_2]_n$, three-dimensional giant molecules. Draw these. Aluminosilicates are a class of compounds based on the three-dimensional silica structure, but in which some of the silicon ions (Si^{4+}) have been replaced by aluminium ions (Al^{3+}). For electro-neutrality, some other cations must be present, such as calcium or sodium ions. The zeolites are a class of aluminosilicates containing water which can be driven off by heating. This leaves a porous open network which results in a very large effective surface area. These can be used in two ways: (a) to absorb water or other small molecules and to separate them from larger molecules—as "molecular sieves"; and (b) as cation-exchange columns in which the cation on the column (sodium for example) can be exchanged with calcium ions in solution, thus softening water. More efficient resins can now be made synthetically.

The major part of the building industry is based upon silicate materials. These include natural silicates, such as sandstone, granite, and slate, and manufactured materials such as cement (made by roasting clay with limestone), concrete (a mixture of cement, sand, silicate rock, and water), and common glass (made by fusing sodium carbonate, limestone, and sand). Clays are used in the ceramics industry (pottery, glazes, tiles, etc.); asbestos for fireproofing; and mica in the electrical industry (in heating elements and condensers) since it is a flexible, elastic, transparent insulator of high dielectric constant and strength. Zeolites are used as ion-exchangers and molecular sieves. Silica is useful for slag formation in metallurgical industries (Section 11.5.2).

The group IVB dioxides (Tables 16.9 and 16.10) are all thermally stable except for lead dioxide (lead(IV) oxide) which decomposes to lead monoxide (lead(II) oxide) and oxygen above 300 °C. Whereas carbon dioxide is soluble, and germanium dioxide sparingly soluble in water, the remaining dioxides are all insoluble. All the dioxides react with hot concentrated or fused alkali (the more acidic carbon dioxide even reacts with dilute alkali) to give solutions containing oxy-anions; the ease of solution decreases down the group as the acidic character of the oxide decreases. Thus, lead dioxide requires molten alkali. These anions are less studied than the silicates, but do not appear to be so readily polymerized. In the solid state germanates GeO_3^{2-}, GeO_4^{4-}, and $Ge(OH)_6^{2-}$ are known, stannates $Sn(OH)_6^{2-}$, and plumbates $Pb(OH)_6^{2-}$. In aqueous solution anions such as $[GeO_2(OH)_2]^{2-}$(aq) exist, and are generally represented as $M(OH)_6^{2-}$; M = Ge, Sn, Pb. No true hydroxides $M(OH)_4$ or $M(OH)_2$ are known (except for silicic acid, $Si(OH)_4$, possibly); these are hydrated oxides. The basic character of the dioxides increases down the group (Table 16.10), and whereas the dioxides of carbon and silicon do not react with acids (except for H—F), the amphoteric dioxides of lead and tin (germanium to a lesser extent) do: tin forms tin(IV) sulphate with concentrated sulphuric acid; and lead forms lead(IV) chloride with cold concentrated hydrochloric acid, which on warming decomposes to lead(II) chloride and chlorine (lead(IV) is an oxidizing state). Silica does react with hydrofluoric acid, HF, mainly owing to the great strength of the Si—F bond.

The stability of the +4 oxidation state relative to the +2 oxidation state decreases down the group, and lead(IV) is an oxidant. Thus if lead dioxide is warmed with concentrated hydrochloric acid, the acid is oxidized to chlorine. This is reflected in the reducing nature of the +2 state. Thus germanium(II) is highly reducing, tin(II) fairly reducing, and lead(II) is stable. The monoxides can be prepared by heating the hydrated oxides and getting rid of the water. This reaction must be performed in nitrogen for germanium and tin, whose monoxides are reducing agents and react with oxygen to yield the dioxide. The monoxides of lead, tin, and germanium are amphoteric, reacting with acids to yield salts, and with alkalis to yield stannites, germanites, and plumbites, written as $M(OH)_4^{2-}$(aq). Germanites and stannites are reducing agents. Tin(II) oxide smoulders in air to form tin dioxide. Lead(II) oxide, litharge, is yellow when cold, orange when hot, and on heating to 470 °C forms, not lead dioxide, but the mixed oxide Pb_3O_4, triplumbic tetroxide or red lead. Above 550 °C this compound decomposes to lead(II) oxide. It is a scarlet insoluble solid. Its structure consists of $[Pb(IV)O_6]^{8-}$ octahedra linked in chains by lead(II) ions, themselves surrounded pyramidally so they form PbO_3 units. The compound therefore contains both lead(II) and lead(IV), and behaves chemically as a mixture of lead(II) oxide and lead dioxide, $PbO_2 \cdot 2PbO$. Thus, with hot concentrated hydrochloric acid it forms lead(II) chloride and chlorine ($PbO \rightarrow PbCl_2$; $PbO_2 \rightarrow PbCl_2 + Cl_2$); with hot dilute nitric acid it forms lead nitrate and lead dioxide ($PbO \rightarrow PbNO_3$; PbO_2 does not react); and with hot concentrated sulphuric acid it forms lead sulphate and oxygen ($PbO \rightarrow PbSO_4$; $PbO_2 \rightarrow PbSO_4 + O_2$). Write out the equations for these reactions by writing out separately those for lead monoxide and lead dioxide, and adding them in the correct proportions to give the reaction for Pb_3O_4.

16.7 HYDRIDES OF GROUP IVB

Corresponding to the alkanes, C_nH_{2n+2}, are the silanes and germanes M_nH_{2n+2} (M = Si, Ge), where n goes up to about ten. Tin and lead form only the mono-compounds, MH_4. Two general methods of preparation are:

1. The action of dilute acid on magnesium silicide (or germanide), or the magnesium alloy of tin (or lead). In the former case a mixture of silanes (germanes) is produced which can be separated by low temperature distillation or gas chromatography:

$$2Mg + Si \xrightarrow[\text{absence of air}]{\text{heat}} Mg_2Si \xrightarrow[\text{acid}]{\text{dilute}} Si_nH_{2n+2} \text{ mixture.}$$

2. The action of lithium aluminium hydride on a suitable chloride:

$$2Si_2Cl_6 + 3LiAlH_4 \rightarrow 2Si_2H_6 + 3LiCl + 3AlCl_3.$$

Mixed chlorohydrides, SiH_3Cl, can also be formed. The general reactivity of the hydrides increases as the group is descended, and this has already been discussed (Section 16.4.1). For example, the silanes are spontaneously inflammable in air at 25 °C, are strong reducing agents, are readily hydrolysed by bases to hydrated silica, and dissociate to the elements at temperatures above 400° (cf. methane).

16.8 HALIDES OF GROUP IVB

There are three general methods of preparation, though not all work for each halide: (a) heating the elements together; (b) heating the dioxides with anhydrous or concentrated halogen acid (for tetrahalides); and (c) halogenation of hydrides, oxides, sulphides, and so on:

$$SiO_2 + 2CaF_2 + 2H_2SO_4 \longrightarrow 2CaSO_4 + SiF_4 + 2H_2O \text{ (via HF)}$$
$$C_2H_6 + 6Cl_2 \xrightarrow{\text{heat or light}} C_2Cl_6 + 6HCl.$$

The free energies of formation of the tetrahalides are given in Table 16.11. All four tetrahalides are known for all the elements, except $PbBr_4$ and PbI_4. In general, the stability with respect to dissociation (and other reactions) decreases as the halogen changes from fluorine to iodine, and also as the group is descended from silicon to lead (carbon is anomalous). This correlates well with the bond strengths. The absence of $PbBr_4$ and PbI_4 is not unexpected (Section 15.1). Even the tetrachloride of lead is unstable and decomposes at room temperature to lead(II)

Table 16.11. The free energies of formation of some compounds, MX_4, $\Delta G_f^\ominus$(g) (kJ mol^{-1}).

Compound	MF$_4$	MCl$_4$	MBr$_4$	MI$_4$	MH$_4$
C	−879	−61	+67		−50.8
Si	−1506	−573	−431		+56.9
Ge	−1188	−456	−318	−105	+71
Sn		−440	−350		+188

Table 16.12. Properties of the carbon tetrahalides, CX_4.

	CF$_4$	CCl$_4$	CBr$_4$	CI$_4$
Bond strength, C—X (kJ mol^{-1})	489	326	272	238
Melting point (°C)	−185	−22.9	93.7	171
Boiling point (°C)	−128	76.4	decomposes	decomposes
Nature at 25 °C	colourless gas	colourless liquid	pale yellow solid	bright red crystals
Thermal stability	very stable	moderately stable; mild chlorinating agent		$\longrightarrow C_2I_4 + I_2$ decomposed by heat or light

chloride and chlorine. All the tetrahalides are relatively volatile covalent compounds, except SnF$_4$ and PbF$_4$ (involatile polymeric partly ionic structures). They are more stable with respect to dissociation into elements than the hydrides. All the halides are thermodynamically unstable with respect to hydrolysis, but whereas the carbon compounds are inert, the halides of the other elements are not, and are all hydrolysed in solution, the tendency decreasing as the group is ascended. Unlike the other halides, silicon tetrafluoride hydrolyses to the complex anion SiF_6^{2-}, owing to the strength of the Si—F bond:

$$SiF_4(l) + 2H_2O(l) \to SiO_2(s) + 4HF(aq)$$
$$2SiF_4(l) + 4HF(aq) \to 2SiF_6^{2-}(aq) + 4H^+(aq)$$
$$\overline{3SiF_4(l) + 2H_2O(l) \to SiO_2(s) + 2SiF_6^{2-}(aq) + 4H^+(aq)}$$

Apart from this very stable complex anion, the tendency for the tetrahalides to form halo-complexes increases as the metallic character of the group IVB element increases.

Lead(II) chloride, PbCl$_2$, can be prepared either by the action of chlorine on lead, or by the addition of chloride ions to an aqueous solution of lead ions. Tin(II) chloride, SnCl$_2$, is prepared either by boiling concentrated hydrochloric acid with tin, when the dihydrate SnCl$_2 \cdot 2H_2O$ is obtained, or by the action of hydrogen chloride gas on tin, when the anhydrous salt is obtained. Both are white crystalline solids.

The dihalides increase in stability as the group is descended: the polymeric compounds GeX$_2$ are strongly reducing and easily disproportionate; the compounds SnX$_2$ are mild reducing agents; but the compounds PbX$_2$ are stable. All form complex ions: $GeCl_3^-$, $PbCl_4^{2-}$, SnF_3^-, etc. The vapours of the lead(II) and tin(II) halides contain the monomers MX_2 (M = Sn, Pb; X = I, Br, Cl). (Are these linear or bent? Problem 15.) This indicates that the halide salts possess a large degree of covalent character, though they are crystalline solids, and this is generally true for all tin and lead salts, particularly those of tin. This is expected, since both metals are only weakly electropositive, and tin less than lead. Tin(II) chloride, $SnCl_2$, is in fact soluble in organic solvents, whereas lead(II) chloride is only sparingly soluble in alcohol. Moreover, molten $PbCl_2$ is a good conductor (m.p. 500 °C) but molten $SnCl_2$ is not (m.p. 247 °C). Thus, tin(II) chloride is best regarded as mainly covalent, lead(II) chloride as mainly ionic.

Tin(II) salts are generally soluble in water, despite their covalent character, owing to the large hydration enthalpy of the tin(II) ion (cf. aluminium chloride). Several hydrolysed complex species exist in solution. In perchlorate solutions (non-co-ordinating anions) the aquated species $[Sn_3(OH)_4]^{2+}$, $[Sn_2(OH)_2]^{2+}$, $[Sn(OH)]^+$, etc., exist. In chloride solutions (and those of other co-ordinating anions), as well as the above cationic species, several aquated anionic species also occur: $[Sn(OH)_2Cl]^-$, etc. This is illustrative of the amphoteric nature of tin (forms cations and anions). Dissolution of tin(II) chloride in water produces a white basic precipitate, $Sn(OH)Cl$, which can be dissolved in acid. Addition of alkali to tin(II) aqueous solutions yields a precipitate of the hydrated oxide, $SnO \cdot xH_2O$, often written as $Sn(OH)_2$, which dissolves in an excess of hydroxide ions to yield stannites, $Sn(OH)_6^{4-}$. These, like all tin(II) compounds, are powerful reductants.

Lead(II) salts are usually insoluble, exceptions being the acetate and nitrate, which dissolve in water to yield hydrated lead(II) ions, $Pb^{2+}(aq)$. These are much less extensively hydrolysed than the tin(II) salts. All lead(II) salts are soluble in an excess of alkali to yield plumbites, $Pb(OH)_4^{2-}$. Thus, aqueous solutions of lead(II) salts on the addition of hydroxyl ions first form the non-stoichiometric hydrated oxide, $PbO \cdot xH_2O$, which dissolves on the addition of an excess of alkali.

16.9 SULPHIDES OF GROUP IVB

The disulphides MS_2, are all known (except the lead compound—Section 15.1) and are prepared by heating the elements together. All are covalent compounds: carbon disulphide forms discrete molecules and is a pale-yellow liquid; the others are colourless solids (except tin(IV) sulphide—yellow) with giant covalent structures. All are insoluble in water, and those of germanium and silicon are readily hydrolysed to the hydrated dioxide. The sulphides of carbon, germanium, and tin, are soluble in an excess of alkali or an excess of sulphide ions, yielding stannates and thiostannates, SiS_3^{2-}, etc.

The monosulphides of tin and lead, MS, are mainly-covalent compounds which are semi-conductors. The black lead(II) sulphide (found native as the ore galena) has the rock salt structure, and the dark-brown tin(II) sulphide has a distorted rock salt structure. Both are precipitated by the action of H_2S on an aqueous solution of $M^{2+}(aq)$ ions; both are insoluble and high-melting; both are insoluble in dilute hydrochloric acid but are soluble in the concentrated acid owing to the formation of the $MCl_4^{2-}(aq)$ complex ion. Lead sulphide dissolves in nitric acid, since the sulphide ion is oxidized to sulphur (Section 18.10). The more acidic nature of tin(II) compared to the basic lead(II) is demonstrated by the reaction of hydroxyl or sulphide ions on the sulphides. Tin(II) sulphide is dissolved by hot concentrated alkali, yielding a mixture of stannite and thiostannite ions, SnS_2^{2-}, $Sn(OH)_4^{2-}$, etc.; lead(II) sulphide is not. Similarly tin(II) sulphide dissolves in yellow ammonium sulphide to form the complex anions, SnS_3^{2-}, etc.; lead(II) sulphide does not. Lead(II) sulphide is slowly oxidized by air to lead oxide and lead sulphate.

16.10 PROPERTIES OF LEAD(II) AND TIN(II) IONS

Lead(II) salts are colourless, unless either the anion is coloured as in chromate, or the anion is relatively easily oxidized when a charge transfer transition (Section A.9) may cause colour as in PbS and PbI_2. They are also poisonous and generally insoluble in water (except the acetate and nitrate), and therefore many characteristic precipitates are formed by the addition of appropriate ions: $PbCl_2$ (white), $PbBr_2$ (yellow), PbI_2 (golden), all three of which are soluble in hot water; PbS (black); $PbSO_4$ (white); $PbO \cdot xH_2O$ (white, and soluble in an excess of alkali, but not in ammonia since lead does not form complex ammines); and $PbCrO_4$ (yellow).

Tin(II) salts are colourless and generally soluble in water, but are extensively hydrolysed:

$$SnCl_2 + H_2O \underset{HCl}{\overset{H_2O}{\rightleftharpoons}} Sn(OH)Cl(s) + HCl.$$

Solutions of tin(II) salts react with alkalis to yield a white precipitate of hydrated tin(II) oxide, $SnO \cdot xH_2O$, which then dissolves in an excess of alkali to form stannites. Hydrogen sulphide precipitates brown tin(II) sulphide from tin(II) solutions. Tin(II) solutions are reducing agents, and will reduce: iron(III) ions to iron(II) ions; nitric acid to hydroxylamine; mercury(II) ions to mercury(I) salts, and then to mercury; permanganate to manganese(II); and dichromate to chromium(III) ions.

16.11 ORGANOMETALLIC COMPOUNDS

The group IVB elements form many organometallic compounds (Chapter 27). The most important of these are lead tetraethyl and silicones. Lead tetraethyl is prepared by heating a sodium/lead alloy with ethyl chloride (cf. Section 16.7, preparation of silanes, etc.). Despite the fact that all lead compounds (and particularly lead tetraethyl) are poisonous, it is used as an anti-knock agent in petrol.

Silicon tetrachloride reacts with Grignard reagents to form alkylchlorosilanes such as mono-, di-, and trimethylchlorosilane, $(CH_3)SiCl_3$, $(CH_3)_2SiCl_2$, and $(CH_3)_3SiCl$:

$$RMgCl + SiCl_4 \longrightarrow MgCl_2 + RSiCl_3 \xrightarrow{RMgCl} R_2SiCl_2 \text{ etc.}$$

These hydrolyse and polymerize to yield polymeric molecules called silicones (Fig. 16.9). The hydrolysis and polymerization of methyltrichlorosilane yields a complex sheet polymer. The chain size is limited by the size of the alkyl group, and the amount of cross linking is limited by the relative amounts of trimethyl and dimethyl compounds present. These are regulated to give silicones with the desired properties. These compounds can be liquids or solids, and have many

$$SiCl_4 + H_2O \rightarrow SiO_2 \text{ (giant polymer)}$$

$$(CH_3)_3SiCl + H_2O \xrightarrow{\text{hydrolysis}} HCl + (CH_3)_3Si\text{—}OH \xrightarrow{\text{polymerization}} (CH_3)_3Si\text{—}O\text{—}Si(CH_3)_3$$

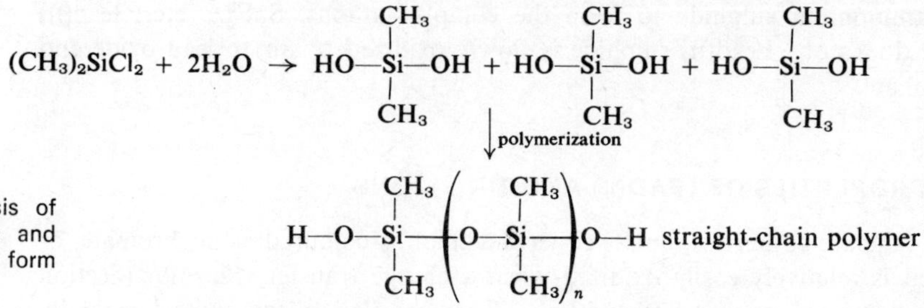

$$(CH_3)_2SiCl_2 + 2H_2O \rightarrow HO\text{—}\underset{\underset{CH_3}{|}}{\overset{\overset{CH_3}{|}}{Si}}\text{—}OH + HO\text{—}\underset{\underset{CH_3}{|}}{\overset{\overset{CH_3}{|}}{Si}}\text{—}OH + HO\text{—}\underset{\underset{CH_3}{|}}{\overset{\overset{CH_3}{|}}{Si}}\text{—}OH$$

$$\downarrow \text{polymerization}$$

$$H\text{—}O\text{—}\underset{\underset{CH_3}{|}}{\overset{\overset{CH_3}{|}}{Si}}\text{—}\left(O\text{—}\underset{\underset{CH_3}{|}}{\overset{\overset{CH_3}{|}}{Si}}\right)_n\text{—}O\text{—}H \quad \text{straight-chain polymer}$$

Fig. 16.9 Hydrolysis of silicon tetrachloride and alkylchlorosilanes to form polymers.

useful properties; they are chemically inert, heat resistant up to about 200°, water repellent, good electrical insulators, and show little viscosity change with temperature. They are used as lubricants, insulators, protective coatings, greases, and low temperature oils.

16.12 OXALATES

Oxalates are the salts of oxalic acid, $(CO_2H)_2$. Their properties are:

1. All oxalates are insoluble except those of the group IA metals (except lithium), iron(II), and ammonium.
2. All are soluble in solutions of strong acids:

$$BaC_2O_4(s) + 2H^+(aq) \rightleftharpoons Ba^{2+}(aq) + H_2C_2O_4(aq).$$

3. All oxalates are decomposed by heat; the products which are obtained depend upon the thermal stability of the carbonate and oxide of the metal (Section 11.2):

$$BaC_2O_4(s) \rightarrow BaCO_3(s) + CO(g)$$
$$Ag_2C_2O_4(s) \rightarrow 2Ag(s) + 2CO_2(g).$$

4. Oxalates react with warm concentrated sulphuric acid to yield carbon dioxide and carbon monoxide:

$$Na_2C_2O_4 + H_2SO_4 \longrightarrow Na_2SO_4 + H_2C_2O_4 \xrightarrow{-H_2O} CO + CO_2.$$

5. Solutions of soluble oxalates in water:
   a) react with barium chloride solution to precipitate white barium oxalate, which is soluble in strong acids but insoluble in oxalic acid, ammonium oxalate, and acetic acid.
   b) react with silver nitrate solution to yield white silver oxalate which is soluble in dilute nitric acid and ammonia solution (cf. AgCl):

$$2Ag^+(aq) + C_2O_4^{2-}(aq) \rightarrow Ag_2C_2O_4(s).$$

   c) decolorize permanganate solution on heating, giving off carbon dioxide:

$$2MnO_4^-(aq) + 5C_2O_4^{2-}(aq) + 16H^+(aq) \rightarrow$$
$$2Mn^{2+}(aq) + 10CO_2(g) + 8H_2O(l).$$

6. Oxalates and oxalic acid are highly poisonous. They occur in the leaves of rhubarb and beet.

PROBLEMS

1. Why is the melting point of diamond much higher than that of silicon, despite their similar structures?

2. Why are the properties of carbon dioxide so different from those of silica?

3. The temperatures (°C) at which the group IVB hydrides, MH_4, begin to decompose rapidly are: CH_4, 600°; SiH_4, 450°; GeH_4, 300°; SnH_4, 100°; and PbH_4, 0°. Account for this trend.

4. The compounds $M_{2n}H_{2n+2}$ where M is a group IVB element are known for carbon (n up to 10^6), silicon and germanium (n up to eight), tin (n up to two), and lead (n is one only). The thermal stability of corresponding hydrides, M_2H_6 for example, decreases as the group is descended. Account for these facts.

5. Compare the properties of the following pairs of compounds (empirical formulae given):
   a) CO_2 and SiO_2;
   b) CCl_4 and $SiCl_4$;
   c) $(CH_3)_2CO$ and $[(CH_3)_2SiO]_n$.
   Can you explain these differences?

6. Why is elemental carbon a better reductant than either boron or silicon?

7. The melting points of the group IVB dioxides, MO_2, are (°C): CO_2, $-57°$; SiO_2, 1720°; GeO_2, 1120°; and SnO_2, 1130°. What does this tell us about the structures of these compounds?

8. By reference to the elements of group IVB show how the chemical character of the elements in an s- or p-block group changes as the atomic number increases.

9. The tendency to form catenated compounds in group IVB decreases as the atomic number increases. Comment.

10. Comment on the stability of germanium(II) compounds in aqueous solution at pH = 0 (25 °C).

11. Comment on the relative stability of the corresponding compounds (e.g. chlorides and hydrides) of silicon and carbon with respect to hydrolysis, oxidation, and thermal decomposition.

12. Draw on graph paper the free energy diagram for the carbon group elements at pH = 14 from the data in Table 16.2. From this deduce whether the +4 oxidation state is generally more oxidizing in acid or alkali solution.

13. Would the free energy change for the hypothetical reaction: CO_2 (silica form) → $CO_2(g)$, be greater or less than -169 kJ mol^{-1}? (Section 16.4.3, Fig. 16.4.)

14. Relate the types of carbide formed by the elements to their positions in the Periodic Table.

15. What is the shape of the following species: $Cr(CO)_6$, $Fe(CO)_5$, $Ni(CO)_4$, CO_3^{2-}, C_2H_4, CO_2, CS_2, HCN, C_2H_2, CH_4, $SiCl_4$, $SnCl_2$, and $PbBr_2$?

BIBLIOGRAPHY

Most of the books on inorganic chemistry in the general list contain material on the chemistry of group IVB.

A selection is:

Johnson, R. C. *Introductory Descriptive Chemistry*, Benjamin (1966).

Cottrell, T. L. *Chemistry*, Oxford University Press (1970).

Duffy, J. A. *General Inorganic Chemistry*, Longman (1966).

Douglas, B. E., and D. H. McDaniel. *Concepts and Models of Inorganic Chemistry*, Blaisdell (1965).

Cotton, F. A., and G. A. Wilkinson. *Advanced Inorganic Chemistry*, Interscience (1966).

Heslop, R. B., and P. L. Robinson. *Inorganic Chemistry*, Elsevier (1967).

Pauling, L. *The Nature of the Chemical Bond*, Oxford University Press (1967).

Philips, C. S. G., and R. J. P. Williams. *Inorganic Chemistry*, Volumes I and II, Oxford University Press (1965)

CHAPTER 17

The Nitrogen Group: Group VB

17.1 GENERAL CHEMISTRY

In this group, like group IVB, it is the differences in the properties of the elements and their compounds which is the most striking feature. The variation in properties is vast: the elements range from the electronegative non-metal nitrogen, to the very weakly electropositive metal bismuth, via the semi-metals arsenic and antimony. The usual trends in properties are observed in descending the group however (Table 17.1): increasing metallicity, radius, and so on. The greatest change in properties occurs between nitrogen and phosphorus, between which elements there is little resemblance other than the similar stoichiometries of some simple compounds—NX_3, PX_3, etc. Arsenic is more electronegative than "expected" by simple extrapolation from nitrogen and phosphorus—the usual middle row anomaly (Section 10.3.2). Frequently the chemistry of arsenic and antimony is similar, and often that of antimony and bismuth.

Only fluorine and oxygen are more electronegative than nitrogen. As usual, for first row elements the octet rule is useful. Nitrogen can form stable compounds in which it achieves the noble gas configuration in several ways:

1. Electron gain to form nitrides (of the electropositive group IA elements).
2. Covalent bond formation by the sharing of electrons. There are several types:

$$-\overset{|}{\underset{|}{N}}\overset{x}{_{x}}\quad -N{=}N-\quad {^{x}_{x}}N{\equiv}N{^{x}_{x}}\quad -N\overset{\displaystyle O}{\underset{\displaystyle O}{\diagup}}$$

3. Electron gain and covalent bond formation, e.g. the amides, NH_2^-, and the imides, NH^{2-}, (again, formed by the group IA elements only).
4. Electron loss and covalent bond formation, e.g. NH_4^+, $N_2H_5^+$. Alternatively these can be regarded as being formed by lone pair donation to acceptor atoms such as H^+ (Lewis acids).

The oxygen compounds of nitrogen frequently do not obey the octet rule, e.g. nitric oxide.

The oxidation number of nitrogen varies from -3 to $+5$ owing to covalent bond formation of the element to atoms of very different electronegativity; therefore, a classification of compounds in terms of oxidation number is less useful than

17.1	General chemistry
17.2	Oxidation state diagrams: stability of oxidation states
17.3	Group trends
17.4	The anomalous nature of nitrogen
17.5	Occurrence, preparation, and uses
17.6	Allotropes
17.7	Reactions of the elements
17.8	Nitrides and phosphides
17.9	Hydrides
17.10	Halides
17.11	Oxides
17.12	Sulphides
17.13	The oxyacids of nitrogen
17.14	The oxyacids of phosphorus
17.15	Oxyacids, oxyanions, and solution chemistry of arsenic, etc.

Table 17.1. Group VB properties.

Property	Nitrogen N	Phosphorus P	Arsenic As	Antimony Sb	Bismuth Bi
Atomic number	7	15	33	51	83
Electron configuration (outer)	$2s^2 2p^3$	$3s^2 3p^3$	$3d^{10}4s^2 4p^3$	$4d^{10}5s^2 5p^3$	$4f^{14}5d^{10}6s^2 6p^3$
Isotopes (in order of abundance)	14, 15	31	75	121, 123	209
Atomic weight	14.0067	30.9738	74.9216	121.75	208.980
Metallic radius (Å)		1.28	1.39	1.59	1.70
Ionic radius (Å)	1.71(−3)[a]	2.12(−3)[a]	2.22(−3)[a]	0.62(+5)[a]	1.20(+3)[a]
Covalent radius (Å)	0.70	1.10	1.21	1.41	1.48
Van der Waals radius (Å)	1.5	1.8	2.0	2.2	
Atomic volume ($cm^3 mol^{-1}$)	17.3	17.0	13.1	18.2	21.4
Boiling point (°C)	−196	280[w]	610[s,gr]	1380[gr]	1560
Melting point (°C)	−210	44[w]	817[p,gr]	630[gr]	271
Enthalpy of fusion (kJ mol^{-1})	0.36	0.628	27.7	19.9	10.9
Enthalpy of vaporization (kJ mol^{-1})	2.8	12.9	32.4	195	172
Enthalpy of atomization or sublimation (kJ mol^{-1})	473	315	287	262	207
Density (g cm^{-3})	0.81	2.34[r] 1.82[w] 2.7[b]	5.73[gr]	6.7[gr]	9.8
Electronegativity (A/R)	3.05	2.05	2.2	1.8	1.65
Electron affinity (kJ mol^{-1})	−31	60			
Ionization energy (kJ mol^{-1}):					
1 + 2 + 3	8839	5864	5478	4866	4766
1 + 2 + 3 + 4 + 5	25,755	17,090	16,357	14,469	14490
Common co-ordination numbers	3, 4	3, 4, 5, 6	3, 4, (5), 6	3, 4, (5), 6	3, 6
Common oxidation states	−3, −2, −1, 0, 1, 2, 3, 4, 5	−3, (1), 3, 5	−3, 3, 5	(−3), 3, 5	3, (5)
Abundance, p.p.m.		1200	5	1	0.2
Atomic conductance ($ohm^{-1} cm^{-4}$)			2.3	1.4	0.4
Element structures	colourless gas N_2	red-polymer; black-layer; white-P_4	grey-metallic layer	grey-metallic layer	metallic layer

[a] ionic charge; [w] white; [gr] grey; [p] 36 atmospheres; [s] sublimes; [r] red; [b] black. Oxidation states: (2) less stable state, 2 very stable state.

one in terms of types of compound (oxides, hydrides, etc.). The lower elements of the group show fewer stable oxidation states than nitrogen, but still more than two each; consequently, there is a vast chemistry of this group.

For phosphorus and the remaining elements, as the electronegativity of the elements decreases: electron gain to form anions becomes less favourable; lone pair donation from the molecules MX_3 (M = P, etc.) becomes less common (e.g. PH_3 is hardly basic); the tendency to form cations increases slowly; and the tendency for covalent bond formation decreases slowly. However, other types of compound are formed: expansion of the octet to form complex anions like PF_6^-, and five-valent compounds such as PF_5, $Cl_3P{=}O$, etc.; and the formation of cations in the +3 oxidation state by antimony and bismuth, e.g. SbO^+, BiO^+ (typical basic salts of weakly electropositive metals), $Sb_2(SO_4)_3$, and $Bi(NO_3)_3 \cdot 3H_2O$.

A wide range of shapes and co-ordination numbers are found in the group (Table 17.1). For nitrogen, the maximum co-ordination number is four (Section 16.4.4); the remaining elements show co-ordination numbers up to six (expansion of octet). The shapes of the molecules can be obtained by application of the Gillespie–Nyholm rules.

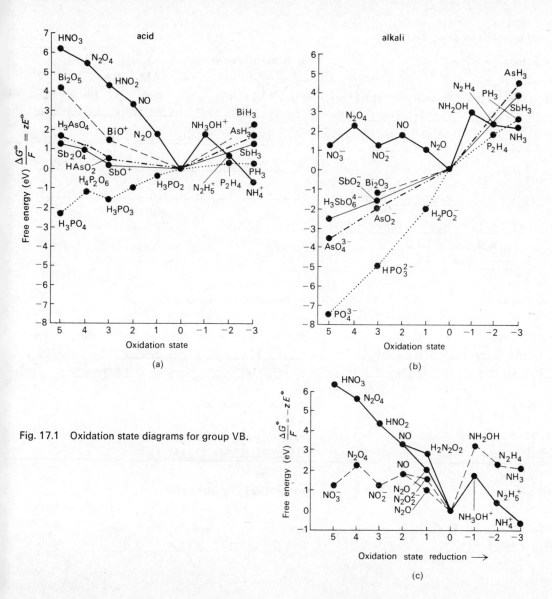

Fig. 17.1 Oxidation state diagrams for group VB.

17.2 OXIDATION STATE DIAGRAMS: STABILITY OF OXIDATION STATES

Group VB has the most complicated chemistry of all the main groups. The oxidation state diagrams, Figs. 17.1 (a), (b), and (c), illustrate much of the chemistry succinctly. The student is advised to reconstruct these diagrams himself from the reduction potential data given in Table 17.2, and to draw a separate diagram for each element containing both the acid and alkali data, as in Fig. 17.1(c) for nitrogen.

These diagrams illustrate that (see Section 7.12.3):

1. Nitrogen(0), as the nitrogen molecule, is very stable in both acid and alkali (deep minimum). It is thermodynamically unstable with respect to reduction

17.2 OXIDATION STATE DIAGRAMS: STABILITY OF OXIDATION STATES   405

Table 17.2. Reduction potentials for group VB elements.

(a) Nitrogen in acid solution.

```
  +5     +4     +3     +2     +1     0     -1     -2     -3
  NO3⁻   N2O4   HNO2   NO     N2O    N2    NH3OH⁺ N2H5⁺  NH4⁺
  |+0.794|+1.07 |+1.0  |+1.59 |+1.77 |-1.87|+1.41 |+1.275|

  |     +0.94       |               |    -0.23      |

  |     +0.96       |

  |            +1.11            |   -0.05   |  +1.35  |

              |       +1.29          |    +0.27      |
                              H2N2O2
                              | 0.71 | 2.65 |
```

(b) Nitrogen in alkali solution.

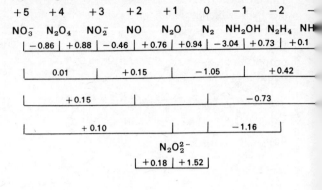

(c) Phosphorus in acid solution.

```
  +5       +4       +3      +1      0       -2      -3
  H3PO4   H4P2O6   H3PO3   H3PO2   P4     P2H4    PH3
  | -0.94 | 0.38   | -0.5  | -0.51 |(-0.10)|(-0.0) |
  |    -0.28       |  -0.5         |  -0.06        |
```

(d) Phosphorus in alkali solution.

```
  +5       +4        +3       +1       0       -2     -3
  PO4³⁻   H2P2O6²⁻  HPO3²⁻   H2PO2⁻   P4     P2H4   PH3
  |                 | -1.57  | -2.05  |(-0.9) |(-0.8)|
  |       -1.12             |  -1.73          | -0.89|
```

(e) Arsenic in acid solution.
```
  +5       +3      0      -3
  H3AsO4  HAsO2   As    AsH3
  | +0.56 | +0.25 | -0.6 |
```

(f) Arsenic in alkali solution.
```
  +5       +3      0      -3
  AsO4³⁻  AsO2⁻   As    AsH3
  | -0.67 | -0.68 | -1.43 |
```

(g) Antimony in acid solution.
```
  +5       +4       +3     0      -3
  Sb2O5   Sb2O4   SbO⁺    Sb    SbH3
  | +0.48 | +0.68 | +0.21 | -0.5 |

  |    +0.58       |
```

(h) Antimony in alkali solution.
```
  +5              +3       0       -3
  H3SbO6⁴⁻       SbO2⁻    Sb    SbH3
  |   (-0.4)     | -0.66  |(-1.34)|
```

(i) Bismuth in acid solution.
```
  +5       +3     0      -3
  Bi2O5   BiO⁺   Bi    BiH3
  | (+1.6)| +0.32 | (-0.8) |
```

(j) Bismuth in alkali solution.
```
  +3      0
  Bi2O3   Bi
  | -0.46 |
```

to the NH_4^+ ion in acid solution (below it on diagram), but this reaction does not occur readily, owing to the inertness of nitrogen (kinetic control). Phosphorus is unstable with respect to disproportionation, especially in alkaline solution, and in fact the element reacts with alkali to yield phosphine and phosphoric acid. The other elements are stable with respect to disproportionation in acid solution, and become increasingly stable in alkaline solution

as the group is descended. In fact, only arsenic of the heavier elements reacts (in fused alkali), to yield, not the unstable arsine, AsH_3, but hydrogen.
2. The nitrogen (-1) state is unstable, both to oxidation and reduction, and in fact hydroxylamine can act as both oxidant and reductant. This state does not exist for the other elements.
3. For nitrogen, all oxidation states between $+1$ and $+4$ (inclusive) tend to disproportionate in acid solution, generally to the nitrate ion and nitrogen, but not always (kinetic factors); for example,

$$3HNO_2(aq) \rightarrow H^+(aq) + NO_3^-(aq) + 2NO(g) + H_2O(l).$$

However, the nitrite ion is stable in alkali (and neutral) solution with respect to disproportionation. Similarly, for phosphorus, nearly all intermediate oxidation states both in acid and alkali, tend to disproportionate into phosphorus(5) (phosphoric acid or phosphates) and phosphorus (-3), (phosphine, PH_3). By comparison, the $+3$ states of arsenic, antimony, and bismuth, become increasingly stable with respect to disproportionation (inert pair effect). This is a general trend for the $+3$ state from nitrogen to bismuth.
4. The oxidation state -3 becomes less stable as the group is descended; and the compound bismuthine, BiH_3, is very unstable.
5. In alkali the higher positive oxidation states become much poorer oxidants (much lower on the diagram). This is generally true for oxyanions, and consequently they are prepared by oxidation in alkaline solution (Section 7.12.4) when it is necessary, e.g. preparation of permanganate (Section 20.16).
6. The stability of the $+5$ oxidation state relative to the $+3$ oxidation state decreases as the group is descended (relative heights of points on diagram) except that nitrogen(v) is a better oxidant than expected. This is more easily seen by reference to the M^{5+}/M^{3+} reduction potentials, and is discussed in Section 17.3.4. The oxidizing power of the $+5$ state decreases in the order $Bi \gg N > Sb \geqslant As > P$. The reducing power of the $+3$ state is in the reverse order, and phosphorus(III) is by far the best reductant. In fact, phosphorus(v) is so stable, especially in alkali, that all the oxidation states of phosphorus are reducing except phosphorus(v). The reduction potentials are, of course, dependent on pH. Thus, in acid, arsenic acid oxidizes iodide ions to iodine, but in alkali arsenite reduces iodine to iodide:

$$H_3AsO_4(aq) + 2I^-(aq) + 2H^+(aq) \xrightarrow{acid} HAsO_2(aq) + I_2 + 2H_2O(aq);$$
$$AsO_4^{3-}(aq) + 2I^-(aq) + 2H_2O(l) \xleftarrow{alkali} AsO_2^-(aq) + I_2 + 4OH^-(aq).$$

17.3 GROUP TRENDS

Table 17.1 illustrates the trends in physical and atomic properties. Overall these are similar to those of other p-block groups. Some important points are listed below.

1. The electropositive character of the elements increases as the group is descended. Thus the stability of cationic states increases on descending the group. The cation Bi^{3+} probably exists in salts of strong acids (of weakly co-ordinating anions) e.g. BiF_3, $Bi(NO_3)_3$. The cation Sb^{3+} probably exists in $Sb_2(SO_4)_3$. In solution these cations are extensively hydrolysed, and are best represented

Table 17.3. The relative stabilities of the +3 and +5 oxidation states.

Couple	Reduction potential	Nature of (V) state	Nature of (III) state	Product of element + oxygen
$N(V)/N(III)$	+0.94	oxyacids, oxidizing	faintly reducing	NO_2
$P(V)/P(III)$	−0.28	oxyacids, stable	reducing	P_2O_5
$As(V)/As(III)$	+0.56	oxyacids, oxidizing	mildly reducing	As_2O_3
$Sb(V)/Sb(III)$	+0.58	no oxyacid, oxidizing	mildly reducing	Sb_2O_3
$Bi(V)/Bi(III)$	+1.6	few compounds, strongly oxidizing, no oxyacid	stable	Bi_2O_3

as the oxycations, $BiO^+(aq)$ and $SbO^+(aq)$ (typical of very weakly electropositive metals), which probably also exist in some salts such as BiOCl. There is a corresponding decrease in the stability of oxyanions down the group: nitrogen forms nitrates, NO_3^-, etc.; phosphorus forms phosphates, PO_4^{3-}, etc.; arsenic and antimony form only the complex hydroxy anions, $Sb(OH)_6^-$, $Sb(OH)_4^-$, and $As(OH)_4^-$; and bismuth forms none at all.

2. The acidic character of the oxides decreases as the group is descended. The pentoxides, as expected (Section 11.1.1), are all acidic, and the acidity decreases as the group is descended. The trioxides change from the acidic N_2O_3, P_2O_3, and As_2O_3 to the amphoteric Sb_2O_3, and the basic B_2O_3. This change is shown nicely by the nature of the product of the reaction of the element with nitric acid, where phosphoric acid, arsenic acid, hydrated antimony-(V) oxide, and the salt bismuth(III) nitrate are formed. This is also illustrated by the hydrolysis of the halides: phosphorus trichloride yields phosphorus acid; arsenic trichloride yields arsenic(III) oxide (less acidic than phosphorus(III) oxide); but antimony and bismuth yield the oxychlorides, MOCl. On the other hand, nitrogen trichloride yields ammonia, since the polarity of the $N^{\delta-}$—$Cl^{\delta+}$ bond differs from the polarity of the others, $M^{\delta+}$—$Cl^{\delta-}$ (Section 16.4.7).

3. The hydrides, MH_3, decrease in stability (which correlates with bond energies) and basicity down the group (Section 11.1.3).

4. The stability of the +5 oxidation state relative to the +3 oxidation state decreases as the group is descended, except that nitrogen(V) is a good oxidant (Fig. 17.1). This is illustrated, in Table 17.3, by the reduction potentials and the stabilities of compounds in the +5 oxidation state. As the stability of the +5 oxidation state decreases, that of the +3 state increases (inert pair effect). This elegantly illustrates both the general trend, and the first and middle row anomalies (both arsenic(V) and nitrogen(V) are more oxidizing than "expected").

17.4 THE ANOMALOUS NATURE OF NITROGEN

This is a typical list of differences between a first row s- or p-block element and subsequent elements in the same group (especially the second element). The differences are due to nitrogen being much smaller and more electronegative than phosphorus, capable of forming stable multiple bonds with itself (carbon and oxygen), and not being able to expand the octet.

1. The elemental form of nitrogen is the multiple-bonded stable molecule, N≡N, whereas phosphorus exists as a solid in various forms containing single-bonded phosphorus. This is due to the fact that the bond energy of triply bonded nitrogen is much greater than that for phosphorus, $B_{N\equiv N} \gg B_{P\equiv P}$, but the reverse is true for the single bonds, $B_{P-P} > B_{N-N}$. This is easily illustrated using the bond energies (Table A.12) to calculate the heats of reaction for the hypothetical reactions: $P_4(s) \rightarrow 2\text{"}P_2\text{"}(g)$ and $\text{"}N_4\text{"}(s) \rightarrow 2N_2(g)$; where the solid "$N_4$" is assumed to have the same structure as white phosphorus (Section 17.6) and the gas "P_2" is assumed to have the same structure as nitrogen gas. To break P_4(white) into four gaseous phosphorus atoms, six single P—P bonds must be broken:

		M is phosphorus	M is nitrogen
(i)	$M_4(g) \rightarrow 4M(g)$	$\Delta H^\ominus = +6 \times 209 = +1254$	$6 \times 159 = +954$
(ii)	$4M(g) \rightarrow 2M_2(g)$	$\Delta H^\ominus = -2 \times 523 = -1046$	$-2 \times 946 = -1892$
(i) + (ii)	$M_4(g) \rightarrow 2M_2(g)$	$\Delta H^\ominus = +208 \text{ kJ}$	$\Delta H^\ominus = -938 \text{ kJ}$

Clearly the stable forms are P_4 and N_2. Since the enthalpy of vaporization of white phosphorus ($P_4(s) \rightarrow P_4(g)$; $\Delta H = +59$ kJ) is so small (the molecules are held together only by weak van der Waals forces), the fact that we have considered the stability of gaseous instead of solid P_4 does not matter. Nor does the fact that we have used enthalpy data (Section 10.3.18). A slightly easier method of calculating reaction enthalpies from bond energies is merely to subtract the enthalpies of formation of the substances (relative to gaseous atoms) from one another (Section 4.12); for example:

$$P_4(g) \rightarrow 2P_2(g); \quad \Delta H = -2 \times 523 - (-6 \times 209) = +208 \text{ kJ}.$$
$\Delta H_f^\ominus \quad -6 \times 209 \quad -2 \times 523$

2. Catenation is more common in phosphorus compounds than in nitrogen compounds. This is related to the relative weakness of the N—N single bond, compared to the P—P single bond. Compounds containing N—N bonds are unstable with respect to decomposition into the elements, and also to oxidation. Hydrazine, for example, $NH_2 \cdot NH_2$, decomposes into ammonia, nitrogen, and hydrogen at 250°, and its aqueous solution decomposes at 25 °C in the presence of platinum black to ammonia and hydrogen. Similarly, it burns in air to nitrogen and water, and will explode in oxygen at 25 °C in the presence of platinum. The weakness of the N—N bond (and the O—O, N—O, N—F, O—F, and F—F bonds) is attributed to repulsion of the non-bonding electrons (Section 19.6). Actually, all the group VB elements form compounds of the type R_2—M—M—R_2 (R = H, alkyl), but in general the catenating tendency is less than that in group IVB, and decreases in the order P > N > As ⩾ Sb ⋙ Bi. This differs from the corresponding order in group IVB, where the first row element forms the most stable catenated compounds.
3. The oxides of nitrogen are gaseous and contain multiple bonds between nitrogen and oxygen, e.g. NO; whereas the oxides of phosphorus, etc., are solids containing single P—O bonds. This is due to the relatively high N=O and N≡O bond strengths, and the relatively low N—O bond strength; whereas with phosphorus it is the other way around. Consider the heats of

formation of P$_4$O$_6$ (Fig. 17.6) and the hypothetical "N$_4$O$_6$" (which would similarly contain 12N—O bonds):

$$2N_2(g) + 3O_2(g) \rightarrow N_4O_6(s); \quad \Delta H_f^\ominus = +1430 \text{ kJ};$$
$$\Delta H_f^\ominus \quad -2 \times 946 \quad -3 \times 498 \quad -12 \times 163$$

$$P_4(g) + 3O_2(g) \rightarrow P_4O_6(s); \quad \Delta H_f^\ominus = -1668 \text{ kJ}$$
$$\Delta H_f^\ominus \quad -6 \times 209 \quad -3 \times 498 \quad -12 \times 368$$

Clearly, the molecule P$_4$O$_6$ is stable with respect to dissociation into the elements, but the molecule "N$_4$O$_6$" is not. In fact, all compounds containing N—O bonds tend to be unstable with respect to dissociation either into elements, or into a more stable compound; whereas those containing P—O bonds tend to be stable. For example, hydroxylamine, NH$_2$OH, decomposes at room temperature into nitrogen, nitrous oxide, ammonia, and water. Also, whereas the compounds P(OR)$_3$ are known, there are no nitrogen analogues; instead, the compound R—O—N=O is formed. The acid "N(OH)$_3$" does not exist (calculate its enthalpy of formation—Problem 16), whereas "P(OH)$_3$" does, though it does not have the expected structure (Section 17.14). Condensed phosphorous acids and anions containing P—O bonds are stable whereas the nitrogen analogues are much less stable. Conversely, there are many stable compounds containing multiple-bonded nitrogen compounds which have no analogues in the chemistry of phosphorus, such as cyanides, and isocyanides, for example.

4. Nitrogen cannot expand the octet, whereas phosphorus and the other elements can, and the latter elements can therefore have co-ordination numbers greater than four. Thus they form compounds (PF$_5$, etc.) and complexes (PF$_6^-$, etc.) which have no nitrogen analogues. This has other effects on the chemistry. For example, the compound P(OH)$_3$ is less stable than the compound H(OH)$_2$PO, since the latter contains more bonds (Fig. 17.8). (Calculate the enthalpy of formation of these from their bond energies—Problem 17.) Quite often, therefore, low valent phosphorous acids contain —P(=O)(H) fragments where —P—OH might have been expected, since phosphorus can expand the octet. Phosphonitrilic compounds (Section 17.10) contain five-co-ordinate phosphorus, and have no nitrogen analogues. The hydrolysis of phosphorus trichloride is fast and yields phosphorous and hydrochloric acids; that of nitrogen trichloride is slow and yields ammonia and hypochlorous acid, despite the fact that the decomposition to nitrous acid is thermodynamically more favourable. This is clearly a kinetic effect. The mechanism of the hydrolysis of phosphorus trichloride is similar to that of silicon tetrachloride (Fig. 16.5) and involves the formation of an intermediate four co-ordinate species. This is impossible for nitrogen, which therefore hydrolyses by a different mechanism, probably involving hydroxyl ion attack on the partly negative chlorine atom, which is therefore slower and yields different products.

5. Hydrogen bonding. As a result of the high electronegativity of nitrogen relative to carbon, and the existence of a lone pair of electrons on nitrogen, ammonia is very different from methane in that it is hydrogen-bonded (anoma-

lous boiling point, melting point, etc.) and shows basic properties. In contrast, phosphine (PH_3) is hardly basic, and stibine (SbH_3) not at all; nor is hydrogen-bonding noticeable in these molecules. The decreasing electronegativity also affects the bond angles in the hydrides (Section 17.9).

6. Nitrogen is very inert. Sometimes this lack of reactivity is thermodynamic, since the high bond energy of the nitrogen molecule tends to make the free energy of its reactions positive. For example, all the oxides of nitrogen are thermodynamically unstable with respect to dissociation into the elements, mainly as a consequence of the high bond strength of $N{\equiv}N$, and the low bond strength of N—O. However, frequently, reactions are observed not to take place when the free energy change is negative. For example, nitrogen does not react with hydrogen at room temperature despite the negative free energy change. Therefore this is a kinetic effect.

$$N_2(g) + 3H_2(g) \rightleftharpoons 2NH_3(g); \qquad \Delta G^\ominus_{25°} = -33.4, \, K_{25°} = 8 \times 10^5.$$

To increase the rate of a reaction involving a contraction in volume, it is necessary (Le Chatelier, Section 7.3) to increase the pressure and also to increase the temperature and add a catalyst. But the reaction is exothermic, and although increasing the temperature to 500 °C obtains a satisfactory rate of reaction, the equilibrium constant decreases to $K_{500°} = 10^{-4}$ (Le Chatelier). Nevertheless, the conditions for the production of ammonia are optimized under these conditions: the equilibrium yield of ammonia is smaller, but the rate at which it is produced is much greater. The increased pressure does not affect the equilibrium constant, but does increase the yield of ammonia (Haber process).

Nitrogen is generally more reactive at high temperatures where, presumably, nitrogen atoms are involved in the reactions. Thus the high bond energy of nitrogen is a factor tending to make it inert, since it excludes mechanisms involving nitrogen atoms at low temperatures. The only processes which occur at room temperature are: (a) the reaction with lithium to yield lithium nitride (Section 14.2.7); (b) the fixation (combination) of nitrogen by bacteria on root nodules of plants such as beans, peas, clover, etc. (important in the nitrogen cycle); and (c) the fixation of nitrogen by some transition metal compounds. At higher temperatures, nitrogen becomes more reactive yielding nitrides with most elements. Some important reactions of molecular nitrogen are those with hydrogen to give ammonia (500 °C), oxygen to give nitric oxide (1000 °C), and calcium carbide to give calcium cyanamide (1000 °C). Calcium cyanamide is used as a fertilizer, since it is hydrolysed in the soil to urea and ammonium carbonate successively:

$$CaC_2 + N_2 \xrightarrow{1000\,°C} C + CaN{\cdot}CN \xrightarrow{3H_2O} Ca(OH)_2 + NH_2CONH_2 \xrightarrow{2H_2O} (NH_4)_2CO_3.$$

When an electric discharge is passed through nitrogen gas at low pressure, nitrogen atoms are formed. The recombination to form nitrogen is slower than expected, taking several minutes, and is accompanied by a yellow glow. The gas is very reactive—it is called "active nitrogen"—and forms nitrides with most elements.

17.5 OCCURRENCE, PREPARATION, AND USES

Nitrogen occurs as an inert diatomic gas, 78 percent by volume in the atmosphere. Inorganic nitrogen compounds are usually soluble and are rarely found in nature except that huge quantities of sodium nitrate are found in the North Chile desert (Chile saltpetre). Both nitrogen and phosphorus are essential constituents of all plant and animal tissue: nitrogen is present in proteins; and phosphorus is present as calcium phosphate in bones and teeth, and in organic esters (adenosine triphosphate, Section 28.2.4) which are thought by many biochemists to provide energy for cell reactions by hydrolysis of the phosphate bond.

Nitrogen is manufactured by fractional distillation of liquid air, and contains about 1 percent argon and oxygen. The oxygen is removed by passing the mixture either over heated copper or through strong reductants such as V^{2+}(aq) or Cr^{2+}(aq).

Very pure nitrogen is obtained by heating sodium azide (**danger**—explosive):

$$2NaN_3(s) \rightarrow 2Na(l) + 3N_2(g).$$

Several methods of preparation can be used in the laboratory (although of course it is seldom prepared since it is generally available from cylinders):

1. The removal of carbon dioxide (by alkali) and oxygen (by hot copper) from air.
2. The thermal decomposition of ammonium dichromate, during which heat, light, and steam (and green chromium(III) oxide) are evolved ("volcano reaction"):

$$(NH_4)_2Cr_2O_7(s) \rightarrow 4H_2O(g) + Cr_2O_3(s) + N_2(g).$$

3. The thermal decomposition of unstable ammonium nitrite solution, made by mixing ammonium chloride and sodium nitrite in solution:

$$NH_4NO_2(aq) \rightarrow 4H_2O(l) + N_2(g).$$

Both the above reactions typify the unstable nature of salts which contain an oxidizable cation (NH_4^+) and an oxidizing anion, and (4) is similar.

4. The action of concentrated ammonia on oxidants such as chlorine, bromine, or hypochlorite (bleaching powder) solutions:

$$3OCl^-(aq) + 2NH_3(aq) \rightarrow 3Cl^-(aq) + 3H_2O(l) + N_2(g).$$

Phosphorus is found in phosphate ores such as apatite, $Ca_3(PO_4)_2$, and fluoro-apatite $3Ca_3(PO_4)_2 \cdot CaF_2$. Arsenic, antimony, and bismuth occur mainly as sulphide ores: arsenical pyrites, FeAsS; stibnite, Sb_2S_3; and bismuth glance, Bi_2S_3. About 90 percent of the phosphate ores mined are used in the production of fertilizer. Apatite, for example, is ground and heated with 70 percent sulphuric acid, producing calcium sulphate and the soluble fertilizer calcium dihydrogen phosphate ("superphosphate"):

$$Ca_3(PO_4)_2(s) + 2H_2SO_4 \rightarrow 2CaSO_4(s) + Ca(H_2PO_4)_2(s).$$

White phosphorus is manufactured by heating phosphate ore with sand and coke in an electric furnace to about 1500 °C, when the volatile pentoxide formed initially, in an inert atmosphere of carbon monoxide, is reduced by the coke (Section 11.5.1) to phosphorus, and is distilled out and condensed under water as white phosphorus:

$$2Ca_3(PO_4)_2(s) + 6SiO_2(s) \rightarrow 6CaSiO_3(slag) + P_4O_{10} \xrightarrow{10C} P_4(w) + 10CO(g).$$

Antimony, arsenic, and bismuth are obtained by roasting the sulphide ore and reducing the oxide obtained with carbon:

$$Sb_2S_3(s) + \tfrac{9}{2}O_2(g) \rightarrow 3SO_2(g) + Sb_2O_3(s) \xrightarrow{3C} 2Sb(s) + 3CO(g).$$

Nitrogen is used in the manufacture of ammonia, and thence nitric acid and nitrates (fertilizer). Phosphates are used as fertilizers and as detergents, phosphorus in matches, fireworks, and rat poisons. Arsenic, antimony, and bismuth are used for various pharmaceutical purposes, and as alloys. Many of these expand on solidification, and are used in castings and metal type. Bismuth forms low-melting alloys, used in safety valves and fire sprinklers.

17.6 ALLOTROPES

Black phosphorus resembles graphite in structure and properties (grey solid, metallic lustre, flaky, conductor of heat and electricity). It consists of polymeric layers of phosphorus atoms held together by strong covalent bonds, and weakly bonded to adjacent layers, just as in graphite. However, the layers are not planar, but corrugated (Fig. 17.2), since clearly the arrangement of atoms around trivalent phosphorus must be pyramidal (Section 12.7). This is the most metallic of the allotropes of phosphorus, and is often called "metallic phosphorus". **White phosphorus** is a waxy translucent solid which is transparent when freshly sublimed. It is a molecular solid consisting of tetrahedral molecules, P_4, held together by weak van der Waals forces. As expected, therefore, it is soft, soluble in organic solvents but not water (in fact it is stored in water to protect it from oxygen), is appreciably volatile, even at 25 °C, and is poisonous. The vapour glows in the dark in oxygen (faint green) owing to a complex oxidation reaction (phosphorescence). White phosphorus is metastable, and changes to red phosphorus on prolonged exposure to light, or more quickly on heating in the presence of iodine as catalyst. It is reactive, probably because of angular strain in the P_4 molecule where the angles are only 60°. Thus, on heating, one bond is broken and a polymeric form of phosphorus is formed, **red phosphorus**, in which the angles are less strained (Fig. 17.2). This is a dark-red opaque powder, insoluble in organic solvents, involatile and therefore non-toxic. It is less reactive than white phosphorus. Black, red, and white phosphorus are each polymorphic. **Brown phosphorus** is stable only at temperatures less than −196 °C (Fig. 17.3). It is prepared by rapidly condensing the vapour, $P_2(g)$, and probably contains P_2 molecules. White phosphorus is always obtained on cooling the liquid or vapour, or by sublimation. The reactivity towards other substances decreases in the order: brown > white > red > black (inert).

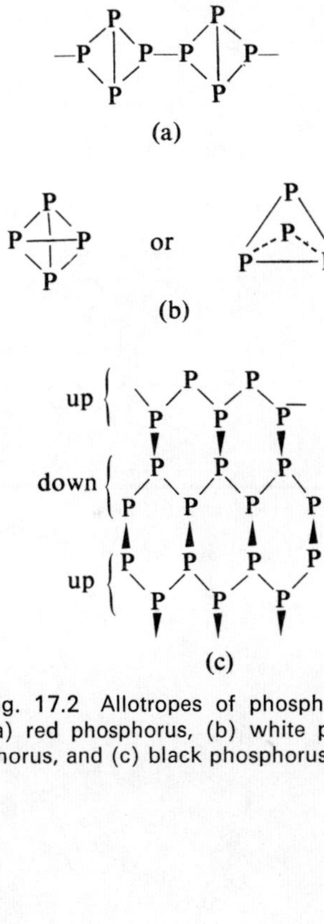

In general, on descending a p-block group, the metallic allotropes become more stable. Nitrogen shows only a non-metallic form. Phosphorus has several allotropes, one of which is partly metallic in nature, but of which the red non-metallic

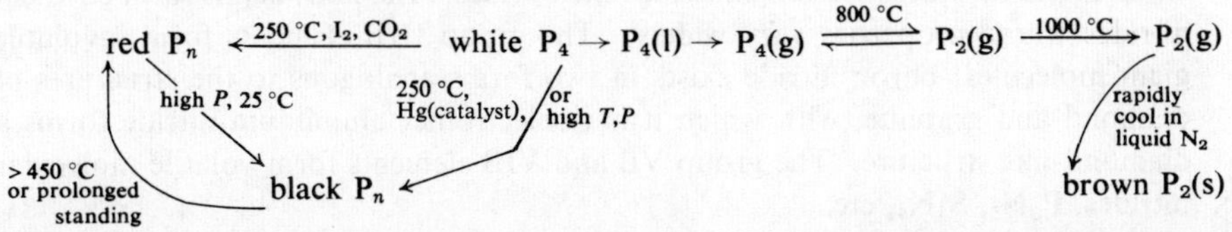

Fig. 17.2 Allotropes of phosphorus: (a) red phosphorus, (b) white phosphorus, and (c) black phosphorus.

Fig. 17.3 Interconversions of the allotropes of phosphorus.

form appears to be the most stable (thermodynamically). Rapid condensation of the vapours of arsenic and antimony yields reactive yellow solids (cf. white phosphorus), but these non-metallic allotropes rapidly change to the grey metallic allotropes of the elements. These (and bismuth) are stable lustrous metallic crystalline solids, which are fairly good conductors, but which are brittle. They have layer structures similar to that of black phosphorus.

The chemistry of the elements can be related to the structures. For example, white phosphorus is more reactive than nitrogen because the P—P bond is more easily broken than the N≡N bond.

17.7 REACTIONS OF THE ELEMENTS

Nitrogen is reactive only at high temperatures. White phosphorus is reactive but the red and black allotropes are not. With oxygen: nitrogen combines only at high temperatures to yield nitric oxide; white phosphorus combines so readily that it is stored under water, the products being the "trioxide" P_2O_6, and "pentoxide" P_4O_{10}; and arsenic, antimony, and bismuth on strong heating yield the trioxides. Only nitrogen reacts with hydrogen. All except nitrogen react with chlorine to form the chlorides: PCl_5 and PCl_3; $AsCl_3$; $SbCl_3$ and $SbCl_5$; and $BiCl_3$ (inert pair effect). All react with sulphur, except nitrogen, to form the sulphides. All react with some metals, but the products differ: nitrogen forms ionic and interstitial nitrides; the phosphides formed are less ionic; and the arsenides, etc., are more like alloys. Only white phosphorus is soluble in organic solvents. Arsenic, antimony, and bismuth react with hot concentrated sulphuric acid to form arsenious acid H_2AsO_3, antimony(III) sulphate $Sb_2(SO_4)_3$, and bismuth(III) sulphate. With nitric acid: red or white phosphorus forms phosphoric acid (moderately concentrated nitric); arsenic forms arsenious acid (hot dilute nitric) and arsenic acid, H_3AsO_4 (hot concentrated nitric); antimony forms antimony(III) oxide (moderately concentrated nitric) and antimony(V) oxide (concentrated nitric); and bismuth forms bismuth(III) oxide, Bi_2O_3. White phosphorus disproportionates in aqueous alkali to form phosphates and phosphine. Arsenic reacts with molten alkali to form arsenite, AsO_2^-, and hydrogen (not the unstable arsine AsH_3).

17.8 NITRIDES AND PHOSPHIDES

An extensive series of nitrides is formed by the elements and these can be classified in a similar manner to the carbides (Section 16.5). Ionic nitrides are formed by the group IIA metals and zinc and lithium (those of the other group IA metals are explosive). They are obtained by heating the metal in nitrogen or ammonia. They contain the nitride ion, N^{3-}, and decompose readily in water, forming the metal oxide or hydroxide, and ammonia. The transition metal nitrides are similarly prepared, and their properties and classification are similar to those of the transition metal carbides (Section 16.5) and the borides. The non-metals form covalent nitrides whose properties vary widely. The group IIIB elements form involatile giant molecules: boron nitride exists in two forms analogous to the structures of diamond and graphite, with which it is isoelectronic; aluminium nitride forms a diamond-like structure. The group VB and VIB elements form volatile molecular nitrides, P_3N_5, S_4N_4, etc.

Phosphorus forms a similar range of compounds, but these are generally less ionic than the corresponding nitrogen compounds. The tendency to form covalent molecular solids decreases down the group in the order P > As ⩾ Sb ≫ Bi. With metals, the compounds formed by arsenic, antimony, and bismuth become increasingly like alloys.

17.9 HYDRIDES

17.9.1 Preparation

Ammonia is prepared industrially by direct combination of nitrogen and hydrogen in the Haber process. The nitrogen and hydrogen are made by passing the correct amounts of air and steam over coke to obtain the three to one ratio, and the other gases are removed (CO, O_2, CO_2, H_2S). A gas containing 10–20 percent ammonia is obtained from the reaction, and is condensed or dissolved in water; the unreacted nitrogen and hydrogen are recycled through the process. This synthesis is the first step in the commercial fixation of nitrogen; most of the ammonia is converted to nitric acid. Ammonia is used in the manufacture of fertilizer, nitric acid, rubber, and nylon, and as an inert or reducing atmosphere in metallurgy.

The laboratory preparations of the hydrides include:

1. The displacement of the hydrides from ammonium (and phosphonium) salts by alkali. Equal parts of an intimate mixture of ammonium chloride and calcium hydroxide are heated, and the gaseous mixture obtained is dried over calcium oxide and collected by upward delivery in air:

$$2NH_4Cl(s) + Ca(OH)_2(s) \rightarrow CaCl_2(s) + 2H_2O(g) + 2NH_3(g).$$

2. The hydrolysis of binary compounds or alloys (nitrides, phosphides, arsenides, antimonides, and the magnesium alloy of bismuth) by water or dilute acids:

$$AlP(s) + 3H_2O(l) \rightarrow Al(OH)_3(s) + PH_3(g);$$
$$Li_3N(s) + 3H_2O(l) \rightarrow NH_3(g) + 3LiOH(s).$$

3. The reduction of compounds by zinc and hydrochloric acid.
4. The reaction of alkali on white phosphorus:

$$P_4(s) + 3OH^-(aq) + 3H_2O(l) \rightarrow PH_3(g) + 3H_2PO_2^-(aq).$$

17.9.2 Physical Properties

Ammonia is a colourless pungent gas, lighter than air, and is the most soluble of all gases in water. Commercial concentrated ammonia is 35 percent w/w, and its density is 0.88 (dissolution of ammonia causes expansion). When boiled, concentrated ammonia gives off ammonia gas. The other hydrides, MH_3, are also colourless gases at 25 °C, but are heavier than air and sparingly soluble in water. All the molecules are pyramidal (Section 12.7) and the bond angle decreases (Table 17.4) as the group is descended and as the electronegativity of the group V element decreases. This decrease in the electronegativity causes the bonding electron pairs to be found closer to the hydrogen atoms (on average) as the group is descended; therefore the repulsion between bonding electron pairs in the vicinity of the central

Table 17.4. The properties of the group VB hydrides, MH_3.

Property	Ammonia NH_3	Phosphine PH_3	Arsine AsH_3	Stibine SbH_3	Bismuthine BiH_3
Melting point (°C)	−78	−134	−116	−88	
Boiling point (°C)	−33	−87	−62	−17	22
Solubility, v/v, 20 °C	739	0.26	0.20	0.20	
Odour	pungent	decaying fish	garlic-like		
Bond energy, M—H	389 kJ mol^{-1}	322	297	255	
$\Delta G_f^\ominus$ (kJ mol^{-1})	−16.7	+13.4	+66.9	+146	
Bond angle	106° 47′	93° 30′	92°	91° 30′	
Dipole moment (D)	1.44	0.55	0.15		
Decomposition to elements	500 °C	on gentle heating	at 20 °C	at 20 °C	very unstable

atom decreases, and the bond angle decreases. Only in the case of nitrogen is the difference in electronegativity between the group VB element and hydrogen sufficiently large to allow hydrogen bonding. Many of the anomalous properties of ammonia can be related to its hydrogen-bonding capacity: high melting and boiling points; large enthalpy of vaporization (therefore the liquid is easily handled); high dielectric constant (therefore liquid ammonia is a good ionizing solvent); high solubility; easy liquefaction; and high polarity. Liquid ammonia can be obtained either by cooling the gas at atmospheric pressure, or by compression (nine atmospheres) at 25 °C. Its physical properties are similar to those of water; it is neutral to litmus, and it is a good ionizing solvent (but not as good as water). The thermal stability of the gases, MH_3, decreases as the group is descended (Table 17.4) and this is the basis of the Marsh test for arsenic.

17.9.3 Chemical Properties

All the gases are reducing agents (Fig. 17.1) (except for ammonia in acid solution), their reducing power increasing as the group is descended. Ammonia is surprisingly inert to oxidation (typical first row kinetic effect) and does not burn in air unless heated to high temperatures. In oxygen, at a somewhat lower temperature, it burns with a yellow flame:

$$4NH_3(g) + 3O_2(g) \rightarrow 2N_2(g) + 6H_2O(g); \quad \Delta G^\ominus = -1305 \text{ kJ}; K_{25°} = 10^{228}.$$

In the presence of a platinum catalyst at 800 °C, an alternative reaction occurs, despite the lower $\Delta G^\ominus$ value (kinetic effect), to produce nitric oxide. This is used industrially in the manufacture of nitric acid.

$$4NH_3(g) + 5O_2(g) \rightarrow 4NO(g) + 6H_2O(g); \quad \Delta G^\ominus = -1132 \text{ kJ}; K_{25°} = 10^{198}.$$

The remaining hydrides are much more reactive (lower activation energies), and phosphine ignites spontaneously at 150 °C in air to form phosphoric acid (reflecting the great stability of phosphorus(v), Fig. 17.1). Actually, the phosphine normally prepared contains diphosphine, P_2H_4, which ignites at room temperature in air. An excess of ammonia reacts with chlorine to form ammonium chloride.

$$2NH_3(g) + 3Cl_2(g) \xrightarrow{} N_2(g) + 6HCl(g) \xrightarrow{6NH_3} 6NH_4Cl(s).$$

In the presence of an excess of chlorine, however, the highly explosive yellow oil, nitrogen trichloride, is formed. Fluorine reacts similarly to form the stable gas, nitrogen trifluoride; and aqueous ammonia reacts with bromine (at low temperatures) and iodine to form the unstable ammonates, $NI_3 \cdot nNH_3$ and $NBr_3 \cdot 6NH_3$:

$$3Cl_2(g) + NH_3(g) \rightarrow 3HCl(g) + NCl_3(l).$$

The other hydrides are more reactive; for example, phosphine and chlorine ignite spontaneously. Gaseous ammonia reduces heated metal oxides such as copper oxide; and the other hydrides reduce easily reduced metal ions such as Ag^+ and Cu^{2+} to the free metal, or the arsenide, etc.

Ammonia acts as a Lewis base and forms numerous addition or co-ordination compounds (ammines), both with metal ions, forming complexes such as $Cu(NH_3)_4^{2+}$, $Ag(NH_3)_2^+$, and $Co(NH_3)_6^{2+}$, and with neutral acceptor molecules, forming addition compounds such as $NH_3 \rightarrow BF_3$. Phosphine and the heavier Group VB hydrides are much less basic and they form only a few unstable addition compounds. A special case of complexes occurs when the Lewis acid (acceptor) involved is the hydrogen ion, $H^+(aq)$ or H_3O^+. Ammonia is basic and forms stable salts:

$$NH_3 + H^+ \rightarrow NH_4^+ \quad \text{ammonium ion.}$$

Ammonia is miscible with water in all proportions, owing to its hydrogen-bonding capacity. Crystalline hydrates, which are hydrogen-bonded polymers, can be obtained at low temperatures—$NH_3 \cdot H_2O$, $2NH_3 \cdot H_2O$, etc. Neither the covalent compound $NH_4OH(s)$ (which would contain five-co-ordinate nitrogen!), nor the ionic compound $NH_4^+OH^-(s)$ are obtained. In aqueous solution the situation is probably similar, and the main species present are hydrogen-bonded species written $NH_3(aq)$, with some ions $NH_4^+(aq)$ and $OH^-(aq)$:

$$NH_3(aq) + H_2O(l) \rightleftharpoons NH_4^+(aq) + OH^-(aq); \quad K = \frac{[NH_4^+][OH^-]}{[NH_3(aq)]} = 1.8 \times 10^{-5}.$$

The degree of ionization is slight and the misleadingly named "ammonium hydroxide" is a weak base (better called "aqueous ammonia"). The solution turns blue litmus red. Ammonia neutralizes acids to form salts, and reacts with carbon dioxide to form ammonium carbamate. Phosphine is only feebly basic, and the remaining hydrides are not basic at all. In non-aqueous conditions strong acids do form phosphonium salts, for example $PH_4^+I^-$, but these are rapidly hydrolysed by water.

Ammonia is less reactive to metals than water is, but the electropositive metals form amides (Section 17.1); thus ammonia behaves as a weak acid:

$$Na(s) + NH_3(g) \xrightarrow{heat} Na^+NH_2^-(s) + \tfrac{1}{2}H_2(g).$$

The other hydrides show no acid character.

17.9.4 Ammonium Salts

The ammonium ion is said to be a "pseudo alkali metal ion" because some of its properties resemble those of the alkali metal ions, particularly K^+ (1.33 Å) and Rb^+ (1.48 Å), since its size (1.43 Å) is intermediate between these. Thus many stable salts are known which are largely ionic, white (unless the anion is coloured), usually anhydrous crystalline solids, which are usually soluble and of similar structures to those of the corresponding potassium salts. One exception is ammonium fluoride which, owing to hydrogen bonding, has the wurtzite structure. Differences between ammonium salts and potassium salts are due to the fact that ammonia is a gas, and that the ammonium ion is oxidizable and partly hydrolysed. Thus:

1. Solutions of ammonium salts of strong acids are slightly acidic owing to hydrolysis. If such solutions are boiled, ammonia gas is evolved, the equilibrium moves to the right, and the solution becomes more acidic:

$$NH_4^+(aq) + H_2O(l) \rightleftharpoons NH_3(aq) + H_3O^+(aq); \quad pK = 5.$$

2. Ammonium salts of volatile acids sublime on heating (chloride, carbonate, and even the nitrate at moderate temperatures):

$$NH_4^+Cl^-(s) \rightleftharpoons NH_3(g) + HCl(g).$$

3. Ammonium salts of oxidizing anions decompose on strong heating by oxidizing the ammonia to (usually) nitrogen, N_2, or nitrous oxide, N_2O. Examples are: the decomposition of the nitrate to nitrous oxide (Section 11.3.3); dichromate to nitrogen; and the nitrite to nitrogen (Section 17.5).

Hydrazine, NH_2—NH_2, is an endothermic (but kinetically fairly inert) fuming colourless liquid, and is the nitrogen analogue of hydrogen peroxide. It burns in air to nitrogen in a highly endothermic reaction, and is a potential rocket fuel. It is less basic than ammonia but forms two series of salts, $N_2H_6^{2+}$ (solid state only) and $N_2H_5^+$. It can act as both an oxidant—forming ammonia—and as a reductant —forming nitrogen (Fig. 17.1).

Hydroxylamine, NH_2OH, is an unstable (Problem 18) deliquescent white solid, which is a weaker base than ammonia, but forms stable salts, such as $NH_3OH^+Cl^-$, hydroxylaminium hydrochloride. It shows an intermediate oxidation state of nitrogen, like hydrazine, and can therefore act as either oxidant or reductant.

17.10 HALIDES

17.10.1 Trihalides

All the trihalides MX_3[M=N, P, As, Sb, Bi; X=F, Cl, Br, I] exist. Those of nitrogen are the least stable (except for NF_3 which is inert, cf. CCl_4): the compound NCl_3 is explosive; and NBr_3 and NI_3 exist only as ammonates, and detonate when the ammonia is removed. General methods of preparation are the action of an excess of the element, M, on the halogen (except nitrogen—too inert, and PF_3), and the action of the hydrogen halide on the trioxide or trisulphide. All the molecules are pyramidal in the gas phase, but not all the solids have discrete molecular lattices; BiF_3, for example, has an ionic lattice. All hydrolyse rapidly (except NF_3) to the hydrated oxide ($As_2O_3 \cdot xH_2O$), to the acid (H_3PO_3), or to the oxycation (SbOCl, BiOCl). Nitrogen trichloride forms ammonia and hypochlorous acid, not nitrous acid and hydrochloric acid (Problem 9). The ease of hydrolysis decreases from phosphorus to bismuth and from iodine to fluorine. Thus, in the case of antimony trichloride an equilibrium is obtained:

$$SbCl_3(aq) + H_2O(l) \underset{acid}{\overset{water}{\rightleftharpoons}} SbOCl(s) + 2HCl(aq).$$

All except NX_3 and PF_3 can act as Lewis acids (acceptors), especially the fluorides and chlorides, for example SbF_4^-, and $BiCl_5^{2-}$. They can also act as Lewis bases (donors) especially PF_3, for example $Ni(\leftarrow PF_3)_4$; cf. $Ni(CO)_4$. The ease of oxidation of the trihalides MX_3 decreases in the order $P > As \geqslant Sb \gg Bi$ (N not at all). They are halogenating agents. Some reactions of phosphorus trichloride, a typical example, are: $H_2O \rightarrow H_3PO_3$; $NH_3(l) \rightarrow P(NH_2)_3$; $X_2(Br_2, Cl_2, F_2) \rightarrow PX_2Cl_3$; $ZnF_2 \rightarrow PF_3$; I_2 (glacial acetic) $\rightarrow P_2I_4$; $HI(g) \rightarrow PI_3$; S(sealed tube, 130 °C) $\rightarrow$ $PSCl_3$; $BaS \rightarrow P_2S_3$; $O_2 \rightarrow POCl_3$; acetic acid $\rightarrow POCl_3 + CH_3COCl$; $RMgX \rightarrow PR_3, PR_2Cl, PRCl_2$; $AgNCO \rightarrow P(NCO)_3$; $Ni(CO)_4 \rightarrow Ni(PCl_3)_4$.

17.10.2 Pentahalides

No nitrogen pentahalides are known (Problem 14), and the stability with respect to dissociation into MX_3 and X_2 (X = halogen) of those of the other group elements decreases in the orders $P > Sb > As > Bi$, and $F > Cl > Br$. The stable pentahalides formed are PF_5, PCl_5, PBr_5, AsF_5, SbF_5, $SbCl_5$, and BiF_5, though PBr_5 and $SbCl_5$ readily lose X_2 and form the trihalide. This is much as expected: the oxidation state $+5$ becomes more oxidizing down the group (inert pair effect), except that arsenic(v) is more oxidizing than expected (middle row anomaly); and fluorine is the least oxidizable, iodine the most oxidizable halogen. All have trigonal bipyramidal structures in the gas phase (Section 12.7) but the structures in the solid state generally differ, for example $PCl_4^+PCl_6^-(s)$ and $PBr_4^+Br^-(s)$. Can you suggest why PBr_6^- and PI_5 do not exist despite the fact that phosphorus(v) is not an oxidizing state (Problem 19)?

These compounds can be prepared by the action of an excess of halogen on the element or the trihalide MX_3. For example, if the liquid PCl_3 is slowly dropped into a flask through which a stream of chlorine is being passed, yellow solid phosphorus pentachloride collects in the bottom of the flask. Owing to easy dissociation into the halogen X_2, these compounds are good halogenating agents, the least stable being the most effective: PBr_5, $SbCl_5$, and BiF_5. With water, they yield the oxyacid (H_3PO_4, H_3AsO_4), the hydrated oxide ($Sb_2O_5 \cdot xH_2O$), or the oxycation (BiO^+)—inert pair effect. All except BiF_5 act as Lewis acids, e.g. PF_6^-, $PF_5 \cdot D$ (D = an oxygen or nitrogen donor ligand, amines, ethers, etc.).

Phosphorus pentachloride is a typical example. It is a yellowish-white solid, which fumes strongly in moist air. When heated, it sublimes, but dissociates on stronger heating. It is an effective chlorinating agent. Some typical reactions are: $H_2O \rightarrow H_3PO_4$; $H_2SO_4 \rightarrow POCl_3 + HO \cdot SO_2 \cdot Cl$; $H_2S \rightarrow P_2S_5$; $SO_2 \rightarrow SO_2Cl_2 + PCl_3$; $NH_4Cl \rightarrow$ phosphonitrilic chlorides (Section 17.10.4); excess $I_2 \rightarrow ICl + PCl_3$; $Na \rightarrow Na_3P$; $ROH \rightarrow RCl + POCl_3$ (R = H, alkyl).

17.10.3 Oxyhalides

The nitrosyl halides, XNO, are covalent gases. Are these molecules linear or bent (Problem 6)? They are very powerful oxidizing agents whose general stability and inertness decreases in the order F > Cl > Br. With water they form a mixture of oxides and oxyacids of nitrogen. Stoichiometric analogues of antimony and bismuth, MOCl, exist. The nitryl halides, XNO_2 (X = F, Cl), are also reactive gases. The oxyhalides of phosphorus, POX_3 (X = F, Cl, Br), are prepared by the hydrolysis of PX_5 with a little water, or by the oxidation of PX_3, or by the action of phosphorus pentoxide (P_4O_{10}) on PX_5. Their reactions are similar to those of the pentahalides, PX_5, but they are milder halogenating agents.

17.10.4 Phosphonitrilic Halides

These are prepared by fusing together phosphorus pentachloride and ammonium chloride, and separating the mixture $[PCl_2N]_n$ by distillation. Rings up to $n = 17$ have been obtained, containing Br, F, NCS, alkyl, and other groups (Fig. 17.4).

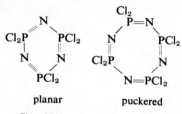

Fig. 17.4 Two phosphonitrilic halides.

17.11 OXIDES

Nitrogen forms the oxides N_2O, NO, N_2O_3, NO_2, N_2O_4, and N_2O_5, as well as the less well-characterized NO_3 and N_2O_6. All the oxides of nitrogen are thermodynamically unstable relative to dissociation into the elements, but are inert towards this reaction (typical first row element property). Compounds containing single N—O bonds tend to be unstable because of the low N—O single bond enthalpy. The existence of gaseous nitrogen oxides was discussed in Section 17.4.

17.11.1 Nitrous Oxide (Dinitrogen Oxide)

Nitrous oxide or dinitrogen oxide (b.p. $-88\ °C$; m.p. $-102\ °C$; $\Delta G_f^{\ominus} = +104$ kJ mol^{-1}) is a colourless gas obtained by heating ammonium nitrate gently:

$$NH_4NO_3(s) \xrightarrow{melt} 2H_2O(g) + N_2O(g).$$

It has a faint sweet smell, is denser than air, and only slightly soluble in water. It is the least reactive and least noxious of the nitrogen oxides and, mixed with oxygen, is used as an anaesthetic by dentists (laughing gas). It is inert at 25 °C, but decomposes to nitrogen, oxygen, and nitric oxide at 500 °C, and therefore supports the combustion of glowing combustible materials better than air (more oxygen!). It is neutral to water, and hyponitrous acid $H_2N_2O_2$ is not formed; nor does nitrous oxide react with acids or alkalis. It is isoelectronic with carbon dioxide, and like it, it is linear but has a small dipole moment: $N^-{=}N^+{=}O \leftrightarrow N{\equiv}N^+{-}O^-$. Despite its position on the oxidation state diagram, it is neither oxidant nor reductant, and it does not disproportionate (kinetically stable, or inert).

17.11.2 Nitric Oxide (Nitrogen Oxide)

Nitric oxide or nitrogen oxide (b.p. $-152\ °C$; m.p. $-164\ °C$; $\Delta G_f^{\ominus} = +86.6$ kJ mol^{-1}) is frequently obtained during the reduction of nitric acid by reductants such as copper, iron(II) ions, and so on, but it is contaminated by nitrogen dioxide:

$$3Cu(s) + 8HNO_3\ (50\%) \rightarrow 3Cu(NO_3)_2(aq) + 4H_2O(l) + 2NO(g).$$

Purer nitric oxide can be prepared by the reduction of nitrous acid by iron(II) or iodide ions, or by the catalytic oxidation of ammonia.

Table 17.5. The oxides of nitrogen.

Property	N_2O	NO	N_2O_3	N_2O_4/NO_2	N_2O_5
Oxidation number	+1	+2	+3	+4	+5
Colour of solid	colourless	colourless	blue	colourless	colourless
Molecular weight	44	30	76	92	108
$\Delta H_f^\ominus$(g) (kJ mol^{-1})	+82	90.4	83.8	9.2/33.2	11 [−43(s)]
$\Delta G_f^\ominus$(g) (kJ mol^{-1})	+104	86.6	139.4	97.8/51.3	115 [114(s)]
Melting point (°C)	−102	−164	−102	−11	30(d)
Boiling point (°C)	−88	−152	+3.5(d)	21(d)	47(d)

The colourless gas is slightly heavier than air, sparingly soluble in water, and (like the other oxides of nitrogen) thermally stable despite its positive free energy of formation (kinetic control), and only decomposes above 1000 °C to nitrogen and oxygen. Thus it supports the combustion of only fiercely burning (highly exothermic) substances such as magnesium and sulphur. It is difficult to write a satisfactory valence-bond structure for nitric oxide. It is best considered as containing $2\frac{1}{2}$ bonds, N≝O, since although it is paramagnetic (Section A.9), its properties are not typical of molecules containing a single electron on one atom: it is colourless; it is relatively unreactive; and it does not polymerize. It is a neutral oxide, like nitrous oxide, and does not react with alkalis or acids. It reacts very quickly with oxygen to form the brown gas nitrogen dioxide. Nitric oxide contains nitrogen in an intermediate oxidation state (Fig. 17.1), and acts as both oxidant and reductant (though it is kinetically inert to disproportionation). Thus with acid permanganate it is quantitatively oxidized to nitric acid (analytical determination of NO), and with halogens it yields the nitrosyl halides XNO; with reductants a variety of products can be obtained, for example, nitrous oxide from sulphur dioxide, and hydroxylamine from acidified tin(II) salts. With iron(II) salts it forms the complex cation $[Fe(NO)(H_2O)_5]^{2+}$ which is the brown product of the "brown ring test".

17.11.3 Nitrogen Dioxide and Nitrogen Tetroxide (Dinitrogen Tetroxide)

$$N_2O_4(s) \underset{-11°C}{\rightleftharpoons} N_2O_4(l) \underset{21°C}{\rightleftharpoons} N_2O_4(g) \rightleftharpoons 2NO_2 \xrightarrow{140°C}$$

colourless solid, diamagnetic | pale-yellow 0.01% NO_2 | deep red-brown 0.1% NO_2 | 90% at 100°C

$$2NO_2 \xrightarrow{600°C} 2NO + O_2$$
100% at 140°C — colourless gases

Nitrogen dioxide ($\Delta G_f^\ominus = +51.3$ kJ mol^{-1}) can be prepared: by the action of various reducing agents on nitric acid, for example copper on concentrated nitric acid,

$$Cu(s) + 4HNO_3(conc.) \rightarrow 2NO_2(g) + Cu^{2+}(aq) + 2NO_3^-(aq) + 2H_2O(l);$$

by the reaction between nitric oxide and oxygen; and by the decomposition of heavy metal nitrates (Section 11.2). It is a very deep-brown gas, denser than air, and reacts with water. It contains a lone electron on the nitrogen (Fig. 17.5), and

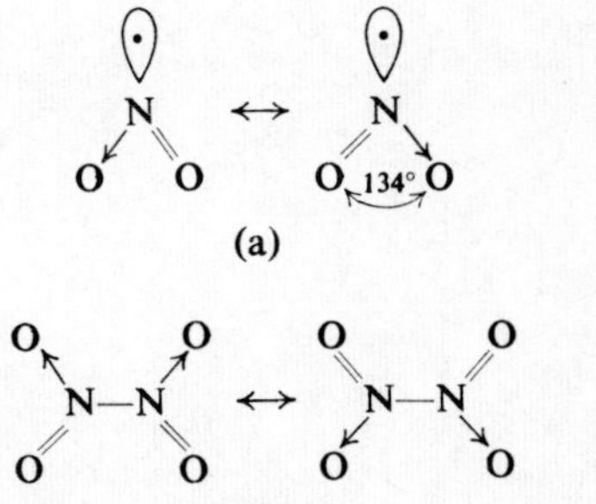

Fig. 17.5 Valence bond structures of (a) nitrogen dioxide, and (b) dinitrogen tetroxide.

is a typical one-electron compound: it polymerizes to colourless dinitrogen tetroxide at low temperatures (m.p. $-10\,°C$; b.p. $21\,°C(d)$; $\Delta G_f^\ominus = +97.8\,kJ$), is coloured, and is paramagnetic. Although nitrogen dioxide is endothermic, its thermal decomposition is not complete until $600\,°C$. Nevertheless, it supports the combustion of phosphorus, sulphur, etc. It is unstable with respect to disproportionation (Fig. 17.1), and in cold water it disproportionates to nitric acid and nitrous acid:

$$H_2O(l) + 2NO_2(g) \xrightarrow{cold} H^+(aq) + NO_3^-(aq) + HNO_2(aq).$$

Nitrous acid is itself unstable with respect to disproportionation, and on warming it forms nitric acid and nitric oxide (synthesis of nitric acid). Consequently the gas is corrosive. In nitrogen dioxide the nitrogen is in an intermediate oxidation state and the compound can act as both oxidant (mainly) and reductant (Fig. 17.1). It is a powerful oxidizing agent; some examples are:

$$I^- \to I_2, \quad SO_2 \to SO_3, \quad Cu \xrightarrow{hot} CuO, \quad \text{and} \quad H_2S \to S.$$

Powerful oxidants such as permanganate oxidize it to nitric acid.

Dinitrogen tetroxide (Fig. 17.5), is planar, colourless, and diamagnetic. It contains a very long, weak, N—N bond (1.75 Å), and the bond energy is small (58 kJ mol^{-1}).

17.11.4 Dinitrogen Pentoxide

N_2O_5 (b.p. $47\,°C(d)$; m.p. $30\,°C(d)$; $\Delta G_f^\ominus = +114\,kJ\,mol^{-1}$) can be prepared from the dehydration of nitric acid by phosphorus pentoxide, and is distilled out in ozone and oxygen, and collected as a white solid in a cooled container. It is deliquescent (forming nitric acid), decomposes slowly at $25\,°C$ to nitrogen dioxide and oxygen, and is a powerful oxidant. The stable form of the solid is not a molecular solid, but is nitronium nitrate, $NO_2^+ NO_3^-$. What shape is the nitronium ion, NO_2^+ (Problem 6)?

$$O_2N-O-NO_2 \xrightarrow[90\,K]{condense\ at} \text{molecular solid} \xrightarrow{on\ standing} NO_2^+ NO_3^-(s)$$

17.11.5 The Pentoxides of Phosphorus, Arsenic, Antimony, and Bismuth

These can be prepared by the oxidation of the elements or the trioxides. Increasingly powerful oxidants are required on descending the group as the stability of the $+5$ oxidation state decreases (inert pair effect). Thus, white phosphorus burns in an excess of oxygen to yield the pentoxide, whereas the pentoxides of arsenic and antimony are obtained by the action of hot concentrated nitric acid on the elements (or trioxides). With bismuth, however, this method yields the trioxide, and this is converted to an insoluble, impure, non-stoichiometric, highly oxidizing, brown compound called sodium bismuthate, "$NaBiO_3$", by the action of alkaline sodium peroxide. All the pentoxides decompose on heating to the trioxides, the ease of decomposition increasing down the group.

Phosphorus pentoxide is a white, deliquescent, polymorphic molecular solid,

P_4O_{10}. The four phosphorus atoms are arranged at the corners of a tetrahedron, with six of the oxygen atoms between the phosphorus atoms, just above the six sides of the tetrahedron, and the other four oxygen atoms bonded one to each phosphorus atom (Fig. 17.6). The molecular structure is the same in the liquid and gas phases. With water, a mixture of phosphorus(v) acids is obtained, the ratio depending upon the conditions, especially the amount of water:

$$P_2O_5(s) + \text{a little } H_2O \rightarrow (HPO_3)_n \xrightarrow{H_2O} H_4P_2O_7 \xrightarrow{H_2O} H_3PO_4.$$
$$\text{meta-} \qquad\qquad \text{pyro-} \qquad \text{ortho-phosphoric acid}$$

Thus, it is an acidic oxide, and reacts with basic oxides to give phosphates. It is also a powerful dehydrating agent, extracting the water from air and from various compounds:

$$H_2SO_4 \rightarrow SO_3; \qquad HNO_3 \rightarrow N_2O_5; \qquad RCONH_2 \rightarrow RCN.$$

The structures of the pentoxides of arsenic, antimony, and bismuth are not known. With water, arsenic pentoxide dissolves to give arsenic acid, and antimony pentoxide dissolves to give an acidic solution of the hydrated oxide. With alkalis, they yield the arsenate oxyanion, AsO_4^{3-}, and the complex anion, $Sb(OH)_6^-$, showing the increasing metallic nature of the elements. The oxidizing power of the +5 oxidation state increases down the group (Fig. 17.1) and sodium bismuthate is one of the most powerful oxidants known.

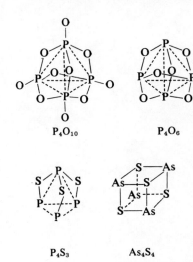

Fig. 17.6 The structures of some group V compounds.

17.11.6 The Trioxides

Dinitrogen trioxide, N_2O_3 (m.p. $-102\,°C$; $\Delta G_f^\ominus = +139.4$ kJ mol^{-1}) is a deep-blue solid which exists only in the solid state. It is prepared by cooling an equimolar mixture of nitric oxide and nitrogen dioxide, and behaves chemically like such a mixture. Phosphorus trioxide is prepared by burning white phosphorus in a limited supply of oxygen. It is a white volatile solid soluble in organic solvents (m.p. 24 °C; b.p. 175 °C; $\Delta H_f^\ominus = -1640$ kJ mol^{-1}), and its structure is that of P_4O_{10} without the apical oxygen atoms (Fig. 17.6). It slowly oxidizes to the pentoxide at 25 °C in air, and inflames on heating. It reacts with cold water to form phosphorus(III) acid, H_3PO_3, which is in an intermediate oxidation state (Fig. 17.1), and disproportionates to phosphoric acid and a mixture of reduction products (PH_3, P, etc.). Thus the oxide is acidic, and reacts with alkalis to form phosphites:

$$P_4O_6(s) + 8OH^-(aq) \rightarrow 4HPO_3^{2-}(aq) + 2H_2O(l).$$

It is a reducing agent (Fig. 17.1), being oxidized by oxidants to the stable phosphorus(v) state, e.g. $X_2 \rightarrow POX_3$ (X = halogen).

The trioxides of arsenic and antimony are white volatile solids and have the same structure as P_4O_6. Bismuth(III) oxide (bismuth trioxide) has an ionic lattice and is yellow. Only that of arsenic dissolves in water, yielding the hydrated oxide, $As_2O_3 \cdot xH_2O$, called "arsenious acid". Both arsenic(III) (arsenious) and antimony(III) (antimonous) oxides, As_2O_3 and Sb_2O_3, dissolve in alkalis to give the arsenite, AsO_2^-(aq), and antimonite, $Sb(OH)_4^-$(aq), ions. Bismuth(III) oxide shows no acidic properties, and is unaffected by alkalis. In contrast to the acidic properties, which decrease down the group, the basic properties of the oxides increase:

arsenic(III) oxide is more soluble in hydrochloric acid than in water; antimony(III) oxide forms a basic nitrate; and bismuth(III) oxide readily dissolves in acids to form bismuth salts.

17.12 SULPHIDES

Arsenic and phosphorus each form a number of sulphides, prepared by heating together the elements, all of which are molecular solids, for example P_4S_3, P_4S_{10}, P_4S_5, P_4S_7 (each of which contains a tetrahedron of phosphorus atoms), As_4S_3, and As_4S_4.

The trisulphides of arsenic (yellow As_4S_6), antimony (an orange polymeric solid which goes black when heated, $(Sb_2S_3)_n$), and bismuth (black polymeric $(Bi_2S_3)_n$), can be prepared either by reaction between the elements, or by the reaction of hydrogen sulphide on acid solutions containing the elements in the +3 oxidation state. The trisulphides of arsenic and antimony are soluble in alkali solutions or in yellow ammonium sulphide, $(NH_4)_2S_x$:

$$2As_2S_3(s) + 4OH^-(aq) \rightarrow AsO_2^-(aq) + 3AsS_2^-(aq) + 2H_2O(l)$$
(arsenite + thioarsenite)
$$As_2S_3(s) + 3S^{2-}(aq) + 2S \rightarrow 2AsS_4^{3-}(aq) \quad \text{(thioarsenate).}$$

Bismuth trisulphide is basic and is not soluble in alkali or ammonium sulphide.

The pentasulphides of arsenic (bright yellow) and antimony (red) can be prepared by direct action of elements; by action of hydrogen sulphide on solutions containing antimony(V) and arsenic(V); or by the action of acid on thioarsenates or thioantimonates:

$$2AsS_4^{3-}(aq) + 6H^+(aq) \rightarrow As_2S_5(s) + 3H_2S(g).$$

These are also soluble in alkali and ammonium sulphide, but insoluble in water and acids. Bismuth pentasulphide does not exist (Section 15.1).

17.13 THE OXYACIDS OF NITROGEN

17.13.1 Nitrous Acid

This cannot be isolated pure. Aqueous solutions of the acid are obtained by acidification of nitrites, or by passing equimolar mixtures of nitric oxide and nitrogen dioxide into water:

$$Ba(NO_2)_2(aq) + H_2SO_4(aq) \rightarrow BaSO_4(s) + 2HNO_2(aq).$$

Nitrous acid contains nitrogen in an unstable intermediate oxidation state (Fig. 17.1), and the compound is unstable in the pure state, disproportionates in solution, and can act as an oxidant (e.g. $I^-(aq) \rightarrow I_2$; $Fe^{2+}(aq) \rightarrow Fe^{3+}(aq)$; $Sn^{2+}(aq) \rightarrow Sn^{4+}(aq)$; $H_2SO_3 \rightarrow H_2SO_4$; $H_2S \rightarrow S$), or as a reductant with strong oxidants (e.g. $MnO_4^-(aq) \rightarrow NO_3^- + Mn^{2+}$). Clearly, acid solutions of nitrites show the same reactions. It is a weak monobasic acid, giving rise to a series of salts, the nitrites (Fig. 17.7). Alkali metal nitrites are prepared by heating nitrates, preferably in the presence of lead or carbon:

$$NaNO_3(s) + C(s) \rightarrow NaNO_2(s) + CO(g).$$

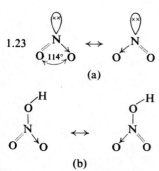

Fig. 17.7 The structures of (a) the nitrite ion NO_2^-, and (b) nitric acid.

Alternatively, a mixture of nitric oxide and nitrogen dioxide can be passed into an alkaline solution:

$$2Na^+(aq) + 2OH^-(aq) + NO(g) + NO_2(g) \rightarrow$$
$$2Na^+(aq) + 2NO_2^-(aq) + H_2O(l).$$

Nitrites are all soluble in water (except Ba^{2+}, Pb^{2+}, and Ag^+), and are decomposed by acids, yielding nitrogen dioxide:

$$2NO_2^-(aq) + 2H^+(aq) \rightarrow H_2O(l) + NO(g) + NO_2(g).$$

They yield a white precipitate when added to a solution of barium, lead, or silver ions. When boiled with ammonium chloride, nitrogen is obtained (Section 17.5). Devarda's alloy reacts with nitrites (and nitrates) to yield ammonia. Sodium nitrite is important for diazotization reactions in the dye industry (Section 24.8.1).

17.13.2 Nitric Acid

Nitric acid is manufactured from ammonia by oxidation to nitric oxide, which is mixed with air and passed into water, yielding nitric acid and nitric oxide, which is then recycled:

$$2NO(g) + O_2(g) \rightarrow 2NO_2(g); \quad 3NO_2(g) + H_2O(l) \rightarrow 2HNO_3(aq) + NO(g).$$

In the laboratory, nitric acid can be prepared by heating equal weights of concentrated sulphuric acid and sodium nitrate, and distilling off and condensing the nitric acid. Concentrated nitric acid is often yellow owing to the presence of nitrogen dioxide, formed by the photochemical decomposition of nitric acid, or by heating:

$$4HNO_3(l) \rightarrow 4NO_2(g) + 2H_2O(l) + O_2(g).$$

Fuming nitric acid is a solution of nitrogen dioxide in nitric acid. Nitric acid (m.p. -41.5 °C, b.p. 86 °C) is a colourless liquid which readily decomposes to nitrogen dioxide, and is soon yellow. The constant boiling point mixture with water contains 68 percent acid ($d = 1.4$ g cm^{-3}), and is called concentrated nitric acid. The vapour contains the planar molecule $HONO_2$ (Fig. 17.7). The pure liquid has no acidic properties, but in aqueous solution it is a monobasic strong acid. Concentrated nitric acid is a strong oxidant; in fact, the only metals which evolve hydrogen from even the dilute acid are manganese and magnesium. The products of the reactions with metals depend upon the conditions, particularly the concentration of the acid, the temperature, and the other reactant. Any of the lower oxidation states of nitrogen can be obtained, and usually a mixture of products results. All equations are over-simplifications, and generally only the major products are included; for example, the reactions with copper to yield nitric oxide and nitrogen dioxide (above) are typical of those of weakly electropositive metals. More electropositive metals tend to reduce the acid further to ammonia and hydroxylamine:

$$10H^+(aq) + NO_3^-(aq) + 4Zn(s) \rightarrow 4Zn^{2+}(aq) + NH_4^+(aq) + 3H_2O(l).$$

Concentrated nitric acid often passivates metals, such as Al, Cr, and Fe. Tin and antimony form the hydrated metal oxides and oxides of nitrogen. Only gold, platinum, iridium, and rhodium are not attacked. Concentrated nitric acid also oxidizes iron(II) to iron(III) ions, being itself reduced to nitric oxide. This is the basis of the brown ring test for nitrates, which in the presence of concentrated sulphuric acid and iron(II) form the brown complex, $[Fe(NO)(H_2O)_5]^{2+}$:

$$4H^+(aq) + 3Fe^{2+}(aq) + NO_3^-(aq) \rightarrow 3Fe^{3+}(aq) + NO(g) + 2H_2O(l)$$
$$NO(g) + Fe^{2+}(aq) \rightarrow [Fe(NO)(H_2O)_5]^{2+} \quad \text{brown.}$$

Non-metals are oxidized to oxides or oxyacids, for example: $C \rightarrow CO_2$; $P \rightarrow H_3PO_4$; $As \rightarrow H_3AsO_4$; $S \rightarrow H_2SO_4$; $I_2 \rightarrow HIO_3$, etc. Organic compounds are frequently oxidized, e.g. toluene → benzoic acid, but are often nitrated especially in the presence of concentrated sulphuric acid (Section 22.2.6).

Nitrates exist for nearly all metals, often hydrated, and usually soluble. The thermal decomposition of nitrates is discussed in Section 11.2. The nitrate ion is trigonal planar (Section 12.7). Nitrates are not good oxidizing agents, unless acidic (Fig. 17.1)—nitric acid! However, Devarda's alloy (Cu/Zn/Al) reduces solutions of alkaline nitrites or nitrates to ammonia:

$$3NO_3^-(aq) + 8Al(s) + 5OH^-(aq) + 2H_2O(l) \rightarrow 3NH_3(g) + 8AlO_2^-(aq).$$

17.14 THE OXYACIDS OF PHOSPHORUS

Generally in chemistry, the most highly hydroxylated acid known for an element in a particular oxidation state is called the ortho-acid. That acid obtained from it by the loss of one water is the meta-acid, and the pyro-acid is an intermediate, usually obtained by the action of heat. The prefix "hypo" generally indicates a lower oxygen content than the parent acid. This is illustrated in Fig. 17.8 for the phosphorus acids.

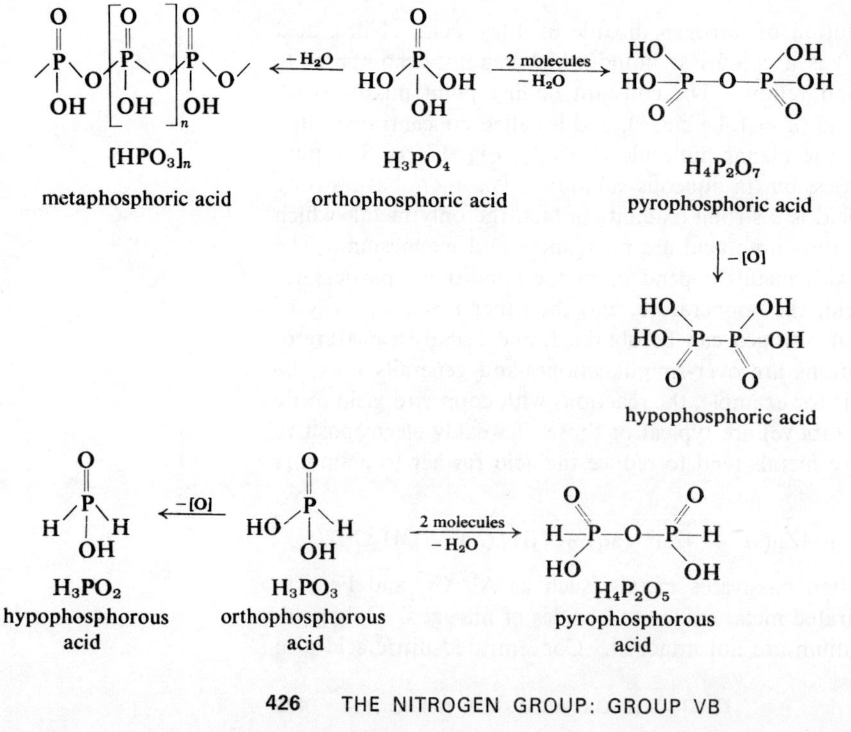

Fig. 17.8 Some oxyacids of phosphorus.

The phosphorus acids and oxyanions all contain approximately tetrahedral four-co-ordinate phosphorus. The phosphorus(v) acids and anions (phosphorus in oxidation state +5, called phosphor*ic*) all contain P=O bonds. The lower phosphorus acids and anions (phosphorus in oxidation states lower than +5, called phosphor*ous*) contain either P—P or P—H bonds, but not both. Oxyanions and acids containing P—H bonds are in low oxidation states and are reducing (though the reactions may be slow) and they tend to disproportionate to phosphate and phosphine. Condensed acids containing P—P bonds are more slowly hydrolysed than those containing P—O—P bonds, which themselves do not hydrolyse quickly to the monomeric acids. Therefore each of the condensed acids has a distinctive aqueous chemistry. The phosphate link P—O—P is very important in the chemistry of biologically active molecules, since many biochemists consider it to be the prime store of energy in biological systems. The energy of the bond is allegedly released to the system (muscle tissue, etc.) by hydrolysis of the phosphate link in adenosine triphosphate, ATP (Section 28.2.4):

$$\text{R}-\text{O}-\overset{\overset{\displaystyle O}{\|}}{\underset{\underset{\displaystyle OH}{|}}{P}}-\text{O}-\overset{\overset{\displaystyle O}{\|}}{\underset{\underset{\displaystyle OH}{|}}{P}}-\text{O}-\overset{\overset{\displaystyle O}{\|}}{\underset{\underset{\displaystyle OH}{|}}{P}}-\text{OH},$$

the so-called "high energy phosphate".

Phosphorus–hydrogen bonds are not ionizable to give H^+, and do not confer acidity. Therefore all phosphorus acids must contain a —P—OH group, and the number of such groups is the basicity of the acid. Why is there no metaphosphorous acid $[HPO_2]_n$—(Problem 15)? The oxyanions of the acids are obtained by removing H^+ from a P—OH group in the acid. All the oxyanions tend to form complexes

Table 17.6. The acids of phosphorus and their properties.

Acid	Nature	Preparation	Anion	
H_3PO_2 or $H_2P(OH)O$ hypophosphorous	crystalline white solid	white P_4 + alkali $\rightarrow H_2PO_2^-$ + H_2	$H_2PO_2^-$ hypophosphite	strongly reducing monobasic $pK = 2$
H_3PO_3 or $HPO(OH)_2$ orthophosphorous	deliquescent colourless solid	P_2O_3 or PCl_3 + H_2O	$H_2PO_3^-$, HPO_3^{2-} phosphite	reducing, but slow dibasic $pK_1 = 2$ $pK_2 = 6$
$H_4P_2O_5$ pyrophosphorous	white solid	PCl_3 + H_3PO_3	$H_2P_2O_5^{2-}$ pyrophosphite	reducing dibasic
$H_4P_2O_6$ hypophosphoric	white solid	red P + alkali	$P_2O_6^{4-}$ hypophosphate	not reducing or oxidizing tetrabasic $pK_1 = 2$
H_3PO_4 orthophosphoric	white solid	P_2O_5 + H_2O	$H_2PO_4^-$, HPO_4^{2-}, PO_4^{3-} phosphate	not oxidizing, tribasic
$H_4P_2O_7$ pyrophosphoric	colourless solid	heat phosphates or phosphoric acid	$P_2O_7^{4-}$ pyrophosphate	tetrabasic $pK_1 = 2$
	linear and cyclic anions	heating phosphates	$[PO_3(PO_3)_n \cdot OPO_3]^{(4+n)-}$ $[PO_3]_n^{n-}$ polyphosphate	

17.14 THE OXYACIDS OF PHOSPHORUS

with metal ions. As well as the acids and corresponding anions of phosphorus given in Fig. 17.8, there is a wide range of condensed phosphorus(v) acids and anions in chains, called polyphosphates $[P_nO_{3n+1}]^{(n+2)-}$, and in rings, called metaphosphates, $[PO_3]_n^{n-}$.

Orthophosphoric acid is prepared in the laboratory by the oxidation of red phosphorus with 1:1 nitric acid by warming together and then concentrating over concentrated sulphuric acid when white prisms of phosphoric acid crystallize out. If the temperature rises above 180 °C, pyrophosphoric acid is obtained.

$$P_4(s) + 10HNO_3(aq) + H_2O(l) \rightarrow 5NO(g) + 5NO_2(g) + 4H_3PO_4(s)$$
$$2H_3PO_4(s) \rightarrow H_4P_2O_7(s) + 2H_2O(g).$$

It is manufactured either by treating calcium phosphate ore (or bone ash) with concentrated sulphuric acid, or by hydrating phosphorus pentoxide. The pure solid is a three-dimensional hydrogen-bonded structure; in concentrated aqueous solution much intermolecular hydrogen bonding persists and the solutions are syrupy. At concentrations below about 50 percent, however, all the hydrogen bonding is to water. It is water soluble, like the meta- and pyro-acids, and the conversions of the three acids and their anions into each other in aqueous solution are slow; therefore each has a distinctive aqueous chemistry.

Orthophosphoric acid is a moderately strong acid and forms three series of salts (triprotic or tribasic): sodium dihydrogen phosphate, NaH_2PO_4, weakly acidic by hydrolysis, pH 3–5; disodium hydrogen phosphate, Na_2HPO_4, weakly alkaline by hydrolysis, pH 8–10; and trisodium phosphate, Na_3PO_4, strongly alkaline by hydrolysis, pH12. Thus, on titrating phosphoric acid with sodium hydroxide, the indicator methyl orange changes colour after the formation of NaH_2PO_4, and phenolphthalein changes colour after the formation of Na_2HPO_4. Neither phosphoric acid nor phosphates are oxidizing agents (Fig. 17.1).

Pyrophosphoric acid is obtained by heating orthophosphoric acid between 200 and 300 °C. It is a stronger acid than phosphoric, $K = 1.4 \times 10^{-1}$, and is tetraprotic, though only two series of salts are commonly known: $Na_4P_2O_7 \cdot 10H_2O$, etc.; and $Na_2H_2P_2O_7$, etc.,

$$H_3PO_4 \underset{\text{boiling water}}{\xrightarrow{200-300\,°C}} H_4P_2O_7 \xrightarrow{300\,°C} (HPO_3)_n \quad \text{(similarly for anions).}$$

Metaphosphoric acid is a mixture of polymeric linear and cyclic chains, and (like its salts) is a glass.

Table 17.7. Tests for phosphates.

Reagent	ortho-	meta-	pyro-
neutral $AgNO_3$	yellow precipitate	white precipitate	white precipitate
egg albumen	no reaction	no reaction	coagulates

Draw the structures of the tripolyphosphate ion, $P_3O_{10}^{5-}$ (linear) and the trimetaphosphate ion, $P_3O_9^{3-}$ (cyclic).

Phosphates of most cations are known, and are very important since they are an essential constituent of living matter: sugar phosphates are important in photosynthesis; nucleic acids contain phosphates; the driving force for many cell reactions is thought by many biochemists to be provided by free energy of phosphate hydrolysis; and phosphates are constituents of bones and teeth. Thus phosphates are important fertilizers. Sodium phosphates are used as water softeners, and in the laboratory phosphates are used as complexing anions for transition metal ions to keep them in solution. The naturally occurring phosphates are all orthophosphates. All the group IA phosphates and ammonium phosphate are soluble. The others are generally insoluble in water, but soluble in mineral acids and even in acetic acid (except for those of titanium, zirconium, and thorium, which are even insoluble in mineral acid). The insoluble double phosphate, $Mg(NH_4)PO_4$, is precipitated in the test for magnesium ions. The test for phosphate ion is the precipitation of the bright-yellow insoluble ammonium phosphomolybdate in the presence of an excess of nitric acid and ammonium molybdate.

17.15 OXYACIDS, OXYANIONS, AND SOLUTION CHEMISTRY OF ARSENIC, ETC.

The oxyanions and oxyacids of arsenic are similar to those of phosphorus (preparations, structures, properties) except that arsenic acid is moderately oxidizing (Fig. 17.1), arsenious acid is probably the hydrated oxide, and the condensed acids and oxyanions are less stable to hydrolysis to the monomers. Antimony and bismuth show the differences expected from the increasing electropositivity down the group: they form no free oxyacids; and their salts are not oxyanions, but are hydroxyanions, with increased co-ordination number. Antimony(v) oxide is acidic, and no antimony(v) cations exist. The anion $Sb(OH)_6^-$ is oxidizing. Antimony(III) oxide is amphoteric and thus the salt $Sb_2(SO_4)_3$ exists, though it is hydrolysed in solution to $SbO^+(aq)$ and the hydroxyanion $Sb(OH)_4^-(aq)$. Bismuth-(v) is very strongly oxidizing, and only the impure insoluble solid "sodium bismuthate" is known. Bismuth(III) oxide is basic and forms salts which are hydrolysed in solution to $BiO^+(aq)$.

PROBLEMS

1. Calculate the enthalpy and free energy changes of the reactions:

    $$NH_4NO_3(s) \rightarrow N_2(g) + \tfrac{1}{2}O_2(g) + 2H_2O(g)$$
    $$NH_4NO_3(s) \rightarrow N_2O(g) + 2H_2O(g).$$

    Comment on the results.

2. Why are the properties of "$N(OH)_3$" and "$P(OH)_3$" so different?

3. Summarize the trends of the group IVB elements with respect to: the first ionization energy; the ionic character of the binary compounds with calcium; the acid–base character of the oxides; and the volatility of the hydrides.

4. Construct the oxidation state diagram for nitrogen from the reduction potential data given in Section 17.1. Using this diagram comment on: the action of water on N_2O_4; and the thermal decomposition of the salts ammonium nitrate and ammonium nitrite.

5. From the reduction potential data given below, state whether chlorine "should" oxidize nitrogen dioxide and nitrogen:

Cl_2/Cl^-, $E^\ominus = +1.36$ V; NO_3^-/NO_2, $E^\ominus = +0.8$ V; N_2O/N_2, $E^\ominus = -1.77$ V.

In fact, chlorine oxidizes nitrogen dioxide but not nitrogen. Comment.

6. What are the shapes of the following species: NH_3, NF_3, $PCl_5(g)$, $PCl_5(s)$, $PBr_5(g)$, $PBr_5(s)$, $ClNO$, $ClNO_2$, $POCl_3$, "P_2O_5", NO_2^+, "P_2O_3", $HNO_3(g)$, NO_2^-?

7. Why are compounds containing nitrogen–nitrogen bonds uncommon?

8. Why are the oxides of nitrogen so different from those of phosphorus?

9. Why is the hydrolysis of NCl_3 slower than the hydrolysis of PCl_3, and a different type of product obtained?

10. Comment on the bond angles of the group VB hydrides, MH_3.

11. Why is white phosphorus so much more reactive than nitrogen?

12. Why does the structure of $NH_4^+F^-$ differ from that of the other ammonium halides?

13. Why does NCl_3 hydrolyse rapidly whereas NF_3 is inert to hydrolysis?

14. Why are the halides NX_5 unknown?

15. Why is metaphosphorous acid, HPO_2, not known?

16. Calculate (from the bond enthalpies) the enthalpy of formation of the (nonexistent) molecular compound "$N(OH)_3$". Discuss the result.

17. Calculate the enthalpies of formation (from the bond enthalpies) of the molecular compounds $P(OH)_3$ and $H(OH)_2PO$. Discuss the result.

18. What are the main causes of the "instability" of hydroxylamine?

19. Why are PBr_6^- and PI_5 unknown, despite the fact that phosphorus(v) is not an oxidizing state?

BIBLIOGRAPHY

Chilton T. H., *Strong Water: Nitric Acid, Its Sources, Methods of Manufacture and Uses*, Massachusetts Institute of Technology (1970).

    A detailed but readable account of nitric acid.

van Wazer, J. R. *Phosphorus and its Compounds*, Interscience (1958).

Guggenheim, E. A. "Odd Molecules," *Journal of Chemical Education*, **43**, 474 (1966).

    A simple discussion of the stability of "odd molecules" (those containing an unpaired electron), using nitric oxide as an example.

Jolly, W. L. *The Inorganic Chemistry of Nitrogen*, Benjamin (1964).

Jolly, W. L. *The Chemistry of the Non-Metals*, Prentice-Hall (1966).

In addition most of the books on inorganic chemistry in the general list contain material on the chemistry of group VA. A selection is listed at the end of Chapter 16.

CHAPTER 18

The Oxygen Group: Group VIB

18.1 FEATURES OF THE OXIDATION STATE DIAGRAM

Before looking at the general chemistry of Group VIB (Section 18.2) it is convenient to examine the oxidation state diagram. The following points should be obvious from Fig. 18.1 and from the reduction potential data in Table 18.1.

1. The -2 oxidation state is well established in this group (cf. the -3 state in group VB), not only for oxygen but also for sulphur, selenium, and even tellurium. However, the stability of this oxidation state decreases down the group as the electronegativity of the elements decreases, as expected. Thus, hydrogen telluride, H_2Te, is thermally unstable at 25 °C and is a strong reducing agent, even in acid conditions. Water, however, is thermally stable and is a poor reducing agent.
2. The $+6$ oxidation state becomes more strongly oxidizing as the group is descended (oxygen is anomalous). Polonium(VI) is so strongly oxidizing that its existence is in doubt. Selenium(VI) is more strongly oxidizing than expected (middle row anomaly, Section 10.3.2). Thus the oxidizing power of higher oxidation states decreases in the order (a typical one for p-block elements, with the first element anomalous and the second and third inverted):

O	≫	Po	⋙	Se	> Te > S.
(first row anomaly —no higher oxidation states exist)		probably no $+6$ oxidation state exists		middle row anomaly	

As usual the higher oxidation states are poorer oxidants in alkali than in acid solutions (Section 7.12.3).
3. The intermediate oxidation state $+4$ becomes increasingly stable with respect to disproportionation as the group is descended. Sulphur(IV) is more stable to disproportionation than appears from the main figure (see inset in figure).
4. Oxygen is a strong oxidant, particularly in acid solution. Hydrogen peroxide is an intermediate oxidation state and is unstable with respect to disproportionation, and can act as either a reductant or oxidant, especially the latter.
5. Sulphur(VI) is a poor oxidant, even in acid solution (even worse in alkali, SO_4^{2-}). Sulphur(IV) is an intermediate oxidation state and in acid can behave as either oxidant (good) or reductant (mild). In acid solution, thiosulphate is easily oxidized to the tetrathionate ion, $S_4O_6^{2-}$, by mild oxidants (e.g. I_2), but

18.1 Features of the oxidation state diagram
18.2 General chemistry
18.3 Group trends
18.4 The anomalous nature of oxygen
18.5 Occurrence, preparation, and allotropes
18.6 Reactivity of the elements
18.7 Oxides
18.8 Hydrogen peroxide
18.9 Hydrides
18.10 Sulphides
18.11 Halogen compounds
18.12 Oxides
18.13 Oxyacids
18.14 Oxidation and reduction

Table 18.1. Reduction potential data for group VIB.

(a) Acid

O.N.	0	−1	−2	+6	+4	+2	0	−2	+6	+4	0	−2	+6	+4	0	−2
					H_2SO_3											
	O_2	H_2O_2	H_2O	SO_4^{2-}	SO_2	$S_2O_3^{2-}$	S	H_2S	SeO_4^{2-}	H_2SeO_3	Se	H_2Se	H_6TeO_6	TeO_2	Te	H_2Te
		+0.68	+1.77		+0.17	+0.40	+0.50	+0.14		+1.15	+0.74	−0.40		+1.02	+0.53	−0.72

(b) Alkali

O.N.	0	−2	+6	+4	+2	0	−2	+6	+4	0	−2	+6	+4	0	−2
	O_2	HO_2^-	OH^-	SO_4^{2-}	SO_3^{2-}	$S_2O_3^{2-}$	S	S^{2-}	SeO_4^{2-}	SeO_3^{2-}	Se	Se^{2-}	TeO_4^{2-}	TeO_3^{2-}	Te	Te^{2-}
		−0.08	+0.87		−0.98	−0.58	−0.74	−0.51		+0.05	−0.37	−0.92		+0.4	−0.57	−1.14
		+0.4														

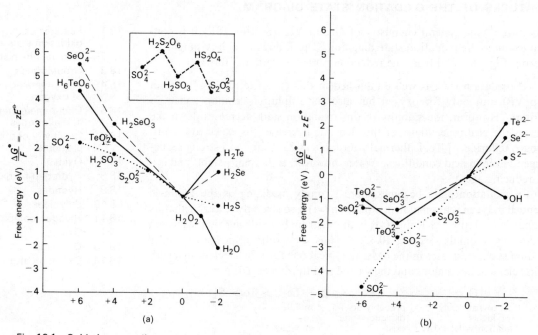

Fig. 18.1 Oxidation state diagrams for group VIB in (a) acid, and (b) base.

to sulphur dioxide and sulphate by stronger oxidants. Moreover, it is unstable with respect to disproportionation in acid solution, but is more stable in alkali. Hydrogen sulphide is a mild reducing agent in acid solution, but more powerful in alkali (as the sulphide ion). All the higher oxidation states of sulphur are poor oxidants in alkali; in fact, all except sulphate are strong reductants.

6. Selenium(VI) is a powerful oxidant in acid solution. Even in alkaline solution selenium(VI) and tellurium(VI) are oxidizing, and consequently selenium(IV) and tellurium (IV) are very stable.

18.2 GENERAL CHEMISTRY

The most important trend in the group is the general increase in electropositivity on descending the group from oxygen to polonium: oxygen is a reactive electronegative non-metal; sulphur is a less electronegative non-metal; selenium and tellurium are semi-metals; and polonium is a weakly electropositive metal, rather like lead, but is rare and radioactive. This classification is demonstrated by application of any of the usual criteria: structure, physical properties, and atomic properties (Tables 10.9 and 10.10); the emergence of cationic properties on the descending of the group; the increasing basic and ionic character of oxides (tellurium dioxide has the rutile structure and reacts with acids to yield $TeCl_4$); the increasing tendency to form complexes, $SeBr_6^{2-}$, $TeBr_6^{2-}$, etc.; and the decreasing stability and increasing acidity of the hydrides. The greatest change in properties occurs between sulphur and oxygen (first row anomaly) (Table 18.2). Selenium is more electronegative than "expected" (Section 10.3.2).

The often quoted resemblances between this group and group VIA (Cr, Mo, W) are restricted to stoichiometric resemblances to compounds in similar oxidation states, especially the state +6, and any inevitable consequences: the oxidizing nature of high oxidation states; and the similar structures (often isomorphous) and solubilities of selected compounds (chromates and sulphates).

Oxygen is a first row element and usually obeys the octet rule. It can form stable compounds in which it achieves the noble gas configuration in several ways (cf. nitrogen):

1. Electron gain to form an oxide ion O^{2-} (solid state only). The remaining elements (except polonium) can also form ions, M^{2-}, but only with the most electropositive elements.
2. Formation of two single covalent bonds, e.g. H—O—H. The other elements also form such compounds, but with decreasing stability down the group.
3. Formation of one double bond, e.g. O=C=O (also S=C=S).
4. Gain of one electron and formation of one single bond, e.g. OH^-. The analogues of the other elements, $S-H^-$ etc., decrease in stability (with respect to hydrolysis for example) as the group is descended.
5. Formation of three bonds, in which the compound acts as a Lewis base, e.g. $H_2O \rightarrow H_3O^+$. Again, the stability of such compounds decreases down the group.
6. Rarely, the oxide ion can form four-co-ordinate bonds to a cation, for example in the basic acetates of beryllium and zinc: $OM_4(O \cdot CO \cdot CH_3)_6$ (M = Be, Zn). In these compounds the oxide ion is at the centre of a tetrahedron of metal ions. The metal ions are joined to each other along the six sides of the tetrahedron by the bridging acetate ligand,

In all of its compounds, except those with fluorine and itself, oxygen is in the oxidation state −2 (by definition, since it is the second most electronegative element). The other elements also form compounds, as above, in the oxidation state −2. However, these compounds are generally covalent. Sulphur can often replace oxygen directly in compounds, for example sulphates, SO_4^{2-}, to form thiosulphates, $S_2O_3^{2-}$. But, in addition, sulphur and the other elements form compounds with other oxidation numbers, especially $+4$ and $+6$. Also, since the heavier elements in the group are not restricted to an outer octet of electrons, a wider variety of co-ordination numbers (two to eight) and shapes are found in the chemistry of sulphur, etc., than in that of oxygen. The application of the Gillespie–Nyholm rules usually gives the correct structure of covalent molecules in this group (Problem 7).

18.3 GROUP TRENDS

Table 18.2 illustrates the trends in physical and atomic properties, which are much as expected. The trends are very similar to those in group V, and are listed below and discussed elsewhere.

1. There is an overall increase in metallicity down the group (Section 11.4).
2. The first element (Section 14.2.7) and the middle row element (Sections 10.3.2 and 15.1) are anomalous.
3. The stability of the -2 oxidation state decreases down the group (Section 15.1).
4. The $+6$ oxidation state decreases in stability down the group, and the stability of the $+4$ state increases correspondingly (Section 17.3.4).
5. The tendency to form compounds containing larger co-ordination numbers increases down the group, e.g. TeF_8^{2-}.
6. The tendency to form condensed acids decreases sharply from sulphur to selenium.
7. Catenation decreases sharply from sulphur to selenium.
8. The acidity of the oxides decreases down the group (Section 11.1.1).
9. The hydrides show decreasing thermal stability and increasing acidity as the group is descended (Section 11.1.3).

18.4 THE ANOMALOUS NATURE OF OXYGEN

As usual, the first member of the group is anomalous. This is very obvious in this group where oxygen differs from sulphur dramatically, and the remaining three elements form a well-graded series (stated in this way since selenium is often anomalous if the series from sulphur downwards is considered, Section 10.3.2).

The anomalous nature of oxygen is, as usual, the result of the following factors:

1. Oxygen is smaller and more electronegative than sulphur and the other elements. Thus, the bonding in stoichiometrically similar compounds is much more ionic in the oxygen compounds than in the others, and the properties and relative stabilities of such compounds often differ considerably. An extreme example is the presence of hydrogen bonding in water, which is almost completely absent in hydrogen sulphide, causing the properties of these compounds to differ considerably (Section 13.6).
2. Oxygen is restricted to eight electrons in its valence electron shell; sulphur is not. This restricts the co-ordination number of oxygen to four, and in practice

Table 18.2. Group VIB properties.

Property	Oxygen O	Sulphur S	Selenium Se	Tellurium Te	Polonium Po
Atomic number	8	16	34	52	84
Electron configuration (outer)	$2s^22p^4$	$3s^23p^4$	$3d^{10}4s^24p^4$	$4d^{10}5s^25p^4$	$4f^{14}5d^{10}6s^26p^4$
Isotopes (in order of abundance)	16, 18, 17	32, 34, 33, 36	80, 78, 82, 76, 77, 74	130, 128, 126, 125, 124, 122, 123, 120	
Atomic weight	15.9994	32.064	78.96	127.6	(210)
Metallic radius (Å)			1.4	1.6	1.76
Ionic radius, M^{2-} (Å)	1.40	1.84	1.98	2.21	
Covalent radius (Å) double bond	0.62	0.94	1.07	1.27	
single bond	0.66	1.04	1.17	1.37	1.46
Van der Waals radius (Å)	1.4	1.9	2.0	2.2	
Atomic volume (cm^3 mol^{-1})	14.0	15.4	16.5	20.4	22.4
Boiling point (°C)	−183	445	685	990	962
Melting point (°C)	−218	119^m	217^g	450	254
Enthalpy of fusion (kJ mol^{-1})	0.22	1.42	5.3	17.7	11
Enthalpy of vaporization (kJ mol^{-1})	3.39	10.5	27.6	50	103
Enthalpy of atomization (kJ mol^{-1})	250	280	207	197	145
Density (g cm^{-3})	1.14	2.07^{rh}	4.79^g	6.25	9.4
Electronegativity (A/R)	3.5	2.45	2.5	2.0	1.75
Ionization energy, 1st (kJ mol^{-1})	1314	1000	941	869	
Electron affinity (kJ mol^{-1}): M^{2-}	−700	−330	−406		
M^-	141	196			
Electrode potential, M/M^{2-} (alkali) (V)		−0.51	−0.92	−1.14	
Common co-ordination numbers	1, 2, (3), (4)	2, 4, 6	2, 4, 6	6, (7), (8)	
Common oxidation numbers	−2, (−1)	−2, 2, 4, 6	−2, 2, 4, 6	2, 4, 6	2, 4
Abundance, p.p.m.	466,000	520	0.09	0.002	

^{m} monoclinic; ^{rh} rhombic; ^{g} grey.

it rarely exceeds two, whereas the co-ordination numbers of sulphur, etc., can exceed four. Therefore, sulphur compounds exist which have no oxygen analogues. Nevertheless, all the elements have similar electron configurations and similar compounds do exist in the oxidation state −2. The expansion of the octet for sulphur, etc., leads to the formation of compounds containing higher oxidation numbers than two for these elements. Thus, whereas oxygen shows mainly the oxidation number −2, and the co-ordination number two in its compounds, sulphur and the other elements show higher co-ordination and oxidation numbers, including positive oxidation numbers, because of their lower electronegativity. Oxygen forms weak single bonds to itself, carbon, nitrogen, and fluorine (Table A.11).

3. The bond energy of the oxygen–oxygen double bond, O=O, is more than three times that of the oxygen–oxygen single bond, which is characteristically low (Section 15.1). By comparison, the sulphur–sulphur double bond, S=S, is less than twice as strong as the sulphur–sulphur single bond. This results in catenated O—O—O chains being unstable relative to oxygen, O_2(g); but catenated S—S—S chains being stable relative to the molecule S=S. Thus, at normal temperatures, O_2(g) and S_8(s) (Fig. 18.2) are the normal forms of the elements (see Problem 9). Moreover, there are many catenated sulphur compounds for which no oxygen analogues exist (except H—O—O—H). However, these are not very stable to sulphur abstraction from the chain. For example,

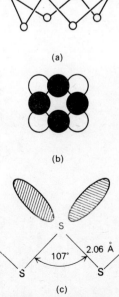

Fig. 18.2 Rhombic sulphur.

consider the chain hydrogen polysulphides H_2S_n ($n = 1$ to 8), and the dissociation reaction below.

$$H_2S_n \rightarrow H_2S + \left(\frac{n-1}{8}\right) S_8.$$

Both the reactants and products contain the same number of H—S and S—S bonds (draw out and see). The enthalpy of reaction is therefore close to zero (actually slightly endothermic) and the free energy of reactions is therefore negative (Section 6.12); thus the compound is unstable, especially on heating. The same is true for the polysulphides of the electropositive metal ions, S_n^{2-} ($n = 1$ to 5 in aqueous solution), made by adding sulphur to sulphides, and an equilibrium containing all species is present in solution.

Oxygen is much more abundant than sulphur. Both elements form allotropes (Section 18.5). Oxygen is slightly more reactive than sulphur with other elements (Section 18.6) but less reactive than the halogens are. From the energy required to form the ion O^{2-}(g) (Table 18.2), it looks at first as if oxygen would not form ionic oxides (cf. the electron affinities of the halogens):

$$\tfrac{1}{2}O_2(g) \rightarrow O(g) \rightarrow O^{2-}(g); \quad \Delta H = +700 \text{ kJ mol}^{-1}.$$

But the very high lattice energy of the small doubly charged anion more than makes up for this, and even the oxides of fairly weakly electropositive metals have a large degree of ionic character, and are more ionic than the corresponding sulphides, etc. Oxygen, like fluorine (Section 19.1), brings out the highest oxidation number of elements in covalent compounds as oxides or oxyanions, such as permanganate for example. Clearly, with the element in a highly oxidizing state, manganese (+7) for example, only highly electronegative anions like F^- and O^{2-} are capable of coexistence (Section 14.1.2). The compounds of oxygen and sulphur show many differences: the acidity, hydrogen bonding, reducing properties, Lewis base behaviour, and bond angles in the hydrides; the lesser ionic character and lower stability of sulphides compared to oxides; and the dissimilarity of the halides (Section 18.11).

18.5 OCCURRENCE, PREPARATION, AND ALLOTROPES

Oxygen is the most widely distributed and abundant (50 percent) element in the earth's crust, and is present in air (20 percent by volume), water (80 percent), silicates, oxides, and oxyanions. Oxygen has three isotopes, and can exist as two allotropes: oxygen, O_2, a colourless gas; and ozone, O_3, a pale-blue gas. Oxygen has a bond order of two (from the bond length and bond enthalpy) and shows paramagnetism equivalent to two unpaired electrons. Thus the electronic structure cannot be written O=O, since although this gives the correct bond order, it has no unpaired electrons. The best valence-bond structure that can be written is $\ce{:O::O:}$, which contains one two-electron bond, and two three-electron bonds each containing one pair of electrons and one unpaired electron. Since a three-electron bond has the same bond strength as a one-electron bond, or half that of a two-electron bond, this representation is correct from the point of view of the bond order and the paramagnetism. The oxygen molecule is atypical for a molecule containing unpaired electrons since it is scarcely coloured and shows little tendency to polymerize.

Oxygen is prepared industrially by the liquefaction of air and subsequent fractional distillation. In the laboratory it can be prepared by the action of heat on: the nitrates of the group IA elements; higher oxides, such as lead dioxide; potassium chlorate in the presence of manganese dioxide catalyst; potassium permanganate; and mercury(II) oxide; or from sodium peroxide.

Oxygen forms a pale blue liquid and solid, and at 25 °C is a colourless, odourless gas, slightly more dense than air, and slightly soluble in water (3 percent by volume at 20 °C). It is more soluble in organic solvents. Certain transition metal complexes form weak complexes with oxygen, and act as oxygen carriers, for example haemoglobin (Section 20.17).

Ozone (m.p. -250; b.p. -112; $\Delta H_f^\ominus = +142$ kJ mol^{-1}) can be prepared by passing a slow stream of oxygen through an electric discharge, condensing the product (up to 10 percent ozone), and purifying by fractional distillation. It is an endothermic compound and its decomposition to oxygen is easily catalysed. Pure ozone is explosive. It forms a pale-blue diamagnetic gas, a deep-blue liquid, and a black-violet solid. It occurs in the upper atmosphere owing to the action of ultraviolet light on oxygen. In the absence of such light and catalysts, its decomposition to oxygen is slow up to about 250 °C (high activation energy, i.e. metastable). Since it is endothermic, the equilibrium mixture at high temperatures contains more than that at 25 °C, at which temperature the equilibrium mixture contains practically no ozone. But at higher temperatures the rate of attainment of equilibrium is quicker. Therefore ozone is prepared at a high temperature, and cooled to room temperature where the equilibrium mixture appropriate to the higher temperature is stabilized (kinetic control). The effect of increasing the temperature of such a mixture is to decompose ozone, despite the fact that it is more stable at higher temperatures! The O—O bond length in ozone is 1.28 Å, intermediate between an O—O bond (1.49 Å) and a double O=O bond (1.21 Å). Draw the valence-bond structure for ozone. Is the molecule linear or bent, and what is the bond angle approximately (see Problem 7)? Ozone is a powerful oxidant (see below). It attacks double bonds to form ozonides (Section 22.2.4), and therefore deteriorates rubber (Section 22.2.4). The oxidizing powers of oxygen and ozone are contrasted in Table 18.3: ozone is a more powerful oxidant under all conditions. Few oxidants are more effective in acid solution. Oxygen itself should be a good oxidant, especially in acid solution. But its reactions are often slow and are under kinetic control. For example, the oxidation of iron(II) to iron(III) by oxygen is quicker in alkaline solution! Oxygen usually requires elevated temperatures in its reactions, whereas ozone is frequently effective at 25° C, for example the "tailing" of mercury owing to oxide formation. Ozone is analysed by oxidation of iodide to iodine. Some other typical oxidations are: iron(II) to iron(III); sulphide to suphate; metals to oxides; chlorine dioxide to the hexoxide; and peroxides to oxygen.

Oxygen is used in oxyacetylene and hydrogen flames for cutting and welding, in steel manufacture, and as a rocket fuel. Ozone is used for bleaching paper and wax, and sterilizing water.

Sulphur is fairly abundant and occurs: as the element; in the gases sulphur dioxide and hydrogen sulphide; in sulphide ores such as pyrites, FeS_2, and galena, PbS; and as sulphates such as anhydrite, $CaSO_4$, and gypsum, $CaSO_4 \cdot 2H_2O$. Sulphur is frequently present in organic matter (body tissue, hair, onions, mustard). Some body sulphur can be replaced by arsenic structurally but not functionally; therefore arsenic is poisonous. Elemental sulphur cannot be mined conventionally

Table 18.3. The relative oxidizing powers of oxygen and ozone at various pH values.

Oxygen	$E^\ominus$	pH	Ozone	$E^\ominus$
$O_2(g) + 4H^+(aq) + 4e^- \rightarrow 2H_2O(l)$	+1.23	0	$O_3(g) + 2H^+(aq) + 2e^- \rightarrow O_2(g) + H_2O(l)$	+2.07
$O_2(g) + 4H^+(aq) + 4e^- \rightarrow 2H_2O(l)$	+0.85	7	$O_3(g) + 2H^+(aq) + 2e^- \rightarrow O_2(g) + H_2O(l)$	+1.65
$O_2(g) + 2H_2O(aq) + 4e^- \rightarrow 4OH^-(aq)$	+0.4	14	$O_3(g) + H_2O(l) + 2e^- \rightarrow O_2(g) + 2OH^-(aq)$	+1.24

Table 18.4. Properties of various oxygen species.

Compound	O_2^+	O_2	O_2^- (K)	O_2^{2-} (Ba)	O_3	O_3^-	HO—OH
Bond length (Å)	1.12	1.21	1.28	1.49	1.28	1.2	1.49
Bond energy (kJ mol^{-1})	628	498			301		142
Bond order		2		1	$1\frac{1}{2}$		1
Bond angle					117°		97° (O—O—H)
Magnetism	P(1)	P(2)	P(1)	D(0)	D(0)	P(1)	D(0)
Colours	red	blue	yellow	colourless	blue	red	colourless

P—paramagnetic; D—diamagnetic. Figures in brackets give number of unpaired electrons.

because of quicksands found above the deposits, and is obtained by the Frasch process ("tapped" like oil). Three concentric pipes, about six feet in diameter, are sunk into the bed of sulphur. Superheated steam is forced down the outer ring, and this melts the sulphur which is forced up the next-to-outer ring by compressed air applied down the central pipe. The sulphur is collected in large vats in which it solidifies, and is more than 99 percent pure. Selenium and tellurium occur in sulphide ores as impurities. They are separated from the flue dust obtained in roasting pyrites, or the anode slime obtained in purifying copper. All the allotropes of polonium are short-lived, highly radioactive, products obtained from the decay of uranium and thorium minerals.

Sulphur is allotropic and shows a wide variety of different forms in each phase. Sulphur can exist in at least six allotropic forms in the solid state (polymorphs). Orthorhombic (usually called rhombic) sulphur is the stable form up to 96 °C, and consists of eight sulphur atoms covalently bonded together in a puckered ring (Fig. 18.2), in which, as expected, there is an approximately tetrahedral arrangement of electron pairs about each sulphur atom. The intermolecular forces between the S_8 molecules in the solid state are the weak van der Waals forces, therefore the melting point is low (113 °C if heated quickly to avoid formation of monoclinic sulphur). As expected, rhombic sulphur is insoluble in water but soluble in organic solvents, and in fact is prepared by dissolving any form of sulphur in carbon disulphide and allowing evaporation to take place below 95.5 °C, when squat crystals of rhombic sulphur are obtained. Monoclinic sulphur is stable between 95.5 and 119 °C (the melting point). It also is composed of S_8 units, but they are arranged differently in the lattice from those in monoclinic sulphur (see Problem 11). Monoclinic sulphur is prepared by crystallizing the element above 95.5 °C: either by melting roll sulphur, allowing it to cool, piercing the crystalline crust with a needle, and pouring out the remaining molten sulphur; or by crystallizing from hot toluene (Problem 12). Monoclinic and rhombic sulphur are examples of what type of allotropy? Purple sulphur is obtained by quickly cooling sulphur vapour at 500 °C and 0.1 mm pressure in liquid nitrogen. A solid, S_2, is obtained which is purple and paramagnetic (cf. O_2). On warming this solid, a mixture of monoclinic and amorphous sulphur is obtained. Engel's sulphur, S_6, is obtained by treating concentrated hydrochloric acid with a thiosulphate (Section 18.13.5) at 0 °C, extracting with toluene, and crystallizing. Within hours it changes to a mixture of rhombic and plastic sulphur. Can you guess the shape of the S_6 molecule (Problem 7)? An unstable S_{10} form of sulphur can be obtained by the reaction of S_4Cl_2 on the acid H_2S_6. Plastic sulphur is an amorphous fibrous form. It is a supercooled liquid consisting of long sulphur chains, not arranged regularly. It is obtained by pouring molten roll sulphur into cold water to cool it quickly, which has the result that the long chains present in the liquid (see below) are given no chance to change to a more stable form. The colour of plastic sulphur depends on the temperature of the molten sulphur, and varies from transparent amber to brown. It is like putty or heated glass, it is insoluble in all solvents, and slowly changes to rhombic sulphur.

The melting point of sulphur depends upon the rate at which the rhombic sulphur is heated. At the melting point it is a transparent amber mobile liquid (S_8 rings); at about 160 °C it darkens and the viscosity suddenly increases (an equilibrium between S_8 rings and long chains is present); at about 190 °C it is still darkening and maximum viscosity is obtained (very long chains) and the liquid will not pour (plastic sulphur can now be obtained on rapid cooling); at 200 °C the liquid is now black (the colour is associated with unpaired electrons at the ends of the sulphur chains), but the viscosity starts to fall slowly (a temperature effect); between 200 and 444.6 °C (the boiling point) the mobility of the liquid increases more quickly owing to the formation of S_8 molecules; and at 444.6 °C the liquid boils and a vapour S_8 is obtained which can be condensed to give a pale yellow powder called flowers of sulphur. This behaviour changes dramatically in the presence of impurities such as iodine, which stabilize the chains by forming terminal S—I bonds.

Sulphur vapour provides an example of dynamic allotropy (Section 10.3.1):

$$S_8(g) \rightleftharpoons 4S_2(g) \rightleftharpoons 8S(g)$$

| 450 ° | 1000 °C | 2000 °C |
| bromine colour | dark red | yellow |

Selenium and tellurium both have several allotropes, which have been studied less than those of sulphur. Rhombic and monoclinic modifications of selenium exist which probably contain Se_8 molecules. These are dark red solids obtained by evaporating solutions of the element in carbon disulphide. Both are thermally unstable relative to a grey semi-metallic form, which is prepared either by heating one of the red allotropes, or by slowly cooling liquid selenium. This form consists of very long chains of covalently bonded selenium atoms spiralling round an axis parallel to the crystal axis. There is a metallic type of interaction between adjacent chains. The conductivity of this grey allotrope is small but is increased five hundred times when light is shone on it. This property is called photoconductivity, and selenium is used in photoelectric cells. This allotrope, unlike the molecular forms, Se_8, is insoluble in all solvents unless it reacts. Several amorphous allotropes of selenium are also known. Tellurium also has a semi-metallic allotrope with the same structure as grey selenium (see Problem 13), but it also has a metallic allotrope, as

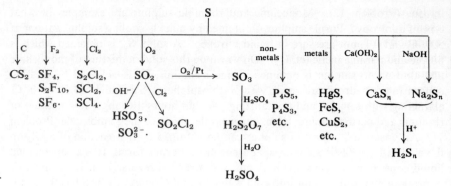

Fig. 18.3 The properties of sulphur.

well as several amorphous forms. In the vapour phase both selenium and tellurium have a number of species in equilibrium (dynamic allotropy), but the molecules are smaller than those of sulphur under the same conditions (less tendency to catenate). Polonium has two metallic allotropes (increasing metallic character).

Sulphur is used in the manufacture of sulphuric acid, gunpowder, matches, and many drugs, and in the vulcanization of rubber. Selenium is used in photocells, vulcanization, rectifiers, and in the manufacture of red glass and some alloys.

18.6 REACTIVITY OF THE ELEMENTS

The general reactivity of the elements decreases in the order $O > S > Se > Te$; both oxygen and sulphur are fairly reactive. Oxygen is slightly less reactive than the halogens, but reacts with all elements (usually at elevated temperatures) except the noble gases, the halogens, and a few noble metals; and it forms compounds with all elements except the lower noble gases, helium, argon, and neon. Selenium and tellurium also react with most elements on heating to form selenides and tellurides. The elements sulphur, selenium, and tellurium do not react with water or dilute acids (except for polonium), but are oxidized by concentrated nitric acid to yield H_2SO_4, H_2SeO_3, and H_2TeO_3 (note the decreasing stability of the higher oxidation state +6). Oxygen does not react with alkali, but the remaining elements disproportionate to M^{2-} and MO_3^{2-} (Fig. 18.1). In the case of sulphur these anions react further in the presence of an excess of sulphur to yield the ions S_n^{2-} and $S_2O_3^{2-}$.

18.7 OXIDES

The oxides of the elements can be classified into: normal oxides (Section 11.1.1); peroxides; superoxides; ozonides (below); suboxides, which contain element–element bonds and are therefore deficient in the amount of oxygen expected from the normal oxidation numbers of the elements (e.g. C_3O_2, $O=C=C=C=O$, carbon suboxide); dioxides, which are the higher oxides of metals and are therefore usually oxidizing (e.g. MnO_2 and PbO_2); and neutral oxides (e.g. N_2O and CO).

Peroxides contain the diamagnetic peroxide ion, O_2^{2-}, and are formed only by the group IA and heavier group IIA metals (Section 14.2.5). In acid or water they yield hydrogen peroxide. They are powerful oxidants: organic material often yields carbonate on mild heating; and many metals yield oxyanions—iron, for example, gives the ferrate ion, FeO_4^{2-}. Since the oxygen is in an intermediate oxidation state, peroxides can also be oxidized by very powerful oxidants (cf. H_2O_2) that is, they act as reductants. Superoxides are formed by the most electropositive metals (Section 14.2.5) and contain the paramagnetic (one unpaired electron) superoxide ion, O_2^-. They are powerful oxidants. Their reaction with carbon dioxide is used to remove carbon dioxide and regenerate oxygen in closed systems:

$$4MO_2(s) + 2CO_2(g) \rightarrow 2M_2CO_3(s) + 3O_2(g); \quad M = Na, \text{ etc.}$$

Ozonides are formed by the most electropositive elements (Rb, Cs, K, and the ammonium ion):

$$2O_3(g) + 2MOH(s) \rightarrow 2MO_3(s) + H_2O(g) + \tfrac{1}{2}O_2(g).$$

They are paramagnetic, coloured, and unstable, slowly decomposing, even at room temperature, to superoxides and oxygen. What shape is the O_3^- ion (Problem 7)?

18.8 HYDROGEN PEROXIDE

Peroxy compounds contain either the —O—O— group or the O_2^{2-} ion (cf. perchloric acid, etc., which do not). Hydrogen peroxide has the open-book structure shown in Fig. 18.4. It is manufactured by the electrolysis of aqueous sulphuric acid to peroxydisulphuric acid, which on distillation hydrolyses to hydrogen peroxide which is removed as it is formed. The dilute solution (about 30 percent) obtained can be fractionally distilled (90–99 percent) and fractionally crystallized at low temperature to yield pure hydrogen peroxide.

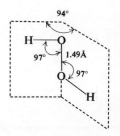

Fig. 18.4 The structure of hydrogen peroxide.

$$\begin{aligned}
\text{Cathode:} &\quad 2H^+(aq) + e^- \rightarrow H_2(g); \\
\text{Anode:} &\quad 2HSO_4^-(aq) \rightarrow 2e^- + 2H^+(aq) + S_2O_8^{2-}(aq); \\
&\quad S_2O_8^{2-}(aq) + 2H_2O \rightarrow H_2O_2 + 2HSO_4^-(aq).
\end{aligned}$$

Hydrogen peroxide (m.p. $-0.89\ °C$; b.p. $150\ °C$; $\Delta H_{vap} = 51.6\ kJ\ mol^{-1}$ —cf. $H_2O = 43.9$; $\Delta H_f^\ominus = -188\ kJ\ mol^{-1}$) is a pale-blue viscous liquid which freezes to a white crystalline solid. This has a three-dimensional hydrogen-bonded structure which persists into the liquid phase and which is therefore viscous. In fact, it is more highly associated than water, with a dielectric constant of 90 (that of water is 80). It is useless as a solvent, however, since it is both highly oxidizing and disproportionates easily. It is miscible in all proportions with water, and, being less volatile than water, its aqueous solutions can be concentrated: even simple evaporation in an open dish concentrates it to about 60 percent, but beyond this it decomposes. It is unstable with respect to disproportionation (Fig. 18.1) both in acid and alkali, but is kinetically stabilized, especially in the presence of either acids (the opposite effect to that expected thermodynamically) or complexing anions, such as phosphate, which remove traces of transition metal ion catalysts from solution. Thus, if kept in a dark smooth glass bottle in the absence of heavy metal ions, metals, dust, and so on, pure hydrogen peroxide keeps for weeks; but any sudden changes (temperature, percussion) may cause it to explode:

$$H_2O_2(l) \rightarrow H_2O(l) + \tfrac{1}{2}O_2(g); \quad \Delta H = -98.7\ kJ.$$

Since the oxygen is in an intermediate oxidation state, hydrogen peroxide can act as oxidant or reductant, in either acid (H_2O_2) or alkali (HO_2^-) solution. Thermodynamically, it ought to be a better oxidant in acid than in alkali:

	acid	$E^{\ominus}$(V)	alkali	$E^{\ominus}$(V)
As oxidant:	$H_2O_2(l) + 2H^+(aq) + 2e^- \to 2H_2O$	+1.77;	$HO_2^-(aq) + H_2O(l) + 2e^- \to 3OH^-(aq)$	+0.87
As reductant:	$O_2(g) + 2H^+(aq) + 2e^- \to H_2O_2$	+0.68;	$O_2(g) + H_2O(l) + 2e^- \to HO_2^-(aq) + OH^-(aq)$	−0.08

In fact, acid oxidations tend to be slow (like the disproportionation)—a kinetic effect—unless catalysts are used. Despite this, acid solutions are used, where possible, to avoid the concurrent disproportionation in the presence of base. Conversely, hydrogen peroxide is a better reductant in the presence of base than in acid, and sometimes reactions are reversed on changing the pH. A few examples of oxidizing reactions are: the restoration of oil paintings where the white lead (basic lead carbonate) used has absorbed hydrogen sulphide and turned to black lead sulphide, which is oxidized by peroxide to white lead sulphate; iodide to iodine (quantitative in acid); sulphite and thiosulphate to sulphate; and the oxidation of organic materials (e.g. dark hair to light, and the destruction of coagulated blood resulting in cleaning of wounds). Oxygen from the hydrogen peroxide is always produced when it acts as a reductant. A few examples are: ammoniacal silver nitrate to silver; acid permanganate to manganese(II) (quantitative, used for peroxide estimation); and chlorine to chloride. It is more acidic than water, and forms a series of salts—the peroxides:

$$H_2O_2(aq) + H_2O(l) \rightleftharpoons H_3O^+(aq) + HO_2^-(aq); \qquad K = 1.5 \times 10^{-12}.$$

Tests for hydrogen peroxide include: the formation of the blue perchromic acid, CrO_5, which is stable in ether and formed by the action of peroxide on acid dichromate; and the decoloration of lead sulphide on filter paper. Pure hydrogen peroxide is used as a rocket fuel, and in aqueous solution it is used as a bleach for wool, straw, hair, and wood pulp, as an antiseptic, and in the manufacture of porous concrete and foam rubber (leavened by the oxygen from the decomposition).

18.9 HYDRIDES

(Compare water, Section 13.6.) These are evil-smelling, colourless, poisonous gases. Hydrogen sulphide is as poisonous as hydrogen cyanide, but is so strong smelling that it is rarely dangerous. The bad-egg smell of hydrogen sulphide is actually due to impurities, and the pure gas has a sweetish, though still offensive, odour. The gases are prepared by the action of acids on metal chalconides such as FeS, Al_2Se_3, and Al_2Te_3 (Section 11.1.5), or a magnesium–polonium alloy (cf. Section 11.1.3).

The volatility of the hydrides changes in the order $H_2O \ll H_2S > H_2Se > H_2Te$ (see Problem 14). The thermal stability of the hydrides with respect to dissociation to the elements decreases (see Problem 15) in the order $H_2O \gg H_2S$ (stable up to 280 °C) > H_2Se > $H_2Te \gg H_2Po$ (unstable at 0 °C).

$$H_2S(aq) + H_2O(l) \rightleftharpoons H_3O^+(aq) + HS^-(aq); \qquad K_1 = 10^{-7}.$$
$$HS^-(aq) + H_2O(l) \rightleftharpoons H_3O^+(aq) + S^{2-}(aq); \qquad K_2 = 10^{-14}.$$
$$K_1 = 1.3 \times 10^{-4}(Se); K_1 = 3.7 \times 10^{-3}(Te).$$

All are weak acids, the acidity increasing as the group is descended. Since at 25 °C the concentration of a saturated aqueous solution of hydrogen sulphide is only 10^{-1}M, the concentration of sulphide ions in an aqueous solution is very small. Thus only very insoluble metal sulphides can be precipitated from aqueous solution by hydrogen sulphide, especially in acid solution, and this fact is used in schemes of qualitative analysis (see below) to recognize and separate such metal ions from those forming more soluble sulphides. In alkaline solution, relatively more soluble (but still "conventionally insoluble") metal sulphides can be precipitated (Section 18.10). The reducing power of the hydrides increases down the group. This is illustrated by the standard electrode potential data $E^\ominus$ for the couples S/H_2S etc., in acid, and S/S^{2-}, etc., in base (Table 18.1):

Acid: O_2/H_2O, $+1.23$ V; S/H_2S, $+0.14$ V; Se/H_2Se, -0.4 V; Te/H_2Te, -0.72 V.

Base: O_2/OH^-, $+0.4$ V; S/S^{2-}, -0.48 V; Se/Se^{2-}, -0.92 V; Te/Te^{2-}, -1.14 V.

or more simply in Fig. 18.1. Thus the oxidation of water to oxygen is very difficult and requires a strong oxidizing agent. In other words, water is a poor reductant.

$$H^+ + e^- \rightleftharpoons \tfrac{1}{2}H_2; \qquad E^\ominus = 0 \quad \text{(acid)}.$$
$$H_2O + e^- \rightleftharpoons \tfrac{1}{2}H_2 + OH^-; \qquad E^\ominus = -0.8 \quad \text{(base)}.$$

The remainder are successively stronger reductants, and in fact the hydrides of selenium and tellurium "should" reduce water to hydrogen (thermodynamic argument) but the reaction is very slow (kinetic argument), and their aqueous solutions are stable. Hydrogen sulphide reduces even mild oxidants, such as iron(III) solutions, and is itself usually oxidized to sulphur. Some of its reactions are: $Fe^{3+}(aq) \to Fe^{2+}(aq)$; $MnO_4^-(aq) \to Mn^{2+}(aq)$; $H_2O_2(aq) \to H_2O$; $SO_2 \to S(s)$; $HNO_3 \to NO_2(g)$; $Cl_2(aq) \to Cl^-(aq)$; $Cr_2O_7^{2-}(aq) \to Cr^{3+}(aq)$; $H_2SO_4 \to SO_2(g)$; and $O_2(aq) \to H_2O$ (slowly). All the hydrides burn in air with a blue flame to yield either the dioxide, MO_2, or the element M (M = S, Se, Te), depending upon the amount of oxygen available. The bond angles change regularly from H_2O to H_2Te (Fig. 13.5).

Sulphanes (Section 18.4.) (or hydrogen polysulphides) are compounds of general formula H_2S_n (n, from 2 to 6) obtained when concentrated aqueous solutions of metal polysulphides are carefully treated with acids. Polysulphides are compounds of sulphur and another element which contain more sulphur atoms than would be expected from a consideration of the normal valencies of the other element and bivalent sulphur. The ions of the more electropositive metals (Group IA except lithium, and the heavier Group IIA metals) form polysulphides when sulphur is added to concentrated aqueous or liquid ammonia solutions of the metal sulphides. They contain chains of sulphur atoms S_n^{2-}.

18.10 SULPHIDES

(See Section 11.1.5.) Most metals and many non-metals react with sulphur directly to form binary sulphides, usually on heating but often very readily, as in the case of mercury. A variety of compounds is obtained, often depending upon the conditions since many elements form more than one compound. Alternatively, insoluble metal sulphides can be prepared by addition of hydrogen sulphide to a solution of metal ions; and the group IA sulphides and hydrosulphides (HS⁻) can be prepared by addition of hydrogen sulphide to alkali:

$$\text{excess alkali:} \quad 2NaOH + H_2S \rightarrow Na_2S + 2H_2O;$$
$$\text{excess sulphide:} \quad NaOH + H_2S \rightarrow NaHS + H_2O.$$

All sulphides are insoluble except those of groups IA and IIA, and the sulphides of aluminium, chromium, and the rare earths. All the soluble sulphides are hydrolysed in water,

$$S^{2-}(aq) + 2H_2O(l) \rightleftharpoons 2OH^-(aq) + H_2S(s),$$

the ease of hydrolysis increasing in the order given above: the group IA sulphides only slightly (but they do smell of hydrogen sulphide); the heavier group IIA sulphides more so; and those of magnesium, chromium, aluminium, and the rare earths are almost completely hydrolysed. Only the group IA and, to a lesser extent, IIA metals form hydrosulphides and polysulphides (Section 11.2), such as NaHS, CsS_6, and BaS_4.

Many sulphides are highly coloured. When insoluble sulphides are precipitated they form powders, but when fused and crystallized they often have a metallic appearance (Section 11.1.5). On heating in air, most sulphides are converted to the sulphate, oxide, or the metal, depending on the electropositivity of the metal (Section 11.2); some examples are $BaS \rightarrow BaSO_4$, $CuS \rightarrow CuO$, $HgS \rightarrow Hg$.

If there is any reaction with non-oxidizing acids, hydrogen sulphide is evolved. But some sulphides are so insoluble that, far from reacting with acids, the sulphides can be precipitated by hydrogen sulphide from an aqueous acid solution of the metal ions. Consider the solution of a metal sulphide, MS, in an aqueous acid solution:

$$MS(s) \rightleftharpoons M^{2+}(aq) + S^{2-}(aq); \tag{18.1}$$

$$S^{2-}(aq) + 2H^+(aq) \rightleftharpoons H_2S(g). \tag{18.2}$$

Clearly (by applying Le Chatelier's principle, Section 7.3), as the concentration of hydrogen ions is increased (or as the solution is made more acidic) the sulphide becomes more soluble. But some sulphides are so insoluble that even concentrated hydrochloric acid will not dissolve them. Thus two classes of water-insoluble sulphides exist:

1. Those that are soluble in dilute hydrochloric acid (2M). These include most of the sulphides of the first row of the d-block elements (manganese, iron, cobalt, nickel, and zinc), which have solubility products (see below) between 10^{-20} and 10^{-30}. These sulphides precipitate only in non-acid solution.
2. Those that are insoluble in dilute hydrochloric acid. These include the sulphides of most of the least electropositive metals in the Periodic Table (copper, silver, gold, mercury(II), cadmium, tin, lead, arsenic, antimony, and bismuth), which have solubility products less than 10^{-30}. These are precipitated by hydrogen sulphide in dilute acid. The really insoluble sulphides of copper, bismuth, and mercury, (solubility products about 10^{-50}) will even precipitate in a solution of concentrated hydrochloric acid, but those of antimony, arsenic, and cadmium are soluble in the concentrated acid. To dissolve the sulphides of copper, bismuth, mercury, tin, silver, and gold, oxidizing con-

ditions (e.g. nitric acid) are necessary to remove the sulphide ion from solution and displace the equilibria to the right:

$$CuS(s) \rightleftharpoons CuS(aq) \rightleftharpoons Cu^{2+}(aq) + S^{2-}(aq)$$
$$S^{2-}(aq) \xrightarrow{oxidation} SO_4^{2-}, S, \text{etc.}$$

These phenomena can be discussed by reference to equilibrium constants. From Eq. (18.1), following the discussion in Section 9.1.2, we can write an equilibrium constant $K = [M^{2+}][S^{2-}]/[MS(s)]$; or $K_{sp} = [M^{2+}][S^{2-}]$, since $[MS(s)]$ is constant. This new equilibrium constant, K_{sp}, is called the solubility product of the particular compound. When the product of the concentrations of the ions in solution $[M^{2+}]$ and $[S^{2-}]$ exceeds this value, K_{sp}, then solid MS is precipitated. Consider saturating a solution of various metallic salts of about equal molar concentrations, say all $[M^{2+}] = 1$. To precipitate any sulphide, of solubility product K_{sp}, $1 \times [S^{2-}] = K_{sp}$. In other words, the sulphide ion concentration has to become greater than the solubility product K_{sp} to precipitate the sulphide. Clearly, from Eqs. (18.1) and (18.2), the sulphide ion concentration can be regulated to some extent by the acidity or alkalinity of the solution; therefore sulphides can be precipitated somewhat selectively. The greater the concentration of hydrogen ions, the smaller the concentration of sulphide ions and the more insoluble a sulphide must be to precipitate. The sulphides of zinc, etc., with relatively high solubilities and solubility products, must be precipitated in alkali solutions, where the sulphide ion concentration is increased.

Some insoluble sulphides, such as SnS, SnS_2, Sb_2S_3, and As_2S_3, are soluble in alkali (i.e. amphoteric), forming mixtures of thio ions and oxyanions:

$$3SnS_2(s) + 6OH^-(aq) \rightarrow 2SnS_3^{2-}(aq) + SnO_3^{2-}(aq) + 3H_2O(l).$$

These reactions are used to separate them from the insoluble sulphides which are also insoluble in alkalis (copper, etc.).

There are many general tests for sulphides. Sulphides often give hydrogen sulphide when treated with hydrochloric acid, and coloured polysulphides when treated with sulphur. Aqueous solutions of sulphides yield black lead sulphide when treated with lead acetate, black silver sulphide when treated with neutral silver nitrate, but no precipitate when treated with barium chloride. They yield sulphur when treated with either acid permanganate or acid dichromate; and a violet colour with alkaline nitroprusside (what is the correct name for this ion? Section 20.7):

$$S^{2-}(aq) + [Fe(CN)_5NO]^{2-}(aq) \rightarrow \text{violet } [Fe(CN)_5 \cdot NOS]^{4-}(aq).$$

18.11 HALOGEN COMPOUNDS

Like oxygen (Section 19.12), sulphur and the other group elements form many compounds with the halogens (M_2X_2, MX_2, MX_4, MF_6). There is little similarity between the oxygen compounds and those of the rest of the group (except in the case of the fluorine compounds), however, since sulphur and the other group VI elements are formally in positive oxidation states, and oxygen is in a negative oxidation state in these compounds. Generally, the order of thermal stability of such compounds decreases in the order $F > Cl > Br > I$. The compounds are generally reactive (halogenating agents), they tend to hydrolyse readily, and those of intermediate oxidation number disproportionate. The compounds MF_6 are unexpectedly inert, with respect to hydrolysis for example:

$$SF_6(g) + 4H_2O(l) \rightarrow H_2SO_4(aq) + 6HF(aq); \qquad \Delta G^\ominus = -460 \text{ kJ}.$$

Comment on this (Problem 2).

Oxyhalides and halosulphonic acids exist. The sulphuryl halides have molecular formulae MO_2X_2 (M = S; X = F or Cl); also (M = Se, X = F). Sulphuryl chloride is a fuming hydrolysable liquid. What is its shape (Problem 7)? The thionyl halides, MOX_2 (M = S or Se; X = F, Cl, or Br) are also fuming liquids, and are used to replace OH by X in organic chemistry. What is their molecular shape (Problem 7)? The halosulphonic acids have four groups around a central sulphur atom, $X(OH) \cdot SO_2$ (X = F, Cl, or Br). What is their shape, and are they strong or weak acids (Problems 7 and 16)?

18.12 OXIDES

Sulphur, selenium, tellurium, and polonium each form two oxides, MO_2 and MO_3.

1) Dioxides. As the group is descended and the group VI elements become more metallic, the dioxides become more ionic, weaker reducing agents and stronger oxidizing agents, and less acidic. Thus sulphur dioxide, SO_2, is a gas with a covalent molecular structure (Fig. 18.5); it is a strong acid and a good reductant. Selenium dioxide has a non-planar chain structure (Fig. 18.5), and tellurium dioxide has the mainly ionic rutile structure; both are amphoteric (mainly acidic) white solids and are mild oxidants. Polonium dioxide is an oxidizing agent and is amphoteric.

The dioxides can be prepared by burning the element in air. Sulphur dioxide is obtained as a side-product from many reactions: the action of dilute acids on sulphites and thiosulphates; and the reaction of concentrated sulphuric acid on many metals. Industrially it is obtained: by burning sulphur; as a by-product from the roasting of sulphur ores, e.g. FeS_2; and from calcium sulphate, by heating with coke, $2CaSO_4 + C \rightarrow 2CaO + CO_2 + 2SO_2$. The calcium oxide is heated with clay to obtain cement:

$$4CaO + \underset{\text{clay}}{Al_2Si_2O_7} \xrightarrow{\text{heat}} \underbrace{2CaSiO_3 + Ca_2Al_2O_5}_{\text{cement}}.$$

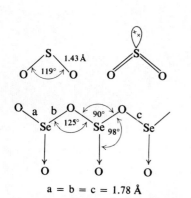

Fig. 18.5 The structures of sulphur and selenium dioxides.

The structure of sulphur dioxide is given in Fig. 18.5. The short bond distance indicates the presence of double bond character. At low temperatures sulphur dioxide forms a molecular lattice. At normal temperatures sulphur dioxide is a colourless dense gas with a pungent odour. It is poisonous, particularly to lower life, and is used as a disinfectant. It is easily liquefied, (-10 °C, 3 atmospheres) and some self-ionization takes place, as in the case of water: $2SO_2(l) \rightleftharpoons SO^{2+}(l) + SO_3^{2-}(l)$. It is an ionizing agent and dissolves many inorganic (and organic) compounds; thus it is a useful non-aqueous solvent. The gas is incombustible, but some fiercely burning metals can dissociate it and therefore continue burning (e.g. Mg, K, Na). It reacts with oxygen (Section 18.13.3), with chlorine to yield sulphuryl chloride, and with phosphorus pentachloride to give thionyl chloride and phosphorus oxychloride. It is very soluble in water, yielding an acid solution (Section

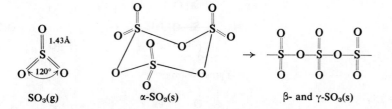

Fig. 18.6 The structure of sulphur trioxide.

18.13.1) and reacts with alkalis to give sulphites or bisulphites, depending upon which reactant is in excess. The redox behaviour of sulphur dioxide is discussed in Section 18.13.1.

Sulphur dioxide is used: in the manufacture of sulphuric acid; as a solvent; as a reducing agent, for example in bleaching the yellow dye in wool, flour, paper, and straw (the yellow colour may return on prolonged exposure to air); and as an antichlor after bleaching with chlorine.

2) Trioxides. Sulphur trioxide, SO_3, is formed from sulphur dioxide and oxygen (Section 18.13.3). The reaction is very slow except in the presence of a platinum catalyst at 500 °C (98 percent yield), and on freezing the mixture a white solid, sulphur trioxide (m.p. 17 °C) is obtained. In the gas phase (as expected, Section 12.7) the molecule is triangular (Fig. 18.6), and the short bond length indicates the double-bond character of each identical S=O bond. This gas readily polymerizes because of the instability of the S=O double bond relative to two S—O single bonds (Table A.11). If the gas is condensed at −80 °C, a trimer is formed (Fig. 18.6), called the α-form, (cf. the chair form of cyclohexane, in Fig. 21.11). These transparent ice-like crystals polymerize further in the presence of moisture to the β-form (Fig. 18.6). This is composed of infinite helical chains, like asbestos (Section 16.6). Another form, the γ-form, is similar to the β-form but the chains may be cross-linked. What structural feature do all the solid forms have in common (Problem 17)?

Sulphur trioxide dissolves in water to form sulphuric acid. This is not how sulphuric acid is prepared because the trioxide does not dissolve easily. The gaseous trioxide tends to pass through water and form a mist. Instead, the gas is passed through concentrated sulphuric acid, in which it dissolves nicely, forming "oleum" or "fuming sulphuric acid." This is diluted with water to give sulphuric acid:

$$H_2SO_4(l) + SO_3(g) \rightarrow H_2S_2O_7(l) \xrightarrow{H_2O} 2H_2SO_4(aq).$$

Thus sulphur trioxide is a highly acidic oxide. It is also a powerful oxidant: for example, $P_4 \rightarrow P_4O_{10}$, $HBr \rightarrow Br_2$. Like sulphuric acid, it is a sulphonating agent (Section 22.2.6). The relative instability of the S=O double bonds is shown in the tendency of the trioxide to form addition compounds: $H_2O \rightarrow H_2SO_4$; $BaO \rightarrow BaSO_4$; and $HCl \rightarrow$ chlorosulphonic acid, $Cl(OH) \cdot SO_2$.

18.13 OXYACIDS

Sulphur, selenium, and tellurium all form oxyacids. As usual, the co-ordination number of the anions and acids of the more metallic members is higher than that of the less metallic members, for example $SO_2(OH)_2$ and $Te(OH)_6$. Just as in the case

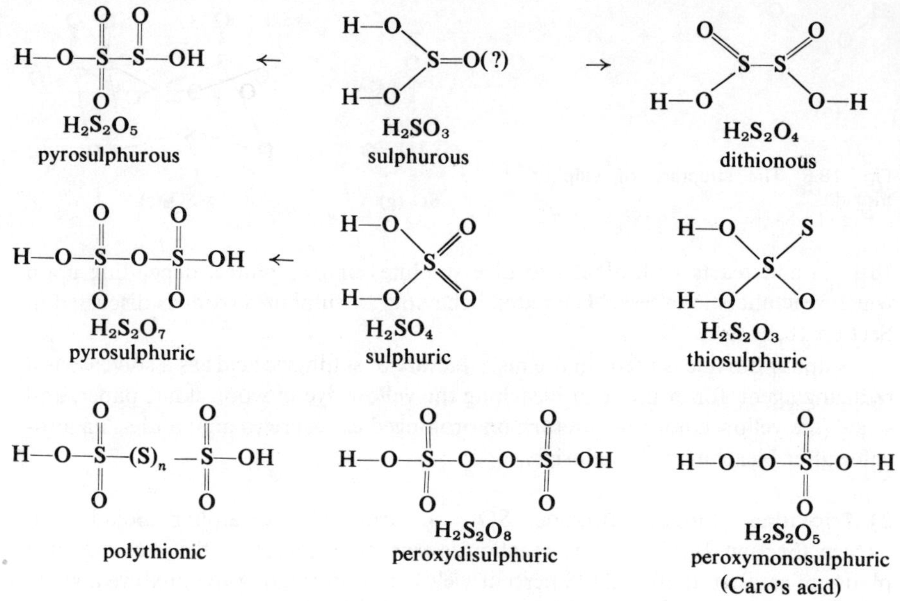

Fig. 18.7 Some oxyacids of sulphur. The peroxyacids are oxidants; the lower acids are reductants.

$H_2S_2O_5$ pyrosulphurous; H_2SO_3 sulphurous; $H_2S_2O_4$ dithionous;
$H_2S_2O_7$ pyrosulphuric; H_2SO_4 sulphuric; $H_2S_2O_3$ thiosulphuric;
polythionic; $H_2S_2O_8$ peroxydisulphuric; $H_2S_2O_5$ peroxymonosulphuric (Caro's acid)

of phosphorus, there are many sulphur acids, for some of which the free acids are unstable and only the corresponding oxyanions and salts are known pure. Acids containing S—S bonds are known which have no analogues among the other members of the group. Again, like phosphorus, the nomenclature of these acids is not strictly systematic. Sulphur*ic* acids are those based on the higher oxidation state (VI); sulphur*ous* acids are those based on the lower oxidation state (IV). The term pyro- is used in the same way as in phosphorus chemistry (Section 17.14). Thio-acids contain sulphur–sulphur bonds: dithio-acids contain one such bond, S—S; trithio-acids contain two such bonds, S—S—S, etc. These are known up to hexathio-acids, and like the sulphanes and polysulphides (Section 18.9) they readily form equilibrium mixtures. Peroxy-acids contain the peroxy group. Examples of all of these are shown in Fig. 18.7. Their shapes can be deduced from the Gillespie–Nyholm rules.

18.13.1 Sulphurous Acid [SO_2(aq) or $SO_2 \cdot xH_2O$]

Sulphur dioxide dissolves in water to yield an acid solution called "sulphurous acid", "H_2SO_3". In fact, the free acid has never been prepared. When aqueous solutions of the gas are frozen, a "compound", $H_2SO_3 \cdot 6H_2O$, is formed; this is now known to be a gas clathrate (Section 12.5) of approximate stoichiometry $SO_2 \cdot 7H_2O$. Nor is there any evidence for the existence of the free acid, "H_2SO_3", in solution:

$$SO_2(g) + xH_2O(l) \rightarrow SO_2 \cdot xH_2O \rightleftharpoons HSO_3^-(aq) + H_3O^+ + (x-2)H_2O$$

$$K_1 = \frac{[HSO_3^-][H_3O^+]}{[SO_2 \cdot xH_2O]} = 1.3 \times 10^{-2} \ (25\,°C); \quad K_2 = 6.2 \times 10^{-8}.$$

Three series of solid salts exist: sulphites (SO_3^{2-}), bisulphites (HSO_3^-), and metabisulphites ($S_2O_5^{2-}$). These ions also exist in solution in an aquated form. If one of

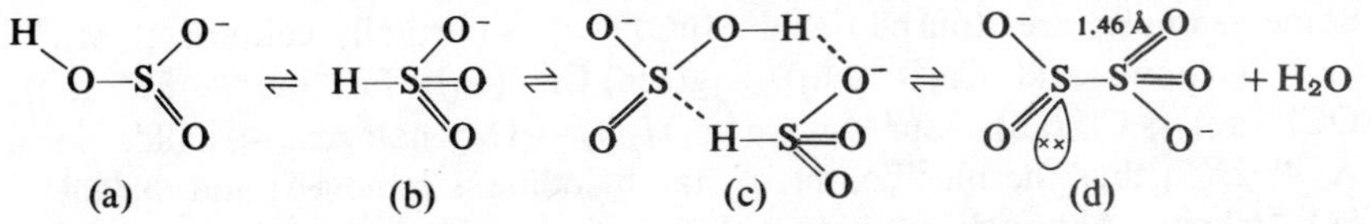

Fig. 18.8 The tautomers of the bisulphite species (a and b), and the formation of the metabisulphite ion (d).

the latter two is dissolved in water, an equilibrium exists between four species (Fig. 18.8).

Thus the bisulphite species (Figs. 18.8a and b) exist as tautomers (Section 28.3) which react via a hydrogen-bonded species (Fig. 18.8c) to form the metabisulphite (Fig. 18.8d). Clearly (Le Chatelier), the bisulphite species predominate in dilute solution, the metabisulphite ion in concentrated solution. Thus, if solid bisulphites or their solutions are heated, metabisulphites are obtained. Conversely, since metabisulphites are non-hygroscopic and more stable than bisulphites, they are useful sources of bisulphite ions in solution. Bisulphites are obtained by passing an excess of sulphur dioxide into a metal hydroxide or carbonate solution. If an equal volume of the original hydroxide solution is added to this solution, the corresponding sulphite solution is formed, from which the solid can be obtained by evaporation:

$$NaOH(aq) + SO_2(aq) \longrightarrow NaHSO_3(aq)$$
$$NaOH(aq) + NaHSO_3(aq) \longrightarrow Na_2SO_3(aq) + H_2O(l)$$
$$Na_2SO_3(aq) \xrightarrow{evaporate} Na_2SO_3 \cdot 7H_2O(s).$$

The alkali metal bisulphites, sulphites, and metabisulphites are known as solids as well as in solution, the metabisulphites being the least soluble. Sulphites of many other metals exist in the solid state, but their bisulphites exist only in solution (cf. carbonates, Section 16.6).

18.13.2 Reactions of Sulphites

1. With acids, sulphur dioxide is evolved (also bisulphites and metabisulphites).
2. Most heavy metal ions precipitate their insoluble sulphites from aqueous sulphite solutions. Some examples are: $Pb^{2+}(aq) \rightarrow PbSO_3(s)$; $Ba^{2+}(aq) \rightarrow BaSO_3(s)$; $Ca^{2+}(aq) \rightarrow CaSO_3(s)$; and $Ag^+(aq) \rightarrow Ag_2SO_3(s)$. These are all soluble in dilute acid (cf. sulphates), and some are soluble in an excess of sulphite ions owing to the formation of soluble complex anions, such as $[AgSO_3]^-(aq)$. Silver sulphite is soluble in hydrochloric acid, in ammonia (Section 17.9.3), and in an excess of sulphite ion. Sulphur dioxide gives a "positive" lime-water test (Section 16.6) since the initial precipitate of calcium sulphite dissolves in an excess of sulphur dioxide to form a solution of the bisulphite!
3. Sulphur dioxide and sulphites contain sulphur in an intermediate oxidation state (+4) (Fig. 18.1). Therefore, it is not unexpected that they can act as both oxidant and reductant, and can undergo disproportionation reactions. They are good reducing agents, especially in alkaline solution, being themselves oxidized to sulphate ions (Fig. 18.1 or Table 18.1):

Acid: $SO_4^{2-}(aq) + 4H^+(aq) + 2e^- \rightleftharpoons SO_2(aq) + 2H_2O(l)$; $E^\ominus = 0.17$ V;

Alkali: $SO_4^{2-}(aq) + H_2O(l) + 2e^- \rightleftharpoons SO_3^{2-}(aq) + 2OH^-(aq)$; $E^\ominus = -0.93$ V.

Some examples are: (purple) acid MnO_4^-(aq) → (virtually colourless) Mn^{2+} (aq); (yellow) acid $Cr_2O_7^{2-}$(aq) → (green) Cr^{3+}(aq); Fe^{3+}(aq) → Fe^{2+}(aq); OCl^-(aq) → Cl^-(aq); Cu^{2+} → Cu^+; H_2O_2 → O_2; halogens → halide ions, X_2 → $2X^-$, thus the blue colour of starch–iodine is removed; and O_2(aq) → SO_4^{2-} (slow). Although sulphate is the common oxidation product of sulphur dioxide and sulphites, some oxidants (e.g. Fe^{3+}, MnO_2) react to yield the dithionate ion, $S_2O_6^{2-}$. What is the oxidation number of sulphur in this compound? Give equations for these reactions (Section 18.14). In neutral conditions or acid solution, but not alkaline (Fig. 18.1), sulphites and sulphur dioxide can act as oxidants in the presence of powerful reducing agents: H_2S(aq) → S(s); some metals $\xrightarrow{SO_2}$ oxides; and C $\xrightarrow[1000\ °C]{SO_2}$ CO_2(g). The reactions are often complex because of the formation of many different products, usually mainly sulphur but often tetrathionate ($S_4O_6^{2-}$), thiosulphate ($S_2O_3^{2-}$), dithionite ($S_2O_4^{2-}$), and hydrogen sulphide. Zinc gives mainly dithionite, and sulphur gives mainly thiosulphate. Sulphites disproportionate on heating to yield sulphate and a sulphur product of lower oxidation number; for example, solid sulphites give sulphides:

$$4SO_3^{2-}(s) \rightarrow 3SO_4^{2-}(s) + S^{2-}(s).$$

Bisulphites give addition compounds with aldehydes (Section 23.3.2).

18.13.3 Sulphuric Acid

This is the most important chemical in any industrial country. There are several methods of manufacture, all of which involve the three steps: preparation of sulphur dioxide (Section 18.12); SO_2 → SO_3; and SO_3 → H_2SO_4. Most sulphuric acid is now manufactured by the "contact process". In this, sulphur trioxide is produced by passing sulphur dioxide with oxygen over a vanadium pentoxide catalyst. This reaction,

$$2SO_2(g) + O_2(g) \rightarrow 2SO_3(g); \quad \Delta H = -98\ kJ, \Delta G = -70\ kJ,$$

is slow to attain equilibrium. The rate of reaction can be increased by the presence of a suitable catalyst, by raising the temperature, by using an excess of oxygen, and by increasing the pressure. Platinum is a very efficient catalyst here, but is readily poisoned by the unavoidable presence of traces of sulphur and arsenic. Vanadium pentoxide, V_2O_5, is less efficient but is less readily poisoned and is much cheaper, and is therefore used. For an exothermic reaction, raising the temperature also decreases the equilibrium constant (Section 7.4). Thus, at 500 °C a 98 percent conversion of sulphur dioxide is achieved, whereas at 1000 °C practically no sulphur trioxide is produced (though the little that is formed is produced very rapidly!). An increase in pressure not only increases the rate of reaction but also increases the yield of the trioxide. Thus from the point of view of obtaining the maximum amount of product, we require a low temperature, a high pressure, and an excess of oxygen. From the point of view of a rapid reaction we require a high temperature, a catalyst, and a high pressure. The optimum conditions (balancing cost, yield, and rate) are found to be: vanadium pentoxide catalyst; 500 °C; an excess of air (cheap oxygen); and a pressure of one atmosphere only. These conditions produce a 98 percent yield quickly. The mixture of sulphur trioxide and excess air produced is passed through concentrated sulphuric acid to yield oleum which can be diluted as required (Section 18.12).

Properties. Below its melting point, 10.4 °C, anhydrous sulphuric acid is a white crystalline solid consisting of a three-dimensional hydrogen-bonded network, which persists in the liquid state and even in concentrated aqueous solutions, making such solutions viscous. The covalent nature of the bonding (Fig. 18.7), rationalizes the low melting point and the poor conductivity of the compound, though some self-ionization occurs and it finds some use as an ionizing solvent.

$$2H_2SO_4(l) \rightleftharpoons H_3SO_4^+ + HSO_4^-.$$

The pure acid decomposes slightly on standing (and particularly on warming) to evolve sulphur trioxide, and on heating it dilutes itself until the maximum boiling point mixture (Section 5.4) is obtained at 330 °C (98.3 percent H_2SO_4). Similarly, if dilute sulphuric acid is distilled, it concentrates itself to the maximum boiling point mixture.

Concentrated and pure sulphuric acid have a great affinity for water, and are used as dehydrating agents. The reaction with water is highly exothermic, and various hydrates are formed on freezing the solution:

$$H_2O + H_2SO_4 \rightarrow H_3O^+ + HSO_4^-.$$

The acid dries gases which do not react with it (e.g. SO_2, Cl_2, N_2, O_2), dehydrates many crystalline hydrates (e.g. $CuSO_4 \cdot 5H_2O$), and removes the elements of water from many organic compounds (e.g. $C_2H_5OH \rightarrow C_2H_4$, oxalic acid $H_2C_2O_4 \rightarrow CO + CO_2$, formic acid $HCO_2H \rightarrow CO$, carbohydrates $C_{12}H_{22}O_{11} \rightarrow 12C$).

Concentrated sulphuric acid is a powerful oxidant, especially when hot. The oxidizing power decreases rapidly on dilution. Mixtures of reduction products of the acid are usually produced during such reactions, often containing mainly sulphur dioxide, but also sulphur and hydrogen sulphide. Some examples are: $C \rightarrow CO_2$; $S \rightarrow SO_2$; $P \rightarrow H_3PO_4$; $I^- \rightarrow I_2$, etc. Organic material is often oxidized vigorously. Frequently, however, the anhydrous acid sulphonates organic material (Section 22.2.6). The acid is also used to assist nitration (Section 26.5).

In aqueous solution sulphuric acid is a strong dibasic acid (Section 9.3), forming two series of salts (sulphates and bisulphates) on neutralization with bases:

$H_2SO_4(aq) + NaOH(aq) \rightarrow H_2O(l) + NaHSO_4(aq)$, bisulphate or hydrogen sulphate

$H_2SO_4(aq) + 2NaOH(aq) \rightarrow 2H_2O(l) + Na_2SO_4(aq)$, sulphate

The dilute acid reacts with electropositive metals to evolve hydrogen (Section 11.4.2), and with carbonates to liberate carbon dioxide. Since it is less volatile than nitric and hydrochloric acids, concentrated sulphuric acid liberates these acids from their salts:

$$H_2SO_4(l) + NaCl(s) \xrightarrow{heat} HCl(g) + NaHSO_4(l).$$

There are various species present in water / sulphur trioxide systems (H_2SO_4, $S_2O_7^{2-}$, $H_2S_2O_7$, $H_3SO_4^+$, etc.) depending mainly upon the concentration, but the main species present in aqueous solutions are represented by the equilibria:

$$H_2SO_4(l) + H_2O(l) \rightleftharpoons H_3O^+(aq) + HSO_4^-(aq); \quad K \text{ very large.}$$
$$HSO_4^-(aq) + H_2O(l) \rightleftharpoons H_3O^+(aq) + SO_4^{2-}(aq); \quad K = 10^{-2} \text{ at } 25 \text{ °C.}$$

Thus the relative concentration of sulphate ions to bisulphate ions is small, except in very dilute solutions.

18.13.4 Sulphates

There are three classes of compound: sulphates; bisulphates; and pyrosulphates. All sulphates are soluble (Section 10.3.17) except for those of barium, strontium, and lead (very insoluble), and silver and calcium (sparingly soluble). All are colourless (unless the cation is coloured), crystalline solids, often containing water of crystallization, e.g. $CuSO_4 \cdot 5H_2O$ (Fig. 11.8). The thermal stability of sulphates depends upon the electropositivity of the metal (Section 11.2). Unlike the sulphites (Section 18.13.2), the insoluble sulphates do not dissolve in dilute hydrochloric acid. Solid bisulphates are known only for the alkali metals and the ammonium ion. These are anhydrous, stable, colourless salts which decompose on heating to yield pyrosulphates, such as $Na_2S_2O_7$. These, in turn, form the sulphate on further heating:

$$2NaHSO_4 \xrightarrow{300 \text{ °C}} Na_2S_2O_7 \xrightarrow{heat} Na_2SO_4 + SO_3.$$

Aqueous solutions of bisulphates are acidic and stable, but pyrosulphates are rapidly hydrolysed to bisulphates.

18.13.5 Thiosulphates, $S_2O_3^{2-}$

These compounds contain sulphur in an intermediate oxidation state, and they are prone to disproportionation reactions (Fig. 18.1). Thiosulphuric acid, for example, is unstable above -9 °C, and aqueous solutions of the acid rapidly disproportionate at room temperature to sulphur and sulphur dioxide. Thus, the addition of hydrochloric acid to thiosulphates precipitates sulphur, slowly at 25 °C, more rapidly on warming:

$$S_2O_3^{2-}(s) + 2H^+(aq) \rightarrow \text{"}H_2S_2O_3\text{"} \rightarrow H_2O(l) + S(s) + SO_2(g).$$

The salts of the acid can be prepared either by: boiling sulphur with sulphite solutions, when, for example, sodium thiosulphate, $Na_2S_2O_3 \cdot 5H_2O$ ("hypo"), can be crystallized; or by boiling sulphur with alkaline solutions, causing disproportionation of the element (Fig. 18.1):

$$SO_3^{2-}(aq) + S(s) \rightarrow S_2O_3^{2-}(aq);$$
$$6OH^-(aq) + 4S(s) \rightarrow 3H_2O(l) + S_2O_3^{2-}(aq) + 2S^{2-}(aq).$$

Many metals give rise to thiosulphates, all of which are stable, soluble (except those of lead, silver, and barium), crystalline solids. Many cations form complex ions (Section 20.4) with the thiosulphate ion; the silver ion, Ag^+, for example, forms the soluble anion $[Ag(S_2O_3)_2]^{3-}$, and silver salts are therefore soluble in thiosulphate solutions:

$$AgBr(s) + 2S_2O_3^{2-}(aq) \rightarrow [Ag(S_2O_3)_2]^{3-}(aq) + Br^-(aq).$$

This reaction is used in photography to remove the unreacted light-sensitive silver bromide from photographic plates to avoid further reaction with light. Thiosulphates disproportionate on heating. Hypo, for example, when heated first dissolves in its water of crystallization, then steam is evolved, solidification occurs, and sulphur is deposited:

$$4Na_2S_2O_3(s) \rightarrow 3Na_2SO_4(s) + Na_2S(s) + 4S(s).$$

The thiosulphates of the less electropositive metals are less stable (Section 11.2), and some are even decomposed by boiling in solution. Solutions of lead, silver, and barium ions each precipitate their insoluble thiosulphates from thiosulphate solutions; but whereas those of the weakly electropositive lead and silver are soluble in an excess of thiosulphate, forming stable complex anions, that of barium is not (Table 10.10). Sulphur is in a low oxidation state in thiosulphates and they are mild reductants, e.g. $I_2 \rightarrow 2I^-$; and $Fe^{3+}(aq) \rightarrow Fe^{2+}(aq)$. The reaction with iodine to form the tetrathionate ion, $S_4O_6^{2-}$, is rapid and quantitative, and is a very useful reaction in volumetric analysis:

$$I_2 + 2S_2O_3^{2-}(aq) \rightarrow 2I^-(aq) + S_4O_6^{2-}(aq).$$

Iron(III) is reduced to iron(II) via the dark-violet unstable complex anion, $[Fe(S_2O_3)_2]^-$:

$$2Fe^{3+}(aq) + 2S_2O_3^{2-}(aq) \rightarrow 2Fe^{2+}(aq) + S_4O_6^{2-}(aq).$$

18.14 OXIDATION AND REDUCTION

The concepts of oxidation and reduction have gone through several stages of (further) generalization, during which time the names have lost their original meaning, but have been retained as labels for the widened concepts (Section 10.3.19). These two related concepts are excellent examples of how a large body of material has been systematized, and attention drawn to the similarities between many reactions (correlation).

Oxidation was originally defined as *the addition of oxygen to a compound*. In the reaction below, the zinc is said to be oxidized:

$$2Zn(s) + O_2(g) \rightarrow 2ZnO(s).$$

Reduction was originally defined as *the removal of oxygen from a compound*, since during this process the weight of the original compound was reduced. In the reaction below, the iron oxide is said to be reduced:

$$Fe_2O_3(s) + 3C(s) \rightarrow 2Fe(s) + 3CO(g).$$

Consider the two reactions below. The first is clearly an oxidation. Surely the second reaction must also be classified similarly?

$$H_2S(aq) + 2O_2(g) \rightarrow H_2SO_4(aq)$$
$$2H_2S(aq) + O_2(g) \rightarrow 2S(s) + 2H_2O(l)$$

Often, when oxygen is reacted with a compound containing hydrogen, instead of the oxygen being added to the compound, hydrogen is removed as water, mainly owing to the great stability of water. Thus the concept of oxidation was extended to cover *the removal of hydrogen* from a compound (dehydrogenation); and, conversely, the concept of reduction was extended to cover *the addition of hydrogen* to a compound or element (hydrogenation).

Manganese dioxide, for example, oxidizes hydrogen chloride to chlorine:

$$MnO_2(s) + 4HCl(aq) \rightarrow MnCl_2(aq) + 2HCl(aq) + Cl_2(g).$$

Manganese dioxide is an example of an **oxidizing agent** or **oxidant**, a compound which causes oxidation of another substance. Carbon is an example of a **reducing agent** or **reductant,** a compound which causes reduction of another substance (e.g. Fe_2O_3, above).

Consider the two reactions:

$$4FeO(s) + O_2(g) \rightarrow 2Fe_2O_3(s); \tag{18.3}$$

$$4FeS(s) + 2S(l) \rightarrow 2Fe_2S_3(s). \tag{18.4}$$

The similarity of these reactions pleads that they should be classified similarly. Since Eq. (18.3) is an oxidation, then the concept of oxidation must be widened to include reactions such as Eq. (18.4). Therefore, oxidation was defined as a process in which the ratio of the non-metallic to metallic constituent is increased, and reduction as a process in which the above ratio is decreased. Clearly, the dependence of the concepts on oxygen, hydrogen, or reduction in weight has now been lost.

18.14.1 Electronic Theory of Oxidation and Reduction

Once reactions had been classified in this way, scientists turned their attention to an explanation of *why* these reactions were similar, in terms of electronic theory. What "really" happens in these reactions is identical as far as the iron ions are concerned: they each "lose" another electron to the non-metal:

$$4Fe^{2+} + O_2 \rightarrow 4Fe^{3+} + 2O^{2-}$$
$$4Fe^{2+} + 2S \rightarrow 4Fe^{3+} + 2S^{2-}.$$

Therefore, **oxidation** is defined as a process in which *electrons are removed from a species*, and, conversely, **reduction** is defined as a process in which *electrons are added*. Thus, elemental sulphur and oxygen are reduced to their ions, in the processes above. The complementary nature of the definitions is obvious. Since electrons cannot exist independently under normal conditions, then *oxidation and reduction must always occur together:* if electrons are gained by one species, then they must have been lost by another. This is *always* true (how often can this be said?). Thus in any reduction/oxidation (usually abbreviated to "redox") process, one species is an oxidant, another a reductant. While it is not strictly justified to classify chemical substances as oxidants and reductants without a reference to a particular process (since many species can act in either capacity depending upon the conditions, especially upon the nature of the other reactant), it is often convenient to do so. The classification is a good approximation but it must be remembered that there are substances which are oxidants in one process, but reductants in another (hydrogen peroxide for example). Part of such a classification is given in Table 18.5. One danger of this treatment is that it implies that the mechanism of all redox reactions is by simple electron transfer. This is not true.

The redox character of any species can easily be investigated in a qualitative manner. Oxidants will react with known reducing agents: $H_2S \rightarrow S$ (yellow); acidified $KI \rightarrow I_2$ (brown solution in an excess of KI); and $Fe^{2+}(aq) \rightarrow Fe^{3+}(aq)$

Table 18.5. A classification of compounds as reductants and oxidants.

	Oxidants	Reductants
Gases	O_2; Cl_2	H_2, CO, H_2S, SO_2
Solutions	HNO_3(aq); OCl^-(aq); H_2O_2 metals in high oxidation numbers (aq)	SO_2(aq); H_2O_2
Solids	higher oxides e.g. PbO_2; $KMnO_4$	metals, carbon

(red colour in presence of CNS^-). Reductants will react with known oxidizing agents: acidified dichromate (yellow) → Cr^{3+}(aq) (green); acidified permanganate (purple) → Mn^{2+}(aq) (almost colourless); Fe^{3+}(aq) → Fe^{2+}(aq); and a little I_2 in KI (brown) → colourless I^-.

Balancing redox equations. The concept of electron transfer is useful in balancing equations. The rules given below apply to acid solutions only, and are illustrated with reference to the reaction between the dichromate and the iron(II) ions in aqueous acid solution:

$$Cr_2O_7^{2-}(aq) + Fe^{2+}(aq) \rightarrow Cr^{3+}(aq) + Fe^{3+}(aq).$$

Notice that the reactants and products must be known.

1. Write down the product and reactant for each redox half-reaction (Section 7.8) separately, and treat them independently, balancing any oxygen atoms with water, and making sure that atoms other than hydrogen are balanced.

   reduction *oxidation*
   $Cr_2O_7^{2-} \rightarrow 2Cr^{3+} + 7H_2O$ $Fe^{2+} \rightarrow Fe^{3+}$

2. Balance the hydrogen atoms by adding H^+:

   $Cr_2O_7^{2-} + 14H^+ \rightarrow 2Cr^{3+} + 7H_2O$ $Fe^{2+} \rightarrow Fe^{3+}$.

3. Balance the half-reactions electrically by adding the appropriate number of electrons:

   $6e^- + Cr_2O_7^{2-} + 14H^+ \rightarrow 2Cr^{3+} + 7H_2O$ $Fe^{2+} \rightarrow Fe^{3+} + e^-$.

   These are now the half-cell reactions.

4. Multiply each half-cell reaction by the number of electrons appearing in the other, and add the equations to eliminate the electrons. Simplify the final equation if possible.

   $6e^- + Cr_2O_7^{2-} + 14H^+ \rightarrow 2Cr^{3+} + 7H_2O$
   $6Fe^{2+} \rightarrow 6Fe^{3+} + 6e^-$

   $Cr_2O_7^{2-}(aq) + 6Fe^{2+}(aq) + 14H^+(aq) \rightarrow 2Cr^{3+}(aq) + 6Fe^{3+}(aq) + 7H_2O(l)$

The "molecular" equation, if required, can be obtained by adding equal numbers of K^+ (two) and SO_4^{2-} (thirteen) to each side of the equation:

$K_2Cr_2O_7 + 6FeSO_4 + 7H_2SO_4 \rightarrow Cr_2(SO_4)_3 + 3Fe_2(SO_4)_3 + K_2SO_4 + 7H_2O.$

Using this method, equations which balance electrically (law of conservation of charge) and with respect to the number of atoms (law of conservation of mass) on each side of the equation (this is what "balancing" equations means), can be written for all redox reactions, and also for all half-cell reactions which are often abbreviated as couples, e.g. $Cr_2O_7^{2-}/Cr^{3+}$. One important point: just because an equation can be written for a particular reaction, it does not mean that the reaction takes place.

For reactions in alkaline solutions, which contain virtually no hydrogen ions, the oxygen and hydrogen atoms (rules 1 and 2) are balanced by adding the species H_2O and OH^- instead of H_2O and H^+. Actually, it is usually simpler to proceed *as if* the solution were acidic, and remove any H^+ ions appearing in the completed equation by adding the same number of OH^- ions to both sides of the equation. Neutral solutions can be treated similarly, but in this case only the H^+ ions appearing on the left hand (reactant) side of the equation need be removed at the end. The same procedures can be adopted in formulating half-cell reactions.

18.14.2 Oxidation Number (O.N.)

The most comprehensive definition of oxidation (reduction) involves the concept of oxidation number, often abbreviated to O.N. In using this idea, all compounds (and this includes all covalent species except for elements) are treated as if they are completely ionic, despite the fact that no compounds are "completely ionic". From this apparently unpromising start a concept develops which is of great use in chemistry, though this will only become apparent later. As usual in science, the value of any definition or concept lies in its usefulness, not in the aesthetic, moral, or emotional reaction it generates (this does not always appear to be true in a number of other fields of human activity). In covalent compounds, therefore, the bonding electrons are imagined to be "owned" by the more electronegative atom, and the compound treated as if it is ionic. The oxidation number of each atom of each element in the compound is the charge left on that atom after this imaginary exercise has been performed. In the case of ionic compounds, the oxidation number of each atom is the charge already on the atom (ion). In the case of elements, since all the atoms are the same and no one is more electronegative than any other, the charge on each atom after distributing the bonding electrons fairly is equal to zero. Thus the atoms in all elements in the elemental state have an oxidation number of zero. Some examples are shown in Table 18.6.

In practice, it is found simpler to derive a series of simple rules from the ideas above and from considerations of relative electronegativities and maximum bond-forming capacities of atoms, and to use these in deriving the oxidation numbers of atoms in their compounds. Thus, the **oxidation number** of an atom in a given species is defined as *the formal (i.e. not real) charge on the atom in the compound or ion in question, after the bonding electrons have been allocated according to a particular set of rules:*

1. The O.N. of the atoms of an uncombined element is zero.
2. The O.N. of monatomic ions is the charge on the ion.
3. The O.N. of fluorine is always minus one (-1), since it is the most electronegative element, and never forms more than one bond (Section 19.6). The

Table 18.6. The allocation of oxidation numbers.

Compound	Structure	"Imagined ionic structure"	Oxidation numbers
Sodium chloride	Na^+ Cl^-	Na^+ Cl^-	$Na = +1$; $Cl = -1$.
Hydrogen chloride	$H{-}Cl$	H^+ Cl^-	$H = +1$; $Cl = -1$.
Water	$H{-}O{-}H$	H^+ O^{2-} H^+	$H = +1$; $O = -2$.
Carbon tetrachloride (tetrachloromethane)	$Cl{-}C(Cl)(Cl){-}Cl$	Cl^- C^{4+} Cl^- Cl^- Cl^-	$C = +4$; $Cl = -1$.
Chloroform (trichloromethane)	$Cl{-}C(H)(Cl){-}Cl$	H^+ Cl^- C^{2+} Cl^- Cl^-	$C = +2$; $Cl = -1$; $H = +1$.
Permanganate ion	$[O{=}Mn(={O})(={O}){=}O]^-$	$[O^{2-}\ O^{2-}\ Mn^{7+}\ O^{2-}\ O^{2-}]^-$	$Mn = +7$; $O = -2$.
Hydrogen peroxide	$H{-}O{-}O{-}H$ or	$(O{-}O)^{2-}$ H^+ H^+ or O^- O^- H^+ H^+	$H = +1$; $O = -1$.

O.N. of the other halogens is usually minus one (-1), except when they are combined with more electronegative elements.

4. The O.N. of oxygen is almost always minus two (-2), except when bonded to fluorine (the only element that is more electronegative than oxygen) or in peroxides, when the O.N. is minus one (-1) owing to the existence of an oxygen–oxygen bond.

5. The O.N. of metals is usually positive, since they are usually combined with the more electronegative non-metals.

6. The O.N. of hydrogen bonded to a metal is minus one (-1), and hydrogen bonded to a non-metal, plus one ($+1$). This is due to the electronegativity of hydrogen being intermediate between metals and non-metals, and its maximum covalency being one.

7. The sum of the oxidation numbers of each atom (represented by $\sum$ O.N. in shorthand) in an ion or compound is equal to the charge on the ion or compound. This rule enables the unknown oxidation number of a particular atom in a given species to be determined. For example, consider the manganese in the permanganate ion, MnO_4^-. We know (rule 7) that $\sum$ O.N. $= -1$, and

that the O.N. of each of the four oxygen atoms is equal to -2. If we let M stand for the unknown oxidation number of the manganese atom in the permanganate ion, then $\sum \text{O.N.} = M + 4(-2) = -1$; therefore $M = +7$, and the oxidation number of manganese is $+7$ (Table 18.6). For ammonia, NH_3, if we let N stand for the O.N. of nitrogen in ammonia, then since the O.N. of hydrogen is $+1$ (rule 6), and since $\sum \text{O.N.} = 0$ (rule 7), therefore: $3(+1) + N = 0$, and $N = -3$. Thus the oxidation number of nitrogen in ammonia is minus three (-3).

Two important points must be understood. First, an element may have different oxidation numbers in different compounds. For example, calculate the oxidation numbers of carbon in the compounds CH_4, CH_3F, CH_2F_2, CHF_3, and CF_4. This merely reflects the different electronegativities of the other atoms. Secondly, in compounds containing more than one atom of a given element, for example carbon in propan-1,2-diol, fractional oxidation numbers can occur. This happens because the overall oxidation number of carbon is an average of three

different oxidation numbers: $(-1 + 0 - 3) \div 3 = -\frac{4}{3}$. This can be obtained from the rules in the normal way. Let C be the unknown (average) oxidation number of carbon in this compound:

$$\sum \text{O.N.} = 8(+1) + 2(-2) + 3C = 0; \quad \therefore 3C = -4; \quad \therefore C = -\tfrac{4}{3}.$$

What is the average oxidation number of sulphur in the tetrathionate ion, $S_4O_6^{2-}$?

The most comprehensive definitions of oxidation and reduction are: *oxidation is a process in which the oxidation number of an element is increased algebraically; reduction is a process in which the oxidation number of an element is decreased algebraically*. (N.B. a change from -2 to -3 is a decrease, and a change from -3 to -2 is an increase in oxidation number.) For example, the change $CH_4 \to CBr_4$ is an oxidation of the carbon, since its oxidation number changes from minus four (-4) in CH_4 to plus four $(+4)$ in CBr_4.

Use in balancing redox equations. The oxidation number method provides a simpler method for balancing equations than that previously described, since one set of rules is adequate for all situations. Again the method is illustrated for the dichromate–iron(II) reaction in acid solution.

1. Write down the products and reactants for each half-reaction separately, and treat them independently:

   reduction oxidation
   $Cr_2O_7^{2-} \to Cr^{3+}$ $Fe^{2+} \to Fe^{3+}$.

2. Balance each partial equation (half-reaction) with respect to the mass (number

of atoms) of the elements being oxidized and reduced (chromium and iron in this example):

$$Cr_2O_7^{2-} \rightarrow 2Cr^{3+} \qquad Fe^{2+} \rightarrow Fe^{3+}.$$

3. Write down the total oxidation numbers (abbreviated to T.O.N.), that is the oxidation number of the element being oxidized or reduced, multiplied by the number of such atoms present:

$$Cr_2O_7^{2-} \rightarrow 2Cr^{3+} \qquad Fe^{2+} \rightarrow Fe^{3+}.$$
T.O.N. $\quad +12 \qquad +6 \qquad\quad +2 \qquad +3$

4. Write down the change in total oxidation numbers:

$$Cr_2O_7^{2-} \rightarrow 2Cr^{3+} \qquad Fe^{2+} \rightarrow Fe^{3+}.$$
T.O.N. $\quad +12 \qquad +6 \qquad\quad +2 \qquad +3$
Change $\quad\underline{\quad -6 \quad} \qquad\quad \underline{\quad +1 \quad}$

5. Balance the half-cell reactions by multiplication such that the decrease in the total oxidation number in one is equal to the increase in the total oxidation number in the other. This is done by multiplying each half-cell reaction by the numerical value of the total oxidation number change in the other ("six" and "one" in this example) and simplifying if necessary (not in this example):

$$Cr_2O_7^{2-} \rightarrow 2Cr^{3+} \qquad 6Fe^{2+} \rightarrow 6Fe^{3+}.$$

6. Add the "balanced" half-reactions together:

$$Cr_2O_7^{2-} + 6Fe^{2+} \rightarrow 2Cr^{3+} + 6Fe^{3+}.$$

The equations are now balanced with respect to the masses of the species undergoing oxidation and reduction, but must now be balanced with respect to oxygen, hydrogen, and electrical charge.

7. The ionic charges are balanced by adding H^+ or OH^-, depending on the conditions of the reaction. In this example the solution is acidic, and there are total ionic charges of plus ten ($+10$) and plus twenty four ($+24$) on the left and right hand sides respectively, therefore $14H^+$ ions are added to the left hand side:

$$Cr_2O_7^{2-} + 6Fe^{2+} + 14H^+ \rightarrow 2Cr^{3+} + 6Fe^{3+}.$$

8. Balance the hydrogen and oxygen atoms by adding water:

$$Cr_2O_7^{2-}(aq) + 6Fe^{2+}(aq) + 14H^+(aq) \rightarrow$$
$$2Cr^{3+}(aq) + 6Fe^{3+}(aq) + 7H_2O(l).$$

This is the complete equation.

Another example, in alkaline solution this time, is the reaction of Devarda's alloy (50 percent Cu, 45 percent Al, 5 percent Zn) with the nitrate ion:

$$Al + NO_3^- \rightarrow AlO_2^- + NH_3.$$

Rules
1 and 2. $NO_3^- \rightarrow NH_3 \qquad Al \rightarrow AlO_2^-.$
3. $\quad +5 \qquad -3 \qquad\quad 0 \qquad +3$
4. $\quad \underline{\quad -8 \quad} \qquad\quad \underline{\quad +3 \quad}$

5. $3NO_3^- \to 3NH_3 \qquad 8Al \to 8AlO_2^-$.
6. $3NO_3^- + 8Al \to 8AlO_2^- + 3NH_3$.
7. $3NO_3^- + 8Al + 5OH^- \to 8AlO_2^- + 3NH_3$.
8. $3NO_3^-(aq) + 8Al(s) + 5OH^-(aq) + 2H_2O(l) \to$
$$8AlO_2^-(aq) + 3NH_3(l).$$

Organic reactions can be written down similarly, for example the oxidation of ethanol to acetaldehyde by acid dichromate (Section 23.1.2):

Rules

1 and 2. $\qquad Cr_2O_7^{2-} \to 2Cr^{3+} \qquad C_2H_5OH \to C_2H_4O$.

3 and 4. T.O.N. $\quad +12 \qquad +6 \qquad\quad -4 \qquad\quad -2$

$\qquad\qquad\qquad\qquad \underbrace{\qquad -6 \qquad} \qquad \underbrace{\qquad +2 \qquad}$

5 and 6. $2CrO_7^{2-} + 6C_2H_5OH \to 4Cr^{3+} + 6C_2H_4O$.
   or $\quad Cr_2O_7^{2-} + 3C_2H_5OH \to 2Cr^{3+} + 3C_2H_4O$.
7. $Cr_2O_7^{2-} + 3C_2H_5OH + 8H^+ \to 2Cr^{3+} + 3C_2H_4O$.
8. $Cr_2O_7^{2-}(aq) + 3C_2H_5OH(l) + 8H^+(aq) \to$
$$2Cr^{3+}(aq) + 3C_2H_4O(l) + 7H_2O(l).$$

With practice, this can be carried out almost by inspection; for example, consider a totally unfamiliar example, the reduction of nitric acid by zinc to give hydrazine, N_2H_4.

$\qquad\qquad\qquad\qquad Zn + 2NO_3^- \to Zn^{2+} + N_2H_4$
T.O.N. $\qquad 0 \qquad +10 \qquad +2 \qquad -4$
Change $\qquad \underbrace{\qquad\qquad | \quad +2 \quad |}$
$\qquad\qquad\qquad\qquad\qquad \underbrace{\qquad -14 \qquad}$

Therefore $7Zn + 2NO_3^- \to 7Zn^{2+} + N_2H_4$;
$7Zn + 2NO_3^- + 16H^+ \to 7Zn^{2+} + N_2H_4 + 6H_2O$;

or if required in "molecular" form, add $14NO_3^-$ to each side of the equation:

$$7Zn(s) + 14HNO_3(aq) \to 7Zn(NO_3)_2(aq) + N_2H_4(g) + 6H_2O(l).$$

The student is advised to practise his technique on redox reactions as they appear on his reading of this book.

Disproportionation reactions (Section 7.12.2) are treated in exactly the same way except that both half-reactions have the same reactant (but different products). As an example, consider the disproportionation of the manganate ion $MnO_4^{2-}(aq)$, in neutral solution, to form manganese dioxide and permanganate (Section 20.16):

Rules

1 to 4. $\qquad\qquad MnO_4^{2-} \to MnO_2 \qquad MnO_4^{2-} \to MnO_4^-$.
T.O.N. $\qquad +6 \qquad +4 \qquad\qquad +6 \qquad +7$
Change $\qquad \underbrace{\qquad -2 \qquad} \qquad \underbrace{\qquad -1 \qquad}$

5. $MnO_4^{2-} \to MnO_2 \qquad 2MnO_4^{2-} \to 2MnO_4^-$.
6. $MnO_4^{2-} + 2MnO_4^{2-} \to MnO_2 + 2MnO_4^-$.
7. $MnO_4^{2-} + 2MnO_4^{2-} \to MnO_2 + 2MnO_4^- + 4OH^-$.
   (N.B. the charges cannot be balanced by adding $2H^+$ ions to the left hand side unless the solution is acidic.)
8. $3MnO_4^{2-}(aq) + 2H_2O(l) \to MnO_2(s) + 2MnO_4^-(aq) + 4OH^-(aq)$.

PROBLEMS

1. Why is oxygen a gas, whereas sulphur is a solid?
2. Tellurium hexafluoride, TeF_6, is rapidly hydrolysed by water to telluric acid, $Te(OH)_6$, and hydrochloric acid; whereas sulphur hexafluoride is unattacked, even by boiling alkali. Comment.
3. How do (a) the boiling points, (b) the acid strengths, and (c) the bond angles of the hydrides of the group VIB elements change as the group is descended? Comment.
4. Plot the melting points and boiling points of the group VIB elements against atomic number, and compare these with similar curves for the group IA elements. Comment.
5. Which are the most acidic oxides in each of the following pairs:
   a) CaO and CO;
   b) MnO and Mn_2O_7;
   c) N_2O and N_2O_5;
   d) Cr_2O_3 and CrO_3?
6. Classify the following oxides in terms of their structure and acid–base characteristics: Cl_2O_7, Fe_2O_3, CaO, and SiO_2.
7. What are the shapes of the following species: $O_3(g)$, S_6, S_8, O_3^-, R_3O^+, SiF_6^{2-}, SF_6, SF_4, SO_2Cl_2, $SOCl_2$, SOF_4, $Cl(OH)SO_2$, SO_2, SO_3, SO_4^{2-}, $S_2O_3^{2-}$, and SO_3^{2-}?
8. Draw the structure of the basic acetate of zinc, $OZn_4(O \cdot CO \cdot CH_3)_6$.
9. Using the bond energies in Table A.11, calculate the enthalpies of reaction for the changes: $S_8(g) \rightarrow 4S_2(g)$; and '$O_8$'$(g) \rightarrow 4O_2(g)$.
10. Why are the reactions of oxygen often slow at 25 °C, whereas those of ozone are quicker?
11. Do you think that the enthalpy of reaction and the activation energy for the change rhombic $\rightarrow$ monoclinic sulphur are large or small?
12. Why is toluene used for the recrystallization of monoclinic sulphur instead of benzene?
13. Is the conductivity of semi-metallic tellurium greater or less than that of grey selenium (same structure)?
14. The volatility of the group VIB hydrides changes in the order $H_2O \ll H_2S > H_2Se > H_2Te$. Rationalize this order.
15. Rationalize (a) the acidity, and (b) the thermal stability with respect to dissociation into the constituent elements, of the group VIB hydrides.
16. Are the halosulphonic acids, $X(OH)SO_2$, strong or weak acids?
17. What important structural feature do the α, β, and γ forms of SO_3 have in common?
18. What is the main reason why compounds containing S—S bonds are more stable than compounds containing O—O, Se—Se, or Te—Te bonds?

BIBLIOGRAPHY

Sharpe, A. G. *Principles of Oxidation and Reduction*, Royal Institute of Chemistry (1968).

Sowerby, D. B., and M. F. A. Dove. "Oxidation States in Inorganic Compounds," *Education in Chemistry*, **1**, No. 2, 83–90 (1964).

Mahan, B. H. *University Chemistry* (second edition), Addison-Wesley (1969).

In addition most of the books on inorganic chemistry in the general list contain material on the chemistry of group VIB. A selection is given at the end of Chapter 16.

CHAPTER 19

The Halogens: Group VIIB

19.1 GENERAL CHEMISTRY AND GROUP TRENDS

As expected from the general trend of increasing electronegativity across the *p*-block group (Section 11.4), this is a group of electronegative elements whose similar electron configuration, one short of the noble gas configuration, leads to the elements showing similar chemistries. Imposed upon this marked similarity of the elements (the most important single feature of the chemistry in this group) is the usual increase in metallicity down the group. This trend is much less important here than in the middle of the *p*-block elements, and only iodine shows cationic properties to any degree, and then only slightly. As usual, the first member of the series is anomalous (Section 14.2.7), as is the middle row element, bromine, though the "middle row anomaly" effect (Section 10.3.2) fades towards the end of the table. The latter effect is illustrated well by the reduction potential data (Table 19.3 and Fig. 19.1). In general the order of decreasing oxidizing power of any positive oxidation state, $M(+n)$, is given by $F(+n) \gg Br(+n) \geqslant Cl(+n) > I(+n)$. Positive oxidation states of fluorine do not exist (by definition), and those of bromine are more highly oxidizing than expected. The recently synthesized perbromate ion, BrO_4^-, is probably particularly highly oxidizing, which is why it has proved difficult to prepare.

The electron configuration of all the elements is one electron short of the noble gas configuration, and the chemistry is dominated by the tendency of these elements to complete the octet and to form: (a) ionic compounds with metals, containing the ion X^-; and (b) covalent compounds with non-metals (or metals in high oxidation states), containing the single covalent bond $-X$. In addition they can form bridged compounds in which they are two-co-ordinate, such as Al_2Br_6 and $H-F$. The elements other than fluorine can form compounds in which they display positive oxidation numbers up to $+7$, in which the co-ordination number is often greater than two. Typically for *p*-block elements, the oxidation numbers differ in steps of two (cf. transition elements) as Fig. 19.1 shows. As expected (Table 19.1) the negative oxidation state of -1 decreases in stability relative to the element as the group is descended and the electronegativity decreases. All the positive oxidation states of these elements are oxidizing (higher in Fig. 19.1 than X^-), even in alkaline solution, though as usual the higher oxidation states are less strongly oxidizing in basic conditions than in acidic (Section 17.2). In both alkali and acid, the $+3$ oxidation state is generally unstable with respect to both disproportionation and reduction (Fig. 19.1).

19.1 General chemistry and group trends
19.2 Occurrence and preparation of the elements
19.3 Properties of the elements
19.4 Chemical properties of the elements
19.5 Uses
19.6 The anomalous nature of fluorine
19.7 Preparation of hydrogen halides
19.8 Hydrogen fluoride
19.9 Properties of hydrogen chloride, bromide, and iodide
19.10 Halides
19.11 Interhalogen compounds and polyhalides
19.12 The oxides and oxyacids of the halogens

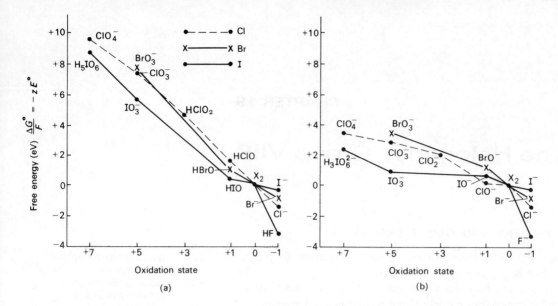

Fig. 19.1 Oxidation state diagrams of the halogens: (a) acid, and (b) alkaline.

The halogens exist in all states as non-polar diatomic molecules. The electronegativity of each element decreases down the group. The trends in other properties (Table 19.1) are much as expected. The decreasing volatility down the group is related to the increasing strength of van der Waals forces (Section 4.9), and this is reflected in the increasing enthalpies of vaporization, etc., and the increasing melting and boiling points. The electrode potentials X_2/X^- become less positive down the group, reflecting the fact that the free energy change of the reaction $\frac{1}{2}X_2(\text{stand.}) \to X^-(aq)$ becomes less negative on descending the group. In other words, fluorine is the best oxidizing agent in aqueous solution (actually this is true under any conditions). The factors involved in this change are reproduced in Fig. 19.2. Clearly the order of oxidizing power F > Cl > Br > I is not due to the relative electron affinities, but is mainly due to the relative hydration energies, and to a lesser extent to the weak dissociation energy of the fluorine–fluorine bond. This treatment, as usual, assumes that enthalpy changes mainly determine free energy changes, in other words that we can ignore small differences in entropy between the elements (Section 10.3.18). The reduction potential data for the oxidation of water to oxygen,

$$O_2(g) + 4H^+(aq) + 4e^- \to 2H_2O(l); \quad E^\ominus = +1.23 \text{ (in acid solution)}$$

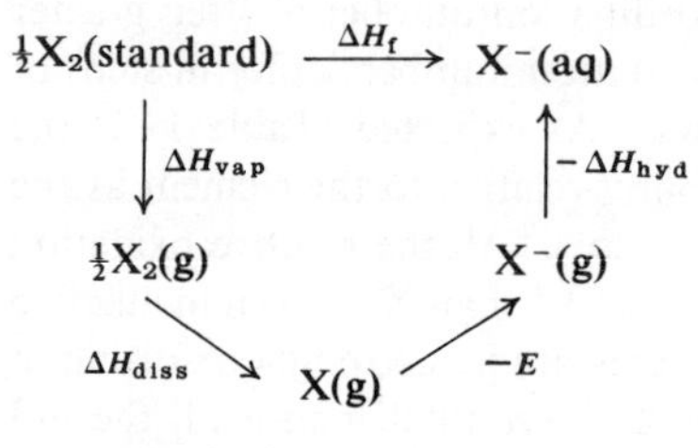

Fig. 19.2 Halogens as oxidants in aqueous solutions (see Fig. 10.14).

	F	Cl	Br	I
ΔH_{vap}	0	0	15	30
ΔH_{diss}	79.1	122	96	76
ΔE	−333	−348	−340	−297
$-\Delta H_{hyd}$	−460	−385	−351	−305
$\frac{1}{2}X_2 \to X^-(aq); \Delta H_f(\text{sum})$	−714	−611	−580	−496

Table 19.1. Properties of the halogens, group VIIB.

Property	Fluorine F	Chlorine Cl	Bromine Br	Iodine I
Atomic number	9	17	35	53
Electron configuration (outer)	$2s^2 2p^5$	$3s^2 3p^5$	$3p^6 3d^{10} 4s^2 4p^5$	$4p^6 4d^{10} 5s^2 5p^5$
Isotopes (in order of abundance)	19	35, 37	79, 81	127
Atomic weight	18.9984	35.453	79.909	126.904
Atomic radius (in molecule) (Å)	0.72	0.99	1.14	1.33
Ionic radius, X^- (Å)	1.36	1.81	1.95	2.16
Covalent radius (Å)	0.64	0.99	1.14	1.33
Van der Waals radius (Å)	1.35	1.85	1.95	2.15
Atomic volume ($cm^3\ mol^{-1}$)	17.1	22.7	25.6	25.7
Melting point (°C)	−220	−101	−7.3	113
Boiling point (°C)	−188	−34.5	59	183
Enthalpy of fusion ($kJ\ mol^{-1}$)	0.26	3.2	5.27	7.8
Enthalpy of vaporization ($kJ\ mol^{-1}$)	3.27	10.2	15	30
Enthalpy of atomization ($kJ\ mol^{-1}$)	79.1	122	111	106
Enthalpy of hydration, X^- (g) ($kJ\ mol^{-1}$)	460	385	351	305
Density (liquid) ($g\ cm^{-3}$)	1.11 (b.p.)	1.56 (b.p.)	3.12	(4.94) solid
Electronegativity (A/R)	4.10	2.85	2.75	2.20
Ionization energy ($kJ\ mol^{-1}$)	1681	1255	1142	1007
Electron affinity ($kJ\ mol^{-1}$)	333	348	340	297
Electrode potential, $\frac{1}{2}X_2(g) + e^- \rightarrow X^-(aq)$ (V)	+2.87	+1.36	+1.07(l)	+0.54(s)
Co-ordination numbers	1, (2)	1, 2, 3, 4	1, 2, 3, 5	1, 2, 3, 4, 5, 6, 7
Oxidation states	−I	−I, I, III, V, VII	−I, I, III, V	−I, I, III, V, VII
Abundance, p.p.m.	800	480	2	0.3
Solubility (g/100 g of water) at 20 °C	decomposes	0.59	3.6	0.018
Lattice energy, potassium salt KX ($kJ\ mol^{-1}$)	817	718	656	615
Lattice energy, calcium salt CaX_2 ($kJ\ mol^{-1}$)	2581	2247		
Colour and physical state at 20 °C	pale-yellow gas	greenish-yellow gas	red-brown liquid	black solid, metallic lustre, violet vapour

show that whereas iodine and bromine cannot, chlorine and fluorine should (thermodynamically), oxidize water to oxygen. Fluorine, in fact, does oxidize water to oxygen (and ozone), but the reaction of chlorine with water is very slow and in fact hypochlorous acid, HOCl, is formed, which slowly decomposes to hydrochloric acid and oxygen. Since the lattice energies of ionic compounds also decrease from the fluoride to the iodide (Table 19.1), in the same way as the hydration energies, then the order of oxidizing power is the same for solid state reactions (Section 10.3.18). Moreover, the bond energies of covalent compounds generally decrease from fluorine to iodine (Table 19.2), except for the F—F bond which is exceptionally weak (attributed to non-bonding electron repulsion, Section 19.6).

Table 19.2. Bond energies for the halogens ($kJ\ mol^{-1}$).

	B	C	N	O	F	Si	P	S	Cl	As	Br	Ge	Sn	Sb	H
F	644	489	280	213	158	598	498	285	251	485	251	473		452	565
Cl	443	326	192	205	251	402	331	272	244	310	209	339	314	314	431
Br	368	272		201	251	331	268	213	209	255	192	280	268	264	364
I	272	238			243	234	184		209	188	176	213	197	184	299

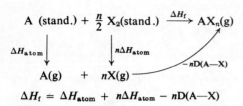

$$A \text{ (stand.)} + \frac{n}{2} X_2 \text{(stand.)} \xrightarrow{\Delta H_f} AX_n(g)$$

$$\Delta H_{atom} \downarrow \quad \downarrow n\Delta H_{atom} \quad \diagup -nD(A\!-\!X)$$

$$A(g) \quad + \quad nX(g)$$

$$\Delta H_f = \Delta H_{atom} + n\Delta H_{atom} - nD(A\!-\!X)$$

Fig. 19.3 Enthalpy of formation of covalent halide, $AX_n(g)$, $(-\Delta H_f)$.

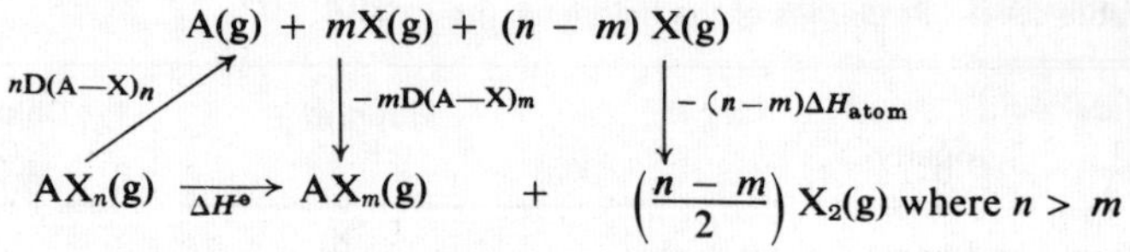

$$A(g) + mX(g) + (n-m)X(g)$$

$$nD(A\!-\!X)_n \diagup \quad \big| -mD(A\!-\!X)_m \quad \big| -(n-m)\Delta H_{atom}$$

$$AX_n(g) \xrightarrow{\Delta H^\ominus} AX_m(g) \quad + \quad \left(\frac{n-m}{2}\right) X_2(g) \text{ where } n > m$$

e.g. $PCl_5(g) \rightarrow PCl_3(g) + Cl_2(g)$

$$\Delta H^\ominus = nD(A\!-\!X)_n - mD(A\!-\!X)_m - (n-m)\Delta H_{atom}$$

Fig. 19.4 Dissociation of a higher halide compound to a lower.

This combination of high lattice energies, high bond energies with other atoms, and high hydration energies, together with the weakness of the F—F bond, makes fluorine one of the strongest oxidants known from a thermodynamic point of view. In fact it is also very reactive (fast reactions) and this is presumably also related to the low dissociation energy of the F_2 molecule, since the molecule must dissociate at one stage during the reaction. Compare fluorine and the other halogens, which are also fairly reactive, with hydrogen, oxygen, and nitrogen, which are much less reactive. This idea of associating reactivity with bond strength appears useful.

There are two other results of the high bond energies of element–fluorine bonds. First, fluorine compounds tend to be more stable with respect to dissociation into the constituent elements; in general the order of stability of halogen compounds is $F > Cl > Br > I$. The enthalpy of formation of a covalent compound AX_n, ΔH_f, is given in Fig. 19.3, where $D(A\!-\!X)$ is the bond enthalpy of the A—X bond. Such a compound would be PF_5, for example. The more negative the enthalpy of formation, ΔH_f, the more stable the compound is. Clearly, the greater bond enthalpies of A—F bonds and the lower dissociation enthalpy of the fluorine molecule make fluorine compounds more stable than those of the other halogens. Almost invariably the order of stability of the compounds AX_n relative to elemental dissociation is $F > Cl > Br > I$. Examples are provided by the compounds NX_3 (Section 17.10), OX_2 (Section 19.12.1), H—X (Section 19.9), and the van Arkel method of purifying elements via iodides (Section 11.5.2). The fact that the lighter halogens replace the heavier from their compounds is also related to this.

The second result is that the relative capacity of the halogens to bring out the highest oxidation numbers of other elements decreases in the order $F > Cl > Br > I$. This is difficult to discuss briefly since many factors are involved. Usually, the higher halide yields on dissociation, not the elements, but a lower halide and the halogen as in Fig. 19.4. Clearly the enthalpy of dissociation $\Delta H^\ominus$ (restricting ourselves to enthalpy data as usual, Section 10.3.18), depends upon three terms, involving the bond energies $D(A\!-\!X)_n$ of the A—X bond in the molecule AX_n, $D(A\!-\!X)_m$ of the A—X bond in the molecule AX_m (these two are usually different), and the dissociation energy of the halogens. For the higher halide to be stable with respect to dissociation into the lower halide, $\Delta H^\ominus$ must be positive. Factors favouring this are low dissociation energies of the halogens, and high bond energies, since $n > m$ (think about this). The result of these two factors leads to the order given above; the low bond dissociation energy of F_2, and the high element–fluorine bond enthalpies, lead to the stabilization of AF_n relative to ACl_n, etc. Examples of higher fluorides which have no chloro-analogues are: AsF_5, BiF_5, BrF_5, CrF_6, IF_7, IrF_6, SiF_6, SF_6, and XeF_6. Iodine, for example, forms

Table 19.3. Electrode potentials for the halogens.

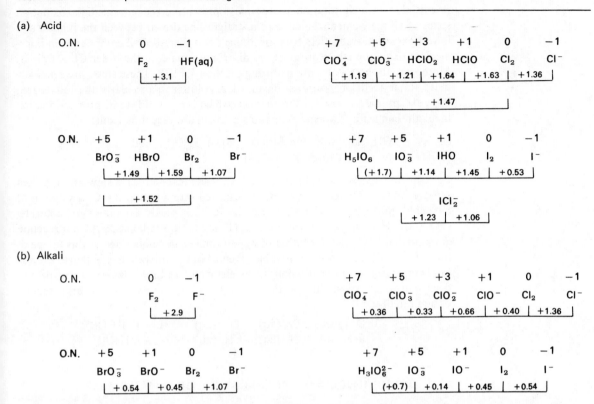

the compounds IF$_7$, ICl$_3$, IBr, and I$_2$. The same treatment could be given for ionic compounds, leading to the same result, where lattice energies would replace bond energies in the discussion.

The ionization energies of the halogens are very high. No cation F$^+$ is possible under normal conditions, even solvated. Note that no positive oxidation states of fluorine are found, by definition, since it is the most electronegative element (Section 8.14.2). The thermodynamic possibility of the existence of the ion X$^+$ increases down the group, and free iodine cations, I$^+$, probably exist; for example, iodine monochloride, ICl, in concentrated sulphuric acid solution probably contains I$^+$. Conversely, the ionic character of the halides, X$^-$, decreases in the order F$^-$ > Cl$^-$ > Br$^-$ > I$^-$. For example, compare the halides of aluminium (Section 15.2.6); the increased covalent character of the iodides correlates with their greater solubility in organic solvents, their lower melting point, and so on. The increasing polarizability of the halogens correlates well with their increasing tendency to form charge transfer complexes (Section A.9): F$_2$ < Cl$_2$ < Br$_2$ < I$_2$. In general, the tendency to form stable complexes, MX$_6^{2-}$ etc., decreases in the order F > Cl > Br > I, though this does depend upon the nature of the central atom or ion. The later transition metal ions, for example those of Pt(II), Pd(II), Cu(I), Ag(I) etc., show the reverse order. The acids H—X show increasing acid strength and reducing power down the group (Section 13.4).

Disproportionation reactions are important in halogen chemistry. Figure 19.1 reveals that in acid solution the points representing the intermediate oxidation states of all the elements lie close to a straight line drawn between the highest and lowest oxidation states (less true for iodine), and therefore all compounds in intermediate oxidation states are liable to disproportionate, in the order Cl > Br > I. Moreover, all these states are oxidizing (Section 7.12.3). These states are relatively more stable to disproportionation in alkali solution, particularly the iodate ion, IO_3^-, except in each case for the element, which is less stable to disproportionation in alkali than acid. Two such reactions in alkali are very important:

(1) $X_2 + 2OH^-(aq) \rightarrow X^-(aq) + OX^-(aq) + H_2O(l)$ and
(2) $3OX^-(aq) \rightarrow 2X^-(aq) + XO_3^-(aq)$.

For chlorine, reaction (1) is rapid at 20 °C, while reaction (2) is slow at 20 °C, but rapid at 70 °C. Therefore, if chlorine is passed into alkali at 20 °C, a solution of hypochlorite, OCl^-, is obtained; but at 70 °C it yields a solution of chlorate, $ClO_3^-(aq)$. For bromine, both reactions (1) and (2) are rapid at 20 °C, but reaction (2) is slow at 0 °C, and a solution of hypobromite can be obtained at this temperature. For iodine, reaction (2) is rapid, even at 0 °C, and iodate is obtained even at low temperatures. In acid solution the elements are more stable with respect to disproportionation, and in fact with iodine and bromine in acid the reverse reaction occurs rapidly:

$$BrO_3^-(aq) + 5Br^-(aq) + 6H^+(aq) \rightarrow 3Br_2 + 3H_2O(l);$$
$$IO_3^-(aq) + 5I^-(aq) + 6H^+(aq) \rightarrow 3I_2 + 3H_2O(l).$$

19.2 OCCURRENCE AND PREPARATION OF THE ELEMENTS

Fluorine and chlorine are fairly abundant, bromine and iodine less so, (Table 19.1). Fluorine is present mainly in the insoluble fluorides of calcium: calcium fluoride (fluospar, U.S.A.); cryolite, Na_3AlF_6 (Greenland); and fluoroapatite, $3Ca_3(PO_4)_2 \cdot CaF_2$. Sea-water contains the chlorides, bromides, and iodides, of sodium, potassium, magnesium, and calcium, but mainly sodium chloride (2.5 percent by weight). The deposits of dried-up seas contain these compounds, mainly the chlorides, for example sodium chloride and carnallite, $KCl \cdot MgCl_2$. Certain forms of marine life concentrate iodine in their systems; various seaweeds, for example, contain up to 0.5 percent of iodine. In addition, Chile saltpetre (Section 14.2.1), contains up to 0.2 percent of sodium iodate.

Fluorine was first prepared by Moissan in 1886. Aqueous electrolysis of fluoride solutions yields oxygen, and in any case fluorine reacts with water. To prepare fluorine by electrolysis no other anion must be present (fluorine is the most electronegative element), and not even a trace of water, which would react with the fluorine to yield oxygen, which would burn away the carbon anode at the temperature of the reaction. Anhydrous pure hydrogen fluoride is used, but it is a non-conductor, and therefore potassium fluoride is added to make the solution conducting. A mixture with a molar proportion of one potassium fluoride to about two and one-half hydrogen fluoride melts at about 90 °C, and this is used in the preparation. During the electrolysis, fluorine is evolved at the anode, hydrogen at the cathode, and they are collected separately. The gases must not be allowed to mix. As the hydrogen fluoride is used up, more is added to prevent the melting point of the mixture from rising. The electrolysis is conducted in a steel, copper, or nickel vessel, with a carbon anode impregnated with copper to render it inert, and a steel cathode.

Chlorine, bromine, and iodine can all be prepared by heating the appropriate alkali metal halide with manganese dioxide, MnO_2, and sulphuric acid:

$$2NaX + MnO_2 + 3H_2SO_4 \rightarrow 2NaHSO_4 + 2H_2O + X_2.$$

Chlorine can be prepared by the action of concentrated hydrochloric acid on a strong oxidizing agent such as manganese dioxide (requires heating) or potassium permanganate:

$$MnO_2(s) + 4HCl(aq) \longrightarrow MnCl_2(aq) + Cl_2(g) + 2H_2O(l);$$
$$2KMnO_4(s) + 16HCl(aq) \longrightarrow 2MnCl_2(aq) + 5Cl_2(g) + 2KCl(aq) + 8H_2O(l).$$

Chlorine can also be prepared by the action of dilute acids on bleaching powder (Section 19.12.3).

Chlorine is manufactured in three ways: by the electrolysis of fused sodium chloride (Section 11.5.2); by the electrolysis of concentrated solutions of brine, yielding chlorine at the anode, hydrogen at the cathode, and leaving sodium hydroxide in solution; and by the copper-catalysed oxidation of hydrogen chloride gas to chlorine and water. Bromine is obtained from either the bromide in seawater, or the mother liquor from the extraction of carnallite. The aqueous solution is acidified to a pH of 3.5, to avoid hydrolysis of the bromine produced, and a mixture of air and chlorine is passed through the solution. The chlorine displaces the aqueous bromide and the air transports it out of solution.

$$\tfrac{1}{2}Cl_2(g) + Br^-(aq) \rightarrow Cl^-(aq) + \tfrac{1}{2}Br_2(g).$$

Iodine is obtained from the iodate left in the mother liquor after the crystallization of sodium nitrate from Chile saltpetre, by treatment with bisulphite:

$$2IO_3^-(aq) + 5HSO_3^-(aq) \rightarrow 3HSO_4^-(aq) + 2SO_4^{2-}(aq) + H_2O(l) + I_2.$$

In another method, the soluble salts in dried seaweed are extracted with water, the filtrate is concentrated and is then treated with chlorine, which displaces the iodine.

19.3 PROPERTIES OF THE ELEMENTS

The halogens are coloured (the depth of colour increasing down the group), volatile, diatomic molecules with typical penetrating odours. There is no special significance in the facts that fluorine and chlorine are gases, bromine is a liquid, and iodine a solid, at 25 °C. On a planet on which the ambient temperature was 25 K, all would be solids, and the inert gases would span the phase range: He(g); Ne(l); Ar, etc. (s). All the halogens dissolve slightly in water and colour it, but fluorine reacts to give oxygen and ozone. They are more soluble in organic solvents, the resulting colour depending upon the halogen and the solvent. In non-polar solvents such as carbon disulphide, carbon tetrachloride, paraffins, etc., iodine is violet, bromine is red, and chlorine is yellow. Presumably these are present as free molecules as in the gas phase. But in nucleophilic (electron-donating) solvents, N, such as benzene, alkenes, alcohols, ketones, liquid sulphur dioxide, etc., the halogens give brown solutions (especially iodine, though chlorine and bromine do to a lesser

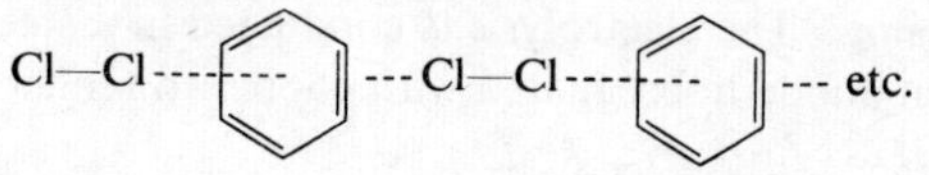

Fig. 19.5 Charge-transfer complex between benzene and chlorine.

extent). These colours are due to complex compound formation, $N \rightarrow I_2$ in solution. Such compounds are called charge-transfer compounds since the electronic transition which causes the colour results in a transfer of charge from the solvent molecule to the iodine (Section A.9). In other words, an electron located mainly in an orbital on the solvent molecule jumps into an orbital located mainly on the iodine, and absorbs energy from light, causing the compound to look coloured (Section A.9). Such compounds have been isolated at low temperatures (Fig. 19.5). These complexes form before halogens react with benzene, alkenes, etc. (Chapters 22 to 26).

The halogens are stable relative to dissociation into atoms at normal temperatures, but are very reactive in the presence of other molecules. Astatine (Greek—unstable) exists only as short-lived radioactive isotopes and is not found in nature. Tracer studies show that it is more electropositive than iodine, as expected.

By no means can iodine be described as a metal, or even as a semi-metal. However, it is more metallic than the other halogens, as is obvious from Table 19.1. Further evidence is: the slight metallic lustre of solid iodine and the small conductivity of the molten solid; the "amphoteric" nature of the hydroxide, HOI, in that compounds Na^+OI^- (acid behaviour) and ICl (base behaviour, Section 11.1.2) exist; and the existence of compounds containing positive iodine, I^+, e.g. $[I \cdot py_2]^+$, and ICl in concentrated sulphuric acid (py = pyridine, Fig. 20.2).

19.4 CHEMICAL PROPERTIES OF THE ELEMENTS

The reactivity of the elements decreases on descending the group. Fluorine is the most reactive of all elements. This is because it has a very low dissociation energy (Table 19.1) and therefore a low activation energy for most reactions (since reactions must involve bond-breaking), and it forms stable bonds with other elements, both ionic and covalent (Tables 19.1 and 19.2). The reactivity of fluorine prevented its isolation until 1886. The element bonds to all elements except helium, neon, and argon, and combines directly with all elements, except nitrogen and the inert gases, often at ordinary temperatures, and often very vigorously. Organic compounds often inflame in fluorine, inorganic compounds are fluorinated (but more controlled methods of fluorination are now available, for example using bromine trifluoride, BrF_3). Chlorine and bromine also combine with most elements directly (but not oxygen or carbon, for example) but less vigorously than does fluorine. Iodine does not combine with some elements, sulphur for example, but is still fairly reactive, for example with phosphorus, iron, and mercury. Obviously, none of the halogens is found in the uncombined state, but they are found as compounds. Metals generally yield ionic solids in their reactions with the halogens; non-metals give covalent halides. In both cases the stoichiometry of the products with different halogens may differ: for example, phosphorus yields the compounds PCl_5 and PI_3, antimony yields $SbCl_5$ and SbI_3 (Section 17.10). Fluorine reacts with water, and in fact it fumes in moist air:

$$2F_2(g) + 2H_2O(g) \rightarrow 4HF(g) + O_2(g).$$

The other halogens are sparingly soluble in water ($Br_2 > Cl_2 > I_2$), and partly react:

$$X_2 + 2H_2O(l) \rightleftharpoons HOX(aq) + H_3O^+(aq) + X^-(aq).$$

In a saturated solution at 25 °C about one-third of the chlorine, one percent of the bromine, and very little of the iodine has reacted. In sunlight the hypohalous acid formed, HOX (Section 19.12), decomposes to oxygen and the hydrogen halide, and therefore both chlorine water and bromine water evolve oxygen in sunlight. When chlorine is passed into cold water, crystals of the water clathrate (Section 12.5), chlorine hydrate ($Cl_2 \cdot 7.3\ H_2O$), separate. Similarly, the clathrates $Br_2 \cdot 1.0\ H_2O$ and $I_2 \cdot xH_2O$ have been reported. Iodine is very soluble in an aqueous solution of potassium iodide, owing to the formation of the deep-brown triiodide ion I_3^-. This ion dissociates readily to iodine and iodide, and therefore the solution behaves as if it contained free iodine. This is a useful method of obtaining an aqueous "solution" of iodine. Fluorine reacts with alkalis to yield the oxide, F_2O:

$$2F_2(g) + 2OH^-(2\%\ aq.\ solution) \rightarrow F_2O(g) + 2F^-(aq) + H_2O(l).$$

The other halogens react with alkalis to yield a solution of the hypohalite ion, $OX^-(aq)$, which may disproportionate (Section 19.12.3).

All the halogens are oxidizing agents, their power as oxidants decreasing in the order $F_2 > Cl_2 > Br_2 > I_2$: fluorine, for example, explodes with hydrogen in the dark; chlorine does so only in bright sunlight; bromine reacts with hydrogen only in the presence of a platinum catalyst at 200 °C; and the reaction with iodine is reversible. Whereas chlorine rapidly oxidizes compounds containing hydrogen, for example warm turpentine ($C_{10}H_{16}$) inflames in chlorine,

$$C_{10}H_{16}(l) + 8Cl_2(g) \rightarrow 10C(s) + 16HCl(g),$$

bromine and iodine are much less reactive. Iodine is easily the weakest oxidant of the four. The element will not oxidize ammonia to nitrogen. The reduction potential for the couple I_2/I^- is only $+0.53$ V, and even mild oxidizing agents oxidize aqueous iodides to iodine, for example the oxygen of the air. This reaction, I_2/I^-, is very useful since it is easily reversible, it is quantitative in many cases, an indicator (starch) exists, and iodine can be kept in solution as the ion $I_3^-(aq)$ in the presence of an excess of potassium iodide. Iodine is usually estimated using sodium thiosulphate solution, which is itself oxidized to sodium tetrathionate:

$$2S_2O_3^{2-}(aq) + I_2(aq) \rightarrow S_4O_6^{2-}(aq) + 2I^-(aq).$$

Iodine in the presence of starch is deep blue. Starch contains a helical chain of glucose molecules. The iodine molecules are absorbed reversibly in the helical chain to form a deep-blue compound.

In general, the electrode potentials are useful for predicting the feasibility of reactions (Problem 5). Moreover, the speed of reaction usually decreases in the order $Cl_2 > Br_2 > I_2$.

19.5 USES

Chlorine is used as a cheap industrial oxidant: in the manufacture of bromine (Section 19.2); as a bleaching agent (either as chlorine or bleaching powder); and as a germicide (water sterilization). It is used in the manufacture of non-inflammable solvents (carbon tetrachloride, etc.), hydrogen chloride, and plastics such as PVC (Section 22.2.4). Bromine is used in the manufacture of various organic compounds including ethylene dibromide which is added to lead tetraethyl (an anti-knock agent) in petrols to prevent accumulation of lead in the engine. It is considered preferable to accumulate the lead in our bodies. Silver bromide and iodide are used in black-and-white photography. Photographic plates contain silver bromide which decomposes to silver on exposure to light. The silver forms the black opaque shadow on the "negative" (why is it given this name?), and the unreacted silver bromide is removed from the plate as a soluble complex anion, $[Ag(S_2O_3)_2]^{3-}$, formed by reaction with sodium thiosulphate ("hypo").

19.6 THE ANOMALOUS NATURE OF FLUORINE

Fluorine differs from the other elements in the group because of:

1. Its "unexpectedly" low dissociation energy (Table 19.1), attributed to non-bonding electron repulsion (Section 15.1). This results in its compounds being slightly more stable with respect to dissociation into the elements than expected —a thermodynamic effect. Also, more importantly, it results in a low activation energy for most reactions, therefore its reactions are quicker than those of other halogens under the same conditions (for example with hydrogen, Section 19.4), and it is said to be more "reactive" (a kinetic result). However, not only does fluorine react more quickly than the other halogens, but its compounds are generally more stable. This is rationalized below.
2. The small size of the fluorine atom. This results in its covalent bonds to elements other than oxygen, nitrogen, and itself (Section 15.1) being stronger than those of the other halogens (Table 19.2). Consequently, covalent fluorides have larger negative free energies and enthalpies of formation, and are more stable with respect to dissociation into both the constituent elements and into compounds of lower oxidation number (Figs. 19.3 and 19.4) than the corresponding chlorides, etc. In other words, fluorine brings out higher oxidation numbers of other atoms than the other halogens (Section 19.1).
3. The small size of the fluoride ion. This causes ionic fluorides to have higher lattice energies than the corresponding compounds of the other halides, with similar results as in the case of covalent fluorides, above.
4. Its restriction to an octet of outer electrons. Two results of this are: the inertness of many fluoro compounds, for example, carbon tetrafluoride and other fluorohydrocarbons, and silicon hexafluoride; and the restriction of fluorine to monocovalency.

The result of these differences is that fluorine is the most electronegative element in the Periodic Table, and is a powerful oxidant, despite the fact that its electron affinity is less than that of chlorine (Section 10.3.4).

Some differences between fluorine and the other halogens are listed below:

1. Fluorine is more reactive and its compounds are more stable.
2. Fluorine is almost invariably monoco-ordinate (co-ordination number = 1) and is never more than monocovalent.
3. The solubility of fluorine compounds often differs from that of the corre-

sponding chlorine compounds, for example silver fluoride is soluble, calcium and magnesium fluorides are insoluble.
4. Fluorides are more ionic.
5. Point (4) may result in fluorides having different crystal structures, for example

| Compound: | $AlF_3(s)$, | $AlCl_3(s)$, | $Al_2Br_6(s)$, | $Al_2I_6(s)$, | $CdF_2(s)$, | $CdI_2(s)$, |
| Structure: | 3-D ionic | layer | dimeric | dimeric | fluorite | layer |

6. Fluorine forms strong hydrogen bonds (Section 4.8), resulting in the properties of hydrogen fluoride, like water, being anomalous (Section 19.8).
7. Fluorine brings out the highest oxidation (and co-ordination) numbers from other elements (Section 19.1).
8. Fluorine compounds are often more inert (e.g. SF_6).
9. Trifluoroacetic acid is a strong acid (Section 28.5.1), and the substituted amine $(CF_3)_3N$ is not basic.
10. Fluorine forms stronger bonds, both ionic and covalent.
11. The reactions of fluorine with oxygen-containing compounds, and the compounds of fluorine containing oxygen, often differ from those of the other halogens. For example, the reaction of fluorine with water yields a mixture of fluoride, fluorine monoxide, oxygen, ozone, and hydrogen peroxide. With alkali, fluorine yields mainly:

cold dilute alkali: $2F_2(g) + 2OH^-(aq) \rightarrow$
$$F_2O(g) + 2F^-(aq) + H_2O(l);$$
hot concentrated alkali: $2F_2(g) + 4OH^-(aq) \rightarrow$
$$4F^-(aq) + O_2(g) + 2H_2O(l).$$

The recently prepared oxyacid of fluorine, HOF, is very unstable; and the oxides of fluorine are not acidic!

19.7 PREPARATION OF HYDROGEN HALIDES

1. *Direct combination*: $H_2(g) + X_2(g) \rightarrow 2HX(g)$. Hydrogen fluoride cannot be prepared in this way (Section 13.2), but hydrogen chloride is manufactured by burning chlorine in an excess of hydrogen, and hydrogen bromide and iodide by passing the gaseous mixture of elements over a heated platinum catalyst.
2. *Reaction of concentrated sulphuric acid on a halide* (fluoride and chloride):

$$NaX(s) + H_2SO_4(l) \longrightarrow HX(g) + NaHSO_4 \xrightarrow[\text{excess NaX}]{\text{high temp.}} Na_2SO_4 + HX(g).$$

Hydrogen fluoride is manufactured from calcium fluoride, and hydrogen chloride was formerly manufactured from rock salt by this method (Leblanc process). Hydrogen bromide and iodide cannot be prepared using concentrated sulphuric acid (mainly the halogen is obtained), but can be prepared using a non-volatile non-oxidizing acid such as syrupy phosphoric acid.
3. *Hydrolysis of covalent halides* (Section 11.1.4); especially for the bromide and iodide. Phosphorus tribromide and triiodide are prepared *in situ* and hydrolysed immediately. For example, liquid bromine is conveniently dropped on to a mixture of red phosphorus and water; the mixture of gases evolved is passed

through glass beads coated with moist red phosphorus to remove bromine, over calcium chloride to remove water, and collected by downward delivery in air:

$$2P(s) + 3Br_2(l) \longrightarrow 2PBr_3(l) \xrightarrow{6H_2O} 2H_3PO_3(aq) + 6HX(g).$$

Hydrogen iodide is obtained by dropping water into a mixture of red phosphorus and iodine.

19.8 HYDROGEN FLUORIDE

Hydrogen fluoride is the principal source of fluorine and the fluorides. The pure compound is a colourless fuming liquid at 15 °C, and like fluorine it is extremely dangerous, causing severe burns. In the presence of moisture it attacks glass. The sodium and calcium ions in the glass are converted to fluorides, which frost and etch the glass:

$$4HF(g) + SiO_2(s) \rightarrow SiF_4(g) + 2H_2O(l).$$

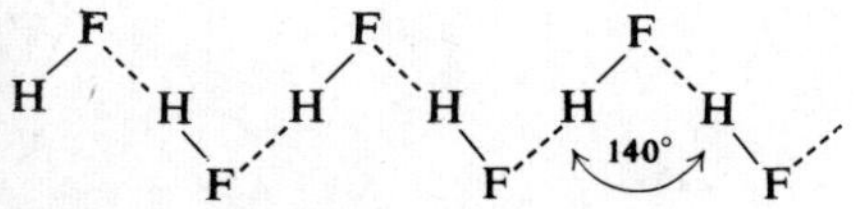

Like water (Section 13.6), hydrogen fluoride is a hydrogen-bonded compound, with a high dielectric constant and high dipole moment. Unlike water, however, solid hydrogen fluoride cannot show a three-dimensional structure and is an infinite chain in the solid state (Fig. 19.6). In the liquid state it is also heavily hydrogen-bonded; and even in the gas phase, up to 60 °C, association is observed, probably involving an equilibrium of chains and rings, $(H-F)_n$, up to $n = 6$. Other results of hydrogen bonding are: the "unexpectedly" low volatility (high melting and boiling points, Table 19.4); and the formation of the hydrogen-bonded anion HF_2^- manifesting itself in the formation of acid salts, solvent action, and the increasing strength of the acid as the concentration increases (see below). Fluorides react with hydrogen fluoride to form acid salts containing the symmetrical anion HF_2^-, in which the hydrogens around the fluoride are in equivalent positions. This can only be written in terms of two canonical structures (Section 4.12.1):

$$[F\text{-----}H-F]^- \leftrightarrow [F-H\text{-----}F]^-$$

Fig. 19.6 The structure of hydrogen fluoride.

Table 19.4. Properties of hydrogen halides.

Property	HF	HCl	HBr	HI
Melting point (°C)	−83	−114	−87	−51
Boiling point (°C)	+20	−85	−67	−35
Solubility in water (g/100 g)	miscible	51	120	154
Constant boiling point mixture: temperature	120 °C	110 °C	125 °C	127 °C
composition	35% HF	20% HCl	47% HBr	57% HI
$\frac{1}{2}X_2 + \frac{1}{2}H_2(g) \rightarrow HX(g)$; ΔH_f (kJ mol^{-1})	−271	−92	−36	+26
Bond length (Å)	0.86	1.28	1.42	1.60
Dipole moment (D)	1.82	1.12	0.83	0.44
Percent ionic bonding	43	17	17	5
Apparent degree of dissociation, 0.1 M	0.085	0.92	0.93	0.95
Percent dissociation (gas) at 1000 °C	—	3×10^{-7}	0.003	19
Dielectric constant (liquid)	66	9	6	3
Density at boiling point (g cm^{-3})	0.99	1.187	2.160	2.799

The bond strength of the hydrogen–fluorine bonds is intermediate between that of a hydrogen bond (represented by a dashed line) and a covalent bond—about 100 kJ mol^{-1}. After water, it is the most successful solvent known, for both organic and inorganic compounds. Very pure hydrogen fluoride self-ionizes as does water:

$$3HF \rightleftharpoons H_2F^+ + HF_2^-.$$

Aqueous solutions of hydrogen fluoride conduct electricity:

$$(H\text{---}F)_n + H_2O(l) \rightleftharpoons H_3O^+(aq) + F^-(aq); \quad K = 7 \times 10^{-4} \, (25 \, °C).$$

Thus it is a weak acid (Section 9.3). However, in concentrated solutions the reaction

$$F^-(aq) + HF(aq) \rightleftharpoons HF_2^-(aq); \quad K = 5.1$$

displaces the above equilibrium to the right and increases the hydrogen ion concentration (or as is said "increases the acid strength"). This is the opposite effect to the usual result of increasing the concentration.

19.9 PROPERTIES OF HYDROGEN CHLORIDE, BROMIDE, AND IODIDE

These are colourless pungent gases at 20 °C. They fume in moist air and are very soluble in water, from which hydrates can be frozen out: $HX \cdot 3H_2O$, etc. Hydrogen chloride is used as a catalyst in esterification (Section 23.1.2). They form constant boiling point mixtures (Table 19.4). The regular gradation of properties among these three compounds, and the anomalous nature of the fluoride, is obvious from the table. The hydrogen-bonding capacity of these three compounds is much less than that of hydrogen fluoride. Thus the pure liquids do not self-ionize, and are non-conducting. However, it is thought that the solid phases of the bromide and chloride may have the same hydrogen-bonded structure as solid hydrogen fluoride.

The hydrogen halides all dissolve in water to form hydroxonium ions, that is, they are acidic:

$$H\text{---}X(g) + H_2O(l) \rightarrow H^+(aq) + X^-(aq).$$

This is mainly due to the fact that the hydration energies are greater than the bond strengths involved (Section 15.2.6). Thus the aqueous solutions are good conductors of electricity, and all undergo the typical reactions of acids with metals, metal oxides, carbonates, and hydroxides. The relative strengths of these acids are discussed in Section 13.4. In water, hydrogen fluoride behaves as a weak acid, and the others are all strong acids; but in solvents which are poorer proton acceptors than is water, such as formic acid, it can be seen that the order of increasing acid strength is $HF \ll HCl < HBr < HI$. Aqua regia is a mixture of three parts of concentrated hydrochloric acid to one part of concentrated nitric acid. It dissolves platinum and gold (Section 10.4.2.). The power of the hydrogen halides to act as reducing agents decreases in the order HF (not at all) $\ll HCl \ll HBr < HI$. Thus hydrogen chloride reacts only with the strongest oxidizing agents such as metal peroxides, permanganates, dichromates, lead dioxide, etc.; hydrogen bromide reacts with moderately strong oxidants such as concentrated sulphuric acid, chlorine, and hydrogen peroxide; and hydrogen iodide reacts with even mild oxidants such as oxygen, aqueous iron(III) salts, dilute nitric acid, etc. Solutions of hydrogen bromide and hydrogen iodide are slowly oxidized by air to the halogens.

19.10 HALIDES

There are several methods of preparation (Section 11.3.2). The reaction of the dry hydrogen halide or the halogen on the heated metal yields the metal halide. If the metal exhibits more than one stable oxidation state (Section 10.3.20), the former method yields the halide of the metal in a lower oxidation state, the latter method yields the metal in a higher oxidation state:

$$2Fe(s) + 3Cl_2(g) \rightarrow 2FeCl_3(s);$$
$$Fe(s) + 2HCl(g) \rightarrow FeCl_2(s) + H_2(g).$$

The reaction of an aqueous solution of the hydrogen halide on the metal oxide, carbonate, or hydroxide, yields the corresponding salt on crystallization, frequently in hydrated form (Section 11.3.1).

Metathetical reactions (Section 11.3.2) can be used for the insoluble halides of silver, lead, and mercury(I). Metallurgical processes often use the chlorination of oxides in the presence of carbon (Section 11.5.2). Hydrated chlorides frequently form basic chlorides on heating (Section 11.3.3), and can be converted to the anhydrous chlorides by heating with thionyl chloride or 2,2-dimethoxypropane, a ketal (Section 23.3.2):

$$CrCl_3 \cdot 6H_2O(s) + 6SOCl_2(l) \rightarrow 6SO_2(g) + 12HCl(g) + CrCl_3(s);$$
$$CrCl_3 \cdot 6H_2O(s) + 6(CH_3)_2C(OMe)_2(l) \rightarrow CrCl_3(s) + 6(CH_3)_2CO(g) + 12CH_3OH.$$

The halides of non-metals are formed by direct action, often at normal temperatures. Where more than one stable oxidation state of the non-metal exists, both can frequently be obtained separately, by controlling the conditions:

$$2P(s) + 3Cl_2(g) \rightarrow 2PCl_3(l); \quad 2P(s) + 5Cl_2(g) \rightarrow 2PCl_5(s).$$
$$\text{excess} \qquad\qquad\qquad\qquad\qquad \text{excess}$$

Halogen exchange (a lighter halogen for a heavier) using the hydrogen halide or halogen can frequently be employed.

The relative stabilities of the halides were discussed in Section 19.1. Fluorides tend to be the most stable halides, and fluorine tends to stabilize other elements in higher oxidation states than do the other halogens. Such compounds, for example tungsten hexafluoride (WF_6), are covalent volatile compounds, often hydrolysed by water. Where the other element is exerting its maximum covalence, e.g. CF_4, SF_6, the compounds may be very inert compared to other halogen compounds, owing to the inability of fluorine to expand the octet. In other cases, hydrolysis to fluoro-anions, acids, or oxides occurs e.g. $TeF_6 \rightarrow Te(OH)_6$. All ionic fluorides, MF, have the rock-salt structure, except ammonium fluoride (Section 17.9). Fluorides are more ionic than the corresponding chlorides, and may have different structures. For example all fluorides, MF_2, have the fluorite or rutile (ionic) structures, whereas many chlorides have the more covalent layer structures (Table 19.5). The solubilities of ionic fluorides often differ from those of the other halides: AgF is soluble; the group IIA fluorides and lithium fluoride are sparingly soluble. The covalent fluorides are often inert, have high bond energies, and often have high ionic character.

The chlorides of the elements are discussed in Section 11.1.4. The reactions of ionic chlorides, bromides, and iodides, are summarized below (see Table 19.6). All common halides (Cl^-, Br^-, I^-) are soluble except for those of the ions Pb^{2+},

Table 19.5. Structures of halides.

		(a)	(b)	
Three dimensional	caesium chloride	MX	8:8	CsCl, CsBr, CsI, TlCl, TlBr, KHF$_2$
ionic character increasing ↑	sodium chloride	MX	6:6	CsF, MX (M = Li, Na, K, Rb; X = F, Cl, Br, I); NH$_4$X (X = Cl, Br, I); AgX (X = F, Cl, Br)
	fluorite	MX$_2$	8:4	CaF$_2$, SrF$_2$, BaF$_2$, PbF$_2$, CdF$_2$, HgF$_2$
	rutile	MX$_2$	6:3	CaBr$_2$, CaCl$_2$, SrCl$_2$, CuF$_2$, PdF$_2$, ZnF$_2$, MgF$_2$, CoF$_2$, NiF$_2$, FeF$_2$
	zinc blende	MX	4:4	AgI, CuCl, CuBr, CuI
	wurtzite	MX	4:4	NH$_4$F
Partly covalent layer structures	cadmium chloride	MX$_2$	ccp $\frac{1}{2}$ oct.	CdCl$_2$, ZnBr$_2$, MgCl$_2$, CoCl$_2$, NiCl$_2$, FeCl$_2$
	cadmium iodide	MX$_2$	hcp $\frac{1}{2}$ oct.	CdBr$_2$, CdI$_2$, ZnI$_2$, NiBr$_2$, PbI$_2$, CaI$_2$, MgBr$_2$, MgI$_2$, CoBr$_2$, CoI$_2$, FeBr$_2$, FeI$_2$
	chromium(III) chloride	MX$_3$	ccp $\frac{1}{3}$ oct.	CrCl$_3$
	iron(III) chloride	MX$_3$	hcp $\frac{1}{3}$ oct.	MCl$_3$, (M = Fe, Sc, Ti, V); MBr$_3$ (M = Fe, Ti, Cr); MI$_3$ (M = Bi, Sb, As)
Mainly covalent chain structures				CuBr$_2$, CuCl$_2$, PdCl$_2$, BeCl$_2$
Covalent molecular structures				Al$_2$Br$_6$, Al$_2$I$_6$, Fe$_2$Cl$_6$(g)

(a) stoichiometry; (b) type of co-ordination (Section 3.3) and classification of structure (Table 3.2).

Hg_2^{2+}, Ag^+, and Cu^+. Sodium chloride reacts with concentrated sulphuric acid to give hydrogen chloride gas, but the reactions of this acid with bromides and iodides give mainly the halogen. The same reaction in the presence of a more powerful oxidizing agent (manganese dioxide for example) yields the halogen in each case. Sodium chloride and potassium dichromate when treated with concentrated sulphuric acid yield the volatile red compound, chromyl chloride; this reacts with sodium hydroxide to form sodium chromate, which can be recognized by the formation of the yellow insoluble lead chromate on the addition of lead ions:

$$K_2Cr_2O_7 + 6H_2SO_4 + 4NaCl \longrightarrow$$
$$2KHSO_4 + 4NaHSO_4 + 3H_2O + 2CrO_2Cl_2 \xrightarrow{OH^-} CrO_4^{2-}.$$

Bromides and iodides yield bromine and iodine respectively when treated in the same way (cf. MnO$_2$). Aqueous solutions of halide ions precipitate insoluble silver halides from aqueous silver ion solutions. They are all insoluble in dilute nitric acid, but the white silver chloride is soluble in dilute ammonia, the less soluble cream silver bromide is insoluble in dilute but soluble in concentrated ammonia, and the least soluble yellow silver iodide is even insoluble in concentrated ammonia. Aqueous solutions of lead acetate precipitate white lead halides from solutions of halide ions (lead iodide is yellow), which are soluble in hot water. Acid hypochlorite solutions oxidize aqueous solutions of bromide ions to bromine, and iodide ions to iodine. In the presence of a layer of carbon tetrachloride, the

Table 19.6. Reactions of the halides.

Reaction	F⁻(aq)	Cl⁻(aq)	Br⁻(aq)	I⁻(aq)
Silver nitrate (aq)	no precipitate	white precipitate	cream precipitate	yellow precipitate
Solubility of AgX in				
a) water	soluble	insoluble	insoluble	insoluble
b) conc. HNO_3(aq)	soluble	insoluble	insoluble	insoluble
c) dilute NH_3(aq)	soluble	soluble	insoluble	insoluble
d) conc. NH_3(aq)	soluble	soluble	soluble	insoluble
Lead acetate	white precipitate	white precipitate	white precipitate	white precipitate
Chlorine (aq) or hypochlorite	no reaction	no reaction	yields bromine (orange-red in CCl_4)	yields iodine (violet in CCl_4)
Copper(II) (aq)	no reaction	no reaction	no reaction	cream precipitate

Reaction	F⁻(s)	Cl⁻(s)	Br⁻(s)	I⁻(s)
Conc. H_2SO_4	HF(g)	HCl(g)	Br_2(+HBr)	I_2(+HI)
Conc. H_2SO_4 + MnO_2(s)	HF	Cl_2	Br_2	I_2
Conc. H_2SO_4 + $K_2Cr_2O_7$(s)	HF	CrO_2Cl_2 chromyl chloride	Br_2	I_2

organic layer is coloured red and violet respectively. Solutions of iodide ions precipitate copper(I) iodide from aqueous solutions of copper(II) ions, and red mercury(II) iodide from solutions of mercury(II). This iodide is soluble in an excess of iodide ions owing to complex formation, HgI_4^{2-} (Section 20.4).

19.11 INTERHALOGEN COMPOUNDS AND POLYHALIDES

The interest in the interhalogens is mainly in (a) their structures, which can be "forecast" from the Gillespie–Nyholm rules, (b) their use as fluorinating agents, and (c) their possible use as non-aqueous solvents. The compounds are of the type AB_n, where A and B are halogens, A is heavier than B, and $n = 1, 3, 5,$ or 7. Ternary interhalogens, e.g. AB_nX_2, have not yet been prepared (though they exist as polyhalide anions, Fig. 19.7), since they decompose to a halogen X_2 and a binary interhalogen, AB_n. Most of the interhalogens known are fluorides. All can be prepared by direct combination of the constituent elements, or by metathesis (lower halogen on interhalogen):

$$Cl_2(g) + F_2(g) \xrightarrow{200\,°C} 2ClF(g); \quad Cl_2(g) + 3F_2(g) \xrightarrow{300\,°C} 2ClF_3.$$

They are all covalent molecules which exist as diamagnetic volatile solids or liquids, except chlorine monofluoride which is a gas at 25 °C. Their physical properties (colour, for example) are intermediate between those of the constituent halogens, except that their melting and boiling points are a little higher than

AB	AB$_3$	AB$_5$	AB$_7$	ABC$^-$		AB$_4^-$	
ClF	ClF$_3$	(ClF$_5$)	IF$_7$	Br$_3^-$	ClBr$_2^-$	ClF$_4^-$	IF$_6^-$, I$_5^-$
ClI	BrF$_3$	BrF$_5$		I$_3^-$	IBr$_2^-$	BrF$_4^-$	I$_7^-$
BrI	(IF$_3$)	IF$_5$		BrCl$_2^-$	IBrF$^-$	IF$_4^-$	I$_9^-$
BrCl	I$_2$Cl$_6$			ICl$_2^-$	IBrCl$^-$	ICl$_4^-$	
BrF						ICl$_3$F$^-$	I$_8^{2-}$
(IF)							irregular chains

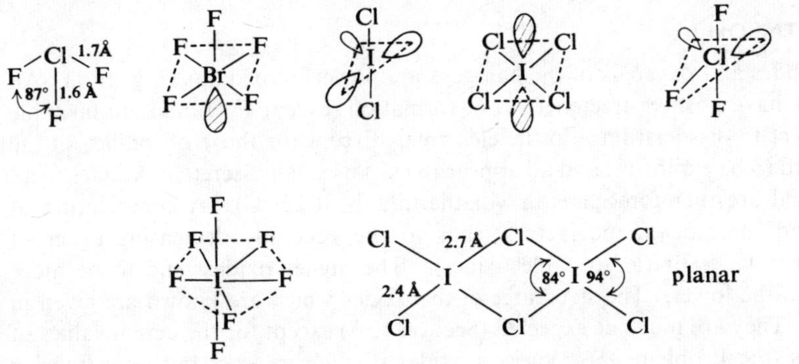

Fig. 19.7 Interhalogen compounds and polyhalides.

expected by interpolation. They are more reactive than the halogens, and all have negative or small positive enthalpies of formation (e.g. IBr, $\Delta H_f^\ominus = +40.8$ kJ mol^{-1}). The interhalogens are hydrolysed by water or alkali to the halide ion of the lighter halogen, B$^-$, and the hypohalite ion OA$^-$, which may react further. The interhalogens add across alkene double bonds, and iodine monochloride is used as a measure of unsaturation of fats and oils. They are strong oxidants, and can be used to halogenate metals and metal oxides for example, except in the case of the most electropositive metals which form polyhalides (below). Lower halides are displaced from their salts e.g. MI$_2$ + Cl$_2$ → MCl$_2$ + I$_2$. The fluorine compounds are very useful fluorinating agents; bromine trifluoride for example is widely used. All the interhalogens are potentially useful non-aqueous solvents, showing similar solvent properties to water. Self-ionization of liquid interhalogens produces polyhalide ions, such as ICl$_4^-$:

$$2ICl_3(l) \rightleftharpoons ICl_2^+ + ICl_4^-;$$
$$2IF_5 \rightleftharpoons IF_4^+ + IF_6^-.$$

These can be regarded as halogens (or interhalogens) associated with halide ions:

$$ICl_3 + Cl^- \rightarrow ICl_4^-.$$

They have been known for some time. The solubility of iodine in water is low, but is greatly increased by the presence of an excess of iodide ions:

$$I_2(s) \rightarrow I_2(aq); \quad I_2(aq) + I^-(aq) \rightarrow I_3^-(aq), \quad \text{deep brown.}$$

The large I$_3^-$ ion is very soluble as expected, (Section 10.3.17). Few polyhalide ions are stable in aqueous solution however, and they are normally isolated as salts containing large cations (more stable), usually by the reaction of a halogen or interhalogen with a halide, often in the absence of a solvent:

$$K^+F^-(s) + BrF_3(l) \rightarrow K^+BrF_4^-(s).$$

Apart from the polyiodides, I_5^- etc., which are irregular chains, the shapes of the polyhalides can be "predicted" from the Gillespie–Nyholm rules. They all decompose on heating to the halide and halogen (or interhalogen):

$$Rb^+ICl_2^-(s) \rightarrow RbCl(s) + ICl(l).$$

19.12 THE OXIDES AND OXYACIDS OF THE HALOGENS

19.12.1 The Oxides

All the oxides and oxyacids of the halogens are powerful oxidants (Fig. 19.1). All the oxides have positive free energies of formation (except F_2O) and are unstable with respect to dissociation into the elements. Except for those of iodine, all the oxides tend to be explosive, and all appear to exist as small discrete molecules in all phases (and are therefore gases or volatile liquids at 25 °C). A combination of kinetic and thermodynamic factors leads to the generally decreasing order of stability I > Cl > Br (least stable oxides). The higher oxides tend to be more stable than the lower. The structures of the oxides which are known are given in Fig. 19.8. They are much as expected (Section 12.7) except for the actual values of the angles (see Problem 22). Various oxides of fluorine exist but only fluorine monoxide, F_2O, is thermally stable at 25 °C. It is non-explosive (negative $\Delta G_f^\ominus$) and is prepared by bubbling fluorine through an aqueous 2 percent solution of sodium hydroxide:

$$2F_2(g) + 2OH^-(aq) \rightarrow 2F^-(aq) + F_2O(g) + H_2O(l).$$

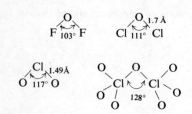

Fig. 19.8 Structures of oxides of halogens.

It is a pungent gas, slowly hydrolysed by alkali (and even by water in time) to oxygen; it is a strong oxidant; it is not acidic; and its solution in water is neutral.

All the oxides of chlorine are unstable relative to dissociation into the elements (positive $\Delta G_f^\ominus$) and tend to explode. They are all highly reactive oxidizing agents; and are all molecular, covalent, volatile compounds. Chlorine monoxide, Cl_2O, is a yellowish-red gas, prepared by passing chlorine over fresh yellow mercury(II) oxide:

$$2Cl_2(g) + 2HgO(s) \rightarrow HgO \cdot HgCl_2(s) + Cl_2O(g)$$

It reacts with water to form a solution of hypochlorous acid. Chlorine dioxide is a yellow gas, which explodes unless diluted with another gas. It is therefore best prepared, not by the reaction of concentrated sulphuric acid on a chlorate, but by the similar reaction using oxalic acid instead, when the product is diluted by carbon dioxide. The equations are given below.

$$H_2SO_4 + KClO_3 \rightarrow KHSO_4 + HClO_3.$$

The chloric acid then disproportionates.

$$3HClO_3 \xrightarrow{\text{disprop}^n} HClO_4 + H_2O + 2ClO_2.$$

Some carbon dioxide is formed when oxalic acid is used:

$$2HClO_3 + H_2C_2O_4 \rightarrow 2CO_2 + 2H_2O + 2ClO_2.$$

It is an odd-electron molecule (draw a valence-bond structure for it) and is therefore paramagnetic, but is not associated. The gas dissolves in water to yield an aqueous solution, which is stable in the dark (the hydrate $ClO_2 \cdot 8H_2O$ can be obtained), but which disproportionates in sunlight or in alkali (rapidly):

$$6ClO_2(g) + 3H_2O(l) \rightarrow HCl(aq) + 5HClO_3(aq);$$
$$2ClO_2(g) + 2OH^-(aq) \rightarrow ClO_2^-(aq) + ClO_3^-(aq) + H_2O(l).$$

It is a powerful oxidant and is used commercially as a bleach and a germicide. Chlorine hexoxide, Cl_2O_6, is a red oily liquid, prepared by treating ozone with chlorine dioxide:

$$2O_3(g) + 2ClO_2(g) \rightarrow Cl_2O_6(l) + 2O_2(g).$$

It disproportionates readily in water or alkali to chlorate and perchlorate. Chlorine heptoxide, Cl_2O_7, is a colourless oil, and is the most stable of the chlorine oxides, but it still explodes on shock. It is prepared by dehydrating perchloric acid with phosphorus pentoxide at $-10\ °C$. It is acidic and is a strong oxidant.

The bromine oxides are the least stable halogen oxides ("middle row anomaly") and exist only at low temperatures. They are prepared from bromine and ozone at low temperatures, or from oxygen in an electric discharge. They are very powerful oxidants, and react with water to yield mixtures of acids. The iodine oxides are insoluble solids which decompose on heating. Iodine pentoxide is a white solid prepared by heating iodic acid at 200 °C. It is stable at 25 °C and is used as an oxidant, for example in the estimation of carbon monoxide:

$$I_2O_5(s) + 5CO(g) \rightarrow 5CO_2(g) + I_2(s).$$

The other oxides are less stable. The oxides and oxyacids of the halogens are summarized in Table 19.7.

19.12.2 Species Present in Aqueous Solutions of the Halogens

All the halogens are soluble in water to some extent (fluorine reacts, Section 19.4). At pH = 1 the species present are:

$$X_2(aq) + H_2O(l) \rightleftharpoons H^+(aq) + X^-(aq) + HOX(aq);$$
$$K = 10^{-4}(Cl);\ 10^{-8}(Br);\ 10^{-13}(I).$$

Table 19.7. The oxides and oxyacids of the halogens.

Oxidation state	Fluorine[a]	Chlorine		Bromine		Iodine	
+1	F_2O, F_2O_2, F_2O_3, F_2O_4	$Cl_2O(g)$	HOCl*	$Br_2O(l)$	HOBr*		HOI
+3	(F—O—H)		$HClO_2$*		$HBrO_2$*?		
+4		$ClO_2(g)$		$BrO_2(d)$		$I_2O_4(s)$	
+5			$HClO_3$*		$HBrO_3$	$I_2O_5(s)$	HIO_3
+6		$Cl_2O_6(l)$		$BrO_3(s)$ (or Br_3O_8)			
+7		$Cl_2O_7(l)$	$HClO_4$	$Br_2O_7(?)$	$HBrO_4$*	$I_2O_7(s)$ $I_4O_9(s)$	HIO_4 etc.

* In solution only.
[a] Oxidation states do not apply to fluorine.

N.B. This is a poor preparation of the acids HOX, even for HOCl, but the equilibrium can be moved to the right in the presence of mercury(II) oxide, yielding the basic chloride $HgCl_2 \cdot HgO$. The acids themselves are unstable with respect to disproportionation, especially in alkaline solution (Section 19.1). The result is that in alkaline solution the species formed are: $NaOCl(20°)$ and $NaClO_3(100°)$; $NaOBr(0°)$ and $NaBrO_3(25°)$; and $NaIO_3$ (even at 0 °C).

Fluorine reacts rapidly with water to liberate oxygen; chlorine and bromine react similarly but much more slowly via the decomposition of the hypohalous acid, HOX. Iodine is more stable, and in fact in acid solution the reverse reaction occurs, and iodide ions are oxidized by oxygen from the air to iodine.

19.12.3 Oxyacids

In general, for the oxyacids of the halogens, as the oxidation number increases from $+1$ to $+7$: (a) the thermal stability increases; (b) the acid strength increases (Section 13.4); and (c) the oxidizing power generally decreases in the order (for a given acid HXO_n) $Cl > Br > I$.

Hypohalous acids, HOX(aq), can be prepared in aqueous solution by shaking the halogen in water with mercury(II) oxide:

$$2X_2(aq) + 2HgO(s) + H_2O(l) \rightarrow HgX_2 \cdot HgO(s) + 2HOX(aq).$$

The compounds are known only in aqueous solution, and are all weak acids. All decompose slowly in solution:

$$HOX(aq) \rightleftharpoons XO^-(aq) + H^+(aq); \quad K_{298} = 3 \times 10^{-8}(Cl); 2 \times 10^{-9}(Br);$$
$$10^{-11}(I).$$

The fluorine analogue, HOF, has been prepared only recently, and is very unstable. Solutions of alkali metal hypochlorites and hypobromites can be prepared by passing chlorine and bromine into cold alkali solutions (Section 19.1). These ions readily disproportionate (Fig. 19.1). The alkali metal hypochlorites can be isolated as hydrated solids, such as $NaOCl \cdot 7H_2O$. The hypohalous acids and hypohalites are oxidizing agents, and are used as bleaches. Oxidation can occur by oxygen transfer:

$$*OCl^-(aq) + NO_2^-(aq) \rightarrow Cl^-(aq) + [*ONO_2]^-(aq).$$

Sodium hypochlorite solution is prepared by the electrolysis of cold brine (Section 19.1) in such a way that the chlorine evolved at the anode mixes with the alkaline solution left around the cathode after the liberation of hydrogen. Usually the chloride ions left in solution do not matter. Sodium hypochlorite is strongly hydrolysed in solution (is the solution acid or alkali?). It can decompose in one of two ways: to oxygen, catalysed by sunlight, and various transition metal ions, $Ni^{2+}(aq)$, $Fe^{2+}(aq)$, $Co^{2+}(aq)$, (Section 20.11); and to chlorate by disproportionation, especially when heated:

$$2OCl^-(aq) \rightarrow 2Cl^-(aq) + O_2(g);$$
$$3OCl^-(aq) \rightarrow 2Cl^-(aq) + ClO_3^-(aq).$$

Hypochlorites yield chlorine when treated with dilute acids. They are strong oxidizing agents, and are used as bleaches.

$$OCl^-(aq) + 2H^+(aq) + Cl^-(aq) \rightarrow H_2O(l) + Cl_2(g);$$
$$HOCl(aq) + 2Fe^{2+}(aq) + H^+(aq) \rightarrow H_2O(l) + Cl^-(aq) + 2Fe^{3+}(aq);$$
$$\text{(coloured) dye} + ClO^-(aq) \rightarrow Cl^-(aq) + dye \cdot O \text{ (colourless)}.$$

Bleaching powder is an off-white solid prepared by passing chlorine through a counter-current of agitated slaked lime, $Ca(OH)_2(s)$. A mixture of calcium, chloride, hydroxide, and hypochlorite ions is formed, in the form of a fairly stable (slight smell of chlorine) non-deliquescent solid which is sparingly soluble in water. It shows the properties of the hypochlorite ion, above, and is used as a bleach for fabrics and wood pulp, and as a disinfectant.

Draw the structures of the acids HOX, HOXO, $HOXO_2$, and $HOXO_3$ (see Problem 7). Halous acids, $HO \cdot XO$, and halites, XO_2^-, are known (for certain) only for chlorine. The acid HOClO readily disproportionates to hypochlorous and chloric acid (Fig. 19.1). The chlorite ion is slightly more stable in alkaline solution.

Halic acids, $HOXO_2$, and halates, XO_3^-, are known for chlorine, bromine, and iodine, but only iodic acid has been obtained pure. The salts of all three acids are known in the solid state. Solutions of chloric and bromic acids are obtained by treating the halogen with hot concentrated barium hydroxide, and treating the barium halate obtained with dilute sulphuric acid:

$$6X_2 + 6Ba(OH)_2(aq) \rightarrow 5BaX_2(aq) + Ba(XO_3)_2(s) + 6H_2O(l);$$
$$Ba(XO_3)_2(s) + H_2SO_4(aq) \rightarrow BaSO_4(s) + 2HXO_3(aq).$$

Iodic acid is obtained by heating concentrated nitric acid with iodine:

$$3I_2(s) + 10HNO_3 \rightarrow 6HIO_3(s) + 10NO(g) + 2H_2O(l)$$

On cooling the mixture, white deliquescent crystals of iodic acid are obtained. All the compounds HXO_3 are strong acids and strong oxidizing agents, and have a pyramidal structure. Unlike the others, chloric acid is unstable with respect to disproportionation to chloride and perchlorate ions (Fig. 19.1), but the reaction is very slow (also in alkali). Solutions of each acid decompose on heating to give oxygen and the halogen. The halates are prepared either by dissolving the halogen in the appropriate hot concentrated alkali, or by electrolysis of hot concentrated halide solutions. All halates are soluble in water, except those of the ions Pb^{2+}, Ba^{2+}, and Ag^+. On heating, solid halates give a variety of products. Chlorates disproportionate initially, and the perchlorate produced then decomposes:

$$4KClO_3(s) \xrightarrow{<370\,°C} KCl(s) + 3KClO_4(s) \xrightarrow{>370\,°C} 3KCl(s) + O_2(g)$$
$$\text{(catalysed by } MnO_2\text{).}$$

Bromates and iodates, particularly those of the less electropositive metals, tend to give the free halogen and the oxide:

$$2KIO_3(s) \rightarrow 2KI(s) + 3O_2(g); \text{ but } 2Zn(BrO_3)_2(s) \rightarrow 2ZnO(s) + 2Br_2(g) + 5O_2(g).$$

Potassium chlorate is used as an oxidant in matches and fireworks. Sodium chlorate is used as a weed-killer. Potassium bromate and iodate are used as primary standards (volumetric oxidants). Iodate is used to standardize thiosulphate:

$$IO_3^-(aq) + 6H^+(aq) + 5I^-(aq) \rightarrow 3I_2 + 3H_2O(l).$$

In the presence of concentrated hydrochloric acid a different reaction occurs,

$$Cl^-(aq) + IO_3^-(aq) + 6H^+(aq) + 4e^- \rightarrow ICl + 3H_2O(l),$$

since the iodine is oxidized further. This is useful for the standardization of many reducing agents in the presence of carbon tetrachloride. Since the reaction goes via iodine, the end point is the disappearance of the initially formed iodine (violet in the organic layer). It has been used to estimate: arsenic(III), antimony(III), tin(II), iodide, hydrogen peroxide, mercury(I), hydrazine, etc. Potassium bromate can be used similarly.

Perbromates and perbromic acid have recently been prepared by the oxidation of bromates with xenon difluoride. Probably they are very powerful oxidants. Perchloric acid and perchlorates cannot be prepared by the disproportionation of chlorates since, though it is thermodynamically favourable, the reaction is very slow. Instead anodic oxidation of chlorates is used to form perchlorates, and perchloric acid is formed by the action of concentrated sulphuric acid on the salts. Perchloric acid is a colourless oily liquid, difficult to obtain pure, thermally unstable, a powerful oxidant (dangerously explosive with organic material), and it combines vigorously with water. It is one of the strongest acids known (Section 13.4). This is indicative of the poor complexing power of the perchlorate anion for the ion H_3O^+ and other cations. Perchlorates are known for all electropositive metal ions. They are usually soluble in water (Section 10.3.17) except for a few salts of large cations with very low hydration energies (e.g. Cs^+, Rb^+, K^+). The salts are often isomorphous with those of other large tetrahedral anions (e.g. BF_4^-, SO_4^{2-}, MnO_4^-).

In solution periodic acid is present in several forms, and several different types of salt are known, for example Ag_5IO_6, $AgIO_4$, $Ag_2(HIO_5)$, and Ag_3IO_5. Periodic acid and periodates are powerful oxidants, for example they oxidize manganese(II) to permanganate. In organic chemistry periodic acid is used to determine 1,2-glycols:

$$\underset{\underset{OH}{|}}{R-CH}-\underset{\underset{OH}{|}}{CH-R'} \rightarrow RCHO + R'CHO.$$

PROBLEMS

1. Explain the differences in their reactions with water between:
   a) NCl_3 and PCl_3;
   b) CCl_4 and $SiCl_4$.

2. Classify the structures of the following compounds as molecular, ionic, or giant-molecular: $NaCl$, $AlBr_3$, $SiCl_4$, $PdCl_2$, BCl_3, $CdCl_2$, $TiCl_2$, $TiCl_4$, CCl_4, ICl, and $BaCl_2$. What *simple* experiments could you perform to confirm your judgements? How do the above compounds react with water?

3. What are the probable structural types of the compounds given below?

	NaCl	$MgCl_2$	$AlCl_3$	$SiCl_4$	PCl_5
m.p. °C	800	712	190	−68	148
b.p. °C	1465	1418	180	57	164

4. Predict the following properties of astatine: atomic weight; molecular formula; boiling and melting point; atomic volume; the ionic and covalent radii; possible stability of At_3^-; solubility of $AgAt$; the reactions of the element with sodium and hydrogen; the reactions of $NaAt$ with concentrated sulphuric acid and dichromate; the acid strength of HAt; and the reaction of H_5AtO_6 with magnesium.

$$H_5AtO_6 \xrightarrow{1.6V} AtO_3^- \xrightarrow{1.5V} HOAt \xrightarrow{1.0V} At \xrightarrow{0.3V} At^-$$

5. Show how the reduction potentials of the couples X_2/X^- determine:
   a) the reactions of the halogens (X_2) with water;
   b) the methods of preparation of the elements.

6. Chlorine can be prepared by the reaction between hot concentrated hydrochloric acid and manganese dioxide. Does this strike you as surprising?

$$MnO_2/Mn^{2+}, \quad E^\ominus = 1.23 \text{ V}; \quad Cl_2/Cl^-, \quad E^\ominus = 1.36 \text{ V}.$$

7. What are the shapes of the following species: ICl_2^-, ClF_3, ICl_4^-, IF_5, IF_7, $TeCl_4$, ClO_4^-, HOX, HOXO, $HOXO_2$, and $HOXO_3$?

8. Why do elements show their highest oxidation states in their compounds with (oxygen and) fluorine?

9. What are the hydrolysis products of the compounds IBr and BrCl? Comment.

10. The brown colour of an acidified dilute solution of iodine in aqueous potassium iodide is intensified by the addition of nitrite, but decolorized by the addition of phosphite. Comment.

11. Comment on the low bond strength of the fluorine molecule.

12. Why is fluorine more reactive than the other halogens?

13. Calculate the equilibrium constants for the dissociation of the hydrogen halides (H—X) in aqueous solution:

	F	Cl	Br	I
$H—X + H_2O \rightleftharpoons H_3O^+(aq) + X^-(aq)$, $\Delta G^\ominus =$	17.7	−40.8	−54.2	−57 kJ

Comment on the relative acid strengths.

14. Ammonium fluoride has the wurtzite structure, whereas ammonium chloride has the sodium chloride structure. Comment.

15. Nitrogen pentafluoride is unknown. Comment.

16. After the sodium nitrate has been extracted from caliche, a quantity of aqueous iodate containing about 10 g l^{-1} of iodine remains. How much sodium bisulphite is required to extract it?

17. Compare the trends in atomic volume, boiling point, and melting point of the halogens with those in
    a) group IA,
    b) group 0 (noble gases).

18. Calculate the equilibrium constants for the disproportionation reactions:
    a) $ClO^-(aq) \rightarrow Cl^-(aq) + ClO_3^-(aq)$ in alkali;
    b) $Br_2 \rightarrow Br^-(aq) + BrO_3^-(aq)$ in alkali;
    c) $I_2 \rightarrow IO_3^-(aq) + I^-(aq)$ in alkali.

19. Discuss the trends in (a) reducing action, (b) thermal stability, and (c) acid strength, of the hydrogen halides.

20. Which is the strongest acid of the following:
    HClO, $HClO_2$, $HClO_3$, or $HClO_4$? Comment on this, and draw the structures of these acids.

21. Which of the halogens (excluding astatine) provides the group VIIB example of: the weakest acid HX; the largest atom; the smallest ionization energy; the strongest reductant; the best hydrogen-bonder.
22. Are any of the bond angles in Fig. 19.8 surprising?

BIBLIOGRAPHY

Most of the books on inorganic chemistry in the general list contain material on the chemistry of group VIIB. A selection is given at the end of Chapter 16.

CHAPTER 20

d-Block Elements

20.1 DEFINITION OF A TRANSITION METAL ION

In the long form of the Periodic Table (Section 10.1) the transition elements occur in the d-block between groups IIIA and IIB. In the "thought experiment" of building up the elements by successively adding one proton (and perhaps some neutrons) to the nucleus, and one electron to the orbital electrons, they are formed during the filling of electrons into the d-orbitals. This chapter is mainly about the chemistry of the first transition series, titanium to copper, though brief mention will be made of the second (zirconium to silver) and third series (hafnium to gold) "formed" when the $4d$- and $5d$-orbitals, respectively, are filled with electrons.

Transition metals can be defined as: *those elements at least one of whose simple ions has an incomplete outer shell of d-electrons* (*that is, contains between one and nine electrons*). This is a conceptual definition, and is useful because a transition metal ion can be recognized immediately by looking at its electron configuration. An operational definition would be: an element whose compounds show certain characteristic properties; these are defined separately in Section 20.3 and summarized in Section 20.11. Note that since it is the chemistry of the *compounds* of the transition metals with which we are mainly concerned, they are defined in terms of their compounds. Scandium, for example, invariably forms scandium(III) compounds in which the ion Sc^{3+} has the electron configuration $1s^2 2s^2 2p^6 3s^2 3p^6 3d^0$. Therefore, scandium is not classified as a transition metal, nor do its compounds show any transitional characteristics. The same is true for zinc, which forms the ion Zn^{2+} exclusively ($1s^2 2s^2 2p^6 3s^2 3p^6 3d^{10}$). Copper shows intermediate behaviour since it forms compounds in two oxidation states: copper(I) and copper(II). The electron configuration of the copper(I) ion, Cu^+, is $1s^2 2s^2 2p^6 3s^2 3p^6 3d^{10}$, and consequently it is not a transition metal ion and shows no transitional characteristics. The copper(II) ion, however, has the configuration, $1s^2 2s^2 2p^6 3s^2 3p^6 3d^9$, and is a transition metal ion; and copper(II) compounds are coloured, paramagnetic, etc. Thus not all members of the d-block are classified as transition metals, and not all the oxidation states of transition metals show transitional properties.

20.2 PERIODIC TRENDS

The transition elements show a horizontal similarity in their physical and chemical properties (Table 20.1), as well as the usual vertical relationship. This horizontal similarity is so marked that the chemistry of the first transition series, Ti to Cu, is

20.1 Definition of a transition metal ion
20.2 Periodic trends
20.3 General properties
20.4 Transition metal complexes
20.5 Stereochemistry
20.6 Isomerism
20.7 Rules for the nomenclature of complexes
20.8 The uses of complex compounds
20.9 Trends among transition elements
20.10 Differences between the first and the last two transition series
20.11 Summary of general properties of transition elements
20.12 The scandium group and the lanthanides
20.13 The titanium group, group IVA
20.14 The vanadium group, group VA
20.15 The chromium group, group VIA
20.16 The manganese group, group VIIA
20.17 The iron group, group VIII
20.18 The cobalt group, group VIII
20.19 The nickel group, group VIII
20.20 The copper group, group IB
20.21 Zinc, cadmium, and mercury, group IIB

Table 20.1. Properties of the d-block elements and calcium.

Property		Calcium Ca		Scandium Sc		Titanium Ti		Vanadium V	
Atomic number (weight)		20 (40.08)		21 (44.956)		22 (47.90)		23 (50.942)	
Electron configuration (outer)		$3d^0 4s^2$		$3d^1 4s^2$		$3d^2 4s^2$		$3d^3 4s^2$	
Isotopes (in order of abundance)		40, 44, 42, 48, 43, 46		45		48, 46, 47, 49, 50		51, 50	
Metallic radius (Å)		1.97		1.64		1.47		1.35	
Ionic radius (Å)		2+		3+		3+	4+	3+	4+
		0.99		0.81		0.76	0.68	0.74	0.60
Covalent radius (Å)		1.74		1.44		1.32		1.22	
Atomic volume (cm³ mol⁻¹)		26		15		10.6		8.3	
Boiling point (°C)		1487		2727		3260		3400	
Melting point (°C)		845		1539		1675		1900	
*Enthalpies fus vap		8.7	161	16	335	15.5	443	17.6	443
(kJ mol⁻¹) atom hyd		177	1565	390	3915	469		502	
Density (g cm⁻³)		1.54		3.0		4.5		6.11	
Electronegativity (A/R)		1.05		1.2		1.3		1.45	
Ionization energy (kJ mol⁻¹)	1st 2nd	590	1146	633	1235	659	1309	650	1414
	3rd 4th	4941	6464	2388	7130	2648	4171	2866	4631
	5th 6th	8142	10500	8874	10720	9627	11577	6293	12435
†Electrode potential (V)		(II)		(III)		(III)	(IV)	(II)	(III)
		−2.87		−2.1		−1.2	−1.63	−1.2	−0.86
Structure		ccp, hcp		ccp, hcp		hcp, bcc		bcc	
Abundance, p.p.m.		36,300		1.5		4400		150	
Colour, M²⁺ (aq)		colourless		—		brown		lavender	
Highest oxide		CaO		Sc_2O_3		TiO_2		V_2O_5	
Highest chloride		$CaCl_2$		$ScCl_3$		$TiCl_4$		VCl_4	
Highest fluoride		CaF_2		ScF_3		TiF_4		VF_5	
Configuration, M²⁺		$3d^0 4s^0$		—		$3d^2$		$3d^3$	
Configuration, M³⁺		—		$3d^0 4s^0$		$3d^1$		$3d^2$	

* Enthalpies of hydration are all for M^{2+} except Sc^{3+}.
† (III) refers to couple M^{3+}/M, etc.
‡ Distorted.

often discussed separately from that of the second and third series, which are more similar to one another than to the first series (Section 10.3.2). This horizontal similarity contrasts sharply with the trend in traversing a row of the s- and p-block elements, lithium to fluorine for example, where it is the difference between the elements that is the most striking characteristic (Section 11.4). This difference is due to the fact that, whereas in building up the elements from lithium to fluorine etc. the nuclear charge is being increased and an electron is being added to the valence shell of electrons, in the transition element build-up the nuclear charge is also being increased, but an inner d-electron is being added. In the former case the added s- and p-electrons shield each other badly from the extra nuclear charge, and along the series the atomic radii decrease sharply and the electronegativity and ionization energies increase sharply. These cause major differences in valency and chemical properties from element to element. The effect of adding electrons to an inner shell, however, in the case of the transition elements, is that the difference from element to element is much less than that in the representative elements. The atomic radii do decrease along the series, since no electrons can completely shield the effect of increasing nuclear charge, but only slightly. Similarly, the ionization energies and electronegativities tend to increase along the series, but only slightly

	Chromium Cr		Manganese Mn		Iron Fe		Cobalt Co		Nickel Ni		Copper Cu		Zinc Zn	
	24 (51.996)		25 (54.938)		26 (55.847)		27 (58.933)		28 (58.710)		29 (63.54)		30 (65.37)	
	$3d^54s^1$		$3d^54s^2$		$3d^64s^2$		$3d^74s^2$		$3d^84s^2$		$3d^{10}4s^1$		$3d^{10}4s^2$	
	52, 53, 50, 54		55		56, 54, 57, 58		59		58, 60, 62, 61, 64		63, 65		64, 66, 68, 67, 70	
	1.30		1.35		1.26		1.25		1.25		1.28		1.37	
	2+	3+	2+	3+	2+	3+	2+	3+	2+	3+	1+	2+	2+	
	0.84	0.69	0.80	0.66	0.76	0.64	0.74	0.63	0.72	0.62	0.96	0.69	0.74	
	1.18		1.17		1.17		1.16		1.15		1.17		1.25	
	7.2		7.4		7.1		6.7		6.6		7.1		9.2	
	2480		2097		3000		2900		2732		2595		907	
	1890		1244		1535		1495		1453		1083		419	
	13.8	305	14.6	225	15.4	354	15.2	389	17.6	379	13.1	305	7.4	115
	397	1820	284	1815	406	1890	439	2025	427	2075	341	2075	130	2017
	7.2		7.44		7.86		8.86		8.90		8.92		7.13	
	1.55		1.6		1.65		1.7		1.75		1.75		1.65	
	653	1591	717	1509	762	1561	759	1644	736	1751	745	1958	906	1732
	2992	4861	3259	5021	2958	5502	3230	5104	3391	5400	3556	5681	3828	5983
	7050	8745	7322	9874	7531	10250	8054	10230	7531	10627	7908	10230	8284	10795
	(II)	(III)	(II)	(III)	(II)	(III)	(II)	(III)	(II)		(I)	(II)	(II)	
	−0.91	−0.74	−1.18	−0.28	−0.44	−0.04	−0.28	+0.4	−0.25		+0.52	+0.34	−0.76	
	hcp, bcc		complex		ccp, bcc		ccp, hcp		ccp, hcp		ccp		hcp‡	
	200		1000		50,000		40		100		70		80	
	blue		pale pink		pale green		pink		green		blue		colourless	
	CrO_3		Mn_2O_7		Fe_2O_3		Co_2O_3		Ni_2O_3		CuO		ZnO	
	$CrCl_4$		$MnCl_3$		$FeCl_3$		$CoCl_2$		$NiCl_2$		$CuCl_2$		$ZnCl_2$	
	CrF_6		MnF_4		FeF_3		CoF_3		NiF_2		CuF_2		ZnF_2	
	$3d^4$		$3d^5$		$3d^6$		$3d^7$		$3d^8$		$3d^9$		$3d^{10}$	
	$3d^3$		$3d^4$		$3d^5$		$3d^6$		$3d^7$		$3d^8$		—	

compared to the representative elements. Thus the transition elements become less basic or less electropositive along the series (see, for example, electrode potentials in Table 20.1) just as in a row of representative elements, but the change from element to element is much less marked.

20.3 GENERAL PROPERTIES

All the transition elements are typical metals of moderate to weak electropositivity (e.g. electrode potentials, Table 20.1). They are all less electropositive than the s-block metals and although the difference from one element to another is comparatively small, since there are seven such differences the span of character from the moderately electropositive titanium to the noble metal copper is quite large. Except for copper, all the first row transition elements are white and show the properties of moderate to weakly electropositive metals (Section 10.5). The good cohesive properties compared with those of the s-block elements (high melting and boiling points, large enthalpies of vaporization etc., high densities, and low atomic volumes) are clearly associated with the presence of d-electrons, which must be involved in metallic bonding (Section 4.7). The good mechanical properties reflect this strong interatomic bonding, and they are widely used structural materials. Their electropositivity and atomic radii are smaller than those of the *s*-block elements in the same row (Table 20.1), owing to the increased nuclear charge. But the change is less than that in going across an *s*- or *p*-block row (see above). As expected from Table 10.10, the properties of the transition metals compared to the *s*-block metals are: they are better conductors and less reactive with acids and water (and with oxidizing agents generally); their oxides and hydroxides of low oxidation number (2 or 3) are less basic and less soluble; their salts are less ionic (though M^{2+} and M^{3+} are still largely ionic), less thermally stable, more hydrated, more hydrolysed, and more easily reduced; and they form few salts with large polarizable anions, but more "complex compounds" (Section 20.4). The electrode potentials are lower than those of the *s*-block elements (Section 10.3.18) but, nevertheless, they would be expected to be moderate reducing agents (from a thermodynamic point of view). For example, all the first row series (except copper), from a thermodynamic point of view, "should" displace hydrogen gas from acids and even water. In fact, inertness to water and acids is a characteristic of almost all the transition elements except manganese; that is, the rate of reaction is very small. The high activation energies for this reaction are due either to coherent oxide layers which protect most of the metals, or to high hydrogen overpotentials (Section 8.9) or both. Cobalt and nickel do not even react with concentrated hydrochloric acid. Even stronger oxidizing agents than the ion $H^+(aq)$ often react slowly with transition metals. The high enthalpy of vaporization, which is a factor contributing to the low electrode potential of these metals (Section 10.3.18) may also be a factor contributing to the high activation energy (Section 8.3). Thus, according to their electronegativities and reduction potentials, the first row transition series ought to be moderately reactive metals. But owing to the coherent oxide layer, the reactions are slow. At high temperatures, as usual, the reactions come under thermodynamic control, however, and the metals react with most non-metals.

The ionization energies are greater than those of the *s*-block elements in the same row of the Periodic Table, because of the increased nuclear charge, but the increase along the row is less than that in the *s*- and *p*-blocks (see above). The changes in atomic radii reflect the changes in ionization energies. Exceptions to the general trend occur in the second ionization energies of chromium and copper which are anomalously high. Interestingly, this occurs when an electron is being removed from a filled (d^{10}) or half-filled (d^5) shell. This also occurs for nitrogen and the inert gases (Section 10.3.3). The extra difficulty in removing an electron from such a configuration is referred to as the "stability of the half-filled or filled shell". This concept crops up often in rationalizations of the chemistry of manganese, chromium, iron, zinc, and copper. Examples are: the stability of the manganese(II), iron(III), zinc(II), and copper(I) ions; the anomalous ground state configurations of the chromium and copper atoms; and the facts that the second ionization energy of zinc is lower than, and the third ionization energy of zinc is greater than, those of copper. That the ionization energies of these elements are greater than those of the *s*-block elements, but less than those of the *p*-block elements, leads to the formation of both ionic and covalent compounds. In forming the ions M^+, M^{2+} etc., of a particular element, the increase in successive ionization energies is not great. This leads to the occurrence of various stable oxidation states for each element (cf. group IA, Section 14.2.4). In the formation of ions it is the lower energy *s*-electrons which are lost. The *s*- and *d*-electrons do not differ much in energy, however, which is why the difference in ionization energies is relatively small. Although the compounds of the transition elements containing higher oxidation states are mainly covalent, the ionization energies are still a useful guide to the possibility of their formation.

Despite the weakly electropositive character of these elements, the compounds in the oxidation states $+2$ and $+3$ are usually best regarded as mainly ionic. Thus the lower oxides are basic, though the aqueous solutions of ions are often hydrolysed. But many oxides and chlorides in the $+3$ oxidation state are mainly covalent; for example Fe_2Cl_6 is molecular and Mn_2O_3 is amphoteric.

Most of the metals crystallize in more than one form. There are more glide planes in the face-centred cubic structure (Section 3.3) than in the hexagonal close-packed or body-centred cubic structures, and those metals which usually crystallize in this structure are more easily worked. Thus copper and iron are relatively soft, malleable, and ductile, while chromium and vanadium are harder.

The similar sizes of the metals and their ions result in the formation of substitutional alloys, and many isomorphous salts, such as $M^{2+}SO_4^{2-}(NH_4^+)_2SO_4^{2-} \cdot 6H_2O$, $M^{3+}K^+(SO_4^{2-})_2 \cdot 12H_2O$, and $M^{2+}(NH_4^+)PO_4^{3-}$.

The hydroxides precipitated by adding hydroxyl ions to aqueous solutions of the metal ions are true hydroxides (i.e. form an ionic lattice of metal and hydroxyl ions) in the case of the $+2$ oxidation state only. Higher hydroxides in the $+3$ etc. oxidation states are in fact hydrated oxides, e.g. $Fe_2O_3 \cdot xH_2O$. This is a characteristic of weakly electropositive metals (Section 10.5).

20.4 TRANSITION METAL COMPLEXES

A completely satisfactory definition of complex compounds is difficult to formulate; it is attempted below. Complex compounds (which may be neutral or ionic) are compounds which contain a central atom or ion (usually a metal, often referred to as the "nuclear atom") closely surrounded by a cluster of other ions or molecules (called *ligands*). The ligands are usually bonded to the nuclear atom by what are classically described as co-ordinate bonds, and complexes or complex compounds are often referred to as "co-ordination compounds". The number of nearest neighbours (ligands) to the nuclear atom is referred to as the "co-ordination number" of the central atom (C.N.), and these nearest neighbours constitute what is known as the "first co-ordination sphere". Complex ions tend to retain their identity even in solution, though partial dissociation may occur. Another distinguishing characteristic is that both the nuclear atom and ligands are usually capable of independent existence as stable chemical species. As usual, cases occur in which it is not clear whether the compounds should be considered as complexes or not. As always with definitions, it is the usefulness that matters. The critical question is whether it is useful to consider a particular compound as a complex. Compounds such as AlF_6^{3-}, $Ti(H_2O)_6^{3+}$, and $CoCl_4^{2-}$ are usefully considered as complexes; CH_4, ClO_4^-, etc., are not (e.g. C^{4+} has no independent existence). An intermediate type is that formed by ligands being added to molecules, as opposed to atoms or ions, e.g. $SiF_4 + 2F^- \rightarrow SiF_6^{2-}$. These are normally considered as complexes. A sub-group called adducts is formed by the reaction of two neutral molecules, e.g. $BF_3 + NH_3 \rightarrow BF_3 \cdot NH_3$. Thus, no perfect definition exists; it is a matter of judgement (Fig. 20.1). Using the above model of complexes it is expected that the most stable co-ordination compounds would be formed by

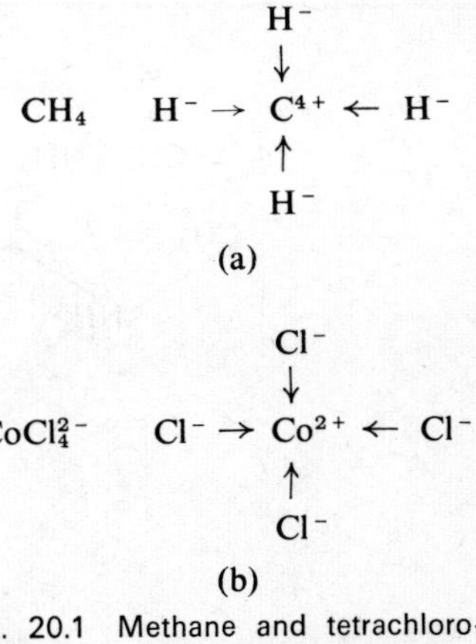

$$CH_4 \quad H^- \begin{matrix} H^- \\ \downarrow \\ \rightarrow C^{4+} \leftarrow H^- \\ \uparrow \\ H^- \end{matrix}$$
(a)

$$CoCl_4^{2-} \quad Cl^- \begin{matrix} Cl^- \\ \downarrow \\ \rightarrow Co^{2+} \leftarrow Cl^- \\ \uparrow \\ Cl^- \end{matrix}$$
(b)

Fig. 20.1 Methane and tetrachlorocobalt(II). (a) This is not a useful way to look at the electronic structure of methane. Therefore it is not considered as a complex. (b) This is a useful way of looking at the structure of the tetrahedral $CoCl_4^{2-}$ ion (Section 20.18). Therefore it is considered to be a complex compound.

the interaction of highly polarizing cations (those of weakly electropositive metals) with stable donor ligands (those with lone pairs capable of forming co-ordinate bonds) e.g. NH_3, H_2O, CN^-, Cl^-, NO_2^-, etc. This is roughly correct, but the situation is much more complicated.

20.5 STEREOCHEMISTRY

The spatial arrangement or stereochemistry of complexes cannot be predicted from the Gillespie–Nyholm rules, that is, *d-electrons differ from s- and p-electrons in their influence on the spatial structure of compounds.* The stereochemistry of simple inorganic molecules is more difficult than the stereochemistry of correspondingly simple organic molecules since: (a) inorganic central atoms can have co-ordination numbers from two to nine; (b) different stereochemistries are possible for some co-ordination numbers; and (c) various types of isomerism exist (below). The most common co-ordination numbers are two (linear), four (tetrahedral or square planar), and six (octahedral). Examples are:

1. Two-fold co-ordination, especially gold(I), silver(I), and copper(I); for example: $Au(CN)_2^-$, $Ag(NH_3)_2^+$, and $CuCl_2^-$ (all linear).
2. Four-fold co-ordination. Square planar geometry is more common than tetrahedral (cf. *s-* and *p-*block elements) and is found for palladium(II), nickel(II), copper(II), and gold(III); for example $AuCl_4^-$.
3. Six-fold co-ordination is by far the most common, and is always octahedral; for example $Co(H_2O)_6^{2+}$.

Ligands that occupy only one co-ordination position are called **monodentate ligands**, e.g. Cl^-, NO_2^-, NH_3, H_2O, etc. Those that occupy more than one position are called **multidentate** or **chelate ligands,** e.g. ethylenediamine, $NH_2CH_2CH_2NH_2$, abbreviated "en", and the resulting compound is called a **chelate compound,** e.g. $[Cu\ en_2]^{2+}\ SO_3^{2-}$ (Fig. 20.2a).

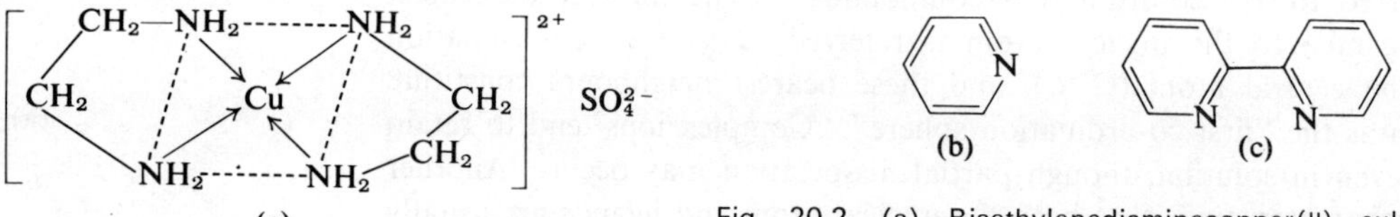

Fig. 20.2 (a) Bisethylenediaminecopper(II) sulphate (square-planar). (b) Pyridine. (c) Dipyridyl (abbreviated as dipy).

Ethylenediamine, which can occupy two co-ordination positions, is said to be bidentate; ligands which can occupy three, four, five, and six positions are said to be ter-, quadri-, quinqui-, and sexadentate respectively. Ethylenediaminetetra-acetic acid (EDTA), for example, is sexadentate (Section 14.3.4).

20.6 ISOMERISM

Compounds which have the same molecular formula, but different structural formulae (Section 21.1) are said to be isomers. There are many types of isomerism possible (about a dozen). Those important in transition metal chemistry are:

1) **Structural isomerism.** This occurs in complex ions when the composition of the first co-ordination sphere may differ from one compound to another. The class may be subdivided into sub-classes; for example:

a) Ionization isomerism, where the compounds yield different ions in solution, e.g. $[Co(NH_3)_5Br]^{2+}SO_4^{2-}$ and $[Co(NH_3)_5SO_4]^+Br^-$.
b) Hydration isomerism, e.g. $[Cr(H_2O)_6]Cl_3$ (violet), $[Cr(H_2O)_5Cl]Cl_2 \cdot H_2O$ (light green), and $[Cr(H_2O)_4Cl_2]Cl \cdot 2H_2O$ (dark green). These compounds show different conductivities in solution, and they precipitate different amounts of silver chloride from cold silver nitrate solution, in the ratio $3:2:1$ respectively.
c) Co-ordination isomerism, e.g. $[Co(NH_3)_6]^{3+}[Cr(C_2O_4)_3]^{3-}$ and $[Cr(NH_3)_6]^{3+}[Co(C_2O_4)_3]^{3-}$.

2) **Geometrical isomerism.** These are isomers in which the composition of the first co-ordination sphere is the same, but the geometrical arrangement of the ligands varies. This is often called *cis–trans* isomerism, referring to the relative positions of two selected ligands. This isomerism is only possible for co-ordination numbers greater than or equal to four, e.g. *cis* and *trans* $PtCl_2(NH_3)_2$ (Fig. 20.3). Compare *cis* and *trans* butenedioic acid, Section 21.4.

3) **Optical isomerism** (Section 21.3). Any molecule which contains no plane or centre of symmetry may exist in two forms which are non-superposable mirror images of one another. These have identical chemical and physical properties except that they rotate the plane of plane-polarized light equally but in opposite directions, and they react differently with other optically active compounds. Such compounds are normally obtained as a $50:50$ mixture (called a racemic mixture) of the two optical isomers (called enantiomers, or enantiomorphs), which is therefore optically inactive, such as $[Co(en)_3]^{3+}3Cl^-$ (Fig. 20.4).

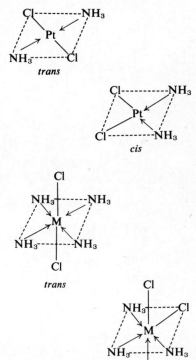

Fig. 20.3 Two examples of *cis–trans* isomerism.

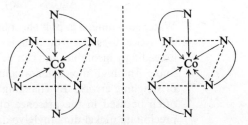

Fig. 20.4 The enantiomers of $[Co(en)_3]^{3+}$ (en = ethylenediamine), also called 1,2-diaminoethane.

20.7 RULES FOR THE NOMENCLATURE OF COMPLEXES

1. Non-ionic complexes are given a one-word name. In ionic complexes the cation is named first, then the anion separately.
2. Neutral ligands are named as the molecule, negative ligands end in *-o*, and positive ligands (rare) end in *-ium*. Historical exceptions to this rule are water (aquo), ammonia (ammine), carbon monoxide (carbonyl), and nitric oxide (nitrosyl). Some examples are: ethylenediamine, $NH_2CH_2CH_2NH_2$; chloro, Cl^-; hydroxo, OH^-; oxalato, $C_2O_4^{2-}$; and hydrazinium, $[NH_2 \cdot NH_3]^+$.

3. The ligands in a complex are named first and are placed in the order: (a) anionic ligands; (b) neutral ligands; and (c) cationic ligands; they are run together as one word. Within each sub-class, the ligands are written in order of increasing structural complexity (e.g. chloro before oxalato). The number of identical ligands is given by the Greek prefixes di-, tri-, tetra-, penta-, and hexa-, in the case of simple ligands; but the prefixes bis-, tris-, tetrakis-, etc. are used for complex ligands which themselves often contain the former prefixes; e.g. $[Pt(NH_3)_3(NO_2)Cl_2]^+Br^-$ is dichloronitro*tri*ammineplatinum(IV) bromide; but $(Ph_3P)_3RhCl$ is *tris*(triphenylphosphine)rhodium(I) chloride.
4. The name of the central atom is given after the ligands, together with its oxidation state designated by a Roman numeral in brackets. When the complex is a cation or a neutral molecule, the name of the atom is unchanged, but when it is an anion, the suffix -*ate* is added, e.g. potassium hexacyanoferrate-(III), $K_3Fe(CN)_6$.

20.8 THE USES OF COMPLEX COMPOUNDS

Complex compounds find many uses in industry, the laboratory, and the home. Some of these are classified below:

1. Extraction of metals; for example aluminium (Section 11.5.2), and gold (Section 20).
2. Electroplating, e.g. silver plating. One important factor in obtaining a uniform coherent layer instead of a spongey loose layer of the plated metal is a low constant concentration of aqueous silver ions. This is obtained by using aqueous potassium dicyanoargentate(I), $K[Ag(CN)_2]$, as the electrolyte. The small concentration of aqueous silver ions is maintained by dissociation of the complex anion:

$$Ag(CN)_2^-(aq) \rightleftharpoons Ag^+(aq) + 2CN^-(aq).$$

3. Water softening, e.g. polyphosphates (Section 17.14).
4. Analytical uses. Neutral complexes of metal ions with anionic ligands are often exceedingly insoluble, and can be used in gravimetric analysis to estimate metal ions. The high molecular weight of such complexes reduces the significance of weighing errors. By changing the conditions, such as pH, the method can often be used in the presence of other metal ions. One example is the red precipitate nickel dimethylglyoxime. Industrial determinations of metal ions, in which speed of determination is more important than *very high* accuracy, are often colorimetric. Metal ions in the presence of a given ligand, such as water, often absorb radiation at specific wavelengths at which other metal ions absorb only weakly, if at all (Section A.9). Since the amount of absorption depends upon the concentration of the metal ion, its measurement can be used to determine the metal ion concentrations; for example iron as the iron(III) complex thiocyanate cation $[Fe(SCN)]^{2+}(aq)$, which is an intense red colour. This can often be used together with simultaneous solvent extraction. The metal ion can be extracted from aqueous solution by shaking with an excess of ligand in an organic solvent, and the concentration of the complex determined in the organic layer, e.g. colorimetrically. Solutions of metal ions can also be titrated with suitable ligands, of which the disodium salt of EDTA (Section

14.3.4) is the most widely used. The complexes formed by EDTA and the metal ions are very stable, and at the end point, when sufficient EDTA equivalent to the amount of metal ion present has been added, the concentration of the metal ion present rapidly diminishes (the same principle as in pH titrations). Other ligands which form moderately stable coloured compounds with the metal ions involved are used as indicators. At the end point, where there are no longer any aqueous metal ions or coloured complex present, owing to the formation of the more stable EDTA complex, the colour of such complex indicators disappears.

5. The insoluble coloured neutral complexes are used as pigments (colouring agents) in industry for paints, and so on.
6. Enzymes are often very important complexes in biological systems; the ligands are usually large proteins. See also chlorophyll (Section 14.3.4) and haemoglobin (Section 20.17). Iron, magnesium, and copper have long been known to be essential for life. The function of other essential metal ions—"trace elements"—is less well understood.

20.9 TRENDS AMONG TRANSITION ELEMENTS

Trends in the stability of the compounds of transition elements are difficult to discuss and all generalizations are rough and ready, partly because the stability of a particular oxidation state depends upon the conditions (for example the type of ligand present (Section A.9) and partly because it depends upon what is meant by "stability" (stable to which reaction? kinetic or thermodynamic?) (Section 16.4.1). For example, is the compound $TiCl_2$ stable? It is stable with respect to dissociation into its elements up to about 400 °C in the absence of air, but it is immediately oxidized by air. Some of the factors influencing the stability of various oxidation states are:

1) Ionization energies. The second ionization energy of copper is unusually high; this fact rationalizes the existence of the unusual $+1$ oxidation state under normal conditions, present in copper(I) compounds. The unusually high third ionization energies of nickel, copper, and zinc "explain" the instability to reduction of the oxidation states copper(III), etc. Similarly, much of the comparative chemistry of nickel and platinum is rationalized by the data in Table 20.2. Thus the stability trend is Ni(II) > Pt(II); but Pt(IV) > Ni(IV).

Table 20.2. Ionization energies of nickel and platinum (kJ mol^{-1}).

	Nickel	Platinum
Sum of first two ionization energies	2487	2661
Sum of first four ionization energies	11,278	9364

Table 20.3. The relative stability of the $+2$ and $+3$ aqueous ions of the first row d-block elements.

Cr	Mn	Fe	Co	Ni	Cu	Zn
$\underline{3}$	(3)	$\underline{3}$	(3)	((3))	((3))	
(2)	$\underline{\underline{2}}$	$\underline{\underline{2}}$	$\underline{\underline{2}}$	$\underline{\underline{2}}$	$\underline{\underline{2}}$	$\underline{\underline{2}}$

For meanings of symbols see Table 20.4.

2) Electronic structure. Table 20.3 shows the relative stabilities of the $+2$ and $+3$ oxidation states of the first row of the d-block elements. Manganese(II) and iron(III) are much more stable than expected—"stability of half-filled shell" (Section 20.3).

3) **Nature of ligands.** Certain types of ligand stabilize certain oxidation states (Section A.9). Moreover, chelating ligands form more stable complexes than similar monodentate ligands, even though the bond energies are similar. This is called the *chelate effect*, and is clearly an entropy effect. Can you explain it?

4) **Ambient conditions: solvent, temperature, pH, etc.**

Summary of Trends

Despite the objections to talking about "stability", there are a number of trends apparent within the transition series:

a) Within each group (e.g. Cr, Mo, W) the elements show a range of oxidation states but they differ in relative stability. In general, the last two elements in a particular group show higher co-ordination numbers, and high oxidation states which are more stable relative to their lower oxidation states than those of the first row elements (Table 20.7). Thus, whereas chromium(VI) is an oxidizing state, molybdenum(VI) is non-oxidizing. Correspondingly, the lower oxidation states of the last two series are relatively less stable. Thus zirconium(III) is more strongly reducing than titanium(III) (Table 20.5). There is rarely any simple aqueous chemistry for the $+2$ and $+3$ ions of the last two series (cf. Ni^{2+}(aq), Cu^{2+}(aq) etc.).

b) For the first transition series the maximum oxidation number for any element can be obtained by adding the number of unpaired d-electrons (Table 2.5) to the two s-electrons in the "ground" state of the free atom, written as $d^m s^2$ (i.e. cheating in the case of copper and chromium). Thus the highest valency state found from titanium to manganese (usually as the oxy-compound or the fluoride only) is the group number (the total number of d- and s-electrons), since up to the configuration d^5, all d-electrons are unpaired. Beyond the configuration d^5 the number of unpaired d-electrons decreases by one from element to element, and so does the maximum oxidation state attained (Table 20.4).

c) In general, the stability of the possible higher oxidation states decreases in crossing the series from left to right (Table 20.4 and Fig. 20.5). Thus, titanium(IV) is stable whereas manganese(VII) is oxidizing. After manganese, the higher oxidation states rapidly become unstable. The stability of the half-filled shell upsets a tidy trend in the relative stability of the aqueous M^{3+}(aq) and M^{2+}(aq) ions (Table 20.3). From chromium to zinc the stability of the $+3$ ion relative to the $+2$ decreases in stability, except that manganese(II)(d^5) and iron(III)(d^5) are more stable than "expected".

d) Negative and zero oxidation numbers can exist, but only in the presence of ligands which stabilize low oxidation states, e.g. carbon monoxide, dipyridyl, etc. (Section A.9). The existence of such states is indicative of the weakly electropositive nature of the metals. Copper is unusual in that it forms compounds in the unusual oxidation state $+1$ with common ligands (Cl^-, etc.).

e) Oxidation states between the highest oxidation state and the oxidation states $+2$ and $+3$, tend to disproportionate, as in the p-block elements (Section

Table 20.4. The oxidation states of the first row d-block elements and calcium.

	Ca	Sc	Ti	V	Cr	Mn	Fe	Co	Ni	Cu	Zn
Number of unpaired d-electrons + s-electrons	2	3	4	5	6	7	6	5	4	3	2
*Configuration	d^0s^1	d^1s^2	d^2s^2	d^3s^2	"d^4s^2"	d^5s^2	d^6s^2	d^7s^2	d^8s^2	"d^9s^2"	$d^{10}s^2$
Maximum possible oxidation state	2	3	4	5	6	7	6	5	4	3	2
Actual oxidation states	$\underline{2}$	$\underline{3}$	4	5°	6°	7°	6°	5	$(4)^O$	$(3)^O$	$\underline{2}$
		$(2)^R$	3^R	4	$(5)^d$	$(6)^d$	$(5)^O$	$(4)^O$	$(3)^O$	2	
			$(2)^R$	3^R	$(4)^d$	$(5)^d$	$(4)^O$	3°	$\underline{2}$	1	
			(1)	$(2)^R$	3	4^d	3	$\underline{2}$	0		
			(0)	(1)	2^R	$(3)^O$	$\underline{2}$	0			
			(−1)	(0)	(1)	2	0				
				(−1)	(0)	(1)					
					(−1)	(0)					
					(−2)	(−1)					

(2) less stable state; 2 very stable; $\underline{2}$ stable state; ^{R} reducing; ^{O} oxidizing; ^{d} disproportionates.
* These are "incorrect" for Cr and Cu.

15.1). Examples are (Fig. 20.5) manganese(III) and (VI). Copper(I) also tends to disproportionate (Section 20.20).

Other trends worth mentioning are:

f) As the oxidation state increases, the compounds become more covalent, the oxides more acidic (Section 11.1.1), and the halides more susceptible to hydrolysis (Section 11.1.4).
g) Oxygen ligands form complexes which are tetrahedral for metals in oxidation states (IV) to (VII), and octahedral for lower oxidation states.

20.10 DIFFERENCES BETWEEN THE FIRST AND THE LAST TWO TRANSITION SERIES

The insertion of the lanthanides between barium and hafnium has an effect on the chemistry of the succeeding elements similar to, but more marked than, that produced by the lesser known transition insertion or contraction (Section 10.3.2). The lanthanide insertion upsets the "expected" regular trend in properties down groups IVA, VA, etc., for example Ti to Zr to Hf. Since a new shell of electrons has been added between zirconium and hafnium, an increase in atomic radii and other changes in properties similar to those occurring between titanium and zirconium would be expected. This does not occur. Hafnium, unlike zirconium and titanium, has an underlying filled 4f-shell of electrons. These diffuse orbitals do not protect the outer electrons well from the increased nuclear charge, and the effective nuclear charge (Section 2.4.3) is greater than expected. Consequently, the atomic radii of the third row series are smaller than expected by simple extrapolation down the series, and the electronegativity is greater. What do you think the trend in cohesive properties (Section 10.3.13, etc.) will be? The effect of the lanthanide insertion (or contraction as it is often called) is greater than that of the transition insertion (or contraction) since: (a) more elements are involved, and the increase in nuclear charge is therefore greater; and (b) the f-orbitals are less protecting than the

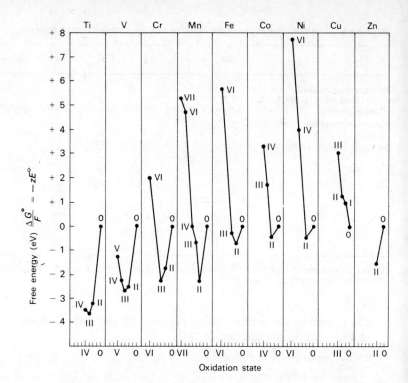

Fig. 20.5 First row transition series oxidation state diagram. Main points: (a) decreasing stability of the higher (group) oxidation states along the series; (b) increasing stability of the +2 state relative to the +3 state.

d-orbitals. In fact, the resulting increased effective nuclear charge cancels the usual group increase in size (owing to the extra shell of electrons) almost exactly, and the radii and many other properties of the second and third series are almost identical. This results in the typical group characteristic trend down the transition series: *the first element differs from the last two, which are very similar to one another.* The effect gradually dies out along the series, and in fact zinc and cadmium are more similar to one another than to mercury. Compared to the first series, the elements of the last two series differ as follows:

1. They form higher oxidation states, which are more stable relative to the lower oxidation states (two or three): zirconium(III) is more strongly reducing than titanium(III); molybdenum(VI) is a non-oxidizing state, whereas chromium(VI) is highly oxidizing; osmium(VIII) exists, but not iron(VIII); chromium(II) exists, but not tungsten(II); and no simple aquo ions M^{2+}(aq) or M^{3+}(aq) exist for the second and third series, but they are common in the first series (N.B. cationic complexes with other ligands may be formed, e.g. $Pt(NH_3)_4^{2+}$).
2. They are less abundant. An extreme example is that of technetium which is radioactive and not found naturally.
3. They form compounds with larger co-ordination numbers, such as ZrF_7^{3-}, cf. TiF_6^{2-}; the oxyanions of molybdenum are based on octahedral MoO_6 units, whereas those of chromium are based on tetrahedral CrO_4 units.
4. They give larger crystal field splitting energies (Section A.9). This is partly due to the greater extension in space of the 4d- and 5d-orbitals compared to the 3d, which are therefore able to interact more strongly with the ligand orbitals. The

result is that low spin complexes are the type invariably found in the latter two series.
5. They have a greater tendency to form polymers. One type of polymer is that containing metal–metal bonds, and this is certainly related to the higher cohesive properties of such elements.
6. They are much less reactive, e.g. have lower reduction potentials. These elements are more noble than their first row counterparts.

Despite this impressive list of differences, there are sufficient similarities in any transition metal group to justify their classification by groups in the Periodic Table. Like the first row, the general reactivity decreases from left to right across the series, so that the noble metals are found together at the bottom right hand side of the d-block elements. Moreover, the general properties of transition elements, listed below, (coloured compounds, etc.) are common to all the elements.

20.11 SUMMARY OF GENERAL PROPERTIES OF TRANSITION ELEMENTS

1. Transition elements show a horizontal similarity of properties (Section 20.2), in contrast to the representative elements (Section 11.4), as well as a typical vertical group relationship (Section 10.1).
2. They are all metals, varying from moderately electropositive (e.g. titanium) to noble (e.g. copper). Their properties reflect this (Section 10.5) except that most are less reactive than expected under normal conditions owing to a protective coating of oxide (Section 20.3).
3. They form compounds in a range of oxidation states (often called "variable valency") which differ by units of one (cf. p-block elements, Section 10.3.20). There is very little difference in energy between the $(n - 1)d$ and the ns electrons. This is reflected in the values of the ionization energies for these elements. These show only a gradual increase as the overall positive charge on the ion increases, and no severe increase until an inert gas configuration is broken into (Section 10.3.3). Since these energies are comparable to bond energies, a range of compounds in different oxidation states is possible. Note that although compounds in oxidation states greater than $+3$ are largely covalent and not ionic, the ionization energies are nevertheless a useful indication of the energy required to attain the formal oxidation state. Zero and even negative oxidation states are possible with certain ligands (Section A.9) and this is indicative of the weakly electropositive character of the elements concerned. The trends in relative stability of various oxidation states were discussed in Sections 20.9 and 20.10. Obviously, the character of the compounds is very dependent upon the oxidation state of the element (Section 11.1.1). The compounds of manganese are a good example (Section 20.16) of the increasing oxidizing nature and covalent character of the compounds, and the increasing acidic character of the oxides, as the oxidation state increases.
4. The elements and their compounds show marked catalytic activity. For example, nickel and platinum metals are good hydrogenation catalysts (Section 22.2.4), catalysing the hydrogenation of unsaturated hydrocarbons to saturated compounds. Other examples are iron in the Haber Process (Section 17.9), and vanadium pentoxide in the contact process. This property must be

associated with the ability to form compounds of variable oxidation number. One possible example is the decomposition of bleaching powder by cobalt salts:

$$Co^{2+}(aq) + OCl^-(aq) + H_2O \rightarrow Co^{3+}(aq) + Cl^-(aq) + 2OH^-(aq);$$
$$2Co^{3+}(aq) + 2OH^-(aq) \rightarrow 2Co^{2+}(aq) + \tfrac{1}{2}O_2(g) + H_2O(l) \text{ etc.}$$

5. The formation of coloured compounds is discussed in Section A.9.
6. The paramagnetism of some compounds is discussed in Section A.9.
7. Complex ion formation (Section 20.4) is often quoted as a typical property of transition elements. Actually, all metal ions, even those of the most electropositive elements, form complex ions, but the ions of the more weakly electropositive metals form more stable complexes with a greater range of ligands than those of the s-block elements. It is thus a property of weakly electropositive metals which the transition elements show to a marked degree. This is at least partly due (simple model) to the high effective nuclear charge which the transition metal ions exert at bonding distances from the nuclei, maximizing their attraction for ligands (Section 10.3.16). Not only are the transition metal ions relatively small, but owing to the poorly shielding d-electrons present, the nuclear charge attracting the ligands is greater than expected (Section 2.4.3) and they attract ligands more than non-transition metal ions of the same size and overall charge, and therefore form more stable complexes. Frequently, but not always, these complexes have the inert gas configuration, for example $Fe(CN)_6^{4-}$, $Ni(CO)_4$, etc.
8. They form many "interstitial compounds". These are prepared by heating the metal with a non-metal or a compound of a non-metal (such as ammonia, or a hydrocarbon), when a binary compound of the transition metal and the non-metal is formed. In these compounds the atoms of the metal occupy some of the interstices (Section 3.3) in a metallic lattice. Only small non-metals show this property: hydrogen, carbon, boron, and nitrogen. Oxygen and fluorine are too electronegative and form ionic compounds; most of the others are too large. Carbon and nitrogen always occupy octahedral holes; hydrogen is smaller and always occupies tetrahedral holes. Such compounds are usually confined to the earlier transition elements (Problem 28). The name "interstitial" refers to an earlier model of the bonding when it was thought that the non-metals merely "popped" into the vacant holes in the lattice. There are clearly strong bonding forces, however, since: the structure of the metal often changes during the formation of such compounds, say from hexagonal close-packed to an expanded face-centred cubic structure; the heat of formation is often high; the properties of the metal are often considerably altered, e.g. harder, loss of conductivity; and only transition metals form such compounds, therefore the d-electrons are presumably involved in a bonding capacity. A range of properties exists, but the compounds are often hard (especially carbides and nitrides, e.g. tungsten carbide) they often have metallic appearance, high melting and boiling points, are good conductors, and are generally chemically inert except to oxidizing agents. The composition is generally non-stoichiometric, e.g. $TiH_{1.73}$, $PdH_{0.6}$, $VH_{0.56}$, but may approach regular stoichiometry and a regular structure, e.g. TiC and VN (sodium chloride structure), and ZrH (zinc blende structure). Carbon steels are interstitial iron–carbon compounds in which the interstitial carbon prevents the iron atoms sliding

easily over one another, making the iron harder and stronger, but more brittle. The transition metals form carbides so easily that only copper and nickel can be extracted by conventional carbon reduction methods. Some, like titanium, react with nitrogen so readily at high temperatures that they must be refined in argon. Palladium, platinum, and iron are permeable to hydrogen at high temperatures. Palladium can absorb (reversibly) eight hundred times its own volume of hydrogen (Section 20.19).

The later transition elements of the first series form non-stoichiometric carbides (etc.) with irregular structures (Section 16.5), such as Cr_7C_3, which are much more reactive than the interstitial carbides of the earlier elements.

The elements also form a wide range of alloys with each other, owing to their similar size and characteristics. Particularly important are the alloys of iron with the other elements, which give a wide range of properties. For example, hardness and toughness can be achieved without consequent brittleness. Alloy formation is convenient since the preparation of the pure metals is usually difficult (Problem 29), but to form alloys it is not necessary to purify the metal first. The metals tend to confer their (desirable) properties on the particular steel: titanium steels are light, hard, strong, workable, and chemically inert (used in engines for aircraft, ships, and rockets, and in chemical plants); chromium and vanadium steels are hard, strong, workable, and corrosion resistant (stainless steels); molybdenum and tungsten steels, in addition, have high melting points; and manganese steels are extremely tough.

Similarities between the transition elements (the A groups) and the *p*-block elements (the B groups) are often emphasized. For example, since the group IVA element titanium shows a non-transition oxidation state of $+4$, it was bound to show some resemblances to at least one of the group IVB elements, since the latter group also shows an oxidation state of $+4$, and spans a wide range of electropositivity. Tin, in fact, turns out to resemble titanium the most, but this is only significant in that both are fairly weakly electropositive metals with a non-transition oxidation state $+4$. This results in bond type and stoichiometric similarities between corresponding compounds, and in the similar acid–base character of the oxides: the compounds SnO_2 and TiO_2 are both amphoteric. Similar relationships exist in groups V (oxidation state $+5$), VI (oxidation state $+6$), and VII (oxidation state $+7$).

20.12 THE SCANDIUM GROUP AND THE LANTHANIDES

Scandium, yttrium, and lanthanum form compounds only in the oxidation state $+3$, which has the configuration ns^0, $(n-1)d^0$. Thus they are *not* transition elements and show no transitional characteristics. The group relationship is similar to that observed in the *s*-block groups (Section 11.4.1.): the size and electropositivity (metallicity) of the elements increase as the group is descended. For example, the aquo-ion $M^{3+}(aq)$ is appreciably hydrolysed in the case of scandium but not for the heavier two elements (Section 10.5). Similarly, the normal hydroxides, $M(OH)_3$, are precipitated from solution by hydroxyl ions for the heavier two elements; in the case of scandium the hydrous oxide $Sc_2O_3 \cdot xH_2O$ is produced. Scandium is slightly more metallic than aluminium (compare electrode potentials) but resembles it in many ways: it reacts vigorously with water when the protective oxide layer is removed; the insoluble hydrous oxide is amphoteric; and only compounds in the +3 oxidation state are formed. Yttrium and lanthanum are similar to scandium but are more reactive. The lanthanides are the fourteen elements that follow lanthanum and they are produced in the "thought experiment" of forming the elements (Section 2.5) during the filling of the $4f$-subshell. These elements, cerium to lutetium, were called the "rare earth elements", but since they are not particularly rare this name has been abandoned, and they are now called the lanthanides. The main feature of their chemistry is their great similarity to one another, and this caused difficulties of separation at one time, since they are always found in nature together. This is now achieved easily by column chromatography. The metals are produced by reducing the trihalides with calcium. All the fourteen elements are very similar to one another since they have similar outer electron configurations. They are fairly reactive metals (for example they decompose water slowly) whose reactivity and metallicity decrease only slightly along the series (Section 20.2). For example, the reduction potential M^{3+}/M decreases from -2.48 (cerium) to -2.25 Å (lutetium), and the ionic radii (M^{3+}) decrease from 1.03 (cerium) to 0.85 Å (lutetium). The most common oxidation state is +3, though most of the elements also form compounds in other oxidation states. The presence of the lanthanides affects the chemistry of the following elements to a marked degree (Section 10.3.2) and the effect is known as the **lanthanide contraction.** Thus, in groups IVA, VA, VIA, VIIA, and VIII the chemistry of the last two elements is similar and differs slightly from that of the first element, which is less electropositive. For example, in group IVA, the dioxides of zirconium and hafnium are more basic (less soluble in alkali, more soluble in acid), and the ions M^{4+}(aq) (M = Zr, Hf) are less extensively hydrolysed in solution (Section 13.4) than those of titanium.

20.13 THE TITANIUM GROUP, GROUP IVA (Table 20.5)

Titanium, zirconium, and hafnium

Titanium is a typical transition element (Section 20.11). It is a silvery-white, hard, strong, ductile, metal, and is the least dense of all the transition elements (Problem 26). It is surprisingly inert at normal temperatures (Section 20.3), but above 500 °C it reacts vigorously with non-metals to form the compounds $TiCl_4$, TiO_2, etc., and with steam to yield the dioxide and hydrogen. The high strength to

Table 20.5

Ti	Zr	Hf	Ti		Nature of oxide	Nature of compounds of titanium
$\underline{\underline{4}}$	$\underline{\underline{4}}$	$\underline{\underline{4}}^a$	Ti(IV)	d^0	amphoteric, mainly acidic	non-transitional, covalent, stable, titanyl ion, complex anions
3^R	$(3)^R$	$(3)^R$	Ti(III)	d^1	basic, ionic	transitional, ionic, $Ti(H_2O)_6^{3+}$, reducing
$(2)^R$	$((2))^R$	$((2))^R$	Ti(II)	d^2	basic, ionic non-stoichiometric	transitional, ionic, very strongly reducing
0	0		Ti(0)			no carbonyl, $[Ti(dipy)_3]^0$, purple
(−1)			Ti(−1)			no carbonyl, $Li[Ti(dipy)_3]$, black

form complexes with ligands which stabilize low oxidation states (Section A.9)

^{a} increasing stability of higher oxidation states (Section 20.10); dipy = dipyridyl (Fig. 20.2); for meanings of other symbols see Table 20.4.

weight ratio, its resistance to corrosion, and the high abundance of titanium have not led to the replacement of iron as a structural material by titanium. This is due to its reactivity at high temperatures at which it forms brittle interstitial compounds with carbon and nitrogen, and thus precludes normal metallurgical processes and makes pure titanium relatively expensive. The important ores are rutile, TiO_2, and ilmenite, $FeTiO_3$. Industrially, the pure metal is prepared by magnesium reduction of the chloride in an inert atmosphere (Section 11.5.2) or by the van Arkel method (for small amounts).

At room temperature dilute sulphuric, hydrofluoric, and hot concentrated hydrochloric acids each react to yield hydrogen and stable complex anions of titanium(III) (why not titanium(IV)?), such as TiF_6^{3-}. Oxidizing acids yield titanium-(IV) compounds such as the basic sulphate $TiO \cdot SO_4$ (concentrated sulphuric) and the hydrated dioxide $TiO_2 \cdot 4H_2O$ (hot nitric). Alkalis do not react.

Titanium shows the general characteristics of moderately electropositive metals (Section 10.5), and also those of transition elements in its compounds in the +2 and +3 oxidation states (Section 20.11) (why not in the +4 state?). It forms both anionic and cationic complexes. In its highest oxidation state, +4, the compounds are largely covalent (e.g. $TiCl_4$), and its salts may be basic (e.g. $TiO \cdot SO_4$). In its lower oxidation states, +2 and +3, the salts are largely ionic, and in this case are typically transitional (coloured, etc.).

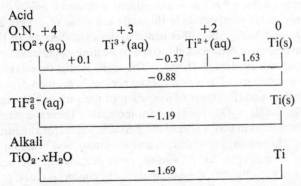

The +4 oxidation state is the most stable one in the group, and in this group it is a non-transition state, d^0. Titanium dioxide is a white solid used as a pigment, which does not darken in air (cf. basic lead carbonate). It is found naturally in one of three modifications, one of which is rutile. The form obtained by heating to a high temperature is inert and refractory, dissolving only in fused alkali or concentrated sulphuric acid (cf. alumina). But the hydrous oxide $TiO_2 \cdot xH_2O$, prepared by adding hydroxyl ions to a solution of titanium(IV), is amphoteric. It dissolves in alkalis to give "titanates" which, in solution, are mixtures of anions such as $TiO_2(OH)_2(aq)^{2-}$ and $TiO(OH)_4(aq)^{2-}$. In strong acids it dissolves to give mixtures of complex anions (e.g. $TiCl_6^{2-}$) and cations (e.g. $[Ti(H_2O)_3(OH)_3]^+$, etc.), depending upon the nature of acid (Section 16.8). No cation, $Ti(H_2O)_6^{4+}$, is formed (Problem 30). This cationic mixture is referred to as the "titanyl" ion and is written $TiO(aq)^{2+}$ for short. Acid solutions of titanium(IV) react quantitatively with hydrogen peroxide to yield the intensely coloured orange species, $Ti(O_2)^{2+}(aq)$.

Titanium tetrachloride is a colourless covalent low-boiling liquid, rapidly hydrolysed by water to the hydrous dioxide and hydrochloric acid via the basic salt $TiOCl_2$, which can be isolated. The tetrachloride finds use as an important catalyst.

The ions Ti^{2+} and Ti^{3+} are transition metal ions, and are coloured and paramagnetic. Their salts are formed by the reaction of titanium(IV) compounds with strong reducing agents. For example: zinc and hydrochloric acid reduce titanium tetrachloride to titanium(III) trichloride solution (purple); and the heterogeneous reaction of titanium on titanium tetrachloride at elevated temperatures yields titanium(II) dichloride. Solutions of titanium(III) in water are reducing but are stable in the absence of oxygen, and are slightly acidic (Section 13.4). They form the purple hydrous oxide ($Ti_2O_3 \cdot xH_2O$) with alkalis, which is basic, ionic, and insoluble in water. The oxidation state +2 exists only in the solid state and is powerfully reducing. For example, it reduces water to hydrogen, and the dichloride inflames in air when heated.

20.14 THE VANADIUM GROUP, GROUP VA (Table 20.6)

Vanadium, niobium, and tantalum

The reactions (Fig. 20.6) and general chemistry of vanadium are very similar to those of titanium, except that vanadium is less electropositive and less abundant. The pure metal is prepared by the reduction of either the chloride, VCl_4, by magnesium, or the pentoxide by aluminium, or alternatively by the van Arkel method (VI_3). No concentrated ores have been found.

The oxidation state +5 is a non-transitional state and is mildly oxidizing (Fig. 20.5). The only stable compounds in this state are those of oxygen and fluorine, e.g. VF_5, V_2O_5, and VOF_3. No other pentahalides exist (Section 15.1). Vanadium pentoxide is an orange or brown solid formed by heating vanadium in oxygen, or by heating ammonium vanadate, NH_4VO_3. The oxide is slightly soluble in water to give pale-yellow solutions. The actual species present in solution depends upon the pH of the solution, and the extent of hydration of the various polymeric species is not known (Fig. 20.7). V_2O_5 itself is a giant molecule. There is a general tendency to form polymeric ions in this region of the Periodic Table: for vanadium (niobium and tantalum), chromium (molybdenum and tungsten), and titanium. Both rings and chains are formed (Fig. 20.7). Various types of vanadate can be precipitated from solutions of vanadium(V), depending upon the conditions: pH, concentration, temperature, and the nature of the precipitating metal ion. Vanadium pentoxide is amphoteric. It dissolves in alkali to form "vanadates" (mixtures of the ions in Fig. 20.7), and in acids to give cationic mixtures containing mainly the hydrated so-called pervanadyl ion, $VO_2^+(aq)$. This does not contain the peroxy ion O_2^{2-}. Acidified vanadium(V) solutions react with hydrogen peroxide to form the red ether-insoluble peroxy-cation, $V(O_2)^{3+}(aq)$, which is not bleached by fluoride ions (compare titanium), and this reaction is used as a qualitative test.

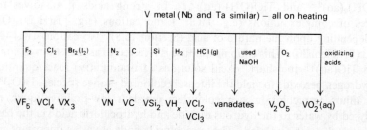

Fig. 20.6 The reactions of vanadium.

Table 20.6

V	Nb	Ta	V		Nature of oxide	Nature of compounds of vanadium
5°	5	5^a	V(v)	d^0	amphoteric, mainly acidic	non-transitional, covalent, mildly oxidizing, VO_2^+ (aq)
4	$(4)^d$	$(4)^d$	V(IV)	d^1	amphoteric, mainly basic	transitional, covalent, stable, VO^{2+} (aq)
3^R	3^R	3^R	V(III)	d^2	basic, ionic, non-stoichiometric	transitional, ionic, reducing, $V(H_2O)_6^{3+}$
$(2)^R$	$(2)^R$	$(2)^R$	V(II)	d^3	basic, ionic, non-stoichiometric	transitional, ionic, strongly reducing, $V(H_2O)_6^{2+}$
(1)			V(I)			$[V(dipy)_3]^+$ ⎫ form complexes with
(0)			V(0)			$[V(dipy)_3]$ and $V(CO)_6$ ⎬ ligands which stabi-
(−1)			V(−1)			$V(CO)_6^-$ ⎭ lize low oxidation states (Section A.9)

For meanings of symbols see Tables 20.4 and 20.5.

Mild reducing agents (H_2S, SO_2, Fe^{2+}, oxalates, Sn^{2+}, I^-, etc.) reduce vanadium(v) salts and solutions to vanadium(IV) compounds. For example, the dark-blue solid vanadium dioxide, VO_2, can be prepared by heating the pentoxide with oxalic acid (reversed by heating in air). This oxide, VO_2, is amphoteric, reacting with fused alkalis to yield a mixture of anions called vanadites (VO_3^{2-}(aq), VO_4^{4-}(aq), etc.), and with aqueous acids to yield blue solutions containing mainly the hydrated VO^{2+}(aq) cation. The molecular tetrahalides, VF_4, VCl_4, and VBr_4 (unstable), are known but not VI_4 (Section 19.1). Ordinarily this is the most stable state of vanadium. This could not be deduced from Fig. 20.5 (a similar problem exists for titanium). Vanadium(III) *is* the most stable state under the conditions appropriate to the diagram (Section 7.12.3) but under normal laboratory conditions in a slightly oxidizing atmosphere (air), the +4 oxidation state is the most stable state. Diagrams such as Fig. 20.5 always exaggerate slightly the stability of "reducing" states relative to normal atmospheric conditions.

Vanadium(III) compounds are formed by more powerful reduction of vanadium(V) (or IV) compounds. For example, the reaction of hydrogen or carbon

pH	>12	12 to 9	9 to 7	7 to 6.5	6.5 to 2.2	<2.2	
Number of V atoms	1	2	3 or 4	5 to 8	8	10	1
Approximate formulae	VO_4^{3-}	$V_2O_6(OH)^{3-}$	$V_3O_9^{3-}$	$V_5O_{14}^{3-}$	$V_2O_5 \cdot xH_2O$	$V_{10}O_{28}^{6-}$	⇌ VO_2^+
	colour-less	to -----------→ red brown			orange brown	pale yellow	yellow

Some suggested species:

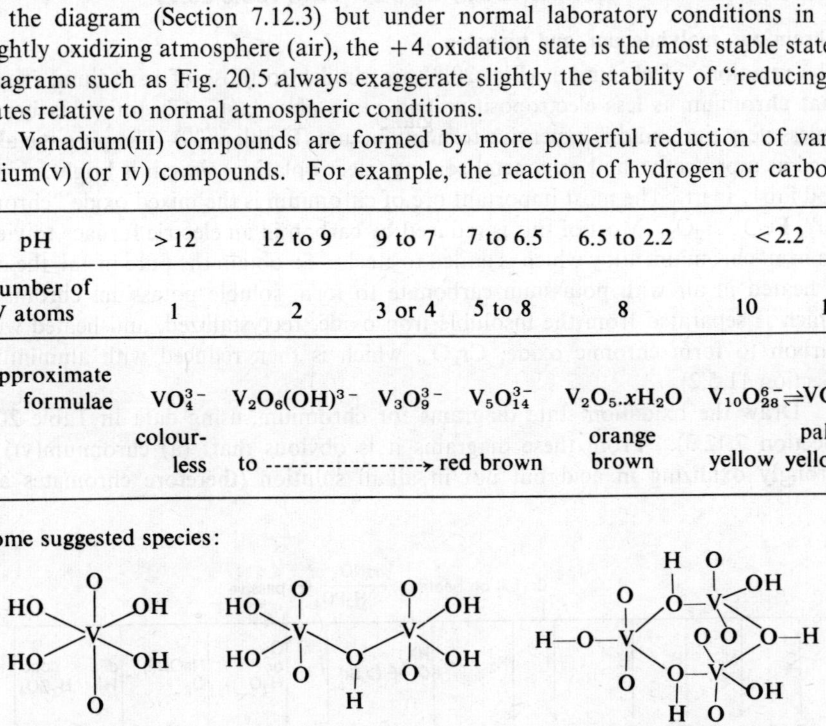

$[VO_4 \cdot 2H_2O]^{3-}$ $[V_2O_6 \cdot OH \cdot 2H_2O]^{3-}$ $[V_3O_9 \cdot 3H_2O]^{3-}$

Fig. 20.7 Species present in vanadium(v) solutions at various pH values.

monoxide with vanadium pentoxide yields the black non-stoichiometric basic oxide, V_2O_3. This reacts with acids to yield the green reducing aquo-cation, $V(H_2O)_6^{3+}$, or anionic complexes such as VF_6^{3-}. The green hydrous oxide, $V_2O_3 \cdot xH_2O$, can be precipitated from such solutions by the addition of alkali. Vanadium(III) compounds are strong reductants, and aqueous solutions are readily oxidized by atmospheric oxygen. Apart from its reducing properties, vanadium(III) resembles chromium(III); for example it forms alums and all four trihalides.

Vanadium(II) compounds are formed by further reduction of higher oxidation states; for example, the black non-stoichiometric oxide, VO, (rock-salt lattice) is formed by heating vanadium with the pentoxide. It is a basic oxide, and dissolves in water to yield the pale-violet or lavender cation, $V(H_2O)_6^{2+}$. This is powerfully reducing, and even reduces water slowly. The addition of alkali yields the violet hydroxide, $V(OH)_2$. A few crystalline salts ($VSO_4 \cdot 7H_2O$, etc.) and all the dihalides, VX_2, exist.

All these oxidation states of vanadium can be observed by treatment of an aqueous solution of vanadium(V) with zinc and hydrochloric acid on heating, when the +4, +3, and +2, states can be observed in turn: the solution turns blue, then green, then violet. This is due to the gradually decreasing electrode potentials:

VO_2^+(aq)		VO^{2+}(aq)		$V(H_2O)_6^{3+}$		$V(H_2O)_6^{2+}$		V^0
	+1.0		+0.36		−0.25		−1.2	

20.15 THE CHROMIUM GROUP, GROUP VIA (Table 20.7)

Chromium, molybdenum, and tungsten

The reactions of chromium (Fig. 20.8) are similar to those of vanadium, except that chromium is less electropositive and more abundant. Chromium is brittle unless very pure, and is expensive to manufacture (Problem 29). The pure metal is used as a protective and ornametal coating (electroplating) since it is bright, shiny, and fairly inert. The most important ore of chromium is the mixed oxide "chromite", $FeO \cdot Cr_2O_3$. Most of this is reduced by carbon in an electric furnace to yield an iron–chromium alloy which is added to steels. To obtain the pure metal, the ore is heated in air with potassium carbonate to form soluble potassium chromate, which is separated from the insoluble iron oxide, recrystallized, and heated with carbon to form chromic oxide, Cr_2O_3, which is then reduced with aluminium (Section 11.5.2).

Draw the oxidation state diagrams for chromium, using data in Table 20.8 (Section 7.12.3). From these diagrams it is obvious that: (a) chromium(VI) is strongly oxidizing in acid but not in alkali solution (therefore chromates are

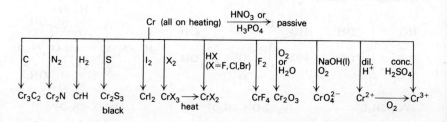

Fig. 20.8 The reactions of chromium.

Table 20.7

Cr	Mo	W	Cr		Nature of oxide	Nature of compounds of chromium
6º	6	6ª	Cr(VI)	d^0	acidic, covalent	non-transitional, covalent, oxidizing
(5)ᵈ	5		Cr(V)	d^1		transitional ⎫ disproportionate, unstable,
(4)ᵈ	4		Cr(IV)	d^2		transitional ⎭ oxidizing
3	3		Cr(III)	d^3	amphoteric, mainly basic	transitional, ionic, stable, many complexes
(2)ᴿ			Cr(II)	d^4	ionic, basic	transitional, ionic, Cr(H$_2$O)$_6^{2+}$, powerful reductant
(1)			Cr(I)	d^5		[Cr(dipy)$_3$]$^+$ClO$_4^-$ ⎫ form complexes with lig-
(0)			Cr(0)	d^6		Cr(CO)$_6$ and [Cr(dipy)$_3$] ⎪ ands which stabilize low
(−1)			Cr(−1)			Cr$_2$(CO)$_{10}^{2-}$ ⎬ oxidation states (Section
(−2)			Cr(−2)			Cr(CO)$_5^{2-}$ ⎭ A.9)

For meanings of symbols see Tables 20.4 and 20.5.

prepared in alkaline conditions (Section 17.2.5); (b) in acid solution chromium(III) is the most stable oxidation state, but not in alkali; and (c) dichromate in acid solution should oxidize water to oxygen (O$_2$/H$_2$O, $E^\ominus = +1.23$ V), but in fact it does not (Section 13.6). Data are not available for the oxidation states +4 and +5. Solid and gaseous compounds in these states do exist, for example, CrX$_4$ (X = F, Cl, Br) and CrF$_5$, but in aqueous solution they disproportionate rapidly (often a characteristic of intermediate oxidation states).

The +6 oxidation state is highly oxidizing in acid solution, and only compounds with electronegative ligands (O, F, Cl) exist (Section 14.1): CrO$_3$, CrF$_6$, CrO$_4^{2-}$, Cr$_2$O$_7^{2-}$, CrO$_2$F$_2$, and CrO$_2$Cl$_2$. In most of these compounds the chromium is tetrahedrally surrounded by the ligands (Fig. 20.9). Potassium dichromate (orange) is used as a primary standard in volumetric analysis for estimating reducing agents (Cr$_2$O$_7^{2-}$(aq) + 14H$^+$(aq) + 6e$^-$ → 2Cr^{3+}(aq) + 7H$_2$O(l)), for example I$^-$ → I$_2$; Fe^{2+} → Fe^{3+}; etc. It has the advantage over potassium permanganate that it is more stable in solution, and since it oxidizes chloride ions in dilute solution much more slowly, it can be used in the presence of dilute hydrochloric acid. However, dichromate titrations require an indicator. When the dichromate is boiled with concentrated hydrochloric acid, one oxygen atom is replaced to form the orange solid potassium chlorochromate K[CrO$_3$Cl]. When dichromate is boiled with concentrated sulphuric acid and sodium chloride, two oxygen atoms are replaced to yield the red-brown vapour (b.p. 117°) chromyl chloride, CrO$_2$Cl$_2$. Both compounds are readily hydrolysed to dichromate. The formation of chromyl chloride is a good test for the chloride ion since the analogous bromide and iodide are unknown (Section 15.1), and the fluoride is not formed under these conditions.

The tendency to polymerize and yield poly acids (acids with more than one atom other than oxygen in the anion) and their salts is less than that of vanadium(V),

Table 20.8. Reduction potentials for chromium.

(a) Acid				(b) Alkali			
Cr$_2$O$_7^{2-}$	Cr^{3+}(aq)	Cr^{2+}(aq)	Cr	CrO$_4^{2-}$(aq)	Cr(OH)$_3$	Cr(OH)$_2$	Cr
+1.33		−0.4	−0.9	−0.13		−1.1	−1.4

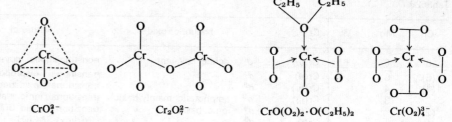

Fig. 20.9 Some tetrahedral chromium compounds.

CrO_4^{2-} $Cr_2O_7^{2-}$ $CrO(O_2)_2 \cdot O(C_2H_5)_2$ $Cr(O_2)_4^{3-}$

molybdenum(VI), or tungsten(VI), but it exists. The chromate (yellow)–dichromate (orange) change is the first stage in this process,

$$2CrO_4^{2-}(aq) + 2H^+(aq) \xrightleftharpoons[\text{alkali}]{\text{acid}} Cr_2O_7^{2-}(aq) + H_2O(l) \quad \text{(about pH = 4)},$$

and at pH values of less than zero, anions such as $Cr_3O_{10}^{2-}$(aq) and $Cr_4O_{13}^{2-}$(aq) exist. In concentrated sulphuric acid, the red acidic oxide, CrO_3, is formed. Many chromates are insoluble, and can be precipitated as yellow solids even from acid solutions; for example lead, barium, and silver. No cationic forms of chromium(VI) such as CrO_2^{2+} exist—cf. vanadium(V). As expected chromium(VI) is more acidic and more highly oxidizing than vanadium(V). Hydrogen peroxide reacts with acid dichromate solutions to form a blue coloration in the presence of ether owing to the formation of the peroxy compound, $CrO(O_2)_2 \cdot (C_2H_5)_2O$ (Fig. 20.9). Other peroxy compounds are known, for example $Cr(O_2)_4^{3-}$ (Fig. 20.9). What is the oxidation state of chromium in this compound?

The +3 oxidation state is outstandingly stable and a wide variety of compounds are formed. The octahedral violet aquo-ion, $Cr(H_2O)_6^{3+}$(aq), exists in aqueous solution and is stable to oxidation and reduction, except in alkaline solution when it is readily oxidized. Many complex compounds of chromium(III) exist with ligands which stabilize moderate oxidation states, such as H_2O, NH_3, Cl^-, etc. (Section A.9). Almost invariably the stereochemistry about chromium is octahedral, and the compounds are thermodynamically stable and kinetically inert, both of which are associated with the "stability" of the half-filled t_{2g} shell (Section A.9). Thus the complex ions in solution are stable to substitution by other ligands, as well as to redox reactions, the symmetrical t_{2g}^3 configuration protecting the nucleus from nucleophilic substitution (Section 26.3). The ion $Cr(H_2O)_6^{3+}$ exists in some solids, such as $Cr_2(SO_4)_3 \cdot 18H_2O$ and $KCr(SO_4)_2 \cdot 12H_2O$, as well as in aqueous solutions in the absence of strongly complexing ligands such as chloride. In the presence of such ligands the water molecules can be replaced, and hydration isomerism is observed: $[Cr(H_2O)_6]Cl_3$ is violet; $[Cr(H_2O)_5Cl]Cl_2 \cdot H_2O$ is pale green; $[Cr(H_2O)_4Cl_2]Cl \cdot 2H_2O$ is dark green; and $[Cr(H_2O)_3Cl_3]$ is brown.

The green oxide Cr_2O_3 can be prepared by heating ammonium dichromate (Section 17.5) and unless it is rendered inert by strong heating, it will dissolve both in acids, to form the violet ion $Cr(H_2O)_6^{3+}$, and in alkalis to form green chromites which are mixtures of ions such as $Cr(OH)_6^{3-}$(aq), $[Cr(OH)_5H_2O]^{2-}$, etc. Chromites are easily oxidized by air or hydrogen peroxide to chromates. On addition of alkalis to solutions of the ion $Cr(H_2O)_6^{3+}$(aq), the green hydrated amphoteric oxide, $Cr_2O_3 \cdot xH_2O$, is formed. Ammonia also precipitates the hydrated oxide, but (unlike alkalis) even concentrated ammonia only slightly redissolves the precipitate to form some complex ammine, $Cr(NH_3)_6^{3+}$. The aqueous ion, $Cr(H_2O)_6^{3+}$(aq), is slightly acidic (Section 13.4).

The +2 oxidation state is strongly reducing, but is stable in water in the absence of oxidants such as oxygen. The sky-blue aquo-ion, $Cr(H_2O)_6^{2+}$, can be prepared either by the action of aqueous acids on chromium in the absence of oxygen, or by the reduction of chromium(III) by strong reducing agents, such as zinc and hydrochloric acid. It rapidly turns violet in air owing to atmospheric oxidation. Many compounds are known and are stable in the solid state or in solution in the absence of oxygen: the black oxide CrO; CrX_2 (X = F, Cl, Br, I); $CrSO_4 \cdot 5H_2O$; and many complexes.

20.16 THE MANGANESE GROUP, GROUP VIIA (Table 20.9)

Manganese, technetium, and rhenium

Manganese is a typical transition element, but its properties often differ slightly from those expected from interpolation from neighbouring elements (Table 20.1). Often these can be associated with the stability of the half-filled shell, d^5 (Section 20.3). For example its cohesive properties (Section 10.3.13 etc.) are less than expected (lower b.p., etc.) since the d-electrons are participating less than expected in metallic bonding, the first and second ionization energies are lower than expected (good shielding of d^5 configuration), and the reduction potential M^{2+}/M is greater than expected. Moreover, since the metal is *not* protected by an oxide layer, it is outstandingly reactive among the transition series, even dissolving in water slowly. On heating, it reacts with most non-metals, except hydrogen, to form the compounds Mn_3N_2, MnS, Mn_3C, Mn_3O_4, and MnX_2 (X = F, Cl, Br, I). It reacts with dilute acids to form Mn^{2+}(aq) and hydrogen, and with oxidizing acids to form Mn^{2+}(aq), and reduction products of the acid (SO_2, NO_2, etc.). The main ore is pyrolusite, MnO_2. This oxide reacts too fiercely with aluminium, so it is heated to form the oxide, Mn_3O_4, purified by distillation, and then reduced with aluminium to the pure metal. The pure metal is brittle and little used.

The oxidation state diagram of manganese in acid solution was discussed in Section 7.12.3. The major points are: the stability of the +2 oxidation state under

Table 20.9

Mn	Tc	Re	Mn		Nature of oxide	Nature of compounds of manganese	
7º	7	7	Mn(VII)	d^0	acidic, covalent	non-transitional, covalent, powerfully oxidizing	
(6)^d	(6)	(6)	Mn(VI)	d^1	(acidic)	transitional, disproportionates, powerfully oxidizing	
(5)^d	(5)	(5)	Mn(V)	d^2	(acidic)	transitional, disproportionates, powerfully oxidizing	
4^d	4	4	Mn(IV)	d^3	amphoteric, mainly acidic	transitional, oxidizing	
(3)º	(3)	3	Mn(III)	d^4	mainly basic and ionic	transitional, disproportionates, oxidizing	
<u>2</u>	(2)	(2)	Mn(II)	d^5	basic	transitional, stable, many compounds and complexes, $Mn(H_2O)_6^{2+}$	
(1)		1	Mn(I)			⎫ complexes of ligands	$Mn(CO)_5Cl$, $Mn(RNC)_6^+$
(0)	0	0	Mn(0)			⎬ which stabilize low	$Mn_2(CO)_{10}$
(−1)		−1	Mn(−1)			⎨ oxidation states, e.g.	$Mn(CO)_5^-$
(−2)			Mn(−2)			⎭ CO, RNC etc.	

For meanings of symbols see Table 20.4.

these (and most other) conditions; the powerfully oxidizing nature of manganese-(VII) (it should oxidize water, but actually the reaction is very slow and it does not); and the oxidizing nature and instability with respect to disproportionation of the $+3$, $+5$, and $+6$ states. In alkaline conditions the situation differs dramatically; draw the oxidation state diagram from the potential data given below:

alkali: MnO_4^- $\quad$ MnO_4^{2-} $\quad$ MnO_4^{3-} $\quad$ MnO_2 $\quad$ Mn(OH)_3 $\quad$ Mn(OH)_2 $\quad$ Mn
$\quad\quad\quad\quad$ | $+0.56$ | $+0.34$ | $+0.84$ | -0.2 | $+0.1$ | -1.55 |

From the diagram it should be obvious that in alkali: manganese dioxide is the most stable state (just), and is the usual reduction product of permanganate in non-acid conditions; the higher oxidizing states are much less oxidizing (Section 7.12.4); and the intermediate oxidation states are more stable with respect to disproportionation.

The $+2$ oxidation state is very stable, especially in acid solution. Most complexes and compounds in this state have octahedral stereochemistry about the manganese and are high spin (Section A.9), showing the electron configuration $t_{2g}^3 e_g^2$. Their stability can be partly associated with this half-filled shell (Section 20.3). The green insoluble oxide, MnO, is entirely basic, reacting with acids to form aqueous solutions of salts containing the stable pale-pink aquo-ion, $\text{Mn(H}_2\text{O)}_6^{2+}$. When manganese(II) oxide is warmed in air, the brown oxide Mn_2O_3 is formed, and on further heating the very stable mixed oxide, Mn_3O_4, results. Ammonia (except in the presence of ammonium chloride, Section 14.3.5) and sodium hydroxide both precipitate the white hydroxide, Mn(OH)_2, from solutions of $\text{Mn(H}_2\text{O)}_6^{2+}$, and the hydroxide does not dissolve in an excess of either reagent (it is not amphoteric, nor are the ammines stable in aqueous solution). The white precipitate is rapidly oxidized in air to the brown hydrous oxide, $\text{Mn}_2\text{O}_3 \cdot x\text{H}_2\text{O}$. Alkaline hydrogen sulphide precipitates the flesh-coloured hydrous sulphide, $\text{MnS} \cdot x\text{H}_2\text{O}$, from solutions of $\text{Mn(H}_2\text{O)}_6^{2+}$, which when heated in nitrogen forms the green anhydrous sulphide. Both sulphides are soluble in acids, releasing hydrogen sulphide and yielding the aquo-ion, $\text{Mn(H}_2\text{O)}_6^{2+}$. The carbonate and basic carbonate can be obtained from the aquo-ion by the addition of aqueous bicarbonate and carbonate solutions respectively (Section 16.6). In acid solution, very powerful oxidants (e.g. bismuthate) are required to oxidize the aquo-ion, $\text{Mn(H}_2\text{O)}_6^{2+}$, to permanganate; less powerful oxidants (e.g. hypochlorite) convert it to manganese dioxide. But in alkaline solution the oxidation is relatively easy. Manganese(II) forms an extensive series of salts with all the common anions, most of which are soluble in water (except for the carbonate and phosphate), and if crystallized from aqueous solution are usually hydrated. Many complexes are known, but they are not very stable in aqueous solution.

There is not an extensive chemistry of manganese(III), since this oxidation state disproportionates in solution to manganese dioxide and Mn^{2+}(aq), and is easily reduced. As usual, it can be stabilized as either an insoluble compound, e.g. the brown oxide Mn_2O_3, or as a stable complex e.g. Mn(CN)_6^{3-} or MnF_6^{3-} (Section A.9). The only stable halide is MnF_3. Manganese(IV) is similar; only the compounds MnF_4, MnO_2, and a few complexes (e.g. MnF_6^{2-}, and MnCl_6^{2-}) are known, but it is familiar owing to the insolubility and fair thermal stability of manganese dioxide (which changes to the stable oxide, Mn_3O_4, only above 900 °C). Manganese dioxide is a grey-black non-stoichiometric (oxygen-deficient) solid. It is insoluble in dilute acids (therefore it is not a peroxide). Cold concentrated hydrochloric acid reacts with the dioxide to form the unstable chloride, $MnCl_4$ (as the dark-brown complex $MnCl_6^{2-}$), which dissociates on heating to yield chlorine, or is hydrolysed by water to yield manganese dioxide. Manganese(V) and (VI) are stable only in alkaline conditions as the blue permanganite ion, MnO_4^{3-}, and the dark-green manganate ion MnO_4^{2-}. Both disproportionate immediately in dilute alkali, neutral, or acid solution, yielding manganese dioxide and permanganate. Write the equations for these reactions (Section 18.14).

Manganese(VII) is highly oxidizing and exists only as the oxygen and fluorine compounds (Section 15.1), Mn_2O_7 (green), MnO_3F, and the deep-purple permanganate ion MnO_4^- (all tetrahedral). How can you account for their deep colours (Problem 22)? Potassium permanganate is manufactured by fusing potassium hydroxide and manganese dioxide together in air, and oxidizing the manganate formed either electrolytically or with chlorine. In the laboratory the green manganate is extracted with water, concentrated by evaporation, and carbon dioxide passed in in order to effect disproportionation:

$$3MnO_4^{2-}(aq) + 2H_2O(l) \underset{OH^-}{\overset{H^+}{\rightleftharpoons}} 2MnO_4^-(aq) + MnO_2(s) + 4OH^-(aq) \longrightarrow$$
$$OH^-(aq) + CO_2(g) \rightarrow HCO_3^-(aq).$$

Clearly this is unsatisfactory in an industrial context, since one-third of the manganese finishes as manganese dioxide. Permanganate is about the most powerful oxidant which is stable in water (Section 13.6). Thus it is very useful in volumetric analysis, particularly since it requires no indicator. In acid solution, it is quantitatively reduced to the manganese(II) ion by most reductants, e.g. oxalate, iodide, iron(II), etc. Only sulphuric acid can be used, since permanganate slowly oxidizes chloride to chlorine. Potassium permanganate is not a primary standard since it is difficult to obtain pure, and it is slowly reduced by water to manganese dioxide, especially in the presence of light or acid.

acid: $\quad MnO_4^-(aq) + 8H^+(aq) + 5e^- \rightarrow Mn^{2+}(aq) + 4H_2O(l);$
$$E^\ominus = +1.51 \text{ V}$$

alkali: $\quad MnO_4^-(aq) + 2H_2O(l) + 3e^- \rightarrow MnO_2(s) + 4OH^-(aq);$
$$E^\ominus = +0.58 \text{ V}$$

$$4MnO_4^-(aq) + 4H^+(aq) \xrightarrow{light} 3O_2(g) + 2H_2O(l) + 4MnO_2(s).$$

It is less useful as an oxidant in alkaline solution. At 250 °C permanganate decomposes to manganate, manganese dioxide, and oxygen. In concentrated alkali it forms oxygen and the manganate ion. Concentrated sulphuric acid reacts to give the covalent, explosive, acidic, green oil, Mn_2O_7. Write the equations for these reactions (Section 18.14).

20.17 THE IRON GROUP, GROUP VIII (Table 20.10)

Iron, ruthenium, and osmium

The chemistry of iron differs very markedly from that of ruthenium and osmium. Iron is very similar to cobalt and nickel; ruthenium and osmium more closely resemble the other "platinum metals" (platinum, palladium, rhodium, and iridium) which are inert, have a wider range of oxidation states, have no simple aquated ions such as $Fe^{2+}(aq)$, are outstanding catalysts, and are not ferromagnetic.

Table 20.10

Fe	Ru	Os	Fe		Nature of oxide	Nature of compounds of iron
(6)⁰	6	6ᵃ	Fe(VI)	d^2		transitional, powerful oxidant, FeO_4^{2-}
(5)⁰	(5)	(5)	Fe(V)	d^3		FeO_4^-
(4)⁰	4	4	Fe(IV)	d^4		$Fe[diarsine_2Cl_2]^{2+}$
<u>3</u>	3	3	Fe(III)	d^5	amphoteric (mainly basic), non-stoichiometric	$Fe(H_2O)_6^{3+}$ mild oxidant, many compounds
<u>2</u>	2	2	Fe(II)	d^6	amphoteric (mainly basic), non-stoichiometric	$Fe(H_2O)_6^{2+}$ stable, many compounds
1	1		Fe(I)	d^7		$[Fe(H_2O)_5NO]^{2+}$
(0)	(0)	(0)	Fe(0)			$Fe(CO)_5$ ⎫ form complexes with ligands which
(−2)	(−2)	(−2)	Fe(−2)			$Fe(CO)_4^{2-}$ ⎬ stabilize low oxidation states (Section A.9)

For meanings of symbols see Tables 20.4 and 20.5.

Iron is abundant, easily obtained, and (even when impure) has relatively desirable mechanical properties: thus it is very important. The most useful ores are: haematite, Fe_2O_3; magnetite, Fe_3O_4; limonite, $FeO(OH)$; and siderite, $FeCO_3$. Pure iron is obtained by heating purified iron(II) oxalate in a vacuum, and reducing the iron(II) oxide obtained with hydrogen. It is a white lustrous metal and is relatively soft. Most iron used contains other elements in order to improve its properties (for example, to harden it), and such alloys are called steels. There are three stages in the production of steels. First, the ore is roasted to the stable iron(III) oxide, Fe_2O_3. Then the oxide is reduced with carbon (coke) in the presence of calcium carbonate (Section 11.5.2) in a blast furnace to form a low-melting form of impure iron known as "pig iron" (m.p. about 1200 °C). This contains many impurities (Mn, Si, P, S, etc.) and up to 4 percent carbon, some as the interstitial compound, Fe_3C. Pig iron is used for gates, pipes, and so on, which are formed from moulds; it is very brittle. The reactions taking place in the furnace are complex, but the most important are:

$$CaCO_3(s) \rightarrow CO_2(g) + CaO(s);$$
$$CO_2(g) + C(s) \rightarrow 2CO(g);$$
$$CaO(s) + SiO_2(s) \rightarrow CaSiO_3(l).$$

Top of furnace (200 °C): $\quad 3Fe_2O_3(s) + CO(g) \rightarrow 2Fe_3O_4(s) + CO_2(g);$
Lower down (300 °C): $\quad Fe_3O_4(s) + CO(g) \rightarrow 3FeO(s) + CO_2(g);$
Bottom of furnace (>1000 °C): $\quad FeO(s) + CO(g) \rightarrow Fe(l) + CO_2(g).$

Finally, the impurities are removed by burning in oxygen ($C \rightarrow CO_2$, $S \rightarrow SO_2$) and by addition of calcium carbonate (as slags, e.g. $CaSiO_3$). Wrought iron is much purer than pig iron (less than 1.5 percent carbon). This is hardly used nowadays; instead steels, obtained by adding suitable alloying agents such as titanium, vanadium, and manganese, generally in the form of an easily prepared concentrated alloy such as ferro-vanadium, are employed. These alloying agents not only impart specific properties to the alloy, but also clear the last traces of carbon, oxygen, etc., which tend to make the iron brittle if left in.

Iron is moderately reactive. In moist air it "rusts" and forms the brown hydrated oxide, $Fe_2O_3 \cdot xH_2O$. This oxide is non-coherent and permeable; consequently it does not protect the metal from further reaction. The necessary con-

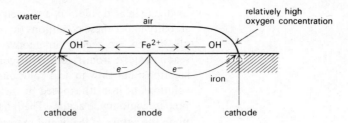

Fig. 20.10 A model of rusting.

ditions for rusting are: (a) impure iron (all iron is, unless specially prepared); (b) the presence of water (very thin invisible layers suffice); and (c) the presence of oxygen. Rusting is accelerated by the presence of electrolytes, especially acids, and is retarded by alkalis. A model of the process is given in Fig. 20.10, which represents a drop of water on iron, surrounded by air. The impurities in the iron (or strains, or junctions with other materials) form the cathode of a cell, the associated anode of which is an area of purer iron. At the cathode, reduction of atmospheric oxygen takes place; and at the anode (inside the drop where the oxygen concentration is lower) oxidation of iron occurs to form iron(II) ions:

cathode: $O_2(aq) + 2H_2O(l) + 4e^- \rightarrow 4OH^-(aq)$
anode: $Fe(s) \rightarrow Fe^{2+}(aq) + 2e^-$.

The hydroxyl and iron(II) ions diffuse away from the electrodes (therefore the electrodes are not coated) and precipitate iron(II) hydroxide, which is oxidized in air to hydrous iron(III) oxide. A low pH assists the dissolution of iron, which is hindered by a high pH. The presence of electrolytes in the water (sodium chloride in sea-side areas for example) increases the conductivity, and impurities in the iron assist cell action. There are two general methods of protection: (a) the application of a protective layer (paint, tin, zinc electroplating, etc.); and (b) the presence of a sacrificial metal. The idea here is that in the presence of a more electropositive metal, iron will not rust because the more reactive metal will be oxidized first. Thus, the galvanizing of iron by zinc is more successful than tin-plating, because once tin-plating is scratched, and the iron exposed, the more electropositive iron rusts. However, even if galvanized iron is scratched, the zinc is oxidized preferentially. Underground piping is protected by regular strips of zinc, etc.

Iron combines with most non-metals (not nitrogen) when heated gently to form the compounds Fe_2O_3 (via Fe_3O_4), FeS and FeS_2, FeF_3, $FeCl_3$, FeI_2, Fe_3C, etc. It reacts with dilute acids in the absence of air to yield the iron(II) ion, Fe^{2+}, and hydrogen (and smelly impurities owing to carbides etc., in the iron). Nitric and concentrated sulphuric acids yield mixtures of Fe^{2+} and Fe^{3+} ions together with the reduction products of the acids (mainly NH_4^+ and SO_2 respectively); but concentrated nitric acid and aqua regia render the metal passive (Section 10.4.2). Fused alkalis yield the ion $Fe(OH)_4^{2-}$ in the absence of air, and ferrates, FeO_4^{2-}, in the presence of oxygen.

The reduction potential data for iron in acid and alkali is:

acid: FeO_4^{2-} $\underset{>1.9}{\quad}$ Fe^{3+} $\underset{+0.77}{\quad}$ Fe^{2+} $\underset{-0.44}{\quad}$ Fe *alkali:* FeO_4^{2-} $\underset{>0.9}{\quad}$ $Fe(OH)_3$ $\underset{-0.56}{\quad}$ $Fe(OH)_2$ $\underset{-0.89}{\quad}$ Fe

Draw the free energy diagrams in acidic and alkaline solution (Section 7.12.3). It is obvious from these diagrams that the +6 oxidation state is very powerfully oxidizing in acid solution, and even in alkali. The +2 and +3 states are of similar stability under these conditions (actually true for most conditions) in both acid and alkali, and are readily interconverted. Thus, even mild oxidants like oxygen readily oxidize iron(II) to iron(III) ions. This reaction is rapid in alkaline or neutral solutions, but slow in acid solutions (N.B. this is a *kinetic* factor). Thus aqueous solutions of iron(II) exposed to the air always contain iron(III) ions unless freshly prepared and made acidic. The relative stabilities of the +2 and +3 states depend upon the nature of the ligand present (Section A.9). For example, the iron(III) state is more stable in the presence of cyanide ion than in water:

$$[Fe(III)(CN)_6]^{3-} + e^- \rightleftharpoons [Fe(II)(CN)_6]^{4-}; \quad E^\ominus = +0.36 \text{ V}$$
$$[Fe(III)(H_2O)_6]^{3+} + e^- \rightleftharpoons [Fe(II)(H_2O)_6]^{2+}; \quad E^\ominus = +0.77 \text{ V}.$$

The +2 oxidation state of iron is very stable and forms salts with all stable anions giving a wide range of hydrated and anhydrous binary salts, oxysalts, and double salts. The double salt known as Mohr's salt, $FeSO_4 \cdot (NH_4)_2SO_4 \cdot 6H_2O$ is fairly stable relative to both oxidation and loss of water, and is used as a primary standard. The black oxide, FeO, burns spontaneously in air to iron(III) oxide. Pure iron(II) hydroxide is white and is rapidly oxidized in air to the brown hydrated iron(III) oxide. It dissolves slowly in hot concentrated alkali to the ferrosite ion $Fe(OH)_6^{4-}$, and rapidly in dilute acids to form solutions of the pale-green aquo-ion $Fe(H_2O)_6^{2+}$, except in the presence of strongly co-ordinating ligands such as cyanide. This aquo-ion is hardly hydrolysed and is barely acidic (Section 13.4). Its oxidation is discussed above, and the reactions of the aquated iron(II) ion are given in Fig. 20.11. Many complex ions are known, most of which have octahedral stereochemistry, but most are unstable in water, yielding the aquo-ion. One water-stable complex is the yellow hexacyanoferrate(II) ion, $[Fe(II)(CN)_6]^{4-}$ (ferrocyanide), present in the compound $K_4[Fe(II)(CN)_6] \cdot 3H_2O$. On treatment with acid this compound evolves the poisonous hydrogen cyanide gas; on oxidation with chlorine it forms the iron(III) compound $K_3[Fe(III)(CN)_6]$ which contains the hexacyanoferrate(III) ion (ferricyanide); it forms the pigment Prussian blue on treatment with aqueous iron(III) ions, $KFe(II)Fe(III)(CN)_6$; and it forms the white

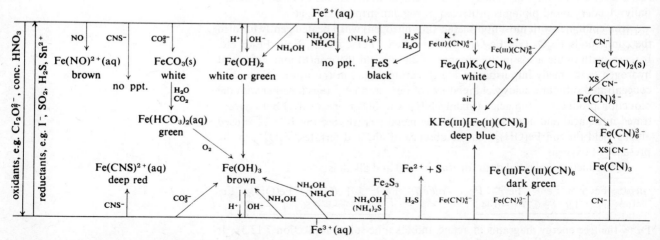

Fig. 20.11 The reactions of the aqueous Fe^{2+} and Fe^{3+} ions.

Everitt's salt on treatment with aquated iron(II) ions, $K_2Fe(II)Fe(II)(CN)_6$. The relationship between Everitt's salt, Prussian blue, and the oxidation product of Prussian blue, a green insoluble compound called Berlin green, is given below:

$$Fe(III)Fe(III)(CN)_6 \underset{\text{e.g. Cl}_2}{\xleftarrow{\text{oxidation}}} KFe(II)Fe(III)(CN)_6 \underset{\text{e.g. O}_2\text{(air)}}{\xleftarrow{\text{oxidation}}} K_2Fe(II)Fe(II)(CN)_6.$$
$$\text{Berlin green} \qquad\qquad\qquad \text{Prussian blue} \qquad\qquad\qquad \text{Everitt's salt (white)}$$

The structures of these solids are relatively simple (draw them): in each case the iron atoms occupy the corners of a cube (simple cubic); the cyanide ions occupy the centres of the cube edges; and the potassium ion occupies the centre of the cubes where necessary (at the centre of every other cube in the case of Prussian blue).

The active part of the large complex oxygen-carrying molecule haemoglobin, is an iron(II) atom co-ordinated to a complex nitrogen ligand, a porphyrin (Fig. 21.10), which occupies four of the six octahedral co-ordination positions (Fig. 20.12). The fifth position is occupied by a nitrogen from another part of the large molecule and the vacant sixth position can be occupied weakly and reversibly by oxygen, which is how oxygen is transported around the body. Unfortunately, the more strongly bonding ligands CN^-, CO, and PF_3, etc., can also occupy this position (but irreversibly) and are poisons. Chlorophyll is similar, but the central metal ion is magnesium not iron.

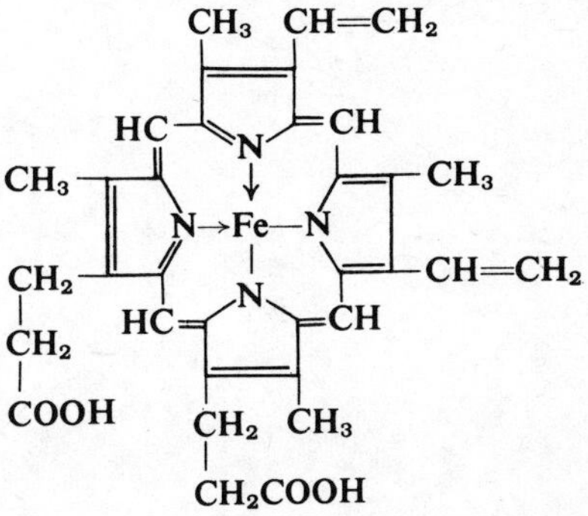

Iron(III) is a mildly oxidizing state, and salts and complexes are known for all anions except those easily oxidized (iodide, sulphide, Section 15.1). The compounds are more covalent than the iron(II) compounds: like aluminium chloride, iron(III) chloride (which is deep red-brown and hygroscopic) has a layer structure in the solid state and is a chlorine-bridged dimer in the vapour phase. In aqueous solution iron(III) exists as the strongly hydrolysed, acidic, hydroxy aquo-ion. The aquo-ion $Fe(H_2O)_6^{3+}$ exists only at pH values below 0. The yellow hydroxy species $[Fe(H_2O)_5OH]^{2+}$ exist at pH values between 0 and 3, and above this hydroxy-bridged species are formed (Section 20.14) and the solutions become colloidal.

$$[Fe(H_2O)_6]^{3+} \underset{H^+}{\rightleftharpoons} [Fe(H_2O)_5OH]^{2+} + H_3O^+ \underset{}{\overset{OH^-}{\rightleftharpoons}} [Fe(H_2O)_4(OH)_2]^+ + H_3O^+$$
very pale purple $\qquad\qquad$ yellow

At even higher pH values, hydrous iron(III) oxide is precipitated. This oxide, $Fe_2O_3 \cdot xH_2O$, is amphoteric: it dissolves on boiling in very concentrated alkalis to form the anion $Fe(OH)_6^{3-}$; and it dissolves in acids to form the mixture of aquo-ions above, mainly the yellow hydroxy species in dilute acid (for convenience it is represented as Fe^{3+}aq). The reactions of the yellow Fe^{3+}(aq) are summarized in Fig. 20.11 and Table 20.11. It is a mild oxidizing agent, reacting with reductants such as H_2S, SO_2, Sn^{2+}(aq), etc. Iron(III) forms many complexes, mainly octahedral

Fig. 20.12 The iron porphyrin complex with the side chains as shown here is called a haem group.

Table 20.11. A comparison of the reactions of the aqueous iron(II) and iron(III) ions.

Test	Fe^{2+}(aq)	Fe^{3+}(aq)
NH_3(aq)	green precipitate	red-brown precipitate
$Fe(III)(CN)_6^{3-}$	Turnbull's blue (precipitate)	brown or green solution
$Fe(II)(CN)_6^{4-}$	Everitt's salt (white precipitate)	Prussian blue (precipitate)
CNS^-(aq)	no colour	intense red colour

Turnbull's blue and Prussian blue are the same compound

and high spin. Quite often these are intensely coloured, owing to charge transfer transitions (Section A.9) not *d–d* transitions: for example, the ligand thiocyanate, SCN^-, forms the deep-red anion $[Fe(SCN)(H_2O)_5]^{2+}$. Oxalate ions form the stable soluble complex anion $[FeOx_3]^{3-}$, and consequently oxalic acid removes rust stains.

20.18 THE COBALT GROUP, GROUP VIII (Table 20.12)

Cobalt, rhodium, and iridium

Draw the oxidation state diagram for cobalt (Section 7.12.3) using the data given below:

$$\text{acid:} \quad CoO_2 \underset{>1.8}{\quad} Co^{3+}(aq) \underset{+1.82}{\quad} Co^{2+}(aq) \underset{-0.28}{\quad} Co$$

$$\text{alkali:} \quad CoO_2 \underset{+0.7}{\quad} Co(OH)_3 \underset{+0.14}{\quad} Co(OH)_2 \underset{-0.72}{\quad} Co$$

In neutral or acidic aqueous solutions containing no strongly complexing ligands, the stable ion present is the pink aquo-ion $Co(H_2O)_6^{2+}$. In the presence of strongly complexing ligands, for example the cyanide ion or nitrogen ligands such as NH_3, NO_2^-, etc., the relative stability of the $+2$ and $+3$ states is reversed (Section A.9).

$$Co(H_2O)_6^{3+} + e^- \rightleftharpoons Co(H_2O)_6^{2+}; \quad E^\ominus = +1.82 \text{ V at pH} = 0$$
$$Co(NH_3)_6^{3+} + e^- \rightleftharpoons Co(NH_3)_6^{2+}; \quad E^\ominus = +0.1 \text{ V at pH} = 0$$
$$Co(CN)_6^{3-} + e^- \rightleftharpoons Co(CN)_6^{4-}; \quad E^\ominus = -0.8 \text{ V at pH} = 0.$$

The ion $Co(III)(H_2O)_6^{3+}$ is powerfully oxidizing, and oxidizes water to oxygen. The ion $Co(NH_3)_6^{3+}$ is much more stable and, in fact, in the mildly oxidizing conditions on this planet (remember the figures refer to the hydrogen electrode) this is the stable state in the presence of ammonia. Thus, a general preparation of cobalt(III) complexes is to pass oxygen through (or add hydrogen peroxide to) a cobalt(II) solution in the presence of the ligand (e.g. ammonia) and some activated carbon catalyst. The cobalt(III) state is even more stable in the presence of the cyanide ligand, and in fact $Co(II)(CN)_6^{4-}$ reduces water to hydrogen! The ammine complexes can be fitted on the diagram using the data below:

$$Co(NH_3)_6^{3+} \underset{+0.1}{\quad} Co(NH_3)_6^{2+} \underset{-0.42}{\quad} Co$$

In general the $+3$ oxidation state is the stable one for small non-oxidizable anions (CoF_3, $Co(OH)_3$) and for most complexes; whereas the $+2$ oxidation state is stable for most simple salts, and in aqueous solution. The higher oxidation states (III) and (IV) are powerfully oxidizing, but are more stable in alkaline solution. Cobalt(III) is an oxidizing state and does not form complexes or salts with oxidizable ligands such as sulphide, iodide, bromide, or chloride.

Cobalt is a bluish-white ferromagnetic metal, harder and stronger than iron, and slightly less reactive. On heating it reacts with many non-metals forming Co_3O_4 (via CoO), CoF_2 and CoF_3, CoX_2(X = Cl, Br, I), CoS, etc. It dissolves slowly in dilute acids yielding hydrogen and solutions of the aquo-ion $Co(H_2O)_6^{2+}$ from which hydrated salts, $CoSO_4 \cdot 7H_2O$, etc., can be crystallized. Concentrated nitric acid renders the metal passive; alkalis do not react.

Table 20.12

Co	Rh	Ir	Co		Nature of oxide	Nature of compounds
	6⁰	6⁰,ᵃ		d^3		transitional, RhF_6 IrF_6
		(5)⁰		d^4		transitional, IrF_6^-
(4)⁰	4	4	Co(IV)	d^5		transitional, CoF_6^{2-}, powerfully oxidizing
3⁰	3	3	Co(III)	d^6		transitional, $Co(H_2O)_6^{3+}$, powerfully oxidizing, many stable complexes
2	2	2	Co(II)	d^7	amphoteric, mainly basic	transitional, $Co(H_2O)_6^{2+}$, stable, many compounds
(1)	(1)	(1)	Co(I)			⎫ complexes of ligands $Co(RNC)_6^+$
(0)	(0)	(0)	Co(0)			⎬ which stabilize low oxida- $Co_2(CO)_8$, $Co(CN)_4^{4-}$
(−1)			Co(−1)			⎭ tion states (Section A.9) $Co(CO)_4^-$

For meanings of symbols see Tables 20.4 and 20.5.

Cobalt is comparatively rare, and the main sources are residues from arsenic ores of nickel, copper, and lead. Consequently, its metallurgy is complicated, but essentially consists of roasting the arsenic ore in air to obtain the mixed oxide Co_3O_4, and the reduction of this to cobalt by carbon or aluminium. It is an essential constituent of fertile soils, and is present in vitamin B_{12} (see below).

The oxidation state +2 is stable, and a wide variety of simple salts exists. When these are hydrated they usually contain the ion $Co(H_2O)_6^{2+}$, and are pink; when anhydrous they are frequently blue. The olive-green oxide CoO is obtained by heating the metal in air below 400 °C; above this temperature the mixed oxide Co_3O_4 is formed. The true hydroxide $Co(OH)_2$ is obtained in one of two forms, blue or pink (the stable form); it is amphoteric and dissolves in dilute acid to form the ion $Co(H_2O)_6^{2+}$, and in very concentrated alkali to form the deep-blue ion $Co(OH)_4^{2-}$. The reactions of the aquo-ion, $Co(H_2O)_6^{2+}$, are given in Fig. 20.13. Numerous complexes of various stereochemistries are found, mainly octahedral and tetrahedral. The tetrahedral complexes are intensely coloured (Section A9): the intense blue ion $CoCl_4^{2-}$ can be formed by concentrating an aqueous solution of cobalt(II) chloride or by adding an excess of chloride ions. In fact, the pale-pink aqueous solution of cobalt chloride can be used as an invisible ink: on drying it is

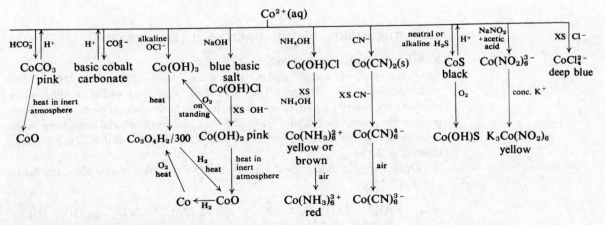

Fig. 20.13 The reactions of the cobalt(II) aqueous ion.

Table 20.13

Ni	Pd	Pt	Ni		Nature of oxide	Nature of compounds
		(6)º				PtF_6, PtO_3
		(5)º				PtF_5, PtF_6^-
(4)º	4	4ª	Ni(IV)	d^6		PtX_4, PtO_2, NiF_6^{2-}, PdX_6^{2-}, PdF_4, $PdO_2 \cdot xH_2O$
(3)º	(3)		Ni(III)	d^7	$Ni_2O_3 \cdot xH_2O$	
2	2	2	Ni(II)	d^8	basic NiO	transitional, $Ni(H_2O)_6^{2+}$, many complexes
	(1)					$Ni_2(CN)_6^{4-}$, $Pd_2(CN)_6^{4-}$ ⎫
0	(0)	(0)	Ni(0)			$Ni(CO)_4$, $Ni(PF_3)_4$, ⎬ complexes of ligands which stabilize low oxidation states
(−1)						$Pt(PPh_3)_4$ ⎪ (Section A.9)
						$Ni_2(CO)_6^{2-}$ ⎭

For meanings of symbols see Tables 20.4 and 20.5.

invisible and, on heating, the water is removed from the compound $CoCl_2 \cdot 6H_2O$ and the intense-blue ion $CoCl_4^{2-}$ is formed. All the tetrahedral complexes are high spin (Section A.9); the octahedral complexes may be either low or high spin, the change-over occurring between the ligands NH_3 and NO_2^- in the spectrochemical series (Section A.9). How many unpaired electrons does the cation $Co(NH_3)_6^{2+}$ have?

Only a few simple salts of the +3 oxidation state are known, and these are strong oxidants, e.g. CoF_3, $Co_2(SO_4)_3$. But a wide variety of stable complexes have been prepared; all are octahedral and low spin (except CoF_6^{3-}). They are the most widely studied of all complexes because they are not only stable (thermodynamically) but are also inert and undergo substitution reactions only slowly, on account of their filled t_{2g} shell of electrons (Section A.9). They are therefore easy to prepare and study. By preparing various isomers, Alfred Werner determined their co-ordination numbers and stereochemistries without the aid of modern techniques, and laid the basis of modern co-ordination chemistry. A biologically important cobalt(III) complex is vitamin B_{12}. This enzyme contains cobalt(III) in a porphyrin-like ring similar to that of haem (Section 20.17). The sixth co-ordination position is again the active site, and is occupied by another ligand.

20.19 THE NICKEL GROUP, GROUP VIII (Table 20.13)

Nickel, palladium, and platinum

The chemistry of nickel differs appreciably from that of palladium and platinum, which are similar. The main differences are: nickel is much more reactive; no binary carbonyls are known for the latter two elements; nickel compounds are much more labile; and compounds of platinum and palladium in the +2 oxidation state are not ionic, e.g. $PtCl_2$ is a polymer, and $PdCl_2$ is an infinite chain (Fig. 11.3).

Draw the oxidation state diagram from the reduction potentials given below (Section 7.11.3):

acid: NiO_4^{2-} NiO_2 Ni^{2+} Ni
          | >1.8 | +1.78 | −0.25 |

alkali: NiO_4^{2-} NiO_2 $Ni(OH)_2$ Ni
          | >0.4 | +0.49 | −0.72 |

From the diagram it is obvious that, especially in acid solution, the higher oxidation states are very unstable relative to the +2 oxidation state, and the chemistry of nickel is largely confined to this state.

Nickel is found combined with arsenic, sulphur, and antimony. One such ore is Millerite, NiS. The method of extraction depends upon the ore, but basically consists of roasting the ore to form nickel oxide, followed by carbon reduction. Nickel of high purity is prepared by the Mond process in which the volatile nickel carbonyl is formed by treating nickel with carbon monoxide at 70 °C, purified, and decomposed to nickel at 200 °C:

$$Ni(s) + 4CO(g) \underset{200°C}{\overset{70°C}{\rightleftharpoons}} Ni(CO)_4(g).$$

Nickel is a ferromagnetic, silvery-white, typical weakly electropositive element. It is less reactive than iron and cobalt are to acids and oxygen, owing to its oxide coating, and is used for electroplating. On heating, it reacts with most non-metals to form: NiO; NiX_2 (X = F, Cl, Br, I); NiS; Ni_3C, etc. It reacts with steam at red heat, it absorbs hydrogen, reacts only slowly with dilute acids (more quickly with nitric), is rendered passive by concentrated nitric acid, and does not react with alkalis even in the presence of oxygen (therefore it is used for crucibles and inert electrodes in alkaline media).

The +2 oxidation state is the only stable state and a great variety of compounds and complexes are formed by nearly all anions (Section 15.1), including highly oxidizing anions (e.g. ClO_2^- but not ClO^-) and easily oxidizable anions (I^-, S^{2-}, etc.). Nickel oxide is a green solid (NaCl structure), water-insoluble, and basic, dissolving in acids to yield solutions of the green aquo-ion, $Ni(H_2O)_6^{2+}$. The addition of hydroxyl ions to these solutions precipitates a green gelatinous precipitate of the basic hydrous nickel hydroxide, $Ni(OH)_2 \cdot xH_2O$. The hydroxide does not dissolve in an excess of alkali, but does in an excess of ammonia, forming the blue ammine, $Ni(NH_3)_6^{2+}$ (note the change of colour—Section A.9). The black nickel sulphide is precipitated in neutral or alkaline solutions by sulphide ions, and is soluble in acid unless it has been left standing when, like cobalt, it oxidizes to the insoluble basic sulphide Ni(OH)S. The halides, oxysalts, and double salts ($NiSO_4 \cdot M_2SO_4 \cdot 7H_2O$) are all hydrated when crystallized from aqueous solution, and usually contain the $Ni(H_2O)_6^{2+}$ ion and are therefore green. In solution these dissolve to give solutions of the aquo-ion, except in the presence of strongly complexing ligands. The reactions of the Ni^{2+}(aq) ion are summarized in Fig. 20.14. Despite the fact that there is only one stable oxidation state, the chemistry of nickel complexes is quite complicated because of the existence of various stereochemistries (and often of equilibria between them in solution) and polymer formation. The octahedral complexes (e.g. $Ni(H_2O)_6^{2+}$ and $Ni(NH_3)_6^{2+}$) all have two unpaired

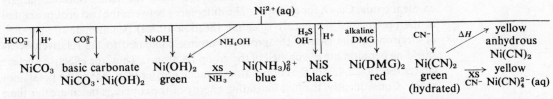

Fig. 20.14 The reactions of the aquo-ion Ni^{2+}(aq).

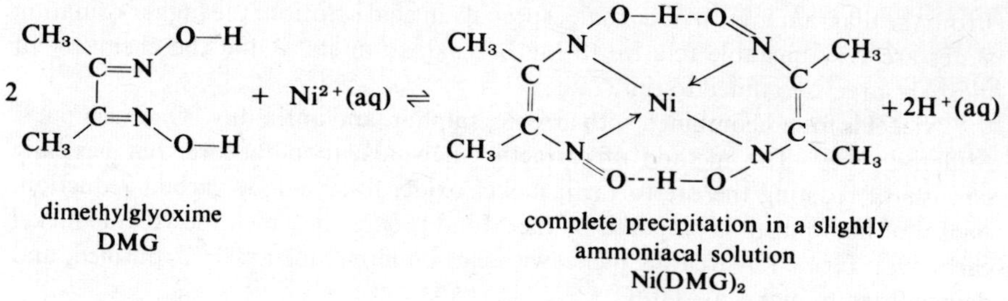

Fig. 20.15 The formation of nickel dimethylglyoxime.

dimethylglyoxime
DMG

complete precipitation in a slightly ammoniacal solution
Ni(DMG)$_2$

electrons (Section A.9). Tetrahedral complexes are formed by the halide ions, e.g. [NiCl$_4$]$^{2-}$, and are often an intense blue colour. Other ligands form square-planar complexes, such as Ni(CN)$_4^{2-}$ and Ni(DMG)$_2$, etc., (Fig. 20.15) which are often red, brown, or yellow. The extreme insolubility of the red compound, nickel dimethylglyoxime, is not only due to the fact that it is a neutral complex, but also because in the solid the molecules are linked in chains to form polymers by nickel–nickel bonding, just like a deck of cards.

20.20 THE COPPER GROUP, GROUP IB (Table 20.14)

Copper, silver, and gold

Table 20. 14

Cu	Ag	Au	Cu		Nature of oxide	Nature of compounds of copper
(3)º	(3)º	3^a	Cu(III)	d^8	acidic	trransitional, oxidizing
2	2º	1	Cu(II)	d^9	basic	tansitional, salts and solution chemistry, Cu^{2+} (aq)
1	1		Cu(I)	d^{10}	basic	non-transitional, simple salts and complexes

For meanings of symbols see Tables 20.4 and 20.5.

20.20.1 Comparison with the Group IA Metals

Draw a table comparing the properties of copper and potassium (Tables 10.10, 14.1, and 20.1). Despite the fact that the elements of groups IA and IB have the same outer electron configuration, s^1, their properties are very different: the alkali metals are reactive, whereas the group IB elements are unreactive (noble). The only similarity between the groups is the existence of non-transitional compounds in which the elements have the oxidation state +1. Even here, though the compounds have necessarily similar stoichiometries, their properties are very different: compare the colourless, soluble, ionic oxide K$_2$O with the red, insoluble, largely covalent oxide, Cu$_2$O, for example. The differences between the two groups are due to the presence of the poorly shielding (Section 10.3.2) penultimate d^{10}-subshell of electrons in the case of the group IB atoms, compared to the relatively well-shielding penultimate p^6-subshell of the group IA atoms. Thus, the effective nuclear charge (Section 2.4.3) of the copper atom is much greater than that of the potassium atom. Consequently, the first ionization energy of the copper is much greater than that of potassium, and therefore copper forms compounds much less readily than potassium, and is said to be noble (unreactive). Also involved in this lack of reactivity are the high cohesive forces in copper (Section 10.3.13, etc.). Moreover, the compounds of copper are more covalent than the corresponding potassium compounds (Section 4.3), and the complexes are more stable (Section 10.5). However, the second ionization energy of copper is much less than that of potassium, and this leads to the formation of stable copper(II) compounds. The differences between the properties of the group IA and IB elements are summarized in Table 10.10, and are those between a strongly and weakly electropositive metal (Section 10.5).

20.20.2 Oxidation State Diagrams

From the data in Table 20.15 draw the oxidation state diagrams (Section 7.12.3) for copper, silver, and gold, each in both acidic and alkaline conditions. As usual, the presence of complexing ligands changes the relative stability of these states (Section A.9) and some relevant data, in acid solution, are:

$$\underset{+1.0}{AuCl_4^- \quad | \quad Au} \qquad \underset{+0.54}{Cu^{2+} + Cl^- \quad | \quad CuCl} \quad \underset{+0.14}{| \quad Cu + Cl^-}$$

The relative stabilities of the various species are obvious from the diagrams. The more obvious points are: the oxidizing nature of all the positive oxidation states in acid solution, and the inertness of the elements; the increased stability of these states in alkaline solution; and the instability of the ions $Cu^+(aq)$ and $Au^+(aq)$ with respect to disproportionation in acid solution (except for copper(I) in the presence of chloride). Thus:

$$2Cu^+(aq) \rightleftharpoons Cu(s) + Cu^{2+}(aq); \quad E^\ominus = +0.18 \text{ V};$$
$$K = [Cu^{2+}]/[Cu^+]^2 = 10^6$$

$$2Ag^+(aq) \rightleftharpoons Ag(s) + Ag^{2+}(aq); \quad E^\ominus = -0.59 \text{ V};$$
$$K = [Ag^{2+}]/[Ag^+]^2 = 10^{-20}$$

$$3Au^+(aq) \rightleftharpoons 2Au(s) + Au^{3+}(aq); \quad E^\ominus = +0.18 \text{ V};$$
$$K = [Au^{3+}]/[Au^+]^3 = 10^9$$

Therefore, in aqueous solution the copper(I) ion, $Cu^+(aq)$, disproportionates unless the concentration of the ion $Cu^+(aq)$ can be reduced relative to that of the ion $Cu^{2+}(aq)$, when the equilibrium is shifted to the left (Le Chatelier) in order to keep

Table 20.15

(a) Acid				(b) Alkali			
+3	+2	+1	0	+3	+2	+1	0
CuO^+	Cu^{2+}	Cu^+	Cu		$Cu(OH)_2$	$Cu(OH)$	Cu
|(+1.8)|	+0.15 |	+0.5 |			|−0.08|	−0.36 |	
AgO^+	Ag^{2+}	Ag^+	Ag	Ag_2O_3	AgO	Ag_2O	Ag
|(+2.1)|	+1.98 |	+0.8 |		|+0.74|	+0.37 |	+0.344 |	
Au^{3+}	Au^{2+}	Au^+	Au	H_2AuO_3			Au
|<1.29|	>1.29 |(+1.68)|				+0.7		
	+1.5						

the equilibrium constant K at the same value. There are two ways to achieve this. First, the presence of ions whose copper(I) salts are much less soluble than their copper(II) salts (for example the ions I^-, Br^-, Cl^-, CN^-, and SCN^-) stabilize the copper(I) state. Indeed, if some of these ions (I^-, CN^-) are added to copper(II) aqueous solutions, the copper(I) salt is precipitated:

$$Cu^{2+}(aq) + 2I^-(aq) \rightarrow CuI(s) + \tfrac{1}{2}I_2.$$

In some cases the stable insoluble copper(I) compound is soluble in an excess of the reagent, forming stable complex anions: $CuCl_4^{3-}$, $Cu(CN)_4^{3-}$, and $Cu(NH_3)_2^+$. These cases are discussed below. Secondly, complexing agents (ligands) which form more stable complexes with copper(I) than with copper(II) also prevent disproportionation, again by preferentially lowering the concentration $[Cu^+(aq)]$. These include nearly all ligands (Cl^-, NH_3, CN^-, $S_2O_3^{2-}$) except for chelating ligands which do not form stable copper(I) complexes, because the most stable copper(I) complexes are linear (e.g. $[Cl—Cu—Cl]^-$) and chelating ligands cannot form linear complexes:

$$2Cu(NH_3)_2^+ \rightleftharpoons Cu(NH_3)_4^{2+} + Cu; \qquad K = 2 \times 10^{-2}$$
$$2CuCl + 2\,en \rightleftharpoons Cu(en)_2^{2+} + 2Cl^- + Cu; \qquad K = 10^5.$$

Ethylenediamine (Section 20.5) is abbreviated as en.

In the absence of anions which form insoluble copper(I) compounds, or ligands which form stable copper(I) complexes (Cl^-, Br^-, I^-, NH_3, etc.), copper(I) compounds disproportionate immediately they are placed in aqueous solution:

$$Cu_2SO_4(s) + H_2O(l) \rightarrow SO_4^{2-}(aq) + Cu(s) + Cu^{2+}(aq).$$

The relative stabilities of the possible oxidation states of each metal are surprisingly different from one another. The $+1$ oxidation state is a non-transitional state (d^{10}), and in it the compounds are generally colourless (except when the anion is coloured, or charge-transfer transitions occur, Section A.9), and are largely covalent and insoluble (except for complexes). Their stereochemistry is usually either linear or trigonal (Fig. 20.16a). The $+1$ state is the most stable for silver compounds, both in solution and in the solid state, and is generally the most stable state for solid compounds (especially at high temperatures) and complexes of gold and copper. The $+2$ oxidation state is the stable state in aqueous solution for copper, but is much less stable for silver (which Ag(II) compounds do you expect to be the most stable?—Problem 37), and very unstable for gold. The $+3$ oxidation

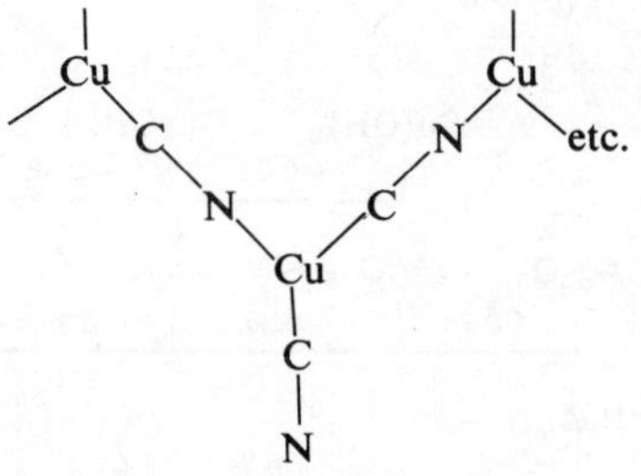

$[Cu(CN)_2^-]_x$ in solid $[KCu(CN)_2]$

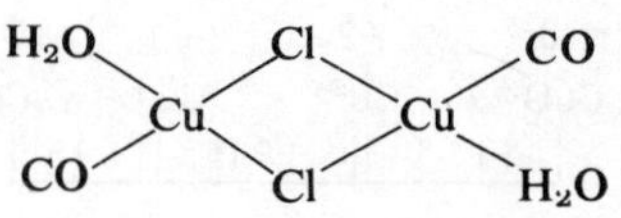

$[Cu(CO)Cl \cdot H_2O]_2$

(a)   (b)

Fig. 20.16 Structures of two copper(I) compounds.

state increases in stability as the group is descended; copper(III) compounds are rare, but gold(III) is the stable oxidation state for this element.

20.20.3 The Occurrence, Extraction, and Properties of Copper

Copper is widely distributed in nature. It is found as the element, the sulphide, copper pyrites $CuFeS_2$, the carbonate, the arsenide, and the chloride. The extraction of the metal (Section 11.5.1) from two sources, is summarized below:

1. $CuCO_3 \cdot 3Cu(OH)_2 \xrightarrow{heat} CuO \xrightarrow{C} Cu$
2. $CuFeS_2 \xrightarrow{roast} Cu_2O + Cu_2S \longrightarrow SO_2 + Cu$.

The refining (Section 11.5.2) which is the final stage of purification, is interesting. Three relevant couples are: Ag^+/Ag, $E^\ominus = 0.8$ V; Cu^{2+}/Cu, $E^\ominus = 0.34$ V; and and Fe^{2+}/Fe, $E^\ominus = -0.44$ V. The potential applied to the cell is such that: copper and iron dissolve as the ions $Cu^{2+}(aq)$ and $Fe^{2+}(aq)$, but the noble metals (Ag, Au, Pt) drop to the bottom as anode sludge, and are recovered; and copper(II) ions are deposited on the cathode (but not iron(II) ions, which stay in solution).

20.20.4 Properties of the Elements

Many of the properties of copper, silver, and gold differ. There is no simple explanation for this. They are all moderately soft metals which are very ductile and malleable (particularly gold), are excellent conductors, have high densities, and fairly high melting and boiling points. They are all fairly inert, and their compounds are easily reduced to the metal. This inertness increases as the group is descended (Table 20.16) and, correspondingly, the thermal stability of the oxides and salts decreases from copper to gold. Thus, copper (on roofs for example) may in time produce the green basic salts: $CuSO_4 \cdot 3Cu(OH)_2$; $CuCO_3 \cdot Cu(OH)_2$; and even $CuCl_2 \cdot Cu(OH)_2$ at the sea-side. Silver may form the sulphide, but gold does not react with the atmosphere. Copper also reacts with most non-metals on heating (Table 20.16).

20.20.5 Copper(I) Compounds

Copper(I) oxide is a red or yellow compound formed: by heating copper(II) oxide above 900 °C; by boiling a copper(I) halide with sodium hydroxide (no copper(I) hydroxide is known); or by the action of glucose (a reductant) on an aqueous alkaline solution of copper(II) ions in the presence of the tartrate ion (which prevents the precipitation of copper(II) hydroxide by complexing of the ion Cu^{2+}). The copper(I) halides, CuX ($X = Cl^-$, Br^-, I^-), are insoluble compounds with the zinc blende structure. The chloride and bromide can be prepared by the sulphur dioxide reduction of copper(II) solutions in the presence of the corresponding hydrogen halide.

20.20.6 Copper(II) Compounds

A large number of salts and many complexes are known for the stable oxidation state +2 of copper. They are all paramagnetic and coloured, and many different stereochemistries are found, particularly the distorted octahedral and square

Table 20.16. The reactions of copper, silver, and gold.

Reactant	Copper	Silver	Gold
Oxygen	$\rightarrow$ CuO $\longrightarrow$ Cu_2O	high pressure, Ag_2O	X
Water	X	X	X
Halogens	$\rightarrow$ CuCl $\rightarrow$ $CuCl_2$ (also Br); CuF_2 CuI	AgF_2, AgCl, AgBr, AgI	Au_2Cl_6 Au_2Br_6
Air	green $CuCO_3 \cdot Cu(OH)_2$ (sea-side) $CuCl_2 \cdot 3Cu(OH)_2$	$\rightarrow Ag_2S$	X
Sulphur	$\rightarrow Cu_2S \rightarrow CuS$	$\rightarrow Ag_2S$	X
Dilute acids in absence of air	X	X	X
Dilute acids in presence of air	Cu^{2+} slowly	X	X
Concentrated HCl, boil	$CuCl_2^- + H_2$	X	X
Concentrated H_2SO_4	$Cu^{2+} + SO_2$ (also CuS etc.)	$Ag^+ + SO_2$	X
HNO_3, dilute	$Cu^{2+} + NO$	$Ag^+ + NO$	X
concentrated	$Cu^{2+} + NO_2$	$Ag^+ + NO_2$	X
Aqua regia	$Cu^{2+} + NO_2$	X	$AuCl_4^- + NO_2$
Alkalis	X	X	X

X = no reaction.

planar structures. Aqueous solutions of copper(II) compounds contain the light-blue ion, $Cu(H_2O)_6^{2+}$. The reactions of this ion are summarized in Fig. 20.17.

20.20.7 Silver Compounds

All silver(I) salts are diamagnetic, and nearly all are insoluble and colourless. The only common ions which do not precipitate silver ions from aqueous solution (Section 10.3.17) are the fluoride ion and the large anions: NO_3^-, ClO_4^-, ClO_3^-, MnO_4^-, SO_4^{2-}. The reactions of the silver ion are summarized in Fig. 20.18. The insoluble silver halides, AgX (X = I^-, Br^-, Cl^-), are insoluble in dilute nitric acid, but are soluble in concentrated ammonia (silver chloride even in dilute ammonia) except for the insoluble iodide. Coloured silver(I) compounds include: the sulphide and oxide (brown-black); the iodide and phosphate (yellow); the bromide (pale-yellow); the chromate (dark-red); and the hexacyanoferrate (orange). The silver halides have the sodium chloride structure except for the iodide (zinc blende structure).

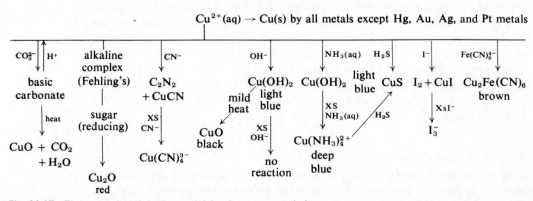

Fig. 20.17 The properties of the copper(II) ion in aqueous solution.

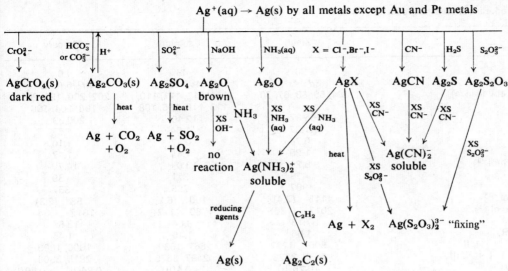

Fig. 20.18 The reactions of the aqueous silver ion, Ag^+(aq).

20.21 ZINC, CADMIUM, AND MERCURY, GROUP IIB

20.21.1 General Chemistry: Trends in the Group

This group does not fit well into the long form of the Periodic Table (Section 10.1). Although it is in the d-block group of elements, the group shows none of the characteristic properties of transition metals (complex formation is best regarded as a characteristic of weakly electropositive metals). Also, there is a wide variation of properties within the group: zinc and cadmium are moderately electropositive elements (see electrode potentials, Table 20.17) and show similar chemistries, but mercury is a noble metal similar to silver. This is an extreme case of the trend found in the transition series, where the third row element is less electropositive than expected (Section 20.2). For example: the hydroxide and pure carbonate of mercury(II) do not exist; mercury is noble; and mercury(II) salts are generally less soluble than those of zinc and cadmium (mercury(I) salts are all insoluble except the nitrate and chloride). As usual, one similarity between all members of the group, owing to the similar electron configuration, is the existence of the common oxidation state $+2$. There is no evidence for oxidation states greater than two in this group, since, as expected, the third ionization energy is very high (Section 10.3.3). The formal $+1$ oxidation state exists for mercury, owing to the existence of metal–metal bonding in the mercury(I) ion Hg_2^{2+}, but not for cadmium and zinc under normal conditions. Thus, like the main p-block group, the lower oxidation states increase in stability (relative to the higher) as the group is descended (cf. transition series, Section 20.2), but unlike the p-block group, the stable states do not differ by units of two (cf. transition series). This is typical intermediate behaviour. The $+1$ oxidation state is not paramagnetic, however, since it is due to metal–metal bonding and it is not the ion Hg^+, but the ion Hg_2^{2+} that is found. In fact, no s^1 (paramagnetic) compounds are known in nature: they either disproportionate or form metal–metal bonds.

Table 20.17. The properties of the zinc group.

	Zinc	Cadmium	Mercury
Atomic number	30	48	80
Electron configuration (outer)	$3d^{10}4s^2$	$4d^{10}5s^2$	$4f^{14}5d^{10}6s^2$
Isotopes (in order of abundance)	64, 66, 68, 67, 70	114, 112, 111, 110, 113, 116, 106, 108	202, 200, 199, 201, 198, 204, 196
Atomic weight	65.37	112.40	200.59
Metallic radius (Å)	1.37	1.54	1.57
Ionic radius (Å)	0.74	0.97	1.10
Covalent radius (Å)	1.25	1.44	1.47
Atomic volume ($cm^3\ mol^{-1}$)	9.2	13.0	14.7
Melting point (°C)	419	321	−39
Boiling point (°C)	907	765	357
$\Delta H_{vap}(\Delta H_{fus})$ (kJ mol^{-1})	115 (7.4)	100 (6.1)	58 (2.3)
$\Delta H_{hyd}(\Delta H_{atom})$ (kJ mol^{-1})	2017 (130)	1780 (112)	1812 (61)
Density (g cm^{-3})	7.13	8.64	13.6
Electronegativity (A/R)	1.65	1.46	1.45
Ionization energies (kJ mol^{-1}) I, II	906.2, 1732	867, 1630	1006, 1809
I + II, III	2638, 3828	2497, 3625	2815, 3309
Electrode potential (V)	−0.76(II)	−0.4(II)	+0.84(II); +0.79(I)
Common co-ordination numbers	(2), 4, 6	4, 6	2, 4, (6)
Common oxidation numbers	+2	+2	+2, +1
Abundance, p.p.m.	80	0.2	0.5

The metallic character decreases down the group. If the cation-forming ability is taken as a criterion of metallicity, then this is illustrated by the electrode potential data (Table 20.17), for example. This trend is reflected in the whole chemistry of the elements. The danger of using just one criterion for this purpose (Section 10.5) is shown by considering the character of the oxides: zinc oxide is amphoteric; cadmium oxide is mainly basic; and mercury(II) oxide is basic—giving the opposite trend! The general reactivity of the metals decreases down the group. The low hydration enthalpy and high ionization energies "cause" the nobility of mercury (the first indication of the inert pair effect in the table!). The ionic character of corresponding compounds decreases as the group is descended: $Zn \geqslant Cd \gg Hg$. Thus mercury compounds tend to be volatile, can be sublimed, and are more soluble in organic solvents than the corresponding zinc and cadmium compounds. Zinc is, nevertheless, only moderately electropositive, and its compounds have appreciable covalent character: for example, zinc and cadmium oxides can be sublimed. The covalent organometallic compounds of mercury, HgR_2, HgRX, etc. (Chapter 27) are more stable than those of cadmium and zinc ($Hg \gg Cd > Zn$), which are unstable in air and water. All three form stable complex compounds, the stability increasing in the order $Zn \leqslant Cd < Hg$, correlating with the tendency to covalent bond formation. The tendency for complexes with lower co-ordination numbers to become more common increases down the group (Table 20.17).

Zinc and cadmium resemble magnesium in many ways. There is nothing fundamental in this. Since they show the oxidation state +2 like the group IIA elements, it is not surprising that there is a weakly electropositive element in the group they resemble.

20.21.2 Occurrence, Extraction, and Uses

Zinc is moderately abundant, cadmium and mercury much less so. But all three elements are well known because of the existence of concentrated deposits, their simple extractive metallurgy (Section 11.5.2), and their widespread use.

Zinc finds use in galvanizing (Section 20.17), for the cases of dry batteries, and as a constituent of alloys such as brass. Zinc oxide is a white pigment which does not blacken in air, and is also used in ointments and cosmetics. A mixture of zinc oxide and chromic oxide catalyses the hydrogenation of carbon monoxide to form methanol. Cadmium alloys are useful, and the metal is also used for electroplating. Mercury is the only liquid metal at normal temperatures, and is useful as a liquid conductor (thermometers, electrical apparatus, etc.). Its alloys with other metals are called amalgams, and many are useful: as tooth fillings (Sn, Hg, Ag); as reducing agents; and as mirror coatings (Sn, Hg). Zinc sulphide and cadmium sulphide are fluorescent and are used on television screens.

20.21.3 Properties

All three metals are silvery-white. Zinc and cadmium tarnish in air to form a protective oxide layer. Pure mercury does not react at 25 °C in dry air, but impurities catalyse the reaction and an oxide layer is formed which "tails" on glass. They all have low tensile strengths. All three metals are fairly volatile, especially mercury; this, together with its surprising solubility in water and toxicity, makes mercury *very* dangerous. All form monatomic vapours. They all react with non-metals, at 25 °C in a few cases, at moderate temperatures in most. Zinc and cadmium burn in air at moderate temperatures to form the oxide MO, zinc with a blue-green flame, cadmium with a red flame. Mercury forms the red oxide, HgO, at 300 °C, which decomposes to the elements at 400 °C. The halogens form the compounds MX_2 (M = Zn, Cd; X = halogen), and Hg_2X_2 and HgX_2. The iodides of mercury can be obtained by rubbing mercury and iodine together with alcohol at 25 °C, when the olive green mercury(I) iodide or the scarlet mercury(II) iodide are formed, depending upon the proportions of reactants. All three metals form the sulphide, MS. The precipitation of the insoluble sulphides of zinc (white), cadmium (yellow), and mercury(II) is discussed in Section 18.10. Only zinc reacts with hot alkali, yielding hydrogen and the zincate ion, $Zn(OH)_4^{2-}$ (Fig. 20.19). None of these

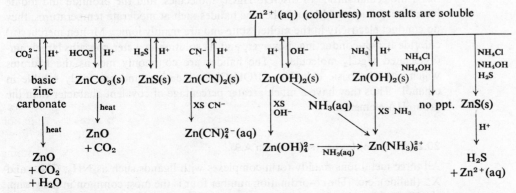

Fig. 20.19 The reactions of the zinc ion.

elements reacts with non-oxidizing acids when pure (kinetic control, Section 8.9), but zinc and cadmium react in the presence of impurities such as copper. All three react with oxidizing acids (Section 10.4.2). They all form alloys with metals, for example, brass (Zn, Cu) and low-melting alloys like Wood's metal (Bi, Pb, Sn, Cd).

20.21.4 Oxides and Hydroxides

These are predominantly covalent in character: even those of zinc are poor conductors in the fused state. All the group VIB binary compounds of these metals have the zinc blende or wurtzite structures except for cadmium oxide (sodium chloride structure). This illustrates the preference for tetrahedral co-ordination in this group. Zinc oxide and cadmium oxide have defect lattices, and are coloured (Section A.9). The oxide colour deepens from zinc to cadmium, and also as the temperature rises and the defects increase. Thus zinc is white at 25 °C, and becomes yellow on heating; cadmium oxide is white at 0 °C, yellow at 25 °C, and black on heating. Both can be sublimed, showing the covalent character.

Aqueous solutions of the ions M^{2+}(aq) react with hydroxyl ions to form white precipitates of zinc and cadmium hydroxide, and a yellow precipitate of mercury(II) oxide. This has a structure consisting of zig-zag chains of Hg—O—Hg—O, etc., whereas the red form has the zinc blende structure. Mercury(II) hydroxide does not exist (Section 11.2). The yellow form of the oxide is somewhat soluble in water, and changes to the red form on heating. Zinc hydroxide is amphoteric and dissolves in an excess of aqueous alkali to form the zincate ion $Zn(OH)_4^{2-}$(aq); cadmium hydroxide does not. Both, however, dissolve in an excess of ammonia to yield ammines, $M(NH_3)_4^{2+}$.

20.21.5 Halides

These illustrate Fajans's rules (Section 4.3) and the concepts of polarization and polarizability nicely. The polarizing power of the cations increases in the order $Zn^{2+} < Cd^{2+} < Hg^{2+}$; the polarizability of the anions increases in the order $F^- < Cl^- < Br^- < I^-$. Therefore, zinc fluoride is the most ionic of the halides, and mercury(II) iodide the most covalent. This is largely borne out by their properties: structures; trends in solubilities; and melting and boiling points. For example, mercury(II) fluoride has the largely ionic fluorite structure, but mercury(II) chloride has a molecular lattice of discrete $HgCl_2$ molecules, and the bromide and iodide have layer lattices. Although the zinc halides melt at moderate temperatures, they do conduct electricity in the molten state and are mainly ionic. Molten mercury(II) chloride is non-conducting, however, and its solution in water contains largely undissociated $HgCl_2$ molecules. The halides are commonly met as the hydrates which give the basic chloride, M(OH)Cl, on heating, and are fairly soluble in ethanol. Thus they have a much greater percentage of covalent character than the group IIA elements.

20.21.6 Complexes (*see* Section A.9)

All three metal ions readily form complexes with ligands such as NH_3, CN^-, and X^- (halide), etc. The co-ordination number four is the most common in the group, e.g. $Zn(CN)_4^{2-}$, $CdBr_4^{2-}$, and HgI_4^{2-}. What shape are these ions (Problem 7, Chapter

12)? Mercury(II) ions, when treated with iodide ions, yield a scarlet precipitate of the dimorphic mercury(II) iodide (via the yellow form) which dissolves in an excess of iodide ions to form the stable complex anion, HgI_4^{2-}. The copper salt, $CuHgI_4$, is dimorphic, the red form existing below 70 °C, a black form above. This is used for heat-sensitive paints.

20.21.7 Mercury(I) Compounds

The conclusive evidence of X-ray analysis exists to show that mercury(I) compounds in the solid state exist only in the form containing the metal–metal bond, Hg—Hg. Of course, independent evidence is needed for solutions since, despite the nature of the species in the solid state, molecules such as Hg—Cl, and ions such as Hg^+(aq) could, but do not, exist in solution. In fact it is found that the equilibrium constant K_1 describing the equilibrium represented by Eq. (20.1), is constant, whereas K_2 is not.

$$Hg(s) + Hg^{2+}(aq) \rightleftharpoons Hg_2^{2+}(aq); \quad K_1 = \frac{[Hg_2^{2+}]}{[Hg^{2+}]} \quad (20.1)$$

$$Hg(s) + Hg^{2+}(aq) \rightleftharpoons 2Hg^+(aq); \quad K_2 = \frac{[Hg^+]^2}{[Hg^{2+}]} \quad (20.2)$$

This verification of the law of chemical equilibrium is powerful evidence. Other evidence includes the diamagnetism of such compounds, and the fact that the electrical conductivities of ionic mercury(I) compounds are typical of electrolytes of the type $M^{2+}2X^-$ (2:1 electrolytes) rather than those of the type M^+X^- (1:1 electrolytes).

The relative stability of mercury and the aqueous mercury(I) and mercury(II) ions are described by the reduction potentials:

$$Hg_2^{2+}(aq) + 2e^- \rightleftharpoons 2Hg(l); \quad E^\ominus = +0.789 \text{ V} \quad (20.3)$$

$$2Hg^{2+}(aq) + 2e^- \rightleftharpoons Hg_2^{2+}(aq); \quad E^\ominus = +0.920 \text{ V} \quad (20.4)$$

$$Hg^{2+}(aq) + 2e^- \rightleftharpoons Hg(l); \quad E^\ominus = +0.854 \text{ V}. \quad (20.5)$$

The electrode potential of Eq. (20.4) is obtained from Eqs. (20.3) and (20.5) (Section 7.12.2). The free energy change and equilibrium constant K for the disproportionation of the ion Hg_2^{2+}(aq) can be obtained similarly:

$$Hg_2^{2+} \rightleftharpoons Hg(l) + Hg^{2+}(aq); \quad E^\ominus = -0.13 \text{ V}; \quad \Delta G^\ominus = 12.5 \text{ kJ}; \quad K = 6 \times 10^{-3}.$$

Therefore, in aqueous solution the disproportionation of the ion Hg_2^{2+}(aq) scarcely occurs. But it "only just doesn't occur" and any reagents which reduce the mercury(II) ion concentration $[Hg^{2+}]$ relative to that of mercury(I), $[Hg_2^{2+}]$ will cause disproportionation to occur (Le Chatelier). Three classes of reagent effect this (and this includes most reagents):

1. Those whose mercury(II) salt is less soluble or more stable than their mercury(I) salt (OH^-, S^{2-}, etc.), for example:

$$Hg_2^{2+}(aq) + 2OH^-(aq) \rightarrow Hg(l) + HgO(s) + H_2O.$$

2. Those forming covalent mercury(II) salts (CN^-):

$$Hg_2^{2+}(aq) + 2CN^-(aq) \rightarrow Hg(l) + Hg(CN)_2(aq).$$

3. Those forming more stable mercury(II) complexes than mercury(I) complexes—all ligands except the weakly co-ordinating ligands perchlorate and nitrate.

Effectively, this means that apart from the soluble nitrate and perchlorate, the only mercury(I) compounds that are stable in the presence of water are: the insoluble compounds Hg_2X_2 (X = Cl, Br, I) and Hg_2SO_4. The sulphide, hydroxide, and oxide of mercury(I) are unknown.

PROBLEMS

1. Plot against atomic number, for the elements from calcium to zinc (inclusive):
   a) the first ionization energy;
   b) the sum of the first two ionization energies.
   c) the sum of the first three ionization energies;
   d) the sum of the first four ionization energies.
   Comment on any trends, maxima, or minima, etc.

2. What is the general trend in the ionic radius of the transition ions in the +2 oxidation state along the series? Discuss.

3. Comment on the large sixth ionization energy for vanadium.

4. Why are the atomic radii of the second and third transition series so similar?

5. Will an acid solution of dichromate oxidize any of the following: F^-, Cl^-, Br^-, I^-, H_2O, H_2S (to S), Hg_2^{2+}, Cu (to Cu^{2+}), Mn^{2+}, HNO_2?

6. Look up the reduction potential data for iron in Section 20.17.
   a) Is the oxidation of iron(II) to iron(III) (thermodynamically) easier in acid or alkaline solution?
   b) Calculate the equilibrium constant for the reaction:
   $$Fe(s) + 2Fe^{3+}(aq) \rightarrow 3Fe^{2+}(aq)$$
   in acid solution at 25 °C.
   c) Is iron(VI) more stable in acid or alkaline solution?
   d) Will acidic Fe^{2+}(aq) oxidize hydrogen peroxide to oxygen in acid solution?
   $$H_2O_2(aq) \rightarrow O_2(g) + 2H^+(aq) + 2e^-; \qquad E^\ominus = -0.7 \text{ V}.$$

7. Calculate the enthalpy of reaction for the reaction:
   $$3Fe(s) + 4H_2O(l) \rightarrow Fe_3O_4(s) + 4H_2(g).$$

8. Calculate the lattice energies of the silver halides, AgX. (What information do you need, and where can you find it?)

9. Name the following complexes: [Pt(ethylenediamine)$_3$]Cl$_4$, [Pt(NH$_3$)$_5$Cl]Br$_3$, K$_4$[Fe(CN)$_6$], cis[Co(NH$_3$)$_4$Cl$_2$]Cl, [Cr(H$_2$O)$_6$]Cl$_3$, and trans[PtCl$_2$(NH$_3$)$_2$].

10. Why are oxyanions such as the permanganate ion usually prepared in alkaline conditions?

11. Why does titanium find relatively little structural use as a metal despite its abundance?

12. Explain why the anion CoF_6^{3-} is paramagnetic, while the similar anion $Co(CN)_6^{3-}$ is diamagnetic.

13. Explain the following changes: when a pink aqueous solution of a cobalt(II) salt is treated with aqueous ammonia a blue precipitate is formed which on the

addition of an excess of ammonia solution forms a yellow-brown solution; on standing in air this solution turns red.

14. Comment on the stability to reduction of the ions: $Co(H_2O)_6^{3+}$, $Co(NH_3)_6^{3+}$, and $Co(CN)_6^{3-}$.

15. Describe and explain the changes that occur when a solution of vanadium(v) (ammonium vanadate for example) is treated with zinc and hydrochloric acid.

16. Explain the following changes:
    a) When a solution of vanadium(v) (ammonium vanadate for example) is treated with sulphur dioxide a deep blue solution is obtained (A).
    b) When a solution of vanadium(v) is treated with acidified zinc amalgam a violet solution is obtained (B).
    c) When solutions (A) and (B) containing equal amounts of vanadium are mixed, a green solution is obtained (C).

17. The ionic radius of the ion V^{2+} is similar to those of the ions Ca^{2+} and Cd^{2+}, yet its properties are very different (it is powerfully reducing). Comment.

18. To which particular properties of the following substances do you attribute their various uses:
    a) tungsten as electric wire (filaments etc.);
    b) tungsten carbide as drill tips;
    c) chromium for coating iron?

19. Consider the reduction potential data for the manganese group (manganese technetium, rhenium). Which oxidation state(s) of these three elements:
    a) disproportionate(s);
    b) is the most powerful oxidant?

20. Does zinc react with an aqueous solution of manganese(II)?

21. Describe and explain the reaction between the green manganate ion and carbon dioxide.

22. Explain the following:
    a) VF_5 is known, but VCl_5 is not.
    b) $FeCl_3$ is known, but FeI_3 is not.
    c) Tungsten carbide is hard and brittle.
    d) Copper(II) compounds are common, but gold(II) compounds are very rare.
    e) No scandium(IV) state exists.
    f) Some vanadium(v) species are coloured (d^0 configuration).
    g) Manganese(VII) compounds are deeply coloured.

23. Explain the general trend in the relative stability of the $+2$ and $+3$ oxidation states of the first row transition elements. Which elements provide exceptions? Why?

24. a) How many isomers are found of the following (octahedral) compounds and ions: $Co(NH_3)_6^{3+}$, $Co(NH_3)_5Cl$, $Co(NH_3)_4Cl_2$, and $Co(NH_3)_3Cl_3$?
    b) The compound $Pt(NH_3)_2Cl_2$ forms two isomers. What can you deduce from this?

25. Which oxide of manganese is most likely to be:
    a) basic and ionic;
    b) acidic and covalent?

26. Which of the first row transition elements has:
    a) the highest second ionization energy;
    b) the highest third ionization energy;
    c) the highest enthalpy of atomization;
    d) the highest boiling point and melting point;
    e) the highest atomic conductance;
    f) the highest abundance;
    g) a carbonyl, $M(CO)_5$;
    h) a colourless ion, M^+;
    i) a compound, MO_3F;
    j) the lowest density?

27. Do you expect manganese(VIII) compounds to be stable?

28. Why do only the early members of the first transition series form interstitial carbides, nitrides, and so on?

29. Why is it rarely possible to extract transition metals from their ores by conventional carbon-reduction methods?

30. Why are no aquated ions found in oxidation states of metals higher than three, for example $Ti(H_2O)_4^{4+}$?

31. Why does calcium oxalate dissolve in EDTA?

32. Describe and explain what happens when an aqueous solution of the dark green isomer, $CrCl_3 \cdot 6H_2O$, is treated with an excess of an aqueous solution of silver nitrate, filtered, and the filtrate warmed.

33. Describe and explain what happens when an aqueous solution of ammonia is added to:
    a) $CuSO_4(aq)$;
    b) $NiSO_4(aq)$.

34. Describe the types of isomerism found in:
    a) $PtCl_2 \cdot 2NH_3$;
    b) $[Co\ en_2Cl_2]^+$;
    c) $CrCl_3 \cdot 6H_2O$.

35. Summarize the main differences between the group IA and group IB elements. Comment on what appears to be the main reason(s) for these differences.

36. Why is copper a much poorer reductant than its neighbours in the Periodic Table:
    a) zinc;
    b) nickel?

37. Which simple silver(II) compounds do you expect to be the most stable?

38. What is the reaction of copper(I) oxide with:
    a) dilute nitric or sulphuric acid;
    b) concentrated hydrochloric acid;
    c) ammonia solution?

39. Which type of metal does zinc resemble most closely: a group IIA metal, a transition metal, or a B-group metal? Give examples.

40. Calculate the equilibrium constant for the disproportionation of the aqueous mercury(I) ion, Hg_2^{2+}, from the following data:

$$Hg^{2+}(aq) + e^- \rightleftharpoons Hg(l); \qquad E^\ominus = +0.854$$
$$Hg_2^{2+}(aq) + 2e^- \rightleftharpoons 2Hg(l); \qquad E^\ominus = +0.789$$
$$Hg_2^{2+}(aq) \rightleftharpoons Hg(l) + Hg^{2+}(aq); \qquad E^\ominus = ?$$

Despite this result the disproportionation of $Hg_2^{2+}(aq)$ often occurs. Under what conditions?

41. Account for the following differences:

	Boiling point	Solubility
HgF_2	650	insoluble in organic solvents
$HgCl_2$	300	soluble in organic solvents

BIBLIOGRAPHY

Basolo, F., and R. Johnson. *Coordination Chemistry*, Benjamin (1964).

    A good introduction to the subject.

Larsen, E. M. *Transitional Elements*, Benjamin (1965).

    Much descriptive chemistry.

Harvey, K. B., and G. B. Porter. *Introduction to Physical Inorganic Chemistry*, Addison-Wesley (1963).

    An undergraduate-level text.

Stark, J. G. (Editor). *Modern Chemistry*, Part 5: "Coordination Chemistry," article by G. G. Schlessinger, Penguin (1970).

Sanderson, R. T. *Inorganic Chemistry*, Chapter 7: "Principles of Coordination Chemistry," Reinhold (1967).

    A qualitative description of the general principles.

Orgel, L. P. *An Introduction to Transition Metal Chemistry–Ligand Field Theory*, Methuen (1966).

    This will take the student beyond the level of treatment in this book up to more sophisticated approaches.

Chilton, J. P. *Principles of Metallic Corrosion* (second edition), Royal Institute of Chemistry (1968)

Bond, G. C. *Principles of Catalysis* (second edition), Royal Institute of Chemistry (1968).

Kauffman, G. B. "A. Werner's Coordination Theory," *Education in Chemistry*, 4, No. 1, 11–18 (1967).

    A brief, easily understandable historical introduction to the work of Alfred Werner.

Kauffman, G. B. *Classics in Coordination Chemistry*, Part I: "The Selected Papers of Alfred Werner," Dover (1968).

In addition most of the books on inorganic chemistry in the general list contain material on the chemistry of the transition metals.

CHAPTER 21

Hydrocarbons: Structure and Geometry

21.1 Chain structures
21.2 Nomenclature
21.3 The geometry of four-co-ordinate carbon: optical isomerism
21.4 Unsaturation and geometrical isomerism
21.5 Ring structures
21.6 Summary

The unique features of carbon have been discussed at some length (Section 16.4). The ability to catenate and to form multiple bonds, and a covalency of four, are the principal features which enable carbon to form such an enormous range of compounds. The formation of so many varied ring and chain structures is not equalled by any other element, and carbon compounds therefore have a special place in chemistry. A further unique feature of carbon, a consequence of its other unique features, is its place in biological systems. Nearly all important biological compounds (and there are very many of these) are compounds of carbon, and this has given rise to the name *organic chemistry* to encompass the chemistry of carbon compounds. Indeed, at one time it was thought that some special, mystical property of living things—"vital force"—was necessary to synthesize organic compounds although many have now been synthesized in the laboratory, including a very large number which have no (known) place in biological systems.

The kinetic stability of organic compounds has also been referred to (Section 16.4) and organic chemists have spent much time and energy in studying the kinetics and other features (Section 8.10) of organic reactions in order to work out feasible reaction pathways and mechanisms. This theme is taken up in Chapter 26, where the mechanisms of some of the more important types of reaction are discussed. The chapters which precede Chapter 26 lay emphasis upon the patterns which emerge from a wide study of organic reactions. These patterns, or relationships, are ultimately a function of the underlying principles—mechanism, structure, thermochemistry, etc.—but the principles are perhaps best appreciated against the background of these patterns. The enquiring reader will wish to relate the mechanistic ideas to the patterns of earlier chapters. He is encouraged to do this and help is provided in the form of certain types of problem. Chapters beyond Chapter 26 make use of these principles, so that there is a gradual increase in sophistication as progress is made through Chapters 21 to 28, although it is not essential to take these chapters in numerical sequence.

The great majority of organic compounds can be considered to be composed of molecules which are built up, basically, of a skeleton formed by a number of carbon atoms bonded together, and to which may be attached various "functional groups". Most of the four valencies of carbon which are not involved in bonding to other carbon atoms are generally satisfied by bonds to hydrogen. If all such bonds are satisfied in this way we have what is known as a hydrocarbon. Propane is an example of this class of compound and is diagrammatically represented as

$$\begin{array}{c} \text{H} \ \text{H} \ \text{H} \\ | \ \ | \ \ | \\ \text{H}-\text{C}-\text{C}-\text{C}-\text{H} \\ | \ \ | \ \ | \\ \text{H} \ \text{H} \ \text{H} \end{array}$$

or, more conveniently, as $CH_3CH_2CH_3$. There is a basic carbon chain of three atoms and there are eight hydrogen atoms satisfying the carbon atom valencies which are not involved in holding the chain together.

Sometimes one or more hydrogen atoms on the carbon skeleton are replaced by other atoms or groups of atoms. Ethanol (ethyl alcohol) is a compound which illustrates this. One of the hydrogen atoms of ethane, CH_3CH_3, has been replaced by an —OH group to give CH_3CH_2OH. We sometimes say that ethanol is a substituted ethane and the —OH group is the substituent, having been substituted for a hydrogen atom. Such replacing groups are known as **functional groups**. Because the hydrocarbon skeleton is chemically rather unreactive, it is the functional group (or groups) in a molecule which give it its chemical characteristics. Thus organic compounds are classified chemically according to the functional groups which are attached to their basic skeletons, rather than according to the skeletons themselves. The class of any organic compound is normally indicated by the word-ending of its name. The suffix, *-ol*, for example, as in ethanol, characterizes the general family of the hydroxyl group containing compounds known as alcohols. We shall be concerning ourselves with the more common functional groups which are to be found in organic molecules, and the chemistry of the compounds which contain them. We shall also be concerned to some extent with the effect one functional group may have on the chemistry of other functional groups in the same molecule. Accordingly, the discussion of organic chemistry is arranged in eight chapters, four of which are devoted to relatively simple functional groups, classified by the most characteristic atom in the group, two to molecules containing no functional groups (hydrocarbons), one to molecules containing several functional groups, and one to the nature of organic reactions.

Before launching into this first chapter on the hydrocarbons, it would perhaps be appropriate to indicate that the skeletons of some organic molecules do not consist of uninterrupted chains of carbon atoms. Sometimes atoms other than carbon are involved in the chain. For example, in protein molecules the basic chain also contains nitrogen atoms arranged with carbon atoms as follows: —C—C—N—C—C—N—C—C—N— etc. (the skeletons of these molecules are very long). Our first generalization, however, that the majority of organic compounds have a carbon skeleton, is still valid and in the chapters which follow we shall confine ourselves to molecules with small carbon skeletons. In this chapter we consider the basic types of skeleton carbon can form; some of the properties of these molecules without functional groups are discussed in Chapter 22.

Note. In writing equations throughout the discussions of physical and inorganic chemistry the use of state symbols has been common practice. For quantitative physico-chemical work the precise definition of state is often necessary and, as far as inorganic reactions are concerned, most of these take place in relatively dilute aqueous solution. It is therefore easy—and accurate—to describe the various ions or molecules as "aqueous". Many organic reactions, however, do not take place in such easily defined conditions. Furthermore, the reagents are themselves often covalent liquids. If one considers, for example, the reaction between acetic acid and ethanol, which yields an equilibrium mixture containing ethyl acetate (an ester) and water,

$$CH_3COOH + C_2H_5OH \rightleftharpoons CH_3COOC_2H_5 + H_2O,$$

it is very difficult to know how to describe the state of any of the reagents. We have a mixture of four liquids, each of which will interact (physically) with all of the others. We certainly could not write, for example, $CH_3COOH(aq)$, to represent the acetic acid which is dissolved in ethyl acetate and ethanol as well as in water. Consequently, we rarely find state symbols useful in writing organic reactions and, in discussing these reactions, state symbols will normally be omitted.

21.1 CHAIN STRUCTURES

Catenation, the formation of chains, is perhaps one of the most characteristic features of carbon chemistry (Section 16.4). No other element can form chains (and rings) which are as large, as stable (Section 16.4), or as geometrically varied as those which carbon can form and which are held together only by bonds between atoms of the same element. Other elements in the same group of the Periodic Table (group IVB), in particular silicon and germanium, do form chains of this type but those known are relatively small and relatively unstable. The mean silicon–silicon bond energy is 226 kJ mol^{-1} as opposed to 347 kJ mol^{-1} for the carbon–carbon bond. Furthermore, the silicon–oxygen bond, for example, is very much stronger (464 kJ mol^{-1}) than the silicon–silicon bond, whereas the carbon–oxygen bond is of similar strength (about 360 kJ mol^{-1}) to the carbon–carbon bond. This leads, in an oxygen-rich environment, to a thermodynamic preference for Si—O—Si—O—Si— chains rather than Si—Si—Si— chains, a preference not paralleled in carbon chemistry. The well-known silicone polymers have long-chain

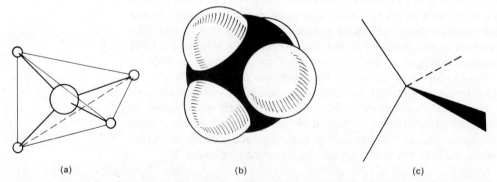

Fig. 21.1 Tetrahedral geometry of methane. (a) The larger, central sphere is the carbon atom, the small spheres hydrogen. Heavy lines represent the bonds, lighter lines the tetrahedron at the corners of which lie the hydrogen atoms. Bond lengths are exaggerated to make the angular relationships clear. (b) Model of methane, with van der Waals and covalent radii approximately to scale. (c) Convention for representing the three-dimensional molecule (and others) on paper (i.e. in two dimensions). Solid lines lie in the plane of the paper, broken lines recede below the paper, away from the reader. The wedge shaped lines (perspective) come out towards the reader. Atoms lie at the intersections of and the ends of lines.

structures based on Si—O—Si—O—Si— chains; quartz, one of the hardest materials known, depends for its strength on a three-dimensional network of Si—O—Si— linkages (Section 16.6):

$$\begin{array}{c} \text{Si} \\ | \\ \text{O} \\ | \\ \text{Si—O—Si—O—Si—O} \\ | \\ \text{O} \\ | \\ \text{Si} \end{array}$$

Catenation is also very limited in elements in the periodic groups III and V. Boron (group III) and phosphorus (group V) catenate more than other members of their groups, but to a very small extent relative to carbon.

The simplest hydrocarbon is methane which contains only one carbon atom and has the formula CH_4. We now know that the four hydrogen atoms are tetrahedrally disposed around the carbon atom—that is, they lie at the corners of a regular tetrahedron with the carbon atom at the centre as in Fig. 21.1. Thus the conventional diagram of methane:

$$\begin{array}{c} \text{H} \\ | \\ \text{H—C—H} \\ | \\ \text{H} \end{array}$$

tells us nothing of the geometry of the molecule. The molecule is certainly not planar, and the inter C—H bond angles are certainly not right angles as the diagram would appear to indicate (the angles are in fact about 109°). It is important to realize that these plane diagrams are very diagrammatic indeed and tell us nothing about the real geometry of the molecules that they are representing.

If one of the hydrogen atoms of methane is replaced by another carbon atom then we arrive at the next simplest hydrocarbon, ethane. We have effectively replaced a hydrogen atom in methane with a —CH_3 group. This group is known as a **methyl group**, the name clearly deriving from methane. Ethane consists of two methyl groups joined together, CH_3CH_3 or C_2H_6. If we replace a hydrogen atom in ethane with another methyl group we obtain propane, $CH_3CH_2CH_3$, and we can continue to make a series of hydrocarbons in this way, each one containing one more carbon atom than its predecessor. Table 21.1 lists the simpler members of this series together with their names and molecular formulae.

These hydrocarbons are referred to as **straight-chain hydrocarbons**, and the usual way of writing their structures, as in Table 21.1, gives the impression that the

Table 21.1

CH_4	CH_4	methane
CH_3CH_3	C_2H_6	ethane
$CH_3CH_2CH_3$	C_3H_8	propane
$CH_3CH_2CH_2CH_3$	C_4H_{10}	butane
$CH_3CH_2CH_2CH_2CH_3$	C_5H_{12}	pentane
$CH_3CH_2CH_2CH_2CH_2CH_3$	C_6H_{14}	hexane
and so on		

molecules are indeed straight. It is important to stress that the angle between any two carbon–carbon bonds is about 109° (the tetrahedral angle), not 180° as these written structures may seem to imply, and their real shapes will be more like zig-zags therefore.

Another important point to be recognized in this connection is that there is almost free rotation about each of the bonds in these systems. This is not very important for the C—H bonds, but it is important in the case of C—C bonds. If you can imagine yourself to be very small and sitting on one of the carbon atoms of ethane, then you would see, if you looked at the other methyl group, that it was able to rotate rather like a propeller. At low temperatures it would probably not rotate very much, but at higher temperatures it would rotate quite rapidly. The situation is illustrated in Fig. 21.2. Clearly, together with the fact that the interbond angles are less than 180°, this means that for larger straight-chain hydrocarbons there is a multitude of shapes which the molecule can adopt without any breaking or making of chemical bonds. The situation as far as the term "straight" is concerned is rather similar to that of a real chain built up from many metallic links. The chain can be rolled up, coiled up, or held in any overall shape that one chooses but the links remain joined together in the same sequence and the chain is fundamentally the same whatever its overall shape. The term "straight chain" in this connection simply refers to the sequence of atoms. Each carbon atom is joined to two neighbouring carbon atoms—no more and no less—except for the two atoms at either end which are joined to only one other carbon atom. *This sequence is known as the structure of the molecule*, and Fig. 21.3 illustrates the limitations of the planar diagrams for representing these structures. Each molecule in the diagram is in fact identical with its fellows. Provided that one knows their limitations, these plane diagrams are still useful of course. They are the most convenient way of writing structures because three-dimensional diagrams of large molecules are very difficult to draw or print.

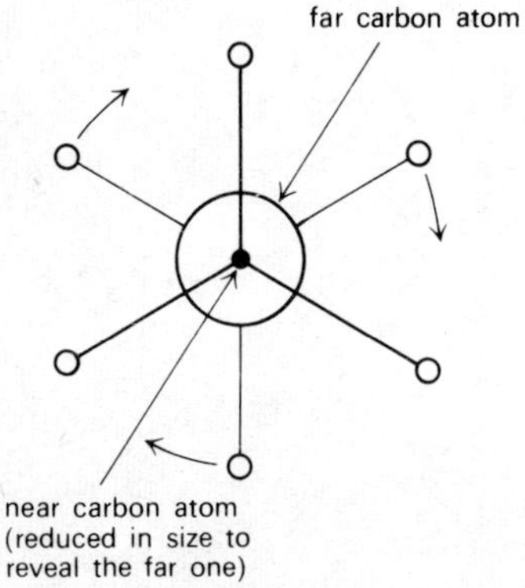

The importance of this sequence, or structure, will perhaps be clearer if we consider **branched-chain hydrocarbons**. There are, for example, two distinct structures we can write for a molecule of molecular formula C_4H_{10}. We can write:

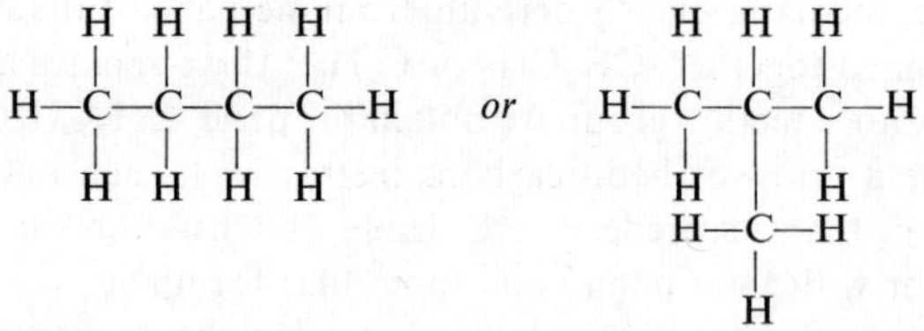

Fig. 21.2 Rotation about carbon–carbon single bonds. This is a molecule of ethane and the two carbon atoms are in line with the reader's eye. The carbon–carbon bond is perpendicular to the plane of the paper.

A is the straight-chain structure (butane) which we have already met but B is an new one. B contains one carbon atom which is joined to three others and there are three ends to the molecule. Such a molecule has what we call a branched chain and has a different sequence of atoms, or structure, from butane. The two have the same molecular formula and are therefore **isomers**. Because they differ in structure they are **structural isomers**. Their chemical and physical properties are different and they are different and quite distinct compounds.

The branched chain which we have just considered "forks" from one branch into two. Clearly, with carbon having a valency of four, it is possible for a branched-chain hydrocarbon to branch out three ways and the carbon atom at the branching

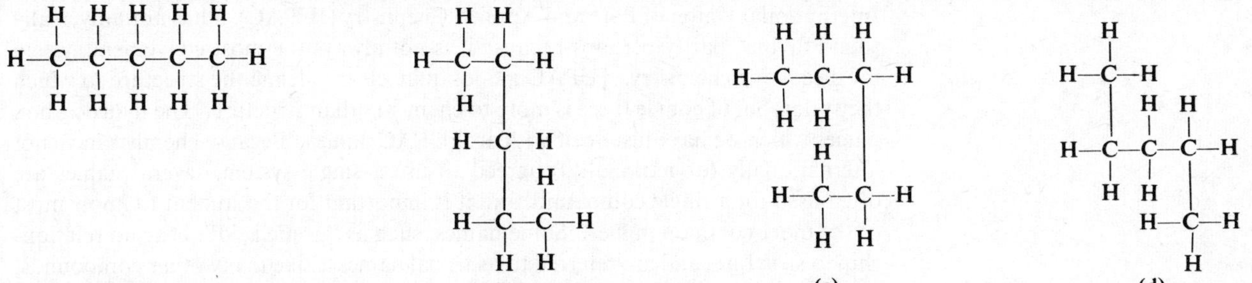

Fig. 21.3 Each of these molecules has the same sequence of atoms (i.e. structure) and is therefore identical with all the rest.

point will then be connected to four others. Figure 21.4c illustrates such a hydrocarbon.

It is often adequate to abbreviate the plane diagrams such as those in Fig. 21.4 in such a way that they can be written more easily. Such abbreviations have been used already, and they are easy to relate to the plane diagrams. For example, the structures in Fig. 21.4 would be written, in the abbreviated form, on a single line as follows:

$CH_3CH_2CH_2CH_2CH_3$ (pentane);
$CH_3CH_2CH(CH_3)_2$ (2-methylbutane);
$C(CH_3)_4$ (2,2-dimethylpropane).

21.2 NOMENCLATURE

A branched-chain hydrocarbon is named as a derivative of the hydrocarbon corresponding to the longest straight chain which it contains. Thus, whilst the first example in Fig. 21.4 is called pentane, the second is called 2-methylbutane (a derivative of butane), and the third, 2,2-dimethylpropane. The numbers indicate the positions of the substituent groups, or side chains, on the main chain which gives the molecule its name. On this basis, of course, the second example could have been called 3-methylbutane if the main chain had been numbered from its other end (left hand end in Fig. 21.4). In practice, we choose to number the main chain from the end which leads to the smallest numbers in the final name. All of these fully saturated (Section 21.4) hydrocarbons are known as **alkanes**, *alk-* being the general term for *eth-*, *prop-*, *but-*, *pent-*, *hex-*, etc. (Table 21.1) and *-ane* being the common ending for all saturated hydrocarbons. The ending of the systematic name of any organic compound normally indicates the class of the compound in question. The ending *-ol* in ethanol, for instance, indicates that ethanol is an alcohol and the ending *-oic acid* in propanoic acid indicates that propanoic acid is a carboxylic acid. The general name for monovalent groups derived from alkanes, such as methyl, ethyl, pentyl, is thus **alkyl**. All such groups are alkyl groups and the general symbol R is sometimes used to represent them in generalized structural formulae.

The systematic naming of organic compounds is a somewhat thorny problem. Although there is a system applicable to virtually all compounds, devised by the

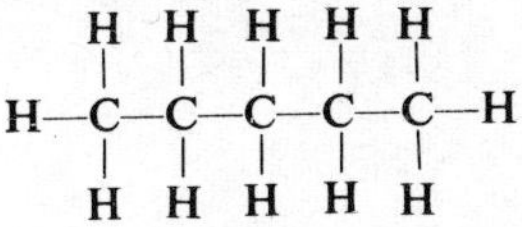

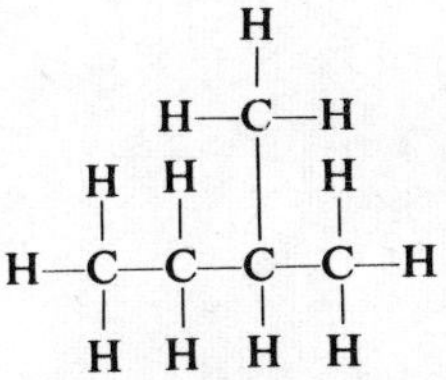

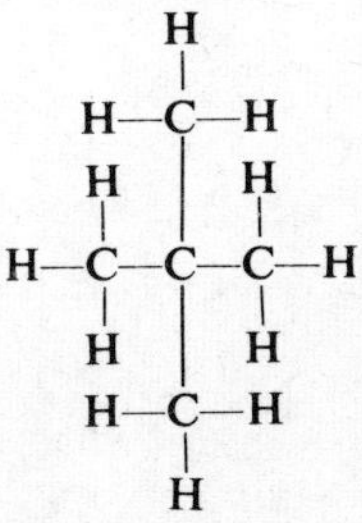

Fig. 21.4 All of these alkane molecules contain five carbon atoms. They are all structural isomers, and illustrate (a) a straight-chain structure, and (b) and (c) both types of branching.

International Union of Pure and Applied Chemistry (IUPAC), it has not universally found favour, partly perhaps because it is not always the most convenient system for discussing chemistry. IUPAC names quite clearly define the structure to which they refer, but of course there is more to chemistry than structure. The hydrocarbon names which we have just dealt with are IUPAC names. Because chemists have not internationally (or nationally!) agreed to use a single system, several names are often used for a single compound, and it is important for the student to know most of the more common names. Some names, such as "acetic acid", bear no relationship to structure, and are referred to as trivial names. Like many other compounds, however, acetic acid has given rise to a variety of other names, deriving from the same root, which are thoroughly worked into chemistry and other disciplines, such as biology and biochemistry. Names like acetyl, acetylation, acetate, acetoacetate are widespread. It is clear, therefore, that the student should know acetic acid by that name as well, perhaps, as the more systematic but rarely used "ethanoic acid". It is our policy in this book to use IUPAC names as far as possible, but where other names are more commonly employed, we use these. Often both are given. Occasionally, the IUPAC system allows more than one type of name to be used for one type of compound. A particular example of this is the case of halogenoalkanes, or alkyl halides (Chapter 25)—hydrocarbon molecules in which one hydrogen atom is replaced by a halogen atom. Naming these compounds as substituted hydrocarbons—halogenoalkanes—is generally preferred, but it is often necessary to lay emphasis on the alkyl group in the molecule in order to make discussion of the chemistry clearer. We have generally, therefore, referred to such molecules as halogenoalkanes (e.g. 2-chloropropane, $CH_3CH(Cl)CH_3$), but where discussion is clarified by calling them alkyl halides (e.g. prop-2-yl chloride for $CH_3CH(Cl)CH_3$) we have done this, as in Chapter 24 for example. As well as clarifying the discussion, we hope that this will help students to become familiar with both names; this is important because both are used in the language of modern chemistry. In several other instances we have departed from "preferred" IUPAC nomenclature if this clarifies the chemistry and is also in accordance with current chemical practice. In Section A.19 we list alternative names for some of the more important compounds referred to in this book, and when new types of organic compounds are introduced, nomenclature is briefly discussed.

21.3 THE GEOMETRY OF FOUR-CO-ORDINATE CARBON: OPTICAL ISOMERISM

(See Section 20.6.) Modern methods of spectroscopy and X-ray crystallography have made it possible to determine the shapes of many molecules which would have been impossible many years ago. The tetrahedral shape of carbon compounds has, however, been known since long before these methods were made available and the principles which were first used to deduce the shape of these compounds still have important applications today. They are fundamental to a proper understanding of present day chemical ideas and therefore merit our early attention.

Certain compounds are known to exist in two forms, each form having chemical and physical properties which are identical with those of the other form. The two forms differ in only one respect. Each form, if plane-polarized light (see below) is shone through a solution of it in a suitable solvent, will rotate the plane

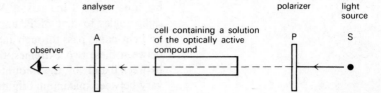

Fig. 21.5 Apparatus for measurement of optical activity.

of polarization of that light. One form will rotate it in one direction and the other form will rotate it in the other direction. The direction in which they rotate this plane of polarization is virtually the only criterion for distinction between the two forms (there are, in fact, some other, more sophisticated methods available now), as in nearly every other respect they appear to be identical. Compounds which have this property are said to be optically active. The value of this rotation, α, divided by the length of the cell, l (in decimetres), and by the concentration, c, of the solution (in grams per cubic centimetre) is known as the specific rotation, $[\alpha]_D$. Each pure form of a given compound has a particular value of $[\alpha]_D$. $[\alpha]_D = \alpha/lc$. The subscript D refers to the wavelength of the light used, which is normally that corresponding to the sodium D-line (yellow).

The kind of apparatus which is used to measure this rotation is, in principle, very simple and is illustrated in Fig. 21.5. The analyser, A, and the polarizer, P, are devices for polarizing light from a normal monochromatic light source, S. Light emanating from S approaches P (solid line) and on passage through P becomes polarized and passes on (dotted line). Before the light reaches P it is made up of waves which vibrate in all directions about the axis of the beam. Figure 21.6(b) is a representation of the wave nature of a single light ray illustrating that the waves (arrows) vibrate in all directions about the axis of the ray. When the ray reaches P the components of the wave vibrations in one direction only are allowed to pass through, the remainder being cut out by P. The material of which A and P are made will allow vibrations in only one plane to pass, and the light which emerges from P is said to be plane-polarized with the plane of polarization determined by P. If P were to be rotated about the axis of the ray, then the plane of polarization of the emergent ray would be rotated by a similar amount. We shall refer to this directional property of P as its axis. Figure 21.6(c) illustrates the nature of the light emerging from P, polarized in a vertical plane because the axis of P is vertical.

In the absence of the cell in Fig. 21.5, the polarized beam would pass through the apparatus until it reached A. It could then pass through A with maximum

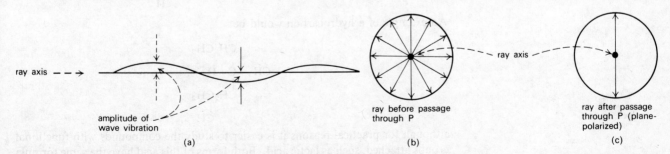

Fig. 21.6 Representation of unpolarized light (b), and polarized light (c).

21.3 GEOMETRY OF FOUR-CO-ORDINATE CARBON: OPTICAL ISOMERISM

brightness only if the axis of A were parallel to that of P. If the axis of A were at right angles to that of P, and also therefore to the plane of polarization, then no light could pass through and the observer would see a dark field. For situations between these two extremes, that is with the angle between the axes of A and P between 0 and 90°, the amount of light passing through to the observer's eye would vary between maximum brightness and darkness according to how close the angle was to 0 or 90°. At 0° the observer would see light of maximum intensity. If A were rotated about the axis of the light beam then the intensity of the light would drop gradually until the angle reached 90° when no light would pass. As the angle went on increasing the light would brighten again until, at an angle of 180°, with the axes of A and P parallel again, the light would be at maximum brightness.

Consider now the case where the apparatus has been so adjusted, in the absence of the cell, that the axes of A and P are parallel, that is with P fixed A has been rotated to a position of maximum brightness. If the cell, containing a solution of optically active material, is now put into position, the plane-polarized light from P will have to pass through it. In so doing, its plane of polarization will be rotated by an angle and in a direction which will be determined by the material in the cell. Thus, on reaching A, the plane of polarization will no longer be parallel to the axis of A and the observer will no longer see maximum brightness. If he rotates A, however, he will see maximum brightness again when the axis of A is parallel to the plane of polarization, and the angle through which A has to be rotated to give brightness again can be measured easily and is equal to the angle through which the plane of polarization has been rotated by the compound in the solution. Compounds which require the observer to rotate the analyser to the right (clockwise) are referred to as **dextrorotatory** and their rotation is defined as positive, and those which require rotation of the analyser in the opposite direction are referred to as **laevorotatory** and have a negative rotation.

To return to our original point—the shape of carbon compounds—it is necessary to consider what types of compound exhibit optical activity, and how one form of an optically active compound differs from the other.

The simpler types of optically active compound contain one carbon atom to which four different groups are attached—an *asymmetric carbon atom*. It does not matter what the groups are, provided that they are different. Such a compound would be lactic acid (2-hydroxypropanoic acid):

$$CH_3CH(OH)CO_2H \quad or \quad CH_3-\overset{\overset{\displaystyle OH}{|}}{\underset{\underset{\displaystyle H}{|}}{C}}-COOH.$$

An example of a hydrocarbon would be

$$CH_3-\overset{\overset{\displaystyle CH_2CH_3}{|}}{\underset{\underset{\displaystyle CH_2CH_2}{|}}{C}}-H$$
$$\underset{\displaystyle CH_3}{|}$$

although for practical reasons it is easier to study the compounds with functional groups attached, such as lactic acid. Both forms of this acid have the same formula and are therefore isomers. They also have the same structure, however (they contain the same atoms in the same order or sequence), and cannot therefore be structural isomers and the question arises as to just how the two forms do differ. To simplify the problem, we shall refer to the four different groups simply as a, b, c, and d. The key to the solution lies in the fact that only if the four groups attached to carbon are different does the molecule show optical activity, and of those molecules that do, two isomers—no more and no less—are known. If these compounds were planar, then with four different groups attached to the carbon atom a total of three different isomers ought to exist:

$$\begin{array}{ccc} \overset{a}{|} & \overset{a}{|} & \overset{a}{|} \\ d-C-b & d-C-c & c-C-b. \\ \overset{|}{c} & \overset{|}{b} & \overset{|}{d} \end{array}$$

Furthermore, if the compounds are really planar, then four different groups would not be an essential requirement for isomerism, as compounds of type a_2Cbc should also exhibit it (see problem section). The only shape which fits the observed requirements for optical activity is the tetrahedron with carbon at the centre and the relationship between two optical isomers (as they are called) is illustrated in Fig. 21.7. It will be seen that one is the mirror image of the other and that one cannot be superimposed upon the other. (You will find building and studying models of the two isomers particularly helpful in appreciating this point.) They are called **enantiomorphs** and such molecules, which lack both a centre and a plane of symmetry, are said to be **chiral**. The fact that one isomer rotates the plane of polarized light by an equal amount in the opposite direction to the other is consistent with this mirror image relationship. It was thus possible for van't Hoff and Le Bel to work out the correct shape of carbon compounds in 1874, long before the sophisticated modern theories and techniques were available. Mixtures of equal amounts of each form for any pair of enantiomorphs have a zero rotation, as each form makes an equal but opposite contribution. Such mixtures, called **racemic mixtures**, are difficult to separate, and special techniques are required.

These isomers differ only in the way in which the various parts of the molecules are arranged in space; they do not differ in structure. Isomerism such as this is therefore known as **stereoisomerism**. We have so far discussed only one type of stereoisomerism—optical isomerism—but there is another type which we shall discuss when carbon–carbon double bonds and ring compounds are considered. This second type of stereoisomerism is known as geometrical isomerism.

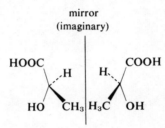

Fig. 21.7 Enantiomorphs of lactic acid.

21.4 UNSATURATION AND GEOMETRICAL ISOMERISM

The hydrocarbons dealt with so far, the alkanes (or paraffins as they used to be called), all contain carbon which is exerting its normal covalency of four. It turns out that all of these molecules conform to the general formula C_nH_{2n+2} in which all of the carbon valencies not concerned in carbon–carbon bonds are satisfied by bonding to hydrogen. The number $2n + 2$, where n is the number of carbon atoms, is the maximum number of hydrogen atoms which can be accommodated in the molecule (containing n carbon atoms), and molecules such as these are regarded as being **saturated**. Thus, the alkanes are saturated hydrocarbons.

Many hydrocarbons are known, however, which contain fewer than this maximum number and these compounds may be thought of as being **unsaturated**.

Amongst the hydrocarbons there are two well-known series of compounds, for example, which conform to the general formulae C_nH_{2n} and C_nH_{2n-2}. Compounds conforming to either of these general formulae again fall into two separate groups according to their chemical behaviour. One group is chemically (somewhat) unreactive, like the saturated hydrocarbons discussed above, and one group is extremely reactive so that we have reactive C_2H_{2n} and C_nH_{2n-2} compounds and unreactive C_2H_{2n} and C_nH_{2n-2} compounds. The group resembling the saturated hydrocarbons is known to be composed of molecules whose carbon skeletons are cyclic in structure. For example cyclohexane, having the molecular formula C_6H_{12} (C_nH_{2n}), has the structure:

$$\begin{array}{c} CH_2 \\ H_2C \quad CH_2 \\ H_2C \quad CH_2 \\ CH_2 \end{array}$$

As would be expected from this structure, which is essentially a variation on the usual alkane theme, the properties of cyclohexane are very similar to those of, say, hexane. We can, therefore, explain the existence and properties of the unreactive alkane-like compounds in terms of cyclic or ring structures which, in the C_nH_{2n} series, contain one more carbon–carbon bond and therefore two less hydrogen atoms than the corresponding C_n open-chain alkane structure. They are known as **cycloalkanes** (the prefix *cyclo-* clearly differentiates the cyclic molecules from the others).

This explanation is supported by the fact that there is no member of the alkane-like C_nH_{2n} series which contains two carbon atoms only. Clearly a hydrocarbon molecule with only two carbon atoms cannot form a ring. The alkane-like C_nH_{2n-2} group is similarly explained in terms of a skeleton containing two rings, and this is supported by the fact that no member of the series is known which contains three carbon atoms only. Decalin is an example of this series of compounds and, having the molecular formula $C_{10}H_{18}$ (C_nH_{2n-2}), it has the structure:

$$\begin{array}{c} CH_2 \quad CH_2 \\ H_2C \quad CH \quad CH_2 \\ H_2C \quad CH \quad CH_2 \\ CH_2 \quad CH_2 \end{array}$$

This still leaves the question of the reactive compounds which conform to either of these two general formulae, and it is these compounds which we shall consider in detail now. A fuller discussion of ring compounds is deferred until the next section.

That these are not ring compounds is indicated by the existence of molecules in both series which contain only two atoms of carbon. Ethylene (systematic name, ethene), C_2H_4, and acetylene (systematic name, ethyne), C_2H_2, are the molecules concerned, and a consideration of these two compounds will adequately serve to indicate the nature of the other compounds which make up the series.

The existence of multiple bonds between carbon atoms was put forward to explain how, in terms of classical covalent bond theory, these molecules could exist. The satisfaction of two valencies of one carbon atom by two of another readily explained the existence and structure of ethylene and other members of the reactive C_nH_{2n} group, whilst the satisfaction of three valencies of one carbon atom by three of another similarly accounted for acetylene and the C_nH_{2n-2} group. The situation can be illustrated with Lewis ("dot and cross") diagrams, which were introduced in Section 4.5 and which show how electrons pair to form bonds. The "lines" in a normal structural diagram represent these electron-pair bonds.

$$\text{H}_2\text{C}::\text{CH}_2 \qquad \text{H}:\text{C}\vdots\vdots\text{C}:\text{H}$$

$$\underset{\text{ethylene}}{\text{H}_2\text{C}=\text{CH}_2} \qquad \underset{\text{acetylene}}{\text{H}-\text{C}\equiv\text{C}-\text{H}}$$

The carbon–carbon linkage in ethylene is known as a double bond and that between the carbon atoms of acetylene as a triple bond. A fuller discussion of the nature of these bonds is given in Section A.2.

The fact that these compounds are very much more reactive than alkanes is clearly due to the presence of a different kind of chemical bonding in the molecules. Chemical reaction is a matter of making and breaking chemical bonds (Chapter 26). Most of the reactions of these compounds lead to products which are essentially adducts, which contain no multiple bonds and which are therefore saturated compounds. For example, both ethylene and acetylene react with hydrogen, in the presence of a suitable catalyst, to form ethane. In this reaction, and many others, the double (or triple) bond has "opened"—without the separation of the two carbon atoms—and the hydrogen has fully satisfied all of the carbon valencies so "released". This type of reaction is an addition reaction and is characteristic of compounds containing multiple bonds.

$$\underset{\substack{\text{ethylene} \\ \text{(ethene)}}}{\text{H}_2\text{C}=\text{CH}_2} + \underset{\text{hydrogen}}{\text{H}_2} \rightarrow \underset{\text{ethane}}{\text{H}_3\text{C}-\text{CH}_3}$$

$$\underset{\substack{\text{acetylene} \\ \text{(ethyne)}}}{\text{H}-\text{C}\equiv\text{C}-\text{H}} + 2\underset{\text{hydrogen}}{\text{H}_2} \rightarrow \underset{\text{ethane}}{\text{H}_3\text{C}-\text{CH}_3}$$

An **addition reaction** may be defined as one in which the reactants combine, or add, together to give a single product: $A + B \rightarrow C$. More examples of such reactions will be given when we deal with the detailed chemical properties of the hydrocarbons.

Just as we built up the series of alkanes from methane to hexane, all of general formula C_nH_{2n+2}, so we can build up the series from ethylene to hexene, all of formula C_nH_{2n} (Table 21.2).

Table 21.2. C_nH_{2n} hydrocarbons.

	IUPAC name	(Common name)
$CH_2{=}CH_2$	ethene	(ethylene)
$CH_2{=}CHCH_3$	propene	(propylene)
$CH_2{=}CHCH_2CH_3$	butene	(butylene)
$CH_2{=}CHCH_2CH_2CH_3$	pentene	
$CH_2{=}CHCH_2CH_2CH_2CH_3$	hexene	

Similarly (Table 21.3), we can build up the series from acetylene to hexyne, C_nH_{2n-2} (note that there are no one-carbon members of either of these series).

Table 21.3. C_nH_{2n-2} hydrocarbons.

	IUPAC name	(Common name)
$CH{\equiv}CH$	ethyne	(acetylene)
$HC{\equiv}CCH_3$	propyne	
$HC{\equiv}CCH_2CH_3$	butyne	
$HC{\equiv}CCH_2CH_2CH_3$	pentyne	
$HC{\equiv}CCH_2CH_2CH_2CH_3$	hexyne	
and so on		

It will be apparent that the cyclic compounds mentioned in this section on unsaturation are very much more akin, both structurally and chemically, to the alkanes discussed in the previous section. Although from their general formulae they would appear to be unsaturated it would clearly be a mistake to classify them with the compounds containing multiple bonds. Accordingly, compounds containing multiple bonds between carbon atoms are defined as unsaturated and cyclohexane, for example, is a saturated compound. One must also point out at this stage that multiple bonds can be formed by atoms other than carbon. We shall consider compounds in this book with double bonds between carbon and oxygen ($\diagdown{C}{=}O \diagup$), between carbon and nitrogen ($\diagdown{C}{=}N{-} \diagup$), and between nitrogen and nitrogen ($-N{=}N-$). We shall also meet compounds with triple bonds between carbon and nitrogen ($-C{\equiv}N$). All of these compounds undergo addition reactions which are characteristic of compounds containing multiple bonds, although, according to general usage, the term unsaturated is confined to those compounds with multiple bonds between carbon atoms.

(*Note.* The use of terms such as "C_nH_{2n-2} series" is somewhat arbitrary. Thus, C_4H_6 may have the structure $CH{\equiv}CCH_2CH_3$ or $CH_2{=}CH{-}CH{=}CH_2$. Clearly these two compounds are different and the usefulness of this "series" concept, based on molecular formulae, is therefore very limited. A series in which each member differs from the members adjacent to it by "CH_2" is known as a homologous series, any one member being a homologue of the others. A non-hydrocarbon homologous series for example would be the saturated alcohol series, $C_nH_{2n+1}OH$. Again, these are terms of very limited usefulness in modern chemistry; they are not employed in this book.)

As we have seen, the names of most saturated hydrocarbons are composed of two parts, the first indicating the number of carbon atoms in the molecule and the second, common to all the compounds, indicating that they are saturated hydrocarbons. Thus, in the case of pentane, *pent-* tells us that the molecule contains five carbon atoms and *-ane* tells us that the molecule is a saturated hydrocarbon. The first four members of the series, however, retain their early names which do not derive from the number of carbon atoms contained in their molecules, although they still convey that information. If the hydrocarbon has a ring structure, then the prefix *cyclo-* is used, as in "cyclohexane" for example. The systematic names of the compounds containing carbon–carbon double bonds are similarly derived. The first part of the names is exactly the same as for the saturated compounds; only the ending is different. Thus the compounds as a whole are referred to as **alkenes**, pentene, $CH_2{=}CH_2CH_2CH_2CH_3$, being an alkene corresponding to pentane. Similarly, the compounds containing triple bonds are known as **alkynes**, ethyne being the correct (though rarely used) systematic name for acetylene. The older general names for these two groups of compounds, which are still very commonly used, are olefines (alkenes) and acetylenes (alkynes).

The naming of alkenes and alkynes raises another very important point. For example, we have so far written pentene as $CH_2{=}CH_2CH_2CH_2CH_3$, although clearly $CH_3CH{=}CHCH_2CH_3$ also merits the name pentene. These two molecules are isomers, having the same formula, and they are further examples of structural isomerism even though their basic skeletons are identical. They are structural isomers because they differ in the position of the double bond in the sequence of atoms as a whole, and we saw earlier that structure was a matter of sequence of atoms. We are able to name the isomers by the adoption of the same simple numbering system that we used in naming branched alkanes (Section 21.2). The carbon atoms in the longest "straight" hydrocarbon chain are numbered consecutively from one end of the molecule to the other. Thus the two pentenes which we have just considered would be called pent-1-ene, $\overset{1}{C}H_2{=}\overset{2}{C}H_2\overset{3}{C}H_2\overset{4}{C}H_2\overset{5}{C}H_3$, and pent-2-ene, $\overset{1}{C}H_3\overset{2}{C}H{=}\overset{3}{C}H\overset{4}{C}H_2\overset{5}{C}H_3$, the number used indicating that the *-ene* part of the molecule lies between the carbon atom of that number and the next one higher. In pent-2-ene the double bond lies between carbon atom 2 and carbon atom 3. Note that pent-3-ene is identical with pent-2-ene. Thus there are two alternative names for the same molecule according to which end of it we start numbering from. As with branched alkanes, we adopt the name containing the lower number, that is, pent-2-ene.

The system for alkynes is exactly the same: hex-3-yne has the structure $CH_3CH_2C{\equiv}CCH_2CH_3$.

This technique of numbering the carbon atoms in a simple basic carbon skeleton finds considerable use in the naming of cyclic and more complicated non-cyclic compounds.

As well as giving rise to this sort of structural isomerism, double bonds also give rise to a form of *stereoisomerism* where isomers having the same structure exhibit different chemical and physical properties. This is quite distinct from optical isomerism, in which the isomers do not differ chemically and show only a very subtle difference in physical properties.

This type of stereoisomerism, known as **geometrical isomerism** (Section 20.6), is well illustrated by reference to two well-known alkenes, maleic and fumaric acids. Both have the structure

$$HO-\underset{\underset{O}{\|}}{C}-CH=CH-\underset{\underset{O}{\|}}{C}-OH$$

and the systematic name butenedioic acid, but have different chemical and physical properties. Maleic acid, for instance, melts at 130 °C whereas fumaric acid has a melting point of 287 °C. Fairly gentle heating of maleic acid results in a chemical reaction in which the elements of water are eliminated from the molecule leaving an acid anhydride of structure

$$\begin{array}{c} CH-C=O \\ \| \qquad \diagdown \\ \qquad \qquad O \\ \| \qquad \diagup \\ CH-C=O \end{array}$$

Similar treatment of fumaric acid results in no chemical reaction. Only very powerful heating of fumaric acid can succeed in bringing about reaction and when this happens the products are the same as those obtained from maleic acid.

The explanation of this chemical difference (and others) is found in the arrangements in space—the stereochemistry—of the groups of atoms constituting maleic and fumaric acids. The double bond, unlike a carbon–carbon single bond, will not allow free rotation about the carbon–carbon axis (the nature of the double bond is discussed in Section A.2). Furthermore, in alkenes the four bonds, and therefore the four substituents, which are found at either end of the double bond, all lie in the same plane. Thus ethylene and substituted ethylenes (maleic and fumaric acids are substituted ethylenes) are flat, planar molecules, and we should expect to find that ethylenes with a substituent at each end exist in two isomeric forms. We can write two isomeric forms, differing in geometry, for maleic and fumaric acids:

A   B

Considering again the nature of the reaction we have just discussed and the products obtained, it is clear that only isomer A above could undergo such a reaction, as the two COOH groups are close enough only in that isomer to permit the cyclic anhydride to form. Isomer A (the *cis* isomer) must be maleic acid therefore, and isomer B (the *trans* isomer) fumaric acid. Full, systematic names are

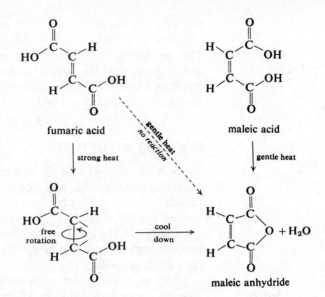

Fig. 21.8 Stereochemistry of fumaric and maleic acids.

cis-butenedioic acid (maleic) and *trans*-butenedioic acid (fumaric). The formation of maleic anhydride from fumaric acid by the application of very strong heat leads us to the conclusion that the extra energy supplied to the molecule is necessary to force a rotation about the double bond. This is discussed more fully in Section A.2. A summary of these reactions is given in Fig. 21.8.

Thus, from the observation of this type of isomerism, the early organic chemists were able to deduce that there was no free rotation about a carbon–carbon double bond and that ethylene and substituted ethylenes were planar. The absence of free rotation about the double bond, as proved by the existence of the isomers, does not in itself prove that the molecules are flat, of course, although by rather more sophisticated arguments the planar nature of the molecules can be deduced from the complete, observed nature of the isomers of substituted ethylenes.

Similar isomerism in the case of simple alkynes has never been observed and it was possible to deduce that either they were completely linear (see Fig. 21.9), or that there was completely free rotation about the triple bond, or both. Modern physical methods reveal that the acetylene molecule is linear and because of this linearity any rotation about the triple bond would be impossible to detect.

As would be expected, double bonds are stronger than single bonds and triple bonds stronger than double bonds (Section 4.12). It takes more energy to pull apart atoms which are held together by multiple bonds than those held together by single bonds. The bond energy of C=C is 611 kJ mol^{-1}, and that of C≡C is 837 kJ mol^{-1}. The single C—C bond energy is 347 kJ mol^{-1}. Further, the higher the bond order the closer together are the atomic nuclei. Thus, in acetylene the carbon–carbon bond length is 1.20 Å, in ethylene it is 1.34 Å, and in ethane it is 1.54 Å. The bond order is just another way of expressing the multiplicity of a bond. Thus in acetylene the carbon–carbon bond order is three, in ethylene two, and in ethane one.

Ethylene is more reactive towards many reagents than is ethane and it is important to realize that this does not mean that the carbon–carbon linkage in ethylene (double bond) is weaker than that in ethane (single bond). As we have

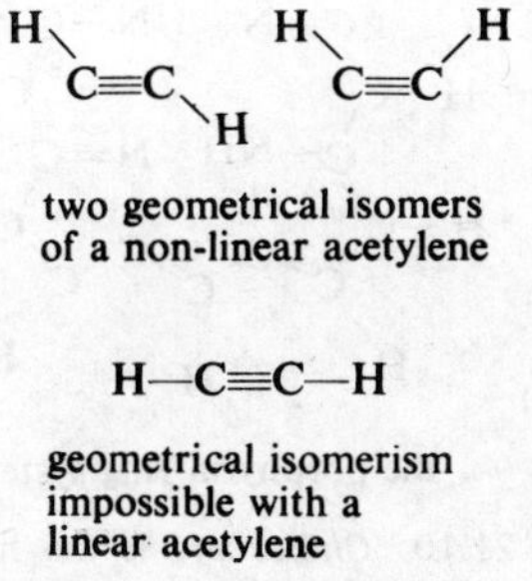

Fig. 21.9 The absence of geometrical isomerism in acetylenes suggests that they are linear.

just seen, this is not the case. The point is that the reactions of ethylene do not involve—at any rate as a first step—the rupture of the carbon–carbon linkage. They involve the saturation of the molecule with the formation of more bonds, and in the process the carbon–carbon linkage is converted to a single bond. In reactions where the carbon–carbon linkage is completely broken, it is probably converted to a single bond first.

21.5 RING STRUCTURES

We have already met with the concept of ring structures and cyclic molecules in discussing unsaturated hydrocarbons. We use the terms "ring" and "cyclic" even though the molecules are far from circular; they are used in rather the same sense that we use the term "straight" in connection with chain structures of alkanes (Section 21.1). The terms simply mean that there is a continuous bonding connection, through other atoms in the molecule, from any one atom in the skeleton back again to itself.

Ring compounds are very common amongst man-made compounds and amongst compounds found in nature. Figure 21.10 illustrates two complex but common types of ring compound found in nature, but we must confine ourselves for now to the study of rather simpler molecules.

The most common and most extensively studied ring systems are perhaps those containing six atoms in the basic skeleton of the ring, although rings of many shapes and sizes are known. The six-membered ring allows the easy accommodation of the tetrahedral angle associated with fully saturated carbon (that is, carbon with four single covalent bonds). It can be shown that the geometry of cyclohexane is as in Fig. 21.11 with the carbon–carbon bond angles almost exactly tetrahedral. Far from being circular, the ring does not even lie in one plane. Larger rings, by virtue of the free rotation which the single bond permits, can buckle and twist into less regular shapes, thus allowing for the accommodation of the tetrahedral bond angle. Rings of fewer than five atoms, however, necessitate severe straining of the bonds as the bond angles required are very much less than the tetrahedral angle. Cyclopropane is known, but because of this strain in the ring it is rather more reactive than most other saturated hydrocarbons. There is a marked tendency for the ring to spring open under the influence of this strain and an attacking reagent.

The concept of ring strain is not merely an unsupported idea arising out of a contemplation of bond angles. It has a sound experimental basis and can be expressed quantitatively. Using the data in Chapters 13 and 16, we can calculate a value which we would expect for the heat of formation of, say, cyclopropane. Two "stages" need to be considered for this calculation:

$$3C(s) + 3H_2(g)$$
$$\downarrow \qquad\qquad\qquad CH_2$$
$$3C(g) + 6H(g) \rightarrow CH_2\!-\!\!-\!\!CH_2(g).$$

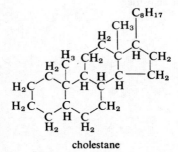

Fig. 21.10 *Cholestane* is the hydrocarbon skeleton of many compounds found in the human body, including the important group of steroidal hormones. The *porphyrin* ring system is the basic skeleton of the active part of haemoglobin and also of an important group of respiratory substances and other important biological molecules.

$\Delta H^\ominus$ for the first stage is three times the heat of atomization of carbon ($+716$ kJ mol^{-1}) plus three times the dissociation energy of the hydrogen molecule ($+436$ kJ mol^{-1}). That is $+2148 + 1308$ kJ $= +3456$ kJ. $\Delta H^\ominus$ for the second stage is six times the C—H bond energy plus three times the C—C bond energy. That is, $-2484 - 1041$ kJ, $= -3525$ kJ. The sum of these two $\Delta H^\ominus$ terms,

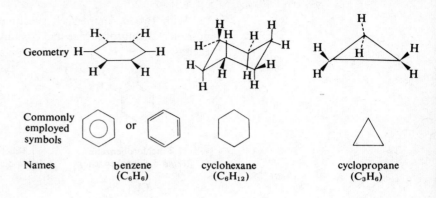

Fig. 21.11 Some important ring systems. Note the C—C—C bond angles. *Benzene* 120° (sp^2 hybridization, Section A.2). *Cyclohexane* (chair form) 109° (sp^3 hybridization, Section A.2). *Cyclopropane* 60°, 49° less than the tetrahedral angle, 109°, of methane and cyclohexane. Carbon atoms are represented as the "corners", in the symbols and hydrogen atoms are not explicitly shown.

−69 kJ, is the predicted heat of formation of cyclopropane. The observed value, however, is +55 kJ mol^{-1} and it would seem that cyclopropane "contains" 124 kJ mol^{-1} *more* than was expected. This, we believe, is a measure of the energy required to constrain the hydrocarbon bonds away from their preferred tetrahedral angles. In point of fact, we believe that our basic assumption that the C—H bond energies in methane and cyclopropane are the same is not fully justified; there is reason to believe that the C—H bond in cyclopropane is about 12.6 kJ stronger than in methane. This leads to a modified value for the predicted heat of formation of cyclopropane of $(-69) - (6 \times 12.6) = -144.6$ kJ mol^{-1}, and a modified value of $(124 + 6 \times 12.6) = 199.6$ kJ for the extra heat content. This figure of almost 200 kJ mol^{-1} is a quantitative measure of the ring strain in cyclopropane—a measure of the energy to be "released" when the ring springs open. In cyclobutane, where bond angles are less far removed from the tetrahedral value, the ring strain is down to about 130 kJ mol^{-1}; in cyclopentane it is less than 30 kJ mol^{-1}; in cyclohexane it is zero.

Benzene, C_6H_6, is another very well-known ring compound with a six-atom skeleton. As well as being cyclic, however, it contains what may formally be represented as three double bonds. A fuller discussion of the nature of the bonding in benzene is given in Sections 4.12 and A.2. Unlike cyclohexane, the atoms of benzene all lie in one plane. Thus benzene is very different in shape from cyclohexane as dictated by the nature of the bonding which also dictates that the chemistry of benzene is very different from that of cyclohexane. Compounds such as benzene and its relatives, showing similar properties, are said to be **aromatic**, in contrast with the other compounds so far considered which are **aliphatic**. Aromatic hydrocarbons are known collectively as **arenes** and corresponding monovalent groups, derived by the removal of one hydrogen atom from an arene, as **aryl groups**. The symbol Ar is often used to represent such groups in generalized structural formulae. Rings containing functional or other groups replacing hydrogen atoms may give rise to the possibility of structural isomerism. With only one such substituent, only one structure can exist. Thus chlorobenzene and chlorocyclohexane have only one possible structure each:

With two or more such substituents several alternative structures may be possible. Dichlorobenzene and dichlorocyclohexane have three possible structures each:

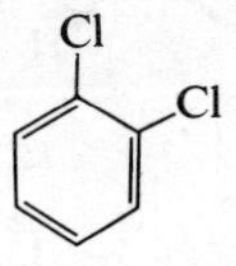

1,2-dichlorobenzene (*ortho*-dichlorobenzene)

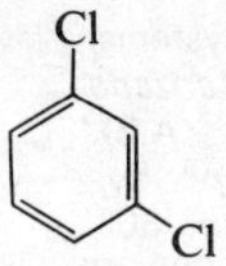

1,3-dichlorobenzene (*meta*-dichlorobenzene)

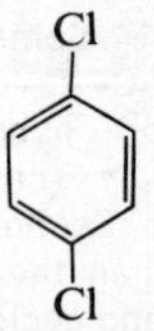

1,4-dichlorobenzene (*para*-dichlorobenzene)

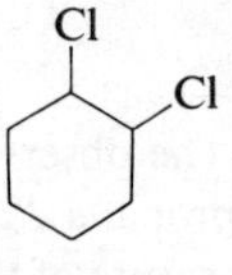

1,2-dichlorocyclohexane

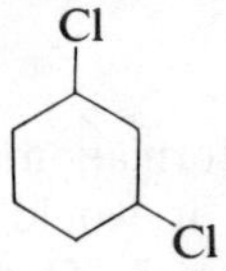

1,3-dichlorocyclohexane

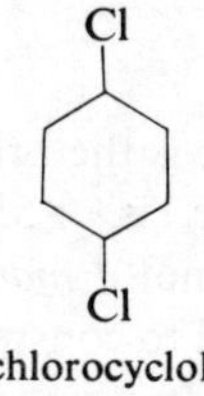

1,4-dichlorocyclohexane

This numbering system of nomenclature is of almost universal applicability for cyclic structures. The names based on the prefixes *ortho-*, *meta-*, and *para-* are used only for compounds derived from benzene (benzene derivatives), however, and, although still in common use, are technically out of date. They are useful only for benzene rings with two substituents.

When the carbon–carbon bond is contained in a ring structure, completely free rotation about that bond is not possible. Complete rotation of 360° about one bond can only occur if other ring bonds break. Thus, with ring compounds, as with the alkenes the possibility of geometrical isomerism arises (Section 21.4). Because all the carbon atoms and all the hydrogen atoms (or other substituents) of a molecule of benzene (or its derivatives) lie in the same plane, geometrical isomerism is not possible, but it is possible in other suitably substituted non-aromatic rings. The simplest case is that of a cyclopropane molecule with two substituents. If we consider such a molecule, say cyclopropane-1,2-dicarboxylic acid, we see that there are two possible isomers, a *cis* isomer and a *trans* isomer, corresponding to maleic

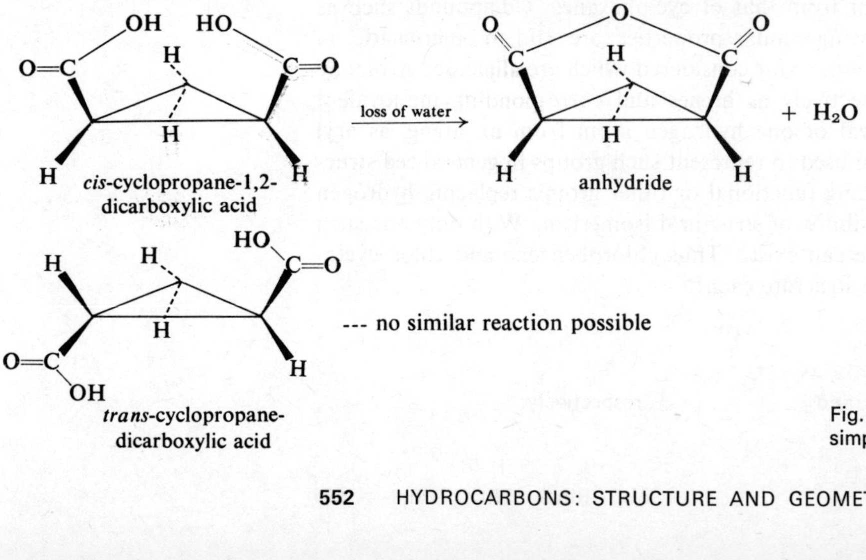

cis-cyclopropane-1,2-dicarboxylic acid

$\xrightarrow{\text{loss of water}}$

anhydride + H_2O

trans-cyclopropane-dicarboxylic acid

--- no similar reaction possible

Fig. 21.12 Geometrical isomerism in simple ring compounds.

and fumaric acids which we chose as examples to illustrate geometrical isomerism about a carbon–carbon double bond. *Cis*-cyclopropane-1,2-dicarboxylic acid, like maleic acid (which is also a *cis* isomer), forms an anhydride readily, whereas *trans*-cyclopropane-1,2-dicarboxylic acid does not (Fig. 21.12).

21.6 SUMMARY

Carbon, by virtue of its ability to form strong, covalent bonds to other carbon atoms, is unique in the size, number, and variety of the molecules to which it can give rise. *Hydrocarbons* are molecules in which the only other type of bond is between carbon and hydrogen, and these molecules most commonly form the basic skeleton of all organic molecules. They usually take the form of *chains* (*alkanes*)—"straight" or "branched"—and *rings* (*cycloalkanes*). The geometry of the molecules is a function of the tetrahedral disposition of the four covalent bonds which a carbon atom can form, and the ease with which rotation can occur about these bonds. In certain circumstances, such as when four different groups are attached to one carbon atom (an *asymmetric* carbon atom), the structures are *chiral* and exist in two isomeric forms, one being the non-superimposable mirror image

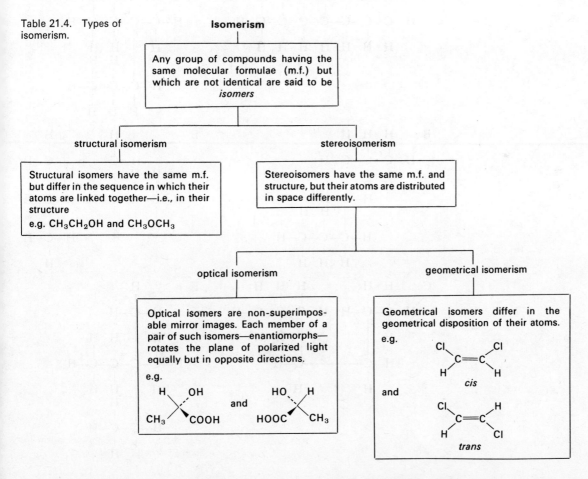

Table 21.4. Types of isomerism.

of the other. This type of isomerism is known as *optical isomerism*—a type of *stereoisomerism*—and one form is described as the *enantiomorph* of the other.

Carbon atoms may also link together by means of multiple bonds—double (*alkenes*) or triple (*alkynes*). Rotation about double bonds is restricted and, in suitable molecules, gives rise to *geometrical isomerism*, another form of stereoisomerism. Compounds containing multiple bonds between carbon atoms are said to be *unsaturated*. The four single bonds emanating from a $\overset{\diagdown}{\diagup}C{=}C\overset{\diagup}{\diagdown}$ unit, and the C=C unit itself, all lie in one plane, with bond angles of about 120°. The triply bound unit, —C≡C—, and the two single bonds emanating from it, are *linear*.

The sequence in which atoms are linked together in a molecule is known as the *structure*.

PROBLEMS

1. Study the following six molecules all of which are of molecular formula C_7H_{16}:

a) Which molecules are identical with—have the same structure as—molecule A which is *heptane*?
b) Work out the systematic names for the molecules which are *not* identical with A.
c) Which, if any, of these molecules contain an asymmetric carbon atom?

2. Write down the structural formulae of:
a) five different, simple alkanes;
b) five different, simple alkenes (only *one* double bond);
c) five different, simple alkynes (only *one* triple bond);
d) five different, simple cycloalkanes (only *one* ring).
Show that they conform to the general formulae: C_nH_{2n+2}, C_nH_{2n}, C_nH_{2n-2}, and C_nH_{2n} respectively.

3. How many structural isomers of formula C_5H_{12} can you write down? Give the systematic name for each one.

4. A compound, A, has the formula C_5H_8.
a) What is the maximum number of rings it could contain?
b) What is the maximum number of double bonds it could contain?
c) What is the maximum number of triple bonds it could contain?
On treatment with an excess of hydrogen over a nickel catalyst, one mole of A absorbed 22.4 litres of hydrogen (at s.t.p).
d) What is the formula of the product of this hydrogenation reaction?
e) How many rings does A contain?
f) How many double bonds does A contain?
g) How many triple bonds does A contain?

5. A hydrocarbon has a molecular weight of approximately 60 and contains 17.2 percent of hydrogen. Work out the molecular formula (Section 29.6) of the hydrocarbon and write down all of the structural isomers which have this formula.

6. A hydrocarbon has a molecular weight of 56.0 and contains 85.7 percent carbon. Work out the molecular formula (Section 29.6) of the hydrocarbon and write down all possible structural and stereoisomers corresponding to this formula.

7. What is the smallest alkane molecule (no rings or multiple bonds) which could exist in two enantiomorphic forms? Give the structure, and mark the asymmetric carbon atom with an asterisk.

8. Which of the following structures could exist in two enantiomorphic forms?

$CH_3CH(NH_2)COOH$ $CH_2(NH_2)COOH$
$CH_2(OH)CH_2OH$ $CHFClBr$
$CH_3CH(Cl)COOH$ $CH_2(Cl)CH_2COOH$

Write down those which do, and indicate asymmetric carbon atoms by marking them with an asterisk.

9. Which of the following structures would exhibit geometrical isomerism? Where appropriate draw a diagram of each pair of geometrical isomers.

ClCH=CHCl

CH₃CH=CHCH₂CH₃

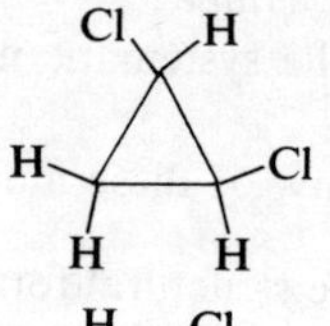

ClCH=CCl₂

(cyclopropane structures with Cl and H substituents)

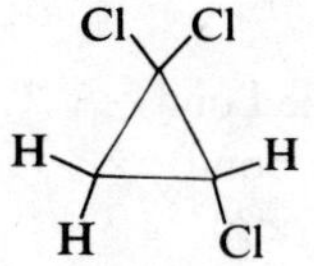

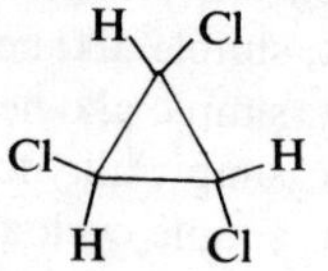

10. How many stereoisomers of the molecule a—C—b can be written, assuming

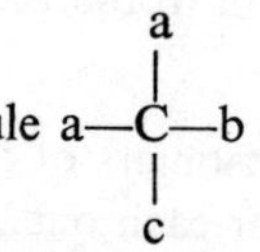

    $$\text{a}-\underset{\underset{c}{|}}{\overset{\overset{a}{|}}{C}}-\text{b}$$

    that the shape of the molecule is:
    a) square planar—carbon at the centre of a square, attached groups at each corner;
    b) tetrahedral;
    c) square-based pyramidal.
    Make a diagram for each isomer.

11. Tartaric acid has the structure indicated below, and contains two asymmetric carbon atoms (marked with asterisks).

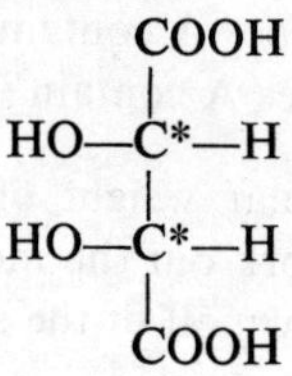

    $$\begin{array}{c} \text{COOH} \\ | \\ \text{HO}-\text{C*}-\text{H} \\ | \\ \text{HO}-\text{C*}-\text{H} \\ | \\ \text{COOH} \end{array}$$

    a) Predict how many optical isomers would exist for this compound. (*Hint*—remember that each asymmetric carbon atom can have *two* different arrangements (configurations) in space.
    b) How many pairs of enantiomorphs—non-superimposable mirror images—can you pick out from your predicted isomers?
    c) Go to your school library and consult a more advanced text on organic chemistry. Read the section on tartaric acid stereochemistry and carefully note down what the isomers are, and how they are related. Compare these notes with your predictions.

12. a) Find out as much as you can, from your school library, about the methods available for separating a racemic mixture into its enantiomorphic components. Make brief notes on the methods.
    b) If you had a mixture of all of the possible stereoisomers of tartaric acid (see Problem 11) how far, in principle, would you expect to be able to separate the mixture into its components by normal methods, such as recrystallization from a common solvent?

BIBLIOGRAPHY

Benfey, O. T. *From Vital Force to Structural Formulas*, Houghton-Mifflin (1964).

> An excellent and very readable account of the historical development of organic chemistry.

Cahn, R. S. *An Introduction to Chemical Nomenclature* (third edition), Butterworth (1968).

Stark, J. G. (Editor). *Modern Chemistry*, Part 2: "Stereochemistry," article by G. Baddeley, Penguin (1970).

Allinger, N. L., and J. Allinger. *Structures of Organic Molecules*, Prentice-Hall (1965).

> Primarily intended for use by undergraduates during their earlier courses but provides good reading for the sixth-former who wishes to expand his understanding of the molecular geometry of organic compounds.

Mislow, K. *Introduction to Stereochemistry*, Benjamin (1965).

> Rather an advanced treatment, at university level, but might provide some insight for the sixth-former whose studies have reached a high level.

All of the books on organic chemistry in the general list cover, at differing levels, the material of this chapter.

CHAPTER 22

Properties of Hydrocarbons

22.1 Physical properties of hydrocarbons
22.2 Chemical behaviour
22.3 Important methods of hydrocarbon synthesis

22.1 PHYSICAL PROPERTIES OF HYDROCARBONS

Note. In reading through the sections on physical properties which appear throughout, readers will find it helpful to ensure that they are familiar with the contents of Chapter 4, in particular, Sections 4.8 and 4.9.

The hydrocarbons are entirely covalent in character, and as such tend, at any rate in the cases of the smaller molecules, to be volatile. The molecular weight in fact largely determines the volatility of the hydrocarbons with the lower members being gases at room temperatures and the higher members progressing through liquids and solids with boiling and melting points rising with increasing molecular weight. Only the relatively weak van der Waals forces (Section 4.9) oppose the vaporization process, tending to prevent the separation of one molecule from others as it passes into the gas phase and acquires independence. When, in other types of compound, other forces (such as hydrogen bonding) come into play then other patterns may be superimposed upon this one. To illustrate this tendency the boiling points and melting points, together with the molecular weights, of some of the compounds we have already met are listed in Table 22.1. Some other compounds are included to widen the range of the table.

The properties of polythene (polyethylene or, more systematically, polyethene) $(CH_2)_n$, where n may be as large as several tens of thousands, are no doubt familiar to all. These very large hydrocarbon molecules are clearly far from being volatile, and some types of polythene are quite high-melting (up to about 135 °C).

Table 22.1. Boiling and melting points of hydrocarbons (°C).

Compound	Molecular weight	Boiling point	Melting point
Methane	16	−161.5	−184
Ethane	30	−88.3	−172
Propane	44	−42.7	−189.9
Benzene	78	80	5
Cyclohexane	84	81.4	6.5
Naphthalene ($C_{10}H_8$)	128	217.9	80.2

Table 22.2. Crude oil fractions.

	Boiling range (°C)
Solvent fractions	20–120
Petrol (motor fuel)	70–200
Kerosene (paraffin oil)	200–300
Lubricating oil, greases, and waxes	300

Crude oil consists largely of hydrocarbon material, and, on refining by distillation, can be separated into many useful fractions which differ in boiling point; you are probably familiar with these. Some of the more important fractions are listed in Table 22.2.

Being covalent in character, the hydrocarbons are soluble in, and, if liquids themselves, will dissolve many other covalent compounds. Thus naphthalene dissolves readily in benzene, and petrol is a good solvent for removing grease stains from fabrics. They are non-conductors of electricity in solution or in the molten or vapour state.

22.2 CHEMICAL BEHAVIOUR

22.2.1 Combustion

One property that all hydrocarbons have in common is that they burn in oxygen or air to give carbon dioxide and water. For example:

$$C_3H_8(\text{propane})(g) + 5O_2(g) \rightarrow 3CO_2(g) + 4H_2O(g).$$

Aromatic compounds such as benzene, and some alkenes and alkynes are very rich in carbon when compared with the saturated hydrocarbons and, when burnt in air, they tend to produce elemental carbon. Thus their combustion results in a good deal of soot and the flame is coloured yellow by the glowing soot particles. This provides the basis of a qualitative test for aromatic compounds which usually burn in this way, although, because some other compounds may also burn with a smoky flame, the result of such a test should be regarded as a guide rather than as a definite proof of aromaticity.

Complete combustion of hydrocarbons in an excess of oxygen also provides the basis for quantitative analysis. It is fairly easy to collect and measure the amounts of water and carbon dioxide produced on combustion of a known weight of hydrocarbon and thus to calculate the composition of the compound. The method is in fact of very wide application as the presence of elements other than carbon or hydrogen in a molecule does not usually interfere with the combustion. Thus it is possible to measure the percentage of carbon and hydrogen in an organic compound simply by burning it, quantitatively, in a suitable apparatus (Section 29.6).

The gaseous hydrocarbons form explosive mixtures with air or oxygen. Acetylene–air mixtures detonate with great violence.

22.2.2 Introduction of Functional Groups

As we have seen, hydrocarbons form the skeletons of most organic molecules; hence the introduction of functional groups into hydrocarbon molecules is an important consideration. It is perhaps not quite so important as one might at first suppose because compounds obtained from natural sources often already contain functional groups. However, they are not always the groups which we require and we often need to introduce new groups or convert the original groups into others. Furthermore, the hydrocarbons produced by the petroleum industry are an important source of organic chemicals and it is necessary to introduce functional groups into some of these hydrocarbons to satisfy the world's requirements for organic chemicals. Methods of introducing functional groups are emphasized in the following discussion of hydrocarbon reactions.

22.2.3 Reactions of Alkanes

The saturated hydrocarbons are often said to be chemically unreactive and to a large extent this is true. There are certain reagents, however, which react very readily with hydrocarbons and these are reagents which can give rise to free radicals —species which contain unpaired electrons (Section 26.2). The halogens, such as chlorine and bromine, are such reagents and these can react rapidly with hydrocarbons if the conditions are right.

If a few drops of bromine are added to a sample of a suitable liquid alkane (e.g. hexane or cyclohexane) in a glass test-tube and strong sunlight is allowed to fall upon the mixture, bubbles of hydrogen bromide gas are soon seen to form and rapidly leave the liquid. Thus a chemical reaction is taking place and taking place quite rapidly. The rate of gas evolution is seen to fall off if the tube is removed from the sunlight and placed in a region of relatively low intensity light.

The effect of the energy of the sunlight is to break up the bromine molecule into bromine atoms, each of which contains one lone electron and is therefore a free radical (homolytic fission, Section 26.2):

$$:\!\ddot{\text{Br}}\!:\!\ddot{\text{Br}}\!: \rightarrow 2:\!\ddot{\text{Br}}.$$

These reactive atoms will then attack a hydrocarbon molecule by "abstracting" a hydrogen atom, forming in the process a molecule of hydrogen bromide and a new radical derived from the hydrocarbon molecule:

$$\text{Br}\cdot + \text{H}:\overset{\overset{\text{H}}{|}}{\underset{\underset{\text{H}}{|}}{\text{C}}}:\overset{|}{\underset{|}{\text{C}}}- \rightarrow \text{Br}:\text{H} + \cdot\overset{\overset{\text{H}}{|}}{\underset{\underset{\text{H}}{|}}{\text{C}}}:\overset{|}{\underset{|}{\text{C}}}-$$

The hydrocarbon radical, like the bromine atom, is also reactive and can attack a bromine molecule to form bromoalkane and a bromine atom:

$$\text{Br}:\text{Br} + \cdot\overset{|}{\underset{|}{\text{C}}}-\overset{|}{\underset{|}{\text{C}}}- \rightarrow \text{Br}\cdot + \text{Br}:\overset{|}{\underset{|}{\text{C}}}-\overset{|}{\underset{|}{\text{C}}}-$$

The latter can attack another hydrocarbon molecule and form hydrogen bromide and a hydrocarbon radical and the process is continuous until all the reactants are consumed or the radicals removed. The spare radicals at the end of the reaction can recombine to form molecules such as Br_2.

Thus we have a process which results in the conversion of halogen and alkane into halogenated alkane and hydrogen halide and which we can represent by a single simple equation, such as:

$$CH_4 + 4Cl_2 \rightarrow CCl_4 + 4HCl.$$

From what we have said about the reaction in the previous paragraph it will be apparent that this equation is very much an oversimplification of what actually takes place, although for practical purposes it accurately represents the products and reactants and their reacting quantities, which is one of the most useful functions of a chemical equation.

As a practicable method for introducing functional groups into an alkane molecule, this method is very limited indeed, largely because it is not easy to control the position of entry of the halogen atom into the hydrocarbon or the number of halogen atoms which enter. The reaction between one mole each of bromine and hexane, for example, would produce a mixture of many compounds, including some unreacted hexane. Some of the halogenated compounds would be structural isomers. Examples of halogenated species would be:

$$CH_3CH_2CH_2CH_2CH_2CH_2Br, \quad CH_3CH_2CH_2CH_2\underset{\underset{Br}{|}}{C}HCH_3,$$

$$CH_3CH_2CH_2\underset{\underset{Br}{|}}{C}HCH_2CH_3, \quad CH_3CH_2CH_2CH_2CH_2CHBr_2,$$

$$CH_3CH_2CH_2CH_2\underset{\underset{Br}{|}}{C}HCH_2Br, \quad \text{and so on.}$$

Only in certain limited cases does the method find useful application. An example would be the chlorination of methylbenzene (toluene) which is a derivative of benzene containing a methyl group, $C_6H_5CH_3$. The benzene part of this molecule is not readily attacked by chlorine atoms (as discussed shortly in connection with benzene itself) so that there is only one site for chlorine to attack under conditions which favour free radical reaction (strong light and high temperatures). Chlorination of methylbenzene, with suitable control over quantities, can produce three useful compounds: (chloromethyl)benzene (benzyl chloride), $C_6H_5CH_2Cl$; (dichloromethyl)benzene, $C_6H_5CHCl_2$; and (trichloromethyl)benzene, $C_6H_5CCl_3$.

A reaction in which an atom, or group of atoms, is replaced by another is known as a **substitution reaction**.

The vigour with which chlorine can react with some hydrocarbons can be illustrated by plunging a lighted taper into a gas-jar of chlorine. The wax (hydrocarbon material) continues to burn with an orange-red flame as the chlorine "strips" off the hydrogen atoms to form hydrogen chloride leaving a black smoke of soot (carbon). In this case the reaction is not a substitution reaction.

22.2.4 Reactions of Alkenes

These compounds are very much more reactive towards many common reagents than are the alkanes. They readily undergo addition reactions to form saturated products and are therefore extremely useful for preparing substituted alkanes—they are much more useful for this purpose than the alkanes themselves. For example, ethylene will react with hydrogen chloride to produce chloroethane:

$$CH_2\!=\!CH_2 + HCl \rightarrow CH_3CH_2Cl.$$

In this reaction there is no possibility of introducing more than one chlorine atom into the molecule and only one product—a halogenoalkane or alkyl halide—can result. Thus this would be a much better way of making chloroethane (ethyl chloride) than chlorinating ethane which could give rise to a large number of different products. The reactions of hydrogen chloride with some more complex alkenes are also useful and lead to predictable products. With propene, for example, only two products are possible, according to the direction in which the hydrogen chloride adds across the double bond:

$$CH_3CH{=}CH_2 + HCl \rightarrow CH_3\underset{\underset{Cl}{|}}{C}HCH_3$$

or

$$CH_3CH{=}CH_2 + HCl \rightarrow CH_3CH_2CH_2Cl.$$

In practice the first reaction predominates and the product is largely $CH_3\underset{\underset{Cl}{|}}{C}HCH_3$.

When the alkene which reacts with hydrogen chloride has alkyl groups adjacent to the double bond, as is the case here (one methyl group), the hydrogen atom from the hydrogen chloride molecule attaches itself to the carbon atom which already carries the greater number of hydrogen atoms—i.e. the least substituted carbon atom. In general, to indicate the predominant product, we can write:

$$RCH{=}CH_2 + HCl \rightarrow R\underset{\underset{Cl}{|}}{C}HCH_3$$

and

$$R_2C{=}CHR + HCl \rightarrow R_2\underset{\underset{Cl}{|}}{C}CH_2R.$$

This is known as Markownikoff's rule and it applies to other important additions of strong acids to alkenes. For alkenes of general formula $RCH{=}CHR'$ and $R_2C{=}CR'_2$, where R and R' are different alkyl groups, the direction of HCl addition is more difficult to predict but large amounts of both products will probably arise.

Both hydrogen bromide, HBr, and sulphuric acid, H_2SO_4, (strong acids) undergo important addition reactions with alkenes as follows (the reactions are illustrated with ethylene but apply to other alkenes as well):

$$CH_2{=}CH_2 + HBr \rightarrow CH_3CH_2Br \qquad \text{bromoethane (ethyl bromide)}$$

$$CH_2{=}CH_2 + \underset{\text{sulphuric acid}}{\overset{HO}{\underset{HO}{{>}}}S\overset{O}{\underset{O}{{<}}}} \rightarrow \underset{HO}{\overset{CH_3CH_2O}{{>}}}S\overset{O}{\underset{O}{{<}}} \qquad \text{ethyl hydrogen sulphate}$$

The importance of the latter reaction lies in the fact that ethyl hydrogen sulphate is very easily hydrolysed to yield ethanol (ethyl alcohol), an important industrial chemical (the term "hydrolysis" is defined and explained in Section 23.2.1):

$$\underset{HO}{\overset{CH_3CH_2O}{{>}}}S\overset{O}{\underset{O}{{<}}} + H_2O \rightarrow CH_3CH_2OH + \underset{HO}{\overset{HO}{{>}}}S\overset{O}{\underset{O}{{<}}}$$

What amounts to the addition of water across an alkene double bond can be achieved by dissolving the alkene in sulphuric acid and then adding water. Addition of sulphuric acid is followed by hydrolysis to yield an alcohol. Cheap and readily available ethylene can thus be converted, on a large scale, to ethanol:

$$CH_2\!\!=\!\!CH_2 + H_2SO_4 \longrightarrow CH_3CH_2OSO_3H \xrightarrow{H_2O} CH_3CH_2OH + H_2SO_4.$$

In some circumstances, in the presence of an acid catalyst, water will add directly to an alkene to yield an alcohol:

$$\mathrm{\underset{/}{\overset{\backslash}{C}}\!\!=\!\!\underset{\backslash}{\overset{/}{C}}} + H_2O \xrightarrow{H^+} H\!-\!\underset{|}{\overset{|}{C}}\!-\!\underset{|}{\overset{|}{C}}\!-\!OH.$$

Large quantities of ethanol are manufactured industrially from ethylene by both methods.

Halogen molecules themselves will add to alkenes and in the process introduce two functional groups into the molecule:

$$CH_3CH\!\!=\!\!CH_2 + Br_2 \rightarrow CH_3\underset{\underset{Br}{|}}{CH}CH_2Br.$$

Iodine and chlorine behave similarly. Treatment of a compound containing a carbon–carbon double (or triple) bond with a solution of bromine, which is yellow-brown in colour, results in rapid decolorization of the reagent as the bromine is consumed in an addition reaction. No hydrogen bromide is produced (compare the reaction of bromine with an alkane). This makes a very good test for unsaturated compounds for the decolorization can be observed very easily in a test-tube.

At this point it will be expedient to consider a number of reactions of alkenes which do not result in the introduction of functional groups into the molecule. We shall deal with alkynes similarly. In this way, by considering a selection of their reactions, we shall be better able to see the characteristics of these compounds in their correct perspective.

Another important reagent which will add to alkenes is hydrogen gas. In the presence of a suitable catalyst—usually a finely divided transition metal such as nickel, palladium, or platinum—hydrogen molecules split apart and add across the double bond of an alkene, converting it to an alkane:

$$CH_3CH\!\!=\!\!CH_2 + H_2 \xrightarrow{\text{Ni catalyst}} CH_3CH_2CH_3.$$

This is often referred to as catalytic hydrogenation. It is the basis of the process for hardening vegetable oils into edible fats. The oils are esters of long-chain fatty (carboxylic) acids (Sections 23.1 and 23.2) which contain double bonds. On hydrogenation over nickel, these oils become fully saturated, solid, and more butter-like. This is the basis of margarine production. This "synthetic butter" has the advantage that it does not contain those factors hazardous to health (such as cholesterol) which are present in most animal fats. There are other clear advantages in situations where there is a shortage of food. Large quantities of these "synthetic" fats are now consumed.

$$CH_3(CH_2)_7CH\!\!=\!\!CH(CH_2)_7COOR + H_2 \rightarrow CH_3(CH_2)_{16}COOR.$$

An important point is illustrated in this reaction, and that is that the carbon–carbon double bond in this ester behaves in exactly the same way as the double bond in the propene which we have just considered. In short, we may note the generalization that a multiple bond (or other functional groups for that matter) behaves in essentially the same way whatever other groups are present in the molecule. The behaviour may be modified if the other groups are suitably situated in the molecule—sometimes very greatly—but by and large their behaviour is based on the behaviour of the group when it is the only group in a simple molecule. It is for this reason that we shall be principally concerned in this book with simple molecules containing only one functional group. We shall meet some examples, however, of molecules with more than one functional group in which one group influences the properties of the other (Chapter 28).

Alkenes undergo a very useful addition reaction with ozone to form ozonides. The reaction is complex and has been difficult to study on account of the explosive nature of the ozonides produced. Undoubtedly, the carbon–carbon double bond is not completely broken in the initial reaction between alkene and ozone, but in the final ozonide it is. The ozonides are usually assigned a cyclic structure:

$$RCH{=}CHR' + O_3 \xrightarrow{\text{several steps}} RCH\underset{O-O}{\overset{O}{\diagup\diagdown}}CHR'.$$

alkene ozone ozonide

The importance of this reaction arises from the ready hydrolysis of ozonides to give carbonyl compounds (Section 23.3) which can be easily identified. Under reducing conditions (say in the presence of zinc and acetic acid), which prevent any aldehyde fragments from being oxidized to carboxylic acids (Section 23.3), the ozonide breaks up to give aldehydes or ketones according to the structure of the original alkene:

$$RCH{=}CHR' \rightarrow R{-}CH\underset{O-O}{\overset{O}{\diagup\diagdown}}CHR' \xrightarrow{H_2O} RCHO + R'CHO + H_2O_2;$$

aldehyde aldehyde

$$\underset{R'}{\overset{R}{\diagdown}}C{=}CH_2 \rightarrow \underset{R'}{\overset{R}{\diagdown}}C\underset{O-O}{\overset{O}{\diagup\diagdown}}CH_2 \rightarrow RCOR' + CH_2O + H_2O_2. \quad (H_2O_2 \xrightarrow{\text{rapid reduction}} H_2O.)$$

ketone formaldehyde

Identification of the aldehyde/ketone fragments enables the alkene structure to be deduced. Thus ozonolysis of an unknown alkene, C_5H_{10}, to ethyl methyl ketone and formaldehyde ($CH_3COCH_2CH_3$ and CH_2O) leads to the conclusion that the original alkene had the structure:

$$\underset{CH_3CH_2}{\overset{CH_3}{\diagdown}}C{=}CH_2.$$

One knows that the carbonyl fragments were linked by a double bond between their carbonyl carbon atoms. This process of ozonolysis is a powerful structural tool and has been extensively employed in structure determinations.

The final addition reaction of alkenes which we shall consider is that in which the unsaturated molecules add to themselves to form very long chain molecules. When a gas like ethylene is subjected to high pressures and temperatures and is brought into contact with the right sort of catalysts, a very involatile, tough material is produced which contains carbon and hydrogen only and has a very high molecular weight. It is formed by the addition of many ethylene molecules, one to the other, to form long chains of CH_2 units:

$CH_2{=}CH_2 \quad CH_2{=}CH_2 \quad CH_2{=}CH_2 \quad CH_2{=}CH_2 \quad CH_2{=}CH_2 \quad CH_2{=}CH_2$
$\downarrow$
$—CH_2—CH_2—CH_2—CH_2—CH_2—CH_2—CH_2—CH_2—CH_2—CH_2—CH_2—CH_2—$

At the very ends of the chains a hydrogen atom may be gained or lost so that the chains terminate as $CH_2—CH_2—CH{=}CH_2$ or as $—CH_2—CH_2—CH_2—CH_3$. The chains are so long, however, that these terminal units do not significantly alter the formula of the material from $(CH_2)_n$. Molecular weights of these chains may be as high as several hundreds of thousands. Any individual sample of the material will contain a variety of non-identical chains so that in that sense it is not a pure compound. This process of linking together many similar simple units (or "bricks") to form larger, more complex molecules is known as **polymerization**, the product being known as a **polymer** and the simple units as **monomers**.

It is the length of these chains which enables the molecules to bind together in the bulk of the material which they go to make up, and to lend it the strength which we associate with modern plastics and certain naturally occurring tough materials (wood, leather, muscle fibre, and so on). In certain materials the strength is increased by "cross linking" between individual molecular chains. Polyethylene, or polythene, which is its usual abbreviation, is the name of the material derived from ethylene. By varying the polymerization conditions, the amount of chain branching and the length of the chains can be varied; hence we are familiar today with a whole range of polythenes with somewhat differing physical properties. Low-density polythene (familiar as washing up bowls, toys, etc.) melts at about 105 °C and has highly branched molecules, whereas high density polythene has straight-chain molecules and melts at about 135 °C. The straight chains can pack more closely together and more regularly than can the branched-chain molecules, and the van der Waals forces can be expected to be greater therefore. By using variously substituted ethylenes a much larger range of polyalkene plastics can be made. Table 22.3 lists a number of alkene monomers which, on polymerization, give the familiar plastics indicated. These vinyl polymers are a large family, although they represent only a small proportion of the vast range of polymeric materials—man-made or naturally occurring—which is made use of in modern materials.

Table 22.3

Monomer			
Systematic name	Common name	Formula	Polymer
chloroethene	vinyl chloride	$CH_2{=}CHCl$	polyvinyl chloride, PVC
ethenyl ethanoate	vinyl acetate	$CH_2{=}CHOCOCH_3$	polyvinyl acetate
phenylethene	styrene	$CH_2{=}CHC_6H_5$	polystyrene
propene	propylene	$CH_2{=}CH—CH_3$	polypropylene
tetrafluoroethene	tetrafluoroethylene	$F_2C{=}CF_2$	Teflon (polytetra-fluoroethylene)

Polymers of dienes (such as buta-1,3-diene, $CH_2=CH-CH=CH_2$) still contain double bonds and, due to the restriction of rotation about these bonds (Section 21.4), the molecules tend to be awkward and "rigidly angular" near them.

$$nCH_2=CH-CH=CH_2 \rightarrow (CH_2-CH=CH-CH_2)_n.$$

Free rotation about single bonds allows a saturated polymer molecule to adopt pretty well any convenient overall shape (Section 21.1). The diene polymers are therefore inefficiently packed together and the van der Waals forces are low. When subjected to tension, the molecules are pulled more nearly straight, but spring back again when released. They are elastic. Such polymers are called **elastomers** and include buna-rubber, a synthetic polymer of 2,3-dimethylbuta-1,3-diene [$CH_2=C(CH_3)C(CH_3)=CH_2$], and polychloropropene (neoprene rubber), a synthetic polymer of 2-chlorobuta-1,3-diene [$CH_2=C(Cl)CH=CH_2$, chloroprene]. Natural rubber is a polymer of 2-methylbuta-1,3-diene, better known as isoprene [$CH_2=C(CH_3)CH=CH_2$]. The five-carbon atom skeleton of isoprene has in fact been identified as the common "building brick" of a large number of naturally occurring biological molecules. Two examples of such molecules are shown with the isoprene units picked out in Fig. 22.1. The emphasis here is simply on the skeletal structure, as double bonds and other functional groups can quite "easily" come and go in biological systems.

22.2.5 Reactions of Alkynes

The alkynes also undergo addition reactions which are very similar to those of the alkenes. Functional groups may therefore be introduced into the molecules by means of these reactions. The same set of acids as for the alkene series will add across the triple bond. So will halogen molecules.

An interesting point arises in this connection because a triple bond is usually saturated in two stages. First it is converted into a double bond and then to a single bond. Thus, with hydrogen chloride, acetylene reacts in two stages:

$$CH \equiv CH + HCl \rightarrow CH_2=CHCl;$$

$$CH_2=CHCl + HCl \rightarrow CH_3CHCl_2.$$

The interesting question is whether the reaction can be stopped after only the first stage has taken place so that the reaction can be used to prepare substituted alkenes. It turns out that, with suitable control of reactant quantities and reaction conditions, this can be done, and the reaction we have just considered is used industrially to produce chloroethylene, $CH_2=CHCl$, better known perhaps as

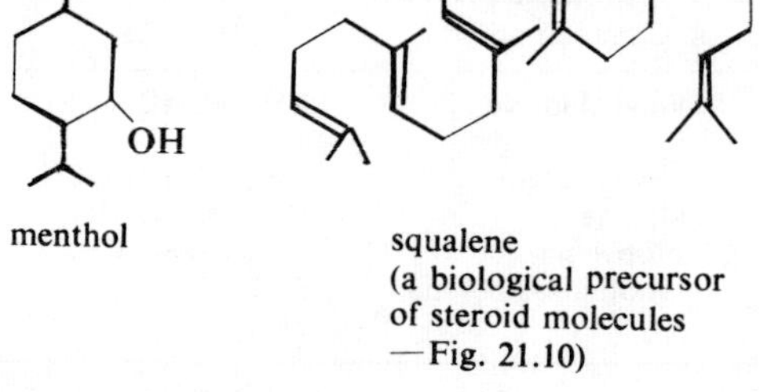

Fig. 22.1 Biological molecules "built" up from isoprene units. The isoprene units are picked out with heavy lines. (Only carbon atoms are shown in these diagrams—either at the angles, or at the ends of bonds.)

menthol

squalene
(a biological precursor of steroid molecules
—Fig. 21.10)

vinyl chloride ("vinyl" is to $CH_2=CH-$ as "ethyl" is to CH_3CH_2-). Vinyl chloride is the raw material for the production of the very important plastic polyvinyl chloride or PVC (Table 22.3), familiar as an insulating material in the electrical industry (cable sheaths) and as a waterproof clothing fabric.

We have seen how treatment of ethylene with sulphuric acid, followed by water, results in the production of ethanol. This two-reaction process can be represented by the use of a single equation which indicates that, essentially, the elements of water have been added across the double bond:

$$CH_2=CH_2 + H_2O \rightarrow CH_3CH_2OH.$$

Treatment of acetylene with aqueous sulphuric acid, in the presence of a mercury(II) salt which catalyses the reaction, results in the direct addition of water across the triple bond. The product we should expect after the addition of only one molecule of water, assuming that the reaction goes in the same way as it does for ethylene, would be vinyl alcohol (ethenol), $CH_2=CH-OH$. This molecule is very unstable however (unlike ethanol), and immediately rearranges its constituent atoms to form acetaldehyde (ethanal), $CH_3\overset{\overset{H}{|}}{C}=O$. Note that the change, from a structural point of view, is very slight. The only thing that has happened is that one hydrogen atom has migrated from oxygen to carbon and the double bond has moved one place along. The basic skeleton of the molecule, C—C—O, is unchanged. In fact we believe that the two structures are in equilibrium with each other, although in this case (as with most simple aldehydes and ketones) there is too little of the enol form present at equilibrium to be directly detected.

$$\underset{\text{enol form}}{CH_2=CH-OH} \rightleftharpoons CH_3CHO.$$

The term "enol" (Section 28.3) is composed of those word-endings characterizing carbon–carbon double bonds (*ene*) and alcohols (*ol*). This phenomenon, involving reversible migration of a proton to give two very similar structural isomers in equilibrium with each other, is known as **tautomerism**. It is more fully discussed in Section 28.3. In this particular case, the result is that the addition of a second molecule of water across what was originally the triple bond is prevented, and we have:

$$HC\equiv CH + H_2O \xrightarrow[\text{HgSO}_4]{\text{H}_2\text{SO}_4} CH_3CHO \quad \text{(acetaldehyde)}.$$

This, too, is an important industrial process, and the economic importance of additions to simple hydrocarbons containing multiple bonds cannot be overemphasized.

The reaction of chlorine with acetylene proceeds with explosive violence, but with careful control of the reaction conditions tetrachloroethane (acetylene tetrachloride) can be produced. This is an important solvent. The reaction with bromine in aqueous solution proceeds more gently, and alkynes will reveal their unsaturated character quite conveniently, by decolorizing bromine solutions.

Hydrogen (with a suitable metal catalyst again), will add across the triple bond and produce a saturated molecule. Hydrogenation, for example, converts acetylene to ethane in the same way as it converts ethylene to ethane:

$$CH\equiv CH + 2H_2 \rightarrow CH_3CH_3.$$

This reaction serves to prove that both ethylene and acetylene have the same two-carbon atom skeleton as ethane. Note also that whereas one molecule of ethylene absorbs one molecule of hydrogen, one molecule of acetylene absorbs two of hydrogen. For each mole of hydrogen that a mole of compound will absorb on catalytic hydrogenation we ascribe to it one "degree of unsaturation". Thus ethylene has one degree of unsaturation and acetylene two. Buta-1,3-diene, $CH_2{=}CH{-}CH{=}CH_2$, having two double bonds, will, like acetylene, absorb two molecules of hydrogen and therefore has two degrees of unsaturation too.

The number of degrees of unsaturation of an unknown compound can be measured quite easily by measuring the amount of hydrogen one mole will absorb on catalytic hydrogenation. This means of course that the molecular weight must also be determined. This experimental measure of unsaturation is most useful in determining how many double or triple bonds a compound contains, although, as we have seen above in connection with acetylene and buta-1,3-diene, it cannot distinguish between one triple bond and two double bonds.

Like alkenes, alkynes show a tendency to polymerize, although they have not made available a wide and useful range of plastics as the alkenes have. Compounds of certain transition metals, or boron trifluoride, will catalyse the trimerization of alkynes to give benzene derivatives:

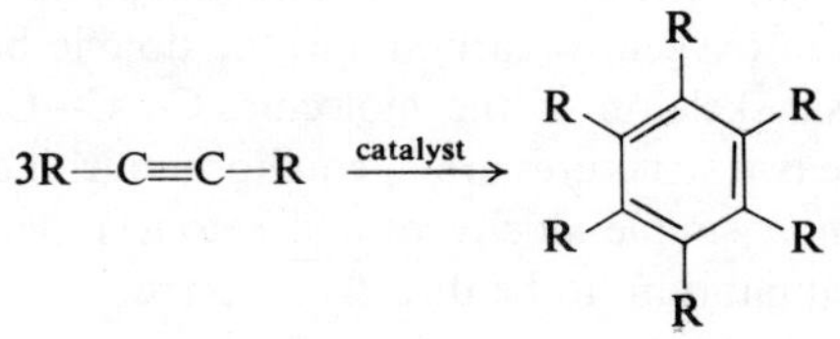

$$3R{-}C{\equiv}C{-}R \xrightarrow{\text{catalyst}} \text{hexasubstituted benzene}$$

One great point of difference between alkenes and alkynes is the relatively high acidity of the acetylenic hydrogen atom, ${\equiv}C{-}H$. Clearly this applies only to acetylene itself and to monosubstituted alkynes, $RC{\equiv}CH$. Such molecules, in which the triple bond extends to the terminal carbon atom, are known as *terminal alkynes*. *Non-terminal alkynes*, such as $R{-}C{\equiv}C{-}R'$, contain no such hydrogen atom. Alkanes and alkenes are inert towards bases. Sodamide, $NaNH_2$, which is a very strong base, although having no effect on alkanes and alkenes, will remove a proton (a hydrogen ion, H^+) from a terminal alkyne to form a salt, $Na^+RC{\equiv}C^-$. The reaction with acetylene can be written:

$$NaNH_2 + HC{\equiv}CH \rightarrow NaC{\equiv}CH + NH_3.$$

These salts of alkynes—acetylides—are useful, largely because they afford a valuable means of synthesizing carbon–carbon bonds. They will react with halogenoalkanes, for example, with the acetylide ion displacing a halide ion:

$$CH_3C{\equiv}CNa + CH_3CH_2Br \rightarrow CH_3CH_2C{\equiv}CCH_3 + NaBr.$$

Treatment of them with water or mineral acids regenerates the free alkyne:

$$NaC{\equiv}CH + H_2O \rightarrow HC{\equiv}CH + NaOH;$$
$$Cu_2C_2 + 2HCl \rightarrow Cu_2Cl_2 + C_2H_2.$$

These compounds and their uses are discussed at somewhat greater length in Chapter 27 (see also Section 16.5).

Metal acetylides can also be prepared by means of simple double decomposition reactions. Passing acetylene gas through ammoniacal copper(I) chloride, for example, produces an immediate red precipitate of copper(I) acetylide. Silver acetylide (white) may be similarly precipitated from ammoniacal silver nitrate by acetylene gas:

$$Cu_2Cl_2 + C_2H_2 \rightarrow Cu_2C_2 + 2HCl;$$
$$2AgNO_3 + C_2H_2 \rightarrow Ag_2C_2 + 2HNO_3.$$

Note that these are reversible reactions as quoted. In the previous paragraph, excess hydrochloric acid and removal of gaseous acetylene ensure that the copper(I) reaction proceeds in the reverse direction. Here excess acetylene and the low solubility of copper(I) acetylide ensure that the latter is the product.

The effect of strongly heating lime and coke together is to produce calcium carbide, CaC_2 (better termed calcium(II) dicarbide or calcium acetylide). Water releases acetylene gas from this readily available compound and this forms the most convenient synthesis of acetylene industrially or in the laboratory:

$$CaC_2 + 2H_2O \rightarrow HC\equiv CH + Ca(OH)_2.$$

The old acetylene lamps, once used on bicycles etc., were charged with calcium(II) dicarbide and water and the acetylene gas produced was burnt for illumination.

In connection with the combustion of hydrocarbons, the great violence with which acetylene–air mixtures exploded was referred to. In fact alkynes are dangerous compounds to handle and store, as they show a marked tendency to explode vio-

Table 22.4. Some important addition reactions of alkenes and alkynes. (Hal = halogen)

	Reagent	Product
Alkenes $\diagup_{C=C}\diagdown$	$H_2 \rightarrow H-C-C-H$	alkanes
	$HHal \rightarrow H-C-C-Hal$	halogenoalkanes
	$Hal_2 \rightarrow Hal-C-C-Hal$	dihalogenoalkanes
	$H_2SO_4 \rightarrow H-C-C-O-SO_3H$	alkyl hydrogen sulphates
	$H_2O \rightarrow H-C-C-OH$	alcohols
	alkene (polymerization) $\rightarrow \left[\begin{array}{c} -C-C- \end{array} \right]_n$	polyalkene
Alkynes $-C\equiv C-$	behave similarly. One mole of reagent may give the corresponding compound containing a double bond; two moles give the fully saturated product. Water (one mole) gives an aldehyde or ketone: $-C\equiv C- + H_2O \rightarrow -CH_2-\overset{O}{\underset{\|}{C}}-.$	

lently whether mixed with an appropriate amount of air or not. The heavy metal acetylides (for example, of copper(I) and silver) are particularly dangerous and will explode, if dry, on warming or on being struck.

22.2.6 Reactions of Aromatic Hydrocarbons: Benzene

The reactions which benzene undergoes and which are useful for the introduction of functional groups are very different in character from those of alkanes, alkenes, and alkynes. Substitution reactions occur but under very different conditions to those effective for the alkanes. Sometimes a catalyst is required but the important difference is that the conditions of heat and light required for the free radical reactions of the alkanes are not required here. The reactions are of a completely different character and are discussed more fully in Section 26.5. Unlike the alkenes and alkynes, benzene does not undergo addition reactions very easily. Rather, it undergoes substitution, thus preserving its aromatic character (Section 26.5). Under certain conditions halogens can be made to add to benzene to produce, for example, 1,2,3,4,5,6-hexachlorocyclohexane (benzene hexachloride, BHC, a well-known insecticide):

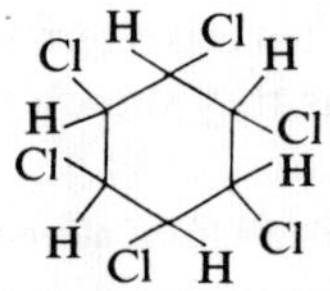

The ring can also be catalytically hydrogenated, under forcing conditions, to give cyclohexane:

$$C_6H_6 + 3H_2 \rightarrow C_6H_{12}.$$

Some of the important reactions of benzene are as follows:

Halogenation

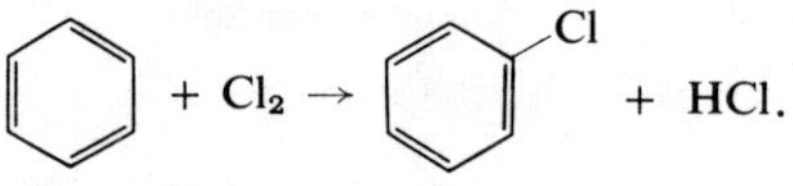

$$C_6H_6 + Cl_2 \rightarrow C_6H_5Cl \text{ (chlorobenzene)} + HCl.$$

$$C_6H_6 + Br_2 \rightarrow C_6H_5Br \text{ (bromobenzene)} + HBr.$$

Nitration

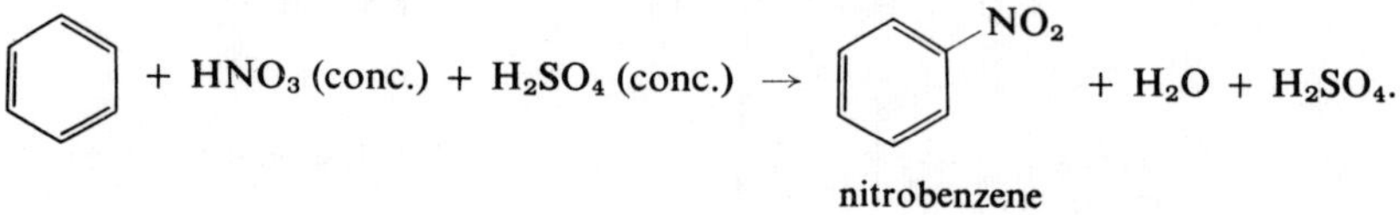

$$C_6H_6 + HNO_3 \text{ (conc.)} + H_2SO_4 \text{ (conc.)} \rightarrow C_6H_5NO_2 \text{ (nitrobenzene)} + H_2O + H_2SO_4.$$

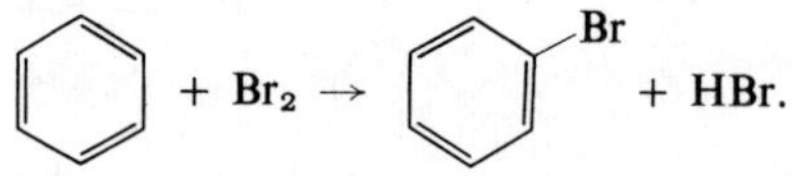

The concentrated sulphuric acid is a catalyst for this reaction.

Friedel–Crafts reaction

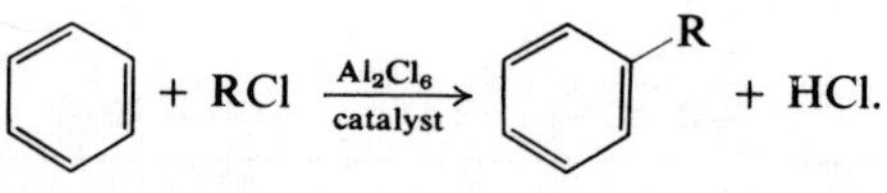

$$C_6H_6 + RCl \xrightarrow[\text{catalyst}]{Al_2Cl_6} C_6H_5R + HCl.$$

RCl may be a halogenoalkane (alkyl halide) or an acid chloride; for example:

$$C_6H_6 + CH_3Cl \xrightarrow[\text{catalyst}]{Al_2Cl_6} C_6H_5CH_3;$$
(chloromethane) → methylbenzene (toluene)

$$C_6H_6 + CH_3COCl \longrightarrow C_6H_5COCH_3.$$
(acetyl chloride) → methyl phenyl ketone (acetophenone)

Sulphonation

C$_6$H$_6$ + fuming H$_2$SO$_4$ (essentially SO$_3$) → C$_6$H$_5$SO$_3$H (benzenesulphonic acid) + H$_2$O

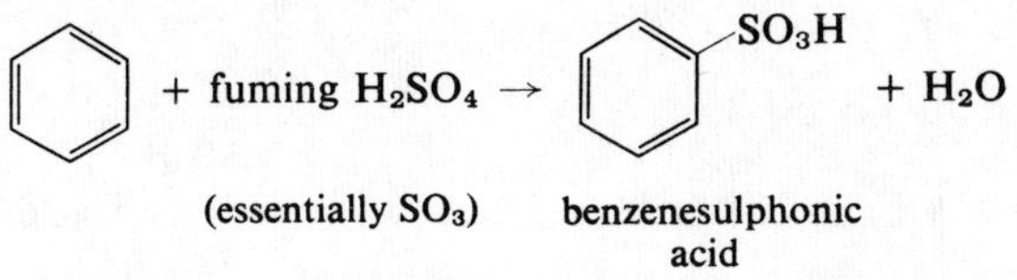

Treatment of benzenesulphonic acid with molten sodium hydroxide cleaves the carbon–sulphur bond and replaces the SO$_3$H group with OH:

$$C_6H_5SO_3H + NaOH \rightarrow C_6H_5OH + Na_2SO_3.$$

The resulting hydroxybenzene is known as phenol, and in combination with this reaction, sulphonation is an important method of introducing hydroxyl groups into the benzene ring. Other groups may be introduced similarly.

One of the remarkable features of many of the above reactions is the ability of the benzene ring itself to remain intact even in the presence of such powerful oxidants as hot concentrated nitric and sulphuric acids. Many aliphatic compounds lacking the peculiar stability (associated with electron delocalization, Section 4.12) of benzene and other aromatic molecules would not survive. Alkylbenzenes, and benzenes with other aliphatic substituent groups, are oxidized by powerful oxidants, such as permanganate, to the corresponding aromatic carboxylic acid. The carbon of the carboxylate group, being at its highest oxidation level short of carbon dioxide, is the sole survivor of the aliphatic part of the molecule:

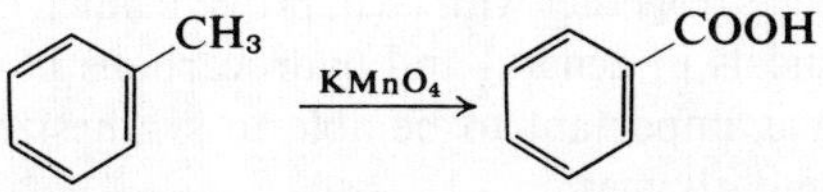

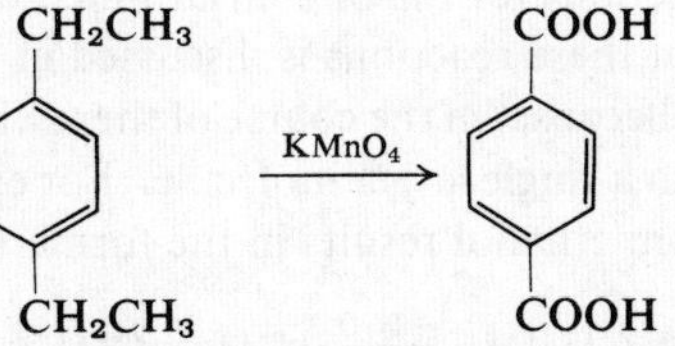

C$_6$H$_5$CH$_3$ $\xrightarrow{KMnO_4}$ C$_6$H$_5$COOH

1,4-(CH$_3$CH$_2$)$_2$C$_6$H$_4$ $\xrightarrow{KMnO_4}$ 1,4-(COOH)$_2$C$_6$H$_4$

This feature is of use in structure determination in that identification of the aromatic acid, remaining after oxidation of a compound containing aliphatic side chains, enables the positions of the side chains to be determined. The identification of benzene-1,4-dicarboxylic acid (terephthalic acid) on oxidation of diethylbenzene, for instance, shows that the side chains of the dialkylbenzene were at positions 1 and 4.

Table 22.5. Important substitution reactions of alkanes and arenes. The character of the alkane reactions is quite different from those of the arene reactions. (Hal = Cl or Br.)

- C$_6$H$_6$ + Hal$_2$ → C$_6$H$_5$Hal — halogenation
- C$_6$H$_6$ + HNO$_3$/H$_2$SO$_4$ (NO$_2^+$) → C$_6$H$_5$NO$_2$ — nitration
- C$_6$H$_6$ + SO$_3$ → C$_6$H$_5$SO$_3$H — sulphonation
- Friedel–Crafts reactions:
  - C$_6$H$_6$ + RHal/Al$_2$Hal$_6$ → C$_6$H$_5$R — alkylation
  - C$_6$H$_6$ + RCOHal/Al$_2$Hal$_6$ → C$_6$H$_5$COR — acylation
- alkane: —C—H + Hal$_2$ (light, heat) → —C—Hal + HHal — halogenation (halogenoalkane)

22.3 IMPORTANT METHODS OF HYDROCARBON SYNTHESIS

Natural petroleum and gas, together with coal, provide much of the world's requirements of organic chemicals in general and hydrocarbons in particular. The only hydrocarbons which it is important to be able to synthesize are the unsaturated ones, that is, alkenes and alkynes.

The most important reactions for making unsaturated compounds are **elimination reactions**. The nature of these reactions is discussed in Section 26.6. They are called elimination reactions because, in the course of them, the elements of a smaller molecule are eliminated from a single larger molecule. For example, the elimination of the elements of water from ethanol results in the formation of ethylene:

$$CH_3CH_2OH \xrightarrow{-H_2O} CH_2{=}CH_2.$$

This reaction is catalysed by strong acids, and the treatment of ethanol with concentrated sulphuric acid at a high temperature (170 °C) is an important laboratory preparation of ethylene. In fact this is an important general method for the laboratory preparation of alkenes.

Two vital elimination reactions, responsible for the industrial preparation of alkenes, are "cracking" reactions. Ethane, at temperatures of over 800 °C, "cracks" to ethylene and hydrogen. Propane, in a similar reaction, yields propene and hydrogen. In a different reaction, involving the elimination of methane, propane also yields ethylene.

$$CH_3CH_3 \rightarrow CH_2{=}CH_2 + H_2$$
$$CH_3CH_2CH_3 \rightarrow CH_3CH{=}CH_2 + H_2$$
$$CH_3CH_2CH_3 \rightarrow CH_2{=}CH_2 + CH_4.$$

Massive quantities of alkenes are produced in this way in North America where natural gas is rich in ethane and propane ("wet"). European natural gas consists mainly of methane (is "dry") and is virtually useless for ethylene production. Cracking of petroleum fractions containing ethane and higher alkanes also yields alkenes, as well as aromatic hydrocarbons, on an industrial scale.

The elimination of the elements of hydrogen chloride, hydrogen bromide, or hydrogen iodide from halogenoalkanes, by treatment with alkali, is another important reaction for the production of double and triple bonds. It is rather less useful than acid-catalysed water elimination, however, as certain halogenoalkanes give very poor yields of alkene. Halogenoethanes (ethyl halides), for example, give very little ethylene.

The use of high concentrations of alkali at high temperatures in non-polar solvents (commonly ethanol) favours the elimination reaction at the expense of the competing, alternative hydrolysis reaction (Section 26.6).

The structure of the halogenoalkane (alkyl halide) is also an important factor in this competition. Thus those containing simple primary alkyl groups, which would give simple alkenes, usually give very little alkene whereas secondary and tertiary halides, which give highly branched alkenes, give good yields. (The terms primary, secondary, and tertiary are explained in Section 23.1.2 in connection with alcohols.)

$$C_2H_5Br \xrightarrow[\text{ethanol}]{\text{KOH}} C_2H_4 \qquad \text{very little;}$$
bromoethane ethylene
(ethyl bromide)
(a primary alkyl halide)

$$CH_3CH_2\underset{\underset{Br}{|}}{C}HCH_3 \xrightarrow[\text{ethanol}]{\text{KOH}} CH_3CH{=}CHCH_3 \qquad \text{good yield;}$$
2-bromobutane but-2-ene
(a secondary alkyl halide)

$$(CH_3)_3CBr \xrightarrow[\text{ethanol}]{\text{KOH}} (CH_3)_2C{=}CH_2 \qquad \text{very good yield.}$$
2-bromo-2-methylpropane 2-methylpropene
(tertiary-butyl bromide)
(a tertiary alkyl halide)

Note that in the case of 2-bromobutane, the major product is but-2-ene. The alternative possible elimination product, but-1-ene, $CH_2{=}CHCH_2CH_3$, is formed only in small amount as the reaction favours the more highly branched but-2-ene.

Treatment of a dihalogenoalkane with alcoholic potassium hydroxide may lead to an alkyne, although, again, yields are not high in the case of the simple compounds such as acetylene itself. The best way of preparing acetylene is from calcium(II) dicarbide and water (Section 22.2.5):

$$\underset{\text{1,2-dibromoethane}}{BrCH_2CH_2Br} \xrightarrow[\text{alcohol}]{KOH} \underset{\text{acetylene}}{CH\equiv CH}.$$

A very important method of synthesizing alkanes, which is not an elimination reaction, is the Wittig reaction. This is a condensation reaction and involves the use of a phosphorus compound and a carbonyl compound:

$$\underset{\substack{\text{aldehyde}\\\text{or ketone}}}{\overset{R'}{\underset{R''}{\diagdown}}C=O} + \underset{\substack{\text{Wittig}\\\text{reagent}}}{Ph_3P=CHR'''} \rightarrow \underset{\text{alkene}}{\overset{R'}{\underset{R''}{\diagdown}}C=CHR'''} + \underset{\substack{\text{triphenylphosphine}\\\text{oxide}}}{Ph_3PO}.$$

For example:

$$\underset{\text{benzaldehyde}}{PhCHO} + \underset{\substack{\text{methylene}\\\text{triphenyl-}\\\text{phosphorane}}}{Ph_3P=CH_2} \rightarrow \underset{\substack{\text{(styrene)}\\\text{(phenylethene)}}}{PhCH=CH_2} + Ph_3PO.$$

Table 22.6. Important alkene synthesizing reactions. All but the lower reaction (Wittig reaction) are elimination reactions and the molecules eliminated are shown over the arrows.

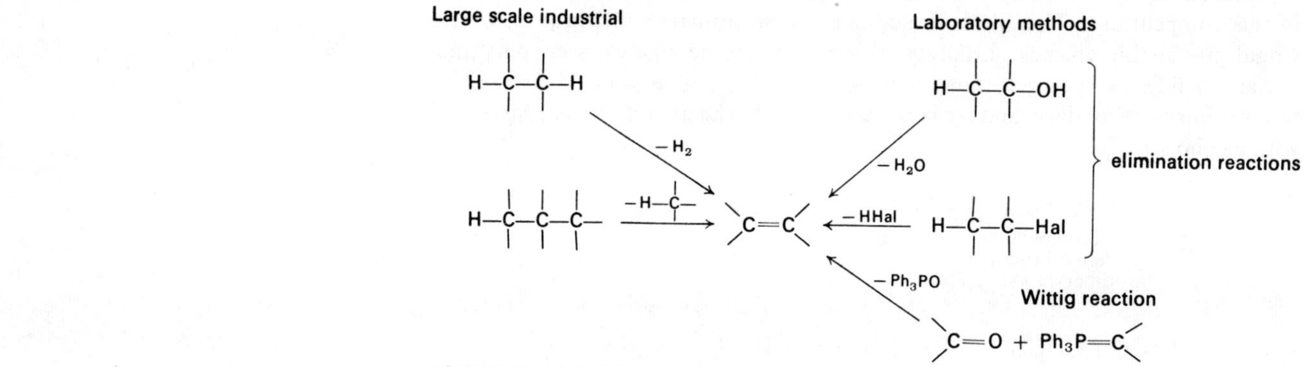

Large scale industrial   **Laboratory methods**

[Reaction scheme showing various elimination reactions converging to C=C, with arrows labeled $-H_2$, $-H-C-$, $-H_2O$, $-HHal$, $-Ph_3PO$, from starting materials H-C-C-H, H-C-C-C-, H-C-C-OH, H-C-C-Hal, and C=O + Ph_3P=C (Wittig reaction).]

elimination reactions

Wittig reaction

PROBLEMS

1. Suggest simple, chemical tests for distinguishing one compound from the other in the following pairs of compounds:

   a) $CH_3CH_2CH_2C\equiv CH$ and $CH_3CH_2C\equiv CCH_3$;

   b) $CH_3CH_2CH=CHCH_2CH_3$ and ⬡ (cyclohexane);

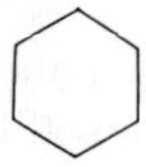

   c)  ⬡ (benzene) and ⬡ (cyclohexane);

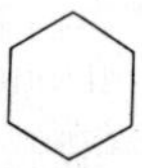

d) 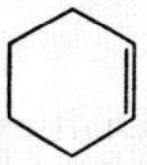 ⬡ (cyclohexene) and $CH_3CH_2C\equiv CCH_2CH_3$;

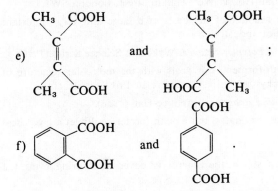

2. Write down the structures for all of the possible products from the chlorination of ethane. Write down an equation expressing the stoichiometry of a reaction between ethane and chlorine (both gaseous) to give hexachloroethane (C_2Cl_6).

3. Study the following list of compounds carefully, and answer the questions set out below:

   i) $C_6H_5C\equiv CH$;

   ii) $C_6H_5CH=CH_2$;

   iii) $CH_3C\equiv CCH_2CH_3$;

   iv) CH_3CH_3;

   v) C_6H_6;

   vi) $C_6H_5CH_3$;

   vii) ⬡ (cyclohexene).

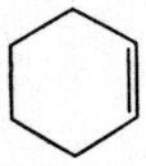

   a) Which of these compounds would decolorize a solution of bromine in tetrachloromethane (carbon tetrachloride) instantly, with no evolution of hydrogen bromide? (Bromine is brown, and gives a yellow-brown solution in tetrachloromethane which is a common, colourless solvent.)
   b) Which compounds would react with chlorine *only* under the influence of light or heat, evolving hydrogen chloride?
   c) Which compound would give a white precipitate with ammoniacal silver nitrate solution?
   d) Which compound would react with CH_3COCl and Al_2Cl_6 (catalyst) to give $C_6H_5COCH_3$?
   e) Which compound would give the aldehyde $\underset{CHO}{\overset{CHO}{|}}(CH_2)_4|$ on ozonolysis?
   f) Which compound would give an alkane on reaction with hydrogen over a nickel catalyst?
   g) Which compound is fully saturated?

BIBLIOGRAPHY

Petroleum Chemicals, available from Education Section, Information Division, BP Chemicals (UK) Ltd., West Halkin House, West Halkin Street, London SW1.

> A clear and informative account of the basis of a large and very important part of the organic chemical industry.

Treloar, L. R. G. *Introduction to Polymer Science*, Wykeham Science Series (1970).

> Primarily written for sixth-formers, this deals with the molecular structure of polymers and how their physical properties are related to their structure.

Stock, L. M. *Aromatic Substitution Reactions*, Prentice-Hall (1968).

> Primarily written for undergraduates, but a sixth-former could get a great deal from this.

All of the books on organic chemistry in the general list cover, at differing levels, the material of this chapter.

CHAPTER 23

Compounds with Functional Groups containing Oxygen

There are three main functional groups containing oxygen which we shall consider in this chapter: the *hydroxyl group*, —OH, found in alcohols, phenols, and carboxylic acids; the *carbon–oxygen–carbon linkage*, found in ethers, carboxylic esters, and carboxylic acid anhydrides; and the *carbonyl group*, $-\overset{\overset{\displaystyle O}{\|}}{C}-$, found in aldehydes, ketones, carboxylic acids, and derivatives of carboxylic acids. The types of compound we shall be considering are illustrated in Table 23.1 where it will be seen that some of the compounds contain two of these groups and some contain a second group which is not based on oxygen. All of these compounds with two functional groups are related to, and derived from, the carboxylic acids, RCOOH, and the two groups in each of these types of compound are so close together that they affect one another profoundly. So much are the properties of the one group modified by the second that the combinations of groups, —COOH, $CONH_2$, and COCl are often considered as single functional groups in their own right. It will, however, be instructive to consider them here as being made up of two groups which are profoundly affecting one another.

23.1 The hydroxyl group
23.2 Carbon–oxygen–carbon linkages
23.3 The carbonyl group

23.1 THE HYDROXYL GROUP

As with all the functional groups we shall examine, we shall study here the properties of hydroxyl groups attached to different simple skeletons. The nature of the skeleton to which a group is attached may well modify the properties of the group, and it is important to know how the properties of groups vary when they are attached to the various different skeletons. Thus, we shall study the compounds:

1. R—OH, alcohols (primary, secondary, and tertiary). R is an alkyl group (primary, secondary, or tertiary).
2. Ar—OH, phenols. Ar is an aryl group e.g. C_6H_5—, derived from benzene.
3. $R-\overset{\overset{\displaystyle O}{\|}}{C}-OH$ and $Ar-\overset{\overset{\displaystyle O}{\|}}{C}-OH$, carboxylic acids. The alkyl (aliphatic) compounds are very similar to the aromatic ones, and we shall not be concerned here with their differences.

Table 23.1. Functional groups based on oxygen.

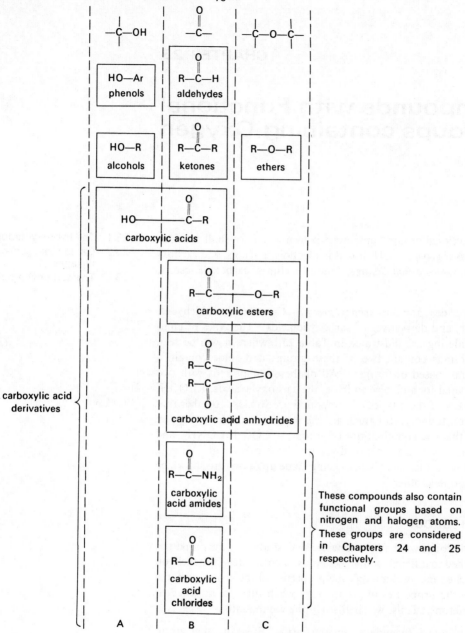

R may be replaced by an aromatic group **Ar**, in any of these compounds except alcohols.

Nomenclature of hydroxy compounds depends upon the type of compound:

Alcohols. The IUPAC name for an alcohol is derived from that of the parent alkane by removing the terminal *e* and adding the ending *-ol*. Thus ethanol, CH_3CH_2OH, is derived from ethane. Propan-1-ol, $CH_3CH_2CH_2OH$, and propan-2-ol, $CH_3CH(OH)CH_3$ are derived from propane, with the numerals indicating the position of the —OH group. Older names still commonly employed include iso-propanol, for propan-2-ol, and tertiary-butanol for 2-methylpropan-2-ol, $(CH_3)_3COH$. In this book we shall use the IUPAC names throughout, but these two older names are used sufficiently often in books and articles on chemistry, that they are worth knowing.

Phenols. The only really important phenol at this level is phenol itself, C_6H_5OH, which gives its name to this whole class. The name derives from an early name, phene, once used for benzene and still firmly entrenched in chemical nomenclature. The group C_6H_5—, the phenyl group (sometimes abbreviated as Ph), also derives its name from phene.

Carboxylic acids. The IUPAC name is derived from the parent hydrocarbon again, which includes the carboxyl group (COOH) carbon atom. CH_3COOH is thus ethanoic acid, and CH_3CH_2COOH propanoic acid. When the chain is numbered, the carboxyl carbon is numbered 1, so that 1-methylpropanoic acid is $(CH_3)_2CHCOOH$. The trivial names "acetic acid" and "formic acid" (ethanoic and methanoic acids, respectively), however, are still in normal use, and they will therefore be used in this book instead of the IUPAC names. IUPAC names will, however, be used for all other non-cyclic acids. Cyclic compounds are named differently, as the carboxylic acid group clearly cannot be part of a ring. Cyclo-$C_6H_{11}COOH$ is named as cyclohexanecarboxylic acid, and C_6H_5COOH as benzenecarboxylic acid. The latter is more commonly called benzoic acid, and this is therefore the name used in this book. Salts and esters are named as alkanoates, or carboxylates, such as sodium propanoate (CH_3CH_2COONa), and ethyl cyclohexanecarboxylate ($C_6H_{11}COOCH_2CH_3$).

23.1.1 Physical Properties

The boiling points of all these compounds are very much higher than one would expect solely on the basis of their molecular weights. They are very much like water in this respect, for they are all capable of forming hydrogen bonds between the hydroxyl hydrogen atom of one molecule and the oxygen atom of another (Section 4.8 and Fig. 23.1). The molecules therefore tend to "stick" together and are more difficult to separate and drive off into the vapour phase than would be corresponding molecules of similar weight which cannot form hydrogen bonds. Thus, for example, butan-1-ol ($CH_3CH_2CH_2CH_2OH$) boils at 118 °C, whereas diethyl ether (ethoxyethane, $CH_3CH_2OCH_2CH_3$), a structural isomer, boils at 35 °C. Molecules of the latter compound are unable to form hydrogen bonds to their neighbours because hydrocarbon type hydrogen atoms are incapable of forming such bonds.

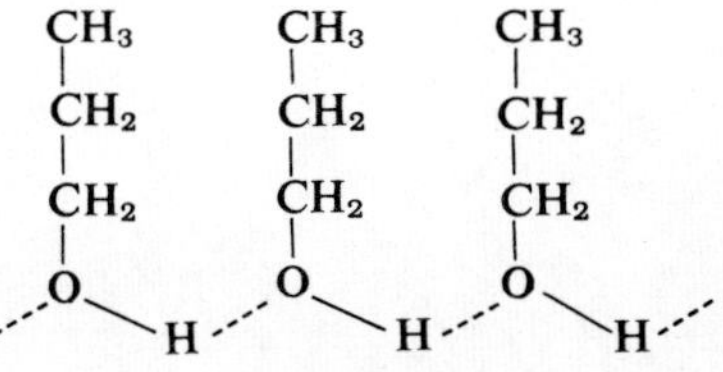

Apart from this effect of hydrogen bonding, the boiling points of hydroxyl compounds vary in the usual fashion in so far as, within groups of related compounds, the boiling points rise steadily with increasing molecular weight.

This ability to form hydrogen bonds also confers upon compounds containing the hydroxyl group a high to moderate solubility in water (hydrogen bonds form between water and the organic hydroxyl compound). In those compounds with large carbon skeletons and a few hydroxyl groups (or only one), however, the effect of the hydrogen bonding hydroxyl group is unimportant. Cholesterol, $C_{27}H_{45}OH$, for example, is quite insoluble in water.

Fig. 23.1 Hydrogen bonding between molecules of propan-1-ol.

23.1.2 Chemical Behaviour

1) Acidity. All hydroxyl groups show acidic properties to a certain degree (Section 9.3 gives a fuller discussion of acidity and basicity). They are acidic by virtue of the fact that the hydrogen atom of the hydroxyl group can be removed as a proton (H^+) by bases. The skeleton to which the group is attached, however, will influence the "ease" with which protons can be removed, that is, the strength of the compounds as acids.

To illustrate this effect we shall consider ethanol (C_2H_5OH), a typical alcohol, phenol (C_6H_5—OH, C_6H_5OH), a typical phenol, and acetic acid ($CH_3\overset{O}{\underset{\|}{C}}$—OH), a typical carboxylic acid.

Sodium hydrogen carbonate is a weak base and requires a moderately strong acid to neutralize it. Of the three compounds under consideration, only acetic acid is a strong enough acid to do this:

$$CH_3COOH + NaHCO_3 \rightarrow CH_3COO^-Na^+ + H_2O + CO_2 \quad \text{(effervescence)};$$
sodium acetate
(a salt)

$C_6H_5OH + NaHCO_3$ no reaction;

$C_2H_5OH + NaHCO_3$ no reaction.

If, instead of using the weak base sodium hydrogen carbonate, we use the stronger base sodium hydroxide, then we observe that acetic acid and phenol are sufficiently strong acids to react:

$$CH_3COOH + NaOH \rightarrow CH_3COO^-Na^+ + H_2O;$$
$$C_6H_5OH + NaOH \rightarrow C_6H_5O^-Na^+ + H_2O;$$
sodium phenoxide

$C_2H_5OH + NaOH$ no detectable reaction.

Sodium metal is an even stronger base than sodium hydroxide. It is able to remove protons from acids by donating an electron to them and releasing them as hydrogen gas:

$$Na + H^+ \rightarrow Na^+ + \tfrac{1}{2}H_2.$$

It is so strong a base as to react dangerously even with water which is a very weak acid. It reacts with all three types of hydroxyl compound—dangerously with acetic acid so that in practice that experiment is never done. We have, however:

$$C_6H_5OH + Na \rightarrow C_6H_5O^-Na^+ + \tfrac{1}{2}H_2;$$
$$C_2H_5OH + Na \rightarrow C_2H_5O^-Na^+ + \tfrac{1}{2}H_2.$$
sodium ethoxide

Thus, from the evidence we have just studied, it is possible to say that acid strengths increase in the sequence ethanol < phenol < acetic acid, and this sequence expresses the general situation that the hydroxyl group is more acidic in carboxylic acids than in phenols and more acidic in phenols than in alcohols. (Both the benzene ring in phenols and the carbonyl group in the carboxylic acids are able to delocalize the negative charge in the resultant anion which, in the case of the alkoxide anion, has to sit solely on the oxygen atom. The carbonyl group is more effective in delocalizing that charge than is the benzene ring; hence, as the name implies, carboxylic acids are stronger acids than phenols (the formate ion is discussed in this context in Section 4.12, and the phenolate anion in Section 28.5.1).

The acidic nature of hydroxyl groups is not peculiar to organic compounds. Many of the well-known inorganic oxyacids, for instance, contain hydroxyl groups which are acidic (Section 11.1.2). For example:

$$(HO)_2SO_2(aq) + 2NaOH(aq) \rightarrow Na_2SO_4(aq) + 2H_2O;$$
sulphuric acid

$$(HO)_3P=O(aq) + 3NaOH(aq) \rightarrow Na_3PO_4(aq) + 3H_2O;$$
phosphoric acid

$$(HO)_3Al(aq) + NaOH(aq) \rightarrow NaAlO_2(aq) + 2H_2O.$$
aluminium hydroxide

2) Basicity. Because the oxygen atom of the hydroxyl group has lone pairs of electrons, not involved in bonding, it is in principle possible for it to accept a proton from a stronger acid—in other words to behave as a base:

$$-\overset{..}{\underset{..}{O}}-H + H^+ \rightarrow -\overset{H}{\underset{|}{\overset{|+}{O}}}-H.$$

This basic character is realized in practice, and strong acids will protonate hydroxy compounds which, being acidic as well, are therefore amphoteric. The carbonyl group oxygen atom is more basic than the hydroxyl oxygen atom, and carboxylic acids will accept protons on the carbonyl group rather more readily than on the hydroxyl. The basicity may be readily demonstrated in the laboratory by immersing a suitable pair of electrodes, connected in series with a torch bulb and battery, in a mixture of, say, pure sulphuric and glacial acetic acids. Whereas the separate liquids are non-conducting, the mixture is conducting and the bulb lights up.

$$R-OH + H^+ \rightarrow R-\overset{+}{O}H_2$$

$$Ar-OH + H^+ \rightarrow Ar-\overset{+}{O}H_2$$

$$R-\underset{\|}{\overset{O}{C}}-OH + H^+ \rightarrow R-\underset{\|}{\overset{\overset{+}{O}H}{C}}-OH.$$

3) Alkylation. Alkylation involves the replacement of a hydrogen atom by an alkyl group. Thus, alkylation of hydroxyl groups may be represented as follows for each of the three types of hydroxyl compounds we are considering:

alcohols → ethers: e.g. C_2H_5OH → $C_2H_5OC_2H_5$;
ethanol   diethyl ether
(ethoxyethane)

phenols → ethers: e.g. C_6H_5OH → $C_6H_5OC_2H_5$;
phenol   ethyl phenyl ether
(ethoxybenzene)

carboxylic acids → esters: e.g. CH_3COOH → $CH_3COOC_2H_5$.
acetic acid   ethyl acetate

Reactions of type:

$$Y\text{—}OH + HOR \xrightarrow{acid} Y\text{—}O\text{—}R + H_2O.$$

This reaction works well only for *alcohols* (Y—OH = R—OH) and *carboxylic acids* (Y = $R\text{—}\overset{\overset{O}{\|}}{C}\text{—}$). *Alcohols* require sufficient concentrated sulphuric acid to dehydrate them, and this dehydration may take place within one molecule of alcohol (leading to alkene formation) or between two, according to the amount of acid used and the precise experimental conditions (the temperature must not rise too much as at 170 °C alkene formation becomes important):

$$C_2H_5OH + \text{excess } H_2SO_4 \xrightarrow{170\,°C} CH_2\text{=}CH_2 + H_2O \quad \text{(largely)};$$
ethanol   ethylene

$$\text{excess } C_2H_5OH + H_2SO_4 \xrightarrow{140\,°C} C_2H_5OC_2H_5 + H_2O \quad \text{(largely)}.$$
ethanol   diethyl ether
(ethoxyethane)

Acids require catalytic quantities of acid (dry hydrogen chloride gas or concentrated sulphuric acid) to reach equilibrium. Although the reaction normally only goes to an equilibrium position it is still useful for preparing esters.

$$CH_3COOH + C_2H_5O\text{*}H \underset{\text{or dry HCl}}{\overset{\text{conc. }H_2SO_4}{\rightleftharpoons}} CH_3COO\text{*}CH_2CH_3 + H_2O.$$
acetic acid   ethanol   ethyl acetate
(the ethyl ester of acetic acid)

This particular reaction is known as *esterification* rather than alkylation and is rather different from alkylation reactions in so far as the oxygen atom to which the alkyl group is attached in the ester has come with the alkyl group from the alcohol (see asterisk). It is not the oxygen atom of the acid hydroxyl group. It is, however, convenient to consider the reaction here, even though the similarity to the other reactions is only formal.

Reactions of type:

$$Y\text{—}O^- + \text{halogen—}R \rightarrow Y\text{—}O\text{—}R + \text{halide ion}^-.$$
(i.e metal halide)

Metal salts of hydroxy compounds (discussed in Section 23.1.2, acidity—very easy to make from the parent hydroxyl compound) are often used in this way as they are more reactive (Section 26.3) than the parent hydroxy compound and the reaction works well (Section 23.2.3) for *alcohols, phenols*, and *carboxylic acids*.

Mainly sodium salts of alcohols and phenols are used whereas silver salts of carboxylic acids are commonly employed. Iodo-, bromo-, or chloroalkanes may be used, but iodoalkanes are best:

$$C_3H_7ONa + IC_2H_5 \rightarrow C_3H_7OC_2H_5 + NaI;$$
sodium pro- iodoethane   ethyl propyl ether
poxide   (1-ethoxypropane)

$$C_6H_5ONa + ICH_3 \rightarrow C_6H_5OCH_3 + NaI;$$
sodium phen- iodo-   methyl phenyl ether
oxide   methane (methoxybenzene)

$$CH_3COOAg + C_4H_9I \rightarrow CH_3COOC_4H_9 + AgI.$$
silver acetate 1-iodo- butyl acetate
                butane   (the butyl ester of acetic acid)

Note that methyl phenyl ether cannot be made from iodobenzene and sodium methoxide. Halogenobenzenes are very unreactive in this type of reaction (Section 28.5.2).

4) Acylation. Acylation involves the replacement of a hydrogen atom with an acyl group, $R-\overset{O}{\underset{\|}{C}}-$. The acetyl group, $CH_3-\overset{O}{\underset{\|}{C}}-$, is an acyl group and replacement of hydrogen by acetyl is known as acetylation (note the derivation of the name from the acetic acid root). We have, in the case of hydroxyl groups:

alcohols → esters: e.g. $C_2H_5OH \rightarrow C_2H_5O\overset{O}{\underset{\|}{C}}CH_3;$
                         ethanol   ethyl acetate

phenols → esters: e.g. $C_6H_5OH \rightarrow C_6H_5O\overset{O}{\underset{\|}{C}}C_2H_5;$
                         phenol   phenyl propanoate

acids → acid anhydrides: e.g. $CH_3COOH \rightarrow CH_3COOCOCH_3.$
                               acetic acid   acetic anhydride

Reactions of type:

$$Y-OH + R-\overset{O}{\underset{\|}{C}}-OH \xrightarrow{acid} YO-\overset{O}{\underset{\|}{C}}-R.$$

This reaction works well only for *alcohols*, for example:

$$C_2H_5OH + CH_3COOH \xrightleftharpoons{H_2SO_4} C_2H_5O\overset{O}{\underset{\|}{O}}CH_3 + H_2O.$$
ethanol   acetic acid   ethyl acetate

We have already considered this esterification reaction in connection with the alkylation of the acid hydroxyl group. Here we consider it in connection with the acylation of the alcohol hydroxyl group.

Reactions of type:

$$Y-OH + R-\overset{O}{\underset{\|}{C}}-Cl \rightarrow Y-O-\overset{O}{\underset{\|}{C}}-R + HCl.$$
                acid
                chloride

Sometimes this reaction is carried out in the presence of a base which may act as a catalyst as well as "mopping" up the hydrogen chloride produced. The reaction works well for *phenols* and *alcohols*.

$$\underset{\text{phenol}}{C_6H_5OH} + \underset{\substack{\text{benzoyl}\\\text{chloride}}}{C_6H_5\overset{\overset{O}{\|}}{C}Cl} \xrightarrow{\text{NaOH}} \underset{\text{phenyl benzoate}}{C_6H_5O\overset{\overset{O}{\|}}{C}C_6H_5} + HCl$$

$$(HCl \xrightarrow{\text{NaOH}} NaCl)$$

$$\underset{\text{ethanol}}{C_2H_5OH} + \underset{\substack{\text{acetyl}\\\text{chloride}}}{CH_3\overset{\overset{O}{\|}}{C}Cl} \longrightarrow \underset{\text{ethyl acetate}}{C_2H_5O\overset{\overset{O}{\|}}{C}CH_3}.$$

Acid anhydrides behave rather like acid chlorides in this respect. A mixture of acetic anhydride ($CH_3\overset{\overset{O}{\|}}{C}$—O—$\overset{\overset{O}{\|}}{C}CH_3$) and acetic acid makes a very convenient acetylating mixture; for example:

$$\underset{\text{phenol}}{C_6H_5OH} + \underset{\text{acetic anhydride}}{CH_3\overset{\overset{O}{\|}}{C}\text{—O—}\overset{\overset{O}{\|}}{C}CH_3} \rightarrow \underset{\text{phenyl acetate}}{C_6H_5O\overset{\overset{O}{\|}}{C}CH_3} + \underset{\text{acetic acid}}{CH_3\overset{\overset{O}{\|}}{C}OH}.$$

Simple carboxylic acids do not easily lose water to form acylated derivatives (acid anhydrides); for example:

$$\underset{\text{acetic acid}}{CH_3COOH} + CH_3COOH \text{ does not yield } \underset{\text{acetic anhydride}}{CH_3COOCOCH_3}.$$

Nor do they react readily with acyl chlorides; for example:

$$\underset{\text{acetic acid}}{CH_3COOH} + \underset{\substack{\text{acetyl}\\\text{chloride}}}{CH_3COCl} \text{ does not yield } \underset{\text{acetic anhydride}}{CH_3COOCOCH_3} + HCl.$$

Their salts, however, are more reactive (Section 26.3) and will react with acid chlorides to produce acid anhydrides; for example:

$$\underset{\substack{\text{sodium}\\\text{acetate}\\\text{(essentially}\\CH_3COO^-)}}{CH_3COONa} + \underset{\substack{\text{acetyl}\\\text{chloride}}}{CH_3COCl} \xrightarrow{\text{heat}} \underset{\substack{\text{acetic}\\\text{anhydride}}}{CH_3COOCOCH_3} + NaCl.$$

5) Dehydration We have already seen (Section 22.3) that alcohols can be dehydrated. We can represent the reaction in general by the equation:

$$\begin{array}{c} H \\ | \\ -\overset{|}{\underset{|}{C^2}}-\overset{|}{\underset{OH}{C^1}}- \end{array} \rightarrow \overset{2}{C}=\overset{1}{C} + H_2O.$$

584 COMPOUNDS WITH FUNCTIONAL GROUPS CONTAINING OXYGEN

Clearly, if there is no hydrogen atom available on carbon atom 2 of the alcohol, then no dehydration can take place. Consider phenol,

$$\text{H}-\underset{\underset{H}{|}}{\underset{|}{\bigcirc}}-\text{OH}.$$

There is certainly a hydrogen atom on carbon atom 2, but if this is lost in a dehydration reaction we should form the compound

$$\text{H}-\bigcirc \quad \text{(benzyne)}.$$

This compound, if formed, would be very strained because normal triple bonded compounds are linear (cf. H—C≡C—H). We cannot readily dehydrate phenols.

It is possible to eliminate a molecule of water from carboxylic acids. Thus, dehydration of acetic acid leads to ketene (systematic name ethenone):

$$CH_3COOH \xrightarrow{-H_2O} CH_2=C=O.$$

This reaction forms the basis of an industrial synthesis of ketene, although it is not a convenient laboratory preparation. In short, dehydration of carboxylic acids is possible but difficult.

Thus, only alcohols can be dehydrated conveniently. The dehydration can be achieved by heating with concentrated sulphuric acid (excess at high temperature—170 °C) *or* by passing the alcohol vapour over hot aluminium oxide. Phosphoric acid is also a good catalyst for this reaction.

$$CH_3CH_2OH \rightarrow CH_2=CH_2 + H_2O.$$

6) Oxidation. The oxidation of alcohols proceeds essentially by the removal of hydrogen to form a double bond between carbon and oxygen:

$$-\underset{|}{\overset{H}{\underset{|}{C}}}-OH \xrightarrow{-2H} -\underset{|}{C}=O.$$

It is, therefore, important that a hydrogen atom should be attached to the same carbon atom as the hydroxyl group. Consider the following types of alcohol:

$$\underset{A}{H-\underset{\underset{H}{|}}{\overset{H}{|}}{C}-OH} \quad \underset{B}{-\underset{\underset{H}{|}}{\overset{H}{|}}{C}-\underset{\underset{H}{|}}{\overset{H}{|}}{C}-OH} \quad \underset{C}{-\underset{\underset{H}{|}}{\overset{|}{C}}-\underset{\underset{|}{|}}{\overset{|}{C}}-OH} \quad \underset{D}{-\underset{\underset{|}{|}}{\overset{|}{C}}-\underset{\underset{|}{|}}{\overset{|}{C}}-OH}$$

The carbon atom to which the hydroxyl group is attached in each case has varying numbers of carbon atoms attached to it. In A it has none and there is only one alcohol—methanol—which fits this requirement. In B the carbon atom in question has *one* other carbon atom attached to it (this in turn may have other groups attached); it is known as a *primary* carbon atom and characterizes a primary alkyl group. For similar reasons, that in C is known as a *secondary* carbon atom, characterizing secondary alkyl groups, and that in D as a *tertiary* carbon atom, characterizing tertiary alkyl groups. Alcohols of type B are known as primary alcohols therefore, and those of types C and D as secondary and tertiary alcohols, respectively. Methanol, A, is a special case.

Oxidation by removal of hydrogen can proceed for A, B, and C but *not* for D as there is no hydrogen atom available on the tertiary carbon atom. The products we would obtain, in each case, contain a carbonyl group, $\diagup\!\!\!\!C=O$. Illustrating this with the simplest member of each group ([O] stands for oxidation), we have:

A $\quad CH_3OH \xrightarrow{[O]} \begin{array}{c} H \\ \diagdown \\ C=O; \\ \diagup \\ H \end{array}$
$\qquad$ methanol $\qquad\qquad$ formaldehyde
$\qquad\qquad\qquad\qquad\qquad$ (methanal)

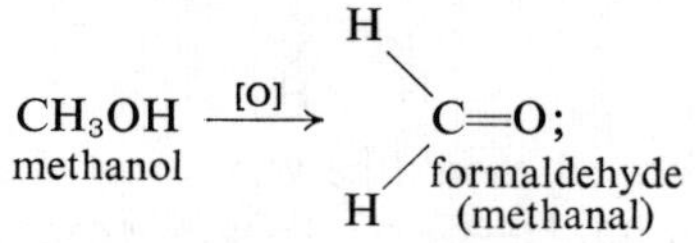

B $\quad CH_3CH_2OH \xrightarrow[-2H]{[O]} CH_3-\overset{\overset{H}{|}}{C}=O;$
$\qquad$ ethanol $\qquad\qquad$ acetaldehyde
$\qquad\qquad\qquad\qquad$ (ethanal)

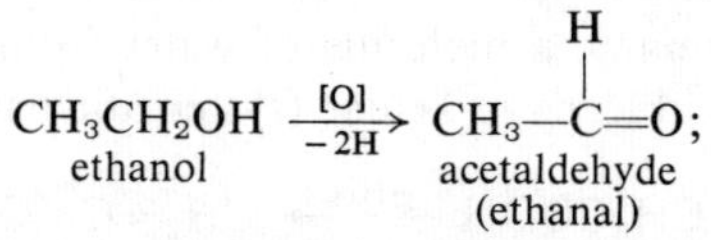

C $\quad \begin{array}{c} CH_3 \\ \diagdown \\ CHOH \\ \diagup \\ CH_3 \end{array} \xrightarrow[-2H]{[O]} \begin{array}{c} CH_3 \\ \diagdown \\ C=O; \\ \diagup \\ CH_3 \end{array}$
$\qquad$ propan-2-ol $\qquad\qquad$ acetone
$\qquad$ (isopropanol) $\qquad\qquad$ (propanone)

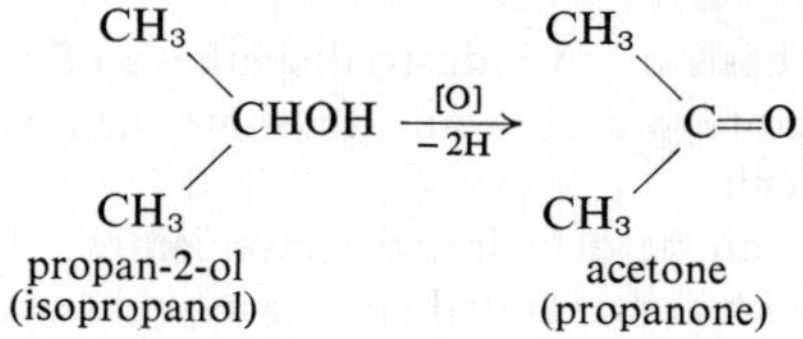

D $\quad \begin{array}{c} CH_3 \\ \diagdown \\ CH_3-C-OH \\ \diagup \\ CH_3 \end{array} \quad$ no easy oxidation.
$\qquad$ 2-methylpropan-2-ol
$\qquad$ (tertiary-butanol)

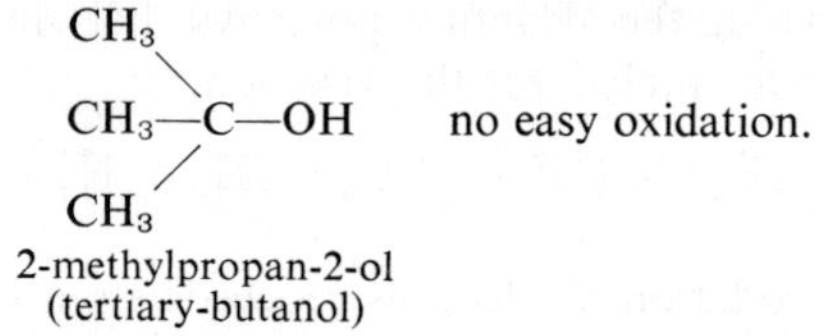

Note that the products from primary alcohols and methanol still have hydrogen atoms attached to the carbon atom which originally carried the hydroxyl group and which now forms part of a carbonyl group. They are therefore capable of further ready oxidation to carboxylic acids:

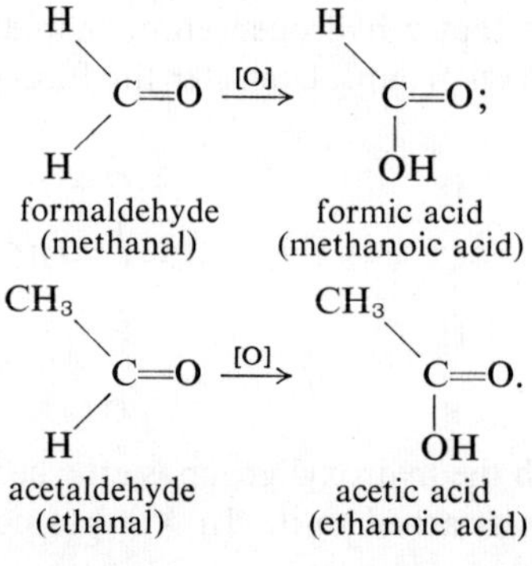

$\begin{array}{c} H \\ \diagdown \\ C=O \\ \diagup \\ H \end{array} \xrightarrow{[O]} \begin{array}{c} H \\ \diagdown \\ C=O; \\ | \\ OH \end{array}$
formaldehyde $\qquad$ formic acid
(methanal) $\qquad\quad$ (methanoic acid)

$\begin{array}{c} CH_3 \\ \diagdown \\ C=O \\ \diagup \\ H \end{array} \xrightarrow{[O]} \begin{array}{c} CH_3 \\ \diagdown \\ C=O. \\ | \\ OH \end{array}$
acetaldehyde $\qquad$ acetic acid
(ethanal) $\qquad\quad$ (ethanoic acid)

Of these acids, only formic acid, derived from methanol, is capable of further ready oxidation, as it is the only acid with a hydrogen atom attached to a carbonyl group. Formic acid yields carbonic acid—carbon dioxide and water:

$$\underset{HO}{\overset{H}{>}}C=O \xrightarrow{[O]} \underset{HO}{\overset{HO}{>}}C=O \longrightarrow CO_2 + H_2O.$$

All of the compounds mentioned so far which have no hydrogen atoms attached to the oxygen-bearing carbon atom are difficult to oxidize. They *can* be oxidized, but only with powerful oxidizing agents which break up the whole carbon skeleton into several smaller fragments. A mixture of sulphuric acid and potassium dichromate is a good oxidizing agent for performing the ready oxidations; for example:

$$\underset{\text{ethanol}}{CH_3CH_2OH} \xrightarrow{H_2SO_4/K_2Cr_2O_7} \underset{\text{acetaldehyde}}{CH_3CHO} \xrightarrow{H_2SO_4/K_2Cr_2O_7} \underset{\text{acetic acid}}{CH_3COOH}.$$

The whole sequence is summarized in Table 23.2.

Although with primary alcohols it is easy to oxidize them all the way to carboxylic acids, it is possible to isolate the product at the aldehyde stage. It is possible to prepare acetaldehyde by the oxidation of ethanol with $K_2Cr_2O_7/H_2SO_4$, for example, in good yield, by distilling the acetaldehyde out of the boiling reaction mixture before it is further oxidized. It has a lower boiling point (no hydroxyl group and no hydrogen bonding) than either ethanol or acetic acid.

A suitable alcohol may also be catalytically converted to a carbonyl compound, in the vapour phase, by a metallic copper surface. Hydrogen is removed from the alcohol and adsorbed onto the surface of the metal (whence it may be removed by oxygen or air to form water). This reaction, referred to as dehydrogenation, only produces the aldehyde (or ketone) and does not proceed to the carboxylic acid stage.

$$\underset{\text{methanol}}{CH_3OH} \xrightarrow{\text{Cu catalyst}} \underset{\text{formaldehyde}}{CH_2O};$$

$$\underset{\text{ethanol}}{CH_3CH_2OH} \longrightarrow \underset{\text{acetaldehyde}}{CH_3CHO}.$$

7) Replacement of hydroxyl by halogen. All simple hydroxyl-containing compounds react with phosphorus pentachloride to give a product containing a chlorine atom where the hydroxyl group was originally situated. In general:

$$X\text{—}OH + PCl_5 \rightarrow X\text{—}Cl + POCl_3 + HCl.$$

This is a very important way of making chlorine compounds although it does not work very well with phenols. With carboxylic acids phosphorus trichloride may be successfully employed in place of phosphorus pentachloride. Thionyl chloride is often a more convenient reagent than either as the other products, which are not required, are all gaseous and easily removed.

$$\underset{\text{ethanol}}{CH_3CH_2OH} + PCl_5 \rightarrow \underset{\text{chloroethane}}{CH_3CH_2Cl} + POCl_3 + HCl$$

$$\underset{\text{acetic acid}}{CH_3COOH} + PCl_5 \rightarrow \underset{\text{acetyl chloride}}{CH_3COCl} + POCl_3 + HCl$$

$$\underset{\text{propan-1-ol}}{CH_3CH_2CH_2OH} + SOCl_2 \rightarrow \underset{\text{1-chloropropane}}{CH_3CH_2CH_2Cl} + SO_2 + HCl.$$

Bromine and iodine compounds may be similarly prepared from hydroxy compounds using phosphorus tribromide or iodide which may be prepared *in situ* from red phosphorus and the halogen. Thus, iodoethane can be prepared from ethanol, red phosphorus, and iodine:

$$\underset{\text{ethanol}}{C_2H_5OH} + P + I_2 \rightarrow C_2H_5OH + PI_3 \rightarrow \underset{\text{iodoethane}}{C_2H_5I}.$$

The hydroxyl group of alcohols may be replaced by a halogen atom by reaction with the appropriate hydrogen halide. This method is unsuitable for phenols and

Table 23.2. Oxidation of alcohols.

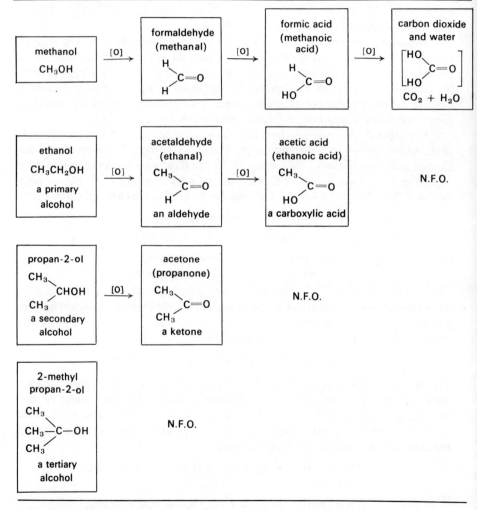

[O]—oxidation (e.g. by $H_2SO_4/K_2Cr_2O_7$ mixture)
N.F.O.—no further oxidation without more vigorous conditions and cleavage of carbon–carbon bonds.

carboxylic acids, and a catalyst (zinc chloride) is required for primary and secondary alcohols.

$$(CH_3)_3COH + HBr \rightarrow (CH_3)_3CBr.$$
2-methylpropan-2-ol   2-bromo-2-methylpropane
(tertiary-butanol)   (tertiary-butyl bromide)

The hydrogen halide used can be generated *in situ* from sulphuric acid and potassium bromide so that in practice we may have:

$$CH_3CH_2OH + KBr + H_2SO_4 \rightarrow CH_3CH_2Br + KHSO_4 + H_2O.$$
ethanol   bromoethane

No zinc chloride is then required as the sulphuric acid acts as catalyst.

8) Decarboxylation. A reaction which is peculiar to carboxylic acids is that in which the —CO_2H group is lost to yield a hydrocarbon. This is not strictly a reaction of the hydroxyl group alone but reflects the influence of the two groups, C=O and OH, one upon the other. If they are further apart in the molecule then there is no corresponding reaction. Upon heating with soda lime (a mixture of calcium and sodium hydroxides) a carboxylic acid is decarboxylated to yield a hydrocarbon and sodium carbonate:

$$C_6H_5COOH \xrightarrow{NaOH} C_6H_5COONa \xrightarrow{NaOH} C_6H_6 + Na_2CO_3;$$
benzoic acid   sodium benzoate   benzene

$$CH_3COOH \xrightarrow{NaOH} CH_3COONa \xrightarrow{NaOH} CH_4 + Na_2CO_3.$$
acetic acid   sodium acetate   methane

With aliphatic acids larger than acetic the products are complex mixtures of hydrocarbons and the yield of methane from acetate is usually pretty low anyway. The reaction is of little preparative value, therefore, in the aliphatic series, although it is often a useful way of removing carboxylic acid groups from aromatic molecules. The ready and useful decarboxylation of certain carboxylic acids containing a second carbonyl group in the molecule is discussed in Section 28.3.

23.1.3 Introduction of the Hydroxyl Group into Molecules

The direct introduction of a hydroxyl group on to a carbon skeleton (perhaps by treating an alkane with some suitable reagent) is not easy. It is therefore necessary to introduce other groups first, or to use compounds which contain functional groups already which can be converted to the hydroxyl group.

1) Hydrolysis reactions. If a molecule A—B (A and B may be atoms or groups of atoms), in the presence of water (usually as solvent), breaks up so that one part of a water molecule is associated with A and the other part with B, then the reaction is termed a hydrolysis reaction:

$$A—B + H_2O \rightarrow AOH + HB.$$

Sometimes, usually if A and B are linked by a multiple bond, hydrolysis may proceed:

$$A=B + H_2O \rightarrow A=O + H_2B.$$

If A were a carbon group then, clearly, a reaction of this type could be used to introduce a hydroxyl group on to that carbon group. A number of such reactions,

Table 23.3. Important reactions of hydroxy compounds.

A. Reactions in which oxygen remains bound to carbon

acidity

$$\diagdown\!\!\!C\!\!-\!\!OH + B \rightleftharpoons \diagdown\!\!\!C\!\!-\!\!O^- + BH^+$$

| B represents a base |

basicity

$$\diagdown\!\!\!C\!\!-\!\!OH + HA \rightleftharpoons \diagdown\!\!\!C\!\!-\!\!\overset{+}{O}H_2 + A^-$$

| HA represents an acid |

alkylation

reactions with other hydroxy compounds

$$\diagdown\!\!\!C\!\!-\!\!OH + HO\!\!-\!\!C\diagdown \rightarrow \diagdown\!\!\!C\!\!-\!\!O\!\!-\!\!C\diagdown + H_2O$$

| usually requires an acid catalyst |

reactions of the oxyanion with halogenoalkanes

$$\diagdown\!\!\!C\!\!-\!\!O^- + \diagdown\!\!\!C\!\!-\!\!Hal \rightarrow \diagdown\!\!\!C\!\!-\!\!O\!\!-\!\!C\diagdown + Hal^-$$

| proton removed from OH to generate the more reactive anion |

acylation

$$\diagdown\!\!\!C\!\!-\!\!OH + RCOOH \rightarrow \diagdown\!\!\!C\!\!-\!\!O\!\!-\!\!COR + H_2O$$

RCOCl + HCl
(RCO)$_2$O + RCOOH

| it may be necessary to remove proton first in some cases to generate anion, e.g. $CH_3CO_2^-$ |

oxidation

$$\begin{matrix}H\\|\\-C\!\!-\!\!OH\\|\end{matrix} \rightarrow \diagdown\!\!\!C\!\!=\!\!O$$

| alcohols yield aldehydes, ketones, or carboxylic acids |

B. Reactions in which oxygen is lost

replacement of —OH by halogen

$$\diagdown\!\!\!C\!\!-\!\!OH + HHal \rightarrow \diagdown\!\!\!C\!\!-\!\!Hal$$

PCl$_5$
SOCl$_2$

| phenols are not susceptible to these reactions |

dehydration

$$H\!\!-\!\!\overset{|}{\underset{|}{C}}\!\!-\!\!\overset{|}{\underset{|}{C}}\!\!-\!\!OH \rightarrow \diagdown\!\!\!C\!\!=\!\!C\diagup + H_2O$$

---

in which the hydroxyl group derives from water, are used to prepare hydroxy compounds. If the carbon skeleton is represented by A, then B normally has to be based on elements from groups VB, VIB, or VIIB of the Periodic Table. This will be clear from the general methods of introducing hydroxyl groups given below.

B is derived from group VB (nitrogen):

$$A\!\!-\!\!NH_2 + H_2O \rightarrow AOH + NH_3.$$

This reaction proceeds only if A is an acyl group, R—C(=O)—, and is therefore of use for the preparation of *carboxylic acids* only. Alcohols and phenols are not prepared in this way as amines, RNH_2 and $ArNH_2$, are not hydrolysed when treated with aqueous acids or bases.

$$RCONH_2 + H_2O \xrightarrow{\text{NaOH (hot aq. soln.)}} RCOOH + NH_3$$
acid amide　　　　　　　　　　　　　carboxylic acid

$$RCOOH + NaOH \longrightarrow RCOONa + H_2O.$$

The free carboxylic acid can be liberated from its salt by addition of mineral acid (carboxylic acids are usually weaker acids than the mineral acids):

$$2CH_3COONa + H_2SO_4 \rightarrow 2CH_3COOH + Na_2SO_4.$$
sodium acetate　　　　　　　acetic acid

The reaction may also be catalysed by acids:

$$2RCONH_2 + 2H_2O \xrightarrow{H_2SO_4} 2RCOOH + 2NH_3;$$

$$2NH_3 + H_2SO_4 \longrightarrow (NH_4)_2SO_4.$$

For example:

$$CH_3CONH_2 + H_2O \xrightarrow{\text{acid or base}} CH_3COOH + NH_3.$$
acetamide　　　　　　　　　acetic acid

Aliphatic or aromatic nitriles (cyanides), R—C≡N, Ar—C≡N, will, with aqueous alkali or acid, react by addition of water to produce an amide. Hydrolysis of this follows and a carboxylic acid is produced. This is one of the most important ways of making carboxylic acids because nitriles are relatively easy to make (Section 24.6). For example:

$$C_6H_5C\equiv N \xrightarrow{H_2O/NaOH} C_6H_5CONH_2 \xrightarrow{H_2O/NaOH} C_6H_5COONa + NH_3;$$
benzonitrile　　　　　　　benzamide　　　　　　　benzoic acid (sodium salt)

$$CH_3C\equiv N \xrightarrow{H_2O/H_2SO_4} CH_3CONH_2 \xrightarrow{H_2O/H_2SO_4} CH_3COOH + (NH_4)_2SO_4.$$
acetonitrile　　　　　　　acetamide　　　　　　　acetic acid

B is derived from group VIB (oxygen):

$$\begin{array}{l} A\text{—}OR \\ \phantom{A\text{—}}\text{—}OAr \end{array} + H_2O \rightarrow AOH + \begin{array}{l} HOR \\ HOAr. \end{array}$$

Again, this method is only of use if A is an acyl group, that is if A—OR is an ester. Again, the reaction may be catalysed by acids or alkalis. A—OR must therefore be an ester and the reaction may be used to prepare *carboxylic acids* (which will be derived from A) or *alcohols* and *phenols* (which will be derived from —OR or —OAr). Ethers are extremely resistant to hydrolysis. Thus, we may have:

$$\underset{\text{ester}}{RC(=O)-OR'} + H_2O \rightarrow \underset{\text{carboxylic acid}}{RC(=O)OH} + \underset{\text{alcohol}}{HOR'};$$

$$\underset{\text{ester}}{RC(=O)-OAr} + H_2O \rightarrow \underset{\text{carboxylic acid}}{RC(=O)OH} + \underset{\text{phenol}}{HOAr}.$$

Note that we may also hydrolyse A—OR if R is an acyl group too. We would then be dealing with an acid anhydride, such as $CH_3COOCOCH_3$, acetic anhydride, and we should obtain two molecules of carboxylic acid:

$$\underset{\text{acetic anhydride}}{CH_3COOCOCH_3} + H_2O \rightarrow \underset{\text{acetic acid}}{2CH_3COOH}.$$

This is not a very important way of making carboxylic acids, though, as the anhydride usually has to be made from the acid in the first place.

Acid hydrolyses of esters (i.e. those carried out in acid solution) usually lead to an equilibrium position. This is simply the reverse of the esterification reaction. Those carried out with alkali proceed to completion as the acid produced in the equilibrium is rapidly removed by the alkali to form a salt.

Fats are naturally occurring esters, usually derived from glycerol (propane-1,2,3-triol) which contains three hydroxyl groups:

$$\begin{array}{c} CH_2OH \\ | \\ CHOH. \\ | \\ CH_2OH \end{array}$$

Each or any of these OH groups may take part in the ester linkage to a carboxylic acid residue. Glycerol tristearate or tristearin (propane-1,2,3-triyl trioctadecanoate) is found in animal fats and has the structure:

$$\begin{array}{c} CH_2O-C(=O)-(CH_2)_{16}CH_3 \\ | \\ CHO-C(=O)-(CH_2)_{16}CH_3 \\ | \\ CH_2O-C(=O)-(CH_2)_{16}CH_3 \end{array}$$

When animal fat containing this is boiled with sodium hydroxide solution the ester is hydrolysed to produce glycerol and sodium stearate. Glycerol has many uses in the chemical industry (explosives, pharmaceuticals, cosmetics etc.) and sodium stearate is a soap. We can write the hydrolysis:

$$\begin{array}{l} \text{CH}_2\text{OC(O)(CH}_2)_{16}\text{CH}_3 \\ | \\ \text{CHOC(O)(CH}_2)_{16}\text{CH}_3 \;+\; 3\text{NaOH} \;\rightarrow\; \\ | \\ \text{CH}_2\text{OC(O)(CH}_2)_{16}\text{CH}_3 \end{array} \quad \begin{array}{l} \text{CH}_2\text{OH} \\ | \\ \text{CHOH} \;+\; 3\text{NaOC(O)(CH}_2)_{16}\text{CH}_3 \\ | \\ \text{CH}_2\text{OH} \end{array}$$

tristearin   glycerol   sodium stearate

Because this type of reaction is the basis of soap production, the general procedure of alkaline hydrolysis of esters is often referred to as *saponification*.

Sulphur, another group VIB element, is often used as a basis for B (in our hydrolysis model A—B) when introducing hydroxyl groups into aromatic compounds. A would then be an aromatic group like phenyl, C_6H_5. Thus molten KOH will then convert benzenesulphonic acid into potassium phenolate (because water is not present during the C—S cleavage, this is not, strictly, a hydrolysis reaction). This potassium salt will readily liberate the free phenol on treatment with acid.

$$C_6H_5\text{-SO}_2\text{-OH} + 3\text{KOH} \xrightarrow{\text{heat}} C_6H_5\text{-OK} + K_2SO_3 + 2H_2O \qquad C_6H_5\text{-OK} + HCl \rightarrow C_6H_5\text{-OH} + KCl.$$

benzenesulphonic acid   potassium phenolate

B is derived from group VIIB (halogens):

$$\text{A—halogen} + H_2O \rightarrow \text{A—OH} + \text{H—halogen}.$$

This reaction works very well if A is an alkyl group or an acyl group. If A is aromatic the reaction usually will not take place. The reaction is therefore of no use for phenol preparation and is of little use for preparing carboxylic acids because the acyl chloride (acid chloride) required would probably have to be made from the acid itself.

$$\text{R—Hal} + H_2O \rightarrow \text{ROH} + \text{HHal}$$

$$\text{RCO—Hal} + H_2O \rightarrow \text{RCOOH} + \text{HHal}$$

$$\text{ArHal} + H_2O \text{ no reaction.}$$

Halogenoalkanes (alkyl halides) react very slowly with water but conveniently and rapidly with aqueous alkali. Acids have no catalytic effect on this reaction. For example:

$$C_3H_7Cl + \text{NaOH(aq)} \rightarrow C_3H_7OH + \text{NaBr}.$$

1-chloropropane   propan-1-ol
(prop-1-yl chloride)

Acid chlorides react rapidly and often violently with water without added catalyst; for example:

$$CH_3CH_2COCl + H_2O \rightarrow CH_3CH_2COOH + HCl.$$

propanoyl chloride   propanoic acid

Table 23.4

Compound	Hydrolysis	Compound	Hydrolysis
$CH_3CH_2CH_2Cl$ 1-chloropropane (prop-1-yl chloride)	hydrolyses *slowly* in water hydrolyses at a *moderate rate* in aqueous alkali	$CH_3CH_2\overset{\underset{\parallel}{O}}{C}Cl$ propanoyl chloride	hydrolyses *rapidly* in water—no catalyst necessary
$CH_3CH_2CH_2OCH_2CH_3$ ethyl prop-1-yl ether (1-ethoxypropane)	virtually *no* hydrolysis	$CH_3CH_2\overset{\underset{\parallel}{O}}{C}OCH_2CH_3$ ethyl propanoate	hydrolyses *slowly* in water hydrolyses at a *moderate rate* in aqueous acid or alkali
$CH_3CH_2CH_2NH_2$ prop-1-ylamine	*no* hydrolysis	$CH_3CH_2\overset{\underset{\parallel}{O}}{C}NH_2$ propanamide	hydrolyses *very slowly* in water hydrolyses at *moderate rate* in aqueous acid or alkali

It will be interesting to look back at this point to compare the last two hydrolysis reactions with four which we considered earlier. In Table 23.4 the hydrolysis of amides, amines, ethers, and esters is contrasted with the hydrolysis of 1-chloropropane and propanoyl chloride. The compounds used to illustrate the comparison are chosen to be, structurally, as nearly similar as possible.

Two general conclusions may be drawn from this table. First, carbon–halogen bonds are more readily hydrolysed than carbon–oxygen or carbon–nitrogen bonds. Second, if the halogen, oxygen, or nitrogen atoms are attached to a carbonyl carbon atom (i.e. the compounds are relatives of the carboxylic acids), then the bond from them to carbon is much more easily hydrolysed. If hydrolysis was not possible in the absence of the carbonyl group, then it becomes so in the presence of the carbonyl group. Thus, the carbonyl group in this position exerts a considerable influence on the reactivity of the adjacent group (Section 26.4).

2) Oxidation and reduction reactions. Reference back to Table 23.2 will show that alcohols and carboxylic acids, both hydroxyl-containing compounds, are at opposite ends of a sequence of oxidation reactions. Thus, oxidation of the H—C=O group (aldehydes) with aqueous sulphuric acid and potassium dichromate converts the H—C part to HO—C. Aldehydes can be oxidized to carboxylic acids. This is an important way of manufacturing carboxylic acids. Acetaldehyde, for example, made by hydrating acetylene (Section 22.2.3), is oxidized by air, in the presence of a manganese(II) acetate catalyst, to acetic acid:

$$CH_3CHO \xrightarrow[Mn^{2+}]{air} CH_3COOH.$$
acetaldehyde   acetic acid

Carboxylic acids are also made by the oxidation of alcohols. Vinegar is made by bacterial oxidation of the alcohol (ethanol) in wine which produces acetic acid, the characteristic, acidic, sour component of vinegar:

$$CH_3CH_2OH \xrightarrow{[O]} CH_3COOH.$$
ethanol   acetic acid

Open bottles of wine quite soon become sour by this process. Acids are also manufactured by direct oxidation of hydrocarbons.

One can traverse the sequences of Table 23.2 from right to left, as well as from left to right, by the employment of suitable reducing agents. Lithium aluminium hydride (LiAlH$_4$) will reduce carboxylic acids (after their conversion to esters), aldehydes, or ketones to the corresponding primary or secondary alcohols. A number of reducing agents suitable for converting aldehydes and ketones to primary and secondary alcohols, is available. They include hydrogen (in the presence of a platinum or palladium catalyst), sodium plus ethanol, and sodium borohydride (NaBH$_4$). These reagents will not reduce carboxylic acids, which are very difficult to reduce. Normally only lithium aluminium hydride will reduce them and even then, it is usual to esterify them first. For example:

$$\underset{\substack{\text{acetone}\\\text{(propanone)}}}{CH_3COCH_3} \xrightarrow{NaBH_4} \underset{\text{propan-2-ol}}{CH_3CH(OH)CH_3};$$

$$\underset{\text{propanoic acid}}{CH_3CH_2COOH} \rightarrow \underset{\text{methyl propanoate}}{CH_3CH_2COOCH_3} \xrightarrow{LiAlH_4} \underset{\text{propan-1-ol}}{CH_3CH_2CH_2OH} + \underset{\text{methanol}}{CH_3OH};$$

cyclohexanone $\xrightarrow{H_2/Pt}$ cyclohexanol

As well as alcohols (primary and secondary only), phenols may also be prepared by reduction of carbonyl compounds. The carbonyl compounds used are *quinones* and these can only yield dihydroxybenzenes (phenol itself cannot be prepared this way); for example:

p-quinone $\xrightarrow{SO_2/H_2O}$ hydroquinone

3) Hydration of alkenes. The addition of water to ethylene has been met with several times already. This is an important general method of preparing *alcohols* industrially and in the laboratory. It is clearly not applicable to the preparation of phenols or carboxylic acids. For example:

$$CH_2{=}CH_2 \xrightarrow[\text{or } H_3PO_4,\, H_2O]{H_2SO_4,\, H_2O,} CH_3CH_2OH.$$

A very useful alternative method of achieving this addition involves the prior addition of the boron hydride, B$_2$H$_6$ (diborane), to the alkene:

$$B_2H_6 + 6RCH{=}CH_2 \rightarrow 2(RCH_2CH_2)_3B.$$

This is known as hydroboration. Treatment of the boron compound produced with hydrogen peroxide and alkali yields the desired alcohol:

$$(RCH_2CH_2)_3B \xrightarrow{H_2O_2/OH^-} 3RCH_2CH_2OH + H_3BO_3.$$

An interesting point about this reaction is that the OH has appeared on the unsubstituted end of the alkene with which we started, which was an ethylene substituted with an R group at one end only. Treatment of this molecule with sulphuric acid/water would have resulted in the addition of water the other way round (Markownikoff rule), giving R—CH—CH$_3$ as product (this is discussed
$$\text{R—CH(OH)—CH}_3$$
further in Sections 27.2 and 27.3).

4) Conversion of the amino function to hydroxyl. Treatment of compounds containing the primary amino function (—NH$_2$) with aqueous nitrous acid usually produces the corresponding hydroxy compound. This general method is applicable to alcohols (sometimes), phenols, and carboxylic acids. It does not involve hydrolysis of the amine which is not possible. For example:

$$\text{CH}_3\text{CH}_2\text{NH}_2 \xrightarrow{\text{HNO}_2/\text{H}_2\text{O}} \text{CH}_3\text{CH}_2\text{OH} + \text{N}_2 + \text{H}_2\text{O}$$
ethylamine → ethanol

(this is not a good general method for alcohols);

$$\text{C}_6\text{H}_5\text{NH}_2 \xrightarrow{\text{HNO}_2/\text{H}_2\text{O}} \text{C}_6\text{H}_5\text{OH}$$
phenylamine (warm) → phenol
(aniline)

$$\text{C}_6\text{H}_5\text{CONH}_2 \xrightarrow{\text{HNO}_2} \text{C}_6\text{H}_5\text{COOH}$$
benzamide → benzoic acid

Table 23.5. Important methods of introducing the hydroxyl group into organic molecules.

hydrolysis

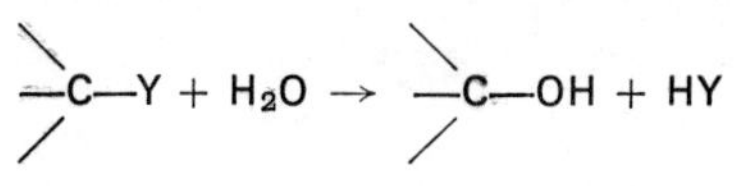

$$\text{>C—Y} + \text{H}_2\text{O} \rightarrow \text{>C—OH} + \text{HY}$$

Y must be an atom or group based on the electronegative elements of groups VB, VIB, or VIIB

oxidation/reduction

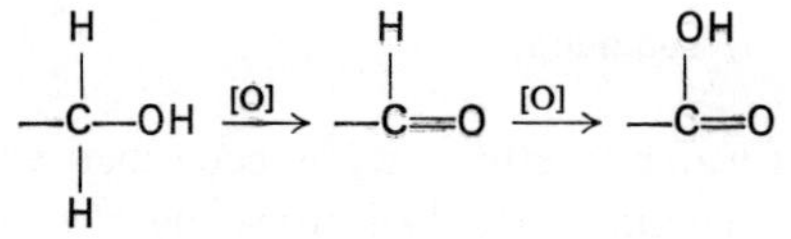

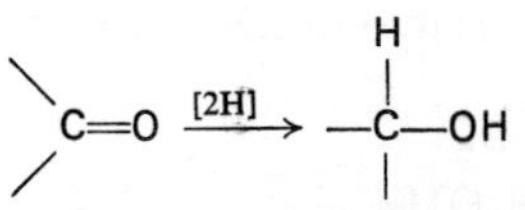

$$\text{>CH—OH} \xrightarrow{[O]} \text{>C=O} \xrightarrow{[O]} \text{HO—C=O}$$

$$\text{>C=O} \xrightarrow{[2H]} \text{>CH—OH}$$

hydration (reverse of dehydration)

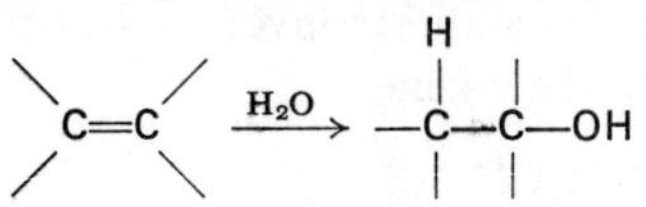

$$\text{>C=C<} \xrightarrow{\text{H}_2\text{O}} \text{>CH—C(OH)<}$$

treatment of primary amino compounds with nitrous acid

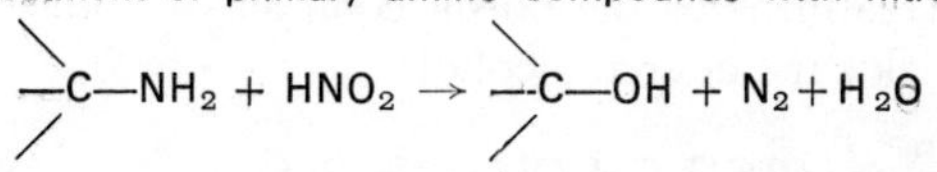

$$\text{>C—NH}_2 + \text{HNO}_2 \rightarrow \text{>C—OH} + \text{N}_2 + \text{H}_2\text{O}$$

rarely of use for alcohols

5) Fermentation. The fermentation of sugars (and other carbohydrate materials) might well be mentioned at this point, although it is not strictly a method of introducing hydroxyl groups into molecules. Sucrose, for example, a complex molecule of formula $C_{12}H_{22}O_{11}$ (Section 28.2), is broken down in a long sequence of reactions by certain enzymes (catalysts) in yeast to alcohols, notably ethanol. Whilst this is still the basis of wine production, it is no longer used to produce alcohols for industrial purposes. A summary of the reaction would be:

$$C_{12}H_{22}O_{11} + H_2O \rightarrow 4C_2H_5OH + 4CO_2.$$

23.2 CARBON–OXYGEN–CARBON LINKAGES

We have just considered, at some length, the properties of the hydroxyl group attached to a carbon skeleton. We may write it as $\diagdown\text{C}-\text{O}-\text{H}\diagup$ where the other atoms or groups attached to carbon are not specified. What we have just written might be regarded as a substituted water molecule in which one of the hydrogen atoms of water has been replaced by a carbon group. If we replace *both* of the hydrogen atoms by carbon groups then we obtain a molecule with the C—O—C linkage which will possess none of the properties of water associated with the water hydrogen atoms. Thus, for example, water reacts with sodium to yield hydrogen and sodium hydroxide in a highly exothermic and rapid reaction. Such rapid release of large quantities of energy can lead to explosion.

$$2HOH + 2Na \rightarrow 2HONa + H_2.$$

Diethyl ether, though, cannot react with sodium as only the "water-type" of hydrogen atom (active, or acidic hydrogen) can react.

$$2C_2H_5OC_2H_5 + 2Na \quad \text{no reaction.}$$
diethyl ether
(ethoxyethane)

Note that an alcohol is like water in this respect, because it contains an active hydrogen atom (attached to the oxygen), although it reacts with sodium with much less violence than does water:

$$2C_2H_5OH + 2Na \rightarrow 2C_2H_5ONa + H_2.$$

We shall consider three types of compound here which contain carbon–oxygen–carbon linkages, each of which has been met with already, at least once, in connection with the hydroxyl group.

1. R—O—R and Ar—O—Ar, ethers. The differences between the C—O—C linkages in R—O—R compounds and Ar—O—Ar compounds are relatively slight. R is an alkyl group, Ar an aromatic (aryl) group.
2. R—C=O esters. These are derivatives of carboxylic acids.
       |
      O—R′
3. R—C=O acid anhydrides. These, too, are derivatives of carboxylic acids.
       |
      O
       |
   R—C=O

Nomenclature of these compounds again depends upon the type of compound in question:

Ethers. There are two commonly used ways of naming ethers on the IUPAC system. Each alkyl (or aryl) group may be specified and followed by the word *ether.* Diethyl ether is thus the name for $CH_3CH_2OCH_2CH_3$, and ethyl methyl ether is the name for $CH_3OCH_2CH_3$. Alternatively, ethers may be named as substituted hydrocarbons, such as methoxybenzene ($CH_3OC_6H_5$), and 1-ethoxypropane ($CH_3CH_2OCH_2CH_2CH_3$). As both methods are often used, we use both in this book.

Esters and anhydrides. These are named simply as derivatives of the parent acids. Ethyl acetate (or ethyl ethanoate) is the name for $CH_3COOCH_2CH_3$, methyl propanoate for $CH_3CH_2COOCH_3$. Ethyl benzoate is $CH_3CH_2OCOC_6H_5$. Acetic anhydride (or ethanoic anhydride) is $CH_3COOCOCH_3$, and propanoic anhydride is $CH_3CH_2COOCOCH_2CH_3$.

23.2.1 Physical Properties

At room temperature most of these compounds are liquids or solids. Only dimethyl and ethyl methyl ethers are gases at room temperature. The boiling point of each of these types of compound rises with increasing molecular weight (Section 22.1) and is not complicated by hydrogen bonding.

23.2.2 Chemical Behaviour

1) Acidity. With no hydrogen atom attached to the oxygen atom the C—O—C grouping is devoid of acidic properties. Thus diethyl ether (ethoxyethane), $C_2H_5OC_2H_5$, does not release hydrogen when treated with sodium.

2) Basicity. Although the absence of hydrogen attached to oxygen means an absence of acidity, the oxygen in the —C—O—C— link is still basic because it possesses lone pairs of electrons not involved in bonding. Thus diethyl ether, which is almost insoluble in water, dissolves to some extent in strong hydrochloric acid. It does so because it forms a salt which, being ionic, is soluble in water.

$$C_2H_5OC_2H_5 + HCl(aq) \rightarrow \underset{\substack{\text{diethyl ether hydrochloride}\\\text{(soluble in water)}}}{\overset{C_2H_5}{\underset{C_2H_5}{\diagdown}}\overset{+}{O}{-}H\ Cl^-}$$

diethyl ether
(insoluble in water)

This salt cannot be isolated as a pure compound as it dissociates into the ether and hydrochloric acid when the water and excess acid are evaporated from its solution.

Acid anhydrides and esters will produce solutions with concentrated sulphuric acid which conduct electricity owing to the formation of ions by protonation of the oxygen functions. Because C=O is more basic than C—O—C, however, it is more extensively protonated than is the C—O—C oxygen. Strong acids, at high concentration, are needed as both carbonyl and ether type oxygens are relatively weakly basic.

$$\text{CH}_3-\overset{\displaystyle O}{\underset{\displaystyle OC_2H_5}{C}} + H_2SO_4 \rightarrow \text{CH}_3-\overset{\displaystyle \overset{+}{O}-H}{\underset{\displaystyle OC_2H_5}{C}} + HSO_4^-.$$

ethyl acetate

$$\begin{matrix} \text{CH}_3-C\!\!\diagup\!\!\overset{O}{} \\ \diagdown O \\ \text{CH}_3-C\!\!\diagup\!\! \\ \overset{}{\underset{O}{}} \end{matrix} + H_2SO_4 \rightarrow \begin{matrix} \text{CH}_3-C\!\!\diagup\!\!\overset{\overset{+}{O}-H}{} \\ \diagdown O \\ \text{CH}_3-C\!\!\diagup\!\! \\ \overset{}{\underset{O}{}} \end{matrix} + HSO_4^-.$$

acetic anhydride

3) Hydrolysis. The hydrolysis of esters and acid anhydrides has already been discussed in connection with the production of hydroxyl-containing compounds. Carbon–oxygen–carbon compounds in general, however, show varying resistance to hydrolysis.

a) *Ethers.* Ethers are essentially inert to hydrolysis. Treatment with strong solutions of strong acids at high temperatures will break the carbon–oxygen bond, however, although the reaction is not strictly a hydrolytic one:

$$\text{ROR} + \text{HI} \rightarrow \text{RI} + \text{ROH};$$

$$\underset{\substack{\text{methyl phenyl ether}\\\text{(methoxy benzene)}}}{C_6H_5OCH_3} + HI \rightarrow \underset{\text{phenol}}{C_6H_5OH} + \underset{\text{iodomethane}}{CH_3I}.$$

Hydriodic acid is often the reagent of choice.

b) *Esters.* These are hydrolysed fairly easily by aqueous solutions of acids or alkalis. They are hydrolysed very slowly by water alone:

$$\underset{\text{ethyl acetate}}{CH_3COOC_2H_5} \xrightarrow{H_2O} \underset{\text{acetic acid}}{CH_3COOH} + \underset{\text{ethanol}}{HOC_2H_5}.$$

c) *Acid anhydrides.* Acid anhydrides are hydrolysed readily by water alone:

$$\underset{\text{acetic anhydride}}{CH_3COOCOCH_3} + H_2O \rightarrow \underset{\text{acetic acid}}{2CH_3COOH}.$$

It should be noted that this oxygen bridge between the residues of two molecules of acid is not peculiar to carbon and organic chemistry:

$$\underbrace{CH_3-\overset{\displaystyle O}{\underset{\displaystyle \|}{C}}}_{\substack{\text{acetic acid}\\\text{residue}}}\underbrace{-O-}_{\substack{\text{oxygen}\\\text{bridge}}}\underbrace{\overset{\displaystyle O}{\underset{\displaystyle \|}{C}}-CH_3}_{\substack{\text{acetic acid}\\\text{residue}}}$$

Thus many of the acidic oxides of inorganic chemistry are acid anhydrides containing oxygen bridges and are derived from residues of the parent acid connected by oxygen bridges (Sections 17.11 and 18.12). With water they yield the parent acids (hence the name, acid anhydride).

$$\underbrace{\begin{array}{c}\text{sulphuric acid residue}\end{array}}_{\text{"sulphur trioxide"}\ (S_3O_9)} + 3H_2O \rightarrow 3 \underset{\text{sulphuric acid}}{\begin{array}{c}HO\\HO\end{array}\!\!S\!\!\begin{array}{c}O\\O\end{array}}$$

$$\underbrace{\begin{array}{c}\text{phosphoric acid residue}\end{array}}_{\text{"phosphorus pentoxide"}\ (P_4O_{10})} + 6H_2O \rightarrow 4\ \underset{\text{phosphoric acid}}{HO\!-\!\!\!\begin{array}{c}HO\\HO\end{array}\!\!\!P\!=\!O}$$

4) Acylation reactions. We have seen that acid anhydrides hydrolyse with water. They also undergo a similar reaction with alcohols and phenols to produce esters:

$$\underset{\text{acetic anhydride}}{\begin{array}{c}CH_3C\overset{O}{\diagup}\\ \diagdown O\\ CH_3C\diagup\\ \diagdown O\end{array}} + \underset{\text{ethanol}}{C_2H_5OH} \rightarrow \underset{\text{ethyl acetate}}{CH_3\!-\!C\overset{O}{\diagdown OC_2H_5}} + \underset{\text{acetic acid}}{CH_3C\overset{OH}{\diagdown O}}$$

$$\underset{\text{acetic anhydride}}{\begin{array}{c}CH_3C\overset{O}{\diagup}\\ \diagdown O\\ CH_3C\diagup\\ \diagdown O\end{array}} + \underset{\text{phenol}}{C_6H_5OH} \rightarrow \underset{\text{phenyl acetate}}{CH_3C\overset{O}{\diagdown OC_6H_5}} + CH_3C\overset{OH}{\diagdown O}$$

compare

$$\begin{array}{c}CH_3C\overset{O}{\diagup}\\ \diagdown O\\ CH_3C\diagup\\ \diagdown O\end{array} + \underset{\text{water}}{HOH} \rightarrow CH_3C\overset{O}{\diagdown OH} + CH_3C\overset{OH}{\diagdown O}$$

In these reactions the "active hydrogens" of ethanol, phenol, and water have been replaced by an acetyl group: they have been acetylated. Acid anhydrides are useful reagents for acylating hydroxy compounds.

Hydrogen atoms attached to nitrogen are also "active" and thus ammonia and primary and secondary amines (Section 24.1 and 24.2) can also be acylated by acid anhydrides:

$$\begin{array}{c}CH_3-C(=O)\\ O\\ CH_3-C(=O)\end{array} + \quad NH_3, \quad C_2H_5NH_2, \quad (C_2H_5)_2NH.$$
$$\text{ammonia} \quad \text{ethylamine} \quad \text{diethylamine}$$

$$\downarrow \qquad\qquad \downarrow \qquad\qquad \downarrow$$

$$CH_3COOH + CH_3CONH_2, \quad CH_3CONHC_2H_5, \quad CH_3CON(C_2H_5)_2.$$
acetic acid   acetamide   N-ethylacetamide   N,N-diethylacetamide

Acid chlorides are useful acylating agents too (Section 23.1.2), for example:

$$CH_3COCl + C_2H_5NH_2 \rightarrow CH_3CONHC_2H_5 + HCl.$$
acetyl chloride   ethylamine   N-ethylacetamide

Esters may also be useful acylating agents, particularly for the nitrogen type of active hydrogen compounds. They are not as commonly used, however, as acid chlorides or acid anhydrides.

$$\begin{array}{l}COOC_2H_5\\ |\\ COOC_2H_5\end{array} + 2NH_3 \rightarrow \begin{array}{l}CONH_2\\ |\\ CONH_2\end{array} + 2C_2H_5OH.$$
diethyl oxalate   oxamide   ethanol
(diethyl ethanedioate) (ethanediamide)

Ethers are clearly unsuitable for use in acylation reactions because they are very unreactive and in any case contain no acyl group.

5) Epoxides and ring strain. Cyclic ethers, such as epoxyethane (ethylene oxide)

$$\underset{CH_2-CH_2}{\overset{O}{\triangle}}$$

where the carbon–oxygen–carbon linkage is incorporated into a very small ring, provide a sharp contrast with the unreactive open-chain ethers. The ring size in this case dictates a C—O—C bond angle of about 60°, more than 40° less than the preferred, approximately tetrahedral angle (cf. H—O—H angle in water, Section 4.10). Prediction of the heat of formation of ethylene oxide, in the same way as a value was predicted for cyclopropane in Section 21.5, leads to a value of -217 kJ mol^{-1} (the C—H bond energy is again taken as being 426 kJ mol^{-1}, i.e. 12.6 kJ stronger than in methane). The experimentally observed value is -53 kJ mol^{-1}, leading to a value of 164 kJ mol^{-1} for the ring strain, very similar in magnitude to the ring strain in cyclopropane (about 200 kJ mol^{-1}). This strain makes the ethylene oxide molecule very ready to "spring" open and hydrolysis is easy. Other ring opening reactions, too, are easy and ethylene oxide

Table 23.6. Important reactions of compounds containing the carbon–oxygen–carbon linkage.

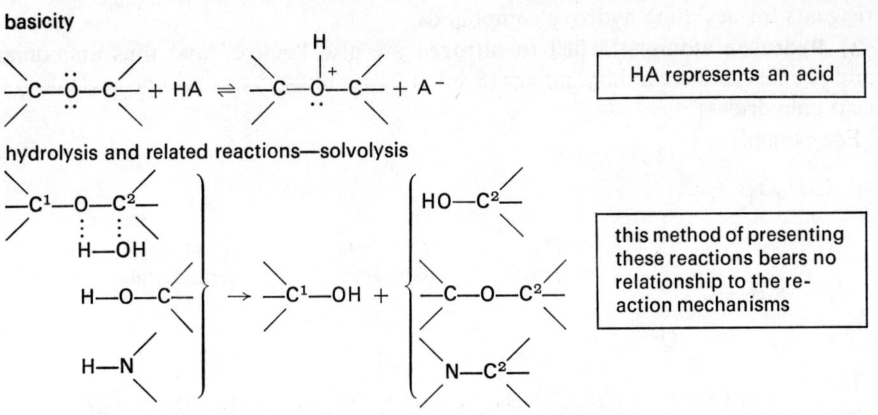

basicity

$$\mathrm{\overset{\diagdown}{\underset{\diagup}{C}}-\overset{..}{\underset{..}{O}}-\overset{\diagdown}{\underset{\diagup}{C}}- + HA \rightleftharpoons \overset{\diagdown}{\underset{\diagup}{C}}-\overset{\overset{H}{|}}{\underset{..}{\overset{+}{O}}}-\overset{\diagdown}{\underset{\diagup}{C}}- + A^-}$$

HA represents an acid

hydrolysis and related reactions—solvolysis

$$\left. \begin{array}{c} -\mathrm{C^1-O-C^2-} \\ \mathrm{H-OH} \\ \mathrm{H-O-C-} \\ \mathrm{H-N} \end{array} \right\} \rightarrow -\mathrm{C^1-OH} + \left\{ \begin{array}{c} \mathrm{HO-C^2-} \\ -\mathrm{C-O-C^2-} \\ \mathrm{N-C^2-} \end{array} \right.$$

this method of presenting these reactions bears no relationship to the reaction mechanisms

---

is an important synthetic intermediate both industrially and in the research laboratory:

$$\underset{CH_2-CH_2}{\overset{O}{\triangle}} + \xrightarrow[\text{aqueous acid}]{\text{dilute}} HOCH_2CH_2OH$$
ethylene glycol (ethane-1,2-diol)

$$\xrightarrow{\text{conc. HCl}} HOCH_2CH_2Cl$$
ethylene chlorohydrin (2-chloroethanol)

$$\xrightarrow{NH_3} HOCH_2CH_2NH_2.$$
ethanolamine (2-aminoethanol)

It is made commercially by the catalytic oxidation of ethylene with air:

$$CH_2{=}CH_2 \xrightarrow[\text{Ag catalyst}]{\text{air at 2–300 °C and about 20 atm}} \underset{CH_2-CH_2}{\overset{O}{\triangle}}$$

Other alkene oxides (epoxides) are made by treating an alkene with a peroxy derivative of a carboxylic acid:

$$\underset{\text{alkene}}{R-CH{=}CH-R'} + \underset{\text{perbenzoic acid}}{C_6H_5\overset{\overset{O}{\|}}{C}OOH} \rightarrow \underset{\text{epoxide}}{R-\overset{O}{\overset{\triangle}{CH-CH}}-R'} + \underset{\text{benzoic acid}}{C_6H_5\overset{\overset{O}{\|}}{C}OH}.$$

23.2.3 Production of the C—O—C Linkage

1) Acid-catalysed reaction between two hydroxy compounds

$$X{-}OH + HO{-}Y \xrightarrow{\text{acid}} X{-}O{-}Y + H_2O.$$

For example:

$$\underset{\text{ethanol}}{2C_2H_5OH} \xrightarrow[140\,°C]{H_2SO_4} \underset{\text{diethyl ether}}{C_2H_5OC_2H_5} + H_2O;$$

$$\underset{\text{acetic acid}}{CH_3COOH} + \underset{\text{ethanol}}{C_2H_5OH} \xrightarrow{H_2SO_4} \underset{\text{ethyl acetate}}{CH_3COOC_2H_5} + H_2O.$$

This type of reaction is suitable for the preparation of ethers and esters respectively and is unsuitable for preparing acid anhydrides (Section 23.1.2).

2) Reaction between hydroxy compounds and halogen compounds

$$X\text{—}OH + Cl\text{—}Y \rightarrow XOY + HCl.$$

For example:

$$\underset{\text{ethanol}}{C_2H_5OH} + \underset{\substack{\text{acetyl} \\ \text{chloride}}}{ClCOCH_3} \longrightarrow \underset{\text{ethyl acetate}}{C_2H_5OCOCH_3} + HCl;$$

$$\underset{\text{phenol}}{C_6H_5OH} + \underset{\substack{\text{benzoyl} \\ \text{chloride}}}{ClCOC_6H_5} \xrightarrow{\text{NaOH}} \underset{\text{phenyl benzoate}}{C_6H_5OCOC_6H_5} + HCl \text{ (removed by NaOH)}.$$

The only suitable C–halogen compounds for this reaction are acid chlorides and the only hydroxy compounds are alcohols and phenols. Thus, the only C—O—C compounds which can be prepared in this way are esters.

3) Reaction between salts of hydroxy compounds and halogen compounds.

This method is suitable for the preparation of all three types of compound, and has been discussed already in connection with the alkylation and acylation of hydroxy compounds (Section 23.1.2).

$$\underset{\text{(X—O metal)}}{X\text{—}O^-} + Cl\text{—}Y \rightarrow X\text{—}O\text{—}Y + \underset{\text{(metal chloride)}}{Cl^-}.$$

We have just seen that the hydroxyl group itself is only sufficiently reactive to react with acid chlorides—the most reactive (in this respect) type of C—Cl compound that we shall consider. When a proton is removed from the hydroxyl group, however, to give the ionic salt (X—OH $\xrightarrow{\text{base}}$ X—O$^-$) the XO$^-$ ion is more reactive than XOH. XO$^-$ will usually react with halogenoalkanes as well as acid chlorides, although it will not react with halogenoarenes (e.g. bromobenzene). (XO$^-$ is more powerfully nucleophilic than the uncharged X—O—H, and is able to displace halide ion from halogenoalkanes, Section 26.3.)

Thus, we have:

a) *Ethers*

$$\underset{\substack{\text{sodium} \\ \text{ethoxide}}}{C_2H_5ONa} + \underset{\text{iodomethane}}{CH_3I} \rightarrow \underset{\substack{\text{ethyl methyl} \\ \text{ether} \\ \text{(methoxyethane)}}}{C_2H_5OCH_3} + NaI$$

$$\underset{\substack{\text{sodium} \\ \text{phenoxide}}}{C_6H_5ONa} + \underset{\text{bromoethane}}{C_2H_5Br} \rightarrow \underset{\substack{\text{phenyl ethyl} \\ \text{ether} \\ \text{(ethoxybenzene)}}}{C_6H_5OC_2H_5} + NaBr.$$

($C_2H_5ONa + C_6H_5Br$ *no* reaction as aryl halides are unreactive in this respect.)

b) *Esters*

$$\underset{\text{silver acetate}}{CH_3COOAg} + \underset{\text{iodoethane}}{C_2H_5I} \rightarrow \underset{\text{ethyl acetate}}{CH_3COOC_2H_5} + AgI.$$

c) *Acid anhydrides*

$$CH_3COONa + CH_3COCl \xrightarrow{heat} CH_3COOCOCH_3 + NaCl.$$
sodium acetate   acetyl chloride   acetic anhydride

4) The reaction between acid anhydrides, which already contain a C—O—C linkage, and alcohols or phenols. This produces a new C—O—C linkage. We have already seen how this reaction produces esters (Section 23.1.2).

$$CH_3COOCOCH_3 + C_2H_5OH \rightarrow CH_3COOC_2H_5 + CH_3COOH.$$
acetic anhydride   ethanol   ethyl acetate   acetic acid

The alkylation and acylation reactions of hydroxy compounds (Table 23.3) summarize important methods of synthesizing carbon–oxygen–carbon linkages.

23.3 THE CARBONYL GROUP

The simplest molecules containing carbonyl groups are aldehydes and ketones. Formaldehyde (methanal) is the smallest such molecule and is, essentially, a carbonyl group attached to two hydrogen atoms: $\begin{array}{c}H\\ \diagdown\\ C=O.\\ \diagup\\ H\end{array}$ If one of these hydrogen atoms is replaced by a hydrocarbon group, aromatic or aliphatic, then we have an aldehyde, $\begin{array}{c}R\\ \diagdown\\ C=O,\\ \diagup\\ H\end{array}$ such as $\begin{array}{c}CH_3\\ \diagdown\\ C=O,\\ \diagup\\ H\end{array}$ acetaldehyde (ethanal), and $\begin{array}{c}C_6H_5\\ \diagdown\\ C=O,\\ \diagup\\ H\end{array}$ benzaldehyde. Replacement of both hydrogen atoms by hydrocarbon groups leads to ketones, $\begin{array}{c}R\\ \diagdown\\ C=O,\\ \diagup\\ R'\end{array}$ such as CH_3COCH_3, dimethyl ketone (better known as acetone but more systematically as propanone), and $CH_3COC_6H_5$, methyl phenyl ketone (better known as acetophenone). The hydrogen atom attached to the carbonyl group of aldehydes is responsible for the differences between the chemical properties of aldehydes and ketones, and is thus part of the aldehyde function. Aldehydes are normally written RCHO, and ketones RCOR'.

We have already come across the carbonyl group in connection with the carboxylic acids where we saw how it affected the character of the hydroxyl group. The effect of the hydroxyl group on the carbonyl group is perhaps more marked since, in carboxylic acids, the chemical reactions of the carbonyl group have apparently little in common with those of aldehydes and ketones (but see Section 26.4). We shall note in the following account that carboxylic acids and their derivatives show very few of the "carbonyl" reactions shown by aldehydes and ketones (the —COOH group is often regarded as a single functional group).

The types of compound containing carbonyl groups which we shall consider are, therefore:

1. R—CHO, aliphatic aldehydes.
2. Ar—CHO, aromatic aldehydes.
3. $\begin{array}{c}R\\ \diagdown\\ \diagup\\ R'\end{array}$C=O, aromatic and aliphatic ketones.
4. R—C$\begin{array}{c}\diagup\!\!\!\!\!\!O\\ \diagdown OH\end{array}$, carboxylic acids.
5. R—C$\begin{array}{c}\diagup\!\!\!\!\!\!O\\ \diagdown Cl\end{array}$, acid chlorides.
6. R—C$\begin{array}{c}\diagup\!\!\!\!\!\!O\\ \diagdown OR'\end{array}$, esters.
7. $\begin{array}{c}\text{R—C}\diagup\!\!\!\!\!\!O\\ \diagdown O\\ \text{R—C}\diagdown\!\!\!\!\!\!O\end{array}$, acid anhydrides.
8. R—C$\begin{array}{c}\diagup\!\!\!\!\!\!O\\ \diagdown NH_2\end{array}$, amides.

(5) to (8) are derivatives of carboxylic acids. (Any of the R groups in (1) to (8) can be replaced with aromatic groups, Ar.)

Nomenclature, for the various types of compound, is as follows:

Aldehydes. The IUPAC name derives from the parent hydrocarbon and the ending -*al* is used to indicate that the compound is an aldehyde. Thus ethanal is CH_3CHO and 2-methylpropanal is $(CH_3)_2CHCHO$. Numbering starts at the aldehyde carbon atom which is always number one. An alternative nomenclature, generally used for many of the simpler and more common aldehydes, uses the name of the acid, to which the aldehyde may be oxidized, as the root, and *aldehyde* as the word-ending. Thus CH_3CHO is normally called acetaldehyde (after acetic acid) and C_6H_5CHO benzaldehyde (after benzoic acid).

Ketones. The IUPAC system uses the hydrocarbon corresponding to the longest straight carbon chain in the molecule to derive the name from, with the ending

-one. Thus propanone is CH_3COCH_3 (no need for numbering here as there is only one possible position for the ketone carbonyl) and pentan-2-one is $CH_3COCH_2CH_2CH_3$. An alternative IUPAC system simply specifies both groups attached to the carbonyl group and uses the term *ketone*, so that CH_3COCH_3 would be dimethyl ketone and $C_6H_5COCH_3$ methyl phenyl ketone. Trivial names, such as acetone and acetophenone for the last two compounds (respectively), are still in very common use, though, and the student should be familiar with these (indicated where appropriate in the text).

Acids and their derivatives. Carboxylic acids, esters, and anhydrides have already been discussed (Section 23.1 and 23.2). Amides are named after their parent acids by using the ending *-amide*. Ethanamide (or, more usually, acetamide) is the name for CH_3CONH_2, propanamide for $CH_3CH_2CONH_2$, and benzamide for $C_6H_5CONH_2$. Acid chlorides are also named from their parent acids, using the ending *-oyl chloride*; thus propanoyl chloride is CH_3CH_2COCl and benzoyl chloride C_6H_5COCl. Acetyl chloride (or ethanoyl chloride) is CH_3COCl, and in this case, the *o* is omitted.

Sometimes, in referring to carbon atoms in a carbonyl-containing molecule, a system of Greek letters is used, which indicates how far away the carbon atom is from the carbonyl group. The system does not correspond to the IUPAC numbering as is clear from the following examples:

$$\overset{5}{C}H_3 - \overset{4}{C}H_2 - \overset{3}{C}H_2 - \overset{2}{C}O - \overset{1}{C}H_3 ;$$
$$\gamma\beta\alpha\alpha$$
pentan-2-one

$$\overset{4}{C}H_3 - \overset{3}{C}H_2 - \overset{2}{C}H_2 - \overset{1}{C}OOH.$$
$$\gamma\beta\alpha$$
butanoic acid

23.3.1 Physical Properties

The boiling points of aldehydes and ketones rise steadily with increasing molecular weight. Formaldehyde is a gas, the rest are liquids and (for larger molecular weights) solids. The boiling point of a given aldehyde or ketone is usually higher than that of a hydrocarbon or ether of similar molecular weight (Table 23.7) due to the electronic nature of the carbonyl group; no hydrogen bonding is involved. (The attraction of the carbonyl dipoles—Section 4.9—makes it difficult to drive molecules into the gas phase.) The boiling points of carboxylic acids are higher than those of aldehydes or ketones of similar molecular weight as hydrogen bonds

Table 23.7

	Molecular weight	Boiling point (°C)
CH_3CHO acetaldehyde	44	21
CH_3OCH_3 dimethyl ether	46	−25
$CH_3CH_2CH_3$ propane	44	−42

Table 23.8

	Molecular weight	Boiling point (°C)
Acetic acid	60	118
Acetone (propanone)	58	56
Propan-2-ol	60	82

usually exist between the hydroxyl hydrogen atom of one molecule and the carbonyl oxygen atom of another. Acetic acid, for example, can exist as a dimer due to this hydrogen bonding, which is generally more effective than dipole attractions in raising boiling points as is revealed by comparing the boiling points of propan-2-ol and acetone (propanone) (Table 23.8):

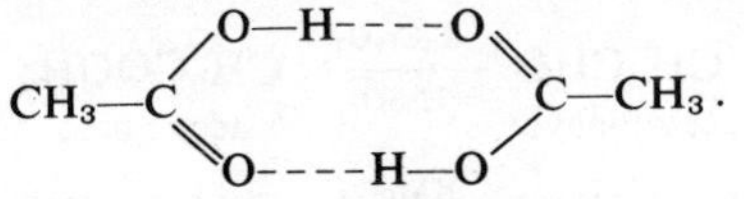

$$CH_3-C{\overset{O-H----O}{\underset{O----H-O}{}}}C-CH_3.$$

acetic acid dimer

23.3.2 Chemical Behaviour

1) Basicity. The carbonyl group is weakly basic in character (although more basic than an ether oxygen atom) and, provided that a sufficiently strong acid is used, will accept a proton on the oxygen atom. This property is common to the carbonyl group of aldehydes, ketones, and carboxylic acids and their derivatives. Sometimes this protonation may be followed by another chemical reaction (see, for example, point 8 below on the acid-catalysed polymerization of aldehydes) but it is usually the first thing that happens to a carbonyl compound on treatment with acid:

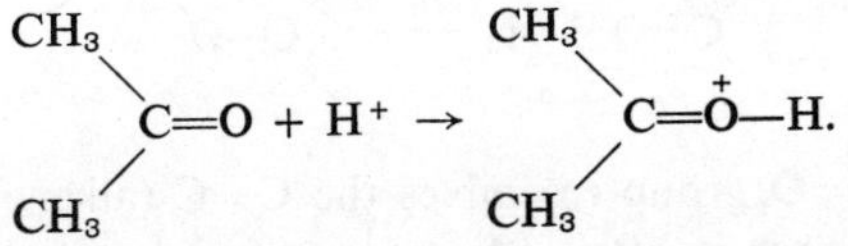

$$(CH_3)_2C=O + H^+ \rightarrow (CH_3)_2C=\overset{+}{O}-H.$$

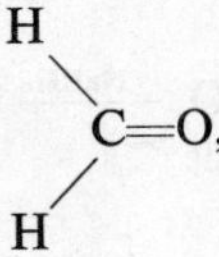

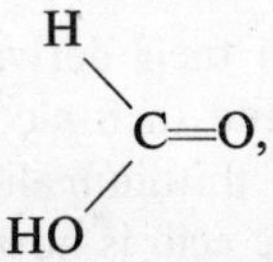

2) Oxidation. Reference to Table 23.2 will be helpful in considering the oxidation and reduction of carbonyl compounds. We first considered Table 23.2 in connection with the oxidation of alcohols and we saw there that carbonyl compounds with an "aldehyde hydrogen atom", that is a hydrogen atom attached to the carbonyl carbon atom, were readily oxidized with cleavage of that carbon–hydrogen bond. Thus formaldehyde, $H_2C=O$, is readily oxidized to formic acid, $\underset{HO}{\overset{H}{\diagdown}}C=O$, which, as it still contains an "aldehyde hydrogen atom", is further oxidized to carbonic acid, $\begin{array}{c}HO\\ \diagdown\\ C{=}O\\ \diagup\\ HO\end{array}$, that is, carbon dioxide and water.

Formic acid is the only simple carboxylic acid which is easily oxidized. The others, containing no "aldehyde hydrogen atom", are oxidized only with difficulty as breakage of a carbon–carbon bond is involved. Thus, only aldehydes are good reducing agents; ketones and carboxylic acid derivatives are not. Simple aldehydes (and formic acid) will reduce ammoniacal silver nitrate to metallic silver; they will reduce alkaline copper(II) salts (Fehling's solution) to copper(I) oxide (red Cu_2O precipitates from the blue solution, Section 20.21). Ketones and carboxylic acids will not reduce these solutions.

The product obtained by oxidizing an aldehyde is a carboxylic acid and this reaction is sometimes used to prepare the latter:

$$CH_3CHO \xrightarrow[H_2SO_4]{K_2Cr_2O_7} CH_3COOH;$$
$$\text{acetaldehyde} \qquad\qquad \text{acetic acid}$$

$$C_6H_5CHO \xrightarrow{KMnO_4} C_6H_5COOH.$$
$$\text{benzaldehyde} \qquad\qquad \text{benzoic acid}$$

(On the basis of the earlier discussion on oxidation reactions—Section 18.14—you should attempt to write balanced equations for these reactions.)

3) Reduction. The reduction of aldehydes and ketones gives rise to alcohols: primary alcohols from aldehydes and secondary alcohols from ketones. The reduction can be brought about by treating the compound with hydrogen and a nickel catalyst when hydrogenation of the double bond occurs:

$$\diagup\!\!\!\diagdown\!\!\!C{=}O + H_2 \xrightarrow{Ni} \diagup\!\!\!\diagdown\!\!\!\overset{H\;\;H}{C{-}O}.$$

In this respect the C=O group resembles the C=C (alkene) group. Reduction of aldehydes and ketones is often effected in the laboratory by treatment with complex metal hydrides, such as lithium aluminium hydride ($LiAlH_4$), or sodium borohydride ($NaBH_4$) (Section 13.5):

$$CH_3COCH_3 \xrightarrow[Ni]{H_2} CH_3CH(OH)CH_3;$$
$$\text{acetone} \qquad\qquad \text{propan-2-ol}$$
$$\text{(propanone)}$$

$$C_6H_5CHO \xrightarrow{NaBH_4} C_6H_5CH_2OH.$$
$$\text{benzaldehyde} \qquad\qquad \text{benzyl alcohol}$$
$$\text{(phenylmethanol)}$$

One might suppose that reduction of carboxylic acids and their derivatives (reading Table 23.2 from right to left) should give aldehydes and then alcohols. In practice, however, carboxylic acids are very difficult to reduce. Lithium aluminium hydride is one of the only reagents which can do this and the acid is normally esterified first. The product is a primary alcohol. Any aldehyde which might be produced would be reduced more readily than any unreacted acid.

$$\underset{\text{propanoic acid}}{CH_3CH_2COOH} \longrightarrow \underset{\text{methyl propanoate}}{CH_3CH_2COOCH_3} \xrightarrow{LiAlH_4} \underset{\text{propan-1-ol}}{CH_3CH_2CH_2OH} + CH_3OH.$$

The carbonyl reductions which we have so far considered involve the addition of hydrogen to a double bond, and the carbonyl oxygen remains in the molecule as part of a hydroxyl group. By choosing the right reducing agent we can remove that oxygen altogether so that we can convert C=O into CH_2. Zinc amalgam and concentrated hydrochloric acid will effect this reduction (Clemmensen reduction):

$$\underset{\substack{\text{methyl phenyl}\\\text{ketone}}}{C_6H_5COCH_3} \xrightarrow[HCl]{Zn/Hg} \underset{\text{ethylbenzene}}{C_6H_5CH_2CH_3}.$$

Carboxylic acids cannot be reduced in this way to hydrocarbons. No ethane is obtained, for instance, on treating acetic acid with these reagents.

4) Addition reactions. Like alkenes, carbonyl compounds contain a double bond and, like alkenes, carbonyl compounds will undergo addition reactions. We have already seen that they will add hydrogen to form hydroxyl compounds,

$$\diagdown\!\!\!\!\diagup C=O \rightarrow \diagdown\!\!\!\!\diagup \overset{H}{\underset{}{C}}\!\!-\!\!\overset{H}{\underset{}{O}},$$

and that this applies only to aldehydes and ketones, not to carboxylic acids.

Aldehydes and ketones undergo a number of addition reactions and a summary of some of the important ones follows.

a) $\diagdown\!\!\!\!\diagup C=O + H_2 \xrightarrow{Ni} \diagdown\!\!\!\!\diagup \overset{H}{\underset{}{C}}\!\!-\!\!\overset{H}{\underset{}{O}}.$

For example:

$$\underset{\substack{\text{acetone}\\\text{(propanone)}}}{CH_3COCH} + H_2 \xrightarrow{Ni} \underset{\text{propan-2-ol}}{CH_3-\underset{H}{\overset{OH}{C}}-CH_3}.$$

b) $\diagdown\!\!\!\!\diagup C=O + HCN \rightarrow \diagdown\!\!\!\!\diagup \overset{CN}{\underset{}{C}}\!\!-\!\!\overset{H}{\underset{}{O}}.$

For example:

$$\underset{\text{acetone}}{CH_3COCH_3} + HCN \rightarrow \underset{\substack{\text{acetone}\\\text{cyanohydrin}}}{CH_3-\underset{CN}{\overset{OH}{C}}-CH_3};$$

$$CH_3CHO + HCN \rightarrow CH_3-\underset{\underset{CN}{|}}{\overset{\overset{OH}{|}}{C}}-H.$$

acetaldehyde   acetaldehyde cyanohydrin

c) $\underset{HO}{\overset{Na^+O^-}{}}C=O + \underset{}{\overset{}{S}}=O \rightarrow$ sodium bisulphite adduct (cyclic structure with S-O-Na$^+$)

For example:

$$CH_3CHO + NaHSO_3 \rightarrow CH_3-\underset{\underset{H}{|}}{\overset{\overset{OH}{|}}{C}}-SO_3Na.$$

acetaldehyde   sodium bisulphite   acetaldehyde bisulphite addition compound

This adduct sometimes separates from aqueous solution and treatment with dilute mineral acid converts it back to the original aldehyde or ketone:

$$CH_3-\underset{\underset{H}{|}}{\overset{\overset{OH}{|}}{C}}-SO_3Na \xrightarrow[HCl]{dilute} CH_3CHO + NaCl + H_2O + SO_2.$$

acetaldehyde bisulphite addition compound   acetaldehyde

The reaction therefore forms the basis of a method for purifying aldehydes and ketones, useful when they must be separated from other materials.

Alcohols, in the presence of a dry, acid catalyst, add to aldehyde carbonyl groups to form hemi-acetals. The reaction, however, is usually followed rapidly by a second reaction, which is not an addition, and the final product is an acetal:

$$\underset{H}{\overset{R}{}}C=O + R'OH \xrightarrow{acid} \underset{\underset{OH}{|}}{\overset{\overset{OR'}{|}}{\underset{H}{\overset{R}{C}}}} \xrightarrow[R'OH]{acid} \underset{\underset{OR'}{|}}{\overset{\overset{OR'}{|}}{\underset{H}{\overset{R}{C}}}} + H_2O.$$

hemi-acetal   acetal

$$CH_3CHO + C_2H_5OH \xrightarrow[HCl]{dry} CH_3\underset{\underset{OH}{|}}{\overset{\overset{OC_2H_5}{|}}{C}}-H \xrightarrow[HCl]{C_2H_5OH} CH_3\underset{\underset{OC_2H_5}{|}}{\overset{\overset{OC_2H_5}{|}}{C}}-H + H_2O;$$

acetaldehyde   ethanol   acetaldehyde hemi-acetal (1-ethoxyethanol)   acetaldehyde diethyl acetal (1,1-diethoxyethane)

Hemi-acetals are usually too unstable to be isolated, decomposing easily to aldehyde and alcohol. Ketones give the corresponding compounds—ketals—only with difficulty except in the case of their reaction with a dihydroxy compound, when the cyclic ketal forms readily. For example:

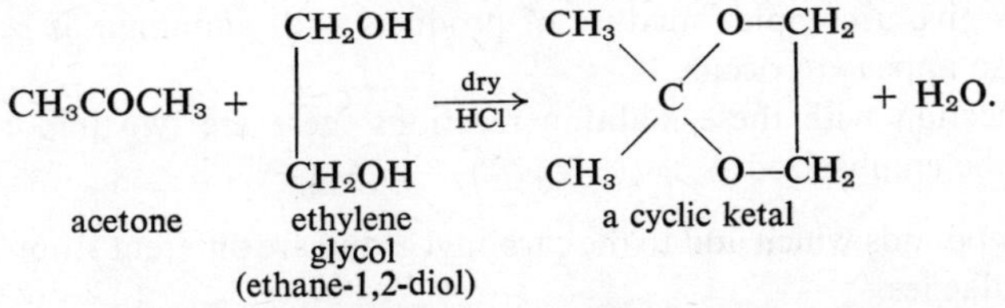

$$\text{CH}_3\text{COCH}_3 + \begin{array}{c}\text{CH}_2\text{OH}\\|\\\text{CH}_2\text{OH}\end{array} \xrightarrow[\text{HCl}]{\text{dry}} \begin{array}{c}\text{CH}_3\quad\;\;\text{O}-\text{CH}_2\\\diagdown\;\;\diagup\\\text{C}\\\diagup\;\;\diagdown\\\text{CH}_3\quad\;\;\text{O}-\text{CH}_2\end{array} + \text{H}_2\text{O}.$$

acetone   ethylene glycol (ethane-1,2-diol)   a cyclic ketal

Treatment with dilute aqueous acid readily regenerates the original carbonyl compound from acetals or ketals by reversing the above reactions. (This reversal is ensured by the presence of an excess of water. *Dry* conditions are necessary for acetal or ketal formation.) The process involves the ready hydrolysis of a carbon–oxygen bond in a carbon–oxygen–carbon linkage (cf. ether hydrolysis in Section 23.2.2). Two alkoxy groups attached to the same carbon atom "labilize" each other with respect to acid-catalysed hydrolysis, and we do not observe the resistance to this type of cleavage reaction shown by simple ethers.

Ammonia will add to simple aldehydes, such as acetaldehyde, to yield an adduct which is unstable and which polymerizes. Treatment of acetaldehyde in ether with ammonia gas results in the precipitation of a white solid whose empirical formula corresponds to a simple adduct. It was called, therefore, "acetaldehyde ammonia" and assigned the structure:

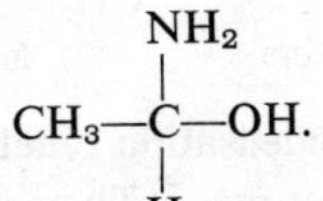

$$\begin{array}{c}\text{NH}_2\\|\\\text{CH}_3-\text{C}-\text{OH}.\\|\\\text{H}\end{array}$$

The correct structure is now known to be

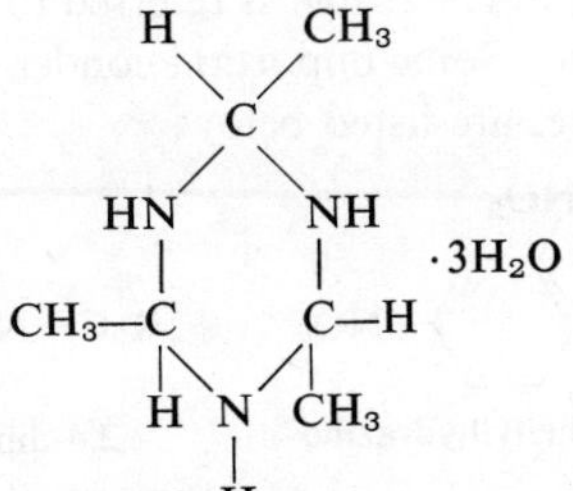

$$\begin{array}{c}\quad\;\;\text{H}\quad\text{CH}_3\\\diagdown\;\;\diagup\\\text{C}\\\diagup\;\;\diagdown\\\text{HN}\quad\text{NH}\\|\qquad\quad|\qquad\cdot 3\text{H}_2\text{O}\\\text{CH}_3-\text{C}\quad\text{C}-\text{H}\\\diagup\;\diagdown\;\diagup\;\diagdown\\\text{H}\quad\text{N}\quad\text{CH}_3\\|\\\text{H}\end{array}$$

and the simple adduct is believed to trimerize, very soon after it is formed, to this final product which separates with three moles of water of crystallization:

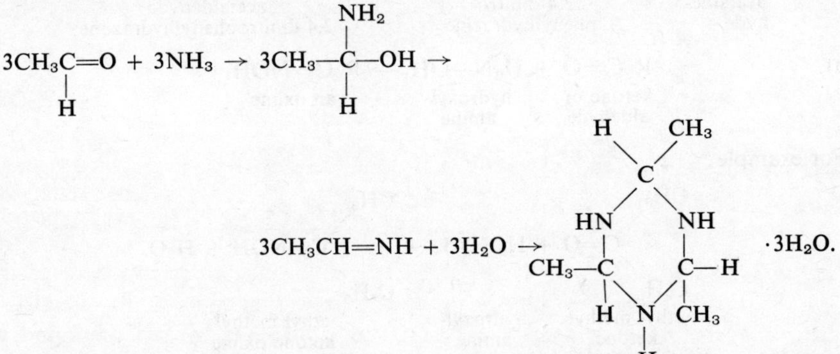

$$3\begin{array}{c}\text{CH}_3\text{C}=\text{O}\\|\\\text{H}\end{array} + 3\text{NH}_3 \rightarrow 3\begin{array}{c}\text{NH}_2\\|\\\text{CH}_3-\text{C}-\text{OH}\\|\\\text{H}\end{array} \rightarrow$$

$$3\text{CH}_3\text{CH}=\text{NH} + 3\text{H}_2\text{O} \rightarrow \begin{array}{c}\quad\;\;\text{H}\quad\text{CH}_3\\\diagdown\;\;\diagup\\\text{C}\\\diagup\;\;\diagdown\\\text{HN}\quad\text{NH}\\|\qquad\quad|\\\text{CH}_3-\text{C}\quad\text{C}-\text{H}\\\diagup\;\diagdown\;\diagup\;\diagdown\\\text{H}\quad\text{N}\quad\text{CH}_3\\|\\\text{H}\end{array} \cdot 3\text{H}_2\text{O}.$$

Ketones give a complex mixture of products with ammonia as several other reactions also appear to occur.

In connection with these addition reactions there are two important points which must be emphasized:

a) The compounds which add to the carbonyl group are different from those which add to alkenes.
b) Carboxylic acids and their derivatives do not undergo these reactions.

5) Condensation reactions. When the elements of a smaller molecule are lost from a single larger molecule, the chemical reaction involved is known as an **elimination reaction** and a multiple (usually double) bond is formed. Sometimes however, the elements of a smaller molecule are eliminated from two larger molecules—part of the smaller molecule from each of the larger ones—and the two larger molecules can link together. Thus, for example, a molecule of water is removed from one molecule of acetone and one molecule of phenylhydrazine when the two react:

$$\begin{array}{c}CH_3\\ \diagdown\\ CH_3\end{array}\!\!C\!\!=\!\!O + H_2N\!-\!NHC_6H_5 \rightarrow \begin{array}{c}CH_3\\ \diagdown\\ CH_3\end{array}\!\!C\!\!=\!\!N\!-\!NHC_6H_5 + H_2O.$$

acetone   phenylhydrazine   acetone phenylhydrazone

Such a reaction is known as a **condensation reaction** and reactions of this type are characteristic of aldehydes and ketones. They do not constitute a fourth fundamental type of organic reaction as they are normally made up of both an addition step and an elimination step. The reader is referred to Sections 26.2 and 26.4 for a fuller discussion of this point. Some important condensation reactions, in addition to that with phenylhydrazine, are listed below.

a) $R_2C\!=\!O + H_2NNH\!-\!\!\!\bigcirc\!\!\!-NO_2 \rightarrow R_2C\!=\!NNH\!-\!\!\!\bigcirc\!\!\!-NO_2 + H_2O.$
      (with NO_2 substituent)

ketone or   2,4-dinitrophenylhydrazine   2,4-dinitrophenylhydrazone
aldehyde

For example:

$$CH_3CHO + H_2NNHC_6H_3(NO_2)_2 \rightarrow CH_3CH\!=\!NNHC_6H_3(NO_2)_2.$$

acetalde-   2,4-dinitro-   acetaldehyde
hyde   phenylhydrazine   2,4-dinitrophenylhydrazone

b) $R_2C\!=\!O + H_2N\!-\!OH \rightarrow R_2C\!=\!NOH.$

             ketone or   hydroxyl-   an oxime
             aldehyde   amine

For example:

$$\begin{array}{c}CH_3\\ \diagdown\\ C_2H_5\end{array}\!\!C\!\!=\!\!O + H_2NOH \rightarrow \begin{array}{c}CH_3\\ \diagdown\\ C_2H_5\end{array}\!\!C\!\!=\!\!NOH + H_2O.$$

ethyl methyl   hydroxyl-   ethyl methyl
ketone   amine   ketone oxime
(butanone)   (butanone oxime)

c) $\quad R_2C{=}O + H_2NR \rightarrow R_2C{=}NR + H_2O$.
            ketone or   amine   a "Schiff's base"
            aldehyde (primary)   or imine

This reaction normally only works well for aromatic compounds (Section 24.5). For example:

$$C_6H_5CHO + H_2NC_6H_5 \rightarrow C_6H_5CH{=}NC_6H_5 + H_2O.$$
  benzaldehyde   phenylamine   N-benzylidene
                 (aniline)   phenylamine
                            (benzylidene aniline)

d) $\quad R_2C{=}O + Ph_3P{=}CHR' \rightarrow R_2C{=}CHR' + Ph_3P{=}O$.
       ketone or   A "Wittig"   alkene   triphenyl-
       aldehyde   reagent   phosphine
                                                        oxide

For example:

$$PhCHO + Ph_3P{=}CHPh \rightarrow PhCH{=}CHPh + Ph_3P{=}O.$$
  benzalde-   benzylidene   1,2-diphenyl-   triphenyl-
   hyde   triphenyl-   ethene   phosphine
          phosphorane   (stilbene)   oxide

This is a very important method of synthesizing alkenes.

The reaction with 2,4-dinitrophenylhydrazine is very easily carried out in a test-tube and whilst many simple ketones and aldehydes are liquids, the product is usually an insoluble crystalline solid. Thus the product forms a yellow-orange precipitate and the reaction provides a useful test for aldehydes and ketones. Carboxylic acids and their derivatives (amides, anhydrides, esters, acid chlorides, etc.) do not undergo these condensation reactions.

6) Substitution reactions. Whilst carboxylic acids and their derivatives do not undergo the addition and condensation reactions which characterize aldehyde and ketone chemistry, they do undergo substitution reactions which are only very rarely a feature of aldehyde or ketone chemistry. Substitution reactions characterize the carboxylic acid family in much the same way that addition and condensation reactions characterize the aldehyde/ketone family. From a mechanistic point of view (Section 26.4) these three types of reaction are quite closely related, although they do seem, at first sight, to be very different.

A typical substitution reaction would be the hydrolysis of an ester, such as ethyl acetate, where the $-OC_2H_5$ group of the ester has been replaced by $-OH$:

$$CH_3COOC_2H_5 + H_2O \rightleftharpoons CH_3COOH + C_2H_5OH.$$

This is a reversible reaction, and the reverse reaction involves the replacement of the $-OH$ group of acetic acid by $-OC_2H_5$. Because these reactions involve the fission of carbon–oxygen single bonds they have been dealt with under that heading (Sections 23.1 and 23.2). That the carbonyl group is intimately involved in the reaction, however, is clear from its influence in rendering the ester carbon–oxygen bond so much more labile (easy to break) than is the carbon–oxygen bond in ethers. The carbonyl group exerts a similar influence on bonds between carbon and other electronegative elements such as the halogens (acid halides, Section 25.2) and nitrogen (acid amides, Section 24.1). This influence is illustrated in Table 23.4. Important reactions of carboxylic acid derivatives are:

a) *Hydrolysis*

$$RCOCl + H_2O \rightarrow RCOOH + HCl$$
$$RCOOCOR + H_2O \rightarrow 2RCOOH$$
$$RCOOR' + H_2O \rightleftharpoons RCOOH + R'OH$$
$$RCONH_2 + H_2O \rightarrow RCOOH + NH_3.$$

direction of decreasing reactivity towards water, ammonia, and alcohols ↓

b) *Ammonolysis*

$$RCOCl + NH_3 \rightarrow RCONH_2 + HCl$$
$$RCOOCOR + NH_3 \rightarrow RCONH_2 + RCOOH$$
$$RCOOR' + NH_3 \rightarrow RCONH_2 + R'OH.$$

(rarely of preparative use)

c) *Alcoholysis*

$$RCOCl + R''OH \rightarrow RCOOR'' + HCl$$
$$RCOOCOR + R''OH \rightarrow RCOOR'' + RCOOH$$
$$RCOOR' + R''OH \rightarrow RCOOR'' + R'OH.$$

The term "hydrolysis" has already been defined (Section 23.1.3). *Ammonolysis* and *alcoholysis* are similarly defined, with ammonia or an alcohol, respectively, assuming the role of water in hydrolysis. The nature of reactions such as these is discussed in Section 26.4 where it is illustrated with particular reference to the ammonolysis of acetyl chloride.

The sequence of reactivity of the carbonyl compounds with respect to hydrolysis is the same for ammonolysis and alcoholysis too. All of these reactions are subject to catalysis by acids and bases and such catalysts are normally necessary to enable all but acid chlorides to react at conveniently high rates. Acetyl chloride reacts vigorously and exothermically with water, for example, without added catalysts, "spitting" and fuming dangerously. Acetamide must be boiled with aqueous alkali (or acid) to effect hydrolysis.

7) The effect of alkali. There are two common types of reaction which aldehydes and ketones undergo in the presence of alkali. Some will undergo an aldol reaction, others will undergo a Cannizzaro reaction. (The base-catalysed hydrolysis of acid derivatives has been considered under point 6 above.)

a) *Aldol reactions.* Treatment of acetaldehyde with dilute sodium hydroxide solution results in the formation of a carbon–carbon bond to yield a dimer of acetaldehyde known as aldol:

$$2CH_3C\underset{H}{\overset{O}{\diagup\!\!\!\diagdown}} \xrightarrow{NaOH} CH_3-\underset{H}{\overset{OH}{C}}-CH_2-C\underset{H}{\overset{O}{\diagup\!\!\!\diagdown}}$$

acetaldehyde → aldol (3-hydroxybutanal)

This compound gives its name to all reactions of this type. One molecule of acetaldehyde has added across the carbonyl double bond of another in the sense:

$$\begin{array}{c} H-CH_2-CHO \\ \vdots \\ O=C-CH_3 \\ | \\ H \end{array} \rightarrow \begin{array}{c} CH_2CHO \\ | \\ HO-C-CH_3 \\ | \\ H \end{array}$$

The nature of this reaction is more fully discussed in Section 28.3.

Similarly, on appropriate treatment with the right alkali (barium hydroxide), acetone dimerizes:

$$2CH_3COCH_3 \xrightarrow{Ba(OH)_2} CH_3-\underset{\underset{CH_3}{|}}{\overset{\overset{OH}{|}}{C}}-CH_2COCH_3.$$

acetone   diacetone alcohol
(4-hydroxy-4-methyl-pentan-2-one)

Only compounds which contain a hydrogen atom on the carbon atom adjacent to the carbonyl group can undergo this reaction and this, of course, is because a hydrogen atom in that position—the α-position—is involved in the addition as outlined above. Thus, the following aldehydes cannot undergo an aldol reaction because they contain no α-hydrogen atom:

C_6H_5-CHO $H_2C=O$ $(CH_3)_3C-CHO$

benzaldehyde   formaldehyde   2,2-dimethylpropanal

Acetaldehyde, on treatment with concentrated caustic soda solution, produces a yellowy-brown gelatinous resin which probably arises from aldol reactions between a number of aldehyde molecules, giving long-chain molecules. This is one method for distinguishing between simple aldehydes and ketones as the latter do not yield these resinous precipitates.

b) *Cannizzaro reactions.* Those aldehydes containing no α-hydrogen atom undergo the Cannizzaro reaction with caustic soda solution. Ketones do not undergo this reaction at all. The reaction is quite different from the aldol reaction and no carbon–carbon bond is formed. It is curious in that, under the influence of alkali, one molecule of aldehyde appears to oxidize another to a carboxylic acid, being itself reduced to a primary alcohol. Thus, benzaldehyde and the other aldehydes listed above react as follows, undergoing disproportionation:

$$2C_6H_5CHO \xrightarrow{NaOH} C_6H_5COOH + C_6H_5CH_2OH;$$

benzaldehyde   benzoic acid   benzyl alcohol
(phenylmethanol)

$$2H_2CO \xrightarrow{NaOH} HCOOH + CH_3OH;$$

formaldehyde (methanal)   formic acid (methanoic acid)   methanol

$$(CH_3)_3C-CHO \xrightarrow{NaOH} (CH_3)_3C-COOH + (CH_3)_3C-CH_2OH.$$

2,2-dimethylpropanal   2,2-dimethylpropanoic acid   2,2-dimethylpropan-1-ol

c) *The haloform reaction.* One very special type of aldehyde or ketone undergoes a rather unusual reaction with alkali in which a carbon–carbon bond is broken. Ketones or aldehydes containing a trihalogenomethyl group attached to the carbonyl carbon atom will lose that group on reaction with alkali. For example, tri-iodoacetaldehyde and trichloroacetone break up as follows:

$$\underbrace{CI_3CHO}_{\substack{\text{tri-iodomethyl} \\ \text{group of tri-} \\ \text{iodoacetaldehyde}}} + NaOH \rightarrow \underbrace{CI_3H}_{\substack{\text{tri-iodo-} \\ \text{methane} \\ \text{(iodoform)}}} + \underbrace{HCOONa}_{\substack{\text{sodium} \\ \text{formate}}};$$

$$\underbrace{CCl_3COCH_3}_{\substack{\text{trichloromethyl} \\ \text{group of tri-} \\ \text{chloroacetone}}} + NaOH \rightarrow \underbrace{CCl_3H}_{\substack{\text{trichloro-} \\ \text{methane} \\ \text{(chloroform)}}} + \underbrace{CH_3COONa}_{\substack{\text{sodium} \\ \text{acetate}}}.$$

Compare these reactions with that of acetyl chloride and sodium hydroxide:

$$\underset{\substack{\text{acetyl} \\ \text{chloride}}}{ClCOCH_3} + NaOH \rightarrow NaCl + CH_3COOH;$$

$$CH_3COOH + NaOH \rightarrow \underset{\text{sodium acetate}}{CH_3COONa} + H_2O.$$

This reaction is very unusual in so far as the aldehydes and ketones involved are behaving as though they were carboxylic acid derivatives. The trihalogenomethyl group is displaced as is the halogen atom of acid halides. In fact the three electronegative halogen atoms have conferred some of their own character upon the carbon atom to which they are attached.

As well as being of interest because of this unusual feature of carbon–carbon bond cleavage, this reaction is important for other reasons. Trichloromethane (chloroform) used to be used as an anaesthetic and is a very useful solvent whilst tri-iodomethane (iodoform) has useful antiseptic properties. The reaction is therefore used to prepare these compounds and the starting materials are very much simpler than one might suppose. Thus, treatment of acetone with iodine, under alkaline conditions, replaces all three hydrogen atoms of *one* methyl group with iodine very rapidly:

$$\underset{\text{acetone}}{CH_3COCH_3} + 3I_2 + 3OH^- \rightarrow \underset{\substack{\text{tri-} \\ \text{iodoacetone}}}{CI_3COCH_3} + 3H_2O + 3I^-.$$

Under the alkaline conditions prevailing, the tri-iodoacetone is broken up into tri-iodomethane and sodium acetate, as indicated earlier, and a yellow precipitate of iodoform is obtained. The same result is achieved by treating acetone with potassium iodide and alkaline sodium hypochlorite solution. The hypochlorite oxidizes potassium iodide to iodine and this, together with the alkali present, produces tri-iodomethane (iodoform) from the acetone:

$$\underset{\substack{\text{sodium} \\ \text{hypochlorite}}}{NaOCl} + 2KI + H_2O \rightarrow NaCl + I_2 + 2KOH.$$

This sequence of reactions, resulting from treatment with iodine and alkali, is known as the iodoform reaction. The general term haloform reaction covers this as well as the parallel reactions involving chlorine and bromine, the name of the reaction deriving from the trivial names of the products.

From what has been said so far it will be clear that any ketone or aldehyde containing a CH_3CO- group (acetaldehyde, CH_3CHO, is the only possible aldehyde containing a CH_3CO- group) will give this yellow precipitate of tri-iodomethane with iodine and sodium carbonate solution or potassium iodide and sodium hypochlorite solution, and the iodoform reaction is therefore important for detecting the presence of CH_3CO- groups in compounds.

By virtue of the oxidizing powers of iodine itself, or of sodium hypochlorite, the reaction will go one stage further. Any alcohol which can be oxidized to a ketone (or aldehyde) containing the CH_3CO group will also undergo the iodoform reaction:

$$CH_3CH_2OH \xrightarrow{\text{oxidation}} CH_3CHO \xrightarrow{\text{substitution}} CI_3CHO \xrightarrow{\text{cleavage}} CI_3H + HCOOH.$$

Summarizing, the iodoform reaction can be used to detect the $CH_3-\overset{\overset{\displaystyle H}{|}}{\underset{\underset{\displaystyle OH}{|}}{C}}-$ group or the $CH_3\overset{\overset{\displaystyle O}{\|}}{C}-$ group when they are attached to carbon or hydrogen (acetic acid or its derivatives, such as acetamide or methyl acetate, will not undergo the iodoform reaction). Treatment of compounds containing either of these groups with iodine and alkali, or sodium iodide and sodium hypochlorite, gives a yellow precipitate of iodoform. Trichloromethane (chloroform) and tribromomethane (bromoform) may be similarly prepared although these reactions do not provide such a convenient test as does the iodoform reaction, since only iodoform is a solid at room temperature, the others being liquids.

8) Other reactions. Simple aldehydes often give, on treatment with certain acids and bases, various polymers. Individual compounds behave somewhat differently, but the polymerization of formaldehyde and acetaldehyde will serve to illustrate this behaviour.

a) *Formaldehyde*, CH_2O

Slow evaporation of aqueous solutions leaves a low molecular weight straight-chain polymer:

$$(n + 2)\underset{\text{formaldehyde}}{CH_2O} \rightarrow \underset{\text{polymer}}{HOCH_2(OCH_2)_nOCH_2OH}.$$

Dilute aqueous acid yields trioxan, a cyclic trimer:

$$3\underset{\text{formaldehyde}}{CH_2O} \rightarrow \underset{\text{trioxan}}{\begin{array}{c}O\\ \diagup \ \diagdown\\ CH_2 \quad CH_2\\ | \qquad |\\ O \qquad O\\ \diagdown \ \diagup\\ CH_2\end{array}}$$

b) *Acetaldehyde*, CH_3CHO

In the presence of sulphuric acid, acetaldehyde polymerizes to produce paraldehyde, a cyclic trimer, which regenerates acetaldehyde when warmed with sulphuric acid:

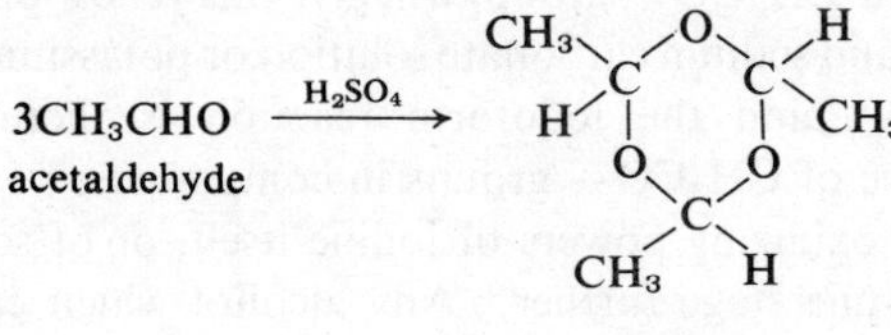

$$3CH_3CHO \xrightarrow{H_2SO_4} \text{paraldehyde}$$

acetaldehyde

At temperatures below zero (centigrade), sulphuric acid catalyses the polymerization of acetaldehyde to metaldehyde, a cyclic polymer containing four units of acetaldehyde:

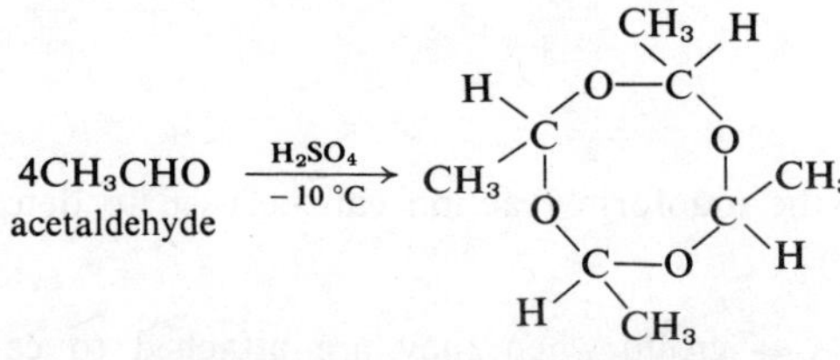

$$4CH_3CHO \xrightarrow[-10\,°C]{H_2SO_4} \text{metaldehyde}$$

acetaldehyde

This polymer (a solid) is well known as meta-fuel or as a slug killer.

Treatment of aldehydes or ketones with phosphorus pentachloride results in the replacement of the oxygen atom with two chlorine atoms:

$$CH_3CHO + PCl_5 \rightarrow CH_3\underset{Cl}{\overset{H}{C}}-Cl + POCl_3;$$

acetaldehyde   phosphorus pentachloride   1,1-dichloroethane   phosphorus oxychloride

$$CH_3COCH_2CH_3 + PCl_5 \rightarrow CH_3-\underset{Cl}{\overset{Cl}{C}}-CH_2CH_3.$$

ethyl methyl ketone (butanone)   2,2-dichlorobutane

This is a useful way of making such dichlorides.

9) Differences between carbonyl compounds. In most of the reactions of carbonyl compounds with which we have dealt, only aldehydes and ketones have been mentioned. The carbonyl group in carboxylic acids and their derivatives shows very few reactions of the normal aldehyde/ketone carbonyl and it is important to note that the presence of a group VB (e.g. nitrogen in amides), group VIB (oxygen in acids and esters), or group VIIB (chlorine in acid chlorides) atom adjacent to the carbonyl group modifies its behaviour considerably. Most of the addition, condensation, and polymerization reactions are simply not known for carboxylic acid derivatives which normally react by substitution (RCOX → RCOY). The behaviour of carboxylic acids towards reducing agents is considered with that of aldehydes and ketones in Section 23.3.2 and some similarity between carboxylic acid derivatives and aldehydes and ketones was noted when we considered the iodoform reaction (although this was largely a question of the aldehyde or ketone

Table 23.9. Important reactions at the carbonyl group. Reactions at adjacent H—C⟨, such as the *aldol reactions*, are also important. These can lead to carbon–carbon bond synthesis and, in this respect, are important in biological systems.

Carboxylic acids and their derivatives (X or Y is a halogen atom or a group based on electronegative atoms such as oxygen or nitrogen)

Aldehydes/ketones (X, Y = hydrogen *or* a carbon group)

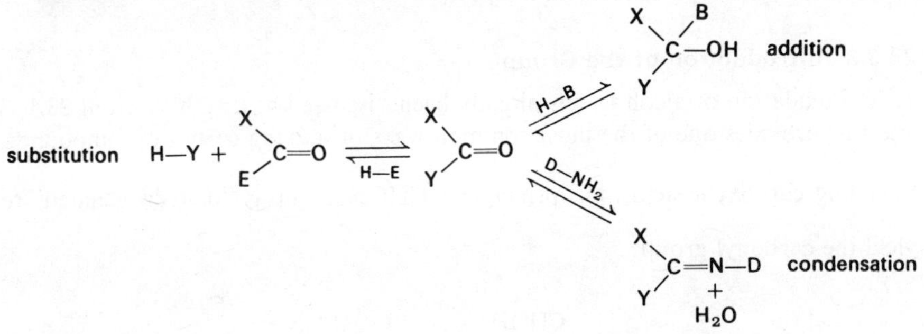

substitution   H—Y + X\C=O / E   ⇌ (H—E)   X\C=O / Y

addition (via H—B):   X\C—OH / Y\B

condensation (via D—NH$_2$):   X\C=N—D / Y   + H$_2$O

E is a group based on an electronegative atom such as oxygen or nitrogen

Note that the addition reagents, H-B, are different from those which add to the carbon–carbon double bond of alkenes

Some important, common, examples of reagents, all of which are nucleophiles, are:

H—E: H—OH, H—OR, H—NH$_2$
H—B: H—H, H—CN, H—SO$_3$Na, H—OR
H$_2$N—D: H$_2$N—OH, H$_2$N—NHC$_6$H$_5$.

basicity

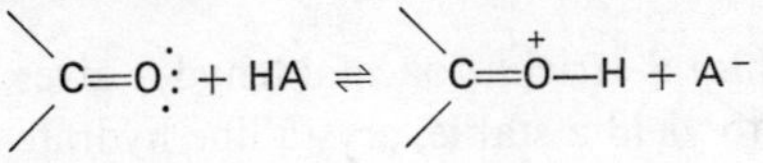

\C=O: + HA ⇌ \C=O$^+$—H + A$^-$ HA represents an acid

oxidation

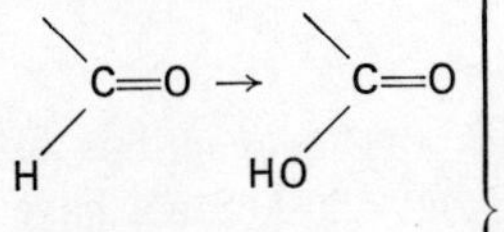

\C=O (H) → \C=O (HO)

⎫
⎬ combined in the Cannizzaro reaction
⎭

reduction

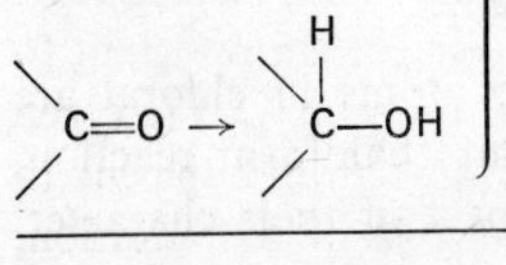

\C=O → \C—OH (H)

23.3 THE CARBONYL GROUP

being modified until it behaved in the carboxylic acid fashion, that is, underwent a substitution reaction).

Differences between aldehydes and ketones are relatively few, but a distinction can be made because aldehydes are reducing agents whereas ketones are not. Aldehydes will reduce Fehling's solution to red copper(I) oxide and ammoniacal silver nitrate to metallic silver; ketones will not do this (Section 23.3.2). A further useful method of distinguishing between the two types of compound involves the use of Schiff's reagent. This is a solution of magenta which has been decolorized by treatment with sulphur dioxide, and aldehydes rapidly restore this colour whereas ketones do so only slowly or not at all. The reaction is a complex one and is not related to the reducing properties of the carbonyl compounds involved. Because the ketonic carbonyl group is shielded by relatively large carbon groups, it is normally less susceptible to reaction at the carbonyl carbon than is the aldehyde carbonyl group which always has the very tiny proton on one side. Aldehydes are more reactive than ketones in this respect.

23.3.3 Introduction of the Group

1. The oxidation of alcohols has already been discussed at length (Section 23.1.2) and this provides one of the most common ways of making carbonyl compounds, including carboxylic acids. In principle a $\diagdown\!\!\!\text{CHOH}$ group is "dehydrogenated" to yield the carbonyl group:

$$\diagdown\!\!\!\text{CHOH} \xrightarrow{-2H} \diagdown\!\!\!\text{C}=O.$$

Large quantities of acetaldehyde and acetone are manufactured by the dehydrogenation of ethanol and propan-2-ol (isopropyl alcohol) over copper catalysts (at about 500 °C).

2. It is interesting that, in most circumstances, two or more hydroxyl groups attached to one carbon atom constitutes an unstable arrangement and a molecule of water is lost with the consequent formation of a carbonyl group:

$$\diagdown\!\!\!\text{C}\!\!\diagup\!\!\!\begin{array}{c}\text{OH}\\\text{OH}\end{array} \rightarrow \diagdown\!\!\!\text{C}=O + H_2O.$$

There are a few exceptions to the rule and chloral (trichloroacetaldehyde) gives rise to one. This compound reacts with water to yield a stable, crystalline hydrate of this type:

$$CCl_3CHO + H_2O \rightarrow CCl_3CH\!\!\diagup\!\!\!\begin{array}{c}\text{OH}\\\text{OH}\end{array}.$$

chloral   chloral hydrate

We have already noted that the electronegative chlorine atoms in chloral are responsible for modifying the normal aldehyde behaviour (haloform reaction, Section 23.3.2) and this is another illustration of this aspect of their character.

Apart from a few exceptions, then, this rule generally holds good for most molecules, and any reaction, therefore, which might be expected to yield such a dihydroxy compound, would in fact yield a carbonyl compound.

To illustrate the point consider the following sequence of reactions which is used to manufacture benzaldehyde from methylbenzene (toluene) (Section 22.2.3).

$$C_6H_5CH_3 \xrightarrow[\text{heat}]{\text{Cl}_2, \text{ bright light}} C_6H_5CHCl_2 \xrightarrow[\text{(hydrolysis)}]{\text{Ca(OH)}_2, \text{H}_2\text{O}} C_6H_5CHO$$

methylbenzene (toluene) → (dichloromethyl)benzene → benzaldehyde

The dihydroxy compound, $C_6H_5CH(OH)_2$, is too unstable to exist and loses water to give benzaldehyde. (Trichloromethyl) benzene is also formed by the chlorination of methylbenzene and hydrolysis of this, instead of leading to the compound $C_6H_5C(OH)_3$, gives benzoic acid.

3. The dry distillation of calcium or barium salts of carboxylic acids often leads to ketones. Thus calcium acetate, on being strongly heated, gives acetone vapour which can be condensed and collected:

$$(CH_3COO)_2Ca \xrightarrow[\text{strongly}]{\text{heat}} CH_3COCH_3 + CaCO_3.$$

calcium acetate → acetone

This method is primarily of use for ketones which are symmetrical. Heating together a mixture of calcium acetate and calcium propanoate would yield *some* ethyl methyl ketone but, clearly, diethyl ketone and acetone would also be formed, as shown schematically:

$$(CH_3COO)_2Ca + (CH_3CH_2COO)_2Ca \xrightarrow[\text{strongly}]{\text{heat}}$$

calcium acetate + calcium propanoate

$$CH_3COCH_2CH_3 + CH_3CH_2COCH_2CH_3 + CH_3COCH_3.$$

ethyl methyl ketone + diethyl ketone + acetone

For similar reasons, aldehydes cannot be conveniently prepared in this way (that is, by using calcium formate, $(HCOO)_2Ca$, as one component of the mixture to be treated).

4. Passage of acetylene through aqueous sulphuric acid, in the presence of a mercury(II) salt (catalyst), results in the hydration of acetylene. The product is acetaldehyde (Section 22.2.3):

$$CH \equiv CH + H_2O \xrightarrow{H^+} [CH_2 = CHOH] \longrightarrow CH_3CHO.$$

acetylene → acetaldehyde

This is an important method for the production of acetaldehyde. Other acetylenes may also be converted to carbonyl compounds in this way:

$$CH_3C \equiv CH + H_2O \rightarrow CH_3COCH_3.$$

propyne → acetone

Acetylenes can also be converted to carbonyl compounds by hydroboration (Section 23.1.3). Addition of diborane to an acetylene yields an alkenyl boron compound which, on treatment with hydrogen peroxide and alkali, gives a carbonyl compound:

$$6R\text{—}C\equiv CH + B_2H_6 \longrightarrow 2(R\text{—}CH\text{=}CH)_3B \xrightarrow[OH^-]{H_2O_2} 6(RCH\text{=}CHOH)$$
$$\longrightarrow 6RCH_2CHO.$$

5. In discussing methods for introducing hydroxyl groups (Section 23.1.3) the hydrolysis of nitriles (cyanides) to give carboxylic acids was quoted:

$$RCN + 2H_2O \xrightarrow[\text{base catalyst}]{\text{acid or}} RCOOH + NH_3.$$

Clearly this also represents a method of introducing a carbonyl group.

Aldehydes can also be produced from nitriles by reduction followed by hydrolysis:

$$\underset{\text{benzonitrile}}{C_6H_5C\equiv N} + [2H] \xrightarrow[\text{in aq. HCl}]{SnCl_2} \underset{\text{benzaldimine}}{C_6H_5CH\text{=}NH};$$

$$C_6H_5CH\text{=}NH + H_2O \longrightarrow \underset{\text{benzaldehyde}}{C_6H_5CHO} + NH_3.$$

This is known as Stephen's method.

Table 23.10. Important methods of introducing the carbonyl group into organic molecules.

oxidation	$-\overset{H}{\underset{\|}{C}}-OH \xrightarrow{-2H} -C\text{=}O$
hydrolysis	$-\overset{Hal}{\underset{\|}{C}}-Hal \xrightarrow{+H_2O} {>}C\text{=}O + 2HHal$
	${>}C\text{=}N\text{—}H \xrightarrow{+H_2O} {>}C\text{=}O + NH_3$
hydration	$-C\equiv C- \xrightarrow{+H_2O} -\overset{O}{\overset{\|\|}{C}}-CH_2-$

from nitriles by related methods

$$-\overset{|}{\underset{|}{C}}-C\equiv N$$

reduction ↙   hydration ↓   Grignard reagent RMgHal ↘

$-\overset{|}{\underset{|}{C}}-CH\text{=}NH$ $-\overset{|}{\underset{|}{C}}-\overset{O}{\overset{\|\|}{C}}-NH_2$ $-\overset{|}{\underset{|}{C}}-\underset{R}{C}\text{=}NMgHal$

↓ hydrolysis   ↓ hydrolysis   ↓ hydrolysis

$-\overset{|}{\underset{|}{C}}-CH\text{=}O + NH_3$ $-\overset{|}{\underset{|}{C}}-\overset{O}{\overset{\|\|}{C}}-OH + NH_3$ $-\overset{|}{\underset{|}{C}}-\underset{R}{C}\text{=}O + NH_3 + Mg(OH)Hal$

aldehyde   carboxylic acid   ketone

Ketones cannot be prepared in this way but nitriles can be converted to ketones by treatment with a Grignard reagent (Section 27.3):

$$R\text{—}C\equiv N + R'MgX \rightarrow R\text{—}C\!=\!N\text{—}MgX \text{ (addition);}$$
$$\qquad\qquad\qquad\text{(Grignard}\qquad\quad |$$
$$\qquad\qquad\qquad\text{ reagent)}\qquad\quad R'$$

$$R\text{—}C\!=\!NMgX + 2HCl(aq) + H_2O \rightarrow R\text{—}C\!=\!O + NH_4Cl + MgXCl.$$
$$\quad |\qquad\qquad\qquad\qquad\qquad\qquad\qquad\quad |\qquad\qquad\quad\text{(subsequent}$$
$$\quad R'\qquad\qquad\qquad\qquad\qquad\qquad\qquad\quad R'\qquad\qquad\qquad\text{hydrolysis)}$$

For example:

$$\begin{array}{c}CH_3 \\ \diagdown \\ \qquad CHCN \\ \diagup \\ CH_3\end{array} + C_2H_5MgBr \xrightarrow[\text{hydrolysis}]{\text{(1) addition}} \begin{array}{c}CH_3 \quad O \\ \diagdown \quad \parallel \\ \qquad CHCCH_2CH_3. \\ \diagup \\ CH_3\end{array}$$

2-methylpro-   ethyl-   ethyl isopropyl
panonitrile   magnesium   ketone
                bromide   (2-methylpentan-3-one)

The nitrile (or cyanide) group is thus seen to be a useful carbonyl group precursor as is the alkyne, or acetylene, group. Both contain a triple bond.

PROBLEMS

1. Study the following list of compounds:
   i) C_6H_5OH;
   ii) $CH_3CH_2COOCH_3$;
   iii) $CH_3COOCOCH_3$;
   iv) [benzene ring with COOH and OH substituents];
   v) $(CH_3)_3COH$;
   vi) $CH_2\text{—}CH\text{—}CH_3$ (with O bridging CH_2 and CH);
   vii) $C_6H_5COOCH_3$;
   viii) CH_3COOH;
   ix) [phthalic anhydride structure];
   x) $CH_3CH(OH)COOH$;
   xi) $C_6H_5COC_6H_5$;
   xii) C_6H_5CHO.

   a) Which of these compounds contain a hydroxyl group?
   b) Which of these compounds contain two hydroxyl groups?
   c) Which of these compounds contain a carbonyl group?
   d) Which of these compounds contain two carbonyl groups?
   e) Which of these compounds contain a carbon–oxygen–carbon linkage?
   f) Which of these compounds are: phenols, alcohols, ethers, carboxylic acids, acid anhydrides, esters, aldehydes, or ketones?

2. Examine the following list of alcohols:
   i) CH_3CH_2OH;
   ii) $(CH_3)_3COH$;
   iii) $CH_3CH_2CH_2OH$;
   iv) [cyclohexanol structure];

v) (CH₃)₂CHOH;
vi) CH₃CH₂CHCH₂CH₃;
         |
        OH

viii)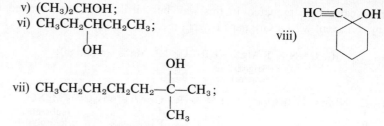

vii) CH₃CH₂CH₂CH₂CH₂—C(OH)(CH₃)—CH₃;

a) Which of these are primary alcohols?
b) Which are secondary alcohols?
c) Which are tertiary alcohols?

3. Arrange the following compounds in the sequence in which you would expect their boiling points to increase. Give your reasons.
   a) CH_3CH_2COOH.
   b) CH_3COOCH_3.
   c) $CH_3CH_2CH_2OH$.
   d) $CH_3CH_2CH_2CH_2OH$.

Look up the known boiling points in a suitable list of physical constants from your library and see how the observed order compares with the one you have written.

4. Which of the following would you expect to dissolve in pure sulphuric acid to give a conducting solution? Give reasons for your choice.
   a) $CH_3CH_2COCH_2CH_3$.
   b) $CH_3CH_2CH_2CH_2CH_3$.
   c) $CH_3CH_2CH(OH)CH_2CH_3$.
   d) 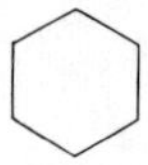 (cyclohexane);
   e) $CH_3CH_2CH_2CH_2COOH$.

5. Arrange the following compounds in the order in which you would expect their acidity to increase:
   a) 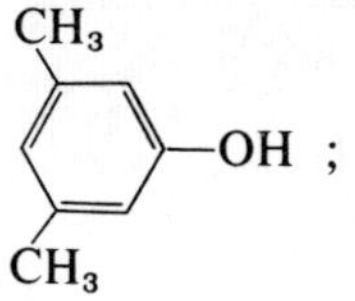 3,5-dimethylphenol (benzene ring with CH₃ at 3,5 positions and OH);
   b) $CH_3CH_2CH_2OH$;
   c) 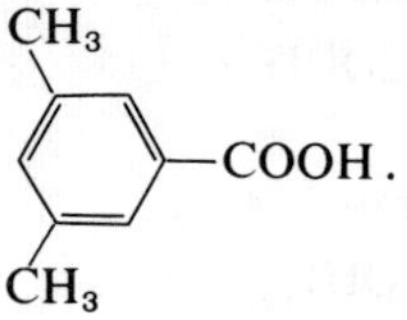 3,5-dimethylbenzoic acid (benzene ring with CH₃ at 3,5 positions and COOH).

6. Study the following list of compounds, and answer the questions below:
   i) CH_3CH_2OH;
   ii) $CH_3CH(OH)CH_3$;
   iii) $CH_3CH_2CH_2COOCH_3$;
   iv) $CH_3COOCH_2CH_2CH_3$;
   v) $CH_3CH_2COCH_3$;
   vi) CH_3CHO;
   vii) $CH_3CH_2COOCOCH_2CH_3$;
   viii) CH_3COOH;
   ix) $CH_3CH_2OCH_2CH_3$.

a) Which compound would give, on oxidation, a ketone (give the structure of the ketone)?
b) Which compound would, on dissolving in water, immediately produce a solution which would release carbon dioxide from sodium hydrogen carbonate solution?
c) Which compound would produce a solution capable of releasing carbon dioxide from sodium hydrogen carbonate solution only after it had been in contact with water for some time?

d) Which compound could be reduced to a secondary alcohol?
e) Which compound would give a primary alcohol on reduction?
f) Which compound (not itself an ester) would react with an alcohol to give an ester?
g) Which compounds, on suitable oxidation, would yield a carboxylic acid?
h) Which compound, on suitable oxidation, would yield an aldehyde?
i) Which compound would, on acid- or base-catalysed hydrolysis, give acetic acid and propan-1-ol?
j) Which ester in this list is a derivative of butanoic acid?
k) Which compounds in this list would give a pale-yellow precipitate of tri-iodomethane (iodoform) if treated with iodine and alkali?
l) Which compounds contain a carbon–oxygen–carbon linkage?
m) Of those compounds containing a carbon–oxygen–carbon linkage, which would be *most* resistant to hydrolysis?
n) Of those compounds mentioned in (l), which would be *least* resistant to hydrolysis—that is *most reactive* towards water?
o) Which compound is a derivative of propanoic acid?

7. Study the following list of compounds:
   i) CH_3COCH_3;
   ii) $C_6H_5CH_2CH_2CHO$;
   iii) $CH_3CH_2CH_2-\underset{\underset{CH_3}{|}}{\overset{\overset{CH_3}{|}}{C}}-CHO$;
   iv) $(CH_3)_3C-\overset{\overset{O}{\|}}{C}-C(CH_3)_3$;
   v) cyclohexyl with CH_3 and CHO substituents.

   a) Which of these compounds would you expect to undergo the aldol reaction?
   b) Which would you expect to undergo the Cannizzaro reaction?

8. Study the following list of compounds:
   i) CH_3CONH_2;
   ii) $CH_3CH_2COCH_2CH_3$;
   iii) $CH_3CH(OH)CH_2CH_2CHO$;
   iv) $C_6H_5COCH_3$;
   v) CH_3COCl;
   vi) $CH_3COCH(CH_3)_2$;
   vii) $CH_3COOCH_2CH_3$.

   a) Which of these compounds would you expect to undergo the iodoform reaction (when treated with iodine and alkali)?
   b) Which would you expect to undergo hydrolysis when treated with warm, aqueous alkali?
   c) Which would you expect to give a yellow-orange precipitate with 2,4-dinitrophenylhydrazine reagent?

9. Define acetylation. How may the following compounds be acetylated:
   a) phenol;
   b) ethanol;
   c) acetic acid;
   d) benzene?

10. Phosphorus oxychloride, $O=PCl_3$, is the acid chloride of phosphoric acid, $O=P(OH)_3$. What products might you expect from reaction between phosphorus oxychloride and ethanol?

11. Define alkylation. How may the compounds listed in Problem 9 be methylated?
12. Suggest simple chemical tests for distinguishing one compound from the other in the following pairs:
    a) CH_3CHO and CH_3COOH;
    b) CH_3CHO and CH_3COCH_3;
    c) $CH_3COOCOCH_3$ and $CH_3CH_2OCH_2CH_3$;
    d) $CH_3COOCH_2CH_3$ and $CH_3COOCOCH_3$;
    e) CH_3COCl and CH_3CH_2Cl;
    f) $CH_3COOCH_2CH_3$ and $CH_3CH_2COCH_2CH_3$;
    g) $CH_3CH_2OCH_2CH_3$ and CH_3COCH_3;
    h) $CH_3COCH_2CH_3$ and $CH_3CH_2COCH_2CH_3$;
    i) $CH_3CH_2CH_2CH_2\underset{\underset{O}{\diagdown\diagup}}{CH-CH_2}$ and (tetrahydropyran).

13. Suggest, in outline, methods for converting propan-1-ol ($CH_3CH_2CH_2OH$) into:
    a) prop-1-yl acetate;
    b) 1-bromopropane (prop-1-yl bromide);
    c) propene;
    d) propan-2-ol (isopropanol—$CH_3CH(OH)CH_3$);
    e) 1-(prop-1-oxy)propane (diprop-1-yl ether);
    f) prop-1-oxybenzene (phenyl prop-1-yl ether);
    g) propanal (CH_3CH_2CHO);
    h) propanoic acid.

14. Suggest, in outline, methods for converting benzene-1,2-dicarboxylic acid (phthalic acid, o-C$_6$H$_4$(COOH)$_2$) into:
    a) its dimethyl ester;
    b) its diphenyl ester;
    c) 1,2-bis(hydroxymethyl) benzene [o-C$_6$H$_4$(CH$_2$OH)$_2$];
    d) benzene.

15. Predict the outcome of reaction, *if any*, between the reagents in the following sets:
    a) $C_6H_5O^- + C_6H_5CH_2Br \rightarrow$;
    b) $C_6H_5CH_2O^- + C_6H_5Br \rightarrow$;
    c) $(CH_3)_2CHCH_2OH + CH_3COCl \rightarrow$;
    d) p-HOC$_6$H$_4$OH + excess $(CH_3CO)_2O \rightarrow$;
    e) $(CH_3)_2CHCH_2OH$ + excess concentrated $H_3PO_4 \xrightarrow{heat}$;
    f) $(CH_3)_3COH$ + excess concentrated $H_3PO_4 \xrightarrow{heat}$;

g) $CH_3CH\underset{O}{-\!\!-\!\!\diagdown\!\!\diagup\!\!-\!\!-}CHCH_3$ + concentrated $NH_3(aq) \rightarrow$;

h) $CH_3COCH_2CH_3 + NH_2OH \rightarrow$;

i) $CH_3CH_2CHO + NaHSO_3 \rightarrow$;

j) [naphthoquinone structure] + $SO_2(aq) \rightarrow$.

16. A compound, of molecular weight about 50, contains 52.2 percent carbon, and 13.0 percent hydrogen. It contains only carbon, hydrogen, and oxygen.
    a) Calculate the molecular formula (Section 29.6) and write down all of the possible structures which satisfy this formula.
    b) The compound reacts with acetyl chloride to yield an acetyl derivative (i.e. is acetylated). What is the structure of the compound?

17. A compound has the molecular formula, $C_3H_6O_2$. How many structures are consistent with this formula? Given that the structure of this compound contains no rings and no double bonds between carbon atoms, suggest how you might determine which of the remaining possible structures is the correct one.

18. A compound X has the molecular formula, C_3H_8O. On treatment with hot, concentrated phosphoric acid X gives a hydrocarbon, Y, which rapidly decolorizes a solution of bromine in carbon tetrachloride. On treatment with acetyl chloride, X is acetylated. Oxidation of X with potassium dichromate and sulphuric acid yields a compound Z which gives a yellow-orange precipitate with 2,4-dinitrophenylhydrazine. Z is not a reducing agent, and will not, for example, give metallic silver with ammoniacal silver nitrate.

    Deduce the structures of X, Y, and Z, explaining carefully what each piece of evidence tells you.

19. An oily liquid, A, is insoluble in water but, on heating with aqueous sodium hydroxide solution for half an hour, dissolves. From the reaction mixture can be distilled a liquid B, which gives a yellow precipitate with iodine and alkali and which, on careful oxidation, yields an aldehyde, C. C gives a yellow precipitate with iodine and alkali. If sulphuric acid is added to the solution obtained by heating A with sodium hydroxide, a white solid, D, separates which liberates carbon dioxide from sodium hydrogen carbonate solution. One mole of D reacts with one mole of sodium hydrogen carbonate to give one mole of carbon dioxide. Heating D with soda lime converts it to benzene.

    Elucidate the structures of compounds A to D, giving all of your reasoning.

20. An aromatic compound W (a derivative of benzene) has the molecular formula $C_9H_{12}O_3$. On heating it strongly with concentrated hydriodic acid, iodomethane (methyl iodide) is evolved and is converted by a suitable reaction to methanol and aqueous iodide ions. Silver nitrate is added and 705 mg of precipitated silver iodide is obtained thus from 168 mg of W. How many methoxy groups (CH_3O- groups) does W contain? Are these joined to the ring directly (as in methoxybenzene, $CH_3OC_6H_5$) or via other carbon atoms

(as in methoxymethylbenzene, $CH_3OCH_2C_6H_5$)? (*Hint*—how is the C—O—C linkage broken in simple ethers?)

21. Find out, by consulting books in your library, as much as you can about the chemistry of one or more of the following:

   a) Oxalic acid (ethanedioic acid):

   $$\begin{array}{c} COOH \\ | \\ COOH \end{array}$$

   b) Ethylene glycol (ethane-1,2-diol):

   $$\begin{array}{c} CH_2OH \\ | \\ CH_2OH \end{array}$$

   c) Glycerol (glycerine, propane-1,2,3-triol):

   $$\begin{array}{c} CH_2OH \\ | \\ CHOH \\ | \\ CH_2OH \end{array}$$

   Note, in particular, properties which would not be easily predicted on the basis of your knowledge of simple carboxylic acids (for a) or alcohols (for b and c).

BIBLIOGRAPHY

Gutsche, C. D. *Chemistry of Carbonyl Compounds*, Prentice-Hall (1967).

> Primarily written for undergraduates, but useful for the enthusiastic sixth-former.

Taylor, R. J. *The Chemistry of Glycerides*, Unilever Education Booklet, Advanced Series No. 4, available from Unilever Education Section, Unilever House, Blackfriars, London EC4.

> A clear account, beautifully and plentifully illustrated, of this aspect of the chemistry of fats.

This chapter covers a wide range of material and one (or more) of the books on organic chemistry from the general list would provide a most helpful form of further reading.

CHAPTER 24

Compounds with Functional Groups containing Nitrogen

Like oxygen, nitrogen is an electronegative element which forms a simple hydride (NH_3, ammonia; cf. H_2O, water). Just as oxygen forms two series of organic compounds which might be regarded as substituted waters, C—OH compounds and compounds containing the C—O—C linkage, so nitrogen forms three series of substituted ammonias and a series of substituted ammonium (NH_4^+) compounds as well (substituted H_3O^+ compounds are not common). Because nitrogen can exert two covalencies (four and three), both of which are greater than that of oxygen (two), the possibilities for forming functional groups, some involving multiple bonds, are much greater for nitrogen than for oxygen (the nature of nitrogen is more fully discussed in Chapter 17).

Thus the table of the nitrogen functions (Table 24.1) which we shall consider in this chapter is more extensive than the corresponding chart of oxygen functions (Table 23.1). The structurally related oxygen analogues, which must be borne in mind when considering the nitrogen compounds, are indicated where appropriate.

24.1 Primary amino groups
24.2 Secondary amino groups
24.3 Tertiary amino groups
24.4 Quaternary ammonium compounds
24.5 Imino groups
24.6 Nitriles and isonitriles
24.7 The nitro group
24.8 Diazonium group

24.1 PRIMARY AMINO GROUPS

As with other functional groups, the nature of the skeleton to which it is attached affects the character of the amino group and we shall consider three different types of primary amino compound, as follows, which will emphasize this generalization:

1. R—NH_2, primary alkylamines (primary aliphatic amines).
2. Ar—NH_2, primary arylamines (primary aromatic amines).
3. R—$CONH_2$ and Ar—$CONH_2$, amides of carboxylic acids.

Nomenclature for *all* amino compounds (Sections 24.1, 24.2, and 24.3) is as follows:

Amines. The alkyl groups are specified, and the word is terminated by *-amine*. Ethylamine is $CH_3CH_2NH_2$, dimethylamine is $(CH_3)_2NH$, and N,N-dimethylphenylamine is $C_6H_5N(CH_3)_2$. Aniline, the trivial name for phenylamine, $C_6H_5NH_2$, is in very common use and is therefore an important name to remember.
Amides. The nomenclature of these compounds has already been fully discussed (Section 23.3).

Table 24.1. Nitrogen functions.

Function	Name of function	Common types of compound with examples		Analogous oxygen function
$C-NH_2$	primary amino	primary amines; $C_6H_5NH_2$ aniline	amides CH_3CONH_2 acetamide	$C-OH$
$C-NHR$	secondary amino	secondary amines; $C_6H_5NHCH_3$ N-methylaniline	amides $CH_3CONHCH_3$ N-methylacetamide	$C-OH$ $C-O-C$
$C-NR_2$	tertiary amino	tertiary amines; $(C_2H_5)_3N$ triethylamine	amides $CH_3CON(CH_3)_2$ N,N-dimethylacetamide	$C-O-C$
$C-\overset{+}{N}R_3$	quaternary ammonium	quaternary ammonium hydroxides and salts $(CH_3)_4\overset{+}{N}\overset{-}{O}H$ tetramethylammonium hydroxide $(CH_3)_4\overset{+}{N}\overset{-}{C}l$ tetramethylammonium chloride		$C-\overset{+}{\underset{\underset{C}{\mid}}{O}}-C$ (rare)
$C=N-H$ $C=N-R$	imino alkyl (or aryl) imino	imines or Schiff's bases $C_6H_5CH=NC_6H_5$ benzylideneaniline		$C=O$
$C-C\equiv N$	cyano	cyanides (nitriles) CH_3CN methyl cyanide (acetonitrile)		none
$C-N\rightleftharpoons C$	isocyano	isocyanides $CH_3-N\rightleftharpoons C$ methyl isocyanide		none
$C-N\overset{\overset{O}{\parallel}}{\searrow_O}$	nitro	nitro compounds CH_3NO_2 nitromethane		none
$C-N\overset{\overset{O}{\parallel}}{}$	nitroso	nitroso compounds C_6H_5NO nitrosobenzene		none
$C-\overset{+}{N}\equiv N$	diazonium	diazonium salts $C_6H_5\overset{+}{N}_2\overset{-}{C}l$ phenyldiazonium chloride		none

24.1.1 Physical Properties

Again, with the larger molecules we observe higher melting and boiling points. The lower molecular weight members are gaseous. Hydrogen bonding is much less important than with hydroxy compounds, hence methylamine, for example, (CH_3NH_2, mol. wt. 31) is a gas (b.p. -7.6 °C), whereas methanol (CH_3OH, mol. wt. 32) is a liquid (b.p. 64 °C). An amide, containing a carbonyl group, has a higher boiling point than an amine of similar molecular weight:

$C_2H_5NH_2$, mol. wt. 45, b.p. 19 °C;
$C_3H_7NH_2$, mol. wt. 59, b.p. 49 °C;
CH_3CONH_2, mol. wt. 59, b.p. 222 °C.

24.1.2 Chemical Behaviour

We have already noted that compounds containing amino groups are substituted ammonias. Primary amino compounds derive from the replacement of one hydrogen atom of ammonia with a carbon group, secondary amino compounds from the replacement of two hydrogen atoms, and tertiary amino compounds from the replacement of all three. Quaternary ammonium salts can be thought of as ammonium salts in which all four hydrogen atoms of the ammonium ion have been replaced by carbon groups.

1) Acidity. Like alcohols, primary amino compounds will lose a proton if they are treated with a strong enough base. They are thus acidic although rather more weakly so than the corresponding hydroxy compounds. They can be arranged in order of increasing acidity as follows: $RNH_2 < ArNH_2 < RCONH_2$. Sodium metal will remove a proton from RNH_2 ($Na + RNH_2 \rightarrow Na^+ R\bar{N}H + \frac{1}{2}H_2$), but even mercury(II) oxide will form a mercury(II) salt with amides ($HgO + 2RCONH_2 \rightarrow (RCONH)_2Hg + H_2O$). A Grignard reagent (Section 27.1) is also a strong enough base to form a magnesium salt with primary amines ($RNH_2 + R'MgBr \rightarrow RNH \cdot MgBr + R'H$). Primary amines are too weakly acidic to react with any aqueous alkalis, hence, in water their acidic properties can be neglected.

2) Basicity. Whilst primary amino compounds are much weaker acids than their oxygen analogues it is not surprising to note that they are stronger bases. They owe their character as bases to the lone pair of electrons on nitrogen, as does ammonia itself, and can form strong bonds to protons (hydrogen ions):

$$\diagdown\!\!\!\!\!\!\underset{\diagup}{N}\!:\ +\ H^+\ \rightarrow\ \diagdown\!\!\!\!\!\!\underset{\diagup}{\overset{+}{N}}\!\!-\!\!H.$$

The ease with which the lone pair can bond to the proton, and hence the base strength of the compound, depends upon the groups attached to nitrogen.

A carbonyl group adjacent to nitrogen reduces the availability of the electrons, and amides are therefore weak bases (acetamide, CH_3CONH_2, is a weaker base than methylamine, CH_3NH_2). Aromatic rings also reduce the availability of the electrons, whereas aliphatic groups tend to increase it. Thus phenylamine (aniline), $C_6H_5NH_2$, is a weaker base than ammonia, whereas methylamine, CH_3NH_2, is stronger (Section 28.5.1).

Typical reactions with acids are:

$CH_3NH_2 + HCl \rightarrow CH_3\overset{+}{N}H_3 + Cl^-$ (sometimes written as $CH_3NH_2 \cdot HCl$);
methylamine   methylammonium chloride
(or methylamine hydrochloride)

$C_6H_5NH_2 + HCl \rightarrow C_6H_5\overset{+}{N}H_3 + Cl^-$.
phenylamine   phenylammonium chloride
(aniline)   (anilinium chloride
or aniline hydrochloride)

All amines are weaker bases than sodium (or potassium) hydroxide, and so are released from their salts (as is ammonia) by treatment with aqueous alkali:

$$CH_3\overset{+}{N}H_3 + Cl^- + Na^+ + OH^- \rightarrow CH_3NH_2 + Na^+ + Cl^- + H_2O.$$

3) Reactions with alkali. Primary amino compounds are too weakly acidic to react with aqueous alkalis to form salts. Amides, being more strongly acidic than amines, may react to a small extent with aqueous alkali, but the equilibrium will lie well over towards the free amide:

$$RCONH_2 + OH^- \rightleftharpoons RCO\overset{-}{N}H + H_2O.$$

In hot aqueous alkali, however, the C—N bond of amides is hydrolysed, liberating ammonia and yielding a carboxylic acid salt. The C—N bond of amines is usually inert in this respect (for an exception, see Section 24.3.2).

$$RCONH_2 + NaOH(aq) \xrightarrow{heat} RCOONa + NH_3$$
$$CH_3CONH_2 + NaOH(aq) \longrightarrow CH_3COONa + NH_3$$
acetamide sodium acetate ammonia

$$C_6H_5CONH_2 + NaOH(aq) \longrightarrow C_6H_5COONa + NH_3$$
benzamide sodium benzoate ammonia

$$RNH_2 + NaOH \text{no reaction.}$$

This reaction is also catalysed by acids in which case the free carboxylic acid is formed together with an ammonium salt:

$$\begin{array}{c} HO\!\mid\! H \\ RCO\!\mid\! NH_2 \end{array} \rightarrow RCOOH + NH_3 \xrightarrow{H_2SO_4} RCOOH + (NH_4)_2SO_4.$$

With pure water, in the absence of acid or base, the reaction proceeds too slowly for easy observation. The effect of the carbonyl group in increasing the reactivity of the C—N bond has already been referred to (Section 23.2, Table 23.4).

4) Alkylation. Most primary amines (aliphatic or aromatic) react with alkyl halides (halogenoalkanes) to produce a new nitrogen–carbon bond. Alkyl iodides are better (more reactive) than chlorides or bromides for this reaction.

$$RNH_2 + R'I \rightarrow \begin{array}{c} R H \\ \diagdown \overset{+}{} \diagup \\ N \\ \diagup \diagdown \\ R' H \end{array} + I^-$$

$$\left[\text{compare} RNH_2 + HI \rightarrow \begin{array}{c} R H \\ \diagdown \overset{+}{} \diagup \\ N \\ \diagup \diagdown \\ H H \end{array} + I^- \right].$$

Examples:

$$C_2H_5NH_2 + C_2H_5I \rightarrow (C_2H_5)_2\overset{+}{N}H_2 + I^-$$
ethylamine ethyl diethylammonium
 iodide iodide

$$C_6H_5NH_2 + CH_3I \rightarrow \underset{\underset{CH_3}{|}}{\overset{\overset{C_6H_5}{|}}{N^+}} \begin{matrix} H \\ \diagdown \\ \diagup \\ H \end{matrix} + I^-.$$

phenylamine   methyl   N-methylphenyl-
(aniline)   iodide   ammonium iodide

Note that the compounds $RR'\overset{+}{N}H_2\, I^-$ are also the products you would expect from the reaction of $RR'NH$ (secondary amines) with hydriodic acid. Thus, the products are salts of secondary amines and on treatment with sodium hydroxide will give the free secondary amine (see 24.1.2(2) above). We have thus a means of converting primary amines into secondary amines:

$$RNH_2 + R'I \rightarrow RR'\overset{+}{N}H_2\, I^- \xrightarrow{NaOH} RR'NH + H_2O.$$

The primary amine has in fact been alkylated (alkylation is defined in Section 23.1.2 in connection with the OH group).

The corresponding alkylation of an alcohol with $R'I$ requires that the alcohol be first converted to its more reactive sodium salt. Amines are thus seen to be more reactive towards alkyl halides than alcohols as they do not need to be converted to their sodium salt first:

$$RNH_2 + R'I \xrightarrow[\text{(2) NaOH}]{\text{(1) heat}} RR'NH;$$
$$ROH + R'I \quad\quad \text{no reaction};$$
$$R\bar{O}\,Na^+ + R'I \longrightarrow ROR' + NaI.$$

Amides are not readily alkylated. They are much less reactive towards alkyl halides than are amines.

5) Acylation. Acylation (as defined in Section 23.1.2) of the primary amino function leads to the replacement of a hydrogen atom by an acyl group:

$$-NH_2 \rightarrow -NHCOR.$$

This proves to be very readily achieved for primary amines—aliphatic or aromatic —and can be brought about by treatment of the amines with a carboxylic acid, acid chloride, or acid anhydride:

$$CH_3NH_2 + CH_3COOH \text{ (excess)} \rightarrow CH_3COO^-\, \overset{+}{N}H_3CH_3;$$
methylamine   acetic acid   methylammonium acetate

$$CH_3COO^-\, \overset{+}{N}H_3CH_3 \xrightarrow[\text{(excess acetic acid)}]{\text{heat}} CH_3CONHCH_3 + H_2O.$$
N-methylacetamide

Reaction proceeds very much more readily with acid chlorides or anhydrides:

$$CH_3NH_2 + (CH_3CO)_2O \rightarrow CH_3NHCOCH_3 + CH_3COOH;$$
methylamine   acetic anhydride   N-methylacetamide   acetic acid

$$C_6H_5NH_2 + C_6H_5COCl \xrightarrow{NaOH} C_6H_5NHCOC_6H_5 + NaCl + H_2O.$$
phenylamine   benzoyl   N-phenylbenzamide
(aniline)   chloride   (benzanilide)

(The sodium hydroxide removes hydrogen chloride from the reaction mixture by neutralizing it to form sodium chloride.)

Thus, like alcohols and phenols, primary amines are readily acylated. Amides, however, are generally difficult to acylate. One exception to this rule is urea—an amide of carbonic acid—which "acylates" itself when heated:

$$\underset{\text{urea}}{NH_2CONH_2} + NH_2CONH_2 \xrightarrow{\text{heat}} \underset{\text{biuret}}{NH_2CONHCONH_2} + NH_3.$$

Biuret, in fact, reacts with more urea to form longer chains, for example:

$$\underset{\text{biuret}}{NH_2CONHCONH_2} + \underset{\text{urea}}{NH_2CONH_2} \rightarrow \underset{\text{triuret}}{NH_2CONHCONHCONH_2} + NH_3.$$

6) Effect of nitrous acid. Aqueous nitrous acid (unstable and usually generated as required *in situ* from sodium nitrite and aqueous mineral acid) attacks the primary amino group with the quantitative liberation of nitrogen gas:

$$\underset{\text{acetamide}}{CH_3CONH_2} + HNO_2 \rightarrow \underset{\text{acetic acid}}{CH_3COOH} + H_2O + N_2.$$

By measuring the volume of gas liberated it is easy to estimate the amount of amino compound originally present.

With aliphatic amines, although the amount of nitrogen liberated is still quantitative (one mole of molecular nitrogen per mole of amine), the organic part of the molecule may suffer several of many possible fates. The amino group in the amide above is replaced, in good yield, by a hydroxyl group but this does not usually happen with alkylamines. Although some alcohol is formed, alkenes and other compounds are formed as well. It is believed that a diazonium salt is formed, by reaction of the amine with nitrous acid, which rapidly decomposes to give nitrogen gas (quantitatively) and a carbonium ion (a positively charged carbon group):

$$\underset{\text{amine}}{RNH_2} + HNO_2 \rightarrow \underset{\substack{\text{diazonium}\\\text{salt}}}{R\overset{+}{N}\!\equiv\!N} + OH^- \xrightarrow{H^+} \underset{\substack{\text{carbonium}\\\text{ion}}}{R^+} + H_2O + N_2.$$

The carbonium ion is unstable and may lose a hydrogen ion to form an alkene (Section 26.6), or react with a hydroxide ion to form an alcohol (Section 26.3), or react with several other things as well. Thus the yield of nitrogen gas is 100 percent, whereas the original amine is converted into a mixture of several compounds. This reaction is therefore rarely of use for preparing alcohols from primary aliphatic amines.

Primary aromatic amines similarly give diazonium salts with nitrous acid, but aromatic diazonium salts are very much more stable than their aliphatic analogues and, provided that the temperature is kept low (below about 5 °C), they react no further and remain in solution. Their reactions are discussed in detail later in this chapter.

24.1.3 Introduction of the Group

1) Alkylation or acylation of ammonia. The alkylation and acylation of primary amines has been discussed in detail (Section 24.1.2). Analogous reactions with ammonia, instead of primary amines, lead to amines and amides respectively:

$$NH_3 \xrightarrow{\text{alkylation}} RNH_2;$$
$$NH_3 \xrightarrow{\text{acylation}} RCONH_2 \text{ (or ArCONH}_2\text{)}.$$

Aromatic amines cannot be prepared by "arylating" ammonia due to the stability of chloro- and bromobenzenes which has been mentioned already (Section 23.1.2; see also Section 28.5.2). We saw that they cannot be hydrolysed by water to form phenols and they cannot react with alcohols (or their sodium salts) to form ethers. All three types of reaction are very similar and, as well as being regarded as alkylations, they can be regarded as "olysis" reactions (cf. "hydrolysis", Sections 23.1.3 and 23.3.2). Thus:

R—Cl + H$_2$O → ROH + HCl; hydrolysis of RCl or alkylation of H$_2$O

R—Cl + HOR → ROR′ + NaCl; alcoholysis of RCl or alkylation of ROH
   (usually as sodium salt)

R—Cl + NH$_3$ → RNH$_2$ + HCl; ammonolysis of RCl or alkylation of NH$_3$
   salt

The aryl–halogen bond is usually not susceptible to hydrolysis, alcoholysis, or ammonolysis, that is, it cannot be hydrolysed, alcoholysed, or ammonolysed.

In conclusion, it should be pointed out that the reaction between alkyl halides and ammonia, usually carried out in solution in alcohol at high temperatures and pressures (sealed tube at about 100 °C), usually leads in practice to mixtures of primary, secondary, and tertiary amines as well as the corresponding quaternary ammonium compounds. Thus it is necessary to separate these products to obtain those which are required:

$$C_2H_5I + NH_3 \rightarrow \text{mixture} \begin{cases} C_2H_5\overset{+}{N}H_3 & I^- \\ (C_2H_5)_2\overset{+}{N}H_2 & I^- \\ (C_2H_5)_3\overset{+}{N}H & I^- \\ (C_2H_5)_4\overset{+}{N} & I^- \end{cases}$$
ethyl iodide

mixture + NaOH → [C$_2$H$_5$NH$_2$ + NaI] + [(C$_2$H$_5$)$_2$NH + NaI] +
                 ethyl amine   diethylamine

[(C$_2$H$_5$)$_3$N + NaI] + [(C$_2$H$_5$)$_4\overset{+}{N}$ OH$^-$ + NaI].
  triethylamine   tetraethylammonium hydroxide

A related method, involving the use of phthalimide, gives pure primary amines with no secondary or tertiary amines as impurities. Phthalimide is closely related to the amides which have already been discussed. With two carbonyl groups adjacent to the N—H group, however, the latter is rendered much more acidic than the normal amide —NH$_2$ group and it reacts with ethanolic potassium hydroxide to give a potassium salt. It is also related to amides in that it can be hydrolysed (acid or base catalyst necessary) to yield ammonia and phthalic acid.

phthalimide + KOH ⟶ potassiophthalimide + H$_2$O

phthalimide + 2H₂O $\xrightarrow[\text{catalyst}]{\text{acid or base}}$ phthalic acid (benzene-1,2-dicarboxylic acid) + NH₃

compare CH₃CO|NH₂ + HO|H $\xrightarrow[\text{acid or base catalyst}]{\text{H}_2\text{O}}$ CH₃COOH + NH₃.

The potassium salt of an N—H compound is much more reactive towards alkyl halides than the free N—H compound, just as the potassium (or sodium) salt of an alcohol is more reactive than the alcohol. Thus it is easy to prepare an "N-alkyl" phthalimide and to hydrolyse this to the corresponding primary alkylamine:

phthalimide + KOH ⟶ potassiophthalimide + H₂O;

potassiophthalimide-NK + RI ⟶ N-alkylphthalimide-NR + KI;

N-alkylphthalimide + 2H₂O $\xrightarrow[\text{catalyst}]{\text{acid or base}}$ phthalic acid + RNH₂.

Whilst this may seem a rather cumbersome way of introducing a primary amino function it is still a very practical one because it obviates the necessity of an even more cumbersome separation procedure which has to follow the ammonia/alkyl halide method.

Acylation of ammonia with acid chlorides or anhydrides usually leads to a single product in a "clean" reaction:

CH₃COCl + NH₃ → CH₃CONH₂ + HCl;
acetyl chloride   acetamide

$$(CH_3CO)_2O + NH_3 \rightarrow CH_3CONH_2 + CH_3COOH;$$
acetic anhydride   acetamide   acetic acid

phthalic anhydride + NH₃ → mono-amide of phthalic acid

Esters may sometimes acylate ammonia in good yields and afford a method of preparing amides. Oxamide, for example, crystallizes readily on treating diethyl oxalate with aqueous ammonia:

$$\begin{array}{l} COOC_2H_5 \\ | \\ COOC_2H_5 \end{array} + 2NH_3 \rightarrow \begin{array}{l} CONH_2 \\ | \\ CONH_2 \end{array} + 2C_2H_5OH$$

diethyl oxalate   oxamide   ethanol

$$\left[\text{oxalic acid (ethanedioic acid) is } \begin{array}{l} COOH \\ | \\ COOH \end{array} \right].$$

2) Reduction of nitro compounds. The nitro group, $-N\begin{subarray}{l}O\\O\end{subarray}$, is readily reduced in acidic conditions to the primary amino group. This allows a very convenient synthesis of primary aromatic amines which are not readily prepared from ammonia. The nitro group is very readily introduced into aromatic hydrocarbons and, by subsequent reduction, primary amines are produced. The reaction is not confined to aromatic amines but finds most application there. A suitable reducing agent is tin in hydrochloric acid.

$$ArNO_2 \xrightarrow{[6H]} ArNH_2 + 2H_2O$$
$$2C_6H_5NO_2 + 3Sn + 12HCl \rightarrow 2C_6H_5NH_2 + 3SnCl_4 + 4H_2O.$$
nitrobenzene   phenylamine
(aniline)

Nitroso compounds (nitroso group is $-N=O$) may also be reduced thus but they are considerably less useful than nitro compounds. The method is only of use for preparing primary amines and is not applicable to amide or secondary or tertiary amine syntheses.

3) Reduction of nitriles (cyanides). Aliphatic or aromatic nitriles can be reduced to primary amines:

$$R-C \equiv N \xrightarrow{4H} R-CH_2-NH_2.$$

This is formally related to the hydrogenation of alkynes to alkanes although the reagents (and reactions) involved are very different. Thus, lithium aluminium hydride or sodium and ethanol are normally used for nitrile reductions, whereas hydrogen and a metal catalyst are employed for the hydrogenation of alkynes:

$$CH_3CH_2C{\equiv}N + 4Na + 4C_2H_5OH \rightarrow CH_3CH_2CH_2NH_2 + 4NaOC_2H_5;$$
propanonitrile (ethyl cyanide) — prop-1-ylamine

$$CH_3C{\equiv}N \xrightarrow{LiAlH_4} CH_3CH_2NH_2.$$
acetonitrile — ethylamine

An important feature of this reaction is that it lengthens the alkyl group of the original molecule. Thus, starting with ethyl cyanide we obtained prop-1-ylamine.

As a preparative method, the reaction is clearly only applicable to primary amines in which the amino function is attached to a primary carbon atom (definition, Section 23.1.2). It is of no use for amides or for secondary and tertiary amines.

4) Conversion of amides to amines. There are two ways of converting amides to amines, one in which the carbonyl group is reduced and one in which it is apparently "nipped out":

$$RCONH_2 \begin{array}{c} \xrightarrow{4H} RCH_2NH_2 \\ \xrightarrow{-CO} RNH_2 \end{array}$$

The latter only works with amides containing primary amino functions, whereas the former works with further substituents on nitrogen. Typical reactions (reagents indicated) are:

$$CH_3CONH_2 \begin{array}{c} \xrightarrow{Na + C_2H_5OH \text{ or } LiAlH_4} CH_3CH_2NH_2 \text{ (ethylamine)} \\ \xrightarrow{Br_2 + 3NaOH} CH_3NH_2 + NaBr + Na_2CO_3 \text{ (methylamine)} \end{array}$$
acetamide

Note that the "nipping out" reaction affords a method for chain shortening as illustrated in the following sequence:

$$CH_3CH_2OH \xrightarrow{Na_2Cr_2O_7} CH_3COOH \xrightarrow{PCl_5} CH_3COCl$$
ethanol — acetic acid — acetyl chloride

$$CH_3COOH \xrightarrow{NH_3, \text{heat}} CH_3CONH_2 \xleftarrow{NH_3} CH_3COCl$$

$$CH_3NH_2 \xleftarrow{Br_2/NaOH} CH_3CONH_2$$
methylamine — acetamide

24.2 SECONDARY AMINO GROUPS

The functions considered under this heading are —NHR, alkylamino, and —NHAr, arylamino, and the types of compound are:

1. R—NHR′ (R′ the same as or different from R)
 R—NHAr
 Ar—NHAr′ (Ar′ the same as or different from Ar) } secondary amines.

638 COMPOUNDS WITH FUNCTIONAL GROUPS CONTAINING NITROGEN

2. $\left.\begin{array}{l}\text{RCONHR}'\\ \text{ArCONHR}\\ \text{RCONHAr}\\ \text{ArCONHAr}'\end{array}\right\}$ N-substituted amides.

The properties of these compounds—physical or chemical—are similar to those of the corresponding primary amino compounds. Only where their chemistries differ will details be given.

24.2.1 Physical Properties

The physical properties of these compounds are parallel to those of their primary amino analogues. Only in so far as they may have a higher molecular weight than their primary analogues will their boiling or melting points also be higher. Hydrogen bonding is again possible but again less important than in alcohols.

24.2.2 Chemical Behaviour

1) Acidity and basicity. These properties also run parallel to those of their primary analogues. With a lone pair of electrons on nitrogen, and a hydrogen atom attached to nitrogen, they are similarly capable of behaving as acids and bases. The carbon skeletons will determine the acid or base strengths in much the same way as they did with the primary compounds.

2) Reactions with alkali. Again, these properties parallel those of the primary compounds. Hydrolysis of the C—N bond occurs only with amides:

$$\underset{\text{N-phenylpropanamide}}{CH_3CH_2CONHC_6H_5} + H_2O \xrightarrow[\text{or base}]{\text{acid}} \underset{\text{propanoic acid}}{CH_3CH_2COOH} + \underset{\substack{\text{phenylamine}\\\text{(aniline)}}}{C_6H_5NH_2}.$$

3) Alkylation and acylation. The hydrogen atom attached to nitrogen is capable of replacement by alkyl or acyl groups and again, therefore, the secondary compounds are similar to the primary. Whereas with the latter, however, there are two such hydrogen atoms capable of replacement, in the former there is only one:

$$\underset{\substack{\text{N-methyl-}\\\text{phenylamine}\\\text{(N-methylaniline)}}}{C_6H_5NHCH_3} + \underset{\substack{\text{methyl}\\\text{iodide}}}{CH_3I} \rightarrow \underset{\substack{\text{N,N-dimethyl-}\\\text{phenylamine}\\\text{(N,N-dimethylaniline)}}}{\overset{+}{C_6H_5NH(CH_3)_2}\ I^- \atop \downarrow\ NaOH \atop C_6H_5N(CH_3)_2;}$$

$$\underset{\substack{\text{N-methyl-}\\\text{phenylamine}}}{C_6H_5NHCH_3} + \underset{\substack{\text{acetic}\\\text{anhydride}}}{(CH_3CO)_2O} \rightarrow \underset{\substack{\text{N-methyl-N-phenyl-}\\\text{acetamide}}}{C_6H_5N(CH_3)COCH_3}.$$

N-alkyl amides are difficult to acylate, but this can be achieved with an excess of acylating agent and high temperature.

4) Effect of nitrous acid. In their reactions with nitrous acid the secondary compounds differ fundamentally from the primary compounds. Here the presence of one rather than two hydrogen atoms has a profound effect on the course of the reaction. The lone hydrogen atom is replaced by a nitroso group yielding an N-nitroso compound.

$$RNHR' + HNO_2 \rightarrow RN\begin{matrix}N=O\\ \\R'\end{matrix} + H_2O$$

$$ArNHAr' + HNO_2 \rightarrow ArN\begin{matrix}N=O\\ \\Ar'\end{matrix} + H_2O$$

$$\underset{\substack{\text{N-methyl-}\\\text{phenylamine}}}{C_6H_5NHCH_3} + HNO_2 \rightarrow \underset{\substack{\text{N-nitroso-N-methyl-}\\\text{phenylamine}}}{C_6H_5N(NO)CH_3} + H_2O.$$

It is also possible to prepare N-nitroso derivatives of N-substituted amides in this way.

24.2.3 Introduction of the Group

1) Alkylation or acylation of primary compounds. This procedure exactly parallels that for the introduction of the primary amino function except that primary amino compounds are used in place of ammonia. Mixtures of products again result from these alkylations.

$$\underset{\substack{\text{methyl-}\\\text{amine}}}{CH_3NH_2} + \underset{\substack{\text{methyl}\\\text{iodide}}}{CH_3I} \rightarrow (CH_3)_2\overset{+}{N}H_2\ I^- \xrightarrow{\text{alkali}} \underset{\text{dimethylamine}}{(CH_3)_2NH}$$

$$\underset{\text{prop-2-ylamine}}{(CH_3)_2CHNH_2} + \underset{\text{acetic anhydride}}{(CH_3CO)_2O} \rightarrow \underset{\text{N-prop-2-ylacetamide}}{(CH_3)_2CHNHCOCH_3}.$$

The phthalimide amine synthesis is only applicable to primary amines.

2) Conversion of amides to amines. This reaction is the precise analogue of that quoted for primary amino compounds:

$$\underset{\text{N-methylacetamide}}{CH_3CONHCH_3} \xrightarrow{Na/C_2H_5OH} \underset{\text{ethylmethylamine}}{CH_3CH_2NHCH_3}.$$

24.3 TERTIARY AMINO GROUPS

The groups considered are —NR_2, dialkylamino, —NAr_2, diarylamino, and —NRAr, alkylarylamino. The types of compound are:

1. R—NR_2 and Ar—NAr_2, and various combinations of Ar or R attached to N. Tertiary amines.
2. RCONXY and ArCONXY, (X and/or Y may be R and/or Ar). N,N-disubstituted amides.

Again, many of the properties of these compounds are closely related to those of their primary or secondary analogues. Only where they differ from these analogues will a detailed discussion be presented.

24.3.1 Physical Properties

The physical properties of these compounds are closely similar to those of their primary and secondary analogues. Hydrogen bonding is impossible, however, as there are no hydrogen atoms attached to nitrogen, and the absence of such hydrogen atoms is chiefly responsible for differences between the chemical properties of these compounds and those of other amino compounds.

24.3.2 Chemical Behaviour

1) Acidity and basicity. In the absence of hydrogen atoms attached to nitrogen, clearly there is no scope for acidic character. The lone pair of electrons, however, is still present on nitrogen and this confers basic character. Base strengths again depend upon the carbon groups attached to nitrogen.

2) Alkylation and acylation. The absence of hydrogen atoms on nitrogen clearly means that acylation and alkylation, as formally defined, cannot take place. Nonetheless, tertiary amines do react with alkyl halides to form new C—N bonds in reactions which are similar to those of the primary and secondary compounds. The difference is, however, that treatment with alkali cannot remove a proton from the product which we call a quaternary ammonium salt. There is no proton to remove.

$$R_3N + R'X \rightarrow R_3\overset{+}{N}R'\ X^-$$

$$(C_2H_5)_3N + C_2H_5I \rightarrow (C_2H_5)_4\overset{+}{N}\ I^-.$$

triethyl-amine   ethyl iodide   tetraethylammonium iodide (a quarternary ammonium salt)

The reaction of tertiary amines with acylating agents probably leads to similar ions but these are relatively unstable and there may only be a small quantity present in equilibrium with reactants:

$$RCOCl + R'_3N \rightleftharpoons RCO\overset{+}{N}R'_3\ Cl^-.$$

3) Effect of nitrous acid. With aliphatic tertiary amines only normal acid–base reactions occur. Amines carrying an aromatic group on nitrogen are normally attacked at a ring position, however, to give a C-nitroso compound:

C$_6$H$_5$N(CH$_3$)$_2$ + HNO$_2$ → 4-NO-C$_6$H$_4$-N(CH$_3$)$_2$ + H$_2$O.

N,N-dimethylphenylamine   4-nitroso-N,N-dimethylphenylamine

Amides do not react.

The introduction of the nitroso group into the ring has an interesting effect on the carbon–nitrogen (amino) bond. It so "activates" it that it becomes susceptible to alkali-catalysed hydrolysis. A secondary amine is produced by hydrolysis, as well as 4-nitrosophenol, and this reaction is sometimes employed to prepare secondary amines:

$$\underset{\text{4-nitroso-N,N-dimethylphenylamine}}{C_6H_4(N(CH_3)_2)(NO)} + H_2O \rightarrow \underset{\text{4-nitrosophenol}}{C_6H_4(OH)(NO)} + \underset{\text{dimethylamine}}{(CH_3)_2NH}.$$

N,N-disubstituted amides are not affected by nitrous acid.

4) C—N bond cleavage in amines. It has already been pointed out that the C—N bond in amines is not susceptible to hydrolysis. It can, however, be broken by treatment with concentrated solutions of strong acids at high temperature, and in this respect amines closely resemble ethers (Section 23.2.2). The reaction is not confined to tertiary amines.

$$\underset{\text{N,N-dimethyl-phenylamine}}{C_6H_5N(CH_3)_2} + HI \xrightarrow{\text{heat}} \underset{\text{N-methyl-phenylamine}}{C_6H_5NHCH_3} + \underset{\text{methyl iodide}}{CH_3I}$$

$$\underset{\text{N-methyl-phenylamine}}{C_6H_5NHCH_3} + HI \xrightarrow{\text{heat}} \underset{\substack{\text{phenylamine}\\\text{(aniline)}}}{C_6H_5NH_2} + \underset{\text{methyl iodide}}{CH_3I}.$$

24.3.3 Introduction of the Group

1) Alkylation or acylation of secondary amines. The procedure is very similar to that described for the alkylation or acylation of primary amines:

$$\underset{\text{dimethylamine}}{(CH_3)_2NH} + \underset{\text{methyl iodide}}{CH_3I} \longrightarrow \underset{\text{trimethylammonium iodide}}{(CH_3)_3\overset{+}{N}H\ I^-} \xrightarrow{\text{NaOH}} \underset{\text{trimethylamine}}{(CH_3)_3N;}$$

$$\underset{\text{N-ethylphenylamine}}{C_6H_5NHC_2H_5} + \underset{\text{acetic anhydride}}{(CH_3CO)_2O} \longrightarrow \underset{\text{N-ethyl-N-phenylacetamide}}{C_6H_5N(C_2H_5)COCH_3}.$$

2) Conversion of amides to amines. This is exactly analogous to the reductions described for primary and secondary amino compounds:

$$\underset{\text{N,N-dimethylpropanamide}}{C_2H_5CON(CH_3)_2} \xrightarrow{\text{Na/C}_2\text{H}_5\text{OH}} \underset{\text{N,N-dimethylprop-1-ylamine}}{C_2H_5CH_2N(CH_3)_2}.$$

24.4 QUATERNARY AMMONIUM COMPOUNDS

These compounds are of type $R_4\overset{+}{N}\ X^-$ where R is usually an alkyl or aryl group and X^- a suitable negative ion (e.g. OH^-, Cl^-, HSO_4^-). With four carbon groups attached to nitrogen, they represent a logical extension of the classes of compound

Table 24.2. Important reactions of amino compounds—amines and amides. Groups symbolized by R may be either alkyl, aryl (Ar), or acyl (RCO). R' is exclusively alkyl and R" may be alkyl or aryl.

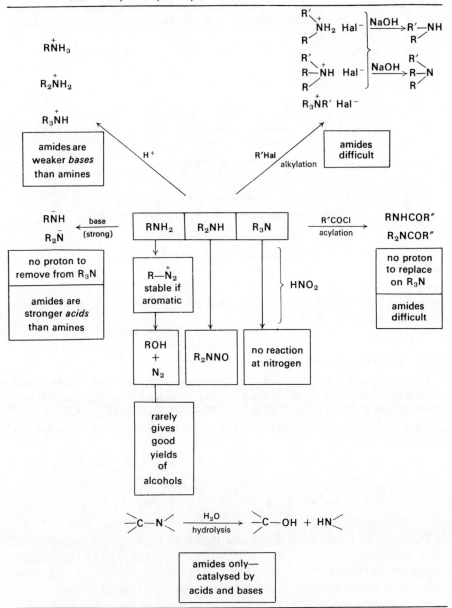

which we have just considered with one, two, or three carbon groups on nitrogen. They are obtained by alkylating tertiary amines (Section 24.3.2).

24.4.1 Physical Properties

Quaternary ammonium compounds are usually crystalline ionic solids having the usual ionic properties (Section 4.1). They are usually soluble in water, although

24.4 QUATERNARY AMMONIUM COMPOUNDS 643

Table 24.3 Important methods for synthesizing amino compounds. The groups symbolized by R′ in this table may be alkyl or aryl (Ar) groups. R is exclusively alkyl.

Important methods of C—N bond formation

Alkylation of $\ce{>NH}$ $\xrightarrow{\text{R—Hal}}$ $\ce{>N+(H)(R)}$ + Hal$^-$ $\xrightarrow{\text{NaOH}}$ $\ce{>N-R}$ + NaHal

Acylation of $\ce{>NH}$ $\xrightarrow{\text{R′COX}}$ $\ce{>NCOR′}$ + HX

acylation and alkylation of amides is difficult

X = OCOR—acid anhydride
= Cl —acid chloride
= OR —ester

Special methods for synthesis of primary amines

R′NO$_2$ $\xrightarrow{\text{Sn/HCl}}$ R′NH$_2$

R′C≡N $\xrightarrow{\text{LiAlH}_4}$ R′CH$_2$NH$_2$

when the carbon groups are very large, and the ionic portion of the molecule therefore relatively small and unimportant, solubility in covalent organic solvents may be favoured.

24.4.2 Chemical Behaviour

1. These compounds are fundamentally different in character from the amino and alkylamino compounds which we have already discussed. They have no lone pair of non-bonding electrons and can therefore be neither nucleophiles (Section 26.3) nor bases. They do not, for example, react with alkyl halides, or acids, nor can they be acylated.

2. Treatment of a quaternary ammonium halide with an aqueous suspension of silver oxide usually results in the precipitation of silver halide and a solution of a quaternary ammonium hydroxide remains. The resulting quaternary ammonium hydroxide is of base strength, in water, comparable with sodium hydroxide.

$$2R_4\overset{+}{N}X^- + Ag_2O + H_2O \rightarrow 2R_4\overset{+}{N}OH^- + 2AgX.$$
$$(\equiv 2AgOH)$$

For example:

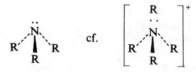

$$2(C_2H_5)_4\overset{+}{N}I^- + Ag_2O + H_2O \rightarrow 2(C_2H_5)_4\overset{+}{N}OH^- + 2AgI.$$

tetraethylam-
monium iodide   tetraethylammonium
                              hydroxide

Why, then, one might ask, is it a so much stronger base than ammonium hydroxide (which is much weaker than sodium hydroxide)? The answer to this question lies in the fact that a proton can be removed from the unsubstituted ammonium ion by the hydroxide ion. Thus ammonium ions and hydroxide ions are in equilibrium, in water, with ammonia and water molecules (Section 17.9.3), and the base strength of ammonia, which is expressed in terms of the equilibrium constant of this reaction, is the controlling factor. No such equilibrium can exist with the quaternary ammonium hydroxides (which contain no such protons), and these are always completely dissociated, like sodium and potassium hydroxides, and their base strength is essentially that of the hydroxide ion.

3. Quaternary ammonium hydroxides, when heated, undergo an interesting and useful elimination reaction in which an alkene is produced; for example:

$$(C_2H_5)_4 \overset{+}{N} OH^- \xrightarrow{\text{heat}} (C_2H_5)_3N + H_2O + CH_2{=}CH_2.$$

tetraethylammonium hydroxide   triethylamine   ethylene

In this example a hydrogen atom and the triethylamino group are removed—eliminated—from the molecule, leaving an alkene:

$$\underset{CH_2 \text{—} CH_2}{H \overset{\text{eliminated}}{\curvearrowleft \curvearrowright} \overset{+}{N}Et_3} \quad OH^-$$

This is a useful reaction for alkene synthesis.

Heating tetra-alkylammonium halides in a vacuum results in dealkylation of a different type. What is essentially the reverse of the amine + alkyl halide reaction occurs:

$$R_4\overset{+}{N} X^- \rightarrow R_3N + RX.$$

Finally, a comment on the oxygen analogues of quaternary ammonium salts might be made. These are much less common than quaternary ammonium salts and are only made by treating ethers with very powerful alkylating agents. They are themselves highly reactive alkylating agents. Alkyl tetrafluoroborates can be used to prepare them and the alkylating agent is essentially a carbonium ion:

$$R^+ BF_4^- + R_2O \rightarrow R_3\overset{+}{O} BF_4^-.$$

24.5 IMINO GROUPS

Here we shall consider the C=N—H (imino), C=N—R (alkylimino), and C=N—Ar (arylimino) groups. They are included because they are the structural analogues of the carbonyl function, proceeding from C=O to C=N—, rather than because their chemistry is particularly important. We have already indicated that there are structural relationships between nitrogen and oxygen compounds and it is logical to ask what the nitrogen analogue of the carbonyl group is, and to enquire as to its chemistry.

In fact the chemistry of the imino groups is much less important than that of the carbonyl group and compounds containing imino groups are probably best seen as derivatives of carbonyl compounds. Thus, they are usually prepared by condensation of a carbonyl compound with a primary amino compound and they usually revert to these parent compounds very readily on hydrolysis. It is usually only in the case of aromatic compounds that stable, crystalline compounds, sometimes known as Schiff's bases, can be isolated. Dry phenylamine (aniline) reacts with dry benzaldehyde when warmed, for example, to yield an arylimino compound, N-benzylidenephenylamine. This compound is very easily hydrolysed back however,

$$C_6H_5CHO + H_2NC_6H_5 \underset{\text{hydrolysis}}{\overset{\text{condensation}}{\rightleftharpoons}} C_6H_5CH{=}NC_6H_5 + H_2O,$$

benzaldehyde phenylamine N-benzylidenephenyl-
(aniline) amine

and a pure sample of N-benzylidenephenylamine is very soon contaminated with benzaldehyde and phenylamine (aniline) unless special precautions are taken to keep the sample very dry. Oximes and hydrazones (Section 23.3.2) are clearly close relatives of imino compounds but are generally much more stable to hydrolysis.

24.6 NITRILES AND ISONITRILES

Reactions of alkyl halides with inorganic cyanides lead to products which depend upon the inorganic cyanide used. Reaction of an alkyl iodide with potassium cyanide leads largely to a nitrile (alkyl cyanide), although a little isonitrile (isocyanide) may also be formed. With silver cyanide, however, the isonitrile predominates.

$$RI + KCN \rightarrow RCN + KI$$
nitrile
(alkyl cyanide)

$$RI + AgCN \rightarrow RNC + AgI.$$
isonitrile
(alkyl isocyanide)

The very different natures of these two inorganic cyanides should be noted: potassium cyanide is ionic, whereas silver cyanide has a covalent, polymeric structure rather like $(CuCN)_x$ (Section 20.21).

Both types of compounds contain multiple bonds. The bonding between nitrogen and carbon in isonitriles is usually written $R{-}N{\equiv}C$. Compare this with the nitrile, $R{-}C{\equiv}N$.

Nitriles usually derive their names from the carboxylic acid which they would yield on hydrolysis (see below). Thus CH_3CN is acetonitrile (hydrolysis yields acetic acid), CH_3CH_2CN propanonitrile (propanoic acid on hydrolysis), and C_6H_5CN benzonitrile. Alternatively, these compounds may be called cyanides—methyl, prop-1-yl, and phenyl cyanide, respectively. Compounds RNC, or ArNC, are alkyl or aryl isonitriles (or isocyanides).

24.6.1 Physical Properties

Nitriles are more soluble in water than isonitriles and both dissolve in organic solvents. Nitriles are not as toxic as compounds containing a free cyanide ion although isonitriles are both toxic and foul smelling. The lower nitriles and isonitriles are all liquids.

24.6.2 Chemical Behaviour

1) Hydrolysis. Both types of compound are hydrolysed by aqueous acid but the products are different and reflect the different structures (sequences of atoms) of the compounds being hydrolysed:

$$RC{\equiv}N + 2H_2O \xrightarrow{acid} RCOOH + NH_3;$$

nitrile   carboxylic (forms a
[alkyl (or aryl)   acid   salt with the
cyanide]   aqueous acid)

$$R{-}N{\equiv}C + 2H_2O \xrightarrow{acid} R{-}NH_2 + HCOOH.$$

isonitrile   primary   formic
(alkyl iso-   alkylamine   acid
cyanide)   (forms a salt
                with the
                aqueous acid)

Note that the nitrile carbon atom remains attached to the main carbon skeleton when the nitrile is hydrolysed, whereas the nitrogen atom remains attached in the isonitrile hydrolysis. It is also important to realize that hydrolysis does *not* lead to removal of the complete CN group and therefore alcohols are not produced:

$$\begin{array}{c}RCN\\(RNC)\end{array} + H_2O \text{ do not react to give } ROH + HCN.$$

This is presumably a kinetic effect (Section 26.1), as alcohols do not react with hydrogen cyanide to give nitriles and water. Examples:

$$CH_3CN + 2H_2O \rightarrow CH_3COOH + NH_3;$$

acetonitrile   acetic acid
(methyl cyanide)

$$C_2H_5N{\equiv}C + 2H_2O \rightarrow C_2H_5NH_2 + HCOOH.$$

ethyl   ethylamine   formic acid
isonitrile

Aqueous alkali (hot) will also bring about the hydrolysis of nitriles but does *not* affect isonitriles.

2) Reduction. Reduction of these compounds has already been referred to and is summarized here by the following general equations:

$$RC{\equiv}N + [4H] \xrightarrow{Na/C_2H_5OH} RCH_2NH_2;$$
$$RC{\equiv}N + [2H] \xrightarrow{SnCl_2} RCH{=}NH \xrightarrow{H_2O} RCHO;$$
$$RN{\equiv}C + [4H] \xrightarrow[Ni]{H_2} RNHCH_3.$$

The third reaction is of little preparative significance, but demonstrates the presence of an R—N linkage in the isonitrile.

24.6.3 Introduction of the Groups

1. Reaction of alkyl halides with inorganic cyanides is probably the most convenient and important method—particularly for nitriles. These reactions, discussed earlier, are summarized:

$$RX + KCN \rightarrow RCN + KX;$$
$$RX + AgCN \rightarrow RNC + AgX.$$

This method is of no use for the preparation of the aromatic compounds.

2. Aryl diazonium salts react with potassium tetracyanocuprate(I), $K_3[Cu(CN)_4]$ (KCN + CuCN), to yield aromatic nitriles; for example:

$$KCN + C_6H_5N_2^+ \; Cl^- \xrightarrow[Cu+]{KCN} C_6H_5CN + N_2 + KCl.$$

benzenedia-   benzonitrile
zonium chloride (phenyl cyanide)

This method is applicable only to aromatic nitriles.

3. Dehydration of amides, by heating them with phosphorus pentoxide, gives nitriles:

$$RCONH_2 \xrightarrow{-H_2O} RC\equiv N;$$
$$CH_3CONH_2 \xrightarrow{-H_2O} CH_3C\equiv N.$$

acetamide   acetonitrile
                  (methyl cyanide)

4. Addition of hydrogen cyanide to aldehydes and ketones leads to nitriles (Section 23.3.2):

$$\diagdown \!\!\!\! C=O + HCN \rightarrow \diagdown \!\!\!\! C(OH)(CN) \diagup$$

For example:

$$CH_3CHO + HCN \rightarrow CH_3\!-\!\underset{CN}{\overset{OH}{\underset{|}{\overset{|}{C}}}}\!-\!H.$$

acetaldehyde   acetaldehyde cyanohydrin

24.7 THE NITRO GROUP

The structure of the nitro group is $-N\!\!\diagup\!\!\diagdown\!\overset{O}{\underset{O}{}}$. Nitroalkanes have the general structure, RNO_2, and aromatic nitro compounds, $ArNO_2$. Nitro compounds should not be confused with the structurally isomeric nitrite esters which contain no C—N bond (e.g. nitroethane has the structure, $CH_3CH_2N\!\!\diagup\!\!\diagdown\!\overset{O}{\underset{O}{}}$, whereas the structure of ethyl

648   COMPOUNDS WITH FUNCTIONAL GROUPS CONTAINING NITROGEN nitrite is $CH_3CH_2-O-N=O$). Simple nitro compounds are named as substituted hydrocarbons, for example, nitromethane CH_3NO_2, nitrobenzene $C_6H_5NO_2$.

24.7.1 Physical Properties

The lower alkyl compounds are volatile liquids of moderately high solubility in water. Aromatic nitro compounds are usually crystalline solids (nitrobenzene itself, $C_6H_5NO_2$, is a high-boiling liquid), insoluble in water.

24.7.2 Chemical Behaviour

1. The nitro group can be reduced to a primary amino group and, particularly with aromatic compounds, is an important precursor of primary amines (aromatic amines cannot be prepared by treating an aromatic halide with ammonia—Section 24.1.3):

$$RNO_2 \xrightarrow{[H]} RNH_2 + H_2O;$$
$$ArNO_2 \longrightarrow ArNH_2.$$

The fact that an amine is so produced demonstrates the presence of a C—N bond in the original nitro compound and differentiates nitro compounds from the isomeric nitrites, reduction of which could not lead to an amine:

$$R-O-N=O \text{ cannot lead to } R-NH_2.$$

2. An interesting feature of the chemistry of nitroalkanes is their ability to form salts when treated with a suitably strong base. Nitromethane, for example, reacts with sodium ethoxide to give a sodium salt which can be isolated quite easily as a white solid:

$$CH_3NO_2 + NaOEt \rightarrow (CH_2NO_2)Na + EtOH.$$

The removal of a hydrogen atom from an alkyl group is quite remarkable and in most circumstances is not possible. The nitro group, however, renders the adjacent C—H bonds more acidic than they could normally be and makes this proton removal possible. This "acid conferring" character of the nitro group is not unique and other groups, such as carbonyl, also show this effect. A fuller discussion of this effect is to be found in Section 28.3, when diethyl malonate and ethyl acetoacetate are discussed.

24.7.3 Introduction of the Group

1) Nitration of hydrocarbons. Treatment of aromatic hydrocarbons with nitrating mixture (a mixture of concentrated nitric and sulphuric acids) usually results in the introduction of a nitro group. Nitrobenzene, for example, is conveniently prepared by heating benzene with nitrating mixture (this reaction, and other related reactions, are discussed more fully in Section 26.5):

$$\underset{\text{benzene}}{C_6H_6} + HNO_3 \xrightarrow{H_2SO_4} \underset{\text{nitrobenzene}}{C_6H_5NO_2} + H_2O.$$

Treatment of alkanes with nitric acid at high temperatures (vapour phase nitration) also yields nitro compounds. The reaction is very different from aromatic nitration, however (it is a complex free radical reaction), and often mixtures of products are obtained.

2) Nitration of halogenoalkane. By treating a halogenoalkane with a metal nitrite (e.g. potassium nitrite) a nitroalkane is obtained, although the isomeric nitrite ester is usually obtained as well:

$$R-Hal + KNO_2 \rightarrow RNO_2 + RONO + KHal.$$

These can usually be separated quite easily, though.

A better synthesis is one employing a halogenated carboxylic acid rather than a simple alkyl halide. The carboxylic acid grouping is lost during the reaction, and the product is a simple nitroalkane, for example:

$$KNO_2 + ClCH_2COOH \rightarrow KCl + CH_3NO_2 + CO_2.$$

24.8 DIAZONIUM GROUP

24.8.1 Production of Diazonium Salts

Treatment of a primary amine with nitrous acid, in the presence of an excess of mineral acid, results in the production of a diazonium salt. In an excess of hydrochloric acid, a diazonium chloride is produced:

$$R-NH_2 + HNO_2 + HCl \rightarrow R-\overset{+}{N}\!\!\equiv\!\!N + Cl^- + 2H_2O;$$

$$Ar-NH_2 + HNO_2 + HCl \rightarrow Ar-\overset{+}{N}\!\!\equiv\!\!N + Cl^- + 2H_2O.$$

Aliphatic diazonium salts are so unstable that they decompose as soon as they are formed and liberate nitrogen and a carbonium ion (Section 24.1.2) but aromatic diazonium salts are much more stable and, provided that the temperature is kept below about 5 °C, will remain in the solution in which they are formed. An example of an aromatic diazonium salt is benzenediazonium chloride, formed by "diazotizing" phenylamine (aniline):

$$C_6H_5NH_2 + HNO_2 + HCl \xrightarrow{5\,°C} C_6H_5\overset{+}{N}\!\!\equiv\!\!N + Cl^- + 2H_2O.$$
phenylamine   benzenediazonium
(aniline)   chloride

Delocalization is possible only in the aromatic salts and probably accounts for their relatively enhanced stability.

24.8.2 Reactions of Diazonium Salts

1. The importance of these compounds lies principally in the fact that, although sufficiently stable to exist in solution at low temperatures, they are still highly reactive. The diazonium group, $-\overset{+}{N}\!\!\equiv\!\!N$, is closely related to a nitrogen molecule,

N≡N, and, as the latter is very stable indeed, it is thermodynamically favourable for the N_2 entity to separate from the phenyl group at higher temperatures, taking with it the bonding pair of electrons. This is, of course, how the alkyl compounds also decompose:

$$\text{Ar}-\overset{+}{\text{N}}\equiv\text{N} \rightarrow \text{Ar}^+ + \text{N}\equiv\text{N}.$$

This means that in the aromatic diazonium salts, the N_2 group can be very readily replaced by a nucleophile (Section 26.3), such as I^- or OH^-, and these compounds undergo a series of such substitution reactions which are of the utmost preparative importance because very few aromatic compounds undergo nucleophilic substitution reactions at all (Sections 26.5 and 28.5). Chlorobenzene, for example, is not readily hydrolysed and does not react readily with amines or potassium cyanide; thus methods of introducing hydroxyl, amino, and cyanide groups, which are useful for the aliphatic analogues, simply do not work.

Some of the groups which can be introduced by replacement of the diazonium group are as follows, using the simple phenyl compounds for illustration:

$$C_6H_5\overset{+}{N_2}\,\overset{-}{Cl} + H_2O \rightarrow C_6H_5OH + N_2 + HCl;$$
phenol

$$C_6H_5N_2Cl + KI \rightarrow C_6H_5I + N_2 + KCl;$$
iodobenzene

$$C_6H_5N_2Cl \rightarrow C_6H_5Cl + N_2;$$
chlorobenzene

$$C_6H_5N_2Br \rightarrow C_6H_5Br + N_2;$$
bromobenzene

$$C_6H_5N_2Cl + KCN \rightarrow C_6H_5CN + N_2 + KCl.$$
benzonitrile

The last three of these reactions require metallic copper, or a copper salt (copper(I) halide plus hydrogen halide, or copper(I) cyanide, as appropriate), as catalyst and are probably not nucleophilic substitution reactions. However, from a preparative point of view, the end result is the same—important functional groups can be introduced into the ring as the stable nitrogen molecule is lost.

2. In addition to substitution reactions involving loss of nitrogen, diazonium salts undergo a series of reactions with certain other aromatic molecules in which the nitrogen is retained.

Addition of a diazonium salt solution to a strongly alkaline solution of a phenol results in such a reaction—termed a coupling reaction. For example:

$$C_6H_5\overset{+}{N_2}\,H\overset{-}{SO_4} + \text{C}_6\text{H}_5\text{—OH} \xrightarrow{\text{NaOH}} C_6H_5\text{—N}=\text{N—}\text{C}_6\text{H}_4\text{—OH};$$

benzenediazonium hydrogen sulphate   phenol   4-hydroxyazobenzene

$$C_6H_5\overset{+}{N_2}\,Cl^- + \text{2-naphthol} \xrightarrow{\text{NaOH}} C_6H_5\text{—N}=\text{N—(naphthol-OH)}$$

benzenediazonium chloride   2-naphthol   2-hydroxynaphthalene-1-azobenzene

Many of these products are highly coloured (2-hydroxynaphthalene-1-azobenzene separates as a crimson precipitate) and are used as dyes (aniline dyes). Similar reactions occur with certain aromatic amines.

3. The reduction of diazonium salts, depending upon conditions and the reducing agent employed, may give products with or without the original nitrogen. In either case, the reaction is important. Reduction with hypophosphorous acid results in loss of nitrogen and its replacement by hydrogen:

$$C_6H_5\overset{+}{N}_2 \ Cl^- \xrightarrow{H_3PO_2} C_6H_6.$$
benzenediazonium chloride → benzene

It is a useful way of removing unwanted amino groups (or groups which can be converted to amino) from aromatics:

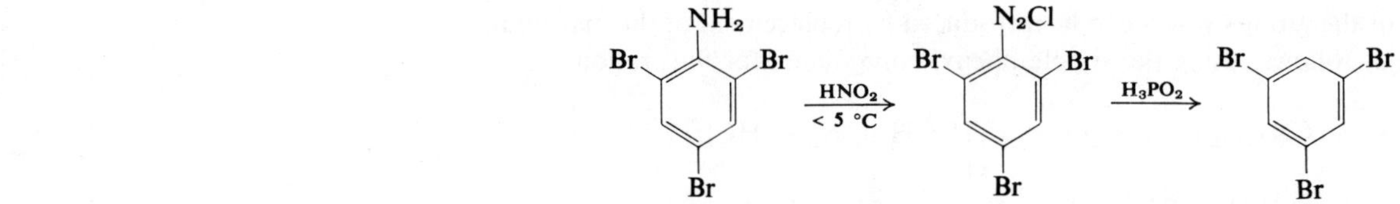

2,4,6-tribromoaniline $\xrightarrow[<5\ °C]{HNO_2}$ 2,4,6-tribromobenzenediazonium chloride $\xrightarrow{H_3PO_2}$ 1,3,5-tribromobenzene

Reduction with tin(II) chloride in hydrochloric acid involves retention of nitrogen and a substituted hydrazine is obtained. Phenylhydrazine (Section 23.3.2) may be prepared in this way:

$$C_6H_5\overset{+}{N}_2 \ \overset{-}{H}SO_4 \xrightarrow[HCl]{SnCl_2} C_6H_5NHNH_2.$$
benzenediazonium hydrogen sulphate → phenylhydrazine

Table 24.4. Important reactions of aromatic diazonium salts.

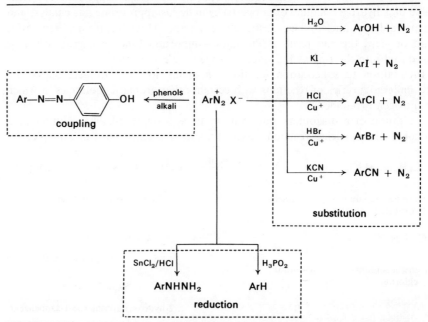

652 COMPOUNDS WITH FUNCTIONAL GROUPS CONTAINING NITROGEN

PROBLEMS

1. Study the following list of compounds:
   i) $C_6H_5NH_2$;
   ii) $C_6H_5CONH_2$;
   iii) $C_6H_5CH_2NHCH_3$;
   iv) $CH_3CON(CH_3)_2$;
   v) $CH_3CH_2NHCOCH_3$;
   vi) ortho-$C_6H_4(CONH_2)(COOH)$.

   a) Which compounds contain a primary amino group?
   b) Which compounds contain a secondary amino group?
   c) Which compounds contain a tertiary amino group?

2. Referring to the compounds listed in Problem 1:
   a) Which compounds would be hydrolysed by hot aqueous acid or alkali? Write equations for these reactions.
   b) Which compounds are primary amines?
   c) Which compounds are secondary amines?
   d) Which compounds are tertiary amines?
   e) Which compounds are amides?

3. Arrange the following three compounds in the order in which you would expect their boiling points to increase, giving your reasons:
   a) $CH_3CH_2CH_2OH$;
   b) $CH_3CH_2CH_2NH_2$;
   c) $CH_3CH_2CONH_2$.

   Look up the known boiling points in a suitable list of physical constants from your library and see how the observed order compares with your prediction.

4. Arrange the following three compounds in the order in which you would expect their basicity to increase:
   a) $C_6H_5CH_2NH_2$;
   b) $C_6H_5CONH_2$;
   c) phthalimide (o-$C_6H_4(CO)_2NH$).

5. Urea (or carbamide), NH_2CONH_2, is an important, naturally occurring compound and is clearly a type of amide (—$CONH_2$). Predict what the products would be:
   a) on boiling urea with aqueous sodium hydroxide solution;
   b) on treating urea with nitrous acid.

   Acetamide (ethanamide) is derived from acetic acid (ethanoic acid). From what acid is urea derived?

6. Go to your library, consult some more advanced texts, and find out as much as you can about urea. Write a summary of the chemistry of urea, paying particular attention to points of difference between urea and other amides, to the biological role of urea, and to the historical significance of the early synthesis of urea (discovered by Wöhler).

7. Study the following list of compounds:
   i) $CH_3CH_2CH_2NH_2$;

   ii)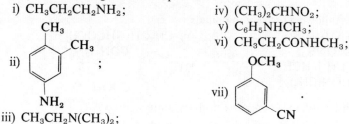

   iii) $CH_3CH_2N(CH_3)_2$;

   iv) $(CH_3)_2CHNO_2$;
   v) $C_6H_5NHCH_3$;
   vi) $CH_3CH_2CONHCH_3$;

   vii) (3-methoxybenzonitrile: benzene ring with OCH_3 and CN substituents)

   a) Which of these compounds can be acetylated?
   b) Which compounds can be methylated with iodomethane (methyl iodide)?
   c) Which compound would give nitrogen when treated with nitrous acid?
   d) Which compound would give a stable diazonium salt with nitrous acid at low temperatures?
   e) Which compounds would give a primary amine on reduction?
   f) Which compounds would give a quaternary ammonium salt if treated with iodomethane?
   g) Which compound would give a secondary amine on reduction?
   h) Which compound would give a carboxylic acid and ammonia on hydrolysis?
   i) Which compound would give a carboxylic acid and a primary amine on hydrolysis?

8. Outline simple chemical tests which would enable you to distinguish one compound from the other in the following pairs:
   a) $CH_3CH_2NH_2$ and $(CH_3)_2NH$;

   b) 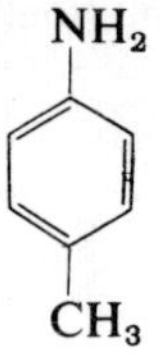 (4-methylaniline) and (4-methylcyclohexylamine);

   c) $CH_3CH_2NH_2$ and CH_3CONH_2;
   d) $CH_3CH_2CH_2C{\equiv}N$ and $CH_3CH_2CH_2CONH_2$;
   e) $CH_3\overset{+}{N}H_3\ Cl^-$ and $(CH_3)_4\overset{+}{N}\ Cl^-$;
   f) $(CH_3)_3\overset{+}{N}H\ HSO_4^-$ and $(CH_3)_3\overset{+}{N}H\ Cl^-$;

   g) 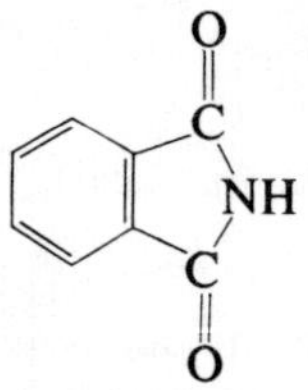 (phthalimide) and (isoindoline);

   h) $\underset{CH_2NH_2}{CH_2COOH}$ and $\underset{CH_2-NO_2}{CH_2-CH_3}$;

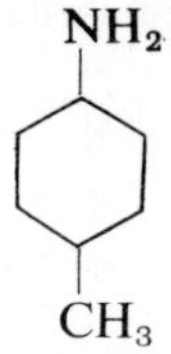

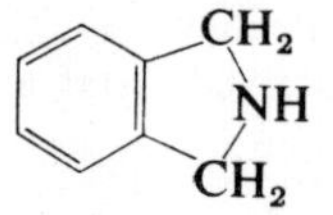

i)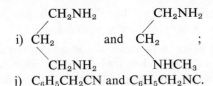

j) $C_6H_5CH_2CN$ and $C_6H_5CH_2NC$.

9. 1.30 g of a compound, A, of molecular formula $C_5H_{12}N_2O_2$, gave 0.448 dm³ of nitrogen gas on treatment with nitrous acid. How many primary amino groups are present in each molecule of A? Similar treatment of 0.65 g of B, $C_5H_{12}N_2O_2$, gave 0.112 dm³ of nitrogen. How many primary amino groups are there in each molecule of B?

Problems 10 to 13. Suggest, in outline, how each of the conversions in Problems 10 to 13 may be brought about.

10. a) benzene → nitrobenzene.

b) nitrobenzene → aniline.

c) aniline → chlorobenzene.

d) 3,5-dimethylbenzene → 2-iodo-1,3-dimethylbenzene (with CH₃ groups at 3,5 and I at the remaining position).

11. a) $CH_3CH_2Br \rightarrow CH_3CH_2CN$.
    b) $CH_3CH_2CN \rightarrow CH_3CH_2CH_2NH_2$.
    c) $CH_3I \rightarrow CH_3CH_2NH_2$.
    d) $C_6H_5CH_2CH_2Br \rightarrow C_6H_5CH_2CH_2CH_2NHCOCH_3$.

12. a) $CH_3CH_2NH_2 \rightarrow CH_3CH_2\overset{+}{N}(CH_3)_3\ I^-$.
    b) $CH_3CH_2\overset{+}{N}(CH_3)_3\ I^- \rightarrow CH_3CH_2\overset{+}{N}(CH_3)_3\ OH^-$.
    c) $CH_3CH_2\overset{+}{N}(CH_3)_3\ OH^- \rightarrow CH_2{=}CH_2 + (CH_3)_3N + H_2O$.
    d) $C_6H_5CH_2CH_2NH_2 \rightarrow C_6H_5CH{=}CH_2$.
    e) $CH_3CH_2CH_2I \rightarrow CH_3CH_2CH{=}CH_2$ (*hint*—consider Problem 11).

13. a) $C_6H_5CH_2Br \rightarrow C_6H_5CH_2COOH$.

b) 2-methylaniline → 2-methyl-4-(2-methyl-5-hydroxyphenylazo)... (o-toluidine → azo dye with o-cresol).

c) $C_6H_5CH_2COOH \rightarrow C_6H_5CH_2NH_2$.

14. A compound of formula C_3H_9N can be acetylated, but gives no nitrogen gas when treated with nitrous acid. What is the structure of this compound?

15. A compound X, $C_5H_{11}NO$, on refluxing with aqueous sodium hydroxide yields a volatile compound, Y, with a "fishy" odour and molecular formula, C_2H_7N. Y liberates nitrogen from nitrous acid. On acidifying the liquid which remains after refluxing X with sodium hydroxide and distilling out Y, an acidic material, Z, is obtained. Z can be synthesized by hydrolysing the product of the reaction between bromoethane (ethyl bromide) and potassium cyanide. What are the structures of X, Y, and Z? Give your reasoning. Why does Z not distil out of the reaction mixture when Y is distilled out?

16. In the form of a brief summary, make a comparison between the chemistries of primary, secondary, and tertiary amines. Emphasize points of similarity and points of difference. From your summary, suggest methods whereby one type of amine could be distinguished from the other.

BIBLIOGRAPHY

This chapter covers a wide range of material and one (or more) of the books on organic chemistry from the general list would provide the most helpful form of further reading.

CHAPTER 25

Compounds with Halogens as Functional Groups

The halogens are predominantly univalent elements and are therefore capable of forming a much narrower range of compounds than either oxygen or nitrogen can. They form simple hydrides (e.g. hydrogen chloride) and their organic derivatives could be looked upon, formally, as being substituted hydrogen halides. The ethyl chloride molecule, for example, C_2H_5Cl, can be regarded as a hydrogen chloride molecule in which the hydrogen atom has been replaced by an ethyl group. Because of their univalency, the halogens, in forming hydrides, bond to only one hydrogen atom; thus, when that is replaced by a carbon group, no other hydrogen atoms remain attached to the halogen. Consequently, those aspects of the chemistry of oxygen and nitrogen compounds (acidity, alkylation, acylation) which depend upon the presence of N—H and O—H bonds have no parallel here.

Most of the halogens are more electronegative than carbon and they all form stable negative ions. Their hydrides, for example, are highly ionized in water, consisting largely of hydrogen ions and halide ions:

$$HCl + H_2O \rightleftharpoons H_3O^+ + Cl^-.$$

Hydrogen fluoride is least inclined to ionize thus, and hydrogen iodide ionizes to the greatest extent (Section 13.4). Most of the reactions of organic halogen compounds result in the production of stable halide ions, although very few of these compounds ionize to any significant extent in the manner of the hydrogen halides; the carbon–halogen bond is more resistant to ionization than is the hydrogen–halogen bond.

For each halogen, therefore, we are confined to a single functional group (carbon–halogen bond) and the compounds which we shall consider vary only in the identity of the halogen atom and in the nature of the organic skeleton to which it is attached. Variations in the identity of the halogen atom result in relatively trivial variations in chemistry, although fluorine is perhaps the exception. We shall, however, be concerned largely with chlorine, bromine, and iodine compounds, which are very similar in character, and the most important variations in

25.1 Physical properties
25.2 Chemical behaviour
25.3 Introduction of the groups

their chemistry arise from variations in the carbon skeleton. The principal types of compound to be considered are as follows:

R—Hal, halogenoalkanes or alkyl halides;
Ar—Hal, halogenoarenes or aryl halides;
RCOHal, acyl halides or acid halides.

Simple aliphatic and aromatic halogen compounds (other than acid chlorides) are most commonly looked upon as substituted (halogenated) hydrocarbons, rather than as substituted hydrogen halides, and this is reflected in the system of nomenclature now most favoured. On this system, as substituted hydrocarbons, CH_3CH_2Cl is chloroethane, $CH_3CH_2CH(Br)CH_3$ is 2-bromobutane, and C_6H_5I is iodobenzene. They are halogenoalkanes or halogenoarenes. Often, however, it is desirable to stress that halogenoalkanes contain an alkyl group and they may then be named as alkyl halides. Names such as ethyl chloride (CH_3CH_2Cl) or but-2-yl bromide ($CH_3CH_2CH(Br)CH_3$) are typical. Older names for alkyl groups—or at any rate, names which do not conform to IUPAC rules—are still commonly used too, and are of some importance. Tertiary-butyl bromide, for example, is the name often given to $(CH_3)_3CBr$, which on the IUPAC system would be 2-methylprop-2-yl bromide or 2-bromo-2-methylpropane. This stresses that the alkyl group is tertiary, a point of considerable chemical importance. Because both types of name are important, we have used both in this book. We have tended to give preference to the halogenoalkane names, but have also used alkyl halide names where they seemed to be more helpful. In some places, both names have been quoted.

25.1 PHYSICAL PROPERTIES

With no complications due to hydrogen bonding, the boiling points of these compounds tend to rise steadily, in the usual way, with increasing molecular weight. The high atomic weight of the halogens, and the polarity of the carbon–halogen bond (Section 4.4) tend to make boiling points fairly high for a given size of carbon chain with the more polar acid halides having the highest. A comparison is given in Table 25.1.

Table 25.1

	Molecular weight	Boiling point (°C)
Chloroethane (ethyl chloride)	64	12.5
1-Chloropropane (prop-1-yl chloride)	78	46
Acetyl chloride	78	55
Ethane	30	−84

Halogens also confer upon their compounds a fairly high density, and most halogen compounds, in the liquid state, are denser than water. Compare the various data in Table 25.2.

Table 25.2

	Molecular weight	Density (g cm^{-3})
Dichloromethane (methylene chloride)	85	1.38
Bromoethane (ethyl bromide)	109	1.46
Trichloromethane (chloroform)	120	1.50
Tetrachloromethane (carbon tetrachloride)	155	1.59
Benzene	78	0.87
Methylbenzene (toluene)	92	0.88
Decane ($C_{10}H_{22}$)	142	0.75
Undecane ($C_{11}H_{24}$)	156	0.75

25.2 CHEMICAL BEHAVIOUR

25.2.1 Substitution Reactions

The readiness of halogen atoms to form stable negative ions makes it possible for species containing lone pairs of electrons to attack the carbon atom to which a halogen may be attached and, with their lone pair of electrons, to displace the electron pair bonding halogen to carbon entirely on to the halogen atom. This type of reaction, nucleophilic substitution (Section 26.3), and the ability to undergo this type of reaction, is a characteristic of most carbon–halogen compounds—chlorides, bromides, and iodides.

Halogenoalkanes undergo these reactions readily, as illustrated by the following general equations (Hal = halogen atom). Iodo compounds are usually more reactive than bromo compounds which are, in turn, more so than chloro compounds.

$$R\text{—Hal} + H_2O \rightarrow ROH + Hal^- + H^+ \quad (1)$$
$$\text{alcohol}$$

$$R\text{—Hal} + OH^- \rightarrow ROH + Hal^- \quad (2)$$
$$(\text{NaOH(aq)}) \quad \text{alcohol}$$

$$R\text{—Hal} + NH_3 \rightarrow R\overset{+}{N}H_3 + Hal^- \quad (3)$$
$$(R\overset{+}{N}H_3\ Hal^-) \xrightarrow{\text{NaOH}} RNH_2 + H_2O + NaHal$$

$$R\text{—Hal} + CN^- \rightarrow RCN + Hal^- \quad (4)$$
$$(\text{KCN}) \quad \text{nitrile}$$

$$R\text{—Hal} + R'O^- \rightarrow R\text{—}OR' + Hal^- \quad (5)$$
$$(\text{NaOR}') \quad \text{ether}$$

We have met each of these reactions before in connection with the synthesis of alcohols, amines, nitriles, and ethers. Reaction (1) is usually very slow whereas Reaction (2) is much more rapid. Thus the hydrolysis (definition in Section 23.1.3) of halogenoalkanes is more conveniently achieved in alkaline solution (aqueous sodium hydroxide) than in neutral solution (water alone).

The corresponding aromatic compounds are characterized by their reluctance to undergo these nucleophilic substitution reactions and are normally inert to water, aqueous alkali, ammonia, potassium cyanide, and so on. None of these reagents will affect chlorobenzene, C_6H_5Cl, for example. A fuller discussion of this effect is given in Section 28.5.

By contrast, acyl chlorides undergo substitution reactions more readily than halogenoalkanes (the nature of the reaction is discussed in Section 26.4) and we can provide several general illustrations, all of which have been mentioned in Chapters 23 and 24, as follows:

$$RCOCl + H_2O \rightarrow RCOOH + HCl;$$
carboxylic acid

$$RCOCl + NH_3 \rightarrow RCONH_2 + HCl;$$
amide

$$RCOCl + R'OH \rightarrow RCOOR' + HCl.$$
ester

In each of these reactions the acyl chloride, itself a derivative of a carboxylic acid, gives rise to another carboxylic acid derivative and is of very high reactivity. Acetyl chloride, for example, gives off fumes of hydrogen chloride in moist air and, when added to water, "spits" and fumes dangerously in a rapid and exothermic reaction.

Other reactions in which the halogen atom may be replaced include reductions and the Friedel–Crafts reaction. Some of these are related, in type, to the nucleophilic substitutions discussed above.

1) Reduction. Halogenoalkanes may be reduced to alkanes by a zinc/copper couple or by lithium aluminium hydride:

$$R\text{—Hal} \xrightarrow{2H} RH + HHal.$$

Hydrogen iodide will usually reduce iodoalkanes to alkanes and, as iodoalkanes are obtained from alcohols and hydrogen iodide, treatment of alcohols with hydrogen iodide (in the presence of red phosphorus) results in the reduction of C—OH to C—H:

$$ROH \xrightarrow[\text{red P}]{HI} RH.$$

Reduction of halogenoarenes is difficult. Reduction of acyl chlorides also poses problems and the only useful reduction is achieved with hydrogen in the presence of a palladium catalyst. The product is an aldehyde.

$$RCOCl \xrightarrow[Pd]{H_2} RCHO.$$

For example:

C₆H₅—COCl → C₆H₅—CHO.

benzoyl chloride   benzaldehyde

2) Friedel–Crafts reaction. Halogenoalkanes and acid halides, in the presence of aluminium chloride as catalyst, will react with aromatic rings to give a product in which the aromatic ring has replaced the halogen atom. The aromatic compound has been acylated or alkylated:

benzene + CH₃COCl $\xrightarrow{AlCl_3}$ C₆H₅—COCH₃ + HCl

methyl phenyl ketone (acetophenone)

methylbenzene (toluene) + CH₃Cl $\xrightarrow{AlCl_3}$ 1,3-dimethylbenzene (*meta*-xylene) + HCl.

25.2.2 Elimination Reactions

When a halogenoalkane is treated with alkali (e.g. sodium hydroxide) as well as hydrolysis, the elimination of the elements of hydrogen halide may also occur to give an alkene:

$$-\underset{|}{\overset{H}{C}}-\underset{|}{\overset{OH}{C}}- \xleftarrow[\text{hydrolysis}]{\text{NaOH}} -\underset{|}{\overset{H}{C}}-\underset{|}{\overset{Hal}{C}}- \xrightarrow[\text{elimination}]{\text{NaOH}} \ \ \mathrm{C=C} \ + \ \mathrm{NaHal} + \mathrm{H_2O}.$$

This competition between elimination and hydrolysis is discussed in Sections 22.3 and 26.6 and these discussions should be referred to at this point. This elimination reaction is often useful for alkene synthesis, although for ethylene itself the reaction is ineffective and is generally poor when applied to primary alkyl halides.

With acyl halides such eliminations are possible but rare. Clearly a non-aqueous solvent would be required which does not react with acyl halides, and this immediately rules out water and alcohols.

$$-\underset{|}{\overset{H}{C}}-\underset{}{\overset{Hal}{C}}=O \ \rightarrow \ \mathrm{C=C=O}.$$
(a ketene)

With aromatic halides, under forcing conditions, eliminations may occur but the species produced (a benzyne) is highly reactive and undergoes an addition reaction as soon as it is formed to yield what is essentially a substitution product:

bromobenzene $\xrightarrow[\text{heat}]{\text{NaNH}_2}$ benzyne + NH₃ + NaBr;

benzyne + NH₃ ⟶ C₆H₅NH₂.

(Na⁺NH₂⁻ is a very strong base, $NH_2^- + H^+ \rightarrow NH_3$.)

25.2.3 The Effect of Halogens on Molecules Containing Them

The highly electronegative chlorine atoms are able to stabilize an ion like Cl_3C^- by polarizing the bonds. Negative charge is drained away from carbon and the charge, which would otherwise be localized on carbon, is effectively delocalized:

$$Cl^{\delta-} \leftarrow \underset{\underset{Cl^{\delta-}}{\uparrow}}{\overset{\overset{Cl^{\delta-}}{\downarrow}}{C^{\delta-}}}$$

This has a profound effect on the chemistry of compounds containing the trichloromethyl group (or its bromo or iodo analogues). We have already seen, for example (Section 23.3.2), that a trihalogenomethyl aldehyde or ketone, unlike a normal methyl ketone, is hydrolysed by alkali. This is largely due to the fact that the Hal_3C^- ion is stable enough (Sections 26.3 and 26.4) to be displaced by OH^-.

Chloroform itself, $CHCl_3$, is rendered weakly acidic and aqueous alkali can remove a hydrogen ion to yield the (relatively) stable Cl_3C^- ion:

$$HCCl_3 + OH^- \rightleftharpoons {}^-CCl_3 + H_2O.$$

The production of $^-CCl_3$, which can undergo further reactions with suitable compounds, is believed to be the basis of a number of base-catalysed reactions of trichloromethane(chloroform).

A final illustration of the influence which halogens may exert on molecules containing them is provided by chloral(trichloroacetaldehyde). This aldehyde, unlike normal aldehydes, forms a stable crystalline hydrate containing two hydroxyl groups on one carbon atom—usually a very unstable arrangement. Although most simple aldehydes (and ketones) form such hydrates in aqueous solution, they are usually too unstable to be isolated. They dissociate back to aldehyde (or ketone) and water.

$$CCl_3CHO + H_2O \rightarrow CCl_3\!-\!\underset{\underset{OH}{|}}{\overset{\overset{OH}{|}}{C}}\!-\!H.$$

chloral (trichloroacetaldehyde)   chloral hydrate

Compare:

$$CH_3CHO + H_2O \rightleftharpoons CH_3\!-\!\underset{\underset{OH}{|}}{\overset{\overset{OH}{|}}{C}}\!-\!H.$$

25.2.4 The Carbon–Chlorine Bond in Perspective

In Section 23.1.2 it was pointed out that covalent hydroxy compounds, organic or inorganic, have something in common. They are all acidic. Compounds containing covalently bound halogen atoms have much in common, too, and, in general, they all behave in much the same way. They can usually undergo nucleophilic substitution reactions for instance. Thus, phosphorus oxychloride ($POCl_3$) will react with alcohols to yield trialkyl phosphates:

$$O\!=\!P\!\underset{Cl}{\overset{Cl}{\diagup\!\!\!\diagdown}}\!Cl + 3ROH \rightarrow O\!=\!P\!\underset{OR}{\overset{OR}{\diagup\!\!\!\diagdown}}\!OR + 3HCl,$$

or with water to yield phosphoric acid:

$$O=P(Cl)_3 + 3H_2O \rightarrow O=P(OH)_3 + 3HCl.$$

Other covalent inorganic chlorides behave similarly:

$$SO_2Cl_2 + 2H_2O \rightarrow SO_2(OH)_2 + 2HCl;$$

$$AlCl_3 + 3H_2O \rightarrow Al(OH)_3 + 3HCl.$$

In so far as all these compounds give rise to acidic hydroxy compounds on hydrolysis, together with hydrogen chloride, they can be regarded as acid chlorides. In much the same way, alkyl and acyl halides give rise to acidic hydroxy compounds on hydrolysis, together with hydrogen chloride. Thus the character of the covalent carbon–chlorine bond must be seen against a background of the characters of all covalent chlorine compounds which have much in common. The common feature stressed here is the ability of nucleophiles to displace halide ions.

Table 25.3. Principal reactions of halogeno compounds. Note that they all involve expulsion of the halogen atom as halide ion.

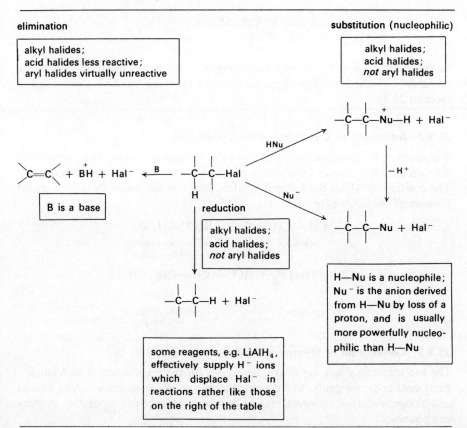

25.2 CHEMICAL BEHAVIOUR

25.3 INTRODUCTION OF THE GROUPS

25.3.1 Direct Halogenation of Hydrocarbons

Reaction between halogens and hydrocarbons results in the replacement of hydrogen with chlorine:

$$-\overset{|}{\underset{|}{C}}-H + Cl_2 \rightarrow -\overset{|}{\underset{|}{C}}-Cl + HCl.$$

Particularly in the case of alkanes, however, this is not usually a good preparative method as it is difficult to direct the halogen on to the required carbon atom and difficult to prevent it from attacking the molecule at several different positions (Section 22.2.3). With aromatic hydrocarbons, however, the reaction has an entirely different character (Section 26.5) and often affords a useful preparation of halogenoarenes.

A carbonyl group, when present in a molecule, has the useful ability to "activate" adjacent C—H groups towards halogenation. Thus, treatment of, say, propanoic acid with chlorine results largely in attack at the C—H adjacent to the carbonyl group (the α-position):

$$CH_3CH_2COOH + Cl_2 \rightarrow CH_3\underset{Cl}{\underset{|}{CH}}COOH + HCl$$

$$\text{very little} \rightarrow \underset{Cl}{\underset{|}{CH_2}}CH_2COOH + HCl.$$

In the case of carbonyl compounds, therefore, direct halogenation is a useful preparative method for introducing a halogen atom on to the α-carbon atoms (Section 28.3).

25.3.2 Additions to Unsaturated Hydrocarbons

The addition of hydrogen halide or elemental halogen to unsaturated compounds (Section 22.2.4) is a more useful way of preparing aliphatic halogen compounds. The position at which the halogen is introduced into the molecule is fixed by the position of the double (or triple) bond.

$$CH_2{=}CH_2 + Br_2 \rightarrow BrCH_2CH_2Br$$
    ethylene   1,2-dibromoethane
                    (ethylene bromide)

$$CH_3CH{=}CH_2 + HCl \rightarrow CH_3-\underset{Cl}{\underset{|}{CH}}-CH_3.$$
    propene   2-chloropropane

25.3.3 Conversion of Hydroxyl Groups (see Section 23.1.2)

The two methods above are clearly unsuitable for the preparation of acid halides, being confined, essentially, to halogenoalkanes and halogenoarenes. Acid halides and halogenoalkanes, however, may be obtained from the corresponding hydroxy compounds.

1. Reaction with phosphorus pentachloride (PCl$_5$) or thionyl chloride (SOCl$_2$). These compounds successfully convert hydroxy compounds to chlorides (although with phenols, the conversion is very poor). Bromo and iodo compounds may be prepared by similar methods (Section 23.1.2).

$$C_2H_5OH + SOCl_2 \rightarrow C_2H_5Cl + HCl + SO_2$$
ethanol chloroethane (ethyl chloride)

$$CH_3COOH + PCl_5 \rightarrow CH_3COCl + POCl_3 + HCl.$$
acetic acid acetyl chloride

2. Hydrogen halides will convert alcohols (only) to halogenoalkanes (Section 23.1.2). Sometimes sulphuric acid or the appropriate zinc halide is used as a catalyst, particularly with primary alcohols which react very slowly. Secondary alcohols react faster, and tertiary alcohols faster still.

$$C_2H_5OH + KBr + \text{excess } H_2SO_4 \rightarrow C_2H_5Br + KHSO_4 + H_2O$$
ethanol bromoethane (ethyl bromide)

$$(CH_3)_3COH + HCl \rightarrow (CH_3)_3CCl + H_2O.$$
2-methylpropan-2-ol (tertiary-butanol) 2-chloro-2-methylpropane (tertiary-butyl chloride)

Table 25.4. Important methods of introducing the carbon–halogen bond into organic molecules.

1. Halogenation of hydrocarbons

—C—H + Hal$_2$ → —C—Hal + HHal

> careful distinction should be made between reactions of aromatic —C—H and aliphatic —C—H

2. Addition to alkenes

C=C + Hal$_2$ → Hal—C—C—Hal

HHal → H—C—C—Hal

3. Conversion of —C—OH compounds

—C—OH + HHal → —C—Hal
PCl$_5$
SOCl$_2$

> gives virtually no product in the case of phenols, aromatic —C—OH

4. Conversion of diazonium salts

Ar—N$_2^+$ $\xrightarrow{\text{metal halide}}$ Ar—Hal + N$_2$ + metal$^+$

> applicable only to aromatic halides, but very important for these as (2) and (3) above are not applicable

25.3 INTRODUCTION OF THE GROUPS

25.3.4 Conversion of Diazonium Salts

By reaction with the appropriate inorganic halide and catalyst, if necessary, diazonium salts may be converted to halogenoarenes. This is one of the most useful ways of introducing halogens into an aromatic ring.

$$C_6H_5\overset{+}{N}_2\ HSO_4^- + KI \rightarrow C_6H_5I + KHSO_4 + N_2$$
benzenediazonium hydrogen sulphate → iodobenzene

$$C_6H_5\overset{+}{N}_2\ Cl^- \xrightarrow[HCl]{CuCl} C_6H_5Cl + N_2$$
benzenediazonium chloride → chlorobenzene

$$C_6H_5\overset{+}{N}_2\ HSO_4^- \xrightarrow[HBr]{CuBr} C_6H_5Br + N_2 + H_2SO_4$$
bromobenzene

PROBLEMS

1. Study the following list of compounds:
   i) CH_3COCl;
   ii) $CH_3CH_2CH_2Br$;
   iii) C_6H_5Cl;
   iv) 3-chlorobenzoyl chloride (benzene ring with Cl and COCl);
   v) 4,5-dimethylbenzene-1,2-dicarbonyl dichloride (benzene ring with two CH_3 and two COCl groups);
   vi) $CH_3CH_2CH(I)CH_3$.

   a) Which compounds are acid chlorides?
   b) Which compounds are halogenoalkanes (alkyl halides)?
   c) Which compounds are halogenoarenes?

2. Arrange the following three compounds in the order in which you would expect their reactivity towards ammonia to increase:
   a) $(CH_3)_2CHCOCl$;
   b) 4-chlorotoluene (benzene ring with Cl and CH_3 para);
   c) $(CH_3)_2CHCH_2Cl$.

3. What *type* of compound would be obtained when acetyl chloride reacts with:
   a) ammonia;
   b) ethanol;

c) water;
d) phenol?

4. What *type* of compound would be obtained when 1-bromobutane reacts with:
   a) ammonia;
   b) sodium phenoxide;
   c) sodium hydroxide (aqueous);
   d) sodium ethoxide?

5. How would you "induce" ethanol to react with 1-bromopropane? Would a similar "inducement" be necessary for ethanol to react with acetyl chloride?

6. There are two possible products which may result from reaction between 2-bromopropane and sodium ethoxide (a strong base). What are they?

7. Suggest possible products which might be expected from reaction between a halogenoalkane (say bromoethane) and:
   a) sodium hydrogen sulphide (NaSH—essentially HS$^-$ ion);
   b) phosphine, PH_3;
   c) sodium acetylide, Na^+ $^-C{\equiv}CH$.
   (*Hint*—examine the electronic structure of these reagents and, if necessary, read Section 25.2.1 again.)

8. Suggest, in outline, how the following conversions may be brought about:
   a) $CH_3CH_2Br \rightarrow CH_3CH_2OH$;
   b) $CH_3CH_2I \rightarrow CH_3CH_2OC_6H_5$;
   c) $CH_3CH_2Br \rightarrow CH_3CH_2COOH$;
   d) $CH_3CH_2COCl \rightarrow (CH_3CH_2CO)_2O$;
   e) $CH_3CH_2COCl \rightarrow CH_3CH_2COOCH_2CH_3$;
   f) $CH_3CH_2COCl \rightarrow CH_3CH_2CONH_2$;
   g) $CH_3CH_2COCl \rightarrow CH_3CH_2CONHCH_3$.

9. Indicate the products, if any, which you would expect to form by reaction between the following combinations of reagents:
   a) $(CH_3CH_2)_3CBr + KOH$
      (KOH used as a hot, concentrated solution in ethanol);
   b) $CH_3CH_2CH_2Cl + KOH(aq)$;
   c) $CH_3CH_2CH_2CH_2I + KCN$;
   d) $CH_3-\!\!\bigcirc\!\!-COCl + NH_3$;
   e) $CH_3-\!\!\bigcirc\!\!-COCl + CH_3CH_2CH_2OH$;
   f) $CH_3-\!\!\bigcirc\!\!-COCl + (CH_3)_2CHCH_2NHCH_3$;
   g) $CH_3-\!\!\bigcirc\!\!-Br + NH_3$;
   h) $C_6H_5COCl + H_2O$;
   i) $CH_3CH_2Br + NaOH(aq)$;
   j) $CH_3CH_2I + NaOCH_2CH_2CH_3$.

10. A compound, W, has molecular formula C_3H_5OCl, and contains a carbonyl group. Given this information, what possible structures could W have? W fumes in moist air and reacts vigorously with cold water to give a strongly acidic solution, and, with ammonia, yields an amide. Suggest which of your possible structures is the correct one.

11. A compound, X, has molecular formula, C_3H_7Cl. On hydrolysis it yields a compound, Y, C_3H_8O, which is oxidized by potassium dichromate and sulphuric acid to Z, C_3H_6O. Z gives a yellow precipitate with 2,4-dinitrophenylhydrazine reagent, but fails to give a precipitate with iodine and alkali. Treatment of Z with ammoniacal silver nitrate gives a silver mirror. Deduce the structures of X, Y, and Z, giving your reasoning.

BIBLIOGRAPHY

This chapter covers a wide range of material and one (or more) of the books on organic chemistry from the general list would provide the most helpful form of further reading.

CHAPTER 26

Organic Reactions as the Making and Breaking of Chemical Bonds

26.1 INTRODUCTION

A good deal of descriptive organic chemistry has been dealt with in the previous four chapters, although the emphasis has been on the important patterns to which the behaviour of organic compounds conforms. The establishment of a wide range of facts, and the discernment of patterns and relationships amongst them, is a vital first step in the scientific development of any area. The patterns and relationships help us to formulate the right questions of the "how?" and "why?" variety, and to set up hypothetical models to account for the observed behaviour, which can be tested and developed into acceptable scientific theory. It is the purpose of this chapter to outline some of the important ideas which are now thought to underly these patterns of behaviour and, because these ideas are inevitably based upon the fundamental principles of chemistry described earlier in this book, to establish a firm link between organic chemistry and the other traditional branches of chemistry.

The two most important factors which determine whether a reaction will occur or not, and if it does occur, what course it will take, are undoubtedly thermodynamics (Chapters 6 and 7) and chemical kinetics (Chapter 8). The use of thermodynamics in predicting the feasibility of any chemical reaction has been explained (Section 7.5). If the products of the process are thermodynamically very unstable with reference to the reactants, then the reaction cannot possibly proceed spontaneously in a closed system. If reactants and products are of similar stability, then an equilibrium mixture may be obtained in which quantities of both reactants and products can be measured. However, even if the thermodynamics of a reaction suggest that it should proceed spontaneously to completion, it may fail to do so in practice, at room temperature, because the activation energy (Section 8.3) of the process is too high. The kinetics of the process then becomes the most important determining factor, and any consideration of that process must concern itself primarily with kinetics and factors which influence activation energies. A very large proportion of organic reactions fall into this category, and the organic chemist frequently has to concern himself, therefore, with kinetics. This point can be very simply illustrated by reference to the combustion of practically any organic compound. Methane, for example, in the presence of oxygen or air, is thermodynamically unstable with reference to carbon dioxide and water:

$$CH_4(g) + 2O_2(g) \rightarrow CO_2(g) + 2H_2O(l); \qquad \Delta G^\ominus = -818 \text{ kJ mol}^{-1}.$$

26.1 Introduction
26.2 Classification of reactions
26.3 Nucleophilic substitution at tetrahedral carbon
26.4 Reactions of nucleophiles with multiply bound carbon
26.5 Reactions of electrophiles with multiply bound carbon
26.6 Elimination reactions

Application of a spark or a flame will excite a number of molecules, raising their energy above the activation energy, so that this small proportion of the reactants may react. Thereafter the reactants are "activated" by the energy released by those molecules which have already undergone this highly exothermic process.

It may be the case that the *direct* reaction of one compound with another has a free energy of activation so high that one would not expect it to proceed at a measurable rate at room temperature. Products may nevertheless be observed to form rapidly if an *indirect* route is available for the conversion of reactants into products. Consider the reaction between A and BC to give AB and C:

$$A + BC \rightarrow AB + C.$$

If the free energy of activation for this process is too high, then it will not occur at room temperature. If, however, A can react with BC to give ABC in a low activation energy reaction, and ABC can subsequently decompose to AB and C in another reaction of low activation energy, then A and BC will indeed react to give AB and C via ABC

$$A + BC \rightarrow ABC$$
$$ABC \rightarrow AB + C.$$

This combination of reactions is referred to as a "reaction pathway" and each component reaction as a "step". The reaction of A with BC would then be a two-step process and it would proceed provided that the free energies of activation of both steps were low enough. The slower reaction of the two (the rate determining step) would determine the rate of the overall reaction. The investigation of an organic reaction very often, therefore, involves the establishment of the various steps which may make up the reaction pathway, and in the paragraphs which follow, we shall be concerned to interpret a number of reactions in terms of multi-step pathways.

The reaction pathway is often referred to as the **mechanism** of a reaction, although in its fullest sense the term mechanism covers rather more than just a reaction pathway. It is concerned not only with identifying which species collide, leading to reaction, but also with the geometry of their collision and with the redistribution of electrons which takes place. It might be said that the mechanism of a reaction is concerned with its pathway and with the nature of the important transition states (Section 8.7) involved.

Another important way in which kinetics exerts a controlling influence on chemical reactions is observed in cases where two (or more) reactants may give rise to more than one product (or set of products) in processes which are thermodynamically irreversible. Consider the reaction of XY with Z, which can lead to two different products, XYZ and ZXY:

$$XY + Z \rightarrow XYZ$$
$$XY + Z \rightarrow ZXY.$$

If both reactions were reversible, then an equilibrium would eventually be reached in which the most stable of all the species involved would predominate. Thus, if XYZ were very much more stable than ZXY and the reactants themselves, XYZ would be the principal reaction product. In this case, the reaction is under thermodynamic control (Section 8.9). If, on the other hand, both reactions were irreversible, then the product obtained would be that of the faster reaction, irrespective of which product was the more stable. Thus, if ZXY were formed more rapidly than XYZ, then, even though XYZ is more stable than ZXY, ZXY would be the predominant product. In this case, the reaction is under kinetic control (Section 8.9).

26.2 CLASSIFICATION OF REACTIONS

Three basic types of reaction have already been defined in the earlier chapters on organic chemistry. These are *substitution reactions*, *addition reactions*, and *elimination reactions* and they constitute the three fundamental reaction types encountered in organic chemistry (a fourth type, rearrangement reactions, is outside the scope of this treatment). Whilst extremely useful, however, this type of classification tells us nothing about the way these reactions occur. In particular it tells us nothing of the way in which electrons behave during reaction.

A different classification, related to the processes of bond breaking and making, is more useful from this point of view. This simply defines two ways in which an electron pair bond can break (or be made) in terms of whether the pair becomes separated or not. Thus, a covalent bond between two atoms A and B may break because the electron pair moves over completely to one atom:

$$A : B \rightarrow A^+ + :B^-$$

Such a process is referred to as **heterolytic fission**, and reactions in which electron pairs move in this way are called **heterolytic** or **ionic** reactions. The small "curly" arrow refers to the movement of the electron pair. The alternative way in which the bond A—B might break is with the separation of the electron pair:

$$A : B \rightarrow A\cdot + B\cdot$$

Note the use of half-arrows to denote the shift of single electrons. The species produced inevitably contain odd, unpaired electrons and are therefore usually highly reactive. They are referred to as free radicals and the bond breaking process is known as **homolytic fission**. Relatively few reactions dealt with in this book are of this type, though, so we shall primarily be concerned in this chapter with heterolytic reactions. One important free radical reaction is dealt with in detail where it arises in Section 22.2.3 (halogenation of alkanes).

By way of an introduction to the nature of heterolytic reactions it would be useful to consider a relatively simple inorganic reaction first. The reaction between hydrogen chloride and water, for example, can be examined. Both of these materials are covalent, the former a gas and the latter a liquid, and with both of them in the pure state, they are very poor conductors of electricity. On mixing the two, however, a good deal of heat is liberated and the resultant solution of hydrogen chloride in water is an extremely good conductor. Evidently a reaction has occurred between these two species which has produced something different which can conduct electricity. From what has been said earlier on electrolytes and conductivity, it will be apparent that there are large numbers of ions in solution and the reaction can be described by the following simple equation:

$$H_2O + HCl \rightarrow H_3O^+ + Cl^-.$$

The bond between hydrogen and chlorine has broken and a new one has formed between hydrogen and oxygen. As the chlorine is now present as a chloride ion and a positive charge is associated with the hydrogen, it is apparent that the bond has broken heterolytically, with the electron pair moving over from hydrogen to chlorine. The high stability of the proton and the chloride ion in aqueous solution, together with the fact that the hydrogen chloride molecule was polarized in the sense $\overset{\delta+}{H}-\overset{\delta-}{Cl}$ (Section 4.4) due to electronegativity differences between hydrogen and chlorine, allow us to understand that the bond breaks in this direction (as opposed to $H-Cl \rightarrow H^- + Cl^+$). However, whilst the chlorine thus departs with a complete octet, the hydrogen atom is not allowed to remain "naked", stripped of all electrons. A lone pair on a water molecule has been donated to the potential hydrogen ion and, whilst maintaining the oxygen octet, fills the hydrogen orbitals.

One very important effect to note at this stage is the nature of the reaction in terms of the movement of electron pairs which, for clarity, are circled here, and in some other instances.*

* In these and the equation which follow, dots represent the electrons of the bond shown by a line.

Both reactants and products have stable electron configurations with full outer shells. The reaction is really just a rearrangement of electrons between the reacting species, and the driving force is not the requirement of atoms to achieve a full outer shell. This is rarely a driving force for chemical reactions since the atoms in most molecules have filled shells already. Rather, the driving force is related to the energies of the bonds broken and the bonds which have been made, together with the solvation energies of the various species present and any entropy changes which may have been involved.

A final point in connection with this reaction, which might be made before launching into a consideration of organic reactions, is that it can be regarded as a substitution reaction. If it is considered from the standpoint of the hydrogen chloride molecule, it can be seen that the chlorine atom has been replaced by the group H_2O^+. We can look at this as a water molecule displacing a chloride ion from hydrogen chloride, and being substituted for the displaced chloride ion. This point will be referred to again when we consider nucleophilic substitution reactions below. The two processes are very closely related.

26.3 NUCLEOPHILIC SUBSTITUTION AT TETRAHEDRAL CARBON

The term "tetrahedral carbon" in this heading simply refers to carbon atoms which are attached to four other atoms or groups and which therefore adopt a tetrahedral disposition of bonds. The carbon atom of methane is a tetrahedral carbon. The heading therefore excludes substitution reactions which take place at carbon atoms which are involved in multiple bonding and which are therefore not tetrahedral. An alternative term might have been "saturated carbon".

We have, in the preceding chapters, met a large number of reactions in which substitution takes place at a tetrahedral carbon atom. Most of these fall into the category which we are now discussing. Some examples are:

1. Reaction of hydrogen chloride with 2-methylpropan-2-ol (tertiary-butanol) to give 2-chloro-2-methylpropane (tertiary-butyl chloride):

$$(CH_3)_3COH + HCl \rightarrow (CH_3)_3CCl + H_2O.$$

In this reaction chlorine has replaced the hydroxyl group of the alcohol and the substitution involves bonds to the tetrahedral central carbon atom of the tertiary-butyl group.

2. Cleavage of the carbon–oxygen bond in methyl phenyl ether (methoxybenzene) with hydrogen iodide:

$$C_6H_5OCH_3 + HI \rightarrow C_6H_5OH + CH_3I.$$

Iodine has replaced the C_6H_5O- group of the ether and the substitution has taken place at the tetrahedral carbon of the methyl group.

3. Hydrolysis of alkyl halides (halogenoalkanes) with aqueous sodium hydroxide:

$$CH_3CH_2CH_2Cl + NaOH \rightarrow CH_3CH_2CH_2OH + NaCl.$$

Again the substitution has occurred at tetrahedral carbon. This time an OH group has replaced a chlorine atom.

The last reaction, hydrolysis of an alkyl halide, will be very suitable as an example which we can look at closely in our study of this type of reaction.

To begin with, knowing that in aqueous solution sodium hydroxide is present as a mixture of sodium and hydroxide ions, and sodium chloride as sodium and chloride ions, we can rewrite the equation more realistically as follows:

$$CH_3CH_2CH_2Cl + Na^+ + OH^- \rightarrow CH_3CH_2CH_2OH + Na^+ + Cl^-.$$

The sodium ion appears on both sides of the equation and seems to take no part in the reaction. We might, therefore, rewrite the equation again:

$$CH_3CH_2CH_2Cl + OH^- \rightarrow CH_3CH_2CH_2OH + Cl^-.$$

There are sound experimental reasons for believing that this is an accurate representation of the reaction, and that we are indeed concerned with reaction between the alkyl halide and a hydroxide ion.

Comparison of this reaction with that between hydrogen chloride and water (Section 26.2) indicates that the two reactions are very similar. In this case, a chloride ion has been displaced from carbon (as opposed to hydrogen) and the species responsible for the displacement, like water, contains lone pairs of electrons. We could, therefore, write down the reaction in much the same way that we wrote down the hydrogen chloride / water reaction:

$$[H\!-\!\ddot{\underset{..}{O}}\!:]^- \quad \underset{H}{\overset{CH_3-CH_2\ \ H}{\diagdown\!\diagup}}C\underset{Cl}{\diagdown} \rightarrow H-O-\underset{H}{\overset{H}{\underset{|}{C}}}-CH_2-CH_3 + Cl^-.$$

Once again the covalent bond has broken with the electron pair moving on to the halogen atom to convert it into a very stable negatively charged chloride ion. Chlorine being of greater electronegativity than carbon, the bond is already (i.e. before reaction) polarized towards chlorine, $\overset{\delta+\ \ \ \delta-}{C-Cl}$, and this will also help to determine the direction of cleavage. The stability of the aqueous chloride ion is probably the most important factor, though, in determining that the chlorine can carry the electrons, and this is demonstrated by the fact that alkyl iodides are also hydrolysed thus (more rapidly than are chlorides). Iodine and carbon have almost equal electronegativities and the polarity of the carbon–iodine bond, being virtually zero, cannot be the important factor. The ability of iodine to accept the bonding electron pair and form the stable solvated iodide ion matters more.

From a thermodynamic point of view, the carbon–iodine bond is more ready to break than is the carbon–chlorine bond. Although the iodide ion is less well solvated than the chloride ion (solvation energies 305 and 385 kJ mol^{-1} respectively), the carbon–iodine bond is much weaker than the carbon–chlorine bond (bond energies 238 and 326 kJ mol^{-1} respectively). Examination of the relative acid strengths of hydrogen chloride and hydrogen iodide (Section 13.4) reveals a similar situation. The weaker hydrogen–iodine bond dictates that hydrogen iodide is a stronger acid than is hydrogen chloride. The similarity between the hydrolysis of a carbon–chlorine bond and the aqueous dissociation of hydrogen chloride has already been noted.

Thermodynamics, however, control the extent to which a reaction goes (equilibrium constant) and not the rate at which it goes (rate constant). We therefore get no obvious help from these thermodynamic arguments in accounting for the fact that alkyl iodides are more reactive, that is react *faster*, than alkyl chlorides. We should rather be considering the free energies of activation for these reactions. Free energy of activation (Section 8.7) is concerned with the difference in free energy between the ground states of the unreacted materials and the transition state for the reaction. The precise nature of the latter is very difficult to determine, as it has virtually no finite period of existence, and any contemplation of a transition state usually has to concern itself with an approximate model. In this particular case, as in some others (Section 28.5.2), we can use the nature of the products as a guide to the nature of the transition state. We envisage the carbon–halogen bond in the process of breaking and, as the bond stretches, negative charge must be building up on the halogen atom as it begins to assume the character of a free halide ion. Solvation of the developing charge must occur in much the same way as for the "free" halide ion. With a higher solvation energy for the chloride ion than for the iodide ion, one might expect that the transition state for fission of the carbon–chlorine bond would be of lower free energy than that for the carbon–iodine bond. However, because the latter bond is very much weaker than the carbon–chlorine bond it is much easier to stretch. The net result, largely because of the weakness of the carbon–iodine bond, is that the transition state for the nucleophilic displacement of iodide is of lower free energy than that for the displacement of chloride. Factors stabilizing products, relative to reactants, also stabilize the transition state. In short, the alkyl iodide reaction has the lower free energy of activation and proceeds faster than the alkyl chloride reaction. We sometimes say, in this context, that the iodide ion is a better *leaving group* than is the chloride ion.

Note that, with respect to the reaction between prop-1-yl chloride and hydroxide ion, we have said nothing about whether the leaving group departs before, after, or simultaneously with the approach of the hydroxide ion. We can rule out the second case, with hydroxide bonding to carbon before chloride has left, on the grounds that this would involve the formation of a compound of pentavalent carbon, $\underset{\underset{\text{Cl}}{|}}{\overset{\overset{\text{OH}}{|}}{\text{CH}_3\text{CH}_2\text{CH}_2}}$. Carbon cannot expand its octet to contain the necessary ten electrons and this cannot therefore form. There is certainly no evidence that it does. The first case, with chloride ion leaving before hydroxide ion bonds to carbon, would involve a two-step process, the first being the ionization of the alkyl halide to alkyl carbonium ion and halide ion:

$$\text{CH}_3\text{CH}_2\text{CH}_2\text{Cl} \rightarrow \text{CH}_3\text{CH}_2\text{CH}_2^+ + \text{Cl}^-;$$
$$\text{CH}_3\text{CH}_2\text{CH}_2^+ + \text{OH}^- \rightarrow \text{CH}_3\text{CH}_2\text{CH}_2\text{OH}.$$

Whilst we do not believe that primary alkyl halides such as prop-1-yl chloride react in this way, there is ample evidence to show that certain others do. The alkyl halide we are discussing here, and many others, react according to the third alternative indicated, that is, with the bond to the new group building up as that to the leaving group breaks down (because carbon cannot expand its octet, the new electron pair makes room for itself by "pushing" the old pair out). However, in connection with nucleophilic substitution at tetrahedral carbon, we shall not pursue this question any further and no distinction will be drawn between the two possibilities in the subsequent paragraphs.

The attacking species, the OH^- ion in this case, is referred to as a **nucleophile**—a "lover of nuclei". The essential feature of a nucleophile is the lone pair of electrons which can react with centres capable of accommodating them. Whilst the OH^- ion is defined thus as **nucleophilic**, the carbon atom, which accommodates the lone pair, may be said to be **electrophilic**—loving electrons. The substitution reaction as a whole is defined as a **nucleophilic substitution** (S_N reaction), the definition being made from the standpoint of the organic reactant which is seen as being attacked by the nucleophile. This is sometimes called a nucleophilic attack. This is a purely arbitrary definition, of course, and one could equally well have said that the hydroxide ion had been attacked by an electrophile, and that the reaction was therefore an electrophilic one. However, the standard practice of defining the reaction from the standpoint of the organic species is universally adopted, and the definition is therefore useful and means the same thing to everyone.

The essential requirement for a nucleophile is a lone pair of electrons, and it is interesting to note that the requirement for a base is the same. The hydroxide ion, OH^-, in bonding to a proton through a lone pair is behaving as a base. In reacting as above with an alkyl halide, however, it is behaving as a nucleophile. The question arises, therefore, are the terms nucleophilicity and basicity synonymous? Basicity—the strength of a base—is expressed as an equilibrium constant (K_b). It is a function of thermodynamics. Further, if we adopt the Brönsted definition of acidity and basicity, a base and basicity are related solely to protonation reactions. Nucleophilicity—the capacity to behave as a nucleophile—is essentially a kinetic term, and nucleophiles are compared according to the relative rates with which they can react with a common compound. The choice of this common compound may be made from a wide range of materials as there is no unique species which characterizes a nucleophile in the way that a proton characterizes a base. Thus, for two nucleophiles A and B, A may be more nucleophilic than B with reference to their reaction with, say, compound X, but B *may* be more nucleophilic than A with reference to their reaction with compound Y. Strictly, it is necessary to specify the reaction in question when nucleophiles are compared.

Thus, with such different definitions of basicity and nucleophilicity, we need not expect the two properties to be parallel. In practice, of course, they often are, and a species which is a powerful base is often also a powerful nucleophile. Hydroxide ion is a strong base and a powerful nucleophile, for example. However, there are exceptions. Basicity increases along the series of alkoxide ions CH_3O^-, $CH_3CH_2O^-$, $(CH_3)_2CHO^-$, and $(CH_3)_3CO^-$, whereas nucleophilicity decreases along the same series.

There is a broad range of nucleophiles which will react with alkyl halides. Some of the important ones, of varying reactivity, are as follows (compare with the list of reagents which react with alkyl halides as listed in Section 25.2.1): HO^-, H_2O, NH_3, $N\equiv C^-$, RO^-, and ROH. In general, if a nucleophile contains a hydrogen atom bound to the lone-pair bearing atom (as is the case with water), removal of that hydrogen as a proton yields an anion which is a more powerful nucleophile than the neutral molecule. HO^- is more nucleophilic than H_2O (Sections 23.1.3 and 25.2.1); RO^- is more nucleophilic than ROH (Section 23.2.3). A negative charge, however, is by no means a necessary feature of a nucleophile.

To conclude this discussion of nucleophilic substitution at tetrahedral carbon, it would be appropriate to mention again examples (1) and (2), which were quoted at the beginning of the discussion, and to relate them to what we have said about this type of reaction.

In example (1), chloride ion, from concentrated aqueous hydrochloric acid, has apparently displaced a hydroxide ion from the alcohol. It is of interest to note that sodium chloride, rich in chloride ions, will not convert an alcohol to an alkyl halide. Although we have the nucleophile present, the reaction will not take place because hydroxide ion itself is a very poor leaving group:

$$(CH_3)_3COH + Cl^- \text{ do not react to give: } (CH_3)_3CCl + OH^-.$$

It will be recalled that alcohols are weakly basic in character (Section 23.1.2) and that when dissolved in strong acids the oxygen lone pairs are protonated. Tertiary-butanol conforms to this behaviour and we could represent the initial reaction between the alcohol and the acid as:

$$(CH_3)_3C-\overset{..}{\underset{..}{O}}\overset{H}{\diagdown} \; + \; H\div Cl \rightarrow (CH_3)_3C-\overset{+}{O}\overset{H}{\underset{H}{\diagdown}} \; + \; Cl^-.$$

This has modified the OH group considerably and we now see that the chloride ion has to displace not a hydroxide ion, but a pre-formed water molecule. The latter makes a very much better leaving group than does a hydroxide ion, and the next stage of the reaction takes place easily and rapidly:

$$:\!\overset{..}{\underset{..}{Cl}}\!:^- \; + \; H_3C-\underset{\underset{CH_3}{|}}{\overset{\overset{CH_3}{|}}{C}}-\overset{+}{O}H_2 \rightarrow Cl-\underset{\underset{CH_3}{|}}{\overset{\overset{CH_3}{|}}{C}}-CH_3 + H_2O.$$

The acid exerts its catalytic influence by making available an alternative pathway for the reaction which is of lower activation energy than that involving direct interaction between a chloride ion and the alcohol. Reaction can therefore proceed at a reasonable rate.

Example (2) can be rationalized in a similar manner. Phenoxide and alkoxide ions, like hydroxide ions, are very poor leaving groups and it is necessary for the acid to protonate the ether oxygen before further reaction can occur. The iodide ion has then to displace a pre-formed phenol molecule, which is much easier than displacing a phenoxide ion, and the process can be described in terms of two steps:

$$CH_3\ddot{O}C_6H_5 + H:\ddot{I} \rightarrow CH_3-\overset{+}{\underset{H}{\ddot{O}}}-C_6H_5 + I^-$$

$$:\ddot{I}:^- + CH_3-\overset{H}{\underset{+}{\ddot{O}}}-C_6H_5 \rightarrow I-CH_3 + :\overset{H}{\ddot{O}}-C_6H_5$$

26.4 REACTIONS OF NUCLEOPHILES WITH MULTIPLY BOUND CARBON

The lack of reactivity of aryl halides towards reagents which we have described in this chapter as nucleophiles has been referred to in Section 25.2.1, and is discussed again, from a mechanistic standpoint, in Section 28.5.2. Vinyl halides (halogen attached to unsaturated carbon, as in $CH_2=CHCl$) and acetylenic halides, of type $RC\equiv CHal$, are also resistant to simple nucleophilic displacement of halide ion. We shall not, therefore, be concerned here with nucleophilic displacements of halide ions (or other leaving groups, which are similarly difficult to displace) from carbon atoms involved in multiple bonds to other carbon atoms.

Carbon atoms multiply bound to different, more electronegative atoms, however, are normally very reactive towards nucleophiles; in Section 23.3.2 a large number of reactions of aldehydes, ketones, and carboxylic acid derivatives were discussed and many of these can be interpreted in terms of nucleophilic attack at the carbonyl carbon atom. Addition and condensation reactions are common with aldehydes and ketones whereas carboxylic acid derivatives are very different, commonly undergoing substitution reactions—Section 23.3.2(6).

The carbonyl group, being composed of atoms of widely differing electronegativity, is polar in character with electrons (of both the σ-bond and the π-bond —see Section A.2) displaced somewhat towards oxygen (the more electronegative element of the two). The C=O group therefore constitutes a dipole (Section 4.4), in terms of which some of the physical properties of carbonyl-containing compounds were discussed in Section 23.3.1. Furthermore, the group is easily polarized further by other species. That is, other species which have themselves an uneven distribution of charge, can influence the electron distribution in the carbonyl group. The approach of a species with a high density of negative charge at one end (e.g. a nucleophile) to the carbonyl carbon, can push (polarize) the electrons still farther over towards oxygen. (Using the multiple-bond model described in Section A.2 we can see that the electrons in a σ-bond have what we might refer to as their "centre of gravity" somewhere on the straight line between the bonded

*See note on page 672.

nuclei, whereas π-electrons have two such "centres of gravity", well to either side of the same straight line. π-electrons are therefore essentially farther away from the positive charges of the nuclei than are the σ-electrons, and are consequently more easily displaced (polarized) by an external field. It is the π-electrons which we think of as moving on to oxygen when a nucleophile attacks the carbon.) A nucleophile can easily attack the carbonyl carbon atom and displace a pair of electrons* entirely over on to oxygen, and the result, in the case, say, of ammonia (a nucleophile) attacking acetaldehyde, is a dipolar adduct. A negative charge is much more easily accommodated on oxygen than on the less electronegative carbon.

$$CH_3:C(H)=O\!:\ + :NH_3 \rightarrow CH_3-\overset{\overset{:\ddot{O}:^-}{|}}{\underset{\underset{H}{|}}{C}}-\overset{+}{N}H_3 \xrightarrow[+H^+]{-H^+} CH_3-\overset{\overset{:\ddot{O}H}{|}}{\underset{\underset{H}{|}}{C}}-\ddot{N}H_2$$

This loses its ionic nature by losing a hydrogen ion from nitrogen and acquiring one on oxygen. In this particular case, further reaction converts the uncharged adduct into something else (Section 23.3.2), but the way in which most carbonyl compound adducts are formed is illustrated and the means provided for a useful comparison with the reaction between the structurally similar acetyl chloride and ammonia.

The other characteristic type of reaction associated with aldehydes and ketones is the *condensation reaction*, although this is not really a fundamentally new type of reaction (Section 26.2). These reactions can usually be shown to involve a reaction pathway which is composed of two basic parts. The first is an addition reaction, like the one we have just been discussing, and the second is an elimination reaction, rather like the elimination of water from alcohols which is fully discussed later. The formation of a phenylhydrazone will serve to illustrate:

$$\underset{CH_3}{\overset{CH_3}{>}}C=O\ +\ NH_2NHPh\ \rightarrow\ \underset{CH_3}{\overset{CH_3}{>}}\underset{NH_2NHPh}{\overset{O^-}{\underset{+}{C}}}\ \rightarrow\ \underset{CH_3}{\overset{CH_3}{>}}\underset{NHNHPh}{\overset{OH}{C}}$$

$$\underset{CH_3}{\overset{CH_3}{>}}\underset{NHNHPh}{\overset{OH}{C}}\ \rightarrow\ \underset{CH_3}{\overset{CH_3}{>}}C=NNHPh\ +\ H_2O.$$

Consider, now, the attack of an ammonia molecule on acetyl chloride. Structurally, the only difference between acetaldehyde and acetyl chloride is that the latter has a chlorine atom where the former has a hydrogen, and we believe that the initial attack of the ammonia molecule is the same in both cases. A dipolar adduct is again formed:

$$CH_3-\overset{\overset{\ddot{O}:}{\|}}{\underset{\underset{Cl}{|}}{C}}\ :NH_3\ \rightarrow\ CH_3-\overset{\overset{O^-}{|}}{\underset{\underset{Cl}{|}}{C}}-\overset{+}{N}H_3.$$

A different reaction course is kinetically open to this species, though, due to the fact that chloride ion is a good leaving group. The electrons displaced on to oxygen can return towards carbon, regenerating the carbonyl group and displacing a chloride ion:

$$CH_3-\underset{Cl}{\underset{|}{\overset{:\ddot{O}:^-}{\overset{|}{C}}}}-\overset{+}{N}H_3 \xrightarrow{-Cl^-} CH_3-\underset{\overset{+}{N}H_3}{\overset{O}{\overset{\|}{C}}} \xrightarrow{-H^+} CH_3-\underset{NH_2}{\overset{O}{\overset{\|}{C}}}$$

The result of this, after loss of a proton, is a neutral acetamide molecule. In the case of the dipolar acetaldehyde adduct, if the electrons had come back towards carbon to regenerate the carbonyl group, they could only have done so by displacing the newly introduced nitrogen group, that is, by reversing the whole reaction. The reaction is undoubtedly reversible, but thermodynamics dictate that at any rate some product should be formed. The reason why acetaldehyde does not form a substitution product is that to do so, the electron pair returning towards carbon would have to displace either a hydride ion, H^-, or a methyl anion, CH_3^-. Neither of these is very stable, and they are therefore very poor leaving groups and cannot be thus displaced (leaving groups are discussed in Section 26.3).

We might note, at this point, that all carboxylic acid derivatives have an electronegative atom attached to the carbonyl carbon atom such that it can form at least a reasonable leaving group. This is basically why the behaviour of the acid derivatives is apparently so different from that of aldehydes and ketones (Table 23.9). The similarity of the addition step in both types of reaction should not be overlooked. We might also note here the similarity between the mechanism of the acetyl chloride / ammonia reaction, and the second and third steps of the ester hydrolysis mechanism outlined in Section 8.10.

One very important point, emerging from the discussion so far, is that the reaction between ammonia and acetyl chloride, a nucleophilic substitution reaction, is in one respect very different from a corresponding nucleophilic substitution reaction at tetrahedral carbon, say the reaction between ammonia and ethyl chloride:

$$CH_3CH_2-Cl + :NH_3 \xrightarrow{-Cl^-} CH_3CH_2\overset{+}{N}H_3 \xrightarrow{-H^+} CH_3CH_2NH_2.$$

In the acid chloride reaction, the new bond between nitrogen and carbon forms before the old bond between carbon and chlorine breaks. In reactions at tetrahedral carbon this is impossible, as we have seen (Section 26.3). The energy released in the formation of the new bond is available to activate the reaction in which the old bond is broken, and this is why, in the particular cases just quoted, the carbon–chlorine bond in acetyl chloride is so very much more reactive than that in ethyl chloride. We have a rationalization of the activating influence of the carbonyl group, with respect to the nucleophilic cleavage of carbon–chlorine, carbon–oxygen, and carbon–nitrogen bonds, referred to in earlier chapters, in particular Table 23.4 (Section 23.1.3).

26.5 REACTIONS OF ELECTROPHILES WITH MULTIPLY BOUND CARBON

We have seen that the lack of electrical symmetry, and the ability of oxygen to carry a negative charge, permitted nucleophiles to attack and bond to the carbonyl carbon. The carbon–carbon double bond in simple alkenes is very much more symmetrical than the double bond between oxygen and carbon, and is normally unreactive towards nucleophiles. However, there is a high density of easily polarizable (Section 26.4) electrons between the carbon nuclei, and species which can accommodate electrons (electrophiles) are likely to interact with these. Most of the simple addition reactions of alkenes, discussed in Section 22.2.4, are initiated in this way in a process we can describe as electrophilic attack.

Consider the approach of a bromine molecule to a molecule of ethylene, such that the bromine molecule approaches the centre of the ethylene molecule end on:

$$\begin{array}{c} CH_2 \\ \| \\ CH_2 \end{array} \quad Br\!-\!Br \;\rightarrow\; \begin{array}{c} CH_2Br \\ | \\ CH_2 \\ + \end{array} + :\!\ddot{Br}\!:^- \;\rightarrow\; \begin{array}{c} CH_2Br \\ | \\ CH_2Br \end{array}$$

There will be many chance collisions of this variety in a mixture of both compounds. The high density of electrons in ethylene will tend to polarize the bond between the bromine atoms such that the bonding pair is displaced towards the bromine atom remote from the ethylene molecule. The atom presented to the ethylene molecule will thus become positively charged, and this will in turn polarize the electrons in the double bond. We believe that reaction occurs by the complete displacement of the bonding pair of bromine on to one of the atoms. In short, a pair of electrons from the ethylene double bond displaces a bromide ion from a bromine molecule. (On the ethylene model described in Section A.2, it is the easily polarized π-electrons which are involved here.) We do not define this as a nucleophilic reaction, though, as such definitions are made from the standpoint of the organic molecule (Section 26.3). We define it as electrophilic attack on the ethylene molecule, with bromine, as the electrophile, accepting a pair of electrons from the alkene. We might envisage this as the transfer of a bromonium ion, $:\!\ddot{Br}\!^+$, from the bromine molecule to the ethylene molecule, in much the same way as a proton is transferred from a hydrogen bromide molecule to a water molecule:

$$H_2O: \;+\; H\!-\!Br \;\rightarrow\; H_3O^+ \;+\; Br^-.$$

The proton does not become free. Neither does the electrophilic bromonium ion. The latter has only six electrons in its outer shell. Bromide ion completes the reaction by attacking the carbonium ion.

Other common addition reactions of alkenes, described in Section 22.2.4 and which are of this mechanistic type, include the addition of hydrogen halides and the addition of sulphuric acid:

$$CH_2\!=\!CH_2 + HCl \rightarrow CH_3CH_2Cl;$$
$$CH_2\!=\!CH_2 + H_2SO_4 \rightarrow CH_3CH_2OSO_3H.$$

In as much as bromine initially transfers a bromonium ion on to the alkene, these reagents initially transfer a proton. All of these reactions are known as electrophilic addition reactions. The reaction of sulphuric acid with ethylene will serve to illustrate:

It will be of interest to consider at this point the effect of bromine on an aromatic system like benzene, which contains the "equivalent" of three double bonds (i.e. six π-electrons—see Section A.2). Bromine certainly attacks benzene, and we believe that it does so in a way which is very similar to that in which it attacks ethylene, with the formation of a positively charged intermediate species. The positive charge on this intermediate ion is delocalized, with canonical forms:

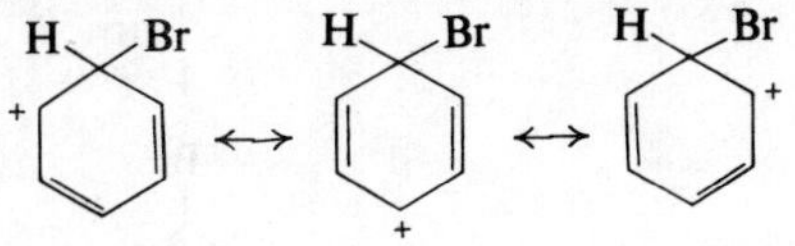

The ion may be represented as

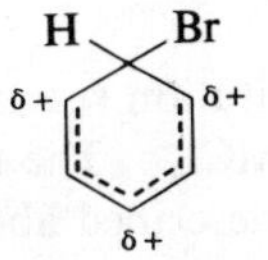

which illustrates that the delocalization is not completely "cyclic" as in benzene itself, and that the ion is not aromatic. It will have a much smaller delocalization energy than benzene or its aromatic derivatives. The formation of the ion, and the two alternative pathways by means of which it could lose its charge, are illustrated below. The positive intermediate formed between ethylene and bromonium ion is also faced with this choice of course, but it chooses the opposite way to lose the charge. This is because more energy is released if it follows this route, the fully saturated product being more stable than the alternative, separate bromoethylene and hydrogen bromide, $CH_2\!=\!CHBr + HBr$. This is a thermodynamic effect. In the case of benzene, however, a hydrogen ion is lost, and the final product contains the same sort of electron system as benzene itself. That is, the product is also aromatic and therefore much more stable than the alternative cyclohexadiene compound. If the latter had been the product, the aromatic delocalization energy of benzene would have been lost, and this dictates that the product is the substituted benzene. Thus, with bromine, benzene undergoes an electrophilic substitution reaction, whereas ethylene undergoes electrophilic addition:

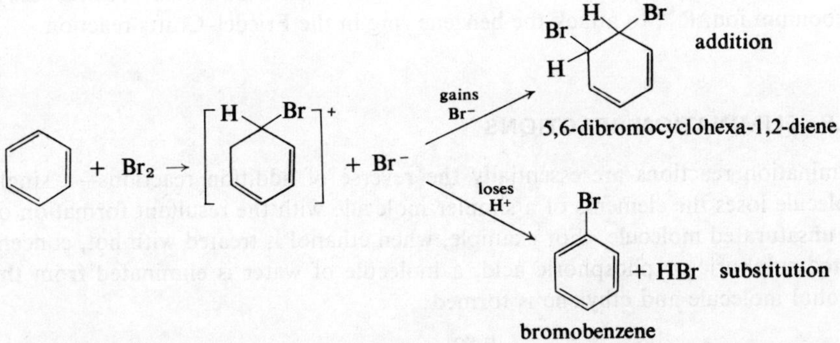

A catalyst which can accept an electron pair (i.e. which is itself an electrophile) is generally used in this reaction. By withdrawing electrons from the bromine molecule and ultimately accepting a bromide ion (Br^-), it helps to break the molecule and promotes the formation of the carbon–bromine bond. Iron(III) bromide, for example, is an effective catalyst and one can envisage it removing a bromide ion to form $FeBr_4^-$:

$$C_6H_6 + Br-Br + FeBr_3 \rightarrow [C_6H_6(H)(Br)]^+ + FeBr_4^-$$

$$\downarrow -H^+$$

$$C_6H_5Br$$

Common electrophiles in substitution reactions with benzene (Section 22.2.6) are Br^+, Cl^+, NO_2^+, SO_3, and $R^+AlCl_4^-$. The first two are most commonly provided by the halogen molecule, as described above for bromine. Only rarely is a free bromonium ion involved in such reactions, although salts such as bromonium perchlorate, $Br^+ClO_4^-$, are known and are powerful brominating agents. The nitronium ion, NO_2^+ (Section 8.5), is present in the mixture of concentrated sulphuric and nitric acids used for nitrating benzene. The sulphuric acid protonates the nitric acid which then decomposes:

$$H_2SO_4 + HO-NO_2 \rightarrow H_2\overset{+}{O}-NO_2 + HSO_4^-;$$

$$H_2\overset{+}{O}-NO_2 \rightarrow H_2O + NO_2^+;$$

$$H_2O + H_2SO_4 \rightarrow H_3\overset{+}{O} + HSO_4^-.$$

Overall:

$$2H_2SO_4 + HNO_3 \rightarrow NO_2^+ + H_3O^+ + 2HSO_4^-.$$

SO_3 is electrophilic because, like the carbonyl group (Section 26.4), the S=O group is strongly polarized, with negative oxygen and positive sulphur. Fuming sulphuric acid is a solution of sulphur trioxide in pure sulphuric acid. $R^+AlCl_4^-$ is the reactive species formed by an alkyl halide and aluminium chloride which provides alkyl carbonium ion, R^+, to attack the benzene ring in the Friedel–Crafts reaction.

26.6 ELIMINATION REACTIONS

Elimination reactions are essentially the reverse of addition reactions—a single molecule loses the elements of a simpler molecule with the resultant formation of an unsaturated molecule. For example, when ethanol is treated with hot, concentrated sulphuric or phosphoric acid, a molecule of water is eliminated from the alcohol molecule and ethylene is formed:

$$CH_3CH_2OH \xrightarrow{H_2SO_4} CH_2=CH_2 + H_2O.$$

Not all elimination reactions involve loss from adjacent atoms to yield multiple bonds as in this case, although the majority do, and these are the more important ones to which we shall confine our attention.

We shall examine here two types of elimination reaction, those which are catalysed by acids and those which are catalysed by bases. The example above is clearly a case of acid-catalysed elimination. This reaction is somewhat complicated by the fact that several other reactions may also occur at the same time, but we shall consider the straightforward elimination of water catalysed by the acid.

Like the oxygen atom of water, the oxygen atom of an alcohol has lone pairs of electrons which can be protonated by acids. As was pointed out, the protonated species carries a pre-formed water molecule. The positively charged oxygen atom draws to itself the electron pair bonding it to carbon, and a water molecule leaves. A positively charged alkyl group (a carbonium ion) remains, which can easily lose a proton to the solvent and form ethylene.

$$H_3C-CH_2-\ddot{O}H + H-A \rightarrow H_3C-CH_2-\overset{+}{O}H_2 + A^-$$
acid

$$H_3C-CH_2-\overset{+}{O}H_2 \rightarrow H_3C-CH_2^+ + :\ddot{O}H_2$$

$$H_2\ddot{O}: + H-CH_2-CH_2^+ \rightarrow CH_2{=}CH_2 + (H_3O)^+$$

A typical base-catalysed elimination reaction is that in which an alkyl halide is treated with a base and loses the elements of the corresponding hydrogen halide (Section 22.3). In general, where X is a halogen atom (chlorine, bromine, or iodine):

$$H-\underset{|}{\overset{|}{C}}-\underset{|}{\overset{|}{C}}-X \xrightarrow{base} \underset{|}{C}{=}\underset{|}{C} + HX$$

HX forms a salt with the base.

As most bases are also nucleophiles, it is apparent that the displacement of halide ion by the base is a possible alternative to the required elimination reaction. For example, the treatment of $H-\underset{|}{\overset{|}{C}}-\underset{|}{\overset{|}{C}}-X$ with an aqueous solution of sodium hydroxide may result in the displacement of X^- by OH^- with the formation of the alcohol $H-\underset{|}{\overset{|}{C}}-\underset{|}{\overset{|}{C}}-OH$ (Section 26.3). For the elimination reaction to predominate, it is necessary, ideally, to carry out the reaction in a solvent of low polarity with a very strong base which is not very effective in the displacement reaction—that is, a strong base which is a weak nucleophile. A high concentration of base and a high temperature also favour elimination. In practice a compromise is reached. Most strong bases are themselves polar in nature and therefore require a solvent of moderate polarity. A solution of potassium hydroxide in ethanol (alcoholic potash) is often the reagent of choice and ethanol, whilst polar, is less so than water. The nature of the alkyl group attached to the halogen also plays an important role in determining the ratio of elimination to substitution. Thus, tertiary alkyl halides (e.g. tertiary-butyl bromide, $(CH_3)_3CBr$) favour elimination whereas primary alkyl halides favour substitution. Ethyl halides yield very little ethylene indeed when treated with alcoholic potash. Secondary halides are of intermediate character in this respect.

The characteristic property of a base is that it will accept a hydrogen ion, or proton, from an acid. In the case of this elimination reaction we believe that the base accepts a proton from the carbon atom adjacent to that to which the halogen is attached. At the same time the electrons from the ruptured carbon–hydrogen bond move into place between the two carbon atoms to form a double bond, and in so doing, expel the halide ion. In so far as the double bond is formed at the same time as the carbon–hydrogen and carbon–halogen bonds are broken, this reaction is similar to the nucleophilic displacement reaction described earlier where the new bond formed as the old bond was breaking (Section 26.3). Instead of attacking the carbon atom to which the halogen is attached, though, the reagent attacks a hydrogen atom on the adjacent carbon atom (it behaves as a base, rather than as a nucleophile).

$$B{:}^- + H{-}C{-}C{-}X \rightarrow [B{-}H]^+ + \mathrm{C{=}C} + :\ddot{X}{:}^-$$

base

As with nucleophilic substitution reactions, there is an alternative pathway in which the alkyl halide ionizes first. Instead of the positive alkyl ion (carbonium ion) reacting with the attacking reagent by addition, to give a substitution product, it loses a hydrogen ion to form a neutral alkene. Again, the reagent has behaved as a base rather than as a nucleophile.

$$H{-}C{-}C{-}\ddot{X}{:} \rightarrow H{-}C{-}C^+ + :\ddot{X}{:}^-$$

$$B{:}^- + H{-}C{-}C^+ \rightarrow [B{-}H]^+ + \mathrm{C{=}C}$$

Table 26.1 Summary chart. This chart is simply a summary of reaction types which have been covered in this chapter. It is not an attempt to cover *all* types of reaction nor even all types of organic reaction.

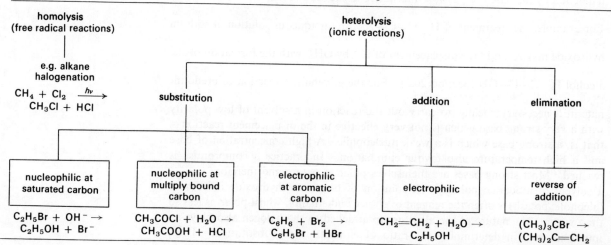

PROBLEMS

(*A cautionary note*—remember that without experimental verification, suggested mechanisms cannot be accepted as valid. It is important, nevertheless, to be able to postulate sensible mechanisms—and other scientific models—based on systems which are known and understood, before they can be tested.)

1. Classify the following as addition, substitution, or elimination reactions:
   a) $CH_2=CHCH_3 + Br_2 \rightarrow BrCH_2CH(Br)CH_3$;
   b) $CH_3CH(OH)CH_3 \rightarrow CH_2=CHCH_3 + H_2O$;
   c) o-nitrotoluene + $Cl_2 \rightarrow$ o-nitrobenzyl chloride + HCl;
   d) p-toluidine + $2Br_2 \rightarrow$ 2,6-dibromo-4-methylaniline + 2HBr;
   e) $CH_3CH=CH_2 + CH_2 \rightarrow CH_3CH\text{—}CH_2$ (cyclopropane with CH₂ bridge);
   f) 1-hydroxy-1-methylcyclopentane $\rightarrow$ 1-methylcyclopentene $+ H_2O$;
   g) $(CH_3)_3COCOCH_3 \rightarrow (CH_3)_2C=CH_2 + CH_3COOH$;
   h) $CH_3CHO + H_2 \rightarrow CH_3CH_2OH$;
   i) $CH_3CH_2Br + C_6H_5O^- \rightarrow CH_3CH_2OC_6H_5 + Br^-$;
   j) $(C_6H_5)_2PCl + 2(C_2H_5)_2NH \rightarrow (C_6H_5)_2PN(C_2H_5)_2 + (C_2H_5)_2\overset{+}{N}H_2\ Cl^-$;
   k) $CH_3CH(Cl)COOH + 3NH_3 \rightarrow CH_3CH(NH_2)COONH_4 + NH_4Cl$;
   l) cyclohexyl-$\overset{+}{N}(CH_3)_3$ + $OH^- \rightarrow$ cyclohexene $+ (CH_3)_3N + H_2O$;
   m) $CH_3COCH_3 + CH_3MgBr \rightarrow CH_3\text{—}\underset{CH_3}{\overset{CH_3}{C}}\text{—}OMgBr$;
   n) $(CH_3)_3COP(OC_6H_5)_2 \rightarrow (CH_3)_2C=CH_2 + HO\text{—}P(OC_6H_5)_2$ (with P=O in both).

2. a) Draw the electronic structures for the nucleophiles, listed in Section 26.3, which react with halogenalkanes (alkyl halides). Show only the electrons in the outer shells of atoms. Indicate clearly the lone pairs of electrons which make each species a nucleophile.*

   (e.g. for water: H:Ö:H — lone pairs)

*See note on page 672.

b) Using curved arrows to indicate electron pair shifts, write equations illustrating feasible mechanisms for the reaction of each of these nucleophiles with ethyl bromide. Assume that they are all nucleophilic substitutions.

3. Draw the electronic structures (outer shells only) of the following electrophiles, clearly indicating why they are electrophiles: Br^+, NO_2^+, H^+, SO_3, and CH_3^+.

4. Classify the following species as free radicals, electrophiles, or nucleophiles, giving your reasons in each case: Cl^+, CH_3, CH_3NH_2, CH_3^-, Cl, I^-, BF_3, H^+, and CH_2.
(The last species is known, but is highly reactive and has a very short life time.)

5. Which of the following substitution reactions would you classify as nucleophilic (S_N) and which as electrophilic (S_E)? Indicate clearly which species are nucleophiles and which electrophiles.

a) $CH_3CH_2Br + KOH(aq) \rightarrow CH_3CH_2OH + KBr(aq)$.

b) $C_6H_5-CH_3 + Br^+ \rightarrow Br-C_6H_4-CH_3 + H^+$.

c) $CH_3CH_2I + (CH_3)_3N \rightarrow CH_3CH_2\overset{+}{N}(CH_3)_3 + I^-$.

d) $C_6H_5CH_2Br + KCN(aq) \rightarrow C_6H_5CH_2CN + KBr$.

e) $C_6H_5NO_2 + HNO_3 \xrightarrow{H_2SO_4} C_6H_4(NO_2)_2 + H_2O$.

f) $CH_3COCl + H_2O \rightarrow CH_3COOH + HCl$.

g) $C_6H_5COCl + 2NH_3 \rightarrow C_6H_5CONH_2 + NH_4Cl$
(remember, $HCl + NH_3 \rightarrow NH_4Cl$).

h) $CH_3COCl + CH_3OH \rightarrow CH_3COOCH_3 + HCl$.

i) $CH_3COCH_2Cl + NaI \rightarrow CH_3COCH_2I + NaCl$.

j) $CH_3CH_2OC_6H_5 + H_2SO_4 \xrightarrow{heat} C_6H_5OH + CH_3CH_2OSO_3H$.
(ethyl hydrogen sulphate)

k) $\begin{array}{l} COOCH_2CH_3 \\ | \\ COOCH_2CH_3 \end{array} + 2NH_3 \rightarrow \begin{array}{l} CONH_2 \\ | \\ CONH_2 \end{array} + 2CH_3CH_2OH$.

6. Curved arrows, denoting the shift of an electron pair, are used in the following equations. Indicate where they are wrongly used and where correctly used. Where incorrect, give the correct usage, such that your new equation adequately accounts for the conversion of the given reactants into the given products.

a)
$$CH_3OH + CH_3\overset{O}{\underset{\|}{C}}-Cl \rightarrow CH_3-\underset{H}{\overset{+}{O}}-\overset{O}{\underset{\|}{C}}-CH_3 + Cl^-$$

$$Cl^- + H-\underset{CH_3}{\overset{+}{O}}-\overset{O}{\underset{\|}{C}}-CH_3 \rightarrow HCl + CH_3O\overset{O}{\underset{\|}{C}}CH_3$$

b) HO⁻ + H—Cl → H₂O + Cl⁻

c) R—Ö—Ö—R → 2 R—Ö·

d) (CH₃)₃C—Br → (CH₃)₃C⁺ + Br⁻

e) (CH₃)₂CHNH₂ + I—CH₃ → (CH₃)₂CHN⁺H₂CH₃ + I⁻

f) (CH₃)₃N + CH₃—I → (CH₃)₄N⁺ + I⁻

g) CH₃CH₂OH + H—X → CH₃CH₂O⁺H₂ + X⁻

H₂O⁺—CH₂CH₂—H + X⁻ → H₂O + CH₂=CH₂ + HX

7. Look up the haloform reaction (Section 23.3.2). Bearing in mind that trichloromethane (chloroform) is moderately acidic (unlike methane) and reacts accordingly with sodium hydroxide (HCCl₃ + OH⁻ ⇌ ⁻CCl₃ + H₂O), suggest a reason why trichloroacetone, Cl₃CCOCH₃, should behave like a carboxylic acid derivative and hydrolyse in aqueous alkali to trichloromethane and acetate.

8. Suggest a simple mechanism for the reaction between an aldehyde and hydrogen cyanide (Section 23.3.2). It takes place in slightly alkaline solution. (You may assume that it is similar in nature to the reaction between acetaldehyde and ammonia, discussed in this chapter.) Can you suggest why there is *no* reaction at very low pH (HCN is a weak acid)?

9. Using the reaction between ammonia and acetyl chloride as a model, suggest a possible, simple mechanism for the hydrolysis of acetamide:

$$CH_3CONH_2 + H_2O \xrightarrow{H^+} CH_3COOH + NH_3$$
$$\downarrow H^+$$
$$NH_4^+.$$

Why does not ethylamine (CH₃CH₂NH₂) also hydrolyse under these conditions?

10. Using the reaction between bromine and ethylene as a model, suggest a possible mechanism for the reaction between ethylene and hypobromous acid:

$$CH_2=CH_2 + HOBr \rightarrow HOCH_2CH_2Br.$$

Oxygen is of greater electronegativity than bromine.

11. Write a simple mechanism which could account for the reaction between an alkyl chloride, aluminium chloride, and benzene (Friedel–Crafts reaction):

$$RCl + AlCl_3 + C_6H_6 \rightarrow C_6H_5R + AlCl_3 + HCl.$$

Use the aromatic bromination mechanism as a guide.

12. Consider the thermal decomposition of tetraethylammonium hydroxide to give ethylene, triethylamine, and water (read the account in Section 24.4.2) and compare it with reactions discussed in this chapter.
   a) What type of reaction is this (substitution, elimination, addition)?
   b) What role do you think the hydroxide ion might play in this reaction?
   c) What reaction, undergone by halogenoalkanes (alkyl halides) does this

reaction resemble? Point out clearly all of the features which the two reactions have in common.

d) The reaction referred to in (c) above requires a base catalyst. Why is it not necessary to add a base catalyst to bring about this decomposition of tetraethylammonium hydroxide?

Problems 13 to 17 are intended to help you to interpret the patterns emphasized in Chapters 23 to 25 and to appreciate some of the chemical relationships which are important in those areas.

13. Examine the reactions outlined in Table 22.4.
    a) What is the most common type of reaction undergone by alkenes and alkynes (substitution, addition, or elimination)?
    b) Ignoring polymerization and the reaction with hydrogen (which are mechanistically different from the rest), to what mechanistic type do these reactions belong (e.g. electrophilic substitution, nucleophilic addition etc.)?
    c) Using but-2-ene as an example, indicate the important steps in the reaction between an alkene and hydrogen bromide. Use curved arrows to indicate electron pair shifts.

14. The (generalized) reactions under the headings of alkylation and acylation in Table 23.3 represent some of the most important reactions of hydroxyl containing compounds. Examine the three generalized reactions and note that each involves the formation of a new bond between the hydroxyl oxygen atom and a carbon atom of the reagent (this is not so obvious in the case of reactions with acids and acid anhydrides, although experiment shows that it is the oxygen atom of, say, alcohols which attaches to the carbon of the acid or anhydride).
    a) Consider the oxygen atoms of the hydroxyl compounds. Do they have lone pairs of electrons? How many lone pairs?
    b) Do the carbon atoms to which the oxygens become attached have lone pairs of electrons (before reaction)?
    c) Which of these two atoms involved in the reaction—carbon or oxygen—could be classified as a nucleophile?

    Granted that all of these reactions are nucleophilic substitutions, write out feasible mechanisms, using equations and curved arrows, which could apply to these reactions. For simplicity, choose one specific example of each of the three general types, such as:

    $$C_2H_5OH + C_2H_5OH \xrightarrow{H^+} C_2H_5OC_2H_5 + H_2O$$

    (*hint*—remember that much of the alcohol will be present, in acid, as the protonated form $C_2H_5\overset{+}{O}H_2$, Section 26.3);

    $$C_2H_5O^- + C_2H_5Br \rightarrow C_2H_5OC_2H_5 + Br^-;$$
    $$C_2H_5OH + CH_3COCl \rightarrow CH_3COOC_2H_5 + HCl.$$

    Make it clear which is the nucleophile and which is the leaving group.

15. The reactions of compounds containing the carbon–oxygen–carbon linkage are summarized in Table 23.6. They may all be classified as nucleophilic substitutions.

a) Identify the atom which is the nucleophile in each of the three generalized reagents, HOH, —COH, and >N—H.

b) How many lone pairs does each of these atoms contain?

c) Refer to the mechanism outlined for the acid-catalysed hydrolysis of esters (Section 8.8) in which a C—O—C link is broken. What role does the acid catalyst play (Section 26.3)? Rewrite the reaction scheme, using curved arrows to represent electron pair shifts, and clearly indicate which species is the leaving group, and which the nucleophile.

16. Examine the chart at the top of Table 23.9 and, in the light of it, indicate how the following reactions, mentioned earlier in the text (Section 23.3.2), fit into the chart (i.e. are they addition, condensation, or substitution reactions?):

a) $C_6H_5NHNH_2 + CH_3COCH_3 \rightarrow (CH_3)_2C=NNHC_6H_5 + H_2O$;

b) $C_6H_5COOCH_2CH_3 + (CH_3)_2CHOH \xrightarrow{base} C_6H_5COOCH(CH_3)_2 + CH_3CH_2OH$

(base will "activate" the alcohol by removing a proton from oxygen, $B + ROH \rightarrow RO^- + BH^+$)

c) $CCl_3CHO + C_2H_5OH \rightarrow CCl_3\underset{OC_2H_5}{\overset{OH}{\underset{|}{\overset{|}{C}}}}H$

On the assumption that all involve nucleophilic attack on the carbonyl carbon atom, as described for the acetaldehyde/ammonia reaction, indicate, for each of the above equations, which species is the nucleophile and (where appropriate) which the leaving group.

Having successfully interpreted the above three reactions, and related them to Table 23.9, pick out more of the addition, condensation, and substitution reactions of Section 23.3.2 and do likewise for these. This should reveal the mechanistic thread which runs through the pattern of behaviour outlined in Section 23.3. (*Hint*—the bisulphite addition reaction is difficult, but remember the structure of the product and that of the bisulphite ion is $:\underset{OH}{\overset{O^-}{\underset{|}{\overset{|}{S}}}}=O$.)

17. Examine Table 24.2.
   a) How many reactions (at nitrogen) of amino compounds depend upon the nitrogen lone pair (the reactions with nitrous acid are complex, but the initial steps resemble the acylation reactions)?
   b) Compare Table 24.2 with Table 23.3. In what ways do —OH compounds resemble, and in what ways do they differ from, NH compounds (qualitatively)?
   c) Consider the possibilities for electron delocalization in amides (e.g. acetamide) and amines (e.g. ethylamine). Suggest a *possible* reason why amides are stronger acids than amines. Suggest a *possible* reason why amides are

more difficult to alkylate (they alkylate at very low rates) with alkyl iodides, relative to amines (it may be helpful in answering this part, to read Section 28.5.1).

Problems 18 to 26—interpretation of evidence. You are asked to interpret experimental observations in the light of the experience and understanding you have so far gained.

18. Hydrolysis of halogenoalkanes (alkyl halides) is catalysed by bases but not by acids. Suggest reasons for this.

19. Sodium chloride and ethylene do not react. A mixture of sodium chloride, bromine, and ethylene give CH_2BrCH_2Br and CH_2BrCH_2Cl but no CH_2ClCH_2Cl. Can you account for these observations?

20. Suggest two different nucleophilic substitution mechanisms for the ester hydrolysis reaction:

$$CH_3COOCH_3 + H_2O \xrightarrow{OH^-} CH_3COOH + CH_3OH.$$

Given that $CH_3-C(=O)-O^{18}CH_3$ produces CH_3COOH and $CH_3{}^{18}OH$ on alkaline hydrolysis, which mechanism is correct? (*Hint*—which C—O bond breaks in the reaction?)

21. The addition of bromine to cyclopentene gives *trans*-1,2-dibromocyclopentane but no *cis* isomer is obtained. What does this tell you about the way in which bromine reacts with alkenes?

22. Shaking trichloromethane (which is not radioactive) with tritium oxide in the presence of a little base yields, after separation and purification, radioactive trichloromethane (tritium is a radioactive isotope of hydrogen). How may this be interpreted? What does this tell you about the nature of the trichloromethane hydrogen atom?

23. Treatment of ethyl acetoacetate with deuterium oxide and base converts it into a deuterated derivative:

$$CH_3COCH_2COOCH_2CH_3 + 2D_2O \rightarrow CH_3COCD_2COOCH_2CH_3 + 2DOH.$$

What can you conclude about the hydrogen atoms on the central methylene group (CH_2) of ethyl acetoacetate?

24. The reaction between 2-bromo-2-methylpropane (tertiary-butyl bromide) and hydroxide ion in aqueous acetone (solvent) was thought to proceed by one of two possible pathways (Section 26.3):

$(CH_3)_3CBr + OH^- \rightarrow (CH_3)_3COH + Br^-$ (one step);

or
(1) $(CH_3)_3CBr \rightarrow (CH_3)_3C^+ + Br^-$
(2) $(CH_3)_3C^+ + OH^- \rightarrow (CH_3)_3COH$ (two steps).

The reaction is observed to follow first order kinetics—rate depends only upon the concentration of the bromoalkane and not upon the concentration of hydroxide ions. With which of these two reaction pathways is the evidence consistent? What order of reaction would you expect for the other pathway (give a rate expression)?

25. When allyl phenyl ether is heated it rearranges:

[Structure: allyl phenyl ether → 2-allylphenol]

Two possible mechanisms are:

a) [Mechanism a showing dissociation into phenoxide and allyl cation with delocalization $CH_2\!-\!CH\!=\!CH_2 \leftrightarrow CH_2\!=\!CH\!-\!CH_2$, then recombination to cyclohexadienone intermediate, then tautomerization to 2-allylphenol]

b) [Mechanism b showing concerted [3,3]-sigmatropic rearrangement via cyclic transition state to cyclohexadienone intermediate, then tautomerization to 2-allylphenol]

It is observed that the product obtained by heating

[Structure: phenyl crotyl ether (O-CH(H)-CH=CH-CH_3? shown as O-CH_2-CH=CH with H and CH_3)] is [Structure: 2-(1-methylallyl)phenol, OH with CH(CH_3)-CH=CH_2 ortho substituent]

Which of the above two mechanisms is most likely to be correct? (Hint—the allyl carbonium ion, by virtue of the delocalization indicated, is entirely symmetrical about the centre carbon atom, $\overset{\delta+}{CH_2}\!-\!-\!-\!CH\!-\!-\!-\!\overset{\delta+}{CH_2}$.)

26. Halogenoarenes are normally inert to nucleophilic substitution. On heating bromobenzene with the very strong base sodamide, $Na^+NH_2^-$, however, phenylamine is produced. If 4-methylbromobenzene is similarly treated, two products are obtained, 4-methylphenylamine and 3-methylphenylamine. What does this suggest about the nature of this "substitution"? (Hint—see Section 25.2.2.)

BIBLIOGRAPHY

Whitfield, R. C. *A Guide to Understanding Basic Organic Reactions*, Longman (1966).

> A treatment specifically aimed at introducing the sixth-former to organic reaction mechanisms.

Sykes, P. *A Guidebook to Mechanism in Organic Chemistry* (third edition), Longman (1970).

> An excellent and very readable treatment which would appeal to sixth-formers and undergraduates.

Stewart, R. *Investigation of Organic Reactions*, Prentice-Hall (1966).

> Concerned with reaction mechanisms.

Saunders, W. H. *Ionic Aliphatic Reactions*, Prentice-Hall (1965).

Stock, L. M. *Aromatic Substitution Reactions*, Prentice-Hall (1968).

> Primarily written for undergraduates, but a sixth-former could get a great deal from this.

Gould, E. S. *Mechanism and Structure in Organic Chemistry*, Holt (1959).

> This is an advanced account, perhaps more useful for reference at sixth-form level.

All of the books in this list are of wide application and will be found helpful in connection with all chapters concerned with organic chemistry. Similarly, the content of this chapter is covered, at differing levels, by all of the books on organic chemistry from the general list.

CHAPTER 27

Organometallic Compounds

27.1 INTRODUCTION

All the functional groups so far dealt with have one important factor in common: they are based upon atoms which are more electronegative than carbon, and the bonds linking these atoms to carbon are therefore all polarized in the same direction (Section 4.4). The bonding electrons are displaced away from carbon which is always the positive pole. Furthermore, these electronegative elements form negative rather than positive ions and when their bond to carbon breaks, the bonding electrons usually depart with the electronegative element.

- 27.1 Introduction
- 27.2 Methods of forming carbon–metal bonds
- 27.3 Chemical character and synthetic uses of organometallic compounds

Alkyl halides $\overset{\delta+}{-}\overset{\delta-}{C}-Cl$

Amines $\overset{\delta+}{-}\overset{\delta-}{C}-NH_2$

Alcohols $\overset{\delta+}{-}\overset{\delta-}{C}-\overset{\delta+}{O}-H$

Ketones $\overset{\delta+}{C}=\overset{\delta-}{O}$

Ethers $\overset{\delta+}{-}\overset{\delta-}{C}-\overset{\delta+}{O}-C-$

Thus most of the reactions between organic compounds containing these functional groups do not result in the formation of carbon–carbon bonds. The departure of the electronegative elements with the bonding electrons would leave no electrons for carbon–carbon bond formation, and, further, interaction between two positive poles would be required. Alkyl halides, for example, react with amines to form a carbon–nitrogen bond rather than a carbon–carbon bond (Section 24.1.2).

Carbon is not, however, restricted to forming bonds with electronegative elements and many compounds are known in which carbon groups are covalently bonded to metals (metals are less electronegative than carbon). These are called organometallic compounds and, until recently, relatively little was known about their chemistry. Research in organometallic chemistry is developing rapidly at the present time in an area which extends into the traditional fields of both organic and inorganic chemistry. Indeed, the whole concept of this somewhat arbitrary division of chemistry into organic, inorganic, and physical compartments is being undermined by developments in areas such as this.

As a result of the low electronegativity of metals, carbon–metal bonds are polarized so that carbon is now the negative pole and we have bonds of type $\overset{\delta-}{-}\text{C}\overset{\delta+}{-}\text{metal}$ rather than $\overset{\delta+}{-}\text{C}\overset{\delta-}{-}\text{non-metal}$. In addition, metals tend to form positive ions rather than negative ions and therefore these bonds to carbon break with the electron pair remaining on carbon. In reactions between organometallic compounds and compounds with electronegative substituents, attractive interaction between two carbon atoms is thus possible and reactions of this type lead to many useful and important carbon–carbon bond syntheses. Ethyl-lithium, for example, reacts with 1-bromoprop-2-ene (allyl bromide) to give pent-1-ene:

$$\text{ethyl-lithium} \quad CH_3CH_2-Li \qquad \qquad CH_3CH_2$$
$$\text{allyl bromide} \quad CH_2=CHCH_2-Br \quad \rightarrow \quad \overset{|}{CH_2=CHCH_2} \quad + \text{ LiBr};$$

and with acetone to give a tertiary alcohol (as its lithium salt):

$$\text{acetone} \quad CH_3-\underset{CH_3-CH_2}{\overset{CH_3}{\underset{|}{\overset{|}{C}}}}=O \quad \rightarrow \quad CH_3-\underset{CH_3-CH_2}{\overset{CH_3}{\underset{|}{\overset{|}{C}}}}-OLi \quad \xrightarrow{H_2O} \quad CH_3-\underset{CH_3-CH_2}{\overset{CH_3}{\underset{|}{\overset{|}{C}}}}-OH \ + \ LiOH.$$
$$\text{ethyl-lithium} \quad CH_3-CH_2-Li$$

It is important to remember that in non-metallic organic compounds carbon may not be the only element at the positive end of a bond dipole. Thus, in alcohols, for example, the hydrogen atom also bears a positive charge as it does in primary amines:

$$\overset{\delta+}{-}\text{C}\overset{\delta-}{-}\text{O}\overset{\delta+}{-}\text{H} \qquad \overset{\delta+}{-}\text{C}\overset{\delta-}{-}\text{N}\overset{\overset{\delta+}{H}}{\underset{\underset{\delta+}{H}}{}}$$

These hydrogen atoms are acidic in character ("active"). With compounds like this, organometallic compounds usually react to form a C—H bond as it is easier to break an oxygen–hydrogen or nitrogen–hydrogen bond than an oxygen–carbon or nitrogen–carbon bond. Thus ethyl-lithium reacts with ethanol to form ethane and lithium ethoxide, rather than butane and lithium hydroxide:

$$\text{ethyl-lithium} \quad CH_3-CH_2-Li \qquad \qquad CH_3CH_3 \ + \ LiOCH_2CH_3;$$
$$\text{ethanol} \quad H-O-CH_2CH_3 \quad \rightarrow \quad \text{ethane} \qquad \text{lithium ethoxide}$$

not
$$CH_3CH_2-Li \qquad \qquad CH_3CH_2$$
$$CH_3CH_2-O-H \quad \rightarrow \quad \underset{\text{butane}}{\overset{|}{CH_3CH_2}} \ + \ \underset{\text{hydroxide}}{\overset{}{\text{LiOH.}}} \ \text{lithium}$$

This reaction emphasizes that compounds like ethyl-lithium are very strong bases and will remove protons from very weak acids. They cannot be expected to form carbon–carbon bonds, therefore, in the presence of even weakly acidic groups.

From the organic chemist's point of view probably the most useful organometallic compounds are those derived from lithium (group IA) and from magnesium (group IIA) and consequently these are the compounds with which we shall be principally concerned. Compounds of more electropositive metals, however, such as sodium, also have some importance as do those of less electropositive metals. Amongst the latter, organoboranes are perhaps of most interest to the organic chemist.

27.2 METHODS OF FORMING CARBON–METAL BONDS

27.2.1 Reaction Between Metals and Alkyl or Aryl Halides

Probably the most commonly employed method is that involving reaction between the metal and an alkyl or aryl halide. Both lithium and magnesium metals will react thus with organic halides, provided that a suitable solvent such as diethyl ether is used:

a) $\qquad$ RHal + 2Li → RLi + LiHal.

For example:

$$C_4H_9Br + 2Li \rightarrow C_4H_9Li + LiBr;$$
but-1-yl bromide → but-1-yl-lithium

$$C_6H_5Br + 2Li \rightarrow C_6H_5Li + LiBr.$$
bromobenzene → phenyl-lithium

b) $\qquad$ RHal + Mg → RMgHal.

For example:

$$C_2H_5Br + Mg \rightarrow C_2H_5MgBr;$$
ethyl bromide → ethylmagnesium bromide

$$C_6H_5Br + Mg \rightarrow C_6H_5MgBr.$$
bromobenzene → phenylmagnesium bromide

The compounds produced are very sensitive to water and thoroughly dried solvent, apparatus, and reagents are necessary. The reaction is normally carried out by adding the halogen compound, dissolved in ether, to the metal under an atmosphere of dry nitrogen (atmospheric oxygen is inclined to oxidize the products). In the case of magnesium a little iodine may be necessary to initiate the reaction, but, once started, with either metal it continues exothermically to completion and the rate must be controlled carefully by adding the halogen compound slowly.

The synthetic applications of the magnesium compounds were first realized and developed by Grignard and consequently these alkyl or arylmagnesium halides are known as Grignard reagents.

27.2.2 Treatment of Acidic Hydrocarbons with a Metallic Base

Some hydrocarbons have the unusual distinction of possessing relatively acidic hydrogen atoms and these can be "metallated" by treating them with a suitable metallic base. Although such acidic hydrocarbons are few, those which there are become very important. Probably the most important are the alkynes with free acetylenic hydrogen atoms (Section 22.2.3), particularly acetylene itself. Treatment of acetylene with sodamide—a very strong base formed by dissolving sodium in liquid ammonia—results in the formation of sodium acetylide:

$$\underset{\substack{\text{acetylene}\\\text{(ethyne)}}}{\text{H—C}\equiv\text{C—H}} + \underset{\text{sodamide}}{\text{Na}^+\text{NH}_2^-} \rightarrow \underset{\text{sodium acetylide}}{\text{H—C}\equiv\text{C}^-\text{Na}^+} + \text{NH}_3.$$

Because sodium is so very electropositive, sodium acetylide is highly ionic.

Organometallic compounds are themselves basic in character (Section 27.1) and Grignard reagents, for example, will also metallate acetylenes:

$$\underset{\substack{\text{acetylene}\\\text{(ethyne)}}}{\text{HC}\equiv\text{CH}} + \underset{\substack{\text{ethyl-}\\\text{magnesium}\\\text{bromide}}}{\text{C}_2\text{H}_5\text{MgBr}} \rightarrow \underset{\substack{\text{ethynyl-}\\\text{magnesium}\\\text{bromide}}}{\text{HC}\equiv\text{CMgBr}} + \underset{\text{ethane}}{\text{C}_2\text{H}_6}.$$

(The difunctional acetylenic Grignard, $\text{BrMgC}\equiv\text{CMgBr}$, may also be obtained. It is possible to choose reaction conditions favouring either.) This is the most convenient way of obtaining acetylenic Grignard reagents.

The reactions of diethyl malonate and ethyl acetoacetate with sodium ethoxide (Section 28.3) may be referred to at this stage:

$$\underset{\text{diethyl malonate}}{\text{CH}_2(\text{COOC}_2\text{H}_5)_2} + \text{NaOC}_2\text{H}_5 \rightarrow \text{NaCH}(\text{COOC}_2\text{H}_5)_2 + \text{C}_2\text{H}_5\text{OH};$$

$$\underset{\text{ethyl acetoacetate}}{\text{CH}_3\text{COCH}_2\text{COOC}_2\text{H}_5} + \text{NaOC}_2\text{H}_5 \rightarrow \underset{\text{Na}}{\text{CH}_3\text{COCHCOOC}_2\text{H}_5} + \text{C}_2\text{H}_5\text{OH}.$$

Although these two compounds are not hydrocarbons, it is a C—H bond which is being broken in both reactions.

27.2.3 Reaction between Metal Halides and other Organometallic Compounds

Many organometallic compounds are most conveniently prepared from metal halides and other more reactive organometallic compounds. Thus, treatment of cadmium(II) chloride with a Grignard reagent yields an alkylcadmium compound:

$$\text{CdCl}_2 + 2\text{RMgBr} \rightarrow \text{CdR}_2 + 2\text{MgBrCl};$$
$$\text{CdCl}_2 + 2\underset{\substack{\text{ethylmagnesium}\\\text{bromide}}}{\text{C}_2\text{H}_5\text{MgBr}} \rightarrow \underset{\text{diethylcadmium}}{(\text{C}_2\text{H}_5)_2\text{Cd}} + 2\text{MgBrCl}.$$

Organoboranes may be similarly obtained:

$$\text{BF}_3 + 3\text{C}_2\text{H}_5\text{MgBr} \rightarrow \underset{\text{triethylborane}}{(\text{C}_2\text{H}_5)_3\text{B}} + 3\text{MgFBr}.$$

27.2.4 Addition of Metal Hydrides to Alkenes

This is another important method of obtaining organometallic compounds. Organoboranes may be readily obtained by this method as we saw in Section 23.1.3 when we considered the hydroboration process. Boron hydride adds to an alkene in an apparently "anti-Markownikoff" direction, with the hydrogen atom attaching to the carbon bearing fewest hydrogen atoms already:

$$B_2H_6 + RCH=CH_2 \rightarrow 2(RCH_2CH_2)_3B.$$

In order to account for this, it will be necessary to consider the alkene hydration reaction in a little more detail first. An understanding of why the Markownikoff rule is obeyed in acid-catalysed hydration reactions will enable us to see why in this case the rule is reversed.

Electrophilic addition reactions of alkenes were discussed in Section 26.5. The initial step, also in fact the rate determining step, is the addition of a proton from the acid to the alkene to form a carbonium ion. Clearly there are two possible ions which could form:

$$RCH=CH_2 + H^+ \rightarrow R\overset{+}{C}HCH_3 \quad \text{or} \quad RCH_2\overset{+}{C}H_2.$$

The first is a secondary carbonium ion (see definitions of primary, secondary, etc., in connection with alkyl groups, Section 23.1.2) and the second is a primary carbonium ion. We have good reason to believe that a positive charge is more stable on a secondary carbon atom than on a primary carbon atom. The stability is even greater on a tertiary carbon atom. In the transition state leading to the first carbonium ion, charge is building up on a secondary carbon atom whereas in the transition state leading to the second ion, charge is building up on a primary carbon atom. As both ions form from the same reactants, the activation energy for the formation of the secondary ion is smaller than that for the formation of the primary ion. The first ion, the secondary ion, therefore forms faster than the second one, the primary one.

Reaction is completed by the addition of a nucleophilic water molecule to the ion, followed by loss of a proton. Each ion gives a different alcohol:

$$R\overset{+}{C}HCH_3 + :OH_2 \rightarrow \underset{\underset{+OH_2}{|}}{RCHCH_3} \xrightarrow{-H^+} \underset{\underset{OH}{|}}{RCHCH_3};$$

$$RCH_2\overset{+}{C}H_2 + :OH_2 \rightarrow RCH_2CH_2\overset{+}{O}H_2 \xrightarrow{-H^+} RCH_2CH_2OH.$$

Because the reaction is kinetically controlled, and the secondary ion forms faster than the primary ion, the principal product observed is the secondary alcohol, $RCH(OH)CH_3$. In other words, Markownikoff's rule is obeyed.

The initiating step, then, is the addition of a proton to the alkene. A boron hydride, however, cannot yield a proton. Boron has a lower electronegativity than hydrogen and the boron–hydrogen bond is polarized in the sense $\overset{\delta+}{B}$—$\overset{\delta-}{H}$. If the bond were to break it would yield a hydride ion, H^-, rather than a proton, together with a positive boron species. It is therefore the boron entity which behaves as the electrophile, not hydrogen, and in the addition to an alkene it is boron which attaches in the place that hydrogen attaches in a Markownikoff addition of water. The hydrogen atom, having the character of a negative ion, attaches in the place that the nucleophilic oxygen attaches in the water addition. The overall reaction is thus:

$$B_2H_6 + 6RCH=CH_2 \rightarrow 2(RCH_2CH_2)_3B.$$

27.3 CHEMICAL CHARACTER AND SYNTHETIC USES OF ORGANOMETALLIC COMPOUNDS

The chemical reactivity of organometallic compounds depends in large measure upon the nature of the metal concerned. For highly electropositive metals, such as sodium, their organic derivatives may be regarded as being highly ionic, that is composed of separate ions. Sodium acetylide in liquid ammonia, for example, will contain (solvated) sodium and acetylide ions, Na^+ and $HC{\equiv}C^-$. The reactions of these compounds with, say halogenoalkanes, are nucleophilic substitution reactions (Section 26.3) in which the acetylide ion displaces a halide ion from the halogenoalkane.

$$HC{\equiv}\bar{C}: + R{-}Hal \rightarrow HC{\equiv}C{-}R + Hal^-.$$

Acetylide ions are powerful nucleophiles.

With less electropositive metals, however, we are dealing with covalent species in which the carbon atom attached to the metal, as the negative end of a bond dipole, bears a partial negative charge, $metal\overset{\delta+}{-}\overset{\delta-}{C}{-}$. With only a partial negative charge, and a covalent bond to break, carbon groups in such compounds are less reactive—less nucleophilic—than in the ionic compounds. Their reactivity will depend partly upon the extent to which the bond is polarized and, consequently, upon the electronegativity of the metal.

The structure of the covalent compounds is a matter of current research interest. That of the Grignard reagent, probably the most used and most familiar organometallic compound, is still something of an enigma. It is normally written R—Mg—Hal and in some cases this is probably a true reflection of its structure. However, the structure appears to vary according to the experimental conditions and seems to depend, in particular, upon the solvent.

27.3.1 Reactions with Acids

Organometallic compounds of the more electropositive metals, such as lithium, sodium, and magnesium, react even with very weak acids to produce a hydrocarbon derived by protonation of the organic group:

$$\overset{\delta-}{R}{-}\overset{\delta+}{metal} + H^+ \rightarrow RH + metal^+.$$

For example:

$$\underset{\text{butyl-lithium}}{C_4H_9Li} + H_2O \rightarrow \underset{\text{butane}}{C_4H_{10}} + LiOH;$$

$$\underset{\substack{\text{phenyl-}\\\text{magnesium}\\\text{bromide}}}{C_6H_5MgBr} + H_2O \rightarrow \underset{\text{benzene}}{C_6H_6} + Mg(OH)Br.$$

Even ethyl acetoacetate (Section 28.3) is a sufficiently strong acid to liberate hydrocarbon from a Grignard or lithium reagent:

$$CH_3MgI + CH_3COCH_2COOC_2H_5 \rightarrow CH_4 + IMg(CH_3COCHCOOC_2H_5).$$

This type of reaction is useful on two principal counts. It sometimes affords a method of hydrocarbon synthesis or a convenient way of removing a halogen function from a molecule. Alternatively, due to the high sensitivity of Grignard reagents to very weak acids,

$$RBr \xrightarrow{Mg} RMgBr \xrightarrow{H_2O} RH,$$

it forms the basis of a method of detecting and estimating "active hydrogens" (i.e. acidic hydrogens). Thus, after reaction of an excess of methylmagnesium iodide with a measured amount of active hydrogen compound, measurement of the volume of methane evolved allows calculation of the number of active hydrogen atoms present. One mole of methane (22.4 litres at s.t.p.) is the equivalent of one mole of active hydrogen atoms.

$$\underset{\text{1 mole}}{C_2H_5OH} + \underset{\text{excess}}{CH_3MgI} \rightarrow \underset{\substack{\text{1 mole} \\ \text{(22.4 litres at s.t.p.)}}}{CH_4} + C_2H_5OMgI;$$

$$\underset{\text{1 mole}}{HC \equiv CH} + \underset{\text{excess}}{2CH_3MgI} \rightarrow \underset{\substack{\text{2 moles} \\ \text{(44.8 litres at s.t.p.)}}}{2CH_4} + IMgC \equiv CMgI.$$

One mole of ethanol contains one mole of active hydrogen—one mole of acetylene contains two. This is the Zerewitinoff method of active hydrogen determination.

Derivatives of the less electropositive metals (which are of higher electronegativity than magnesium and lithium) are much less reactive towards acids. Trimethylborane, for example, is hydrolysed by water only at about 200 °C, and only one methyl group is lost:

$$(CH_3)_3B + H_2O \rightarrow (CH_3)_2BOH + CH_4.$$

Boiling hydrobromic acid will break all the boron–carbon bonds in tributylborane:

$$(CH_3CH_2CH_2CH_2)_3B + 3H_2O \xrightarrow{HBr} 3CH_3CH_2CH_2CH_3 + H_3BO_3.$$

27.3.2 Reaction with Halogenoalkanes

We have already mentioned this reaction several times (Sections 22.2.3 and 27.1); both organolithium compounds and Grignard reagents react to produce hydrocarbons provided that the halide is particularly reactive. Grignard reagents will not, however, react with simple halogenoalkanes such as bromoethane.

$$\underset{\substack{\text{methyl-} \\ \text{lithium}}}{CH_3Li} + \underset{\substack{\text{2-chloro-} \\ \text{2-methylpropane} \\ \text{(tertiary-butyl} \\ \text{chloride)}}}{(CH_3)_3CCl} \rightarrow \underset{\text{2,2-dimethylpropane}}{(CH_3)_4C}$$

$$\underset{\substack{\text{phenyl-} \\ \text{magnesium} \\ \text{bromide}}}{C_6H_5MgBr} + \underset{\substack{\text{bromomethyl-} \\ \text{benzene} \\ \text{(benzyl bromide)}}}{C_6H_5CH_2Br} \rightarrow \underset{\text{diphenylmethane}}{C_6H_5CH_2C_6H_5}.$$

With acetylides the reaction is even more useful because, as well as resulting in carbon–carbon bond formation, the reaction yields products which are capable of further useful transformation by virtue of the reactive acetylene group which they contain.

$$\text{HC}\equiv\text{CNa} + \text{C}_3\text{H}_7\text{I} \rightarrow \text{HC}\equiv\text{CC}_3\text{H}_7$$

sodium acetylide | 1-iodopropane | pent-1-yne
(prop-1-yl iodide)

With H_2/Pd → $CH_3CH_2C_3H_7$ (pentane)

With H_2O/H_2SO_4 → $CH_3COC_3H_7$ pentan-2-one (methyl propyl ketone)

27.3.3 Reactions with Ketones and Aldehydes

Organometallic compounds such as Grignard reagents and organolithium compounds attack aldehydes and ketones at the carbonyl carbon, forming a new carbon–carbon bond. The alkyl (or aryl) group departs from the metal with the bonding electrons and is nucleophilic. We are dealing here with reaction at a multiple bond and addition is observed rather than substitution. The product is a metal alkoxide which is readily hydrolysed by water to give an alcohol (for practical reasons, aqueous acid is usually used).

$$\overset{\delta+}{\text{C}}=\overset{\delta-}{\text{O}} \quad \text{R—Mg—Hal} \rightarrow \text{—C—O}\quad\text{R Mg—Hal}$$

[The mechanism is probably more complex than this.]

$$\downarrow H_2O$$

$$\text{—C—OH} + \text{Mg(Hal)Cl}$$
  |
  R

$$\text{(CH}_3\text{)}_2\text{C=O} + \text{CH}_3\text{MgBr} \longrightarrow \text{(CH}_3\text{)}_3\text{C—OMgBr} \xrightarrow{\text{HCl(aq)}} \text{(CH}_3\text{)}_3\text{COH}$$

acetone (ketone) → 2-methylpropan-2-ol (tertiary-butanol) a tertiary alcohol

$$\text{CH}_3\text{CH=O} + \text{CH}_3\text{MgBr} \longrightarrow \text{(CH}_3\text{)}_2\text{CH—OMgBr} \xrightarrow{\text{HCl(aq)}} \text{(CH}_3\text{)}_2\text{CHOH}$$

acetaldehyde (aldehyde) → propan-2-ol a secondary alcohol

$$\text{H}_2\text{C=O} + \text{CH}_3\text{MgBr} \longrightarrow \text{H}_2\text{CH(CH}_3\text{)—OMgBr} \xrightarrow{\text{HCl(aq)}} \text{CH}_3\text{CH}_2\text{OH}$$

formaldehyde (aldehyde) → ethanol a primary alcohol

Note that ketones lead to tertiary alcohols, and aldehydes to secondary alcohols. Only formaldehyde, which is unique among carbonyl compounds, can give a primary alcohol. Lithium compounds behave like the magnesium compounds:

$$\underset{R-Li}{\overset{+\ -}{\underset{\curvearrowleft}{C=O}}} \rightarrow -\underset{R}{\underset{|}{C}}-\underset{Li}{\overset{|}{O}} \xrightarrow{HCl(aq)} -\underset{R}{\underset{|}{C}}-OH.$$

27.3.4 Reactions with Acid Chlorides

Acid chlorides have both a carbonyl group and a reactive (Section 26.4) carbon–chlorine group. It is not surprising, therefore, that Grignard reagents and organolithium compounds affect both. The chlorine is displaced in the first instance, by the nucleophilic alkyl group, to yield a ketone which then rapidly reacts further to give a tertiary alcohol.

$$RCOCl + R'MgHal \rightarrow RCOR' + Mg(Hal)Cl$$

$$RCOR' + R'MgHal \rightarrow R-\underset{R'}{\underset{|}{\overset{R'}{\overset{|}{C}}}}-OMgHal$$

$$\downarrow HCl(aq)$$

$$R-\underset{R'}{\underset{|}{\overset{R'}{\overset{|}{C}}}}-OH$$

For example:

$$C_6H_5COCl + 2C_6H_5MgBr \longrightarrow C_6H_5\underset{C_6H_5}{\underset{|}{\overset{C_6H_5}{\overset{|}{C}}}}OMgBr \xrightarrow{HCl(aq)} (C_6H_5)_3COH.$$

benzoyl chloride / phenyl-magnesium bromide / triphenylmethanol (a tertiary alcohol)

The ketone is usually too reactive to be isolated, and this reaction is therefore of no use for ketone synthesis. This, however, is where the less reactive organometallic compounds come into their own, for by choosing a suitable one, the chlorine of acid chlorides may be replaced selectively without further attack on the ketone produced. Dialkyl cadmiums, prepared from cadmium(II) chloride and a Grignard reagent, are suitable for ketone preparation.

$$2RCOCl + R'_2Cd \rightarrow 2RCOR' + CdCl_2.$$

For example:

$$2CH_3COCl + (C_4H_9)_2Cd \rightarrow 2CH_3COC_4H_9 + CdCl_2.$$

acetyl chloride / dibutyl-cadmium / hexan-2-one (butyl methyl ketone)

Note that in order to obtain this selective reaction it was necessary to choose a reagent of low reactivity. This illustrates the general principle that where selectivity is required of reagents, it is most likely to be found amongst those of low reactivity.

27.3.5 Reactions with Esters

We have already noted—Section 23.3.2(6)—that the carbon–oxygen bond in esters can be broken by reagents similar to those which break carbon–chlorine bonds (i.e. nucleophiles such as water). As with the carbon–chlorine bond of acid chlorides, so the ester carbon–oxygen bond can be broken by organolithium and Grignard reagents. Again, the product is a tertiary alcohol:

$$RCOOR' + R''Li \longrightarrow R-\underset{\|}{\underset{O}{C}}-R'' + LiOR'$$

$$RCOR'' + R''Li \longrightarrow R-\underset{R''}{\overset{R''}{\underset{|}{\overset{|}{C}}}}-OLi \xrightarrow{HCl} R-\underset{R''}{\overset{R''}{\underset{|}{\overset{|}{C}}}}-OH.$$

For example:

$$\underset{\text{ethyl benzoate}}{C_6H_5COOC_2H_5} + \underset{\substack{\text{phenyl-}\\\text{lithium}}}{C_6H_5Li} \longrightarrow (C_6H_5)_3COLi$$

$$\downarrow HCl$$

$$\underset{\substack{\text{triphenylmethanol}\\\text{(a tertiary alcohol)}}}{(C_6H_5)_3COH.}$$

27.3.6 Oxidations of Organometallic Compounds

Most organometallic reagents, in solution or otherwise, are handled in an inert, oxygen free, atmosphere. Most are relatively easily oxidized. Molecular oxygen, for example, can oxidize Grignard reagents and the product contains an oxygen atom "inserted" between the carbon and the metal atom:

$$RMgHal + \tfrac{1}{2}O_2 \rightarrow ROMgHal.$$

It is, in fact, a magnesium alkoxide (compare sodium alkoxides, formed by reaction between alcohols and sodium metal) and hydrolysis of this yields a hydroxy compound:

$$ROMgHal + H_2O \rightarrow ROH + Mg(OH)Hal.$$

The reaction is relatively unimportant, however, as a means of preparing alcohols and phenols.

Table 27.1. Principal methods of obtaining organometallic compounds.

R—Hal + M → R—M + MHal (e.g. M = Na, Li, Mg)
(or R—M—Hal)

R—H + metallic base → R—Na + NH$_3$ only certain hydrocarbons, such as acetylene, are sufficiently acidic
(e.g. Na$^+$NH$_2^-$)

R—M— + —M′—Hal → R—M′— + —M—Hal (e.g. M = Mg, M′ = Cd)

H—M< + >C=C< → H—C—C—M< (e.g. M = B)

M = metal atom

Oxidation of trialkylboranes is very much more important, however, and this is usually achieved with a solution of hydrogen peroxide:

$$R_3B + 3H_2O_2 \rightarrow (RO)_3B + H_2O.$$

The product is an ester of boric acid (H_3BO_3 or $B(OH)_3$). Like most esters, these are readily hydrolysed by alkali and the products of hydrolysis are boric acid (as its sodium salt) and an alcohol:

$$(RO)_3B + 3H_2O \rightarrow 3ROH + H_3BO_3.$$

This is the final step in the hydroboration process which has been referred to before (Sections 23.1.3 and 27.2.4) and which is so useful in converting alkenes to alcohols. In practice, the oxidation step and the hydrolysis step are carried out together by treating the borane with an alkaline solution of hydrogen peroxide.

Table 27.2. Principal reactions of organometallic compounds.

PROBLEMS

1. Suggest a method, involving an organometallic reagent, by means of which the following might be synthesized. Use whatever reagents you may consider necessary, as well as the specified starting material. Use as few consecutive reactions as possible.

   a) $(CH_3)_3COH$ (from acetone).
   b) $(CH_3)_2CHCH_2OH$ (from 2-bromopropane).

c) $CH_3COCH_2CH_3$ (from acetyl chloride).
d) $CH_3COC_6H_5$ (from acetonitrile). *Hint*—see Section 23.3.3(5).
e) $(C_6H_5)_3COH$ (from ethyl benzoate).
f) $CH_3CH_2COCH_2CH_3$ (from bromoethane as the *only* organic source of carbon). *Hint*—see Problem 1(d).
g) $CH_3CH_2CH_2CH=CH_2$ (from 1-bromobutane).
h) $(CH_3)_2CHCH_2CH_2CH_2OH$ (from $(CH_3)_2CHCH_2OH$).

2. Suggest how bromoethane may be converted to ethane (use lithium metal and water as reagents—Section 27.2.1).

3. A compound of formula $C_4H_6O_2$ (mol. wt. 46) is known to have structure $HOCH_2CH_2OH$ or CH_3OCH_2OH. From 23 g of this compound, on treatment with an excess of methylmagnesium iodide (CH_3MgI), 11.2 dm³ (at s.t.p.) of methane were obtained. What is the correct structure of the compound?

4. Treatment of ethyl bromoacetate with zinc leads to a stable organozinc compound:

$$BrCH_2COOCH_2CH_3 + Zn \rightarrow BrZnCH_2COOCH_2CH_3.$$

a) Predict the structure of the product obtained by reaction of this compound with acetone (assume that the reaction mixture is treated with dilute acid before the product is isolated).
b) Why is an ester of bromoacetic acid used here, instead of the acid itself?
c) Why would magnesium be an inappropriate metal to use in these reactions?

5. Predict, giving reasons, the products of the following reactions:
a) $PCl_3 + C_6H_5MgBr$ (excess);
b) $SiCl_4 + C_6H_5Li$ (excess);
c) $(CH_3CH_2O)_2C=O + CH_3CH_2MgBr$ (excess);
d) $CH_3CONH_2 + CH_3CH_2Li$.

6. Suggest how you might prepare the isotopically labelled compounds indicated:
a) C_6H_5D, from bromobenzene;
b) CH_3CHDCH_3, from propan-2-ol;
c) $CH_3CH_2CH_2D$, from propene.
In each case, use deuterium oxide (D_2O), as deuterium source.

BIBLIOGRAPHY

Simpson, P. *Organometallic Chemistry of the Main Group Elements*, Longman (1970).

A systematic treatment specifically aimed at introducing the sixth-former to this important and rapidly developing field (usually given very little emphasis in elementary courses and texts).

Rochow, E. G. *Organometallic Chemistry*, Chapman and Hall (1964).

An interesting treatment at a suitable level for sixth-form reading.

CHAPTER 28
Molecules Containing Two or More Functional Groups

Depending upon how close together two functional groups are in the same molecule, and upon the nature of that portion of the molecule which separates them, each will exert an influence upon the other. Usually the character of each individual group is basically the same as it would have been in the absence of the other but probably modified in degree. Thus, for example, the carboxylic acid function in fluoroacetic acid is acidic just as it is in acetic acid. It is, however, more strongly so (but see discussion at end of Section 28.5.1). The hydroxyl and carbonyl functions which make up a carboxylic acid group are very close together and influence each other profoundly. This has already been discussed in Sections 23.1 and 23.3 where the differences between the carboxylic carbonyl group and the ketonic carbonyl group, and those between an alcoholic hydroxyl group and a carboxylic acid hydroxyl group, were emphasized. It may be, however, that in special circumstances a functional group takes on a rather different character. Thus, in the case of α,β-unsaturated carbonyl compounds (Section 28.4) the carbon–carbon double bond takes on aspects of carbonyl group character.

A molecule may also exhibit rather special properties if it contains two functional groups which can react chemically with one another. Thus some of the rather special properties of α-amino acids (Section 28.1) arise because the amino group is able to react with the acid group.

Clearly there are very many possible combinations of functional groups which may occur in a single molecule and it is therefore necessary to select just a few of these for consideration in this chapter. A number of important compounds have been chosen and these will serve to illustrate how functional groups may influence each other and the remainder of the molecule containing them.

28.1 α-Amino acids
28.2 Carbohydrates
28.3 Compounds containing two carbonyl groups attached to the same carbon atom
28.4 α,β-Unsaturated aldehydes and ketones
28.5 Interacting functional groups in aromatic molecules

28.1 α-AMINO ACIDS

α-Amino acids are naturally occurring carboxylic acids with an amino group (usually primary) attached to the α-carbon atom (Section 23.3). They are of general formula, $RCH(NH_2)COOH$, where R may be hydrogen, an alkyl group, or an aryl group. Note that, provided R is not hydrogen, the molecules fulfil the requirements for optical activity (Section 21.3) and therefore exist in two enantiomorphic forms. α-Amino acids are particularly important in plants and animals where, as their principal role, they are the bricks out of which large, polymeric protein molecules are built. Hydrolysis of protein material (see below) yields individual amino acids, and those obtained from natural sources are usually optically active, that is, only one enantiomorph is obtained for each acid. Biological systems can distinguish between one enantiomorph and the other and they normally use only one of any given pair. A notable exception is glycine (aminoacetic acid) which is the amino acid in which R = H and which cannot therefore be optically active.

Taking glycine, the simplest amino acid, as representative we can in the first place note that it shows most of the properties of a primary amine and most of the properties of a carboxylic acid. Thus, it forms salts with acids, it can be acetylated and benzoylated, and it liberates nitrogen, quantitatively, when treated with nitrous acid:

$$CH_2(NH_2)COOH \xrightarrow{HCl} CH_2(\overset{+}{N}H_3)COOH\ Cl^-;$$
glycine   glycine hydrochloride

$$CH_2(NH_2)COOH \xrightarrow{CH_3COCl} CH_3CONHCH_2COOH;$$
N-acetylglycine

$$CH_2(NH_2)COOH \xrightarrow{HNO_2} N_2 + CH_2(OH)COOH.$$
hydroxyacetic acid
(glycollic acid)

It also forms salts with bases and can be esterified:

$$CH_2(NH_2)COOH \xrightarrow{NaOH} CH_2(NH_2)COO^-\ Na^+;$$
sodium glycinate

$$CH_2(NH_2)COOH \xrightarrow{C_2H_5OH} CH_2(NH_2)COOC_2H_5.$$
ethyl glycinate

There are, however, certain properties of glycine which depend upon the presence of both functional groups.

Being both an acid (—COOH group) and a base (—NH$_2$ group), glycine is capable of forming salts internally: the acid group can protonate the amino group. Thus, in neutral solutions and in the solid state, it exists as the dipolar ion, $\overset{+}{H_3N}CH_2COO^-$, known as a **zwitterion**, which is essentially an internal salt (compare $CH_3\overset{+}{N}H_3\ CH_3COO^-$, methylammonium acetate). This ionic nature accounts for the high solubility in water and the high melting point (232 °C, with decomposition) of crystalline glycine.

Again, because it contains both an amino group and a carboxylic acid group, glycine is capable of forming amides derived only from glycine molecules. It cannot easily form internal amides which are entirely analogous to the zwitterionic internal salts (i.e. based on a single glycine molecule) because a covalent bond is required and the resultant compound would contain a highly strained ring structure,

CH$_2$—NH
  \ /
   C
   ‖
   O

(cf. cyclopropane, Section 21.5).

On heating glycine, however, or its ethyl ester, a cyclic amide is formed containing two glycine "residues":

$$\begin{array}{c}NH_2 \\ | \\ CH_2 \\ | \\ COOH\end{array} + \begin{array}{c}HOCO \\ | \\ CH_2 \\ | \\ NH_2\end{array} \xrightarrow{heat} \begin{array}{c}NH-CO \\ / \quad \backslash \\ CH_2 \quad CH_2 \\ \backslash \quad / \\ CO-NH\end{array} + 2H_2O;$$

$$\begin{array}{c}NH_2 \\ | \\ CH_2 \\ | \\ COOC_2H_5\end{array} + \begin{array}{c}C_2H_5OCO \\ | \\ CH_2 \\ | \\ NH_2\end{array} \xrightarrow{heat} \begin{array}{c}NH-CO \\ / \quad \backslash \\ CH_2 \quad CH_2 \\ \backslash \quad / \\ CO-NH\end{array} + 2C_2H_5OH.$$

Non-cyclic amides, which may contain numerous glycine residues, may also be synthesized:

$$H_2NCH_2CONHCH_2COOH$$
glycylglycine

$$H_2NCH_2CONHCH_2CONHCH_2COOH$$
glycylglycylglycine

$$H_2NCH_2CO(NHCH_2CO)_nNHCH_2COOH.$$

These larger molecules are known as peptides and, as synthesized in the laboratory, or isolated from biological systems, may be built up from different amino acids:

$$\begin{array}{cccc}R & R' & R'' & R''' \\ | & | & | & | \\ H_2NCHCONHCHCONHCHCONHCHCO\ldots etc.\end{array}$$

Protein molecules are built up similarly from large numbers of various amino acids and these very long chain molecules are coiled and folded in a specific way which is vital to their biological function. The destruction of this secondary structure, even though the molecule remains intact, usually results in loss of biological activity. The protein is then said to have been *denatured*.

Being amides, proteins and peptides can be hydrolysed and are thus broken down to their constituent amino acids:

$$\begin{array}{cccc}R & R' & R'' & R''' \\ | & | & | & | \\ H_2N-CHCO-NHCHCO-NHCHCO-NHCHCOOH\end{array} + 3H_2O$$

$$\downarrow$$

$$\begin{array}{cccc}R & R' & R'' & R''' \\ | & | & | & | \\ H_2NCHCOOH + H_2NCHCOOH + H_2NCHCOOH + H_2NCHCOOH\end{array}$$

Compare:

$$CH_3CONHCH_3 + H_2O$$
N-methylacetamide

$$\downarrow$$

$$CH_3COOH + H_2NCH_3$$
acetic acid   methylamine

The hydrolysis is catalysed by acids and bases, as for simple amides (Sections 24.1 and 24.2), and by a certain group of enzymes which play an important role in protein digestion. About twenty-four different amino acids are associated with proteins, and the same amino acid may appear many times over, at different points, along the length of a protein chain.

Summary

Amino acids are important biological compounds containing both the *basic amino group* and the *acidic carboxyl group*. The two groups may interact with each other to form internal salts—*zwitterions*, such as $\overset{+}{H_3N}CH_2COO^-$—or *amide linkages*. "Linear" aggregates of amino acids connected by amide linkages are the basis of *protein* and *peptide* structures. Materials containing amino acids—largely proteins—are essential dietary requirements.

28.2 CARBOHYDRATES

28.2.1 Introduction

Having covered the chemistry of aldehydes, ketones, and alcohols in Chapter 23 and aspects of molecular geometry in Chapter 21, we are in a good position to discuss some of the basic chemistry of carbohydrates, a most important group of naturally occurring polyfunctional compounds (compounds containing several functional groups). The name carbohydrate arises from their molecular formulae which can be written as though the molecules were simply combinations of carbon and water. Thus glucose has the molecular formula $C_6H_{12}O_6$ and sucrose $C_{12}H_{22}O_{11}$. These could be written as $C_6(H_2O)_6$ and $C_{12}(H_2O)_{11}$, respectively. These, however, are misrepresentations of their real nature. They bear little or no relationship to the structures as we now know them, and represent very early speculations. Some carbohydrates now known cannot, in fact, be written in this way at all (e.g. deoxyribose, $C_5H_{10}O_4$).

Glucose is a typical, crystalline sugar. It is one of the basic units of more complex sugars and as such is known as a monosaccharide. Combinations of several such units are known as di-, tri-, tetrasaccharides etc., and these are still known as sugars. Combinations of large numbers of monosaccharide units are known as polysaccharides and show properties characteristic of high polymers. The term carbohydrate covers both the sugars and the polysaccharides. Fructose is another common, crystalline monosaccharide and, as a structural isomer of glucose, also has the molecular formula $C_6H_{12}O_6$. Sucrose is a crystalline disaccharide, comprising one fructose unit and one glucose unit. It hydrolyses in water, in a reversible reaction, to yield fructose and glucose. Although the equilibrium is heavily in favour of the separate monosaccharide units, the reaction is reversible and, with the consumption of energy, sucrose is synthesized in biological systems from fructose and glucose:

$$C_{12}H_{11}O_6(\text{sucrose}) + H_2O \rightleftharpoons C_6H_{12}O_6(\text{fructose}) + C_6H_{12}O_6(\text{glucose}).$$

28.2.2 Structure and Chemical Nature

The structure of glucose (strictly, α-D-glucose) is as follows (the hydrogen atoms attached to ring carbons are not shown):

$$\text{[α-D-glucose structure: six-membered ring with CH}_2\text{OH at C5, OH groups at C2, C3, C4, and anomeric OH at C1 (down)]}$$

The basic skeleton is a six-membered ring, like cyclohexane (Section 21.5), having one divalent oxygen atom in the ring. The carbon atoms are numbered, for reference and nomenclature purposes, as shown.

A number of important points emerge from this structure. Firstly, there is a large number of hydroxyl groups. These, it is thought, are responsible for the sweet taste of sugars (other polyhydroxy compounds, such as 1,2-dihydroxy-propane—propylene glycol—$CH_3CH(OH)CH_2OH$, are also sweet and are sold as non-fattening sweeteners), and are certainly responsible for their high solubility in water (Section 23.1.1). The hydroxyl groups at positions 2, 3, 4, and 6 are, in fact, typical alcohol groups, 6 being a primary alcohol group and the others secondary. The hydroxyl group at position 1, however, is different. It is, in fact, the hydroxyl group of a hemi-acetal (Section 23.3.2). A typical hemi-acetal has one hydroxy group, and one alkoxy group (e.g. ethoxy) attached to a single carbon atom. In this case the alkoxy group is complex, and the whole alkoxy group, including the oxygen, is part of the ring. Glucose is thus both an alcohol (tetrahydric) and a hemi-acetal.

Another point which emerges is that there are five asymmetric carbon atoms in glucose, atoms 1, 2, 3, 4, and 5. There are many optical isomers of α-D-glucose, therefore, although only one can be the mirror image of it. This is called α-L-glucose. It does not occur naturally. All of the other stereoisomers of glucose, which are not mirror images of α-D-glucose, are given different names, having different properties, except for the isomer which has a different configuration at position 1. This is known as β-D-glucose:

$$\text{[β-D-glucose structure: six-membered ring with CH}_2\text{OH at C5, OH groups at C2, C3, C4, and anomeric OH at C1 (up)]}$$

α- and β-glucose have a special (chemical) relationship which we shall go on to explain. The α-form is described as the anomer of the β-form, and vice versa, and the reader should note that this use of the Greek letters α and β to denote the geometry at $C_{(1)}$ is different from their normal use in organic chemistry to denote *position* in a molecule.

As we saw earlier (Section 23.3.2), hemi-acetals are very readily decomposed to the parent aldehyde and an alcohol:

$$\underset{\underset{OH}{|}}{\overset{\overset{OR'}{|}}{R-C-H}} \rightleftharpoons R-\overset{\overset{O}{\|}}{C}-H + R'OH.$$

This equilibrium is set up when α-D-glucose dissolves in water, so that the ring opens with the ring oxygen becoming part of an alcohol hydroxyl at position 5.

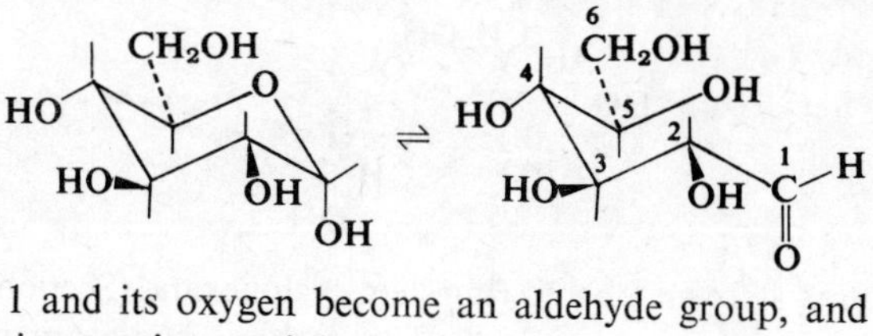

Carbon atom 1 and its oxygen become an aldehyde group, and this open-chain aldehyde form is sometimes written:

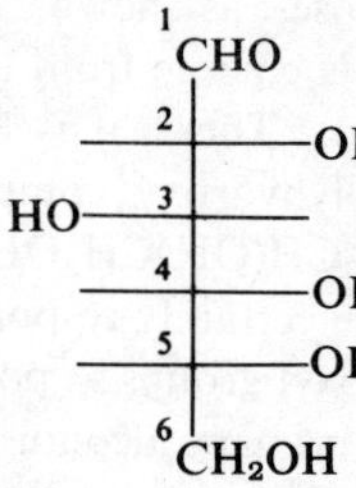

$$\begin{array}{c} \overset{1}{C}HO \\ \overset{2}{|}\!-\!OH \\ HO\!-\!\overset{3}{|} \\ \overset{4}{|}\!-\!OH \\ \overset{5}{|}\!-\!OH \\ \overset{6}{C}H_2OH \end{array}$$

$C_{(1)}$ is no longer asymmetric. Carbon atoms 2, 3, 4, and 5 are shown simply as the intersection of two lines. In the early days of sugar chemistry, this was believed to be the only structure of glucose.

The reverse of this ring opening reaction can be seen as the addition of the $C_{(5)}$ alcohol group to the aldehyde carbonyl group to form the hemi-acetal. In the open-chain form, free rotation about the bond between $C_{(1)}$ and $C_{(2)}$ becomes possible, whereas the ring structure prevented such rotation in the cyclic form. We could envisage such a rotation as follows:

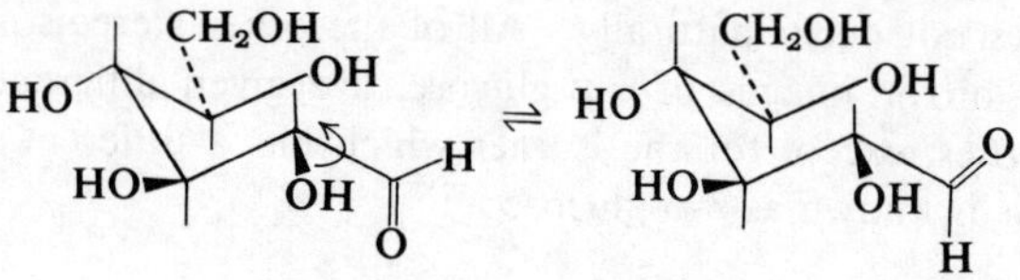

Hemi-acetal formation can now lead to ring closure and $C_{(1)}$ becomes asymmetric again. The result is β-D-glucose, not the original α-form.

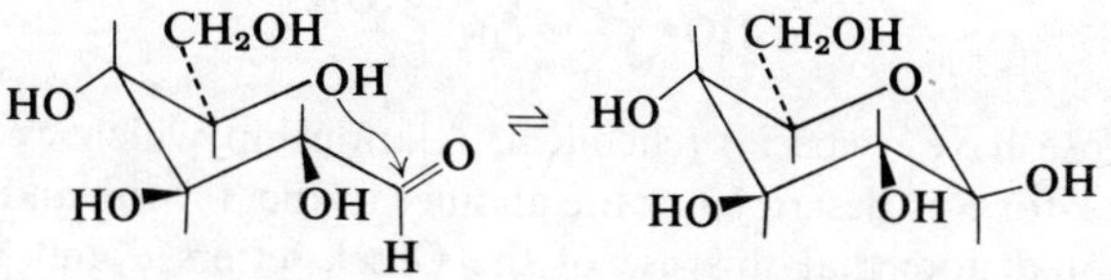

Such cyclic hemi-acetal formation is easy in this case because the geometry of a six-membered ring is exactly suitable for the accommodation of tetrahedral bond angles (Section 21.5). The reacting groups in the open-chain form can—and often must—approach each other frequently and without strain. Furthermore ring closure for glucose does not result in a reduction in the number of molecules in the system, as does hemi-acetal formation between simple aldehydes and alcohols, and so entropy considerations are likely to be more favourable. A similar phenomenon is associated with δ-hydroxy acids, for example, where the alcoholic hydroxyl group can approach the acid group and esterify it. The result is the six-membered ring of a lactone, an internal ester, with an increase in the number of molecules in the system:

$$HOCH_2CH_2CH_2CH_2COOH \rightarrow \begin{array}{c} O \\ CH_2 \quad CO \\ | \quad\quad | \\ CH_2 \quad CH_2 \\ \diagdown \quad \diagup \\ CH_2 \end{array} + H_2O.$$

Five-membered rings, which are nearly flat, also accommodate the tetrahedral bond angle well, hence the five-membered rings so often encountered in carbohydrate chemistry (e.g. in fructose). The $C_{(4)}$ alcohol group of glucose may also add to the aldehyde group to form an acetal and this leads to a five-membered ring structure for glucose. In the case of glucose, however, very little of this form is present in solution.

We thus have, in water, an equilbrium between α-D-glucose and β-D-glucose:

$$\alpha\text{-D-glucose} \rightleftharpoons \text{open-chain aldehyde form} \rightleftharpoons \beta\text{-D-glucose}.$$

The amounts of each form at equilibrium are not equal. Each form, α- and β-, has its own specific rotation ($[\alpha]_D$ values, Section 21.3, $-111°$ and $+19°$ respectively) and when either the pure α- or the pure β-form is dissolved in water the optical rotation slowly changes to the same value, corresponding to the equilibrium mixture of both forms. Note that this is not a racemic mixture (Section 21.3) as α-D-glucose is not the mirror image of β-D-glucose, and this rotation is therefore not zero but is finite ($[\alpha]_D$ of mixture = $+53°$). The phenomenon is known as *mutarotation*.

α-D-Fructose has a structure which is similar to glucose in many respects, but is a hemi-ketal rather than a hemi-acetal and has a five-membered ring rather than a six-membered ring.

cyclic hemi-ketal   open-chain form

The open-chain ketone form is often written:

$$\begin{array}{c} \overset{1}{C}H_2OH \\ \overset{2}{C}=O \\ HO-\overset{3}{C}- \\ -\overset{4}{C}-OH \\ -\overset{5}{C}-OH \\ \overset{6}{C}H_2OH \end{array}$$

The anomer of α-D-fructose is called β-D-fructose and differs from the α-form in the configuration at the hemi-ketal carbon atom, $C_{(2)}$, in the same way that the anomers of glucose differ at the hemi-acetal carbon atom, $C_{(1)}$.

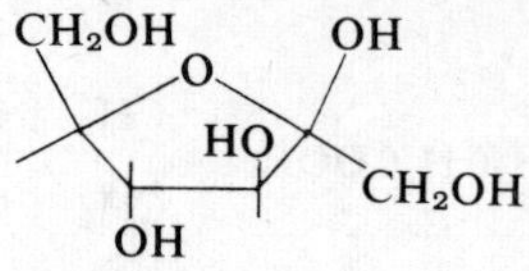

β-D-fructose

28.2.3 Chemistry

Glucose is a good reducing agent and readily reduces Fehling's solution (Sections 20.21 and 23.3.2). This can readily be understood in terms of the oxidation of the open-chain aldehyde form. Removal of the aldehyde form from the ring/chain equilibrium results in conversion of more ring form to the open-chain form, and all of the glucose is ultimately oxidized. This reaction with Fehling's solution is a good test for reducing sugars and is commonly employed in testing urine. Reduction of the reagent implies the presence of glucose which in turn leads to the suspicion that the patient may be suffering from diabetes. Fructose, however, is also a reducing sugar, and this is rather more difficult to understand as ketones do not normally reduce Fehling's solution (Section 23.3.2). However, α-hydroxyketones do reduce Fehling's solution, and the open-chain form of fructose is an α-hydroxyketone. Hydroxyacetone, for example, reduces Fehling's solution, and is oxidized in the process to formic and acetic acids:

$$CH_3COCH_2OH \xrightarrow{[O]} CH_3COOH + HCOOH.$$

Sucrose is a non-reducing sugar and, as we shall see, the ring forms of the component fructose and glucose molecules are "locked" and cannot therefore convert to the reducing open-chain form.

If glucose is treated with methanol and anhydrous hydrogen chloride (the latter is a catalyst) the hemi-acetal hydroxyl at $C_{(1)}$ is methylated, and the molecule becomes an acetal (Section 23.3.2), known as methyl glucoside, with two different alkoxy groups attached to the same carbon atom. A glucoside arises from both α- and β-forms.

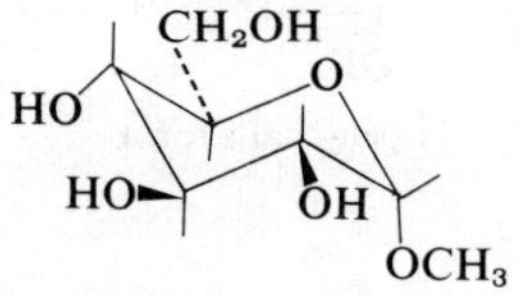

methyl α-D-glucoside 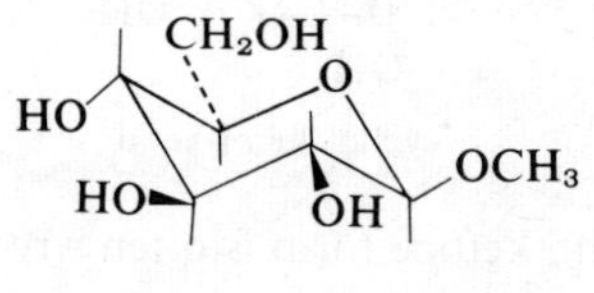   methyl β-D-glucoside

The other hydroxyl groups, typically alcoholic, are not methylated by this reagent. The molecules thus produced are stable at $C_{(1)}$, that is they do not convert to an open-chain form. They are consequently non-reducing and do not affect Fehling's solution. As with other acetals, though (Section 23.3.2), they are hydrolysed by aqueous acid, reverting to normal glucose.

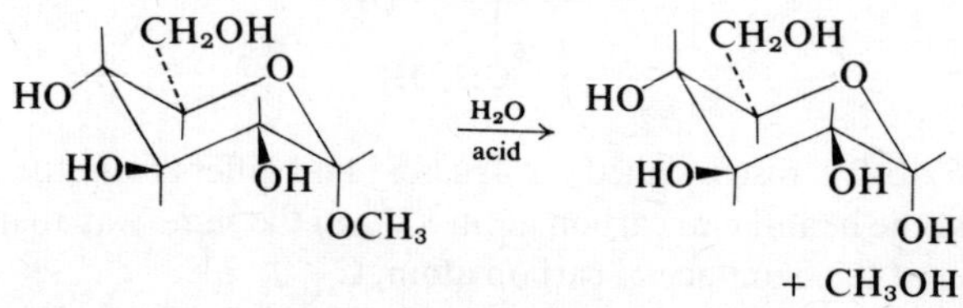

+ CH₃OH

If the methyl glucosides are treated with reagents which normally alkylate alcohols, then all of the other hydroxyl groups are methylated (Section 23.1.2). Methyl iodide and a suitable base may be used, although dimethyl sulphate and aqueous alkali are usually preferred. As acetals are not affected by base, the $C_{(1)}$ methyl group remains in position.

Compare:
$$ROH + base \rightarrow RO^-$$
$$RO^- + CH_3I \rightarrow ROCH_3 + I^-$$
(Section 23.2.3)

The resultant pentamethylglucose, on treatment with aqueous acid, loses the $C_{(1)}$ methyl group, as do the methyl glucosides above, to give a tetramethylglucose. The methoxy groups remaining are normal ether groups, and are not therefore easily hydrolysed (Section 23.2.2).

Derivatives of other monosaccharides, methylated at the hemi-acetal or hemi-ketal hydroxyl group, are similar to the methyl glucosides and all such compounds are known as glycosides. They are essentially non-reducing. The carbon–oxygen–carbon linkage between the acetal (or ketal) carbon and the alcohol residue is known as a *glycoside linkage* and is quite distinct from the ether linkages which attach alkoxy groups to other positions on the ring. The glycoside linkage may be α- or β-, according to whether the parent monosaccharide is the α- or the β-anomer. It is the glycoside linkage which connects monosaccharide units in biological polysaccharides, hence the importance of understanding the chemistry of these simpler methyl derivatives.

28.2.4 Further Biological Aspects

Starch is a polymer of D-glucose and is the form in which plants usually store glucose. It consists of long chains of glucose units held together by α-glycoside linkages between $C_{(1)}$ of one glucose unit and $C_{(4)}$ of another. (The $C_{(4)}$ hydroxyl, an alcohol group, is the analogue of the methanol hydroxyl group in the methyl glycoside synthesis.)

Hence starch is readily hydrolysed by dilute acid to generate D-glucose. Hydrolysis in the plant is enzymic of course, and yields glucose as and when needed. Some of the starch chains are "branched" by glycoside linkages extending from one glucose unit to the hydroxyl group at position 6 of another unit already in a chain. Glucose is an important source of energy in animals (and plants) and is broken down—metabolized—to carbon dioxide and water. The standard free energy change for this process is about 31,500 kJ mol^{-1}. Glucose is the only such metabolite for the human brain. During photosynthesis, in plants, glucose is built up from carbon dioxide under the influence of light energy, and it may then be stored as starch. When released from starch by hydrolysis, it may either be completely broken down again, to provide energy, or be converted into other molecules required by the plant. Glycogen is the analogue of starch in the animal world and has a similar structure to starch.

Cellulose is another important polysaccharide, used by plants for structural purposes, and it forms the basis of many useful fibres obtained from plants as well as of cellophane. Again, it consists of glucose units held together by $C_{(1)}$ to $C_{(4)}$ glycoside linkages, but this time they are β-linkages.

Sucrose, derived from sugar cane and sugar beet, is a disaccharide in which one fructose unit is linked to one glucose unit by a linkage which is glycosidic with respect to *both* molecules. It is α- with respect to glucose, and β- with respect to fructose.

Because the hemi-acetal group of glucose has been converted to acetal and the hemi-ketal group of fructose to ketal, no conversion to open-chain aldehyde or ketone forms can occur and sucrose, like the methyl glucosides, is non-reducing as we have already noted.

A final point on carbohydrate chemistry arises out of their alcoholic nature. They can be acetylated to form acetate esters (Section 23.1.2). More important, from a biological point of view, they can be phosphorylated to form phosphate esters. Adenosine triphosphate (ATP) for example, the so-called energy store of biological systems, is a type of phosphate ester in which the monosaccharide ribose is involved.

[Structure of ATP showing ester linkage, acid anhydride linkages, ribose unit, and adenine unit]

Ester linkages between ribose units hold together the nucleotide units of ribonucleic acid (RNA).

[Structure of RNA showing ester linkages, phosphate diesters, and nucleotide units]

28.2.5 Summary

Carbohydrates are important biological molecules and, like proteins, are essential dietary requirements. They are required by animals as energy sources as well as structural materials. The large number of hydroxyl groups on the molecules are responsible for their solubility in water and probably for their sweet taste as well. Most of the hydroxyl groups of monosaccharides are alcohol groups, but one is associated with a hemi-acetal grouping. Monosaccharide chemistry is essentially the chemistry of these groups, and alcohol, hemi-acetal, and aldehyde properties are a feature. Monosaccharide units can link together, usually through the hemi-acetal hydroxyl group, to form long-chain polymeric structures such as starch, glycogen, and cellulose. The C—O—C linkages holding the units together are part of an acetal grouping.

28.3 COMPOUNDS CONTAINING TWO CARBONYL GROUPS ATTACHED TO THE SAME CARBON ATOM

28.3.1 Malonic Acid

Malonic acid (systematic name propanedioic acid) has the structure
$\begin{array}{c} \text{COOH} \\ | \\ \text{CH}_2 \\ | \\ \text{COOH} \end{array}$
and thus has two carboxylic acid groups (and therefore two carbonyl groups) attached to the central CH_2 carbon atom. In most ways it behaves like a normal carboxylic acid, and thus forms salts, esters, amides, and so on. In one important respect, however, it differs fundamentally from, say, acetic acid (CH_3COOH). Simply on being heated, the compound loses a molecule of CO_2; it is decarboxylated to yield acetic acid:

$$\begin{array}{c} \text{COOH} \\ | \\ \text{CH}_2 \\ | \\ \text{COOH} \end{array} \xrightarrow{\text{heat}} \begin{array}{c} \text{H} \\ | \\ \text{CH}_2 \\ | \\ \text{COOH} \end{array} + CO_2.$$

malonic acid → acetic acid

This is characteristic of all carboxylic acids which contain a carbonyl group separated from the carboxyl group by one carbon atom. A carbonyl group in such a position renders the carboxyl group "labile".

$$\begin{array}{c} R' \quad \text{COOH} \\ \diagdown \diagup \\ C \\ \diagup \diagdown \\ R'' \quad \text{COR}''' \end{array} \xrightarrow{\text{heat}} \begin{array}{c} R' \quad \text{H} \\ \diagdown \diagup \\ C \\ \diagup \diagdown \\ R'' \quad \text{COR}''' \end{array} + CO_2.$$

28.3.2 Diethyl Malonate (Malonic Ester)

The diethyl ester of malonic acid, $CH_2(COOC_2H_5)_2$, is readily prepared and, unlike the free acid, is not easily decarboxylated. It provides a convenient means of examining the effect of the two carbonyl groups on the central CH_2 group.

Treatment of diethyl malonate with sodium ethoxide (a strong base) in ethanol results in the removal of a proton from the CH_2 group and a sodium salt is formed.

$$CH_2(COOC_2H_5)_2 + \bar{O}C_2H_5 \rightarrow \bar{C}H(COOC_2H_5)_2 + C_2H_5OH.$$

The CH_2 group hydrogen atoms are weakly acidic and in this connection contrast strongly with normal hydrocarbon hydrogen atoms. Such a CH_2 group, rendered acidic by neighbouring carbonyl groups, is sometimes referred to as an "active methylene". In a similar manner the carbonyl function enhances the acidity of adjacent —OH groups and N—H groups. The carbonyl group achieves this effect by making it possible for delocalization of the negative charge (Section 4.12) to occur in the malonate anion thus enhancing the stability of the ion and increasing the acid strength of diethyl malonate. There are three canonical forms which can be written for the malonate anion:

$$\begin{array}{c}\text{OC}_2\text{H}_5\\|\\\text{C=O}\\|\\{}^-\text{CH}\\|\\\text{C=O}\\|\\\text{OC}_2\text{H}_5\end{array} \leftrightarrow \begin{array}{c}\text{OC}_2\text{H}_5\\|\\\text{C}-\text{O}^-\\\|\\\text{CH}\\|\\\text{C=O}\\|\\\text{OC}_2\text{H}_5\end{array} \leftrightarrow \begin{array}{c}\text{OC}_2\text{H}_5\\|\\\text{C=O}\\|\\\text{CH}\\\|\\\text{C}-\text{O}^-\\|\\\text{OC}_2\text{H}_5\end{array}$$

Alternatively, the anion may be represented by a single diagram embodying the character of all three forms:

$$\begin{array}{c}\text{OC}_2\text{H}_5\\|\\\text{C}\cdots\text{O}\\\vdots\\\text{CH}-\\\vdots\\\text{C}\cdots\text{O}\\|\\\text{OC}_2\text{H}_5\end{array}$$

By virtue of the pair of electrons which originally bonded the proton to the molecule, and which remain with the anion when the proton is removed, the malonate anion is nucleophilic in character and can displace halide ions from alkyl halides (Section 26.3). In so doing it behaves as though the negative charge were largely localized on the carbon atom and a carbon–carbon bond is formed.

$$^-\text{CH}(\text{COOC}_2\text{H}_5)_2 + \text{R}-\text{Hal} \rightarrow \text{R}-\text{CH}(\text{COOC}_2\text{H}_5)_2 + \text{Hal}^-$$

For example:

$$^-\text{CH}(\text{COOC}_2\text{H}_5)_2 + \underset{\text{but-1-yl bromide}}{\text{CH}_3\text{CH}_2\text{CH}_2\text{CH}_2\text{Br}} \rightarrow$$
$$\underset{\text{diethyl but-1-ylmalonate}}{\text{CH}_3\text{CH}_2\text{CH}_2\text{CH}_2-\text{CH}(\text{COOC}_2\text{H}_5)_2} + \text{Br}^-.$$

This is a very useful way of making carbon–carbon bonds and, in combination with the ready decarboxylation of the free dicarboxylic acid, provides a useful method for synthesizing carboxylic acids of general formula $\text{R}-\text{CH}_2\text{COOH}$:

$$^-\text{CH}(\text{COOC}_2\text{H}_5)_2 + \text{R}-\text{Hal} \rightarrow \text{R}-\text{CH}(\text{COOC}_2\text{H}_5)_2 + \text{Hal}^-;$$
$$\text{RCH}(\text{COOC}_2\text{H}_5)_2 + \text{H}_2\text{O} + \text{NaOH} \xrightarrow{\text{(hydrolysis)}} \text{RCH}(\text{COONa})_2 + \text{C}_2\text{H}_5\text{OH};$$
$$\text{RCH}(\text{COONa})_2 + 2\text{HCl} \longrightarrow \text{RCH}(\text{COOH})_2 \xrightarrow{\text{heat}} \text{RCH}_2\text{COOH} + \text{CO}_2.$$

28.3.3 Ethyl Acetoacetate (Acetoacetic Ester)

Like diethyl malonate, ethyl acetoacetate has an active methylene group situated between two carbonyl groups, $\text{CH}_3\text{COCH}_2\text{COOC}_2\text{H}_5$. One of the carbonyl groups is ketonic, however, as distinct from the other which is part of the carboxylic ester function. The compound forms a sodium salt with sodium ethoxide and the resulting anion is stabilized by delocalization in the same way as in the malonate ion:

$$CH_3COCH_2COOC_2H_5 + C_2H_5O^- \rightarrow CH_3CO\overset{-}{C}HCOOC_2H_5 + C_2H_5OH.$$

$$\begin{array}{c} CH_3 \\ | \\ C=O \\ | \\ ^-CH \\ | \\ C=O \\ | \\ OC_2H_5 \end{array} \leftrightarrow \begin{array}{c} CH_3 \\ | \\ C-O^- \\ | \\ CH \\ | \\ C=O \\ | \\ OC_2H_5 \end{array} \leftrightarrow \begin{array}{c} CH_3 \\ | \\ C=O \\ | \\ CH \\ | \\ C-O^- \\ | \\ OC_2H_5 \end{array}$$

$$\begin{array}{c} CH_3 \\ | \\ C\cdots O \\ | \\ CH\; - \\ | \\ C\cdots O \\ | \\ OC_2H_5 \end{array}$$

The free acid is readily decarboxylated, like malonic acid:

$$CH_3COCH_2COOH \xrightarrow{heat} CH_3COCH_3 + CO_2.$$

For some time the structure of ethyl acetoacetate was a subject of discussion. Whilst it shows many of the characteristic ketone properties it also has properties which indicate that its structure is $CH_3-\underset{\underset{OH}{|}}{C}=CH-COOC_2H_5$ rather than the isomeric ketone structure, $CH_3\underset{\underset{O}{\|}}{C}CH_2COOC_2H_5$. The problem was resolved, however, when it was realized that both of these forms are present in ethyl acetoacetate and are in equilibrium one with the other:

$$\underset{\text{keto form}}{CH_3\underset{\underset{O}{\|}}{C}CH_2COOC_2H_5} \rightleftharpoons \underset{\text{enol form}}{CH_3\underset{\underset{OH}{|}}{C}=CHCOOC_2H_5}.$$

Ethyl acetoacetate, in equilibrium at room temperature, contains about 7.5 percent of the enol form. Conversion of the keto to the enol form merely requires the removal of a proton from the active methylene group and its subsequent return to the oxygen atom.

$$\begin{array}{c} CH_3 \\ | \\ C=O \\ | \\ H-CH \\ | \\ C=O \\ | \\ OC_2H_5 \end{array} \rightleftharpoons \begin{array}{c} CH_3 \\ | \\ C\cdots O \\ | \\ CH\; - \\ | \\ C\cdots O \\ | \\ OC_2H_5 \end{array} +H^+ \rightleftharpoons \begin{array}{c} CH_3 \\ | \\ C-OH \\ \| \\ CH \\ | \\ C=O \\ | \\ OC_2H_5 \end{array}$$

This phenomenon is known as **tautomerism** and the keto and enol forms are known as **tautomers**.

Simple aldehydes and ketones containing α-hydrogens also exhibit keto–enol tautomerism (see discussion of alkyne hydration, Section 22.2.3) although the amount of enol form present at equilibrium is usually too low to be directly detected. Acetone, for example, contains less than one part in 10^5 of its enol form at equilibrium.

$$CH_3COCH_3 \rightleftharpoons CH_3C(OH)=CH_2.$$

The enol form of acetone is none the less important in acetone chemistry. It is believed to be the reactive species in halogenation reactions, (Section 8.5) and accounts for the effect of carbonyl groups in directing halogenation to take place preferentially at α-carbon atoms (Section 25.3.1).

$$CH_3CH_2COC_6H_5 \xrightarrow{H+} CH_3CH=C(OH)C_6H_5 \xrightarrow{Cl_2} CH_3CH(Cl)COC_6H_5 + HCl$$
$$CH_3CH_2COOH \xrightarrow{Cl_2} CH_3CH(Cl)COOH.$$

This control, or "directing influence" of, the carbonyl group renders direct halogenation of this type of aliphatic compound synthetically useful (compare alkane halogenation, Section 22.2).

As an ester, ethyl acetoacetate can be hydrolysed by alkali but the influence of the two carbonyl groups is such that the main acetoacetic acid skeleton breaks up in the process and the free acetoacetic acid (or its salt) is not obtained unless very mild conditions are used. If the ester is heated with dilute aqueous potassium hydroxide, acetone is obtained (ketonic hydrolysis):

$$CH_3COCH_2\text{—}COOC_2H_5 + 2KOH \rightarrow CH_3COCH_3 + K_2CO_3 + C_2H_5OH.$$

With concentrated, alcoholic potassium hydroxide, however, the molecule breaks at the other side of the active methylene group and acetic acid is obtained as its potassium salt (acid hydrolysis):

$$CH_3CO\text{—}CH_2COOC_2H_5 + 2KOH \rightarrow 2CH_3COOK + C_2H_5OH.$$

As in the case of diethyl malonate, the sodium salt of ethyl acetoacetate reacts with alkyl halides to produce carbon–carbon bonds and, with subsequent hydrolysis of the product—ketonic or acid—useful syntheses of ketones and carboxylic acids emerge:

$$CH_3COCH_2COOC_2H_5 + NaOC_2H_5 \rightarrow CH_3CO\bar{C}HCOOC_2H_5 + C_2H_5OH;$$

$$CH_3CO\bar{C}HCOOC_2H_5 + R\text{—}Hal \rightarrow CH_3CO\overset{R}{\underset{|}{C}}HCOOC_2H_5 + Hal^-;$$

$$CH_3CO\overset{R}{\underset{|}{C}}HCOOC_2H_5 \begin{cases} \xrightarrow{\text{dilute KOH}} CH_3COCH_2R + C_2H_5OH + K_2CO_3 \\ \xrightarrow[\text{alcoholic KOH}]{\text{hot conc.}} CH_3COOK + RCH_2COOK + C_2H_5OH. \end{cases}$$

Two principal features, then, due to interaction between the functional groups, are apparent in the chemistry of these compounds which contain a carboxylic acid and a second carbonyl group attached to the same carbon atom. Hydrogen atoms which are also attached to that carbon atom are rendered acidic by the two carbonyl groups, and the carboxylic acid function is readily lost, as CO_2, when heated.

$$\begin{array}{c}|\\ \text{CO}\\ |\\ \text{CH}_2\\ |\\ \text{CO}\\ |\end{array} \xrightarrow{\text{base}} \begin{array}{c}|\\ \text{CO}\\ |\\ \text{CH}^-\\ |\\ \text{CO}\\ |\end{array} \qquad \textit{acidity}$$

$$\begin{array}{c}|\\ \text{COOH}\\ |\\ \text{CH}_2\\ |\\ \text{CO}\\ |\end{array} \xrightarrow{\text{heat}} \begin{array}{c}\text{CH}_3 + \text{CO}_2\\ |\\ \text{CO}\\ |\end{array} \qquad \textit{decarboxylation}$$

If a C—H group is adjacent to only one carbonyl group, although the effect is much weaker, that C—H group is also rendered weakly acidic. Thus, in a simple aldehyde such as acetaldehyde, the hydrogens α- to the carbonyl group can be removed by aqueous alkali. The anion produced is stabilized by delocalization but, as there is only one carbonyl group involved, the delocalization is much less effective than in the case of the malonate or acetoacetate anions, and acetaldehyde is much more weakly acidic than either of the other two compounds. Aqueous alkali can only remove protons from a small proportion of the acetaldehyde molecules present, i.e. the equilibrium lies well over to the left.

$$\text{CH}_3\text{CHO} + \text{OH}^- \rightleftharpoons {}^-\text{CH}_2\text{CHO} + \text{H}_2\text{O}$$

$$^-\text{CH}_2\!\!-\!\!\underset{\text{H}}{\text{C}}\!\!=\!\!\text{O} \leftrightarrow \text{CH}_2\!\!=\!\!\underset{\text{H}}{\text{C}}\!\!-\!\!\text{O}^-.$$

Although aqueous alkali can thus only produce a low concentration of anions from aldehydes and ketones (containing α-hydrogens), these anions are highly reactive nucleophiles and will rapidly attack the carbonyl carbon atoms of undissociated aldehyde (or ketone) molecules. The result is a reaction in which one of the original aldehyde or ketone molecules adds across the carbonyl group of another—the aldol reaction (Section 23.3.2).

$$\text{CH}_3\text{CHO} + \text{OH}^- \rightleftharpoons {}^-\text{CH}_2\text{CHO} + \text{H}_2\text{O}$$

$$\text{CH}_3\!\!-\!\!\underset{\text{H}}{\text{C}}\!\!=\!\!\text{O} + {}^-\text{CH}_2\text{CHO} \rightarrow \text{CH}_3\!\!-\!\!\underset{\text{H}}{\overset{\text{O}^-}{\text{C}}}\!\!-\!\!\text{CH}_2\text{CHO}$$

$$\text{CH}_3\!\!-\!\!\underset{\text{H}}{\overset{\text{O}^-}{\text{C}}}\!\!-\!\!\text{CH}_2\text{CHO} + \text{H}^+ \rightarrow \text{CH}_3\!\!-\!\!\underset{\text{H}}{\overset{\text{OH}}{\text{C}}}\!\!-\!\!\text{CH}_2\text{CHO}.$$

28.3.4 Summary

The *carbonyl group* renders adjacent (α-) C—H groups *acidic* by stabilizing the anion resulting from proton removal. This leads to the phenomenon of *keto–enol tautomerism*. The anions are powerfully nucleophilic, leading to useful *carbon–carbon bond syntheses* in the laboratory (through diethyl malonate, ethyl acetoacetate, and aldol reactions) and in biological systems (aldol reactions). A carbonyl group which is β- to a carboxylic acid group leads to ready loss of the acid group as carbon dioxide—*decarboxylation*.

28.4 α,β-UNSATURATED ALDEHYDES AND KETONES

The use of the Greek letter α to signify position in carbonyl compounds has been explained (Section 23.3). Carbonyl compounds containing a carbon–carbon double bond between the α and β-positions are referred to as α,β-unsaturated carbonyl compounds.

When two separate multiple bonds, present in the same molecule, are separated only by a single bond, we say that they are "conjugated". It has been known for a long time that such an arrangement is rather more stable than any other in which multiple bonds are either immediately adjacent, or separated by more than one single bond. It has also been known for a long time that such an arrangement leads to chemical consequences which would not have been expected solely on the basis of the known chemistries of the isolated multiple bonds. Crotonaldehyde (but-2-enal) is a molecule containing a carbon–carbon double bond conjugated with a carbon–oxygen double bond. It has the structure $CH_3-CH=CH-CH=O$ and is therefore an α,β-unsaturated aldehyde. Aldol and its relatives are very readily dehydrated to yield molecules such as these:

$$CH_3-\overset{\beta}{CH}-\overset{\alpha}{CH_2}CHO \xrightarrow{acid} CH_3-\overset{\beta}{CH}=\overset{\alpha}{CH}-CH=O$$
$$|$$
$$OH$$

aldol crotonaldehyde
(3-hydroxybutanal) (but-2-enal)

$$(CH_3)_2\overset{\beta}{C}-\overset{\alpha}{CH_2}COCH_3 \xrightarrow[heat]{I_2} (CH_3)\overset{\beta}{C}=\overset{\alpha}{CH}COCH_3.$$
$$|$$
$$OH$$

diacetone alcohol mesityl oxide
(4-hydroxy-4- (4-methylpent-3-ene-
methylpentan-2-one) 2-one)

The additional stability of these molecules, relative to others with the same multiple bonds which are not conjugated, can be associated with electron delocalization. There are two arrangements which we can write to indicate electron distribution in, say, the crotonaldehyde molecule:

$$CH_3-CH=CH-CH=O \quad \text{and} \quad CH_3-\overset{+}{CH}-CH=CH-O^-.$$

As a single species we could represent the molecule as $CH_3-\overset{\delta+}{CH}\text{---}CH\text{---}CH\text{---}\overset{\delta-}{O}$.

A third arrangement, $CH_3-\overset{-}{CH}-CH=CH-O^+$, can be rejected because the more electronegative oxygen atom will tend to polarize the electrons towards itself rather than in the opposite direction.

(Using the concept of hybridization of atomic orbitals given in Section A.2, we can construct a molecular-orbital model of the bonding in this molecule. Allocating sp^2 hybridization to the carbon atoms involved in multiple bonding, we see that the unhybridized p-orbitals of the α-carbon and the carbonyl carbon can each overlap in two directions. We suppose that they do in fact do this, and the situation is closely related to that which is described in connection with benzene (Section A.2).

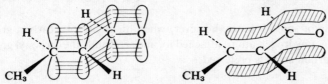

Instead of there being two separate localized π-bonds, there is one long π-orbital extending over four atoms, polarized towards the more electronegative oxygen atom. This model may be compared with the representation indicated above, which was derived from the two canonical forms. It will be apparent from this model, though, that the four atoms involved in the π-bonding must lie in one plane in order that the unhybridized *p*-orbitals can be parallel and thus interact. Any feature of molecules, containing apparently conjugated double bonds, which prevents this coplanarity will reduce the *p*-orbital overlap and the delocalization energy. The double bonds will not be effectively conjugated.)

It is interesting to note that, because of this type of interaction, the polarizing influence of an oxygen atom can be "felt" at considerable distances away from that atom. Thus, in a system such as A below, the carbon atom remote from oxygen would bear a significant positive charge. In similar systems without conjugated multiple bonds, such as B, the polarizing influence of the oxygen atom would not be significantly felt (by transmission through σ-bonds—inductive effect) further away than carbon atom 2.

$$\text{A} \quad \underset{9}{C}=\underset{8}{C}-\underset{7}{C}=\underset{6}{C}-\underset{5}{C}=\underset{4}{C}-\underset{3}{C}=\underset{2}{C}-\underset{1}{C}=O$$

$$\text{B} \quad \underset{9}{C}-\underset{8}{C}-\underset{7}{C}-\underset{6}{C}-\underset{5}{C}-\underset{4}{C}-\underset{3}{C}-\underset{2}{C}-\underset{1}{C}-O-R$$

We have seen (Section 26.5) that an alkene is attacked by electrophilic species which demand a share of the high electron density in the carbon–carbon double bond. By contrast, a carbonyl group is usually attacked at the carbon atom by a nucleophile which contributes electrons to the positive carbon atom and displaces electrons on to oxygen (Section 26.4). With α,β-unsaturated aldehydes and ketones, however, electron density is drawn out of the carbon–carbon double bond by the carbonyl group and the carbon–carbon double bond now assumes some of the chemical character of a carbonyl group. Thus it is susceptible to nucleophilic attack at the positively charged β-carbon atom as well as at the carbonyl carbon atom, and attack may take place at either site according to the structure of the molecule in question. Thus, for example, crotonaldehyde reacts as a normal aldehyde with hydrogen cyanide to yield a cyanohydrin, attack by the nucleophilic cyanide ion taking place preferentially at the carbonyl carbon atom:

$$CH_3CH=CH-\underset{H}{\overset{}{C}}=O \;+\; (:)\bar{C}N: \;\rightarrow$$

$$CH_3-CH=CH-\underset{CN}{\overset{O^-}{C}}-H \;\xrightarrow{H^+}\; CH_3-CH=CH-\underset{CN}{\overset{OH}{C}}-H.$$

In the case of some ketones, however, where there are two carbon groups (each larger than a hydrogen atom) attached to, and screening, the carbonyl group, attack at the β-carbon atom is preferred:

$$:N\equiv\bar{C}(:) \quad CH_2=CH-\underset{\underset{CH_3}{|}}{C}=O \rightarrow N\equiv C-CH_2-CH=\underset{\underset{CH_3}{|}}{C}-O^-$$

$$\downarrow H^+$$

$$N\equiv C-CH_2-CH_2-\underset{\underset{CH_3}{|}}{C}=O.$$

Note that the hydrogen ion attaches to the α-carbon atom and not to oxygen. If it attached to oxygen then an unstable enol (Section 28.3) would result and this would rearrange to the product observed:

$$N\equiv C-CH_2CH=\underset{\underset{CH_3}{|}}{C}-OH \rightarrow N\equiv C-CH_2CH_2-\underset{\underset{CH_3}{|}}{C}=O.$$

Other reagents which add to the carbon group of simple aldehydes and ketones also undergo this type of addition with α,β-unsaturated ketones.

Summary

When two double bonds in the same molecule are separated by *one* single bond they are said to be *conjugated*. When a carbon–carbon double bond is conjugated with a carbonyl group (carbon–oxygen double bond) the interaction is such that the carbon–carbon double bond assumes some of the character of the polar carbonyl group, and may undergo addition reactions normally associated with carbonyl groups.

28.5 INTERACTING FUNCTIONAL GROUPS IN AROMATIC MOLECULES

28.5.1 Nitrophenols

In Chapter 23 (Section 23.1.2) the enhanced acidity of phenols, relative to alcohols, was stressed and it was pointed out that this was due to the ability of the benzene ring to accept and delocalize the negative charge from oxygen in the anion. This charge delocalization may be described in terms of five canonical forms:

The overall representation would be

(The "partial" double bond character, represented by the dotted line, in terms of the molecular-orbital model described in Section A.2, corresponds to *p*-orbital overlap which extends from the ring to an oxygen *p*-orbital (occupied by two electrons) and permits the oxygen *p*-electrons, and consequently the negative charge, to move into the ring.

Note, however, that the negative charge is shared only with the 2- and 4-carbon atoms—*ortho-* and *para-* positions.)

It must be stressed at this point that in phenol itself, a lone pair of electrons on the oxygen is delocalized, which is energetically favourable, although in this case two opposite charges have to be separated and this is energetically unfavourable.

Delocalization in the phenolate ion does not involve separation of opposite charge, and is thus more effective in the phenolate ion than in phenol. It is able to contribute more to the stability of the ion than to the stability of phenol. If it were equally effective in both, then it would not influence the acid strength of phenol.

We are now in a position to consider the effect of the nitro groups in nitrophenols upon the hydroxyl function. A nitro group may be formally written

, with one double bond to oxygen and one dative bond. Both oxygens are equivalent, however, and the nitro group is perhaps best represented as

Because oxygen is more electronegative than carbon and nitrogen, the N=O group is able to withdraw electrons (π-electrons) from systems with which it is conjugated. Thus nitrobenzene, for instance, can be represented as follows:

(It is convenient here, as later, to omit such forms as

and

which do not reflect interaction between the ring and the functional group.)

Provided that the nitro group is in a 2-(*ortho*-) or a 4-(*para*-) position—where the oxygen makes negative charge available—it can spread out the negative charge of a phenolate ion beyond the ring. Canonical forms for the 2-nitrophenolate ion can be written:

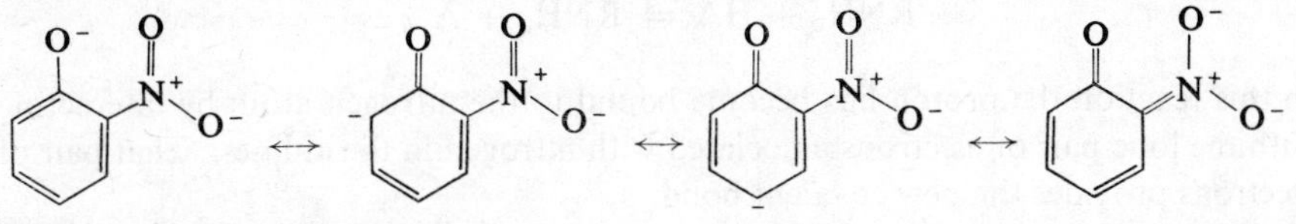

And for the 4-nitrophenolate ion:

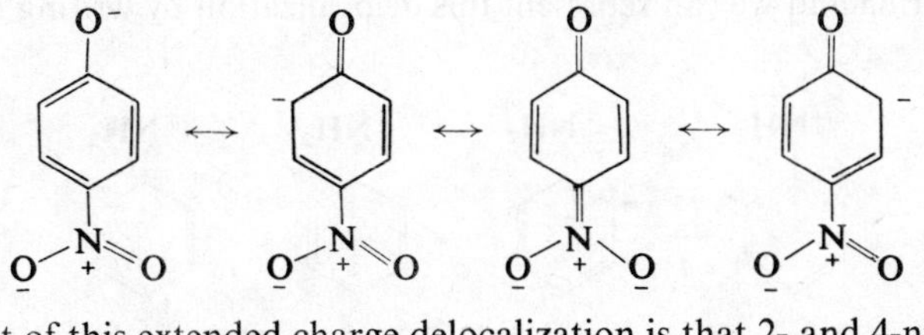

The net result of this extended charge delocalization is that 2- and 4-nitrophenolate ions are stabilized to an even greater extent than is the simpler unsubstituted phenolate ion, and 2- and 4-nitrophenols are consequently stronger acids than phenol. 2,4,6-Trinitrophenol is quite a strong acid and is more commonly known as picric acid (Table 28.1).

Table 28.1

pK_a (Section 9.3.4)	of phenol	9.89
pK_a	of 2-nitrophenol	7.17
pK_a	of 3-nitrophenol	8.28
pK_a	of 4-nitrophenol	7.15
pK_a	of 2,4,6-trinitrophenol (picric acid)	0.38
pK_a	of phosphoric acid	2.12

3-Nitrophenol, as an acid, is strengthened less, relative to 2- and 4-nitrophenols, because canonical forms involving charge delocalization on to the nitro group cannot be written and the stability of the 3-nitrophenolate ion is not enhanced, relative to the simpler phenolate ion, by delocalization.

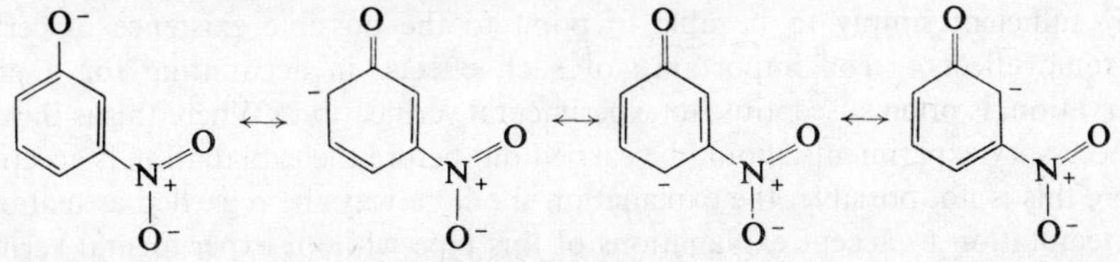

However, 3-nitrophenol is still a stronger acid than phenol, probably because the electronegative nitrogen and oxygen atoms set up a general drift of electrons out of the ring (inductive effect), although this cannot be as effective as the π-electron delocalization in the 2- and 4-nitrophenols.

Base strengths, as well as acid strengths, may also be affected by delocalization, of course. Consider the basic character of the two primary amines, phenylamine (aniline), $C_6H_5NH_2$, and methylamine, CH_3NH_2. Both of these compounds are bases by virtue of the fact that they can accept a proton from an acid, and we can represent their reactions with acids by the general equation:

$$RNH_2 + HA \rightleftharpoons R\overset{+}{N}H_3 + A^-.$$

In this reaction the proton has become bound to the nitrogen atom by interacting with the lone pair of electrons associated with nitrogen in the amines. That pair of electrons provides the new covalent bond.

In phenylamine the lone pair of electrons is delocalized, to some extent, over the aromatic ring and we can represent this delocalization by writing the following forms:

[Resonance structures of aniline showing delocalization of the nitrogen lone pair into the aromatic ring]

A certain delocalization energy is associated with this. When the lone pair becomes involved in bonding a proton to nitrogen, it inevitably has to become localized in that bond and the delocalization energy associated with it in the phenylamine molecule is therefore lost. In methylamine, no such delocalization is possible and there is therefore no delocalization energy to be lost on protonation. Thus, other things being equal, there will be a greater difference in free energy between methylamine and protonated methylamine, than between phenylamine and protonated phenylamine. In reaction with a common acid, HA, the equilibrium for methylamine will lie further over towards the protonated form, therefore, than will be the case for phenylamine. In other words, equilibrium constant K_1 will be greater than equilibrium constant K_2, and methylamine will be a stronger base than phenylamine:

$$CH_3NH_2 + HA \underset{}{\overset{K_1}{\rightleftharpoons}} CH_3\overset{+}{N}H_3 + A^-;$$
$$C_6H_5NH_2 + HA \underset{}{\overset{K_2}{\rightleftharpoons}} C_6H_5\overset{+}{N}H_3 + A^-.$$

Phenylamine, a typical aromatic amine, is indeed observed to be a weaker base than methylamine, a typical aliphatic amine (Section 24.1.2), and we believe that this is largely due to this delocalization effect.

A cautionary word might be appropriate at this point in so far as it is not really sufficient simply to be able to point to the possible existence of certain electronic effects. The importance of such effects, in accounting for a given observation, is often susceptible to experimental verification. Where this is the case the necessary experiments should be carried out before the explanation is accepted. Where this is not possible, the explanation should always be regarded as tentative. The temptation to accept explanations of this type without experimental verification is very strong and there are cases where chemists (who are, after all, human) have yielded to this temptation. A case in point concerns the relative acidities of the halogenated acetic acids. Thus, for example, chloroacetic acid ($CH_2ClCOOH$) is a stronger acid than acetic acid (CH_3COOH). The explanation of this, accepted for many years and quoted in many textbooks, was that the negative charge on the chloroacetate ion was stabilized to some extent by the electronegative halogen atom which, in polarizing the carbon–carbon bond (inductive effect), was able partially to delocalize the negative charge:

$$Cl \leftarrow CH_2 \leftarrow CO_2^-.$$

The direction of the polarization, or "electron pull", is indicated by the arrows. This would mean that the chloroacetate ion was more stable than the acetate ion and that chloroacetic acid would ionize to a greater extent than would acetic acid. In other words the chlorinated compound would be the stronger acid.

A careful study of the temperature dependence of the relevant equilibrium constants, however, carried out quite recently, yielded information about the thermodynamics of these processes which indicated that this could not possibly be the explanation. The important differences are associated with entropy rather than with enthalpy, and the differences in equilibrium constants, and therefore in acid strengths, are probably associated with the solvation of the various species involved. They are apparently not directly related to the stabilization of negative charge within the anions by delocalization effects.

28.5.2 Nitrated Halogenobenzenes

We have already noted (e.g. Section 25.2.1) that halogenobenzenes (e.g. chlorobenzene) are very resistant towards displacement of halide ion by nucleophiles. That is, they do not undergo normal nucleophilic substitution reactions such as hydrolysis:

$$C_6H_5Cl + H_2O \text{ do not react to give } C_6H_5OH + HCl.$$

This can be attributed, at any rate in part, to the development of double bond character between halogen and carbon which helps to prevent hydrolysis from occurring by mechanisms such as are associated with alkyl halides. We can describe this effect in terms of the delocalization of a lone pair of electrons from chlorine:

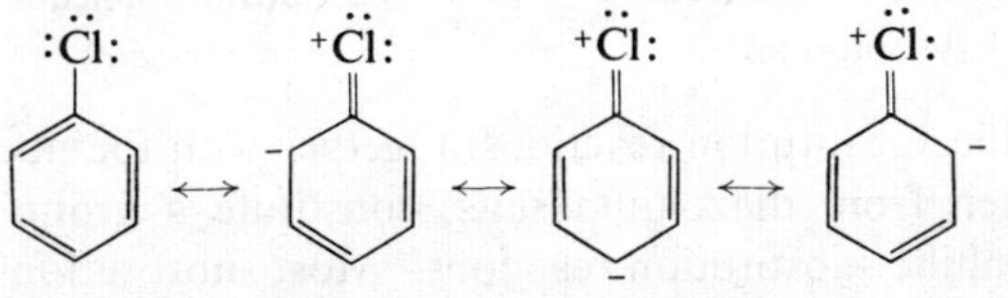

(Using the molecular-orbital model described in the Section A.2, we can build up a picture of this delocalization in much the same way as we did for phenolate anion in Section 28.5.1. The interacting lone pair on chlorine must occupy a *p*-orbital perpendicular to the ring. A similar model can be constructed for vinyl chloride—see below—and the reader could attempt this.)

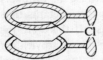

The resistance of vinyl halides to nucleophilic displacement can be explained on a similar model. In vinyl chloride (chloroethene), for example, $CH_2\!=\!CHCl$, we have similar delocalization and strengthening of the carbon–chlorine bond (you could attempt to write canonical forms for vinyl chloride—there are two).

A very interesting change in behaviour is observed, however, when the halogenobenzene contains nitro groups in the 2- position (*ortho*)- or the 4- position (*para*-). A 4-nitro group, for example, enables nucleophiles to displace chlorine from an aromatic ring.

$$\text{4-nitrochlorobenzene} + \text{OH}^- \rightarrow \text{4-nitrophenol} + \text{Cl}^-;$$

$$\text{(4-nitrochlorobenzene)} + \bar{\text{O}}\text{C}_2\text{H}_5 \rightarrow \text{4-nitroethoxybenzene} + \text{Cl}^-.$$

The carbon–chlorine bond in 2,4,6-trinitrochlorobenzene (picryl chloride, the acid chloride derived from 2,4,6-trinitrophenol, picric acid), a compound containing three nitro groups each in an activating position (2-, 4-, and 6-), is very readily broken by nucleophiles.

$$\text{2,4,6-trinitrochlorobenzene (picryl chloride)} + \text{H}_2\text{O} \rightarrow \text{2,4,6-trinitrophenol (picric acid)} + \text{HCl}.$$

These nucleophilic substitution reactions, together with the nucleophilic displacements of nitrogen from diazonium salts, constitute a group of relatively rare aromatic nucleophilic substitution reactions. Most substitutions in aromatic rings are electrophilic (Section 26.5).

This important effect of the nitro group on aromatic halogen compounds, like the effect upon the acidity of phenols (Section 28.5.1), depends upon the ability of the nitro groups to delocalize negative charge. However, this is not a question of thermodynamics and equilibrium constants but rather one of activation energies and rate constants. It is not because the products, phenol and hydrogen chloride, have a higher free energy than the reactants that chlorobenzene fails to undergo hydrolysis in aqueous solutions. If this were the reason, then we should expect hydrogen chloride to react with phenol to yield chlorobenzene and water, and this does not occur either. One must, therefore, conclude that reaction does not take place because the free energy of activation (Section 8.7) is too large, and in seeking an explanation of this effect of the nitro group we must look for factors which would enable the group to lower the free energy of activation of the reaction.

There are good grounds for believing that this type of aromatic substitution proceeds in two steps, with the first of these being the slower and therefore determining the overall rate of reaction (rate determining step):

$$HO^- + \underset{NO_2}{\underset{|}{C_6H_4}}\!-\!Cl \xrightarrow{slow} \underset{NO_2}{\underset{|}{C_6H_4}}(HO)(Cl)^- \xrightarrow{fast} \underset{NO_2}{\underset{|}{C_6H_4}}\!-\!OH + Cl^-.$$

It is the activation energy of this first step which is crucial to the problem. The product of this step bears a negative charge which may be delocalized by a nitro group in the 2- or 4-positions.

[Resonance structures of the intermediate with HO, Cl on one carbon and NO₂ group delocalizing the negative charge, followed by the composite structure with $\delta-$ on the ring carbons and $\delta-$ on the nitro oxygens.]

The resultant stabilization of the ion does not, however, affect the rate at which it is formed. The effect of stabilizing the ion increases the equilibrium constant of its formation reaction by increasing the free energy change, but it cannot affect the free energy of activation, and consequently not the rate.

Consider reaction occurring between a particular hydroxide ion and a particular 4-nitrochlorobenzene molecule. The free energy profile for the reaction (Section 8.7) is shown in Fig. 28.1. Gradually the new carbon–oxygen bond forms,

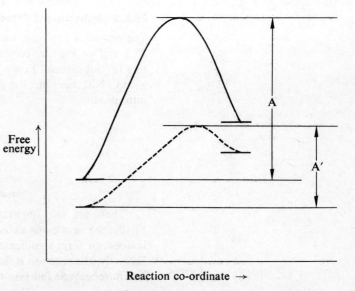

Fig. 28.1 Comparison of free energy profiles for the reactions of hydroxide ion with chlorobenzene (solid line, Y = H) and with 4-nitrochlorobenzene (broken line, Y = NO₂). A is the free energy of activation of the former, A′ that of the latter. A is greater than A′.

28.5 INTERACTING FUNCTIONAL GROUPS IN AROMATIC MOLECULES

and the negative charge builds up in the ring as, in terms of energy, the reacting molecules climb up and over the activation "hump" to form the new ion. At all stages during the formation of the new carbon–oxygen bond there will be some negative charge in the ring which can be delocalized on to the nitro group and therefore the reacting species, at all stages of bond formation, will achieve some degree of stabilization. Thus the nitro group lowers the whole energy profile curve, relative to that for the reaction of chlorobenzene with hydroxide ion (Fig. 28.1), and the free energy of activation is reduced as well as the energy of the resultant ion. The free energy of activation for the chlorobenzene reaction is so high as to prevent hydrolysis from occurring altogether by this mechanism whereas with 4-nitro-chlorobenzene reaction may proceed at a reasonable pace. Neither this hydrolysis mechanism, nor that available to alkyl halides, is available to chlorobenzene.

This effect of the nitro group is not unlike that of the carbonyl group which renders the carbon–chlorine bond of acetyl chloride more reactive towards nucleophiles than that of chloroethane (Section 26.4). The new bond can form before the old one has to break.

The species in the process of bond formation which corresponds to the top of the free energy profile is the transition state (Section 8.7) which, in this case, we may represent diagrammatically with a partially formed bond and a partially developed charge. The reagents before their interaction, are said to be in their ground states.

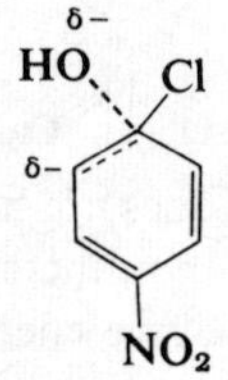

Factors stabilizing the transition state of a given reaction, relative to the ground state, will lower the free energy of activation and increase reaction rate, whereas those raising the free energy of the transition state will have the opposite effect.

28.5.3 Substituent Effects in Aromatic Substitution

The nature of aromatic substitution reactions is discussed in Sections 22.2.5 and 26.5, and we are here concerned with the effect of a group already attached to the ring (a "substituent") on a substitution reaction performed to introduce a second group. Nitrobenzene, for example, may be further nitrated to introduce a second nitro group:

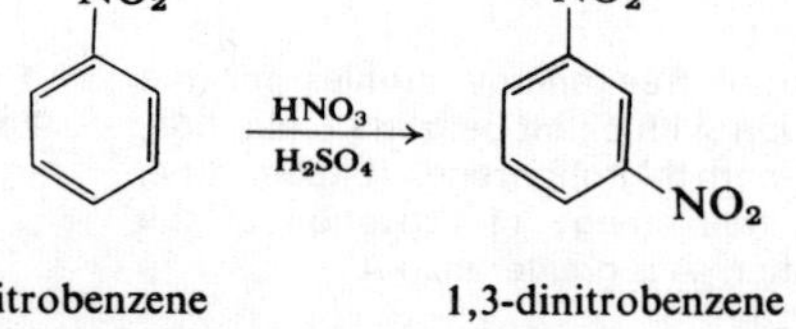

nitrobenzene   1,3-dinitrobenzene

There are two important points which arise from this particular example. Firstly, the new group enters the 3-position to form 1,3- (or *meta*-)dinitrobenzene. It does not form significant amounts of 1,2-(*ortho*-) or 1,4-(*para*-)dinitrobenzene. Secondly, the reaction is slower than is the nitration of benzene itself (both benzene and nitrobenzene fall into the category of less reactive aromatics which nitrate at rates dependent upon the concentration and nature of the aromatic itself, Section 8.5).

To understand the reason for this slow, selective nitration reaction, it is necessary to consider the mechanism of the reaction in a manner similar to that in which we considered the mechanism of nitrochlorobenzene hydrolysis.

The active species in the reaction is believed to be the nitronium ion, NO_2^+, produced by interaction between sulphuric and nitric acids:

$$2H_2SO_4 + HNO_3 \rightarrow NO_2^+ + H_3O^+ + 2HSO_4^-.$$

The nitronium ion is the electrophile which attacks the ring and forms the nitro compound in two steps:

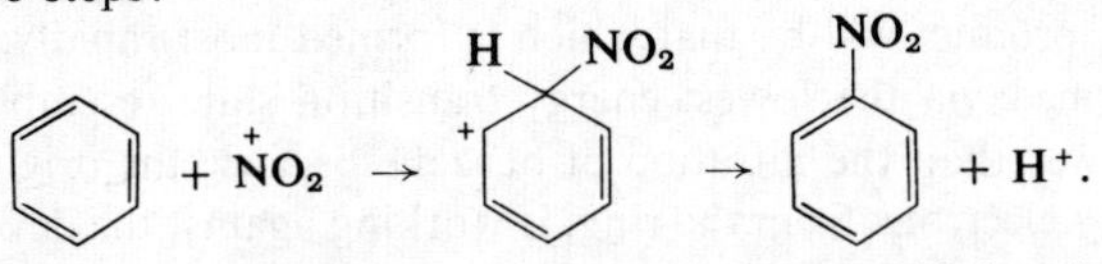

$$\text{benzene} + \overset{+}{NO_2} \rightarrow \text{intermediate} \rightarrow \text{nitrobenzene} + H^+.$$

The first step is the slower and therefore rate determining, and it involves a transition state with developing positive charge, rather like the ion produced. Factors which can stabilize this positive charge will reduce the free energy of the transition state and therefore the free energy of activation of the reaction is lowered and its speed increased. Similarly, factors which "destabilize" the transition state (raise its free energy) will slow down the reaction. As in the nitrochlorobenzene reaction, factors stabilizing the charge in the ion produced by the rate determining step will also stabilize the charge in the transition state, and so we can estimate the stabilizing influences in the transition state by considering those in the ion.

For the nitration of benzene, the positive charge in the intermediate ion is delocalized as follows:

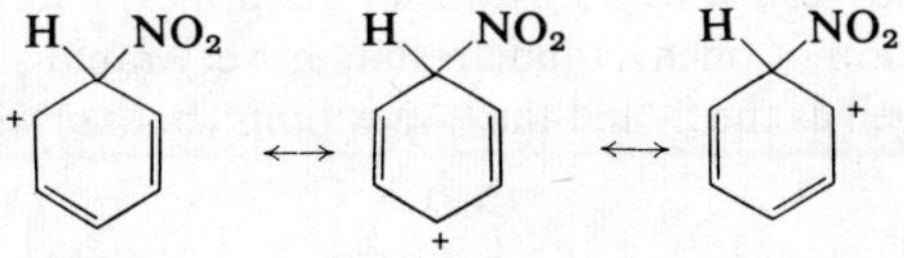

In considering the nitration of nitrobenzene, however, there are three possible intermediate ions according to whether attack has taken place at the 2-, the 3-, or the 4- position:

A

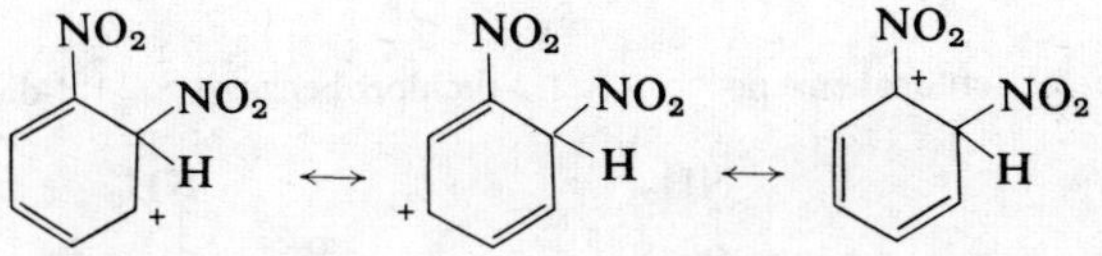

B

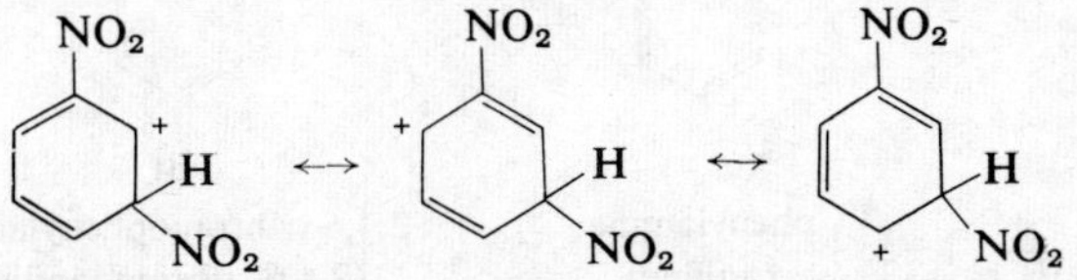

C

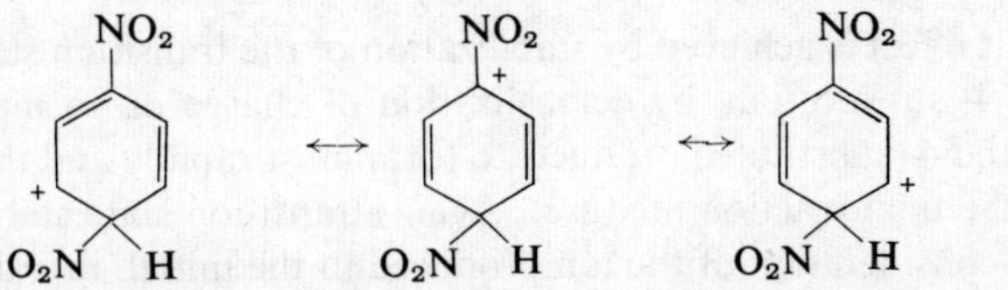

Note that in ions A and C, positive charge is developed at the 1- position which is already being drained of electrons by the original nitro group. Such ions would involve positive charges on adjacent atoms, which is an unstable (high energy) situation,

$$\text{O} \overset{+}{\underset{\underset{+}{|}}{\text{N}}} \bar{\text{O}} \leftarrow \text{mutual repulsion}$$

and therefore reaction via intermediate ions A and C is going to be slower than that via B. Because the overall reaction is irreversible, it is kinetically controlled and the major product will be that which is formed most rapidly, i.e. 1,3-dinitrobenzene is formed via the lowest energy transition state resembling ion B. The reaction is slower than the nitration of benzene because the original nitro group, by withdrawing electrons from the ring, is working against the development of *any* further positive charge in the ring.

The nitro group exerts this influence on all electrophilic substitution reactions (e.g. nitration, bromination, sulphonation) and other electron-withdrawing substituents, such as —$\overset{\text{H}}{\underset{}{\text{C}}}$=O in benzaldehyde, and —$\overset{\text{OH}}{\underset{}{\text{C}}}$=O in benzoic acid, are similarly "*meta-* directing".

Substituents having electrons which can be donated to the ring are able to direct attacking electrophiles to the 2- and 4- positions. Groups with lone pairs, for example, such as —NH_2 in phenylamine (aniline), —OH in phenol, and —Cl in chlorobenzene, direct new substituents to the 2- and 4- positions. Further chlorination of chlorobenzene results in a mixture of 1,2(*ortho*-) and 1,4(*para*-)-dichlorobenzenes. Bromination of phenylamine results in very rapid substitution at the 4- position (*para*-) as well as the 2- and the 6- positions (both *ortho*- positions).

chlorobenzene → 1,2-dichlorobenzene + 1,4-dichlorobenzene (Cl_2 / $FeCl_3$)

phenylamine (aniline) → 2,4,6-tribromophenylamine (2,4,6-tribromoaniline) (Br_2)

This effect is achieved by stabilization of the transition states corresponding to 2- and 4- substitution, by delocalization of charge on to nitrogen, thus enabling the 2- and 4- substituted products to form most rapidly and therefore to predominate in the final reaction mixture. Again, transition state stabilization is suggested from a consideration of the ion produced in the initial, rate determining step.

Summary of common substituents:

Meta- directing (electron withdrawing)
$$-NO_2, -CHO, -COR, -COOR, -SO_3H.$$
Ortho/para- directing (electron donating)
$$-NH_2, -OH, -NR_2, -OR, -Cl, -Br.$$

28.5.4 Summary

Electron-withdrawing groups, such as nitro groups, situated in the 2-, 4-, or 6- (*ortho-* or *para-*) positions of phenols, enhance the acidity of the phenols by stabilizing the anion resulting from loss of the hydroxyl group proton. In the 3- or 5- (*meta-*) positions, electron-withdrawing groups are unable to exert this effect. Nitro groups in the 2-, 4-, or 6- positions of halogenobenzenes render the halogen atom susceptible to ready displacement by nucleophiles. Such nucleophilic substitution of halogen from aromatic rings is relatively rare. Halogenobenzenes with nitro groups in the 3- or 5- positions are inert to nucleophilic substitution, as are the unsubstituted halogenobenzenes.

The presence in a benzene ring of electron-withdrawing substituent groups, like the nitro group (e.g. $-CHO$; $-COR$; $-COOR$; $-SO_3H$), will affect electrophilic substitution reactions by directing the new group into 3- and 5- positions (*meta-*) with respect to the group already there (i.e. the nitro group), and the reaction will be slower than that with benzene itself. Groups which are attached to a benzene ring and which have lone pairs which are delocalized into the ring (e.g. $-NH_2$; $-OH$; $-NR_2$; $-OR$; $-Cl$; $-Br$), however, will direct new groups into 2-, 4-, and 6- positions (*ortho-* and *para-*) with respect to the original group, in reactions which are normally faster than the corresponding substitution in benzene itself.

PROBLEMS

1. Heating ethyl lactate ($CH_3CH(OH)COOCH_2CH_3$) gives rise to ethanol and a neutral compound of molecular weight 144. Suggest a structure for this compound.

2. Predict possible products which could arise simply from heating solutions of $ClCH_2CH_2CH_2CH_2CH_2NH_2$. Do you think that the result of heating a very concentrated solution would be any different from that obtained by heating a very dilute solution?

3. When a pair of electrodes is placed in a suitably arranged solution of glycine, the glycine migrates towards one electrode if the solution is strongly acidic, and towards the other if it is strongly alkaline. Suggest reasons for this, indicating which electrode the glycine would migrate towards in acid solution, and which in alkaline solution.

4. The following is a proposed synthetic method for preparing glycylalanine (alanine is the α-amino acid $CH_3CH(NH_2)COOH$, 2-aminopropanoic acid):
$H_2NCH_2COOH + PCl_5 \rightarrow H_2NCH_2COCl$
$H_2NCH_2COCl + H_2NCH(CH_3)COOH \rightarrow H_2NCH_2CONHCH(CH_3)COOH.$
   glycylalanine

Criticize this proposal as extensively as you can.

5. Examine the following carbohydrate molecule carefully:

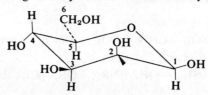

Like glucose, it contains five hydroxyl groups.
   a) Using the numbering system in the diagram, indicate what type of hydroxyl group each one is (alcoholic, primary, secondary, or tertiary; hemi-acetal etc.)
   b) Would you expect this to be a reducing sugar?
   c) Would you expect the optical rotation of this molecule to remain constant if a pure sample of it was dissolved in water?
   d) In the open-chain form, which carbon would form part of the carbonyl group?
   e) In what way does this molecule differ from β-D-glucose?
   f) Write down the structure of the anomer of this molecule.
   g) What would you expect to be the product when this molecule is treated with methanol and hydrogen chloride?

6. Which of the following structures would you expect to reduce Fehling's solution:

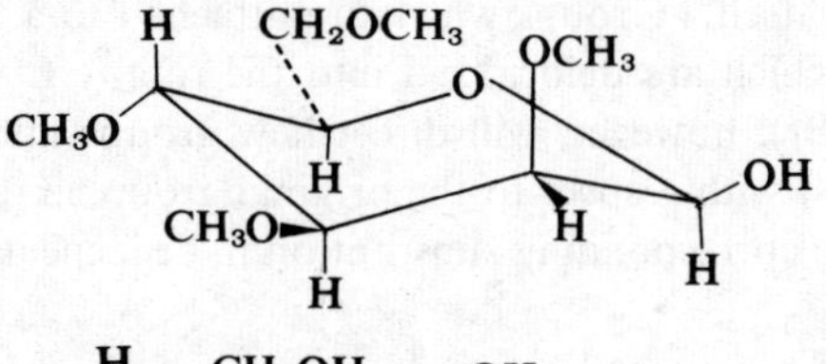

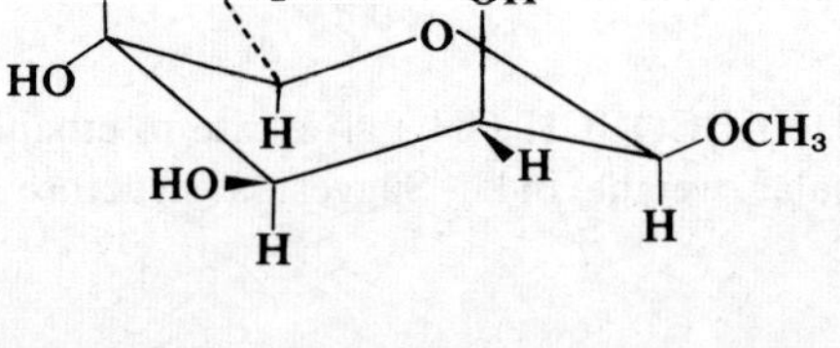

c)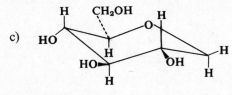

d) 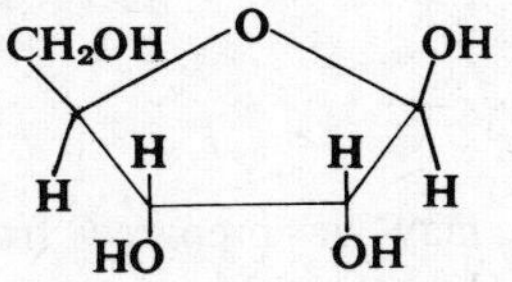 [structure]

e) 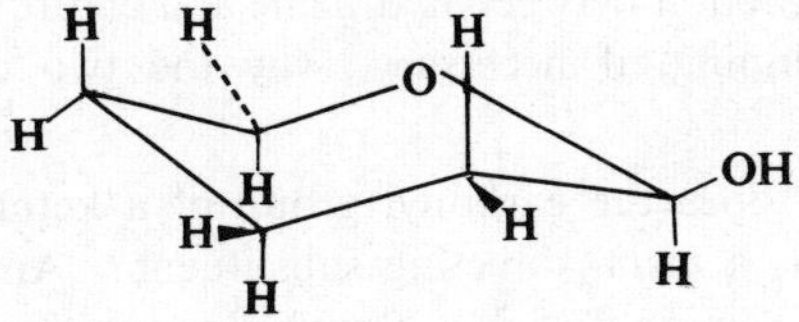 [structure]

f) Cellulose (Section 28.2.4);
g) Cellulose, after having been treated with aqueous acid?

7. a) In using diethyl malonate for the synthesis of carboxlyic acids (Section 28.3.2), why is the ester used, rather than the free acid?
   b) Indicate briefly how the following compounds might be synthesized, using either diethyl malonate or ethyl acetoacetate, whichever is most appropriate:

   i) CH_3CH_2COOH

   ii) $\begin{array}{c}CH_3\\CH_3CH_2\end{array}\!\!\!\diagdown\!\!CHCOOH$

   iii) $CH_3COCH_2CH_3$
   iv) $CH_3COCH(CH_3)CH_2C_6H_5$.

8. Acetylacetone ($CH_3COCH_2COCH_3$) shows properties associated with hydroxy compounds and unsaturated compounds. Why? Would you expect to be able to synthesize the alcohol, $(CH_3)_2C(OH)CH_2COCH_3$, by treating acetylacetone with one mole of methylmagnesium iodide (Section 27.3)?

9. Why does benzaldehyde *not* undergo the aldol reaction? Bearing in mind that aldehyde carbonyl groups are more readily attacked by nucleophiles than are ketone carbonyls (Section 23.3.2), suggest a product which you might reasonably expect to obtain by treatment of an equimolar mixture of benzaldehyde and methyl phenyl ketone (acetophenone), $CH_3COC_6H_5$, with alkali.

10. Which would you expect to be the stronger acid:

a) 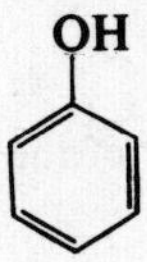 [phenol] or [4-acetylphenol]

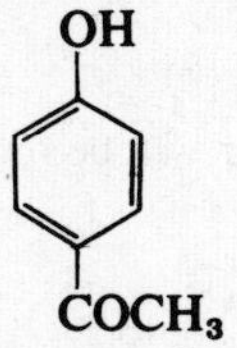

b) [4-hydroxyphenyl COCH₃] or [3-hydroxyphenyl COCH₃] ?

11. Which would you expect to be the stronger base:
    a) phenylamine or 4-nitrophenylamine;
    b) 4-nitrophenylamine or 3-nitrophenylamine?

12. 2,4-Dinitrophenylhydrazine (Section 23.2.2) may be prepared from 2,4-dinitrochlorobenzene and hydrazine ($H_2H—NH_2$). Would you expect phenylhydrazine to be obtainable by reaction between hydrazine and chlorobenzene? Justify your expectation, explaining, if necessary, why the two cases are different.

13. Why, in aromatic substitution, does the carbonyl group of a ketone group (e.g. in $C_6H_5COCH_3$) behave as a *meta-* directing substituent? Analyse the situation as for the nitro group (Section 28.5.3) in terms of possible carbonium ion intermediates.

14. Why does the methoxy group in methoxybenzene behave as an *ortho/para-*directing substituent? Analyse the situation as for the amino group (Section 28.5.3) in terms of carbonium ion intermediates.

BIBLIOGRAPHY

Taylor, R. J. *The Chemistry of Proteins*, Unilever Education Booklet, Advanced Series No. 3, available from Unilever Education Section, Unilever House, Blackfriars, London EC4.

   A beautifully illustrated account of the nature and function of these vitally important compounds.

Guthrie, R. D., and J. Honeyman, *An Introduction to the Chemistry of Carbohydrates* (third edition), Clarendon (1968).

   Primarily written for undergraduates, but still of value to the sixth-former.

Whitfield, R. C. *A Guide to Understanding Basic Organic Reactions*, Longman (1966).

   A treatment specifically aimed at introducing the sixth-former to organic reaction mechanisms.

Sykes, P. *A Guidebook to Mechanism in Organic Chemistry* (third edition), Longman (1970).

   An excellent and very readable treatment which would appeal to sixth-formers and undergraduates.

Stock, L. M. *Aromatic Substitution Reactions*, Prentice-Hall (1968).

   Primarily written for undergraduates, but a sixth-former could get a great deal from this.

Some material on all of the topics in this chapter will be found in all of the books on organic chemistry in the general list.

CHAPTER 29

Separation and Purification of Compounds and the Determination of Molecular Formulae

Most chemical investigations, at some stage, involve the separation of compounds from mixtures, or the purification of compounds, or both. Generally, before the structure and properties of a new compound can be investigated, that compound must first be isolated in a pure state. Before any naturally occurring materials, biological or non-biological, can be understood at the chemical level, they must first be broken down into their component chemical compounds. Before a new chemical reaction can be understood, its products must normally be separated from each other and purified in order that their nature may be determined. There are also other reasons, for example legal, medical, or economic, why it is necessary to be able to detect, isolate, and estimate chemical compounds. Furthermore, before the investigation of the more interesting properties (such as structure) of a new compound, the molecular formula of the compound must be determined. Although these techniques of separation, purification, and analysis may, at first, seem trivial and unimportant they are in fact vital to the continued progress of the understanding of our environment at the chemical level. Further stages of an investigation cannot usually proceed until these preliminary stages are successfully completed. It is the principal aim of this chapter to discuss and describe the techniques involved in these preliminary stages, with particular emphasis on the chemical and physical principles upon which they depend. Most of these principles are discussed elsewhere in this book or are assumed to have been dealt with in more elementary courses.

Although the determination of molecular formulae does not in itself normally present serious problems, the isolation of pure materials frequently does and may, in practice, be the most difficult and time consuming part of a particular piece of work. Perhaps the most difficult field, in this sense, is the investigation of biological systems which are always, from a chemical point of view, mixtures of daunting complexity. It is in this field, too, where the most stringent standards of purity are required, as trace quantities of impurities may have profound biological effects. Successful application of these techniques, then, is, though often difficult, vital. To use a popular misquotation from the eighteenth-century culinary literature, "First catch your hare . . .."

The processes of separation and purification are, of course, very similar. The essential difference between them is that, in the case of purification, one starts with material composed largely of one compound and the small amounts of other

29.1 Recrystallization and related techniques
29.2 Distillation and related techniques
29.3 Chromatography and related techniques
29.4 Chemical methods of separation
29.5 Automation
29.6 Determination of molecular formulae

compounds present must be removed to leave the bulk component pure, whereas with separation one has a mixture containing significant amounts of several compounds (often many) and one may wish to obtain several, or all, of these in a pure state. Inevitably, as both are fundamentally separations, the methods employed in either case are very similar. It is nevertheless still useful to distinguish between them.

Most methods of separation are essentially physical and therefore depend to a large extent upon the physical properties of the compounds involved. The state in which they exist at room temperature (solid, liquid, or gaseous) often has an important bearing upon which methods are appropriate. One would not normally attempt to recrystallize a liquid or a gas, for instance, nor would one normally try to distil a solid. The aim of most methods is to distribute the components between two different phases which can then be separated by a process which is essentially mechanical. Thus, recrystallization depends upon producing crystals of a pure compound in contact with a liquid phase in which impurities (as well as some of the desired compound) are dissolved. The crystals are then separated from the liquid phase (mother liquor) by the simple mechanical process of filtration. Distillation depends upon the separation of a vapour phase from a boiling liquid phase, the two phases being of different composition. The separation of two immiscible liquid phases is also easily achieved by mechanical means and this, too, commonly finds application. It is often convenient to employ chemical means to aid separation processes, but these are usually employed in order to convert one (or more) of the components of a mixture into a compound (or compounds) whose physical properties are further removed from the other, chemically unchanged, components of the mixture. A separation made difficult because the mixture was made up of compounds having similar physical properties may thus be made easier, although if pure samples of the components which have been chemically changed are required, one has to use a chemical reaction which is easily reversed. Acid–base reactions are used in the method described later in this chapter.

One point remains to be considered before we embark upon a consideration of the various methods of separation and purification in detail, and that is the question of criteria of purity. How does one judge the success, or otherwise, of the process used in a particular case?

There are basically two ways of tackling this problem. One of these is to test the homogeneity of the material whose purity is in question by subjecting it to further separation techniques. Thin-layer, paper, and gas chromatography, described later in this chapter, are often thus employed and can very quickly reveal whether a given material is made up of one compound only or of several. The other way depends simply upon measuring some convenient physical constant of the material. Commonly with solids this would be the melting point and with liquids the boiling point and refractive index. Quite how one uses these data depends largely upon whether one is dealing with compounds which are already well known and characterized or whether one is dealing with new compounds, isolated for the first time. If the compounds are already known, then these values will already have been recorded and are characteristic of the compounds in question. Any departure from the correct, recorded value implies the presence of impurities. The melting point of a compound, for example, is depressed by the presence of other, different materials—this is the well-known depression of freezing point (Section 5.4). If the compounds involved are new and not yet characterized, then the usefulness of these data is rather restricted. A liquid which distils over a wide range of temperature is almost certainly impure, although it does not necessarily follow that a liquid which distils over a very narrow temperature range, say one or two degrees, *is* pure. Refractive index is not much help in cases where new compounds are involved, although if the refractive index of the distillate varies during the course of the distillation, the distillate cannot be pure. With unknown solids the situation is easier because, as well as depressing the melting point of solids, impurities broaden it. A pure solid compound usually melts over a very narrow range of one or two degrees, but the presence of other materials, impurities, will both lower its melting point and broaden it to a range of perhaps ten or more. Purification procedures may be repeatedly applied until the melting point is no longer raised and becomes sharp (i.e. the range becomes narrow).

29.1 RECRYSTALLIZATION AND RELATED TECHNIQUES

Recrystallization is a method of purification applicable to crystalline solids. It is probably the oldest, simplest, most effective, and most widely used method of purifying solids and depends essentially upon simple principles of solubility. Sometimes, though, differences in the rate at which one component crystallizes, relative to the others present, may form a satisfactory basis. Usually a solvent is required in which the material to be purified is highly soluble at the boiling point of the solvent, but which will dissolve only very little of the material at room temperature. The impure material is dissolved in the minimum quantity of boiling solvent and, after any insoluble impurities have been filtered off, the resulting solution is allowed to cool down to room temperature. As the temperature drops the solubility drops too and crystalline material begins to separate out from the solution. When the temperature of the mixture reaches that of the room, the amount of material remaining in solution depends upon the solubility at that temperature which, as we have already said, is low. It will be sufficient to saturate the solution at room temperature. Because the impurities were present in only small amounts it is unlikely that they will saturate the solvent and they will therefore remain in solution. In short, the material which has crystallized will be pure. In practice, of course, the procedure may have to be repeated several times before purification is complete, although often a single recrystallization is sufficient. The following illustration will help to clarify the basis of this method.

Suppose we have a 24 g sample of an impure solid, A, which contains 4 g of a contaminant, B, that is the sample contains only 20 g of A and is therefore 83% pure. We will take the solubility of each to be 20 g in 100 cm^3 (200 g l^{-1}) at the boiling point of the chosen solvent and 2 g in 100 cm^3 (20 g l^{-1}) at room temperature, and we will further assume that the presence of one solute will not affect the solubility of the other. 20 g of A requires 100 cm^3 of boiling solvent to dissolve it and this will be more than sufficient to dissolve the 4 g of B which accompanies that amount of A. We therefore dissolve the crude mixture in 100 cm^3 of boiling solvent and, after filtration, allow the solution to cool and crystallize. At room temperature, 2 g of A and 2 g of B will remain in solution and the solid which has crystallized out will therefore contain 18 g of A and 2 g of B—total 20 g. The purity of A is now 18/20, that is 90%, a considerable improvement on the original 83%. The operation is now repeated. 18 g of A will require 90 cm^3 of boiling solvent for complete dissolution and this will be more than sufficient to dissolve the 2 g of B.

The 20 g sample obtained from the first crystallization is therefore dissolved in 90 cm^3 of the boiling solvent and the solution allowed to cool and crystallize. The quantity of A remaining in solution at room temperature will this time be 1.8 g and the quantity of B also 1.8 g. Thus 16.2 g of crystalline A will separate and 0.2 g of B, giving a 16.4 g sample containing 16.2 g of A; the purity has now risen to 16.2/16.4, or 99%, a further considerable improvement. It is fairly easy to see that a third recrystallization should give 100% pure A. 81 cm^3 of boiling solvent will dissolve the 16.2 g of A in the 99% pure sample, and this will be more than is required to hold in solution, at room temperature, all of the 0.2 g of B. 14.6 g of pure A will separate this time, and the "cost" of this operation is the loss of a total of 5.4 g of the original 20 g of A. We have recovered only 73% of the original A, but it is now 100% pure.

Perhaps it should be stressed at this point that this illustration represents an oversimplification. One solute often does affect the solubility of another, for example, and several other complicating factors have not been taken into account. However, it does serve to rationalize the expectation that minor contaminants can be removed by the process and to make qualitative predictions possible on the effect of varying solubility and composition.

The illustration chosen required three operations in order to achieve complete purification. It was perhaps a rather more difficult purification than is normally encountered. However, in practice it might have been made easier if a solvent had been found in which the impurities were less soluble than the compound which was to be purified, rather than of the same solubility. If the solubility of B had been 4 g in 100 cm^3 (40 g l^{-1}), or more, at room temperature, then one operation would have been sufficient. Similarly, if there had been less impurity present, fewer operations would have been necessary. It would be easy to show that if only 2 g (or less) of B had been present with the 20 g of A, that is if A had originally been more than 91% pure, then again only one operation would have been necessary.

It will be very clear from these considerations that the presence of more impurity would have made the matter more difficult. In the example chosen, where A and B were of identical solubility, the composition of a 50/50 mixture of A and B could never be altered by recrystallization (at any rate from that solvent). We now move into the realms of separation rather than of purification and the procedure for effecting this separation is somewhat different from simple recrystallization. It will yield samples of both A and B.

The simplest way (in principle) to achieve this separation is to find a solvent which will dissolve one component but not the other. Sometimes this is possible but more often it is not. It does not have to be 100 percent efficient, of course. If it can produce two samples, one of which is mainly A and the other mainly B, then the operation can be completed by purifying each of these samples by simple recrystallization. Often, however, a more sophisticated approach is necessary. The mixture may be taken up in the appropriate hot solvent, or mixture of solvents, and a small quantity of solid allowed to crystallize. This is collected and, by altering a suitable factor, a little more is allowed to crystallize, and this too is collected. By making continued small alterations to this controlling factor the mixture may be divided up into a number of separate, small quantities. These are known as fractions, and the whole process is referred to as **fractional crystallization**. In this way, relatively small differences in solubility are exploited and are reflected in the differing composition of each fraction. In a particularly difficult case, where a mixture of two components has been divided up by this means into, say, twenty fractions, perhaps the first two fractions contain exclusively one component and the last two exclusively the other, with the intervening sixteen being mixtures of varying composition. An easier case might have given eight or nine fractions each of the pure components, with only three or four "middle fractions". The middle fractions do not, of course, represent entirely lost material. Those which are rich in one component may be purified by simple recrystallization and those which are not may be combined and refractionated. More sophisticated procedures involve recrystallizing various crops of crystals from mother liquors which have already deposited purer crops, in a carefully designed systematic series of operations. The method is often very tedious and time consuming and separations are often more easily and more efficiently achieved by chromatographic methods which are described later. Fractional crystallization is probably, therefore, less used now than once it was.

The factors which one alters to control the fractionation may include concentration, temperature, and solvent composition. If one was using concentration, one would make up the initial solution in a good deal more than the minimum amount of boiling solvent in order to ensure that only a small quantity of solid separated on cooling. After collection of that solid, the mother liquor would be concentrated a little, by boiling or evaporating off some solvent, and allowed to cool and produce a second crop of crystals. Further crops, or fractions, would be produced in the same manner. Temperature can be made use of by collecting the crystals which separate from a solution at successively lower temperatures. With solvent composition as the controlling factor, one would start in much the same way as when using concentration as the factor. One would start with solvent of a particular composition—it may be one pure liquid—and, having collected the first crop of crystals, add to the mother liquor small quantities of a second liquid, thus altering the solvent composition, and collecting the crystals which separate after each successive addition.

Several other methods of separation or purification, related to the crystallization processes just described, may be mentioned briefly at this point. **Fractional precipitation**, for example, is very similar to fractional crystallization in that it depends upon controlling factors affecting the solubility of solids dissolved in a solvent. Precipitation may be brought about chemically or physically and various fractions thus obtained. Precipitation is generally, however, a much less "clean" process than crystallization and precipitates tend to bring with them out of solution other solutes whose solubility may not have been exceeded. In cases where it is difficult to persuade solids to crystallize, methods such as this may have to be employed. Large, complex, biological polymers tend to fall into this category. Proteins, for example, can be crystallized but usually only after they have been obtained in a reasonably pure state. Initial procedures for their isolation from the complex mixtures in which they are found may depend upon methods such as this. They are very polar molecules, as well as being large, and their solubility in water depends very much upon how much other polar material is dissolved in the same solution. Fractional precipitation of proteins from aqueous solutions of biological preparations is often brought about by adding ionic salts to the solution. Precipitated material can be collected after successive additions of a suitable salt. This process is referred to as **salting out**. Before the protein is obtained in a pure state, however, precipitated materials such as these must normally be subjected to a variety of other purification processes as well. Biological molecules usually occur in very complex mixtures alongside numbers of other molecules of very similar structures and properties. The greater the number of impurities present, and the more closely they resemble the molecule which it is required to isolate, the more difficult is the task.

Zone refining, or **zone melting**, is a technique, closely allied to recrystallization, which has been extensively developed in recent years and has found numerous applications particularly in industrial fields. The high-grade germanium metal required by the electronics industry, for example, undergoes zone refining in the final stages of the purification process.

The material in question, already crystalline and of moderately high purity therefore, is packed into a long tube which is suspended vertically. At the top end of the tube a very small, circular, electrical heating element passes right round so that, when switched on, it melts a very thin band of material. It creates a very thin disc of molten material across the tube. By means of a mechanical device, the tube is raised through the heating element (or the element is lowered) at a very low speed. As the element falls with respect to the tube, the molten phase which it leaves behind begins to crystallize again, leaving impurities in the liquid phase. The molten solid effectively takes the place of the solvent in the recrystallization process. Impurities concentrate in the molten phase and drop with it as the element descends, thus ending up concentrated at one end of the tube, whilst the remainder of the tube is packed with the purified, crystalline solid. The process is automatic and the element can make as many passes along the tube as required to achieve the desired standard of purity. Very high standards indeed can be achieved by this method with a minimum of effort. (See Fig. 29.1).

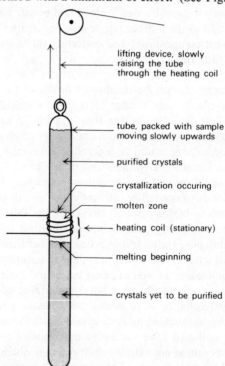

Fig. 29.1 Diagrammatic representation of zone refining apparatus. The direction of movement may be varied to suit the circumstances. Occasionally the tube, and movement, may be horizontal.

29.2 DISTILLATION AND RELATED TECHNIQUES

A liquid—say water—which contains a dissolved involatile material—say sodium chloride—has, in the atmosphere above its surface, pure water vapour, uncontaminated with sodium chloride. If this aqueous solution of sodium chloride is boiled, only water vapour (steam) is evolved and the involatile sodium chloride remains in the liquid phase. It cannot volatilize into the vapour phase. If, therefore, the solution is boiled and the steam evolved is led off and condensed back to the liquid phase by cooling, we obtain pure water as the condensate and, when all the water has boiled off, pure, dry sodium chloride remains. This process is referred to as **distillation**, and has afforded the means of separating the two components of the mixture and enabling them to be purified. It is normally carried out in apparatus such as that illustrated in Fig. 29.2. The thermometer, upon the bulb of which

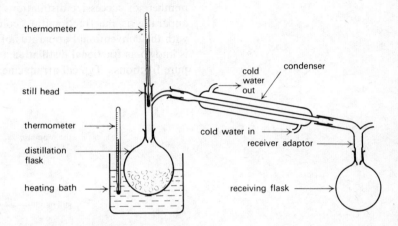

Fig. 29.2 Simple distillation apparatus.

droplets of liquid condense and are in equilibrium with the vapour, records the boiling point of the distilling liquid.

This method is fine for separating liquids from involatile materials, but how useful is it for separating and purifying two liquids? This question is perhaps best answered by studying the phase diagram (Section 5.3) shown in Fig. 29.3. The two liquids, A and B, have boiling points s and t ($t > s$). A has a lower boiling point than B and therefore, at a given temperature, the higher vapour pressure; it is the more volatile liquid. Above a 50/50 mixture of the two liquids the vapour is therefore richer in A than in B. Relative to the liquid phase of any mixture, the vapour phase is always richer in A and deficient with respect to B. For a mixture of composition x, the boiling point can be obtained from the diagram by reading off the temperature corresponding to the point a, on the liquid curve, vertically above x. A horizontal line extended from a until it intersects the vapour curve enables the composition of the vapour, in equilibrium with a liquid mixture of composition x, at its boiling point, to be obtained. The point b represents this intersection and the vapour composition is read off vertically below b. Clearly the vapour is much richer in A than is the liquid. If the mixture of composition x is distilled, the condensing vapour would be represented by b, and the condensate (distillate) must have the same composition (y). The distillation process is represented on the diagram by the path abc. Clearly some sort of separation has been achieved as the (initial) distillate is richer in A and the liquid remaining undistilled

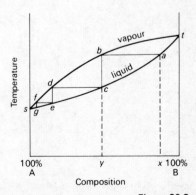

Figure 29.3

must therefore be richer in B. However, before pure A is obtained, the initial distillate would need to go through the process several more times, that is it would have to continue along the path *abcdefg*, etc. In practice the situation is hopeless. It could only be the first drop of distillate which had composition y and, as the distillation proceeds, the composition would gradually move back from y until, when all the liquid had distilled, it would be x again. The boiling point would rise steadily during the operation. Any attempt to collect and redistil the first drop once, let alone several times, would clearly be absurd. For the separation of two liquids by simple distillation, one must be essentially involatile with respect to the other—they must have boiling points which differ by an enormous amount, so that the vapour above the mixture contains only the more volatile component.

Although the employment of numbers of successive, separate distillations is out of the question, apparatus has been in use for many years which enables numbers of successive distillations to take place in one practical operation. This apparatus is the fractionating column, and it can be employed in conjunction with the conventional apparatus of Fig. 29.2. Distillation through such a column is known as **fractional distillation** and enables liquid mixtures to be separated into pure fractions. Typical arrangements are illustrated in Figs. 29.4 and 29.5.

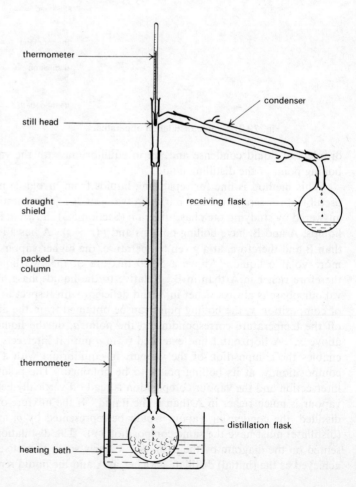

Fig. 29.4 Fractional distillation apparatus.

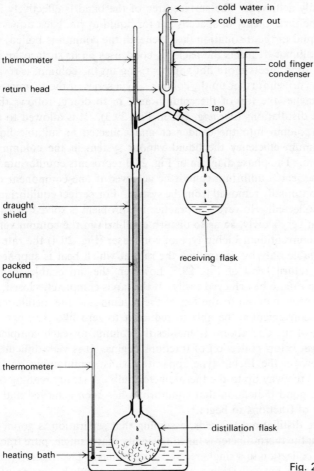

Fig. 29.5 Fractional distillation apparatus with return head.

The column, between the flask and the still head, is packed with a suitable inert material presenting a large surface area to vapour climbing the tube. Small glass helices, small pieces of fine metal mesh, or simply a large number of indentations in the tube itself are commonly employed (an unravelled metal pan scrub gives excellent results in the laboratory). The first two are more efficient as they present a larger surface area than the third. The most effective way of maximizing the surface area of the packing, of course, would be to use the material in the form of a very fine powder, but this is unsuitable as there must be plenty of large gaps in the packing to allow liquid to run back. The most effective method of achieving maximum surface area, whilst retaining plenty of space for liquid to run down, is that employed in a "spinning-band column". A long steel strip, almost as wide as the column and twisted into a spiral shape, is fitted inside the column with its axis coincident with the axis of the column. When stationary, it offers a relatively small surface area to climbing vapour, but, during a distillation, it is kept spinning about its axis at a very high speed by means of a small electric motor. The spinning motion throws condensed liquid off the surface and liquid–vapour equilibrium is very rapidly achieved. The surface area of the band is effectively increased and, at the same time, plenty of space is left for liquid to run back down the column.

As liquid in the distillation flask beneath the column is boiled, vapour climbs into the column. It meets the packing, condenses on its surface, and begins to run down. It soon meets more hot vapour rising up the column, is revaporized by it, and is thus driven further up the column. By a series of these "micro-distillations" vapour reaches the top of the column and, in so doing, follows the pathway of successive distillations (*abcdefg*, etc. in Fig. 29.3). It is allowed to emerge slowly from the column into the condenser and collected in suitable liquid fractions. For maximum efficiency the liquid–vapour system in the column should be in equilibrium. The phase diagram of Fig. 29.3 represents equilibrium compositions. Clearly, perfect equilibrium cannot be achieved if one component of the mixture is being continually removed from the system. For perfect equilibrium, no distillate must be collected. However, in practice the problem is solved by carrying out the distillation very slowly, so as to disturb equilibria in the column very little. With the more conventional Liebig type of condenser (Fig. 29.4) the rate of distillation is controllable only by controlling the rate at which heat is supplied to the flask. With the return head of Fig. 29.5, however, the tap enables a fine control of distillation rate to be achieved easily. If the tap is completely closed, the condenser simply returns material to the top of the column and the distillation rate is zero. It is very convenient to be able to reduce it to zero like this, particularly at the beginning of the operation. It enables the column to reach complete equilibrium before a very slow collection of fractions begins. It is very difficult to reduce the rate to zero in the Liebig type apparatus whilst maintaining hot vapour in the column all the way up to the thermometer bulb. A steady reading on the thermometer is a good indication that equilibrium has been achieved and is the cue for collection of fractions to begin.

As the distillation proceeds, assuming that separation is good, then a steady reading on the thermometer is taken to signify that a single, pure fraction is coming over, and collection in a single receiver is continued. The fraction is not necessarily pure, of course, but collection continues because a steady boiling point is a necessary, though not sufficient, condition for its being pure. When the temperature begins to rise, this is taken to mean that the next component is beginning to distil and collection continues in a new receiver until the temperature is steady again. When it becomes steady, the next, pure component is taken to be coming over and is collected in another receiver. The fraction collected while the temperature was rising is known as a middle fraction, contains both of the components, and is hopefully small.

The course of an efficient fractionation of a mixture of 10 cm^3 of liquid A and 10 cm^3 of liquid B is represented in Fig. 29.6(a). The temperature is plotted against progress as measured by the total volume of liquid distillate collected. The horizontal portions of the curve represent the distillation of pure A and pure B. The size of the middle fraction is defined by the pair of vertical broken lines, and the nearly horizontal alignment of the portions of curve on either side show that the pure fractions distil over a very narrow range of temperature (i.e. the temperature remains steady). A somewhat less efficient distillation is shown in Fig. 29.6(b); the middle fraction is enlarged—relative to the efficient process in (a)—and the boiling range of the "pure" fractions is increased. The curve is less horizontal at its extremes and the quality of the products correspondingly lower. The situation is much

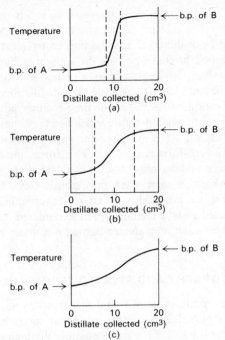

Figure 29.6

worse in Fig. 29.6(c), where separation is very poor indeed. If a more efficient column were not available, then this distillate would be collected in small arbitrary fractions (say 20 of 1 cm³ each) and the course of the distillation followed by measuring the refractive index of each one (refractive index is more reliable than boiling point). The hope would be that the first one (or few) and the last one (or few) would contain pure, or nearly pure, A and B respectively. For a given column, the efficiency is dependent upon the surface area of the packing (the longer the column the better it is in this respect) and upon the rate of distillation or, strictly, the ratio of the distillation rate to the rate at which liquid is returned to the distilling flask from the column. The maintenance of equilibrium (and therefore efficiency) in the column is also helped by shielding the column from external, cooling draughts. Longer columns are often surrounded by electrical heating jackets which not only protect them from irregular cooling by draughts, but also ensure that the total, steady cooling rate is not so great that the vapour cannot climb to the top of the column.

One problem often encountered with liquids, particularly with organic liquids of large molecular weight, is that they decompose at temperatures below their boiling points. This inevitably means that attempts to distil them normally result in extensive, if not complete, decomposition. However, the boiling point of a liquid depends upon atmospheric pressure, or, more accurately, upon the pressure immediately above it. If, by means of a suitable pump, we were to reduce the pressure inside the distillation apparatus, then we would also lower the boiling point of the liquid. At reduced pressure, therefore, such liquids can be distilled without decomposition, and distillation at reduced pressure is in practice a very common procedure.

Finally, in this connection, a brief reference to solids may be made. Some solids, of course, have low melting points and low boiling points and they may, therefore, be purified by distillation. There is the inconvenient possibility that they may crystallize and block the cooler parts of the apparatus, but this can be avoided. None the less, recrystallization is normally to be preferred for solids as it is generally more effective, simpler, and less troublesome than distillation. There are, though, some solids whose boiling points fall *below* their melting points and, as they are heated, when the appropriate temperature is reached they change straight from the solid state into the vapour state. Cooling the vapour transforms it back to the solid state, usually as well-formed crystals. This forms the basis of an excellent means of purifying some solids, and is known as **sublimation**. Relatively few solids have this property which, in some ways, may seem a pity. However, because so few solids can sublime, the probability of impurities subliming with the compound being purified is reduced and the method therefore rendered more likely to succeed. Like distillation, sublimation may also be carried out under reduced pressure.

29.3 CHROMATOGRAPHY AND RELATED TECHNIQUES

The development of chromatography and spectroscopy into techniques which could be usefully employed in a routine way must surely be amongst the most important developments which have made possible the major advances in chemically based science during the last thirty years or so. Whilst chromatography enables small amounts of materials present in complex mixtures to be isolated and purified where at one time this was impossible, spectroscopy gives an enormous amount of information about these very small samples without actually consuming them. Chromatography is now one of the most important, routine procedures in use in practically all chemical and biochemical laboratories, and it therefore occupies an important place in this chapter. More detailed accounts of the method, suitable for sixth form use, are available in specialist monographs such as that by Abbott and Andrews (D. Abbott and R. S. Andrews, *An Introduction to Chromatography*, 2nd Ed., Longman, 1970).

If a mixture is composed of two components, one of which is soluble in solvent A but not in solvent B, and the other soluble in solvent B but not in solvent A, then, provided that the solvents A and B are immiscible, separation of the two components of the mixture is easily achieved by shaking the mixture with a mixture of the two solvents. The less dense solvent will separate out and float on top of the other solvent and there will be one of the components of the original mixture in each solvent layer. A mixture of common salt and naphthalene, for example, could easily be separated by shaking it with a mixture of petrol and water. The salt would dissolve exclusively in the water and the naphthalene in the petrol, with the latter, after the shaken mixture had been allowed to settle, floating on top of the water. Such an operation would normally be carried out in a separating funnel (Fig. 29.7), which has a tap to enable the lower layer to run out into one chosen vessel, after which the upper layer can be poured out through the top of the funnel (to avoid contact with remnants of the lower layer in the stem) into another. Evaporation of solvents then leaves the separated salt and naphthalene.

Such an easy separation as this, however, is relatively rare. Usually both components have some solubility in both liquid phases and complete separation is not,

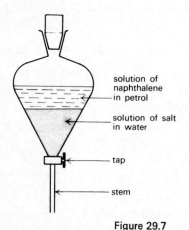

Figure 29.7

therefore, possible by this simple method. The relative amounts of a given component dissolving in the two liquid phases is governed by the distribution or partition coefficient of the component (Section 7.1), and therefore the effectiveness of the separation of a mixture depends upon the partition coefficients of the various components. The partition coefficient of salt, between water and petrol, is infinity whereas that for naphthalene is zero. Only with this enormous difference is the method completely successful.

Consider the case of a mixture of equal amounts of two components, A and B, which is partitioned between equal volumes of two solvents, X and Y. A has partition coefficient 10 (i.e. is more soluble in solvent X, which is the less dense solvent, and therefore will float on Y) and B has coefficient 0.1 (i.e. is more soluble in Y). If 220 g of such a mixture is distributed between equal volumes of X and Y (equal volumes are not essential, but make the calculation easier) then, the 110 g A present will be distributed with 100 g in X, the top solvent layer, and 10 g in Y. Similarly, there will be 10 g of B in the top layer and 100 g in the bottom. Thus, each layer will contain 100 g of one component, contaminated by 10 g of the other, and some degree of separation will have been achieved but it will be far from complete. A technique has been devised, however, for improving this separation to make it the basis of an effective and useful separation method. A whole row of funnels is required, each one charged with the same volume of solvent Y (the denser solvent in this case) and numbered consecutively from 1 upwards. For simplicity, we will first consider the fate of component A only, 110 g of which are placed in funnel 1 along with an equal volume (aliquot) of solvent X. The funnel is thoroughly shaken to establish equilibrium, and allowed to stand while the solvents separate out and settle. From the partition coefficient, we can calculate that there will be 100 g A in the top layer (solvent X) and 10 g in the bottom layer. The top layer is then removed and poured into funnel number 2, where it meets the aliquot of solvent Y already placed there. A fresh aliquot of solvent X is added to funnel number 1, and both funnels are shaken to establish equilibrium. The 100 g of A in funnel 2 will now be distributed with 91 g in the top layer and 9.1 g in the bottom, while the 10 g in funnel 1 will have 9.1 g in the top layer and 0.91 g in the bottom. The top layer from funnel 2 is transferred to funnel 3, that from funnel 1 is transferred to funnel 2, and a fresh aliquot of X is added to funnel 1. All three funnels are shaken and, after settling, the top layers are all transferred to the funnel of next highest number with the further addition of a fresh aliquot of X to funnel 1. The process is continued, bringing a new funnel into play and establishing new equilibria to satisfy the partition coefficient with each operation. After n shaking operations, A is distributed along a row of n funnels. Table 29.1 gives detailed figures for the distribution of A along six funnels after six operations. Each operation is represented by two horizontal rows of fractions, the fractions in the upper row giving the amounts in the upper and lower phases immediately after a transfer, but before shaking, and the fractions in the lower row giving the amounts present in each phase after shaking, that is at equilibrium. The upper and lower phases are represented by the upper and lower parts of the fractions respectively. The reader can clarify this arrangement by correlating operations 1 and 2 in the table with the descriptions in the text. He should also check them to see how they are determined by the distribution law and the partition coefficient. Table 29.2 gives corresponding figures for component B, and, by comparing the two tables, it can be seen that the bulk of A accumulates in funnels of high number, that is at the end

Table 29.1. Distribution of A through six operations.

Operation	Funnel number					
	1	2	3	4	5	6
1	110/0					
	100/10					
2	0/10	100/0				
	9.1/0.91	91/9.1				
3	0/0.91	9.1/9.1	91/0			
	0.83/0.083	16.5/1.6	83/8.3			
4	0/0.083	0.83/1.6	16.5/8.3	83/0		
	0.075/0.0075	2.2/0.22	22.5/2.3	75.5/7.5		
5	0/0.0075	0.075/0.22	2.2/2.3	22.5/7.5	75.5/0	
	0.0068/0.00068	0.27/0.027	4.1/0.41	27.3/2.7	68.6/6.9	
6	0/0.00068	0.0068/0.027	0.27/0.41	4.1/2.7	27.3/6.9	68.6/0
	0.00062/0.000062	0.031/0.0031	0.62/0.062	6.2/0.62	31.1/3.1	62.4/6.2

of the row, whereas B accumulates in funnels of low number, at the beginning of the row. The process can be seen as the passage of solvent X over static solvent Y, with the component most soluble in X moving rapidly with it and the least soluble component moving only very slowly with it. The curves in Fig. 29.8 show the progress of the components along the row, as X moves along it, at various stages. Also included on the curves is the progress of a third component, C, having a partition coefficient of one, intermediate between those of A and B (figures in Table 29.3). The solid line represents A, the broken line B, and the dotted line C. The total amount of the specified component in both phases is plotted against the funnel number, a measure of the distance along the row which the component has moved. Figure 29.8(a) shows the situation after three operations, (b) after six, and (c) after nine. Figures are not given in the tables beyond six operations, but the reader could profitably calculate his own figures for operations 7, 8, and 9 and check that they correspond to the curve in Fig. 29.8(c). After three operations, separation has begun but is clearly not yet very good. After six operations, however, A and B are almost clear of one another. Had the mixture comprised A and

Table 29.2. Distribution of B through six operations.

Operation	Funnel number					
	1	2	3	4	5	6
1	110/0 10/100					
2	0/100 9.1/91	10/0 0.91/9.1				
3	0/91 8.3/83	9.1/9.1 1.6/16.5	0.91/0 0.082/0.82			
4	0/83 7.5/75.5	8.3/16.5 2.3/22.5	1.6/0.82 0.22/2.2	0.082/0 0.0075/0.075		
5	0/75.5 6.9/68.6	7.5/22.5 2.7/27.3	2.3/2.2 0.41/4.1	0.22/0.075 0.027/0.27	0.0075/0 0.00068/0.0068	
6	0/68.6 6.2/62.4	6.9/27.3 3.1/31.1	2.7/4.1 0.62/6.2	0.41/0.27 0.062/0.62	0.027/0.0068 0.0031/0.031	0.00068/0 0.00062/0.000062

B only, the bulk of each component could have been easily recovered in a high state of purity by evaporating the contents of the funnels. Funnels 1, 2, and 3 would yield B, and funnels 4, 5, and 6 would yield A. A further three operations leave A and B virtually completely separated and each is separating well from C. Virtually pure C could be obtained from funnel 5, and good quality C from funnels 4 and 6. B would be obtainable in good quality from funnels 1 and 2, and A from funnels 8 and 9. A series of further operations would give better results still. Separation improves with the number of operations performed. It is also clear that separation is easiest, that is, it requires fewest operations, when the difference in partition coefficients is greatest. A has a coefficient ten times greater than that of C but one hundred times greater than that of B.

With so many funnels to shake, and so many transfers of solvents to perform, this is obviously a very tedious method to operate. Suitable apparatus, however, is available in which the whole procedure can take place mechanically and automatically. A row of wide, parallel glass tubes, in place of separating funnels, suitably interconnected by thinner tubing, is mounted in a rack which can shake

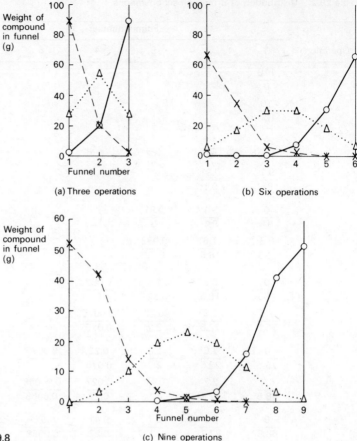

Figure 29.8 (a) Three operations (b) Six operations (c) Nine operations

them and automatically tilt them so that the appropriate solvent transfers can take place after settling. Solvent reservoirs are connected and, after the material has been placed at the beginning of the row, the apparatus can be set up to operate automatically without further attention. In this way 50 or 100 or more shaking and transfer operations can be performed with a minimum of trouble, and excellent separations may be obtained. This method is known as the **counter-current distribution** method, and whilst perhaps not strictly chromatography, depends upon the same principles and serves as a very good illustration of those principles.

The chromatographic technique most closely resembling counter-current distribution is that known as **partition**, or **distribution chromatography**. In this case we consider a liquid—usually water—adsorbed on to the surface of a suitable solid. Finely powdered silica gel is suitable and is commonly employed for this purpose. Consider a glass tube, packed with finely divided silica gel on the surface of which water is adsorbed. The tube, or column, is clamped in a vertical position and the packing is prevented from falling out by a loose glass-wool plug or sintered glass disc. There is a tap at the bottom of the column which is used to control the rate of flow of a second liquid through the column. This liquid is immiscible with water and, being continuously fed into the top of the column, flows down it, over the adsorbed layer of water, under the influence of gravity. Vessels can be placed

Table 29.3. Distribution of C through six operations.

Operation	Funnel number					
	1	2	3	4	5	6
1	110/0 55/55					
2	0/55 27.5/27.5	55/0 27.5/27.5				
3	0/27.5 13.7/13.7	27.5/27.5 27.5/27.5	27.5/0 13.7/13.7			
4	0/13.7 6.9/6.9	13.7/27.5 15.6/15.6	27.5/13.7 15.6/15.6	13.7/0 6.9/6.9		
5	0/6.9 3.4/3.4	6.9/15.6 11.3/11.3	15.6/15.6 15.6/15.6	15.6/6.9 11.3/11.3	6.9/0 3.4/3.4	
6	0/3.4 1.7/1.7	3.4/11.3 7.3/7.3	11.3/15.6 13.4/13.4	15.6/11.3 13.4/13.4	11.3/3.4 7.3/7.3	3.4/0 1.7/1.7

under the column to collect the liquid running through. The arrangement is depicted diagrammatically in Fig. 29.9. As the liquid, known as the eluant, runs down the column over the layer of adsorbed water we have a situation not unlike that in the counter-current distribution method. Any material applied to the top of the column will move down it at a rate determined by its partition coefficient between the eluant, or moving phase, and the water adsorbed on to the silica gel—the stationary phase. Instead of a series of separate equilibria in large vessels, as in the counter-current method, a series of connected equilibria, on a microscopic scale, is set up down the length of the column, provided that the moving phase moves slowly enough. In practice one cannot allow it to move too slowly, or diffusion begins and spreads the applied material out in both directions. When the right rate has been chosen though, the material moves down the column as a narrow band at a speed which depends upon its partition coefficient. Clearly, if several different materials are applied in the form of a mixture to the top of the column, then they will move down the column at different speeds according to their separate, and different, partition coefficients. They emerge from the bottom of the column—are eluted—at different times and are collected in different vessels. If they are

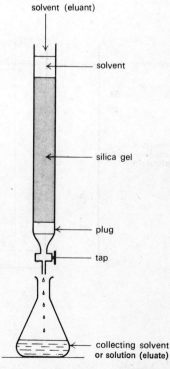

Figure 29.9

coloured, their progress down the column can easily be seen but if this is not the case, then their emergence from the column must be detected by different means. The simplest, though perhaps most tedious method, is simply to collect a whole series of very small fractions of equal volume and to evaporate each to dryness, separately. The weight of residual material indicates the emergence of one of the components. The fractions should be numbered consecutively, in the order of their emergence, and a curve can be plotted of the weight of residual material against fraction number. Eluted components are represented on the curve as peaks, being present, probably, in several consecutive fractions. There are other, notably spectrophotometric methods, although these need not concern us at this point. In much the same way as increasing the number of operations in counter-current distribution increases the separation of components, so increasing the length of the chromatography column improves its separating power.

There are a number of other ways in which this type of chromatography can be set up. Instead of packing the wet silica gel into a column, for instance, it may be spread in a thin film over the surface of a rectangular glass plate (usual size about 20 × 5 cm). The mixture to be examined is then applied to the plate in the form of a very small spot, containing only micrograms of material, near one end of the plate. The position of the spot is marked and the plate is placed in an enclosed glass tank containing a little of the moving phase. The "spot" is at the lower end and, whilst the silica gel should dip below the surface of the moving phase, the spot should be a short distance above it (Fig. 29.10). In this position the moving

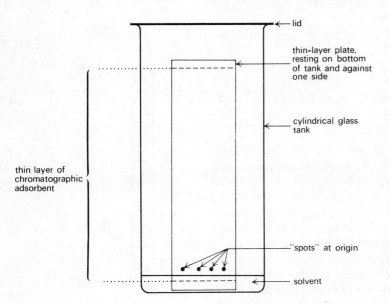

Fig. 29.10 Thin-layer chromatography plate, immediately after being placed in tank, i.e. before chromatography has begun.

phase creeps slowly up the plate under the influence of capillarity and, according to their partition coefficients, carries the components of the mixture, still in the form of spots, up the plate at difference rates. When the moving phase has almost reached the top of the plate, the plate is removed, the position reached (solvent front) by the moving phase is marked, and the plate dried. The positions of the various components may then be seen, as spots, if they are coloured (Fig. 29.11)

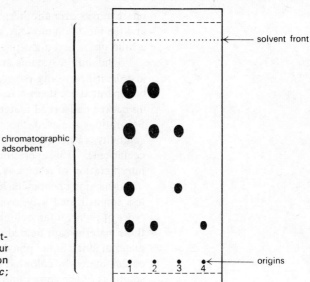

Fig. 29.11 The thin-layer plate from Fig. 29.10 after chromatography has taken place. The spot at origin 1 contained four known components, *a*, *b*, *c*, and *d* (standards). The spots on origins 2, 3, and 4 have been shown to contain *a*, *c*, and *d*; *b* and *c*; and *a*, respectively.

or they may be rendered visible by a variety of techniques. The distance moved by a given spot from the starting point, or origin, divided by the distance moved by the moving phase beyond the origin, is known as the R_f value and, if external conditions are constant, is a function of the partition coefficient of that component. (Note that the solvent properties of the stationary water phase will be modified by its being adsorbed on to the silica gel, and the partition coefficient in question will not, therefore, be that which we would measure by shaking up the material with water and the moving phase.) The R_f value is thus a characteristic of the component and can be used to identify it. This method, known as **thin-layer chromatography**, is therefore of great use in separating and identifying the components of mixtures on a very small scale and very quickly. The various components can also be obtained from the plates if they are required in a pure form for further investigation. The method is routine in most organic chemistry and biochemistry laboratories and is also used by inorganic chemists. It provides a convenient means of analysing mixtures of metal ions, for example, which is much simpler, quicker, and less demanding on skill than the traditional group analysis schemes. With paper chromatography, it has been of particular importance in analysing biological mixtures and the mixtures obtained by chemically breaking down biological molecules. The analysis of the amino acid mixtures obtained from protein or peptide hydrolysis, for example, is often carried out in this way.

Paper chromatography is very similar to thin-layer chromatography and was developed before it. Many of the functions of it have now been taken over by the thin-layer method. It depends upon the use of paper as the solid on which water is adsorbed, and rectangular sheets of paper are used in much the same way as thin-layer plates. The moving phase can descend the strip, though, under the influence of gravity, as well as ascending it through capillary attraction.

All of the chromatographic techniques so far described, as well as the countercurrent method, depend upon distribution between two liquid phases: one liquid phase moves over another. Systems other than liquid–liquid are also used, though, and the most common are solid–liquid and liquid–gas. A liquid phase moves over a solid phase and a gas phase moves over a liquid phase.

Solid–liquid systems are set up in the same ways as the liquid–liquid systems just described, using paper, thin-layer, and column techniques. The materials to be separated are themselves adsorbed on to the solid surface and equilibrium is set up between adsorbed material and material in solution in the moving liquid phase. Separations achieved, therefore, depend upon the balance between adsorption and solubility, rather than on the balance between solubility and solubility (partition coefficient). The operation of this method, **adsorption chromatography,** and the interpretation of results, is much the same as for the liquid–liquid method.

Adsorption phenomena have been employed as a means of purification, in a rather less sophisticated way, over many years. Finely divided charcoal adsorbs certain types of polar, often coloured, material very strongly such that, when solutions of these materials are treated with charcoal, adsorption takes place leaving little or no material in the liquid phase. Organic compounds are often obtained, in crude form, contaminated by coloured materials and these may be readily removed by treating solutions of the compound with charcoal which can then be filtered off along with the impurities. The hot solution obtained during the recrystallization of such a compound is often thus treated with charcoal. Filtration of the hot solution then removes charcoal, on which are adsorbed the coloured impurities, as well as any other insoluble impurities. On cooling, pure "decolorized" crystals may then be expected. This use of decolorizing charcoal is an important stage in the refining of cane sugar, where pure, white crystals of sucrose are required for the food market. A charcoal pad was the essential element in gas masks, issued for possible use by the civilian population during the last war. Gases such as chlorine are strongly adsorbed and are removed from air which has to pass through one of these pads before inhalation.

A variety of other types of solid stationary phase is in use in which the distribution of materials between stationary and moving phase depends upon something other than adsorption or simple partition. Ion exchange resins and gel filtration media are perhaps the most important of these. The latter are solid materials having small holes or voids in their lattice into which solvent can penetrate. The voids are of such a size, however, that large molecules cannot penetrate them, and, because all voids in a given grade of gel are probably not of exactly the same size, but rather vary over a limited range, molecules of intermediate sizes can penetrate limited proportions of the total void volume. The smallest of such molecules can penetrate a large proportion and the largest a small proportion. When a mixture is chromatographed upon such a column, partition equilibria are set up, controlled by molecular size, in which materials are distributed between solvent in the voids and solvent outside. Large molecules which cannot penetrate the voids at all move rapidly down the column at the same rate as the solvent, and no separation between these molecules is achieved. Various grades of gel are available which enable separations to be achieved with molecules of various sizes.

Ion exchange resins are usually made up of organic polymers containing covalently bound acidic or basic groups. The most common acid group is the sulphonic acid group and basic groups are often quaternary ammonium groups. These are represented diagrammatically in Fig. 29.12. Each type contains either an ion or an ionizable group, a proton in the case of the acid resin and a hydroxide

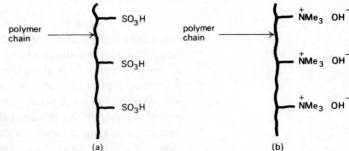

Fig. 29.12 Diagram illustrating portions of synthetic ion exchange resin molecules. (a) Acidic, or cation exchange resin. (b) Basic, or anion exchange resin.

ion in the case of the basic resin illustrated. Either ion can be exchanged for another of the same charge, so that the acid resin is associated with cation exchanges and the basic resin with anion exchanges. When an appropriate mixture is chromatographed over, say, an acid resin, such exchanges are set up between cations dissolved in the solvent phase and cations in the solid phase. These exchanges are usually controlled by the pH of the (normally aqueous) solvent, or other ionic concentrations, and cations may be eluted selectively from the column by controlling these concentrations. This particular chromatographic method finds wide application for organic materials containing ionizable groups and for a variety of inorganic ions. It has, for example, been very useful in effecting the difficult separation of the rare earth metal cations.

A method of analysis superficially resembling paper or thin-layer chromatography is **electrophoresis**. In this process, "spots" of mixture are placed centrally on the plate, or paper, which is placed horizontally. A stationary aqueous phase is used—usually a buffer solution—and, according to pH, ionizable materials adopt their appropriate form. Amino acids, for example, which are zwitterionic (Section 28.1) near neutrality, adopt the form $H_3N^+CHRCOOH$ at high acidities and $H_2NCHRCOO^-$ at low acidities. A high voltage is then applied across the ends of the plate (or paper) and charged species move towards the electrodes, positive ions to the cathode and negative ions to the anode. Rates of movement vary and spots separate. Uncharged materials remain at the starting point. The method is particularly useful in analysis of biological materials, such as blood, containing suitable polar components such as peptides and amino acids. Patterns of spots obtained from biological fluids may be of considerable value in the diagnosis of biochemical disorders.

The liquid–gas system forms the basis of **gas–liquid chromatography** (g.l.c.), alternatively called **vapour-phase chromatography** (v.p.c.). Materials to be separated by this technique must be moderately volatile and separation is achieved by distribution between a stationary, involatile liquid and a moving, inert gas. The equilibrium upon which separations depend here is that between material dissolved in the liquid phase and material in the vapour state above it. The vapour pressure of a given material, above the liquid phase in which it is dissolved, is a measure of this equilibrium and determines how rapidly the material moves with the gas phase. The more volatile materials, having greater vapour pressures, move more rapidly and emerge from the system before the less volatile.

The liquid phase, usually a high molecular weight organic material such as dinonyl phthalate (the ester of phthalic acid, $C_6H_4(CO_2H)_2$, and two moles of nonan-1-ol, $C_9H_{19}OH$), is distributed over the surface of a finely powdered (maximum surface area) inert solid, such as brick dust, packed into a long glass tube. The tube is maintained at a steady temperature and the inert gas, commonly nitrogen or argon, flows along the tube at a steady rate. A micro sample of mixture is injected into the gas stream and is carried through the dispersed liquid phase where separation takes place. The separate components, as they emerge from the end of the tube in the effluent gas, are detected and measured by a device which usually plots (automatically) a curve giving, in essence, the amount of material emerging plotted against time. The emergent materials may also be condensed and collected if pure samples are required. Under given conditions, for a particular rate of flow of the carrier gas, the length of time a compound takes to emerge from the column is a characteristic of that compound and may be used to identify it in much the same way as R_f value is used in paper and thin-layer chromatography. This is, in fact, a method of growing usefulness and application. It is, for example, the method upon which the determination of the alcohol level in blood depends. If indications of alcohol are obtained from a "breathalyser" test on a motorist, a given quantity of blood, from the suspected motorist, after suitable processing, is subjected to g.l.c. analysis in which alcohol, if present, can easily be detected and its amount determined with accuracy. The levels of steroid hormones (sex hormones) present in the urine of pregnant women may also be monitered by g.l.c., although there are additional problems in this case. Steroid hormones often contain polar, hydrogen-bonding groups such as hydroxyl (Section 23.1) which reduce their volatility. A complex alcohol, ROH, may be rendered more volatile, however, if the proton is replaced by a trimethylsilyl group, $(CH_3)_3Si$—, which cannot hydrogen-bond and which reduces the group polarity. The resultant "silylated" molecule, $ROSi(CH_3)_3$, is then sufficiently volatile for g.l.c. analysis.

29.4 CHEMICAL METHODS OF SEPARATION

Mixtures are often met with, in practice, which contain components of very similar physical properties and which are therefore difficult to separate. Sometimes these components may differ in chemical properties in such a way that one (or more) may be subjected to a simple, easily reversed, chemical reaction which converts it to a species of very different physical properties, and which can therefore easily be removed from the mixture. Such a procedure is very often simpler and quicker than, say, a routine fractional crystallization anyway, with components of dissimilar physical properties.

Consider the simple case of a mixture of benzoic acid and naphthalene, both crystalline solids. One of these is an acid and the other a hydrocarbon. Both are insoluble in cold water. If such a mixture is treated with aqueous sodium hydroxide solution, then the acid reacts rapidly to form an ionic sodium salt, sodium benzoate, which is soluble in water and therefore dissolves in the aqueous sodium hydroxide. The hydrocarbon is unaffected by the alkali and does not dissolve, so that the system is neatly and quickly divided into two phases—solid and liquid—with one component in each. The solid naphthalene can be filtered off and the benzoic acid may be recovered by acidifying the alkaline filtrate (with strong mineral acid) whereupon the benzoate ion is protonated to reform benzoic acid which precipitates. In so far as the solid naphthalene phase may contain benzoic acid, intimately mixed in such a way that the alkaline solution cannot easily reach it, it would be more efficient to add a solvent such as ether, which is immiscible with water, rather than to filter off the solid naphthalene. When the naphthalene has dissolved in the ether, the two liquid phases can easily be separated and the naphthalene recovered by evaporating the ether layer to dryness.

What we have done, in this example, is to modify one of the components of the mixture chemically so that separation can be achieved by a solvent extraction process which is simpler, quicker, and easier to perform than the fractional crystallization process which might otherwise have been necessary. We have used an easily reversed reaction so that the modified component (benzoic acid) can easily be regenerated, although this would not have been necessary if only the naphthalene had been required. In cases where the modified component is not required, then reactions may be employed which are not easily reversible.

29.5 AUTOMATION

The successful operation of a number of the techniques described in this chapter by normal manual means would be extremely tedious and very expensive in time. The counter-current distribution method is a case in point, where, say, fifty transfer operations would be virtually impossible by hand. The development of apparatus whereby the technique can be operated automatically, therefore, was a major step forward which allowed full use to be made of the method. Automation has been referred to in other connections, too, notably gas–liquid chromatography and zone refining. Whilst it would be inappropriate in this book to discuss the methods of automation employed, some reference to it is certainly in place as it plays such an important role in increasing the usefulness and application of techniques over a wide range of situations. Details of these methods may be obtained from specialist works such as Abbott and Andrews *An Introduction to Chromatography*.

Outside those cases already mentioned, chromatography has probably been the subject of most developments in automation. Automatically operated machinery is available which enables small fractions of fixed volume to be collected from the bottom of columns over long periods of time (e.g. overnight), with solvent reservoirs continuously supplying eluant to the top of the column. These fractions may be analysed automatically, perhaps by spectroscopic means, so that the different materials emerging from the column can be recovered from the appropriate number of consecutive fractions. Alternatively, devices have been constructed which enable the eluant to be continuously monitored as it emerges from the column, thus enabling the collecting vessel to be changed when a new component begins to emerge. Specialist apparatus, such as amino acid analysers, is commercially available which automates the whole process of analysing mixtures of amino acids by chromatographic techniques.

Even fractional distillation can be automated, at least in part, and commercial apparatus is available, incorporating highly efficient spinning-band columns, in which the distillation temperature is continuously monitored and used to control the collection of fractions.

29.6 DETERMINATION OF MOLECULAR FORMULAE

The molecular formula of a compound is an expression of the numbers of each type of atom present in one molecule of that compound. A compound of formula C_6H_6O contains, in each molecule, six atoms of carbon, six of hydrogen, and one of oxygen. There is no indication of structure or geometry in such a formula, but a knowledge of it is essential before the molecular structure etc. can be investigated.

To work out the molecular formula of any compound, it is necessary to know two things: the elemental composition, usually expressed as the percentage by weight of each element present; and the molecular weight. Although the reader may already be familiar with the calculation of molecular formulae, a simple illustration is given to emphasize the nature of the calculation.

Consider a compound containing 54.5% carbon, 9.1% hydrogen, and no other element but oxygen. The carbon and hydrogen together account for 63.6% (54.5 + 9.1) of the weight of the compound, so that the other 36.4% must be oxygen. We can therefore write down the ratio of the elements present, by weight, as:

$$C : H : O$$
$$54.5 : 9.1 : 36.4$$

Knowing the atomic weights of carbon, hydrogen, and oxygen (12, 1, and 16 respectively), we can convert this weight ratio into the ratio of the numbers of atoms present simply by dividing each number by the appropriate atomic weight. Thus we have the ratio of the numbers of atoms:

$$C : H : O$$
$$\frac{54.4}{12} : 9.1 : \frac{36.4}{16}$$

Simplifying, to give the smallest whole numbers we have:

$$\frac{54.5 \times 16}{12 \times 36.4} : \frac{9.1 \times 16}{36.4} : 1.00$$

i.e. $\quad 1.99 \quad : \quad 4.00 \quad : \quad 1.00$

Clearly, in simple whole numbers, this means that for every oxygen atom present, there are four hydrogen and two carbon atoms. This ratio, known as the **empirical formula**, is written C_2H_4O. It is only a ratio, however, and does not indicate how many atoms there are in a single molecule. It does tell us, though, that the molecular formula must contain an integral number of these empirical formula units, that is it must be $(C_2H_4O)_n$ where n is an integer. The smallest possible value of n is one, and this corresponds to a molecule C_2H_4O having a molecular weight of 44. Whatever the value of n, the molecular weight must be $44n$ (44, 88, 132... etc.) so that even an approximate measure of the molecular weight of the compound will enable us to decide the value of n and to write down the molecular formula. In this case the approximate molecular weight is 90, so that we can see that the correct value is 88, and $n = 2$. The molecular formula is therefore $(C_2H_4O)_2$ or $C_4H_8O_2$.

Although the example chosen here is an organic compound, the method is applicable in exactly the same way for any type of compound, whatever elements, and however many, it may contain. Basically, two experimental measurements must be performed. Firstly the percentage elemental composition (by weight) must be determined (this may involve several different actual experiments) and secondly the molecular weight must be measured. The composition must be accurately known, but an approximate value for the molecular weight is adequate.

Methods employed for molecular weight determination, ranging from the highly accurate methods of mass spectrometry (Section 2.2) to the rather more approximate (and cheaper!) methods depending upon colligative properties, have been discussed in various other parts of the book (Sections 5.4, 5.5, and 5.6) and will not therefore be discussed here. Nor will any attempt be made to cover methods for the estimation of all elements in the Periodic Table. Such methods can be obtained from the standard works on qualitative and quantitative analysis. The approach to the analysis of organic compounds, however, is more general simply because all of these contain carbon and nearly all contain hydrogen. Many also contain nitrogen. Because there are so many organic compounds (far more organic compounds are known than inorganic), and because they all contain this common "core" of elements, a brief account of the analytical methods developed in this field will be given. Because so many "C, H, and N" analyses are required, there has been a high degree of development in this area and a range of commercial instruments is available for their routine execution.

Both detection and estimation of carbon and hydrogen depend essentially upon combustion processes. Simply heating an organic compound with copper oxide (oxygen source) will convert the carbon to carbon dioxide and the hydrogen to water, both of which are easily identified by routine methods. In practice, such a detection is rarely necessary, though, and is of little importance in modern laboratories. Nitrogen, in organic compounds, is usually detected by igniting the compound concerned with molten sodium. Carbon and nitrogen in the compound combine and, under these conditions, are converted to sodium cyanide. The water soluble, ionic cyanide is easily recognized by a standard colour reaction in which it is treated with iron(II) and iron(III) ions. The precipitation of the insoluble, deep-blue, complex cyanide, Prussian blue ($NaFe_2(CN)_6$, Section 20.18) indicates the presence of carbon and nitrogen. Under the conditions of this test—the Lassaigne sodium fusion test—halogens are converted to sodium halides and sulphur to sodium sulphide, all of which can be easily recognized by standard tests for halide or sulphide ions. Again, this method is rarely used in modern laboratories and is passing into chemical history.

All three elements, carbon, hydrogen, and nitrogen are estimated by combustion methods. Carbon and hydrogen are estimated in the same experiment in which a weighed sample of compound is completely burnt in a stream of oxygen gas. All of the carbon is converted to carbon dioxide and all of the hydrogen to water, and both products are collected in tubes containing suitable adsorbents which are weighed before and after the combustion. The weights of water and carbon dioxide produced, and therefore the weights of carbon and hydrogen contained in the original sample, are thus easily obtained. Complete oxidation of nitrogen-containing compounds (mixed with copper oxide as oxygen source, and strongly heated) converts the nitrogen to its gaseous oxides which, on subsequent reduction with metallic copper, are converted to nitrogen gas. This is collected and estimated by measuring its volume at atmospheric pressure.

Fairly sophisticated commercial apparatus is now available in which all three elements are estimated with a single combustion experiment. The whole process is automated with the estimation of carbon dioxide, water vapour, and nitrogen gas carried out by gas–liquid chromatography. The chromatography apparatus is attached to the combustion apparatus so that estimation of the gaseous products follows automatically after the combustion.

PROBLEMS

1. Devise chemical methods for separating the components of the following mixtures:
   a) phenylamine and methylbenzene (both are liquids, of low water solubility);
   b) 4-methylphenylamine and benzoic acid (both of low water solubility);
   c) benzoic acid and methylbenzene;
   d) benzoic acid and 4-methylphenol;
   e) benzoic acid, phenylamine, and naphthalene.
   Assume that all of the compounds listed are of very low water solubility, that the sodium salts of all acidic materials are of high to moderate water solubility, and that the salts of basic materials with hydrochloric or sulphuric acids are of high water solubility. Assume also that all covalent solids in these lists are soluble in ether.

2. Work out the percentage elemental composition of the following compounds (three significant figure accuracy is sufficient—use appropriate values for atomic weights therefore, e.g. C = 12.0, H = 1.00, O = 16.0, N = 14.0, Cl = 35.5, Br = 80.0):
   a) C_6H_5Cl;
   b) $C_6H_5NO_2$;
   c) $CH_3CH_2COOCH(CH_3)_2$;
   d) $BrCH_2CH_2Br$;
   e) $CH_3CON(CH_3)_2$;
   f) $ClCH_2COOCH_2CH_2N(CH_3)_2$;
   g) $[Co(NH_3)_5Br]^{2+}SO_4^{2-}$;
   h) $Cr(H_2O)_6Cl_3$;
   i) $PtCl_2(NH_3)_2$;
   j) $CrO(O_2)_2 \cdot O(CH_2CH_3)_2$.

3. Given the following data, work out the molecular formulae of the compounds referred to (or, where no molecular weight is given, work out the empirical formula).

Compound	%C	%H	%N	Other elements	Approx. mol. wt.
a	52.2	13.0		oxygen only	45
b	53.3	15.5	31.1		45
c	38.9	5.4		oxygen and 38.4% Cl	90
d	43.3	9.04		oxygen and 18.7% P	170
e	4.8	0.4		94.9% Br	250
f	62.1	13.8	24.1		120
g	58.9	9.8		oxygen only	100

h	24.7% K, 34.7% Mn, 40.5% O.
i	26.6% K, 35.4% Cr, 38.1% O.
j	27.0% Si, 72.9% F. Mol. wt. $\approx$ 105.

BIBLIOGRAPHY

Abbott, D., and R. Andrews. *An Introduction to Chromatography* (second edition), Longman (1970).

 Specifically written for the sixth-former, useful for laboratory work.

Mann, F. G., and B. C. Saunders. *Practical Organic Chemistry* (fourth edition), Longman (1960).

 Deals with basic purification techniques as well as a wide variety of organic experiments.

Vogel, A. I. *A Textbook of Practical Organic Chemistry* (third edition), Longman (1966).

 Deals with basic purification techniques as well as a wide variety of organic experiments.

Methods of separation and purification have been traditionally taught in the context of organic chemistry. There is in fact nothing particularly organic about these methods, although many were developed by or for organic chemists. The occurrence of the term "organic chemistry" in the titles of two out of these three books should not be taken therefore as implying that the techniques are solely applicable to organic compounds.

POSTSCRIPT

Organic Synthesis

Throughout the treatment of organic chemistry so far we have stressed methods by means of which the various functional groups can be introduced into molecules, usually involving a single reaction and leaving unanswered the question of where the starting materials come from. Clearly this is an important question. To know that you can make acetic acid by oxidizing ethanol is of little use if you don't know how to obtain ethanol. There are a variety of natural sources of organic chemicals ranging from the particularly important petroleum liquids and natural gases on the one hand, which yield relatively "simple", basic organic chemicals in large quantities, through coal (less important now), to living organisms on the other. The latter yield relatively small quantities of more complex compounds such as penicillin (from moulds) and quinine (from trees). The living sources are usually plants or fungi and can be "cultivated".

The heavy organic chemicals industry is concerned with the transformation of a number of the materials found in oil, natural gas, and so on, into other useful, but still relatively simple, organic molecules which are of direct commercial use in large quantities (e.g. as solvents) or which can be transformed into more complex molecules, perhaps in smaller quantities, in laboratories or in smaller industries. The pharmaceutical industry, however, at the other end of the scale, handles more complex syntheses of more complex molecules, using both the relatively simple molecules from the "heavier" side of the industry and the more complex molecules from living sources.

A fairly complex organic molecule is usually obtained by a series of transformations, starting with a molecule which is "readily available", and not too expensive. A vast range of compounds, of varying complexity, can normally be obtained from a few very simple starting materials, via such series of transformations, and Tables P.1 and P.2 are intended to illustrate this point making use of just a few reactions and a few compounds. It will be clear from the tables that a large number of complex molecules are all related through synthetic sequences and all of the reactions in the tables are simple reactions which we have already met in this book (although some of the compounds we certainly have not met). This generality of reactions is an important principle running through organic chemistry, although it must often be applied cautiously with an eye to exceptions. Growing understanding and experience of organic chemistry leads to improved skills in designing synthetic sequences.

Two different aspects of synthesis are emphasized in the tables. The conversion of one functional group into another is emphasized in Table P.1 where most molecules have only two or three carbon atoms in their skeletons. Table P.2 however, lays emphasis on the building up (and breaking down) of carbon skeletons by reactions involving the making (and breaking) of carbon–carbon bonds. Not all synthetic sequences in the tables, represented by series of arrows, are necessarily viable. Some are there simply to show what can be done, and to emphasize relationships between molecules. Nevertheless, what may be viable today may not be viable tomorrow. Sudden changes in availability or cost of materials may render a commercial synthesis uneconomic and force the manufacturer to adopt another synthesis for his product.

The "heart" of each table is the shaded area in which large-scale industry operates. All compounds in Table P.1 spring ultimately from ethylene and all those in Table P.2 from acetylene or ethanol (which in turn is obtained from ethylene). No organic compounds are included in the tables as reagents unless they have been produced in another part of one of the tables. Syntheses outside the shaded areas are more often carried out on smaller scales, perhaps in laboratories. Space prevents the inclusion of all but the briefest of details for a few of the reactions. Fuller details have been the concern of earlier chapters.

The viability of a given synthesis is a function of a number of factors. The cost of starting materials, reagents, and apparatus ("plant", on an industrial level) is clearly of high importance; more so perhaps in commercial production than in the research laboratory where economic efficiency in the short term is not a primary aim. The natures of the various reactions in a given sequence are also of crucial importance. If every reaction in a sequence were "quantitative", then we should have a highly efficient process. One mole of acetylene, for instance (Table P.2) would lead to one mole of lactic acid, $CH_3CH(OH)COOH$. However, each reaction in each sequence is usually less than perfect, and produces, from a given quantity of starting material, less than the quantity of product predicted by the stoichiometric equation. The actual amount of product obtained in a reaction, relative to the amount predicted by the equation, is known as the *yield* of the reaction, and is usually expressed as a percentage. A quantitative reaction has a yield of 100%. Clearly, the higher the yield the more efficient is the reaction. It is also important to recognize that the greater the number of steps in a synthetic sequence, the more important does the yield of each step become. A sequence of, say, ten reactions, each with a yield of 80%, leads to an overall yield of the final product of $(\frac{80}{100})^{10} \times 100\%, \approx 10\%$. If each step gave a 60% yield, then the overall yield would be less than 1%. A yield of 80% for one reaction is good, and this emphasizes the importance of avoiding low yield reactions in a good synthetic sequence. It also emphasizes the importance of using as few reactions as possible in the sequence from reactants to products. A three-reaction sequence, say, provided it contains no particularly low-yield reactions, is normally preferred relative to a five-reaction sequence.

There are a number of reasons why reactions give the desired product in yields of less than 100 percent. For thermodynamic reasons, the reaction may not go to completion, that is it may reach an equilibrium position. Often, by controlling the conditions, something can be done about this. For example, a product may be distilled out of a reaction mixture as it forms, or precipitated out, thus shifting the equilibrium in favour of products. It may be that a competing reaction—a side-reaction to give unwanted products—proceeds at a significant rate and diverts some of the reactants to unwanted side-products. Again, control of conditions may lead to an improvement. Most industrial reactions are carefully examined, sometimes quite empirically, to discover conditions for maximizing yield before being put into operation. Often a side-product turns out to be marketable, or otherwise useful in its own right, and in an industrial process this might make all the difference as to whether one reaction sequence was favoured, relative to another. In fundamental research, where there is less concern for making a synthesis pay, the overall yield of product may be the most important criterion.

Synthesis is something which runs through the whole of organic chemistry and which most organic chemists—most chemists for that matter—have to indulge in at some time or other. There are a variety of reasons for synthesizing compounds, some more obvious than others. For example, the synthesis of monomers, and ultimately polymers in the form of plastics, fulfils an obvious social need, as do most industrial products. In the research laboratory, however, the need to synthe-

Table P.1

[Reaction scheme showing interconversions among: CH_3CHO, CH_3COOH, CH_3CH_2OH, CH_3CH_2COOH, CH_3CH_2CN, $CH_3CH_2CH_2NH_2$, CH_3CH_2Br, $CH_3CH_2NH_2$, $CH_2=CH_2$, $BrCH_2CH_2Br$, the epoxide CH_2-CH_2 with O, CH_2ClCH_2OH, CH_2OHCH_2OH, CH_2CN/CH_2CN, CH_2NH_2/CH_2NH_2, CH_2COOH/CH_2COOH, $CH_2CH_2NH_2/CH_2CH_2NH_2$, CH_2CN/CH_2OH, CH_2COOH/CH_2OH, $CH_2CH_2NH_2/CH_2OH$, and $CH_2=CHCOOH$. Reagents used: [O], H^+/H_2O, $LiAlH_4$, NaOH, KBr/H_2SO_4, NaCN, NH_3, HBr, HOCl, HCl, air, Br_2, H_2O, [H], $-H_2O$.]

Abbreviations: H^+—acid
[O]—oxidation

size compounds is also very common. If a research worker wishes to investigate a reaction mechanism, say the hydrolysis of esters, he may need a whole range of slightly different esters to examine, some of them quite unusual. He will have to synthesize them, as they probably cannot be bought, or found naturally. Other research problems require unusual molecules, which must be made. Some of the

Table P.2

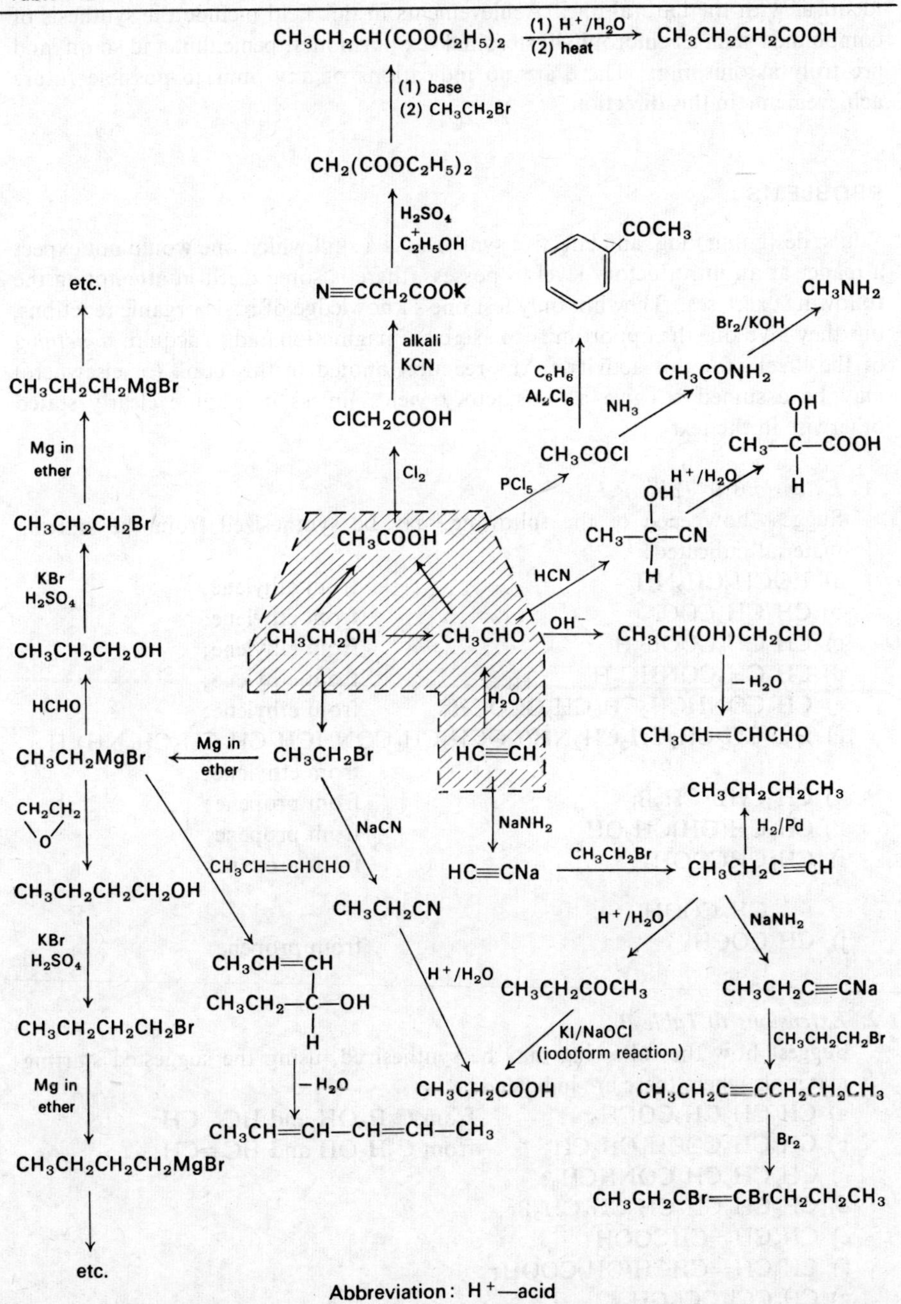

$$CH_3CH_2CH(COOC_2H_5)_2 \xrightarrow[(2) \text{ heat}]{(1) \text{ H}^+/\text{H}_2\text{O}} CH_3CH_2CH_2COOH$$

$$\uparrow (1) \text{ base} \\ (2) \text{ CH}_3\text{CH}_2\text{Br}$$

$$CH_2(COOC_2H_5)_2$$

$$\uparrow \begin{array}{c} H_2SO_4 \\ + \\ C_2H_5OH \end{array}$$

etc.

$$N\equiv CCH_2COOK$$

$$\uparrow \text{alkali} \\ \text{KCN}$$

$$CH_3CH_2CH_2MgBr \qquad ClCH_2COOH$$

Mg in ether $\uparrow \quad Cl_2 \uparrow$

$$CH_3CH_2CH_2Br$$

KBr, H_2SO_4 $\uparrow$

$$CH_3CH_2CH_2OH$$

HCHO $\uparrow$

$$CH_3CH_2MgBr \xleftarrow[\text{ether}]{\text{Mg in}} CH_3CH_2Br$$

$$\begin{array}{c} CH_2CH_2 \\ \diagdown O \diagup \end{array} \updownarrow$$

$$CH_3CH_2CH_2CH_2OH$$

KBr, H_2SO_4 $\downarrow$

$$CH_3CH_2CH_2Br$$

Mg in ether $\downarrow$

$$CH_3CH_2CH_2CH_2MgBr$$

$\downarrow$

etc.

C$_6$H$_6$ / Al$_2$Cl$_6$ → C$_6$H$_5$COCH$_3$

$$CH_3COOH \xrightarrow{PCl_5} CH_3COCl \xrightarrow{NH_3} CH_3CONH_2 \xrightarrow{Br_2/KOH} CH_3NH_2$$

$$CH_3CH_2OH \rightarrow CH_3CHO \xrightarrow{HCN} CH_3-\underset{\underset{H}{|}}{\overset{\overset{OH}{|}}{C}}-CN \xrightarrow{H^+/H_2O} CH_3-\underset{\underset{H}{|}}{\overset{\overset{OH}{|}}{C}}-COOH$$

$$CH_3CHO \xrightarrow{OH^-} CH_3CH(OH)CH_2CHO$$

$$\downarrow -H_2O$$

$$CH_3CH=CHCHO$$

$$HC\equiv CH \xrightarrow{H_2O}$$

$$\downarrow NaNH_2$$

$$CH_3CH=CHCHO \qquad HC\equiv CNa \xrightarrow{CH_3CH_2Br} CH_3CH_2C\equiv CH \xrightarrow{H_2/Pd} CH_3CH_2CH_2CH_3$$

NaCN $\downarrow$

$$CH_3CH_2CN$$

$$CH_3CH=CH \\ | \\ CH_3CH_2-C-OH \\ | \\ H$$

$\downarrow -H_2O$

$$CH_3CH=CH-CH=CH-CH_3$$

$$\xrightarrow{H^+/H_2O} CH_3CH_2COCH_3 \xrightarrow[\text{(iodoform reaction)}]{KI/NaOCl} CH_3CH_2COOH$$

$$\xrightarrow{NaNH_2} CH_3CH_2C\equiv CNa \xrightarrow{CH_3CH_2CH_2Br} CH_3CH_2C\equiv CCH_2CH_2CH_3$$

$$\downarrow Br_2$$

$$CH_3CH_2CBr=CBrCH_2CH_2CH_3$$

Abbreviation: H$^+$—acid

ORGANIC SYNTHESIS 767

longest (in terms both of years of effort, and numbers of reactions in the sequence) and most elegant syntheses, however, are associated with the field of natural products—complex molecules (usually) which occur in living things. A good deal of effort leads the investigators to propose a structure for a given compound, and they then proceed to test their proposals by synthesizing the proposed structures by a sequence of known reactions. The acid test is whether the synthetic material is identical with the natural one. Achievements in this field include the synthesis of compounds such as chlorophyll, steroidal sex hormones, penicillin, and so on, and are truly astonishing. There are no indications of any limit to possible future achievements in this direction.

PROBLEMS

Whilst designing long, and effective syntheses is a skill which one would not expect a reader at an introductory level to possess, there is some merit in attempting the following exercises. They not only test one's knowledge of basic organic reactions, but they give one the opportunity to exercise imagination and to acquire *something* of the "feel" for this activity. Any reaction quoted in this book (or elsewhere) may be assumed to give a satisfactory yield, unless it is quite clearly stated otherwise in the text.

1. *Extensions to Table P.1*
   Suggest how each of the following may be synthesized from the starting material indicated:
   a) $HOCH_2CH_2NH_2$ from ethylene;
   b) CH_3CH_2COCl from ethylene;
   c) $CH_3CH_2COOC_2H_5$ from ethylene;
   d) $CH_3CH_2CONHC_2H_5$ from ethylene;
   e) $CH_3CONHCH_2CH_2CH_2OCOCH_3$ from ethylene;
   f) $H_2NCH_2CH_2CH_2CH_2NH(COCH_2CH_2CONHCH_2CH_2CH_2NH)_nH$
   from ethylene;
   g) $CH_3CHBrCH_2Br$ from propene;
   h) $CH_3CH(OH)CH_2OH$ from propene;
   i) $CH_3CHCOOH$ from propene;
   $|$
   CH_2COOH
   j) CH_3COCH_3 from propene:

2. *Extensions to Table P.2*
   Suggest how the following may be synthesized, using the suggested starting materials where these are indicated:
   a) $CH_3CH_2CH_2COCH_3$ from C_2H_5OH and $HC{\equiv}CH$;
   b) $CH_3CH_2COCH_2CH_2CH_3$ from C_2H_5OH and $HC{\equiv}CH$;
   c) $CH_3CH_2CH_2CONHCH_3$;
   d) $CH_3CH_2CH_2CH_2CH_2CH_2Br$;
   e) $CH_3CH{=}CHCOOH$;
   f) $CH_3CH{=}CHCH(OH)COOH$;
   g) $CH_3CCl_2CCl_2CH_3$;

h) $(CH_3)_2C=CHCOCH_3$ from $CH_3CH(OH)CH_3$ (hint—what is the effect of alkali on acetaldehyde?);

i) $CH_3CH_2CH_2CH_2OCOCH_2Cl$;

j) $CH_2\begin{smallmatrix}CN\\NH_2\end{smallmatrix}$

3. Suggest, in outline, how the following syntheses might be achieved:

a) C₆H₅–NH₂   from benzene;

b) C₆H₅–CONHCH₂CH₃   from benzoic acid and bromoethane;

c) C₆H₅–COCH₃   from benzene and acetic acid;

d) C₆H₅–CH₂OH   from bromobenzene and formaldehyde;

e) C₆H₅–OCH₃   from phenol and methanol.

Appendices

A.1	The covalent bond
A.2	Some further aspects of bonding and structure
A.3	Energy distributions and entropy
A.4	Entropy: molecular interpretation and calorimetric measurement
A.5	Determination and evaluation of entropy changes
A.6	Determination of standard entropies: an example
A.7	Chemical equilibrium
A.8	Transition state theory
A.9	Crystal field theory
A.10	Ionization energies and electron affinities
A.11	Bond energies and bond distances
A.12	Enthalpies of hydration
A.13	Radii, electronegativities, enthalpies of atomization, and atomic volumes of the elements
A.14	Boiling points, melting points, densities, enthalpies of fusion and vaporization, reduction potentials, thermal conductivities, and atomic conductances of the elements
A.15	Thermodynamic properties of selected inorganic substances
A.16	Thermodynamic properties of selected organic substances
A.17	Values of physical constants and conversion factors
A.18	Glossary of symbols
A.19	Glossary of organic nomenclature

A.1 THE COVALENT BOND

If we compare the electron density of the hydrogen molecule with that of the hypothetical situation of two non-interacting hydrogen atoms at the same distance apart, the striking difference is the increased electron density between the nuclei in the case of the molecule. The build up of electron density between the nuclei leads to the formation of a strong bond through electron attraction for the two nuclei. Detailed calculation confirms the essential validity of this qualitative picture of covalent bonding. We can envisage electrons occupying a **molecular orbital** encompassing both nuclei for which there is a significant probability of finding an electron between the nuclei.

In principle, electron densities in molecules can be obtained by solving the Schrödinger equation. In practice this presents major problems except for the simplest of molecules and an alternative approach is required. Chemists naturally think of molecules in terms of their constituent atoms and it does in fact prove convenient to think of molecular orbitals in terms of atomic orbitals. Consider the $1s$ atomic orbitals for two hydrogen atoms centred about two closely placed nuclei as illustrated in Fig. A.1. A molecular orbital for the system can be envisaged as the combination of two atomic orbitals. It must be remembered that the orbitals have wave properties and combination can lead to reinforcement or to destructive interference. In Fig. A.1(a) the orbitals are combined in phase and there is reinforcement whereas in Fig. A.1(b) the orbitals have been combined with opposite phases. As for atoms, two electrons can be assigned to each molecular orbital. For the hydrogen molecule in its lowest energy state both electrons occupy the orbital shown diagramatically in Fig. A.1(a). In this way, by combining atomic orbitals, a molecular orbital is obtained, which leads to a build up of electron density between the nuclei for the electronic ground state of the hydrogen molecule. Excited electronic states exist, as for atoms, and the orbital illustrated in Fig. A.1(b) is of interest in understanding these.

The picture provided for the formation of a covalent bond for the hydrogen molecule can be extended to other molecules. A molecular orbital is regarded as being formed by the overlap of atomic orbitals. Other things being equal the greater the overlap the stronger the bond. The hydrogen molecule is exceptional in that the main source of repulsion is that due to repulsion between the nuclei.

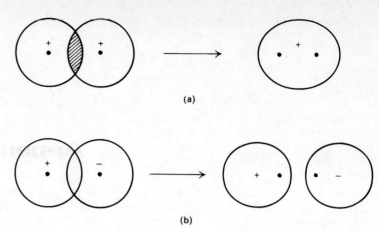

Fig. A.1 Deriving molecular orbitals by combining atomic orbitals.

In other molecules there will be repulsion between other electrons associated with the two atoms. For the hydrogen molecule the molecular orbital is symmetrical with respect to both nuclei. For a bond between two unlike nuclei this will not be the case. The electron density will be greater in the neighbourhood of the more electronegative atom, giving rise to a polar bond. In the extreme case of large electronegativity difference the asymmetry may become so large that an electron is transferred across from one atom to the other leading to an ionic bond. In the covalent bond for the hydrogen molecule the electron density is transferred into the region between the nuclei. This is the opposite extreme from the ionic bond and a continuous range of intermediate bonds between these two may be formed (Section 4.12.2).

A.2 SOME FURTHER ASPECTS OF BONDING AND STRUCTURE

The treatment of atomic orbitals, as we have presented it in this book, has one obvious weakness which may already have occurred to the reader. It bears little relationship to the geometries and structures of the molecules which we have discussed at some length. Thus we have described some orbitals which are non-directional (s-orbitals) and some which are directional and mutually perpendicular (p-orbitals)—Section 2.3.7. We might therefore expect that the only observable bond angles should be right angles. This is manifestly not the case and if the theory were really unable to accommodate the wide spectrum of observed bond angles, then it would have to be abandoned.

In fact, of course, the theory is capable of suitable modification and we describe in this appendix a useful and common treatment of the problem. An understanding of this development is not by any means essential for understanding the chemistry we have covered in the book. However, because it shows how the theory may be reconciled with observation, and because of its place in the language of chemistry, we have thought it wise to include a simple treatment. In placing it thus in an appendix we have avoided an unnecessary clouding of our overall presentation of chemistry with a difficult idea which is interesting and important, but not essential to chemistry at this level.

The principal idea involved in this further development is the **hybridization** of atomic orbitals and, at this early stage, it must be made absolutely clear that this is not a phenomenon. That is, it is not a process which we can observe. It is essentially a mathematical convenience—a way of rewriting the wave functions associated with atomic orbitals such as to reconcile them with the observed geometries of molecules. It has little predictive value in this respect and we are dependent upon experiment to determine molecular geometry and therefore, also, the geometrical distribution of electrons forming bonds. Furthermore, there are other ways of treating the same problem which are just as valid as the hybridization concept. The reason for choosing hybridization as the model to use for discussing certain topics in this book is basically that it is the most convenient one to use in rationalizing a large quantity of data concerned with both structure and reactivity. It is a model commonly employed by working chemists and, indeed, is virtually part of the language of chemistry. One of the dangers of its having become part of that language is that its everyday use seems to imply that it is taken for granted as a factual phenomenon. Life is too short to preface every reference to hybridization with the caution "if we use the hypothetical model . . ." and the concept is therefore often spoken of as though it were indeed a fact. The result of this is that many people have come to believe that it is a phenomenon. Like life, space in this book is also limited, and it is the purpose of this paragraph to justify the use of the idea in the language of subsequent paragraphs and to try to avoid perpetuating the common error that hybridization is something which atoms are observed to do.

From experimental evidence, molecules such as methane are known to be tetrahedral in shape, with carbon at the centre and each of four hydrogen atoms at the corners of the tetrahedron. Each of the four hydrogen atoms is identical with its neighbours in every way, and it is reasonable to conclude, therefore, that the four bonds between carbon and the hydrogen atom are also identical. It is clear that we cannot construct a model of the methane molecule based simply on overlap between the s-orbitals of hydrogen and the separate $2s$- and $2p$-orbitals of carbon. s-Orbitals are different from p-orbitals and both cannot give rise to identical bonds. Neither s- nor p-orbitals are tetrahedrally disposed. It is therefore necessary to modify our mathematical model of the bonding orbitals of carbon in such a way that all four are identical. This can be considered as combining all four orbitals and separating them again into four identical orbitals. This is the (mathematical) process known as hybridization, and, because the new orbitals are made up from one s- and three p-orbitals, they are referred to as "sp^3" hybrids. Each has 25 percent "s-character" and 75 percent "p-character". Each new orbital is identical with the other three and their axes are disposed tetrahedrally from the nucleus. This is illustrated in Fig. A.2.

There is still another problem, however, to be dealt with before we go on to use our hybridization idea to construct a model of the bonding orbitals of methane, and this involves the nature of the ground state configuration of carbon. This configuration is (Section 2.5) $1s^2 2s^2 2p_x^1 2p_y^1 2p_z^0$, and, as there are only two unpaired electrons, both in p-orbitals, we might expect carbon to form only two electron pair bonds. This is only rarely observed, and those species in which it is observed are too unstable to exist for any length of time. Thus, if we are to imagine bonds forming between a carbon and four hydrogen atoms, the carbon atom must first separate its $2s$-electron pair so that we have four unpaired electrons in a new configuration, $1s^2 2s^1 2p_x^1 2p_y^1 2p_z^1$. If we were to bring this process about, then we should have to supply energy to the carbon atom.

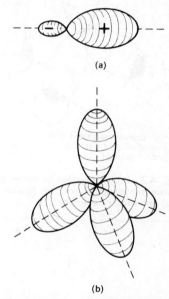

Fig. A.2 (a) A single sp^3 hybrid orbital. (b) Four tetrahedrally disposed sp^3 hybrid orbitals (axes shown as dotted lines). The smaller lobes are not shown.

It will now be helpful to construct a picture of a methane molecule being built up from four separate hydrogen atoms and a carbon atom. Although this process is quite unrelated to the chemical formation of methane, it enables us to understand why these five atoms are more stable when bonded together as methane, than they are as five separate species. If this were not the case, then the molecule would simply fall apart or, rather, would never form. We shall divide up the formation process into several quite arbitrary steps, chosen simply because we can pick out an energy term associated with each step and, when all these are summed, form an impression of which state of affairs is the more stable—that in which we have separate atoms or that in which we have a bonded methane molecule. The steps are:

1. Separation of the carbon 2s-electrons (spin-paired) and promotion of one of them to the p_z-orbital so that each orbital in shell 2 is occupied by a single electron.
2. Hybridization of the 2s- and 2p-orbitals.
3. Bonding by the interaction of the electrons in the hybrid orbital, each with the 1s-electrons of four separate hydrogen atoms, providing high electron density (molecular orbitals) between the nuclei and thus binding them together (Section A.1).

Each of the stages (1) and (2) will clearly require that energy be supplied to the carbon atom. The strength of a carbon–hydrogen bond is high, however (approximately 400 kJ mol^{-1}), and in stage (3) four of them form. The release of so much energy in the final stage more than covers the requirements of the first two and, on summing the requirements of all three stages, it is apparent that the whole process is exothermic. Provided that the loss in entropy is not too great, it is thermodynamically spontaneous. The reverse of this process is therefore *not* spontaneous, and methane does not fall apart. This combination of atomic orbitals to form the molecular orbitals of methane is illustrated in Fig. A.3.

When we make use of the hybridization idea, we are not confined to combinations of all four orbitals at once. Thus we might combine the 2s-orbital with only two of the 2p-orbitals. This gives three new sp^2-orbitals and leaves one p-orbital unhybridized. The hybrid orbitals all lie in one plane and their axes are directed from the nucleus towards the corners of an equilateral triangle with carbon at the centre. Inter-orbital angles are therefore 120°. The unhybridized p-orbital lies perpendicular to this triangle and the whole geometrical arrangement is illustrated in Fig. A.4.

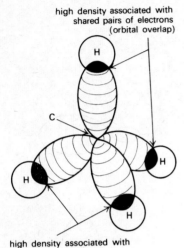

Fig. A.3 Orbital overlaps in methane between hydrogen 1s-orbitals (spherical) and carbon sp^3-orbitals.

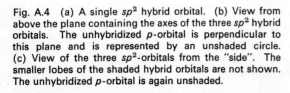

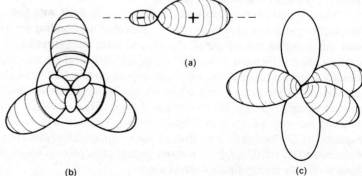

Fig. A.4 (a) A single sp^2 hybrid orbital. (b) View from above the plane containing the axes of the three sp^2 hybrid orbitals. The unhybridized p-orbital is perpendicular to this plane and is represented by an unshaded circle. (c) View of the three sp^2-orbitals from the "side". The smaller lobes of the shaded hybrid orbitals are not shown. The unhybridized p-orbital is again unshaded.

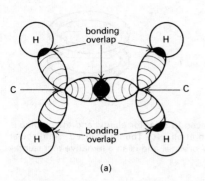 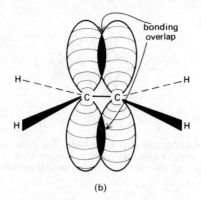 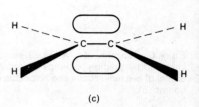

Fig. A.5 (a) Bonding overlap between sp^2 hybrid orbitals of carbon and 1s-orbitals of hydrogen in ethylene, viewed from above the plane of the molecule. These form the σ-bonds in ethylene. Unhybridized p-orbitals are not shown. (b) Sideways bonding overlap between the unhybridized p-orbitals of ethylene, viewed from the edge of the flat molecule. This forms the π-bond. σ-bonds are shown in the conventional stereochemical way, not as combinations of orbitals. (c) This is an alternative representation of (b), but shows the π-molecular orbital, which is the combination of the two p-orbitals shown in (b).

The sp^2 hybrid provides a good model for describing the bonding in molecules which contain carbon bonded to only three other atoms or groups, such as ethylene or benzene. Both of these molecules are known to be flat, with bond angles of 120°, and this correlates well with the sp^2 hybrid geometry.

We can build a model of ethylene by postulating the formation of a bond between the two carbon atoms as a result of overlap between the ends of one hybridized orbital from each atom. Hydrogen atoms can be linked to carbon by bonds formed through overlap of their 1s-orbitals with the ends of the hybridized carbon orbitals not used in bonding the carbon atoms together. This provides the basic skeleton of the molecule, but the description is not yet complete. There remains one electron in each of the two unhybridized p-orbitals, one on each carbon atom. If these two orbitals are parallel, then they can overlap sideways and form a bond in that way. This type of overlap is very different from the type already described and creates two regions of electron density, one on each side of the molecule. The complete picture of the ethylene model is illustrated in Fig. A.5.

The type of bond which is formed by "end-on" overlap is referred to as a σ-bond. It is cylindrically symmetrical about the internuclear axis (the bond axis) and as a consequence of this, rotation of the groups at either end of the bond, relative to one another, does not affect the overlap. Such rotation about single (σ) bonds is a relatively easy process (Section 21.1). The second type of bond, formed by sideways overlap, is called a π-bond. It is not cylindrically symmetrical with respect to the bond axis and rotation about the axis seriously affects the overlap between the two p-orbitals. Maximum overlap, and therefore maximum bond strength, requires the two p-orbitals to be parallel. If the two atomic orbitals are mutually perpendicular, then overlap is at a minimum (effectively zero) and there is essentially no π-bond. Thus, the most stable condition is that in which the two p-orbitals are parallel, and this requires that the two CH_2 units at either end of the molecule should lie in the same plane. The molecule is thus flat and any rotation of the two CH_2 units, relative to one another, requires sufficient energy to break the π-bond. At room temperature, alkene molecules do not have sufficient energy

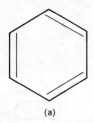
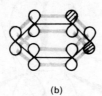
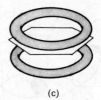

Fig. A.6 Benzene. (a) A Kekulé structure. (b) Overlap between *p*-orbitals. The latter are reduced in size, for clarity, and the overlap is indicated by shading. σ-bonds are represented as straight lines—those to hydrogen are not shown at all. (c) The delocalized π-molecular orbital of benzene, above and below the plane of the ring. This is the combination of the six atomic *p*-orbitals, and is an alternative representation of (b).

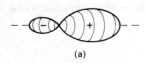

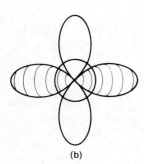

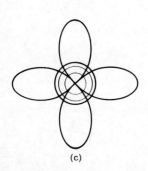

Fig. A.7 (a) A single *sp* hybridized orbital. (b) Both *sp* hybridized orbitals with *p*-orbitals shown (unshaded). The smaller lobes of the hybrid orbitals are not shown. (c) As in (b), but viewed along the common axis of the hybridized orbitals, so that the unhybridized *p*-orbitals are in the plane of the paper.

to allow this rotation to take place, and this results in the restriction of rotation referred to in Section 21.4. The effect of heat in bringing about the conversion of fumaric acid to maleic anhydride is presumably to supply this energy. The σ-bond continues to hold the two carbon atoms together and permits free rotation to occur. The double bond, then, normally represented as a double line (C=C), can be seen to be, on this model, a combination of one σ-bond and one π-bond.

If we use a similar sp^2 model to describe the bonding in benzene (Fig. A.6), with σ-bonds holding the ring carbon atoms together and binding hydrogen to carbon, we note in the first place that all nuclei lie in the same plane. That is, the ring is flat, as observed. We also note an interesting feature of the disposition of the unhybridized *p*-orbitals. There are six of them, of course, and, as with the two in ethylene, they are all parallel. Each holds one electron. If we are to envisage these orbitals overlapping sideways in pairs, to give something equivalent to a Kekulé formula with three isolated double bonds, then we run into a difficulty. Any one of the *p*-orbitals might overlap thus with two others, and it is difficult to decide which one. In fact we do not attempt to envisage such an arrangement, but rather one in which the overlap between *p*-orbitals is continuous round the ring. This gives rise to two circular regions of electron density, one above the plane of the ring and one below it, in much the same way as the ethylene π-orbital has one part above the plane of the molecule and one part below. All six π-electrons are able to become delocalized round the ring and this should be seen in the context of the earlier discussion of delocalization in benzene (Section 4.12).

Molecules in which carbon is bonded to only two other atoms can be rationalized in a similar manner. We need to combine the 2*s*-orbital of carbon with only one of the 2*p*-orbitals and this gives us two *sp* hybrid orbitals together with two unhybridized *p*-orbitals. The two hybrid orbitals lie in the same straight line with the *p*-orbitals perpendicular to this line and remaining, of course, mutually perpendicular (Fig. A.7). Figure A.8 illustrates the description of acetylene, known to be a linear molecule, with one σ-bond and two π-bonds holding the carbon atoms together.

In describing hybridization and its use in building up molecular orbitals to account for the bonding in molecules, we have confined ourselves, so far, to a consideration of carbon. Similar use, though, can be made of the concept with other atoms. Thus sp^3 hybridization for nitrogen in ammonia (Fig. 4.9) (and amines) provides a convenient model for constructing molecular orbitals. The lone

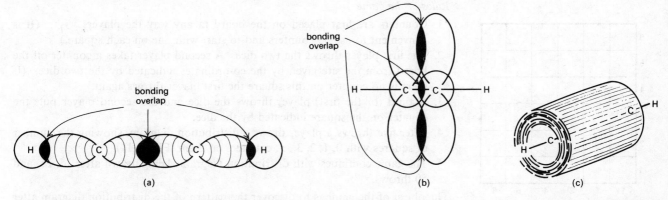

Fig. A.8 (a) Bonding overlap associated with the σ-bonds of acetylene. Unhybridized *p*-orbitals not shown. (b) Bonding overlap between the two sets of unhybridized *p*-orbitals of the acetylene carbon atoms associated with the two π-bonds. The *p*-orbitals on each atom are mutually perpendicular. The σ-bonds are shown as straight lines. (c) The diffuse π-orbitals of acetylene, seen as symmetrically surrounding the σ-bond axis (an alternative representation of (b)).

pair is in an sp^3-orbital and, being closer to the nitrogen nucleus than are the bonding pairs (in the other three sp^3-orbitals), "demands" more "space" than do the latter, thus slightly reducing the H—N—H bond angle below the tetrahedral value. A similar model is helpful for water (Fig. 4.10), with two bonding pairs and two lone pairs. sp^2 hybridization is the appropriate model for boron trifluoride (Section 4.10). More complex molecules, with co-ordination numbers greater than four, may similarly be described by including *d*-orbitals in the hybridization procedure. Thus octahedral geometry would involve d^2sp^3 hybridization. Any observed molecular geometry can be accommodated. (An alternative way of looking at the shapes of molecules is given in Section 12.7.)

A.3 ENERGY DISTRIBUTIONS AND ENTROPY

A Game with Dice and Counters

Consider a planar array of 36 atoms arranged as a square, 6 × 6. Suppose that each atom may accept energy only in quanta ϵ, that is, the energy levels for each atom are equally spaced (Fig. A.9). This is in fact a good model for the vibrational energies of atoms about their mean positions. Suppose we give the system a total of energy 36 ϵ, that is 36 quanta. We are interested in how the energy is likely to be distributed among the atoms.

The game. The game provides a simple model and examines the effect of distributing the energy in a random way. The board on which the game is played is divided into 36 squares as shown in Fig. A.10. The numbers should be distinguished in some way; suppose the horizontal numbers are blue and the vertical ones red. Two dice and a number of counters which fit in the squares are required. One of the dice has blue markings and the other red. By throwing the dice according to the rules of the game the counters are to be distributed on the squares. Each square represents an atom and each counter a quantum of energy.

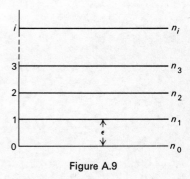

Figure A.9

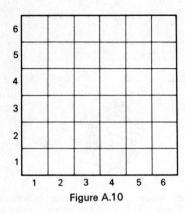

Figure A.10

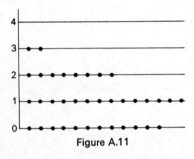

Figure A.11

Rules of the game
1. Counters are first placed on the board in any way the players wish. (It is convenient to use 36 counters and to start with one on each square.)
2. The first player throws the two dice. A second player takes a counter off the board from the site given by the co-ordinates indicated by the two dice. (If there is no counter on this square the first player throws again.)
3. A third (or the first) player throws the dice and the second player puts the counter on the square indicated by the dice.
4. After five throws a player draws a distribution diagram, showing the number of squares with 0, 1, 2, 3 . . . counters on them (Fig. A.11).
5. The game continues with distribution diagrams being drawn after 15, 25, and 50 throws.

The object of the game is to discover the pattern of the distribution diagram after many throws. What would you expect the result to be if much larger numbers were used?

An extension of the game is to calculate

$$W = \frac{n!}{n_0! \, n_1! \, n_2! \cdots n_i!}$$

after 2, 5, 15, 25, and 50 throws. Table A.3.1 will be found helpful. In the expression, n is the total number of squares; and n_0, n_1, n_2, etc. are the numbers of squares

Table A.3.1. Factorials and their logarithms.

n	$n!$	$\log n!$	n	$n!$	$\log n!$
			25	1.5511×10^{25}	25.19065
1	1.0000	0.00000	26	4.0329×10^{26}	26.60562
2	2.0000	0.30103	27	1.0889×10^{28}	28.03698
3	6.0000	0.77815	28	3.0489×10^{29}	29.48414
4	2.4000×10	1.38021	29	8.8418×10^{30}	30.94654
5	1.2000×10^2	2.07918	30	2.6525×10^{32}	32.42366
6	7.2000×10^2	2.85733	31	8.2228×10^{33}	33.91502
7	5.0400×10^3	3.70243	32	2.6313×10^{35}	35.42017
8	4.0320×10^4	4.60552	33	8.6833×10^{36}	36.93869
9	3.6288×10^5	5.55976	34	2.9523×10^{38}	38.47016
10	3.6288×10^6	6.55976	35	1.0333×10^{40}	40.01423
11	3.9917×10^7	7.60116	36	3.7199×10^{41}	41.57054
12	4.7900×10^8	8.68034	37	1.3764×10^{43}	43.13874
13	6.2270×10^9	9.79428	38	5.2302×10^{44}	44.71852
14	8.7178×10^{10}	10.94041	39	2.0398×10^{46}	46.30959
15	1.3077×10^{12}	12.11650	40	8.1592×10^{47}	47.91165
16	2.0923×10^{13}	13.32062	41	3.3453×10^{49}	49.52443
17	3.5569×10^{14}	14.55107	42	1.4050×10^{51}	51.14768
18	6.4024×10^{15}	15.80634	43	6.0415×10^{52}	52.78115
19	1.2165×10^{17}	17.08509	44	2.6583×10^{54}	54.42460
20	2.4329×10^{18}	18.38612	45	1.1962×10^{56}	56.07781
21	5.1091×10^{19}	19.70834	46	5.5026×10^{57}	57.74057
22	1.1240×10^{21}	21.05077	47	2.5862×10^{59}	59.41267
23	2.5852×10^{22}	22.41249	48	1.2414×10^{61}	61.09391
24	6.2045×10^{23}	23.79271	49	6.0828×10^{62}	62.78410
			50	3.0414×10^{64}	64.48307

with no counters, one counter, two counters, etc. W is the number of ways of distributing the counters on the board all having the same distribution diagram. For example it is the number of different drawings that could be made in Fig. 6.8 all for the same distribution diagram of Fig. 6.8(c). It is readily obtained as the number of ways of dividing n objects into groups of $n_0, n_1, n_2, n_3 \ldots$ objects. It is instructive to consider the question: how does W change as the game is played?

Readers interested in extending the ideas to larger numbers are referred to the Nuffield Advanced Physics book, Unit 9, *Change and Chance* (Penguin).

It may be noted that in atomic systems, $n_0, n_1, n_2 \ldots$ are the numbers of atoms with no energy quanta, 1 quantum, 2 quanta, and so on. W is the number of ways of distributing the quanta for the energy distribution with the set of distribution numbers $n_0, n_1, n_2 \ldots n_i$, etc. The most probable distribution of the energy is the one with the maximum value for W, that is the one with the largest number of microstates.

A.4 ENTROPY: MOLECULAR INTERPRETATION AND CALORIMETRIC MEASUREMENT

The molecular interpretation of entropy is summarized in Eq. (6.3):

$$S = k \ln W.$$

The calorimetric measurement of entropy is based on Eq. (6.6):

$$dS = \frac{dq}{T}.$$

In this appendix the second equation is derived from the first for a simple model. The model is that already described in Section 6.8. The model for a crystal is taken to be a set of atoms which may vibrate about their mean positions. The energy levels for each atom are taken to be equally spaced with an energy interval ϵ. Suppose that population numbers of the levels, given by the Boltzmann law (Eq. 6.2), are $n_0, n_1, n_2 \ldots n_i, n_j \ldots$ as illustrated in Fig. A.12. The number of ways W, that is the number of microstates, for the Boltzmann distribution of energy is, from Section A.3, given by

$$W = \frac{N!}{n_0! \, n_1! \, n_2! \cdots n_i! \, n_j! \cdots},$$

where N is the total number of atoms in the system.

Consider a quantity of heat dq, equal to n quanta ϵ, to be transferred to the system, that is

$$dq = n\epsilon.$$

Suppose the quanta are introduced one at a time. Let the first quantum be added to an atom with i quanta. The result is the distribution as before except:

number of atoms with i quanta is $(n_i - 1)$
number of atoms with j quanta is $(n_j + 1)$.

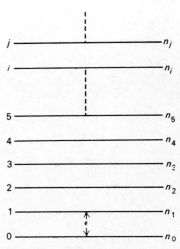

Fig. A.12 Numbers of atoms in each of the equally spaced levels.

As a consequence the number of microstates has changed to W' where

$$W' = \frac{N!}{n_0! \, n_1! \, n_2! \cdots (n_i - 1)! \, (n_j + 1)! \cdots};$$

$$= W \frac{n_i}{n_j + 1} \approx W \frac{n_i}{n_j} \quad \text{since } n_j \gg 1.$$

Hence from the Boltzmann law we obtain

$$W' = W e^{\epsilon/kT},$$

that is $\ln W' = \ln W + \epsilon/kT$.

Quanta may now be added successively, leading, after $n\epsilon$ have been added, to a final value of W'' where

$$\ln W'' = \ln W + n\epsilon/kT.$$

Remembering $n\epsilon = dq$, the change $d \ln W = \ln W'' - \ln W$ is given by

$$d \ln W = dq/kT,$$

and therefore from Eq. (6.3)

$$dS = \frac{dq}{T}.$$

While the relationship has been obtained here for a particular model it is of course a general one.

A.5 DETERMINATION AND EVALUATION OF ENTROPY CHANGES

The basic equation for the experimental determination of entropy changes is (Eq. 6.6):

$$dS = \frac{dq}{T}.$$

For a reversible change, that is one passing through equilibrium states, the entropy change is obtained by integration:

$$\Delta S = \int_1^2 dS = \int_1^2 \frac{dq}{T}. \tag{A.1}$$

Entropy changes encountered in chemistry fall into five categories, which can be indicated by the particular change occurring in the system. They are: (1) change of temperature; (2) change of pressure; (3) change of phase; (4) change on mixing; and (5) chemical change. Each will be considered briefly in turn.

1) Change of temperature. Consider a system of one mole to undergo a reversible change in temperature, T_1 to T_2, at constant pressure. We have from Eq. (6.6) and Eq. (6.8):

$$\Delta S = \int_{T_1}^{T_2} \frac{dq}{T} = \int_{T_1}^{T_2} \frac{C_P \, dT}{T} = \int_{T_1}^{T_2} C_P \, d \ln T. \tag{A.2}$$

The entropy change may be obtained graphically from the plot of C_P against $\ln T$, being given by the area between the ordinates T_1 and T_2.

2) Change of pressure. Consider a system of one mole of ideal gas to be expanded from pressure P_1 to pressure P_2 under conditions of constant temperature. For an ideal gas the internal energy depends only on the pressure and is independent of the temperature. In the present case we have therefore for any infinitesimal step:

$$dU = 0,$$

and so from Eq. (4.1) for an infinitesimal step:

$$dq = P\,dV.$$

It follows that

$$\Delta S = \int_1^2 \frac{dq}{T} = \int_1^2 \frac{P\,dV}{T} = \int_1^2 \frac{RT\,dV}{V\,T} = R\int_1^2 \frac{dV}{V} = R\ln\frac{V_2}{V_1} = R\ln\frac{P_1}{P_2}.$$

That is

$$S_2 - S_1 = R\ln P_1 - R\ln P_2. \tag{A.3}$$

Taking state 2 to be perfectly general with pressure P and entropy S, and state 1 to be the standard state with pressure of 1 atmosphere and entropy $S^\ominus$, we have from the above equation:

$$S = S^\ominus - R\ln P.$$

Hence the entropy may be obtained at any pressure P from the tabulated value of the standard entropy.

3) Change of phase. Consider one mole of substance to undergo phase change (e.g. vaporization) at some temperature T and pressure P with the absorption of heat L. It follows from Eq. (6.6):

$$\Delta S = \frac{L}{T}. \tag{A.4}$$

The equation applies to fusion, vaporization, sublimation, or to change between solid phases.

4) Mixing. Consider one mole of ideal gas A to be mixed with one mole of ideal gas B each of the gases being at temperature T and pressure P:

| A | B |

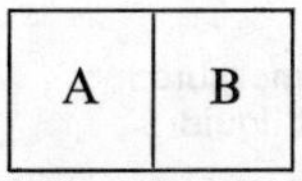

Removal of the barrier in the diagram allows each gas access to the combined volume. Since there is no interaction between molecules of ideal gases, the entropy of each is the same as if it occupied the volume separately. The change in entropy occurs as a result of halving the partial pressures of the two gases. The important equations for most purposes are that the entropy of each component in the mixture is given by

$$S_A = S_A^\ominus - R\ln P_A \tag{A.5}$$

and

$$S_B = S_B^\ominus - R\ln P_B, \tag{A.6}$$

where P_A and P_B are the partial pressures each exerts in the mixture.

5) Chemical change. Consider as an example the chemical change:

$$H_2(g,\ 1\ \text{atm}) + \tfrac{1}{2}O_2(g,\ 1\ \text{atm}) = H_2O(g,\ 1\ \text{atm}).$$

The change is not reversible and $\Delta S^\ominus$ cannot be obtained by application of Eq. (6.6). On the other hand $\Delta S^\ominus$ may be readily evaluated from

$$\Delta S^\ominus = S^\ominus(H_2O) - S^\ominus(H_2) - \tfrac{1}{2}S^\ominus(O_2)$$

using tables of standard entropies. If the entropy change is required for some other pressure and temperature then the equations developed above may be employed, provided the conditions are such that the gases may be assumed to behave ideally.

The equation above for oxygen–hydrogen reaction is an example of the general equation (Eq. 6.10):

$$\Delta S^\ominus = \sum S^\ominus(\text{products}) - \sum S^\ominus(\text{reactants}),$$

in which proper allowance must be made for the number of moles of each product and reactant involved.

A.6 DETERMINATION OF STANDARD ENTROPIES: AN EXAMPLE

The standard entropy of a substance may be obtained by breaking down the change from absolute zero temperature to 298 K and 1 atm into a number of steps which belong to classes 1, 2, or 3 of Section A.5.

The procedure is illustrated for sulphur dioxide in Table A.6.1.

Table A.6.1. Determination of standard entropy for sulphur dioxide. (The numbers refer to the method given in Section A.5.)

Temperature (K)	Step	ΔS (J K^{-1} mol^{-1})
0–15	(1) Graphical, extrapolation of C_P against ln T	0.30
15–197.6	(1) Graphical, using measured heat capacities of solid	20.12
197.6	(3) Fusion	8.95
197.6–263.1	(1) Graphical, using measured heat capacities of liquid	5.96
263.1	(3) Vaporization	22.66
263.1–298.1	(1) Graphical	1.25
		59.24

The method is general. In some cases there may be two or more forms of the solid. At each transition temperature method (3) is applied. If the pressure needs to be changed to give a final state with pressure of 1 atm, then method (2) may be used.

A.7 CHEMICAL EQUILIBRIUM

Consider the equilibrium:

$$N_2O_4(g) \rightleftharpoons 2NO_2(g).$$

Suppose the gas is in a container under conditions of constant temperature and pressure as in Fig. A.13. Suppose the system has reached equilibrium. Then for an infinitesimal displacement in which dn moles of $N_2O_4(g)$ are converted to 2 dn moles of $NO_2(g)$ we may write:

$$dS_{\text{total}} = dS_{\text{surroundings}} + dS_{\text{system}} = 0.$$

The first term (cf. Eq. 6.13)

$$dS_{\text{surroundings}} = \frac{-\Delta H\, dn}{T},$$

which can be rewritten

$$= \frac{-\Delta H^{\ominus}\, dn}{T}$$

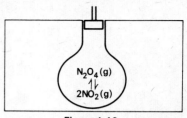

Figure A.13

if the gases are assumed ideal, in which case ΔH is independent of pressure and $\Delta H^{\ominus}$ may be obtained from tabulated values. The second term

$$dS_{\text{system}} = (2S_{NO_2} - S_{N_2O_4})\, dn,$$

where S indicates a molar entropy. From Section A.5 we have:

$$S_{N_2O_4} = S^{\ominus}_{N_2O_4} - R \ln P_{N_2O_4};$$
$$2S_{NO_2} = 2(S^{\ominus}_{NO_2} - R \ln P_{NO_2}).$$

Therefore

$$dS_{\text{system}} = (\Delta S^{\ominus} - 2R \ln P_{NO_2} + R \ln P_{N_2O_4})\, dn,$$

$\Delta S^{\ominus}$ being the standard entropy change, given by

$$\Delta S^{\ominus} = 2S^{\ominus}_{NO_2} - S^{\ominus}_{N_2O_4},$$

the value being obtainable again from tabulated values. Hence we have

$$dS_{\text{total}} = \left(\frac{-\Delta H^{\ominus}}{T} + \Delta S^{\ominus} - 2R \ln P_{NO_2} + R \ln P_{N_2O_4}\right) dn = 0.$$

If the system were not originally at equilibrium, then change continues all the while $dS_{\text{total}} > 0$, and eventually equilibrium is reached when the condition

$$\frac{-\Delta H^{\ominus}}{T} + \Delta S^{\ominus} - 2R \ln P_{NO_2} + R \ln P_{N_2O_4} = 0$$

is satisfied.

The equation simplifies to

$$-R \ln K_P = \frac{\Delta H^{\ominus}}{T} - \Delta S^{\ominus},$$

where

$$K_P = \frac{(P_{NO_2})^2}{(P_{N_2O_4})}.$$

Finally the equation may be rewritten:

$$-RT \ln K_P = \Delta G^{\ominus}.$$

Thus the equilibrium constant is related to the standard free energy of the reaction.

A.7 CHEMICAL EQUILIBRIUM

The equation is quite general for ideal gas reactions. As already indicated, $\Delta G^\ominus$ values may be obtained by combining information from tables of $\Delta H_f^\ominus$ and $S^\ominus$, or alternatively from tabulated values of $\Delta G_f^\ominus$ using Eq. (7.9):

$$\Delta G^\ominus = \sum \Delta G_f^\ominus(\text{products}) - \sum \Delta G_f^\ominus(\text{reactants}).$$

A.8 TRANSITION STATE THEORY

In general we can write for a bimolecular step, in a reaction pathway, the equation

$$A + B \rightleftharpoons (AB)^\ddagger \rightarrow \text{products}.$$

The transition state in this theory is regarded as a sort of compound even though its lifetime is very short. It is assumed that the equilibrium law can be applied to the reaction as written above. This gives an equilibrium constant

$$K^\ddagger = \frac{[(AB)^\ddagger]}{[A][B]}.$$

That is, we may write

$$[(AB)^\ddagger] = K^\ddagger[A][B].$$

The theory supposes next that a unimolecular breakdown of the activated complex occurs. The rate equation for this can be written as

$$v_1 = k_1[(AB)^\ddagger].$$

From the equation written earlier we have therefore

$$v_1 = k_1 K^\ddagger[A][B],$$

where $k_1 K^\ddagger$ is the second order rate constant for the reaction.

We need to look further at k_1 and $K^\ddagger$ because if these could be calculated we could calculate the rate of the chemical reaction which is of course one of the hopes of the chemist. It can be shown by statistical mechanics, rather surprisingly perhaps, that k_1 is the same for all reactions and is given by

$$k_1 = \frac{kT}{h},$$

where T is the thermodynamic temperature, k is the Boltzmann constant, and h is Planck's constant.

Now if by experiment the reaction step is found to be of second order, then we have from experiment:

$$\text{rate} = k_2[A][B];$$

and so the rate constant k_2 should be given by

$$k_2 = \frac{kT}{h} K^\ddagger.$$

We now look further at $K^\ddagger$. There are formidable difficulties in finding $K^\ddagger$ by statistical mechanical methods and so far only a start has been made with simple systems. But there is an alternative method of approaching $K^\ddagger$. Equilibrium constants are related to standard free energy changes (Section 7.5), so following

Eq. (7.6) we may write:

$$-RT \ln K^{\ddagger} = (\Delta G^{\ominus})^{\ddagger},$$

where $(\Delta G^{\ominus})^{\ddagger}$ is the standard free energy difference between the reactants and the transition state. $(\Delta G^{\ominus})^{\ddagger}$ cannot be calculated in terms of the properties of A and B; except for the very simplest of reactions the calculations quickly become too complicated. Nevertheless transition state theory gives a new way of thinking about reactions.

In particular it provides a convenient way of symbolizing the path of a chemical reaction. It provides in fact a very useful diagrammatic way of representing the courses of chemical reactions. The rate of a reaction has been seen to depend on the standard free energy difference between reactants and transition state. Diagrams, for example Fig. 8.6, may be drawn showing how the standard free energy changes in the course of the reaction. Such diagrams are often called reaction profiles (Section 8.7).

A.9 CRYSTAL FIELD THEORY

The thermodynamic stability of *all* chemical compounds relative to dissociation into the constituent parts is due to the lowering of the potential energy of the system when the compound is formed from these constituent parts. Two useful models of this which have been discussed previously (Section 4.1) are:

1. The ionic model, in which the stability of, for example, sodium chloride can be (mainly) attributed to the lattice energy (attraction of oppositely charged ions).
2. The covalent model of bonding (more appropriate for bonding between elements of similar electronegativity), in which the stability of, for example, methane can be attributed to the sharing of electrons between two positive nuclei, causing a lowering in potential energy. A typical (mainly) covalent compound in transition metal chemistry is nickel carbonyl (Fig. A.14). The stability of this compound relative to separated carbon monoxide and nickel metal is (mainly) due to the lowering of the potential energy of the system, which results from the sharing of electrons between the carbon and nickel nuclei, i.e. the formation of bonds:

$$Ni(s) + 4CO(g) \rightarrow Ni(CO)_4(g).$$

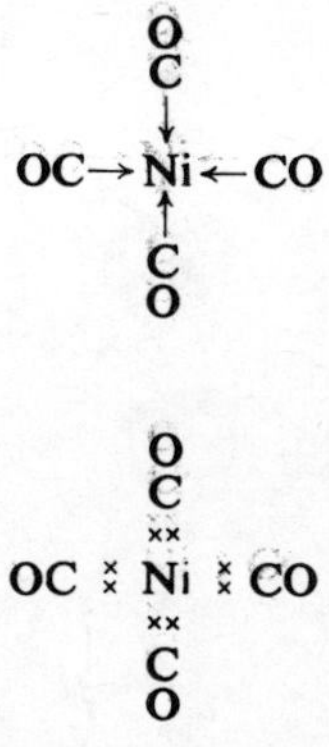

At this level no further explanation about the stability of the compound is necessary. Notice that the *d*-electrons of nickel play no part in this simple and useful model of the bonding (this is not true in more refined models of the bonding, which need not concern us). However, in order to explain some of the special properties of transition metal compounds, such as colour and magnetism, it is necessary to look at the effect of the ligands on these non-bonding (at this level of approximation) *d*-electrons. *Note:* this is not done in order to explain why the compound is stable since we know this (roughly!)—but in order to explain why transition metal compounds are coloured, etc.

Electrostatic crystal field theory is an ionic model. In it the complex is considered as an aggregate of ions or molecules interacting electrostatically (i.e. no covalent bonding), just as in, for example, sodium chloride. The electrostatic

Fig. A.14 Nickel carbonyl.

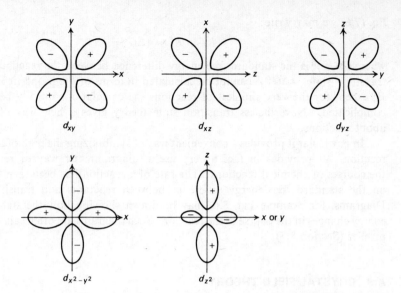

Fig. A.15 The spatial distribution of 3d-orbitals.

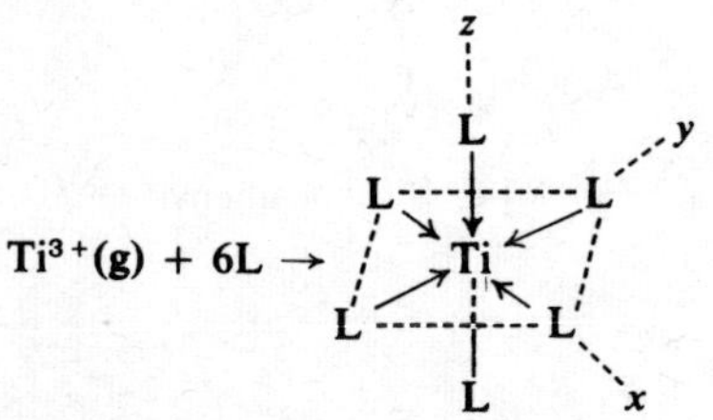

Fig. A.16 The Ti^{3+} cation surrounded octahedrally by six ligands along the co-ordinate axes.

Ti^{3+}(g) + 6L →

attraction is either between oppositely charged ions, as for example Co^{2+} and 4Cl$^-$ in (CoCl$_4$)$^{2-}$, or between the central atom or ion and the negative end of a dipolar molecule, for example water
$$\begin{array}{c} H^{\delta+} \\ \delta- \diagup \\ O \\ \diagdown \\ H^{\delta+} \end{array}$$
, in Co(H$_2$O)$_6^{2+}$. Choosing this end-of-spectrum viewpoint of the bonding is clearly an inadequate model of the bonding generally (e.g. in nickel carbonyl, above, the carbon monoxide molecule is practically non-polar); and, in fact, calculations of bond energies using this model give poor results. The power of the model is not in explaining overall stability of complexes, which we do not require anyway, but in explaining the effect of the ligands on the "non-bonding" d-electrons (shorthand for "electrons in d-orbitals") and thus, in turn, explaining the special properties of transition elements. How do we know that the d-electrons are important in this respect? As usual, the success of the model is the evidence. Most of this is beyond the scope of this discussion. However, bearing in mind that chemists always attempt to relate properties to electron configuration, the clue is given by looking at which cations are coloured (Table 20.1). Clearly the colours of transition metal ions are related in some way to the presence of d-electrons in incompletely filled d-shells. The simplest model might be to attribute the colour to transitions of electrons between filled d- and empty s-orbitals. However: (a) the energies involved in such transitions are too large; and (b) it does not explain why *incomplete* d-shells are necessary.

Effect of Electrostatic Field of Ligands on Non-bonding d-Orbitals

The spatial distribution of the five 3d-orbitals are shown in Fig. A.15. Consider a gaseous titanium ion, Ti^{3+}(g), electron configuration d^1, in free space, that is, with no ligands around it. In this case, the d-orbitals all have the same energy and are said to be degenerate. Now imagine that the ion is surrounded by six ligands, with the co-ordinate system defined in Fig. A.16. The total potential

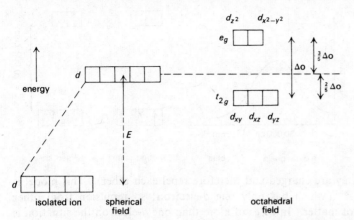

Fig. A.17 The effect of an octahedral field on the *d*-orbitals.

energy of the system relative to infinite separation is lowered owing to electrostatic attraction between the metal ion and the ligands, that is the system is stable relative to dissociation. However, we are more interested in the effect of the electron density of the ligands on the relative energy of the *d*-electrons. There are two main effects: (a) the energy of *all* the *d*-orbitals is raised due to electrostatic repulsion; and (b) the energy of all the *d*-orbitals is not raised equally. The *d*-orbitals (by which is meant "electrons in *d*-orbitals") finish up with different energies, or, as it is said, "their degeneracy is raised". It is obvious why this is so. Two of the orbitals, the $d_{x^2-y^2}$ and the d_{z^2}, have their greatest electron density along the co-ordinate axes on which the ligands are situated. Electrons in these orbitals will be repelled more (raised in energy more) than electrons in the d_{xy}, d_{xz}, and d_{yz} orbitals, which have their maximum electron density between the co-ordinate axes and are therefore repelled less. This is represented in Fig. A.17. For convenience in showing the general increase in energy, it is considered that, first, a spherical field of negative charge (in which the charge is distributed uniformly over a sphere of radius equal to the metal–ligand bond distance) is placed around the metal ion. This results in a *large* energy increase of all the *d*-orbitals. It is then considered that this charge is concentrated into six equal parts at the bonding distance along the co-ordinate axes, as in the bonding situation, when (as above) the orbitals separate into two sets: (a) the d_{xy}, d_{xz}, d_{yz} at lower energy, referred to as the t_{2g} orbitals; and (b) the $d_{x^2-y^2}$, d_{z^2} at higher energy, referred to as the e_g orbitals. It is not obvious from this treatment that the $d_{x^2-y^2}$ and d_{z^2} orbitals (e_g) are still degenerate (have the same energy); but it should be obvious that the t_{2g} orbitals have the same energy. The separation of the e_g and t_{2g} orbitals in the octahedral field provided by the six ligands is called the **crystal field splitting energy**, Δo. (o—for octahedral). This energy is small compared with the general rise in the energy, E. On our ionic model it would be expected that Δo would depend upon: (a) the metal–ligand distance; and (b) the charge or dipole on the ligand. However, the model is too simple to allow accurate energy calculations to be made. In fact, Δo is found by experiment not by calculation.

Consider the spectrum of the aqueous Ti^{3+} ion (Fig. A.18). This ion absorbs electromagnetic radiation in the green part of the visible spectrum, therefore it looks purple (allows red and blue light through). The simplest model for the absorption of light by transition metal compounds is this: electrons are considered to behave as if they are independent of each other's existence (*note:* they clearly

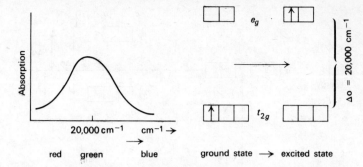

Fig. A.18 The absorption spectrum of the Ti^{3+} (aq) ion, $Ti(H_2O)_6^{3+}$, and the associated electronic transition.

cannot be, since they are charged and therefore repel each other). This model is satisfactory for Ti^{3+}, which has only one d-electron; so repulsion from other d-electrons does not matter. In cases of more than one d-electron the situation is more complicated and more than one absorption band is found. Absorption of electromagnetic radiation is considered to cause promotion of one electron from a lower to a higher energy level. In the case of transition metal compounds, the absorption of light in the visible region of the electromagnetic spectrum corresponds to the transition (or *excitation*) of an electron from one d-orbital to another d-orbital, which are no longer degenerate (of equal energy) in the electrostatic field of the ligands in the complex. Such transitions are called *d–d transitions*, and the resulting spectra are called *d–d spectra*. In the case of tetrahedral molecules, the splitting pattern is the other way around (Fig. A.19), and the crystal field splitting energy, Δt, is less than Δo since: (a) there are only four ligands involved; and (b) in neither case, the t nor the e orbitals, do they point directly at the ligands, therefore the difference in repulsion between the two sets of orbitals is less than in the octahedral case.

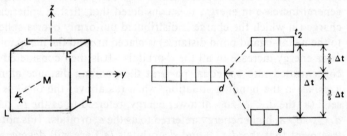

Fig. A.19 The effect of a tetrahedral field on the d-orbitals.

Some Δo values for various ligands and metal ions in octahedral compounds are given in Table A.9.1. Obvious points to notice are:

1. Δo is very similar for metals of the same oxidation number with the same ligand, but Δo increases as the oxidation number increases (a greater interaction is expected between the ligand and the increased charge on the metal).
2. Δo depends upon the ligand. This explains why the colour of transition metal ions changes when the ligand changes. Ligands have been arranged in an order of increasing crystal field splitting energy, Δo, which they induce in a given metal ion. This order is known as the *spectrochemical series*, a small portion of which is:

$$I^- < Br^- < Cl^- < F^- < OH^- < C_2O_4^{2-}$$
$$< H_2O < NH_3 < en < -NO_2^- < CN^-.$$

Table A.9.1. Octahedral crystal field splittings (kcm^{-1})*.

Metal		Ti(III)	V(III)	Cr(III)	Mn(III)	Mn(II)	Fe(III)	Fe(II)	Co(III)	Co(II)	Ni(II)	Cu(II)
						Δo						
Ligands	Cl$^-$	13	13	13.8		7.2					7.2	
	H$_2$O	19	18.6	17.4		8.4	14.2	9.4	18.2	8.2	8.5	11.5
	NH$_3$			21.6					22.9	10.1	10.8	
	CN$^-$			26.7					31.4	33.3		

* kcm^{-1} are the usual units used; 1 kcm^{-1} = 10^3 cm^{-1} ≡ 12 kJ mol^{-1}.

A rough order with reference to the donor atom of the ligand is given by

$$\text{group VIIB} < \text{VIB} < \text{VB} < \text{IVB};$$

for example:

$$\text{Cl}^- < \text{OH}_2 < \text{NH}_3 < \text{CN}^-.$$

This order cannot be explained on an electrostatic model (for example the series has no obvious relation to charge on the ligand), which as usual is little use in calculating actual energy values.

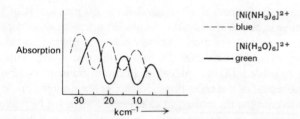

Fig. A.20 The absorption spectra of the ions [Ni(H$_2$O)$_6$]$^{2+}$ and [Ni(NH$_3$)$_6$]$^{2+}$.

Consider the complexes of nickel(II): green [Ni(H$_2$O)$_6$]$^{2+}$ and blue [Ni(NH$_3$)$_6$]$^{2+}$. The ion Ni^{2+}(aq), like all simple ions, exists in aqueous solution as the hydrated cation [Ni(H$_2$O)$_6$]$^{2+}$. The absorption spectra of these ions is given in Fig. A.20. The corresponding bands in [Ni(NH$_3$)$_6$]$^{2+}$ are at a higher energy than in [Ni(H$_2$O)$_6$]$^{2+}$ because Δo is greater in the former compound, and consequently the compounds are different colours. There are many examples of this, for example [Cu(H$_2$O)$_4$]$^{2+}$ is blue, but [Cu(NH$_3$)$_4$]$^{2+}$ is violet (add an excess of concentrated ammonia to an aqueous solution of copper sulphate and see this colour change). A related feature of these ions is that particular ligands may stabilize "unusual" oxidation states (i.e. oxidation states that differ from those normally found in aqueous solutions). Consider the reduction potentials of the Co^{3+}/Co^{2+} system:

$$[\text{Co(NH}_3)_6]^{3+} + e^- \to [\text{Co(NH}_3)_6]^{2+}; \quad E^{\ominus} = +0.1 \text{ V}$$
$$[\text{Co(H}_2\text{O})_6]^{3+} + e^- \to [\text{Co(H}_2\text{O})_6]^{2+}; \quad E^{\ominus} = +1.84 \text{ V}.$$

There is a dramatic difference in the relative stability of the Co^{3+} and Co^{2+} ions in the different environments. Bearing in mind that the above potenials are relative to the hydrogen electrode, but that we live in slightly oxidizing conditions, then it is obvious that in aqueous solutions Co^{2+}(aq) is the stable species, but in ammoniacal solutions Co^{3+} is the stable species. For example, Co^{3+} ammines can be prepared by passing air through ammoniacal solutions of cobalt(II) salts.

This dependence of the stability of oxidation states on the particular ligand present in the complex makes the whole subject difficult to discuss in an exact manner (Section 20.9). For example, in the presence of the ligand carbon monoxide, nickel(0) is a stable oxidation state of nickel; whereas, in the presence of the ligand Cl^-, nickel(II) is the stable oxidation state. It is possible to classify ligands into three categories:

1. Ligands which tend to stabilize "low" oxidation states of transition metals (-1, 0, and $+1$), for example, carbon monoxide and dipyridyl (often called π-bonding ligands).
2. Ligands which tend to stabilize "normal" oxidation states ($+2$ and $+3$), for example H_2O, NH_3, and Cl^-, etc.
3. Ligands which tend to stabilize "high" oxidation states ($+4$ upwards), for example O^{2-} and F^- (non-oxidizable ligands).

In addition, changing the ligand may change the stereochemistry of the complex, with consequent changes in other properties. For example, if an aqueous solution of a cobalt(II) salt is treated with a large excess of chloride ions, the colour changes from pink to a fairly intense blue:

$$[Co(H_2O)_6]^{2+} \xrightarrow{\text{excess } Cl^-} [CoCl_4]^{2-}.$$
$$\text{pale pink} \qquad\qquad \text{fairly deep blue}$$

The blue colour is due to the formation of the tetrahedral complex ion, $[CoCl_4]^{2-}$. Not only is there a change in colour ($\Delta o > \Delta t$), but also an intensification of colour. In general, tetrahedral complexes are more intensely coloured than are octahedral complexes. The theory of all of these effects is beyond this discussion. However, one point which must be made is that not all colour in transition metal compounds can be explained by the model of d–d transitions. Very highly coloured compounds (those with very intense colours, those used in calorimetric analysis, for example) are coloured owing to a **charge transfer spectrum**. In this an electron is transferred from one atom to another, the related absorption of energy occurring in the visible part of the spectrum. Examples are: some metal oxides, sulphides, and halides; some oxyanions such as MnO_4^- and $Cr_2O_7^{2-}$; $[Fe(SCN)]^{2+}$; and compounds containing mixed oxidation states of the same metal, e.g. Prussian blue, which contains Fe^{2+} and Fe^{3+} ions. Defect lattices and non-stoichiometric compounds can be considered as also containing mixed oxidation states (metal oxides, etc.). In the case of permanganate, for example, the colour is due to an electronic transition from an oxygen atom to the manganese.

It is appropriate to summarize three important points about transition metal complexes at this point.

1. In aqueous solution the simple ions are hydrated and are best regarded as complex ions of the ligand H_2O, for example $[Ni(H_2O)_6]^{2+}$, etc. The solid states may be hydrated in the same way, for example, $CuSO_4 \cdot 5H_2O$ (Fig. 11.8).
2. Different ligands can displace water to yield either: (a) soluble complex ions, often of different colour, sometimes of different stereochemistry; or (b) insoluble neutral complexes, if the ligand is negatively charged. These are often used to characterize the ion qualitatively or quantitatively, e.g. nickel dimethylglyoxime (Section 20.19).

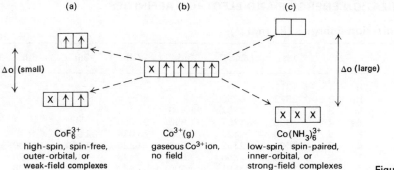

Figure A.21

3. Ligands other than water may stabilize the complex in an unusual oxidation state.

The simple features of the **magnetic properties** of transition metal complexes can also be understood in terms of the crystal field model. Materials which contain *only* paired electrons tend to move (weakly) out of magnetic fields and are said to be *diamagnetic*. Materials which contain unpaired electrons are attracted (strongly) into a magnetic field and are said to be *paramagnetic*. The magnitude of this attraction is taken as a measure of the number of unpaired electrons present in the sample, and is measured fairly easily by a device known as the Gouy balance (other methods are available).

Consider the Co^{3+} ion, configuration d^6. Off-hand, we would expect all compounds containing such ions, but no other paramagnetic material, to show paramagnetism equivalent to four unpaired electrons, since according to the Hund rule the electrons would distribute themselves among the orbitals as in Fig. A.21b. In fact, almost all the complexes of cobalt(III) are found to be diamagnetic, and only a few show the expected paramagnetism. Two major effects determine how the electrons are distributed in the degenerate sets of orbitals (t_{2g} and e_g) which result from the octahedral field of the ligands: (a) the tendency for electrons to spread themselves out (according to the Hund rule) in the maximum number of orbitals and consequently to minimize the repulsive interaction with each other; and (b) the tendency for electrons to fill the lowest energy orbitals available, in this case the t_{2g} set of orbitals, and consequently to minimize the repulsive interaction with the ligand electrons. When the crystal field splitting energy, Δo is sufficiently large by comparison with kT, the electrons occupy the low energy t_{2g} orbitals in preference (the normal situation for cobalt(III) compounds) and in this case the compound is diamagnetic, for example $[Co(NH_3)_6]^{3+}$. Such compounds are called low-spin (spin-paired, inner-orbital, or strong-field) complexes. When the crystal field splitting energy, Δo, is small, the electrons follow the Hund rule, e.g. CoF_6^{3-}, and the complex is paramagnetic (high-spin, spin-free, outer-orbital or weak-field complexes). Clearly, for a given metal ion, M^{n+}, in a given stereochemistry, there will be a point in the spectrochemical series to the left of which the complexes will be high-spin, and to the right of which the complexes will be low-spin. For tetrahedral complexes, since $\Delta t < \Delta o$, it turns out that all known complexes of the first transition series are high-spin. The same principles apply to all transition metal complexes. The critical division in the spectrochemical series is found by experiment.

A.10 IONIZATION ENERGIES AND ELECTRON AFFINITIES

A.10.1 Ionization Energies (kJ mol^{-1})

			1st	2nd	3rd	4th	5th	6th	7th
Hydrogen	H	1	1312						
Helium	He	2	2372	5250					
Lithium	Li	3	5200	7297	11,810				
Beryllium	Be	4	900	1757	14,850	21,000			
Boron	B	5	800	2427	3658	25,024	32,824		
Carbon	C	6	1086	2352	4619	6220	37,820	47,280	
Nitrogen	N	7	1403	2858	4578	7473	9443	53,270	64,360
Oxygen	O	8	1314	3391	5300	7469	10,983	13,326	71,330
Fluorine	F	9	1681	3381	6046	8418	11,017	15,163	17,870
Neon	Ne	10	2081	3964	6150	9376	12,200	15,230	
Sodium	Na	11	495	4563	6912	9540	13,350	16,732	20,110
Magnesium	Mg	12	738	1450	7730	10,550	13,625	18,035	21,735
Aluminium	Al	13	577	1816	2745	11,575	14,840	18,370	23,350
Silicon	Si	14	787	1577	3230	4355	16,090	19,795	19,600
Phosphorus	P	15	1060	1896	2908	4954	6272	21,270	25,413
Sulphur	S	16	1000	2258	3381	4565	6995	8494	27,112
Chlorine	Cl	17	1255	2297	3849	5163	6540	9330	11,029
Argon	Ar	18	1520	2665	3947	5770	7238	8810	11,965
Potassium	K	19	418	3069	4439	5875	7950	9619	11,384
Calcium	Ca	20	590	1146	4941	6464	8142	10,500	12,350
Scandium	Sc	21	633	1235	2388	7130	8874	10,720	13,305
Titanium	Ti	22	659	1309	2648	4171	9627	11,577	13,585
Vanadium	V	23	650	1414	2866	4631	6293	12,435	14,568
Chromium	Cr	24	653	1591	2992	4861	7050	8745	15,525
Manganese	Mn	25	717	1509	3259	5021	7322	9874	11,464
Iron	Fe	26	762	1561	2958	5502	7531	10,250	12,345
Cobalt	Co	27	759	1644	3230	5104	8054	10,230	12,720
Nickel	Ni	28	736	1751	3391	5400	7531	10,627	13,140
Copper	Cu	29	745	1958	3556	5681	7908	10,230	13,514
Zinc	Zn	30	906	1732	3828	5983	8284	10,795	13,725
Gallium	Ga	31	579	1979	2962	6192			
Germanium	Ge	32	760	1537	3301	4410	9013		
Arsenic	As	33	946	1800	2732	4837	6042	12,310	
Selenium	Se	34	941	2075	3096	4121	6602	7853	
Bromine	Br	35	1142	2084	3464	4561	4845	8535	
Krypton	Kr	36	1350	2370	3565	5063	6234	7573	
Rubidium	Rb	37	403	2653	3849	5063			
Strontium	Sr	38	549	1064	4226	5490			
Yttrium	Y	39	615	1180	1975	5983	7431		
Zirconium	Zr	40	659	1268	2217	3305	7991	9540	11,380
Niobium	Nb	41	664	1381	2427	3682	4812	9958	12,050
Molybdenum	Mo	42	688	1557	2619	4477	5990	6945	12,050
Technetium	Tc	43	697	1473	2795	4477	5690	7322	9058
Ruthenium	Ru	44	711	1607	2757	4477	6067	7824	9665
Rhodium	Rh	45	720	1740	2987	4393	6464	8201	10,125
Palladium	Pd	46	804	1874	3222	4707	6360	8661	10,586
Silver	Ag	47	730.9	2072	3368	5012	6736	8577	11,171
Cadmium	Cd	48	867	1630	3625				
Indium	In	49	558	1820	2703	5251			
Tin	Sn	50	707	1412	2941	3929	6987		
Antimony	Sb	51	833	1590	2443	4247	5356		
Tellurium	Te	52	869	1799	2996	3648	5816	6978	
Iodine	I	53	1007	1845	3138				
Xenon	Xe	54	1170	2046	3100	4251	5774		
Caesium	Cs	55	374	2422	3376	4920	5593		
Barium	Ba	56	502	965					
Lanthanum	La	57	541	1103	1858				

			1st	2nd	3rd	4th	5th	6th	7th
Cerium	Ce	58	540	1008	1925	3544			
Praseodymium	Pr	59	527	1017	2238				
Neodymium	Nd	60	527	1033					
Promethium	Pm	61	540						
Samarium	Sm	62	540	1079					
Europium	Eu	63	547	1084					
Gadolinium	Gd	64	594	1155					
Terbium	Tb	65	577	1113					
Dysprosium	Dy	66	573	1125					
Holmium	Ho	67	582	1138					
Erbium	Er	68	586	1151					
Thulium	Tm	69	561	1163					
Ytterbium	Yb	70	598	1167					
Lutetium	Lu	71	594	1381					
Hafnium	Hf	72	674	1435	2029	2992			
Tantalum	Ta	73	745	1569	2155	3192	4351		
Tungsten	W	74	770	1677	2322	3410	4623	5879	
Rhenium	Re	75	761	1598	2510	3632	4916	6171	7615
Osmium	Os	76	837	1640	2410	3850	5188	6569	7991
Iridium	Ir	77	879	1640	2602	3766	5481	6945	8494
Platinum	Pt	78	870	1791	2745	3958	5314	7238	8870
Gold	Au	79	890	1978	2895	4192	5607	7029	9247
Mercury	Hg	80	1006	1809	3309	6945	7908		
Thallium	Tl	81	589	1970	2879	4895			
Lead	Pb	82	715	1452	3079	4084	6723		
Bismuth	Bi	83	703	1611	2452	4351	5373		

A.10.2 Electron Affinities (kJ mol^{-1})

			H 73				He −54
Li 57	Be −66	B 15	C 123	N −31	O 141	F 333	Ne −99
Na 21	Mg −67	Al 26	Si 135	P 60	S 196	Cl 348	
						Br 340	
						I 297	

A.11 BOND ENERGIES AND BOND DISTANCES

Bond energies (kJ mol^{-1}); Bond distances (Å)

Hydrogen																	
Bond H—X	H	C	N	O	F	Si	P	S	Cl	Ge	As	Se	Br	Sn	Sb	Te	I
Bond energy	435	414	389	464	565	318	322	368	431	285	297	305	364	251	255	238	299
Compound	H$_2$	CH$_4$	NH$_3$	H$_2$O	HF	SiH$_4$	PH$_3$	H$_2$S	HCl	GeH$_4$	AsH$_3$	H$_2$Se	HBr	SnH$_4$	SbH$_3$	TeH$_2$	HI
Bond distance	0.74	1.09	1.01	0.96	0.92	1.48	1.44	1.36	1.27	1.59	1.52	1.46	1.41	1.70	1.71		1.61
Carbon																	
Bond C—X	C—C	C≡C	C≡C	C=O													
Bond energy	347	611	837	360													
Compound	organic	alke.	alky.	ethers													
Bond distance	1.54	1.34	1.20	1.43													
Bond C—X					C=O	C≡O	C—N	C=N	C—F	C—Cl	C—Br	C—I	C—Si	C—P	S—C		
Bond energy					736	108	305	305	489	326	272	238	305	264	272		
Compound					alds.	CO	org.	CO$_2$	CF$_4$	CCl$_4$	CBr$_4$	CH$_3$I	SiC(g)	PMe$_3$	SMe$_2$		
Bond distance					1.21	1.13	1.47	1.16	1.37	1.77	1.94	2.14	1.90	1.84	1.82		
Fluorine																	
Bond F—X	F	Cl	Br	I	N	O	Si	P	S	Sb							
Bond energy	158	251	251	243	280	213	598	494	285	452							
Compound	F$_2$	Cl—F	BrF	IF	NF$_3$	F$_2$O	SiF$_4$	PF$_3$	SF$_6$	SbF$_3$							
Bond distance	1.42	1.63	1.75		1.36	1.42	1.56	1.52	1.56	2.03							
Chlorine																	
Bond Cl—X	Cl	Br	I	O	N	Si	P	S	Ge	As	Br	Sn	Sb	Se			
Bond energy	244	209	209	205	192	402	326	272	339	310	218	314	310	243			
Compound	Cl$_2$	ClBr	ICl	Cl$_2$O	NCl$_3$	SiCl$_4$	PCl$_3$	SCl$_2$	GeCl$_4$	AsCl$_3$	BrCl	SnCl$_4$	SbCl$_3$	SeCl$_2$			
Bond distance	1.99	2.14	2.32	1.70	1.75	2.02	2.04	1.99	2.08	2.16	2.14	2.33	2.33	(1.7)			
Nitrogen																	
Bond N—X	N≡N	N=O	N=O$^+$	N—N	N—O	Si—N			N=N	N=C	C≡N	Br	N$_2^+$				
Bond energy	946	632	105	159	163	335			418	615	879						
Compound	N$_2$	NO	NO$^+$	N$_2$H$_4$	NH$_2$OH				N$_2$F$_2$		BrCN	NOBr	N$_2^+$				
Bond distance	1.10	1.15	1.06	1.45	1.42				1.25		1.157	2.14	1.117				
Oxygen																	
Bond O—X	O=O	O—O	P	Br	S=O	Si=O	Si=O	S=O	Ge	As	Cl	O$_3$					
Bond energy	498	142	368	331	523	803	640	293	360	331		301					
Compound	O$_2$	H$_2$O$_2$	P$_4$O$_6$	SiBr$_4$	SO	SiO(g)	SiO$_2$(g)	SiS$_2$	GeO$_2$(s)	As$_4$O$_6$	OCl$^-$	O$_3$					
Bond distance	1.21	1.48	1.66	2.16	1.49	1.51		2.15		1.78	1.56	1.28					
Silicon																	
Bond Si—X	Si	F	Cl		N	C	Si=O	H									
Bond energy	226	598	402		335	305	803	318									
Compound	Si$_2$H$_6$	SiF$_4$	SiCl$_4$		Si$_3$N$_4$(s)	SiC(s)	SiO(g)	SiH$_4$									
Bond distance	2.32	1.56	2.02		1.74		1.51	1.48									
Phosphorus																	
Bond P—X	P—P	P≡P	Cl	Br	F	H	O	N	PSCl$_3$								
Bond energy	209	523	331	268	498	326	368		477								
Compound	P$_4$	P$_2$	PCl$_3$	PBr$_3$	PF$_3$	PH$_3$	P$_4$O$_6$	PN									
Bond distance	2.21	1.89	2.03	2.20	1.52	1.44	1.66	1.49	1.86								
Sulphur																	
Bond S—X	S=S	S—S	S—F	SCl	S$_2$Br$_2$	SH	H$_2$S	S=O	S=O	S=C	S=O	N					
Bond energy	431	264		272	213	368	368	523	477								
Compound	S$_2$	S$_8$	SF$_6$	SCl$_2$	S$_2$Br$_2$	SH	H$_2$S	SO	thiourea		SO$_2$	S$_4$N$_4$H$_4$					
Bond distance	1.89	2.05	1.56	1.99	2.27	1.36		1.49	1.71		1.43	1.67					
Arsenic																	
Bond As—X	As—As	F	Cl	I	H	As—C	S										
Bond energy	180	485	310	188	255												
Compound	As$_4$	AsF$_3$	AsCl$_3$	AsI$_3$	AsH$_3$	AsMe$_3$	As$_4$S$_4$										

Antimony

Bond Sb—X	Sb	F	Cl	Br	I	H
Bond energy	142	452	314	264	184	255
Compound	Sb_4	SbF_3	$SbCl_3$	$SbBr_3$	SbI_3	SbH_3
Bond distance	2.89	2.03	2.33	2.51	2.70	1.71

Boron

Bond B—X	B	F	Cl	Br	I	O	C	H	S
Bond energy	301	644	444	368	272	523			
Compound	B_2F_4	BF_3	BCl_3	BBr_3	BI_3	esters	BMe_3	B_2H_6	BS
Bond distance	1.59	1.29	1.74	1.87		1.36	1.56	1.19	1.61

Selenium

Bond Se—X	Se	F	Cl	Br	I	H
Bond energy	159					305
Compound	Se_2Cl_2	SeF_4	$SeBr_6^{2-}$, $SeCl_2$			H_2Se
Bond distance	2.32	1.76	2.54 (1.7)			1.46

Germanium

Bond Ge—X	Ge	F	Cl	Br	I	H	N	O
Bond energy	188	473	339	280	213	285	255	
Compound	Ge(s)	GeF_4	$GeCl_4$	$GeBr_4$	GeI_4	GeH_4	$Ge_3N_4(s)$	GeO
Bond distance	2.45	1.68	2.08	2.30	2.64	1.59		1.65

Tin

Bond Sn—X	Sn	F	Cl	Br	I	H
Bond energy	150		314	268	197	251
Compound	Sn(s)		$SnCl_4$	$SnBr_4$	SnI_4	SnH_4
Bond distance	2.8		2.33	2.44	2.64	1.7

Bromine

Bond Br—X	Br	I
Bond energy	192	176
Compound	Br_2	IBr
Bond distance	2.28	

Iodine

Bond I—X	I
Bond energy	150
Compound	I_2
Bond distance	2.67

Alke.—alkenes; alky.—alkynes; alds.—aldehydes; org.—organic.

A.12 ENTHALPIES OF HYDRATION

$\Delta H_{hyd}^{\ominus}$ (kJ mol^{-1})

Ion	H^+	Li^+	Na^+	K^+	Rb^+	Cs^+	Tl^+	Ag^+
	1075	498	393	310	284	251	314	460

Ion	Co^{2+}	Ni^{2+}	Cu^{2+}	Zn^{2+}	Cd^{2+}	Hg^{2+}	Pb^{2+}	Sc^{3+}	Y^{3+}	Be^{2+}	Mg^{2+}	Ca^{2+}	Sr^{2+}	Ba^{2+}	Cr^{2+}	Mn^{2+}	Fe^{2+}
	2025	2075	2075	2017	1780	1812	1448	3915	3570	2455	1900	1565	1415	1275	1820	1815	1890
							Tl^{3+}	In^{3+}	Ga^{3+}	La^{3+}	Al^{3+}	Fe^{3+}	Cr^{3+}				
							4140	4060	4645	3240	4630	4340	4350				

Ion	F^-	Cl^-	Br^-	I^-
	460	385	351	305

A.13 RADII, ELECTRONEGATIVITIES, ENTHALPIES OF ATOMIZATION, AND ATOMIC VOLUMES OF THE ELEMENTS

IA	IIA
$_3$Li	$_4$Be
0.60(+1)	0.31(+2)	← Ionic radius (Pauling) (Å) (O.N.)
1.23	0.89	← Covalent radius (Pauling) (Å)
1.55	1.12	← Metallic radius (C.N. = 12) (Å)
1.15	**1.5**	← Electronegativity (Allred/Rochow)
159	322	← Enthalpy of atomization at 25 °C (kJ mol^{-1})
13.1	4.9	← Atomic volume (cm^3 mol^{-1})
		← Van der Waals radius (Å)

H	He
$_1$H	$_2$He
2.08(−1)	—
0.29	—
	—
2.1	—
218	0
14.1	—
1.2	—

IA	IIA	IIIA	IVA	VA	VIA	VIIA	VIII	VIII	VIII
$_{11}$Na	$_{12}$Mg								
0.95(+1)	0.65(+2)								
1.56	1.36								
1.90	1.60								
1.0	**1.25**								
108	150								
23.7	14.0								
$_{19}$K	$_{20}$Ca	$_{21}$Sc	$_{22}$Ti	$_{23}$V	$_{24}$Cr	$_{25}$Mn	$_{26}$Fe	$_{27}$Co	$_{28}$Ni
1.33(+1)	0.99(+2)	0.81(+3)	0.68(+4)	0.60(+4)	0.69(+3)	0.66(+3)	0.64(+3)	0.63(+3)	0.62(+3)
			0.76(+3)	0.74(+3)	0.84(+2)	0.80(+2)	0.76(+2)	0.74(+2)	0.72(+2)
2.03	1.74	1.44	1.32	1.22	1.18	1.17	1.17	1.16	1.15
2.35	1.97	1.64	1.47	1.35	1.30	1.35	1.26	1.25	1.25
0.9	**1.05**	**1.2**	**1.3**	**1.45**	**1.55**	**1.6**	**1.65**	**1.7**	**1.75**
90	177	390	469	502	397	284	406	439	427
95.5	26	15	10.6	8.3	7.2	7.4	7.1	6.7	6.6
$_{37}$Rb	$_{38}$Sr	$_{39}$Y	$_{40}$Zr	$_{41}$Nb	$_{42}$Mo	$_{43}$Tc	$_{44}$Ru	$_{45}$Rh	$_{46}$Pd
1.48(+1)	1.13(+2)	0.93(+3)	0.80(+4)	0.70(+5)	0.62(+6)		0.67(+4)	0.86(+2)	0.86(+2)
					0.68(+4)		0.69(+3)		
2.16	1.91	1.62	1.45	1.34	1.30	1.27	1.25	1.25	1.28
2.48	2.15	1.78	1.60	1.46	1.39	1.36	1.34	1.34	1.37
0.9	**1.0**	**1.1**	**1.2**	**1.25**	**1.3**	**1.35**	**1.4**	**1.45**	**1.35**
85.8	163	410	611	770	653	650	670	577	389
55.8	33.7	19.9	14.1	10.8	9.4	8.6	8.4	8.2	8.9
$_{55}$Cs	$_{56}$Ba	$_{57}$La	$_{72}$Hf	$_{73}$Ta	$_{74}$W	$_{75}$Re	$_{76}$Os	$_{77}$Ir	$_{78}$Pt
1.69(+1)	1.35(+2)	1.15(+3)	0.81(+4)	0.73(+5)	0.64(+6)		0.69(+4)	0.66(+4)	0.96(+2)
					0.68(+4)				
2.35	1.98	1.69	1.44	1.34	1.30	1.28	1.26	1.27	1.30
2.67	2.22	1.88	1.60	1.49	1.41	1.37	1.35	1.36	1.39
0.85	**0.95**	**1.1**	**1.25**	**1.35**	**1.40**	**1.45**	**1.5**	**1.55**	**1.45**
78.2	176	435	703	775	840	790	728	690	510
71	39.3	22.5	13.6	10.9	9.5	8.9	8.4	8.6	9.1

IB	IIB	IIIB	IVB	VB	VIB	VIIB	0
		$_5$B 0.20(+3) 0.82 0.98 2.0 562 4.6	$_6$C 0.15(+4) 2.60(−4) 0.77 0.91 2.5 716 5.3	$_7$N 0.11(+5) 1.71(−3) 0.70 3.05 473 17.3 1.5	$_8$O 0.09(+6) 1.40(−2) 0.66 3.5 250 14.0 1.4	$_9$F 0.07(+7) 1.36(−1) 0.64 4.1 79.1 17.1 1.35	$_{10}$Ne 0 16.8 1.31
		$_{13}$Al 0.50(+3) 1.25 1.43 1.45 326 10.0	$_{14}$Si 0.41(+4) 2.71(−4) 1.17 1.32 1.74 456 12.0	$_{15}$P 0.34(+5) 2.12(−3) 1.10 1.28 2.05 315 17.0 1.8	$_{16}$S 0.29(+6) 1.84(−2) 1.04 2.45 280 15.4 1.9	$_{17}$Cl 0.26(+7) 1.81(−1) 0.99 2.85 122 22.7 1.85	$_{18}$Ar 0 28.5 1.74
$_{29}$Cu 0.96(+1) 0.69(+2) 1.17 1.28 1.75 341 7.1	$_{30}$Zn 0.74(+2) 1.25 1.37 1.65 130 9.2	$_{31}$Ga 0.62(+3) 1.13(+1) 1.25 1.41 1.8 277 11.8	$_{32}$Ge 0.53(+4) 0.93(+2) 1.22 1.37 2.0 376 13.6	$_{33}$As 0.47(+5) 2.22(−3) 1.21 1.39 2.2 287 13.1 2.0	$_{34}$Se 0.42(+6) 1.98(−2) 1.17 1.40 2.5 207 16.5 2.0	$_{35}$Br 0.39(+7) 1.95(−1) 1.14 2.75 111 25.6 1.95	$_{36}$Kr 0 32.3 1.89
$_{47}$Ag 1.26(+1) 1.34 1.44 1.4 289 10.3	$_{48}$Cd 0.97(+2) 1.44 1.54 1.46 112 13.0	$_{49}$In 0.81(+3) 1.32(+1) 1.44 1.66 1.5 243 15.7	$_{50}$Sn 0.71(+4) 1.12(+2) 1.40 1.62 1.7 302 16.3	$_{51}$Sb 0.62(+5) 2.45(−3) 1.41 1.59 1.8 262 18.2 2.2	$_{52}$Te 0.56(+6) 2.21(−2) 1.37 1.60 2.0 197 20.4 2.2	$_{53}$I 0.50(+7) 2.16(−1) 1.33 2.2 106 25.7 2.15	$_{54}$Xe 0 42.9 2.1
$_{79}$Au 1.37(+1) 1.34 1.46 1.40 343 10.2	$_{80}$Hg 1.10(+2) 1.47 1.57 1.45 61 14.7	$_{81}$Tl 0.95(+3) 1.40(+1) 1.55 1.71 1.45 182 17.3	$_{82}$Pb 0.84(+4) 1.20(+2) 1.54 1.75 1.55 195 18.3	$_{83}$Bi 0.74(+5) 1.20(+3) 1.48 1.70 1.65 207 21.4	$_{84}$Po (+6) 1.46 1.76 1.75 145 22.4	$_{85}$At 1.90 92	$_{86}$Rn 0 50.5 2.15

A.13 RADII, ELECTRONEGATIVITIES, ETC., OF THE ELEMENTS

A.14 BOILING POINTS, MELTING POINTS, DENSITIES, ENTHALPIES OF FUSION AND VAPORIZATION, REDUCTION POTENTIALS, THERMAL CONDUCTIVITIES, AND ATOMIC CONDUCTANCES OF THE ELEMENTS

Legend (for each element block):
- Boiling point (°C) Structure
- Melting point (°C)
- Density (g cm^{-3}) (for gases: density of liquid at b.p.)
- Enthalpies of fusion and vaporization (kJ mol^{-1})
- Reduction potential (V)   Oxidation number
- Thermal conductivity (J cm^{-1} s^{-1} K^{-1}) (25°)
- Atomic conductance of elements × 1000 (mol ohm^{-1} cm^{-4})

	H	He
	$_1$H	$_2$He
	−253	−269 H
	−259	−272 P
	0.071	0.126
	0.06 0.45	0.084
	0.0017	0.0013

IA	IIA
$_3$Li 1317 B 180 0.53 2.89 135 −3.04 0.71 8	$_4$Be 2970 H 1280 1.85 9.8 310 −1.85 1.6 51
$_{11}$Na 892 B 97.8 0.97 2.6 98 −2.71 1.34 10	$_{12}$Mg 1107 H 651 1.74 9.0 132 −2.37 1.6 16

IA	IIA	IIIA	IVA	VA	VIA	VIIA	VIII	VIII	VIII
$_{19}$K 774 B 64 0.86 2.32 79 −2.92 0.96 3.5	$_{20}$Ca 1487 CHB 845 1.54 8.7 161 −2.87 1.25 9.6	$_{21}$Sc 2727 HC 1539 3.0 16 335 −2.1 III 0.063 1	$_{22}$Ti 3260 HB 1675 4.5 15.5 429 −1.63 II −1.2 III 0.2 2.2	$_{23}$V 3400 B 1900 6.11 17.6 459 −1.2 II −0.86 III 0.3 4.8	$_{24}$Cr 2480 BH 1890 7.2 13.8 305 −0.91 II −0.74 III 0.67 11	$_{25}$Mn 2097 * 1244 7.44 14.6 225 −1.18 II −0.28 III 1.5 0.7	$_{26}$Fe 3000 BC 1535 7.86 15.4 354 −0.44 II −0.04 III 0.75 24	$_{27}$Co 2900 HC 1495 8.86 15.2 389 −0.28 II +0.4 III 0.67 22	$_{28}$Ni 2732 CH 1453 8.90 17.6 379 −0.25 II III 0.92 22
$_{37}$Rb 688 B 39 1.53 2.3 76 −2.93 1.4	$_{38}$Sr 1334 CHB 769° 2.6 8.7 141 −2.89 1.3	$_{39}$Y 2927 H 1495 4.47 17.1 393 −2.37 III 0.0146 0.9	$_{40}$Zr 3578 HB 1852 6.49 23 544 −1.53 IV 0.26 1.8	$_{41}$Nb 4930 B 2470 8.57 26.8 697 −1.1 III 0.523 7.4	$_{42}$Mo 5560 BH 2610 10.2 27.6 565 −0.2 III 1.46 20	$_{43}$Tc 3927 H 2200 11.5 23 544 +0.4 II	$_{44}$Ru 3900 HC 2250 11.9 25.5 586 +0.45 II	$_{45}$Rh 3700 C 1966 12.48 21.8 531 +0.8 III +0.6 II 0.88 27	$_{46}$Pd 2930 C 1552 12.02 16.7 377 +0.99 II +1.3 IV 0.71 10
$_{55}$Cs 690 B 28.5 1.87 2.1 68 −2.95 0.7	$_{56}$Ba 1140 B 725 3.5 7.7 149 −2.90 0.4	$_{57}$La 3470 HC 920 6.17 11.3 400 −2.52 0.14 0.8	$_{72}$Hf 5400 HB 2150 13.1 21.8 661 −1.7 IV 0.92 2.1	$_{73}$Ta 5425 B 2996 16.6 31.4 753 0.54 7.5	$_{74}$W 5927 B* 3410 19.3 35.2 785 −0.11 III 1.7 19	$_{75}$Re 5640 H 3180 20.9 33.1 670 +0.3 III 0.71	$_{76}$Os 5000 HC 3000 22.5 29.3 650 +0.9 II 5.8	$_{77}$Ir 4530 C 2410 22.4 26.4 636 +1.2 III 0.59 22	$_{78}$Pt 3827 C 1768 21.4 19.7 510 +1.2 II 0.7 10

P—pressure: arsenic (28 atm); helium (26 atm).
S—sublimes.
H—hexagonal close-packed; C—cubic close-packed (f.c.c.); B—body-centred cubic; *complex; † distorted; D—diamond; G—graphite; L—layers; M—metallic; Cov—covalent; Ch—chains; ‖ parallel.
g—grey; w—white.

IB	IIB	IIIB	IVB	VB	VIB	VIIB	0
		$_5$B 2550SB*$_{12}$ 2300 2.34 **22.1 536** 0.01	$_6$C 4327 DG 3550 2.26D **— 718** 0.24 G 0.14 G	$_7$N −196 −210 0.81 **0.36 2.8** 0.00025	$_8$O −183 −218 1.14 **0.22 3.39** 0.00025	$_9$F −188 L −220 1.11 **0.26 3.27** 2.87 	$_{10}$Ne −246 C −249 1.20 **0.33 1.77** 0.00042
		$_{13}$Al 2467 * 660 2.7 **10.7 284** −1.66 2.10 38	$_{14}$Si 2355 D 1410 2.33 **46.5 297** 0.84	$_{15}$P 280 w Cov 44 w 1.82w **0.628 12.9** 0	$_{16}$S 445 Cov 119 2.07 **1.42 10.5** −0.51 0.00029	$_{17}$Cl −34.5 L −101 1.53 **3.2 10.2** 1.36 0.00008	$_{18}$Ar −186 C −189 1.40 **1.18 6.5** 0.00017
$_{29}$Cu 2595 C 1083 8.92 **13.1 305** +0.34 II +0.52 I 3.9 85	$_{30}$Zn 907 H† 419 7.13 **7.4 115** −0.763 1.13 18.4	$_{31}$Ga 2403 * 30 5.91 **5.59 256** −0.53 0.25 5	$_{32}$Ge 2830 D 937 5.32 **31.8 285** 0.0 II 0.59	$_{33}$As 613 S Cov 817 P 5.73 **27.7 32.4** 2.3	$_{34}$Se 685 g ∥Ch 217 4.79 **5.3 27.6** −0.92 0.00004	$_{35}$Br 59 L −7 3.12 **5.27 15** 1.07 	$_{36}$Kr −152 C −157 2.6 **1.64 9** 0.00008
$_{47}$Ag 2212 C 961 10.5 **11.3 254** +0.8 I +1.4 II 4.2 61	$_{48}$Cd 765 H† 321 8.64 **6.1 100** −0.4 0.92 11.5	$_{49}$In 2000 C† 156 7.3 **3.26 225** −0.34 III −0.25 I 0.24 7.8	$_{50}$Sn 2270 DM*w 232 7.3 **7.2 291** −0.136 II 0.01 IV 0.63 5.5	$_{51}$Sb 1380 *L 630 6.7 **19.9 195** 0.21 1.4	$_{52}$Te 990 ∥Ch 450 6.25 **17.7 50** −1.14 0.059	$_{53}$I 183 L 113 4.94 **7.8 30** 0.54 0.0042	$_{54}$Xe −107 C −112 3.06 **2.3 12.6** 0.0004
$_{79}$Au 2966 C 1063 19.3 **12.5 335** +1.7 I +1.4 III 2.9 42	$_{80}$Hg 357 C† −39 13.6 **2.3 58** +0.79 I +0.85 II 0.085 0.7	$_{81}$Tl 1457 * 303 11.8 **4.27 162** −0.34 I +0.72 III 0.39 3.2	$_{82}$Pb 1744 C 327 11.3 **4.78 178** −0.126 II 0.35 2.6	$_{83}$Bi 1560 *L 271 9.8 **10.9 172** 0.085 0.4	$_{84}$Po 962 254 9.4 **11 103** 	$_{85}$At 334 302 **24 —**	$_{86}$Rn −62 C −71 4.4 **2.9 16.4**

A.14 BOILING POINTS, MELTING POINTS, ETC., OF THE ELEMENTS

A.15 THERMODYNAMIC PROPERTIES OF SELECTED INORGANIC SUBSTANCES

$\Delta H_f^\ominus$ standard molar heat of formation
$\Delta G_f^\ominus$ standard molar Gibbs free energy of formation
$S^\ominus$ standard molar entropy.

Values refer to 298 K and to the state indicated in column 2. For aqueous ions data refer to a standard state of unit activity ($a = 1$). Such values are obtained by making use of information derived from extrapolation to infinite dilution. The standard state can be described as a hypothetical ideal solution of molality 1 mol kg^{-1}.

Substance	State	$\Delta H_f^\ominus$ (kJ mol^{-1})	$\Delta G_f^\ominus$ (kJ mol^{-1})	$S^\ominus$ (J mol^{-1} K^{-1})
Aluminium				
Al	s	0	0	27.9
AlCl$_3$	s	−704.2	−628.9	110.7
AlBr$_3$	s	−527.2	−505.0	184.1
Al$_2$O$_3$	s (corundum)	−1675.7	−1582.4	50.9
Al(OH)$_3$	s	−1276.1		
Al$_2$(SO$_4$)$_3$	s	−3440.8	−3100.1	239.3
Al$_2$(SO$_4$)$_3 \cdot$6H$_2$O	s	−5311.7	−4622.6	469.0
Al^{3+}	aq	−524.6	−481.1	−313.3
Antimony				
Sb	s	0	0	45.7
SbCl$_3$	s	−382.0	−324.7	186.2
SbCl$_5$	l	−438.5		
Sb$_4$O$_6$	s	−1409.2	−1213.4	246.0
Sb$_2$S$_3$	s (black)	−174.9	−173.6	182.0
Argon				
Ar	g	0	0	154.7
Arsenic				
As	s (grey)	0	0	35.1
AsH$_3$	g	68.9	22.5	66.4
AsCl$_3$	l	−335.6	−295.0	233.5
AsBr$_3$	s	−195.0		
As$_4$O$_6$	s	−1312.1	−1152.3	214.2
As$_2$O$_5$	s	−915.9	−772.4	105.4
As$_2$S$_3$	s	−169.0	−168.6	163.6
Barium				
Ba	s	0	0	64.9
BaH$_2$	s	−171.1		
BaF$_2$	s	−1200.4	−1148.5	96.2
BaCl$_2$	s	−860.2	−810.9	125.5
BaCl$_2 \cdot$2H$_2$O	s	−1461.9	−1295.8	202.9
Ba(ClO$_3$)$_2 \cdot$H$_2$O	s	−1066.5		
Ba(ClO$_4$)$_2$	s	−806.7		
BaBr$_2$	s	−754.8		
BaBr$_2 \cdot$2H$_2$O	s	−1365.2		
BaI$_2$	s	−602	−597	167
Ba(IO$_3$)$_2$	s	−997.5		
BaO	s	−558.1	−528.4	70.3
BaO$_2$	s	−629.7		
Ba(OH)$_2$	s	−946.4	−856.4 (est)	
BaCO$_3$	s	−1218.8	−1138.9	112.1

Substance	State	$\Delta H_f^\ominus$ (kJ mol^{-1})	$\Delta G_f^\ominus$ (kJ mol^{-1})	$S^\ominus$ (J mol^{-1} K^{-1})
Ba(HCO$_3$)$_2$	aq	−1920.5	−1734.7	200.8
BaS	s	−443.5		
BaSO$_4$	s	−1465.2	−1353.1	132.2
Ba(NO$_3$)$_2$	s	−728.0	−795.0	213.8
BaC$_2$O$_4$·2H$_2$O	s (oxalate)	−1966.5	−1742.2	
BaCrO$_4$	s	−1428.0		
Ba^{2+}	aq	−538.3	−559.7	12.6
Beryllium				
Be	s	0	0	9.5
BeF$_2$	s	−1051.9		
BeCl$_2$	s	−511.7		
BeBr$_2$	s	−369.9		
BeO	s	−610.9	−581.6	14.1
Be(OH)$_2$	s(α)	−907.1		
BeSO$_4$·4H$_2$O	s	−2411.2		
Be(NO$_3$)$_2$·3H$_2$O	s	−787.8		
Be^{2+}	aq	−389.0	−329.2	
Bismuth				
Bi	s	0	0	56.9
BiCl$_3$	s	−379.1	−318.8	177
BiOCl	s	−365.3	−322.2	86.2
Bi$_2$O$_3$	s	−577.0	−496.6	151.5
Bi$_2$S$_3$	s	−143.1	−140.6	200.4
Boron				
B	s	0	0	5.9
B$_2$H$_6$	g (diborane)	31.4	82.8	233.0
BF$_3$	g	−1145.4	−1120.3	254.0
BCl$_3$	l	−427.2	−387.4	206.3
BCl$_3$	g	−395.4	−380.3	290.0
B$_2$O$_3$	s	−1272.8	−1193.7	54.0
Bromine				
Br$_2$	l	0	0	151.6
Br$_2$	g	31.9	3.1	245.4
Br$^-$	aq	−121.4	−103.9	82.4
Br$_3^-$	aq	−125.7	−107.0	215.5
Cadmium				
Cd	s	0	0	51.8
CdBr$_2$	s	−314.6	−293.3	133.5
CdO	s	−254.8	−225.1	54.8
CdS	s	−144.3	−140.6	71.1
CdSO$_4$	s	−820.0	−926.2	137.2
Cd^{2+}	aq	−72.3	−77.6	−61.0
Caesium				
Cs	s	0	0	82.8
CsF	s	−530.9		
CsCl	s	−433.0		
CsBr	s	−394.6	−383.3	121.3
CsI	s	−336.8	−333.5	129.7
Cs$_2$O	s	−317.6		
CsOH	s	−406.7		
Cs$^+$	aq	−247.6	−281.9	133.1
Calcium				
Ca	s	0	0	41.6

A.15 THERMODYNAMIC PROPERTIES OF SELECTED INORGANIC SUBSTANCES

Substance	State	$\Delta H_f^\ominus$ (kJ mol^{-1})	$\Delta G_f^\ominus$ (kJ mol^{-1})	$S^\ominus$ (J mol^{-1} K^{-1})
CaH$_2$	s	−188.7	−149.8	41.8
CaF$_2$	s	−1214.6	−1161.9	69.0
CaCl$_2$	s	−795.0	−750.2	113.8
CaCl$_2 \cdot$H$_2$O	s	−1109.2		
CaCl$_2 \cdot$2H$_2$O	s	−1403.7		
CaCl$_2 \cdot$4H$_2$O	s	−2009.2		
CaCl$_2 \cdot$6H$_2$O	s	−2607.5		
CaBr$_2$	s	−674.9	−656.1	129.7
CaI$_2$	s	−534.7	−529.7	142.3
CaO	s	−635.5	−604.2	39.7
Ca(OH)$_2$	s	−986.6	−896.8	76.1
CaC$_2$	s	−62.8	−67.8	70.3
CaCO$_3$	s (calcite)	−1206.9	−1128.8	92.9
CaCO$_3$	s (aragonite)	−1207.0	−1127.7	88.7
CaS	s	−482.4	−477.4	56.5
CaSO$_4$	s (anhydrite)	−1432.6	−1320.5	106.7
CaSO$_4 \cdot$0.5H$_2$O	s	−1575.3	−1435.1	130.5
CaSO$_4 \cdot$2H$_2$O	s (gypsum)	−2021.3	−1795.8	194.1
Ca(NO$_3$)$_2$	s	−937.2	−741.8	193.3
Ca(NO$_3$)$_2 \cdot$4H$_2$O	s	−2131.3	−1700.8	338.9
Ca$_3$(PO$_4$)$_2$	s(β)	−4137.6	−3899.5	236.0
CaCrO$_4 \cdot$2H$_2$O	s	−1379.0	−1277.4	133.9
CaSi$_2$	s	−150.6		
CaSiO$_3$	s(β)	−1567.3	−1498.7	82.0
Ca$_2$SiO$_4$	s(γ)	−2255.2		
CaC$_2$O$_4 \cdot$H$_2$O	s (oxalate)	−1669.8	−1508.8	156.1
Ca^{2+}	aq	−543.0	−553.0	−55.1

Carbon

Substance	State	$\Delta H_f^\ominus$	$\Delta G_f^\ominus$	$S^\ominus$
C	s (graphite)	0	0	5.7
CO	g	−110.5	−137.3	197.9
CO$_2$	g	−393.7	−394.6	213.8
CS$_2$	l	87.9	63.6	151.0
HCN	g	135.1	124.7	201.7
C$_2$N$_2$	g	307.9	296.3	242.1
CO$_3^{2-}$	aq	−677.0	−527.8	−56.8
HCO$_3^-$	aq	−587.0	−690.6	95.0
CN$^-$	aq	150.6	171.5	94.1
CNO$^-$	aq	−145.9	−97.4	106.7
CNS$^-$	aq	76.4	92.7	144.3
HCO$_2^-$	aq (formate)	−409.9	−334.6	91.6
C$_2$H$_3$O$_2^-$	aq (acetate)	−488.8		
C$_2$O$_4^{2-}$	aq (oxalate)	−824.1	−674.8	51.0
HC$_2$O$_4^-$	aq (hydrogen oxalate)	−817.9	−699.0	153.6

Chlorine

Substance	State	$\Delta H_f^\ominus$	$\Delta G_f^\ominus$	$S^\ominus$
Cl$_2$	g	0	0	223.0
Cl$_2$O	g	80.3	97.9	266.1
ClO$_2$	g	102.5	120.5	256.7
Cl$^-$	aq	−167.1	−131.2	56.5
ClO$^-$	aq	−107.0	−36.7	41.8
ClO$_2^-$	aq	−66.4	17.2	101.3
ClO$_3^-$	aq	−99.1	−3.2	162.3
ClO$_4^-$	aq	−129.2	−8.5	182.0

Chromium

Substance	State	$\Delta H_f^\ominus$	$\Delta G_f^\ominus$	$S^\ominus$
Cr	s	0	0	23.8
CrCl$_3$	s	−563.2	−493.7	125.5
CrO$_2$Cl$_2$	l	−567.8		
Cr$_2$O$_3$	s	−1128.4	−1046.8	81.2

Substance	State	$\Delta H_f^\ominus$ (kJ mol^{-1})	$\Delta G_f^\ominus$ (kJ mol^{-1})	$S^\ominus$ (J mol^{-1} K^{-1})
CrO_3	s	−579.1	−495.8	
$Cr_2(SO_4)_3 \cdot 18H_2O$	s	−8339.5		
Cr^{2+}	aq	−138.8	−164.7	
Cr^{3+}	aq	5481.0	−204.9	
$Cr(H_2O)_6^{3+}$	aq (violet)	−1970.6		
CrO_4^{2-}	aq	−863.1	−706.2	38.5
$HCrO_4^-$	aq	−890.3	−742.6	69.0
$Cr_2O_7^{2-}$	aq	−1460.5	−1257.2	213.8
Cobalt				
Co	s	0	0	30.0
$CoCl_2$	s	−325.5	−282.4	106.3
$CoCl_2 \cdot 6H_2O$	s	−2129.2		
CoO	s	−239.3	−213.4	43.9
Co_3O_4	s	−878.6	−758.9	
$Co(OH)_2$	s	−548.9	−455.6	
$CoSO_4$	s	−868.2	−761.9	113.4
$CoSO_4 \cdot 7H_2O$	s	−2986.5		
$Co(NO_3)_2 \cdot 6H_2O$	s	−2216.3		
Co^{2+}	aq	−67.3	−51.4	−155.1
Co^{3+}	aq		123.8	
Copper				
Cu	s	0	0	33.3
CuCl	s	−136.0	−118.0	84.5
$CuCl_2$	s	−205.9		
Cu_2O	s	−166.7	−146.4	100.8
CuO	s	−155.2	−127.2	43.5
$Cu(OH)_2$	s	−448.5		
Cu_2S	s	−79.5	−86.2	120.9
CuS	s	−48.5	−49.0	66.5
$CuSO_4$	s	−769.9	−661.9	113.4
$CuSO_4 \cdot 5H_2O$	s	−2278.2	−1879.9	305.4
Cu^+	aq	51.9	50.4	−26.3
Cu^{2+}	aq	64.4	65.0	−98.6
Fluorine				
F_2	g	0	0	202.8
F_2O	g	−21.8	−4.6	247.3
F^-	aq	−332.5	−278.7	−13.7
Germanium				
Ge	s	0	0	31.1
$GeCl_4$	l	−543.9		
GeO	s			50.2
GeO_2	s	−551.0	−497.1	55.3
Gold				
Au	s	0	0	47.4
Au_2O_3	s	80.0	163.2	125.5
$AuCl_3$	s	−118.4		
Au^+	aq		161.9	
Au^{3+}	aq		410.9	
$AuCl_4^-$	aq	−325.4	653.5	255.2
$Au(CN)_2^-$	aq	244.3	215.5	414.2
Hydrogen				
H_2	g	0	0	130.6
HF	g	−271.1	−273.2	173.7
HCl	g	−92.3	−95.3	186.7
HBr	g	−36.2	−53.2	198.5

Substance	State	$\Delta H_f^\ominus$ (kJ mol^{-1})	$\Delta G_f^\ominus$ (kJ mol^{-1})	$S^\ominus$ (J mol^{-1} K^{-1})
HI	g	26.5	2.1	206.5
HIO$_3$	s		−238.6	
H$_2$O	l	−285.9	−237.2	70.0
H$_2$O	g	−241.8	−228.6	188.7
H$_2$O$_2$	l	−187.6	−118.0	109.6
H$_2$S	g	−20.6	−33.6	205.7
H$_2$S$_2$	l	−23.1		
H$_2$SO$_4$	l	−814.0	−690.1	156.9
HNO$_3$	l	−173.2	−79.9	155.6
H$_3$PO$_4$	s	−1279.0	−1119.2	110.5
H$_3$BO$_3$	s	−1094.3	−969.0	88.8
H$^+$	aq	0	0	0

Helium

He	g	0	0	126.0

Iodine

I$_2$	s	0	0	116.8
I$_2$	g	62.3	19.4	260.6
ICl	g	17.6	−5.5	247.3
ICl$_3$	s	−88.3	−22.4	172.0
IBr	g	40.8	3.8	258.6
I$_2$O$_5$	s	−158.1	−177.2	
I$^-$	aq	−55.2	−51.5	111.3
I$_3^-$	aq	−51.4	−51.4	239.3
IO$^-$	aq	−107.4	−38.4	−5.3
IO$_3^-$	aq	−221.2	−127.9	118.4
IO$_4^-$	aq	−147.2		
ICl$^-$	aq	0.0	−161.0	
I$_2$Cl$^-$	aq		−132.5	
IBr$_2^-$	aq		−122.9	
I$_2$Br$^-$	aq	−127.9	−109.9	197.5

Iron

Fe	s	0	0	27.2
FeCl$_2$	s	−341.0	−302.1	119.7
FeCl$_3$	s	−405.0		
FeO	s	−266.5	−244.3	54.0
Fe$_2$O$_3$	s	−822.2	−741.0	90.0
Fe$_3$O$_4$	s	−1117.1	−1014.2	146.4
Fe(OH)$_2$	s	−568.2	−483.7	79.5
Fe(OH)$_3$	s	−824.2		
FeCO$_3$	s	−747.7	−674.0	92.9
Fe(CO)$_5$	l	−785.8		
FeS	s(α)	−95.1	−97.6	67.4
FeSO$_4$	s	−922.6		
FeSO$_4\cdot$7H$_2$O	s	3007.0		
Fe^{2+}	aq	−87.8	−84.9	−113.3
Fe^{3+}	aq	−47.6	−9.7	−305.8
Fe(CN)$_6^{3-}$	aq	640.2	719.6	
Fe(CN)$_6^{4-}$	aq	530.1	686.2	

Krypton

Kr	g	0	0	164.0

Lead

Pb	s	0	0	64.8
PbF$_2$	s	−664.0	−617.1	110.5
PbCl$_2$	s	−359.4	−314.1	136.0
PbBr$_2$	s	−278.7	−261.9	161.5

Substance	State	$\Delta H_f^\ominus$ (kJ mol^{-1})	$\Delta G_f^\ominus$ (kJ mol^{-1})	$S^\ominus$ (J mol^{-1} K^{-1})
PbI$_2$	s	−175.5	−173.6	174.8
PbO	s	−217.3	−187.9	68.7
Pb$_3$O$_4$	s	−718.4	−601.2	211.3
PbO$_2$	s	−277.4	−217.4	68.6
PbCO$_3$	s	−700.0	−626.3	131.0
PbS	s	−100.4	−98.7	91.2
PbSO$_4$	s	−919.9	−813.2	148.6
Pb(NO$_3$)$_2$	s	−451.9		
PbCrO$_4$	s	−911		
Pb(C$_2$H$_3$O$_2$)$_2\cdot$3H$_2$O	s (acetate)	−1853.9		
Pb(C$_2$H$_5$)$_4$	l	217.6		
Pb^{2+}	aq	1.6	−24.2	21.3

Lithium

Substance	State	$\Delta H_f^\ominus$	$\Delta G_f^\ominus$	$S^\ominus$
Li	s	0	0	28.0
LiH	s	−90.4	−70.0	24.7
Li$_3$H$_4$	s	−186.6		
LiF	s	−612.1	−583.7	35.9
LiCl	s	−408.8		
LiBr	s	−350.3		
LiI	s	−271.1		
Li$_2$O	s	−595.8		
LiOH	s	−487.2	−443.9	50.2
Li$_2$CO$_3$	s	−1215.5	−1132.6	90.4
LiHCO$_3$	aq	−969.6	−880.9	123.4
LiNO$_3$	s	−482.3		
LiAlH$_4$	s	−101.3		
Li$^+$	aq	−278.3	−293.7	14.2

Magnesium

Substance	State	$\Delta H_f^\ominus$	$\Delta G_f^\ominus$	$S^\ominus$
Mg	s	0	0	32.7
MgCl$_2$	s	−641.8	−592.3	89.5
MgCl$_2\cdot$6H$_2$O	s	−2499.6	−2115.6	366.1
Mg(ClO$_4$)$_2$	s	−588.3		
MgBr$_2$	s	−517.6		
MgF$_2$	s	−1121	−1068	57
MgO	s	−601.7	−569.4	26.8
Mg(OH)$_2$	s	−924.7	−833.7	63.1
MgCO$_3$	s	−1112.9	−1029.3	65.7
MgS	s	−347.3		
MgSO$_4$	s	−1278.2	−1173.6	91.6
MgSO$_4\cdot$7H$_2$O	s	−3383.6		
Mg$_3$N$_2$	s	−461.2		
Mg(NO$_3$)$_2\cdot$6H$_2$O	s	−2612.3		
Mg$_3$(PO$_4$)$_2\cdot$4H$_2$O	s	−4022.9		
Mg$_2$Si	s	−77.8		
MgSiO$_3$	s	−840.3	−763.5	83.3
Mg$_2$SiO$_2$	s	−2042.6	−1923.8	95.0
Mg^{2+}	aq	−461.8	−455.9	−117.9

Manganese

Substance	State	$\Delta H_f^\ominus$	$\Delta G_f^\ominus$	$S^\ominus$
Mn	s	0	0	32.0
MnCl$_2$	s	−482.4	−441.4	117.2
MnCl$_2\cdot$4H$_2$O	s	−1702.9		
MnO	s	−384.9	−363.2	60.2
Mn$_3$O$_4$	s	−1386.6	−1280.3	148.5
Mn$_2$O$_3$	s	−971.1	−893.3	
MnO$_2$	s	−520.9	−466.1	53.1
MnCO$_3$	s	−895.0	−817.6	85.8
MnS	s	−204.2	−208.8	78.2

Substance	State	$\Delta H_f^\ominus$ (kJ mol^{-1})	$\Delta G_f^\ominus$ (kJ mol^{-1})	$S^\ominus$ (J mol^{-1} K^{-1})
MnSO$_4$	s	−1063.6	−956.0	112.1
MnSO$_4 \cdot$H$_2$O	s	−1374.4		
MnSO$_4 \cdot$4H$_2$O	s	−2256.4		
Mn(NO$_3$)$_2 \cdot$6H$_2$O	s	−2370.0		
Mn^{2+}	aq	−218.7	−223.3	−83.6
MnO$_4^-$	aq	−518.3	−425.0	190.0
Mercury				
Hg	l	0	0	76.1
Hg$_2$Cl$_2$	s	−264.8	−210.9	195.8
HgCl$_2$	s		−176.6	
Hg$_2$Br$_2$	s	−206.7	−178.7	213.0
Hg$_2$I$_2$	s	−120.9	−111.3	239.3
HgI$_2$	s (red)	−105.4		
HgO	s (red)	−90.7	−58.5	72.0
HgS	s (black)	−54.0	−46.4	83.3
HgS	s (red)	−58.2	−49.0	77.8
Hg$_2$SO$_4$	s	−742.0	−623.9	200.7
HgSO$_4$	s	−704.2		
Hg$_2^{2+}$	aq		153.9	
Hg^{2+}	aq		164.8	
Neon				
Ne	g	0	0	146.2
Nickel				
Ni	s	0	0	29.9
NiF$_2$	s		−1589.5	
NiCl$_2$	s	−315.9	−272.4	107.1
NiO	s	−244.3	−216.3	38.6
Ni(OH)$_2$	s	−678.2	−453.1	79.5
Ni(CO)$_4$	l			309.6
NiSO$_4 \cdot$7H$_2$O	s	−2688.2	−2221.7	305.9
Ni(NO$_3$)$_2 \cdot$6H$_2$O	s	−2223.4		
Ni^{2+}	aq	−63.9	−46.3	−159.3
Nitrogen				
N$_2$	g	0	0	191.4
N$_2$H$_4$	l	50.6	149.2	121.2
NF$_3$	g	−124.7	−83.7	260.6
NCl$_3$	l	230.1		
N$_2$O	g	82.0	104.2	219.7
NO	g	90.4	86.6	210.5
N$_2$O$_3$	g	83.8	139.4	312.2
NO$_2$	g	33.2	51.3	240.0
N$_2$O$_4$	g	9.2	97.8	304.2
N$_2$O$_5$	s	−43.1	113.8	178.2
NH$_3$	g	−46.0	−16.7	192.5
NH$_4$Cl	s	−315.5	−203.8	94.6
NH$_4$Br	s	−270.3	−175.3	113.0
NH$_4$I	s	−201.4	−112.5	117.2
(NH$_4$)$_2$SO$_4$	s	−1180.9	−901.9	220.1
NH$_4$NO$_3$	s	−365.6	−184.0	151.1
NH$_4^+$	aq	−132.4	−79.3	113.4
N$_2$H$_5^+$	aq	−7.4	82.4	150.6
NO$_2^-$	aq	−104.5	−37.1	140.2
NO$_3^-$	aq	−207.3	−111.2	146.4

Substance	State	$\Delta H_f^\ominus$ (kJ mol^{-1})	$\Delta G_f^\ominus$ (kJ mol^{-1})	$S^\ominus$ (J mol^{-1} K^{-1})
Oxygen				
O_2	g	0	0	205.0
O_3	g	142.3	163.4	237.7
OH^-	aq	−229.9	−157.2	−10.7
Phosphorus				
P	s (white)	0	0	41.0
PH_3	g	5.4	13.4	210.1
PH_4I	s	−69.9	0.8	123.0
PF_3	g	−918.8	−897.5	273.1
PF_5	g	−1595.8		
PCl_3	l	−319.7	−272.4	217.1
PCl_5	s	−463.2		
$POCl_3$	l	−597.1	−520.9	222.5
PBr_3	l	−184.5	−175.7	240.2
PBr_5	s	−269.9		
$POBr_3$	g	−458.6		359.7
P_4O_6	s	−1640.1		
P_4O_{10}	s	−2984.0	−2697.8	228.9
PO_3^-	aq	−976.9	0	0
PO_4^{3-}	aq	−1279.8	−1020.8	−221.7
$P_2O_7^{4-}$	aq	179.1	−1923.7	−104.5
HPO_4^{2-}	aq	−1294.3	−1091.5	−33.4
$H_2PO_4^-$	aq	−1298.5	−1132.6	90.4
HPO_3^{2-}	aq	−968.9		
$H_2PO_3^{2-}$	aq	−969.3		
$H_2PO_2^-$	aq	−613.7		
PH_4^+	aq		67.8	
Potassium				
K	s	0	0	64.2
KF	s	−562.6	−533.1	66.6
KCl	s	−435.9	−408.3	82.7
$KClO_3$	s	−391.2	−289.9	143.0
$KClO_4$	s	−433.5	−304.2	151.0
KBr	s	−392.2	−379.2	96.4
$KBrO_3$	s	−332.2	−243.5	149.2
KI	s	−327.6	−322.3	104.3
KIO_3	s	−508.4	−425.5	151.5
KIO_4	s	−408.4		
K_2O	s	−361.5		
KO_2	s	−560.7		
KOH	s	−425.8		
$KOH \cdot 2H_2O$	s	−1051.0		
K_2CO_3	s	−1146.0		
$KHCO_3$	s	−959.4		
K_2S	s	−418.4		
K_2SO_4	s	−1433.9	−1316.3	175.7
$KHSO_4$	s	−1158.1		
$KAl(SO_4)_2 \cdot 12H_2O$	s	−6057.2	−5137.1	687.4
$KCr(SO_4)_2 \cdot 12H_2O$	s	−5786.9		
KNO_2	s	−370.3		
KNO_3	s	−492.7	−393.1	132.9
$KMnO_4$	s	−813.4	−713.8	171.5
KH_2PO_4	s	−1569		
K_2CrO_4	s	−382.8		
$K_2Cr_2O_7$	s	−2033.0		
KCN	s	−112.5		
KCNS	s	−203.4		
$K_3Fe(CN)_6$	s	−173.2		

A.15 THERMODYNAMIC PROPERTIES OF SELECTED INORGANIC SUBSTANCES

Substance	State	$\Delta H_f^\ominus$ (kJ mol^{-1})	$\Delta G_f^\ominus$ (kJ mol^{-1})	$S^\ominus$ (J mol^{-1} K^{-1})
K$_4$Fe(CN)$_6$	s	−523.4		
K$_4$Fe(CN)$_6$·3H$_2$O	s	−1398.7		
K$^+$	aq	−252.3	−283.2	102.5
Radon				
Rn	g	0	0	176.1
Rubidium				
Rb	s	0	0	76.2
RbF	s	−549.3		
RbCl	s	−430.6		
RbBr	s	−389.2	−378.1	108.3
RbI	s	−328.4	−325.5	118.0
RbOH	s	−476.6	−439.5	113.8
Rb$_2$CO$_3$	s	−1128.0		
RbHCO$_3$	s	−956.0		
Rb$_2$SO$_4$	s	−1424.7		
RbHSO$_4$	s	−1145.2		
RbNO$_3$	s	−454.0	−392.7	270.7
Rb$^+$	aq	−250.1	−282.1	124.3
Silicon				
Si	s	0	0	19.0
SiH$_4$	g	34.3	56.9	204.5
SiF$_4$	g	−1548.1	−1506.2	284.5
SiCl$_4$	l	−640.2	−572.8	239.3
SiCl$_4$	g	−609.6	−569.9	331.4
SiO$_2$	s (quartz)	−910.9	−856.7	41.8
SiO$_2$	s (cristobalite)	−909.5	−855.9	42.7
SiO$_2$	s (tridymite)	−909.1	−855.3	43.5
SiC	s	−65.3	−62.8	16.6
SiS$_2$	s	−145.2		
Silver				
Ag	s	0	0	42.7
AgF	s	−202.9	−184.9	83.7
AgCl	s	−127.0	−109.6	95.8
AgBr	s	−99.5	−95.9	107.1
AgI	s	−62.4	−66.3	114.2
Ag$_2$O	s	−30.6	−10.8	121.8
Ag$_2$CO$_3$	s	−505.8	−437.2	167.4
Ag$_2$S	s(α)	−31.8	−40.3	145.6
Ag$_2$S	s(β)	−29.3	−39.2	150.2
AgNO$_3$	s	−123.1	−32.2	140.9
Ag$_2$CrO$_4$	s	−712.1	−621.7	216.7
AgCN	s	146.2	164.0	83.7
Ag$^+$	aq	105.6	77.1	72.8
Ag(NH$_3$)$_2^+$	aq	−111.7	−16.9	
Ag^{2+}	aq		264.2	
Ag(CN)$_2^-$	aq	269.9	301.5	205.0
Sodium				
Na	s	0	0	51.0
NaH	s	−57.3		
NaF	s	−569.0	−541.0	58.6
NaCl	s	−411.0	−384.0	72.4
NaBr	s	−359.9		
NaI	s	−288.0		
Na$_2$O	s	−415.9	−376.6	72.8
Na$_2$O$_2$	s	−504.6		

Substance	State	$\Delta H_f^\ominus$ (kJ mol^{-1})	$\Delta G_f^\ominus$ (kJ mol^{-1})	$S^\ominus$ (J mol^{-1} K^{-1})
NaOH	s	−426.8		
NaOH·H$_2$O	s	−732.9	−623.4	84.5
Na$_2$CO$_3$	s	−1130.9	−1047.7	136.0
Na$_2$CO$_3$·10H$_2$O	s	−4081.9		
NaHCO$_3$	s	−947.7	−851.9	102.1
Na$_2$S	s	−373.2		
Na$_2$SO$_4$	s	−1384.5	−1266.8	149.5
Na$_2$SO$_4$·10H$_2$O	s	−4324.1	−3644.0	591.9
NaHSO$_4$	s	−1126.3		
Na$_2$S$_2$O$_3$	s	−1117.1		
Na$_2$S$_2$O$_3$·5H$_2$O	s	−2602.0		
NaNO$_2$	s	−359.4	−365.9	120.5
NaNO$_3$	s	−466.7	−365.9	116.3
Na$_2$B$_4$O$_7$·10H$_2$O	s	−6264.3		
Na$_2$SiO$_3$	s	−1518.8	−1426.7	113.8
NaCN	s	−90.0		
NaNH$_2$	s	−118.8		
Na$^+$	aq	−240.0	−261.8	59.0

Strontium

Substance	State	$\Delta H_f^\ominus$	$\Delta G_f^\ominus$	$S^\ominus$
Sr	s	0	0	52.3
SrF$_2$	s	−1214.6		
SrCl$_2$	s	−828.4	−781.2	117.2
SrBr$_2$	s	−715.9		
SrO	s	−590.4	−559.8	54.4
Sr(OH)$_2$	s	−959.4		
Sr(OH)$_2$·8H$_2$O	s	−3352.2		
SrCO$_3$	s	−1218.4	−1137.6	97.1
Sr(HCO$_3$)$_2$	aq	−1927.6	−1731.3	150.6
SrS	s	−452.3		
SrSO$_4$	s	−1444.7	−1334.3	121.8
Sr(NO$_3$)$_2$	s	−975.9		
Sr^{2+}	aq	−545.5	−557.2	−39.2

Sulphur

Substance	State	$\Delta H_f^\ominus$	$\Delta G_f^\ominus$	$S^\ominus$
S	s (rhombic)	0	0	31.9
SF$_4$	g	−774.9	−731.4	291.9
SF$_6$	g	−1209.2	−1105.0	291.2
S$_2$Cl$_2$	l	−60.2		
SOCl$_2$	l	−245.6	−203.4	215.7
SO$_2$Cl$_2$	l	−389.1		
SO$_2$	g	−296.9	−300.4	248.5
SO$_3$	g	−395.4	−370.3	256.1
S^{2-}	aq	33.1	85.8	−14.5
S$_2^{2-}$	aq	30.1	79.5	28.5
S$_3^{2-}$	aq	25.9	73.6	66.1
S$_4^{2-}$	aq	23.0	69.0	103.3
S$_5^{2-}$	aq	21.3	65.7	140.6
HS$^-$	aq	−17.5	−11.9	62.8
SO$_3^{2-}$	aq	−635.4	−486.5	−29.2
HSO$_3^-$	aq	−626.1	−527.7	139.7
SO$_4^{2-}$	aq	−909.2	−744.5	20.1
HSO$_4^-$	aq	−887.2	−755.9	131.8
S$_2$O$_3^{2-}$	aq	−652.2	−518.7	121.3
S$_4$O$_6^{2-}$	aq	−1224.1	−1030.4	259.4

Tin

Substance	State	$\Delta H_f^\ominus$	$\Delta G_f^\ominus$	$S^\ominus$
Sn	s (white)	0	0	51.4
Sn	s (grey)	2.5	4.6	44.8
SnCl$_2$·2H$_2$O	s	−945.2		

Substance	State	$\Delta H_f^\ominus$ (kJ mol^{-1})	$\Delta G_f^\ominus$ (kJ mol^{-1})	$S^\ominus$ (J mol^{-1} K^{-1})
SnCl$_4$	l	−511.5	−440.2	258.6
SnBr$_4$	s	−377.4	−350.2	264.4
SnO	s	−285.8	−256.9	56.5
SnO$_2$	s	−580.7	−519.7	52.3
Sn^{2+}	aq	−8.7	−26.2	
Sn^{4+}	aq	30.5	15.3	
Tungsten				
W	s	0	0	33.5
WO$_3$	s	−1497.5	−1410.8	67.8
WCl$_6$	s	−413.0		
WC	s	−38.0		
WS$_2$	s	−193.7	−193.3	96.2
Uranium				
U	s	0	0	50.3
UF$_6$	g	−2112.9	−2029.2	379.7
UO$_2$	s	−1129.7	−1075.3	77.8
UO$_3$	s	−1263.6	−1184.1	98.6
UO$_2$(NO$_3$)$_2$	s	−1377.4	−1142.7	276.1
UO$_2$(NO$_3$)$_2$·6H$_2$O	s	−3197.8	−2615.0	505.6
U^{2+}	aq		−292.8	
U^{3+}	aq		−312.0	
Xenon				
Xe	g	0	0	169.6
XeF$_2$	s	−133.9	−62.8	133.9
XeF$_4$	s	−261.5	−121.3	146.4
XeF$_6$	s	−380.7		
XeO$_3$	s	401.7		
Zinc				
Zn	s	0	0	41.6
ZnCl$_2$	s	−415.9	−369.4	108.4
ZnBr$_2$	s	−327.2	−310.0	137.2
ZnO	s	−348.0	−318.2	43.9
ZnCO$_3$	s	−812.5	−731.4	82.4
ZnS	s (wurtzite)	−189.5		
ZnS	s (blende)	−202.9	−198.3	57.7
ZnSO$_4$	s	−978.6	−871.5	124.7
ZnSO$_4$·7H$_2$O	s	−3075.7	−2560.2	386.6
Zn^{2+}	aq	−152.3	−147.1	−106.4
Zn(OH)$_4^{2-}$	aq		−863.5	
Zn(NH$_3$)$_4^{2+}$	aq		−304.1	

A.16 THERMODYNAMIC PROPERTIES OF SELECTED ORGANIC SUBSTANCES

$\Delta H_f^\ominus$ standard molar heat of formation
$\Delta G_f^\ominus$ standard molar Gibbs free energy of formation
$S^\ominus$ standard molar entropy.

Values refer to 298 K and to the state indicated in column 3.

Substance	Formula	State	$\Delta H_f^\ominus$ (kJ mol^{-1})	$\Delta G_f^\ominus$ (kJ mol^{-1})	$S^\ominus$ (J mol^{-1} K^{-1})
Straight-chain alkanes					
Methane	CH_4	g	−74.8	−50.8	186.2
Ethane	CH_3CH_3	g	−84.6	−32.8	229.5
Propane	$CH_3CH_2CH_3$	g	−103.8	−23.5	269.9
Butane	$CH_3(CH_2)_2CH_3$	g	−126.1	−17.1	310.1
Pentane	$CH_3(CH_2)_3CH_3$	l	−173.2	−9.6	261.2
Hexane	$CH_3(CH_2)_4CH_3$	l	−198.8	−4.4	295.9
Heptane	$CH_3(CH_2)_5CH_3$	l	−224.4	+1.0	328.5
Octane	$CH_3(CH_2)_6CH_3$	l	−249.9	+6.4	361.1
Nonane	$CH_3(CH_2)_7CH_3$	l	−275.5	+11.8	393.7
Decane	$CH_3(CH_2)_8CH_3$	l	−301.0	+17.2	425.9
Branched alkanes					
2-Methylpropane	$(CH_3)_2CHCH_3$	g	−134.5	−20.9	294.6
2-Methylbutane (isopentane)	$(CH_3)_2CHCH_2CH_3$	l	−179.7	−15.1	260.4
2,2-Dimethylpropane (neopentane)	$C(CH_3)_4$	g	−165.9	−15.2	306.4
Cyclo-alkanes					
Cyclopropane	$(CH_2)_3$	g	+55.2		
Cyclobutane	$(CH_2)_4$	g	−7.1		
Cyclopentane	$CH_2(CH_2)_3CH_2$	l	−105.9	+36.4	204.3
Cyclohexane	$CH_2(CH_2)_4CH_2$	l	−156.2	+26.7	204.4
Alkenes (olefins)					
Ethylene (ethene)	$CH_2{=}CH_2$	g	+52.3	+68.1	219.5
Propene	$CH_2{=}CHCH_3$	g	+20.4	+62.7	266.9
But-1-ene	$CH_2{=}CHCH_2CH_3$	g	−0.1	+71.5	305.6
trans-But-2-ene	$CH_3CH{=}CHCH_3$	g	−11.9	+62.9	296.4
cis-But-2-ene	$CH_3CH{=}CHCH_3$	g	−7.0	+65.9	300.8
Hex-1-ene	$CH_2{=}CH(CH_2)_3CH_3$	l	−41.7	+87.6	384.6
Buta-1,2-diene	$CH_2{=}C{=}CHCH_3$	g	+162.2	+198.4	293.0
Buta-1,3-diene	$CH_2{=}CHCH{=}CH_2$	g	+110.1	+150.6	278.7
Alkynes					
Acetylene (ethyne)	$CH{\equiv}CH$	g	+229.4	+211.7	203.2
Methylacetylene (propyne)	$CH_3C{\equiv}CH$	g	+185.4	+193.8	248.1
Arenes					
Benzene	C_6H_6	l	+49.0	+124.5	172.8
Naphthalene	$C_{10}H_8$	s	+151	+224	336
Toluene (methylbenzene)	$C_6H_5CH_3$	l	+12.1	+115.5	319.7
Ethylbenzene	$C_6H_5CH_2CH_3$	l	−12.5	+119.7	255.2
Propylbenzene	$C_6H_5(CH_2)_2CH_3$	l	−34.4	+123.8	290.5
1,2-Dimethylbenzene	$C_6H_4(CH_3)_2$	l	−24.4	+110.3	246.5
1,3-Dimethylbenzene	$C_6H_4(CH_3)_2$	l	−25.4	+107.6	252.1
1,4-Dimethylbenzene	$C_6H_4(CH_3)_2$	l	−24.4	+110.1	247.4
Styrene (ethenylbenzene)	$C_6H_5CH{=}CH_2$		+103.9(g)	+213.8(g)	345.1(l)
Amines (aminoalkanes, etc.)					
Methylamine	CH_3NH_2	g	−28.0	+27.6	
Dimethylamine	$(CH_3)_2NH$	g	−27.6	−59.0	
Trimethylamine	$(CH_3)_3N$	g	−46.0	−76.6	

Substance	Formula	State	$\Delta H_f^\ominus$ (kJ mol^{-1})	$\Delta G_f^\ominus$ (kJ mol^{-1})	$S^\ominus$ (J mol^{-1} K^{-1})
Ethylamine	CH$_3$CH$_2$NH$_2$	g	−48.5	−41.9	
1-Aminopropane (prop-1-ylamine)	CH$_3$CH$_2$CH$_2$NH$_2$	l	−68.8	−60.2	
1-Aminobutane (but-1-ylamine)	CH$_3$(CH$_2$)$_3$NH$_2$	l	+173.2	−81.8	
2-Aminobutane (but-2-ylamine)	CH$_3$CH$_2$CH(NH$_2$)CH$_3$	l	+160.7		
Aniline (phenylamine)	C$_6$H$_5$NH$_2$	l	+87	+167	319
Organic halogen compounds					
Chloromethane	CH$_3$Cl	g	−82.0		
Bromomethane	CH$_3$Br	g	−35.6		
Iodomethane	CH$_3$I	l	−8.4		
Dichloromethane	CH$_2$Cl$_2$	l	−117.2	−63.2	178.7
Trichloromethane	CHCl$_3$	l	−131.8	−71.5	202.9
Tetrachloromethane	CCl$_4$	l	−316.7		
Tetrabromomethane	CBr$_4$	s	+50.2		
Tetraiodomethane	CI$_4$	s			392
Chloroethane	CH$_3$CH$_2$Cl	g	−105.0		
Bromoethane	CH$_3$CH$_2$Br	l	−85.4		
Chlorobenzene	C$_6$H$_5$Cl	l	+52	+99	314
Iodobenzene	C$_6$H$_5$I	l	+147.4	+213.8	
Benzyl chloride (chloromethylbenzene)	C$_6$H$_5$CH$_2$Cl	l	−70		
Alcohols					
Methanol	CH$_3$OH	l	−238.9	−166.7	127.2
Ethanol	CH$_3$CH$_2$OH	l	−277.0	−174.3	161.2
Propan-1-ol	CH$_3$CH$_2$CH$_2$OH	l	−304.0	−171.3	196.6
Propan-2-ol	CH$_3$CH(OH)CH$_3$	l	−317.9	−180.3	180.5
Butan-1-ol	CH$_3$(CH$_2$)$_2$CH$_2$OH	l	−327.1	−168.9	228.0
Pentan-1-ol	CH$_3$(CH$_2$)$_3$CH$_2$OH	l	−357.1	−161.6	259.0
Hexan-1-ol	CH$_3$(CH$_2$)$_4$CH$_2$OH	l	−379.5	−152.3	289.5
Heptan-1-ol	CH$_3$(CH$_2$)$_5$CH$_2$OH	l	−398.7	−141.8	325.9
Octan-1-ol	CH$_3$(CH$_2$)$_6$CH$_2$OH	l	−425.1	−136.4	354.4
Ethane-1,2-diol	CH$_2$(OH)CH$_2$OH	l	−454.4		
Cyclohexanol	CH$_2$(CH$_2$)$_4$CHOH	s	−358.2		7.9
Ethers					
Dimethyl ether	CH$_3$OCH$_3$	g	−184.1	−112.8	266.7
Diethyl ether	CH$_3$CH$_2$OCH$_2$CH$_3$	l	−279.6	−122.7	251.9
Ethylene oxide	CH$_2$OCH$_2$	g	−53.0	−13.0	242.0
Aldehydes					
Formaldehyde (methanal)	HCHO	g	−115.9		9.8
Acetaldehyde (ethanal)	CH$_3$CHO	g	−166.4		
Propanal	CH$_3$CH$_2$CHO	l	−221.3	−142.1	
Butanal	CH$_3$CH$_2$CH$_2$CHO	l	−219.2	−306.4	
3-Methylpropanal	(CH$_3$)$_2$CHCH$_2$CHO	l	−220.5		
Benzaldehyde	C$_6$H$_5$CHO	l	−77		
Ketones					
Propanone (acetone)	CH$_3$COCH$_3$	l	−216.7	−152.4	
Butanone	CH$_3$CH$_2$COCH$_3$	l	−279.0		
Pentan-3-one	CH$_3$CH$_2$COCH$_2$CH$_3$	l	−308.8		
Methylphenyl ketone	C$_6$H$_5$COCH$_3$	l	−87	+2	373
Carboxylic acids					
Formic acid (methanoic)	HCO$_2$H	l	−409.2		
Acetic acid (ethanoic)	CH$_3$CO$_2$H	l	−488.3		
Propanoic acid	CH$_3$CH$_2$CO$_2$H	l	−509.2	−383.5	
Butanoic acid	CH$_3$CH$_2$CH$_2$CO$_2$H	l	−538.9		
Benzoic acid	C$_6$H$_5$CO$_2$H	s	−394.1		

A.17 VALUES OF PHYSICAL CONSTANTS AND CONVERSION FACTORS

Values of Physical Constants

Physical constant		Value with estimated uncertainty
speed of light in a vacuum	c	$(2.997\,925 \pm 0.000\,003) \times 10^8$ m s^{-1}
mass of hydrogen atom	m_H	$(1.673\,43 \pm 0.000\,08) \times 10^{-27}$ kg
mass of proton	m_p	$(1.672\,52 \pm 0.000\,08) \times 10^{-27}$ kg
mass of neutron	m_n	$(1.674\,82 \pm 0.000\,08) \times 10^{-27}$ kg
mass of electron	m_e	$(9.109\,1 \pm 0.000\,4) \times 10^{-31}$ kg
charge of proton	e	$(1.602\,10 \pm 0.000\,07) \times 10^{-19}$ C
Boltzmann constant	k	$(1.380\,54 \pm 0.000\,18) \times 10^{-23}$ J K^{-1}
Planck constant	h	$(6.625\,6 \pm 0.000\,5) \times 10^{-34}$ J s
Bohr radius	a_0	$(5.291\,67 \pm 0.000\,07) \times 10^{-11}$ m
Avogadro constant	N	$(6.022\,52 \pm 0.000\,28) \times 10^{23}$ mol^{-1}
gas constant	R	$(8.314\,3 \pm 0.001\,2)$ J K^{-1} mol^{-1}
"ice point" temperature	T_{ice}	$(273.150\,0 \pm 0.000\,1)$ K
Faraday constant	F	$(9.648\,70 \pm 0.000\,16) \times 10^4$ C mol^{-1}

Conversion Factors

	erg molecule^{-1}	kcal mol^{-1}	eV mol^{-1}	cm^{-1}	J mol^{-1}
1 erg molecule^{-1}	1	1.44×10^{13}	6.24×10^{11}	5.03×10^{15}	6.02×10^{16}
1 kcal mol^{-1}	6.95×10^{-14}	1	4.3×10^{-2}	350	4.18×10^3
1 eV mol^{-1}	1.60×10^{-12}	23.1	1	8068	9.65×10^4
1 cm^{-1}	1.99×10^{-16}	2.86×10^{-3}	1.24×10^{-4}	1	11.9
1 J mol^{-1}	1.66×10^{-17}	2.39×10^{-4}	1.04×10^{-5}	8.35×10^{-2}	1

1 electrostatic unit of charge $= 3.336 \times 10^{-10}$ C
(One electrostatic unit as a point charge exerts a force of 1 dyne (10^{-5} N) on a similar charge 1 cm away.)
1 electrostatic unit of potential $= 3.00 \times 10^2$ V
1 debye $= 3.336$ C m
1 atomic unit of length $= a_0$ (Bohr radius)
1 litre $= 1$ dm^3

A.18 GLOSSARY OF SYMBOLS

Roman letter symbols are set out alphabetically, followed by Greek letter symbols and a list of subscripts and superscripts. The page on which the symbol is defined or introduced is indicated.

Symbol	Term	Page
a_0	Bohr radius	34
(aq)	aqueous solution	14
A	mass number	24
A_r	relative atomic mass of an element (atomic weight)	13
Ar	aryl group	551
Å	ångström	30
A/R	electronegativity (Allred/Rochow)	80
bcc	body-centred cubic	66
b.p.	boiling point	237
c	speed of light	28
c_X	concentration of X	
ccp	cubic close-packed	65
C	coulomb	15
C_P	heat capacity at constant pressure	126, 138
C_V	heat capacity at constant volume	126, 138
C.N.	co-ordination number	65
d	density	234
(d)	decomposes	
dyn	dyne	813
D	debye	82
$D(X-Y)$	dissociation energy of X—Y	97
e^-	electronic charge	14
e.s.u.	electrostatic unit	813
E	energy	
E	electron affinity	44
E	electromotive force (e.m.f.)	157
$E^\ominus$	standard electromotive force of a cell	162
$E^\ominus$	standard electrode potential	160
E_A	activation energy	180
F	Faraday constant	15
g	gram	
(g)	gaseous state	14
G	Gibbs free energy of system	142
ΔG	Gibbs free energy change for a process	142
$\Delta G_f(X)$	Gibbs free energy of formation of X	155
$\Delta G_{hyd}(X)$	Gibbs free energy of hydration of X	247
$\Delta G_{Ion}(X)$	Gibbs free energy of ionization (refers to gaseous atoms X) [see also under ΔH for further examples of subscripts]	247
h	Planck constant	29

Symbol	Term	Page
hcp	hexagonal close-packed	65
Hal	halogen	
H	enthalpy of system	91
Hz	hertz	30
ΔH	enthalpy change for a process	91
$\Delta H_{atom}(X)$	enthalpy of atomization of X	92
$\Delta H_{diss}(X-Y)$	enthalpy of dissociation of X—Y bond	310
$\Delta H_f(X)$	enthalpy of formation of X	92
$\Delta H_{fus}(X)$	enthalpy of fusion of X	113
$\Delta H_{hyd}(X)$	enthalpy of hydration of X	107
$\Delta H_{latt}(X)$	"lattice energy" of X	103
$\Delta H_{soln}(X)$	enthalpy of solution of X	107
$\Delta H_{sub}(X)$	enthalpy of sublimation of X	113
$\Delta H_{vap}(X)$	enthalpy of vaporization of X	113
I	current	202
I_1	first ionization energy	43
I_2	second ionization energy	43
$I(X)$	ionization energy of X	41
J	joule	90
k	kilo (10^3)	
k	Boltzmann constant	134
k	rate coefficient (constant)	177
K	kelvin	126
K	equilibrium constant	148
$K_1, K_2, \ldots$	successive dissociation constants	391
K_a	acid dissociation constant	206
K_b	base dissociation constant	417
K_c	equilibrium constant (concentration units)	148
K_f	freezing point depression constant	119
K_P	equilibrium constant (pressure units)	149
K_{sp}	solubility product	200
K_w	ionic product for water	206
l	angular momentum quantum number	38
log X	logarithm to the base 10 of X	
ln X	natural logarithm of X	
m	mass of a particle (atom, molecule, ion, electron)	
m	metre	
m_l	magnetic quantum number	38
m_s	spin quantum number	38
m.p.	melting point	236
mol	mole	13
M	relative molecular mass of a substance (molecular weight)	13
M	molarity	16
M	Madelung constant	105
n	principal quantum number	38

Symbol	Term	Page
N	Newton	90
N	Avogadro constant	15
oct.	octahedral	315
O.N.	oxidation number	456
P	pressure	72
P	probability	129
P_X	partial pressure of X	115
P_X°	vapour pressure of pure X	115
pH	$-\log(H_3O^+)$	207
pK	$-\log K$	210
p.p.m.	parts per million	
q	heat absorbed by the system	90
Q	electric charge	
r	radius	
$r(X)$	radius of X	223
$r_+(r_-)$	radius of cation (anion)	105
$r(X{-}Y)$	bond length of X—Y	70
R	Rydberg constant	27
R	ideal gas constant	72
R	resistance	202
R	alkyl group	539
s	second	
stand.	standard state	92
(s)	solid state	14
s.t.p.	standard temperature and pressure	12
S	solubility	200
S	entropy of system	135
ΔS	entropy change for a process [see under ΔH for examples of use of subscripts]	138
t	time	
tet.	tetrahedral	88
T	temperature (thermodynamic; absolute)	
T.O.N.	total oxidation number	459
U	internal energy of system	90
ΔU	internal energy change for a process	90
v	velocity	
v/v	volume %	
V	volume	
V	volt	157
V	potential difference	202
V	potential energy	
w	work; work done by system	90
w/w	weight %	
W	number of microstates	135
W	weight	
x_A	mole fraction of A	115
xs	excess	

Symbol	Term	Page
X	halogen	
Z	atomic number	23
$z_+(z_-)$	charge number of cation (anion)	105
Z_{eff}	effective nuclear charge	47

Greek letters

Symbol	Term	Page
$\alpha, \beta, \ldots$	position of groups in some organic compounds	606
∂	partial differential	
Δ	crystal field splitting	787
Δo	octahedral crystal field splitting	787
Δt	tetrahedral crystal field splitting	788
ϵ	quantum of energy	29
κ	conductivity (specific conductance)	202
λ	wavelength	27
Λ	molar conductance	203
μ	dipole moment	83
ν	frequency	27
Π	osmotic pressure	121
ρ	resistivity (specific resistance)	202
σ	wave number	28
Σ	sum of	
ψ	wave function	35

Subscripts

Symbol	Term	Page
$A, B \ldots X$	refer to substances A, B ... X	115
P, V, T, etc.	indicate constant pressure, volume, temperature etc.	138
$+, -$	refer to positive or negative ion	105
sp	solubility product	200

The following subscripts refer to processes:

Symbol	Term
atom	atomization
comb	combustion
diss	dissociation
f	formation
fus	fusion
hyd	hydration
ion	ionization
soln	solution
solv	solvation
sub	sublimation
vap	vaporization

[see under ΔH for examples of use]

Superscripts

Symbol	Term	Page
$\ominus$	indicates a standard value of a property of a system, or a change between standard states	94
$\ddagger$	refers to transition state (activated complex)	188
$\circ$	indicates a pure substance	115

A.19 GLOSSARY OF ORGANIC NOMENCLATURE

A selection of important organic compounds, often known by different names, is given here. No attempt has been made to make this a comprehensive list—rather it is a guide to help the reader through a confusing situation. It should be used in conjunction with the paragraphs on nomenclature in the text (particularly Section 21.2).

Commonly employed names, including trivial and semi-systematic names, are listed alphabetically in the left hand column. In the right hand column, one particular IUPAC systematic name is given, normally that which would be preferred if a systematic name were used. However, IUPAC rules allow most of the names in the left hand column to be used, including the trivial names, and in practice these are, in fact, in common use. Use of only those names in the right hand column would lead to consistency, but may leave the student out of touch with normal chemical usage.

IUPAC rules often specify alternative systems for naming certain types of compound. Some names in the left hand column are fully systematic in that they represent such an alternative. These are marked with an asterisk.

Common names	Systematic names
acetaldehyde	ethanal
acetaldehyde cyanohydrin	2-hydroxypropanonitrile
acetaldehyde diethyl acetal (acetal)	1,1-diethoxyethane
acetaldehyde hemiacetal	1-ethoxyethanol
acetamide	ethanamide
acetic acid	ethanoic acid
acetic anhydride	ethanoic anhydride
acetoacetic acid	3-oxobutanoic acid
acetone (dimethyl ketone*)	propanone
acetone cyanohydrin	2-hydroxy-2-methylpropanonitrile
acetophenone	methyl phenyl ketone
acetylacetone	pentan-2,4-dione
acetyl chloride	ethanoyl chloride
acetylene	ethyne
acetylene tetrachloride	tetrachloroethane
aldol	3-hydroxybutanal
alkyl halide*	halogenoalkane
aniline	phenylamine
anisole (methyl phenyl ether)*	methoxybenzene
benzaldehyde	benzenecarbaldehyde
benzamide	benzenecarboxamide
benzal dichloride	(dichloromethyl) benzene
benzoic acid	benzenecarboxylic acid
benzonitrile	benzenecarbonitrile
benzoyl chloride	benzenecarbonyl chloride
benzyl alcohol	phenylmethanol
benzyl chloride	(chloromethyl) benzene
carbon tetrachloride	tetrachloromethane
chloroform	trichloromethane

Common names	Systematic names
crotonaldehyde	but-2-enal
diacetone alcohol	4-hydroxy-4-methylpentan-2-one
diethyl ether*	ethoxyethane
diethyl malonate (malonic ester)	diethyl propanedioate
ethanolamine	2-aminoethanol
ethyl acetate	ethyl ethanoate
ethyl acetoacetate (acetoacetic ester)	ethyl 3-oxobutanoate
ethyl alcohol	ethanol
ethyl chloride*	chloroethane
ethyl methyl ether*	methoxyethane
ethyl methyl ketone*	butanone
ethyl phenyl ether*	ethoxybenzene
ethylene	ethene
ethylene dibromide	1,2-dibromoethane
ethylene glycol	ethane-1,2-diol
ethylene chlorohydrin	2-chloroethanol
formaldehyde	methanal
formic acid	methanoic acid
fumaric acid	*trans*-butenedioic acid
glycerol	propane-1,2,3-triol
glycerol tristearate	propane-1,2,3-triyl trioctadecanoate
glycine (aminoacetic acid)	aminoethanoic acid
isopropanol (isopropyl alcohol)	propan-2-ol
lactic acid	2-hydroxypropanoic acid
maleic acid	*cis*-butenedioic acid
malonic acid	propanedioic acid
mesityl oxide	4-methylpent-3-en-2-one
methyl alcohol	methanol
methyl phenyl ether*	methoxybenzene
N-ethylacetamide	N-ethylethanamide
oxalic acid	ethanedioic acid
oxamide	ethanediamide
perbenzoic acid	benzeneperoxycarboxylic acid
phenetole	ethoxybenzene
phthalic acid	benzene-1,2-dicarboxylic acid
phthalimide	benzene-1,2-dicarboximide
stearic acid	octadecanoic acid
stilbene	1,2-diphenylethene
styrene	phenylethene
terephthalic acid	benzene-1,4-dicarboxylic acid
tertiary-butanol (tertiary-butyl alcohol)	2-methylpropan-2-ol
tertiary-butyl bromide	2-bromo-2-methylpropane
toluene	methylbenzene
vinyl alcohol (enol form of acetaldehyde)	ethenol
vinyl choride (chloroethylene)	chloroethene

General Bibliography

SUGGESTIONS FOR FURTHER READING

The task of selecting further reading references has not been an altogether easy one. On this basis, specialist books are called for and very few such works have been produced for school consumption; most are of a more advanced nature. The student who wants to expand on his reading of chemistry would probably be well advised, at any rate in the first instance, to dip into a more advanced textbook (or read it more fully).

A general selection of books is listed below. An attempt has been made to indicate the level at which they may be found helpful by dividing them into groups, but this must be taken as only an approximate indication.

Further important sources of interesting reading are articles published in journals concerned with scientific education, and one would particularly mention those in *Education in Chemistry*, *The School Science Review*, *The Journal of Chemical Education*, and *The Scientific American*. No attempt is made to catalogue these articles here, however. In addition, some of the larger companies in the chemical industry publish useful booklets and pamphlets. Some of these are quoted in the appropriate chapters but others exist and still more are likely to be published in the future. An enquiry to any of the larger chemical companies will probably produce some interesting material.

A. *For sixth-form level work*

Nuffield Advanced Chemistry. *Students' Book I* and *II*, Penguin (1970).

Chemical Bond Approach Project. *Chemical Systems*, McGraw-Hill (1964).

Chemical Education Material Study. *Chemistry: An Experimental Science*, Freeman (1964).

Sienko, M. J., and R. A. Plane. *Chemistry*, McGraw-Hill (1971).

    Makes little assumption of previous knowledge.

Johnson, R. C. *Introductory Descriptive Chemistry*, Benjamin (1966).

Cottrell, T. L. *Chemistry*, Oxford University Press (1970).

    A sophisticated discussion of chemistry at an introductory level.

Hutton, K. *Chemistry*, Penguin (1969).

    Presents an overall picture of chemistry.

Duffy, J. A. *General Inorganic Chemistry*, Longman (1966).

> A good example of an inorganic chemistry core-book which can be used from A-level upwards.

Dickerson, R. E., H. B. Gray, and G. P. Haight. *Chemical Principles*, Benjamin (1970).

> As the title suggests, this book is more concerned with principles than with descriptive chemistry.

B. *Books at a more advanced level, much of which will be understandable to the reader of this book*

Mahan, B. H. *University Chemistry*, Addison-Wesley (1969).

> Introduces fundamental principles and provides systematic treatment of inorganic and organic chemistry.

Plane, R. A., and R. E. Hester. *Elements of Inorganic Chemistry*, Benjamin (1965).

Douglas, B. E., and D. H. McDaniel. *Concepts and Models of Inorganic Chemistry*, Blaisdell (1965).

Jolly, W. L. *The Chemistry of the Non-Metals*, Prentice-Hall (1966).

Stark, J. G. (Editor). *Modern Chemistry*, Penguin (1970).

> A series of articles on some of the main themes in modern chemistry.

Dewar, M. J. S. *An Introduction to Modern Chemistry*, Athlone (1967).

> A general outline of the principles of modern chemistry written for first-year undergraduates from a slightly different point of view than that adopted in this book.

Samuel, D. M. *Industrial Chemistry—Inorganic*, Royal Institute of Chemistry (1966).

Samuel, D. M. *Industrial Chemistry—Organic*, Royal Institute of Chemistry (1966).

De Puy, C., and K. Rinehart. *Introduction to Organic Chemistry*, Wiley (1967).

Tedder, J. M., and A. Nechvatal. *Basic Organic Chemistry—A Mechanistic Approach*, Wiley. Part 1 (1966); Part II (1967).

Bonner, W. A., and A. J. Castro. *Essentials of Modern Organic Chemistry*, Reinhold (1965).

Conrow, K., and R. N. McDonald. *Deductive Organic Chemistry*, Addison-Wesley (1966).

Geissman, T. A. *Principles of Organic Chemistry* (third edition), Freeman (1968).

C. *Books written for undergraduates which may be useful for reference and in which the more descriptive sections could be read in conjunction with this book*

Barrow, G. M. *Physical Chemistry*, McGraw-Hill (1966).

Daniels, F., and R. A. Alberty. *Physical Chemistry*, Wiley (1961).

Moore, W. J. *Physical Chemistry*, Longman (1963).

Cotton, F. A., and G. A. Wilkinson. *Advanced Inorganic Chemistry* (second edition), Interscience (1966).

> A comprehensive university text, mainly non-mathematical.

Heslop, R. B., and P. L. Robinson. *Inorganic Chemistry*, Elsevier (1967).

Pauling L. *The Nature of the Chemical Bond*, Oxford University Press (1967).

Philips, C. S. G., and R. J. P. Williams. *Inorganic Chemistry*, Volumes I and II, Oxford University Press (1965).

Hendrickson J. B., D. J. Cram, and G. S. Hammond. *Organic Chemistry* (third edition), McGraw-Hill (1970).

Morrison, R. T., and R. N. Boyd. *Organic Chemistry* (second edition), Allyn and Bacon (1966).

Corwin, A. H., and M. M. Bursey. *Elements of Organic Chemistry*, Addison-Wesley (1966).

Roberts, J. D., and M. C. Caserio. *Basic Principles of Organic Chemistry*, Benjamin (1965).

> A similar but somewhat less comprehensive text by the same authors and publishers is also available—*Modern Organic Chemistry* (1967).

Roberts, J. D., R. Stewart, and M. C. Caserio. *Organic Chemistry—Methane to Macromolecules*, Benjamin (1971).

D. *Other books*

Hayes, M. *Projects in Chemistry*, Batsford (1967).
> Some useful project possibilities for schools.

Burman, C. R. *How to Find Out in Chemistry*, Pergamon (1966).
> How to use the chemical literature.

Sidgwick, N. V. *The Chemical Elements and Their Compounds*, Volumes I and II, Oxford University Press (1960).
> A description of the properties of the elements and their compounds.

Bassow, H. *The Construction and Use of Atomic and Molecular Models*, Pergamon (1968).

Wells, A. F. *Models in Structural Inorganic Chemistry*, Clarendon (1970).

Stark, J. G., and H. G. Wallace. *Chemistry Data Book*, Murray (1969).

Keller, R. *Basic Tables in Chemistry*, McGraw-Hill (1967).

Aylward, G. H., and T. J. V. Findlay. *S.I. Chemical Data*, Wiley (1971).

Nuffield Advanced Science. *Book of Data*, Penguin (1972).

Answers to Selected Problems

CHAPTER 1

2. a) 69 g; b) 0.04 g; c) 115 g.
3. a) 2; b) 1.2×10^{23}; c) 0.594 g (see Section 9.2).
6. (c).
7. a) 11.2 litres; b) 44.8 litres; c) 168 litres.

CHAPTER 2

4. b) 109,678 cm^{-1}, 1312 kJ mol^{-1}.
5. 495 kJ mol^{-1}.
6. 0.4 Å; a) 6.1×10^{-5} Å; b) 1.3 Å.
7. (b), (c), and (d).
9. (d) and (e).
11. a) $1s^2 2s^2 2p^6 3s$;
    b) $1s^2 2s^2 2p^6 3s^2 3p^6 4s^2$;
    c) $1s^2 2s^2 2p^6 3s^2 3p^6 3d 4s^2$;
    d) $1s^2 2s^2 2p^6$;
    e) $1s^2 2s^2 2p^6 3s^2 3p^6$.

CHAPTER 3

4. 0.707 Å; 9° 35′; 13° 35′.
5. a) i) 0.225 Å; ii) 0.414 Å.
8. a) 0.5; b) 44; c) 3.01×10^{23}; d) 2.7×10^{19}; e) 44.8 litres.
9. a) 8×10^{14}; b) 1.6 kg; c) 65 kg.
10. a) 276 litres; b) 1244 litres; c) 1411 litres.

CHAPTER 4

5. a) H_2S, HOCl, OCl_2.
   b) PH_3, NH_3, PF_3.
   c) Pyramidal, PCl_3; tetrahedral, $SiCl_4$; trigonal bipyramidal, PCl_5.
   d) Polar: NF_3, HCl, CO, H_2O; non-polar: BF_3, CF_4, H_2, Cl_2, CO_2.

8. a) -285.9 kJ; b) -46.15 kJ; c) -1198.6 kJ; d) 2973 kJ; e) 2979 kJ.
10. a) -983 kJ; b) -631.7 kJ; c) -65 kJ; d) -65 kJ; e) -152.3 kJ; f) 4 kJ; g) -281.2 kJ; h) -232.7 kJ.
12. a) -726.6 kJ; b) 244 kJ; c) 122 kJ; d) -56.5 kJ; e) 33 kJ; f) 18 kJ.

CHAPTER 5

1. See Section 5.1. Explore the effect of increasing or decreasing pressure on vapour. Introduce isotopic species of the substance into vapour and after some time examine to see if the isotopic species is present in the liquid—or vice versa.

3. Similar to H_2O. Draw in line for one atmosphere. Remember ammonia boils at -33 °C under this pressure and carbon dioxide sublimes at -78 °C.

6. Make a log plot and use Eqs. (5.2) and (5.3). $\Delta H_{vap}(H_2O) = 41.5$ kJ mol^{-1}.

7. $\Pi V = x_B RT$ (Eq. 5.16). For dilute solutions $x_B = \dfrac{n_B}{n_A + n_B} \approx \dfrac{n_B}{n_A}$.

   V is the volume of 1 mole of A.
   The volume of n_A moles of A is $n_A V$.
   The volume to contain 1 mole of B is $\dfrac{n_A V}{n_B}$.
   Therefore, $V' = \dfrac{n_A V}{n_B}$; hence $\Pi V' = RT$.

8. Use the equation derived in Problem 7. For dilute solution V' is the volume to contain 1 mole of solute. For a solution of molarity M, V' is given in litres by $1/M$. Hence for an ideal solution of non-electrolyte, $M = 7/(0.082 \times 303) = 0.282$ (concentration as moles per litre, osmotic pressure in atmospheres, and R in litre atmospheres per mole per degree). For a solute which ionizes to give two particles and which behaves ideally the molarity required would be 0.141. When a factor of 1.9 arises because of non-ideality, concentration required is 0.148 M.

CHAPTER 6

1. a) 0.25 (1 in 4); b) 0.125 (1 in 8); c) 0.0625 (1 in 16); d) 1 in 2^n.

2. a) Put energy of atom at X equal to 0, 1, 2, and 3 in turn and put in possible energy values for atoms at Y and Z;
   b) 10; c) 3; d) 1, 3, and 6; e) the one with 6 microstates.

3. $n_1/n_0 = e^{-\epsilon/kt}$ (Eq. 6.1). $\epsilon/kT = (4.14 \times 10^{-21})/(1.38 \times 10^{-23} \times 300) = 1$.
   Therefore $n_1/n_0 = 0.368$. Similarly $n_2/n_1 = 0.368$.
   Therefore, $n_2/n_0 = (0.368)^2$. Similarly for n_3 etc.
   Hence the relative population numbers are:

n_0	n_1	n_2	n_3	n_4	n_5
1	0.368	0.135	0.050	0.018	0.007

5. 22 J mol^{-1} and 111 J mol^{-1}.

6. a) $H_2O(s)$, $H_2O(l)$, $H_2O(g)$;
   b) He(g), Ne(g), Ar(g), Xe(g);
   c) $F_2(g)$, $Cl_2(g)$, $Br_2(g)$, $I_2(g)$.

CHAPTER 7

3. See Fig. 7.2. $\Delta H^\ominus = 57$ kJ; $\Delta S^\ominus = 175$ J K^{-1} at both temperatures.

5. $2HI(g) \rightleftharpoons H_2(g) + I_2(g)$.
   $(1 - 2\alpha)$ α α moles

$$K_c = \frac{\alpha^2}{(1 - 2\alpha)^2}; \quad \text{therefore} \quad K_c = \left(\frac{0.1}{0.8}\right)^2 = 1.56 \times 10^{-2}.$$

6. From Eq. (7.2), $K_P = RTK_c$.

$$K_c = \frac{1.74 \times 10^{-1}}{0.082 \times 298} \text{ mol l}^{-1} = 7.2 \times 10^{-3} \text{ mol l}^{-1}.$$

7. a) 3.1 kJ; b) -95.3 kJ; c) -237.2 kJ; d) -100.9 kJ;
   e) -139.8 kJ; f) -840.6 kJ; g) 115.4 kJ.

10. a) 0.34 V; b) 1.10 V; c) 1.08 V.

11. See Fig. 11.11 for comparison.
    At the intersection, $\Delta G^\ominus$ for the two processes,
    $$2C(\text{graphite}) + O_2(g) \rightarrow 2CO(g)$$
    $$\text{and} \quad 2Fe(s) + O_2(g) \quad \rightarrow 2FeO(s),$$
are equal.
Hence for
$$2FeO(s) + 2C(s) \rightarrow 2CO(g) + 2Fe(s),$$
$$\Delta G^\ominus = 0 \quad \text{at the temperature of the intersection.}$$

CHAPTER 8

2. $k = 6.23 \times 10^{-4}$ sec^{-1}.
3. a) Second; b) first; c) third.
4. a) First; b) zero; c) rate $\equiv k[A]$.
   For example:
   $$A \rightarrow X \text{ (slow)}$$
   $$X + B \rightarrow C + D \text{ (fast)}$$
   $k = 2.0 \times 10^{-3}$ sec^{-1}.
6. Use Eq. (8.8); 103 kJ mol^{-1}.
8. a) $k_2[A][B]$, first and first;
   b) $k_1[P]$, first and zero;
   c) $k_3[M]^2[L]$, second and first.

CHAPTER 9

1. See the end of Section 9.1 and also Section 18.10.
   a) mol^2 l^{-2}; b) mol^2 l^{-2}; c) mol^4 l^{-4}.
2. a) 1.64×10^{-11} mol^3 l^{-3}; b) 6.7×10^{-9} mol l^{-1}.
3. a) 1.7×10^{-5} mol l^{-1}; b) 2.8×10^{-8} mol l^{-1}.
5. 114.3 ohm^{-1} mol^{-1} cm^2.

6. i) (b), (d), (e);
   ii) (d) and (f), unlikely under normal conditions, require formation of BF_3H^+ and $H_4O_2^{2+}$.

11. a) pH = 3.
    b) $CH_3CO_2^-(aq) + H_2O(l) \rightleftharpoons CH_3CO_2H(aq) + OH^-(aq)$

$$\frac{[CH_3CO_2H][OH^-]}{[CH_3CO_2^-]} = \frac{[CH_3CO_2H]}{[CH_3CO_2^-][H^+]} \times [H^+][OH^-]$$

$$= \frac{K_w}{K_a} = \frac{1.0 \times 10^{-14}}{1.8 \times 10^{-5}} = 5.6 \times 10^{-10}$$

$[CH_3CO_2H] = [OH^-]$.
Hence $[OH^-] = 1.06 \times 10^{-5}$ and pH = 9.0.

c) From Section 9.3.5

$$[H_3O^+] = \frac{K_a[CH_3CO_2H]}{[CH_3CO_2^-]}.$$

Hence $pH = pK_a - \log \frac{[CH_3CO_2H]}{[CH_3CO_2^-]}$; pH = 5.34.

CHAPTER 10

1. Use electronegativity = $k(I + E)$, where k can be determined from the iodine data and used to determine the other electronegativities (F = 3.9; Cl = 3.1; Br = 2.8). Too few electron affinity data are available.

5. See Section 14.2.4; 672 kJ mol^{-1}.

7. See Section 14.2.4; 436 kJ mol^{-1}.

8. By comparison with the lattice energy of $Mg^{2+}Cl_2^-$ (Section 14.2.4).

9. See Sections 10.3.7, 10.3.4, and 14.2.4.

12. See Section 10.3.1 and Table A.14.

13. a) 0; b) VB; c) VIB.

17. 288 kJ mol^{-1}.

19. See Section 10.3.10.

21. Look at the numbers of neutrons (n) and protons (p) in each nucleus, and at the sum (n + p); beryllium.

22. See Section 10.3.1.

25. See Section 10.3.18.

29. Lanthanide contraction (Section 10.3.2).

30. See Sections 10.3.20, 14.2.4, and 20.3.

CHAPTER 11

4. See Sections:
   a) 11.1.4 and 16.8;
   b) 11.1.3 and 17.9;
   c) 11.1.3, 13.4, 19.8, and 19.9;
   d) 11.1.2, 15.2.4, and 15.2.6.

7. See Section 14.2.7.
8. See Sections 16.1, 12.3, 14.2.4, and 18.1.
10. See Sections:
    a) 10.3.18;
    b) 10.3.18 and 19.9;
    c) 13.4;
    d) 11.1.2;
    e) 10.5.
11. See Sections 3.3, 11.1.1, 19.10, and 4.1, etc.
12. See Sections 11.1.3 and 17.9.
13. See Section 14.2.4.
14. See Table A.11.
15. Decrease in entropy.
16. Compare the slope of the CS_2 line (Fig. 11.12) with the CO line (Fig. 11.11). The CO line slopes down and cuts "all" oxide lines at sufficiently high temperatures.
17. Compare the CO and H_2O lines in Fig. 11.11 with the CCl_4 and HCl lines in Fig. 11.13.
18. Kinetically yes, thermodynamically only slightly better (HCl line in Fig. 11.13.)
19. See Section 11.5.1.
    i) a) 2200 °C; b) 700 °C; c) 1600 °C.
    ii) a) Very high temperature; b) low yields at any temperature; c) very very high temperature.
20. See Section 11.5.1.
    a) Yes; b) yes (1200 °C); c) yes (1100 °C); d) yes.

CHAPTER 12

1. Helium is less soluble in water than is nitrogen.
3. See Sections 10.3.10 and 12.1.
5. See Section 12.5.
7. Consider molecules of the type AB_mC_n etc., where A is a central atom and B, C, etc. are atoms (or groups of atoms) bonded to A. Such molecules, which have the same number of electrons around the central atom and the same number of peripheral atoms bonded directly to A, have the same shape.
9. a) CH_2Cl_2; b) NF_3; c) IF_5; d) SO_2.
12. AlF_6^{3-}, SiF_6^{2-}, PF_6^-; octahedral.

CHAPTER 13

2. Write down the equation for the preparation of the corresponding hydrogen compound using water, and substitute D_2O for H_2O:
$$SO_3 + H_2O \rightarrow H_2SO_4; \quad SO_3 + D_2O \rightarrow D_2SO_4.$$
3. See Section 13.4; $Te(OH)_6$ and $HPO(OH)_2$.

4. See Sections 12.6, 13.4, and 17.9.

5. See Section 11.5.

6. See Section 11.1.3.

11. Less.

12. a) and b) Water > methanol > ether.

13. See Sections 10.3.16, 10.5, 20.3, and A.9, and Table A.12.

CHAPTER 14

1. See Section 10.3.2.

2. If charge density is defined as (charge on ion divided by radius of ion) then it is found that:
   a) that of Li^+ most closely resembles that of Sr^{2+} (poor chemical correlation); and
   b) that of Al^{3+} most closely resembles that of Be^{2+} (good chemical correlation).
   If charge density is defined as (charge on ion divided by volume of ion) then it is found that:
   a) that of Li^+ most closely resembles that of Mg^{2+} (good chemical correlation); and
   b) that of Al^{3+} most closely resembles that of Be^{2+}.

3. $\Delta H_f^\ominus = \Delta H_{atom}(Ca) + \Delta H_{atom}(X) + E(X) + I(Ca) - \Delta H_{latt}(KX)$
   $\Delta H_f^\ominus(CaF) = 177 \quad + 79 \quad - 333 \quad + 590 \quad - 817 = -304$ kJ
   $\Delta H_f^\ominus(CaCl) = 177 \quad + 122 \quad - 348 \quad + 590 \quad - 718 = -177$ kJ
   $\Delta H_f^\ominus(CaBr) = 177 \quad + 111 \quad - 340 \quad + 590 \quad - 656 = -118$ kJ
   $\Delta H_f^\ominus(CaI) \; = 177 \quad + 106 \quad - 297 \quad + 590 \quad - 615 = -39$ kJ.
   All have negative enthalpies of formation. Although the free energies of formation will be less negative in each case (see Table A.15), only in the case of the iodide is $\Delta G_f^\ominus$ likely to be positive. Thus the fluoride, chloride, and bromide are thermodynamically unstable, not with respect to dissociation into the elements, but with respect to disproportionation into the metal and the compound CaX_2 (Section 14.2.4):

$$2CaX \rightarrow CaX_2 + Ca; \quad \Delta G \text{ is negative.}$$

4. a) Solubility correlates with covalent character;
   b) Li^+ is too small;
   c) see Sections 14.1.2 and 11.2;
   d) see Section 10.3.18.

5. The main driving force is the greater bond strength of R—F compared to R—Cl. The difference in the lattice energy between CsF and CsCl is less than that between NaF and NaCl (Sections 10.3.7 and 11.2).

6. By extrapolation of values of other group IA elements *or* interpolation of values from neighbouring elements where appropriate.

7. See Section 10.3.10.

8. Lower ionization energies.

11. See Section 10.3.

12. 2:1. There is an approximate doubling of the effective nuclear charge for the second electron removed (Section 2.4.3).

15. See Section 10.3.3.

16. See Section 10.3.17.

17. $BaCO_3$ (Section 11.2).

CHAPTER 15

2. Both! Tl_2O.

3. Giant polymer.

4. B^{3+} is too small to pack six groups around it.

5. Al_2Br_6(s) is covalent, but in water it hydrolyses to $Al(H_2O)_6^{3+}$.

6. Middle row anomaly (Sections 15.1 and 10.3.2).

7. Reduction potential (Sections 7.7 and 7.12):

$$
\begin{array}{llll}
 & E^\ominus & \Delta G^\ominus & = -zFE^\ominus \\
(1)\ Tl^{3+} + 3e^- \rightleftharpoons Tl; & +0.72; & -3F(+0.72) & = -2.16F \\
(2)\ Tl^+ + e^- \rightleftharpoons Tl; & -0.34; & -F(-0.34) & = +0.34F \\
\hline
(1) - (2)\ Tl^{3+} + 2e^- \rightleftharpoons Tl^+; & & & = -2.50F.
\end{array}
$$

By subtracting equation (2) from equation (1), $\Delta G^\ominus = -2.50F$.
But $\Delta G^\ominus = -zFE^\ominus$, therefore $-2.50F = -zFE^\ominus$;
$Tl^{3+} + 2e^- \rightleftharpoons Tl^+$, therefore $E^\ominus = +1.25$ V.

Equilibrium constant:

$$
\begin{array}{llll}
 & E^\ominus & \Delta G^\ominus & = -zFE^\ominus \\
(1)\ Tl^+ + e^- \rightleftharpoons Tl; & -0.34; & -F(-0.34) & = 0.34F \\
(2)\ Tl^{3+} + 3e^- \rightleftharpoons Tl; & +0.72; & -3F(+0.72) & = -2.16F \\
\hline
[3 \times (1)] - (2)\ 3Tl^+ \rightleftharpoons Tl^{3+} + 2Tl;\ 3 \times 0.34F - (-2.16F) & & & = 3.18F
\end{array}
$$

by subtracting equation (2) from three times equation (1).
Therefore $\Delta G^\ominus = 3.18$ eV $= 3.18 \times 96.5$ kJ.
But $\Delta G^\ominus = -RT \ln K = -5.7 \log K$ kJ;
therefore $-5.7 \log K = 3.18 \times 96.5$, so $\log K = -54$ and $K = 10^{-54}$.

8. See Sections 15.1 and 11.2.

11. Middle row anomaly (Sections 15.1 and 10.3.2).

CHAPTER 16

1. The Si—Si bond energy is lower than the C—C bond energy.

3. See Section 11.1.3.

8. See Sections 14.1 and 15.1.

10. Disproportionate.

12. See Section 7.12.3; acid.

13. See Section 16.4.3. Clearly the reaction solid → gas occurs with increase in entropy; therefore ΔG will be more negative than ΔH (Sections 6.12 and 6.14).

15. See Section 12.7.

CHAPTER 17

1.

	ΔH	ΔG
$NH_4NO_3(s) \rightarrow N_2(g) + \frac{1}{2}O_2(g) + 2H_2O(g)$	-273	-118 kJ
$NH_4NO_3(s) \rightarrow N_2O(g) + 2H_2O(g)$	-269	-36 kJ

(Section 4.18.) Despite these figures nitrous oxide (N_2O) is formed; thus the reaction is under kinetic control (Section 8.9).

2. $N(OH)_3 \rightarrow H\text{---}ONO_2 + H_2O$;
   $P(OH)_3$ is $HPO(OH)_2$ (Section 17.4).

4. $N_2O_4(aq)$ disproportionates. Both ammonium nitrate and ammonium nitrite contain nitrogen in an oxidizing and reducing state; therefore each can "self-react" to form nitrogen in a stable intermediate oxidation state.

5. NO_2, yes; N_2, no (Section 7.12).

9. See Sections 17.4.4 and 11.1.4.

12. See Section 17.9. Because of hydrogen bonding the fluoride ion must be in a tetrahedral site.

14. Octet rule.

15. It would have no P—O—H group, therefore not an acid!

16/17. See Fig. 16.4 and Section 17.4.1.

19. Presumably the phosphorus is too small to pack these large atoms around it.

CHAPTER 18

2. See Section 16.4.5.

3. See Sections 11.1.3, 13.4, and 17.9.

4. See Sections 10.3.10 and 10.3.11.

5. a) CO; b) Mn_2O_7; c) N_2O_5; d) CrO_3.

6. See Section 11.1.1.

8. See Section 18.2.

9. See Section 17.4.

10. The lower bond energy of the O—O bond in ozone (compared to that in oxygen) correlates with lower activation energies for the reactions of ozone.

11. Small. No strong bonds are broken during the change (Section 10.3.1).

12. *Hint*—what are the boiling points of benzene and toluene?

13. Greater (Section 10.5).

16. Strong (Section 13.4).

CHAPTER 19

1. See Sections 11.1.4, 17.4.4, and 16.4.5.
2. See Section 11.1.4.
4. The properties can be "predicted" either by extrapolation from those of the lighter halogens, or by averaging the property of the elements polonium and radon, as appropriate:
   b.p., 290–320 °C; m.p., 220–250 °C; atomic volume, 26 cm^3 mol^{-1};
   ionic radius, 2.2–2.5 Å; covalent radius, 1.4–1.6 Å;
   At_3^- fairly stable; AgAt, very insoluble; Na$^+$At$^-$; HAt; etc.
6. See Section 10.3.19.
9. See Section 19.11. $I-Br \xrightarrow{H_2O} IO^- + Br^-$; $Br-Cl \xrightarrow{H_2O} BrO^- + Cl^-$. The negative oxygen end of the polar water molecule attacks the positive end ($I^{\delta+}-Br^{\delta-}$; $Br^{\delta+}-Cl^{\delta-}$) of the polar interhalogen; compare $SiCl_4$ (Section 16.4.5).
13. See Section 13.4; compare Problem 18 (below).
14. Hydrogen bonding leading to tetrahedral co-ordination.
17. See Sections 10.3.10, 10.3.11, and 10.3.9.
18. See Sections 18.14.1, 7.7, and 7.12.

	$E^\ominus$	$\Delta G^\ominus = -zFE^\ominus$
(1) $ClO^- + H_2O + 2e^- \to Cl^- + 2OH^-$;	+0.89;	$-2F(+0.89)$
(2) $ClO_3^- + 2H_2O + 4e^- \to ClO^- + 4OH^-$;	+0.50;	$-4F(+0.50)$
$2 \times (1) - (2)$ $3ClO^- \to 2Cl^- + ClO_3^-$;		$-4F(0.89 - 0.5)$

by subtracting equation (2) from two times equation (1).
But $\Delta G^\ominus = -RT \ln K = -5.7 \log K$ kJ,
therefore $-4(0.39)F = -4 \times 0.39 \times 96.5 = -5.7 \log K$ kJ,
and $\log K = (4 \times 0.39 \times 96.5)/5.7 = 26.41$.
Therefore $K = 10^{26.4}$.
So for $3ClO^- \to 2Cl^- + ClO_3^-$, $K = 10^{26.4}$.
Similarly for $3Br_2 + 6OH^- \to 5Br^- + BrO_3^- + 3H_2O$, $K = 10^{46.7}$,
and for $3I_2 + 6OH^- \to 5I^- + IO_3^- + 3H_2O$, $K = 10^{25.4}$.

19. See Sections 11.1.3 and 13.4.
20. See Sections 13.4 and 12.7.
22. Yes; greater than the tetrahedral angle (Section 12.7).

CHAPTER 20

1. See Section 10.3.3.
2. See Section 10.3.2.
3. See Section 10.3.3.
4. Lanthanide contraction (Section 10.3.2).
5. Look up reduction potentials of $Cr_2O_7^{2-}/Cr^{3+}$ etc.
   Yes: Br^-, I^-, H_2S, Hg_2^{2+}, Cu, HNO_2.
6. a) Alkaline.
   b) 10^{41}. See Sections 18.14.1, 7.7, and 7.12.
   c) Alkaline.

d) No:

	$E^\ominus$	$\Delta G^\ominus = -zFE^\ominus$
(1) $Fe^{2+} + 2e^- \rightleftharpoons Fe$;	-0.44;	$-2F(-0.44)$
(2) $Fe^{3+} + e^- \rightleftharpoons Fe^{2+}$;	$+0.77$;	$-F(0.77)$
$2 \times (2) - (1)$ $2Fe^{3+} + Fe \rightleftharpoons 3Fe^{2+}$;		$-2F(0.77 + 0.44) = -2(1.21)F$

by subtracting equation (1) from two times equation (2).
Now $\Delta G^\ominus = -zFE^\ominus = -5.7 \log K$ kJ,
therefore $-2 \times 1.21 \times 96.5 = -5.7 \log K$
and $\log K = (2 \times 1.21 \times 96.5)/5.7 = 41$.
So for $2Fe^{3+} + Fe \rightleftharpoons 3Fe^{2+}$, $K = 10^{41}$.
c) Alkaline.
d) No.

7. See Section 4.11.4; 27 kJ.
8. See Sections 10.3.7 and 14.2.4.
9. *Tris*(ethylenediamine)platinum(IV) chloride; chloropentaammineplatinum(IV) bromide; potassium hexacyanoferrate(II); *cis*-dichlorotetraamminecobalt(III) chloride; hexaaquochromium(III) chloride; *trans*-dichlorobisammineplatinum(II).
12. See Section A.9.
16. a) A is the vanadium(IV) species VO^{2+}(aq);
    b) B is the vanadium(II) species $V(H_2O)_6^{2+}$;
    c) The green species is $V(H_2O)_6^{3+}$.
19. See Section 7.12.
20. See Section 10.3.18.
22. a) See Section 15.1; b) see Section 15.1; c) see Section 16.5; g) charge-transfer (Section A.9).
23. See Section 10.3.20.
24. a) One, one, two, two;
    b) must be square planar, not tetrahedral (no isomers).
25. See Section 11.1.1.
27. See Section 10.3.3. No, requires breaking into a complete shell of electrons.
28. See Sections 16.5 and 20.11.
29. See Section 20.11.
31. Complex formation (Section 14.3.4).
36. See Section 10.3.18.
37. Oxide and fluoride.
39. See Sections 10.5 and 20.21.
40. See Sections 7.7 and 7.12.

	$E^\ominus$	$\Delta G^\ominus = -zFE^\ominus$
(1) Hg^{2+}(aq) $+ 2e^- \rightleftharpoons Hg(l)$;	$+0.854$;	$-2F(0.854)$
(2) Hg^{2+}(aq) $+ 2e^- \rightleftharpoons 2Hg(l)$;	$+0.789$;	$-2F(0.789)$
(2) $-$ (1) Hg_2^{2+}(aq) $\rightleftharpoons Hg(l) + Hg^{2+}$;	?	$-2F(-0.065)$

by subtracting equation (1) from equation (2).
$\Delta G^\ominus = -2F(-0.065) = 2 \times 0.065 \times F$
But $\Delta G^\ominus = -RT \ln K = -5.7 \log K$ kJ,
therefore $2 \times 0.065 \times 96.5 = -5.7 \log K$
and $\log K = (-0.13 \times 96.5)/5.7 = -2.2$.
So $K = 10^{-2.2}$.

CHAPTER 21

1. a) B, C, and E correspond to heptane, A.
   b) D is 3-methylhexane; F is 3,3-dimethylpentane.
   c) Carbon atom 3 in 3-methylhexane, D, is asymmetric.
3. Pentane, 2-methylbutane, 2,2-dimethylpropane.
4. a) Two; b) two; c) one; d) C_5H_{10}; e) one; f) one; g) none.
5. There are two structures of formula C_4H_{10}, $(CH_3)_3CH$ and $CH_3CH_2CH_2CH_3$.
6. But-1-ene, 2-methylpropene, *cis*- and *trans*-but-2-ene.
10. a) Three; b) two; c) six.
11. Those who predict four, rather than three, isomers for tartaric acid should, if they complete part (c), realize that two of these are identical, corresponding to the meso-form.

CHAPTER 22

1. The following would form the basis of acceptable solutions (others are also possible):
   a) aqueous ammoniacal silver or copper(I) ions would give a precipitate with the alk-1-yne;
   b) the decolorizing effect of the alkene on bromine water, or its effect on alkaline permanganate, would reveal its identity;
   c) benzene can be nitrated—cyclohexane reacts rapidly with bromine in sunlight (evolution of HBr gas);
   d) ozonolysis [hexanedial (adipaldehyde) from cyclohexene], or quantitative hydrogenation, or halogen addition; cyclohexene would absorb one mole of hydrogen or halogen, the alkyne would absorb two moles;
   e) loss of water from the *cis* acid, on heating, to give the anhydride (cf. maleic acid) would identify that acid;
   f) loss of water from the 1,2-acid to give phthalic anhydride—extrapolation from (e)—would identify that acid.
3. a) (i), (ii), (iii), and (vii);
   b) (iv);
   c) (i);
   d) (v);
   e) (vii);
   f) (iii); (i, ii, v, and vi would give *cyclo*alkanes if conditions were such as to hydrogenate a benzene ring; vii will also give a cycloalkane or hydrogenation);
   g) (iv).

CHAPTER 23

5. $CH_3CH_2CH_2OH$ < [3,5-dimethylphenol] < [3,5-dimethylbenzoic acid].

7. a) (i) and (ii);
   b) (iii) and (v).

10. $(C_2H_5O)_3PO$ (triethyl phosphate) + HCl (or, possibly, the mono or diethyl ester).

12. The following would be acceptable solutions (others are also possible):
    a) acetic acid liberates carbon dioxide from sodium bicarbonate;
    b) acetaldehyde will reduce ammoniacal silver nitrate to silver;
    c) acetic anhydride hydrolyses in water to give an acid solution;
    d) acetic anhydride will give acetic acid and ethyl acetate with ethanol;
    e) acetyl chloride fumes in moist air, and hydrolyses rapidly and exothermically in water;
    f) diethyl ketone will give a yellow precipitate with 2,4-dinitrophenylhydrazine reagent;
    g) acetone will give a yellow precipitate with 2,4-dinitrophenylhydrazine reagent;
    h) ethyl methyl ketone will undergo the iodoform reaction;
    i) the epoxide will react readily with dilute acid to give the 1,2-diol.

15. a) $C_6H_5OCH_2C_6H_5$;
    b) no reaction;
    c) $(CH_3)_2CHCH_2OCOCH_3$;

    d) 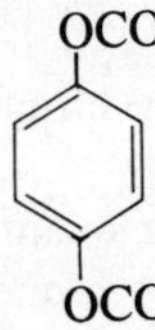
    $$\underset{OCOCH_3}{\overset{OCOCH_3}{\text{C}_6\text{H}_4}} + 2CH_3COOH;$$

    e) $(CH_3)_2C=CH_2$;
    f) $(CH_3)_2C=CH_2$;
    g) $CH_3CH(OH)CH(NH_2)CH_3$;
    h) $\underset{\underset{NOH}{\|}}{CH_3CCH_2CH_3}$;

    i) $CH_3-\underset{\underset{SO_3^-\ Na^+}{|}}{\overset{\overset{OH}{|}}{CH}}$;

    j) 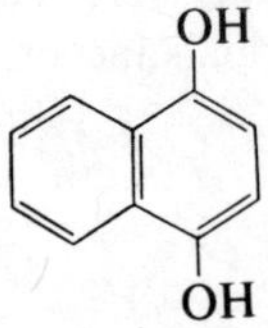 1,4-dihydroxynaphthalene (naphthalene ring with OH at positions 1 and 4).

16. a) C_2H_6O: CH_3OCH_3 or CH_3CH_2OH;
    b) CH_3CH_2OH.

17. There are six open-chain structures and a number of cyclic ones. Some structures containing double bonds between carbon atoms are also "possible", but are unstable.

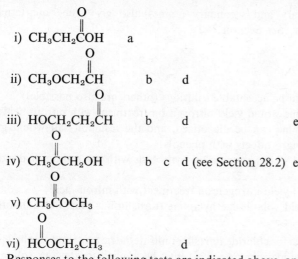

i) CH₃CH₂COOH   a

ii) CH₃OCH₂CHO   b   d

iii) HOCH₂CH₂CHO   b   d   e

iv) CH₃COCH₂OH   b   c   d (see Section 28.2)   e

v) CH₃COCH₃

vi) HCOCH₂CH₃   d

Responses to the following tests are indicated above, and form one satisfactory and systematic method of distinguishing between them:
a) CO_2 with sodium carbonate;
b) yellow precipitate with 2,4-dinitrophenylhydrazine;
c) yellow precipitate of iodoform with iodine and alkali;
d) reduces ammoniacal silver nitrate to silver;
e) can be acetylated.

18. X: $CH_3CH(OH)CH_3$;
    Y: $CH_3CH=CH_2$;
    Z: CH_3COCH_3.

19. A: $C_6H_5COOCH_2CH_3$;
    B: CH_3CH_2OH;
    C: CH_3CHO;
    D: C_6H_5COOH.

20. Three methoxy groups which must be joined directly to the ring as the molecule contains only the structural units $1 \times C_6H_3$ and $3 \times CH_3O$, total molecular, formula $C_9H_{12}O_3$.

CHAPTER 24

4. 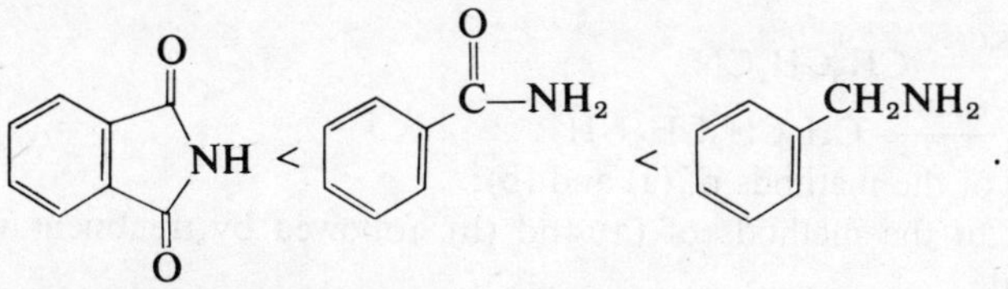 (phthalimide) < (benzamide) < (benzylamine)

7. a) (i), (ii), and (v);
b) (i), (ii), (iii), and (v);
c) only (i) if the temperature is held below 5 °C; nitrogen will also be obtained from (ii) if the temperature becomes higher;
d) (ii);
e) (iv) and (vii);

f) (iii), note primary and secondary amines also give some quarternary ammonium salts. *See* Section 24.1.3;
g) (vi);
h) (vii);
i) (vi).

8. The following would be acceptable solutions (others are also possible):
   a) the primary amine would yield nitrogen on treatment with nitrous acid;
   b) the aromatic amine can be diazotized, and the resultant salt would give coloured coupling products with phenols;
   c) the amide would yield ammonia on hydrolysis with hot sodium hydroxide solution;
   d) the amide would yield nitrogen on treatment with nitrous acid;
   e) alkali would yield volatile methylamine (turns litmus blue, like ammonia) with $CH_3\overset{+}{N}H_3\ Cl^-$;
   f) the standard tests for chloride ion (silver nitrate/nitric acid) and sulphate ion (barium chloride/hydrochloric acid) may be applied to aqueous solutions of these salts;
   g) hot, aqueous sodium hydroxide would hydrolyse the imide, liberating ammonia;
   h) the carboxylic acid would liberate carbon dioxide gas from sodium bicarbonate;
   i) the compound (one mole) with two primary amino groups would give two moles of nitrogen, whereas the compound with only one such group would yield only one mole of nitrogen when treated with aqueous nitrous acid;
   j) the nitrile will yield ammonia with hot, aqueous sodium hydroxide.

9. a) Two primary amino groups;
   b) one primary amino group.

10. a) $C_6H_6 \xrightarrow[\text{conc. } H_2SO_4]{\text{conc. } HNO_3} C_6H_5NO_2$;
    b) $C_6H_5NO_2 \xrightarrow{\text{Sn/HCl}} C_6H_5NH_2$;
    c) $C_6H_5NH_2 \xrightarrow[\text{cold}]{\text{HCl/HNO}_2} C_6H_5N_2^+Cl^- \xrightarrow{\text{Cu(I)Cl}} C_6H_5Cl$;
    d) combined methods of (a), (b), and (c), using KI(aq) instead of Cu(I)Cl for the final stage.

11. a) $CH_3CH_2Br \xrightarrow{\text{KCN}} CH_3CH_2CN$;
    b) $CH_3CH_2CN \xrightarrow{\text{LiAlH}_4} CH_3CH_2CH_2NH_2$;
    c) combination of the methods of (a) and (b);
    d) combination of the methods of (a) and (b), followed by treatment with acetyl chloride.

12. a) $CH_3CH_2NH_2 \xrightarrow[\text{CH}_3\text{I}]{\text{excess}} CH_3CH_2N^+(CH_3)_3I^-$;
    b) $CH_3CH_2\overset{+}{N}(CH_3)_3I^- \xrightarrow[\text{Ag}_2\text{O}]{\text{moist}} CH_3CH_2N^+(CH_3)_3OH^-$;
    c) $CH_3CH_2\overset{+}{N}(CH_3)_3OH^- \xrightarrow{\text{heat}} CH_2{=}CH_2 + (CH_3)_3N + H_2O$;
    d) combination of the methods of (a), (b), and (c);

e) conversion of $CH_3CH_2CH_2I$ to $CH_3CH_2CH_2CH_2NH_2$ (Problem 11) followed by a combination of the methods of (a), (b), and (c).

13. a) $C_6H_5CH_2Br \xrightarrow{KCN} C_6H_5CH_2CN \xrightarrow[\text{warm}]{H^+/H_2O} C_6H_5CH_2COOH$;

b)

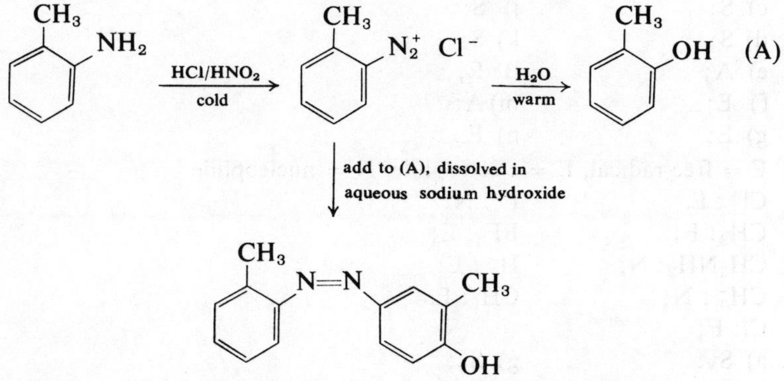

o-methylaniline $\xrightarrow[\text{cold}]{HCl/HNO_2}$ o-methylbenzenediazonium chloride $\xrightarrow[\text{warm}]{H_2O}$ o-cresol (A)

add to (A), dissolved in aqueous sodium hydroxide

↓

(azo coupling product: 2-methylphenyl-N=N-(3-methyl-4-hydroxyphenyl))

c) $C_6H_5CH_2COOH \xrightarrow{PCl_5} C_6H_5CH_2COCl \xrightarrow{NH_3} C_6H_5CH_2CONH_2$

$\xrightarrow{Br_2/KOH(aq)} C_6H_5CH_2NH_2$

14. CH_3CH_2
    $\diagdown$
    NH
    $\diagup$
    CH_3

15. X: $CH_3CH_2CONHCH_2CH_3$;
    Y: $CH_3CH_2NH_2$;
    Z: CH_3CH_2COOH.

CHAPTER 25

5. Convert ethanol to the more reactive sodium ethoxide, by treating it with sodium metal, before introducing the bromopropane. No.

6. $CH_3CH=CH_2$ and CH_3CHCH_3.
   $\phantom{CH_3CH=CH_2 \text{ and } }\qquad\qquad\;\;|$
   $\phantom{CH_3CH=CH_2 \text{ and } }\qquad\;\;OCH_2CH_3$

7. Acceptable suggestions should be based upon the identification of the reagents as nucleophiles—for example (from bromoethane):
   a) CH_3CH_2SH (or $CH_3CH_2SCH_2CH_3$);
   b) $CH_3CH_2PH_2$ (or $CH_3CH_2\overset{+}{P}H_3\ Br^-$);
   c) $CH_3CH_2C\equiv CH$.

10. $CH_3CH_2\overset{O}{\overset{\|}{C}}Cl$; $CH_3\overset{O}{\overset{\|}{C}}CH_2Cl$; $H\overset{O}{\overset{\|}{C}}CH_2CH_2Cl$; $CH_3\overset{Cl}{\underset{|}{C}}H\overset{O}{\overset{\|}{C}}H$.
    The first structure is the correct one.

11. X: $CH_3CH_2CH_2Cl$;
    Y: $CH_3CH_2CH_2OH$;
    Z: CH_3CH_2CHO.

CHAPTER 26

1. A = addition, S = substitution, E = elimination
   - a) A;
   - b) E;
   - c) S;
   - d) S;
   - e) A;
   - f) E;
   - g) E;
   - h) A;
   - i) S;
   - j) S;
   - k) S;
   - l) E;
   - m) A;
   - n) E.

4. F = free radical, E = electrophile, N = nucleophile
   - Cl^+: E,
   - CH_3: F;
   - CH_3NH_2: N;
   - CH_3^-: N;
   - Cl: F;
   - I^-: N;
   - BF_3: E;
   - H^+: E;
   - CH_2: E.

5.
   - a) S_N;
   - b) S_E;
   - c) S_N;
   - d) S_N;
   - e) S_E;
   - f) S_N;
   - g) S_N;
   - h) S_N;
   - i) S_N;
   - j) S_N;
   - k) S_N.

6.
   a) Correct;

   b) $HO^- + H\text{—}Cl \rightarrow H_2O + Cl^-$;

   c) $R\text{—}O\text{—}O\text{—}R \rightarrow 2\ RO\cdot$;

   d) correct;

   e) $(CH_3)_2CHNH_2 + CH_3\text{—}I \rightarrow (CH_3)_2CH\overset{+}{N}H_2CH_3 I^-$;

   f) $(CH_3)_3N + CH_3\text{—}I \rightarrow (CH_3)_4N^+I^-$;

   g) first step correct; second step:

   $H_2O^+\text{—}CH_2\text{—}CH_2\text{—}H + X^- \rightarrow H_2O + CH_2\text{=}CH_2 + HX$.

9. Suggestions such as the following would be reasonable:

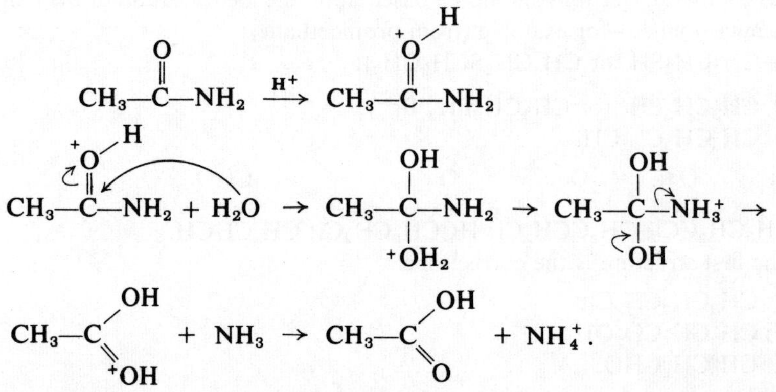

$$CH_3\text{—}\underset{\displaystyle \|}{\overset{\displaystyle O}{C}}\text{—}NH_2 \xrightarrow{H^+} CH_3\text{—}\underset{\displaystyle \|}{\overset{\displaystyle \overset{+}{O}\text{—}H}{C}}\text{—}NH_2$$

$$CH_3\text{—}\underset{\displaystyle \|}{\overset{\displaystyle \overset{+}{O}\text{—}H}{C}}\text{—}NH_2 + H_2O \rightarrow CH_3\text{—}\underset{\displaystyle {}^+OH_2}{\overset{\displaystyle OH}{\underset{\displaystyle |}{C}}}\text{—}NH_2 \rightarrow CH_3\text{—}\overset{\displaystyle OH}{\underset{\displaystyle OH}{C}}\text{—}\overset{+}{N}H_3 \rightarrow$$

$$CH_3\text{—}\overset{\displaystyle OH}{\underset{\displaystyle {}^+OH}{C}} + NH_3 \rightarrow CH_3\text{—}\overset{\displaystyle OH}{\underset{\displaystyle O}{C}} + NH_4^+.$$

Two points should be made:
a) All of the above steps are reversible.
b) It would be very difficult for a student to predict the details of the proton transfer steps. The important points that should be emphasized in the answer are:
   i) nucleophilic attack by water to form a tetrahedral intermediate;
   ii) expulsion of NH_2 from that intermediate to give the acid (or a protonated form of the acid);
   iii) some reasonable role for the acid catalyst, such as protonation of the amide oxygen or nitrogen atom.

10. $CH_2=CH_2 + Br-OH \rightarrow \overset{+}{C}H_2-CH_2-Br + {}^-OH \rightarrow HO-CH_2-CH_2-Br$.

11. $RCl + AlCl_3 \rightarrow R^+ AlCl_4^-$
    $R^+ + C_6H_6 \rightarrow C_6H_5R + H^+$.

14. c) i) $CH_3CH_2-OH \rightleftharpoons CH_3CH_2-\overset{+}{O}H_2$

$CH_3CH_2OH + CH_3CH_2-\overset{+}{O}H_2 \rightleftharpoons CH_3CH_2-\underset{+}{\overset{H}{O}}-CH_2CH_3 + H_2O$

$CH_3CH_2-\underset{+}{\overset{H}{O}}-CH_2CH_3 + H_2O \rightleftharpoons CH_3CH_2OCH_2CH_3 + H_3O^+$;

ii) $CH_3CH_2O^- + CH_3CH_2-Br \rightarrow CH_3CH_2OCH_2CH_3 + Br^-$;

iii) $CH_3CH_2OH + CH_3-\overset{O}{\underset{\|}{C}}-Cl \rightarrow CH_3CH_2-\overset{+}{\underset{H}{O}}-\overset{O^-}{\underset{CH_3}{C}}-Cl$

$\downarrow$

$CH_3CH_2-O-\overset{O}{\underset{\|}{C}}-CH_3 + HCl \leftarrow CH_3CH_2-\overset{+}{\underset{H}{O}}-\overset{O}{\underset{\|}{C}}-CH_3 + Cl^-$

19. Bromine molecules act as sources of "Br^+" and Br^-. Sodium chloride is a source of Cl^-. The only electrophilic species which can transfer to the alkene is therefore Br^+ and the resultant ion $BrCH_2CH_2^+$ can combine with Cl^- or Br^-. As there is no source of Cl^+, there can be no $ClCH_2CH_2^+$ ion involved, and no possibility of producing, directly, $ClCH_2CH_2Cl$.

20. The two mechanisms sought are:

a) $CH_3-\overset{O}{\underset{\|}{C}}-O-CH_3 \quad \bar{O}H \rightarrow CH_3-\overset{O}{\underset{\|}{C}}-O^- + CH_3OH$; and

b) $CH_3-\overset{O}{\underset{\|}{C}}-O-CH_3 + \bar{O}H \rightleftharpoons CH_3-\underset{OH}{\overset{O^-}{\underset{|}{\overset{|}{C}}}}-OCH_3$

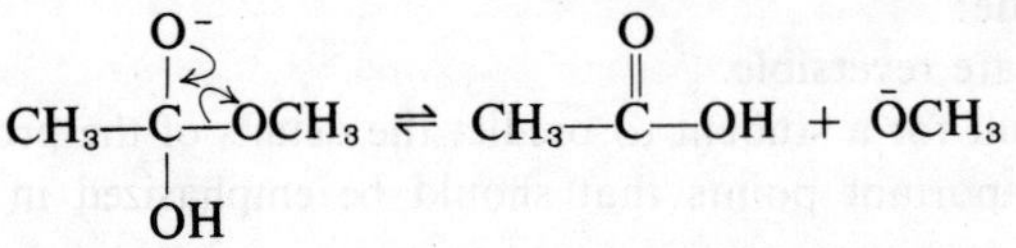

$$CH_3-\underset{\underset{OH}{|}}{\overset{\overset{O^-}{|}}{C}}-OCH_3 \rightleftharpoons CH_3-\overset{\overset{O}{\|}}{C}-OH + \bar{O}CH_3$$

$$CH_3-\overset{\overset{O}{\|}}{C}-OH + \bar{O}CH_3 \rightarrow CH_3-\overset{\overset{O}{\|}}{C}-O^- + HOCH_3.$$

The tracer experiment indicates that it is the bond between carbonyl and oxygen which breaks, not that between the methyl group and oxygen. Mechanism (b) is consistent with this, whereas mechanism (a) is not.

21. One bromine atom approaches from one side of the alkene, and the other bromine atom approaches from the other side. The intermediate ion  has been postulated to account for this with the first bromine "blocking" one side of the alkene molecule.

22. Trichloromethane is acidic—the hydrogen atom can be donated, as a proton, reversibly, to a suitable base.

23. As in Problem 22, the hydrogen atoms are acidic.

24. The two-step mechanism, with the first step being rate determining. The rate expression for the one-step mechanism would be:
$$\text{rate} = k[(CH_3)_3CBr][OH^-].$$

25. Mechanism (b). Mechanism (a) would result in a mixture of

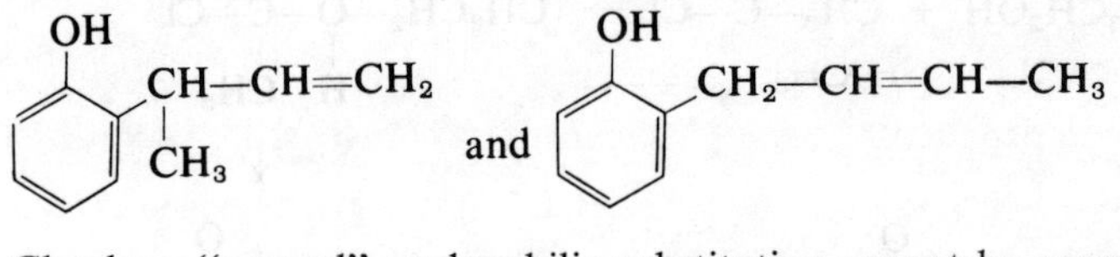

[structure: 2-(1-methylallyl)phenol with OH and CH-CH=CH₂ with CH₃] and [structure: 2-(but-2-enyl)phenol with OH and CH₂-CH=CH-CH₃]

26. Clearly, a "normal" nucleophilic substitution cannot be operating, otherwise only 4-methylphenylamine would have been produced. The evidence suggests that carbon atoms 1 and 2 in the original bromoarene become equivalent in the sense that both can be attacked in a similar way by NH_2^- or NH_3. A possible mechanism which could account for this would be an elimination/addition mechanism (Section 25.2.2):

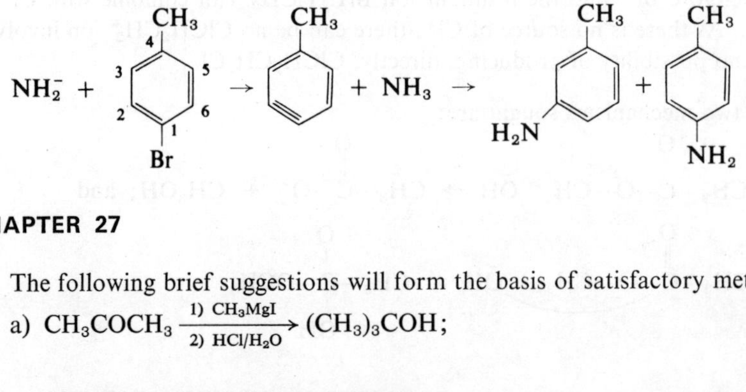

$NH_2^- +$ [4-methyl-1-bromobenzene with numbered positions] $\rightarrow$ [benzyne-like intermediate with CH₃] $+ NH_3 \rightarrow$ [3-amino-toluene with H_2N] $+$ [4-amino-toluene with NH_2]

CHAPTER 27

1. The following brief suggestions will form the basis of satisfactory methods:

   a) $CH_3COCH_3 \xrightarrow[\text{2) HCl/H}_2\text{O}]{\text{1) CH}_3\text{MgI}} (CH_3)_3COH$;

840 ANSWERS TO SELECTED PROBLEMS

b) $CH_3CH(Br)CH_3 \xrightarrow{Mg}{\text{ether}} (CH_3)_2CHMgBr \xrightarrow{\text{1) HCHO (formaldehyde)}}{\text{2) HCl/H}_2\text{O}} (CH_3)_2CHCH_2OH$;

c) $CH_3COCl + (CH_3CH_2)_2Cd \rightarrow CH_3COCH_2CH_3$;

d) $CH_3C\equiv N \xrightarrow{\text{1) C}_6\text{H}_5\text{MgBr}}{\text{2) HCl/H}_2\text{O}} CH_3COC_6H_5$;

e) $C_6H_5COOCH_2CH_3 \xrightarrow{\text{1) excess C}_6\text{H}_5\text{MgBr}}{\text{2) HCl/H}_2\text{O}} (C_6H_5)_3COH$;

f) $CH_3CH_2Br + KCN \rightarrow CH_3CH_2CN$

$\qquad\qquad\qquad\qquad\qquad\qquad\searrow$
$\qquad\qquad\qquad\qquad\qquad\qquad\quad CH_3CH_2COCH_2CH_3$;
$\qquad\qquad\xrightarrow{Mg}{\text{ether}} CH_3CH_2MgBr \nearrow\quad$ (after treatment with aqueous acid)

g) $CH_3CH_2CH_2CH_2Br \xrightarrow{Mg}{\text{ether}} CH_3CH_2CH_2CH_2MgBr$
$\qquad\qquad\qquad\qquad\qquad\qquad\downarrow \text{HCHO (formaldehyde)}$
$CH_3CH_2CH_2CH=CH_2 \xleftarrow{\text{heat}}{H_3/PO_4} CH_3CH_2CH_2CH_2CH_2OH$;

h) $(CH_3)_2CHCH_2OH \xrightarrow{PBr_3} (CH_3)_2CHCH_2Br$
$\qquad\qquad\qquad\qquad\qquad\qquad\qquad\downarrow \begin{array}{c}Mg\\ \text{ether}\end{array}$
$(CH_3)_2CHCH_2CH_2CH_2OH \xleftarrow{\text{1) CH}_2\text{CH}_2\text{ O}}{\text{2) HCl/H}_2\text{O}} (CH_3)_2CHCH_2MgBr$

3. CH_3OCH_2OH.

5. a) $(C_6H_5)_3P$ (triphenylphosphine);
   b) $(C_6H_5)_4Si$ (tetraphenylsilane);
   c) $(CH_3CH_2)_3COH$;
   d) $CH_3CONHLi + CH_3CH_3$.

6. The following suggestions are based on organometallic reactions:

   a) $C_6H_5Br \xrightarrow{Mg}{\text{ether}} C_6H_5MgBr \xrightarrow{D_2O} C_6H_5D + Mg(OD)Br$;

   b) $CH_3CH(OH)CH_3 \xrightarrow{HBr} (CH_3)_2CHBr \xrightarrow{Mg}{\text{ether}} (CH_3)_2CHMgBr$
   $\qquad\qquad\qquad\qquad\qquad\qquad\qquad\qquad\downarrow D_2O$
   $\qquad\qquad\qquad\qquad\qquad\qquad\qquad\quad (CH_3)_2CHD$;

   c) $CH_3CH=CH_2 \xrightarrow{\text{hydroboration}} CH_3CH_2CH_2OH \xrightarrow{KBr}{H_2SO_4} CH_3CH_2CH_2Br$
   $\qquad\qquad\qquad\qquad\qquad\qquad\qquad\qquad\qquad\qquad\downarrow \begin{array}{c}Li\\ \text{ether}\end{array}$
   $\qquad\qquad\qquad\qquad CH_3CH_2CH_2D \xleftarrow{D_2O} CH_3CH_2CH_2Li$

CHAPTER 28

2. [piperidine structure] and $Cl(CH_2)_5NH[(CH_2)_5NH]_n$—H (polymer); the proportion of the cyclic compound would probably be greater in dilute solution and less in concentrated solution.

6. (a), (d), (e), and (g).

9. Benzaldehyde contains no α-hydrogen atom. In a mixture of benzaldehyde and acetophenone, only the latter has α-hydrogens and that only on one side of the carbonyl group. Thus, the only anion which can form is that derived from acetophenone:

$$\overset{-}{C}H_2-\underset{\underset{\|}{O}}{C}-C_6H_5 \leftrightarrow CH_2=\underset{\underset{|}{O^-}}{C}-C_6H_5.$$

This will attack the more accessible aldehyde carbonyl of benzaldehyde, rather than the ketone carbonyl of undissociated acetophenone, to give

$$C_6H_5\underset{\underset{|}{OH}}{C}H-CH_2COC_6H_5$$ which, on loss of water, yields $C_6H_5CH=CHCOC_6H_5$.

10. a) The ketone;
    b) the 1,4-compound (para-compound).

11. a) Phenylamine (aniline);
    b) 3-nitrophenylamine.

CHAPTER 29

2.

	Percentage Compositions						
	C	H	N	O	Halogen	S	Metal
a)	64.0	4.4	—	—	31.6	—	—
b)	58.5	4.1	11.4	26.0	—	—	—
c)	62.1	10.3	—	27.6	—	—	—
d)	12.7	2.1	—	—	85.2	—	—
e)	55.2	10.3	16.1	18.4	—	—	—
f)	43.5	7.3	8.5	19.3	21.4	—	—
g)	—	4.7	21.9	25.0	20.0	10.0	18.4
h)	—	4.5	—	36.0	40.0	—	19.5
i)	—	2.0	9.3	—	23.7	—	65.0
j)	23.3	4.9	—	46.6	—	—	25.2

3. a) C_2H_6O; b) C_2H_7N; c) C_3H_5OCl; d) $C_6H_{15}O_3P$; e) $CHBr_3$; f) $C_6H_{16}N_2$; g) $C_5H_{10}O_2$; h) $KMnO_4$; i) $K_2Cr_2O_7$; j) SiF_4.

Index of Named Substances

This index lists only named substances. Thus it includes sodium chloride and acetaldehyde but alkali halides and aldehydes will be found in the General Index. Frequently trivial as well as systematic names are given for common substances, but it may be helpful in some cases to refer to the glossary on nomenclature (Appendix 19).

Acetaldehyde   567, 586, 587, 594, 604, 606, 610, 614, 615, 617, 618, 620, 621, 678, 679, 700, 720
Acetaldehyde ammonia   611
Acetaldehyde bisulphite addition compound   610
Acetaldehyde cyanohydrin   610, 648
Acetaldehyde diethyl acetal   610
Acetaldehyde 2,4-dinitrophenyl-hydrazone   612
Acetaldehyde hemi-acetal   610
Acetamide   591, 601, 614, 631, 637, 648, 679
Acetic acid   579, 580, 584–587, 589, 591, 594, 599, 601, 604, 607, 726
Acetic acid dimer   607
Acetic anhydride   584, 592, 599, 604
Acetoacetic acid   718
Acetoacetic ester (see Ethyl acetoacetate)
Acetone   586, 595, 607–609, 620, 621, 700
Acetone cyanohydrin   609
Acetonitrile   591, 638, 648
Acetophenone (see Methyl phenyl ketone)
Acetylacetone   345
Acetyl chloride   584, 587, 603, 604, 614, 658, 660, 665, 701
Acetylene   544–546, 567, 568, 574, 621, 696, 765
   hydration   621

Acetylene tetrachloride (see Tetrachloroethane)
N-Acetylglycine   706
Adenosine triphosphate (ATP)   412, 427, 714
Aldol   614, 721
Alum   287
Alumina   364
Aluminium   363, 370
   hydration energy of ion   371
   ores   364
   uses   365
Aluminium bromide   371
Aluminium carbide   366
Aluminium chloride   363, 371
Aluminium fluoride   370, 371
Aluminium hydride   370
Aluminium hydroxide   370, 372, 581
Aluminium iodide   371
Aluminium nitride   365, 366
Aluminium oxide   363, 364, 370
Aluminium oxinate   372
Aminoacetic acid   706
2-Aminoethanol   602
Ammonia   409, 415, 678
   bonding in   88, 776
   hydrate   417
Ammonium carbamate   392, 417
Ammonium dichromate   508
Ammonium nitrite   412
Ammonium oxalate   400
Ammonium phosphomolybdate   429

Ammonium vanadate   504
Anhydrite   349, 437
Aniline   596, 637, 726, 732
Aniline hydrochloride   631
Antimony, structure   414
Antimony(III) oxide   414, 423, 429
Antimony(V) oxide   414, 422, 429
Antimony oxychloride   408
Antimony pentasulphide   424
Antimony trisulphide   424
Apatite   412
Apatite, fluoro-   412
Aqua regia   260
Aragonite   349
Argon   308
Arsenic, structure   414
Arsenic acid   408, 429
Arsenical pyrites   412
Arsenic pentasulphide   424
Arsenic pentoxide   422
Arsenic trioxide   408, 423
Arsenic trisulphide   424
Arsenious acid   429
Arsine   407, 416
Asbestos   394
Astatine   470

Barium   355
Barium carbonate   356, 393
Barium chromate   356
Barium oxalate   356, 400
Barium peroxide   344

Barium sulphate 356
Bauxite 364
Benzaldehyde 574, 608, 615, 622, 732
 manufacture 621
Benzaldimine 622
Benzamide 591, 596, 632
Benzanilide (see N-Phenylbenzamide)
Benzene 551, 558, 569–571, 589, 659
 bonding in 99, 776
 bromination 681
 delocalization in 100, 776
 nitration of 570, 731
Benzenecarboxylic acid (see Benzoic acid)
Benzenediazonium bromide 651, 666
Benzenediazonium chloride 648, 650, 651
Benzene-1,4-dicarboxylic acid 571
Benzene hexachloride, (BHC) 569
Benzenesulphonic acid 571, 593
Benzoic acid 579, 589, 591, 596, 602, 615, 732
Benzonitrile 591, 622, 648
Benzoyl chloride 584, 603
Benzyl alcohol 608, 615
Benzyl bromide 699
Benzyl chloride 561
Benzylidene aniline (see N-Benzylidenephenylamine)
N-Benzylidenephenylamine 613, 646
Benzylidene triphenylphosphorane 613
Benzyne 585, 661
Berlin green 515
Beryl 349
Beryllium,
 electrode potential 357
 properties 356
Beryllium acetate 433
Beryllium acetyl acetonate 357
Beryllium carbide 357, 388
Beryllium carbonate 357
Beryllium dichloride 357
Beryllium oxide 357
Beryllium sulphate 357
Bismuth,
 structure 414
 uses 413
Bismuth glance 412
Bismuthine 416
Bismuth(III) oxide 414, 423, 429
Bismuth(V) oxide 422

Bismuth oxychloride 408
Bismuth trisulphide 424
Biuret 634
Bleaching powder 412, 483, 500
Borax (see Sodium tetraborate)
Borazole 369, 370
Boric acid 364, 366–368, 703
Boric oxide 367
Boron,
 preparation 364
 structure 365
 uses 365
Boron carbide 366, 388
Boron nitride 365, 414
Boron trifluoride 89
Bromic acid 483
Bromine,
 manufacture 469
 preparation 469
 uses 472
Bromobenzene 695
2-Bromobutane 573
Bromoethane 562, 573, 589, 603, 659, 665
Bromomethylbenzene 699
2-Bromo-2-methylpropane, 573, 589
Buta-1,3-diene 566, 568
Butane 537, 698
Butanol, tertiary 579, 586, 589, 665, 700
Butanone 612, 618
But-2-enal 721, 722
But-1-ene 546, 573
But-2-ene 573
Butenedioic acid, cis- 549
Butenedioic acid, trans- 549
Butyl acetate 583
Butyl bromide, tertiary- 573, 589
Butyl chloride, tertiary- 665, 699
Butylene (see Butene)
Butyl-lithium 695, 698
Butyl methyl ketone 701
But-1-yne 546

Cadmium 525
 occurrence 527
 uses 527
Cadmium chloride 69, 701
Cadmium oxide 527, 528
Cadmium sulphide 527
Caesium 337, 340
Caesium chloride 67
Calcite 349, 393

Calcium, volumetric determination of 355
Calcium acetate 621
Calcium acetylide 569
Calcium carbide 569
Calcium carbonate 356
 forms of 349
Calcium cyanamide 411
Calcium dihydrogen phosphate 412
Calcium fluoride 68
Calcium formate 621
Calcium oxalate 356
Calcium propanoate 621
Calcium sulphate (anhydrite) 349
Calcium sulphate (gypsum) 349
Calomel electrode 207
Cane sugar, refining of 756
Carbene 377
Carbon 165, 299
 diamond 69, 378
 graphite 378
Carbon dioxide 361, 385, 389, 391, 395
 crystal structure 70
Carbonic acid 391
Carbon monoxide 389, 390
 uses 391
Carbon suboxide 389
Carbon tetrachloride 386, 659
Carbonyl sulphide 391
Carnallite 339
Cellulose 714
Cement 395, 446
Chalk 349
Charcoal, adsorption by 756
Chile saltpetre 339, 412, 468
Chloral 620, 662
Chloral hydrate 620, 662
Chloric acid 483
Chlorine 251
 manufacture 469
 preparation 469
 structure 225
 uses 471
Chlorine dioxide 480
Chlorine hexoxide 481
Chlorine monoxide 480
Chlorine trifluoride 316
Chloroacetic acid 726
Chlorobenzene 659, 728, 730, 732
2-Chloro-buta-1,3-diene 566
Chloroethane 562, 587, 658, 665
2-Chloroethanol 602
Chloroethene 565, 567, 727

Chloroform 616, 659, 662
Chloromethyl benzene 561
2-Chloro-2-methylpropane 665, 699
Chlorophyll 355, 515
1-Chloropropane 588, 593, 594, 658
2-Chloropropane 664
2-Chloropropanoic acid 664
3-Chloropropanoic acid 664
Chlorosulphonic acid 447
Chromite 506
Chromium(III) chloride 69, 508
Chromium hexacarbonyl 390
Chromium(III) oxide 412, 506
Chromyl chloride 507
Cinnabar 302
Clay 395
Coal 572
Coal gas 389
Cobalt 516
  isotopes 214
Cobalt(II) chloride 517
Concrete 395
Copper,
  extraction 302, 523
  properties 520, 521, 523
  reactions of 524
  refining 523
Copper(I) acetylide 569
Copper(I) carbide 388
Copper(I) chloride 523
Copper(II) fluoride 277
Copper(I) oxide 302, 523
Copper pyrites 523
Copper(II) sulphate, structure 288
Corundum 370
Cristobalite 393
Crotonaldehyde 721, 722
Crude oil fractions 559
Cryolite 364
Cyclobutane, ring strain in 551
Cyclohexane 99, 544, 558, 560
  ring strain in 551
  1,2,3,4,5,6-hexachloro- 569
Cyclohexanecarboxylic acid 579
Cyclohexanol 595
Cyclohexanone 595
Cyclopentane, ring strain in 551
Cyclopropane, ring strain in 551
Cyclopropane-1,2-dicarboxylic acid,
  cis- 553
Cyclopropane-1,2-dicarboxylic acid,
  trans- 553

Decalin 544

Decane 659
Deoxy-ribose 708
Deuterium 24, 319
Deuterium oxide 320
Devarda's alloy 425, 459
Diacetone alcohol 615, 721
1,2-Diaminoethane 493
Diamond 378
  structure 69, 378
Diborane 595
  bonding in 369
1,2-Dibromoethane 472, 574, 664
Dibutylcadmium 701
1,2-Dichlorobenzene (*ortho*) 732
1,4-Dichlorobenzene (*para*) 732
2,2-Dichlorobutane 618
1,1-Dichloroethane 618
Dichloromethane 659
1,1-Diethoxyethane 610
N,N-Diethylacetamide 601
Diethylamine 601, 635
Diethylbenzene 571
Diethylcadmium 696
Diethyl ethanedioate 601
Diethyl ether 582, 597, 598, 602
Diethyl ether hydrochloride 598
Diethyl malonate 696, 716
Diethyl oxalate 601
Difluorocarbene 377
1,2-Dihydroxypropane 709
Dimethylamine 640, 642
Dimethyl ether 606
N,N-Dimethylphenylamine 639
2,2-Dimethylpropanal 615
N,N-Dimethylpropanamide 642
2,2-Dimethylpropane 539, 699
2,2-Dimethylpropanoic acid 615
2,2-Dimethylpropan-1-ol 615
N,N-Dimethylprop-1-ylamine 642
Dimethyl sulphate 713
1,2-Dinitrobenzene (*ortho*) 730
1,3-Dinitrobenzene (*meta*) 730, 732
1,4-Dinitrobenzene (*para*) 730
Dinitrogen oxide 291
Dinitrogen pentoxide 422
Dinitrogen tetroxide 150, 151, 153,
  421, 422
Dinitrogen trioxide 423
Dioxygenyl hexafluoroplatinate(V)
  311
1,2-Diphenylethene 613
Diphenylmethane 699
Diphosphine 416
Dipyridyl (dipy) 492

Disilane 383, 384
Dolomite 300
Dry ice 392

Epoxyethane 601
Ethanal (see Acetaldehyde)
Ethane 383, 384, 537, 549, 558, 572
Ethanediamide 601
Ethanedioic acid 637
Ethane-1,2-diol 602, 611
Ethanoic acid (see Acetic acid)
Ethanol 460, 562, 580, 584, 586,
  587, 589, 594, 596, 597, 599,
  604, 620, 700, 765
  manufacture 563
Ethanolamine (see 2-Aminoethanol)
Ethene (see Ethylene)
Ethenol 567
Ethenone 585
Ethenyl ethanoate 565
Ethoxybenzene 582, 603
Ethoxyethane 582, 597, 598, 602
1-Ethoxyethanol 610
1-Ethoxypropane 583, 594
N-Ethylacetamide 601
Ethyl acetate 582–584, 599,
  602–604
Ethyl acetoacetate 696, 698, 717,
  719
Ethylamine 596, 601, 638
Ethylbenzene 609
Ethyl benzoate 702
Ethyl bromide 562, 573, 589, 603,
  659, 665
Ethyl chloride 562, 587, 658, 665
Ethyl cyanide 638
Ethylene 544–546, 548, 549, 573,
  582, 680, 765
  bonding in 775
Ethylene chlorohydrin 602
Ethylenediamine 493
Ethylenediaminetetra-acetic acid
  (EDTA) 355
Ethylene dibromide 472, 574, 664
Ethylene glycol 602, 611
Ethylene oxide 601
Ethyl glycinate 706
Ethyl hydrogen sulphate 562
Ethyl isonitrile 647
Ethyl isopropyl ketone 623
Ethyl-lithium 694
Ethylmagnesium bromide 623, 696
Ethylmethylamine 640
Ethyl methyl ether 603

Ethyl methyl ketone   612, 618
Ethyl methyl ketone oxime   612
N-Ethylphenylamine   642
N-Ethyl-N-phenylacetamide   642
Ethyl phenyl ether   582, 603
Ethyl propanoate   594
Ethyl propyl ether   583, 594
Ethyne (see Acetylene)
Ethynylmagnesium bromide   696
Everitt's salt   515

Flint   393
Fluorine,
  as oxidizing agent   464
  nature of   472
  preparation   468
  reactions with oxygen compounds   473
Fluorine monoxide   480
Formaldehyde   586, 587, 604, 607, 615, 700
  polymer   617
Formic acid   390, 579, 586, 587, 607, 608, 615
Fumaric acid   548
Fructose, α-D-   708, 711
Fructose, β-D-   711

Galena   302, 380, 398, 437
Gallium,
  metal properties of   363, 365
  oxides   364
Germanium, zone refining of   742
Germanium dioxide   395
Germanium monoxide   395
Glass   395
Glucose, α-D-,
  structure of   708, 709
Glucose, α-L-   709
Glucose, β-D-   709
Glycerol   592
Glycerol tristearate   592
Glycine   706
Glycine hydrochloride   706
Glycollic acid   706
Glyclglycine   707
Glyclglyclglycine   707
Gold,
  properties   520, 521, 523
  reactions   524
Graphite, structure   378
Graphitic fluoride   387
Graphitic oxide   387
Gypsum   437

Haematite   303, 512
Haemoglobin   355, 390, 437, 515
Hafnium   502
Heavy water   320
Helium   26, 308
Hexane   537, 560
Hexan-2-one   701
Hexene   546
Hexyne   546
Hydrazine   409, 418
Hydrogen   147, 165, 218
  atomic   319–321
  electrode   160
  molecule   77, 78
  spectrum   26
  uses   322
Hydrogen bromide,
  properties   475
Hydrogen chloride, 473–475
Hydrogen cyanide   473
Hydrogen fluoride   473, 474
Hydrogen iodide,
  decomposition   179
  properties   475
Hydrogen peroxide   441, 442
Hydrogen sulphide   442
Hydroquinone   595
Hydroxyacetic acid   706
Hydroxyacetone   712
3-Hydroxybutanal   614, 721
Hydroxylamine   410, 418, 421
Hydroxylammonium hydrochloride   418
4-Hydroxy-4-methylpentan-2-one   615, 721
2-Hydroxynaphthalene-1-azobenzene   652
Hypo   452
Hypochlorite   412, 483, 500
Hypochlorous acid   480
Hyponitrous acid   420

Ilmenite   503
Indium, metal properties   365
Invisible ink   517
Iodic acid   483
Iodine   147
  crystal structure   70
  manufacture   469
  preparation   469
Iodine pentoxide   481
1-Iodobutane   583
Iodoethane   583, 588, 603
Iodoform   616

Iodomethane   583, 599, 603
Iridium   516
Iron   458, 511
  pig   512
  wrought   512
Iron(III) bromide   681
Iron(III) chloride   289
Iron(II) hydroxide   513
Iron(III) hydroxide   289, 514
Iron(II) oxide   512
Iron(III) oxide   512
Iron pentacarbonyl   390
Iron pyrites   280
Isoprene,
  polymer   566
  units   566
Iso-propanol   579, 586, 620

Kerosene   559
Ketene   585
Krypton   308
Krypton fluoride   311

Lactic acid   542, 765
Lead,
  extraction   380
  ore   380
  purification   380
  uses   380
Lead(II) chloride   277, 397
Lead(IV) chloride   277, 396
Lead(II) oxide   395
Lead suphide   302, 380
Lead, tetraethyl-   340, 399
Limestone   349
Lime water   392
Limonite   512
Litharge   395
Lithium   346, 347
  nature of   283
Lithium aluminium hydride   274, 328, 595, 608, 637
Lithium borohydride   328, 608
Lithium ethoxide   694
Lithium iodide   245

Magenta   620
Magnesium,
  manufacture   349
  occurrence   349
  uses   349
  volumetric determination   355
Magnesium boride   366, 369
Magnesium carbonate   355

Magnesium chloride,
  basic 286
  Born–Haber cycle 351
  relative stability 351
  thermodynamic stability 352
Magnesium ammonium
  phosphate 355
Magnesium germanide 396
Magnesium silicide 396
Magnetite 512
Maleic acid 548
Malonic acid 716
Malonic ester 716
Manganese 509
  oxidation state diagram 170, 509
Manganese hydroxide 510
Manganese(II) oxide 510
Manganese(III) oxide 510
Manganese(IV) oxide 251, 510, 511
Manganese sulphide 510
Marble 349
Margarine, production 563
Mercury 525
  occurrence 527
  "tails" 527
  uses 527
Mercury(II) chloride 528
Mercury(II) fluoride 528
Mercury(I) iodide 527
Mercury(II) iodide 527, 528
Mercury oxide 527
Mesityl oxide 721
Meta-fuel 618
Metaldehyde 618
Methanal 586, 604, 615
Methane 537, 558, 589
  bonding in 774
Methanoic acid (see Formic acid)
Methanol 391, 585–587, 595, 630
Methoxybenzene 583, 599
Methoxyethane 603
N-Methylacetamide 633, 640, 707
Methylamine 630, 631, 726
Methylammonium acetate 633
Methylbenzene 561, 571, 659
  chloro- 561
  dichloro- 561
  trichloro- 561
2-Methyl-buta-1,3-diene 566
2-Methylbutane 539
Methyl cyanide 648
Methylene chloride 659
Methylene triphenylphosphorane 574

Methyl glucoside 712, 714
Methyl-lithium 699
2-Methylpentan-3-one 623
4-Methylpent-3-en-2-one 721
N-Methyl-N-phenyl acetamide 639
N-Methylphenylamine 639, 640
Methyl phenyl ether 583, 599
Methyl phenyl ketone 571, 609
Methyl propanoate 595
2-Methylpropan-2-ol 579, 586, 589, 665, 700
2-Methylpropanonitrile 623
2-Methylpropene 573
Mica 394
Mohr's salt 514
Molybdenum 506

Naphthalene 193, 558
Natural gas 764
Neon 308
Nickel 518
  ionization energy 495
  isotopes 214
  purification 304
Nickel carbonyl 519, 785
Nickel dimethylglyoxime 494, 520
Niobium 504
Nitric acid 422, 425
Nitric oxide (see Nitrogen oxide)
Nitrobenzene 649
  nitration of 649, 731
4-Nitrochlorobenzene 730
Nitrogen,
  active 411
  oxidation states 407, 408
  oxides 409, 420, 421
Nitrogen oxide 420
Nitrogen dioxide 151, 153
Nitrogen pentoxide (see Dinitrogen pentoxide)
Nitrogen tetroxide (see Dinitrogen tetroxide)
Nitrogen trichloride 410, 417, 418
Nitrogen trifluoride 417
Nitrogen trioxide (see Dinitrogen trioxide)
Nitromethane 649
Nitronium nitrate 422
2-Nitrophenol 725
3-Nitrophenol 725
4-Nitrophenol 725
N-Nitroso-N-methylphenylamine 640
4-Nitrosophenol 642

Nitrous acid 422, 424
Nitrous oxide (see Dinitrogen oxide)

Oleum 447
Orthoboric acid (see Boric acid)
Osmium 511
Oxalic acid 637
Oxamide 601, 637
Oxygen 438
  anomalous nature of 434
  occurrence 436
  uses 437
Ozone 437, 564
  oxidizing power 438

Palladium 501, 518
Palladium chloride, structure 276
Paraldehyde 618
Pentamethylglucose 713
Pentane 537, 539
Pentene 546
Pent-1-ene 546, 547
Pent-2-ene 547
Pentyne 546
Perbenzoic acid 602
Perbromic acid 484
Perchloric acid 484
Perchromic acid 442
Periodic acid 484
Petroleum 572, 764
Phenol 571, 579, 584, 596, 599, 603, 725, 732
Phenyl acetate 584
Phenylamine 596, 637, 725, 732
  bromination of 732
Phenylammonium chloride 631
N-Phenylbenzamide 633
Phenyl benzoate 584, 603
Phenyl cyanide 591, 622, 648
Phenylethene 565, 574
Phenylhydrazine 652
Phenyl-lithium 695, 702
Phenylmagnesium bromide 698, 699, 701
Phenylmethanol 608, 615
N-Phenylpropanamide 639
Phosgene 391
Phosphine 406, 411, 414, 416, 417
Phosphorus
  black 413
  brown 413
  oxidation states 407
  red 222, 413
  white 222, 412, 413

INDEX OF NAMED SUBSTANCES   847

Phosphoric acid,
 hypo 427
 meta 422, 428
 ortho 422, 427, 428
 pyro 422, 427, 428
Phosphorous acid,
 hypo 427
 ortho 410, 423, 427
 pyro 427
Phosphorus pentachloride 419
Phosphorus pentafluoride 89
Phosphorus pentoxide 414, 422
Phosphorus trichloride 410, 419
Phosphorus trioxide 414, 423
Phthalimide 635
Picric acid 725
Picryl chloride 728
Platinum 518
Polonium dioxide 446
Polypropylene 565
Polystyrene 565
Polytetrafluoroethylene 565
Polythene 565
Polyvinyl acetate 565
Polyvinylchloride, (PVC) 565, 567
Potassium aluminium sulphate 287
Potassium bromate 483, 484
Potassium chlorate 483
Potassium chloroplatinate 345
Potassium cobaltinitrite 345
Potassium cyanide 646
Potassium dichromate 507
Potassium dicyanoargentate 494
Potassium ferricyanide 494, 514
Potassium ferrocyanide 287, 514
Potassium hexacyanoferrate(II) 287, 514
Potassium hexacyanoferrate(III) 494, 514
Potassium iodate 483
Potassium perchlorate 345
Potassium permanganate 511
Potassium phenolate 593
Potassium thiocarbonate 279
Producer gas 389
Propanamide 594
Propane 534, 537, 558, 573, 606
Propane, 2,2-dimethyl 539, 699
Propanedioic acid 716
Propane-1,2,3-triol 592
Propane-1,2,3-triyl
 trioctadecanoate 592
Propanoic acid 579, 593, 609, 639
Propan-1-ol 578, 579, 588, 593, 595

Propan-2-ol 579, 586, 595, 606, 608, 620, 700
Propanone 586, 607, 608
Propanonitrile 638
Propanoyl chloride 593, 594
Propene 562, 565, 573
N-Prop-2-ylacetamide 640
Prop-1-ylamine 594, 638
Prop-2-ylamine 640
Prop-1-yl chloride 658
Propylene 562, 565, 573
Propylene glycol 709
Propyne 546, 621
Protium 24, 319
Prussian Blue 515, 761
Pyridine 492
Pyrites 280, 437
Pyrolusite 509

Quartz 393
Quinone 595

Radon 308
Radon fluoride 311
Red lead 395
Rhenium 509
Rhodium 516
Ribonucleic acid (RNA) 715
Ribose 715
Rubber,
 buna- 566
 natural 566
 neoprene 566
Ruby 370
Ruthenium 511

Salicaldehyde 345
Sapphire 370
Selenium 432
 allotropes 439
Selenium dioxide 446
Selenium hydride 443
Semi-water gas 390
Siderite 512
Silica 394
Silicon,
 pure 380
 semi-metal 292
Silicon carbide 388
Silicon dioxide 361
Silicon disulphide,
 structure 280
Silicon tetrachloride 361, 386
Silicon tetrafluoride 396

Silver,
 properties 520, 521, 523
 reactions 524
Silver acetate 603
Silver acetylide 569
Silver carbonate 393
Silver chloride 244
Silver cyanide 646
Silver sulphite 449
Slag, formation of 300
Soap 592, 593
Sodium, spectrum 26, 57
Sodium acetate 580, 584, 589, 591, 604
Sodium acetylide 696, 698
Sodium amide 346, 417
Sodium azide 412
Sodium benzoate 589
Sodium bicarbonate 392
Sodium bismuthate 422
Sodium borohydride 328, 595
Sodium carbonate 392
Sodium chlorate 483
Sodium chloride 67, 342
Sodium ethoxide 580, 603
Sodium glycinate 706
Sodium hypochlorite 482
Sodium iodate 468
Sodium magnesium uranyl acetate 345
Sodium phenoxide 580, 583, 603
Sodium propoxide 583
Sodium stearate 592
Sodium tetraborate 364, 367, 368
Sodium tetrathionate 471
Sodium thiosulphate 452
Starch 713
Steel 501
 production of 512
Stibine 411, 416
Stibnite 412
Stilbene 613
Styrene 565, 574
Sucrose 597, 708, 712, 714
Sulphur 431, 432, 437
 allotropic 438
 Engel's 439
 monoclinic 438
 plastic 439
 properties 440
 purple 439
 rhombic 438
Sulphur dioxide 446, 447
Sulphuric acid 447, 450, 581

Sulphurous acid 448
Sulphur trioxide 447, 450
Sulphuryl chloride 446
Superphosphate 412

Tantalum 504
Technetium 509
Teflon 565
Tellurium 432
   allotropes 439
Tellurium dioxide 446
Tellurium hydride 443
Terephthalic acid 571
Tertiary-butanol 579, 586, 589, 665, 700
1,1,2,2-Tetrachloroethane 567
Tetrachloromethane 386, 659
Tetraethylammonium hydroxide 635, 644, 645
Tetraethylammonium iodide 641, 644
Tetrafluoroethene 565
Tetrafluoroethylene 565
Tetramethylglucose 713
Thallium, metal properties 363–365
Thiocarbonic acid 279
Thionyl chloride 446, 587
Tin,
   allotropes 379
   enantiotropes 379
   extraction 380
   grey 379
Tin(II) chloride 397
Tin(II) oxide 395

Tin(II) sulphide 398
   structure 280
Titanium 502
   in metallurgy 503
   isolation 303, 502
Titanium dioxide 503
Titanium tetrachloride 503
Toluene 561, 571, 659
   nitration 189
Tributylborane 699
Trichloroacetaldehyde 620, 662
Trichloromethane 616, 659, 662
Tridymite 393
Triethylamine 635, 645
Tri-iodoacetone 616
Tri-iodomethane 616
Trimethylamine 642
Trimethylammonium iodide 642
Trimethylborane 699
2,4,6-Trinitrochlorobenzene 728
2,4,6-Trinitrophenol 725
Triphenylmethanol 701, 702
Triphenylphosphine oxide 574, 613
Trioxan 617
Trioxan 617
Trioxan 617
Trioxan 617
Trioxan 617
Trioxan 617
Trioxan 617

Undecane 659

Uranium 20
   urea 392, 634

Vanadium 504
Vanadium pentoxide 504
Vinyl acetate 565
Vinyl alcohol 567
Vinyl chloride 565, 567, 727
Vitamin B$_{12}$ 517, 518

Water 86, 89, 206, 258, 329–333
   enthalpy of 126
   phase diagram 113
   vapour pressure 112
Water gas 390
Wolframite 302
Wood's metal 528

Xenic acid 313
Xenon 308, 311
Xenon fluoride 311
Xenon trioxide 313

Yeast 597

Zinc 525
   occurrence 527
   uses 527
Zinc acetate 433
Zinc blende 303
Zinc fluoride 528
Zinc oxide 527, 528
Zinc sulphide 69, 527
Zirconium 502

General Index

Named substances are not included in this index, but will be found in the Index of Named Substances. Thus aldehydes and alkali halides are included in the General Index but acetaldehyde and sodium chloride are not. Where there are several references the more important are given in boldface.

Absorption spectra   26
Acetals   610
Acetyl group   583, 601
Acetylating mixtures   584
Acetylation   583
Acetylenes (see Alkynes)
Acetylides   377, 388, **568**, 698, 699
Acheson process   379
Acid(s)   204–206
  action of   290
  classification of   323
  conjugate   205
  dissociation constants of   206
  hydro-   323
  oxidizing   259
  strength of   205, 323, 325
Acid anhydrides   597, 598, 604
  alcoholysis   614
  ammonolysis   614
  hydrolysis   599, 614
Acid chlorides   601, 605, 701
  alcoholysis   614
  ammonolysis   614
  hydrolysis   614
Acid halides   658, 660
  from hydroxy compounds,
    reduction of   660, 664
  substitution reactions   613, 659
Acid salts   286
Acidic oxides   219
Acidity   203
  binary hydrides   324

Actinides   217
Activated complex   **180**, 187, 188, 784
Activation energy   176, **180**, 185, 195, 301, 382, 669
Active hydrogen   601
  determination   699
  estimation   699
Active methylene   716, 717
Activity series   **251**, 252, 261, 272, 280, 286, 301
Acyl group   583
Acyl halides (see Acid halides)
Acylation   583, 600, 660
Addition reactions   **545**, 671, 678
Adducts   491, 678
Air   256
Alcohol, determination in blood   758
Alcohols,
  acidity   694
  acylation   584
  dehydration   582, 584, 585
  halogen introduction   588
  hydrogen bond formation   579
  oxidation   585, 588, 594
  preparation   592, **595**, 596
  primary   577, **586–589**, 595, 700
  reactions   610, 676
  secondary   577, **586–589**, 595, 700
  tertiary   577, 586, **700**, 702
Alcoholysis   614, 635
Aldehydes   587, **619–622**, 700, 718

  addition reactions   609
  basicity   607
  condensation reactions   612, 678
  effect of alkali   614, 720
  hydrates   662
  nomenclature   605
  oxidation   594, 607
  physical properties   606
  reduction   595, 608
  test for   613
  α,β-unsaturated   **721**, 722
Aldol reactions   614
Aliphatic compounds   551
Alkali metal halides   243, 244, 281, **344**
  percentage ionic character   345
  stability   103, 281, **341**
Alkali metals   338–340
  order of reactivity   343
  organo-metallic compounds   346
  stability of salts   343
Alkaline earth metals   **348–357**
  carbonates   282
  chlorides   354
  complexes   355
  electrode potentials   353
  solubility of salts   350
  stabilities of salts   354
Alkalinity   203
Alkalis
  action of   291
  aqueous   260

Alkanes 384, 396, **539**, 543, 545
  cyclo 544
  halogenation 664, 671
  reactions of 560, 572
  substituted 561
Alkanoates 579
Alkenes 547
  addition reactions of 570, 664, 680
  electrophilic addition 697
  hydration of **563**, 595, 697
  oxides 602
  reactions 561, 572, 602
  synthesis 613, 645, 661
Alkyl chlorides 674
Alkyl group 539
  primary 586
  secondary 586
  tertiary 586
Alkyl halides 561, 573, 593, **657**, 675, 682, 684, 719
  elimination reactions 683
  hydrolysis 654, 659, **673**
  ionization 675
  nomenclature 658
  physical properties 658
  substitution reactions 659
Alkyl hydrogen sulphates 570
Alkyl iodides 674
Alkyl magnesium halides 695
Alkyl tetrafluoroborates 645
Alkylbenzenes, oxidation 571
Alkylcadmium compounds 696
Alkynes **547**, 573, 696
  non-terminal 568
  reactions 566, 570
  terminal 568
Allotropes 220, **221**, 235, 379, 413
Allotropy 221
  dynamic 223, 439
Alloy formation 501
Alloy phase 275, 280
Allred 80, 232
Alpha decay 25
Alpha rays 20
Aluminates 302, 372
Aluminium,
  alloys 365
  compounds 370–372
  halides 277, 371
  ion 288, **371**, 372
Alumino-silicates 394
Aluminothermic process 303
Alums 371

Amalgams 527
Amides 578, 591, **629–633**, 643
  acylation 633
  alcoholysis 614
  alkylation 632
  ammonolysis 614
  basicity 631
  conversion to amines 638, 640
  cyclic 707
  dehydration 648
  N,N-disubstituted 640
  hydrolysis 594, 614
  physical properties 630
  reactions with alkali 632
  N-substituted 639
Amines,
  C—N bond cleavage 642
  hydrolysis 594
  physical properties 630
Amines,
  primary **629**, 630, 650
    acylation 633, 640
    alkylation 632, 640
    basicity 631, 725
    effect of nitrous acid on aliphatic 634
    introduction of group 634
  secondary 630, 633, **638**, 639
    acylation 632, 639, 642
    alkylation 632, 639, 642
    effect of nitrous acid 639
    introduction of group 640
  tertiary 630, 641, 642
    acylation 641
    alkylation 641
    effect of nitrous acid 641
α-Amino acid 705
Amino acid analysers 759
Amino compounds, reactions 643
Amino groups,
  primary 629
  secondary 638
  tertiary 640
Ammines 417, 528
Ammonia solutions 346
Ammonium salts structure 418
Ammonolysis 614, 635
Amount of substance 15
Amphoteric hydroxide 272
Amphoteric oxide 219, 269
Amphoteric sulphides 279
Amplitude 27
Analytical methods 761
Analyses for C, H and N 761

Angular dependence of electron distribution 41
Aniline dyes 652
Anions 82, 281, 293
  polarizable 283, 338
Anion-stabilizers 285, 338
Anodization 365
Anomalous behaviour of first member 215, 262, 360
Anomalous nature,
  of beryllium 356
  of fluorine 472
  of lithium 283, 339, **346**
  of nitrogen 408
  of oxygen 434
Anomer 709
Antifluorite structure 70
Anti-knock 380
Anti-Markownikoff addition 697
Antimonite ion 423
Apical position 316
Arenes 551, 664
Aristotle 18
Aromatic compounds,
  bonding 98, 551
  nitration 186, 682, 731
Aromatic hydrocarbons,
  production 573
  reactions 569–572
Aromatic substitution,
  electrophilic 681
  nucleophilic 728
  substituents effects 730–733
Arrhenius 181, 204
Arsenic, oxyanions 429
Arsenite ion 414, 423, 424
Aryl groups (Ar) 551
Aryl diazonium salts 648
Aryl halides 658, 677
Arylmagnesium halides 695
Aston 23
Atom, nuclear 21
  structure of 19
Atomic conductance 219, 798, 799
Atomic heats 14
Atomic hydrogen 321
Atomic number 24
Atomic properties 215, 221
Atomic radii 223, 796
Atomic structure 18
Atomic volume **235**, 292, 796
Atomic weights 11
Aufbau approach **49**, 55, 56
Aufbau principle 51

GENERAL INDEX 851

Auto-ionization 332
Automation 759, 762
Avogadro 11, 12
Avogadro constant 15, 73
Avogadro's hypothesis 12

Balancing equations 456
Balancing redox equations 455, 458
Balmer series 27, 33
Banana bonds 369
Barium ion reactions 355
Bartlett 311
Bases 204–205
Basic acetate 433
Basic copper salts 257
Basic oxides 219, 269
Basic salt 286
Benzene derivatives, ortho, meta, and para 552
Beryllates 353
Beta decay 25
Beta rays 20
Bicarbonates 392–393
  action of heat on 285, 291
Biological molecules 412, 427, 566, 714
Bisulphates 451–452
Bisulphites 448
Black-body radiation 28
Blagden's law 119
Blast furnace 512
Bohr 31
Bohr atom 31–35, 39
Bohr radius **34**, 40, 45, 813
Boiling point 116, 117, **237**, 738, 798
  elevation of 119
Boltzmann constant **74**, 134
Boltzmann distribution 136
Boltzmann's law 29, **134**, 780
Bond,
  banana 369
  carbon–carbon 717–720, 765
  carbon–chlorine 662
  carbon–metal 695
  co-ordinate 85
  covalent 40, **79**, 102, 240, 310, 544, 771
  dative 85
  double 545, 549, 723, 775
  hydrogen 86, 116, 117, 271, 274, 330–331
  ionic 77–79, 102, 310
  metallic **85**, 240, 340
  multiple **384**, 544

π– 722, 775
polar 82, 85
rotation about 538, 566, 775
σ– 775
single 549
triple 545, 549, 777
Bond dissociation energy 97
Bond energy 80, 95–97, 101, 274, 549, 794
Bond length 70, 82, 794
Bond order **99**, 549
Bond polarity 102
Bond strength 98, 361
Bond types 79, 224, **270**
Bonding, history 309
Boranes 366, 369
Borate esters 703
Borate glasses 367
Borates 367
Borides 363, 366
Born–Haber cycle **103**, 106, 240, 243, 247, 249, 281, 282, 312, 324, 341, 352, 372
Borohydrides 328, 369
Boron compounds 366, 372
Boron trihalides 367
Boyle's law 72
Bragg 59
Bragg's law **60**, 71
Breathalyser test 758
Bridging halogen atom 276, 357, 371
Bridging hydrogen atom 369–370
Bromates 483
Bromides 277, 477
Bromine oxides 481
Bromonium ion 682
Brönsted theory 204, 205, 675
Brown ring test 421, 426
Buffer solutions 210

Cadmium chloride structure 69
Cadmium compounds 526–528
Caesium chloride structure 67
Calcium ion reactions 355
Cannizzaro 12
  reactions 615, 619
Canonical forms 101, 378
Carbides 387–388
Carbohydrates 708
Carbon,
  analysis for 761
  as reducing agent 165, 299
  peculiarities 381, 384–387
Carbon atom,

  asymmetric 542
  primary 586
  secondary 586
  tertiary 586
  tetrahedral 536, 540
Carbonates,
  action of heat 285, 291
  properties 392
  relative stabilities 283
  solubility 243, 246
  thermal decomposition 283
Carbon–carbon bonds 717–720, 765
Carbon dating 13, 24
Carbon group 374
Carbon–metal bonds 695
Carbon tetrahalides 397
Carbonium ions 634, **675**, 697
Carbonyl compounds,
  addition reactions 609
  condensation reactions 612
  oxidation 607
  physical properties 606
  substitution reactions 613
Carbonyl group 577, 604, 720
  activating influence of 679
  basicity 607
  chemical behaviour 607
  directing influence of 719
  important reactions 619
  introduction 620, 622
Carboxylic acids 582–587, **604–609**, 619
  derivatives 578, 606, 679
  manufacture 594
  nomenclature 579, 606
  oxidation 607
  preparation 591, **594–596**, 622
  reduction 595, 608
  substitution reactions 613
  synthesis 719
Catalysis **189**, 191, 450
  acid 676
Catalytic hydrogenation 563
Catenation 381, 383, 409, **536**
Cathodic reduction 303
Cation-exchange 394
Cations, size of **66**, 82, 245, 265, 281, 293
Cells 156–163
  conductance 202
  e.m.f. of 160, 161
  fuel 156, 158
  silver–copper 162

standard half- 159
zinc–copper 156
Celsius 126
Chain reaction 321
Chain structures,
    inorganic 222, 276, **277**
    organic 536
Change,
    direction of 135, 137
    endothermic **92**, 127
    entropy and 135, 779–782
    exothermic **92**, 127
    in isolated systems 137, 143
    of mixing 124
    reversible 138
Charge density **82**, 228, 283, 284, 288, 294
Charge-transfer compounds 470
Charge-transfer spectrum 790
Charles's law 72
Chelate effect 355, 496
Chelating ligands 355, **492**, 522
Chemical change 111, 127
Chemical equilibrium 148
    thermodynamic treatment 782, 783
Chemical reactions of elements, classification 256
Chemistry,
    definition 1
    inorganic 2
    organic 2, 534
    physical 2
    pure and applied 8
Chiral molecules 543
Chlorates 481
Chloride(s),
    basic 278, 290
    bond type 276, 277
    properties 275–278
Chlorination 561
Chromates 508
Chromatography **748**, 755–757
Chromites 508
Chromium(III) chloride structure 69, 70
Chromium group 506
Chromium steels 501
Clathrate compounds 313
Clemmensen reduction 609
Close-packed structures **64–65**, 235, 308
Cobalt(III) complexes 516, 789, 791
Cobalt(II) compounds 517, 789, 791

Cobalt group 516
Cobaltinitrites 345, 517
Colligative properties 14, 120
Collision theory 186
Collisions 73
Collisions and encounters 186
Complex compounds 287, 491, 500
    uses 494
Complex formation 116
Complex hydrides 328
Complex ion formation 500
Complexes 491
    octahedral 493, 787–791
    paramagnetic 791
    rules for nomenclature 493
    spin character 790
    square-planar 492
    strong and weak field 791
    tetrahedral 492, 788
Condensation reaction 612, 678
Conductance of solutions 201–203, 207
Conductivity,
    electrical 219, **237**
    thermal 237, 798
Conjugation 722, 723
Conservation of charge 456
Conservation of energy 90, 127
Constant boiling mixture 116
Contact process 450
Convergence limit 41
Co-ordinate bond 85
Co-ordination compounds 491
Co-ordination number **65**, 374, 491, 492
Co-ordination sphere 491
Copper(I) compounds 522, 523
Copper(II) compounds 523
Copper group 520
Core electrons 47
Coulomb's law 104
Counter-current distribution 752
Coupling reaction 651
Covalent bond 40, **79**, 102, 310, 771
Covalent radii 70, 224, 228, 235
Cracking (hydrocarbons) 191, 572, 573
Crystal energy 233
Crystal field theory 785–791
Crystal structure 66, 70
Crystals, mixed 287
Cyanides 646
    complex 302, 514

hydrolysis 622
reduction 637
Cyanohydrin 722
Cyclic molecules 544

Dalton 11, 20
Dalton's law of partial pressures 149
Dative bond 85
Davy 203
d-Block elements 217, **487**
    electronic structure 495
    oxidation states of first row 497
de Broglie 30, 35
Debye unit 83
Decarboxylation **589**, 716, 720
Dehydration of alcohols 582, 584, 585
Dehydrogenation 191
Deliquescent hydroxides 257
Delocalization 85, **99**, 100, 237, 722, 725, 776
Democritus 18
Density 234
Deuterium 24, 319, 320
Diabetes 712
Diagonal relationship 284, **348**, 349, 356, 374, 381
Dialkyl cadmiums 701
Diamagnetism, 791
Diamond crystal counter 238
Diamond structure 69
Diazonium group 650–652
Diazonium salts 630, 634, 666
Diazotization 650
Dielectric constant 83
Differential wetting 302
Diffraction,
    electron 62
    neutron 64
    X-ray 59
Dipole–dipole attraction **87**, 116
Dipole moment 82, 83, 101
Disaccharides 708
Disorder 137
Displacement reactions 257
Disproportionation 168, **169**, **171**, 172, 260, 313, 352, 406, 422, 441, 449, 460, 468, 480, 482, 496, 511, 521, 529, 615
Dissociation constant for acids 206
Dissociation constant for water 206
Dissociation of dinitrogen tetroxide 150

GENERAL INDEX 853

Distillation   743
  at reduced pressure   747
Distribution chromatography   752
Distribution coefficient   **146**, 749
Distribution of molecular speeds   75, 130
Dithionites   450
Dobereiner   267
Dot and cross diagrams   84, 545
Double bond character   723
Double decomposition   289
Downs process   339
Dulong and Petit's law   14

Edible facts   563
Effective nuclear charge   47, 56, 225, 229, 234, 241, 283
Eka-silicon   214
Elastomers   566
Electrical conduction   80, 86
Electrical mobilities   228, 331
Electrical work   157
Electrochemical series   **251**, 257, 258, 261
Electrode potential   246
  factors governing   249
Electrolysis   13, 303
Electromagnetic waves   28
Electromotive force (see E.m.f.)
Electron   55
  angular distribution   46
  angular momentum   33
  charge   24
  delocalized   85
  distribution   39, 41
  nonbonding pair   85, 86, 314
  radial distribution   40, 47
  valence   48, 215
Electron affinity   44, 103, **231**, 793
Electron configuration   49, 53, 54, 215
Electron counting   84
Electron deficient compounds   364, 367, 369
Electron density   36, 78, 79
Electron density maps   61, 77–79
Electron diffraction   62
Electron impact   42
Electron penetration   **44**, 46, 48, 56, 229
Electron shells   44, **48**
Electron subshells   48
Electronegativity   80, 81, 101, 103, 215, 232, 265, 796

Electronic theory of oxidation and reduction   454
Electrophiles   682, 731
Electrophilic reactions   675, 680, 681
Electrophoresis   757
Electropositivity   **233**, 261
Electrostatic repulsion theory   314
Elements,
  chemical properties   256
  classification   261
  d-block   487, 488
  group characteristics   215
  p-block   292
  physical properties   221
  s-block   292
Elimination reactions   572, 574, 612, 671, 678, **682**, 683
Ellingham diagrams   164, **165**, 173, 299, 302
  for chlorides   298, 301
  for oxides   295–296
  for sulphides   297, 301
E.m.f.,
  concentration dependence   162
  measurement of   157
  sign convention   160
  standard   162
  temperature dependence   163
Emission spectra   26
Empirical formula   760
Enantiomers   493
Enantiomorphs   493, 543
Enantiotropy   222
Endothermic change   **92**, 127
Energy,
  activation   176, **180**, 185, 195
  band   86
  conservation   90
  distribution   **132**, 134, 135, 179, 777, 778
  hydration   240
  internal   90
  ionization   103
  kinetic   125
  lattice   103
  levels   **41**, 44
  potential   104, 125
  rotational   130
  solvation   186
  thermal   125
  transfer   130
  translational   130
  vibrational   130

Enols   567
Enthalpy   90, 91, 94
  atomization   249, 250, 796
  fusion   113, 126, **238**, 331, 798
  hydration   **240**, **249**, 323, 795
  measurement of changes   92
  of reactions   92, 97, 99, 142
  solution   106, 107, 243, 244
  standard of reaction   153
  sublimation   113
  vaporization   19, 113, 126, **239**, 275, 331, 798
Entropy   134, 135, 139, 143, 152, 777–781
  at absolute zero   138, 143
  molecular interpretation   143, 777–780
  standard molar   139
Entropy changes   140–142, 297
  chemical reactions   140, 141, 781
  measurement   139, 779–782
  mixing   781
  phase change   138, 781
  solution   242
  standard determination   782
  total   142, 158
Enzymes   707
Epoxides   601, 602
Equatorial position   316
Equilibrium   111–113, **146–148**
  acid-base   203
  and free energy change   152, 783
  chemical   147–151
  effect of pressure on   150
  effect of temperature on   151
  phase   111–116
Equilibrium constant   **148**, 150, 153, 155, 173
Equivalence point   207
Esterification   **582**, 583, 598, 601, 603, 605, 613, 614
Esters,
  alcoholysis   614
  ammonolysis   614
  borate   703
  hydrolysis   189, 592, 594, **599**, 703
  phosphate   714
Ethers   597, 598, 603
  hydrolysis   594, 599
  nomenclature   598
Excited state, configurations   49
Exothermic change   **92**, 127
Extraction by partition   114

Face-centred cubic structure **65**, 70, 104
Fajans's rules **81**, 233, 277, 528
Faraday 13
Faraday's laws 201
Fehling's solution **608**, 620, 712
Fermentation 597
Ferricyanide ion (see Hexacyanoferrate(III) ion)
Ferrocyanide ion (see Hexacyanoferrate(II) ion)
Filled shell, stability of 231, 490
Filled subshell 232
First row anomaly 283, 339, **346**, 356, 408, 434, 472
Flame colours 340
Flotation 302
Fluorescence 527
Fluorides 277, 281
  higher 466
  noble gas 311
  stability 312
  structures 476
Fluorite structure 68
Fractional crystallization 740
Fractional distillation 116, 744
Fractional precipitation 741
Fractionating column 116, 744
Frasch process 438
Fraunhofer 26
Free electrons 238
Free energy (see Gibbs free energy)
Free radicals 560, 671
Free rotation 548, 549, 550
Freezing point, depression of 119
Friedel–Crafts reaction 372, 570, **660**, 682
Functional groups 534, 535, 564
  interacting 723
  introduction of 559
  molecules containing two 705

Galvanizing 513
Gamma rays 20
Gangue 302
Gas 572
  kinetic theory 73
  partial pressure 148
Gas constant 72
Gas laws 72
Gas masks 756
Geiger 21
Gel filtration 756
Germanates 395

Germanes 396
Germanites 395
Giant molecules 222, 236
Giant structures 19, **64**, 79, 237
Gibbs free energy **142**, 152, 155
  activation **188**, 674, 728, 730, 785
  atomization 258
  diagrams 169
  profile **188–190**, 729, 785
  solution 243
  standard of formation 155
Gibbs free energy change 156
  and e.m.f. 157
  standard 153
  temperature coefficient 164, 295
Gillespie–Nyholm theory **313**, 377, 404, 434
Glass electrode 207
Glycosides 713
Gold compounds 524
Goniometer 60
Gouy balance 791
Gram-atom 13
Gram-equation 14
Gram-formula 13
Gram-ion 13
Gram-molecule 13
Graphitic compounds 387
Grignard reagent 623, **695**, 698, 700, 701, 702
Ground state **32**, 49
Group 0 216, **308**, 309
Group IA 249, 336, **337**
  ammonia solutions 346
  bicarbonates 292
  cations 288
  chemical properties 338, 340
  hydroxides 291, 344
  oxides 343
  peroxides 343
  physical properties 336–339
  reactions 342
Group IB 520
Group IIA 249, 336, **348**
  ammonia solutions 346
  chemical properties 338, 350
  chlorides 354
  hydroxides 353
  oxides 343, 353
  peroxides 344
  physical properties 337, 348, 350
  reactions 349
  thermal stability of salts 354
Group IIB 525

Group IIIB **360**, 363
  halides 371
  properties 363, 365
  reactions 366
  trends 363
Group IVA 502–503
Group IVB 374
  bond energies 383
  dioxides 389
  electrode potentials 375
  halides 396
  hydrides 396
  oxides 388, 389
  properties 375, **380**, 381
  stability of oxidation states 395
  sulphides 398
Group VA 504
Group VB 403
  halides 418
  hydrides 415, 416
  oxidation state diagrams 405
  oxides 420
  reactions of elements 414
  reduction potentials 406
  sulphides 424
  trends 407
Group VIA 506
Group VIB 431
  hydrides 442
  oxidation state diagrams 431
  oxides 446
  reduction potentials 432
  trends 434
Group VIIA 509
Group VIIB 463
  bond energies 465
  electrode potentials 467
  oxidation state diagrams 464
  trends 463
  uses 471
Group VIII 511, 516
  cobalt compounds 516
  iron compounds 511
  nickel compounds 518
Group characteristics 215
Group oxidation number 253
Group relationship 293
Group valency 294

Haber process 411, **415**, 499
Half-filled shell, stability 231, 255, 490, 509
Half-filled *d*-electron shell 237
Half-filled subshell 232

Halides 275, 476–478
  action of heat on **292**, 298
  solubilities 245
Haloform reaction 616
Halogen acids 102, 480–482
Halogen bridges 367
Halogen groups 663–665
Halogenation 570, 719
Halogenoalkanes 561, 593, **658**, 660, 661, 664, 698
  hydrolysis 659, 673, 674
  substitution reactions 659, 660
Halogenoarenes **660**, 661, 664, 666, 685
Halogenobenzenes, nitration 727
Halogens 463–472
  bond energies 465
  disproportionation 468
  electrode potentials 465
  oxides 480, 481
Halosulphonic acids 446
Heat **90**, 125, 136
Heat capacity 126
Heats of reactions (see Enthalpies)
Heisenberg 40
Hemi-acetals 610, 709
  cyclic 710
Henry's law 198
Hess's law 93
Heterolytic fission 671
Hexacyanoferrate(II) ion 514
Hexacyanoferrate(III) ion 514
Hole(s) 66, 71
  octahedral 66, 274, 280
  tetrahedral 66, 274
Homologous series 547
Homolytic fission 560, 671
Horizontal relationship 294
Hund's rule 52, 53, 791
Hybridization 721, 773–777
Hydration,
  acetylenes 621, 622
  degree of 200
  enthalpy of 249, 795
  ion 200
  water of crystallization and 332
Hydration energy 107
Hydrazones 646
Hydrides 273–275
  acidities 325
  boiling points (Group VIB) 86
  complex 328
  covalent 274
  ionic 273, 322

metal 322
metallic 275
non-metals 321
preparation 328
reducing agents 274, 328
structures 330
Hydroboration 595, 621, **703**
Hydrocarbons 534–547
  combustion 559
  production 573
  properties 558
  reactions of aromatic 569
Hydrogen 319
  active 601
  analysis for 761
  bond energies 331
  reducing agent 165, 301, 321
Hydrogen atom 31, 37, 55
Hydrogen bonding **86**, 116, 117, 271, 274, 410, 579, 630
Hydrogen carbonates 392
Hydrogen electrode 207
Hydrogen halides 324, 473–475
Hydrogen-like systems 45–46
Hydrogen molecule 77, 771
Hydrogen polysulphides 436
Hydrogen sulphates 451
Hydrogenation 563, 568
Hydrolysis **589**, 594, 614, 622, 719
  acid anhydrides 599, 614
  acid chlorides 594, 614
  amides 594, 614
  amines 594
  chlorides 278, 594
  cyanides 622
  esters, 189, 592, 594, **599**, 614
  ethers 594, 599
  ketones 719
  nitriles 622, 647
  partial 278
  reactions of water 333
Hydroxides **272**, 285, 291, 344, 675
Hydroxonium ion 205, 332
Hydroxyketones 712
Hydroxyl compounds 590
  solubility 579
Hydroxyl group 577–581
  introduction 589, 596
Hypochlorites 477, 482

Ice 330
Ideal gas equation 72
Imines 613, 630, **646**
Imino group 645

Indicators 207, **209**, 210
Inductive effect **722**, 725, 727
Inert gases (see Noble gases)
Inert pair effect 254, **255**, 361, 374, 377, 386, 408
Insulators 86
Interatomic distances 223
Interference between waves 60
Interhalogen compounds 478
Interionic forces 105
Intermolecular forces 87
Internal energy 90
International Union of Pure & Applied Chemistry (IUPAC), rules 540, (see also Alcohols, Aldehydes, etc.)
Internuclear distances 223
Interstices 67
Interstitial compounds 275, 500
Iodates 483
Iodides 276, 277, **477**
Iodine oxides 481
Iodoalkanes 660
Iodoform reaction 616
Ion exchange 395, **756**, 757
Ionic bonding 77, **79**, 102, 103, 310
Ionic character 101, 102
Ionic crystal 67, 69, 241
Ionic model 103, 106
Ionic radii 66, 224, **227**, 241
Ionization energies 41, 103, 228, 792
  first 43, 231, 310
  second 495
  successive 43
  third 351, 495
Ion pairs 72
Iron group 511
Iron(II) and Iron(III) compounds 515
Iron(II) and Iron(III) ion reactions 514
Isocyanides 630
Isoelectronic ions 341
Isoelectronic rule 317
Isoelectronic species 267
Isolated system 137
Isomerism 492, 538
  cis-trans 548, 552
  co-ordination 493
  geometrical 493, 548, 552
  hydration 493
  ionization 493
  optical 493, 540, 543
  stereo 543, 548, 790
  structural 493, 538, 547

Isonitriles 646, 647
Isotopes 23, 24
Isotopic tracers 194, 319

Jeans 28
Joule 90

Kekulé 100
Kelvin 126
Ketals 610
Ketene 661
Ketones **604–614**, 619, 623, 700, 718
 addition reactions 609
 basicity 607
 condensation reactions 612, 678
 enol form 718
 effect of alkali 614, 720
 from calcium salts 621
 hydrates 662
 hydrolysis 719
 nomenclature 605
 oxidation 607
 physical properties 606
 reduction 595, 608
 synthesis 623, 719
 test for 613
 α,β-unsaturated **721**, 722
Kinetic control 164, 172, **192**, 193, 251, 256, 258, 322, 382, 421, 437, 442, 514, 528, 670
Kinetic energy of molecules 74
Kinetically stable 222, 385
Kinetics and mechanism **194**, 670
Kinetics of reaction **177**, 178, 194
Kipp's apparatus 391
Kroll process 303

Lability 382
Lactone 710
Lamellar compounds 387
Lanthanide contraction 227, 229, 240, 497, **502**
Lanthanides 217, 226, 233, 243, **501**, 757
Lassaigne sodium fusion test 761
Latent heats (see Enthalpies)
Lattice energy **103**, **233**, 234, 245, 281
Lattice water 288
Lavoisier 203
Law(s) 4, 214
Law of chemical equilibrium 148
Law of conservation of energy 127
Layer structures **69**, 222, 387

Lead halides 396
Lead oxides 395
Lead(II) salts 398, 399
Leaving group 674
Leblanc process 473
Le Chatelier's principle 112, **150**, 198, 290, 332, 411, 449, 521, 529
Lewis 205, 403, 545
Lewis acid **205**, 378, 417, 419
Lewis base **205**, 390, 417, 419, 433
Ligands 491–492
 stabilization of oxidation states 496, 790
Liquids 71
 miscibility **114**, 146, 149
Lithium complexes 345
Lithium compounds 347
Lone pair repulsions 86, 314
Lucretius 18
Lyman series 27

Madelung constant 105
Magnesium compounds 350–354
Magnesium ion,
 reactions 355
 test for 429
Major elements 216
Manganates 511
Manganese couples, reduction potentials 167
Manganese oxides and oxyacids 510
Markownikoff's rule **562**, 596, 697
Marsden 21
Marsh test 416
Mass number 24
Mass spectrometer 23
Maximum oxidation number 255
Maxwell 130
Melting point 738, 798
Mendeléèv 214
Mercury complexes **528**, 530
Mercury(I) compounds 525, **529**
Mercury(II) compounds 525, 528, 529
Metabisulphites 448
Metaborates 367
Metallation 695
Metallic bond **85**, 223, 236, 340
Metallic radius 224, **228**, 235
Metallicity 261, 262
Metalloids 219, 261
Metallurgy **294**, 302
Metals 85, **218**–222, 261–265
 comparison with non-metals 263

electropositivity 264
 extraction 294, **302**
 passive 513, 516
 properties 85, 284, 293
 reactions 261
 refining 294, **304**
 superconducting 238
Metaphosphates 428
Metastable 222
Metathesis 289
Methanides 377, 388
Methyl group 537
Microstate **133**, 135, 140, 780
Middle row anomaly **226**, 229, 233, 360, 362, 403, 419, 431, 463
Mirror image 543
Mixing of gases 128, 781
Mnemonics 52
Models 4, 5, 20, 215, 310
Moissan 468
Molality 16
Molarity 16
Mole 13
Molecular collisions 179
Molecular crystals **70**, 236, 262
Molecular formula 737, 760
Molecular orbital theory 101, 771, 774, 776
Molecular shape 88–89, 313–317
Molecular sieves 394
Molecular structure 64, 315
Molecular weight 11, **118**, 119, 120, 121, 122, 761
Molecularity 182, 183, 194
Molecules,
 chiral 543
 giant 222, 236
 infinite chain 222
 infinite layer 222
 polar 82, 87
Molybdenum compounds 507
Molybdenum steels 501
Monomer 565
Monoprotic acids 286
Monosaccharides 708, 715
Monotropy 222
Moseley 21
Mulliken 80, 232, 265
Mutarotation 711
Multi-electron atoms **46**, 49, 56, 229

Napoleon 379
Natural gases 9, 572, 764
Negative oxidation number 266

GENERAL INDEX 857

Nernst equation 163, 251
Neutron 19, **24**
Neutron diffraction 64
Newton 90
Nickel group 518–520
  square–planar complexes 520
  tetrahedral complexes 520
Nitrates 426
  action of heat 285, 291, 421
Nitration,
  aliphatic 649, 650
  aromatic 189, 570, 649, 731
Nitrides 414
Nitriles 630, **646**, 648
  and ketone synthesis 623
  hydrolysis 622, 647
  reduction 622, 637, 647
Nitrite esters 648, 650
Nitrites 424
Nitro group 648–650
Nitroalkanes 649, 650
Nitrocompounds 630, 648–650
  reduction 637, 649
Nitrogen,
  analysis for 761
  anomalous nature 408
Nitrogen cycle 411
Nitrogen functional groups 630
Nitrogen group 403
  oxidation states 405
  trends 407
Nitrogen oxides 421
Nitrogen oxyacids 424
Nitronium ion 186, 189, 682, 731
Nitrophenols 723, 724
Nitroso compounds 630, 642
  reduction of 637
Nitrosyl halides 421
Noble gas rule 309, 341
Noble gases 83, 216, **308**
  compounds 311–313
  configurations 254
  properties 309–314
Nomenclature 540, 818
  alkyl halides 658
  amino compounds 629
  anhydrides 598
  carbonyl compounds 605
  esters 598
  ethers 598
  halogeno-compounds 658
  hydrocarbons 539
  hydroxy compounds 578
  metal complexes 493

Non-metals 218, 219, 222, 237, 261, 293
Nuclear atom 21, 491
Nucleon 24
Nucleophiles 619, **675**, 676
Nucleophilic substitution 508, 659, 662, **672**, 675
  aromatic 728
Nucleophilicity 675
Nucleotide units 715

Octet rule **83**, 88, 309, 433
  exceptions to 84
Oils,
  lubricating 559
  vegetable 563
Olefins (see Alkenes)
Optical activity 541, 542
Orbitals **36**, 38, 41, 48, 55
  d-, separation 787
  delocalized 101
  half-filled 228
  hybrid 773–777
  molecular 101, 771, 774, 776
  relative energies 51
Order of reaction 177, 178
Ores 294
  treatment 302, 303
Organic synthesis 764–768
Organoboranes 695, 696
Organolithium compounds 699–702
Organo–metallic compounds 346, 372, 399, **693**
  alkali metals 346
  oxidation 702
  preparation 702
  principal reactions 703
  reactivity 698
Osomotic pressure **120**, 122
Overvoltage **193**, 194, 259, 332
Oxalates 400
Oxidant 454
Oxidation 453
  electronic theory 454
Oxidation number (O.N.) 256, 374, 466
  allocation 456, 457
  definition 456
  total (T.O.N.) 459
Oxidation states (see also under each group) 170, **253**, 374, 456
  diagram 171
  relative stabilities 167, **169**, 254, 255, 361, 496, 789–791

s- and p-block 270
  transition metals 496
Oxides 440
  acidic 219
  amphoteric 219, 269
  basic 219, 269
  classification 440
  enthalpy of formation 253, 436
  mixed 269
  normal 269, 291
  reduction by carbon 294, 299, 301
  structure 271
Oxidizing agent 454
Oximes 612, 646
Oxoacids (see Oxyacids)
Oxyacids 325
  arsenic 429
  halogens 480, 481
  inorganic 326
  nitrogen 424
  phosphorus 426
  selenium 447
  sulphur 447–448
  tellurium 447
Oxyanions 283, 425, 426, 448, 482
Oxyhalides 408, 419
Oxygen group 431
Oxygen, types of compounds 433
Ozonides 441, 564
Ozonolysis 564

Packing of spheres 66
Paraffins (see Alkanes)
Paramagnetism 791
Particle model 27, 30
Particles 18–19
Partington 19
Partition chromatography 752
Partition coefficient **146**, 749
Paschen series 27, 33
Pauli exclusion principle 38, **50**, 53
Pauling 80, 232
p-Block elements 292, **360**, 362
  second row 293
Penetration **44**, 46, 48, 56, 229, 230
Peptides 707
Perborates 369
Perbromates 484
Perchlorates 481, 484
Period 216
Periodates 484
Periodic law 214
Periodic properties 269

Periodic table 214, 215
  summary of relationships 292–294
Periodicity 49, **215**
Permanganates 510
Permanganites 511
Peroxides 343, 440, 442
Perxenates 313
Petroleum 559, 572, 764
pH (puissance) 206
pH meter 207
Pharmaceutical industry 764
Phase changes 113
Phase of a wave 60
Phenolate ions 725
Phenols 577, 579, **580**, 592, 595, 596, 651, 723
  acidity 580, 723, 725
  acylation 584
Phenylhydrazones 678
Phosphate esters 714
Phosphates 428
  poly- 427, 428
  solubility 243, 246
  uses 413
Phosphides 414
Phosphites 423
Phosphonitrilic halides 420
Phosphonium salts 417
Photoconductivity 439
Photoelectric effect 28
Photography 452, 472
Photo-ionization 238
Photon 31
Photosynthesis 392, 714
Pidgeon process 300
pK_a 210, 324
pK_a values 326, 327
Planck 28, 31
Planck constant 29
Plane-polarized light 540, 541
Plastics 565
Plumbates 395
Plumbites 395, 398
Polar bonds 82, 85, 102, 673, 693
Polar groups 83
Polar molecules 82, 87, 329
Polarizability 82
Polarization, model of 345
Polarizing power 82, **283**, 284, 322
Pollution 9
Polyalkenes **565**, 570
Polyfunctional compounds 708
Polyhalides 478, 479, 480
Polymerization 565

Polymorphism 221
Polyphosphates 427, 428
Polyprotic acids 286, 326
Polysaccharides 708
Polysulphides 436, **443**, 444
Porphyrins 515
Potassium salts 345
Potential difference 157
Potential energy 104
Potential, standard oxidation 159
Potential, standard reduction 159, 246
Powell · 314
Pregnancy testing 758
Probability 55, 128, 129
Proteins 535, 707
Proton 19, 24
  charge 24
  enthalpy of hydration 323
  transfer 190
Purification 737, 759
Purity, criteria of 738
Pyrosulphates 452

Quanta 28
Quantum numbers 38, 46, 55
Quantum theory 25, 28, 31
Quaternary ammonium compounds 630, **641–644**, 645
Quinones 595

Racemic mixtures 543
Radial distribution 40
Radial probability curves 44
Radioactivity 20, 25
Radius ratio rules 353
Ramsay 26, 308
Raoult's law 115–119
Rare earths (see Lanthanides)
Rate constant **177**, 179
Rate determining step **182**, 670
Rate equation 177
Rate of reaction **176**, 177, 670
  temperature dependence 179
Rayleigh 28
Reaction co-ordinate 188
Reaction mechanism 176, **180**, 195, 670
Reaction molecularity 182–184
Reaction order 177, 178
Reaction pathway 180, 194, 320, 670
Reaction profile 180, 188–190, 729, 785
Reaction rate **176**, 177, 179, 670

Reaction yield 765
Reactions,
  acid-base 205
  addition **545**, 671, 678
  chain 321
  classification **671**, 684
  condensation 671, 684
  coupling 651
  elimination 572, 574, 612, 671, 678, **682**
  isomerization 184
  redox 156, 161
  substitution 561, 613, **672**, 730
Reactivity 382
Recrystallization 739
Redox equilibria 173
Redox equations 455, 458–460
Redox reactions **156**, 161, 454
Reductant 454
Reduction 302, 321, **453–454**
  by carbon 299–301, 453
  by electrolysis 300, 302, 303
  by hydrides 274, 328
  by hydrogen 301
  by metals 303
Reduction potential 159, 246
  Born–Haber cycle for 247
  of water 333
  table of 248, 798
Refining 304
R_f value 755
Refractive index 739
Relative atomic and molecular masses 13
Relative stabilities of oxidation states 167, 169, **254**, 255, 361, 496, 790
Representative elements 216
Repulsive forces 105
Resins 395
Resistivity 202
Resonance 101, 102, 378
Restricted rotation 776
Reversible change 138
Ring compounds 544, 550, 551
Ring strain 98, 550, 551
Roasting in air 302
Rochow 80, 232
Rock–salt structure 67
Rotation about single bonds 538, 775
Rule of eight 309, 385
Rust 257, 513
Rutherford 21
Rydberg constant 27

GENERAL INDEX 859

Saccharides 708–715
Salt bridge 159
Salting out 741
Salts,
    action of heat on 285, 291
    classification 286–289
    preparation 289
Saponification 593
Saturated compounds 543
s-Block elements 292, **336**, 362
    second row 293
Scandium group 501
Schiff's base 613, 630, **646**
Schiff's reagent 620
Schrödinger equation **35**, 39, 771
Screening effect (see Shielding)
Second law of thermodynamics **135**, 163
Selenides 279
Selenium compounds 440, 442, 446
Semi-metals **219**, 222, 261, 262, 293
Semi-permeable membrane 120
Separation,
    chemical methods 758, 759
    physical methods 737–758
Sex hormones 758
Shapes of molecules 88, 314, 315
Shielding effect **47**, 48, 56, 226, 229, 230
    poor value for nd^{10}-shell 226
Sidgwick 314
Silanes 396
Silicates 394
Silicones 399, **400**, 536, 537
Silver–copper cell 162
Silver–ion reactions 525
Silver salts 524
Single-electron atom 44
Six-membered rings 550
Sodium fusion test 761
Sodium salts 345
Solubility **200**, 246, 739
    common compounds 246
    hydroxy compounds 579
    ionic crystals 242–246
Solubility product **200**, 445
Solubility trends 245
Solutions 198–199
Solvation 141, **200**, 228, 327
Solvation energy 106
Specific heats 14
Spectra **25**, 26, 33
Spectrochemical series 788
Spin 38, 55

Spin character of complexes 790
Spinning-band column 745
Stable configurations 311
Stability 382
Stabilities,
    ionic compounds 342
    oxidation states 169, 789–791
    oxides and salts 164
Standard electrode potential 159, 247
Standard enthalpy of formation 94
Standard entropy change 140
Standard free energy change **153**, 173
Standard free energy of activation 188
Standard free energy of formation 155
Standard half-cell 159
Standard molar entropy 139
Standard oxidation potential 159
Standard reduction potential 159
Standard state 92
Standard temperature and pressure 12
Stannates 395, 398
Stannites 260, 395, 398
State function 91
State symbols 14, 535
Stephen's method 622
Stereochemistry 492, 548, 790
Steroid homones 758
Structure,
    antifluorite 70, 271
    body-centred cubic 66, 339
    cadmium chloride 69
    cadmium iodide 271
    caesium chloride 67
    chain 222, 276, 277, 536
    chromium(III) chloride 69
    close-packed 64, 235
    cubic close-packed 65
    diamond 69
    face-centred cubic 65, 70
    fluorite 68
    giant **19**, 64, 79, 237
    hexagonal close-packed 65
    layer **69**, 224, 387
    nickel arsenide 271
    open 66
    pyrites 271
    ring 544, 550, 552
    rock salt 67, 271
    sodium chloride 271

    wurtzite 70, 271
    zinc blende 69, 271
Structure and bonding 77
Structure, determination 59–64
Structure of,
    elements 221
    hydrides 273–275, 322
    metals **85**, 221, 222, 228, 339
    oxides 270, 271
    sulphides 271
Suboxides 440
Sublimation 113, 748
Substitution reactions **561**, 572, 613, 671, 730
Substituents 730
    directing character 732, 733
Sugars 708–715
    fermentation 597
    reducing 712
    taste 709
Sulphanes 443, 448
Sulphates,
    action of heat on 285, 286, 291
    solubility 243, 246
Sulphide ores 302
Sulphides 279, 280, **443**, 444
Sulphites 448, 449
Sulphonation,
    benzene 571
    naphthalene 193
Sulphur,
    hydrides, 422
    oxides 446
    oxyacids 447
Superconductivity 309
Superfluidity 309
Superoxides 343, 441
Sweeteners 709
System,
    definition 90
    isolated **137**, 143
System and its environment 90, 141

Tautomerism 567, 718
    keto-enol 718, 720
Television screens 527
Tellurides 279
Tetra-alkylammonium halides 645
Tetrafluoroborates, alkyl 645
Tetrahedral angle 89, 538
Tetrasaccharides 708
Tetrathionates 450, 453
Theories 4, 214

Thermal conductivity 237, 798
Thermal decomposition,
  of carbonates 164, 282, 283, **285**, 291, 296
  of oxides 296
  of salts 280, 285
Thermal energy (see Heat)
Thermite process 303, 370
Thermodynamic control 164, **192**, 193, 251, 256, 258, 295, 490, 670
Thermodynamically stable 385
Thermodynamically unstable 222
Thermodynamics,
  first law 90
  second law 125, **135**, 158, 163
Thioarsenates 424
Thioarsenites 424
Thiocarbonates 279
Thiocyanates 516
Thionyl halides 446
Thiostannates 280, 398
Thiosulphates 452
Third row anomaly 227
Thomson 23
Tin-plating 513
Tin(II) salts 399
Titanates 503
Titanium group 502
Titanium steels 501
Titanyl ion 503
Titration curves 207, 208
Trace elements 495
Transition contraction 226, 362
Transition elements 217, 218, 227, 233, 270, 294, **487–523**
  complexes 491
  definition 487
  general properties 489, 499
  oxidation states 497
  trends among 487
Transition state **187**, 670, 674, 697, 730
Transition state theory 187–189, 784

Transition temperature 222
Trialkylboranes, oxidation 703
Trialkyl phosphates 662
Tri-iodides 283
Trisaccharides 708
Tritium 24, 319
Trivial names 540, 818
Trouton's rule 239
Tungsten compounds 507
Tungsten steels 501

Uncertain principle 36, 40
Unit cell 65
Unsaturated compounds 543, 546
  test for 563
Unsaturation, degree of 568

Valence-bond theory 101
Valence electrons 48
Valency 374
  variable 499
Valencies,
  common 294
  elements 218
  group 218, 294
Vanadium compounds 505, 506
Vanadium group 504
Vanadium oxides 270
Vanadium steels 501
Van Arkel process 304, 466, 503
Van der Waals forces 19, 70, 71, 72, **87**, 221, 236, 237, 239, 271, 274, 276, 292, 308
Van der Waals radii **70**, 224, 235, 292
Vapour pressure 112, 113, 115
  lowering of 118
  partial 115
Velocities,
  distribution of 75, 130
  root mean square 74
Vibrational energy levels 130
Vinyl halides 727
Vital force 534
Volcano reaction 412

Water 258, **329–333**
  anomalous properties 86, 329
  dissociation constant 206
  enthalpies of phase changes 126
  heavy 320
  lattice 288
  self-ionization 206, 332
  structure 89, 329
Water as a solvent **199**, 242, 246, 332
Water of crystallization 288, 332
Wave description 35
Wave model 27, 30, 31
Wavefunction 55
Wavelengths 27
Wavenumber 28
Weakly electronegative elements 233
Weakly electropositive elements 264
Welding 321
Werner 518
Wittig reaction 574
Wittig reagent 574, 613
Wöhler 653
Work 90
Wurtzite structure 70

Xenates 313
X-ray crystallography 59
X-ray diffraction 59, 61
X-ray goniometer 60
X-ray powder photograph 61
X-rays 21

Yield of reaction 765

Zeolites 394
Zerewitinoff method 699
Zero order kinetics 186
Zincate ion 260, 288
Zincates 260
Zinc blend structure 69
Zinc complexes 528
Zinc compounds 526
Zinc ion reactions 527
Zone refining 380, 742
Zwitterion 706

Printed in Malta by St Paul's Press Ltd

E7987654